INFORMATION, COMPUTER AND APPLICATION ENGINEERING

PROCEEDINGS OF THE 2014 INTERNATIONAL CONFERENCE ON INFORMATION TECHNOLOGY AND COMPUTER APPLICATION ENGINEERING (ITCAE 2014), HONG KONG, P.R. CHINA, DECEMBER 11–12 2014

Information, Computer and Application Engineering

Editors

Hsiang-Chuan Liu
Asia University, Taiwan

Wen-Pei Sung
National Chin-Yi University of Technology, Taiwan

Wenli-Yao
Control Engineering and Information Science Research Association, Hong Kong

CRC Press
Taylor & Francis Group
Boca Raton London New York Leiden

CRC Press is an imprint of the
Taylor & Francis Group, an **informa** business

A BALKEMA BOOK

CRC Press/Balkema is an imprint of the Taylor & Francis Group, an informa business

© 2015 Taylor & Francis Group, London, UK

Typeset by diacriTech, Chennai, India

Published by: CRC Press/Balkema
P.O. Box 11320, 2301 EH Leiden, The Netherlands
e-mail: Pub.NL@taylorandfrancis.com
www.crcpress.com – www.taylorandfrancis.com

ISBN: 978-1-138-02717-6 (Hardback)
ISBN: 978-1-315-73555-9 (eBook PDF)

Table of contents

Session 2: Information, automotive and education engineering

xiv

xv

Information, Computer and Application Engineering – Liu, Sung & Yao (Eds)
© 2015 Taylor & Francis Group, London, ISBN 978-1-138-02717-6

Preface

The 2014 International Conference on Information Technology and Computer Application Engineering (ITCAE 2014) was held August 11-12, 2014 in Hong Kong. The aim was to provide a platform for researchers, engineers and academics as well as industry professionals from all over the world to present their research results and development activities in Computer Application Engineering and Information Science.

Information Technology and Computer Technology have developed rapidly in recent years. Nowadays, these technologies are gradually entering into people's daily lives. In this proceedings volume you will find papers regarding Computational Intelligence, Computer Science and its Applications, Intelligent Information Processing and Knowledge Engineering, Intelligent Networks and Instruments, Multimedia Signal Processing and Analysis, Intelligent Computer-Aided Design Systems, and so on. For this conference, we received more than 800 submissions by email and via the electronic submission system, which were reviewed by international experts, and about 300 papers have been selected for presentation.

On behalf of the guest editors of this book, we would like to thank the conference chairs, organization staff, the authors and the members of International Technological Committee for their hard work. Thanks are also given to CRC Press/Balkema.

We hope that ITCAE 2014 was successful and enjoyable for all participants. We look forward to seeing all of you next year at ITCAE 2015.

December, 2014

Wen-Pei Sung
National Chin-Yi University of Technology

Information, Computer and Application Engineering – Liu, Sung & Yao (Eds)
© 2015 Taylor & Francis Group, London, ISBN 978-1-138-02717-6

ITCAE 2014 committee

CONFERENCE CHAIRMAN

Prof. Wen-Pei Sung, *National Chin-Yi University of Technology, Taiwan*

PROGRAM COMMITTEE

Ghamgeen Izat Rashed, *Wuhan University, China*
Andrey Nikolaevich Belousov, *Head Laboratory of Applied Nanotechnology, Ukraine*
Ramezan ali Mahdavinejad, *University of Tehran, Iran*
Krupa Ranjeet Rasane, *Department of Electronic Communications Engineering KLE Society's College of Engineering, India*
Hao-En Chueh, *Yuanpei University, Taiwan*
Sajjad Jafari, *Semnan University, Iran*
But Adrian, *ELECTROMOTOR company, Timisoara, Bulevardul, Romania*
Prof. Yu-Kuang Zhao, *National Chin-Yi University of Technology, Taiwan*
Yi-Ying Chang, *National Chin-Yi University of Technology, Taiwan*
Darius Bacinskas, *Vilnius Gediminas Technical University, Lithuania*
Viranjay M. Srivastava, *Jaypee University of Information Technology, India*
Anita Kovač Kralj, *University of Maribor, Slovenia*
Chenggui Zhao, *Yunnan Normal University, China*
Wen-Sheng Ou, *National Chin-Yi University of Technology*
Hsiang-Chuan Liu, *Asia University, Taiwan*
Zhou Liang, *Donghua University, China*
Liu Yunan, *University of Michigan, USA*
Wang Liying, *Institute of Water Conservancy and Hydroelectric Power, China*
Chenggui Zhao, *Yunnan University of Finance and Economics, China*
Rahim Jamian, *Universiti Kuala Lumpur Malaysian Spanish Institute, Malaysia*
Wei Fu, *Chongqing University, China*
Lixin Guo, *Northeastern University, China*
Mostafa Shokshok, *National University of Malaysia, Malaysia*
Tjamme Wiegers, *Delft University of Technology, Netherlands*
Bhagavathi Tarigoppula, *Bradley University, USA*
Gang Shi, *Inha University, South Korea*

CO-SPONSORS

International Frontiers of Science and Technology Research Association
Hong Kong Control Engineering and Information Science Research Association

Session 1: Computer science and application engineering

Information, Computer and Application Engineering – Liu, Sung & Yao (Eds)
© 2015 Taylor & Francis Group, London, ISBN 978-1-138-02717-6

Design private cloud of SCADA system

Miao Liu, Man Cang Yuan & Chao Sun
China Petroleum Longhui Automation Engineering Co. Ltd., China
China Petroleum Pipeline Bureau, Langfang, China

Fu Bin Wang
Technology Center of China Petroleum Pipeline Bureau, Langfang, China

ABSTRACT: Cloud computing is fundamentally altering the expectations for how and when computing, storage, and networking resources should be allocated, managed, and consumed. The SCADA (Supervisory Control and Data Acquisition) system is very important to oil and gas pipeline engineering. For increasing resource utilization, reliability, and availability of oil and gas pipeline SCADA system, the private cloud of SCADA system is proposed in this article. This article introduces the realization details about the private cloud of oil and gas pipeline SCADA system.

KEYWORDS: SCADA system; cloud computing; private cloud

1 INTRODUCTION

Cloud Computing refers to both the applications delivered as services over the Internet and the hardware and systems software in the datacenters that provide those services [1]. The services themselves have long been referred to as Software as a Service (SaaS). The datacenter hardware and software is what we will call a Cloud. When a Cloud is made available in a pay-as-you-go manner to the general public, it is a Public Cloud; the service being sold is Utility Computing. The term "Private Cloud" refers to internal datacenters of a business or other organizations that are not made available to the general public[2–3].

Amazon's cloud computing is based on server virtualization technology. Amazon released Xen-based Elastic Compute Cloud™ (EC2), object storage service (S3), and structure data storage service (SimpleDB) [4–5]. AWS becomes the pioneer of Infrastructure as a Service (IaaS) provider [6].

Google's style is based on a technique-specific sandbox. Google published several research papers from 2003 to 2006, which outline a kind of Platform as a Service (PaaS) cloud computing [7–8].

Microsoft Azure™ [9] was released in October 2008, and it uses Windows Azure Hypervisor (WAH) as the underlying cloud infrastructure and .NET as the application container. Azure also offers services, including BLOB object storage and SQL service.

The SCADA system of oil and gas pipeline can manage sequential control transmission of petroleum pipeline, equipment monitoring, data synchronization transmission record, and monitoring operation conditions of every station control system. In addition, the SCADA system has more features, such as leaking detection, system simulation, water hammer protection in advance, and so on. The SCADA system can continuously monitor equipment that are scattered over wide regions, and it can operate remote devices from the control center so that the operating efficiency is improved, energy is saved, and cost is reduced. This system can also guarantee the integrity of pipeline by continuously monitoring the key parameters of the system, such as pressure, flow, oil tank liquid level, and so on [10]. So, the SCADA system is very important to the oil and gas pipeline engineering.

In order to increase resource utilization, reliability, and availability of oil and gas SCADA system, the private cloud of oil and gas pipeline SCADA system is proposed in this article.

The remaining part of this article is organized as follows. In Sect. 2, the architecture of cloud computing and the traditional structure of the SCADA system is described. The private cloud of oil and gas pipeline SCADA system is proposed in Sect. 3. Section 4 describes the scheduling algorithm of the cloud SCADA system. Finally, the article concludes in Sect. 5.

2 ARCHITECTURE OF CLOUD COMPUTING AND TRADITIONAL SCADA SYSTEM

The architecture of cloud computing is shown in Fig. 1. The whole cloud computer system can be divided into the core stack and the management. In the core stack, there are three layers: (1) Resource, (2) Platform, and (3) Application. The resource layer is the infrastructure layer that is composed of physical and virtualized computing, storage, and networking resources. The platform layer is the most complex part that could be divided into many sub-layers; for example, a computing framework manages the transaction dispatching and/or task scheduling. A storage sub-layer provides unlimited storage and caching capability. The application server and other components support the same general application logic as earlier with either on-demand capability or flexible management, such that no components will be the bottle neck of the whole system. Based on the underlying resource and components, the application could support large and distributed transactions and management of huge volume of data. All the layers provide external service through web service or other open interfaces.

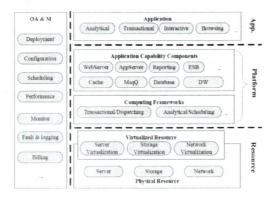

Figure 1.   Architecture of cloud computing.

The SCADA system adopts distributed supervisory control and centralized management. The traditional structure of SCADA system is described in Fig. 2. The system is organized by the supervision center, a lot of site control systems, and communication medium. The supervision center (master station) is the core of system and in charge of controlling and managing system running. It is organized by a lot of data processing servers and database servers. The outside site is the intelligent measure and control module by microprocessor or DSP. It can collect and process the data of remote sites, control local sites, and communicate with the remote supervision center.

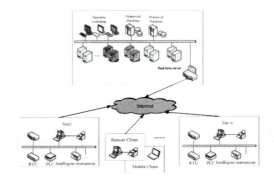

Figure 2.   Structure of SCADA system.

The current SCADA system has some problems as follows:

Servers of supervision center are the command center of the system. They can gather the pipeline running data, state and warning information and provide the real-time database for all operating sites. Nowadays, the SCADA system of oil and gas pipeline adopts Hot Standby as the way of redundant configuration to ensure the reliability of the system. When the master server is working online, the backup server monitors work state of the master server and get the data from the master server for keeping the data consistency between the master server and the backup server. Once the master server breaks down, the backup server takes over the work from the master server immediately, the backup server becomes the master server, and the repaired master server becomes the backup server. As the controlling core of the SCADA system, the central servers are responsible for data collection and controlling of all lines. It is important to ensure the data transmission between the servers and site control system for all SCADA system network normal operation. The single backup server cannot completely ensure the reliability of the system.

Hot Standby is used for ensuring the reliability of the system. With the system scaling out constantly, the number of servers increases. Redundant configuration can cause wasting of server source and an increase in the workload of operation administration and maintenance. So, it is necessary to find a more advanced resource configuration mode to increase resource utilization rate.

With the SCADA system of oil and gas pipeline scaling out constantly, the number of server increases. Lots of operation and data are processed on different servers. However, CPU load is different with different site data. So, different servers have a different load; some servers are overloaded but some servers are just idle. It is necessary to study new load balancing strategies for improving the performance of the system.

4

3 PRIVATE CLOUD DESIGN OF OIL AND GAS PIPELINE SCADA SYSTEM

The private cloud system of the oil and gas pipeline SCADA can be designed as shown in Fig. 3:

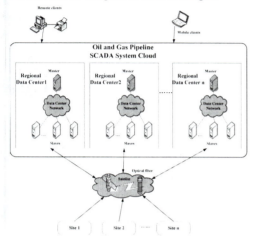

Figure 3. Private cloud system of SCADA.

The Master–Slave structure is used in the oil and gas pipeline SCADA system based on cloud computing. The master server is responsible for maintaining raw data, including name space, access control, mapping information of point configuration defining, and the load information of Slave. In addition, it takes charge of system resource scheduling. In the oil and gas pipeline SCADA system, the cloud structure is used in the data center. All point configuration information is divided into fixed packets. Each point configuration information packet is carried out in n_1 local servers and n_2 remote servers. When real-time data are collected to the nearest regional data center, the master of the regional data center distributes the data to several servers to carry out the task and the primary/backup technology is used to ensure the reliability of the system.

The SCADA system model of the oil and gas pipeline is created as follows:

The real-time data processing tasks of SCADA system are expressed by $T = \{T_1, T_2, ..., T_n\}$. Each task of the SCADA system is independent, and the task is non-preemptive.

Each real-time task can be defined as

$$T_i = (d_i, T_i^1, T_i^2, T_i^3) \quad (1)$$

d_i is the deadline of the real-time task; T_i^1, T_i^2, T_i^3 are the one primary copy and two backup copies, respectively; and the code of the three tasks is entirely same.

$$T_i^1 = (C_i, s_i^1, \rho_i^1) \quad (2)$$

$$T_i^2 = (C_i, s_i^2, \rho_i^2) \quad (3)$$

$$T_i^3 = (C_i, s_i^3, \rho_i^3) \quad (4)$$

Therefore, s_i^1, s_i^2, and s_i^3 are the time for tasks to begin running; ρ_i^1, ρ_i^2, and ρ_i^3 are processors assigned to three task copies.

There are different execution periods for different processors, so defining a computing time vector for each task T_i is as follows:

$$C_i = [c(i,1), ..., c(i,m)] \quad (5)$$

There, c_{ij} represents the execution time for T_i^1, T_i^2, T_i^3 on processor P_j.

If the SCADA system can tolerate 2 processors being disabled, the sum of primary copy execution time and backup copy execution time should be less than or equal to the deadline.

$$\forall i \in [1,n], (c(i, \rho_i^1) + c(i, \rho_i^2)) \leq d_i \wedge (c(i, \rho_i^1) + c(i, \rho_i^3)) \leq d_i \quad (6)$$

For each processor P_i of the SCADA system,

$$\forall i \in [1,m], \sum_{T_i^1 \in \Lambda_i} c(i,j) + \sum_{T_i^2 \in \Lambda_i} c(i,j) \leq d_i, \sum_{T_i^1 \in \Lambda_i} c(i,j) + \sum_{T_i^3 \in \Lambda_i} c(i,j) \leq d_i \quad (7)$$

The data center physical architecture of private cloud includes controller node, compute nodes, NFS server, and network devices. It is shown in Fig. 4.

The function of the Controller node includes showing, monitoring, management, and scheduling. The controller node provides a database and web server to support the management interface of the system. The users issue control commands and operate the system by the management interface of the controller node. Information of other nodes must register to the

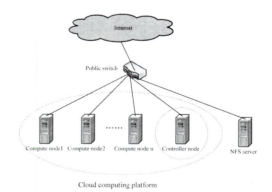

Figure 4. The physical architecture of private cloud.

controller node and the controller node must schedule the resource constantly.

Compute nodes are the operating foundation of the virtual machine. They are in charge of the specific operating work of the private cloud system. Compute nodes compose the resource pool together and provide the resource with a form of virtualization. The virtual machines that the private cloud system provides for outside users are run in compute nodes. Compute nodes are dominated and managed by the controller node.

NFS server provides NFS service, which supports dynamic extension function of the private cloud system.

Network devices include routers and switches. They connect every device of the private cloud system and provide access to the network. In addtion, they take charge of the network address assignment.

4 SCHEDULING ALGORITHM FOR PRIVATE CLOUD SYSTEM

The algorithm is described as follows:

1 Initializing processor set ψ, task set T, single processor scheduling length L_p, real-time task deadline DL, the number of local Slave m, the number of remote Slave n, the number of local Slave with tag configuration information packet n_1 and the number of remote Slave n_2, optimal processor weighting coefficient k_1, k_2, k_3 which is the weighting coefficient of Δ_i, ξ_i, and λ_i, respectively.
2 Master preloads all tag information of all pipelines in the memory of Slave. Each tag configuration information is loaded in local n_1 different servers and remote n_2 different servers.

For each frame SCADA system real-time data{
Find the processors with preload tag configuration, including local processors P_{j_1}, P_{j_2}, and P_{j_3}, remote processor P_{j_4} and P_{j_5}.

Set $Z_{j_i} = k_1 \xi_{j_i} + k_2 \xi_{j_i} + k_3 \xi_{j_i}$, computing P_{j_1}, P_{j_2}, P_{j_3}, P_{j_4}, P_{j_5}, and Z_{j_i}, and place them in ascending order.

Choose a processor with minimal Z_{j_i} in local processor P_{j_1}, P_{j_2}, and P_{j_3} to execute T_i^1.

Choose a processor with sub-minimal Z_{j_i} in local processor P_{j_1}, P_{j_2}, and P_{j_3} to execute T_i^1.

Choose a processor with minimal Z_{j_i} in remote processors P_{j_4} and P_{j_5} to execute T_i^1.
}

5 CONCLUSIONS

SCADA systems are widely used in industries for Supervisory Control and Data Acquisition of industrial processes. For the oil and gas pipeline field, the running data of pipelines can be obtained in real time, processed scientifically and decisions can be made by the SCADA system. For improving the reliability and resource utilization of the traditional oil and gas pipeline SCADA system, the private cloud of the oil and gas pipeline SCADA system is proposed and the real-time scheduling algorithm for the SCADA system based on cloud computing has to be researched. This private cloud of SCADA can realize real-time data multi-copy execution and history data multi-restore. So, the new SCADA system based on cloud computing has more reliability and higher resource utilization.

REFERENCES

[1] Dinh H T, Lee C, Niyato D, et al. A survey of mobile cloud computing: architecture, applications, and approaches. Wireless communications and mobile computing, 2013, 13(18):1587–1611.
[2] Erl T, Puttini R, Mahmood Z. Cloud Computing: Concepts, Technology and Design. 2013,12–18.
[3] ComPUtING C. Cloud computing privacy concerns on our doorstep. Communications of the ACM, 2011, 54(1).
[4] Agmon Ben-Yehuda O, Ben-Yehuda M, Schuster A, et al. Deconstructing Amazon EC2 spot instance pricing. ACM Transactions on Economics and Computation, 2013, 1(3):16.
[5] Juve G, Rynge M, Deelman E, et al. Comparing future-grid, amazon ec2, and open science grid for scientific workflows. Computing in Science & Engineering, 2013, 15(4):20–29.
[6] Kokkinos P, Varvarigou T A, Kretsis A, et al. SuMo: Analysis and Optimization of Amazon EC2 Instances. Journal of Grid Computing, 2014: 1–20.
[7] Zhang Q, Cheng L, Boutaba R. Cloud computing: state-of-the-art and research challenges. Journal of internet services and applications, 2010, 1(1): 7–18.
[8] Marston S, Li Z, Bandyopadhyay S, et al. Cloud computing-The business perspective. Decision Support Systems, 2011, 51(1):176–189.
[9] Feinleib D. The Intersection of Big Data, Mobile, and Cloud Computing[M]//Big Data Bootcamp. Apress, 2014:85–101.
[10] Li Yang, Xiedong Cao, Research on fnn-based security defence architecture model of scada network, Proceedings of IEEE CCIS. 2012. pp.1829–1833.

Information, Computer and Application Engineering – Liu, Sung & Yao (Eds)
© 2015 Taylor & Francis Group, London, ISBN 978-1-138-02717-6

Application of the genetic algorithm in image processing

Y.L. Wang, J.X. Wang & H.B. Kang
Hebei Institute of Architecture and Civil Engineering, Zhang Jiakou, Hebei, China

ABSTRACT: Genetic Algorithm (GA) is a random search and optimization method based on natural selection and genetic mechanism of living beings. In recent years, GA is successfully used in solving the complex optimization and the industrial engineering problems. Therefore, the research on GA has attracted a lot of attention. This article discusses and surveys the status and advances in GA research. The basic algorithms, theory, and implementation techniques of GA are outlined first. Then, many applications of GA in the image processing field are reviewed, such as image segmentation and edge detection, image compression, image restoration, image matching, image enhancement, and image reconstruction. At last, several key problems in this field as well as their development in the future are discussed.

KEYWORDS: genetic algorithm; digital image processing; image segmentation

1 INTRODUCTION

Genetic algorithm (GA) is an adaptive global heuristic search algorithm converges. The basic idea comes from biological evolution and population genetics and reflects the survival of the fittest, evolutionary principle of survival of the fittest. Using the genetic algorithm in scientific research and engineering technology, this basic idea was proposed by professor Holland at the University of Michigan in the early 1960s. As the overall search strategy and optimization, GA calculation does not rely on gradient information, and its very wide range of applications, especially suitable for processing traditional methods, are difficult to solve highly complex nonlinear problems. It is in the adaptive control, combinatorial optimization, pattern recognition, machine learning, planning strategy, information processing, and artificial life field application that it shows more superiority. Image processing is an important research domain of computer vision. In the process of image processing, such as scanning, feature extraction, and image segmentation, there will be some error. It will affect the image effect. How to minimize these errors is an important requirement for the computer vision to achieve practical, GA optimization in these aspects of image processing to find applications. Currently, image segmentation, image restoration, image reconstruction, image retrieval, and image matching have been widely used.

2 PRINCIPLE, BASIC PROPERTIES, AND IMPROVEMENT OF GENETIC ALGORITHM

GA attempted to solve the problem expressed as a chromosome, also known as "string." The solving steps of GA are as follows:

1 Coding: The definition of the mapping from the solution space to the chromosome coding space space.
2 Initialization of population: Under certain restrictions to the initial population, the population is a subspace of the solution space.
3 Design fitness function: The population of each chromosome is decoded into the fitness function to be suitable for the computer, and then, the numerical is calculated.
4 Select: Select the best individual reproductive fitness of the next generation based on size; the higher the fitness, the greater the probability of selection.
5 Cross: The same position of the two randomly selected individuals for the next generation of breeding, the implementation of the exchange on the selected position.
6 Variation: On the one string, gene mutation probability by flipping.
7 Repeat from step 4 until you meet certain performance indicators, hereditary algebra, or regulations.

3 APPLICATION OF GENETIC ALGORITHM IN IMAGE PROCESSING

3.1 *Image restoration based on genetic algorithm*

Image restoration is used to restore a degenerate (or deterioration) image back to its original appearance. It is an important branch of digital image processing. Currently, we have proposed many effective image recovery methods, such as inverse filtering, Wiener filtering method, singular value decomposition pseudo inverse method, and maximum entropy restoration method. Because the function of image degradation is either caused or not caused by the expression of an unknown cause, to make the method face more constraints is too much of a computation problem. Since it is difficult to determine the degradation function, it is limiting the effectiveness of its practical applications.

GA is used to restore the gray image, generally a two-dimensional matrix chromosome encoding gray-scale value of each pixel element. That is, a chromosome represents an image, and each pixel corresponds to a gene, using natural number coding. Each individual's fitness function is shown in Equation 1:

$$F(f_i) = \left\| g - h * f_i \right\|^2 \quad (1)$$

where f_i is representative of the individual i speculating the restored image, g is the observed degradation image, and h is the degradation process. The larger the value, the better the individual functions. In the cross-operation, we generally use the window cross, which selects the same size window in the paternal chromosome matrix to be the exchange. Mutation using the average replacement needs within a small range near a gene variation value.

GA-based image restoration methods, breaking the original theory, and its open architecture are easy to integrate with other methods, such as the combination of fuzzy logic and fuzzy GA and so on. The use of GA to restore the image not only overcomes the effect of noise but also makes the image smoother, the edges without fringe effect, and the visual effect good. Powerful global search capability is a genetic algorithm image restoration method that is effective the main reason.

3.2 *Image enhancement based on genetic algorithm*

Image enhancement is intended to make the image clearer or to emphasize certain features. This can improve the visual effect of the image or facilitate additional processing on the image. Image enhancement techniques mainly include the frequency domain method and the spatial domain method. Frequency domain method is a transformation of the original image, and the transform domain is processed to achieve enhanced purposes. The spatial domain method is image processing, including direct gray-scale transformation, histogram equalization, filtering, and so on.

GA image enhancement technology implementation process is actually a process of finding optimal control parameters or suboptimal solutions. Therefore, we must first select a parameter model.

Munteanu proposed an automatic local image to enhance technology. In order to reduce the amount, we calculate the histogram equalization method, using GA to search for four parameters of its improved algorithm. We use an objective evaluation of the fitness function, without human intervention. The GA is applied as PCA and mutation improvement strategy. Experimental results show that this method achieves automatic, robust, and good results.

GA for image enhancement can obtain a better effect but it takes a longer time to consider its image optimization. We may consider using a parallel GA, which is a subject worthy of study.

3.3 *Genetic algorithm in image reconstruction*

Image reconstruction and image information are carried out by the data in some way for the observed recovery of the original image obtained in the process. Different methods are used to obtain the target projection data and to determine the method of image reconstruction problem solving learning. Reconstruction generated by this methodology consists of mainly the algebraic method, iterative method, and Fourier transform method.

GA used in image reconstruction is mainly used to solve the restored image projection data with noise. The chromosome is defined with the help of voxel values. The initial population of randomly selected chromosomes [1,255] value range. The fitness function is shown in Equation 2:

$$E_{Ray} = \frac{1}{n_x n_y n_z} \sum_{i,j,k=1}^{n_x n_y n_z} \sum_{\theta=1}^{\Theta} \sqrt{(Z_{ijk\theta} - S_{ijk\theta})^2} \quad (2)$$

Among them, i, j, and k were measured target x, y, and z coordinates. The angular position of the image forming apparatus is expressed as θ; n_x, n_y, and n_z represent the measured three-dimensional matrix size targets. Ray sum $S_{ijk\theta}$ is calculated according to the following Equation 3:

$$If \quad voxel \quad V_{ijk} \in Ray \rightarrow S_{ijk\theta} = S_{ijk\theta} + vV_{ijk}$$
$$else \quad S_{ijk\theta} = S_{ijk\theta} \quad (3)$$

In order to increase the search speed and parallel genetic algorithms to be achieved for the noise test data set, the effect is particularly good.

Susumu proposed a genetic relaxation iterative Fourier transform algorithm. This is used to solve the problems of stagnation during image reconstruction, namely the impact of the effect of the reconstructed image selected by the initial set.

The GA that is applied to the reconstructed image is an ongoing exploration of the topic. Currently, the method given in this article is a useful attempt, but there are some problems. For example, speed being generally slow to rebuild the image of the edge is unclear. This needs to be further improved.

3.4 Genetic algorithms in content-based image retrieval

Content-based image retrieval method is based on the image containing color, texture, shape, and space relationship information based on an object. It does this by establishing a feature vector of the image, and as an index of the image for image retrieval technologies. Its effects, image retrieval feature vector encoding method, and the specific image retrieval methods are closely related.

The user only for fitness evaluation, the searching process does not participate in the individual, easy to cause the user fatigue and low speed of convergence. The document presents a visual IGA model (VIGA). The user use the whole search process with a pilot signal, and the subjective emotion direction is toward the development of the genetic process user. The experimental results show that VIGA has a faster convergence speed than the general IGA, and it is very effective to alleviate user fatigue.

GA in the content-based image retrieval is mainly used to extract feature vectors of the image. We use GA global optimization features, dig out the most essential image information, and, ultimately, improve the retrieval accuracy. Simultaneously, the GA method is used for the visualization of high fitness regions of the individual, in order to accelerate the convergence speed of genetic algorithm, and it has a wide application foreground in the perceptual retrieval.

3.5 Image matching based on genetic algorithm

Many image matching algorithms are generally divided into gray scale, correlation based, and feature based, based on a template and a matching transform domain. Among them, the template that matches is the most common method. The method has high precision, the algorithm is stable, but the large amount of calculation is difficult for the high real-time demands of the occasion. GA is introduced to solve the problems related to matching the speed by reducing the number of search locations to reduce the total amount of related computations. Simultaneously, it is in terms of matching that accuracy also achieved good results,

and the algorithm is more stable. Application of GA chromosome is generally related to matching 16 long bits, which are used to indicate the position of the parameter associated with 8-bit matching. The fitness function can be used to similarly measure calculation. Image matching techniques introduced GA related mainly to solve problems related to matching speed, GA by reducing the number of search positions to reduce the total calculation amount dependent. Meanwhile, in terms of matching, accuracy is also achieved along with good results, and the algorithm is more stable.

3.6 Image segmentation based on genetic algorithm

Image segmentation is an important issue in image processing, and computer vision research is a classic problem. Image segmentation is the target and background separation, providing preparation for subsequent image classification, recognition, and so on. Commonly used methods include thresholding segmentation, edge detection, and region tracking method. The threshold method is the most commonly used method. Currently, a large number of threshold selection methods, such as Otsu being among the best histogram entropy method, maintain the minimum error threshold and keep the method of moments.

GA-based image segmentation technology leverages the parallel processing capabilities of GA and uneven nature of the search space, with no special requirements and other characteristics. GA with traditional thresholding segmentation algorithm can speed up the rate of the combination and, to some extent, abandon the traditional method of application; good segmentation results are also achieved. In addition, it can further improve genetic manipulation to improve the search performance of GA. The results of the study show that, relative to other application areas of the image, the application of GA in image segmentation field is the most mature, is effect is remarkable, and it has great potential.

3.7 Fractal image compression based on genetic algorithm

The main theoretical basis of fractal image compression is iterated function systems theory and collage theorem. It is going to solve the problem is when the compressed image as attract midnight how to get IFS parameters. The application of GA in fractal image compression improves the compression ratio and compression accuracy. Since a high compression ratio has greatly improved signal-to-noise ratio, it can also be used for low-bit rate image compression. Moreover, GA has the characteristics of parallel computing and can be reduced in computing time fractals; thus, we quickly need to find the optimal solution.

GA is applied to fractal image compression, improved compression ratio, and compression accuracy. Since GA at high compression ratio has greatly improved signal-to-noise ratio, it can be used for low-bit rate image compression. Moreover, GA has the characteristics of parallel computing, which can reduce the calculation time of fractal and quickly find the optimal solution. However, there are many control parameters of the test, and most of them are dependent on the experience obtained. So, how to control these parameters adaptively to further improve the compression ratio and decoding quality still needs to be studied and explored. Because of the good characteristics of the GA, the prospect will be very broad in combination of fractal.

4 CONCLUSION

In summary, GA is an instructive, robust randomization technique that is ideal for searching the optimal value within such a huge amount of space for image processing. In addition, because of its group operation nature, GA has very good parallel distributed processing characteristics. In particular, each individual GA in the fitness calculation can exist with each other without any independent communication, so the parallel efficiency is very high. It has achieved good results in the field of image processing and has attracted more and more attention. But it also is exposed to many shortcomings and deficiencies in the theory and application of technology. First, although the GA is more robust than other traditional search methods, it is better at global search and local search capability is inadequate. The study found that GA can very quickly reach 90 percent of the optimal solution, but to achieve a true optimal solution we will need to spend a very long time. Therefore, we must take into account the quality of the two indicators as well as the convergence rate reconciliation. In addition to the improvement of the basic theory and methods, we should adopt a neural network, wavelet analysis combined with annealing simulation, and other methods or strategies to improve the local search ability of GA. So, the performance of the algorithm is improved. Second, its convergence speed is decreased

in the variable, ranging from large or no time given range. Moreover, GA selects parameters no quantitative methods most often rely on experience to get. The fitness calibration has a variety of ways, without a simple, generic method, which is not conducive to the use of the GA. Therefore, the mathematical basis of the GA is extremely weak, especially the lack of a profound and universal significance of the theoretical analysis. In addition, GA's "premature" phenomenon quickly converges with a local optimal solution rather than the problem finding it extremely difficult to deal with a global optimal solution. These are the key problems in GA that so far have been the most difficult to solve and that need to be solved urgently. Otherwise, they will limit the further application of GA.

The application of the majority of GA in image processing in the theory of the simulation phase, applied in the actual system is less. How to aim at the aspect of image processing to select the GA structure used in image analysis and processing or the appropriate parameters requires further research content.

With an in-depth theoretical study, to be sure, GA with its unique features enables the algorithm used in image processing problems to become increasingly widespread. Simultaneously, the emergence of a wide range of mathematical methods and powerful computer simulation tools will make considerable progress in GA research. It attempts to make perfect the role of GA in image processing.

REFERENCES

Zhang Yu-jin. Image Processing and Analysis[M]. Beijing: TsinghHa University Press, 1999.

LinLei, Wang Xiaolong, Liu Jiafeng, Research on optimization of a hand written Chinese character recognition system based on genetic algorithm[J] Journal of Computer Research & Development, 2001, 38(6):658–661.

Saupe D, Ruhl M. Evolutionary factal imagecompressio [A]. In Proceedings of IEEE International Conference onImage Processing[C], Lausanne, Switzerland 2006, 1:129–132.

Munteanu C, Rosa A. Towards automatic image enhancement using genetc algorithms [A]. In:Proceedings of the 2000 Congress on Evolutionary Computation CEC00[C], California, USA, 2000:1535–1542.

Lee SU, Chung SY, Park R H. Acomparative performance study of several global thresholding techniques for segmentation[J]. Computer Vision, Graphics and Image Process, 2010, 52(2):191–199.

Information, Computer and Application Engineering – Liu, Sung & Yao (Eds)
© 2015 Taylor & Francis Group, London, ISBN 978-1-138-02717-6

Context-aware ubiquitous learning environment framework: Under the capacity model of attention

D. Fu
Yunnan Minzu University, Kunming, P. R. China

Q. T. Liu
Central China Normal University, Wuhan, P. R. China

ABSTRACT: Most of known research focuses on the technical means rather than the psychological characteristics of the learners in constructing a Ubiquitous Learning Environment (ULE) in the field of ubiquitous learning. Learners probably find it difficult to concentrate and choose appropriate learning contents because of the large amount of open learning resources in ULE. Motivated by the Capability Model of Attention (CMA), which is a model of the allocation of capacity, a novel framework of the Context-aware Ubiquitous Learning Environment (CULE) that gives more consideration to learners' attention is proposed in this article. Specifically, the framework of CULE is designed by analyzing the components related to the allocation of attention according to the CMA. The framework presumably contributes to design and develops a more efficient CULE, because the learner's psychological characteristics are considered.

KEYWORDS: Ubiquitous Learning; Capacity Model of Attention; Context-aware

1 INTRODUCTION

Recently, ubiquitous learning (UL) has emerged as one of the most popular research issues. UL originates from the ideas and technology of ubiquitous computing. The basic technology of UL includes embedded computing, sensing technology, detection technology, and so on. The physical facilities in the real environment comprise sensing devices that are capable of computing, as well as enabling people, information, and the environment to be in harmony.

People can obtain useful information quickly and easily in a ubiquitous learning environment (ULE) while computers are out of their sight (Weiser, 1991). So learners can obtain learning information at any time, any place, and focus on the learning tasks with less consciousness about the technical operation in ULE.

Context-aware ubiquitous learning environment (CULE) attracts much attention in the field of ubiquitous computing. Various learning equipment is interoperable in the learning environment, which results in highly humanized and flexible interaction for learners. Context awareness is one of the fundamental requirements for achieving user-oriented ubiquity. An approach of middleware solution expediting context awareness is presented by some researchers (Ngo et al, 2004). Moreover, Na-songkhla (Na-songkhla, 2011) designs a set of interactive media that can support social awareness of ubiquitous learning. Learning activities such as game-based learning, virtual learning, and digital storytelling can be supported based on the media.

In addition, a CUL platform that uses low-cost cell phones with embedded cameras and internet service to support UL is proposed for enabling users to control their learning activities in the ULE (Hwang et al, 2011). Chen et al (Chen et al, 2012) develop a context-aware ubiquitous learning system (CULS) to improve the learning strategies and acquire knowledge efficiently based on radio-frequency identification, wireless network, and so on. A context-aware application composition system that aims at helping learners utilize the resources dynamically and adapting to the situation within the contexts is proposed (Davidyuk et al, 2011).

To improve the degree of the users' unconsciousness during learning, a learning framework and a blended navigation algorithm for developing a navigation support mechanism is introduced (Chiou et al, 2014). In order to provide the content to the learners appropriately, Temdee (Temdee, 2014) develop a ULE that consists of several leaning objects having multi-agent architecture. These agents include personal agent, content agent, and representation agent. In addition, a ubiquitous learning model is developed (Atifet et al, 2014). The model enables the physical structures to unleash their instructional value and provide a personalized learning agenda within a pervasive smart campus environment.

Most of recent research focuses on improving various technical means (e.g., embedded computing and sensing technology) and implementing related practice projects, rather than paying attention to the learners' psychological characteristics in constructing a ULE. The learner's needs and cognitive characteristics are the important factors that should be considered when any learning environment is constructed. Specially, users learn unconsciously in ULE where a large amount of open and freely information exists. It is assumed that learners may find it difficult to concentrate and choose appropriate learning contents. This involves cognitive psychological factor that is referred to as ATTENTION.

Motivated by earlier observations, this study aims at contributing to present a CULE framework from the perspective of cognitive psychology and particularly focusing on the Capacity Model of Attention (CMA) (Kahneman, 1973).

2 THEORETICAL BACKGROUND: CAPACITY MODEL OF ATTENTION

Attention is the central concept of information processing theory in modern cognitive psychology. It is a problem in the field of psychology and aims at exploring how a person's attention is distributed because the attention is shared by many events occurring simultaneously in a complicated environment.

Several models are presented to explore the selective mechanism of attention. For example, Filter Model (Broadbent, 1958) and Filter-Attenuation Model (Treisman, 1960, 1964) called "perception selection model" consider the location of the filter is after the response. However, Response Selection Model (Deutsch et al, 1963) argues that the filter is after the perception.

Capacity Model of Attention (CMA) (Kahneman, 1973) offers a policy of allocating attention according to the capacity theory (Moray, 1967). This allocation policy is controlled by four factors, including enduring dispositions, momentary intentions, evaluation of demands of capacity, and the effects of arousal. Enduring dispositions are related to the involuntary attention. Momentary intentions involve the current task and purpose. Evaluation of demands is a governor system that guides the supply of capacity (or effort) when the activities are selected by the allocation policy. Arousal that forms miscellaneous determinants is closely related to available capacity.

Principle of attention is used to improve the effectiveness of learning in the field of education. Research is carried out by Couperus (Couperus, 2009) to unveil how selection attention at sensory levels is modulated by implicit learning of perceptual load conditions. Inspired by the premotor theory of visual attention, the model for integrative learning of proactive visual attention and sensory-motor control is proposed by Jeong et al (Jeong et al, 2012).

In this article, the components in CULE related to the allocation of attention are analyzed, and a framework of CULE is proposed according to the model of the allocation of capacity.

3 CONSTRUCTION OF CULE

3.1 *Analysis of related components*

Learning environment includes physical environment and social psychological environment, which are the external conditions for learners. Physical environment means physical devices, which includes computer, handheld device, learning tools, learning resources, and so on. Social psychological environment consists of conception of learning, study ethos, belonging, and so on (see the 1st column of Table 1).

However, the psychological characteristics of learners should be emphasized in constructing a learner-centered learning environment. In this article, the learner's attention and the allocation of attention are discussed when a context-aware learning environment is developed. According to the theory of CMA, allocation of attention is affected by enduring dispositions, momentary intentions, evaluation of demands on capacity, and effects of arousal. The effect of distracter inhibition (LaBerge, 2002) is a factor that is also considered in this article.

The mentioned factors that affect the allocation of attention corresponding to the components of the learning environment are shown in the 2nd column of table 1.

Generally, CULE provides a learning resource that adapts to the learners' needs anywhere and at anytime. Based on the discussion about the components of the learning environment and the allocation of attention, the components in CULE related the allocation of attention are presented in the 3rd column of Table 1. The components include conception of life-long learning, physical environment, design of the tools, stimulation of the learners' motivation, learning resources, and support from human resources.

Conception of life-long learning. According to the theory of life-long learning, it is necessary that learning society and learning organizations be constructed. So the individual needs and wishes are creeping into the learner's consciousness for a long time, which should, in a degree, arouse the learner's attention when they are in the ULE.

Physical environment. Physical environment includes various devices and learning tools that have

Table 1. Components in CULE related the allocation of attention.

Components of the learning environment		Allocation of attention	Components of CULE related to the allocation of attention
Conception of learning		Enduring dispositions, individual needs, and wishes in long term	Conception of life-long learning: Learning society, leaning organization
Physical devices	Physical equipment	Arousal, Involuntary attention, Evaluation of demands on capacity	Physical environment: various devices that have the capacity of ubiquitous computing
	Learning tools	Involuntary attention, Evaluation of demands on capacity, Distracter inhibition	Design of the tools: Selectivity, Controllability, Convenience, Support for learning activities
Learning resources	learning target	Momentary intentions	Stimulation of the learners' motivation: The current needs of learners, learning attitude
	Learning content, Learning activity	Evaluation of demands on capacity, Distracter inhibition	Learning resources: design, development, management, and evaluation
Human resources		Distracter inhibition	Support from human resources: learning companion and others

the capacity of ubiquitous computing in a CULE. Learners will obtain the learning resources anywhere and at anytime, which will contribute to the learner's concentration on the learning activities because the demands on capacity are less.

Design of the tools. It is very important whether the design of the learning tools and other relevant hardware are available or not. A selective, controllable, and convenient learning tool would inhibit the distracters in CULE. The learner would evaluate less demand on capacity and concentrate on the learning tasks.

Stimulation of the learners' motivation. Contrary to the long-term needs and wishes of the learner, the current needs and target would directly impact the learning motivation, learning attitude, and learning consciousness. The design of the learning target of the learning resource should adapt to the learners' goal.

Learning resources or materials. The design of learning content and learning activity would determine whether the learning is efficient in CULE. A good design enables learners to spend less effort on the things unrelated to learning. Learners assess that the learning task costs less demands of capacity, so they choose it to learn. Moreover, interference will be eliminated easily.

Support from human resources. The development of the learner's sociality is important in the CULE. Support from others such as a learning companion will enable learners to gain a sense of belonging and pay attention to their learning tasks.

3.2 Framework of CULE

The framework of CULE based on CMA is shown in figure 1. The figure displays the relationship between the components in CULE and the elements impacting the allocation of a learner's attention. The relationship presumably contributes to design and develops more efficient CULE, because the learner's psychological characteristics are considered, and, consequently, the ubiquitous learning effect is improved.

The learner's action is guided by their consciousness. For example, the conception of life-long learning can impact a person whether pay efforts to learn new knowledge continually in his life. Enduring dispositions are related to the individual needs and

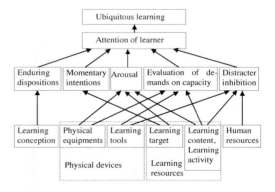

Figure 1. Framework of CULE based on CMA.

wishes in the long term. In other words, the learner's idea about learning determines what and how they should learn. We should continue to carry forward the idea of life-long learning and build a learning community and a learning organization.

Physical devices such as physical equipment and learning tools supporting the learning with the capacity of ubiquitous computing are the infrastructure of the CULE. Their design is related to the evaluation of demands on capacity and the degree of the arousal that impacts the learners to allocate their attention. Consequently, distracter inhibition would be better when the physical devices are selectable, controllable, and convenient. For example, the size of the screen of the handheld devices should be designed larger so that the learner's cognitive load is alleviated. Moreover, the embedded devices should be everywhere and the link is efficient in CULE.

Learning resource loading the learning target, learning content, learning activity, and so on are important components in a learning environment. Momentary intentions involve the learner's current needs, that is, learning target, which could promote the learning motivation and inspire the learner to pay attention to the learning contents and learning activities. In addition, good design of the learning content and activities will also improve the learner's performance when more consideration is given to the learner's attention, such as evaluation of demands on capacity in CULE. For instance, one of the available strategies is to develop a micro-course in which the learning content is divided into small units so that the learner can select the unit they needs and learn it in their fragment time. According to Kahneman (Kahneman, 1973), high arousal causes a person to pay attention to the dominant aspects of the situation. Reasonable utilization of the multimedia in the design of the learning resource would help learners focus on the learning content and learning activities. Consequently, the distracter inhibition would increase.

Moreover, the human resources supporting the learner's cognitive activity from the learning companions are necessary. Interaction between the learners would also contribute to the distracter inhibition, because the learners would not feel lonely and would concentrate on the learning tasks. So a smart human net is needed for helping learners communicate with each other conveniently.

4 CONCLUSION AND FUTURE WORK

This article aims at presenting a novel framework of CULE based on the CMA. Specifically, factors including enduring dispositions, momentary intentions, evaluation of demands on capacity, effects of arousal, and distracter inhibition are considered when CULE is designed. In the future, more experiments should be done for showing the effectiveness of the framework. Moreover, there is also a need to more deeply analyze the learners' attention besides other psychological characteristics when a learning environment is designed.

ACKNOWLEDGMENT

This work is supported by a grant from the Science Research Fund of Education Department of Yunnan Province (NO. 2014Y273).

REFERENCES

Atif, Y., Mathew, S. S. & Lakas, A. 2014. Building a smart campus to support ubiquitous learning. *Journal of Ambient Intelligence and Humanized Computing.* Published online: 27 Feb 2014.

Broadbent, D. E. 1958. *Perception and communication.* London: Pergamon Press Ltd.

Chen, C. C. & Huang, T. C. 2012. Learning in a u-Museum: Developing a Context-Aware Ubiquitous Learning Environment. *Computers & Education* 59(3): 873–883.

Chiou, C., Tseng, J. & Hsu, T. 2014. Navigation Mechanism in Blended Context-Aware Ubiquitous Learning Environment. *Multimedia and Ubiquitous Engineering. Lecture Notes in Electrical Engineering* 308: 199–204. Berlin: springer.

Couperus, J. W. 2009. Implicit learning modulates selective attention at sensory levels of perceptual processing. *Attention, Perception, & Psychophysics* 71(2): 342–351.

Davidyuk, O., Gilman, E., Milara, I. S., Mäkipelto, J., Pyykkönen, M. & Riekki, J. 2011. iCompose: context-aware physical user interface for application composition. *Central European Journal of Computer Science* 1(4): 442–465.

Deutsch, J. A. & Deutsch, D. 1963. Attention: Some theoretical considerations. *Psychological Review* 70: 80–90.

Hwang, G., Wu, C., Tseng, J. & Huang, I. 2011. Development of a ubiquitous learning platform based on a real-time help-seeking mechanism. *British Journal of Educational Technology* 42(6): 992–1002.

Jeong, S., Arie, H., Lee, M. & Tani, J. 2012. Neuro-robotics study on integrative learning of proactive visual attention and motor behaviors. *Cognitive Neurodynamics* 6(1): 43–59.

Kahneman, D. 1973. *Attention and Effort.* Englewood Cliffs, New Jersey: Prentice-Hall.

LaBerge, D. 2002. Attentional control: brief and prolonged. *Psychological Research* 66(4): 220–233.

Moray, N. 1967. Where is capacity limited? A survey and a model. *Acta Psychological* 27: 84–92.

Na-songkhla, J. 2011. An Effect of Interactive Media in a Social Awareness Ubiquitous Learning Community. *The International Conference on Lifelong*

Learning 2011 (ICLLL2011). Kualalumpur, Malaysia, Nov 14–15, 2011.

Ngo, H., Shehzad, A., Liaquat, S., Riaz, M. & Lee, S. 2004. Developing Context-Aware Ubiquitous Computing Systems with a Unified Middleware Framework. *Embedded and Ubiquitous Computing. Proc. intern. EUC 2004., Aizu-Wakamatsu, Japan, August 25–27 2004.* 3207: 672–681. Berlin: Springer.

Temdee, P. 2014. Ubiquitous Learning Environment: Smart Learning Platform with Multi-Agent Architecture. *Wireless Personal Communications* 76(3): 627–641.

Treisman, A. M. 1960. Contextual cues in selective listening. *Quarterly Journal of Experimental Psychology* 12: 242–248.

Treisman, A. M. 1964. Verbal cues, language and meaning in selective attention. *American Journal of Psychology* 77: 206–219.

Weiser, M. 1991. The Computer in the 21st Century. *Scientific American* 265(3), 94–104.

15

A simplified method of clutter covariance matrix of 3DT-SAP STAP algorithm based on distributed memory system

Xiao Qing Cheng, Li Ning Gao & Wen Li
School of Information and Electronics, Beijing Institute of Technology, Beijing, China

ABSTRACT: STAP (Space Time Adaptive Processing) is the key technology to suppress clutter in an airborne phased array radar. The research of STAP mainly focuses on two aspects: One is to research the Reduced-Dimension STAP algorithm, to reduce the dimension of the STAP processor under an acceptable clutter suppression performance and the other is to research the parallel processing system with high efficiency, to accelerate the processing speed with the same hardware. The research shows that 3DT-SAP is the most suitable algorithm to be realized in practice, and the multi-DSP environment based on distributed storage structure is more suitable for 3DT-SAP algorithm mapping researching than other environments. In this research, the modified clutter covariance matrix algorithm can significantly improve efficiency, effectively shorten the time consumed, and does not affect other properties.

KEYWORDS: 3DT-SAP; distributed storage structure; clutter covariance matrix algorithm

1 INTRODUCTION

In 1973, the famous American scientists, Professor Brennan and Professor Reed, proposed STAP (Space Time Adaptive Processing). Our researchers began carrying out studies on STAP at an early stage. In 1991, Bao Zheng proposed space-after-time adaptive cascade processing algorithm (T$A filtering). Wu Renbiao proposed two-dimensional sub-array Capon algorithm, and this method could reduce both the amount of computation and the robustness of error. Based on the principle of sub-space conversion, in 1993, Liu Qingguang proposed joint channel processing method and this method is not sensitive to various errors. Professor Wang Yongliang of Air Force Radar Academy systematically summarized previous work and proposed STAP unified theory and implementation model creatively. Professor Wang Yongliang's achievements make a lot of sense, because researchers could research STAP based on a unified model. More about STAP algorithm theory is included in his monograph Space-Time Adaptive Signal Processing. This paper is based on the research of mapping method of 3DT-SAP STAP algorithm in a distributed storage system, in which clutter covariance matrix algorithm is modified and this modified algorithm could reduce the amount of computation and time consumption.

2 ESTIMATION OF CALCULATION IN 3DT-SAP MAPPING METHOD BASED ON DISTRIBUTED STORAGE SYSTEM

Assuming the number of spatial channels is N, the number of time-domain pulses is K and the number of sampling points of the total distance is L, which are shown in Figure 1.

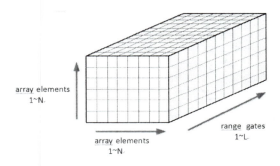

Figure 1. Three-dimensional schematic of the input data of STAP processing.

We divide these L range gates into M parts, and each part has Q range gates ($L=M*Q$ or using overlapping manner). In every Doppler detection channel, we use P samples ($P<Q$) in every distance part to estimate covariance matrix.

The process can be divided into 7 parts:

1 N array elements, L pulses of K PRTs.
2 P range gates, N array elements of K-point FFT weighting, and every operation has $\frac{K}{2}\log_2 K$ complex multiplications. So this part needs $P \times 3N \times 3N$ complex multiplications.
3 In the average covariance matrix operations of P range gates, each has $3N \times 3N$ complex additions. This part needs $P \times 3N \times 3N$ complex additions.
4 K inversions of $3N \times 3N$ dimensional covariance matrix: Each has $O(3^3 N^3)$ complex multiplications. So this part needs $O(3^3 N^3)$ complex multiplications.
5 K multiplications of $3N \times 3N$ dimensional matrix and $3N \times 1$ steering vector: Each has $3N \times 3N$ complex multiplications. This part needs $K \times 3N \times 3N$ complex multiplications.
6 Q range gates, K multiplications of $1 \times 3N$ optimal vector with $3N \times 1$ dimensional vector data: Each has $3N$ complex multiplications. This part needs $Q \times K \times 3N$ complex multiplications.

In summary, in a range gate, the computations of target detection related to all these Doppler channels are as follows. The total complex multiplications are $P \times N \times \frac{K}{2}\log_2 K + P \times 3N \times 3N + K \times O(3^3 N^3) + K \times 3N \times 3N + Q \times K \times 3N$. The total complex additions are $P \times 3N \times 3N$. In addition, the computation of part(3) could be significantly reduced by using modified clutter covariance matrix algorithm.

3 MODIFIED CLUTTER COVARIANCE MATRIX ALGORITHM

When computing the clutter covariance matrix of each range gate, the traditional method involves arranging the data of three adjacent Doppler channels on a fixed range gate in a column vector and calculating them. For example, we detect the second Doppler channel (the first Doppler channel does not need to be detected). It is shown as follows:

$$X(2) = \begin{bmatrix} X_1 \\ X_2 \\ X_3 \end{bmatrix}$$

In the earlier formula, X_1, X_2, X_3 are column vectors with the length of N; is the second Doppler channel's detection snapshot, which is a vector with the length of 3N. Then, the covariance matrix of this Doppler channel is calculated on the range gate. It is shown as follows:

$$R(2) = X(2)X(2)^H = \begin{bmatrix} X_1 \\ X_2 \\ X_3 \end{bmatrix} [X_1^H \quad X_2^H \quad X_3^H]$$

$$= \begin{bmatrix} X_1 X_1^H & X_1 X_2^H & X_1 X_3^H \\ X_2 X_1^H & X_2 X_2^H & X_2 X_3^H \\ X_3 X_1^H & X_3 X_2^H & X_3 X_3^H \end{bmatrix}$$

In the earlier formula, because of the covariance matrix R's Hermite symmetry property, we could calculate $X_1 X_1^H$, $X_2 X_2^H$, $X_3 X_3^H$, $X_1 X_2^H$, $X_1 X_3^H$, $X_2 X_3^H$ and transpose them; then, we can get the entire values of matrix R.

If these matrixes are multiplied without transposition to calculate the matrix R, this computation needs $3N \times 3N = 9N^2$ complex multiplications. And the former method with transposition needs $6N^2$ complex multiplications, which reduce the amount of computation by 2/3.

We detect the third Doppler channel, and it is shown as follows.

$$R(3) = X(3)X(3)^H = \begin{bmatrix} X_2 \\ X_3 \\ X_4 \end{bmatrix} [X_2^H \quad X_3^H \quad X_4^H]$$

$$= \begin{bmatrix} X_2 X_2^H & X_2 X_3^H & X_2 X_4^H \\ X_3 X_2^H & X_3 X_3^H & X_3 X_4^H \\ X_4 X_2^H & X_4 X_3^H & X_4 X_4^H \end{bmatrix}$$

Observing the earlier formula, we have already calculated R(2) by transposition, and the new computations are $X_2 X_4^H, X_3 X_4^H, X_4 X_4^H$ which needs $3N^2$ complex multiplication. So, if we calculate all K Doppler channels' covariance matrixes of the current range gate, we need to calculate the following matrixes:

$X_1 X_1^H, \quad X_1 X_2^H, \quad X_1 X_3^H$

$X_2 X_2^H, \quad X_2 X_3^H, \quad X_2 X_4^H$

$\cdots\cdots$

$X_k X_k^H, \quad X_k X_{k+1}^H, \quad X_k X_{k+2}^H$

$\cdots\cdots$

$X_{K-2} X_{K-2}^H, \quad X_{K-2} X_{K-1}^H, \quad X_{K-2} X_K^H$

$X_{K-1} X_{K-1}^H, \quad X_{K-1} X_K^H$

$X_K X_K^H$

Table 1.

times	1	2	3	4	5	6	7	8	9	10
Time consumed	0.0814	0.0819	0.0830	0.0794	0.0804	0.0761	0.0814	0.0775	0.0815	0.0824
times	11	12	13	14	15	16	17	18	19	20
Time consumed	0.0797	0.0806	0.0815	0.0813	0.0817	0.0798	0.0794	0.0797	0.0799	0.0817

Table 2.

times	1	2	3	4	5	6	7	8	9	10
Time consumed	0.0654	0.0620	0.0622	0.0630	0.0647	0.0651	0.0621	0.0638	0.0638	0.0616
Times	11	12	13	14	15	16	17	18	19	20
Time consumed	0.0618	0.0645	0.0631	0.0645	0.0630	0.0640	0.0645	0.0627	0.0630	0.0636

All these computations are $3(K-1)N^2$ complex multiplications, and if we use the traditional method, it needs $9(K-2)N^2$ complex multiplications. So, the computations of a range gate that are reduced are $(6K-15)N^2$ complex multiplications.

In consideration of actual application, both the number of array elements N and the number of range gates needed to be detected are large, so this method could reduce quite a lot of calculation.

4 MATLAB SIMULATION OF THE CLUTTER COVARIANCE MODIFIED ALGORITHM

4.1 *Comparison of time consumption of traditional and modified method*

We use MATLAB to simulate the clutter covariance modified algorithm.

Suppose there are K=16 Doppler channels; X1, X2...X16 are column vectors with the length of N=128. We choose chirp signal as test data, and calculate R(2), R(3), R(4)...R(14), R(15). We calculate 20 times by traditional method and the time consumed is shown as Table 1. We calculate 20 times by modified method and the time consumed is shown as Table 2 (the unit of time consumed is second).

The comparison of these two methods is shown in Figure 2.

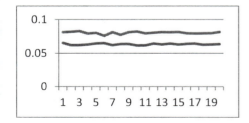

Figure 2.

4.2 *Output simulation waves of the traditional and modified method*

The comparison of 3DT-SAP STAP algorithm's Distance-Doppler wave by the traditional and modified method is shown as follows:

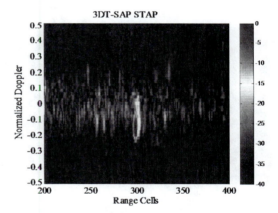

Traditional method

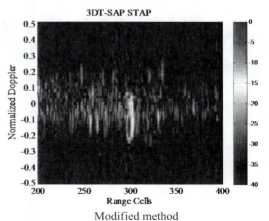

Modified method

Figure 3.

The comparison of 3DT-SAP STAP algorithm's space-time two-dimensional beam verification by the two methods is shown as follows:

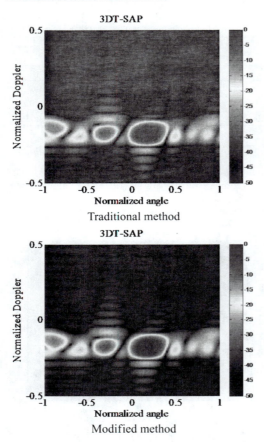

Traditional method

Modified method

Figure 4.

On comparing these waves of 3DT-SAP STAP algorithm by the traditional and modified method, we find that the modified method has no influence on 3DT-SAP STAP algorithm.

5 CONCLUSION

This article introduces a modified method of clutter covariance matrix algorithm in 3DT-SAP STAP algorithm, which is based on distributed storage system. This modified method could significantly reduce computation and time consumed and increase computing speed, which makes sense in practical application.

REFERENCES

[1] Bao Zheng, Liao Guisheng, Wu Renbiao, Zhang Yuhong, Wang Yongliang. Time-Space two-dimensional adaptive filtering of phased array airborne radar during clutter suppression[J]. Journal of Electronics, 1993, 21(9):1–7.
[2] Wu Renbiao, Bao Zheng, Zhang Yuhong. Subarray-level Adaptive Spatial- Temporal Processing for Airborne Radars. Proc. of ICSP, Beijing, Oct. 1993: 391–395.
[3] Liu Qinghuang, Peng Yingning etc. Joint channel conversion method of airborne radar adaptive clutter suppression. Journal of Electronics, 1994, 22(6):1–9.
[4] Wang Yongliang, Peng yingning. Space-Time adaptive signal processing. Beijing: Tsinghua University Press, 2000.

Design and implementation of the fixed-point BAQ decompression algorithm based on FPGA

Wen Li, Liang Chen & Xiao Qing Cheng
School of Information and Electronics, Beijing Institute of Technology, Beijing, China

ABSTRACT: As space-borne SAR (Synthetic Aperture Radar) echo data received by the stationary ground equipment through feedback channel have been compressed based on the BAQ (Block Adaptive Quantization) compression algorithm, in the reverse process, decompression should be implemented beforehand. Based on the fact that quantization loss is inevitable, we introduce an improved method, the fixed-point BAQ decompression algorithm and we implement it in XILINX Virtex5 XC5VLX110T FPGA. It proves that SAR echo data can be restored correctly without making sacrifices to the resource utilization, and the whole system can work stably at high clock frequency.

KEYWORDS: SAR; BAQ; FPGA

1   INTRODUCTION

SAR has priority over optical remote sensor, because SAR can work all day under any kind of weather and it is capable of penetrating obstacles to make images, which therefore makes SAR imaging technology play an important role in a range of scientific fields.

The quantity of remote sensing data collected onboard is so huge that the SAR system cannot make real-time processing and, consequently, it cannot work all day [1].

Something should be done to solve this problem. If we lower PRF (pulse recurrence frequency), we may make azimuth ambiguity worse. If we lower system bandwidth, we may reduce range resolution. If we lower the quantization bit in A/D converter, we may increase quantization noise [2].

Hence, it is more advisable and feasible to make the SAR system meet requirements of real-time performance by way of SAR echo data compression on the premise that the losses after compression are within an acceptable range. It has been proved time and again that BAQ algorithm is extraordinarily applicable in the SAR imaging field [3,4] since it was first used in the Magellan mission [5].

2   THEORY OF BAQ ALGORITHM

That the BAQ algorithm is widely used in the SAR imaging field is based on the following fact: SAR echo data meets the time-varying variance and zero-mean Gauss distribution in range and azimuth as a whole [6].

Through the division of large blocks of data into smaller ones, but not that small because certain quantities of data in a sub-block are necessary to determine the standard deviation, on the principle of a smaller dynamic range compared with that of the whole block [6], we may now assume that each divided block of data meets constant variance, zero-mean Gauss distribution, which makes it possible to compress separated block data with different variances adaptively based on least mean-square error rule. The procedure of BAQ algorithm is shown as follows [7]:

2.1   *Procedure of BAQ compression*

First, the whole block of original data is split into smaller parts with the same size in range and azimuth, the variance of which should be estimated, respectively.

Second, the separated data are normalized according to the variance, making the normalized data meet the unit variance, zero-mean Gauss distribution.

Third, the value of each data in a block is compared with prepared judgment levels so that the quantization level as well as corresponding code word is obtained, where the two types of levels mentioned earlier are determined by some criteria.

2.2 Procedure of BAQ decompression

First, withdraw the standard deviation and the coding word separately.

Second, multiply the two parameters we have in the first step.

Third, quantify the results of the multiplication to certain number of bits.

2.3 Determination of parameters for BAQ decompression

As we can see, it counts a great deal to make the estimation of the standard deviation within each block and the determination of both levels, the judgment level and the quantization level, in the BAQ algorithm.

The standard deviation can be obtained from the statistic mean value of the sample data in the look-up table, which can reduce the computational complexity to a great extent; in addition, the manner in which to derive the judgment level and the quantization level based on least mean-square error rule can be optimum[8].

Since there is no relation between the standard deviation and the mean value in Gauss distribution, we need to construct another distribution that can bridge the standard deviation and another mean value. It is not hard to derive the mapping relationship between the mean value of the absolute value |x| and the standard deviation of the original value x under the condition of the zero-mean Gauss distribution in x. The connection discussed earlier is given in equations set (1), where N is the number of points in a sub-block and x is the value of the SAR echo data. Because we can only obtain certain quantities of points, we replace continuous integral by the sum of a series of piecewise integral[1].

$$E\left[\|X\|\right]= \sum_{k=1}^{N} \frac{|x_k|}{\sqrt{2\pi}\sigma}\int_{x_k}^{x_k+1} \exp\left(-x^2/2\sigma^2\right)dx$$

$$E\left[\|X\|\right]= \sum_{k=1}^{N} |x_k| \tag{1}$$

The judgment level s and the quantization level y can be derived based on east mean-square error rule from equations set (2).

$$E\left[\varepsilon^2\right]= E\left[(x-y)^2\right]= \sum_{k=1}^{M} \int_{s_k}^{s_{k+1}}(x-y_k)^2 P(x)dx$$

$$\frac{\partial E\left[\varepsilon^2\right]}{\partial s_k}=0 \Rightarrow s_k = \frac{y_k+y_{k-1}}{2}$$

$$\frac{\partial E\left[\varepsilon^2\right]}{\partial y_k}=0 \Rightarrow y_k = \frac{\int_{s_k}^{s_{k+1}} xP(x)dx}{\int_{s_k}^{s_{k+1}} P(x)dx} \tag{2}$$

3 DESIGN AND IMPLEMENTATION OF MULTIPLE BAQ DECOMPRESSION MODES BASED ON FPGA

The physical implementation of the BAQ decompression algorithm mainly includes operations of multiplications and looking up tables. To improve the parallelism in this algorithm to some extent, we usually replicate resources, which requires that large block storage and huge number of multiplication units are available.

Block storage as large as 1.6Mb plus and as many as 330,000 logic units are used to set up distributed memory; when necessary, they are integrated in the XILINX VIRTEX 5 FPGA. Besides, fully functioning, configurable IP cores for multiplication and 64 high-frequency application-specific integrated digital signal processors named DSP48es are included in it. It is therefore available to implement the BAQ decompression algorithm in FPGA; meanwhile, it can save a large number of area resources and reduce the cost of equipment by integrating the BAQ decompression module and other logic control circuits altogether in the same FPGA[9].

3.4 Arithmetic Flow of BAQ decompression

Although there are different BAQ compression modes, of which the 8:3, 8:4, and 8:6 modes are the most commonly used, the process of BAQ decompression is almost the same. The arithmetic flow of BAQ decompression can be shown in figure 1.

Here, we take 8:3 BAQ decompression, for example, to specifically introduce the arithmetic process of

Figure 1. Arithmetic flow of BAQ decompression.

BAQ decompression implemented in the FPGA used in this article.

The data format of the three-bit data after decompression and repackage is shown in figure 2. This kind of data format is what the user finally receives.

The size of the BAQ compression block is 512 in range and 8 in azimuth. As for the maximum number of sampling points that is 20k in range, the whole block is separated into 40 smaller parts, which demands that 40 standard deviation codes be recorded.

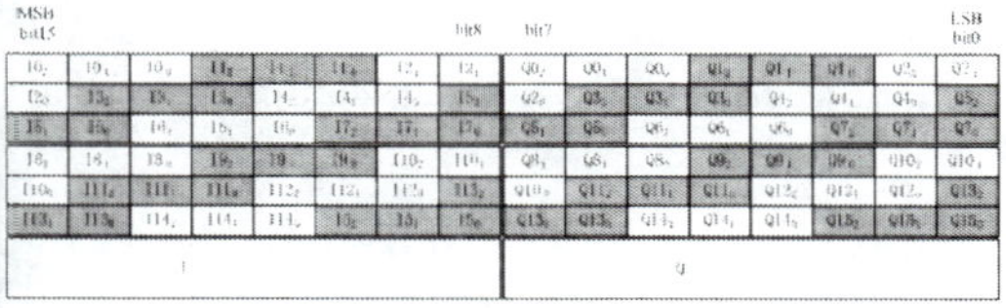

Figure 2.   User data format.

Take channel I for example (it is the same with channel Q): I0~I7 are 3-bit data after BAQ compression, where the subscripts 2, 1, and 0 of the corresponding $I0_2$, $I0_1$, and $I0_0$ represent MSB to LSB, respectively.

The process of BAQ decompression includes three steps: unpacking, decoding, and removal of normalization.

Step1: Unpacking. Unpack data of which the format is shown as follows, that is, to derive independent 3-bit codes.

Step2: Decoding. Decoding is the process by which code words are mapped to the quantization level, and the detailed map relationship is shown as follows.

Step3: Removal of normalization. This process involves mapping the standard deviation code located in the auxiliary segment of the corresponding data under operation to the standard deviation, and then multiplying it by the outcome in Step 2.

3.5   Fixed-point design in FPGA

The standard deviation and the quantization level determined in the process of BAQ compression are necessary in the BAQ decompression process, and the earlier two values should be stored in the look-up table beforehand, which makes it feasible to get access to these values at any time.

Generally, when multiplying the standard deviation by the quantization level, it is, in principle, required to use multipliers that support decimal multiplication such as the fixed-point multiplier and the floating-point multiplier, because the earlier two values are decimal. The fixed-point representation with poor accuracy compared with that of the floating-point representation typically requires more bits in order to approximately represent a number within a certain accuracy. However, the floating-point multiplier requires more resources, which, consequently, makes problems regarding the FPGA resources being consumed by a large sum of cost if multiple multipliers work simultaneously.

Whatever type of multiplier we may choose, its output is ultimately to be quantified to 8bits. We derive high accuracy at the cost of resources, whereas we lose a part of it in the process of quantization.

An approach that we introduce here is to reduce the resources, but making no sacrifices to the accuracy of the final output after quantization is to allow accuracy loss to some extent in the process of multiplication. Experimental results show that the final results in this method are almost identical to those of the floating-point multiplier. The method mentioned earlier is named "fixed-point implementation" and its detailed steps are given next.

In accordance with the requirement for the accuracy, the standard deviation and the quantization level should be multiplied by the integer power of two, respectively, and then the fractional parts of which are to be discarded, and, consequently, the remaining integer parts can be stored in the look-up table. "In accordance with requirements for the accuracy" that we introduced earlier refers to the fact that if we want to keep M bits in fractional parts, we need to multiply by N power of two, where the number of bits in integer parts of the N power of two should be no less than the value of M.

For example, if 4 bits in fractional parts need to be kept, we should multiply by at least 1024. Multipliers that support signed integer representation multiplication are needed to multiply the standard deviation and the quantization level.

As mentioned earlier, reasons that we use signed integer supporting multiplier here by enlarging the factor of the multiplication rather than directly calling the fixed-point decimal supporting multiplier are not that the structure of the fixed-point decimal representation supporting multiplier is more complicated than that of the signed integer representation supporting multiplier (actually the structure of the fixed-point representation multipliers is the same, and it is, in summary, a kind of shift adder multiplier, which means that the complexity to implement such kind of a multiplier differs only in the width of the input to the multiplier), but that the accuracy of the fixed-point decimal representation is not good enough.

In order to meet the needs of the accuracy under the condition of adaptive quantization, the width of the input to each multiplier operating in parallel should be equal to the largest width that the fixed-point number system can represent at worst, and sometimes this width can be very large, which, consequently, results in a serious waste of resources. Fixed number of lower bits in the output of the multiplier should be truncated in the process of quantization to make right shifts, cutting short the result to its true value. We should limit the results after truncation in the range of -128 ~ +127, that is to let the value higher than +127 be +127 and let the value less than -128 be -128.

Here, we multiply the standard deviation and the quantization level by 1024, respectively, and then put them in the look-up table. In consideration of the dynamic range of the standard deviation and the quantization level after enlargement, we choose signed integer multiplier with an input width of 13 bits and 24 bits. Each multiplication unit consumes one piece XTREME DSP slices, so in total we need 22 such kind of units to operate in parallel. We create multiplication units from IP core for multiplication, and the process of place and route can make optimum the specific implementation of these 22 units. The actual number of resources we may use should therefore be less than 22.

3.6 Modular architecture of BAQ decompression

The main function of the BAQ_INTF module is to retrieve SAR echo data, and there are four types of data formats at the receiver, namely, the express mode (the original SAR information is passed on without processing), the first four bits in MSB truncation mode (only the first four most significant bits are sent), the 4 to 8 BAQ decompression mode (eight-bit original data are compressed into four bits by BAQ algorithm), and the 3 to 8 BAQ decompression mode (eight-bit original data are compressed into three bits by BAQ algorithm). The compression format control character of each data packet is included in the auxiliary segment, which involves the first 512 bytes in a frame. According to the structure of the BAQ decompression module, it can be divided into the following four sub-functional modules: BAQ_FRONT_END, BAQ, BAQ_BACK_END, and SRIO_FIFO. The modular architecture is shown in figure 3.

SRIO_RXBUF is responsible for caching SAR echo data coming from RAPID IO terminal, and similarly, SRIO_RCBUF is responsible for caching command information coming from RAPID IO terminal.

BAQ_FRONT_END can read commands from RIO_RCBUF and simultaneously continuously read data from SRIO_RXFIFO, adding frame synchronous bits, valid signal, and so on, some information used to implement BAQ compression in the next module.

The process of BAQ decompression is implemented based on the decompression format bits in the auxiliary segment, and because of the fact that the output bits are larger than the input bits, the output bits need to be enlarged to 512 before being sent to the next processing module.

BAQ_BACK_END is responsible for writing SRIO_RCFIFO and SRIO_RXFIFO with the principle that on pouring some data into SRIO_RXFIFO, it simultaneously stores a command in the SRIO_RCFIFO, which can improve the working efficiency at the most, and even in that case, FIFO will not be empty.

4 SIMULATION AND RESULTS

The main idea behind the BAQ decompression simulation in FPGA is that we decompress sine wave signal based on the fixed-point method using FPGA, where the sine wave signal has been compressed in 8 to 3 mode and 8 to 4 mode, respectively, in advance, and then we compare the outcome with the results from the MATLAB where the results are obtained by way of the floating-point method realized in MATLAB. The results after comparison are shown in figure 4.

The signal CNT_0 denotes the total number of correct double words. We say it is correct if the outcome in ISIM is similar to that in MATLAB. Similarly, the signal CNT_1 denotes the total number of double words where only one byte is wrong and the signal CNT_2 denotes the total number of double words where two bytes are not correct. Finally, CNT_over2 denotes the total number of double words where more than two bytes are erroneous. As we can see from the variant

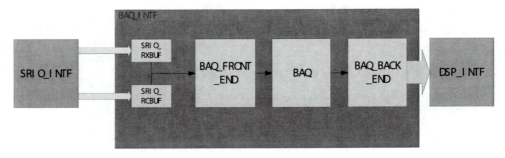

Figure 3.   Modular architecture of BAQ decompression.

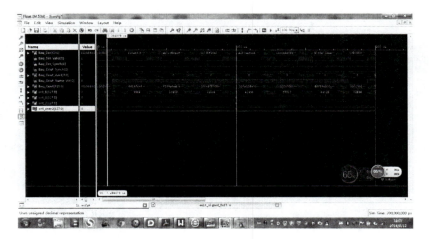

Figure 4.   Simulation results in ISIM.

values of these signals, the results based on the fixed-point method are similar to those in floating-point method. The original SAR echo data after compression are depicted in figure 5, and the corresponding decompressed data are shown in figure 6.

After BAQ decompression, we realize the SAR strip imaging algorithm in figure 7 and SAR scanning imaging algorithm in figure 8.

5   CONCLUSIONS

After implementing the fixed-point BAQ decompression in the XILINX FPGA and comparing the results with those of the floating-point algorithm based on MATLAB, we may safely arrive at the conclusion that the fixed-point method can retrieve SAR echo data without mistakes and optimize the utilization of the area resources as much as possible.

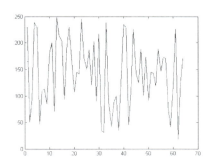

Figure 5.   Sine wave after compression.

Figure 7.   Strip imaging algorithm.

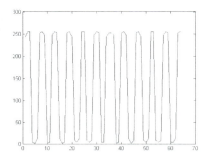

Figure 6.   Sine wave after decompression.

Figure 8.   Scanning imaging algorithm.

REFERENCES

[1] CUI Wei, LI Cheng-shu, TONG Zhi-yong. FPGA implementation of 3 bit Block Adaptive Quantization Algorithm[J]. Journal of Beijing Institute of Technology, 2005, 25(2): 139–142.

[2] Ye Shaohua, Sha Nansheng, Zhu Zhaoda. Study on SAR Raw Data Compression Using Block Adaptive Quantization[J]. Journal of Data Acquisition & Processing, 2001, 16(2): 182–184.

[3] Welch T A. A technique of high performance data compression[J]. IEEE Trans on Computer, 1984, 176(1): 8–19.

[4] Abousleman G P. Compression of hyper spectral imagery using the 3-D DCT and hybrid DPCM/DCT[J]. IEEE trans on Geoscience and Remote Sensing, 1995, 33(1): 26–33.

[5] Kwok R, Johnson W. Block adaptive quantization of magellan SAR data[J]. IEEE Trans on Geosci Remote-Sensing, 1989, 27(4): 375~383.

[6] Curlander J C, McDonough R. Synthetic Aperture Radar, systems and signal processing[M]. Wiley Sciences in remote sensing, John Wiley & Sons, Inc, New York, 1991.

[7] Benz U. A comparison of several algorithms for SAR raw data compression[J]. IEEE Trans GE, 1995, 33(5): 1266–1275.

[8] Habibi A, Wintz P A. Image coding of linear transformation and block quantization[J]. IEEE Trans on Communication Technique, 1991, 19(1): 50–63.

[9] Cui Wei, Han Yueqiu. The ASIC design for synthesizing dual channel gauss white noise[J]. Chinese Journal of Electronics, 2002, 11(3): 327–331.

Information, Computer and Application Engineering – Liu, Sung & Yao (Eds)
© 2015 Taylor & Francis Group, London, ISBN 978-1-138-02717-6

A new multiple search particle swarm optimization

Qing Wei Li
School of Energy and Environment, Southeast University, Nanjing, China

Gui Huan Yao
School of Mechanical and Power Engineering, Nan Jing University of Technology, Nan Jing, China

Tao Zhang
Power China Hubei Electric Engineering Corporation, Hubei, China

ABSTRACT: Particle Swarm Optimization (PSO) algorithm has been a topic of hot research in recent years. In this article, the original PSO with Multiple Adaptive Methods (PSO-MAM) was improved by replacing the sub-gradient method with simulated annealing algorithm. To show the performance of the improved PSO, it was tested in the experiment of benchmark functions, including unimodal functions and multimodal functions together with other state-of-art PSOs. To ensure the conclusions, the tests were implemented independently 30 times. Compared with other algorithms, the improved PSO performs better in the aspects of solution quality and convergence speed.

KEYWORDS: PSO; APSO-MAM; swarm diversity; convergence

1 INTRODUCTION

The particle swarm optimization (PSO) developed by Dr. Eberhart and Dr. Kennedy in 1995 is an evolutionary computation technique that is inspired by social behavior of bird flocking [1]. As it outperforms other intelligent algorithms in many aspects, PSO has been actively studied and applied for many academic and engineering problems such as parameter learning of neural networks[2], control optimization[3], and power station optimization[4].

However, some variants of PSO can only solve some problems and their robustness is poor. Hu[5] provided a PSO with multiple adaptive methods (APSO-MAM), but this method needs sub-gradient information, which is impossible for some problems.

In this article, PSO-MAM is improved with simulated annealing algorithm.

2 A SELF-ADAPTIVE ADJUSTMENT INERTIA-WEIGHTED PARTICLE SWARM OPTIMIZATION

2.1 Standard particle swarm optimizer

PSO algorithm is a swarm intelligence algorithm, and the swarm is composed of a set of particles. The position of particle i is defined as X(xi1, xi2, ... xid), where d is the dimension of the searched space. Particle changes itself as given next:

$$v_{id}^{t+1} = wv_{id}^{t} + c_1 \, rand_1 \times (p_{id} - x_{id}^{t}) + c_2 \, rand_2 \times (p_{gd} - x_{id}^{t}) \tag{1}$$

$$x_{id}^{t+1} = x_{id}^{t} + v_{id}^{t}, 1 \leq i \leq N, 1 \leq d \tag{2}$$

where Pid is the dth dimension of optimal position of particle i; Pg is the dth dimension of optimal position of the whole group; Vi is the dth dimension of position change rate of particle i; w is inertia weight; D is the dimension of the value function; N is particle number; c1 represents reflection of the particle's best position to its velocity; c2 represents reflection of the whole group's best position to velocity; and rand1 and rand2 are two random values in the range(0,1).

2.2 Simulated annealing algorithm

Though the PSO-MAM provided by Hu[5] can give satisfying solutions, the search progress needs the gradient information of the fitness function. But some problems are non-differentiable. So we introduce the simulated annealing algorithm to replace the sub-gradient method.

Simulated annealing algorithm is a probabilistic method proposed for finding the global minimum of a cost function. It works by emulating the physical progress by which a solid is slowly cooled[6].The details are described as follows:

1. Initialization: Produce some initial solutions in the feasible space. Give the high temperature, low temperature, and annealing coefficient.
2. Evaluate the solutions through fitness function.
3. Produce new solutions through variation function.
4. Evaluate the new solutions through fitness function. Update the temperature through Tnew = Told × α, where Tnew is the new temperature, Told is the old temperature, and α is the annealing coefficient. where is the new temperature, is the old temperature, and is the annealing coefficient.
5. If the evaluation value of the new solution is better than that of the old solution, the new solution replaces the old one. Otherwise, the new solution replace the old solution with the probability of exp(D/Tnew), where D is the difference between the new solution evaluation and the old one.
6. If the temperature becomes low, the procedure ends.

3 NUMERICAL SIMULATION

3.1 *Details of the test*

Many parameters should be determined before the search, and these are listed in Table1.

Apart from the average fitness (FV), success rate (SR) and success performance (SP) are also employed to evaluate the performances of PSOs. SR reflects the reliability, and SP reflects the efficiency. They are calculated as follows[7]:

$$SR = \frac{Success\ Times}{Run\ Times}$$

PSO searches successfully only if it can find the solution within the maximum error before the evaluation number reaches the permissible value.

$$SP = \frac{1\text{-}SR}{SR} FE_{max}$$

$+$ mean(evaluation times in the successful search),

where FE_{max} is the maximum evaluation time.

3.2 *Comparisons between AIPSO-MAM and other state-of-the-art PSO algorithms*

In this section, tests are conducted to compare AIPSO-MAM with other state-of-the-art PSO algorithms on unimodal functions, multimodal functions, rotated functions, and noisy functions.

Test results on unimodal functions and multimodal functions are listed in table2. It can be seen that AIPSO-MAM is generally the most outstanding one in the aspect of FV, SP, and SR. AIPSO enhances the convergence speed on Sphere function and Schwefel P2.22 function. Compared with APSO-MAM, SP is reduced from 18138 to 167. The best result provided by other PSOs on Schwefel P1.2 is 9.09E-27. However, the result found by AIPSO-MAM is 0.00E+00.AIPSO-MAM, and it also shows good performance with regard to solution quality and convergence speed dealing with multimodal functions. Though it is slightly worse than APSO-MAM with regard to Ackley function, the convergence speed is greatly enhanced, which is reflected by SP.

4 CONCLUSIONS AND PERSPECTIVES

The original APSO-MAM is improved by replacing the sub-gradient method with the simulated annealing algorithm. AIPSO-MAM can give the most satisfying solutions among the state-of-art PSOs in the aspect of solution quality and convergence speed.

Table 1. Parameters of the test.

Item	Value
particle number	30
maximum evaluation number	300000
significance level of T-test	0.05
maximum permissible error	$10\text{--}5$
trials	30

Table 2. Results of PSOs on unimodal and multimodal functions.

Algorithms	FV	SP	SR(%)	h	FV	SP	SR(%)	h	FV	SP	SR(%)	h
	Sphere				Schwefel p2.22				Schwefel P1.2			
CLPSO	0.00E+00	92887	100	+	0.00E+00	107976	100	+	2.17E+02		0	+
UPSO	0.00E+00	15708	100	+	0.00E+00	20 956	100	+	5.47E-11	183693	100	+
wFIPS	5.21E-27	79 319	100	+	2.46E-14	107719	100	+	1.91E+00		0	+
CPSO-H	2.42E-12	95 478	100	+	2.71E-07	201064	100	+	4.96E+03		0	+
DMS-PSO	7.15E-30	24 082	100	+	0.00E+00	30 186	100	+	1.10E+00		0	+
APSO-MAM	0.00E+00	1505	100	+	0.00E+00	18138	100	+	9.09E-27	35094	100	+
AIPSO-MAM	0.00E+00	118	100		0.00E+00	167	100		0.00E+00	207	100	
	Rastrigin				Ackley				Griewank			
CLPSO	0.00E+00	159419	100	+	8.05E-15	118336	100	+	0.00E+00	121255	100	+
UPSO	6.87E+01		0	+	3.55E-15	22997	100		1.89E-03	85077	83.3	+
wFIPS	2.80E+01		0	+	2.52E-14	117343	100		0.00E+00	98 894	100	+
CPSO-H	9.95E-02		90	+	3.03E-07	199111	100		2.07E-02	621219	36.7	+
DMS-PSO	7.16E+00		0	+	3.55E-15	34 502	100		0.00E+00	29 281	100	+
APSO-MAM	0.00E+00	60842	100	+	0.00E+00	20 012	100	+	0.00E+00	3000	100	+
AIPSO-MAM	0.00E+00	120	100		4.4409E-15	189	100		0.00E+00	152	100	

ACKNOWLEDGMENTS

The authors would like to thank Professor Zhou KeYi of Southeast University for his guidance during the writing of this article and the support of the Natural Science Foundation of Jiangsu Province, China (12KJB470008).

REFERENCES

[1] M.G. Epitropakis, V.P. Plagianakos, M.N. Vrahatis: Inform. Sciences Vol.216 (2012), p.50.
[2] Ping Luo, Peihong Ni, Lihai Yao, S.L. Ho, Guang Zheng Ni, Haixia Xia:Int. J. Appl. Electrom. Vol25(1–4) (2007), p.705.
[3] Song Y., Chen Z., and Yuan Z. New chaotic PSO-based neural network predictive control for nonlinear process. IEEE Transactions on Neural Networks, 18(2), 595–601, 2007.
[4] Vlachogiannis, J.G., Lee K.Y: IEEE T. on Power Syst. 21(4) (2006), p.1718.
[5] Han Huang, Hu Qin, Zhifeng Hao, Andrew Lim: Inform. Sciences Vol.182 (2012), p.125.
[6] Emile H.L. Aarts, Jan H.M. Korst, Peter J.M. van Laarhoven:Simulated Annealing (Springer, German, 1997).
[7] A. Auger, N. Hansen: The 2005 IEEE Congress on Evolutionary Computation, 2005, p.1777.

Information, Computer and Application Engineering – Liu, Sung & Yao (Eds)
© 2015 Taylor & Francis Group, London, ISBN 978-1-138-02717-6

Present situation and analysis of energy saving and emission reduction in electric power industry

Jing Cheng Zou
North China Electric Power University, Baoding, China

ABSTRACT: Authors of papers to proceedings have to type these in a suitable form for direct photographic reproduction by the publisher. In order to ensure uniform style throughout the volume, all the papers have to be prepared strictly according to the instructions given next. A laser printer should be used to print the text. The publisher will reduce the camera-ready copy to 75% and print it in black only. For the convenience of the authors, template files for MS Word 6.0 (and higher) are provided.

KEYWORDS: In order to; MS Word

1 PRESENT SITUATION OF THE ENERGY SAVING AND EMISSION REDUCTION IN ELECTRIC POWER INDUSTRY

Energy-saving emission reduction effectiveness in the electricity industry can be reflected in the power structure, power supply coal consumption, sulfur dioxide / nitrogen oxide emissions of pollutants, the unit-generating capacity of water, wastewater emissions, solid waste comprehensive utilization rate, and so on.

1.1 Development of renewable energy and the adjustment of power structure

In recent years, the scale of investment of the national electric power construction is growing faster, the power investment structure has constantly become more majorizing, and non fossil energy has risen to a higher percentage. Thermal power investment ratio sharply declined, whereas nuclear power and wind power investment ratio greatly increased. Table 1 (2008 ~2013 years, the total installed capacity of power industry and various generator assembly machine capacity) shows that[1] by the end of 2013, the national installed capacity had reached 1332560000 kilowatts. Among them, the power reaches 1006720000 kilowatts, accounting for a total capacity of 75.55%, 1.12% lower than the previous year. Hydropower installed capacity reached 234820000 kilowatts, accounting for a total capacity of 17.62%. Grid connected wind power reached 30980000 kilowatts, a little higher than the previous year.

Table 1. 2013. In 2008, the total installed capacity of electric power industry and various power capacity.

Particular year	2008	2009	2010	2011	2012	2013
The total installed capacity (MW)	92407	94508	97329	114590	123850	133256
The installed capacity of thernal power (k w)	66316	68166	70019	85613	94955	100672
Thernal power installed capacity accounted of the proportion (%)	71.77	72.12	71.94	74.71	76.67	75.55
Hydropower installed capacity (MW)	19879	19885	20636	21459	22163	23482
Wind power (k w)	1648	1850	1925	2510	2876	3098
Nuclear power (k w)	4564	4607	4749	5008	3856	6004

1.2 Promoting the policy of "developing large units and suppressing small ones," adjusting the thermal power structure

China's short-term thermal structure dominated by coal also may not change. Therefore, reducing energy consumption and pollution is an important task in the sustainable development of our power industry to adjust the thermal power structure. According to the statistics, at the beginning of 2009, the capacity of the national small thermal power units was 120000000 kilowatts, accounting for 30% of the thermal power installed capacity. Due to the poor performance of small units, the consumption per kWh is 450 grams, 150 grams more than supercritical thermal power generating units of 600000 kilowatts, and 110 grams more than the thermal power unit of 300000 kW. In the Ten Session of the National People's Congress five conference, Premier Wen Jiabao proposed the target of shutting down 50000000 kilowatts of small thermal power units during the "Eleven Five Year period." Some Opinions on "Quickly Shutting Down Small Thermal Power Units " and related supporting policies made by The State Council made specific provision for the main principle, division of responsibilities, incentive measures, and safeguard measures. These policies not only embody the principles of the market but also play the role of administrative action, organically combining the economic benefits and social benefits, the current interests and long-term interests, the survival of the fittest and optimization of the development in the electric power industry according to actual reality. These have a strong operation, making a good foundation of the policy for shutting down the work to be carried out smoothly.

1.3 Power supply coal consumption decreasing significantly

As an important index of electric power energy-saving emission reduction, the power supply coal consumption reflects energy efficiency of the generator. All the time, coal consumption of power supply in China has a large gap in developed countries. In 2008, China's power industry average coal consumption of power supplies was 374 grams / kWh, higher than the advanced international level of 40~50 grams. This is similar to consuming more than 0.16 billion tons of coal in a year, which caused a great waste of energy. Since the "Twelfth Five Year Plan," the policy of developing large units and suppressing small ones eliminated high-energy consumption units and pollution of small thermal power units. The order of scheduling rules made by the energy consumption accelerates the enterprise's change from high energy consumption to low energy consumption.

1.4 Sulfur dioxide emission of power industry under control

"The Twelfth Five Year Plan" period the national total emissions of major pollutants control plan put forward, to 2015 the total sulfur dioxide emission should be controlled on the target of 24944000 tons, 38.6 percent of which are the power of sulfur dioxide emissions. The realization of this objective is closely related to the whole sustainable economic development. According to the "Existing Coal-fired Power Plant Sulfur Dioxide Governance of 'Twelfth Five Year' Plan," about 198000000 kilowatts of existing coal-fired units are arranged to implement flue gas desulphurization during "Twelfth Five Year Plan" period. Because of the large coal-fired power plant, flue gas desulphurization facilities are completed and put into operation, and the ability of electric power enterprises to control sulfur dioxide emissions has been greatly enhanced.

2 ANALYSIS OF ENERGY CONSERVATION AND EMISSIONS REDUCTION POLICY IN THE POWER INDUSTRY

Currently, the achievements of energy conservation are mainly due to the effective implementation of the relevant state policies. To achieve this purpose, the "Twelfth Five Year" plan will set energy-saving emission reduction targets as legally binding targets and will change the index into the veto condition based on performance evaluation at all levels of government and business effects via the responsibility system. Simultaneously, through the improvement and the supporting laws, policies, planning, and a series of standards, we can make energy-saving quantified emissions reduction targets as various measures that we may operate. We should consider the total sulfur dioxide reduction that targets the responsibility of books during the "Twelfth Five Year Plan" signed between the central government, local government, and the power companies; letters of responsibility for energy-saving targets; and so on. These measures speed up the energy saving and emission reduction in the electric power industry to enhance resource conservation and enable the environmental protection work achieve unprecedented progress[2].

3 SUGGESTIONS

China's energy-saving emission reduction has achieved initial results, but some deficiencies still require to be made perfect. Some suggestions are as follows:

3.1 Continuing to promote the price reform, implementing the price of desulfurization denitration electricity

In the unity of desulfurization price policy encouragement, sulfur dioxide emission reduction in China's electric power industry has achieved obvious effect. But what we cannot deny is that area differences cause some problems of the price rationality to still exist. As another power plant's main emissions, nitrogen oxides can fully use market mechanism and control nitrogen oxide emissions by the price lever. But in the concrete implementation process, denitration price should be set in view of the production situation in various regions of the country. For example, governments are allowed to consider per kilowatt hour for 1.5 cents to 2 cents within the scope according to the actual situation, so that it can be a more reasonable and effective propulsion power plant. Simultaneously, actively promoting the development of flue gas denitration industrialization makes a good foundation for the large-scale construction of the flue gas denitration device.

3.2 Handling the relationship between energy saving and emission reduction correctly, achieving economic and environmental benefits

With the pollution control requirements increasing, the cost of power generation of the power industry increases further. Investment and operation of pollution treatment facilities have increased the cost of power generation. The choice of the technology route of the current power plant pollution control is too single with extensive process design. The pursuit of efficiency is not judged by the actual situation of the standard, the coal quality, the site, and the comprehensive utilization of the by-product plant preferred treatment process. It is wasteful and it also affects the application of circular economy technology and the technology of the independent intellectual property rights. When the power industry makes the plan of energy-saving emission reduction, we should correctly handle the relationship between the standard and the law of cost control, that between energy saving and emission reduction, and that between energy saving and emission reduction and resource conservation; strengthen management of design and operation to pollution control facilities; optimize the pollution control of the facilities; and achieve energy-saving emission reduction and economic benefits of a win–win situation.

3.3 Need to improve the comprehensive utilization rate of the desulfurization by-products

According to statistics, the country has more than 90% of the flue gas desulfurization using limestone–gypsum wet FGD process. With the large-scale operation of the desulfurization unit, desulfurization gypsum production has increased year by year. In 2013, the national desulfurization gypsum production had reached 35000000 tons, but the comprehensive utilization rate is only about 45%. Most of the thermal power plants' desulfurization gypsum mainly undergo the ash field stacking or reclamation treatment. This not only takes up a lot of land and increases the ash field of investment but also may cause secondary pollution to the surrounding environment. The power industry needs to strengthen the operation management of the desulfurization unit, improve the desulfurization gypsum quality, strengthen the cooperation, and actively build the materials sector simultaneously to create conditions for comprehensive utilization of FGD gypsum. The relevant government departments should take measures to increase the comprehensive utilization of desulfurization gypsum support and to improve the comprehensive utilization of desulfurization by-products through economic policies.

3.4 Reducing the unit-generating capacity of water, achieving zero discharge of wastewater

In recent years, with the power industry increasing the water-saving efforts, more and more power plants start using air cooling technology and using municipal reclaimed water as circulating water. Through the water management, achieving wastewater reuse rate has improved reuse of generating units of water consumption and has reduced wastewater. In the implementation of the energy-efficient scheduling or the difference of consumption, they generally set the unit coal consumption rate of the power supply, desulfurization conditions as the evaluation index, and failed to consider the generating units of water consumption, wastewater reuse rate as a rigid index[3]. We can consider making the base on the power consumption of water and water consumption of the power units, rewarding the good and fining the bad[4] to promote the realization of zero discharge wastewater.

4 CONCLUSIONS

Energy saving and emission reduction is a long-term and arduous task for the power industry. Through the formulation of relevant policies and the measures to mobilize and promote energy-saving emission reduction work in the power industry, we can simultaneously attain the goal of reducing the power plant wastewater, waste gas, and waste residue emissions in energy saving. Thus, saving resources and protecting the environment promotes the sustainable development of our national economy.

Author's Introduction: Zou Jingcheng Birthday: 1994.1.22 Sex: Male, Nationality: Han Undergraduate student, Yichang in Hubei Province

Major: Electrical engineering and its automation

REFERENCES

[1] Chinese Federation of electric power enterprises. The national electric power industry statistics Express (2008–2013 year).

[2] Wang Zhixuan. Technology of [2], Huadian electric power industry energy saving emission reduction problem and Counter measure of [J]. 2008, (5): 1–5.

[3] FU Shuti, WANG Haining. On coordination of energy saving and reduction of pollution policy with elect ricity market reformin China. Automation of Electric Power Systems, 2008, 32 (6): 31234.

[4] GENG Jian, GAO Zonghe, ZHANG Xian, et al. A preliminary investigation on power market design considering social energy efficiency. Automation of Electric Power Systems, 2007, 31 (19):18221.

Information, Computer and Application Engineering – Liu, Sung & Yao (Eds)
© 2015 Taylor & Francis Group, London, ISBN 978-1-138-02717-6

Design and application of the patterns of PBL to database curriculum in vocational schools

Hua Yan Li
LangFang Polytechnic Institute, China

ABSTRACT: This study is based on systematic analysis of the database curriculum, students in vocational schools, learning elements of the project, and network collaborative learning platform. The learning patterns of the database course project in vocational schools have been designed with regard to five aspects, including learning situations, the subject of the project, collaborative procedures, learning guidance, and learning assessment. The research findings have been applied timely in the teaching practice, which has made a beneficial attempt to improve the learning effect of the database curriculum in vocational schools.

KEYWORDS: Vocational schools; Database curriculum; Learning patterns of the project; Network platform

1 ANALYSIS OF THE NETWORK PBL IN DATABASE CURRICULUM IN VOCATIONAL SCHOOLS

Database being theoretical, practical, and applicable is a specialized elementary course in computer-related specialty. Almost every computer information management system is related to the application of the database and usually the backend applications. Attention should be paid not only to learning the theory of this curriculum but also to exercising the database design practice.

"Project based learning" (PBL) refers to the learning based on special subjects and projects. It is a series of learning activities carried out usually by grouping the students who make a survey, observation, study, and expression of the new knowledge, as well as display and share the results by centering on the selected subjects. It considers the students as the main part and greatly stimulates their learning enthusiasm. By fulfilling specific project tasks, students greatly improve their ability to analyze and solve problems, ability to cooperate with others, ability to speak and express, ability to cooperate and explore, and so on.

PBL can improve the students' overall quality; however, it also put forward new requirements for them in terms of the learning initiative, the original cognitive level, and the learning style and attitude. Before carrying out PBL, students majoring in e-commerce at Langfang Polytechnic Institute have been chosen as respondents to fill in questionnaires, which include whether having any learning experience of projects, having any interest in project-based learning, approving of the cooperative learning effect, preferring to participate in project-based learning, their proficiency in curriculum knowledge, and so on. The statistical

analysis of the questionnaire shows that most students have a desire to learn and grasp curriculum knowledge, but they lack enthusiasm toward the learning methods based on the traditional structure of knowledge. So they long to have a passion for learning, but they have an internal need for the change of the learning styles; simultaneously, they also worry about whether they can adapt to the new learning style or they can solve the difficulties in their learning and they lack confidence about whether they can achieve success in their studies. According to the analysis of the students' academic records of specialized elementary courses and the introduction given by the teacher on the students' cognitive level, learning style, and learning attitude, it has been found that most of the students have a good learning ability and attitude, enjoy cooperating with other people, and have a certain ability to solve problems. Therefore, it is feasible to add new learning styles and carry out project-based learning.

2 DESIGN OF THE PATTERN OF PBL IN DATABASE CURRICULUM IN VOCATIONAL SCHOOLS

Through the previous analysis of its feasibility, based on the basic idea "to learn by doing," which serves as the guidance, the patterns of the project-based learning in database curriculum in vocational schools have been designed in five aspects as shown next:

2.1 Situation design of PBL in database curriculum

Situation design in database curriculum should combine with the characteristics of vocational education and specialties. The prominent characteristic of the vocational education is to develop the students' practical ability in their respective posts. The training objective of e-commerce is to cultivate high-quality practical and technical talents who have good professional ethics, basic ability of business communication, computer, marketing, and business management. Setting situations related to the students' future occupation can stimulate their interest and encourage them to indulge in further studies such as automobile sales management system, cosmetics management system, library management system, supermarket management system, and salary management system. The real project in the situation as the learning materials for students can make them experience a course of work right from the customer needs to the completion of design.

2.2 Subject design of PBL in database curriculum

In different situations, the subject of the project needs to be designed by a comprehensive analysis of the enterprises' demand for talents, the curriculum characteristics and structure of knowledge, the students' occupational characteristics and interests, and their future posts. Regardless of the specific situation or the subject of a project being carried out for PBL, as long as the students finish the project, they can eventually combine it with practice and fulfill the requirements for curriculum objectives in specific situations to enhance the students' quality and ability.

Taking the design and development of the automobile sales management system as an example, the following five projects can be designed:

First, "the research and planning of the automobile sales management system" includes market research, data model, and database system design. The project requires the students to carry out social research mainly based on the user's data requirements, processing requirements, and requirements for security and integrity. The research findings are used to make a planning book of the automobile sales management system by combining them with the database knowledge. Second, "the conceptual design of the automobile sales management system" mainly covers conceptual model of the system and basic theories of the relational database. The students design the conceptual model of the system according to the results of the research and planning, which will lay a foundation for the subsequent system design. Third, "the structural design and manufacture of the automobile sales management system" mainly includes the design and making of the database and data sheet, the design and making of the form, the design and making of the report forms, and the design and making of the query. The project can be completed by dividing it into four small modules, including "the system's operation interface," "the customers' interface," "the personnel

interface," and "the product interface." Fourth, "the operation and usage of the automobile sales management system" mainly consists of entering, editing, and maintenance of the information on customers, personnel, and products, as well as marketing. It is mainly related to the use and maintenance of the table data, Structured Query Language (SQL), and VBA programming. Fifth, "the summary of developing the automobile sales management system" uses the slide to explain the development of this project. By summing up the experience to find out the shortcomings, the students' ability of making and using the Power Point and their ability of speaking and expressing will be developed.

2.3 Design of the project cooperative learning in database curriculum

PBL is carried out by using the pattern of group collaborative learning. In order to make the cooperative learning effective, it has been designed as grouping control and collaborative processes.

2.3.1 Division of work by grouping

In the early stages of a project, the primary work is to group the students. The groups are divided by following the principle of the students' own free will and the teachers' guidance. There are about 4 people in each group. Each group needs to pick out a person who will be in charge of the project, divide the work among group members to have a clear assignment of their responsibilities, and work collaboratively. In the process of completing the task, each member should indulge in constant communication and provide feedback to improve the design.

2.3.2 Process of the collaborative learning

According to the basic idea of PBL, the implementation of collaborative learning can be divided into the following six aspects:

I. Selecting the subject of the project: According to the characteristics of the members in each group, teachers illustrate the subject and the students can select the project according to their own interest. II. Making a plan for the project: Each group can make plans for learning and for project activities based on the subject. It mainly includes the arrangement of time required for the project and the arrangement for activity process, such as the source of the material and the division of work among the members. III. The planning of project research: It is a process of collecting material by students to solve the problems in accordance with the plan. It requires the students to actively use a variety of knowledge and skills and to construct knowledge. IV. The design and manufacture of the project: It is an important stage of learning and is also a main stage of implementing deep learning, such as research learning, enquiry learning, and cooperative learning. Besides, it is the reason that PBL is different from traditional teaching. Students need to process and reconstruct the collected material by using knowledge and skills that have been acquired during the learning process and they will eventually form their own works. V. The interchange of project results: When the project is completed, there will be a seminar, inviting some leading experts to interchange the results of each group and sharing the successful learning experience and happiness. VI. The assessment of the project activity: It makes PBL different from the traditional teaching. It mainly adopts the method of a combined assessment made by experts, teachers, and students to accomplish the assessment of the learning results and the learning process.

2.4 Guidance design of PBL in database curriculum

The guidance of PBL in database curriculum is mainly arranged at the stage of pre-learning, determining the subject, forming groups, dividing the work among group members, exploring self-study, discussing and negotiating, presenting the results, and summarizing assessment.

The guidance at different stages has a different emphasis, such as at the stage of pre-learning and determining the subject; it requires the guidance of analyzing the subject and interrelating the curriculum knowledge. At the stage of forming groups and dividing the work among group members, it requires the guidance of forming groups scientifically and implementing effective management; at the stage of exploring self-study and learning collaboratively, it requires the guidance of the ability of processing the collected material, further processing the knowledge, and cooperating collaboratively; and at the stage of presenting the results and summarizing assessment, it requires the guidance of developing the students' ability of expression, reflection, and assessment. Besides face-to-face teaching, it can make full use of the network learning platform, QQ, and some other ways to achieve real-time and non–real-time communication between teachers and students.

2.5 Assessment design of PBL in database curriculum

The assessment design of PBL in database curriculum is divided into comprehensive assessment and sub-item assessment. The former refers to an overall assessment of the PBL results. It emphasizes the test of the students' comprehensive quality and focuses not only on the students' academic achievement but

also on their high-level thinking ability, creative spirit, and practical ability. Simultaneously, it requires special attention toward developing the students' good professional psychological quality, their learning interest, and positive emotional experience. Sub-item assessment is the assessment on the completion of the database project emphasizing the learning process. It enables teachers to give the students an objective assessment, which will help the students discover their own strengths. It can not only enable the teacher to educate the students according to their natural ability but also develop the students' self-confidence, which will have a positive effect on their mastering the skills.

Through the earlier mentioned five aspects of design, namely the project situation, the subject of the project, collaborative learning processes, PBL guidance, and assessment, the pattern of the network PBL in database curriculum in vocational schools has taken an initial shape, which is a new effective learning pattern. It can play a positive role of exploiting the students' potential and enhancing their ability of lifelong learning.

3 APPLICATION OF THE NETWORK PBL PATTERN IN DATABASE CURRICULUM IN VOCATIONAL SCHOOLS

3.1 Case of the application of the network PBL in database curriculum

In order to improve the research results, this article illustrates this point by taking chapter ten in the textbook on database application system design as an example. The content of the chapter is a comprehensive application of the knowledge points in the book. By finishing a simple database application system, students can improve their knowledge and quality.

I. Determining the task of the project: Making an assignment book for the project to offer the students some alternative project tasks such as designing the automobile sales management system, cosmetics management system, library management system, supermarket management system, and salary management system. It should include the completion time of the project, requirements for the material submitted, and methods of assessing the results. II. Creating collaborative groups: The whole class is divided into 15 collaborative learning groups. Each group has its own subject and defines the division of work. III. Carrying out collaborative exploration: Teachers instruct each group to learn from the handbook on PBL, which provides instructions for the completion of the project. The subject will be divided into several items, and each item needs to be completed according to the instruction of the project. Group members discuss plans of cooperation together, fully carry out the research, and have an effective research-based learning. IV. Monitoring and guiding the learning process: It requires providing learning resources, guiding learning motivation and learning methods, guiding cooperative exploration and group discussion skills, and checking learning process and results according to the learning plan. For example, Moodle network platform can be used to build a website for PBL in the database curriculum in order to carry out PBL on the database. V. The interchange of learning results: The works of each group will be presented publicly and orally.

3.2 Assessment of the network PBL in database curriculum in vocational schools

The assessment of the network PBL in database curriculum should be based on the comprehensive assessment and sub-item assessment. It includes the assessment of the students' works and the assessment of the completion process of the project. By means of the assessment made by the groups and the teachers, the group works have been graded according to the level of excellence. The students fill in the completion report of the project, final report of practical training, comprehensive ability test, and the questionnaires on the project learning effect, which will assess the students' project results from multiple perspectives.

To sum up, the research and practice carried out in vocational schools on the application of the pattern of the network PBL in database curriculum have shown that the pattern of PBL is an effective new one in vocational schools and is worth being put into practice.

REFERENCES

[1] Pin Liansheng. The psychology of teaching and learning [M]. Shanghai: East China Normal University Press, 1998.
[2] Huang Chunguo, Yin Changhong. The study of PBL under the environment of information technology [J]. Chinese Audio-visual Education, 2007, (5).
[3] Lian Qianping. How to form an effective operating mechanism of PBL [J]. Education Research, 2009.
[4] Yue Caiyi. The study of the patterns of PBL on specialized courses [J]. Vocational Education Research, 2008 (10).
[5] Li Huayan. The study of the optimization of network teaching [J]. Science & Technology Information.

Note: The research has been approved by "the Eleventh Five-year Planning" of Hebei province for education and science in 2010.

Information, Computer and Application Engineering – Liu, Sung & Yao (Eds)
© 2015 Taylor & Francis Group, London, ISBN 978-1-138-02717-6

Design of performance probe architecture for distributed computing system

B. Zeng
Information Management Research Center, Department of management, Naval University of Engineering, Wuhan, Hubei, China

R. Wang
University Library, Training Department, Naval University of Engineering, Wuhan, Hubei, China

ABSTRACT: Monitoring the behavior of a system is the first and key step in tuning the performance of the system. The monitor is the primary tool used to collect, process, and present information on the activity of a system for this purpose. In this paper we describe the design and implementation of the Distributed Probe System (DPS). The key design aspect of DPS is service-oriented. The modular structure of DPS allows it to easily adapt to changes in the underlying platform, thus making DPS portable. By defining uniform interfaces to the modules of DPS in a consistent way, DPS becomes very flexible. Moreover, by using a service-oriented approach in defining the modules it is possible to easily customize existing modules. The flexibility of DPS is also aided by a clear division between the data collecting and data processing modules. DPS has been successfully used to performance tune distributed applications. Moreover, by monitoring numerous applications we gained valuable insight into the operation of the distributed computing such as Grid and Cloud computing.

KEYWORDS: Distributed Computing System; Performance Probe; Service Oriented Design

1 INTRODUCTION

Distributed computing systems are complex systems consisting of many modules, such as processing units, memories, and I/O devices, and many layers of software. For this reason, users often find it difficult to exploit the performance potential of those systems when they first run an application (Wang, W.Y. 2011). The complexity of the system makes it nearly impossible to guess the cause of poor performance, in part, because it is due to subtle interactions between the different components of the system.

A monitoring tool that allows the users to monitor the behavior of the system is therefore essential (Aceto, G. 2013). Such a tool should be capable of recording performance data from all layers of the system, including the hardware, operating system, and application layers, and it must be capable of processing, storing, and displaying this data.

The biggest contribution of Cloud Services Trace (Iosup, A. 2011) is a logic-analyzer-like data visualization tool that displays program activities over a time interval from a trace previously collected by system logs. It has been used for monitoring and performance debugging Amazon Web Services and Google App Engine. A multi-layered monitoring framework for Cloud is presented in (Katsaros, G 2012). The four basic components are the probes, the data pool, the analyst, and the user interface. Agent architecture is

designed that runs in the Root partition to drive the Clouds resource management control and QoS monitoring (Ripal, N. 2010). It captures performance interference interactions, and uses it to perform closed loop resource management.

2 ARCHITECTURE DESIGN

Our goal in designing the Distributed Probe System was to provide system and application programmers with a monitoring tool that is scalable, portable, flexible, and non-intrusive. Scalability means that DPS must not extensively perturb a system with a few processors and yet be able to process and store the large amounts of performance data gathered from systems with a few hundred processors (Xue, C.Y.2013). Portability means that DPS can easily adapt to changes in the underlying hardware, in the operating system, and In the system configuration. Flexibility means that it should be possible to easily incorporate in the monitoring tool a variety of data collection methods or data display formats. Non-intrusiveness means that data and control flow within DPS should perturb the system as little as possible (Sun, W. 2010).

The structure of the Distributed Probe System is shown in Figure 1. The components of the monitor are the probes which are modules that collect data, the Probe Managers that manage the probes in an

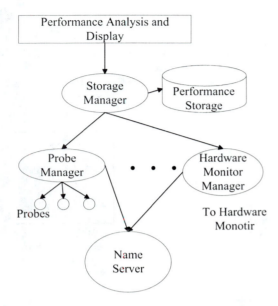

Figure 1. The architecture of the distributed probe system.

cluster, the Storage Managers that manage the permanent storage devices, the Performance Analysis and Display modules, and a Name Server that keeps evidence of all modules in the system.

Probes are data collecting modules inserted into application programs, into servers, or into the OS kernel. Each Probe Manager (PM) manages all probes within one cluster. Upon request, the Probe Manager collects data from the probes and transfers it to the requesting clients, which can be DPS modules or other programs running under Independent. It also relays commands from the client down to the probes, to, for example, manipulate an on/off switch, or to set data collection parameters, such as histogram thresholds. A monitored application program or server must have a Probe Manager if it has a probe.

The Storage Manager is a module that manages secondary storage devices. It supports two types of operations. First, it collects data from the probes specified in the request and writes it to secondary storage. Second, it reads the previously stored data from the secondary storage and passes it to the requesting client.

The Performance Analysis and Display (PAD) modules display and present data collected from the system, and they provide a menu-driven interface through which users can control the operation of DPS. Multiple independent PAD modules, serving different users and monitoring different parts of the system, can run at the same time. The PAD modules can run on remote workstations through a LAN. If a PAD module runs on a remote workstation, a Connection Server module on local system maintains the network connection and issues requests to other DPS modules on the PAD modules' behalf.

The Name Server keeps evidence of all Probe Managers and Storage Managers in the system. The Probe Managers and the Storage Managers register with the Name Server when they are started by sending a registration message with their identification number to the Name Server. Before a client can read data from, or issue a command to an DPS module, it must obtain from the Name Server the identification number of the module. Having obtained a PM's or Storage Manager's identification number, the client can start issuing requests directly to the corresponding Probe Manager or Storage Manager.

The structure of DPS can scale to large systems because the number of data processing and storage modules can increase with the size of the system, and because the modules can be distributed so that they are close to where the data is collected.

The structure is flexible, because it allows display of sampled data in real-time, or the storage of collected data traces to permanent storage. It is portable because the monitor modules are easy to replace when different functionality is required, and because they can be mapped to the processors in various ways. For example, DPS modules can share processors with target computational tasks, they can be assigned to special processors dedicated to monitoring, or they can be off-loaded to remote workstations. Properly mapping the monitor onto the multiprocessor is important, because it can significantly decrease the monitor's intrusiveness.

3 SYSTEM WORKFLOW

The activities of DPS modules are demand driven. All commands arid requests for data are initiated in the PAD module by the user. The user starts the monitoring session by creating a PAD module which connects to the rest of DPS by obtaining the information regarding the current system configuration from the Name Server. The Name Server identifies ail Storage Managers, system PMs, and the PMs of applications belonging to the user, together with the probes they manage. The configuration of the monitor is then presented to the user, who selects the probes that he is interested in and specifies how the data coming from these probes should be processed and displayed as shown in Figure 2.

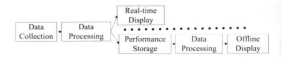

Figure 2. The data flow within DPS.

After the set-up, monitoring of the target program and system is started. The PAD module periodically requests data from those PMs that manage at least one of the user-selected probes. The frequency at which this occurs is set by the user at the beginning of the monitoring session. Sampling probes return data for immediate display by the PAD module. Tracing probes return status information (for example, how much data they have collected and whether the data should be flushed to permanent storage). If a tracing probe has data available, then the PAD module will request a Storage Manager to dump the trace. The PM identifier and the probe are identified in the request. The Storage Manager then requests the trace data from the appropriate PM and writes it to permanent storage. Upon a request from a user, data from permanent storage can then be read, processed, and displayed by the PAD module. The PAD module terminates, when the monitoring session is ended by the user.

4 SYSTEM IMPLEMENTATION

The current implementation of DPS provides the users with a simple mechanism to instrument programs. To instrument a target application, the programmer must add probes to the application code and he must add filter functions for each probe to the PAD module. Programmers can use predefined basic probes that are provided in the form of java classes, or they can customize these probes by deriving their own subclasses from the predefined probes. The filter function takes data obtained from the probe and converts it a format which can be displayed by the PAD module. Again, predefined and default filter functions are available, which the user can use directly or from which the user can derive customized functions.

4.1 *Probe structure*

Probes are modules that collect data. They include one or more Data Collection services, a Data Buffer, and interfaces to the Probe Manager. The basic structure of a probe is shown in Figure 3.

The Data Collection Service is a service that records data and stores it in the probe's Data Buffer. A probe can have multiple Data Collection services.

The Data Set is the basic unit of data collected by the probe. When invoked, the Data Collection Routine updates the entire Data Set. For example, the Data Set can be an event counter, a histogram, or a trace record containing event identification, a timestamp, and event data. A Data Set is a structure of dataSetSize integers. Integers are used as the smallest data entity in DPS in order to speed up data transfers between DPS modules.

The Data Buffer is a two-dimensional integer array of setCount Data Sets. For event counters, the Data Buffer is typically only large enough to contain a single Data Set, namely the counter, but it could be large enough to hold tens of thousands of Data Sets in the case of a trace buffer.

The Data Interface and the Control Interface are procedural interfaces between the probe and the Probe Manager. The Data Interface service is accessed by the Probe Manager each time the PM requests data from the probe. The Control Interface service is called each time the PM issues a command to the probe.

4.2 *Probe manager*

The Probe Manager is a module service that manages within one cluster the data flow from the probes and the control flow to the probes. The Probe Manager consists of three parts: the Probe Manager Server (PM Server), the Probe Information Table, and the data buffer to store data collected from the probes before it is sent to higher-level modules. The structure of the Probe Manager is shown in Figure 4.

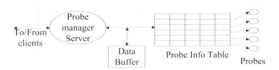

Figure 4. The probe manager.

The Probe Manager Server (PM Server) serves as the interface to other DPS service modules. The Data Buffer holds data from probes before it is sent to higher-level modules. The Probe Information Table contains information, about the probes.

The PM Server is an ordinary Independent service which accepts requests from clients and carries out operations on their behalf. On startup, each PM Server registers its identification number with the Monitor Manager.

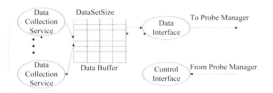

Figure 3. The structure of a probe. The solid lines show the flow of data, the dotted lines show the flow of control.

The data buffer stores data collected from the probes before it is passed on to other DPS modules. The reasons for having a data buffer in the Probe Manager are speed and scalability.

4.3 *Monitor manager*

In our implementation, the Name Server, the Storage Server, and the Connection Servers servicing remote PAD modules have been combined into a single module, which we refer to as the Monitor Manager. The structure of the Monitor Manager with two Connection Servers servicing two remote PAD modules is shown in Figure 5.

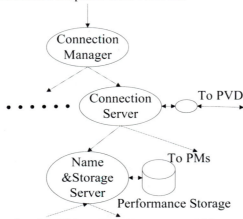

Figure 5.   The structure of the monitor manager.

The Connection Manager creates a Connection Server for each remote PAD module, which then acts as a switch between a remote PAD module and DPS modules on Local cluster, passing commands and data between the PAD module and the DPS modules on Local cluster.

The main components of the Monitor Manager are the Connection Manager and the Name & Storage Server. The Connection Manager is a process that first exports a port to the network and then accepts on this port connection requests from remote PAD modules. For each successful connection request it creates a Connection Server and a socket that provide the connection between the remote PAD module and the DPS modules on Local cluster. The Connection Server is an independent service that passes commands and data requests from the remote PAD module to PMs

and to the Name & Storage Server on Local cluster, and data and status information from the PMs and the Name & Storage Server back to the PAD module.

4.4 *Performance analysis and display module*

The current implementation of the Performance Analysis and Display module runs on a remote workstation connected to local cluster. The PAD module graphically displays data from DPS in real-time and it provides a menu-driven interface that is easy to use to control DPS. Displaying measured data in real-time has proven to be extremely useful in the initial stages of performance tuning parallel programs which usually yield the largest performance improvements. The usefulness of a real-time visualization tool is further increased if users can interactively control the operation of the tool to, for example, switch on or off data streams from different probes, select from various data display formats, display performance data both from user programs and from the operating system, or set data collection parameters for the probes.

5   CONCLUSIONS

In this paper we presented the Distributed Probe System, a software monitoring tool for the distributed computing system. In comparison with other monitoring architecture, DPS is novel in the following aspects.

1 It is scalable. The data collection modules (i. e. the probes and the PMs) ran be replicated as often as needed. The storage modules ran also be replicated, and they can be placed close to data collecting modules in order to minimize the flow of collected data through the system. Filtering of data (in the data collection modules) can be used to reduce the data through the system.
2 It is flexible. We were able to incorporate into the same architectural framework two inherently different data collection methods, namely tracing and sampling.
3 It is little-intrusive. By clearly partitioning the data collection modules from the data processing and storage modules, it is possible to off-load the data processing modules to dedicated workstations.

In the long-term, it would be interesting to extend DPS to new areas such as incorporating hardware monitor events into the DPS framework, to explore new display formats and techniques, or to extend DPS to a full environment for application and system development.

REFERENCES

Aceto, G., Botta, A. & Donato W. 2013. Cloud monitoring: A survey. *Computer Networks* 57(9), 2093–2115.

Iosup, A., Yigitbasi & N., Epema, D. 2011. On the Performance Variability of Production Cloud Services. *Proc of 2011 11th IEEE/ACM International Symposium on Cluster, Cloud and Grid Computing (CCGrid), Newport Beach, 23–26 May 2011.* USA:IEEE.

Katsaros, G, Kousiouris & G. Gogouvitis, S.V. 2012. A Self-adaptive hierarchical monitoring mechanism for Clouds. *Journal of Systems and Software* 85(5), 1029–1041.

Ripal, N., Aman, K. & Alireza, G. 2010. Q-clouds: managing performance interference effects for QoS-aware clouds. *Proceedings of the 5th European conference on Computer systems. NY, USA, 15–16 September 2010.* USA:IEEE.

Sun W, Yuan X.J. & Wang J.P. 2010. Quality of Service Networking for Smart Grid Distribution Monitoring. *Proc of 2010 First IEEE International Conference on Smart Grid Communications (SmartGridComm), Gaithersburg, MD, 4–6 Oct. 2010.* USA:IEEE.

Wang, W.Y., Xu & Y., Khanna, M. 2011. A survey on the communication architectures in smart grid. *Computer Networks* 55(15), 3604–3629.

Xue, C.Y., Liu, F. & Li, H.H. 2014. Research and Design of Performance Monitoring Tool for Hadoop Clusters. *Proceedings of International Conference on Computer Science and Information Technology APADnces in Intelligent Systems and Computing, Kunming, 21–23 September 2013.* German:Springer.

Information, Computer and Application Engineering – Liu, Sung & Yao (Eds)
© 2015 Taylor & Francis Group, London, ISBN 978-1-138-02717-6

An optimal throughput model for multicast routing network

L. Yao, F. Li & W. Hu
Department of Management, Naval University of Engineering, Wuhan, Hubei, China

ABSTRACT: We propose a throughput model for establishing a network that may ensure the minimum throughput requirement for each multicast group is satisfied. This model is especially suitable for supporting Constant Bit Rate (CBR) services or traffic types with given Peak Cell Rate (PCR). By formulating the problem as a mathematical programming problem, we intend to solve this mathematical problem optimally for obtaining a network fitting into our goal, that is, which makes sure the throughput of each multicast group reaches a given level. In terms of performance, our solution has more significant improvement than simple heuristics. The improvement on the total revenue can reach 19% on an average in the throughput model.

KEYWORDS: Throughput Model; Multicast Routing; Heuristic Algorithm; Mesh Network

1 INTRODUCTION

IP Multicast traffic for a particular (source, destination group) pair is transmitted from the source to the receivers via a spanning tree that connects all the hosts in the group. Different IP Multicast routing protocols use different techniques to construct these multicast spanning trees; once a tree is constructed, all multicast traffic is distributed over it.

IP Multicast routing protocols generally follow one of two basic approaches, depending on the expected distribution of multicast group members throughout the network. The first approach is based on assumptions that the multicast group members are densely distributed throughout the network (i.e., many of the subnets contain at least one group member) and that bandwidth is plentiful (Karande, SS. 2011). The so-called dense-mode multicast routing protocols rely on a technique called "flooding" to propagate information to all network routers. Dense-mode routing protocols include Distance Vector Multicast Routing Protocol (DVMRP), Multicast Open Shortest Path First (MOSPF), and Protocol-Independent Multicast - Dense Mode (PIM-DM) (Lim, WS., 2011).

The second approach to multicast routing basically assumes the multicast group members are sparsely distributed throughout the network and bandwidth is not necessarily widely available, for example, across many regions of the Internet or if users are connected via ISDN lines (Wang, C. 2011). Sparse mode does not imply that the group has a few members, but just that they are widely dispersed. In this case, flooding would unnecessarily waste network bandwidth and, hence, could cause serious performance problems. Hence, "sparse-mode" multicast routing protocols must rely on more selective techniques to set up and maintain multicast trees. Sparse-mode routing protocols include Core-Based Trees (CBT) and Protocol-Independent Multicast - Sparse Mode (PIM-SM) (Zhao, X. 2011).

The objective of admission control is to regulate the operation of a network in such a way to ensure the uninterrupted service provision to the existing connections and simultaneously accommodate the new connection requests in an optimum manner. This is done by managing the available network resources and allocating them in an optimum manner among the system users.

The decision is based on (1) Does the new connection affect the QoS of the connections currently being carried out by the network? (2) Can the network provide the QoS requested by the new connection? Once a request is accepted, the required resources must be guaranteed (Cogill, R., 2011).

2 PROBLEM DESCRIPTION

The network is modeled as a graph where each node in the graph represents a router and each arc in the graph represents a link. A user group is an application requesting for transmission in this network, which has one source and one or more destinations. Given the network topology, the capacity of links, and the QoS requirements of every user group, we want to jointly determine the following decision variables: (1) the admission determination of each user group; (2) the routing assignment (a tree for multicasting or path for unicasting) of each user group; and (3) the link set metrics of each link (Song, X. 2011).

Table 1. Notation of given parameters for problem I.

Given Parameters

Notation	Description
G	The set of user groups requesting for connection.
e_g	Revenue generated from admitting multicast group g.
V	The set of nodes in the graph (network)
L	The set of real links in the graph (network)
T_g	The set of trees in the network for multicast/unicast group g.
α_g (packets/sec)	The mean arrival rate of multicast traffic for each multicast group g.
t'_g	An artificial tree for group g with zero cost/revenue, and the link capacity of the tree is infinite
T'_g	$T_g \cup \{t'_g\}$.
D_g	The set of destination of multicast group g.
N_g	The number of multicast group g.
P_{gd}	The set of paths that destination d of multicast group g may use.
P'_{gd}	The set of paths that destination d of multicast group g of the artificial tree t'_g may use.
P'_{gd}	$P_{gd} \cup \{p'_{gd}\}$.
δ_{pl}	1 if link l is on path p, and 0 otherwise.
C_l (packets/sec)	The capacity of link $l \in L$.
h_g	The minimum number of hops to the farthest destination node in multicast group g.
r_g	The multicast root of multicast group g.
I_{r_g}	The incoming links to node r_g.
I_v	The incoming links to node v.

Objective function:

$$\min \sum_{g \in G} -e_g z_g$$

subject to:

$$\sum_{g \in G} y_{gl} \alpha_g \leq C_l \qquad \forall l \in L \qquad (1)$$

$$\sum_{l \in L} y_{gl} \geq \max\{h_g, |D_g|\} \qquad \forall g \in G \qquad (2)$$

$$y_{gl} = 0 \text{ or } 1 \qquad \forall l \in L, g \in G \qquad (3)$$

$$\sum_{d \in D_g} \sum_{p \in P_{gd}} \delta_{pl} x_{gpd} \leq (N_g - 1) y_{gl} \qquad \forall g \in G, l \in L \qquad (4)$$

$$\sum_{p \in P'_{gd}} x_{gpd} = 1 \qquad \forall d \in D_g, g \in G \qquad (5)$$

$$x_{gpd} = 0 \text{ or } 1 \qquad \forall d \in D_g, g \in G, p \in P'_{gd} \qquad (6)$$

Table 2. Notation of problem decision variables.

Notation	Descriptions
α_l	The link set metrics for link $l \in L$.
Z_g	1 if the multicast group g is admitted to the network and 0 otherwise.
Y_{gl}	1 if link l is on the subtree adopted by multicast group g and 0 otherwise.
x_{gpd}	1 if path is selected for group g destined for destination d, and 0 otherwise.

$$\sum_{q \in P_{gd}} \sum_{l \in L} a_l x_{gqd} \delta_{ql} \leq \sum_{l \in L} a_l \delta_{pl}$$

$$\forall d \in D_g, g \in G, p \in P_{gd} \qquad (7)$$

$$a_l \geq 0 \qquad \forall l \in L \qquad (8)$$

$$z_g = \sum_{p \in P_{gd}} x_{gpd} \qquad \forall g \in G, d \in D_g \qquad (9)$$

$$z_g = 0 \text{ or } 1 \qquad \forall g \in G \qquad (10)$$

$$\sum_{l \in I_v} y_{gl} \leq 1 \qquad \forall g \in G, v \in V - \{r_g\} \qquad (11)$$

$$\sum_{l \in I_{r_g}} y_{gl} = 0 \qquad \forall g \in G. \qquad (12)$$

The objective function is to maximize the total "revenue" e_g of servicing the admitted multicast groups g, where $g \in G$ and G are the set of user groups requesting for connection. Revenue e_g can be viewed to reflect the priority of user group g, whereas different choices of e_g may provide different physical meanings of the objective function. For example, if e_g is chosen to be the mean traffic requirement of user group g, then the objective function is to maximize the total system throughput. If e_g is chosen to be the earnings of servicing user group g, then the objective function is to maximize the total system revenue. In general, if user group g is to be given a higher priority, then the corresponding e_g may be assigned a larger value A user group would be rejected whenever its the throughput requirement cannot be satisfied.

Constraint (1) is referred to as the capacity constraint, which requires that the aggregate flows on each link l do not exceed its physical capacity C_l. Constraints (2) and (3) require that the number of links on the multicast tree adopted by the multicast group g be at least the maximum of h_g and the cardinality of D_g. The h_g and the cardinality of D_g are the legitimate lower bounds of the number of links on the multicast tree adopted by the multicast group g. Constraint (4) is referred to as the tree constraint, which requires that the union of the selected path(s) for the destination(s) of user group g forms a tree. Constraints (5) and (6) require that exactly one path is selected for each multicast source/destination pair. The left hand side of (7) is the routing cost for multicast source/destination pair of group g. The right hand side of (7) is the cost of path $p \in P_{gd}$. Constraint (7) requires that for each multicast source/destination pair the shortest path be used to carry the individual addressed traffic where the link set metrics of link l is a_l. Constraint (8) requires that the link set metrics be non-negative. Constraint (9) and (10) require that in the case of a group g that is not admitted to the network, the flag of group g: z_g must be 0 and no paths should be selected. Both constraints (11) and (12) are redundant. Constraint (11) requires the number of selected incoming links y_{gl} to node to be 1 or 0. Constraint (12) requires no selected incoming links y_{gl} to the node, which is the root of multicast group g. Therefore, the links we select can form a tree.

3 HEURISTICS FOR THROUGHPUT MODEL

To calculate primal feasible solutions for the throughput model, solutions to the Lagrangean Relaxation problems are considered (Cogill, R.,2011). The set of $\{x_{gpd}\}$ may not be a valid solution to problems, because the capacity constraints are relaxed. Capacity constraint may be violated for some links. The set of $\{y_{gl}\}$ obtained by solving subproblems may not be a valid solution. It is because of the link capacity constraint and the union of $\{y_{gl}\}$ may not be a tree.

Thus, we need additional heuristics to obtain a primal feasible solution. In this section, we describe the details of the heuristics.

3.1 Sort heuristics

We need to sort all user groups $g \in G$ by their revenue e_g divided by their cost. The cost of user group is the traffic demand of group $g*$ the number of destination of user group g. Pick the group with the largest value and try the next group.

3.2 Link set metrics

In each iteration of solving dual problems, a set of multiplier $\{\beta_l, \varepsilon_{gl}, \lambda_{gd}\}$ is used. A heuristics for determining the link set metrics is to let a_l be $1*\beta_l$, if β_l is not zero. Otherwise, a_l should be 1. Dijkstra shortest path algorithm is then applied to determine the shortest path spanning tree for each group root to deliver the multicast traffic.

3.3 Drop heuristics

Each group constructs its multicast tree, but there is no guarantee that link capacity constraint is not violated. Then, we check traffic flow of each links. If the capacity constraint is satisfied, we get a feasible solution. If it is not, we will reject some user groups and they are assigned to an artificial tree t'_g. The steps of drop heuristics are as follows:

1 Sort all user groups by the value of their revenue divided by the resource they used. The resource they used is their traffic demand α_g* the number of links they selected.
2 Pick the smallest value one and remove it from the network. Next, assign the group to the artificial tree.
3 Check all links. If the capacity constraint is satisfied, stop drop user group.

If it is not satisfied, repeat steps 1 and 2.

3.4 Add heuristics

After drop heuristics, we get a feasible solution that some user groups were dropped. Those groups need another chance for being admitted and for improving the total revenue of the network. The steps of add heuristics are as follows:

1 Sort all user groups that were not admitted based on their traffic demand α_g.
2 Pick the smallest traffic demand one and try to place it in the network.
3 Check all links. If the capacity constraint is satisfied, repeat steps 1 and 2. Otherwise, stop the add heuristics.

4 COMPUTATIONAL EXPERIMENTS

In order to prove that our heuristics is good enough, we also implement one simple algorithm to be compared with our heuristics.

Let a_l to be the inverse of link capacity. According to the link set metrics, every user group constructs its multicast tree. Check all links. If the capacity constraint is satisfied, stop it. Otherwise, drop the user group as the sequence in their group ID.

The model and algorithm were coded in Python and run on a Pentium 4 2.0 G PC with 2 GB RAM. The maximum number of iterations was set to 1000 iterations, but it is flexible to reduce the number of iterations in programs in some special cases (Jahanshahi, M. 2012).

We have tested the algorithms on Mesh networks with 9 nodes. Representative results have been selected for the purpose of demonstration. For each test network, several distinct cases are considered that have a different pre-determined capacity of links and traffic requirement of users. The traffic demand for each user group is drawn from a random variable that is uniformly distributed in a pre-specified range, which is shown in the second column of Table 3. The third and fourth columns show the number and the total traffic demand of new user groups, respectively. Both light and heavy loads are considered in the network, which one can tell by jointly considering the second, third, fourth, and fifth columns. The fifth column is the number of users admitted after applying our model.

Table 3. Summary of computational results of throughput model by using proposed method on mesh network.

Case No.	Traffic Range (Mbps)	No. of Requested User Groups	Total Traffic Of Requested User Groups (Mbps)	No. of Admitted user groups
1	2.0	6	12	6
2	3.0	6	18	6
3	2.0~4.0	6	18.229	6
4	3.0~5.0	6	23.892	6
5	2.0	9	18	9
6	3.0	9	27	9
7	2.0~4.0	9	26.971	9
8	3.0~5.0	9	35.888	7
9	2.0	12	24	12
10	3.0	12	36	11
11	2.0~4.0	12	35.839	11
12	3.0~5.0	12	48.899	8
13	2.0	15	30	15
14	3.0	15	45	12
15	2.0~4.0	15	46.762	12
16	3.0~5.0	15	60.212	10

From the computational results, it is observed that excellent results can be obtained by the throughput model for the Mesh network. For the tested networks, the average error differences are, respectively, 0.026, which means that the solutions of using the proposed method are near optimal.

Table 4. Comparison of proposed method, heuristics, and simple method on mesh network in the throughput model.

Case No.	Traffic Range (Mbps)	No. of Requested User Groups	S1	H1 Improve to S1 (%)
1.	3.0	6	23	4.35
2.	3.0~5.0	6	19	21.05
3.	3.0	9	32	9.38
4.	2.0~4.0	9	33	6.06
5.	3.0~5.0	9	25	8.00
6.	3.0	12	40	15.00
7.	2.0~4.0	12	40	12.50
8.	3.0~5.0	12	31	9.68
9.	2.0	15	56	7.14
10.	3.0	15	40	22.50
11.	2.0~4.0	15	45	4.44
12.	3.0~5.0	15	32	25.00

In Table 4, our algorithm performs better than the S1 heuristic for the throughput model. For the test network, our algorithm achieves approximately 4.35% to 36.73% (average 18.66%) improvement in the total revenue compared with S1 heuristics.

The revenue of each user group is 3, 4, or 5. The number of destination of the user group is 2 or 3 in the throughput model. The root and the destinations of groups are randomly selected in the network.

5 CONCLUSION

At first, we formulate the problem into mathematical formulations. In this solution procedure, we relax some complicated constraints and decompose the primal problems into several sub-problems. Primal feasible solutions are obtained by some heuristics, and we propose a link set metrics adjust method by the aggregate flow on links. We implement the algorithm and test mesh network topologies. In computational experiments, the proposed algorithm determines solutions that are within a few percent of an optimal solution with 9 nodes in the throughput model (error difference in the throughput model is 0.047 on average). In terms of performance, our solution has more significant improvement than simple heuristics. The improvement in the total revenue can reach 19% on an average in the throughput model.

The contribution of this research would be with regard to both practical and academy value. In practice, we implement an algorithm for constructing a QoS-constrained MOSPF network; whereas in academy, our algorithm is a realization of QoS-constrained multicast routing by optimization-based technique and the effectiveness of the algorithms is quite good.

In this article, we only consider fixed-link capacity and whether it can expand dynamically by adding the cost user groups admitted in the network. In other words, one user group can reduce the revenue it takes to expand some bottleneck link capacity. In this manner, the group can be admitted to the network.

REFERENCES

Wang, C., Li, X.Y. & C Jiang. 2011. Multicast Throughput for Hybrid Wireless Networks under Gaussian Channel Model. *IEEE Transactions on Mobile Computing* 10(6):839–852.

Karande, SS., Wang, Z. & Sadjadpour, HR. 2011. Multicast throughput order of network coding in wireless ad-hoc networks. *IEEE Transactions on Communications* 59(2):497–506.

Zhao, X., Guo, J. & Chou, CT. 2011. A high-throughput routing metric for reliable multicast in multi-rate wireless mesh networks. *Proc of 2011 Proceedings IEEE INFOCOM, Shanghai, 10-15 April 2011*. USA: IEEE CFP.

Song, X, Ji, L. & Ning, C. 2011. MANET Multicast Model with Poisson Distribution and Its Performance for Network Coding. *IEICE transactions on communications* E94-B (3):823–826.

Cogill, R. & Shrader, B. 2011. Stable Throughput for Multicast With Random Linear Coding. *IEEE Transactions on Information Theory* 57 (1): 267–281.

Lim, WS., Kim, DW. & Suh, YJ. 2012. Design of efficient multicast protocol for IEEE 802.11 n WLANs and cross-layer optimization for scalable video streaming. *IEEE Transactions on Mobile Computing* 11(5): 780–792.

Jahanshahi, M., Dehghan, M. & Meybodi, MR. 2011. A mathematical formulation for joint channel assignment and multicast routing in multi-channel multi-radio wireless mesh networks. *Journal of Network and Computer Applications* 34(6): 1869–1882.

Information, Computer and Application Engineering – Liu, Sung & Yao (Eds)
© 2015 Taylor & Francis Group, London, ISBN 978-1-138-02717-6

An improved flooding protocol based on network coding

Li Na Zhang, Deng Yin Zhang & Ying Tian Ji
Key Lab of Broadband Wireless Communication and Sensor Network Technology,
Nanjing University of Posts and Telecommunications, Ministry of Education, Nanjing, China.

ABSTRACT: Network coding allows network node to process data by the traditional method of data forwarding, which can effectively improve network throughput, robustness and reliability. In this paper, we propose an improved flooding protocol NCFD (network coding of flooding), which combines network coding technology with flooding protocol, adopting linear network coding technology based on multipath propagation. Computer simulation results show that NCFD could significantly reduce network energy consumption and data redundancy.

KEYWORDS: WSN; Flooding protocol; Network coding; Transmission algorithm;

1 INTRODUCTION

Wireless sensor network (WSN) has the advantages of low cost, high flexibility and the advantages of large-scale self-organized network, which has a broad prospect of application. Flooding protocol (Yan, Zhao & Zhang 2008) is one of the original protocols in WSN. The flooding protocol is necessary for many operations in WSN such as data forwarding, time synchronization, node localization and forming a routing tree. Moreover, in almost all the routing protocols, the flooding strategy is utilized in the stage of data transmission and route discovery. Consequently, the improvement of flooding protocol is of important significance for the performance of itself and other protocols.

Flooding protocol prescribes that when a node completes the initialization of broadcast message, it sends the message to all its neighborhood nodes. When a node receives the flooding algorithm message (FAM) for the first time, it broadcasts this message to its neighborhood nodes. If the node has already received FAM, it discards the message, sends the information data to the destination node and eventually each node in the network receives FAM. The defect of this protocol is the overload of the network and the waste of resource. However, the flood protocol is still a simple and reliable routing algorithm. The flooding in the whole network is the most effective way with excellent robustness in the scene where the nodes movement is intense and the network changes frequently.

Gossiping protocol (Intanagonwiwat, Govindan & Estrin 1998) is one of the improved flooding protocols. It selects a forwarder randomly from its neighbor nodes while forwarding data and this progress is repeated by the selected forwarder. The Gossiping protocol saves the energy but it extends the time consuming and reduces the reliability. Heizelman (Xiao, Wei & Zhou 2006) proposes the SPIN (Sensor Protocol for Information via Negotiation) protocol which focuses on the data and reduces the data size in the way of negotiation between nodes. Some researchers (Govindan, Estrin & Heideman 2000) put forward the DD (Directed Diffusion) protocol in which the Sink node broadcasts a kind of data packet called interest periodically. In this method, the other nodes are informed what information the current node needs which reduces the blindness in flooding progress, saves the node energy and improves the bandwidth utilization. However, for a long time, the improvement direction of the flooding protocol is still limited to the forward method modification.

Network coding is a significant breakthrough in the area of network communication. It allows the network nodes to process the data which is able to increase the network throughput, improve the robust and reliability. The network coding offers a new train of thought to overcome the problem of heavy load and the resource waste in the flooding protocol.

In this paper, we design a network coding scheme with XOR algorithm and we apply this scheme in flooding protocol which forms an innovative routing protocol based on network coding called NCFD. It demonstrates via simulation that our method can effectively reduce the energy consumption and redundancy of network transmission with the increase of nodes in WSN.

2 PRINCIPLE OF FLOODING PROTOCOL

The specific idea of flooding routing algorithm is that each node in WSN is able to receive and forward data packets through broadcast and the received duplicate packets are discarded. The flooding protocol results in the diffusion of data packets centering the source node. In order to prevent the diffusion from occupying too much network resource and to converge the diffusion, it is necessary to set an appropriate TTL (time to live) value to ensure only limited routes are taken in data transmission. Furthermore, to check duplicate packets, each node has to maintain the data packet sequence number SEQ and the neighbor information table NIT. When the source node sends every single data packet, it records the corresponding SEQ into NIT based on which the duplicate packet check is carried out. The realization of flooding algorithm includes three stages which are initialization, route establishment and data forwarding.

2.1 Initialization

Every node broadcast its node information message and the node which receives the message stores related information into NIT. In this stage, each node knows which nodes are directly connected and the connection delay between every two nodes.

2.2 Route establishment

Step1: The source node check its Rendezvous Point Tree (RPT) table, if it is one of the nodes in Route Reply Packet (RREP) message, it forwards the information along a definitive path in a certain RREP. Otherwise, it broadcasts a new Route Request (RREQ) to forward the specific information to next node.

Step2: Node A checks its RPT table when it receives RREQ. Node A replies the RREP if it goes to the upper layer along a definitive path determined by the RREP or reduces the message TTL by 1 which means the message has already taken a hop. When the TTL is reduced to 0, the message will be discarded automatically.

Step3: Node A waits for a certain time. If there is any RREQ costing less energy from the same source node, node A saves the most energy-saving message and wait to forward it until the RREQ arrives at sink.

Step4: Node A discards the message and stores the message information into its own array meanwhile for check whether the message has already been received later.

3 NCFD ALGORITHM DESIGN

The network coding technology is adopted in our NCFD protocol to process data packets which saves the bandwidth resource and energy effectively. The protocol realizes the network coding with the simple XOR operation which means even low computer power is enough. The NCFD protocol add a data processing module between the route establishment and data forwarding based on COPE (Dong, Qian & Chen) coding mechanism.

3.1 Data processing

As Figure 1 shows, we suppose a node sends out data packet p, its neighbor node receives p and the steps of received data processing are as following.

Step1: If p is an original and new data packet, the node updates its NIT and backups the message p into its receiving sink. If p is original but the data is not new, then the node discards it. Then judge whether all the neighbor nodes receive this data packet. If didn't, wait for the coding opportunities and encode the data packet, then send it. Once there's no coding opportunities, send the packet to the output sequence and wait for a random length. If all the neighbor nodes receive this data packet, this node doesn't need to forward it.

Step2: If p is a coded data packet, firstly need to decode this data packet. Then return to step1.

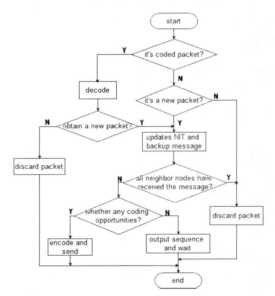

Figure 1. The process of receiving message.

3.2 Coding progress

Step1: Firstly, the node check the data packet whether it is a source data packet. If so, the node sends it out directly. Otherwise, the node calls coding algorithm to look for the coding chance and encodes the packet if the chance exists.

Step2: The node checks its NIT to judge whether its neighbor nodes are able to decode the coded packet. If the packet contains all n source data packets and the NIT has at least n-1 packets, it means the neighbor node is able to decode the coded packet.

Step3: If the conditions are fulfilled, the node conducts XOR operation.

Step4: The node updates the data packet header based on the coded packet set after the whole output queue is traversed through.

3.3 Decoding progress

Step1: Obtain the coded packet information. The node gets the information about all the data packets participating in the coding progress, include the source node address, SEQ, the send time and the total of source data packets in the coded packet.

Step2: Check whether the coded packet can be decoded. The node compares the coded packet with its data sink and gets the number of the different packets. If the number is 1, it means the coded packet can be decoded or it means decoding fails. At the same time, the node records the source address and SEQ of the data packet which is not in the data sink. The information recorded is the information of the decoded original data packet when decoding successes.

Step3: Restore the original data packet. The node conducts XOR operation with the coded packet and known original packet to get the new data packet. Meanwhile, the node updates the packet header by writing the corresponding information in the coded packet head into the new one.

3.4 Example

The node A has data packet Q1, the node B has Q2 and the node C has Q3. We assume that all A, B and C know it own neighbor 2-hop state. Once A, B and C broadcast their own data packet, the node E will receive Q1, Q2 and Q3. The node E updates its NIT with the last hop and 2-hop neighbor node table as shown in Table 1.

Table 1. NIT of node E.

Node	Q_1	Q_2	Q_3
E	1	1	1
A	1	0	1
B	0	1	0
C	1	0	1
D	0	1	0

* An example of new algorithm.

The node E looks for the data packet that can be coded from the output queue before it sends out Q1. Firstly, the node E checks whether its neighbor nodes

can get the missing packet by decoding if $Q1 \oplus Q2$ is sent out. For example, due to all the node A, B, C and D have at least one of Q1 and Q2, they are able to get the missing packet by calculating the OXR of the coded packet and their own packet. Then, the node E checks if it is possible to add another original data packet, and to send out $Q1 \oplus Q2 \oplus Q3$. In this situation, in the view of the node A, it only has the packet Q1. Therefore, it cannot get the missing Q2 and Q3 from $Q1 \oplus Q2 \oplus Q3$ using XOR operation. Finally, the result is that the node E can only code Q1, Q2 and send $Q1 \oplus Q2$. When its four neighbor nodes receive this coded packet, the node B and C can get the missing packet Q1 by calculating the XOR of its own Q2 and the received coded packet. The node A and C can also get the missing packet Q2 in the same way.

4 SIMULATION ANALYSIS

4.1 Simulation environment parameters

In this paper, Network Simulator (NS) is applied as the network simulation platform. The evaluation indexes are energy efficiency and data redundancy.

The main parameters of WSN simulation environment in tcl file are set up as Table 2. The output queue is IFQ queue and the queue length ifqlen is 20, the number of wireless nodes nn is 50. The variable rp represents route protocol and it is the proposed NCFD here.

Table 2. Simulation parameters.

Variable Name	Meaning/value
ifq	Queue/DropTail/PriQueue
ifqlen	20
nn	50
rp	NCFD
chan	Channel/WirelessChannel
prop	Propagation/TwoRayGround
netif	Phy/WirelessPhy
ant	Antenna/OmniAntenna
mac	Mac/802_11
ll	LL
x	100
y	100
stop	100

* The main parameters of WSN simulation environment in tcl file

The wireless transmission model is TwoRayGround. WirelessPhy means the network interface is wireless physical layer. OmniAntenna means the Antenna model is omnidirectional antenna. The DCF mechanism in IEEE802.11 is used in MAC layer. The channel capacity is 2Mbps. The node transmission radius is 250m. The variable ll is the link layer parameter meaning the link layer type is the logic link layer LL. The topological range is set to 100m long and 100m wide. The simulation duration is 100s.

4.2 Energy performance analysis

The total energy consumption comparison of the flooding protocol and NCFD is shown in Figure 2. The consumption of flooding protocol tends to increase by a large margin as the node amount increases while the consumption of NCFD changes slowly. When the node amount is below 50, the consumption of NCFD is very little approach to 0. When it is above 50, though the NCFD consumption increases, it is still apparently less than that of flooding protocol and the consumption of NCFD decreases when the node amount is over 100.

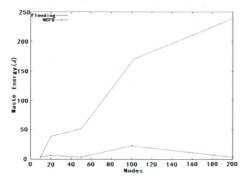

Figure 2.   Energy consumption comparison.

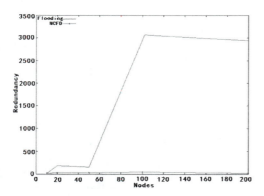

Figure 3.   Redundancy comparison.

4.3 Redundancy performance analysis

The data redundancy comparison of the flooding protocol and NCFD is shown in Figure 3. It is obvious in the Figure 3 that the data redundancy of flooding protocol is showing a rising trend with the increase of node amount while that of NCFD is much less.

5   CONCLUSION

In this paper, based on the characteristic of flooding protocol working in WSN, we introduce the network coding mechanism into the flooding protocol and propose an innovative route protocol based on network coding called NCFD with its detailed implementation scheme. It demonstrates via simulation that our method can effectively reduce the energy consumption and redundancy of network transmission with the increase of nodes in WSN.

ACKNOWLEDGMENTS

This research work is supported by the National Natural Science Foundations of P. R. China (NSFC) under Grant No. 61071093, National 863 Program No.2010AA701202, Returned Overseas Project, Jiangsu Province Major Technology Support Program No.BE2012849, Graduate Research and Innovation project No. KYLX_0812.

REFERENCES

Yan Y & Zhao Z. 2008. Rate-adaptive coding-aware multiple path routing for wireless mesh networks. *IEEE Global Telecommunications Conference*: 543–547.

Sze-Yao Ni & Yuh-Shyan Chen. 1999. The broadcast storm problem in a mobile ad hoc network. *IN: Proc of the 5th ACM/IEEE Int'l Conf on Mobile Computing and Networking (MOBICOM). New York: ACM Press*: 151–162.

Intanagonwiwat, C. & Govindan, R. & Estrin. D. 1998. Directed Diffusion: A Scalable and Robust Communication Paradigm for Sensor Networks. *Proceeding of the 2nd international conference on information processing insensor network*: 326–342.

Debao Xiao & Meijuan Wei. 2006. Secure-SPIN: Secure Sensor Protocol for Information Via negotiation for Wireless Sensor Networks. *Proceedings of the 1st IEEE conference on Industrial Electronics and Applications*: 105–113.

Ahlswede, R. & Cai, N. 2000. Network information flow. *IEEE Trans on Information Theory*: 1204–1216.

Chao Dong & Rui Qian. 2011. Research of network coding COPE in ad hoc. *Journal on communications* 10(12).

Kohvakka, M. & Suhonen, J. 2009. Energy-efficient neighbor discovery protocol for mobile wireless sensor networks. *Ad Hoc Networks* 7(1): 24–41.

Park, J.S. & Lun, D.S. 2006. Performance of network coding in ad hoc networks. *The 25th Military Communications Conf (MILCOM 2006)*.

Deb, B. & Bhatnagar, S. 2003. Nath B. Reinform: reliable information forwarding using multiple paths in sensor networks. *IEEE Int'l Conf. on Local Computer Networks (LCN)*: 406–415.

Katti, S. & Rahul, H. 2008. XORs inthe air: Practical wireless network coding. *IEEE/ACM Transactions on Networking*: 497–510.

Information, Computer and Application Engineering – Liu, Sung & Yao (Eds)
© 2015 Taylor & Francis Group, London, ISBN 978-1-138-02717-6

A pedestrian detection method based on target tracking

Jun Wei Zhao
College of Computer Science, North China Institute of Science and Technology, Beijing, China

Qing Tao Hou
Shandong Huiruo E-commerce Co., LTD, Jinan, China

Jing Gang Zhang
College of Safety Engineering, North China Institute of Science and Technology, Beijing, China

Sen Yang
Shandong Luneng Intelligence Technology CO., Ltd, China

ABSTRACT: Due to the storage time and space constraints, existing pedestrian detection algorithms cannot meet the real-time needs of intelligent video surveillance. An approach based on motion analysis is proposed to solve this problem. The method only extracts gradient histogram features from the objects containing motion information. It avoids window scan on the whole image and saves the processing of Pyramid transformation. So the efficiency of image processing is greatly improved. Compared with conventional pedestrian detection, experimental results show that the approach is fast and effective and it can meet the real-time needs of intelligent video surveillance.

KEYWORDS: Video-analyze; Moving target detection; Gradient histogram; Pedestrian detection

1 INTRODUCTION

Alongside the development of security industry, video surveillance has become more and more intelligent. Simultaneously, image processing, pattern recognition, and artificial intelligence technology have been widely used in the fields of the security video surveillance. On the one hand, the growing prosperity of the surveillance industry brings safety; on the other hand, it brings a huge amount of video data. So, how to retrieve specific video segments that containing pedestrian movement in such a huge amount of data has become the focus of concern.

Currently, the pedestrian detection and retrieval methods are mainly used for contour features of the human body, namely gradient histogram (Histograms of Oriented Gradient, referred to as HOG). Then, they carry out the scale transformation on the image and extract the HOG features on different scales. Finally, they support the vector machine (Support Vector Machine, referred to as SVM) that is used to feature training and classification[1].But in the conventional scheme, the image scale transformation, such as wavelet transform and Pyramid transformation, needs a lot of time and space overhead[2]. As we know, HOG feature extraction needs much time and space overhead, so HOG feature extraction in each scale would enable a higher complexity to be achieved[3]. If this scheme is used to analyze video images in real time, the video may cause a serious delay. In this article, a pedestrian detection method based on motion feature analysis has been proposed. This method is improved based on real-time video monitoring and the accuracy of pedestrian detection. Experimental results show that the approach is fast and effective and it can meet the real-time needs of intelligent video surveillance.

2 PEDESTRIAN DETECTION SYSTEM BASED ON VIDEO ANALYSIS

In the products of the intelligent video surveillance industry, pedestrian detection is a necessary method of video retrieval that has broad application prospects. But real time is a big constraint in intelligent video surveillance. Many mature image detection and recognition algorithms may meet the efficiency bottleneck problem in the real-time video monitoring. This system mainly includes three parts: First, moving target detection on real-time video stream is used to segment moving objects in each frame. Then, the moving target segmentation image is normalized and the HOG feature in the normalized image is extracted. At last, the HOG features as a feature vector has been input SVM feature classifier in order to judge whether the moving target is a pedestrian, and the result is finally output.

The concrete structure is as shown in Figure 1.

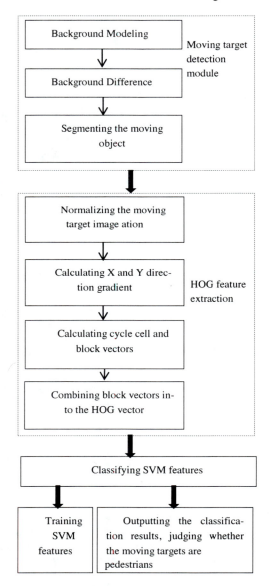

Figure 1.　System structure diagram.

3　DESIGN OF MAIN MODULES IN THE SYSTEM

3.1　*Introduction of the moving target detection*

Generally, the method of moving target detection is background extraction and then background difference. The background modeling is commonly used with camshift, Mixed Gauss (Mixture of Gaussian, referred to as MOG), and codebook. Camshift algorithm is applied to a color video image, but it is not suitable for this method as the video image is almost gray at night. There were no significant differences between both the MOG and the codebook in terms of effect. Seen from the principle, they are both background extraction in the RGB color space[4]. But in the leaves swaying or rippling wave interference in the environment, the MOG stability is better. So the MOG is used for background modeling in this article. The image containing moving objects could be obtained by making a difference between the background image and each frame in a video[5]. In the follow-up of the pedestrian detection step, we only need to carry out the qualitative analysis of these image blocks, and we would be able to judge whether an image contains a pedestrian, avoids scaling, progressive scan high complexity operations on the whole image.

3.2　*Introduction of HOG feature extraction*

HOG is a feature descriptor that is used to describe the target contour, which was proposed by Dalal and Triggs in 2005, researchers of the French national computer technology and control (INRIA)[6].

The first step is used to normalize the target segmentation of image blocks. The normalized size should be similar to that of the SVM training sample. The sample size is 64 x 128 pixels in this article.

Then, the image gradient is calculated using the gradient operator: (-1,0,1). This operator has the advantages of simple calculation, strong practicability, and it can effectively save the test time.

The image is divided into cells. The gradient of the image is projected into each cell, and we can obtain a series of histograms. Then, the image is divided into blocks according to the cell. Each block comprises a plurality of cells. The histogram in each cell is normalized in the block unit. Then, we connect the normalized vector for each block, forming a large vector, which is the HOG feature vector of the target image.

3.3　*Introduction of SVM feature training and classification algorithm*

SVM is a solution that was proposed by V. N. Vapnik, in order to solve nonlinear problems in the 1960s. It constituted a complete theory in the 1990s and was successfully applied in the pattern recognition and artificial intelligence field. It takes attention wildly due to its strong generalization ability.

This method is used to train the sample feature in this article. Manual intercept samples that include 1000 positive images and 1000 negative images are normalized to 64 x 128 pixels in size. Positive and negative samples from each of the 800 pairs are used for training; whereas the other samples are used for testing. In order to ensure the diversity of samples, the positive samples of pedestrians include clothing in a variety of colors, all kinds of shapes, and so on. The negative samples randomly include various environmental backgrounds. The stable feature weights were obtained after SVM training stores in a local file in order to use classification [7].

The HOG feature vectors are input to the classifier. They could make judgment quickly on the input vector by loading the feature weights file that was saved in the local file. The results of pedestrian detection in video image are shown in Figure 2.

4 PERFORMANCE TESTING AND ANALYSIS OF THE EXPERIMENTAL RESULTS OF THE SYSTEM

This system does not affect the accuracy of pedestrian detection algorithms; rather, it improves and optimizes the video processing algorithms, whose aim is mainly to meet the real-time intelligent video surveillance. In Figure 2, (a) is the 1st frame image of the video; (b) is the 129th frame image of the video, and it has more pedestrians and cars compared with (a); (c) is the result image of the video motion detection, in which it can be seen that there are two separate movement target

(a) The first frame

(b) The 129th frame

(c) Motion detection result

(d) Pedestrian detection result

Figure 2. System detection result.

masses of the pedestrians and vehicles; and (d) is the result of extracting the HOG features from the original images and classifying the features by SVM according to the mass position, and the target of the pedestrian is in the white box, whereas the target of the non-pedestrian is in the black box.

Second, different solution images before and after the improvement were tested, respectively, and it can be seen that the processing speed of every frame data is significantly improved. Table 1 shows the data processing time before and after the improvement.

Table 1. Data processing time before and after the improvement.

Image resolution	The average time before the improvement/ms	The average time after the improvement/ms
352×288	121.951	52.632
704×576	337.295	145.571
1280×720	960.943	414.727

5 CONCLUSION

This system introduces the motion detection information, combines with the original HOG feature extraction, and omits the high complexity of the process such as scale transformation in the static images of the human detection method. So, it greatly improves the efficiency of intelligent video surveillance in pedestrian detection; runs truly fast and efficiently; and fully meets the real-time requirements of the video surveillance. Currently, the method has been put into use in the actual product and in the long run, intelligent video analysis has a huge market space and development prospects.

ACKNOWLEDGMENT

This work was financially supported by "the Fundamental Research Funds for the Central Universities of China" (3142013072, 3142013103).

REFERENCES

[1] Qu YongYu, Liu Qing. Pedestrian detection based on HOG and color features [J]. Journal of Wuhan University of Technology, 2011, 33(4):137–141.
[2] JiaHuiXing, Zhang YuJin. A review of pedestrian detection based on computer vision research vehicle auxiliary driving system [J]. Journal of automation, 2007, 33(1):84–90.
[3] Huang Xi, Gu JieFang. Pedestrian detection based on the gradient vector histogram [J]. Science technology and Engineering, 2009, 9(13):3446–3451.
[4] Huo DongHai, Yang Dan, Zhang XiaoHong. A Codebook algorithm for background modeling based on principal component analysis [J]. Journal of automation, 2012, 38(4):591–600.
[5] Huang YongLi, Cao DanHua, Wu YuBin. Segmenting the motion human body image in real-time monitoring system [J]. Opto electronic engineering, 2002, 39(1): 69–72.
[6] NavneetDalal, Bill Triggs. Histograms of Oriented Gradients for Human Detection. [J] Proceedings of the 2005 IEEE Computer Society Conference on Computer Vision and Pattern Recognition (CVPR'05), 2005, 886–893.
[7] Tian Guang, Qi Fei Hu. Hierarchical pedestrian feature transformation and SVM detection algorithm based on mobile camera environment [J]. Chinese Journal of Electronics, 2008, 36(5):1024–1028.

Information, Computer and Application Engineering – Liu, Sung & Yao (Eds)
© 2015 Taylor & Francis Group, London, ISBN 978-1-138-02717-6

Enterprise human resources management system platform optimization design in the information technology era

Yan Zi Wang & Lei Yun Zhang

Jiangxi Science & Technology Normal University, China

ABSTRACT: The computer, network, and other modern information technology has been widely applied in the management field with the advent of the information technology era. The human resources management is an important part of enterprise management, and the management efficiency has great significance for the sustainable development of enterprises. In the information technology era, enterprises should actively and fully make use of information technology, constantly optimize the human resource management system platform, in order to improve the human resources management level, and promote a good and healthy development of enterprises.

KEYWORDS: Information technology; enterprise; human resource management system; optimization

1 INTRODUCTION

Under the market environment of information technology era, the traditional human resource management model is unable to meet the requirements of personnel training and personal incentives, talent evaluation, and technology innovation. Nowadays, the enterprise human resources management is gradually becoming more and more networking and informationized, and it presents the features of adapting to the enterprises' internationalization development and information management and so on. Based on continuously improving the information system and the application of new technology, the enterprise human resources management system just emerged because of the opportunity.

2 PROBLEMS IN THE CURRENT DEVELOPMENT OF THE HUMAN RESOURCE MANAGEMENT SYSTEM

2.1 Human resources management control and localization problem

Many enterprises in the middle of rapid development, due to the continuous expansion of the business scale, undergo a series of human resource management issues of the subsidiary enterprise and parent company. For an enterprise, the enterprise human resources management is its new management field; the enterprise human resources management reform should not only change the traditional mode of human resource management but also require enterprises to

have a good interaction with the subsidiary company in the human resources management aspect, to meet the needs of enterprise expansion. However, for some new enterprises, their development time is not so long, so there are a lot blind spots for them in the human resources management, which will pose a great challenge to the development of enterprise economy.

2.2 The business has gone adrift with the system demands

Currently, some enterprises rely on the manual mode for maintenance and management organizational changing information; this mode enables the company to have a larger workload and a low accuracy, the superior department summary information just according to the organizational level reporting, it is extremely easy to cause information distortion with this reporting information to leaders mode. The companies have not yet specified the department name of the different invested enterprises' same business attributes and same business level, and the invested company always determines their own department name; for instance, the human resources department has so many names: the labor and personnel department, human resources department, the ministry of human resource development, labor wage department, and so on. Because of the different principalship systems between units, the headquarters of the enterprises have great difficulty in the office staffs statistics. In addition, due to the frequent enterprise transfer of personnel, it will be hard to maintain the staff information in time, thus resulting in the data analysis and statistical errors. There also exists the repeated maintenance of the enterprise personnel information by the phenomenon of different departments; statistic data are very difficult, time consuming, and labor intensive.

2.3 *The human resources system problems*

For the enterprise human resources management, the human resources data statistics is extremely important, whether can clearly count the number of staffs and control the total amount of wages or not are both need timely, accurate, and effective statistical data. The existing system of some enterprises is unable to meet the demand of the related statistical data. The different needs of enterprises are specifically manifested in the following aspects: Enhance the query efficiency of the enterprise information and implement the complex statistical functions. For example, reflect all the information of the staff in the corresponding basic information data, and the information data should also be conductive to statistic the same information for simplifying the information query function. Currently, a lot of enterprises always query information on staff through the cooperation between some departments, and the cumbersome procedures cause a waste of the work time. In addition, because the enterprises are using the Excel form to implement management and statistics of the personnel information, this form enables personnel to have a larger workload and lack of data accuracy; therefore, the company must use the special human resources management software. Because of the frequent enterprise personnel labor transfer, if we adopt the manual management mode, it must will lead to the untimely maintaining of personnel information data and will also result in statistics data error.

3 OPTIMIZATION DESIGN OF THE HUMAN RESOURCES MANAGEMENT SYSTEM PLATFORM

3.1 *Optimization design of personnel information module*

As the basic data of the human resources management system, the staff information module has a high rate that is inquired by the enterprise superintendents, that is, to say the personnel information really has the central importance for an enterprise. All employees have a large amount of personnel information, and a lot of the information is very important for the enterprises. Combined with the enterprise business, the staff information such as job tenure, degree and the time of entering the industry, the time of entering the company, the political orientation, and so on should be integrated in the "personal information of

staff." We should code the system according to the field selection and option: Some start with the letter, some start with the numeral, and a lot of information exists in the code-breaking phenomenon, which should commence from the unified standard aspect, combined with the system option number. We should take 001 as the beginning to standardize the employee information.

3.2 *Optimization of personnel transaction processing module*

Personnel transaction process is also an important way for the enterprise to control and standardize the everyday work, strictly control the personnel transaction process, personnel transference, personnel entry, or any other examination and approval work. There will always be a strict division of power about the personnel transaction process for most enterprises. Usually, the staff submit the application and the relevant documents and then wait for approval; it should be according to a different company situation to implement the division of power, and then we should let the branch administrators implement the operation and retake the duties beyond the scope of the operation right.

We need to audit the personnel transaction process via the mode of submitting the personnel documents step by step; according to the actual situation of enterprises, we should let the parent company administrator audit the branch administrator. If the enterprise administrator position exists in the job vacancy phenomenon, at this time, the system will automatically identify, for example, whether the principal role at the higher position exists in the job vacancy phenomenon; then, the system will automatically turn to the deputy position. The deputy manager will instead audit, so as to successfully complete the personnel transaction process.

After completing the personnel transaction process, in accordance with the enterprise business information to correspondingly change the personal information, the relevant information in the personnel transaction process is correlative with each function module of the human resources management system, and the correlative information should be adjusted by the system synchronously, to guarantee consistency of the system data.

Some large enterprises have a large number of staff: Whether it is the staff position adjustment, new employees join, or the turnover processing, and so on, there is a long business process. Considering the business process recording operation and the convenience of checking the increase in the enterprise annual staff situation, it must have a document value system process, implement the effective conditions screening, and then select the year and the enterprise that need to be inquired. Then, the system will automatically produce the personnel transaction process, which records the personnel transaction processing situation in detail. Subdirectories of the personnel transaction process include the new staff introduction schedule, re-entry work after relieving the employer–employee relationship schedule, transregional decruitment schedule, transregional taking office, maintaining employer–employee relationship and back to the position schedule, dismission staff schedule, maintaining employer–employee relationship staff schedule, retiree schedule, the position adjustment of the cadre schedule, the cadre exchange and transference particulars in a district, and so on. Taking the post-adjustment schedule as an example, we may include employee code, name, gender, units, and departments before adjustment; the units, department, and position after adjustment; and the highest degree, employee access channels, and effective date.

3.3 Optimization design of personnel report forms module

The personnel report forms filling is daily work for an enterprise: quarterly labor contract, labor and capital report form at the beginning of the month, and employment report. Each kind of report form has many forms of styles: The information in the form also has complex logic relations; the report form of style will make some corresponding adjustment according to the changes in the enterprise; the general business system administrator is familiar with the Excel form but does not understand the report system development process; and the enterprise personnel report should look from the user perspective, effectively dock the Excel form, support the forms introduced by systems, and ensure that the Excel form is flexible and practically used.

The management level of the enterprise generally has the needs of diverse staff rosters, so it is unreasonable to set the roster as a fixed style; for example, the general settings include the serial number, name, employee, gender, date of birth, time of participating in the work, the time of entering the company,

department, position, labor contract type, ID number, and degree. In actual operation, the user can also act according to their own needs to set the related fields, hide the useless field, adjust the field order, select the appropriate departments or enterprises according to the user set to extract effective information from the employee information, and finally display the information in the form mode. This allows the users to independently and flexibly master the roster and make the roster.

4 CONCLUSION

The twenty-first century is an informationization and globalization era. Along with the rapid development of economy and society, enterprises are facing increasingly fierce competition. Amid the fierce market competition, how can the enterprise gain competitive advantage? The competition of talent is fundamental; as long as the enterprise fully masters the human resources, will the enterprise grasp the initiative in the market war? Human resources management system is a modern management model that functions by optimizing the allocation of human resource of the enterprise by information technology. It has important meanings in improving enterprise management efficiency. In the era of information technology, enterprises need to constantly optimize the human resources management system to make it fit in with the actual needs of enterprises; enable the human resources management to fully play its role; and promote the level of corporate human resources management.

REFERENCES

[1] Li Juan: Design and application of the enterprise human resources management system in the new era [J]. goods and the quality goods, construction and development, 5,718(2014).
[2] Chen Zuchuan: The enterprise human resources management system construction research and practice [J]. communication 2,131(2014).

Information, Computer and Application Engineering – Liu, Sung & Yao (Eds)
© 2015 Taylor & Francis Group, London, ISBN 978-1-138-02717-6

High-frequency signal spectrum sensing method via stochastic resonance

W. Mao, D.Y. Zhang & H. Sun
Nanjing University of Posts and Telecommunications, NanJing, China

ABSTRACT: A spectrum sensing method that uses a stochastic resonance system to detect high-frequency signals is depicted. First of all, the method uses the principle of scale transformation to make the stochastic resonance system subject to the detection of high-frequency primary user signal. Then, by two or more cascaded stochastic resonance systems, it eliminates the effects of high frequency of glitches. The result of theoretical analysis and simulation shows the algorithm can make good use of the stochastic resonance system and improve the detection probability of primary user signal.

KEYWORDS: Cognitive radio; Spectrum sensing; Scale transformation; Cascading stochastic resonance

1 INTRODUCTION

The cognitive ratio achieves dynamic spectrum access. It provides efficient solutions for spectrum allocation and increases the flexibility of networks and terminals. Spectrum sensing is the key technology of cognitive ratio. Previous research mainly focused on energy detection (Kun&Hushen2010), variance matrix detection (Shaowen&Jun&Wei2011), and collaborative detection (Di&Lingge2011). The most difficult challenge is how to detect weak signal effectively. In the low noise ratio, the spectrum sensing performance will be substantially reduced. Stochastic resonance is one of the nonlinear phenomena in physics (Benzi&Sutera&Vulpiani1981). In recent years, stochastic resonance is widely used in signal processing fields. There is little research on the application of spectrum sensing of the cognitive spectrum.

Literature (Liangbin2010) proposes a method that achieves the stochastic resonance subject to high-frequency signal detection by increasing sampling frequency. The best linear approximation between cognitive sampling frequencies f_s and primary user signal f is that $f_s = 500f$. When the primary signal exceeds the range, it is hard for the stochastic resonance system to be applied for detecting the high-frequency signal.

Literature (Yonggang&Taiyong&Yan2011) analyzes the stochastic resonance behaviors of bitable systems connected in series. As to the cascading stochastic resonance system, the first stochastic resonance system has the greatest effect, which transforms environmental noise in the cognitive users sampling mixed signal into colored noise. From the beginning of the second stage of the stochastic resonance system, energy of high-frequency noise is transferred to low-frequency noise ceaselessly. It makes the time-domain waveform become smoother and reduces glitch.

Introducing the stochastic resonance in the spectrum sense can solve the detection problems of weak signal. Literature (Di&Yingpei&Chen2010) brings stochastic resonance into collaborative spectrum sense to improve the performance of collaborative spectrum sensing.

Currently, the research that uses stochastic resonance to improve the cognitive radio spectrum sensing performance still remains at the theoretical analysis and research stage. This literature proposes a method to achieve high-frequency primary user spectrum sensing based on cascading stochastic resonance. On the one hand, the method uses the principle of scale transformation to apply stochastic resonance to detect high-frequency primary user signal. On the other hand, by two or more cascading stochastic resonance, we can make the output signal smoother, reduce the influence of high noise glitch, and improve the probability of signal detection.

2 SCALE TRANSFORMATION PRINCIPLE

Suppose $s(t) = A\cos\omega t$ represents weak periodic primary user signal, A represents amplitude, ω represents signal frequency, and a and b are two variable parameters of the stochastic resonance system. Stochastic resonance system model is defined as

$$\begin{cases} \dfrac{dx}{dt} = ax - bx^3 + A\cos\omega t + \Gamma(t) \\ <\Gamma(t)> = 0, <\Gamma(t), \Gamma(0)> = 2D\delta(t) \end{cases} \quad (1)$$

where $\Gamma(t)$ is the Gaussian white noise surrounding cognitive user. Mean of the noise is zero, and strength is D. $\delta(t)$ acts as unit impulse function. For both cases $a>0$ and $b>0$, we can order $z = x\sqrt{b/a}$, $\tau = at$ and computer normalized transformation to equation (1). We obtain

$$a\sqrt{\frac{a}{b}}\frac{dz}{d\tau} = a\sqrt{\frac{a}{b}}z - a\sqrt{\frac{a}{b}}z^3 + A\cos(\frac{\omega}{a}\tau) + \Gamma(\frac{\tau}{a}) \quad (2)$$

Using equation $\Gamma(\frac{\tau}{a}) = \sqrt{2Da}\xi(\tau)$, we have a transition to equation (2) and obtain

$$\frac{dz}{d\tau} = z - z^3 + \sqrt{\frac{b}{a^3}}A\cos(\frac{\omega}{a}\tau) + \sqrt{\frac{2Db}{a^2}}\xi(\tau) \quad (3)$$

Primary user signal frequency is normalized to ω/a from original ω. According to normalized transform principle, the original stochastic resonance system parameters of a and b can be adjusted and magnified in proportion. High-frequency primary user signal can be normalized to low-frequency signal, which enables the stochastic resonance principle to be applied to detect high-frequency signal.

Suppose the high-frequency user signal is f, then the mixed sampling signal can be computed with fast Fourier Transformation. Using the following formula to estimate the possible existing primary user signal frequency f:

$$f = \frac{1}{L}\left[\frac{r|G_{\max+r}|}{|G_{\max}| + |G_{\max+r}|} + k_{\max}\right] \quad (4)$$

where L is the total sampling time by cognitive users. $k_{\max}$ represents the location of the maximum amplitude spectrum after fast Fourier Transform of sampling mixed signal. $G_{\max}$ represents the amplitude of amplitude spectrum in the location of $k_{\max}$. For $|G_{\max+1}| \geq |G_{\max-1}|$, $r=1$. Let us select a set of best stochastic resonance system parameters $\{A_0, f_0, E=0, \sigma_0^2, a_0, b_0, h_0\}$ as reference of scale transformation. In other words, we adjust parameters a, b, and sampling step h. It follows that

$$a = \frac{f}{f_0}a_0 \quad b = \frac{f}{f_0}b_0 \quad h = \frac{f_0}{f}h_0 \quad (5)$$

Noise variable $\sigma_0^2 = 2D_0/h$ in the environment is in the relationship with strength of actual noise and sampling step h. Noise intensity is $D=2D_0\,b/a^2$ after the scale transformation. In addition, because frequency of the main user signal reduces value after normalization, the sampling step should enlarge to a

times accordingly as many as the original. So after scale transformation, the variance of Gaussian noise is $\sigma^2=4D_0b/(a^3h)$; the ratio of the variance before and after scale transformation is given by

$$\frac{\sigma_0^2}{\sigma^2} = \frac{a^3}{2b} \quad (6)$$

Infer that normalized scale transformation equals the value of primary signal amplitude enlarged to $\sqrt{2b/a^3}$ times as many as the original. However, the value of environmental noise enlarged to $2b/a^3$ times as many as the original, we can obtain that

$$A = \sqrt{\frac{2b}{a^3}}A_0 \quad \sigma^2 = \frac{2b}{a^3}\sigma_0^2 \quad (7)$$

According to the principle of scale transformation, we can make good use of the inverse transformation of scale transformation to make parameters adjust to high-frequency primary user signal and stochastic resonance system parameters. Then, the stochastic resonance system is applied for the detection of high-frequency primary signal.

3 CASCADING STOCHASTIC RESONANCE SYSTEM MODEL

Scale transformation shortens the step of the stochastic resonance system. Owing to the existence of high-frequency component, on the occasion when sensing time is shorter, we use spectrum sensing algorithm to computer further detection. The results are influenced by high-frequency component, which will reduce the probability of successful detection.

N stage cascading of stochastic resonance system can be viewed as a series of n traditional stochastic resonance system. The output from the upper stochastic resonance system will be the input of the next stochastic resonance system. Every stage of the stochastic resonance system is optimized for the last stage. The stochastic resonance of every stage has the same characteristics. Two-stage cascading stochastic resonance system structure is relatively simple and achieves optimization results (Yonggang&TaiyongYan2011). Two-stage cascading stochastic resonance system model is depicted as follows:

$$\begin{cases} \dfrac{dx_1}{dt} = ax_1 - bx_1^3 + s(t) + \Gamma(t) \\ \dfrac{dx_2}{dt} = ax_2 - bx_2^3 + kx_1 \end{cases} \quad (8)$$

Let us call x_1 the sampling mixed signal after processing from the first stochastic resonance system and x_2 the sampling mixed signal after processing from the second stochastic resonance system. Let us define a and b, two variable parameters, in the stochastic resonance system. k is the magnification of the cascading system.

Coefficient of intermediate amplifier often is set to $K = 1$. If the input sampling mixed signal is very weak, it is hard for the second stochastic resonance system to test the primary user signal. We can set K to a larger value (Jinzhao&Guojun&Xiaona2008).

Mixed sampling signal of cognitive user is processed by the first stage of stochastic resonance system. The FFT amplitude spectrum of output mixed signal consists of two parts. One is $S_1(f)$, which is related to cycle primary user signal, and $S_1(f)$ has the same frequency with the primary user signal of input mixed signal. Another is $S_2(f)$, which is related with Gaussian noise and satisfies Lorenz distribution (Jinzhao&Guojun&Xiaona2008):

$$S_1(f) = \frac{2a^2u^1 \exp(-u^2 / 2D) / (\pi D^2)}{(2\pi f_0)^2 + (2u^2 \exp(-u^2 / 2D) / \pi^2)} \delta(f_0 - f) \quad (9)$$

$$S_2(f) = [1 - \frac{a^2u^3 \exp(-u^2 / 2D) / (\pi D^2)}{(2\pi f_0)^2 + (2u^2 \exp(-u^2 / 2D) / \pi^2)}] \quad (10)$$

Among them, u is related with two parameters of a, b and satisfies $u = a / b$. Let us define f_0 as the frequency of primary user signal in the cognitive user sampling signal and D as the environmental noise intensity.

By cascading two traditional stochastic resonance systems, the energy of output high-frequency signal in the first-stage stochastic resonance system is transferred to the low-frequency signal, which means that high frequency is greatly reduced. This algorithm makes output signal smoother and further reduces the influence of high noise glitch. What is more, it improves the probability of signal detection and makes the low-frequency region of intensive noise energy narrower.

4 SIMULATIONS AND ANALYSIS OF EXPERIMENTAL RESULTS

EXPERIMENTAL 1: simulations of scaling transformation performance

Suppose there is a high-frequency primary user $f = 1GHz$. Let us define amplitude $A = 3 \times 10^{10}$, expectation of environmental noise $E = 0$, variance of environmental noise $\sigma^2 = 1 \times 10^{22}$, parameters of stochastic resonance system $a = 1 \times 10^{11}$, $b = 1 \times 10^{11}$, and sampling step $h = 2 \times 10^{-12}s$. According to steps of scaling transformation, let us call $f_0 = 0.01Hz$ as fundamental

frequency. $\Delta = f / f_0 = 1 \times 10^6$ is considered magnification of frequency. System parameters of a and b are multiplied Δ times. The sampling steps are shortened Δ times. Primary user signal amplitude is enlarged $\sqrt{2a^3 / b}$ times. The variance of noise is enlarged $2a^3 / b$ times. Sampling points are unchanged. The output waveform of mixed signal after processing by the stochastic resonance system is as follows:

By the method of scale transformation, sampling

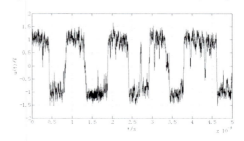

Figure 1. Output time-domain diagram after scale transformation.

steps are shortened to $2 \times 10^{-12}s$. Sampling frequency reaches $500GHz$. Existing sampler AD can support a sampling frequency of $1000GHz$ (Guojun &xiaoping&xiaona2010). By scaling transformation, the stochastic resonance system not only detects high frequency in an actual wireless communication environment but also needs shorter perception time in detecting high-frequency signal.

EXPERIMENTAL 2: simulations of cascading stochastic resonance system

From Figure 2, we observe that there is a high-frequency, low-amplitude glitch in scale transformation. It originates from the high-frequency noise component in cognitive user sampling signal, which reduces probability of detection to a great extent. Now, we can cascade two stochastic resonance systems and take further steps for the output signal from the first-stage stochastic resonance system. The algorithm reduces the high-frequency noise component further. The output signal becomes smoother. As a result, it can improve detection probability.

Let us define the amplitude $A = 0.3$, frequency $f_0 = 0.01\,Hz$; noise expectation $E = 0$ and variance $\sigma^2 = 1$; the parameters of stochastic resonance system $a = 1$ and $b = 1$; and sampling step $h = 0.2s$.

The output signal time-domain waveform of the two-stage stochastic resonance system is shown in Figures 2 and 3. After processing by two-stage stochastic resonance, the glitch of output signal is greatly reduced, and the high-frequency component is greatly restrained. Otherwise, the output signal of second stochastic resonance is fluctuated around 1.4V but the output signal of the first stochastic resonance is fluctuated around 1V, which shows that the second stage

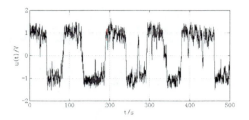

Figure 2. Output time-domain of the first stochastic resonance system.

constitutes a part of the high-frequency noise component transfer to primary user signal energy.

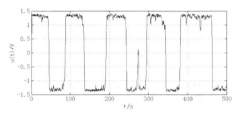

Figure 3. Output time-domain of the second stochastic resonance system.

The high-frequency part (0.05Hz~0.5Hz) of output signal FFT spectrum amplitude by two-stage stochastic resonance system is depicted in Figures 4 and 5. The amplitude of the second stage is obviously reduced, which can minimize the disturbance to the primary user signal detection and improve the probability of detection.

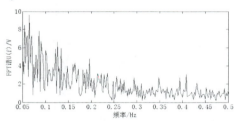

Figure 4. Output FFT spectrum of the first stochastic resonance system.

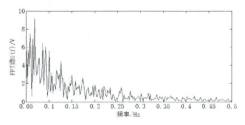

Figure 5. Output FFT spectrum of the second stochastic resonance system.

5 CONCLUSIONS

The literature applies the method of scale transformation to make the stochastic resonance principle suitable for detection of high-frequency primary user signal. As to the existing component of high-frequency noise in the output signal, the first-stage output signal is further processed by two or more cascading stochastic resonance systems. The high-frequency noise component is further reduced. The output signal becomes smoother. So, the algorithm can reduce the influence by high frequency and improve the detection probability of spectrum sensing.

ACKNOWLEDGMENTS

This research work is supported by the National Natural Science Foundations of P. R. China (NSFC) under Grant No. 61071093, National 863 Program No. 2010AA701202, Returned Overseas Project, Jiangsu Province Major Technology Support Program No. BE2012849, and Graduate Research and Innovation project No. SJLK_0383.

REFERENCES

Benzi, R. & Sutera, A. & Vulpiani, A. 1981. The mechanism of stochastic resonance. *Journal of Physics A:Mathematical and General* 14(11):453–457

Di, HE. & Yingpei, LIN. & Chen, HE. 2010. A Novel Spectrum-Sensing Technique in cognitive Radion Based on Stochastic Resonance. *IEEE Transaction on Vehicular Technology* 59(4):1680–1688.

Di, HE. & Lingge, JIANG. 2010. Cooperative Spectrum Sensing Approach Based on Stochastic Resonance Energy Detectors Fusion. *Proceedings of 2011 International Conference on Communications. Kyoto:IEEE*:1–5

Guojun, LI. & Xiaoping, ZENG & Xiaona, ZHOU. 2010. Research based on weak high frequency CW signal detection of stochastic resonance. *Journal of University of Electronic Science and Technology* 39(5):737–741.

Jinzha, LIN. & Guojun, LI. & Xiaona, ZHOU. 2008. Automatic detection of weak high-frequency CW telegraph signal. *Journal of Chongqing University of Posts and Telecommunications* 20(5):6–10

Kun, ZHENG. & Husheng, LI. 2010. Spectrum Sensing in Low SNR Regime Via Stochastic Resonance. *Proceedings of 2010 44th Annual Conference on Information Sciences and Systems. Pr, NJ: IEEE*:1–5

Lingbin, ZHANG. 2010. Improve the sampling frequency to achieve high frequency signal detection by stochastic resonance. *XiangFan College Journal* 31(2):39–41.

Shaowen, ZHANG. & Jun, WANG. & Wei CHEN. 2011. A Spectrum Sensing Algorithm Based on Stochastic Resonance Enhanced Covariance Matrix Detection. *Signal Processing* 27(11):1633–1639

Yonggang, LENG. & Taiyong, WANG. & Yan, GUO. 2011. Stochastic resonance characteristics of cascaded bistable system. *Physics Journal* 54(3):1118–1125.

Information, Computer and Application Engineering – Liu, Sung & Yao (Eds)
© 2015 Taylor & Francis Group, London, ISBN 978-1-138-02717-6

Feedback free distributed video coding rate control algorithm

Deng Yin Zhang
Sci-tech Park of Internet of Things, Nanjing University of Posts and Telecommunication, Nanjing, China

Teng Ma
Key Lab of Broadband Wireless Communication and Sensor Network Technology, Nanjing University of Posts and Telecommunication, Ministry of Education. Nanjing, China

ABSTRACT: In this paper, based on the feedback free distributed video coding system, a rate control algorithm was proposed for Wyner-Ziv frames. First, calculate the maximum correlation of coefficient band level between the source WZ frames and adjacent key frames in DCT domain, and the quantization matrix was determined based on maximum correlation and the QP of the key-frames; then, according to the generated low-complexity side information and source WZ frame with DCT transformation and quantization, allocate the bit plane rate. Experimental results show, compared with the existing rate control algorithm without feedback channel, the proposed algorithm can effectively improve the Rate-Distortion performance of the feedback free system with little additional complexity.

KEYWORDS: Feedback free; Rate control; Transform domain

1 INTRODUCTION

The traditional video coding use the complex motion estimation to remove temporal redundancy of adjacent frames, the computational complexity of the encoder is much higher than the decoder, which is not conducive to reduce power consumption of the encoding device. Therefore, the DVC (Distributed Video Coding) provides a new solution which is resulted from the Slapian-Wolf lossless coding theorem[1], and Wyner-Ziv lossy coding theorem [2]. The typical DVC systems are the transform domain Wyner-Ziv video coding system [3] and the transform domain PRISM video coding system [4]. Compared with the PRISM system, the proposed transform domain Wyner-Ziv coding system by DISCOVER group [5] is less complex, and can take advantage of temporal and spatial correlation of video frames. Currently, the DVC researches are mostly concentrated on Wyner-Ziv video coding system.

In the DVC system, the rate control affects the performance of the entire system, excessive parity bits transmission would cause the waste of resources, reduce the RD(Rate-Distortion) performance of the system; if not enough, the decoder would not decode WZ (Wyner-Ziv) frames correctly. In the DISCOVER coding scheme with feedback channel, the decoder repeated requests parity bits from the encoder, to decode the WZ frame correctly. This rate control method can improve the efficiency of coding, but also greatly increases the complexity of the system,

limiting the application of DVC video system. In [6], an encoder rate control algorithm was proposed in which the rate of bit plane was allocated depended on the source WZ frame and the low-complexity SI (side information). The approach achieves a good allocation of rate, but it ignores the difference of coefficient bands in DCT domain. In [7], the rate control was divided into three layers, and for the WZ frame, calculate the CNM (correlation-noise model) parameter using the coefficient correlation, then the feedback free bit plane rate control algorithm is proposed which takes into account the correlation between the frames, but it cannot calculate the parity bits of each bit plane correctly.

This paper proposes a feedback free rate control algorithm of Wyner-Ziv video coding. First select the quantization matrix based on the maximum correlation of coefficient band level between the source WZ frames and adjacent key frames in DCT domain; then according to the generated low-complexity SI and the source WZ frame with DCT transformation and quantization, allocate the bit plane rate to improve the RD performance of the DVC system.

2 FEEDBACK FREE DVC SYSTEM

The transform domain feedback free DVC architecture is shown in Figure1. At the encoder, the input sequence is divided into key frames (odd frames) and WZ frames (even frames), then allocate WZ frame rate by calculating the relationship between the key frames and WZ frames. The key frames are encoded and decoded using H.264 intra codec, and the WZ frames are encoded and decoded as following.

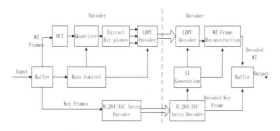

Figure 1.   Transform domain DVC architecture.

Encoding process: first each WZ frame X_t is applied a 4×4 DCT and scanned by Zigzag, the DCT coefficients of the entire frame are grouped together in coefficient bands X_k (k=1, 2…16). Then each coefficient band X_k is uniformly quantized q_k and all of the quantization bits M_k formed bit planes from high to low of the weight of bit. And, each bit plane is sent to the LDPC (Low Density Parity Check) encoder. Only the parity bits generated by the LDPC encoder for each bit plane are sent to the decoder.

Decoding process: first the SI of WZ frame Y_t is generated by applying extrapolation and interpolation on decoded key frames; then WZ frame Y_t is applied a 4×4 DCT, scanned by Zigzag and quantized; the residual statistics of corresponding coefficients between WZ frame and SI is defined as the virtual noise error which can be modeled by a Laplacian distribution, and the Laplacian parameter is calculated by CNM [8]; once the DCT bands of SI Y_k and the CNM parameter are known, the LDPC decoder decode the bit plane with the parity bits received , and output the coefficients q'_k; final the WZ frame X'_k can be reconstructed using Y_k and q_k, and X'_t can be generated by applying 4×4 IDCT on X_k.

3   FEEDBACK FREE DVC RATE CONTROL ALGORITHM

Based on the previous discussion, this paper proposes a feedback free rate control algorithm of WZ frame. The algorithm includes two parts: (1) calculate the correlation of the source WZ frames and adjacent key frames in DCT domain for each coefficient, and select the quantization matrix; (2) according to the generated low-complexity SI to allocate the bit plane rate.

3.1   Selecting quantization matrix

At the encoder of transform domain DVC architecture, each WZ frame X_t is applied a 4×4 DCT (B=4),

the DCT coefficients of the entire frame are grouped together in DCT bands X_k (k=1, 2…16), where X_1 is the DC coefficient band, and the others are the AC bands. In general, the DC coefficient band represents the average energy of the 4×4 blocks. Meanwhile, the QP of the key frames determines the quality of the decoded key frame, and the SI is generated by applying extrapolation and interpolation on decoded key frames, therefore, the QP of the key frames affects the quantization bits of WZ frame.

Assuming that each coefficient band is uniform quantization to 2^{M_k} intervals, then the quantization symbol q_k is represented in binary form M_k. If $M_k = 0$, indicates the coefficient band without coding, and the band is replaced by the SI in the decoder. According to the experiments, the correlation between key frame and WZ frame is different in different coefficient bands. Therefore, this paper calculates the correlation mse_k in each coefficient band to select the quantization matrix.

Use MSE (mean square error) as the parameter of correlation, the correlation between WZ frame and key frame on the k-th coefficient band is

$$mse'_k = \frac{B^2}{MN} \sum_{m,n} ((X_k^{WZ}(m,n) - X_k^{Key}(m,n))^2 \qquad (1)$$

Here, $X_k^{WZ}(m,n)$ and $X_k^{Key}(m,n)$ represent the value of WZ frame and key frame on the k-th coefficient band (k=1, 2…16), M and N denote the length and width of the frame.

According to equation (1), calculate the maximum correlation steps : ① computing the correlation mse_k^f (k=1,2…16) between the WZ frame X_t and previous adjacent key frame X_f on coefficient bands; ② computing the correlation mse_k^b (k=1,2…16) between the WZ frame X_t and the next adjacent key frame X_b on coefficient bands; ③ computing the maximum correlation $mse_k = \min\{mse_k^b, mse_k^f\}$ (k=1,2…16), and select the appropriate quantization matrix.

The QP of key frames decides the quantization bits of WZ frame on DC coefficient band to ensure the fluency of decoded sequence. There are 8 quantization levels for 4×4 block in DCT domain, and Q8 is finest while Q1 the coarsest, as show in Figure 2.

4	3	0	0
3	0	0	0
0	0	0	0
0	0	0	0

Q1

7	6	5	4
6	5	4	3
5	4	3	2
4	3	2	0

Q8

Figure 2.   Quantization matrix.

For AC coefficient bands, based on the maximum correlation mse_k (k=2, 3…16) to select the quantization bits from Q8 or Q1.

For DC coefficient band, the quantization bits M_1 is determined by the QP of key frames; for each AC coefficient band, the quantization bits M_k is determined using equation (2):

$$M_k = \begin{cases} M_1 - (M_1^{Q1} - M_k^{Q})^1 & mse_k < T_k \\ M_1 - (M_1^{Q8} - M_k^{Q})^8 & mse_k \geq T_k \end{cases} \quad (2)$$

Here, M_k^{Q1} and M_k^{Q8} are the quantization bits on k-th coefficient band (k=2, 3…16) of Q1 and Q8, T_k is the threshold to divide mse_k into 2 intervals. Then the quantization matrix Q in this paper is constructed by M_k (k=1, 2…16). As can be seen from Q1 and Q8, for X_{16}, the quantization bits is zero; for X_2 and X_3 ,the quantization bits have no relation on the correlation; for the others, when $mse_k <$ T_k, the quantization is zero, meaning this coefficient band is replaced by SI in decoder without encoding at encoder.

3.2 Allocation bit plane rate

In DVC system, each DCT band X_k of WZ frame is uniformly quantized by Q to q_k and all of the quantization bits M_k formed bit planes from high to low of the weight of bit. And, each bit plane is sent to the LDPC encoder. Only the parity bits generated by the LDPC encoder for each bit plane are sent to the decoder for reconstructing WZ frame. In this paper, the coefficient bands without encoding of WZ frame were filtered using the quantization matrix Q, and the remaining coefficient bands encoded using LDPC encoder need to be estimated the rate of each bit plane. Also, the SI Y_t can be seen as the output of source WZ frame X_t through a "virtual channel". Therefore, the encoder should generate the low-complexity SI Y_t, and calculate the conditional entropy $H(X|Y)$ between X_t and Y_t to achieve optimal encoding rate for each bit plane of WZ frame. Steps are as follows:

① According to the adjacent key frames X_b and X_f, using AI(average interpolation) and FMCI (fast motion compensated interpolation)[6] algorithm generate low-complexity SI Y_t.

② Y_t is applied a 4×4 DCT to Y_k (k=1, 2…16) and quantized by Q to filter the coefficient bands without encoding.

③ The quantized WZ frame and low-complexity SI are formed bit planes from high to low of the weight of bit. And for each couple of the same weight of bit planes using equation (3) to calculate conditional entropy $H(X|Y)$[9] and error probability P_{cros}.

$$H(X|Y) = -P_{cros} \times lbP_{cros} - (1-P_{cros}) \times lb(1-P_{cros})$$
$$P_{cros} = p(Y_i^b = 0, X_i^b = 1) + p(Y_i^b = 1, X_i^b = 0) \quad (3)$$

Here, X_i^b and Y_i^b denote the i-th bit of b-th bit plane of the WZ frame and low-complexity SI.

④ [6] gives a bit plane parity rate estimation algorithm to allocate the parity rate for each bit plane. For the b-th bit plane parity rate R_b, is obtained from equation (4):

$$R_b = \frac{1}{2}H(X|Y) \times \exp(H(X|Y)) + \sqrt{P_{cros}} \quad (4)$$

Here, $H(X|Y)$ is conditional entropy of the b-th bit plane and P_{cros} is error probability of the b-th bit plane. Then, select the appropriate LDPC check matrix to encode the bit plane of WZ frame.

3.3 Algorithm description

① The source WZ frame X_t and adjacent key frames X_b, X_f are applied a 4×4 DCT respectively, and generated 16 coefficient bands.

②According to equation (1), calculate the maximum correlation of coefficient band level between the source WZ frame X_t and adjacent key frames X_b, X_f in DCT domain.

③ According to the QP of key frames and the maximum correlation mse_k, select the appropriate quantization matrix.

④ According to [6], generate low-complexity SI, and calculate the conditional entropy $H(X|Y)$ and error probability P_{cros} for each bit plane, and allocate the rate of each bit plane

4 EXPERMENTNAL RESULTS

To evaluate the RD performance of the proposed algorithm, two QCIF sequences are selected: Hall Monitor@15Hz and Coastguard@30Hz, and the threshold T_k accords to [10]. The sequence is divided into key frames (odd frames) and WZ frames (even frames). The RD performance can be evaluated by PSNR. Figure3 shows the RD performance compares with the encoder rate control algorithm in [6], H.264/AVC intra codec and the DISCOVER [11] codec.

Figure3 shows that the proposed algorithm in this paper has a better RD performance than H.264/AVC intra and [6], but compared with DISCOVER, it is also a gap. For sequence Hall Monitor@15Hz, the proposed algorithm gains up to 0.8dB in PSNR compared to [6], 2.5dB compared to H.264/AVC intra, as show in (a); for sequence Coastguard@30Hz, the proposed algorithm gains up to 0.5dB in PSNR compared to [6], 1dB compared to H.264/AVC intra, as show in (b). On the other hand, compared with DISCOVER which repeated request parity bits until the WZ frame decoded correctly, the proposed algorithm is still a gap of 1dB.

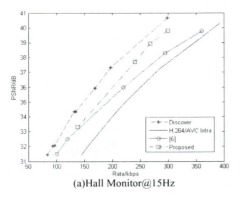

(a)Hall Monitor@15Hz

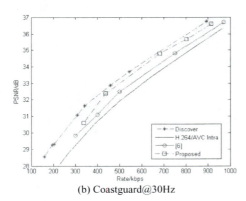

(b) Coastguard@30Hz

Figure 3.   RD performance for different algorithm.

Compared with the existing feedback free rate control algorithm, the proposed algorithm calculates the correlation of coefficient band level between the WZ frames and adjacent key frames, and selects the quantization matrix. The additional computational complexity is small in terms of the encoder. Also, there are some coefficient bands with large correlation are not encoded can reduce computational complexity of the encoder. Therefore, the additional calculation of proposed algorithm is negligible.

5   CONCLUSION

For the feedback free DVC system, a rate control algorithm is proposed based on the correlation of coefficient band level between the WZ frames and adjacent key frames. First, calculate the correlation between the source WZ frames and adjacent key frames in DCT domain, select the quantization matrix; then, according to the generated low-complexity SI and the source WZ frame with DCT transformation and quantization, allocate the bit plane rate. Experimental results show, the proposed algorithm can effectively improve the Rate-Distortion performance of the feedback free system with little additional complexity.

ACKNOWLEDGMENTS

This research work is supported by the National Natural Science Foundations of P. R. China (NSFC) under Grant No. 61071093, National 863 Program No.2010AA701202, Returned Overseas Project, Jiangsu Province Major Technology Support Program No.BE2012849, Graduate Research and Innovation project No. KYLX_0812.

REFERENCES

[1] Slepian, D., & Wolf, J. K. (1973). Noiseless coding of correlated information sources. *Information Theory, IEEE Transactions on, 19*(4), 471–480.
[2] Wyner, A. D., & Ziv, J. (1976). The rate-distortion function for source coding with side information at the decoder. *Information Theory, IEEE Transactions on, 22*(1), 1–10.
[3] Aaron, A., Rane, S. D., Setton, E., & Girod, B. (2004, January). Transform-domain Wyner-Ziv codec for video. In *Electronic Imaging 2004* (pp. 520–528). International Society for Optics and Photonics.
[4] Puri, R., & Ramchandran, K. (2002, October). PRISM: A new robust video coding architecture based on distributed compression principles. In *Proceedings of the annual allerton conference on communication control and computing* (Vol. 40, No. 1, pp. 586–595). The University; 1998.
[5] Artigas, X., Ascenso, J., Dalai, M., Klomp, S., Kubasov, D., & Ouaret, M. (2007, November). The DISCOVER codec: architecture, techniques and evaluation. In *Picture Coding Symposium* (Vol. 17, No. 9, pp. 1103–1120). Lisbon, Portugal.
[6] Brites, C., & Pereira, F. (2007, September). Encoder rate control for transform domain Wyner-Ziv video

coding. In *Image Processing, 2007. ICIP 2007. IEEE International Conference on* (Vol. 2, pp. II-5). IEEE.

[7] Song, B., Yang, M. M., Qin, H., & He, H. (2011). No-feedback rate control algorithm for Wyner-Ziv video coding. *Journal on Communications*, *12*, 002.

[8] Brites, C., & Pereira, F. (2008). Correlation noise modeling for efficient pixel and transform domain Wyner–Ziv video coding. *Circuits and Systems for Video Technology, IEEE Transactions on*, *18*(9), 1177–1190.

[9] Dragotti, P. L., & Gastpar, M. (2009). *Distributed Source Coding: Theory, Algorithms and Applications*. Academic Press.

[10] Mo W. W., Yang C. L. (2011). Quantization Based on Characteristics of DCT Coefficients for Distributed Video Coding. *Computer Engineering*, 38(01): 273–275.

[11] DISCOVER project. The DISCOVER Code Evaluation [R/OL]. http://www.img.lx.it.pt/~discover/home.html.

Information, Computer and Application Engineering – Liu, Sung & Yao (Eds)
© *2015 Taylor & Francis Group, London, ISBN 978-1-138-02717-6*

Coordination strategies of the cloud computing service supply chain based on the service level agreement

Ling Yun Wei & Xiao Han Yang

School of Automation, Beijing University of Posts and Telecommunications, Beijing, China

ABSTRACT: Cloud computing has been expanded so rapidly in recent years. In this paper, we study a cloud computing service supply chain consisting of one Application Infrastructure Provider (AIP) and multiple competing Application Service Providers (ASPs) based on the Service Level Agreement (SLA). A SLA serves as the foundation for the expected level of services. As a result of longitudinal and transverse competitions, the efficiency of integration is always difficult to achieve in the distributed cloud computing service supply chain. To the cloud computing industry, how to coordinate the cloud computing service supply chain based on the SLA to achieve the efficiency of integration is of great importance. We develop and evaluate four situations. We analyze the reasons that the wholesale price contract and revenue sharing contract can't achieve coordination effectively, and propose all quantity discount contract.

KEYWORDS: Cloud Computing; Service Level Agreement; Service Supply Chain; Coordination Strategies

1 INTRODUCTION

With the rise of new technologies and the development of services economic around the world rapidly, the new service model- cloud computing service has caused serious concern of IT industry. According to NIST, cloud computing is a model for enabling ubiquitous, convenient, on-demand network access to a shared pool of configurable computing resources (e.g., networks, servers, storage, applications, and services) that can be rapidly provisioned and released with minimal management effort or service provider interaction [1]. Based on the above definition, cloud computing can be composed of three service models- Infrastructure as a Service (IaaS), that is, raw infrastructure and associated middleware, Platform as a Service (PaaS), that is, developing applications on an abstract platform, and Software as a Service (SaaS),that is, support for running software services remotely [1].

Cloud computing service utilizes the cloud to deliver services [2]. Customers just cost very little upfront investment. Cloud computing eliminates the need for maintaining expensive computing hardware. In China, medium and small-sized enterprises are inclined to purchase cloud computing services that make up for their current shortcomings in the informatization level, funding and talents. Apart from the cost, since the cloud advocates better management of resources, cloud computing also supports the growing concerns of carbon emissions and environmental impact.

Cloud computing is a rising type of service system in IT industry. There are more and more consumers delegating their tasks to cloud providers, so Service Level Agreements (SLA) [3] between consumers and providers emerge as a key aspect. A Service Level Agreement (SLA) is a contract between the provider and customer to ensure the Quality of Service (QoS) which is achieved through a negotiation process. At the same time, Kandukuri, Paturi and Rakshit [4] point out that the SLA is the only legal agreement between the service provider and customer. The paper deems that the provider can gain trust of customer is through the SLA. So agreements in the cloud computing services are essential and necessary because participating parties are independent entities with different objectives and QoS requirements However, it's not possible to fulfill all consumer expectations from the service provider perspective and hence a balance needs to be made via a negotiation process [5]. In the end, they will commit to the service level agreement, see in Figure 1. The SLA serves as the foundation for the expected level of service between the consumer and the provider. There are many quantifiable indicators about the SLA. For these indicators need to be closely monitored [6], we choose the response time as the key indicator during the enforcement the agreement.

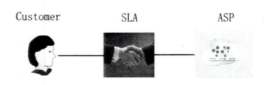

Figure 1. The negotiation process.

Many scholars have done lots of research about the cloud computing. However, there has been relatively little research conducted that studies the coordination strategies in a cloud computing service supply chain. Demirkan and Cheng [7] are the first group of scholars who study the Application Service Provider (ASP) strategies. They discuss the supply chain coordination strategies, and analyze a monopolistic ASP's optimal pricing and capacity policies considering the delay cost of users. Kar and Rakshit [8] evaluate the cloud computing service pricing models and contract terms, and they point out that per transaction charge is the most common way of charging the cloud computing service in industry. We also charge the the ASP services by per transaction. However, neither of them has a discussion about the other actor of the cloud computing service supply chain—Application Infrastructure Provider (AIP). Moreover, previous studies about the supply chain coordination are more based on one AIP and one ASP. Demirkan and Cheng [9] study coordination strategies in an SaaS supply chain consisted of one AIP and one ASP. They propose that it is possible to create the right incentives so that the economically efficient outcome is also the Nash equilibrium. However, their works mentioned above don't involve the horizontal completion between two ASPs. In the contrary, we study the cloud computing service supply chain with one AIP and multiple competing ASPs. We analyze the price competition of multiple ASPs by considering the impact of congestion cost. Our model is more in line with the reality of cloud computing service market.

In this paper, we design a cloud computing service supply chain consisting of one AIP and multiple competing ASPs based on SLA. AIP provide packages the computer capacity as services and supplies them to two ASPs, and in turn ASPs sell the same value-added application services to the common market via the Internet. We consider ASPs who sell their same value-added services to the common market and compete for customers. In order to ensure the quality and reliability of the cloud computing services, the customer and the ASP will sign the SLA.

2 THE MODEL

This section considers a two-stage cloud computing service supply chain which consists of an AIP and multiple competing ASPs based on SLA.

AIP and ASPs are risk neutral, so each player pursues for the expected profit maximization. We assume that there is full information, which means that all of them get the same information at the beginning of the game, i.e., each player is fully aware of all revenues, costs, parameters and rules.

AIP provides the IaaS to multiple ASPs at the wholesale price w per unit of capacity, e.g. Amazon

Web Service (AWS). In turn ASP_i sells his or her SaaS to the market at price p_i per transaction of processing, e.g. on-line ERP service. The ASP_i service system is modeled as an M/M/1 queuing system with processing capacities μ_i in transactions per unit of time, and λ_i is a Poisson rate of transactions per unit of time arriving at the ASP_i system. We assume that the market demand, λ, is fixed and exogenous. In other words, the pricing decisions of the ASPs affect only the resulting market share of each firm, but not the overall market size. We let λ_i denote the market demand of ASP_i, and let μ be the total capacity. Hence, one has

$$\lambda = \sum_{i=1}^{N} \lambda_i , \mu = \sum_{i=1}^{N} \mu_i \qquad (1)$$

Moreover, a stable queuing system requires:

$$\lambda_i \geq 0 , \mu_i \geq \lambda_i + \varepsilon \qquad (2)$$

where $\varepsilon > 0$ is fixed and arbitrarily small. Figure 2 shows the cloud market consisting of multiple

ASP_i faces a random market demand constituted by a large number of customers. ASPs sell the same value-added application services to customers via the Internet. Bertrand competition model is used to characterize the competitive relationship between ASPs under the common market. In this model, price competition influences the ASP_i demand:

$$\lambda_i = \alpha - \beta p_i + \sum_{j=1}^{N} m p_j , i = 1,2 \dots N, j \neq i \qquad (3)$$

where α, β and m are constants; α is the market scale of ASP_i (i.e., the maximum possible demand) ; β is the linear demand distribution parameter ;m is the substitutability coefficient of the services provided by ASPs.

As a result of the queuing delay [10], consumers will wait in the system. Given the effects of congestion cost, let $T_{sj}(\lambda_i, \mu_i)$ denote the expected time each transaction stays in the service system, including actual processing and waiting time. The total expected marginal cost per transaction to a customer is thus the sum of the price and the expected delay cost per transaction. Then the expected cost per transaction is

$$p_i + v T_{si}(\lambda_i, \mu_i) \qquad (4)$$

The value of cloud computing services is represented by the value function $V(\lambda_i)$, and the marginal value function $V'(\lambda_i)$ represents a relationship between the cost per transaction and the number of transactions per unit of time, which depicts the demand curve. Let the demand isoelastic function equal

$$V'(\lambda_i) = p_i = \frac{\alpha - \lambda_i}{\beta + m} + \frac{m(N\alpha - \sum_{j=1}^{N} \lambda_j)}{(\beta + m)(\beta - Nm + m)} , i = 1, \dots N \qquad (5)$$

The market equilibrium is achieved when the marginal value is equated with the marginal cost per transaction. Thus, one has

$$V'(\lambda_i) = p_i + vT_{si} \qquad (6)$$

Hence, we get the maket clearing price

$$p_i = \frac{\alpha - \lambda_i}{\beta + m} + \frac{m(Na - \sum_{j=1}^{N}\lambda_j)}{(\beta+m)(\beta-Nm+m)} - \frac{v}{\mu_i - \lambda_i} \quad i = 1,2\dots N \;(7)$$

It is necessary to enforce SLAs, and we mainly monitor the response time . T_{si} is refer to the respond time in a M/M/1 system. We let s_i denote the service level agreed by the customer and ASPi. Thus, one have

$$s_i = P(T_{si} \leq \tau) = 1 - e^{-(\mu_i - \lambda_i)\tau} \qquad (8)$$

where τ is the limit respond time according to the SLA.

When ASPi can't fulfill all consumer expectations according to the SLA agreed, ASPi will compensate customers for loss. The compensation can be denoted by

$$k\lambda_i(1 - s_i) \qquad (9)$$

where k $k(k > 0)$ is a constant.

The cost to ASPi is the computer capacity paid to AIP in order to acquire capacity, μ_i, and the compensation for failing to fulfill all consumer expectations. The revenue of ASPi equals the quantity of the services purchased by customers, represented by the market arrival rate, λ_i, multiplied by the per transaction price, p_i. Hence, the profit function of the ASPi is described by

$$AP_i = p_i\lambda_i - w\mu_i - k\lambda_i(1 - s_i) \qquad (10)$$

The total cost of AIP includes two components: the marginal cost of computer capacity denoted by parameter c, and a diseconomy of scale cost parameter, $g(g>0)$, presented by Mendelson [11], resulting from the increasing costs of managing capacity and customer access and rising complexity of the business model of ASPs. The revenue of AIP is the computer capacity purchased by ASPi .Based on the above assumptions the profit function of the AIP can be defined by

$$HP = \sum_{i=1}^{N}[(w - c_i)\mu_i - g_i\mu_i^2] \qquad (11)$$

3 COORDINATION STRATEGIES OF THE CLOUD COMPUTING SERVICE SUPPLY CHAIN

In this section, we discuss four coordination strategies: (1) centralized control, (2) wholesale price contract, (3) revenue-sharing contract and (4) all quantity discount contract. As private agents, the target of AIP and ASPs is to maximize their expected profit individually.

3.1 Centralized control

In this situation, a single firm plays an integrated role of one AIP and multiple competing ASPs. We let the number of ASPs to be N (N>1). The supply chain can be coordinated and obtain the maximal profit. The per unit time expected profit of the whole supply chain SP_1 is

$$SP_1 = HP_1 + \sum_{i=1}^{N} AP_{1i} \qquad (12)$$

To find the whole optimal profit

$$max\ SP_1 \qquad (13)$$

such that

$$V'(\lambda_i) = p_i + vT_{si}, 0 \leq \lambda_i \leq \mu_i \qquad (14)$$

The first-order conditions lead to

$$\frac{dSP_1}{d\mu_i} = \frac{v\lambda_i}{(\mu_i - \lambda_i)^2} - c_i - 2g_i\mu_i + \tau k\lambda_i e^{-(\mu_i - \lambda_i)\tau} = 0 \quad (15)$$

$$\frac{dSP_1}{d\lambda_i} = \frac{\alpha - 2\lambda_i}{\beta + m} + \frac{m(Na - 2\sum_{j=1}^{N}\lambda_j)}{(\beta+m)(\beta-Nm+m)} - \frac{v\mu_i}{(\mu_i - \lambda_i)^2}$$
$$- k(\lambda_i\tau + 1)e^{-(\mu_i - \lambda_i)\tau} = 0 \qquad (16)$$

We assume that, $c_1 = c_2 = \cdots c_N$, $g_1 = g_2 = \cdots g_N$. After some algebra, we find that $\lambda_1 = \lambda_2 = \cdots = \lambda_N = \frac{\lambda}{N}$, $\mu_1 = \mu_2 = \cdots = \mu_N = \frac{\mu}{N}$, $p_1 = p_2 = \cdots p_N = p_0$.

Hence, solutions to expressions above provide us the optimal arrival rate , λ^*, the optimal capacity, μ^* ,and the optimal whole profit, $SP_1{}^*$.

3.2 Wholesale price contract

With a wholesale price contract, AIP and ASPi independently make decisions to pursue optimal profits. AIP determine to charge ASPi w per unit computer capacity purchased. Let $HP_2(w)$, $AP_{2i}(p_i, \lambda_i, \mu_1)$ be the per unit time expected profits of AIP and ASPi respectively, and they can be described by

$$HP_2 = \sum_{i=1}^{N}[(w - c_i)\mu_i - g_i\mu_i^2] \qquad (17)$$

$$AP_{2i} = p_i\lambda_i - w\mu_i - k\lambda_i(1 - s_i) \qquad (18)$$

In this case, the ASPi solves the following problem to find his or her own optimal profit

$$max\ AP_{2i} \qquad (19)$$

such that

$$V'(\lambda_i) = p_i + vT_{si}, 0 \leq \lambda_i \leq \mu_i \qquad (20)$$

The first-order conditions lead to

$$\frac{dAP_{2i}}{d\lambda_i} = \frac{\alpha-2\lambda_i}{\beta+m} + \frac{m\left(Na-\lambda_i-\sum_{j=1}^{N}\lambda_j\right)}{(\beta+m)(\beta-Nm+m)} - \frac{v\mu_i}{(\mu_i-\lambda_i)^2} - k(\lambda_i\tau+1)e^{-(\mu_i-\lambda_i)\tau} = 0 \tag{21}$$

$$\frac{dAP_{2i}}{d\mu_i} = \frac{v\lambda_i}{(\mu_i-\lambda_i)^2} - w + \tau k\lambda_i e^{-(\mu_i-\lambda_i)\tau} = 0 \tag{22}$$

After some algebra, we find that $\frac{dAP_{2i}}{d\lambda_i} \neq \frac{dSP_1}{d\lambda_i}$. Hence, the wholesale price contract can't coordinate the cloud computing service supply chain. We also illustrate the results by numerical examples.

3.3 Revenue-sharing contract

In this situation, AIP charges w per unit of capacity purchased plus the ASPi shares AIP a percentage of his revenue. Let $(1-\delta)$ be the fraction of supply chain revenue the AIP earns, so δ is the fraction of supply chain revenue the ASPi keeps.

The profit function of AIP is

$$HP_3 = \sum_{i=1}^{N}[(w - c_i)\mu_i - g_i\mu_i^2 + (1-\delta)p_i\lambda_i] \tag{23}$$

The profit function of ASPi is

$$AP_{3i} = \delta p_i\lambda_i - w\mu_i - k\lambda_i(1 - s_i) \tag{24}$$

In this case, the ASPi solves the following problem to find his or her optimal profit

$$max\ AP_{3i} \tag{25}$$

such that

$$V'(\lambda_i) = p_i + vT_{si}, 0 \le \lambda_i \le \mu_i \tag{26}$$

The first-order conditions lead to

$$\frac{dAP_{3i}}{d\lambda_i} = \delta\left[\frac{\alpha-2\lambda_i}{\beta+m} + \frac{m\left(Na-\lambda_i-\sum_{j=1}^{N}\lambda_j\right)}{(\beta+m)(\beta-Nm+m)} - \frac{v\mu_i}{(\mu_i-\lambda_i)^2}\right] - k(\lambda_i\tau+1)e^{-(\mu_i-\lambda_i)\tau} = 0 \tag{27}$$

$$\frac{dAP_{3i}}{d\mu_i} = \frac{\delta v\lambda_i}{(\mu_i-\lambda_i)^2} - w + \tau k\lambda_i e^{-(\mu_i-\lambda_i)\tau} = 0 \tag{28}$$

We assume that $c_1 = c_2 = \cdots c_N$, $g_1 = g_2 = \cdots g_N$. After some algebra, we find that $\lambda_1 = \lambda_2 = \cdots = \lambda_N = \frac{\lambda}{N}$, $\mu_1 = \mu_2 = \cdots = \mu_N = \frac{\mu}{N}$, $p_1 = p_2 = \cdots p_N = p_0$.

However, after illustrating the results by numerical examples, we find that only when δ tends to zero can the supply chain be coordinated. In other words, ASPi shares so little profit that it refuses this revenue-sharing contract. So the revenue-sharing contract can't coordinate the cloud computing service supply chain.

3.4 All quantity discount contract

In this situation, we investigate whether an all quantity discount strategy coordinates a supply chain with one AIP and two competing ASPs. Let $w(\mu_i)$ be the AIP's wholesale price:

$$w(\mu_i) = \begin{cases} w_0, & \mu_i < \mu^* \\ w_4^*, & \mu_i \ge \mu^* \end{cases} \tag{29}$$

The profit function of ASP$_i$ is

$$HP_4 = \sum_{i=1}^{N}[(w(\mu_i) - c_i)\mu_i - g_i\mu_i^2] \tag{30}$$

The profit function of ASP$_i$ is

$$AP_{4i} = p_i\lambda_i - w(\mu_i)\mu_i - k\lambda_i(1 - s_i) \tag{31}$$

The profit function of the whole supply chain is

$$SP_4 = HP_4 + \sum_{i=1}^{N} AP_{4i} \tag{32}$$

Assuming that the revenue of ASP$_i$ is a constant proportion of SP_4, we have

$$AP_{4i} = \eta_i SP_4, 0 < \sum_{j=1}^{N}\eta_j < 1 \tag{33}$$

We assume that $c_1 = c_2 = \cdots c_N$, $g_1 = g_2 = \cdots g_N$. After some algebra, we find that $\lambda_1 = \lambda_2 = \cdots = \lambda_N = \frac{\lambda}{N}$, $\mu_1 = \mu_2 = \cdots = \mu_N = \frac{\mu}{N}$, $p_1 = p_2 = \cdots p_N = p_0$. Finally, we can get that

$$\eta_1 = \eta_2 = \cdots = \eta_N = \eta \tag{34}$$

$$w(\mu_i) = w_0, \ \mu_i < \frac{\mu^*}{N} \tag{35}$$

$$w(\mu_i) = \{\left[\frac{\alpha-\lambda_i}{\beta+m} + \frac{m\left(Na-\sum_{j=1}^{N}\lambda_j\right)}{(\beta+m)(\beta-Nm+m)} - \frac{v}{\mu_i-\lambda_i}\right]\lambda_i - k\lambda_i e^{-(\mu_i-\lambda_i)\tau} - \eta\sum_{i=1}^{N}\{[\frac{\alpha-\lambda_i}{\beta+m} + \frac{m\left(Na-\sum_{j=1}^{N}\lambda_j\right)}{(\beta+m)(\beta-Nm+m)} - \frac{v}{\mu_i-\lambda_i}]\lambda_i - k\lambda_i e^{-(\mu_i-\lambda_i)\tau} - c_i\mu_i - g_i\mu_i^2\}\}/\mu_i, \ \mu_i \ge \frac{\mu^*}{N} \tag{36}$$

$$\eta SP_4 \ge AP_{11} = AP_{12} \tag{37}$$

As a result of the flexibility in the profit distribution, the all quantity discount contract ensures that participation constraints of AIP and ASPs can be satisfied when the system performance achieves the optimal. So, the supply chain can be coordinated.

4 NUMERICAL EXAMPLES

In this part, we select a set of parameters (see Table 1) to illustrate whether the wholesale price contract, revenue-sharing contract and all quantity discount contract can coordinate the cloud computing service supply chain. Multiple ASPs face the same situation as two ASPs, for the aim of simplicity we assume that there are only two ASPs. We choose the baseline value of the parameter g following Mendelson [10]. To ensure the diseconomy of scale cost parameter doesn't affect the ASP market scale, α, we set it to be 3.00. In order to make the market demand of ASP be positive, we set the linear demand distribution parameter, β, at 1.00, and we set the substitutability coefficient of the services provided by two ASPs, m, at 0.50. When it comes to parameters about the cost, what we are really interested in is their relative effect on the profit function, so we set the marginal capacity of AIP, c, at 1.00. Given the effect of the delay, we assume the delay cost of customers, v, would be about of magnitude lesser, and we set it at 0.025.

Table 1. Baseline parameters.

c	α	v	g	β	m	k	τ
1	3	0.025	1	1	0.5	2	3

After some calculation, we can get the computing results in Table 2. According to the table, we can find that the maximal supply chain expected profit is 1.4304. In other words, when the expected profit can reach 1.4304 with one supply chain contract, we can deem that this contract can coordinate the cloud computing service supply chain. Obviously, the revenue-sharing contract and all quantity discount contract can satisfy the above conditions. However, we can find that only when δ tends to zero can it make the supply chain be coordinated under revenue-sharing contract. This implies that ASPi rarely shares the profit of the supply chain, so in ASPi's case this revenue-sharing contract can't be accepted. Hence, we can draw a conclusion that the revenue-sharing contract can't coordinate the cloud computing service supply chain in fact. In the other hand, we assume that w is 2.1194, and we can find that the whole supply chain expected profit is 1.1982. It shows the all quantity discount contract can coordinate the cloud computing service supply chain. Moreover, the profits of AIP and the two ASPs are not lower than the profits when they don't accept this contract. We illustrate the results by numerical examples.

Table 2. Computing results.

	centralized control	wholesale price contract	revenue-sharing contract	all quantity discount contract
λ_1	0.5527	0.3462	0.7557	0.5527
λ_2	0.5527	0.3462	0.7557	0.5527
λ	1.1054	0.6924	1.5114	1.1054
μ_1	0.8397	0.4698	0.8397	0.7245
μ_2	0.8397	0.4698	0.8397	0.7245
μ	1.4490	0.9397	1.6794	1.4490
p_1	4.7490	5.1053	4.3430	4.7490
p_2	4.7490	5.1053	4.3430	4.7490
p_0	4.7490	5.1053	4.3430	4.7490
δ	–	–	0	–
η	–	–	–	0.3000
w	–	2.0000	0	2.1194
HP	–	0.4982	1.4304	0.5722
AP_1	–	0.3500	0	0.4291
AP_2	–	0.3500	0	0.4291
SP	1.4304	1.1982	1.4304	1.4304

5 CONCLUSIONS

In this article, we aim to coordinate a two-staged cloud computing service supply chain consisting of an AIP and multiple competing ASPs based on SLA. ASPi faces a market which is characterized by a price-sensitive random demand. The ASPi service system is modeled as an M/M/1 queueing system while the effects of congestion are considered. We consider multiple ASPs who sell their value-added services to the common market and compete for customers. We examine four coordination strategies above: namely, centralized control, wholesale price contract, revenue-sharing contract and all quantity discount contract. After analyzing the reasons that the wholesale price contract and revenue sharing contract can't achieve coordination effectively, we propose all quantity discount contract. As a result of the flexibility in the supply chain profit distribution, the all quantity discount contract ensures that participation constraints of one AIP and two competing ASPs can be satisfied when the system performance achieves the optimal. Ultimately, we reach the conclusion that the all quantity discount contract can coordinate the cloud computing service supply chain.

ACKNOWLEDGMENT

This work was supported by National Science Foundation of China (Grant No. 61174167).

REFERENCES

[1] Mell P, Grance T. The NIST definition of cloud computing[J]. National Institute of Standards and Technology, 2009, 53(6):50.

[2] Latif, Rabia, et al. "Cloud Computing Risk Assessment: A Systematic Literature Review." Future Information Technology. Springer Berlin Heidelberg, 2014. 285–295.

[3] Patel P, Ranabahu A H, Sheth A P. Service level agreement in cloud computing[J]. 2009.

[4] Kandukuri B R, Paturi V R, Rakshit A. Cloud security issues[C]//Services Computing, 2009. SCC'09. IEEE International Conference on. IEEE, 2009:517–520.

[5] Wu L, Garg S K, Buyya R, et al. Automated SLA negotiation framework for cloud computing[C]//Cluster, Cloud and Grid Computing (CCGrid), 2013 13th IEEE/ACM International Symposium on. IEEE, 2013:235–244.

[6] Keller, A., Ludwig, H.: The wsla framework: Specifying and monitoring service level agreements for web services. J. Netw. Syst. Manage. 11(1) (2003) 57–81.

[7] Demirkan, Haluk, and Hsing Kenneth Cheng. "The risk and information sharing of application services supply chain." European Journal of Operational Research 187.3 (2008):765–784.

[8] Kar, Arpan Kumar, and Atanu Rakshit. "Pricing of Cloud IaaS Based on Feature Prioritization-A Value Based Approach." Recent Advances in Intelligent Informatics. Springer International Publishing, 2014. 321–330.

[9] Demirkan, Haluk, Hsing Kenneth Cheng, and Subhajyoti Bandyopadhyay. "Coordination strategies in an SaaS supply chain." Journal of Management Information Systems 26.4 (2010):119–143.

[10] Mendelson, Haim. "Pricing computer services: queueing effects." Communications of the ACM 28.3 (1985): 312–321.

[11] Mendelson, Haim. Economies of scale in computing: Grosch's law revisited[J]. Communications of the ACM, 1987, 30(12):148–164.

Information, Computer and Application Engineering – Liu, Sung & Yao (Eds)
© 2015 Taylor & Francis Group, London, ISBN 978-1-138-02717-6

Design and implementation of brush-less DC motor control system

Long Chen, Zhi Hui Zhang & Jian Jian Luo
School of Electronic Information, Hangzhou Dianzi University Hangzhou, Zhejiang Province, China

ABSTRACT: A design of brush-less DC motor control system is presented in this article. It is based on dsPICFJ12MC202 single chip microcomputer. The hardware circuit design includes intelligent power module SD05M50D drive circuit, detection circuit of the Hall sensor, over-current protection circuit, peripheral circuit, SCM download circuit, the motor and the system supply circuit, and so on. In the software, we output 6 channels through micro-controller programming; then, we control drive circuit; last, we realize the phase control of the brush-less motor. We change the external magnetic field direction by controlling the current flow direction of the brush-less motor's three phases. Then, we can drive the motor. This system realizes the control of the square wave and indulges in more precise SPWM controlling of the brush-less DC motor.

KEYWORDS: The single chip processor; BLDC motor; Hall sensor; Back EMF Detection; Over-Current Protection; SPWM

1 INTRODUCTION

With the rapid development of high-power semiconductor devices and the power electronics technology, the brush-less DC motor develops rapidly. It is a mechanic product that uses electronic commutation technique instead of the brush commutation of the DC motor. The brush-less DC motor exhibits a simple structure, reliable operation, easy maintenance, and such series of advantages. It is also similar to the operation of the DC motor with a high efficiency; it is non-excited and has many other advantages. So in today's national economy applications, it has become increasingly popular. This article presents an approach of controlling the brush-less DC motor based on back-EMF zero-crossing detection method. The hardware required is simple, and its software is powerful. Through the related software programming algorithm, we can accurately control the brush-less motor with large torque rotation. The Hall sensor can readily detect the position of the rotor and control the switching on and off through the position signal of the rotor. Lastly, it can achieve electronic commutation.

2 OVERALL DESIGN SCHEME AND OPERATING PRINCIPLE OF THE SYSTEM

A design of the brush-less DC motor control system is presented in this article. It is based on a single chip microcomputer called dsPIC-FJ12MC202. The intelligent power module SD05M50D is selected as the drive.

The control system of the brush-less DC motor mainly includes the following modules: SCM module, intelligent power, Hall sensor detection circuit, Zero-crossing detection circuit, over-current protection circuit, button adjustment, and potent meter control circuit. Figure 1 is about the brush-less DC motor system.

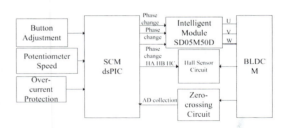

Figure 1. Brush-less DC motor control system.

The working principle of the brush-less DC motor is similar to that of the DC motor in essence. Brushed DC motor commutates the current of the winding with mechanical brush and commutation, whereas the brush-less DC motor adopts the electronic method.

3 HARDWARE DESIGN OF THE SYSTEM

The hardware part uses dsPIC33FJ12MC202 as the control core and SD05M50D as the drive core. The hardware part is designed based on the normal work of the BLDCM.

3.1 Peripheral single-chip circuit design

The single-chip dsPIC33FJ12MC202 is produced by the Microchip Company. It has the control characteristics of a 16-bit single-chip microcomputer and the merits of high-speed DSP. It becomes one of the preferred choices in today's embedded system design schemes.

The MCPWM controller of dsPIC33FJ12MC202 is one of its main features. This device greatly simplifies the control of software and external hardware. The software can generate PWM waveform. By programming, it can generate a three-phase 6-channel PWM waveform. They are independent and work with the same frequency. The intelligent power module SD05M50D is driven through this 6-channel PWM signal. Then, the three-phase input and electronic phase change of the BLDCM are controlled. Each PWM output pin drives a current of approximately 25mA.

Based on these characteristics of the SCM, its external operating circuit is shown in Figure 2.

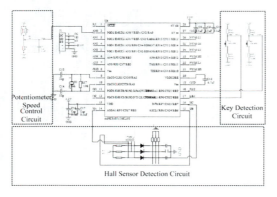

Figure 2. External circuit of dsPIC33FJ12MC202.

In the potent-meter speed control section part, we use rotate button slippage resistance. When it rotates, its output port will detect the voltage changes and the internal AD of the micro-controller reads the value of the voltage. The upper left corner dashed box of figure 1 shows the potent-meter speed control circuit. In order to increase the accuracy of AD collection, we can add a 10kΩ and a 0Ω resistor in parallel for filtering. Then, we can add a 0Ω in series connection to ensure the correctness of the voltage. At last, we achieve the key control functions by detecting changes in the size of AD and performing different actions.

3.2 Design of SMD05M50D circuit

SMD05M50D is a highly integrated intelligent power module, and it is primarily used as a small power motor drive of a fan. It has 6 power MOS and 3 gate drive circuits. It also integrates the under-voltage protection and anti-dv/dt circuits. The circuits provide excellent protection and a broad range of safety work in its internal. The design of SMD05M50D is with a high insulation, easy heat conductivity, and low electromagnetic interference. It provides a very compact package. So it is easy to use, especially for built-in applications of the motor. This is also one of the main reasons that we select it in this system. As the drive part of this system, it can change the current flow the motor to realize the phase change.

SMD05M50D integrates 3 gate drive circuits in its internal, and each includes a gate drive chip and two power FETS. Two PWM inputs control the power FETS on and off. PWM2 is the gate drive chip selection signal, and PWM1 is the circuit operation signal. When PWM1 is high, HO output is high and LO output is low; when PWM1 is low, HO output is low and LO output is high. If the chip select signal PWM2 is off, the upper and lower arms will fully be closed off, as shown in Figure 3.

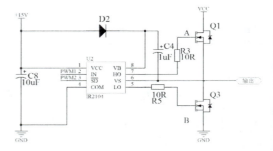

Figure 3. Gate drive circuit.

3.3 Hall sensor circuit design

Most conventional brush-less DC motors use position sensor to determine the rotor positions and realize the phase change by controlling the drive circuit. Due to the presence of the position sensor, the motor size and cost increase, reducing the reliability of the motor. It limits certain occasions of motor application. The hall sensor is often used to detect the position of the rotor inside the motor or the speed deviation in BLDCM drive control system. Because of the hall IC sensor applications, the control system can be greatly simplified. The dashed box below figure 1 shows the hall signal detection circuit of this system. The position signal detected by the hall sensor will be inputted to the MCU after shaping.

3.4 EMF zero-crossing detection circuit

During the rotation of the motor, the stator winding cuts the magnetic induction line of the rotating permanent magnet rotor. According to the law of Electromagnetic Induction, it will generate Back EMF. This method is commonly used among all the control methods without a position sensor in the brush-less DC motor.

No sense of brush-less DC motor needs to use the induced EMF of the third phase to detect the position of the rotor. The method of EMF zero-crossing detection is accomplished by detecting the motor terminal voltage of the three-phase winding. It is calculated by software to get back EMF zero-crossing and delayed 30 electrical angle to give the rotor position signal. So we can control the motor to realize the electronic phase change. It can also be called "terminal voltage detection method."

Figure 4.   EMF zero-crossing detection circuit.

In the circuit of Figure 4, U, V, and W are connected to A, B, and C, which are three lines of the motor. Then, we connect them to AN2, AN3, and AN4 pins of the MCU through a voltage divider network. AN5 pin, which is connected to the MCU, is the estimated midpoint voltage after deformation. During the UV energizer period, open the comparison between AN2 and AN5; during the UW energizer period, open the comparison between AN3 and AN5; and during the VW energizer period, open the comparison between AN5 and AN5. Then, we can successfully detect zero-crossing time of each phase.

4   SOFTWARE DESIGN OF THE SYSTEM

This system achieves the purpose of controlling the brush-less DC motor. The first thing is through PWM and SPWM driving the motor rotation. Furthermore,

we can adjust the rotation of the motor and the size of speed through the buttons and sliding rheostat. According to the tasks described earlier, the system selects dspic33FJ32MC202 micro-controller, leg buttons, sliding rheostat, and brush-less DC motor for programming. Depending on the task, the program flow chart is shown in Figure 6.

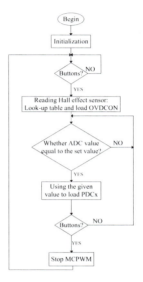

Figure 6.   Program flow chart.

4.1   Design of control part of PWM software

To make sure the speed of the brush-less DC motor is adjustable, variable voltage at the ends of the two phases winding is added. In digital language, variable voltage is obtained by adding different duty cycles of the PWM signal to the BLDCM winding.

By using six MOSFETS, three-phase winding can be driven at a high and low level, either with or without power. For example, when one end of the winding is connected to a high-side driver, the low-side driver can be applied at a variable duty cycle of the PWM signal.

PWM signal is provided by the dedicated PWM module of the dsPIC33FJ12MC202 motor control system. MCPWM module is specially designed for brush-less DC motor control applications.

MCPWM has a dedicated 16-bit PTMR time base register (PTMR). Users can set the PWM output frequency and period through the defined time interval, with the timer counts once per cycle. By selecting a value and loading it into the PTPER register, users can determine the required PWM cycle. We can begin a new cycle by comparing TCY and PTMR with the PTPER. We can set the period and frequency in this manner.

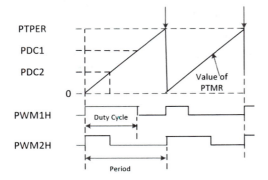

Figure 7.   Edge-aligned PWM.

As shown in Figure 7, the method of controlling duty cycle is similar to the earlier method; we just need to load the three duty cycle registers with a value. On comparison with the cycle, the comparison frequency is twice that of the period. If the value of PTMR matches that of PDCx, then the corresponding output pin of the duty cycle will drive it at a low or high level based on the selected PWM mode.

When we drive the complementary output by controlling a constant bridge arm, these two output pins can also be configured as a separate output mode. We can insert a dead time when the high level becomes low and the low level becomes high to prevent straight leg accident phenomenon of output drive circuit bridge arm.

4.2   Design of control part of SPWM software

Because of brush-less DC motor adopting no sinusoidal stator winding, it is difficult to adapt on some occasions. The occasions require low torque ripple and low noise operation. So using SPWM control mode at low speeds is relatively good.

Sine wave control of the brush-less motor does not actually apply the sine wave to the three-phase input terminal of the motor. It just sets the duty cycle time to be arranged by sine law through adjusting the duty cycle of the PWM waveform. An important conclusion of the sampling control theory is as follows: When a narrow pulse with the same impulse and different shapes are applied to the inertia of the link, the effects are basically the same.

SPWM method uses this conclusion, as it is basically theory. The pulse width of PWM waveform

varies by sine law and it is equal to the sine wave. Using PWM waveform to control the intelligent power module bridge arm off, the area of the output pulse voltage is equal to that of the corresponding interval of the desired output sine wave. We can adjust the output voltage of the invert circuit by changing the frequency and amplitude of the modulation wave.

The concrete realization method is as follows: At first, we assign the sine table to the array, then set each PWM cycle of PWM waveform generation module into interruption status, and lastly change the value of the PWM so that it is comparable with the sine table in the ISR. Then, the cycle of the steps are as described earlier.

5   TEST RESULTS AND ANALYSIS OF THE SYSTEM

The BLDCM runs when it is powered. Then, we can detect the rotor position through internal Hall element and the signal of the Hall element after detecting the circuit. Figure 8 shows the oscilloscope reads the Hall element signal; then through the three channels of the oscilloscope, we display the waveform inputted to the micro-controller.

Commissioning results: Through testing the circuit by the Hall signal, we measured the signal to the micro-controller as a periodic square wave. The signal of the Hall element after conversion by the circuit is neater.

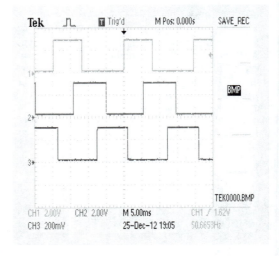

Figure 8.   Oscilloscope reads the Hall element signal.

No sense of blush-less DC motor needs to use the induced EMF of the third phase to detect the position of the rotor.

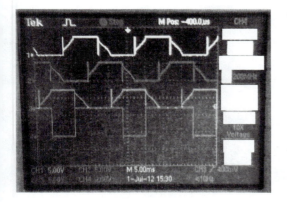

Figure 9. Back-EMF when controlled by the PWM.

When the motor is drive controlled by SPWM, its Back-EMF is shown in Figure 10:

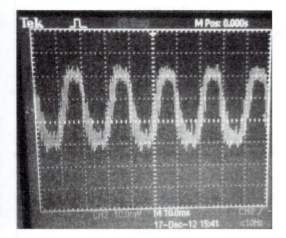

Figure 10. Back-EMF when controlled by the SPWM.

6 CONCLUSIONS

The design of the brush-less DC motor system is based on dsPICFJ12MC202 single-chip microcomputer. After the actual test, the system can accurately control the brush-less DC motor with a larger moment forward, reverse, and speed adjustment and it has stable performance and high reliability. These make us achieve good control of the brush-less DC motor. Practice has proved the following: First, using the back-EMF zero-crossing detection method and hall element to optimize the position of the sensor control method has higher sensitivity and achieves a stable operation state; second, using the three-phase back-EMF zero-crossing signal to determine each extreme point of commutation logic can accurately give commutation time; third, we can avoid the deviation caused by the traditional method of back-EMF in the phase-shifting process; and finally, the back-EMF detection does not need the extra power and the PWM signal can be changed at any duty ratio.

ACKNOWLEDGMENTS

This work was supported in part by the Zhejiang Provincial Natural Science Foundation of China under Grant No. LQ12F01001, and the Scientific Research Fund of Zhejiang Provincial Education Department (No. Y200803237).

REFERENCES

[1] P. Kaur, S. Chatterji. AVR Microcontroller-based automated technique for analysis of DC motors. International Journal of Electronics, 2014, Vol.101 (1), pp.1–9.
[2] Liu Yuhang. The brush-less DC motor control of back electromotive force superposition[J]. China University of Metrology, 2013, 24(2):156–161.
[3] Meng Yanjin, Chang Jie, Zhu Yuguo. Design of brush-less DC motor control system based on DSP[J]. Journal of Shaanxi University of Science and Technology, 2013:3195–3198.
[4] Zhang Wensheng, Hu Qingeng, Wang Wenfeng, Zhang Xin, Shi Zewen, Chen Hao. The research of brush-less DC motor intelligent control system[J], Computing Technology and Automation, 2012, 31(3):69–76.
[5] Shi Zhaolin. DSPIC Digital Signal Controllers Getting Started with the Actual Combat-Getting Started Guide[M]. Beijing University of Aeronalltics and Astronautics Press, 2009.

Forming of the structure of the investment potential of sustainable development of the city

M.L. Moshkevich & L.V. Sevrukova
South West State University, Kursk, Russia

ABSTRACT: The proposed structure of the investment potential of the city, including, according to the author's approach, four main elements: economic, financial, innovation, infrastructure and savings potentials, helped clarify the concept of investment potential of the city. This can be interpreted as functional system resources, identifying opportunities of investment, which can aim at the satisfaction of needs of the population and ensure sustainability of the development of the city.

KEYWORDS: City, sustainability, sustainable development, investment potential

1 INTRODUCTION

In the current economic situation, because of the power of recurring economic crises and their effects, the sustainability of socioeconomic systems remains relevant. Researchers are interested in the national economy and the economy of the territorial entities, industries, and enterprises.

There seems to be a number of changes in the direction of development in the theory of stability of the socioeconomic systems in recent years. Initially, the sustainability of the system was a necessity for survival and normal functioning of the territorial systems. Nowadays, sustainability increases the efficiency of the system, that is, the reserve of development.

The formation of the concept of sustainable development began with the club of Rome report "Limits to growth" in 1972. The report first indicated the fact that the stock of natural resources is limited. The rapid increase of the industry is beginning to threaten the preservation of normal environmental conditions. Therefore, there was a necessity to create a new concept of sustainable development. In consequence, many countries have enacted laws on environmental protection. They began to carry out the transfer of manufacturing from large cities, and some harmful areas of production were closed.

In order to develop proposals for the solution of environmental problems, the United Nations formed the Brundtland Commission. In the report of this Commission, the term "sustainable development" was used for the first time.

After a few years, the Commission concluded that the achievement of sustainable development by solving just environmental problems is impossible without consideration of the socioeconomic issues and it is necessary to speak about sustainable development in a broad sense. In the report, the following definition was given: "Sustainable development is development that satisfied the needs of the present day without compromising the ability of future generations to satisfy their needs."

Thus, issues of sustainability and sustainable development have acquired a new direction of research in connection with the development and effective functioning of the socioeconomic systems.

In Russia, the term "sustainable development" was first officially defined in the Decree of President of the Russian Federation as follows: "On the state strategy of the Russian Federation on environmental protection and sustainable development," as well as in the decree of the President as follows: "On the concept of transition of the Russian Federation to sustainable development."

Initially, the majority of Russian research regarding the theory of sustainable development has been directed to the regions. In Europe, special importance of cities as centers of implementation of principles of sustainable development was highlighted at the European conference held on the sustainable development of cities and towns in Aalborg in 1994. Here, the "Charter of European cities for sustainability (Aalborg Charter) was adopted.

In Russia, the study of sustainable urban development began much later. Nowadays, the studies related to this question are not systematic. Almost every Russian city has a strategic plan with the main goal of achieving sustainable development.

Nowadays, the term "sustainability" is used in many branches of scientific knowledge. In a general sense, sustainability is understood as the system's ability to save the current state in the presence of influences.

A sustainable city should aim not only at the simple growth of quantitative indicators but also at sustainable socioeconomic development, which is based on the rational use of available resources.

From positions of the theory of sustainable development, an urban system is divided into a number of subsystems. The system's stable and efficient development may lead to the achievement of sustainable development of the city. The main subsystems are: social, economic, and environmental. Selection of other subsystems in most cases depends on the author's approach. Any of these subsystems can be considered from the viewpoint of characteristics of a particular capacity that is necessary for the functioning and development of the city.

The potential, in a broad sense, is usually understood as funds, reserves, and sources that are available and can be mobilized, powered, and used for a particular purpose, planning, and solving of any problem. It proposes to define the potential of sustainable development as a strategically important supply source, requiring constant replenishment for its future use for the development.

The main principle of sustainable development is the balance of economic, social, and environmental interests of society. A sustainable development of the city involves always aiming at improving the level and quality of life of the population on subject to the principle of balance of interests. In the current economic conditions, such development is impossible without attraction of additional investment resources. Therefore, one of the main tools of achievement sustainable development of the city can be the investment potential of the city.

Therefore, the sustainability of the city as a territorial socioeconomic system can be understood by its ability for stable functioning and developing in the long term amid constantly changing economic conditions. A sustainable city will be characterized as a process aimed at achieving sustainable functioning and development of the economic subsystem of the city through the efficient use of the existing system of urban potentials.

According to the most common approach used for determining the nature of the investment potential, this concept is characterized by a range of its private potentials' inherent territory. The number of potentials and their types vary depending on the level of consideration of the object along with the potential and the methodology.

The bases of the regional investment potential are items such as natural resource, labor, and industrial,

innovative, institutional, infrastructural, financial, consumer, and tourist potential. Given the structure of the investment potential of the region that is given as not being final, it can be changed.

There is not much information about the structure of the investment potential of the city. It is used as a set of private potentials that characterize the regional system to describe investment potential of the city. But, in our opinion, the composition of private potentials must be adjusted to the peculiarities of the city as a territorial socioeconomic system.

The developed socioeconomic and industrial infrastructure of the city is determined by the infrastructure capacity. Concentration of the institutes of education and science in the city determines the availability of innovative capacity. The shopping center provides consumer potential. High concentrations of population with a higher level of education leads determine the city's labor potential. The relative political and economic independence of the city is possible if there are financial and saving potentials.

It is proposed to determine the investment potential of the city (IPG) as a functional resources system, identifying opportunities to invest, which can be designed to satisfy the needs of the population and provision of sustainable development of the city.

Similar to regional investment potential, investment potential of the city is characterized by a set of specific potentials that define its structure and affect its sustainable growth. The structure of the investment potential of the city is determined by the characteristics of the city as a territorial socioeconomic system.

After the analysis, four private potentials of sustainable city development were selected, thus defining its structure (table 1).

1 Economic and financial potential: This reflects the state of the city's financial resources (budgetary funds, finance, banking and insurance sectors of the city, and funds of enterprises), as well as a cumulative result of economic activity of the population of the city and characterizes the human resources and their educational level;
2 Innovative potential: This is characterized by the level of development of science and the introduction of scientific-technical progress;
3 Infrastructure potential: This reflects infrastructure security;
4 Saving potential: This characterizes the aggregate purchasing power and investment power of the population of the city or a set of features characterizing the level of consumption and accumulation.

This structure can be used for researching investment potential of the city.

Table 1. Structure and indicators for assessing investment potential of the city.

Private potential	Components of potential	Indicators
economic and financial	financial	1. tax base; 2. surplus (deficit) city budget; 3. the profitability of main enterprises located in the territory of the city; 4. the profitability of the banking sector of the city; 5. the profitability of the insurance sector of the city; 6. the number of implemented investment projects in the territory of the city
	industrial	1. gross city product (GCP) per capita; 2. the amount of fixed assets
	labor	1. the number of people employed; 2. the number of officially registered unemployed; 3. life expectancy; 4. the educational level of the population (the population with higher education)
	innovative	1. the number of people employed in research and development; 2. the number of organizations with the main activity of "research and development"; 3. the number of candidates and doctors of Sciences; 4. the number of patents, licenses for utility models 5. the number of Universities, including the state, branches of Universities;
	infrastructure	1. the length of roads with hard surface; 2. departure passenger train (air) transport of general use; 3. the number of registered subscribers of the mobile communication terminal; 4. the number of organizations using information and communication technologies, the global information network; 5. the provision of water supply and gas supply, sewage networks
	saving	1. the cost of a fixed set of consumer goods and services meadow; 2. average accrued wages; 3. volume of paid services to the population; 4. the volume of retail trade; 5. the number of outlets per 1 inhabitant; 6. the number of institutions of culture and leisure, per 1 inhabitant 7. living space per 1 inhabitant; 8. the number of private hire; 9. the amount of savings of the population; 10. the level of real incomes of the population; 11. the ratio between minimum and average wages; 12. the share of wages in (GCP)

2 CONCLUSIONS

The sustainability of the city can be characterized using the indicator of investment potential. Cities have their own peculiarities of functioning and development. Therefore, indicators for assessing the investment potential of the city differ from those used for the regions.

REFERENCES

[1] Davydova, L.V., Ilminsky, S.V. 2007. Formation of strategy of development of the investment potential of the region based on the assessment of investment processes. *Economic analysis: theory and practice* 13(94): 12–22.
[2] Davydova, L.V., Markina S.A. 2007. Investment potential as the basis for economic growth. *Finance and credit* №31(271): 54–57.

Information, Computer and Application Engineering – Liu, Sung & Yao (Eds)
© 2015 Taylor & Francis Group, London, ISBN 978-1-138-02717-6

Optimization of ramjet fuel control system based on GA-PSO

Hui Xian Huang, Sha Zeng, Wei Juan Zhao & Ren Chen
College of Information Engineering, Xiangtan University, Xiangtan, Hunan, China

ABSTRACT: A hybrid optimization algorithm combined with Particle Swarm Optimization (PSO) algorithm and Genetic Algorithm (GA) is proposed for the performance requirements of ramjet fuel control system. The new algorithm inherits the better local search capability of PSOA as well as the fast convergence of GA in such a manner that selection, crossover, and mutation operation are introduced into PSOA in each iteration. Weaknesses such as PSO algorithm being easy to fall into local optimum and of GA being easy to converge have been overcome. This new algorithm is applied to the ramjet fuel control system. The simulation results show that the proposed method can obtain a faster response and better stability as well as restrained overshoot. The comprehensive control performance of the system has been improved a lot.

KEYWORDS: Particle swarm algorithm; Genetic algorithm; Hybrid optimization algorithm; Ramjet fuel control system

1 INTRODUCTION

Fuel control system is the main part of ramjet, and its performance and reliability judge the entire control system. Nowadays, with the continuous improvement of aero-engine technology, requirements for fuel control system become higher. Therefore, the optimization of the ramjet fuel control system is an urgent necessity.

Proportional–Integral–Derivative (PID) control is widely used in industries due to its simplicity and satisfactory control performance[1][2]. Its optimal control performance can be achieved through tuning and optimizing of parameters. Hence, it is of very important theoretical and practical significance to study the PID parameter setting and optimal technology. Along with the advancement of the PID controller parameter setting method, algorithms such as genetic algorithm, particle swarm optimization are applied to PID control parameter setting. Karam M. Elbayomy[3] proposed PID controller optimization by GA for the electro-hydraulic servo control system, and the simulation results verify its effectiveness by comparing it with the classical PID controller and the compensator controller. D. Nangru and D. K. Bairwa[4] presented a modified particle swarm optimization (MPSO)-based PID controller for stable processes, which gives better performance of the MPSO-PID controller than by the traditional method. Ali Tarique and Hossam A. Gabbar[5] suggested the method of PSO for determining the optimal PID controller parameters for steam turbine control and made a comparison with GA.

Since PSO has low convergence rate and is easy to trap in local minima [6], a new algorithm combining PSO algorithm with genetic algorithm was presented to find the optimum PID controller parameters for ramjet fuel control system. After introducing selection, crossover, and mutation operations of GA into PSO, this new hybrid algorithm is of local search capability and global search capability, and it can simultaneously guarantee the population diversity. By optimizing ramjet fuel control system based on the hybrid optimization algorithm, better performance of this control system will be achieved.

2 MATHEMATICAL MODEL OF RAMJET FUEL CONTROL SYSTEM

Ramjet fuel control loop is mainly intended to realize ramjet fuel flow control. The ideal control methods are geometric control and thermal control; by their coordination, the desired operating state of the combustion chamber can be obtained. Figure 1 shows the ideal function module, which includes controller, actuators, turbo pump, and sensors.

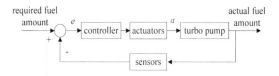

Figure1.   Function block diagram of ramjet fuel control.

When there is inconsistency between the actual and required fuel amount, error signals generated, after the process of the controller, different control signals arise; these signals control the actuator to change the angle of the turbine vanes α and adjust the fuel supply until it reaches the desired fuel amount. In this way, the adjustment of fuel control is achieved.

The initial condition for ramjet fuel control system in this article is as follows: Flight velocity is at 3.0Ma, flight height is at 1800km, and the angle of attack is 0^0. As the mathematical model of ramjet fuel control system is complex and nonlinear, to get intuitive simulation results, the mathematical model is linearized and the transfer function of fuel control system is obtained as given next:

$$G(s) = \frac{165e^{-0.5s}}{(s+20)(s+9.03)} \qquad (1)$$

In this transfer function, a strategy of all-pole approximation is used for the portion of pure delay.

$e^{-\tau s} = \dfrac{1}{\tau^2 + \tau s + 1}$. Compared with other approxima-

tions, all-pole approximation has a shorter adjustment time, no overshoot, and better step response characteristics.

3 CONTROLLER DESIGN

The PID controller transfer function is assumed as

$$G_c(s) = k_p + \frac{k_i}{s} + k_d s \qquad (2)$$

where k_p represents proportional gain, k_i represents integral gain, and k_d represents derivative gain. Adjusting the PID controller parameters effectively improves system performance.

A new GA-PSO method is proposed for tuning PID parameters in order to achieve the optimization of ramjet fuel control system. The block diagram of the proposed GA-PSO-PID controller is shown next:

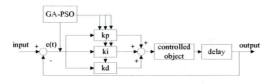

Figure 2. Block diagram of proposed GA-PSO-PID controller.

4 ALGORITHM

4.1 PSO algorithm

PSO algorithm is a swarm intelligence optimization algorithm developed by Kennedy and Eberhart(1995). Its basic principle is as follows: Particles search the optimal position similar to birds in solution space, by tracking the individual extreme P_{best} and group extreme G_{best}. Then, individual position is updated. Three components are contained in each particle, which are position, velocity, and fitness value; they indicate the particle's characters and represent k_p, k_i, k_d, respectively. In each iteration, particles update their position and velocity through formulae (3) and (4) that are given next:

$$v_{id}^{k+1} = \omega v_{id}^k + c_1 r_1(p_{id}^k - x_{id}^k) + c_2 r_2(p_{ad}^k - x_{id}^k) \quad (3)$$

$$x_{id}^{k+1} = x_{id}^k + v_{id}^{k+1} \qquad (4)$$

As the formula given earlier shows, ω indicates inertia weight, $d = 1, 2, \ldots, D$; $i = 1, 2, \ldots, n$; k stands for the current iteration number; v_{id} is particle velocity; c_1 c_1 and c_2 represent non-negative constants, which are called the "acceleration factors"; c_2 and r_1 and r_2 are random numbers that are distributed in [0 1].

For inertia weight factor ω, this article adopts a strategy of decreasing linear weights for the purpose of improving conventional PSO algorithm. The formula is stated as follows:

$$\omega(k) = \omega_{start} + (\omega_{start} - \omega_{end})(\frac{k}{T_{max}})^2$$

$$+ \left(\omega_{end} - \omega_{start}\right)(\frac{2k}{T_{max}}) \qquad (5)$$

In the formula, ω_{start} stands for maximum weighting factor, ω_{end} represents minimum weighting factor, k is the current iteration, and T_{max} states the maximum number of iterations. As this linear decreasing strategy makes the algorithm into a local search at its early stage by accelerating the decline rate of inertia weight, it is considered in formula(3).

4.2 GA

GA was proposed by professor Holland from Michigan university of American in 1969, and it became a kind of simulated evolutionary algorithm when it was summarized by De Jong and Goldberg.

The basic idea is to imitate the evolutionary law of "natural selection, survival of the fittest," and its essential operations are selection, crossover, and mutation. Selection is the step used for selecting the best individual from the current population, crossover can get a new generation of individuals, and mutation involves selecting an individual and changing the value of the string randomly.

4.3 GA-PSO algorithm

Considering the good diversity, global search capability of GA algorithm, fast search speed, and better local search capability of PSO algorithm, a new hybrid algorithm that combined PSO with GA is proposed in this article. Based on PSO algorithm, the selection, crossover, and mutation operations of GA are introduced. In this way, the drawback is that PSO is easy to fall into the local extreme and there is a premature situation in which GA is eliminated. Importantly, the new algorithm not only helps ensure the diversity and quality of particles but also enables the particles to have a better search direction.

The detailed steps of the hybrid algorithm are as follows (*gen*: iteration number of GA; *maxgen*: maximum iteration number of GA; *iter*: iteration number of PSO; *maxiter*: maximum iteration number of PSO):

1 Initialization of related parameters and PID parameters k_p, k_i, and k_d;
2 To calculate the fitness value of particles according to the formulas;
3 Let *gen*=1;
4 Go to step 5 when *gen≤maxgen*; otherwise, produce output results;
5 Let *iter*=1;
6 Update the position and velocity of particles under formulas (2) and (3) when *iter≤maxiter*, or else, go to step 9;
7 Search the individual optimal and group optimal in line with fitness value and assign them to k_p, k_i, k_d;
8 Make *iter*=*iter*+1, repeat step 6;
9 Insert operations of selection, crossover, and mutation to iteration process, making *gen*=*gen*+1 and go back to step 4;
10 Find the best fitness value, then obtain optimal solution of PID parameters;

Figure 3 shows the simplified flow chart of GA-PSO algorithm.

For better performance of ramjet fuel control system, the fitness function is adopted as given next:

$$J = \int_0^{+\infty} (\omega_1 |e(t)| + \omega_2 |y(t)|) dt \qquad (6)$$

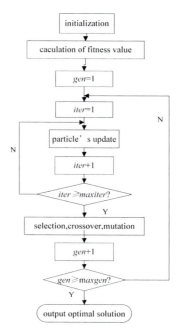

Figure 3. Flow chart of GA-PSO algorithm.

Where $e(t)$ indicates the system error, $y(t)$ refers to the output of the controlled object, and ω_1 and ω_2 are the constants that are used to adjust the system error and overshoot, respectively.

5 SIMULATION RESULTS

To research and analyze the optimize performance of different algorithms, a comparison is made with PSO, GA, and GA-PSO. Parameters related to the simulation process are as follows: Population is *swarm size* = 40; the number of iterations is *ma iter*=40; acceleration constants are $c_1=c_2=2$, dimension D=3; and inertia weights are ω_{start}=0.9, ω_{end}=0.4. Genetic iterations *maxgen*=20 and crossover probability P_c=0.6, mutation probability P_m=0.01. In fitness function, ω_1=1, ω_2=10.

The unit step response curves of different control strategies are presented in figures 4 and 5:

As shown in the illustration, figure 4 presents the unit step response curve by PSO and GA-PSO algorithm, respectively; figure 5 depicts the unit step response curve by GA and GA-PSO separately. Through comparative analysis, we know that the unit step response curve obtained by PSO has oscillation, larger overshoot, and longer stable time. Similarly, the unit step response of GA algorithm has a slow response, longer rising time, and stable time.

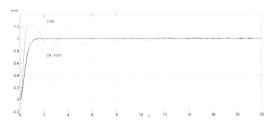

Figure 4.   Unit step response curve by PSO and GA-PSO.

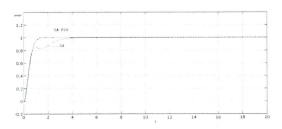

Figure 5.   Unit step response curve by GA and GA-PSO.

However, the new hybrid algorithm GA-PSO has a significant improvement compared with separate PSO or GA; it has a faster response, shorter stable time, and no overshoot.

Furthermore, to demonstrate the immunity of GA-PSO algorithm, a disturbance is added in the simulation at t=10s. Figure 6 is the waveform of PSO and GA-PSO algorithm on disturbance; figure 7 shows the waveform of GA and GA-PSO algorithm on disturbance.

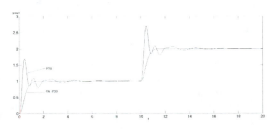

Figure 6.   Waveform of PSO and GA-PSO algorithm on disturbance.

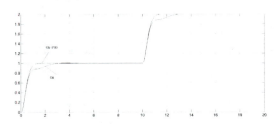

Figure 7.   Waveform of GA and GA-PSO algorithm on disturbance.

According to figures 6 and 7, the robustness of the proposed GA-PSO method is verified. It has better immunity compared with PSO or GA.

6   CONCLUSION

To improve the optimized performance of PSO and thus to enhance the performance of ramjet fuel control system, a new hybrid algorithm GA-PSO is proposed in this article. The hybrid algorithm not only strengthens the information exchange between particles, leading the particles to a better search direction, but also enhances the particles' ability out of the local extreme. Hence, GA-PSO has strong global search ability as well as strong local search ability. The simulation results indicate that, compared with PSO and GA algorithm, the hybrid algorithm GA-PSO possesses better optimization performance with no overshoot, no oscillation of simulation waveforms, and significantly shorter rising time and settling time. It is also of good robustness. All these make a meaningful improvement with regard to the ramjet fuel control system's performance.

REFERENCES

[1] Yang lang. 2006. Study of ramjet thrust control system[D]. Heilongjiang: Harbin institute of technology.
[2] Gao song. 2010. Control system design of scramjet thrust[D]. Heilongjiang: Harbin institute of technology.
[3] Karam M & Elbayomy & Zongxia Jiao & Huaqing Zhang. 2008. PID Controller Optimization by GA and Its Performances on the Electro-hydraulic Servo Control System[J]. Chinese Journal of Aeronautics, 21: 378–384.
[4] D. Nangru & D.K. Bairwa & K. Singh & S. Nema & P.K. Padhy. 2013. "Modified PSO based PID controller for stable process", 2013 International Conference on Control, Automation, Robotics and Embedded Systems (CARE), 10:1–5.
[5] Ali Tarique & Hossam A. Gabbar. 2013. Particle Swarm Optimization (PSO) Based Turbine Control[J]. Intelligent Control and Automation, 04(02): 126–137.
[6] Hanhong Zhu & Yi Wang & Kesheng Wang & Yun Chen. 2011. Particle Swarm Optimization (PSO) for the constrained portfolio optimization problem.[J]. Expert Systems with Applications, 38(8): 10161–10169.
[7] S.M. GirirajKumar & Deepak Jayaraj & Anoop. R.Kishan. 2010. PSO based Tuning of a PID Controller for a High Performance Drilling Machine.[J]. International Journal of Computer Applications, 1(19): 12–18.
[8] Liu wei & Zhou yu ren. 2009. Modified inertia weight particle swarm optimizer[J]. Computer engineering and application. 45(7): 46–48.
[9] Chen gui min & Jiajianyuan & Han qi. 2006. Study on the strategy of decreasing inertia weight in particle swarm optimizational gorithm[J]. Journal of xi'an jiaotong university, 01(40): 53–56+61.
[10] Gejike & Qiuyuhui & Wu chunming & Pu guo lin. 2008. Summary of genetic algorithms research[J]. Application research of computers, 10(25): 2911–2916.

Information, Computer and Application Engineering – Liu, Sung & Yao (Eds)
© 2015 Taylor & Francis Group, London, ISBN 978-1-138-02717-6

An approach of group decision making for multi-granularity linguistic term sets based on 2-tuple linguistic aggregation operators

Zhong Min Zhang
Xi'an Research Institute of High-tech, Xi'an, China
Xi'an Communications Institute, Xi'an, China

Jun Shan Li
Xi'an Research Institute of High-tech, Xi'an, China

Ping Song
Xi'an Communications Institute, Xi'an, China

ABSTRACT: According to the 2-tuple linguistic theory, this article proposes an approach of group decision for the aggregation of multi-granularity linguistic term sets. First, the Language Hybrid Aggregate (LHA) arithmetic operator was extended to 2T-LHA operator and then the different granularity linguistic terms were mapped to the same linguistic term set, which transformed the different granularity linguistic judgment matrixes to being consistent. Furthermore, the group preferences could be drawn from the consistent judgment matrixes with 2T-LHA operators. Finally, an example was given to illustrate the use of the approach. The result shows that the research work can aggregate the multi-granularity linguistic terms with high accuracy and reliability and simultaneously avoid the loss and distortion of information during the decision-making process.

KEYWORDS: group decision making; the Language Hybrid Aggregate (LHA) arithmetic operators; 2-Tuple linguistic; multi-granularity linguistic term sets

1 INTRODUCTION

It is difficult for decision makers to describe the judgment information of decision objects quantitatively due to the complexity of the objective world and ambiguity of human thinking, and human beings are more accustomed to describing the result of the operation of thinking through natural language. Currently, there are two kinds of group decision-making methods based on natural language: One is the transformation of the linguistic information into other forms such as fuzzy number; the other is the symbolic transferring by "subscript operator" to deal with linguistic assessment information [1]. However, there are some limitations in both kinds of methods: The former rests on the premise of certain membership functions, which are difficult to be specified in practice; besides, the computing results of the latter are discrete. So it is necessary to seek the approximate values close to the standard evaluation scales of the initial linguistic term set, which will lead to loss and distortion issues. In the year 2000, Spanish scholar Herera Professor [2] proposed 2-tuple analysis

method to aggregate linguistic information, and the method well makes up for insufficiency of the earlier methods. The studies express the aggregation of the linguistic information as the form of 2-tuple, which consists of a phrase from the linguistic term set and a real number and that can avoid the loss and distortion of information during the decision-making process while achieving high accuracy and reliability.

The research literature on aggregation of linguistic information pay more attention to the situation with the same linguistic term set, and there is less research on different granularity linguistic term sets. In this article, we extend the Language Hybrid Aggregate (LHA) arithmetic operators based on 2-tuple linguistic theory, and propose a group decision-making method of multi-granularity linguistic judgment matrices. The research work can aggregate the multi-granularity linguistic terms and avoid the linguistic loss effectively during the decision-making process; so it can be ensured that the decision results are reasonable and accurate. Finally, a case study on service status evaluation of ECM equipment based on the research work is given.

2 THE 2-TUPLE LINGUISTIC THEORY

2.1 Linguistic evaluation scale

Definition 1[3]. For a linguistic term set: $S^T = \{ s_i | i=0,...,T\text{-}1 \}$, the element s_i is known as the linguistic evaluation scale, and the number of the elements T is said to be the granularity of the linguistic term set, where T is generally odd. Assuming $f : S^T \to T$, $f(s_i) = i$, the function f is said to be the subscript function corresponding to the linguistic evaluation scale; besides, the inverse function $f^{-1} : T \to S^T, f^{-1}(i) = s_i$.

The following properties hold good for the linguistic evaluation scale:

① *Orderliness:* $s_a \succ s_b \Leftrightarrow a > b$,

② *Symmetry:* There exists inverse operation *neg*, which lets neg $(s_b) = s_a$ if $a + b = 0$,

③ *The extreme value:* If $a \geq b$, $\max\{s_a, s_b\} = s_a$ and $\min\{s_a, s_b\} = s_b$.

The following algorithms hold good for the linguistic evaluation scale [4]:
$s_a \oplus s_b = s_{a \oplus b}$, besides,

① *The commutative law* $s_a \oplus s_b = s_b \oplus s_a$,

② *The multiplicative law* $\lambda \cdot s_a = s_{\lambda a}, \lambda \in R$,

③ *The distributive law* $\lambda \cdot (s_a \oplus s_b) = \lambda \cdot s_a \oplus \lambda \cdot s_b$
$\& (\lambda_1 + \lambda_2) \cdot s_a = \lambda_1 \cdot s_a \oplus \lambda_2 \cdot s_a$.

2.2 The 2-tuple linguistic representation

Definition 2[5]. The 2-tuple linguistic representation is in the form of two tuple (s_k, d_k), which is used to represent the linguistic evaluation scales, where s_k denotes the standard linguistic evaluation scale in the linguistic term set S, and d_k is said to be symbolic transfer value that expresses the real offset between the actual evaluation value and s_k. In general, $d_k \in [-0.5, 0.5)$.

Any linguistic evaluation scale can be mapped to the corresponding 2-tuple linguistic representation by the conversion function.

Definition 3[6]. *The conversion function*

$$T : [0, T-1] \circledR S \times [-0.5, 0.5) ,$$

$$T(\beta) = (s_k, d_k). \tag{1}$$

where T denotes the granularity of the linguistic term set S^T, and $\beta \in [0, T-1]$ is the converted real value; then, subscript $k = \text{Round}(\beta)$ and offset $d_k = \beta - k$, Round($\cdot$) is the round operator.

Definition 4[7]. *The comparative law of 2-tuple linguistic representation: for 2-tuple linguistic representation (s_i, d_i) and (s_j, d_j) with the same linguistic term set, if $s_i \succ s_j$, then $(s_i, d_i) \succ (s_j, d_j)$; if $s_i = s_j$, then ①$(s_i, d_i) \succ (s_j, d_j)$ when $d_i > d_j$; ② $(s_i, d_i) = (s_j, d_j)$ when $d_i = d_j$.*

Definition 5. *The multiplicative law of 2-tuple linguistic representation:* $\forall \lambda \in R, \lambda \cdot (s_k, d_k) = (s_\gamma, d_\gamma)$, where $\gamma = \text{Round}[\lambda \cdot (k + d_k)]$, $s_\gamma = f^{-1}(\gamma)$, $d_\gamma = \lambda \cdot (k + d_k) - \gamma$, f^{-1} is known as the inverse function of the subscript function.

3 LHA OPERATOR WITH 2-TUPLE LINGUISTIC REPRESENTATION

Now, we will extended LHA operator to 2T-LHA operator based on 2-tuple linguistic theory.

3.1 LHA operators

The Language Hybrid Aggregate (LHA) arithmetic operators $\text{LHA}_{w,\omega}[] : S_n \to S$,

$$\text{LHA}_{w,\omega}(s_{\alpha_1}, s_{\alpha_2}, \cdots, s_{\alpha_n}) =$$
$$w_1 \cdot s_{\beta_1} \oplus w_2 \cdot s_{\beta_2} \oplus \cdots \oplus w_n \cdot s_{\beta_n} = S_\beta .$$

According to the linguistic evaluation scale algorithm, $\beta = \sum_{i=1}^{n} w_i \beta_i$, where $w = (w_1, w_2, \cdots, w_n)$ is the weight vector associated with the LHA operator, $w_i \in [0,1]$ and $\sum_{i=1}^{n} w_i = 1$; s_{β_i} represents the i-th largest element in the weighted data set $(\bar{s}_{\alpha_1}, \bar{s}_{\alpha_2}, \cdots, \bar{s}_{\alpha_n})$, $\bar{s}_{\alpha_i} = n \omega_i s_{\alpha_i}$, $\omega = (\omega_1, \omega_2, \cdots, \omega_n)$ is the weight vector of the initial data set $(s_{\alpha_1}, s_{\alpha_2}, \cdots, s_{\alpha_n})$, $\omega_i \in [0,1]$, and $\sum_{i=1}^{n} \omega_i = 1$, n is said to be the balancing factor. LHA operator will be degenerated into EOWA operator when $\omega = (\frac{1}{n}, \frac{1}{n}, \cdots, \frac{1}{n})$.

3.2 2T-LHA operators

For the judgment term with 2-tuple linguistic representation $(s_{\alpha_1}, d_{\alpha_1}), (s_{\alpha_2}, d_{\alpha_2}), \cdots, (s_{\alpha_n}, d_{\alpha_n})$, LHA operator based on 2-tuple linguistic theory 2T-LHA$_{w,\omega}$: $(S_n, d_n) \to (S, d)$,

$$\text{2T-LHA}_{w,\omega}[(s_{\alpha_1}, d_{\alpha_1}), (s_{\alpha_2}, d_{\alpha_2}), \cdots, (s_{\alpha_n}, d_{\alpha_n})] =$$
$$w_1 \cdot (s_{\gamma_1}, d_{\gamma_1}) \oplus w_2 \cdot (s_{\gamma_2}, d_{\gamma_2}) \oplus \cdots \oplus w_n \cdot (s_{\gamma_n}, d_{\gamma_n}) = (S_r, d_r).$$

According to Definitions 2 and 3, the aggregated subscript $\gamma = \text{Round}\left\{ \sum_{i=1}^{n} w_i [f(s_{\gamma_i}) + d_{\gamma_i}] \right\}$, the offset $d_r = \sum_{i=1}^{n} w_i [f(s_{\gamma_i}) + d_{\gamma_i}] - \gamma$, where

94

$w = (w_1, w_2, \cdots, w_n)$ is the weight vector associated with the 2T-LHA operator, $w_i \in [0,1]$ and $\sum_{i=1}^{n} w_i = 1$; $(s_{\gamma_i}, d_{\gamma_i})$ represents the i-th largest element in the weighted data set $((\overline{s}_{\alpha_1}, \overline{d}_{\alpha_1}), (\overline{s}_{\alpha_2}, \overline{d}_{\alpha_2}), \cdots, (\overline{s}_{\alpha_n}, \overline{d}_{\alpha_n}))$,

$(\overline{s}_{\alpha_i}, \overline{d}_{\alpha_i}) = n\omega_i \cdot (s_{\alpha_i}, d_{\alpha_i})$, $f(\overline{s}_{\alpha_i}) = Round[\omega_i(f(s_{\alpha_i}) + d_{\alpha_i})]$,

$\overline{d}_{\alpha_i} = \omega_i(f(s_{\alpha_i}) + d_{\alpha_i}) - f(\overline{s}_{\alpha_i})$, $\omega = (\omega_1, \omega_2, \cdots, \omega_n)$ is the weight vector of the initial data set $(s_{\alpha_1}, d_{\alpha_1}), (s_{\alpha_2}, d_{\alpha_2}), \cdots, (s_{\alpha_n}, d_{\alpha_n})$, $\omega_i \in [0,1]$ and $\sum_{i=1}^{n} \omega_i = 1$, $\sum_{i=1}^{n} \omega_i = 1$, n is said to be the balancing factor.

4 GROUP DECISION-MAKING OF MULTI-GRANULARITY LINGUISTIC TERM

Considering the case of a multi-criteria group decision-making, the alternatives set $X = \{x_i | i = 1, 2, \cdots, n\}$, experts set $D = \{d_j | j = 1, 2, \cdots, m\}$, and the expert weight vector $\lambda = (\lambda_1, \lambda_2, \cdots, \lambda_m)$.

Due to the difference of decision makers' knowledge level, thinking habits, and other factors, the granularity of linguistic term sets used may be not the same. In this case, let us assume that the granularities set of the different linguistic term sets corresponding to each decision maker are $\{T_i | i = 1, 2, \cdots, m\}$, and $A^{T_i}(l) = (a_{ij}^{T_i}(l))_{n \times n}$ represent the linguistic judgment matrix of m for the alternatives of n, where $l = 1, \cdots, m$. The linguistic terms with different granularity will be mapped to the same linguistic term set before aggregation. The 2-tuple linguistic representation can effectively avoid information loss and distortion during aggregation and operation with high calculation accuracy and reliability, so the consistent linguistic judgment matrixes with different granularity could be given based on 2-tuple linguistic theory. To this end, we propose the following steps for aggregation of linguistic terms:

Step 1. Establishment of the consistent linguistic terms of judgment matrixes.

According to the conversion function shown as definition 3, the multi-granularity linguistic judgment matrixes $A^{T_i}(l)$ of m should be turned into the consistent 2-tuple linguistic representation $A^T(l)$ based on the same linguistic term set with granularity of T, where $l = 1, \cdots, m$.

The function for consistency:

$TC : S_{[0,T_i-1]}^{T_i} \rightarrow S_{[0,T-1]}^T$,

$TC(S_\alpha^{T_i}) = f^{-1}\left[\dfrac{f(S_\alpha^{T_i})(T-1)}{T_i - 1}\right] = f^{-1}\left[\dfrac{\alpha(T-1)}{T_i - 1}\right] = s_\beta^T$ (2)

where f is a subscript function, f^{-1} is the inverse function of f, and $\beta = \dfrac{\alpha(T-1)}{T_i - 1}$.

The 2-tuple linguistic representations of the evaluation scale $S_\alpha^{T_i}$ corresponding to the consistent linguistic term set S^T are (s_k^T, d_k), where $k = Round(\beta)$, $s_k^T = f^{-1}(k)$, and $d_k = \beta - k$.

Step 2. Aggregation of the consistent 2-tuple linguistic judgment matrixes $A^T(l) = (a_{ij}(l))_{n \times n}$ of m by 2T-LHA operator to obtain the group preference matrix $A^T = (a_{ij})_{n \times n}$, where $l = 1, \cdots, m$ and $a_{ij} = 2T\text{-LHA}_{w,\lambda}(a_{ij}(1), a_{ij}(2), \cdots, a_{ij}(m))$.

The weight vector w associated with 2T-LHA operator can be calculated as follows:

$w_l = Q(l/m) - Q((l-1)/m)$, $l = 1, \cdots, m$. (3),

where $Q(t)$ is said to be the fuzzy linguistic quantized operator [8], $\forall a, b, t \in [0,1]$.

$$Q(t) = \begin{cases} 0 & t < a \\ \dfrac{t-a}{b-a} & a \leq t \leq b \\ 1 & t > b \end{cases}$$ (4)

The values of parameters a and b are shown in Table 1 according to the fuzzy linguistic quantized criteria.

Table 1. Parameters of the fuzzy linguistic quantized operator.

criteria / Parameters	At least half	great majority	Overwhelming majority
a	0	0.3	0.5
b	0.5	0.8	1

Step 3. Calculation of the ordering vector $\theta = (\theta_1, \theta_2, \cdots, \theta_n)^T$ of the group preference matrix by 2T-LHA operator, where

$\theta_i = 2T\text{-LHA}_{w,\lambda}[(s_{\alpha_1}^T, d_{\alpha_1}), (s_{\alpha_2}^T, d_{\alpha_2}), \cdots, (s_{\alpha_n}^T, d_{\alpha_n})] =$
$w_1 = (s_{\gamma_1}^T, d_{\gamma_1}), w_2 = (s_{\gamma_2}^T, d_{\gamma_2}), \cdots, w_n = (s_{\gamma_n}^T, d_{\gamma_n})$.

The component wl of the weight vector w associated with 2T-LHA operator can be calculated as follows:

$w_l = Q(l/n) - Q((l-1)/n)$, $l = 1, \cdots, n$. (5),

where Q(t) is the fuzzy linguistic quantized operator.

Step 4. Sorting of the 2-tuple linguistic results of the ordering vector corresponding to alternatives by the comparative law of 2-tuple linguistic representation.

5 A CASE STUDY

Now, we study a case on service status evaluation of ECM equipment. There are 4 equipments to be evaluated by 3 decision makers who establish linguistic judgment matrixes by pairwise comparison. Besides, the decision makers give the linguistic terms with multi-granularity linguistic term sets.

The alternatives set $X = \{x_1, x_2, x_3, x_4\}$, experts set $D = \{d_1, d_2, d_3\}$, the expert weight vector $\lambda = (1/3, 1/3, 1/3)$, and the linguistic judgment matrixes $A^5(1)$, $A^7(2)$, $A^9(3)$ corresponding to the linguistic term set used, respectively, $S^5 = \{s_0^5, s_1^5, \cdots, s_4^5\}$, $S^7 = \{s_0^7, s_1^7, \cdots, s_6^7\}$, $S^9 = \{s_0^9, s_1^9, \cdots, s_8^9\}$ are given as follows:

$$A^5(1) = \begin{bmatrix} s_2^5 & s_1^5 & s_4^5 & s_0^5 \\ s_3^5 & s_2^5 & s_1^5 & s_3^5 \\ s_0^5 & s_3^5 & s_2^5 & s_1^5 \\ s_4^5 & s_1^5 & s_3^5 & s_2^5 \end{bmatrix},$$

$$A^7(2) = \begin{bmatrix} s_3^7 & s_3^7 & s_5^7 & s_0^7 \\ s_3^7 & s_3^7 & s_0^7 & s_5^7 \\ s_1^7 & s_6^7 & s_3^7 & s_2^7 \\ s_6^7 & s_1^7 & s_4^7 & s_3^7 \end{bmatrix},$$

$$A^9(3) = \begin{bmatrix} s_4^9 & s_4^9 & s_8^9 & s_1^9 \\ s_4^9 & s_4^9 & s_1^9 & s_8^9 \\ s_0^9 & s_7^9 & s_4^9 & s_2^9 \\ s_7^9 & s_0^9 & s_6^9 & s_4^9 \end{bmatrix}.$$

The decision process of this case is as follows according to the aggregation steps shown earlier:

Step 1. Given the consistent linguistic terms of judgment matrixes based on the linguistic term set, S^5 by formula (2) is as follows:

$$A^5(1) = \begin{bmatrix} (s_2^5,0) & (s_1^5,0) & (s_4^5,0) & (s_0^5,0) \\ (s_3^5,0) & (s_2^5,0) & (s_1^5,0) & (s_3^5,0) \\ (s_0^5,0) & (s_3^5,0) & (s_2^5,0) & (s_1^5,0) \\ (s_4^5,0) & (s_1^5,0) & (s_3^5,0) & (s_2^5,0) \end{bmatrix},$$

$$A^5(2) = \begin{bmatrix} (s_2^5,0) & (s_2^5,0) & (s_3^5,0.33) & (s_0^5,0) \\ (s_2^5,0) & (s_2^5,0) & (s_0^5,0) & (s_3^5,0.33) \\ (s_1^5,-0.33) & (s_4^5,0) & (s_2^5,0) & (s_1^5,0.33) \\ (s_4^5,0) & (s_1^5,-0.33) & (s_3^5,-0.33) & (s_2^5,0) \end{bmatrix},$$

$$A^5(3) = \begin{bmatrix} (s_2^5,0) & (s_2^5,0) & (s_4^5,0) & (s_1^5,-0.5) \\ (s_2^5,0) & (s_2^5,0) & (s_1^5,-0.5) & (s_4^5,0) \\ (s_0^5,0) & (s_4^5,-0.5) & (s_2^5,0) & (s_1^5,0) \\ (s_4^5,-0.5) & (s_0^5,0) & (s_3^5,0) & (s_2^5,0) \end{bmatrix}.$$

Step 2. The weight vector, $w = (0, 1/3, 2/3)$, associated with 2T-LHA operator can be calculated with the criterion of overwhelming majority by formulas (3) and (4), and the group preference matrix, A^5, can be drawn by 2T-LHA operator:

$$A^5 = \begin{bmatrix} (s_2^5,0) & (s_1^5,0) & (s_2^5,0.44) & (s_0^5,0) \\ (s_1^5,0.33) & (s_2^5,0) & (s_0^5,0.17) & (s_2^5,0.11) \\ (s_0^5,0) & (s_2^5,0.17) & (s_2^5,0) & (s_1^5,-0.33) \\ (s_3^5,-0.5) & (s_0^5,0.22) & (s_2^5,-0.11) & (s_2^5,0) \end{bmatrix}.$$

Step 3. The ordering vector of the group preference matrix will be calculated by 2T-LHA operator.

If the criterion is of great majority, the weight vector associated with 2T-LHA operator by formulas (4) and (5) will be $w' = (0, 0.4, 0.5, 0.1)$, and the ordering vector of the group preference matrix $W = [(s_1^7,0.3), (s_1^7,0.48), (s_1^7,0.13), (s_2^7,-0.23)]$.

Step 4. The sorting result of ECM equipments service status in this case is $x_4 \succ x_2 \succ x_1 \succ x_3$ according to the ordering vector.

6 CONCLUSIONS

Aiming at solving the problem related to group decision-making of linguistic terms, based on 2-tuple linguistic theory, we extend the LHA operator to 2T-LHA operator and then aggregate the multi-granularity linguistic term sets by making the linguistic judgment matrixes with a different granularity form a decision-making consistency and finally denote their validity and rationality by a case study. The approach can aggregate the multi-granularity linguistic terms and simultaneously avoid the loss and distortion of information during the decision-making process, which will be of good practical value and instructive to a great extent.

REFERENCES

[1] Wei G W: Two-Tuple Linguistic Multiple Attribute Group Decision Making with Incomplete Weight Information, Systems Engineering and Electronics, Vol. 30(2) (2008), p. 273–277.

[2] Herrera F, Martinez L: A 2-Tuple Fuzzy Linguistic Representation Model for Computing with Words, IEEE Trans on Fuzzy Systems, Vol. 8(6) (2000), p. 746–752.

[3] Herrera F, Herrera-Viedma E: Choice functions and mechanisms for linguistic preference relations, European Journal of Operational Research, Vol. 120 (2000), p. 144–161.

[4] Xu Z S. The methods of uncertainty multi-criteria decision making with its application, Beijing: Tsinghua University Press, 2004: 170–171.

[5] Herrera F, Martinez L, Sanchez PJ: Managing non-homogeneous Information in group decision making, European Journal of Operational Research, Vol. 166(l) (2005), p. 115–132.

[6] Xu Z S: A Direct Approach to Group Decision Making with Uncertain Additive Linguistic Preference Relations, Fuzzy Optimization and Decision Making, Vol. 5(1) (2006), p. 21–32.

[7] Jiang Y P, Xing Y N: Consistency Analysis of Two-Tuple Linguistic Judgment Matrix, Journal of Northeastern University (Natural Science), Vol. 28(1) (2007), p 129–132.

[8] Gong Zaiwu, Liu Sifeng: Group decision making method based on two-tuple linguistic for judgment matrices with different fuzzy preferences, Journal of Systems Engineering, Vol. 22(2) (2007), p 185–189.

Information, Computer and Application Engineering – Liu, Sung & Yao (Eds)
© 2015 Taylor & Francis Group, London, ISBN 978-1-138-02717-6

An improved algorithm to generate frequency-hopping sequences

Ping Lai, Hang Bai, Yun Lai Huang & Zhen Zhu

ABSTRACT: Due to the inadequacy of periodic extraction algorithm in the balance and bits recomposition algorithm in the stabilization, an improved algorithm combining bits recomposition algorithm with sequences perturbation method to generate Frequency Hopping (FH) sequences is proposed. The simulation results demonstrate that the FH sequences generated using this method have good properties in balance, stabilization and hamming correlation.

KEYWORDS: FH sequence; perturbation method; bits recomposition; hamming correlation

1 INTRODUCTION

High quality of frequency hopping (FH) sequences is of great importance to FHSS communication system, for it makes FH sequences difficult to analyze[1]. Therefore, families of FH sequences, which are used to specify the transmission frequency at a given time, play an important role in FH communication.

To begin with, a concept called Time of Date (TOD)[2] should be introduced in the FH communication system. To realize real time synchronization, the initial information should be firstly sent. Generating FH sequence based on chaotic sequence is widely studied due to its sensitivity to the initial value, better performance in security and easy realization. Early studies have proposed many methods in generating needed sequences such as mid multi-bit algorithm [3], perturbation method [4] and so on. At the present time, the above methods have some deficiencies in balance, in stabilization [5], etc.

Logistic map is the mostly used chaotic map to study the capabilities of chaotic sequences[6-8]. In this paper, the character of improved Logistic map is studied together with bits recomposition algorithm and sequences perturbation method. Using current TOD as the initial value of the improved Logistic map[9], the properties of the generated sequences like uniformity, balance, stabilization and hamming correlations are simulated

The outline of this paper is as follows. In Section II, we construct an FH sequence based on the Logistic map. In Section III, We test the properties of the generated sequence and compare them with the reference values and an existed algorithm. Finally, some concluding remarks are given in Section IV.

2 IMPROVED ALGORITHM TO GENERATE CHAOTIC FH SEQUENCE

The conclusion draw from the literature [1] is that, the initial value of TOD used in the improved Logistic map must fall in [-1, -0.500000004], [-0.499999998, -0.0000000053], [0.0000000053, 0.499999998] or [0.500000004, 1]. The initial value of TOD is usually converted to binary denotation. The first bit bn-1 is "0" (or "1") if the value of TOD is bigger (or less) than zero.

$$TOD_0 = b_{n-1}b_{n-2}\cdots b_1b_0, b_i \in \{0,1\}, i=0,1,...,n-1. \quad (1)$$

Transform the initial value TOD0 into the origin value of the Logistic map. Then the starting value of the improved Logistic map can be expressed as

$$x_0 = (b_{n-1}\times 2^{-1}+b_{n-2}\times 2^{-2}+...+b_2\times 2^{-n+2})\times L_i+T_i \quad (2)$$
$$i = b_1\times 2+b_0 \, and \, i \in \{0,1,2,3\}$$

The expression of the improved Logistic map is as follows

$$x_{k+1} = 1-2(x_k)^2, \, x_k \in (-1,1) \, and \, k=0,1,2,\cdots \quad (3)$$

The probability density function of the improved Logistic map is

$$\rho = 1/\pi\sqrt{(1-x^2)}, x \in (-1,1) \quad (4)$$

Then, the algorithm used in this paper can be put forward while having known the basics in generating the FH sequence.

Step 1: Use the Eq. (2) and the current value of TOD to obtain the initial value x0for the iterative Eq. (3) to produce a series of TOD sequence. The sequence generated can be described as $X= \{x_i \,|x_i \in (-1,1), i=0,1,...,N-1\}$.

Step 2: Each element of X can be represented using binary system. When n+1 bits are used to represent x_i (a single element belongs to X), the matrix X got through step 1 become a matrix full of "0"and"1", which means:

The first bit in each row is "0" (or "1") if x_i is bigger (or less) than zero.

$$X = \begin{bmatrix} x_{0,0} & x_{0,1} & x_{0,2} & \cdots & x_{0,n} \\ x_{1,0} & x_{1,1} & x_{1,2} & \cdots & x_{1,n} \\ x_{2,0} & x_{2,1} & x_{2,2} & \cdots & x_{2,n} \\ \vdots & \vdots & \vdots & \ddots & \vdots \\ x_{N-1,0} & x_{N-1,1} & x_{N-1,2} & \cdots & x_{N-1,n} \end{bmatrix}, x_{i,j} = 0 \ or \ 1. \quad (5)$$

Step 3: Using the above matrix, divide n+1 bits of one factor into three segments n_1, n_2 and n_3 ($n_1=n_2=$ precision, $n_3=n+1-n_1-n_2$).

The three segments are called virtual value part, perturbation value part and subarea value part respectively. The first two parts will be used to generate binary numbers through modular operations, and the last parts decides which subzone the generated binary numbers should be placed.

Step 4: Choose m (m=log₂q) bits from the above operated binary sequences to construct a new matrix named $X_{N \times m}$, videlicet, the newly formed matrix $X_{N \times m}$ will be:

$$X_{N \times m} = \begin{bmatrix} x_{0,k} & x_{0,k+1} & x_{0,k+2} & \cdots & x_{0,k+m-1} \\ x_{1,k} & x_{1,k+1} & x_{1,k+2} & \cdots & x_{1,k+m-1} \\ x_{2,k} & x_{2,k+1} & x_{2,k+2} & \cdots & x_{2,k+m-1} \\ \vdots & \vdots & \vdots & \ddots & \vdots \\ x_{N-1,k} & x_{N-1,k+1} & x_{N-1,k+2} & \cdots & x_{N-1,k+m-1} \end{bmatrix}$$

$$k+i = (k+i) - \left\lfloor \frac{k+i}{n_1} \right\rfloor * n_1. \quad (6)$$

The operation returns the biggest integer less than (k+i)/n_1.

Step 5: Reshape the N×m dimension matrix X_n to a m×N dimension matrix Q_n to produce the needed FHSS sequence $q_n=\{ q_n(i) \mid q_n(i) \in \{0, 1,...,q-1\}$and $i=0,1,...,N-1\}$

3 PERFORMANCE ANALYSIS

The mostly used properties to test the usability of a sequence are uniformity, balance, auto- and cross hamming correlation, etc. A sequence of good properties is propitious to its anti-jamming, secure, and multiple access properties. Based on the above fact, some properties of the generated sequence are tested to validate its improvement. The simulation results are as follows:

3.1 The performance in uniformity

In reference [10], the results demonstrated that $\{X_n=Q(x_{\delta n})\}$ will be q-array FH sequences (q=2ᵏ, k is a positive integer) when $\{x_n\}$ are sequences generated using Logistic map. In addition, $\{X_n\}$ will be Bernoulli stochastic sequences if $\delta=\log_2 q$. Mi's research demonstrated that matrix reshape have the same property in theoretic [9], while simulation results showed its shortages.

First, to confirm the effectiveness of the proposed algorithm in the uniformity, the simulations are performed with the observation length N=16384, the vector dimension m=6 then compare the results with the initial value distribution of the improved Logistic map and conclusions in reference [5]. The result is shown in Figure 1.

Through Fig. 1, a conclusion can be drawn that, the algorithm put forward in this paper improved the performance greatly in uniformity compared with the initial improved Logistic map and is also better than the method defined in the literature [5]. The frequency distribution of the generated sequence using the proposed algorithm is close to uniform distribution.

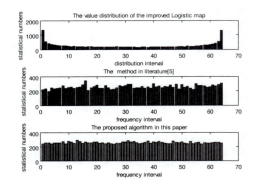

Figure 1. Distribution of three different sequences.

3.2 The performance on balance

Chi-test is adopted to verify the performance on balance of the sequences yielded using the suggested algorithm. The expression of Chi-test is shown below:

$$\chi^2 = \frac{q^2}{N(q-1)} \sum_{i=1}^{q} (N_i - N/q)^2 \tag{7}$$

When q is big enough, the expression above will approach to $\chi^2(q-1)$. The parameter N_i (i=0, 1, 2,...,q-1) in the above formula indicates the ith FH slot appears totally during the whole observation length N.

Choose N=1024, q=64, k=5 and 100 different initial values to verity the effectiveness of proposed algorithm. The simulation results are given in Figure 2.

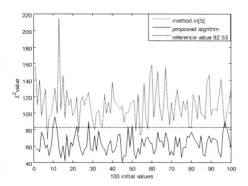

Figure 2. Balance property.

Fig. 2 shows that when the choice of k is not proper, the algorithm proposed has a visible advantage. Furthermore, choose 1000 different initial values of zone1 to test the stabilization property compared with the method in literature [5]. The value of k ranges from 4 to 9.the results are shown in Table 1.

It's obvious that the newly generated sequences have a better fluctuation property especially when |k-8|>0. Then 4 groups of initial values chosen from the intervals mentioned in the preamble to calculate the passing ratio of each zone, the results are shown in table 2. As it's shown, the proposed algorithm has

Table 1. The passing ratio of Chi-test on stabilization (%).

k	4	5	6	7	8	9
Algorithm in [5]	0	7.37	36.63	69.53	82.64	0
Proposed algorithm	95.06	92.94	94.05	95.66	95.56	95.46

Table 2. The passing ratio of chi-testin uniformity for k=8 (%).

	Zone1	Zone2	Zone3	Zone4
Algorithm in [5]	82.64	82.74	81.84	84.76
New algorithm	95.56	95.76	96.17	95.76

better property than the method in [5] (Choose k=8 to insure the best capability of the method related in this literature) in all the 4 zones.

3.3 The hamming correlation property

In FH communication system, collision happens when different users are transmitting information at a same carrier. An important measurement to weigh the capability of FH sequences in this event is the hamming correlation. Its definition is given below:

$$H_{XY}(\tau) = \sum_{i=0}^{N-1} h(X_i, Y_{i+\tau}) \ , \ 0 \leq \tau \leq N-1,$$

$$i+\tau = (i+\tau) - \left\lfloor \frac{i+\tau}{N} \right\rfloor * N \tag{8}$$

X, Y are two FH sequences and

$$h(x,y) = \begin{cases} 1 & ,x=y \\ 0 & ,x \neq y \end{cases} \tag{9}$$

The literature [9] established the mean value of the autocorrelation side lobe $H_{XX}(\tau)$, the cross correlation $H_{XY}(\tau)$, the max autocorrelation side lobe H_{XX} and the max cross correlation H_{XY}, they are shown beneath respectively:

$$\begin{cases} \overline{H}_{XX}(\tau) = \frac{1}{N-1} \sum_{i=0}^{N-1} H_{XX}(\tau) \approx \frac{N}{q} \\ \overline{H}_{XY}(\tau) = \frac{1}{N} \sum_{i=0}^{N-1} H_{XY}(\tau) = \frac{N}{q} \end{cases} \tag{10}$$

$$\begin{cases} H_{XX} = \max_{1 \leq \tau < N} (H_{XX}(\tau))/N = (1 + \sqrt{2q \ln(N)/N})/q \\ H_{XY} = \max_{0 \leq \tau < N} (H_{XY}(\tau))/N = (1 + \sqrt{2q \ln(N)/N})/q \end{cases} \tag{11}$$

In this paper, 50 different initial values are chosen in the condition of q=64 to calculate the mean value of correlation values and compare the results with the method proposed in literature [5].

As it's shown in Figure 3 and Figure 4, the proposed algorithm can provide better hamming auto-correlation properties; the value of the max cross correlation is also below the one got through method in literature [5] and overlaps with the reference value.

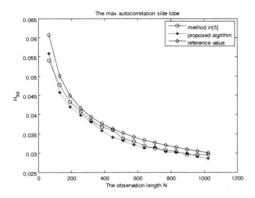

Figure 3.   The max autocorrelation side lobe.

4   CONCLUDING REMARKS

To generate FH sequences, an improved algorithm based on improved Logistic map combining bits recomposition algorithm with sequences perturbation method is proposed. The simulation results demonstrate that under the same conditions, this method can improve the properties of uniformity and the capability of balance. On the other hand, it needn't to consider the synchronization problem with another map

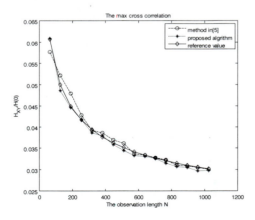

Figure 4.   The max cross correlation.

or sequence, and can greatly decrease the iteration times. To conclude, the newly proposed algorithm is propitious to generate FH sequences for FHSS communication system. Further work should be done to validate other capabilities and seek the algorithm's implement by hardware.

REFERENCES

[1] Zhang Shenru, Den Xiaoyan, Wang Zeyan, Wang Haiyang. Implementation of Hopping Code Generator Using DES Algorithm[J]. Journal of Circuits and System, Vol. 7, No. 3, pp. 35–38, 2002.
[2] Zhang Shenru, Mei Wenhua, Wang Tingchang. Construction of Algorithm for Frequency Hopping Code Generator Using Counting TOD[J]. Journal of Electronics and Information, Vol. 24, No. 8, pp. 1096–1011, 2002.
[3] Wei Jincheng, Wei Wei. Analysis of improved Logistic-Map chaotic sequence[J]. Electronic Design Engineering, Vol. 19, No. 4, pp. 20–23, 2011.
[4] X. Niu, D. Peng, and Z. Zhou, "New classes of optimal frequency hopping sequences with low hit zone with new parameters," in Proc.5th Int. Workshop Signal Design Appl. Commun., Guilin, China, Oct.10–14, pp. 111–114, 2011.
[5] Chen Yongjun, Wu Jie, Xu Hua, Long Min, Bao Yifeng. A Novel Method for Designing Chaotic Frequency Hopping Sequence [J]. Telecommunication Engineering, Vol. 50, No. 9, pp. 24–27, Sep. 2010.
[6] Liu Xiangdong, Zhang Jinhai, LI Zhijie, HE Xiqin. A Generation Method for FH Sequences With Given Minimum Gap Based on Chaotic Dynamic Quantification[J]. Journal of Circuits and System, VoL. 15, No. 4, pp. 96–100, Aug. 2010.
[7] Liu xiangdong, Yan dejun, Duan xiaodong, Wang guangxing. A Chaotic Frequency Hopping Sequences by Mid Multi-bit Quantified and Its Properties[J]. micro-electronics & computer, Vol. 21, No. 8, pp. 5–9, 2004.
[8] Mi Liang, Tang Gang. Design of Frequency-Hopping Sequences Based on Chaotic Map[J]. Journal on Communications, Vol. 26, No. 12, pp. 69–74, p.80, Dec. 2005.
[9] Ling Cong, Sun Songgeng. Frequency-Hopping Sequences by Chaotic Maps for FH/CDMA Communications[J]. ACTA Electronica SINICA, Vol. 27, No. 1, pp. 67–69, Jan. 1999.
[10] Zhang Shuo, Zhang Wei, Gao Kai. A Method to Generate Chaotic DFH Sequence and Its Performance Analysis[J]. Informatization Research, Vol. 36, No. 2, pp. 16–18, Feb. 2010.

Information, Computer and Application Engineering – Liu, Sung & Yao (Eds)
© 2015 Taylor & Francis Group, London, ISBN 978-1-138-02717-6

A study on the strategy of tourism public service system

Yan Zheng

College of Tourism, Dalian University, Dalian, China

ABSTRACT: Currently, there are still some problems in Liaoning tourism public service. Tourism information's propaganda is not in place. Tourists' awareness of tourism public service is low. There are contradictions between tourism public transport supply and demand. The quality of transportation service needs to be improved. Lack of formal tourism centers and standardized management is also a major problem. Tourism insurance mechanisms are inadequate. The free and preferential tourist attractions cannot meet the tourists' needs. Besides, tourist attractions cannot provide tourists with enough professional volunteers. The management and supervision of tourism administrative departments is not sufficient. The government should strengthen propaganda of the tourism public information service and make tourists actively seek tourism public service. In order to improve the quality of public transportation service, the government should make a scientific plan and strengthen its management and supervision. The government should also set up the tourism insurance system for visitors, increase free tourist attractions, and provide volunteers with systematic training of tourism. Tourism departments should strengthen the management and supervision on tourism.

KEYWORDS: Tourism public service; Service quality; Visitors' satisfaction

1 INTRODUCTION

With the development of the popularization of tourism industry, people focus more on the flexibility, the autonomy, and the diversity of tourism, as well as on the quality and security of tourism and the requirement of the public service in those tourism destinations becomes stricter. Travel public service means the process by which the government and some other socioeconomic organizations provide basic and public services for the tourists to meet their public demands, mainly including tourism public information service, travel security service, tourism convenient transportation service, tourism convenience service, tourism administration service, and so on. As an important public tool, travel public service is an essential element in the development of modern tourism activities. The degree of perfection of a city's tourism public service system is an important indicator that is used to measure the maturity of the city's tourism industry. As one of China's best tourism cities, Liaoning is becoming more and more popular in the world. To greet the arrival of the era of national tourism and leisure and meet the growing tourism demand from the tourists, it is an urgent task to strengthen the tourism public service system and is of great significance to improve the quality of the tourism public service.

2 NECESSITY OF LIAONING TOURISM PUBLIC SERVICE SYSTEM CONSTRUCTION

2.1 *Need to balance the contradiction between supply and demand of tourism public service*

The supply of tourism public service means the capacity of the tourism destinations that can meet the demand for tourism public service by the tourists.[1] With the development of the socioeconomic level and the improvement of the people's consumption level, tourists' demands for the tourism public service show a trend of diversification. The diversification, individuation, and differentiation of the tourism public demands result in a shortage situation of tourism public service. The construction of Liaoning Tourism Public Service System is beneficial in order to improve tourism public service functions, expand the supply of tourism public service, increase the diversification of the supply of tourism public service, and alleviate and eliminate the contradictions between the lack of supply and diversification of demands in Liaoning Tourism Public Service System, which can fully meet the diverse needs of the tourism public service by the tourists in Liaoning.

2.2 Need to compensate for deficiencies in the tourism market

Generally speaking, tourism enterprises pursue the maximized profit and provide services for profit.[2] Due to the weaknesses of spontaneity and blindness of the market, tour operators could increase the prices of the tourism services and ignore providing complete and accurate information to the consumers. As a matter of fact, the subject of tourism public service is the government and the biggest feature is nonprofit, whose main purpose is to provide the tourists with public and sharing services and even free services. Therefore, to strengthen the Liaoning Tourism Public Service System construction, we can compensate for its own insurmountable shortcomings, realize the complementarity of the tourism public service and the tourism marketing service, promote the improvement of the tourism service system, and improve the quality of tourism service.

2.3 Need to shape a good image of liaoning tourism

The construction of the quality of tourism public service covers all aspects of tourism development in economics, society, administration, and ecology, throughout the food, housing, transportation, travel, shopping, and entertainment in every aspect.[3] Tourism public service can not only meet the demand of the local residents and improve the standard of living and quality of life for the residents in the tourism destinations but also meet the growing tourism demand, provide high-quality services, increase the satisfaction of both Liaoning Tourism Public Service and the whole tourism industry, and improve the retention and the loyalty of tourists. Therefore, the aim is to strengthen the quality of tourism public service benefits to promote the standardized, orderly, and healthy development of Liaoning Tourism and establish a good brand image of Liaoning Tourism.

2.4 Need to promote the construction of liaoning service-oriented government

Tourism public service is a major task in the tourism administrative departments during the "12th Five-Year Plan." Currently, the construction work of the tourism public service is being valued by relevant departments gradually. The functions of the government are gradually shifted to being service oriented, and tourism authorities at all levels are gradually building the tourism public service as a part of its core functions. Tourism public service has the property of nonprofit and shareability, which means whether the locals or the tourists should enjoy the service provided by the local government. As a result, Liaoning Tourism Administrative Department should accelerate the pace of transformation of the government functions, strengthen public service functions, perform the obligations of serving people, and promote the construction of Liaoning service-oriented government.

2.5 Need to thoroughly implement national tourism legislation

In order to meet the consumers' demand of tourists better, China National Tourism Administration stressed the importance of tourism public service system in 2009 and 2011. In 2013, the promulgated "Travel Act" clearly required the governments at all levels to strengthen and improve public services, strengthen the construction of tourism infrastructure, promote the image of public service, and establish the platform of tourism public information and consultation, which provides the tourists with important information and consulting service about the tourist attractions, routes, traffic, weather, accommodation, security, and medical aid free of charge. To clear this, the tourism destinations have the obligation to remind the tourists of the security risk, indulge in safety supervision and rescue, and clear the fact that the scenic spots that utilize public resources should reflect the nonprofit and gradually become free. The content of Liaoning Tourism Public Service System involves aspects of information, transportation, security, convenience services, and administrative services, which not only satisfy the requirement of the "Travel Act" but also promote the implementation of the relevant tourism regulations.

3 STRATEGY OF LIAONING TOURISM PUBLIC SERVICE SYSTEM CONSTRUCTION

3.1 Expansion of tourism public information service

The government should enhance to propagate the tourism public information service and strengthen the awareness of the tourists to seek tourism public service initiatively.[4] The information that the tourists have about tourism public information service is strongly connected with the propaganda, so if we want the tourists to get more information, we must strengthen it to ensure it is propagated. Therefore, we can establish some specific websites about tourism public information service, open online consulting services for tourism, and publish tourism information at train and bus stations, docks, and some other places of high population concentration that are convenient for the tourists to know and use.

On the other hand, the government should increase the number of electronic facilities to make tourism information diversified. Simultaneously, the scenic spots must keep pace with the information age, increase electronic touch screens, make Wechat two-dimensional code scanning, put in place automatic ticketing, establish scenic intelligent monitoring security, and provide free WiFi so that tourists can get information at anytime and anywhere.

3.2 Promotion of tourism public transportation service

The government should increase the supply of tourism public transportation and ease the contradiction of the supply and demand.[5] The main means of transport are buses, light railways, and taxis, but in the tourist season the tourists' demand cannot be satisfied. Given this situation, the government may increase the number of buses or open some tourist bus lines and increase the fast-track routes and schedules to reduce the waiting time for the tourists. Simultaneously, the government should open the tourism loop bus from Liaoning North Railway Station, which passes some main scenic spots and areas, to realize the "seamless connection" between the tourists and the scenic spots.

The government should accelerate the construction of Liaoning Tourism Hub. We should learn from the experience of Shanghai, Beijing, Hangzhou, and Chengdu; build the government-leading tourism centers; and establish some related official websites. We also need to increase traffic guidance logos, set orientation maps and navigation maps at important road junctions or scenic spots, and set direction signs within the range of 500 meters of scenic areas in case of going in the wrong direction. In addition, the government should increase subsidies, reduce the price of public transportation, and reduce the travel cost of the citizens and tourists.

3.3 Improvement of tourism security and safety service

The instructions of the safety infrastructures and emergency rescue tools in the scenic spots should be clear and accurate; besides the infrastructures, there should be instructions and electronic demonstrators to make sure that the tourists understand the process of operation of those tools. [6] We should improve the travel insurance system and offer tourists a variety of types of travel insurance, especially for the individual tourists. Simultaneously, it is essential to improve Liaoning Travel Insurance System, increase ways to insure, clarify insurance terms, and simplify claims procedures.

3.4 Promotion of tourism convenience service

The government should increase some free scenic spots and reduce the ticket prices of the scenic areas; promote the free opening of the museums and memorials; start no other activities at the scenic spots; vigorously promote the tourism preferential policies to the elderly, children, disabled, low-income groups, and other vulnerable groups; and reduce the ticket prices within reasonable limits. The government should increase the types and number of convenience facilities; increase the number of public restrooms; and set up free drinking fountains at open-air scenic spots such as beaches and squares. Simultaneously, we need to make strict selection criteria for the tourism volunteers, select some volunteers with professional knowledge and students in the tourism management department, and start training in the security-assistance basic common sense and basic knowledge of tourism.

3.5 Strengthening of tourism administration service

As the administration departments for tourism public service, Liaoning Municipal Bureau of Tourism Industry Management Department and Liaoning Tourism Quality Supervision and Management Department should do a good job in the management and supervision, strengthen quality management of tourism service, especially strengthen the supervision and inspection of the tourism food hygiene situation, and establish mechanisms of the quality of tourism service. The government should optimize the shopping environment, establish tourism service reputation security system, and safeguard tourists' legitimate rights and interests. We also need to improve the tourism complaints mechanism, handle the complaining problems fairly and efficiently, and establish tourism public service evaluation mechanisms based on tourist satisfaction.

4 CONCLUSION

With the rapid development of China's tourism industry, Liaoning gradually increases the investment of the tourism public service system, strengthens the construction of tourism public service system, establishes the tourism public information service system, forms the tourism security system, and improves the tourism administrative functions. However, compared with the development of Liaoning's tourism public service and growing tourism public demand, some problems still exist. Therefore, in order to improve the quality of tourism public services, Liaoning municipal government should take appropriate measures.

REFERENCES

[1] YunpengHuang, *Public Service Regulatory Research*: Economic Science Press, 2008, pp. 25–26.
[2] Xinzhang Wang, "Construction of Tourism Public Service System and Tourism Destination," *Tourism Tribune*, vol.27, pp. 40, Jan 2012.
[3] Junpeng Li, "Accelerate the Improvement of Tourism Public Service System," *Tourism Tribune*, vol.27 no. 1, pp. 4, Jan 2012.
[4] Shuang Li, Fucai Huang, Jianzhong Li, "Tourism Public Service: Connotation, Characteristics and Classification Framework," *Tourism Tribune*, vol.4, pp. 20–26, April 2010.
[5] RongHao, "Tourism Public Service and Related Concepts," *Modern Business Trade Industry*, vol.8, pp. 96, Aug 2012.
[6] Jufeng Xu, *Tourism Public Service: Theory and Practice*: China Tourism Press, 2012, pp. 21.

Information, Computer and Application Engineering – Liu, Sung & Yao (Eds)
© 2015 Taylor & Francis Group, London, ISBN 978-1-138-02717-6

Optimization of low-voltage load configuration based on minimizing line loss

Xiao Xiao Cheng & Yu Sheng Zheng
State Grid Henan Electric Research Institute, Zhengzhou, China

Pei Nan Xu
Yexian Power Supply Company, Yexian, China

ABSTRACT: To solve the increase in line loss that is caused by unbalanced low-voltage three-phase load, an optimized model is proposed, which is suitable for optimization of low-voltage load configuration with random low-voltage load distribution or random network structure. By introducing the logic variable, the line loss optimization is deduced to an integer one that can be easily solved by heuristic algorithm. An optimized algorithm is also given, which is suitable for computer programming. For illustration, an application of 10kv distribution area is utilized to show the feasibility of the optimized model. Empirical results show that the model can give the optimized result for different phases of low-voltage load configuration, and the result with the highest line loss decrement will be used as the final decision. This research provides scientific theoretical basis for low-voltage load configuration and plays a vital role in line loss reduction management.

KEYWORDS: Optimization of Low-voltage Load Configuration; Distribution Network Loss; Three-phase Unbalance; Strategies of Loss Reducing Loss

1 INTRODUCTION

There are a large number of distribution transformers in the modern power grid, and the low-voltage grid is also very huge. So, reducing low-voltage grid loss has been valued highly in loss management. Our low-voltage power grids generally use a three-phase four-wire power supply, and then a three-phase unbalanced load has been a common occurrence, which is caused by the single-phase load accessing. For the presence of the three-phase unbalanced load, there is a big current in the neutral line. All of this increased line loss. Optimization of low-voltage load configuration is a simple and economical method of reducing losses. Adjusting only the user access phase, at no additional investment in the case, can effectively reduce the three-phase unbalanced load, which the line loss decreased significantly. In addition, the optimization results can improve power quality and enhance the utilization ratio of equipment.

Currently, solving the problem of the three-phase unbalanced load in practical work is largely empirical, and the effect depends on the professional quality of the technical staff and familiarity with the specific grid. There is no rigorous mathematical description and theoretical guidance and also lack of problem-solving supported software platform entirely by artificial methods. So, the loss reduction effect is difficult to guarantee. In this article, a low-voltage grid line loss calculation method that introduced three-phase unbalanced load has been studied; the mathematical model of three-phase low-voltage load optimization, a suitable computer programming algorithm, and graphical low-voltage grid load optimization computing tools are proposed and developed, to meet the needs of low-voltage grid line loss calculation, analysis, and loss reduction measures.

2 LINE LOSS COMPUTATION AND MATHEMATICAL MODEL OF OPTIMAL ALLOCATION OF LOAD

2.1 Model of line loss introduced unbalanced calculation

First, we define the unbalance of load current as λ, and phases A, B, and C are defined, respectively, as λ_A, λ_B, and λ_C.

$$\lambda = I_{max\varphi} / I_{av} \times 100\% \tag{1}$$

$$\lambda_A = (I_A - I_{av})/I_{av} \times 100\% \qquad (2)$$

$$\lambda_B = (I_B - I_{av})/I_{av} \times 100\% \qquad (3)$$

$$\lambda_C = (I_C - I_{av})/I_{av} \times 100\% \qquad (4)$$

where $I_{\max j} = \max(|I_A - I_{av}|, |I_B - I_{av}|, |I_C - I_{av}|)$ is the largest deviation phase load current; I_A, I_B, and I_C are the current of phases A, B, and C; and I_{av} is the average current, $I_{av} = (I_A + I_B + I_C)/3$.

The active loss of low-voltage line l can be described as

$$\Delta P_l = (3 + \lambda_{Al}^2 + \lambda_{Bl}^2 + \lambda_{Cl}^2) I_{avl}^2 R_l$$
$$+ (\lambda_{Al}^2 + \lambda_{Bl}^2 + \lambda_{Cl}^2 - \lambda_{Al}\lambda_{Bl} - \lambda_{Al}\lambda_{Cl} - \lambda_{Bl}\lambda_{Cl}) I_{avl}^2 R_{0l} \qquad (5)$$

where I_{avl} is the average current of section l. λ_{Al}, λ_{Bl}, and λ_{Cl} are the unbalance of phases A, B, and C. R_l is the phase line resistance of section l. R_{0l} is the neutral line resistance of section l.

Assume that the total number of line sections is L, as shown in Fig. 1. Therefore, the total line loss can be described as

$$\Delta P_\Sigma = \sum_{l=1}^{L} \Delta P_l. \qquad (6)$$

2.2 Model of optimal load allocation

This article introduced logic variable and converted the line loss problem to optimization of low-voltage load configuration.

2.2.1 Introduction of logic variable

In fact, the low-voltage load accesses the network in a single phase or three phases. Then, the symmetry is broken, as shown in Fig. 1. To make the program easy to implement, we need to make equivalent processing (the dashed part). Thus, the network structure becomes symmetrical.

With the dissymmetrical low-voltage grid, to get the branch-node incidence matrix M, we introduced 0-1 logic variable, which can be expressed in the

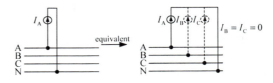

Figure 1. Load equivalent diagram.

connected phase. Assume that the vector of load connected is discarded by 0-1 logic variable.

$$\begin{cases} \mathbf{x}_A = \begin{bmatrix} x_{A1} & x_{A2} & \cdots & x_{AN} \end{bmatrix}^T \\ \mathbf{x}_B = \begin{bmatrix} x_{B1} & x_{B2} & \cdots & x_{BN} \end{bmatrix}^T \\ \mathbf{x}_C = \begin{bmatrix} x_{C1} & x_{C2} & \cdots & x_{CN} \end{bmatrix}^T \end{cases} \qquad (7)$$

where $x_i (i = 1,2,\ldots,N)$ is 0-1 logic variable. If a single phase load i accesses phase A, $x_{Ai} = 1, x_B = 0, x_{Bi} = 0$. If a three-phase load i accesses phase A, B, and C, $x_{Ai} = x_{Bi} = x_{Ci} = 1$.

$$\begin{cases} \mathbf{MI}_{Al} = \mathbf{I}_{An} \\ \mathbf{MI}_{Bl} = \mathbf{I}_{Bn} \\ \mathbf{MI}_{Cl} = \mathbf{I}_{Cn} \end{cases} \qquad (8)$$

$$\begin{cases} \mathbf{I}_{An} = \begin{bmatrix} x_{A1}I_1 & x_{A2}I_2 & \cdots & x_{AN}I_N \end{bmatrix}^T \\ \mathbf{I}_{Bn} = \begin{bmatrix} x_{B1}I_1 & x_{B2}I_2 & \cdots & x_{BN}I_N \end{bmatrix}^T \\ \mathbf{I}_{Cn} = \begin{bmatrix} x_{C1}I_1 & x_{C2}I_2 & \cdots & x_{CN}I_N \end{bmatrix}^T \\ \mathbf{I}_n = \begin{bmatrix} I_1 & I_2 & \cdots & I_N \end{bmatrix}^T \end{cases} \qquad (9)$$

where $\mathbf{I}_{An}$, $\mathbf{I}_{Bn}$, and $\mathbf{I}_{Cn}$ are current vectors from node n into A, B, and C phase. I_{Al}, I_{Bl}, and I_{Cl} are current vectors of A, B, and C phases. $\mathbf{I}_n$ is a note current vector. By using the current results solved by equations (8) and (9), the phase unbalance can be obtained by the solution of equations (1), (2), (3), and (4). The section line loss can be obtained by the solution of equation (5); total line loss can be obtained by the solution of equation (6).

2.2.2 Model of optimal load allocation

In this article, the entity of load allocation optimization is described as follows: With wire type, length, load distribution, and power parameters have been determined; finding the optimal adjustment of phase sequence that the low-voltage load connected, under the premise of single-phase load connected. The objective of this optimal model is accessing the load in optimization result, reducing phase unbalance and minimizing line loss.

Using the analysis from equations (1) to (9), line loss ΔP_l can be described as a function of logic variable, signed $\Delta P_l\left(x_{Ai},x_{Bi},x_{Ci}\right)$. Therefore, the objective function and constraint condition of optimal load configuration is converted to the following:

$$obj. \quad \min_{\mathbf{x}_A,\mathbf{x}_B,\mathbf{x}_C} \Delta P_\Sigma = \sum_{l=1}^{L}\Delta P_l\left(x_{Ai},x_{Bi},x_{Ci}\right) \quad (10)$$

$$s.t. \begin{cases} x_{Ai}+x_{Bi}+x_{Ci}=1, \quad i=1,2,\cdots,N \\ x_{Ai},x_{Bi},x_{Ci}=1 \; or \; 0 \end{cases} \quad (11)$$

where $\min\Delta P$ is low-voltage network loss. From equations (10) and (11), through changing the sequence of the single-phase load, we should calculate the corresponding line loss, comparison, and optimization of line loss. This problem is actually a 0-1 combinatorial optimization problem, and this article uses the heuristic method to solve such a non-linear combinatorial optimization problem, to calculate the optimal line loss and the optimal allocation scheme.

3 APPLICATION EXAMPLES

Based on the optimization mathematical model and algorithm, the corresponding graphical calculation software has been developed. With the optimization results, some low-voltage loads changed the accessed phase in 130 transformer districts that had higher phase unbalance and higher line loss and the actual operation showed good effect. As shown in Fig 2, it is the optimization of a transformer district named YU WANG TAI. The total loss was 890 kWh. The parameters, including length and conductor type of each section, nameplate parameters of transformers, and the quantity of each load, were set in the graphical software. However, the graph of the feeder had also been drawn. Then, the results were marked on the feeder graph automatically, such as in Fig 2.

From the results, before and after optimization loss, reduction effect is obvious, as well as reduced loss of about 192 kWh. Theoretical calculation results and the measured values do not differ much, and they meet the precision requirement, as shown in table 1.

According to the optimized scheme, single-phase loads have been adjusted. After adjustment, with the same monthly supply, line loss decreased by 161 kWh, as shown in table 1. This year, with one area loss of about 1932 kWh, electricity according to 0.5 Yuan/kWh meter, and the loss reduction to save RMB 966 a year, benefits are considerable.

Table 1. Line loss and optimization calculation results.

Items	Loss/kWh	Loss rate/%
Optimal calculation results	739	5.98
The measured results of previous month	1030	8.33
The measured results after optimization implemented	869	7.03

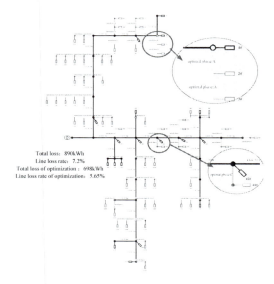

Total loss: 890kWh
Line loss rate: 7.2%
Total loss of optimization: 698kWh
Line loss rate of optimization: 5.65%

Figure 2. Calculation results.

4 CONCLUSIONS

In this article, the author studied on the low-voltage power grid line loss, the unbalanced degree calculation mathematical model, fully considering the characteristic of the low-voltage power grid calculation accuracy being high, easy-to-program implementation, and small computational complexity. The problem of optimal allocation of load had been converted into 0–1 variables combinatorial optimization problem, using the heuristic algorithm. The engineering practice of the optimal solution is given. Through the calculation of the actual power grid, the model is verified, the algorithm is correct, and the result is good.

REFERENCES

Khator S K, Leung L C. Power Distribution Planning A Review of Models and Issues. IEEE Transon Power Systems, 1997, 12(3).

Feng Guihong, Zeng Jianbin, Gong shuqiu, Zhang Bingyi. Line Losses Analysis and Management System of Distribution Networks Based on Loeal Area Network. IEEE/PES Transmission and Distribution Conference & Exhibition: Asia and Pacific Dalian, China, 2005.

MaJin, Xu Jianyuan, Wang shenghui, LinXin. Calculation and Analysis for line losses in Distribution Networks. IEEE Power Delivery Transactions, 2002.

Joanicjusz Naxarko, Zbigniew Styezyski. The fuzZy approach distribution to energy losses ealeulations in low networks. Power Engineering Soeiety voltage Winter Meeting, 2000.

Wen Fu shuan. A refined enetic algorithm for fault section estimation in power systems using the time sequence information of circuit breakers. Electric Machines and Power Systems, 1996.

Wen Fu shuan, Chang C S.A probabilistic approach to alarm processing in power systems using a refined genetic algorithm. InProceedings of International Conference on Intelligent Systems Applications to Power Systems, Orlando(USA), 1996.

M. El-Sobki, M.M. El-Metwally, M.A. Farrang. New Approach for Planning High-Voltage Transmission Network. IEEE Proceedings: PartC, 1986.

N.N. SehraudolP and P.K. Belew. Dynamic Parameter Encoding for Genetie Algorithms. Machine Learning, 1992.

Information, Computer and Application Engineering – Liu, Sung & Yao (Eds)
© *2015 Taylor & Francis Group, London, ISBN 978-1-138-02717-6*

Lattice reduction-aided precoding and detection designs in MIMO two-way relay systems

Lin Wang, Guo Sheng Rui & Hai Bo Zhang
Department of Electronic and Information Engineering; Naval Aeronautical and Astronautical University
Yantai, P. R. China

ABSTRACT: Multiple Input Multiple Output (MIMO) two-way relay system has demonstrated significant gain in spectral efficiency when source nodes exchange information via a relay node. This article considers precoding and detection algorithm when all the nodes are equipped with multiple antennas. According to complex lattice reduction algorithm, we propose a joint Tomlinson-Harashima Precoding and liner detection algorithm. The complexity of the algorithm is very low for reducing the channel matrix only once. The simulation evaluations show that the proposed algorithm can significantly improve the bit error ratio and achieve full diversity A novel algorithm is proposed to solve some problems. Simulation results show the effectiveness of the proposed algorithm

KEYWORDS: lattice reduction; two-way relay system; precoding; detection

1 INTRODUCTION

Cooperative communication can offer significant benefits, including throughput enhancement, coverage extension, and power reduction in wireless communications. As a new cooperative communication scheme, two-way relay systems have received great attention from both academia and industry. It only needs two time slots for two users to exchange information with the help of relay when there is no direct link. This notion is the relay mix that the signals receive from two source nodes for forwarding, and then they apply the self-interference cancellation at each destination to extract the desired information. In comparison with the traditional one-way relaying, two-way relay systems can save two time slots to exchange information between two users. Amplify-and-Forward (AF) is one of the most commonly used relaying protocols. In this protocol, the relay simply amplifies the received signals and forwards them to the destinations. In contrast to DF relay protocol, the AF relay protocol is more attractive for its simplicity of implementation.

When the nodes are equipped with multiple antennas and the channel state information (CSI) is available, two-way relay system is an MIMO relay system that can apply the precoding. But only a few studies of precoding design for MIMO two-way relay systems have been investigated. Note that existing precoding designs in AF MIMO two-way relay systems mostly consider linear precoding. However, nonlinear precoding has been shown to have better

performance than linear precoding in conventional MIMO point-to-point systems and MIMO one-way relay systems. So in future, nonlinear precoding is expected to improve MIMO two-way relay system performance. Tomlinson-Harashima (THP) is one of the common nonlinear precoding techniques, and it includes Successive Interference Cancellation (SIC) and modulo operation. THP algorithm can use Minimum Mean-Square-Error (MMSE) approach and Zero-Forcing (ZF) approach. But the ZF THP algorithm BER performance is not satisfactory with the increment of SNR and cannot achieve the full diversity because of the poor condition number of channel gain matrix. The typical ZF THP algorithm mainly focused on the processing of signals and ignored improving the condition of MIMO channel. The lattice reduction-aided THP (LR-THP) algorithm can achieve the full diversity in conventional MIMO point-to-point systems.

In this article, we will introduce the lattice reduction-aided THP scheme in the two-way MIMO relay systems whose computational power in relay node is very poor. At two sources, the new channel condition is improved with the lattice reduction algorithm for better properties such as better orthogonality and shorter Euclid length. Then, the classical THP structure is used with the new channel gain matrix to transmit signals in the first time slot; at the relay, the received mixed signal is processed by simple modulo operation. Next, the mixed signal is transmitted in the second time slot. The two destinations receive the mixed signal by LR-ZF detection; the reason of adopting the LR-ZF detection is that LR permutation is only performed once. So the useful signal can be decoded by Self-Interference Cancellation. The signal processing is mostly achieved by the two users, and the relay node is the only completed modulo operation. Compared with the existing ZF THP, the complexity in users is applicable, and the processing in relay is quite simple.

2 PRELIMINARIES

The configuration of an MIMO two-way relay system is shown in Fig. 1. Users A and B exchange information with the aid of one relay R, where there is no direct link between the two users. We assume that both users A and B are equipped with N antennas. The relay node has M antennas ($M, N \geq 2$). The entries of the involved channels are independent and identically distributed (i.i.d.) complex Gaussian with zero mean and unit variance. We assume that all MIMO channels are quasi-static, that is, the channel coefficients remain the same during two time slots of information exchange. The contiguous channel matrices are known by users and relay. Thus, matrix H and G are known to relay nodes, but user A only knows matrix H, and user B only know matrix G. This instance corresponds closely to reality. The users and the relay operate in a half-duplex mode. They cannot transmit and receive simultaneously.

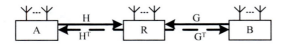

Figure 1.   Configuration of two-way relay system.

The two-way communications take place in two time slots. In the first time slot (multiple access control (MAC) phase), two source nodes simultaneously transmit their signals to relay; the received signal at the relay can be expressed as

$$y_R = HF_A x_A + GF_B x_B + z_R \qquad (1)$$

where $H, G \in \mathbb{C}^{M \times N}$ are full-rank channel matrices from nodes A and B to R. The elements of H and G are independent identical distribution (i.i.d) complex Gaussian with zero mean and unit variance. For $m \in \{A, B\}$, the Fm represents the transmit precoding matrix of m. xm denotes the transmit signal from m. The noise vector zR is the additive white zero-mean complex Gaussian noise with variance σ_R^2. The maximum transmission power at m is assumed to be Pm, and thus the power constraint at two sources can be written as

$$tr(F_m F_m^H) \leq P_m \qquad m \in \{A, B\} \qquad (2)$$

On receiving yR, the relay generates a signal vector xR with simple modulo operation and amplification. So, we have

$$x_R = \alpha \cdot \mathrm{mod}(y_R) \qquad (3)$$

Here, α is an amplification factor. When the maximum transmission power at the relay R is assumed to be PR, the boundary constraint of α can be written as

$$\alpha \leq \sqrt{\frac{P_R}{tr(HF_A F_A^H H^H + GF_B F_B^H G^H + \sigma_R^2 I_M)}} \qquad (4)$$

Then, xR is broadcast to the two users at the second time slot (downlink BC phase). The two users received

$$y_A = H^T x_R + z_A$$
$$y_B = G^T x_R + z_B \qquad (5)$$

This is a point-to-point MIMO channel, so we can detect the signal by detection algorithm. At users A and B, the ZF-DFE receivers are implemented by using a feedforward matrix filter W_A, $W_B \in \mathbb{C}^{n \times n}$, so we have

$$\hat{x}_{RA} = W_A y_A$$
$$\hat{x}_{RB} = W_B y_B \qquad (6)$$

Since users A and B have their own packets, after decoded signal, they could derive what they want to receive by

$$\hat{x}_A = \hat{x}_{RA} - x_B$$
$$\hat{x}_B = \hat{x}_{RB} - x_A \qquad (7)$$

From (7), the total mean squared error (MSE) for both A and B can be denoted as

$$TMSE = E\{|\hat{x}_A - x_A|^2\} + E\{|\hat{x}_B - x_B|^2\} \qquad (8)$$

In the next section, we will design the precoding based on lattice reduction to achieve this objective.

The complex-valued lattice reduction algorithm is proposed, because it greatly reduces the complexity by nearly 50% than the real-valued lattice reduction without sacrificing any performance. A complex-valued lattice of rank m is defined as

$$L \triangleq \left\{ a \mid a = \sum_{i=1}^{m} z_l b_l, \quad z_l \in \mathbb{Z}_j \right\} \quad (9)$$

where $b_l \in \mathbb{C}^n$, and b_l is a complex vector. $\mathbb{Z}_j = \mathbb{Z} + j\mathbb{Z}$ denotes the set of complex integers. Following the CLLL algorithm, we find a "better" channel matrix H=HT from the original channel matrix H.

3 PRECODING AND DETECTION DESIGN

3.1 LR-THP algorithm

In this section, we design precoding under the assumption of the contiguous channel matrices. Tomlinson-Harashima precoding has been used in MIMO channels for a long time. As aforementioned, THP can be combined with LR to enhance the performance of LR-aided linear precoding.

The unitary matrix F at the transmitter works as a feedforward filter that will not change the power of $\tilde{a}$. B is a unit-diagonal lower triangular matrix and serves as a feedback matrix. In the ZF THP algorithm, QR decomposition is performed on HH to generate

$$H = SF^H \quad (10)$$

Here, S is a lower triangular matrix. $U = diag\{1/s_{11},...,1/s_{KK},\}$, where s_{kk} is the diagonal elements of S, k=1,2,...,K and K=min(Nt, Nr). The precoding matrices computation of ZF-THP algorithm can be summarized as follows:

$$H^H = QR$$
$$F = Q^H$$
$$B = R^H U \quad (11)$$

where Q is an Nt × K matrix, such that QQH = I, and R is a K×K upper triangular matrix with diagonal elements {rii}.

According to lattice theories, the process of lattice reduction will make the bases shorter and more orthogonal. With the application of CLLL, we get a modified conjugate transpose of channel gain matrix H^T and G^T with better properties:

$$\hat{H} = H^T T_A$$
$$\hat{G} = G^T T_B \quad (12)$$

where T_m, $m \in \{A, B\}$ is unimodular matrix and contains only complex integers. This can also be rewritten as

$$\hat{H}^T = T_A^T H$$
$$\hat{G}^T = T_B^T G \quad (13)$$

Then, we apply the QR decomposition to the transformed more orthogonal matrix $\hat{H}^T$ and $\hat{G}^T$ to obtain

$$\hat{H}^T = Q_A R_A$$
$$\hat{G}^T = Q_B R_B \quad (14)$$

where Q_m is an $N \times M$ matrix with orthonormal columns, R_m is an $M \times M$ upper triangular matrix with complex diagonal elements, when $m \in \{A, B\}$. According to the structure of THP[12][13], we have

$$B_m = R_m^H U_m$$
$$F_m = Q_m U_m , \quad m \in \{A, B\} \quad (15)$$

Here, U_m is a diagonal matrix with the inverse of diagonal elements of R_m, so $U_m = diag(R_m^{-1})$. Feedback filter B_m and feed forward matrix F_m can be found in the structure of THP. The structure of LR-THP for MIMO two-way relay systems is shown in Fig. 3.

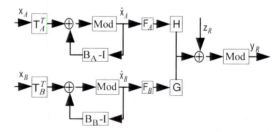

Figure 3. Schematic diagram of the LR -THP algorithm.

In Fig. 3, "Mod" represents modulo operation to reduce the peak or average power. It is a nonlinear process. In the receiver node R, the received signal is processed only by this "Mod."

3.2 LR-MMSE detection

After receiving yR, node R amplifies it as shown in equation (3). Then, a signal vector xR is the transmitted signal at R by constraint of (4). This is a Multi-User MIMO channel. In order to achieve the detection performance, a lot of detection algorithms are proposed. The detector based on maximum likelihood (ML) criterion is best but too complex. ZF detector has a much lower complexity, but its performance is far from that of ML. So we adopt the LR-ZF detection that can do well tradeoff between complexity and performance. LR-ZF detection can reach the full diversity of MIMO channel with a linear structure whose complexity is much lower than the complexity of ML [17]. Since the LR structure has been used in precoding, we can significantly decrease the complexity of the algorithm by reusing for detection.

According to [18], the channel gain matrix H^T and G^T for detection can also be modified as (12) with the application of CLLL. The nulling matrixes W_A and W_B shown in (6) can be computed as

$$W_A = \hat{H}^\dagger = \left(\hat{H}^H \hat{H}\right)^{-1} \hat{H}^H$$
$$W_B = \hat{G}^\dagger = \left(\hat{G}^H \hat{G}\right)^{-1} \hat{G}^H \qquad (16)$$

Then, we can obtain matrixes S_{RA} and S_{RB}, which are composed of the received signal vectors over the H and G.

$$S_{RA} = W_A y_A$$
$$S_{RB} = W_B y_B \qquad (17)$$

The signal vectors in S_{RA} and S_{RB} are shifted and the scaled version from original signal x_R. So, we need to shift the entries back to the original constellation, that is,

$$S_{RA} = T_A^{-1} \hat{x}_{RA}$$
$$S_{RB} = T_B^{-1} \hat{x}_{RB} \qquad (18)$$

From (18), we can shift the entries back to signals $\hat{x}_{RA}$ and $\hat{x}_{RB}$. So, according to (7), users can obtain the signals from the other node.

4 COMPLEXITY AND PERFORMANCE

The purpose of this article is to provide a new precoding and detection algorithm for the two-way relay system; we will consider the complexity and performance. Compared with the traditional ZF-THP and ZF detection, the complexity of the proposed is mostly increasing at LR algorithm. According to [20], the complex LLL algorithm in (12) requires the complex addition $3/2 N^2 (M+N) - 1/2 N^2 - N(M+N)$, and the complex multiplication $3/2 N^2 (M+N) - 1/2 N(M+N)$. When the big O notation is used to describe the

computational complexity of the proposed algorithm, the complex LLL algorithm requires $O[N^3 M \log(N)]$ flop operations (including multiplication and addition). If the real LLL algorithm is used in the proposed algorithm, the complexity will become $O[(2N)^3 (2M) \log(2N)]$. Because the precoding and detection algorithms used LR aid only once, the total complexity is significantly decreased.

Then, we discuss the performance of the algorithm by computer simulations. The proposed LR-THP algorithm can effectively reduce the impact of noise. Since the lattice reduction-aided approaches can achieve full diversity of N, the system performance will be noticeably improved. The LR-ZF detection algorithm can also improve the detection performance.

The simulation is set for a typical MIMO two-way relay system. We assume that all nodes, including A B R, are equipped with four antennas ($M = N = 4$). and all three nodes are transmitted at the same average power P. For comparison, the performance of various algorithms is evaluated with numeric results. The performance results are those of lattice-aided ZF labeled as "LR ZF," MMSE THP without lattice reduction algorithm labeled as "THP," and AF, which is a traditional two-slot MIMO two-way relay scheme labeled as "AF." "LR THP" denotes the proposed LR-THP algorithm in this article. Two kinds of modulation, 4QAM and 16QAM with Gray encoding are used. For fairness, we adopt the LR-ZF detection algorithm in the two users for all schemes.

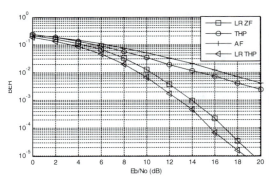

Figure 4. Performance comparison of precoding with 4QAM.

As shown in Fig. 4, LR-THP algorithm shows the better performance by contrast with three other algorithms when the modulation technique adopts 4QAM. At BER = 10^{-3}, a gain of approximately 9dB the proposed LR-THP algorithm is achieved over the classical THP. At BER = 10^{-4}, a gain of approximately 2dB is achieved over the LR-ZF algorithm. We also can detect that the BER performance of classical THP without lattice reduction is even worse than LR-ZF algorithm. The reason is that Lattice reduction process makes the channel matrices shorter and orthogonal.

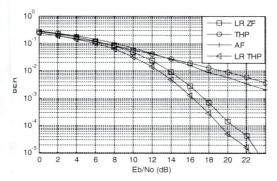

Figure 5. Performance comparison of precoding with 16QAM.

Figure 5 shows the BER performance of the proposed algorithm with 16QAM. We can also conclude that our proposed algorithm achieves the best BER performance. For a 16QAM MIMO two-way relay system, at BER = 10^{-3}, the proposed LR-THP algorithm achieves a gain of approximately 8dB over the classical THP. And at BER = 10^{-4}, a gain of approximately 1.5dB is achieved over the LR-ZF algorithm. Besides, we can also observe that AF algorithm shows the worst performance at low SNR and is the result of the amplification of noise at node R. The LR-THP amplification shows the best performance, because the advantages of both THP and lattice reduction are combined.

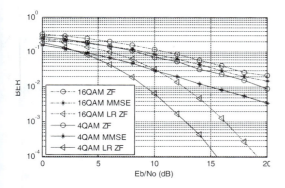

Figure 6. Performance comparison of detection.

Then, we consider the LR-ZF detection algorithm. For comparison, the traditional ZF and MMSE detection is shown in Fig. 6. We assume that the precoding adopts the LR-THP algorithm in three detection. Figure 3 displays the BER performance of LR-ZF detection, ZF detection, and MMSE detection ($M = N = 4$) for 16QAM and 64QAM. We can also conclude that the LR-ZF detection improves the BER performance at the same SNR. The computational complexity of LR-ZF detection is the same as the ZF detection, and it is lower than MMSE detection. For this reason, the LR algorithm has be completed in the precoding algorithm.

5 CONCLUSION

A low-complexity precoding and detection algorithm is proposed for the MIMO two-way relay systems. The proposed algorithm modified the channel gain matrix assisted by lattice reduction. It can better achieve the bit error ratio than the existing THP and ZF detection algorithms. The simulation results for a 4×4 MIMO system show that the proposed algorithm outperforms other algorithms. We achieved a small increase in the complexity of the algorithm. The proposed low-complexity precoding and detection algorithm is promising for future application of MIMO two-way systems.

REFERENCES

[1] Arti M. K., Ranjan K. M, Robert Schober. 2013. Beamforming and Combining in Two-Way AF MIMO Relay Networks. *IEEE Communications Letters* 17(7): 1400–1403.
[2] Sanguinetti L., D'Amico A. A., Rong Y. 2012. A Tutorial on the Optimization of Amplify-and-Forward MIMO Relay Systems. *IEEE Journal on Selected Areas in Communications* 30(8): 1331–1346.
[3] Song C., Lee K.-J. and Lee I. 2012. MMSE-Based MIMO Cooperative Relaying Systems: Closed-Form Designs and Outage Behavior[J]. *IEEE Journal on Selected Areas in Communications* 30(8): 1390–1401.

Information, Computer and Application Engineering – Liu, Sung & Yao (Eds)
© *2015 Taylor & Francis Group, London, ISBN 978-1-138-02717-6*

Countermeasures of ideological education work for college graduates in the new period

Yi Ning Lin

Guangxi Vocational and Technical College of Communications, Guangxi, Nanning, China

ABSTRACT: In recent years, due to pressure on the employment situation, many universities in graduate education pay attention to the practicality of graduate education, carrying out more tasks for the targeted employment guidance service, while ignoring the important position of ideological education of graduate education. In the new period employment system of university graduates, more attention should be paid to the understanding and research thought status and existing problems of the graduates, as well as to the work of ideological education to expand graduates.

KEYWORDS: college graduates, ideological education, countermeasures, thinking

1 IDEOLOGICAL CHARACTERISTICS OF TODAY'S COLLEGE GRADUATES

First, we should make self-accomplishment the center and place emphasis on the realization of self-value. The contemporary university graduates pay more attention to the post regardless of whether the job is beneficial to their own development; too much of emphasis is placed on the implementation and development of their own value and they often ignore the social value. The post with high income, good benefit, more interest, and fast development is becoming the employment hot spot. Along with the call of jobs in the grass-root service, the remote and hard area service has a standoff response. It also exposed the lack of understanding in graduates' occupation values, without really combining the personal development with the national social needs closely.

Second, the sense of competition is enhanced, and the moral consciousness is weakened. With the further establishment of market economy, competition consciousness is constantly strengthened in the talent selection mechanism; "supply and demand to meet" and "self-employment" also increase graduates' competitive awareness. In all kinds of recruitment meetings, they dare to sell themselves, continue to accumulate advantages, hoping that through their own efforts they would obtain the position. But there are also some graduates who lose their principles in the competition, playing down the moral concept, and using improper means to achieve their career goals. This approach shows more important moral quality defects in the quality of the occupation,

and occupation career development is extremely disadvantageous.

Third, the self-ideological confusion and all kinds of psychological problems are easily created. College graduates face severe employment situations and their parents' social expectations. On the one hand, they have to deal with all kinds of school matters in their last year of study as well as the problems of life. On the other hand, they have to enter the talent market and participate in the employment competition. They are often caught in the fierce collision between ideal and the reality, and most people are hovering between wanting to choose their own jobs without taking the risk. Their employment expectations are too high, but they cannot find the combination point, not with social demand. They hope to succeed at an early date, but too much of focus on regional and material treatment, and appear in many complicated and contradictory psychological states. If this state of mind fails to timely adjust, various psychological problems will cause days and months to multiply.

Fourth, innovation consciousness is strengthened gradually, but the actual ability is insufficient. Contemporary college students are active in thinking, and are good at accepting new things, for now advocate building innovation oriented society, innovative campus call a positive response, to advocate in employment in the process of "entrepreneurship" also showed a strong interest, but often for their ability to estimate the lack of understanding, lack of entrepreneurial environment, is easy to appear in the venture problems lead to failure, the real success and the number of graduates is also less, there is also a part of graduates of the lack of confidence in the business, in the face of entrepreneurship and step back.

2 COUNTERMEASURES OF STRENGTHENING AND IMPROVING THE IDEOLOGICAL EDUCATION WORK OF COLLEGE GRADUATES

2.1 Do a good job of college graduates' ideological education work closely combined with the four "tight coupling"

We need to combine the ideological and political education with employment guidance closely. In recent years, many universities have set up employment guiding courses, guiding the students through a comprehensive system of employment; understanding the students' ideological state through the course; strengthening students' situation, publicity, and education of employment policies; and establishing correct occupation ideal and social ideal, which carry out the work of ideological education. In the course of employment guidance, it should guide the students to improve their skills of fair competition, strengthen practice, contract, and honesty education of responsibility to perform their duties. Through education and guidance, the students can have lofty ideals to devote themselves to the national social good; not only have the correct understanding of occupation but also have good occupation morale; not only have a strong sense of talent but also have become the correct way; not only master the national employment policy but also have the development trends of solution for the society; not only strive toward the pursuit of personal value realization but also attempt to contribute more to society.

Ideological education must be combined with psychological health tightly. Choosing a career is a new starting point for the college students after completing school education, to achieve the ideal of life and the value of the brand. Due to differences in their own reality and psychological quality being insufficient, some psychological problems easily produce graduates in the employment process. Without timely adjustment, it will hinder the normal employment and cause more serious psychological problems. So the requirement for university educators is to produce these mental analyses of subjective reasons, help students know themselves, have a good job career ready, especially in the recommendation letter, interview, evaluation, negotiation, and other employment with regard to the part of the frustration psychology preparation. In addition, a variety of problems such as the coordination problem of life education and employment appearing in graduation are also worthy of attention.

Ideological education should be combined with solving the practical difficulties of graduates. For the graduates, the actual problems and difficulties encountered before graduation will be even more than usual, and the actual different problems and difficulties students encounter in employment are not the same. It requires universities graduates ideological education in-depth individual students, to groups with special difficulties to attach great importance to the. The employment problem of graduates such as poverty, it is highly concerned by the nation and the society, for the rural and urban poor families, solve a college students employment, can make one family out of poverty. Therefore, helping poor students successfully complete their education and employment is of great significance. It requires universities to strengthen all aspects of impoverished university students' care and guidance, improve their employment skills, and help them establish confidence, for it provides employment opportunities. In addition, there is still a small part of education and employment "double storm" students, graduation, and the pressure they face are greater than that faced by the average students; therefore, they are more prone to ideological problems and there is also a need to be concerned.

Ideological education must be combined with the graduates' work. Under normal circumstances, before leaving the school, students determine the basic industry, studies have been completed, and there will be a relatively relaxed, free state. Therefore, we must guide the graduates to make good use of this period, to find their own inadequacy to compensate, and to strictly abide by the rules of the school discipline, thus preventing all kinds of things from happening. Simultaneously, we should strengthen education, civilization consciousness of leavers link of thanksgiving education, honesty education, the college ideological education work deepening and sublimation in the last link to the graduates, on the last day the thought course, lay the foundation for its smooth integration with society.

2.2 Ideological education, college graduates in 2 to focus on training students to establish three concepts

Establish a correct concept of employment. The establishment of the concept of employment requires students to first establish a correct concept of values and a self-evaluation system and to clarify their career orientation. Requirements in the careers of graduates must be from the quality, ability, education, and interests of reality, to determine the starting point, good success starting line, and occupation to face reality, to find a suitable position, a positive response to the national "service to the west, service grass-roots" call, to consciously place the personal values that the country needs to combine the most to countries where build up establishment.

Establish the concept of honesty. Integrity in the job market of college students, on the one hand, on the employment management system compliance and enforcement, on the other hand is also reflected in the employment market behavior authenticity and in the performance of protocol on commitment, conscientiously perform their duties. For some graduates, their idea of honesty is indifferent, leading to the opportunistic phenomenon of seeking the position. They should be educated timely in colleges and universities to deal with serious adverse consequences.

Establish the concept of innovation and entrepreneurship. In the development of entrepreneurship education, make students realize that to meet the demands of a new age, they must strengthen their sense of innovation and entrepreneurship. Guide students to enhance innovation and entrepreneurial confidence and courage; encourage and support more independent business conditions of university students to come to the fore; foster innovative advanced models; and develop innovation competition, entrepreneurship enlightenment education. The conditions can open the entrepreneurship curriculum system and further enhance the university students' entrepreneurial ideas and entrepreneurial ability, to create a good business atmosphere. Education enables students to truly possess the quality of entrepreneurship, entrepreneurial skills, courage, and good entrepreneurial character and it pioneers the concept of an active combat professional market, for the future to create and lay a solid foundation for a career.

2.3 Innovative graduates' ideological education mode, enhance their effect

Ideological education of college graduates is a practical and informative task. In practical work, the actual situation should be considered in different periods of the working policy. Graduate students of problems continue to enrich the content of ideological education work, the law of ideological education work for college graduates of master under the new situation problems, methods of improvement, and innovation of ideological education work; improve the vitality and effectiveness of education through various ways. In specific work, we can invite celebrities, entrepreneurs, and successful alumni held report, lectures, interviews, salon, was about the life, career, occupation development, topics such as communication with students, their own will serve as a guide and a strong role model. Simultaneously, we should also pay attention to and give play to the educational function of campus culture, combined with the actual development of graduates' positive campus culture activities, the content and the form of elaborate design of graduation ceremony, and rich civilization of school content, thus deepening the connotation of education. To strengthen the construction and graduates of the information exchange platform, we should fully establish online communication, services, guidance, and a feedback platform using modern educational means and resources, so that ideological education receives more music for graduates and has an excellent effect.

REFERENCES

[1] Wang Jiabin, Zhang Xu. The value of value management in Colleges and Universities[J]. Digest of Management Science, 2007(6).
[2] Li Jing, Li Xiaoan. How to do the work of Ideological and political education of Higher Vocational Graduates[J]. Education and Vocation.

Information, Computer and Application Engineering – Liu, Sung & Yao (Eds)
© 2015 Taylor & Francis Group, London, ISBN 978-1-138-02717-6

Software reliability estimation evaluation method based on Markov model

Xiao Zong Zheng, Ying Gou & Gui Xing Wang
Chongqing College of Engineering, Chongqing, China

ABSTRACT: Markov model is an effective statistical model; this article used Markov model by heuristic iterative to adjust the state transition probability. Then, software reliability estimation heuristic iterative algorithm is given. Through the simulation test analysis, the estimation evaluation method can effectively improve the measurement precision of software reliability.

KEYWORDS: Markov Model; Software Reliability; Estimates; Methods

1 INTRODUCTION

Software quality evaluation and the study of software reliability evaluation has been a hot research topic in the field of software problems. Domestic and foreign scholars have engaged in a great deal of research related to this aspect and have obtained certain research results. Software evaluation method is a more common statistical evaluation, and it is a kind of stochastic assessment method as well as a basic test principle: Use the model to generate test cases by the software, then determine the result of the test statistics, analysis, and, finally, evaluate software quality. Therefore, statistics of test adequacy criteria are generally based on the test using degree of difference and the actual use, in order to ensure the reliability of the estimate according to the test result to represent the reliability of the software actually used[1-3].

In order to ensure the reliability of the software without deviation estimates that need statistics, the evaluation price is very high for a large number of test cases. Many statistical test methods are applied to the software reliability evaluation, and most of them control the statistical test method[4-7].

2 MARKOV MEASUREMENT MODEL

Markov models of initial state and final state are unique, expressed by directed graph and matrix, and described as follows: Directed graph G = (V, E), among them:

V = {1, 2,... ., n} is a node set, that is, the use of the software system.

E={$e_1, e_2, ..., e_m$} is edge set and transfer between states say that when an operation software. From the state i transferred to the state j is a state of one-way transfer. We can use the orderly team < i, j >.

Matrix $P : V \times V -> [0,1]$, the system state transition matrix; among them, P is the state transition probability, the program is in the running stage, and from the current state i is transferred to the state j probability to satisfy the following conditions:

$$P_{<i,j>} \in [0,1] \qquad (1)$$

The use of the software cycle corresponding markov model for:traverse a from initial state to the end state (including the final state and failure state) of the path, and to record is: $y = (y_1, y_2, ..., y_N)$; among them, N is the length of the path, that is, the total number of operations in use in a process at a time. A traverse in the process of the operation i expressed as y_i. y_N is the failure state or final state.

Operation y_i after transfer to the next operation on y_j needs state matrix $P_{<y_i,y_j>}$ random selection. In the process of the software being used in a corresponding path execution, probability is as follows:

$$Pro(y \mid P) = P_{<y_1,y_2>} \times ... \times P_{<y_i,y_j>} \times ... \times P_{<y_k,y_N>} \qquad (2)$$

Shorthand for, $\prod^{sum} Pro(y_{<i,j>} \mid P)$.

Path of execution probability should meet the following constraints:

$$P_{<y_1,y_2>} + ... + P_{<y_i,y_j>} + ... + P_{<y_k,y_{sum}>} = 1 \qquad (3)$$

Shorthand for $\sum^{sum} Pro = 1$.

In the software model based on Markov model, where test cases correspond to the traversal path, for a given software system S, the size of the system is given. Therefore, the reliability of the software system S expectations is as follows:

$$E_S(\prod^{sum} Pro(y_{<i,j>} \mid P)) \tag{4}$$

Among them, E_S is the probability distribution of $Pro(y \mid P)$ that is expected.

3 IMPLEMENTATION OF THE ITERATIVE ALGORITHM

In practice, the first iteration in order to get the optimal solution $\tilde{P}_1$, P by random simulation, the approximate solution of the formula (4) is as follows:

$$\tilde{P}_1 \approx \frac{1}{N} \sum^N (\prod^N Pro(y_{<i,j>} \mid P))) \tag{5}$$

In the second iteration, N path test samples according to the first approximate solution of $\tilde{P}_1$, we get the second approximate optimal solution $\tilde{P}_2$.

$$\tilde{P}_2 \approx \frac{1}{N} \sum^N (\prod^N Pro(y_{<i,j>} \mid \tilde{P}_1))) \tag{6}$$

By the known, we get the k iteration of approximate optimal solutions for

$$\tilde{P}_k \approx \frac{1}{N} \sum^N (\prod^N Pro(y_{<i,j>} \mid \tilde{P}_{k-1}))) \tag{7}$$

Set the software reliability of R in the software in the process of using a failure probability as $\phi_s(y)$; in the time K when the operating system fails, no failure occurs; the N-1 of the traversal path is successful, and the expectation of system failure is as follows:

$$E_\phi(\phi_s(y)) = \int \frac{\phi_s(y_k)}{\tilde{P}_{k-1}} dy \tag{8}$$

$\phi_s(y_k)$ is a random variable and is a linear function of the y_k, solving formula (8) :

$$E_\phi(\phi_s(y)) = \frac{1}{2\tilde{P}_{k-1}} \phi_s(y_k)^2 + \alpha \tag{9}$$

Among them, α is a constant, according to the correction coefficient of failure probability.

When the time k operation failure occurs, then the front k - 1 state is shifted successfully, and the system failure probability of the first k times to remember $\prod^k \phi_s$, derivation formula is as follows:

$$\prod^k_1 \phi_s = P_{<y_1,y_2>} \times ... \times P_{<y_{k-2},y_{k-1}>} \times (1 - P_{<y_{k-1},y_k>}) \tag{10}$$

Actual failure probability of the system is calculated as expected probability; therefore, the following formula is established:

$$\prod^k_1 \phi_s(y) = E_\phi(\phi_s(y)) \tag{11}$$

Using the formulas (9) and (10) in the formula (11), we can calculate the probability of system failure as follows:

$$\phi_s(y) = (2 \times (\prod^k_1 \phi_s - \alpha))^{\frac{1}{2}} \tag{12}$$

For a given software system S, its reliability probability estimation value is

$$\hat{r}_S = 1 - \phi_s(y) \tag{13}$$

Algorithm 1. The Heuristic Iterative Algorithm
 (1). The initialization of the first iteration, approximate solution $\tilde{P}_1$ is obtained.
 (2). Generate test path $(y_1, y_2, ..., y_N)$ and observe whether the software undergoes failure.
 (3). The generated test path and the formula (12) for a solution of $\phi_s(y)$.
 (4). To k = k + 1, repeat iteration, until the failure probability of the correction coefficient variation range is relatively stable, solving the correction coefficient values range as follows: $\alpha \in [0.33, 0.62]$
 (5). Application estimate formula (13) to solve software reliability probability.

4 APPLICATION EXAMPLE ANALYSIS

In order to verify the reliability assessment evaluation method in this article, we use the Gutjahr [5] train dispatching software instance validation experiments. As shown in figure 1, Gutjahr of Markov model, the

122

software contains 12 basic operations, with operation 1 for the initial state and operation 12 for the final state. To the edge (4, 6), (8, 9), (8, 10) leads to execution of key operations 6, 9, and 10; earlier probability of each operation using vector t is expressed as follows:

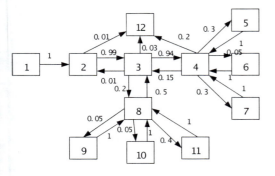

Figure 1.   Markov usage model from Gutjahr.

t=[0, 0.001, 0.001, 0.001, 0.001, 0.0001, 0.001, 0.001, 0.0001, 0.0001, 0.001, 0.001]

To gain a comparative experiment, the simulated annealing (SA) algorithm[7] and the iteration algorithm of this article (iterated algorithm) are compared; the test path sum = 500, 50 times simulation; and the software reliability estimation value of 50 is tested. An experimental comparison chart is shown in figure 2.

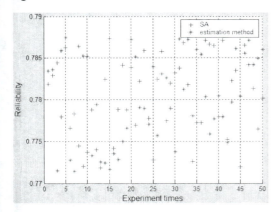

Figure 2.   Reliability estimation of two different profiles.

In 50 times of simulation, the reliability of the SA algorithm to estimate the average is 0.7812, the average variance of 1.77×10^{-4}; in this article, the estimation method of the mean value is 0.7759, and the variance of the average is 8.55×10^{-5}. Through the analysis of test data, in this article, the evaluation of the estimation method can significantly reduce the variance of the estimate, and the precision of estimation is effectively improved.

5   CONCLUSION

In this article, a heuristic iterative algorithm based on Markov model was designed, through the sampling test, and to estimate the reliability of the software. The method using the iterative correction iteration was reduced to an optimization problem, the verification experiment was carried out by Gutjahr instance, and an approximate optimal solution of reliability estimation was obtained.

REFERENCES

[1] Whittacker JA, Poore JH. Markov analysis of software specifications. ACM Trans. on Software Engineering and Method, 1994, 2(1):93–106.
[2] Walton GH, Poore JH. Measuring complexity and converage of software specifications. Information and Software Technology, 2000, 42(12):859–872.
[3] Feng H, Xu XS, Wang J. The determination of resemblance between the usage chain and the test chain in statistical testing. Computer Engineering & Science, 2003, 25(1):17–19 (in Chinese with English abstract).
[4] Gutjahr WJ. Failure risk estimation via Markov Software usage models. In: Schoitsch E, ed. Proc. of the 15th Int'l Conf. on Computer Safety, Reliability and Security (SAFECOMP'96). Springer-Verlag, 1997. 183–192. http://citeseerx.ist.psu.edu/viewdoc/download.
[5] Gutjahr WJ. Importance sampling of test cases in Markovian software usage models. Probability in the Engineering and Informational Sciences, 1997, 11: 19–36. http://citeseerx.ist.psu.edu/viewdoc/download.
[6] Doerner K, Laure E. High performance computing in the optimization of software test plans. Optimization and Engineering, 2002, 3(1):67–87.
[7] Yan J, Wang J, Chen HW. Software statistical test acceleration based on importance sampling. Computer Engineering & Science, 2005, 27(3):64-66 (in Chinses with English abstract).

Design and implementation of web page calendar based on CSS and JavaScript

Tao Liu

College of Information Science and Technology, Zhengzhou Normal University, China

ABSTRACT: With the continuous development of the web development technology, dynamic web pages with interaction ability will be the mainstream. JavaScript combined with CSS is used to design web pages with dynamic effects, which becomes a preferred scheme of many web designers. This article put forward a design method of web page calendar based on CSS and JavaScript. First of all, use CSS to achieve the appearance style effect of the calendar; then, use JavaScript to achieve the date display function of the calendar. Experimental results show that according to different system dates, a calendar of the corresponding month is displayed.

KEYWORDS: CSS; JavaScript; DOM; Web

1 INTRODUCTION

JavaScript is a scripting language based on objects and is event driven. JavaScript can develop interactive web pages; in order to realize the JavaScript code execution, it needs to be embedded into HTML[1]. Using JavaScript, we can reconstruct the entire HTML document. JavaScript set up the document object model(DOM); the Document Object Model, or DOM, is the fundamental API for representing and manipulating the content of HTML documents. First, you should understand that the nested elements of an HTML document are represented in the DOM as a tree of objects. The tree representation of an HTML document contains nodes representing HTML tags or elements, such as <body> and<p>, and nodes representing strings of text. An HTML document may also contain nodes representing HTML comments. Through the DOM, we can create, add, delete, and change the elements in the HTML document as well as its properties and methods.

DOM provides a hierarchical object model that is used to access the HTML page; the hierarchical object model is based on the HTML document structure. Using HTML parser based on DOM, we can transform an HTML page into a collection of object model(A DOM node tree); through the DOM node tree, we can receive access to elements in an HTML page[2]. The DOM can be seen as an objectification HTML data interface that defines the logic structure of the HTML document as well as gives a method to access and handle the HTML document. We can use the DOM to dynamically create HTML document elements, traverse the entire HTML document, add, delete, and modify the HTML document elements, and so on.

Cascading Style Sheets (CSS) is a standard for specifying the visual presentation of HTML documents. CSS is intended for use by graphic designers: It allows a designer to precisely specify fonts, colors, margins, indentation, borders, and even the position of document elements. But CSS is also of interest to client-side JavaScript programmers, because CSS styles can be scripted. Scripted CSS enables a variety of interesting visual effects: You can create animated transitions for document content "slides in" from the right, for example, or create an expanding and collapsing outline list in which the user can control the amount of information that is displayed. When first introduced, scripted visual effects such as these were revolutionary. The JavaScript and CSS techniques that produced them were loosely referred to as Dynamic HTML or DHTML, a term that has ever since fallen out of favor.

2 DESIGN AND ANALYSIS

When people browse the web, they often use the calendar to show what day it is and what date it is to prompt visitors; with CSS + JavaScript, we can design a friendly interface and a convenient, practical calendar. First of all, CSS is used to beautify the appearance of the calendar interface[3]; then, by using the DOM method, we create a calendar in the HTML page, which can display the current month and today's date in red font. How to write JavaScript code to implement the calendar function is the core part of the design. In order to realize the page calendar, two problems must be solved urgently. The first involves using getFullYear(), getMonth(), getDay(), and getDate() method function to get information of the current month, including the current month, the date of the last day of the current month, what day is the first day of the current month, and what is the date today. The second, first of all, according to the data provided by the Date object using the DOM methods creates a calendar table and then a calendar function "CreateCalendar (Monthone, nDays)" is defined, CreateCalendar can use "Monthone" and "nDays," these two parameters, through the functional operation to create a calendar. At last, an empty array is created and named "myarray"; using "myarray.push()" method function, computing results of the "CreateCalendar(Monthone,nDays)" are added to the "myarray" order by date. Then through the "myarray.shift()" method function, loop statements (myarray.length>0) are used in turn to extract the calendar information in the "myarray"; with seven lines of calendar information to one line, fill in the calendar information into its corresponding position in the table. The effect after the design program is running, as shown in Figure 1.

2014-11-16						
SUN	MON	TUE	WED	THU	FRI	SAT
						1
2	3	4	5	6	7	8
9	10	11	12	13	14	15
16	17	18	19	20	21	22
23	24	25	26	27	28	29
30	31					

Figure 1. Calendar.

3 REALIZATION OF THE CALENDAR DESIGN

3.1 *Set the CSS styles*

Using CSS style, change the appearance of the calendar. CSS uses inline citation[4]; then, write CSS codes and put them in between <style> and </style> of the head. The CSS codes are as follows:

```
<style type="text/css">
table {
border-collapse:collapse;
width:212px;
height:250px;
text-align:center;
vertical-align:center;
background-color:#EAF2D3;
color:#000000;
font-size:1.2em;
font-family:"Trebuchet MS", Arial, Helvetica, sans-serif;
}
table,td,tr{
border: 1px solid #98bf21;
}
td {
padding:10px;
}
</style>
```

3.2 *JavaScript code is designed and embedded in the HTML page*

Step1: First, an empty Date object instance is created by using "new Date()"; it was named myDate, and myDate initial value is the current system date. Then, using getFullYear(), getMonth(), and getDate(), these method functions get the current year, month, and day. Here, getMonth() function returns 0 to 11 numbers, 0 is to point to January, and 11 is to point to December[5]. Therefore, when the month is displayed in the calendar, it should be used getMonth() +1. At last, through getDay(), getDate() functions calculate what day is the first day of every month and what date is the last day of every month. JavaScript codes are as follows:

```
var myDate = new Date();
var thisYear = myDate.getFullYear();
var thisMonth = myDate.getMonth();
var thisDate = myDate.getDate();
var Monthone=
new Date(thisYear,thisMonth,1).getDay();
var nDays=
new Date(thisYear,thisMonth,0).getDate();
```

Step2: First, the way of using the DOM to create the calendar table of elements: myTable means created table element node object, mytbody means created the main body of table element node object, titleTr means create header row element node object, and titleTd means create cell element node object in header row. Second, the table header element properties are set: Using titleTd.setAttribute ("colspan", "7") to set the header row of cells across the column numbers are seven; using titleTd.InnerHTML so it shows the current date in titleTd. Third, using appendChild() method, the function will create the element nodes and be added to the HTML page. JavaScript codes are as follows:

```
var myTable = document.createElement("table");
var mytbody = document.createElement("tbody");
var titleTr = document.createElement("tr");
```

```
var titleTd = document.createElement("td");
titleTd.setAttribute("colspan", "7");
titleTd.innerHTML = thisYear + "-" + (thisMonth
+ 1) + "-"+thisDate;
titleTr.appendChild(titleTd);
mytbody.appendChild(titleTr);
myTable.appendChild(mytbody);
document.body.appendChild(myTable);
```

Step3: Custom two functions are used to create the td and tr tags. Because in the calendar every row has seven cells, the appendChild() method function is to be repeated 7 times so that cell element nodes are appended to the row element node. JavaScript codes are as follows:

```
function CreateTd(x)
{
var myTable_Td = document.createElement("td");
myTable_Td.innerHTML = x;
return myTable_Td;
}
function CreateTr(x1, x2, x3, x4, x5, x6, x7)
{
var myTable_Tr = document.createElement("tr");
myTable_Tr.appendChild(CreateTd(x1));
myTable_Tr.appendChild(CreateTd(x2));
myTable_Tr.appendChild(CreateTd(x3));
myTable_Tr.appendChild(CreateTd(x4));
myTable_Tr.appendChild(CreateTd(x5));
myTable_Tr.appendChild(CreateTd(x6));
myTable_Tr.appendChild(CreateTd(x7));
return myTable_Tr;
}
```

Step4: According to these data of what day is the first day of every month and what date is the last day of every month, we create a calendar. JavaScript codes are as follows:

```
function CreateCalendar(Monthone, nDays)
{
var myarray = new Array();
// According to what day is the first day of this
month, we can judge how many empty cells are in
front of it. All these empty cells fill in empty strings;
every empty string, in turn, is appended to myarray
and is defined as an "empty array."
for (var i = 0; i < Monthone; i++)
{
myarray.push("");
}
// According to what date is the last day of this
month, we can judge how many days this month has
and then convert these dates to strings. These, in turn,
are appended at the end of myarray.
for (var j = 1; j <= nDays; j++)
{
```

```
if (thisDate == j)
{ // Today's date are shown in red.
myarray.push("<div onclick=
'alert(titleTd.innerHTML);'><font
color='red'>"+ j.toString() +"</font></div>");
}
else
{
myarray.push(j.toString());
}
}
// Judgment after calendar is added into the myar-
ray; the spaces that are not filled in the calendar table
should be filled in empty strings and are appended at
the end of the myarray
if (myarray.length >= 28 && myarray.length <= 35)
{
var nums = 35 - myarray.length;
for (var k = 0; k < nums; k++)
{
myarray.push("");
}
}
if (myarray.length >35 && myarray.length <= 42)
{
var nums = 42 - myarray.length;
for (var k = 0; k < nums; k++)
{
myarray.push("");
}
}
// Take out every element in the array and put them
in the calendar table corresponding to their position.
Each one should take one element, and the length of
the array is minus one.
while (myarray.length > 0)
{
mytbody.appendChild(CreateTr(myarray.shift(),
myarray.shift(),myarray.shift(),myarray.shift(), myar-
ray.shift(), myarray.shift(), myarray.shift()));
}
}
```

Step5: Create a calendar header. JavaScript codes are as follows:

```
mytbody.appendChild(CreateTr("SUN", "MON",
" TUE", " WED", " THU", " FRI", " SAT"));
```

Step6: Create a calendar. JavaScript codes are as follows:

```
CreateCalendar(Monthone, nDays);
```

4 CONCLUSION

The emergence of CSS and JavaScript technology brings vitality and development to the website design[6]. In this article, the JavaScript was introduced briefly and through a calendar design example, the JavaScript codes were implemented. In the implementation process, this design uses the DOM model, Date object, and each custom function as well as its call relations between them. The method to design the calendar function is relatively simple. Introducing JavaScript event handler in the design can achieve more complex web pages related to the calendar function.

REFERENCES

[1] Xing-wei Hao, "Web development technology", BeiJing: Higher Education Press, 2006.12.

[2] Ning Dong, "JavaScript and Ajax applications", BeiJing: China WaterPower Press, 2014.01.

[3] Chong Li, Shu-hua Xiong, Ying-ying Wei, "Design and Implementation of Tab Panel Based on CSS and JavaScript", Computer Technology and Development, 2011, 21(3):28–30.

[4] Di Xiong, "Realization of dynamic Tabbed menu by CSS and JavaScript", Journal of HuBei TV University, 2009, 29(2): 155–156.

[5] Li-rong Zhou, "Making dynamic web page based on JavaScript", Computer learning, 2005(3): 11–13.

[6] Jensen S H, Moller A, "Type Analysis for JavaScript", STAT IC ANALYSIS, 2009, (5673): 238–255.

Information, Computer and Application Engineering – Liu, Sung & Yao (Eds)
© 2015 Taylor & Francis Group, London, ISBN 978-1-138-02717-6

Recognition and location of circular objects based on machine vision

J. Zhou & D.R. Zhu
College of Mechanical and Electrical Engineering, Anhui Jianzhu University, Hefei, China

ABSTRACT: Extracting the coordinates at the center of the circular objects accurately is of great significance in the actual image processing and analysis system. After studying the classical recognition and location based on machine vision and the Matlab programming that the disc movement has achieved, this article presents a method by adding external devices to improve the center coordinates precision. The experimental results show that the double capturing devices can improve coordinate precision and verify the feasibility of this approach.

KEYWORDS: Machine vision; Matlab, Image processing; Center location; External capture devices

1 INTRODUCTION

Currently, in terms of disk location technology, many mature methods are existing. Among them, the Hough Transform, which can run in Matlab, is the most common approach. But because of its large computation, massive data occupy a lot of memory, and it is difficult to meet the requirements in the center location of the real-time system test. Therefore, Zhang Hongmin put forward an improved algorithm on this basis. The study by Wang Jian and Wang Xiaotong, which is based on gradient, improved Hough Transform. Wang Shubin and others present right triangle method, mid-perpendicular, method and equidistance method: These obtain better positioning action of standard and fragmentary circle. Jin Tao and others study the outline and the center of the circle based on optics measurement. Ma Wenluo, Huang Shilei, and Zhang Peng present the analysis of moving objects, which is based on the vision-based tracking algorithm. Circle detection and positioning technology are closely linked.

These articles are mostly from the perspective of algorithm; this article will study the improvements of external devices that can be positioned more precisely.

2 THEORETICAL PROOF

The method proposed in this article aims at improving the positioning accuracy and at increasing the number of external devices.

2.1 Camera imaging principle

Camera imaging and pin-hole imaging have the same principle. In ideal conditions, imaging principle is shown in Figure 1.

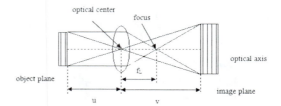

Figure 1. Lens imaging principle.

In Figure 1, f_L stands for focal length, v stands for image distance, and u stands for object distance. The three elements are taken into Gauss formula with Eq.(1)

$$\frac{1}{f_L} = \frac{1}{u} + \frac{1}{v} \qquad (1)$$

It must be pointed out that although it is established in theory, camera imaging and pin-hole imaging have a significant difference in terms of practice. The main reason is that focal length and image distance are two independent parameters in the lens system, but in the pin-hole system, they are exactly the same value. However, this does not affect the use of the principle of pin-hole imaging for analyzing the system, because the relationship between imaging units is consistent.

2.2 *Theoretical analysis*

In this experiment, three same model shooting equipment are located in the left, middle, and right, respectively, and pictures are taken simultaneously. The schematic diagram is shown in Figure 2.

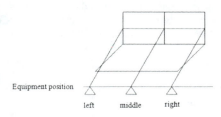

Figure 2.   Schematic diagram of equipment position.

According to the principle of pin-hole imaging, the equipment takes the pictures of an ellipse whose long axis and short axis are not equal (circle is a special case of the ellipse).

3   EXPERIMENTAL EQUIPMENT AND PROCEDURE

3.1 *Introduction of experimental equipment*

3.1.1 *Programming software*

In this study, researchers use Matlab software to program.

Matlab is a combination of two English words, "Matrix" and "Laboratory." It can achieve numerical analysis, matrix computation, image processing, and other powerful features. The researchers use version Matlab R2011b, which can directly invoke C/C++.

3.1.2 *Shooting equipment*

The experiment uses Sony DSC-T70 to shoot and sets the shooting interval to be 0.5s.

3.2 *Experimental procedure*

1 In Matlab, the procedure can be programmed to achieve a uniformly accelerated linear motion and precise coordinates of every shooting point can be sorted out. The result is shown in Table 1.
2 The three cameras are located in the left, middle, and right position to shoot simultaneously and the shooting interval should be set to 0.5s; each camera shoots 9 pictures.
3 Using Matlab to analyze pictures

1 Machine vision positioning system
   The machine vision positioning system has its input, processing, and output structure; a brief system chart is shown in Figure 3.
2 Choose one picture shown in Figure 4 as an example that is shot in the 2.0s by the camera located in the middle; it can describe the localization process in Matlab. Other pictures are analyzed in a similar manner.

Steps are as follows:
Import Figure 4 to Matlab, take Hough Transform to locate circular object, and obtain the test results as follows:
Center 115 129 radius 20
Center 115 130 radius 20
Center 116 130 radius 20

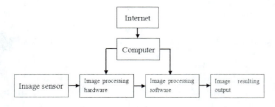

Figure 3.   Vision localization system chart.

Table 1.   Exact coordinates of each time point.

Test time(s)	0	0.5	1.0	1.5	2.0	2.5	3.0	3.5	4.0
Figure Movement distance	0	4.0625	15	38.0625	80	149.0625	255	409.0625	624
Real movement distance	0	1.2500	3	5.2500	8	11.2500	15	19.2500	24

Figure 4. Picture shot in the 2.0s by camera located in the middle.

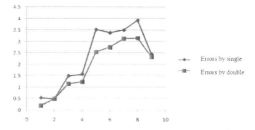

Figure 5. Error comparison.

Center 115 129 radius 21
Center 115 130 radius 21
Center 116 130 radius 21,

In this experiment, only X coordinate is needed. The peak point is X=115, and we can know the real movement distance is 11.5 based on Table 1. Other data can be analyzed in a similar manner.

Use the average method to integrate the data by the left and right cameras; the data are shown in Table 3.

A comparison of single and double shot devices is conducted to check whether it has a significant effect with regard to improving the positioning accuracy or not.

The errors are shown in Table 4; error comparison is shown in Figure 5.

4 EXPERIMENTAL RESULTS

The earlier data show that increasing the number of external equipment to improve the accuracy of coordinates is feasible; it can directly be seen from the figure that there is a larger error by a single camera whereas the error is reduced by a double camera. However, due to various factors, there still exists a certain degree of error that needs to be improved in future work.

ACKNOWLEDGMENT

The content of this article belongs to the project supported by the Anhui Province Key Technology R&D Program (No. 1301022066).

Table 2. Data by the middle camera.

Test time(s)	0	0.5	1.0	1.5	2.0	2.5	3.0	3.5	4.0
Data by middle	0.53	1.75	4.50	6.83	11.5	14.63	18.50	23.18	26.45

Table 3. Data by the left and right cameras.

Test time(s)	0	0.5	1.0	1.5	2.0	2.5	3.0	3.5	4.0
Data by left	0.30	1.75	4.63	7.25	11.85	14.53	19.18	21.30	27.53
Data by right	0.10	1.75	3.67	5.73	9.25	13.47	17.08	23.5	25.13
Data by average method	0.20	1.75	4.15	6.49	10.55	14.00	18.13	22.40	26.33

Table 4. Comparison of two methods of error.

Errors by single	0.53	0.50	1.50	1.58	3.50	3.38	3.50	3.93	2.45
Errors by double	0.20	0.50	1.15	1.24	2.55	2.75	3.13	3.15	2.33

REFERENCES

Zhang, H.M & Zhang, Y.J. 2000. The rapid extraction method of center coordinates based on generalized hough transform. Journal of Jianghan Petroleum Institute.

Wang, J & Wang, X.T. 2006. Fast circle detection using randomized hough transform based on gradient. Application Research of Computers.

Wang, S.B & Yang, J.J. 2013. Research of center positioning method based on the geometric characteristics. Electronic design engineering.

Jin, T & Jia, H.Z. 2010. The ectraction method of moving circle and circle center research. Laser Journal.

Ma, W.L & Hu, J.X. 2012. Research on servo tracking of moving object based on visual. Computer Engineering.

Huang, S.L. & Chen, S.L. 2013. Moving object tracking algorithm based on visual perception mechanism. Application research of Computers.

Zhang, P. & Wang, R.S. 2005. An approach to the remote sensing image analysis based on visual attention. Journal of electronics & information technology.

Hu, J. & Lin, X.R. 2005. Fast Algorithm of Real-time Image Locating and its Application. Computer Engineering and Applications.

Liu, Q.X. & Zhu, C.P. 2003. Image Location Arithmetic in the Intelligent Vehicle Plate Recognization System. Computer Engineering.

He, Y.L. 2011. Research on Centroid Localization Algorithm for Wireless Sensor Network Based RSSI. Computer Simulation.

Zhang, P. & Zhu, Z.H. 2007. Machine vision technique and its application to automation of mechanical manufacture. Journal of Hefei University of Technology.

Information, Computer and Application Engineering – Liu, Sung & Yao (Eds)
© 2015 Taylor & Francis Group, London, ISBN 978-1-138-02717-6

Intelligent traffic management platform research and design based on big data

Ling Ling Hu, Xu Xu & Ning Li
The 3rd Research Institute, Ministry of Public Security, Shanghai, China

ABSTRACT: With the growth of vehicles and traffic flow density, massive traffic management data emerge sharply, and the complexity of transportation systems increases rapidly, thus big data appears as an innovative technical method. This paper analyzes present situations of intelligent traffic management systems, and proposes an intelligent traffic management platform based on big data technology. Discussions are presented from four aspects, interface, application, data and system, and research technical route focuses on storage, analysis, and security. Improvement on public security traffic management can be expected.

KEYWORDS: big data; Intelligent Traffic Management; Platform

1 INTRODUCTION

With the rapid development of social economy, the number of vehicles continues to grow. By the end of 2013, the number of vehicles in China exceeded 250 million, and the number of vehicle drivers was nearly 280 million[1]. With the increasing of vehicles and traffic flow, the scale and the complexity of traffic system has rapidly increased, which causes traffic management facing enormous pressure. In the new era of traffic environment, how to organize traffic management safely, orderly, and flexibly, service people travel, improve the green transportation service quality effectively, reduce exhaust emissions, earnestly safeguard the traffic safety, and so on have brought new challenges to traffic science and technology work. "Big data," a characteristic that includes volume, velocity, variety, and veracity [2], provides a perfect technology for traffic management innovation [3].

In this article, we will first analyze the status of China's intelligent traffic management information system, next propose an intelligent traffic management system based on the technology of large data, the integration design method, and study the technical routes, which is the final conclusion.

2 INTELLIGENT TRAFFIC MANAGEMENT INFORMATION SYSTEM APPLICATION STATUS

In the process of intelligent transportation information construction, big cities such as Beijing, Shanghai, and other places whose information systems have covered almost all the businesses such as traffic management, traffic police, traffic police service examination system, vehicle administration files management system, traffic wardens put information management system, license plate recognition system, ground road traffic information collection and dissemination system, video event detection system, video monitoring system, and the GIS application platform and other dozens of information systems. Using these systems, we can improve work efficiency and benefit, relieve traffic congestion, and improve the level of traffic management of the city.

Simultaneously, the traffic data are rich in resources of all kinds of systems, such as vehicle information, driver information, traffic flow, object identification, and other structured data, also including electronic documents, voice, images, video and other semi-structured and unstructured data, and traffic data from scarce to abundant. The public security traffic has officially entered the new "data" era as the unit of structured, semi-structured, and unstructured polymorphism coexistence. But simultaneously, questions are quite obvious:

1 There is not enough sharing of information resources between the various business units. Information isolated island, long acquisition, and repeated acquisition phenomenon exist generally, which is a serious influence on the development of traffic management information. For example, the vehicle data are lacking in the depth of integration, and information of public security traffic management mainly focuses on the existing vehicle static information, such as basic information management, car information, rendering service limited separation, multiple portals landing, and

artificial Association records. There is a lack of comprehensive information service based on dynamic information service.

2 Overall dynamic data collection and application development is not balanced. In only one day, more than 60,000 taxis will generate hundreds of millions of GPS data, the greater the amount of data plate recognition and traffic surveillance video. Traditional traffic data analysis method has found it difficult to effectively support such large data. Unstructured data such as audio and video graphics occupy a lot of system resources but rely heavily on artificial intelligence transformation judged, intelligence information to generate time-consuming, high-strength, poor service. There is a serious lack of intelligence unstructured data storage, management, and use of tools.

3 The overall level of development and utilization of information resources is lagging behind. The ability to dig useful information to guide transportation planning decisions from big data and to improve the management efficiency is limited. Through public information resources, enhancing the efficiency of road traffic capacity is limited, and information resource utilization inefficient seriously affects the level of traffic management command.

3 INTELLIGENT TRAFFIC MANAGEMENT PLATFORM BASED ON THE TECHNOLOGY OF BIG DATA

The intelligent traffic management platform based on big data integrate business systems for traffic management, control systems and video monitoring systems and other application systems, also static and dynamic, structured and unstructured data. We should deepen the application, dig the role of data in the traffic management decision-making, and enhance traffic control intelligence coordination and ability to induce traffic, and improve traffic management level. We should carry out the integration of the interface, applications, resources (data and systems), and other aspects of making full use of the multi-level framework that has been set up by integration.

3.1 Interface integration

Compliance with unified access entrance, the principle of unified user and rights management, unified interface style and mode of operation, integration of all applications, information websites and information resources, application portal technology integrated display. The main achievements of the following functions should be completed: (1) Unified application access portal; (2) Unified information display and query retrieves entrance; (3) Unified user rights management; (4) Cross-application business flow and information integration; (5) Interdepartmental site information search mining and concentrated expression; and (6) Comprehensive display of traffic information resources.

3.2 Application integration

According to the business nature of police system, we should divide the traffic police application system into business management, intelligent transportation, dispatching, decision making, party government, and police security. We should integrate the application system according to the following category:

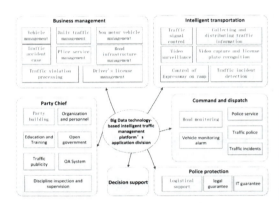

Figure 1. Big data technology-based intelligent traffic management platform's application division.

3.3 Data integration

Data integration is an important part of resource integration. Data integration is divided into business data integration and analysis data integration. Business data integration is carried out with business systems integration; the purpose of analysis data integration is to decision support, merge static, dynamic, structured, and unstructured data, consisting of multidimensional data. We should implement a full range of analysis and mining and provide in-depth access to the content and application. Simultaneously, there should be access to video, images, and other media information, combined with the use of GIS and GPS. They are displayed in the portal in a new, intuitive, and practical way.

3.4 System integration

System integration is designed to take full advantage of the hardware equipment and other infrastructure resources. Using virtualization technology, fully tap the server and storage potential and utilization rate; reduce the number of servers and energy consumption. Establish a unified platform to supporting the upper application services. System application support platform consists of PGIS [4], platform-based GIS application platform, statistical analysis, query, retrieval, and so on.

Using the concept and technology of cloud computing, we should establish public security traffic management and intelligent transport cloud. Using Hadoop [5] technology, we should build computer equipment and other shared resource pool and achieve intelligent traffic clouds. We should meet the real-time traffic flow speed operation and response requirements. Using real application cluster [6] and other technology, we should establish a shared resource pool of computer and storage devices to achieve computing cloud, storage cloud, and application cloud for security traffic management; to meet the large number of users for accessing, analyzing, and mining large amounts of data requirements.

4 PLATFORM TECHNOLOGY

According to the life cycle of management of traffic data processing based on large data, the intelligent traffic management platform can be divided into big data acquisition and preprocessing, big data storage management, big data analysis, security, and big data service, such as shown in figure 2.

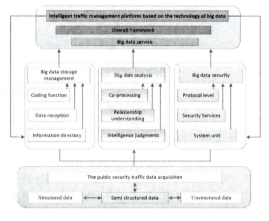

Figure 2. Platform technology.

4.1 Big data storage management

in view of the existing data source in traffic management information system, we should analyze the characteristics and applications of the basic information, statistical data, dynamic data, and so on, which are scattered in different subsystems. Combined with the typical business environment of the traffic management, we should determine its storage management strategy, according to big data resource planning and application requirements.

4.1.1 Coding function

Vehicle entity encoding uses the form of "rear section of front end code + code" as a code. We should establish registration and inquiries encoding mechanism. Entity encoding is mainly intended to solve the question of the uniqueness of the entity tag for the vehicle. To prevent excessive expansion of the code length, we should not consider the entity attributes much when we code.

Coding enables the information that exists to be collected by various types of devices that are aggregated, integration, and transformation, so that it can be recognized by the computer, and it is easy for storage in the form of a two-dimensional code. It is the foundation for the establishment of the database.

4.1.2 Basic information function

The basic information module receives the basic information provided by public security police. The information is updated real timely on the information platform.

Basic information: name, code, category (categories and sub-categories), work (fixed and mobile), perceived scope, purpose, object monitored, collection frequency, and each datum;

Location information: X coordinates, Y coordinates, latitude, longitude, description, and county;

Managers' Information: Responsibility unit type code, the unit name responsible, organization code, contact name, contact telephone number, and contact address of the responsible units;

Device information: Device type, device status, supporting industries, installation (time, unit), and production (time, unit).

4.1.3 Information directory

Information directory mainly records vehicle information structure and information attributes. Information structure through the directory tree structure shows the relationship between vehicle information. Information attribute describes the management attribute, including the source, destination, and so on, for the control and management, reflecting the internal correlation data, and it makes the depth of data mining easy.

4.1.4 Resource management

Traffic big data resource management has three functions of data access in a way of virtual disk, personal storage and community data sharing, and user/community management. It provides disk access interface seamless integration of local operating system.

The user can use the storage service as a local disk. It provides data community storage service, sharing data between multiple users conveniently.

4.2 Big data security

4.2.1 Overall security model

Security architecture includes the protocol hierarchy, security services, system element, and the security mechanism. Figure 3 shows the relationship between protocol, service, and unit. It forms three different angles of view and clearly depicts the relationship between the basic structure of safety system and each part. It reflects the security requirements and the common platform architecture.

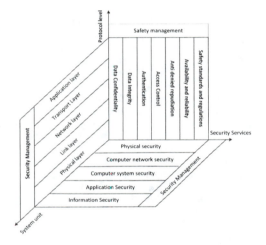

Figure 3.   Security architecture.

4.2.2 Security service model

Security architecture includes five important security services: authentication, access control, data integrity, data confidentiality, and non-repudiation. These security services reflect the security requirements of information systems. The security service such as entity identification, authentication, and access control is the most important.

4.2.3 Protocol layer model

Investigate security architecture from the angle of network architecture and protocol layers. Get the network security protocol layer model and the realization mechanism of seven kinds of basic security services in of TCP/IP protocol for each level. Give them the protocol-level position.

4.2.4 System unit model

In the phase of demonstration implementation, a variety of security services and security mechanisms to the various agreements should be implemented ultimately to physical entity units, including physical platform security, network platform security, application security, and system security.

4.3 Traffic data mining analysis function.

The mining analysis enables large amounts of real-time data and historical data that are saved in the car industry to be widely used. And the data are converted into useful information and knowledge. Finally, relevant information and knowledge are converted into useful information, thus servicing the society through each vehicle industry.

4.3.1 Global centralized control

In order to facilitate the implementation of distributed data mining, design a centralized control for the entire system for solving communication overhead and how to undertake global decision-making in the global scope of the problem.

4.3.2 Parallel and distributed data mining

Through the parallel algorithm to divide the data into subsets, reduce the time complexity of the whole data mining so that the performance is improved.

4.3.3 Knowledge sharing and distributed design

For distributed mining among the sites, use the form of knowledge that can be understood. Advantages of distribution are the biggest support software reuse. System designers can use the soft component existing. They can optimize the division of labor, greatly reduce the coding workload, improve efficiency, and reduce the costs.

4.3.4 Unstructured semantic extraction

The key technological achievements, such as knowledge of the unstructured data about cars, understanding and description of surveillance video content, and unstructured data management, need to be integrated into application-oriented management. According to the application demand, unstructured information processing and customized services of the public security work flow should be analyzed. On the basis of this, design service-oriented software architecture. Develop service package interface, realizing the information-related services based on unstructured data.

5 CONCLUSION

In the background of the information society, focus on vehicle elements of public security vehicles, carry out research and design on an intelligent traffic management platform based on big data technology, and strengthen integration and sharing between the various types of shared IT systems. Through high-quality big data application service levels enhance public security and traffic management capabilities to serve the people; promote the intelligent traffic management mechanism innovation based on big data technology. This has important academic value and practical significance.

ACKNOWLEDGMENT

This article is supported by Soft Science Research Program of Shanghai Science and Technology No 14692101800.

REFERENCES

[1] Chinanews, "By the end of last year the number of vehicles break through the 250 million driving close to 280 million people in our country", http://www.chinanews.com/gn/2014/01-28/5794423.shtml.
[2] Zhao G. Big Data Technology and Application Practice. Beijing: publishing house of electronics industry, 2014:4–5.
[3] Zhou F, Shi J J, Cui L. Initial Probe into Police Work in the Big Data Era. Journal of Shanghai Police College. 2013(23):34–37.
[4] PGIS.http://baike.baidu.com/view/4740930 .htm?fr=aladdin.
[5] Cui J, Lin T, Lan H, Design and Development of the Mass Data Storage Platform Based on Hadoop, Journal of Computer Research and Development, 2012(49):12–18.
[6] Hu M L, Luo J. Dynamic real-time scheduling with task migration for handling bag-of-tasks applications on clusters. The 18th IEEE International Conference on Networks (ICON). Singapore. IEEE.2012:222–226.

Information, Computer and Application Engineering – Liu, Sung & Yao (Eds)
© 2015 Taylor & Francis Group, London, ISBN 978-1-138-02717-6

Construction and application of multitask-driven teaching model aiming at developing computational thinking—taking Visual Basic.Net programming course as an example

Qing Liu & Ji Ming Zheng
School of Business Information, Shanghai University of International Business and Economics, China

ABSTRACT: Computational thinking is not only an important concept in the international computer community at present but also an important topic to be studied in the current computer education. Based on methods of computational thinking, this article classifies tasks into simulation tasks, key tasks, and independent tasks. Considering the characteristics of the task-driven teaching model, this article proposes a multitask-driven teaching model aiming at developing computational thinking. By using Visual Basic.NET programming course as an example, it explains, from aspects of a teacher's teaching process, students' learning process as well as communication, reflection, and evaluation; how to use the model to improve students' computational thinking ability and to encourage students to use the ability to solve problems.

KEYWORDS: computational thinking; task-driven; teaching model; programming

1 INTRODUCTION

In March 2006, Jeannette M. Wing, Department Head and Professor of Computer Science of Carnegie Mellon University in the United States, defined computational thinking in *Communications of the ACM*, which is an authoritative computer magazine in the United States. Professor Wing believes that computational thinking involves solving problems, designing systems, and understanding human behavior, by drawing on the concepts fundamental to computer sciences. Computational thinking includes a range of mental tools that reflect the breadth of the field of computer science (JEANNETTE M.2006).

When computational thinking was put forward, it caused a general response in the international computer community and received widespread support from the American education circle. In June 2008, Computer Science Teachers Association (CSTA) published the online report titled "Computational Thinking: A problem solving tool for every classroom supported by Microsoft." The report points out that the essence of computational thinking is thinking about data and ideas, and using and combining these resources to solve problems. Computational thinking is a required skill for 21st-century success that teachers can foster using subject-specific simulations and modeling. (Philips.2008). In July of the same year, in "Computational Thinking and Thinking about Computing," Professor Wing pointed out that the essence of computational thinking is abstraction.

Deeper computational thinking will help us not only to model more and more complex systems but also to analyze the massive amounts of data we collect and generate (Wing. 1989).

Currently, computational thinking is not only an important concept in the international computer community but also an important topic to be studied in the current computer education. In China, Professor G.L. Chen, academician of Chinese Academy of Sciences, pointed out that in the universities computational thinking not only revitalizes the computer education but also helps make revolutionary research achievements in the science and engineering field. (G.L. & R.S. (2011)). Professor Y.Z. Zhu regarded computational thinking, experiment thinking, and theoretical thinking as three scientific thinking modes of human beings (Y.Z.2009). Professor D.C. Zhan put forward the computation tree, which is a multi-dimensional observation framework of computational thinking, and discussed its effect on software engineering education (D.C. et al. 2014). Combined with anchored instruction, cognitive apprenticeship, and computational thinking methods, L. & H.W (2012) constructed the Two-level Project-Based Instruction Model (TPBIM) and applied it to the web page design and production courses. Y.S. et al. (2014) applied the concept of computational thinking to the multimedia teaching, so as to improve students' ability of computational thinking. In spite of this, how to cultivate students' ability of computational thinking in the specific teaching and learning process is still in the exploratory stage. It is an important direction worth our study. Based on this, according to the methods of computational thinking and the characteristics of the task-driven teaching model, this article proposed a multitask-driven teaching model aiming at developing computational thinking. This teaching model is the further improvement and sublimation of task-driven teaching model. Integrating various application methods of computational thinking, it can help teachers improve teaching efficiency and help students develop thinking abilities.

2 MULTITASK-DRIVEN TEACHING MODEL AIMING AT DEVELOPING COMPUTATIONAL THINKING

Task-driven method is a teaching model based on the constructivist teaching theory. In this teaching model, teachers design teaching content into one or more specific tasks, and they let students master teaching contents to achieve the target of this course by completing some specific tasks. Students take initiatives in learning, and teachers give guidance in this method. Tasks as the main line, teacher as the leadership, and students as the main body are the basic characteristics of this teaching method (X. et al. 2011).

In the traditional task-driven teaching model, the teacher designs tasks and evaluates the tasks completed by the learners, but there is no requirement of the methods used by the learners.

The multitask-driven teaching model aiming at developing computational thinking has meaning in two aspects. On one hand, the concept of computational thinking runs through the teaching process. The teacher applies the method of computational thinking to design tasks; on the other hand, learners are encouraged to implement computational thinking methods to solve the problems when completing the tasks, which improves the learners' thinking ability. The schematic diagram of the multitask-driven teaching model aiming at developing computational thinking is shown in Figure 1.

With multitask as the main line, the model connects teacher's teaching process to students' learning process, designs tasks with the method of computational thinking, and guides the students to use the method of computational thinking to complete the tasks. According to the syllabus and knowledge, combined with a series of methods of computational thinking (problem reduction, separation of concerns, simulation, heuristic reasoning, recursive thinking, pushback, etc.), the teacher designs and puts forward three kinds of tasks, which are simulation tasks, key tasks, and independent tasks. In the simulation tasks, the teacher uses the method of problem reduction in computational thinking to simplify the real items as necessary. The relevant procedures and statements are provided directly by the teacher. Students only need to imitate to build a system framework; then, the teacher uses the method of separation of concerns to decompose a complex and huge system

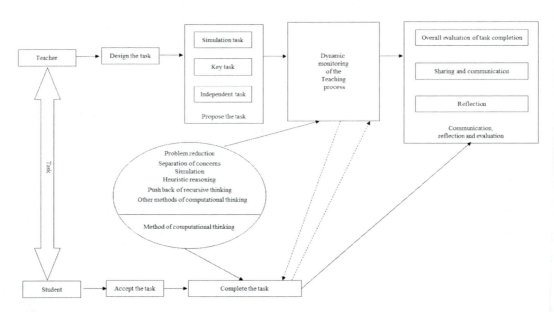

Figure 1.　Schematic diagram of the multitask-driven teaching model aiming at developing computational thinking.

into multiple modules, solves each module by stage and task, and presents it to students in the form of simulation. According to the basic process of development system, this kind of tasks allows students to know the creation process of the whole system. It is the key task for the teacher to use the embedding, recursion, abstraction and decomposition methods of knowledge to embed the real and reasonable code into the tasks for students' study and discussion. This stage is mainly the analysis and explanation of key knowledge points. In this type of tasks, students will understand the diversity of all kinds of algorithm and independently analyze the code in the first kind of tasks, so as to cultivate the students' ability of computational thinking and improve their understanding of the algorithm. In the independent tasks, after the completion of the earlier two kinds of tasks, the students independently select topics in small groups and use the method of computational thinking to complete the task.

Through these three kinds of tasks, the teacher simulates the whole process of project development, and helps and guides students to solve problems effectively. With the teacher's dynamic monitoring and help, students also use the method of computational thinking to accomplish tasks. After the completion of the tasks, there is sharing and communication between the teacher and students and between students and students. The teacher summarizes and evaluates the task completion. The students reflect on their learning process and methods. We need to emphatically point out that the tasks should be chosen from the real projects. The tasks familiar and often applied by the students are the best. In these tasks, the teacher plays the role of an expert and a leader to guide students. When they really master the knowledge points and know how to use the method of computational thinking, the students can build their own learning framework and methods independently, and they can unconsciously integrate the method of computational thinking. As Professor Wing said, if computational thinking is used everywhere, then it will touch everyone directly or indirectly (Wing. 1989).

3 PROBLEMS EXISTING IN THE TRADITIONAL TEACHING PROCESS

3.1 *Introduction to Visual Basic.NET programming courses*

Visual Basic.NET programming is an important professional foundation course in our university. It is theoretical and practical. The main task of this course is to make the students master the basic method and design ideas of program design, as well as cultivate students' ability to analyze and solve problems. The

content of the course mainly includes the following aspects. First, we have the basic methods of program design. This part is mainly intended to teach students the basic methods and techniques of sequential, selection, loop, array, function, database, and other programming. Second is the design idea and method. This part is mainly intended to help the students master the basic process and design idea of project development based on practical projects. Third, we have the content of practical teaching. This part is mainly intended to complete the construction of the practical project based on the selected design ideas.

3.2 *Problems existing in the teaching process*

In terms of the traditional teaching methods and organization of course content, the author mainly found the following problems.

1 In teaching, the teacher explains the knowledge points alone. The students face the examples without logic correlation. After completing the whole course, students do not know how to develop the system or how to complete a specific task. They cannot put the knowledge they have acquired into practice.
2 The examples in the teaching are specially designed for teaching, rather than from the actual project. After learning, students do not know how to draw inferences about other cases from one instance or to apply them in practice.
3 The theory is apart from practice. After learning about the basic process of project development, the students cannot use it in the actual project development, and they do not know which ideas and methods of program design they can use in the actual project development.

In order to solve these problems, in the process of the teaching of this course, we use the task-driven method as the main teaching method, integrate the theory of computational thinking, build the multi-task-driven teaching model aiming at developing computational thinking, and improve students' ability to use the knowledge they have acquired to solve practical problems and ability of design innovation.

4 SPECIFIC APPLICATION IN VISUAL BASIC. NET PROGRAMMING COURSE

We consider computational thinking as the process thinking for problem solving. It includes finding the problem, determining the problem, and solving the problem. It can clearly describe the operation planning, problem constraints, and range restriction of problem solving (C.J. & Y. (2012)). With Visual

Basic.NET programming course as an example, the author explained the specific operation method of the multitask-driven teaching model aiming at developing computational thinking.

4.3 Teachers' teaching process

4.3.1 Teachers designs the task

According to the syllabus, the teacher timely supplements and updates the ideas and methods of program design, allows students to know the frontier theory, transfers the latest computer technology, uses separation of concerns and other methods to establish the corresponding problem base in the corresponding project combined with his/her own project development experience, integrates the knowledge points into the design problem, and builds the bridge to solve the problem with project, task, and problem as the unit.

4.3.2 Propose the task

We propose a total of three tasks for students.

The first task requires students to complete the construction of the universal examination platform. This task belongs to the simulation task of the first kind. Because this task has been simplified, the large system is decomposed into multiple modules, and related programs are provided directly by the teacher, it is convenient for students to complete this task. Through the task, the students can master the basic process of building a practical system.

The second task requires students to complete the construction of the student information management system. This task belongs to key tasks of the second kind. The students have completed the first task and have certain knowledge and understanding of development of the system, so at this stage students can extrapolate according to the knowledge acquired in the first task and build the basic framework of the student information management system. Because the programs and statements in the first task are provided directly by the teacher, students know little about NET language. Therefore, in this task, we explain the code embedded into the system, such as login with different permission, inquiring of various qualifications, inserting into database operation, and so on. We will explain a part of it when we use it, so as to make the knowledge difficult to understand and learning interesting and practical. These codes not only combine the basic method and classic algorithm of program design but also guide students to discuss about the code, encourage students to provide a variety of solutions and to analyze advantages and disadvantages, and conversely analyze the code in the first task of the universal examination platform on this basis. Thus, students can acquire basic knowledge in the process of programming and expand computational thinking in the process of learning (J.H. & L.J. (2011)).

The third task is the independent task. Students choose their own familiar projects to complete the system creation in small groups. The students have completed two tasks under the guidance of the teacher and basically mastered the method and language of system development; so in this task, students are required to use the method of computational thinking to design, analyze, and program to complete the construction of the system by themselves.

4.3.3 Dynamic monitoring of the teaching process

In the first task, the students are quite familiar with the process of examination; so with the participation of students, the universal examination platform is divided into the login and verification module, examination and marking module, score inquiry module, test database management module, and so on. The students only need to construct every module to complete the construction of the whole system. This stage is mainly imitation. The teacher supervises students' completion. Students may ask the questions that we will talk about in the latter part of this article. At this time, the teacher will encourage students and give a brief description.

Urging students to finish the second task is the key to learning. The second task requires students to complete this task on the basis of the completion and understanding of the first task. Because various algorithms and operations are involved in this task, the teacher should encourage students to design different algorithms and understand the diversity of the algorithms.

In the third task, students independently develop the system in small groups. The teacher mainly adopts the heuristic method in computational thinking to guide students to think more and improves students' ability to use computational thinking.

In the completion of the three tasks, in addition to knowing students' completion in the classroom and providing guidance to the students, we especially open the online classroom and provide a platform for students to ask questions, study by themselves, and discuss. In the process of completing each task, students may directly post a message in this area when they encounter problems. When the students ask and answer questions, the teacher mainly uses the heuristic method to give guidance, provides some information to students, encourages students to use Internet to search for the answer and solve problems through multiple channels, and promotes students to complete the learning tasks. In this process, the teacher monitors students to complete the process with tasks as the core, aiming at cultivating students' ability of computational thinking.

4.4 Student's learning process

4.4.1 Accept the task

The students first preview before class, know about the general teaching content, analyze problems one by one according to the tasks provided by the teacher, and define the tasks. For example, in the first task, students should know the function of each module on the universal examination platform; in the second task, in addition to the analysis of functions of each module, students should know the classical algorithm; in the third task, students should determine the topic by themselves, define the division of labor in each group, and prepare a variety of materials.

4.4.2 Complete the task

Under the teacher's guidance and with the teacher's help, with tasks as the core, students use various methods in computational thinking to accomplish tasks. The first task is mainly imitation. Students can easily complete it under the teacher's guidance. The second task is more difficult than the first one and involves more content. In the process of imitation, students should have their own understanding and innovation. Structure and effectiveness, two important characteristics of computer science, are algorithm diversity and machine realization mechanism in the experiment teaching of program design. Both of them are very effective for the training of computational thinking (J.H. & L.J. (2011)). Computational thinking has a strong ability to innovate. The highest goal of cultivating computational thinking is innovation (Q. & L. (.2011.)). When completing the tasks, students can put forward their different points of view, algorithms, and methods. Both inside and outside the classroom, the teacher can initiate discussion and review, and students also can evaluate each other, which will encourage students to use the method of computational thinking to solve problems in the process of completing the tasks, and it will encourage the students to feel recognized and become more interested in learning.

4.4.3 Communication, reflection, and evaluation

Communication, reflection, and evaluation constitute a key part in the mode. Both the teacher and students need to participate. The teacher will comprehensively evaluate the tasks completed by students in terms of knowledge system, result innovation, ability of computational thinking, and so on; summarize the cooperation of the whole team and the training of computational thinking. Students sum up and internalize the knowledge according to the teacher's comments, deepen their understanding of the knowledge system, and improve their own ability to extrapolate. Students reflect on their own creation process. Students and groups communicate, comment on and help each other, and construct their own knowledge system.

5 CONCLUSION

Seen from the theory of modern study psychology, the development of thinking is an emotional process. Didactic education separated from the actual situation finds it hard to realize the migration of students' thinking ability (F. & J.Q. (2013)). The thinking development level is the key to students' success. The mental teaching has an important influence on students' future. Through the multitask-driven teaching model aiming at developing computational thinking, students become their own masters.

In Visual Basic.NET program design course, we teach through the development tasks of multiple actual projects and solve problems for students who cannot really develop the system after learning the programming language. In the whole teaching process, we use computational thinking approach to both design and present the tasks, teach through heuristic reasoning for the problems that students encounter in the learning process, provide useful information, and encourage students to complete the learning tasks. In the classroom teaching, through the design task, the teacher simulates the whole process of the development of a project, uses the method of separation of concerns (SOC) in computational thinking to change the complex and huge development project into various tasks and to decompose every task into small problems, and uses heuristic reasoning and other methods to help students solve problems effectively. In the link of the simulation project, through constantly solving problems, students make progress in being able to find the problems by themselves, and they consciously use the computational thinking method to solve the problems.

Of course, we have only used a part of the newly emerging and booming computational thinking. There are more methods worth thinking and exploring. We will appreciate the charm of computational thinking that is put into practice.

REFERENCES

Johnson, H.L. 1965. Artistic development in autistic children. *Child Development* 65(1):13–16.

JEANNETTE M.WANG. 2006. Computational Thinking. *Communications of ACM.* 49(3):33–35.

Philips P. (2008). Comptional Thinking : Aproblem_solving tool for every classroom[EB/OL]. *http://www.csta .acm.org/Resources/sub/ResourceFiles/Computational Thinking.pdf.*

Wing JM. 2008. Computational Thinking and Thinking About Computing. *Philosophical Transactions of the Royal Society.* 366:3717–3725.

G.L. Chen & R.S. Dong. 2011. Computational Thinking and Basic Education of University Computer. *China University Teaching.* (1):7–11,32.

Y.Z. Zhu. 2009. Discussion about Computational Thinking-Scientific Orientation, Basic Principle and Innovation Path of Computational Thinking. *Journal of Computer Science.* 36(4):53–55.

D.C. Zhan, L.S. Nie & X.F. Xu. 2014. Computatinal Thinking and Its Inpact on Software Engineering Education. *Software Engineering Education for a Global E-Service Economy Progress in IS 2014*:29–40.

L.Guo & H.W. Ye. 2012, Construction and Application of TPBIM in the Pilot Course of Teaching Reform of Computer Common Course in the Universities of Guangdong Province-Take Web Design and Production Course as an Example. *China Educational Technology.* (12):111–114.

Y.S. Zhang, Y. Gao, J.S. Zou & A.Q. Bao. 2014. Research on Multimedia Teaching and Cultivation of Capacity for Computational Thinking. *Frontier and Future Development of Information Technology in Medicine and Education Lecture Notes in Electrical Engineering* Volume 269:2779–2783.

X. Li, Yanfei.F. Peng & J.G. Sun. 2011. Task-Driven and Cooperative-Working Based Compiler Principle Teaching Reform. *Advances in Computer Science, Environment, Ecoinformatics, and Education Communications in Computer and Information Science* Volume 218:448–451.

C.J. Gan & Y. Zhou. 2012. Basic Computer Teaching and Cultivation of Computational Thinking of Arts Students. *Journal of Computer Education.* (19):20–23.

J.H. Chen & L.J. Dai. 2011. Program Design Experiment Teaching Aiming to Develop Computational Thinking. *Journal of Experimental Technology and Management.* (1):125–127.

Q. Mou & L. Tan. 2011. Research and Progress of Computational Thinking. *Journal of Computer Science.* (3):10–15.

F. Li & J.Q. Wang. 2013. Computational Thinking: A Kind ofIntrinsic Value in the Information Technology Courses. *China Educational Technology.* (8):19–23.

Information, Computer and Application Engineering – Liu, Sung & Yao (Eds)
© 2015 Taylor & Francis Group, London, ISBN 978-1-138-02717-6

Initial alignment of large azimuth misalignment angle in SINS based on MEP-CKF

Hui Li & Jie Wei Ruan & Fang Zhao & Ji Lin
Shenyang University of Technology, Shenyang, Liaoning, China

ABSTRACT: In view of initial alignment of large azimuth misalignment angle in SINS, in order to improve the alignment accuracy and reduce the amount of calculation, a new nonlinear filtering method named MEF-CKF is applied. This algorithm combines error prediction filter model (Model Error Prediction, MEP) and cubature kalman filter (Cubature Kalman Filter, CKF); MEP takes measurement error of the inertial measurement unit as the model error and inserts the value of the model error to correct the state that is predicted in CKF. In the nonlinear model of initial alignment, we compare CKF, MEP EKF, and MEP - CKF in terms of algorithm. The experimental result shows that the MEP-CKF algorithm can effectively solve the problem of nonlinear initial alignment; it performs better not only in filter performance but also in alignment accuracy.

KEYWORDS: SINS; MEP-CKF; Initial alignment; Large azimuth misalignment

1 INTRODUCTION

Initial alignment is one of the key technologies in SINS[1]. In the condition of alignment of large azimuth misalignment angle, calculation errors based on Euler angle[2] and quaternion[3] are usually used to describe the nonlinear error model of SINS in the misalignment angle. EKF algorithm is the earliest method used to process the nonlinear error. EKF uses the first order of Taylor expansion to approximate the nonlinear expression, which will inevitably lead to the model error. In order to overcome the disadvantage of EKF, Unscented kalman filter[4] and central Difference kalman filter[5] are presented by scholars. In 2009, Simon Haykin proposed a new nonlinear method called Cubature Kalman Filter[6] (CKF). CKF does not need to make linear the nonlinear model and it improves the initial alignment accuracy. But CKF still contains model error in strong nonlinear condition, which will lead to low filter accuracy. According to the real-time filtering algorithm for arbitrary nonlinear system model, MP-Model[7] predictive is proposed by Crassidis. It can estimate model error in real time and correct the error effectively.

A new algorithm called MEP-CKF is proposed by combining the model prediction[8] with CKF algorithm. Based on the foundation of derivative large azimuth, we compare MEP-CKF, MEP-EKF with CKF and confirm that MEP-CKF exhibits accuracy improvement for initial alignment.

2 NONLINEAR ERROR MODEL OF LARGE AZIMUTH MISALIGNMENT ANGLE IN SINS

Navigation geographic coordinate system can be divided into East (E), north (N), and day (U) coordinates. The influence of various error sources, deviation between the real geographic coordinate system, and calculating the geographic coordinate system result in the navigation calculation error. Using Euler Angle $\delta\phi = (\phi_e, \phi_n, \phi_u)^T$ for the misalignment angle between the navigation system n and system c, ϕ_e ϕ_n ϕ_u represent the roll angle error, pitch angle error, and yaw angle error. The error transformation matrix from c to n is as follows:

$$C_c^n = \begin{bmatrix} c\phi_e c\phi_u + s\phi_e s\phi_n s\phi_u & c\phi_n s\phi_u & s\phi_e c\phi_u - c\phi_e s\phi_n s\phi_u \\ -c\phi_e s\phi_u + s\phi_e s\phi_n c\phi_u & c\phi_n c\phi_u & -s\phi_e c\phi_u - c\phi_e s\phi_n c\phi_u \\ -c\phi_n s\phi_u & s\phi_n & c\phi_e c\phi_n \end{bmatrix}$$

(1)

In the formula (1), c represents cos and s represents sin.

Transfer equation of attitude error of A matrix can be written as

$$C_c^n = C_c^n \left[\omega_{nc}^c \times \right]$$

(2)

In the formula (2), $[\omega\times]$ is the angular velocity vector of the anti symmetric matrix form,, ω_{nc}^{c} is the rotating angular velocity projection in c system based on n system, and $\omega_{nc}^{c} = C_{b}^{c}\omega_{nc}^{b}$, ω_{nc}^{b} is the rotating angular velocity projection in the carrier coordinates based on n system.

[1]Namely, for SINS rotation angular velocity error $\omega_{nc}^{b} = \omega_{nb}^{b} - \omega_{cb}^{b}$, and ω_{nb}^{b} is the angular velocity of the navigation calculation, ω_{nb}^{b} is the rotating angular velocity projection in the carrier coordinates based on n system

The attitude error matrix relation between n system, c system, and b system is

$$C_{c}^{n} = C_{b}^{n} - C_{b}^{c} = (I - C_{n}^{c})C_{b}^{n} \tag{3}$$

In the formula (3), C_{b}^{c} is the attitude matrix of the navigation calculation system with $C_{b}^{c} = [C_{c}^{b}]^{T}$; C_{b}^{n} is the direction cosine attitude matrix in n system based on b system.

The transfer equation of C_{b}^{n} is

$$\dot{C}_{b}^{n} = C_{b}^{n}[\bar{\omega}_{ib}^{b}\times] - [\bar{\omega}_{in}^{n}\times]C_{b}^{n} \tag{4}$$

ω_{ib}^{b} is the rotating angular velocity projection in the b system based on i system. It is the real output value of Gyro. ω_{in}^{n} is the rotating angular velocity projection in the n system based on i system. $\bar{\omega}_{ib}^{b}$ is the theory output value of Gyro, and $\omega_{ib}^{b} = \bar{\omega}_{ib}^{b} + \delta\omega_{ib}^{b}$, $\delta\omega_{ib}^{b}$ is the real output value of Gyro. It generally consists of constant drift and random drift.

Suppose angular velocity in c referred to n is ω_{nc}^{c}, we can have relationships between the ω_{nc}^{c} and Euler misalignment angle ϕ_e, ϕ_n, ϕ_u::

$$\omega_{nc}^{c} = \begin{bmatrix} -s\phi_n & 0 & c\phi_n \\ 0 & 1 & 0 \\ c\phi_n & 0 & s\phi_n \end{bmatrix} \begin{bmatrix} 1 & 0 & 0 \\ 0 & c\phi_n & s\phi_n \\ 0 & -s\phi_n & c\phi_n \end{bmatrix} \begin{bmatrix} \phi_e \\ 0 \\ 0 \end{bmatrix}$$

$$+ \begin{bmatrix} -s\phi_n & 0 & c\phi_n \\ 0 & 1 & 0 \\ c\phi_n & 0 & s\phi_n \end{bmatrix} \begin{bmatrix} 0 \\ \phi_n \\ 0 \end{bmatrix} + C_{n}^{c} \begin{bmatrix} 0 \\ 0 \\ \phi_u \end{bmatrix} \tag{5}$$

Differential on both sides of formal (3) and substituting formulas (2), (4), and (5), we can get::

$$\omega_{nc}^{c} = C_{nc}\delta\dot{\phi} \tag{6}$$

In the formula (6), $C_{nc} = \begin{bmatrix} -s\phi_n & 0 & -c\phi_n s\phi_e \\ 0 & 1 & 0 \\ c\phi_n & 0 & s\phi_n s\phi_e \end{bmatrix}$

The dynamic transmission equation of attitude error is

$$\delta\dot{\phi} = C_{nc}^{-1}\omega_{nc}^{n}$$

$$= C_{nc}^{-1}\left[(I - C_{nc}^{c})\omega_{in}^{c} + \delta\omega_{in}^{n} - C_{b}^{c}\delta\omega_{ib}^{b}\right] \tag{7}$$

The dynamic transmission equation of speed error is

$$\delta\dot{V} = C_{n}^{c}C_{b}^{n}f^{b} - C_{b}^{n}f^{b} + C_{b}^{n}\delta f^{b}$$

$$- (2\omega_{ie}^{n} + \omega_{en}^{n})\times\delta V^{n}$$

$$- (2\delta\omega_{ie}^{n} + \delta\omega_{en}^{n})\times V^{n} + \delta g^{n} \tag{8}$$

In the formula (8), ω_{ie}^{n} is the projection of earth rotation angular velocity in a geographic coordinate system. f^{b} is the output values of accelerometer, ω_{en}^{n} is the angle velocity projection caused by carrier motion relative to the earth in the navigation system, and g^{n} is a projection of the gravity vector based on the navigation system.

δf^{b} is measurement error of the accelerometer, which is componented by constant drift and random error, namely, $\delta f^{b} = \nabla + w_{a}$, δg^{n} is gravity measurement error.

Formula (8) shows that the dynamic transmission error rate of the model is nonlinear, and under quiescent conditions with speed at zero, the equation (8) can be written as follows:

$$\delta\dot{V} = C_{n}^{c}C_{b}^{n}f^{b} - C_{b}^{n}f^{b} + C_{b}^{n}\delta f^{b}$$

$$- (2\omega_{ie}^{n} + \omega_{en}^{n})\times\delta V^{n} + \delta g^{n} \tag{9}$$

3 ALGORITHM OF INITIAL ALIGNMENT BASED ON MEPF-CKF

3.1 Basic algorithm of MEPF filter

Error prediction filter (MEP) is a kind of filtering method based on nonlinear system model, and its principle of work is to estimate model error in the system by tracking forecast output and the measured output, so as to continuously revise the dynamic model.

Assuming a nonlinear system as follows:

$$\dot{x}(t) = f(x(t),t) + G(x(t),t)\,d(t) \tag{10}$$

$$z(t) = H(x(t),t) + v(t) \tag{11}$$

146

In the formula (11): $f \in R^n$ is nonlinear function that is continuously differentiable; $x(t) \in R^n$ is state variable; G is error vector in the model; $G \in R^{n \times q}$ is error distribution matrix in the model; H is measurement vector; and $v(t)$ is measurement noise vector, and assuming the Gaussian white noise with zero mean and covariance is $E\{v(t)v^T(t)\} = R$.

We establish a state estimation equation and measurement equation as follows:

$d(t) \in R^q$

$$\hat{x}(t) = f(\hat{x}(t),t) + G(\hat{x}(t),t)\,\hat{d}(t) \qquad (12)$$

$$\hat{z}(t) = H(\hat{x}(t),t) \qquad (13)$$

We differentiate the output of measurement equation (13), forecast consecutively, place the $\hat{x}$ of state (12) at the equality right end, take a small time interval Δt, expand the i component that is in $\hat{z}(t + \Delta t)$ to Taylor series of P_i order, and ignore the higher-order terms, according to Lee derivative definitions:

$\hat{z}_i(t + \Delta t)$

$$\approx \hat{z}_i(t) + \Delta t L_f^1(h_i) + \frac{\Delta t^2}{2!}L_f^2(h_i) + \cdots \qquad (14)$$

$$+ \frac{\Delta t^{p_i}}{p_i!}L_f^{p_i}(h_i) + \frac{\Delta t^{p_i}}{p_i!}\frac{\partial L_f^{p_i-1}(h_i)}{\partial \hat{x}}G(\hat{x}(t),t)d(t)$$

In the formula (14), h_i is the i turn of component in the $H(\hat{x}(t),t)$, p_i is the component in $d(t)$, which is the lowest order in h_i that first appeared, $i = 1,2,...,m$. $L_f^k(h_i)$ is Lee derivative definitions with k order:

$$\begin{aligned} L_f^0(h_i) &= h_i & k = 0 \\ L_f^k(h_i) &= \frac{\partial L_f^{k-1}(h_i)}{\partial \hat{x}}f(\hat{x}(t),t) & k \geq 1 \end{aligned} \qquad (15)$$

Substituting formula (14) in the matrix form, we can get formula

$$\hat{z}_i(t+\Delta t) \approx \hat{z}_i(t) + S(\hat{x}(t),t) + \Lambda(\Delta t)U(\hat{x}(t))d(t) \quad (16)$$

In the formula (16), $\Lambda(\Delta t) \in R^{m \times m}$ is diagonal matrix and the diagonal elements are

$$\lambda_{ii} = \Delta t^{p_i}/p_i! \qquad i = 1,2,\cdots,m \qquad (17)$$

$U(\hat{x}(t)) \in R^{m \times q}$ is sensitivity matrix, and it can be expressed as

$$U(\hat{x}(t)) = \begin{bmatrix} L_{g_1}L_f^{p_1-1}(h_1) & \cdots & L_{g_q}L_f^{p_1-1}(h_1) \\ L_{g_1}L_f^{p_2-1}(h_2) & \cdots & L_{g_q}L_f^{p_2-1}(h_2) \\ \vdots & \vdots & \vdots \\ L_{g_1}L_f^{p_m-1}(h_m) & \cdots & L_{g_q}L_f^{p_m-1}(h_m) \end{bmatrix} \qquad (18)$$

In the formula (18): g_j is the j row in G, $j=1, 2, ..., q$. $L_{g_j}L_f^k(h_i)$ is the Lee derivative definition with one order in $L_f^k(h_i)$ function about $g_j(\hat{x}(t), t)$, and it can be expressed as

$$L_{g_j}L_f^k(h_i) = \frac{\partial L_f^k(h_i)}{\partial \hat{x}}g_j(\hat{x}(t),t) \qquad (19)$$

$S(\hat{x}(t), \Delta t)$ is q dimension column vector, and each component is

$$S(\hat{x}(t),\Delta t) = \sum_{k=1}^{p_i}\frac{\Delta t^k}{k!}L_f^k(h_i) \quad i = 1,2,\cdots q \quad (20)$$

Assuming that the small time interval is constant, then, $z(t + \Delta t) = z_{k+1}$ and $z(t) = z_k$. According to the working principle of predictive filter, we can define the performance index function $J(d_k)$. It consists of a weighted sum of squares that is the difference between the measuring output and the predicted output of the weighted sum of squares with the model error, which is corrected as follows:

$$J(d_k) = \frac{1}{2}[z_{k+1} - \hat{z}_{k+1}]^T \cdot R^{-1}[z_{k+1} - \hat{z}_{k+1}] + \frac{1}{2}d_k^T W d_k \quad (21)$$

In the formula (21), $W \in R^{q \times q}$ is weighed matrix of model error.

In order to make the J minimum, it needs to meet the condition that if $\partial J/\partial d_k = 0$, we can get error estimation modal between the time interval $[t_k, t_{k-1}]$:

$$\hat{d}_k = -M_1 \cdot M_2[S(\hat{x},\Delta t) - z_{k+1} + \hat{z}_k] \quad (22)$$

In the formula (22)

$$M_1 = \{[\Lambda(\Delta t)U(\hat{x}_k)]^T R^{-1}[\Lambda(\Delta t)U(\hat{x}_k)] + W\}^{-1},$$

$$M_2 = [\Lambda(\Delta t)U(\hat{x}_k)]^T R^{-1}$$

We can conclude the filtering process as follows: According to the formulas (13) and (22), we can calculate the predicted output vector and the tracking measurements by using the state estimation $\hat{x}_k$ at the time t_k; then, we can get error estimation modal $\hat{d}_k$ between the time interval $[t_k, t_{k-1}]$; next, according to the formula (12), we can get the state of estimation $\hat{x}_{k+1}$ at the time t_{k+1}.

3.2 Algorithm based on CKF

The algorithm based on CKF uses the spherical-radial cubature rule to select the 2n point with the same weight to approximate the integral value:

$$I(f) = \sum_{i=1}^{2n} w_i f(\xi_i) \quad (i=1..2n) \tag{23}$$

$$\xi_i = \sqrt{2n/2}[1]_i \quad \omega_i = 1/2n \tag{24}$$

In the formulas (23) and (24), n is the state of the system dimension, $i=1, 2, \ldots, 2n$ $[1]_i$ is the i row $[1]$ in the set of points that is completely symmetrical; we can calculate Cubature (ω_i, ξ_i) by formula (24):

Assuming the posterior density function is $p(x_{k-1}) = N(x_{k-1/k-1}, p_{k-1/k-1})$, at the time k-1, the CKF algorithm is expressed as follows:

1 Time Update
①The calculation of Cubature;

$$x_{i,k-1} = \sqrt{P_{k-1}}\xi_i + \hat{x}_{k-1} \tag{25}$$

②Spreading Cubature by the state equation

$$x_{i,k|k-1}^* = f\left(x_{i,k-1}\right) \tag{26}$$

③ Estimate the predicted value of the state at time k

$$\hat{x}_{k|k-1} = \frac{1}{2n}\sum_{i=1}^{2n} x_{i,k|k-1}^* \tag{27}$$

④Estimate the state error covariance that is predicted at time k

$$\hat{P}_{k|k-1} = \frac{1}{2n}\sum_{i=1}^{2n} x_{i,k|k-1}^* x_{i,k|k-1}^{*T} - \hat{x}_{k|k-1}\hat{x}_{k|k-1}^T + Q_{k-1} \tag{28}$$

2 Measurement Update
①The calculation of Cubature

$$x_{i,k} = \sqrt{P_k}\xi_i + \hat{x}_{k,k-1} \tag{29}$$

②Spreading Cubature by the measurement equation

$$z_{i,k} = h(x_{i,k}) \tag{30}$$

③Estimate the predictive value, innovation variance, and covariance at time k

$$\hat{z}_k = \frac{1}{2n}\sum_{i=1}^{2n} z_{i,k} \tag{31}$$

$$P_{xz,k} = \sum_{i=1}^{2n} \omega_i x_{i,k} x_{i,k}^T - \hat{x}_{k-1}\hat{z}_k^T \tag{32}$$

$$P_{zz,k} = \sum_{i=1}^{2n} \omega_i z_{i,k} z_{i,k}^T - \hat{z}_k \hat{z}_k^T + R_k \tag{33}$$

④Estimate the Kalman state and the state covariance at time k

$$W_k = P_{xz,k} P_{zz,k}^{-1} \tag{34}$$

$$\hat{x}_k = \hat{x}_{k|k-1} + W_k (z_k - \hat{z}_k) \tag{35}$$

$$\hat{P}_k = \hat{P}_{k|k-1} - W_k P_{zz,k} W_k^T \tag{36}$$

3.3 Algorithm of initial alignment based on MEPF-CKF

Substituting attitude, speed as the state variable of the system, we can get the nonlinear system state equation as follows:

$$\dot{x}(t) = f(x(t)) + G(t)d(t) + M(t)w(t) \tag{37}$$

In the formula (37), $x = [\phi_e\ \phi_n\ \phi_u\ \delta v_e\ \delta v_n\ \delta v_u]^T$ is the state vector; is the distribution matrix; and is input matrix of the process noise; $d = [\varepsilon_x, \varepsilon_y, \varepsilon_z, \nabla_x, \nabla_y, \nabla_z]$, which is model error vector with the accelerometer bias $\nabla = (\nabla_x, \nabla_y, \nabla_z)^T$ and gyro drift ; $\varepsilon = (\varepsilon_x, \varepsilon_y, \varepsilon_z)^T$ is noise vector of the system, including constant accelerometer, gyro drift noise, and driving noise of random walk.

In the measurement equation (23), $z = [\delta v_e, \delta v_n, \delta v_u]$ is the observation vector, $\delta v_e, \delta v_n, \delta v_u$ is the difference between SINS algorithm speed with mileage wheel speed, and H is a measurement matrix as follows:

$$H = [0.(-V_l^n \times)] \tag{38}$$

In the formula (38), $V_l^n \times$ is the speed of the product for SINS

The algorithm of MEP-CKF is as follows:

① Substituting the system state value $\hat{x}_k$ at time t_k into the formulas (17), (18), and (19), we can get the MEP model error parameter of, and $\Lambda(\Delta t)$, $U(\hat{x}_k)$, $S(\hat{x}_k, \Delta t)$

② Then, substitute $\hat{x}_k$ into the formula (13) and obtain the predicted output $\hat{z}_k$

$$\hat{z}_k = H\hat{x}_k \tag{39}$$

148

③ Substitute the observed values z_{k+1} at time t_{k+1} and the parameters obtained in step ①② into equation (22); then, we get the model error prediction value

④ Use $\hat{d}_k$ to modify the Cubature at time t_k in the CKF.

$$x_{i,k+1|k}^{*} = f(x_{i,k+1|k}, \hat{d}_k) \quad (40)$$

⑤ Modify the state value of CKF

$$\hat{x}_{k+1|k} = \frac{1}{2n} \sum_{i=1}^{2n} f(x_{i,k+1|k}^{*}, \hat{d}_k) \quad (41)$$

⑥ Predict variance at the time k.

$$\hat{P}_{k+1|k} = \frac{1}{2n} \sum_{i=1}^{2n} x_{i,k+1|k}^{*} x_{i,k+1|k}^{*T} - \hat{x}_{k+1|k} \hat{x}_{k+1|k}^{T} + Q_k \quad (42)$$

According to formulas (28) to (36), we should keep updating measuring to obtain the state estimation value $\hat{x}_{k+1}$ and the variance $\hat{P}_{k+1}$ in a timely manner.

4 EXPERIMENT AND RESULT

4.4 Experimental conditions

At the start of the experiment, the latitude is 41.7377°, longitude is 123.2447°, and height is 0m. Assuming the Gyro constant drift is 0.01°/h, random drift is 0.05°/h, the constant drift of the accelerometer is 1e-4*g, the random drift is 5e-5*g, and the error is 10°.

Assuming the initial error angles ϕ_e ϕ_n ϕ_u of the model is 1°,1°,10°, the initial system state X is 0, the state of motion is as follows: 0~20s, the motion is uniformly accelerated; then, it is static for 20s; 40–50s, the motion is uniformly accelerated again; 50~300s, it maintains uniform motion and is static.

4.5 Analysis of experiment

Based on Figures 1 to 3, we can conclude that in the roll angle, pitch angle, and yaw angle, the MEP-CKF's convergence speed is optimal, MEP-EKF times and CKF is slower. In the aspect of numerical stability, MEP-CKF exhibits better performance than CKF.

As can be seen from Table 1, in terms of roll angle, the estimation accuracy of MEP-CKF is better than CKF, MEP-EKF less. Due to the initial alignment of large azimuth misalignment angle, the estimation accuracy of CKF is better than EKF[9]. Compared with CKF, MEP-EKF still has a lower accuracy, though MEP-EKF algorithm has a higher accuracy than EKF algorithm. In terms of pitch angle, value

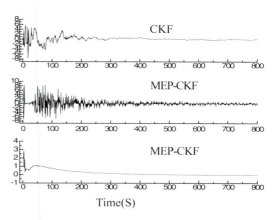

Figure 1.   Roll alignment comparison.

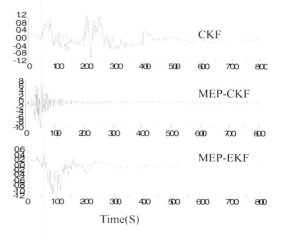

Figure 2.   Pitch alignment comparison.

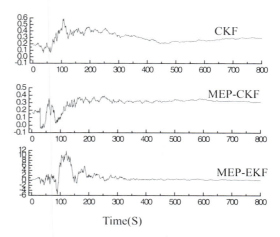

Figure 3.   Yaw alignment comparison.

149

Table 1. Estimation and average comparison of misalignment angle.

	CKF	MEP-CKF	MEP-EKF
Roll angle/ (rad)	-0.005198	-0.012836	-0.084139
Pitch angle/ (rad)	-0.089598	-0.022713	-0.024183
Yaw angle/ (rad)	0.705075	0.692192	0.700603

that is calculated by MEP-EKF and MEP-CKF is close and compared with MEP-EKF and MEP-CKF, the deviation difference of CKF is about 0.06 radian. In terms of yaw angle, MEP-CKF, MEP-EKF, and CKF have a similar estimation accuracy.

5 CONCLUSIONS

This article combines the MEP algorithm with CKF, and it reaches a nonlinear filter algorithm MEP-CKF based on the foundation of CKF. The experiment result shows that MEP-CKF and MEP-EKF have a similar estimation accuracy, but in the aspect of convergence speed and numerical stability, MEP-CKF algorithm shows higher performance. Compared with CKF, MEP-CKF exhibits much better performance than CKF in the aspect of estimation accuracy, convergence speed, and numerical stability; it confirms that MEP-CKF is a better nonlinear filter than CKF.

REFERENCES

[1] Liu G H, LI Q X, Shi W, et al. Application of dynamic Kalman filtering instate estimate of navigation test[J]. Chinese Journal of Scientific Instrument, 2009, 30(2); 396–400.
[2] Tan Hong-li, Huang Xin-sheng, Yue Dong-xue. Rapid transfer alignment based on large misaugnment SINS error model[J]. Journal of National University of Defense Technology. 2008, 30 (6):19–23.
[3] Wang J, Guo X S, ZHou ZH F. Establishmengt of errors model for SINS on a stationary base with large azimuth misalignment angle based on quaternion[J]. Piezoelectrics Acoustooptics, 2014, 10, 36(5).
[4] Xia Jia He, Qin Yong Yuan, Zhao Chang Shan. Study on nonlinear alignment method for low precision INS[J]. Chinese Journal of Scientific Instrument. 2009, 30 (8):1618–1622.
[5] Norgaard M. Poulsen N K. Ravn O. eral. New developments in state estimation for nonlinear systems[J]. Automatica, 2000. 36(11); 1627–1638.
[6] Arasaratnam I, Haykin S. Cubature kalman filter [J]. IEEE Transactions on automatic control. 2009, 54 (6):1254–1269.
[7] Crassidis J L, Markley F L. Predictive filtering for nonlinear systems[J]. Journal of Guidance, Control and Dynamics. 1997, 20(3):566–572.
[8] ZHang Hong-mei, Deng Zheng-long, Lin Yu-rong. UKF method based on model Error Prediction[J]. ACTA AERONAUTICA ET ASTRONAUT ICA SINICA. 2004, 25(6):598–601.
[9] Sun Feng,Tang LiJun.Initial alignment of large azimuth misalignment angle in SINS based on CKF[J]. Chinese Journal of Scientific Instrument. 2012, 33 (2):327–333.

Information, Computer and Application Engineering – Liu, Sung & Yao (Eds)
© 2015 Taylor & Francis Group, London, ISBN 978-1-138-02717-6

Modeling on bus production planning system based on assembly to order

R.J. Wu

Henan Information and Statistics Vocational College, Zhengzhou, Henan, China

ABSTRACT: A model and design associated with bus production is disclosed. The model, which is related to the bus production planning system, includes constructing a three-level system model of the system based on ATO, namely plant level, shop-floor level, and shift-group level. Based on the model, the functions of the system are designed and implemented under the Visual Studio 2008 integrated development environment. The system's validity thus lies in a bus manufacturing plant.

KEYWORDS: Assembly to order; Production planning; Modeling

1 INTRODUCTION

For customers, Bus or Coach is an important asset. As a tool, the customers depend on it to run the business, so they have different demands of the vehicle equipment. Bus/coach production includes component processing, purchasing, and assembling in accordance with the requirement of the order. This is known as the Assembly to Order (ATO) production.

From the supply chain point of view, ATO is a production method by which the firms produce customized products as per clients' request using the components in the inventory. Through the standardized and modular design of products, they get several series of products. According to the market forecast, they produce or purchase components in advance and provide various options. The assembly of the production is driven by the specific sale, so it is a multi-product and small-batch production, and it is mixed and flexible. In the enterprise information system, the ERP system helps firms deal with the material requirements planning, by which they can undertake the production planning and purchasing planning of the components. But in the production planning field, the ERP system cannot meet the requirements of the workshop production planning and production control because of the restrictions of the enterprise production capacity, production batch, scheduling rules, and scheduling algorithm. Wang, Z.Q. et al. built a production planning model for the enterprises in the supply chain alliance aiming at the profit maximization for controlling the storage cost, ordering cost, and shortage cost of components. Reference shows a production management system that generates assembly planning and production planning according to the orders and the BOM. Xiao, P. et al. built a flexible production planning model that is oriented

to capacity to minimize the production costs and inventory costs. These models are used to solve the problems in the collaborative production planning of the firms that are in the supply chain or the internal master production schedule, but the execution of the planning is seldom involved.

Bus production belongs to the discrete manufacturing, the production cycle of vehicle assembly process is long, and the production process control is complex. In ATO mode, it is the problem to be currently solved in the assembly process and how to arrange production schedule, vehicle online order, stage production planning, and detailed work plan to conduct the workshop material transportation and workshop production coordination, and then to arrange production equipment and work shift for operators in different workshops. Facing the complex manufacturing process, the capacity balance and process equilibrium should be considered. In different working environments, the production planning should be arranged in the concept of multi-level design to get three-level production planning, namely plant level, shop-floor level, and shift-group level.

2 MODELING ON BUS PRODUCTION SCHEDULING SYSTEM

2.1 *Plant-level production planning: order rollout plan*

In the ATO operation mode, order is the source of production activities of the enterprise. The production order process is as follows. communicating with the customers, creating tentative orders, transforming them into firm orders after auditing, making material requirements planning according to the firm orders, producing and assembling, finished goods check-in, and delivering and settling accounts. The whole process is shown in figure 1.

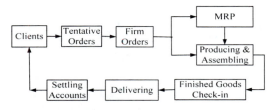

Figure 1.   Production order process model.

In the process of assembly, it is a matter of cardinal significance to develop an Order Rollout Plan according to the model, amount, priority, and delivery of the order to meet a client's requirement under the constraints of production capacity, production situation, and scheduling rule. Order Rollout Plan includes confirming production line, production sequence, and production cycle, and it is the basis of the production preparation.

Production capacity is affected by the amount of assembly firm, the quantity of production line, the capacity of production line, and tact time.

Production situation includes the production planning of each line in assembly firms, the schedule of each order, equipment condition, the status of inventory, and customer order.

Scheduling rule involves the model priority, order priority, production continuity, delivery date, and rules for the order split.

2.2 Shop-floor-level production planning: phased production plan

The bus consists of a chassis and a body that are welded together. Chassis production includes three stages, namely production of work in process (chassis), frame welding, and chassis assembly. Body production has five stages: production of work in process (body), body welding, body painting, vehicle assembly, and quality inspection. The bus assembly stage model is shown in figure 2.

In order to coordinate the works in plant-level scheduling and shop-floor-level fulfilling, we should develop the Phased Production Plan based on the Order Rollout Plan. The chassis production plan is assigned according to the body production plan.

2.3 Shift-group-level production planning: detailed operation plan

There are huge differences in organization and production environment among different shop floors. For example, production of work in process consists of job shop, frame welding, and body welding; chassis assembly and vehicle assembly are flow shops; body painting is a flexible flow shop; and quality inspection is an open shop. In this article, we take body painting as an example.

2.3.1 Processes
Body painting includes ten processes: surface treatment, under coating, Poly-Putty Base, PVC coating, shrinkage hole repairing, finish coating, varnishing, finishing touches, and so on. Each process has 2–5 sub-processes, so there are 30 sub-processes in body painting.

2.3.2 Rated time
Rated times are different among different processes. For the production smoothing, we should arrange shifts to keep the processes stay on the same beat.

2.3.3 Characteristics
Different processes have different characteristics. The capacity of pretreatment is concerned with the amount of shifts for assembly. The capacities of under coating and under coat baking are restricted by the amount of painting room and drying room. The body painting is a flexible flow shop shown in figure 3.

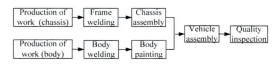

Figure 2.   Bus assembly stage model.

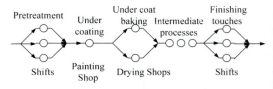

Figure 3.   Body painting model.

2.3.4 *Constraints*

Constraints include delivery of welding process, requirement of assembly process, priority of order, position purpose, and so on.

2.4 *Three-hierarchical structure model of production planning*

In general, based on the requirement of bus production process controlling, we built a hierarchy structure model of production planning system shown in figure 4. In this model, we developed a production plan according to the vehicle model, amount, and delivery date of the firm order.

3 SYSTEM'S DESIGN AND IMPLEMENTATION

3.1 *System's function design*

There are nine functions in the Production Planning System: Organization Management, Calendar Management, Process Management, Constraint Management, Order management, On-Site Data Management, Production Plan fulfilling, Plan Releasing, and Rights Management, as shown in figure 5.

Organization Management: Plant Definition, Shop Floor Definition, Production Line Definition, and Shift Group Definition.

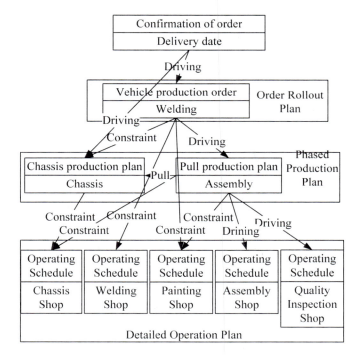

Figure 4.   Hierarchical structure model of production planning system.

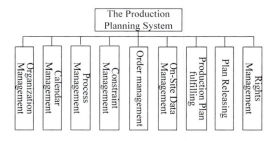

Figure 5.   Function structure in the production planning system.

Calendar Management: Plant timetable Management, Shop Floor Timetable Management, and Shift Group Timetable Management.

Process Management: Stage Management, Working Hours Management.

Constraint Management: Order Rollout Plan Constraint, Phased Production Plan Constraint, and Detailed Operation Plan Constraint.

Order Management: Order Data Capture, Order Data Check (include the information for chassis, body, and frame).

On-Site Data Management: On-Site Data Capture, On-Site Data Check.

Production Plan fulfilling: Order Rollout Plan Check, Order Rollout Plan Confirmation, Order Rollout Plan Fulfilling, Phased Production Plan Check, Phased Production Plan Confirmation, Phased Production Plan Fulfilling, Detailed Operation Plan Check, Detailed Operation Plan Confirmation, and Detailed Operation Plan Fulfilling.

Plan Releasing: Order Rollout Plan Inquiry, Phased Production Plan Inquiry, and Detailed Operation Plan Inquiry.

Rights Management: Role Management, User Management.

3.2 *Process of the production planning system*

the process of the Production Planning System is divided into three sections, static data management, dynamic data management, and Plan Releasing. The section of static data includes organization management, calendar management, process management, and constraint management; the scheduling algorithm and scheduling objectives are determined by this section. The section of dynamic data includes order management, on-site data management, and production plan fulfilling, and they are the operands of the system. Plan Releasing has different levels, such as plant level, shop-floor level, and shift-group level, which will make the whole production process orderly and coordinated. The process of the production planning system is shown in figure 6.

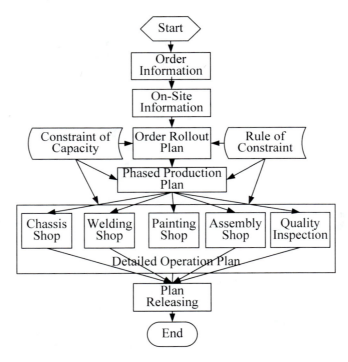

Figure 6.   Process of the production planning system.

3.3 The system's implementation

The system's implementation is based on the Microsoft.NET Framework3.0, under the Visual Studio 2008 integrated development environment. It is a combination structure of B/S and C/S. C/S structure is used for the algorithm implementation and data maintenance, and B/S structure is used for the release and inquiry of the results. The system is developed in object-oriented programming, which benefits the upgrading and maintenance of the system.

The data of organization management, calendar management, process management, and constraint management are stored in SQL Server 2005 databases. Order information and on-site information are from the databases for ERP and MES database, from which they are Extracted, Transformed, and Loaded (ETL) into the database of the production planning system.

4 SYSTEMS IMPLEMENTATION

This system was used in a bus manufacturing plant, deployed in 2 welding shops, 2 painting shops, 2 assembly shops, and 1 quality inspection shop. It enhanced the shop floor controlling and improved the production efficiency.

REFERENCES

Chen, X.W. 2009. Network Order Management System for Automobile Assembly-to-order Manufacturing Enterprises. Computer Knowledge and Technology 18: 1532–1535.

Cui, S.J. 2008. Development Trend of Bus Painting. Modern Paint & Finishing 11(1):27–36.

Liang, L. et al. 2007. Integrated Decision Model for AOT Supply Chain Design with Coordination of Strategic and Operational Planning. Journal of Systems & Management 16(2):144–149.

Liu, M.& Wu, C. 2008. Intelligent Optimization Algorithm and the Application of Manufacturing Process. Beijing: National Defence Industry Press.

Li, Z. 2005. Development of the ATO Production Management System. Manufacture Information Engineering of China 34(9):88–92.

Lu, Y.J. et al. 2008. Research on configuration design of assemble-to-order product. Me-chemical & Electrical engineering Magazine 25(5):24–28.

Pinedo, M. 2005. Scheduling: Theory, Algorithms, and System. Beijing: Tsinghua University Press.

Wen, F. & Liu, P.L. 2009. Strengthen the Bus Production Process Quality Control. Bus Technology and Research 4:59–61.

Xiao, P. et al. 2007. Research on an Aggregate Production Planning model Objection to Volume Flexibility. Mechanical Engineering & Automation 6(3):79–82.

Yan, B.C. 2008. Coating Process of Passenger Car Body. Electroplating & Finishing 27(2):52–54.

Information, Computer and Application Engineering – Liu, Sung & Yao (Eds)
© 2015 Taylor & Francis Group, London, ISBN 978-1-138-02717-6

Light scattering characteristic from particles located on a surface

L. Gong & H. L. Hou
School of Photoelectric Engineering, Xi'an Technological University, Xi'an, China
Science and Technology on Electromechanical Dynamic Control Laboratory, Xi'an, China

J. L. Zou
Science and Technology on Electromechanical Dynamic Control Laboratory, Xi'an, China

ABSTRACT: Based on the Finite Difference Time Domain (FDTD) method, the Generalized Perfectly Matched Lays (GPML) can work very well against the half-space problem about the optical surface and defect particles. The connect boundary condition is reduced by the three-wave method. The reciprocity theorem is applied to the near-far field extrapolation. The angle distributions of double particles with different ratios are shown. Some selected calculations on the effects of the shape are described.

KEYWORDS: scattering characteristic; FDTD method; particle; surface

1 INTRODUCTION

Light scattering is a powerful tool in optical surface quality control. The finite-difference time-domain (FDTD) technique has proved to be a simple yet powerful tool for solving problems related to radio-wave propagation, light scattering, and electromagnetic scattering[1].Investigation of light scattered by optical surface by means of the computer simulation seems to be a reliable tool for investigation of the properties and functional abilities of the optical system by defect particles with the optical surface[2].

It helps reduce manufacturing and experimental costs. The scattering sources particles above the surface give rise to polarizations and scattering intensity that differ from those predicted by microroughness. It can help derive some information about the particle such as sizes, shapes, and so on[3,4]. With this goal in mind, a number of theories have been developed to predict the scattering by particles on the surfaces[5-7]. Validation of those theories, however, have been carried out in a limited number of cases; for these reasons, scholars from our group studied with the Finite Difference Time Domain(FDTD) method[8,9]. Because the FDTD is applied directly to Maxwell equations, there has been a strong trend in the past few decades. FDTD is a simple technique, since it does not require profound knowledge of Maxwell theory. It is based on simple mathematical operations, which can be handled even by very simple computers for nanoparticles.

This article is arranged as follows. In Section 2, the theoretical treatment is given for the determination of the scattering problem of particles and surface. In Section 3, we present numerical results about spheres and optical surface with different conditions. In the last section, the conclusions are given.

2 HALF SPACE FDTD METHOD

In the optical system manufacture industry, many steps have been experienced, such as deposit, polish, and so on. During the procedures, multi-body defect particles at different positions may be leaded in. The multi-body defect particles are regarded as spheres in this article due to their limited length (see Figure 1.).

Figure 1. Schematic view of defect particle on surface.

There are some half-space problems in optics and electromagnatics about the upon and down spaces with two different mediums. In the half-space problem of the optical system manufacture, the incident wave will produce reflectance and transmission wave. The incident field and reflectance field is in the upon space, whereas the transmission filed is in the down space. The method described earlier is called the "three-wave method."

Figure 2 shows the measuring geometry, the schematic of half-space scattering zone, and the field about s polarization light used in this article. Plane wave polarized light of wavelength λ irradiates the surface and the particles at an incident angle of θ_i and an azimuth angle of ϕ. The incident field can be written as $\overline{E}_{inc}$ or $\overline{H}_{inc}$, the reflectance field as $\overline{E}_{ref}$ or $\overline{H}_{ref}$, the transmission field as $\overline{E}_t$ or $\overline{H}_t$, and the scattering electric field as $\overline{E}_s$ or $\overline{H}_s$. The upon-surface field can be obtained as $\overline{E} = \overline{E}_{inc} + \overline{E}_{ref} + \overline{E}_s$ and $\overline{H} = \overline{H}_{inc} + \overline{H}_{ref} + \overline{H}_s$.

The down-surface field can be obtained as $\overline{E} = \overline{E}_t + \overline{E}_s$ and $\overline{H} = \overline{H}_t + \overline{H}_s$.

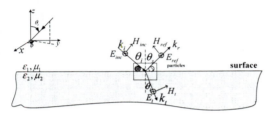

Figure 2. Measuring geometry and the schematic of half-space scattering zone and the field about s polarization light.

Because the simulated space is limited to FDTD of half space, the perfectly matched layer(PML) needs to locate the outside of the simulated zone in order to simulate the infinite space. This article describes a generalized perfectly matched layer (GPML) that extends the original PML to absorb both the propagating and evanescent waves in lossless and lossy media. In the GPML area, the split field E_{xy} and H_{xy} satisfies the following equations:

$$\varepsilon \frac{\partial E_{xy}}{\partial t} + \left(\sigma + \sigma_y\right) E_{xy} + \frac{\sigma\sigma_y}{\varepsilon_r} E_{xy(1)} = \frac{\partial H_z}{s_{y0}(y)\partial y} \quad (1)$$

$$\mu \frac{\partial H_{xy}}{\partial t} + \left(\sigma^* + \sigma_y^*\right) H_{xy} + \frac{\sigma^*\sigma_y^*}{\mu_r} H_{xy(1)} = -\frac{\partial E_z}{s_{y0}(y)\partial y} \quad (2)$$

where $\left(\varepsilon, \mu, \sigma, \sigma^*\right)$ are the electromagnetic parameters. The equation is simplified to nonmagnetic material as

$$\mu_0 \frac{\partial H_{xy}}{\partial t} + \sigma_y^* H_{xy} = -\frac{\partial E_z}{s_{y0}(y)\partial y} \quad (3)$$

where other parameters are defined as

$$\begin{cases} s_{y0}(y) = 1 + s_m \left(\frac{y}{\delta}\right)^2 \\ \sigma_y(y) = \sigma_m \sin^2\left(\frac{\pi y}{2\delta}\right) \\ \sigma_y^*(y) = \frac{\mu_0}{\varepsilon}\sigma_y(y) \end{cases} \quad (4)$$

where δ is the thickness of GPML; R_{th} is the theoretical reflectance ratio with vertical incidence wave. The s_m and σ_m are selected as follows:

$$\frac{\lambda}{(1+s_m)} > (2 \sim 3)\Delta y \quad (5)$$

$$\sigma_m = -\frac{\varepsilon c/\delta}{1+s_m\left(1/3 + 2/\pi^2\right)} \ln R_{th} \quad (6)$$

R_{th} is the reflection coefficient of GPML with the vertical incidence wave. From the theoretical standard, the smaller the numerical value, the better the situations. However, the value is too small to lead to an increase of the reflection coefficient. It is better to select R_{th} between 10^{-8} and 10^{-10}.

In order to get the differential scattering cross-section (DSCS), the numerical data of the near field need to be extrapolated to the far field. For the half-space problem, the reflectance, transmission, and refraction procedures of wafer are so complex that the reciprocity theorem is applied to near-far field extrapolation to solve the half-space Green function.

Because the distance of two points is infinitely far, the far field extrapolation is defined as

$$E_\theta(J,\hat{q},r_a;r_\infty) = E_q(J,\hat{\theta},r_\infty;r_a)$$
$$q = x,y,z \quad (7)$$

The equation means the radiated electric field $E_\theta(J,\hat{q},r_a;r_\infty)$ of current J with q direction that is located at a equal to the radiated electric field $E_p(J,\theta,r_\infty;r_a)$ of current J with θ direction that is located at ∞.

3 NUMERICAL RESULTS AND DISCUSSION

To validate the theory, some results are compared with the Method of Moment (MOM) first, both of which are very well identical. It proves the reliability of the method in this article. The wavelength of the incident wave through this article is 0.633 μm, and the size of the particle is about r = 0.4 μm with the incident angle of 0°. The reflective index is n_{K9} = 1.52, n_{cu} = 0.25 + 3.41i, respectively.

The schematic views of particles are described in in detail in Figure 4, which shows one's radius is 0.4 μm; whereas the radius of the others is 0.2 μm, 0.4μm, 0.6 μm, and 0.8 μm, respectively.

It can be observed from Figure 4 that (1) the radius of spheres affects the scattering intensity in the lateral direction (0°–30°) and the larger the radius r_2 is, the stronger the interaction in the lateral direction (0°–30°) is. (2) The larger the radius r_2, the more are the extremum points in the lateral direction (30°–90°).

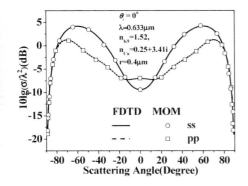

Figure 3. Results of this article compared with MOM method.

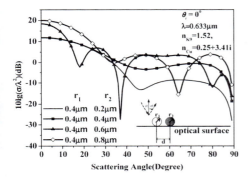

Figure 4. Angular distributions of scattering intensity of spheres for different radius.

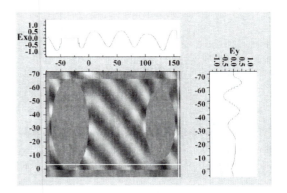

Figure 5a. Electric field distribution of double sphere particles on a surface.

Figure 5b. Electric field image of double sphere particles on a surface with 1:1 ratio.

In the project, the particle distribution can be received by the analysis of the angular distributions of scattering intensity. (3) The angular distribution is symmetry to 0° for the incident angle that is 0°. Then, the scattering angle scale is (0°–90°).

The effects of the sphere radius are described by some selected calculations. And the electric field distribution of double sphere particle on a surface will be described in Figure 5a and Figure 5b.

From Figure 5, we can see that the electric field inside the sphere is zero, because the material of the sphere is Cu. The profiles show that the field with a difference value is more obvious with p polarization incident light. The electric field distribution shape of sphere is not the sphere, because the ratio of Figure 5a is not 1:1. The 1:1 ratio case is given in Figure 5b.

To sum up, with the development of super-large-scale integrated circuits, the calculation of light scattering is widely used in the semiconductor industry for the detection of wafers and equipment quality. One of the challenges is to identify the information about the contaminants above the wafers. The scattering properties provide an effective measurement of the location and the shape in the semiconductor industry for surface qualities.

4 CONCLUSIONS

In summary, the multiple scattering fields are derived based on an improved analysis of the half-space problem of surfaces with particles located on a surface. The accuracy of the theory is verified by comparing the numerical results of some special cases of a particle located on a surface with those given by MOM. The effects of the sphere radius are numerically analyzed. The radius of spheres affects the scattering intensity in the lateral direction (0°–30°) and the larger the radius r_2 is, the stronger the interaction is. The larger the radius r2, the more the extremum points in the lateral direction (30°–90°). In the project, the particle distribution can be received by the analysis of the angular distributions of scattering intensity. The field with a difference value is more obvious with p polarization incident light, which makes it easy to inverse the characteristic of defect particles. In the project, p polarization light is suggested to the nondestructive examination.

ACKNOWLEDGMENTS

This work was supported by the General Armament Department Key Laboratory Foundation under Grant No. 9140C360302120C36136. In addition, the authors acknowledge support from the Natural Science Foundation of China (61308071), the Natural Science Foundation in Shaanxi Province of China (2013JQ8018), and the Special Natural Science Fund Projects in Shaanxi Province of China (2013JK0633). The program is financed by the open foundation of Shaanxi Key Laboratory of Photoelectric Measurement and Instrument Technology.

REFERENCES

[1] Yee, K.S. 1966. Numerical solution of initial boundary value problvolving Maxwell's equation in istropic media, *IEEE Trans*:302–307.

[2] Eremina, E. et al. 2011. Computational nano-optic technology based on discrete sources method, *Journal of Modern Optics* 58(5–6):384–399.

[3] Schmidt, V. et al.2011. Preconditioning techniques for iterative solvers in the Discrete Sources Method, *Journal of Quantitative Spectroscopy & Radiative Transfer* 112(11):1705–1710.

[4] Grishina, N. et al. 2011. Modelling of different TIRM setups by the Discrete Sources Method, *Journal of Quantitative Spectroscopy & Radiative Transfer* 112(11):1825–1832.

[5] Karlsson, A. et al. 2005. Numerical simulations of light scattering by red blood cells, *IEEE Trans. Biomed. Eng.*52(1):13–18.

[6] Karamehmedovic, M. et al. 2011. Comparison of numerical methods in near-field computation for metallic nanoparticles. *Optics Express*, 19(9):8939–8953.

[7] Eremina, E. 2009.Light scattering by an erythrocyte based on Discrete Sources Method: shape and refractive index influence, *Journal of Quantitative Spectroscopy and Radiative Transfer* 110(14–16):1526–1534.

[8] Gong, L. & Wu Z.S.2012. Analysis of composite light scattering properties between wafers and many shapes of particles with different positions. *Acta Optica Sinica* 32(6):0629003-1-0629003-6.

[9] Gong, L. et al. 2012.Light scattering by wafers and multi shaped defect particles. *High Power Laser and Particle Beams* 24(11):2731–2734.

Information, Computer and Application Engineering – Liu, Sung & Yao (Eds)
© *2015 Taylor & Francis Group, London, ISBN 978-1-138-02717-6*

Technology-based private enterprises technological innovation case investigation and policy research

Jun Hua Yan

Handan Polytechnic College, Hebei, Handan, China

ABSTRACT: Based on technology-based private enterprises making a definition and theory of technology innovation, on the basis of elaborating the combination of science and technology, this article analyzes in detail the restriction factors of technology-based private enterprises technological innovation in the present situation, put forward enhancement of technology-based private enterprises technological innovation capability of countermeasures and suggestions, so as to promote sound and rapid economic development in our country and, eventually, improved the comprehensive national strength and competitiveness of our country.

KEYWORDS: Technology-based private enterprises; Technical innovation; Innovation strategy and policy

1 TECHNOLOGY-BASED PRIVATE ENTERPRISES

1.1 *Technology-based private enterprises are defined*

Technology-based private enterprises refers to workers as the main body of science and technology, the nongovernmental business, being mainly engaged in the industrialization of scientific and technological achievements, as well as technology development, technology transfer, technical consulting, technology services, and other business activities of enterprises. Basic characteristics of technology-based private enterprises are "voluntary self-raised funds, portfolio, operate independently, self-financing, self-discipline and self development" of the management mechanism, its products and services that have high technology, and knowledge content of the enterprise. Following the principle of the market to its development, we should implement an independent decision-making mechanism, the market-oriented operation mechanism, and the evolution mechanism of talents, according to the contribution payment incentive mechanism. Then, we should adopt a market oriented to a product, in order to benefit as the core, the raw material, production-demand-sales chain of the enterprise technology innovation mode.

1.2 *Characteristics of technology-based private enterprises*

1.2.1 In organizations, technical personnel account for the proportion of high, generally more than 30%; in some knowledge-intensive, technology-intensive areas, the proportion of science and technology personnel increases by more than 50%. On a subject or professional direction has strong ability to research and develop high-tech products, generally in the same industry advanced level or has its unique feature.

1.2.2 In the aspect of management, technology-based private enterprises not only should have quite advanced and new technology products and development capacity but also should have strong ability to market development and management. They should adopt scientific methods to develop the market. Because judging whether an enterprise science and technology enterprises, not only look at whether to adopt high technology means, and should see whether the areas of its business in the technology progress rapidly, the field of high and new technology to see whether the master degree theses of master of high technology enterprise into realistic productivity, the productivity is mainly embodied in the product.

1.2.3 Organization structure is a simple, flexible management mechanism. Technology-based private enterprises are mostly in accordance with self-raised funds; the principles of voluntariness combination operate independently, as in the case of self-financing. Organizational structure is relatively simple, some indulge in only production and sales of two departments, and some do not have a formal organizational structure, only to be responsible for the individual or temporary organization according to the production and operation situation. Simple organization structure, on the one hand, means flexible operational mechanism, no absolute boundaries between the development and marketing, and research and development personnel can have direct contact with the user. Communication is conducive to being timely according to the market demand and in encouraging entrepreneurs to be innovative in their decision-making. On the other hand, it can save operating costs effectively and reduce the agent risk. But with the expansion of enterprises, under the impact of the market economy, its shortcomings are also gradually exposed, such as unclear has not clear, labor relations, nepotism, and so on can be affected by the enterprise, which serves as the important obstacle in the development of the future.

2 OBSTACLE FACTORS OF TECHNOLOGY INNOVATION IN TECHNOLOGY-BASED PRIVATE ENTERPRISES

With strong science and technology private enterprise development in our country, the industrial growth rate is far more than the size of the general industry. The profitability of enterprises in growing and economic efficiency has also improved; more and more technology-based private enterprises exhibit core competitiveness, and their ability to compete in the fierce market competition continues to be enhanced. In the process of its rapid development, however, enterprises still face some problems and difficulties in reality.

2.1 Less research and developmental ability of enterprises

Currently, private technology enterprises in our country can be roughly classified into three types: One is the research and development driven. An important feature of this kind of enterprise is that it is market oriented combined with technical drawing, insists for a long time on independent research, development, and independent innovation, relies on its own research and development of new technologies, new products, and new technology occupying and expanding the market, gains a competitive advantage, and continues to develop and expand. Investment in research and development is more than 5% and sometimes even more than 10%. The second is "technology-industry-trade" or "trade -industry- technology" combination model. Even general companies have a strong research and development strength and new technologies, new products, new technology, and so on, but the development of the enterprise growth is not primarily based on research and development but on "technology-industry-trade" or "trade -industry-technology" combined with patterns and mechanisms of private and business philosophy, and so on. Most of the time, companies lack core technology and original innovation technology, and their investment in research and development intensity is generally less than 3%. The third is a technical trade or logistics. This kind of company mainly consists of trade

and logistics, its research and development strength is relatively thin, research and development is the focus of the marketing and logistics portfolio allocation, and generally do not involve product research and development; thus, investment in research and development spending is generally less than 2%.

2.2 Personnel lack scientific and technological acumen

Technology-based private enterprises technological innovation is the lack of technology development talent, technology-based private enterprises' less professional and technical personnel, and the core technical personnel's more than part-time nature. Technical personnel's professional level is uneven, and in solving some key technical problems of technological innovation, there will be a lack of some necessary interdisciplinary technical personnel in terms of help. Innovation personnel's own development ability is limited, thus the intellectual support of technology innovation cannot be guaranteed, which makes the technology innovation of technology-based private enterprises resort to foreign technology, to promote the enterprise's technical innovation. In some technology-based private enterprises, on the other hand, because they do not pay attention to the road of the development of science and technology, there is no real play with regard to the role of the science and technology being the first productive force, not for enterprise development of human resource management planning, lack of effective talent incentive mechanism, make the technology-based private enterprises keep talents, the phenomenon of brain drain frequent increasingly serious.

2.3 Lack of the strategic management in technology innovation

Practice has proved that using strategic management will directly affect the enterprise's business performance, especially the high failure rate of technology-based private enterprises. However, many technology-based private enterprises are not aware of this; in practice, few technology-based private enterprises exist due to using the method of strategic management. As a result, in some technology-based private enterprises, many problems have been existing in the management of technology innovation, such as not being guided by market demand for technical innovation, blindly fixing the quotas for marketing in order, and ignoring the market research of technology innovation and future development trends of the market judgment. This leads to upgrading of new products, which is difficult to keep up with the changes of market demand, technological innovation, and the market as an entity. Some technology-based private enterprises, because of the lack of consciousness of strategic management, turn their attention to technical innovation while simultaneously ignoring the market development of new products. The cause of the lack of effective marketing strategy formulation is not known to market innovative products, there is a large amount of backlog, and the technology innovation of technology-based private enterprises poses serious setbacks.

2.4 Enterprise cannot solve the financing problem effectively; technology innovation cycle is longer than the other cycles

Financing channels in diversity, wide coverage, such as capital market financing, government, bank loans, financial loans, risk investment, angel investment, and so on can provide financial support for science and technology private enterprise technology innovation. However, when applying for funds all kinds of barriers exist, such as enterprise credit rating, guarantees, and financing formalities to complexity in technology-based private enterprises face the financing channels, financing difficult predicament. Technology innovation cycle is longer, and technology innovation and development mode change. The cash uses track deviation, the serious influence enterprise development for a long time, as a result of the science and technology private enterprise personnel, funds, equipment and so on the strength of the weak, and lack support for the guidance of the government or related services, with a long cycle combined with technology innovation, enterprise to get due to technical innovation of enterprise benefit period is longer, longer cycle will make the enterprise the management difficulty, therefore, more and more small and medium-sized enterprises using the limited funds to purchase directly through innovative products to equipment production.

3 COUNTERMEASURE TO PROMOTE SCIENCE AND TECHNOLOGY PRIVATE ENTERPRISE TECHNOLOGY INNOVATION

3.1 Government involvement in guarantee, optimizing financing channels

First, credit financing is provided in technology-based private enterprises; the existing resources should make full use of the government; the government should be reviewed for enterprise credit, for potential, reputable science; and technology private enterprises should provide guarantee financing, lower the difficulty of small and medium-sized enterprise financing, and promote the small and medium-sized enterprise technology innovation activities to ensure they run

smoothly; Second, the government formulated related policies, optimized the technology innovation enterprise financing channels, improved the plight of more than one channel while financing difficulties at this time, helped open up new financing channels of private enterprises, and provided the necessary means of preferential subsidies. Third, for the development of science and technology private enterprise product or process, such as technology innovation cycle being long, effective, or slow, the bank can deal with the enterprise funds, finance companies, credit departments such as communication and coordination, appropriate relax repayment period, the normal operation of technology-based private enterprises technology innovation management.

3.2 Perfect the enterprise technology innovation system of organization management mechanism and management system

Perfecting organization management mechanism and management system involves improving enterprise technology innovation ability with a strong security guarantee. The enterprise is composed of a series of functions of the economic organization; regardless of whether enterprise organization structure and pipe system is reasonable or not, it will affect the efficiency and the success or failure of technological innovation. A reasonable organizational structure can guarantee the information communication channel being unobstructed and coordinate with various departments. An effective management system can promote enterprise technology innovation work. Through the establishment of the management system for enterprises, we can provide the foundation of innovation, provide technical innovation talent reserves, strengthen technology innovation potential, personnel training, and strengthen the technical personnel of innovation consciousness. We can ensure the innovation of technical personnel and the innovation of material and spiritual double incentive, mobilize the innovation consciousness and innovation power, and make technical innovation of enterprise culture. To strengthen enterprise innovation ability, we should shorten the cycle of technological innovation, increase the enterprise competitiveness of science and technology, and increase the ability to grasp the opportunity, thus getting an unexpected harvest.

3.3 Strengthen self-management, improve the ability of innovation management

Our aim should be to strengthen the level of science and technology innovation management of private enterprises vigorously, in management innovation to ensure the efficiency of resource allocation, and to use management innovation to adapt and promote the innovation of the product, process, and equipment. We should strategically build innovation management mechanism, strengthen the innovation of idea generation, research and development, technology management and organization, design and manufacturing, the user participation and marketing, and a series related to enterprise technology innovation process management, from the system, to make use of the direction of science and technology resources, the configuration mechanism, the effective management of efficiency, setting up within the industry, enterprise innovation resources sharing and migration mechanism, to maximize the technology service efficiency of funds. Third, in order to promote science organization enterprise technological innovation activities, we should improve the project management of technology innovation ability; to perform a good job of innovation projects of the review, we should pay attention to key support strategic product development and major, key technology research. We should also pay attention to economic benefit while simultaneously paying attention to the increase of the stock of knowledge as well as technology accumulation. Third, technology-based private enterprises within each strategic business units are formulated as departments, thus furthering horizontal technical development. Each department should be contacted by a professional technical meeting platform, on the basis of the new topics in technology strategy research, to predict the direction of the market and strengthen the company's technical creativity. Fourth, we should attach importance to the cultivation of innovation of enterprise culture. Entrepreneurs should first have the courage to take responsibility and risk, to promote a high-value, corporate culture. In order to strengthen learning about foreign enterprises' spirit of innovation and innovation management idea, we should vigorously promote technological innovation, system innovation, service innovation, and continuous innovation to better serve customers. We should organize various activities to fully arouse the enthusiasm of staff innovation and let the innovative ideology be deeply rooted in the hearts of the people and the innovation spirit and the culture in the brand construction of science and technology. Finally, in order to strengthen and improve the company's incentive mechanism, by formulating a special innovation incentive policy, we could further

enhance the forming ability and the quality of patents, proprietary technology center into an influential.

3.4 Promote the effective combination

We should establish an enterprise as the main body of technological innovation, but we cannot ignore other departments of energy, universities, research institutes, and enterprises while conducting conduct scientific experiments and development. To establish technical innovation alliance, we should integrate all sorts of innovation resources. An effective combination of production needs to carry out the following two tasks: First, the production, the three parties is established on the basis of the same interests, strengthen cooperation, enterprises as the main body of technology innovation cannot be changed, the science and technology into productivity, the enterprise must be play a leading role, colleges and universities as the main, mutual benefit and common development. Second, establishing the technology platform of tripartite joint innovation study, invention patent in China for a long time, and the development of productive forces is an uncoordinated phenomenon. These are the disadvantages of the legacy of planned economy; the scientific research institutions only patented inventions, don't ask, don't production transformation, enterprises need to new technology, but it can't get, the intermediate links have serious disconnectedness. To establish a joint innovation technology platform, efforts for the development of the relationship between, can be better for technical innovation, reduce the lag problem of science and technology into production.

REFERENCES

[1] Bai-Zhou Li. The several factors of technology innovation of small and medium enterprises [J]. Journal of academic exchanges, 2010.
[2] KuangHui. Technology innovation and sustainable development of small and medium-sized enterprise [J]. Journal of popular science and technology, 2013.
[3] Rui-Qing Yu. Small and mid-sized enterprise financing difficulties and improve the way-in baoding city, hebei province, for example [J]. China economic studies, 2013.

Information, Computer and Application Engineering – Liu, Sung & Yao (Eds)
© 2015 Taylor & Francis Group, London, ISBN 978-1-138-02717-6

Discussion on the relationship between development of science and technology and vocational education

Zhen Shang
Beijing Polytechnic, Beijing, China

ABSTRACT: Vocational education is not only the need for economic development but also the need for education development. This article begins from the several major vocational education models in the international sphere. Then, a brief introduction to the present situation of China's vocational education development is made. Finally, the effects of the development of science and technology on vocational education are discussed. The study provides some ideas to explore the reform of Chinese vocational education.

KEYWORDS: Vocational Education; Development of Science and Technology; Reform

1   INTRODUCTION

High-quality talent with a pioneering spirit, the courage to explore, and innovation are needed in the 21st century. Education must adapt to socioeconomic development. Vocational education should keep pace with the times, reflecting the spirit of the times, to effectively promote socioeconomic development and progress.

Competition is the manifestation of the market economy. Vocational education trained workers are urgently needed in the production and service sector. The quality of graduates will directly affect not only the quality of employees but also the overall image of the work force at the entrepreneurial level.

2   MAIN VOCATIONAL EDUCATION MODEL IN INTERNATIONAL SPHERE

Internationally, the degree of development and the status of popularization of vocational education is an important symbol that measures the degree of modernization and the degree of social civilization [1]. Vocational education system is more perfect in major developed countries in the world. It can be said that there is not a single developed country that does not pay attention to vocational education, not a single developed country that does not need to be supported by strong vocational education. Currently vocational education models in the international sphere are as follows:

1 "Dual system" vocational education model in Germany.

"Dual system" vocational education model is a successful model, which is the enterprise-based division of labor cooperation between schools and businesses. There is a close integration of theory and practice, with a focus on practice.

2   Competence-Based Vocational Education in the United States and Canada

The core of Competence-Based Vocational Education, abbreviated as CBE, is translated into Chinese as competency-based education or ability-based education. In summary, CBE theory is based on competency, and it is the teaching ideology that emphasizes ability and capability training.

3   TAFE in Australia

TAFE stands for technical and further education colleges. This education and training model is Australia's unique vocational education and training, the largest education and training organization that is compulsory in Australia, and the main provider of vocational education and training (VET). TAFE is equivalent to the synthesis of vocational school, technical school, secondary school, and college in China.

4   Complementary Vocational Education Model in Japan

The vocational education model in Japan is a complementary vocational education model that is based on the national school, supplemented by business school and a private school. This educational model is based on an indivisible contractual relationship between education and economics [2].

5   Models of Initial Vocational Education and Training in Austria

Full-time vocational schools and apprenticeship are the main models of initial vocational education and training in Austria. The full-time vocational schools model includes VET schools and colleges, university of applied science, and apprenticeship with more emphasis on hands-on training in the enterprise, business premises, and part-time apprenticeship training schools simultaneously. Then, it introduced the operation of these two models in detail through some examples, including management finance, teaching method, teacher training, learning, and so on. Based on these facts, this article analyzed the mechanisms that pushed these two models toward achieving success. These mechanisms included management mechanism of clear responsibility, the mechanism of stable funds devotion, the sound law mechanism, and the unimpeded information mechanism. This article still analyzed the deep reasons that form the two models in Austria.

3 DOMESTIC VOCATIONAL EDUCATION DEVELOPMENT

3.1 Several major vocational education models in China

Since reform and opening, the development of vocational education shows a rising trend in general. In recent years, the development of vocational education attaches great importance to both the party and the country [3]. With the continuous introduction of favorable policies, vocational education is facing unprecedented opportunities for development. Currently, the full-time vocational school models mainly include the following types in our country:

1 (1) the East and West, urban and rural Joint School-running; (2) order-based HR training model; (3) HR training model of work-school curriculum system; (4) "the all-dimensional" integration educational pattern; (5) school-running model of comprehensive cooperation combined with the local economy; (6) "Combining Learning with Working" model; and (7) "Teaching Factory" Model.

3.2 Problems of vocational education in China

Currently, China's full-time vocational school has occupied half of vocational education, but the quality of development is not satisfactory. Based on the percentage of school number and demanded population percentage, vocational schools cannot meet the needs of the vocational education population, and many schools do not recruit students. The main reasons are as follows:

1 Many local high schools also restrict the development of vocational education; most students and parents think that going to high school, as well as admitting to ordinary colleges and universities is more promising. This ideas also restricts the development of vocational education. And employers would give priority to graduates of ordinary colleges and universities. States must completely break the old concept of talent. Simultaneously, states must increase status and treatment of skilled personnel, particularly highly skilled personnel. And these need to be reflected in a certain form of regulatory documents.

2 Vocational education models are outdated, many are a replica of general education, and there is no innovation. Students are trained in the mode poor hands and lack academic ability. Employment prospects are not optimistic. Affected by general education, our full-time vocational school education mainly uses the transfer-received teaching model. Other types of teaching models are assisted. Organizational form of teaching is mainly intended to imitate patterns of general education, to take the class system, with the theme of classroom teaching. Some organizational forms are rarely used, such as collaborative teaching, on-site teaching, ability grouping system, open teaching, and so on.

3 The education system needs to be reformed. Faculties should be encouraged to give full play to the initiative. Colleges should actively explore new modes of vocational education; they should not only provide support in terms of policy but also give them protection in the capital.

4 EFFECTS OF THE RAPID DEVELOPMENT OF SCIENCE AND TECHNOLOGY DEVELOPMENT ON VOCATIONAL EDUCATION

With the rapid development of science and technology, market, industry, occupation, and job are constantly changing [4]. On the one hand, the traditional industries and professional positions continue to decrease, and some professional positions even disappear. With increased automation, extensive use of automated production lines and robots, production line workers are freed from heavy labor, which requires highly qualified workers. They can not only manipulate the machine to complete production but also need to know the machine problems that may occur, in order to facilitate timely daily production for repair and maintenance. On the other hand, with human resources being marketable, the rational flow of talent has become the normal form of the modern market economy. Therefore, vocational education should be proactive, to pay attention to the development trend of new technology and new industries at home and abroad, in particular, to understand the domestic industrial policy and industrial structure positions trends, and to train students to adapt to the rapid development of science and technology.

Effects of the rapid development of science and technology on vocational education are as follows:

1 Students' initiative and innovation education

In the modern enterprise, a large number of foreign advanced technologies are introduced, the degree of automation in the workshop is constantly increasing, demand for workers is also rising, and it is no longer a mere apprenticeship. Apprentices also need to have an independent ability to learn, which requires, according to the idea of lifelong education, comprehensively improving the quality of students, especially training the ability of autonomous learning, constantly updating knowledge, and cultivating students' innovative spirit and ability, which will lay a solid foundation for their life's work, continuous training, and learning.

2 International communication

With the coming of the 21st century, economic participation in international competition continues to expand throughout the country. A large number of companies are heavily involved in international trade and international competition, which are not only from a professional, but also from the quality. Due to technical update speed, production technology of business and industry, management procedures, and product quality are carried out in line with international standards to increase competitiveness. It is becoming increasingly common for the product quality management system to be certified by the international quality management system standard. Many companies use it not only as a marketing tool but also as an internal workflow system. For this reason, these industries and companies need a lot of international talent. These people should have the awareness to participate in international competition and should have the skills to adapt to international competition.

3 Corporate training (technical training)

Modern enterprises pay more attention to training, which not only enables enterprises to continuously develop but also instills hope in young people for their work, to ensure that the young people can be more long-term services for enterprises. On the one hand, corporate training needs of staff should have an appropriate knowledge base, so that employees have room for improvement; on the other hand, it also requires companies grading training for staff. However, most technical training of new employees in enterprises is a unified whole, in one part of which employees are enrolled from the community; the other part is the school–enterprise cooperation, the students of "orders class" culture, which results in some of the staff repetition forming a resource waste.

5 CONCLUSIONS

China's economic situation is good in the future. The vocational education model will be doomed to be innovative, because professional skilled personnel are urgently needed. The reform of vocational education needs to keep pace with the development and to be consistent with the pulse of the world. Based on the several major vocational education models that are presented in the international sphere, this article briefly describes the status of development of vocational education in China. Then, the author discusses and studies the effects of the rapid development of science and technology on vocational education. This article provides some ideas for exploring the reform of vocational education in our country.

REFERENCES

[1] Xu Xin. A Study on Models of Initial Vocational Education and Training in Austria [D]. Southwest University, 2010.
[2] Professor HARTON (Japan), Translated by Liang Qi. Situation and reform of vocational education in Japan [M]. Beijing: Higher Education Press, 2005 (4).
[3] Li Hongwei. Case Study of Attractiveness of Vocational Education: Students' Visual Angle [J]. Vocational and Technical Education, 2010(31):10.
[4] An Jie. The Research on Enterprise Education in Vocational Schools [D]. Soochow University, 2008.

Information, Computer and Application Engineering – Liu, Sung & Yao (Eds)
© 2015 Taylor & Francis Group, London, ISBN 978-1-138-02717-6

Education technology, information technology, and course integration

Ying Song & Jing Sun
Harbin University of Science and Technology, China

ABSTRACT: Considering the fact of people's awareness of information technology and modern education technology, this article elaborates the requirement of education technology, information technology, and the requirement gist and method of curriculum integration.

KEYWORDS: The comprehensive concept of education technology information technology and curriculum integration

1 DEFINITION OF EDUCATION TECHNOLOGY, INFORMATION TECHNOLOGY, AND CURRICULUM INTEGRATION

1.1 Curriculum integration

The intention is to test the system factor such as the setting of the course, the structure of the course, the goal of teaching, the strategies of teaching, and the commitment of teaching.

We can see that we ought to use the holistic connected and dialectic viewpoint to learn and research education and the relationship between the factors.

1.2 Integrating information technology into the curriculum

Professor Li Ke dong defines it as one of the new teaching methods that is used to complete the task of education teaching that combines information technology, information resource, information method manpower resource, and course details together.

1.3 Integrating education technology into the curriculum

It means that education technology combines with course structure, course details, course resources, and implementation of course, becoming a part of the course, the organic section of the course details, and the implementation of the course. The curriculum between education technology and the course is not meant to be brought in passively, but to be brought forward to accept and promote the course. The curriculum of Integrating Education Technology will affect each sector related to itself, moreover, then influence and punch it.

2 REQUIREMENT OF EDUCATION TECHNOLOGY, INFORMATION TECHNOLOGY, AND CURRICULUM

2.1 Requirement of the teachers

Information age asks the teachers to gain a higher quality. It will be efficient for those who know how to use modern education technology to serve their own teaching and activity by managing to promote self-quality. Meanwhile, in the environment of information technology, the teachers are requested not only to use the means of modern information technology but also to build a new sense and theory to reexamine and guide every single part of the teaching activities.

2.2 Requirement of the students

The information technology and education technology combined with the curriculum is the course that strongly requests theory and practice, which means that the students ought to pay attention to the theory related to the actual theory by studying seriously and understanding the theory and method to bring the information technology and education technology into the practice of education. Therefore, students ought to switch the mode in order to become the ones that own knowledge autonomously, becoming the structure and the participant instead of doing things passively.

2.3 Requirement of the media

For the sake of the curriculum of the information technology and the modern education technology, we must establish education, scientific research, development, training, management, and information service in the integration of the software and hardware environment. Simultaneously, we are supposed to arrange those reasonably, including hardware, equipment, net environment, and so on. Setting up the system based on the computer network (network) as the center, symbolized by the multi-media audio-visual education equipment construction, we should promote the advantages in teaching. What is more, we should improve the qualities of the teachers and the communication while training becomes vital to the whole system.

3 ACCORDING TO INTEGRATION

"Leading - Subject" teaching theory is the main theoretical basis of the construction of education technology and curriculum integration teaching. "Leading - Subject" teaching theory absorbs the constructivist teaching theory of "to learn as the center" and the Ausubel "to teach as the center," the theory of learning and teaching, to avoid being short and to absorb the merits of learning and teaching theory. In the process of the teaching development, we should fully respect the principal position of students' learning, encourage the students' autonomous learning, and promote the content of the teaching of independent thinking. In the teaching process, teachers should choose the learning content and plan the learning process, dominant role of help and guidance, and so on. We should make learning and teaching be organically unified and reflect on people's all-round development as the ultimate goal of educational thought.

4 METHOD OF EDUCATION TECHNOLOGY, INFORMATION TECHNOLOGY, AND CURRICULUM INTEGRATION

4.1 Basis of resource integration

Resource integration should involve a kind of education resources as the goal to optimize the form of an organization. Educators and especially managers, through the rational allocation of hardware resources, make full use of network resources, train human resources development and organic reconstruction of curriculum resources, to achieve an organizational form of "integration." This is the guarantee of curriculum integration and foundation.

The ways to streamline integration - integration process.

"Integration process" involves educators and learners in the form of a specific operation for the purpose of teaching and learning. It involves applying the learning of information technology and discipline courses of study to explore the organic fusion in the process of the same activity and to achieve the purpose of "assimilation" and "adaptation" in the process of constructivism advocated. This process includes three basic stages:

4.1.1 *Preparation before class, class implementation stage, and development stage after class. The main task of the three stages is as follows:*
Preparation before class, mainly for the teachers.

In view of the teaching content that has a carefully designed teaching scheme, seek a breakthrough point; browse relevant website home pages to find information and capture point; order the information from the main, link building, and scaffolding teaching platform; detect network security and interactivity; and use moral education for network civilization.

4.1.2 *Emphasizes the interaction between teachers and students in classroom implementation stage*
According to the teaching plan, we should begin to learn the following:

Teacher learning objectives are put forward (including necessary network knowledge learning), as well as the corresponding learning methods (individual or group learning). Students, on the basis of teacher's import problem, relatively independently decide to study methods by which they could try to solve the problems of teachers. We should start discussions between students or between teachers and students and teachers should explain the learning process, to summarize the common understanding and to pay attention to individual problems.

4.1.3 *Development stage after class*
We should give priority to student activities. Generally, these can be divided into four kinds:

Consolidate the application of network knowledge; through the project learning website of curriculum knowledge; by applying the website for broadening curriculum knowledge; and using methods such as E-mail with teachers, classmates, and even with other online learners.

4.2 To focus on the purpose of the integration–integration ability

The ultimate purpose of education is to cultivate students' ability, and the purpose of modern education is to cultivate citizens' lifelong learning ability. "Curriculum integration," as a means of teaching and learning, is no exception. "The ability to integrate" refers to a study on "resource integration" and "process integration" of this kind of curriculum integration, more effectively with the aid of increasingly advanced education technology, to cultivate both having higher information literacy and using the modern information technology for learning with regard to the learners of the new century.

5 SUMMARY AND REFLECTION

Education workers must know clearly that education faces the service object is a rich emotional, full of vigor and vitality of the individual, not simply to technology application to promote the development of related disciplines as the goal. It should consider the educational goal of cultivating people's sound personality. Education technology is constantly practicing the theory of technology in a big social and education background. Amid the background of quality education, education technology must be people oriented, and it should not blindly make technology and ignore the educated spiritual culture. At this stage, what we call education technology is to realize the education and the teaching process optimization. From the literal understanding, it is easy to fall into a pure technical level. From solving problems of education and teaching practice and from the visible problems to considering the education technology, it is necessary to deepen the understanding, not just retaining the simple understanding of technology. In facing the education, technology is also to be highlighted. It has both the teacher's control and students' interaction. In the application of education, technology plays an important role.

In a word, new curriculum reform advocates to build to meet the requirements of quality education in the 21st-century basic education curriculum system. It develops the cause of education technology with the aim of providing a lot of lessons for the present day. Education technology in the position and role of the information technology and curriculum integration is a theory of other technologies that cannot be replaced. Only fully aware of this, information technology and curriculum integration to a more healthy development, the new curriculum reform goals can be achieved.

REFERENCES

[1] Kedong Li, The Objectives And Methods Of Information Technology And Curriculum Integration [J]. Electrochemical Education Research, 2001, (4).
[2] Kekang, He, Wenguang Li, Education Technology [M]. Beijing:Beijing Normal University Press, 2002. 10.
[3] Kekang He, The Theory And Method Of Deep Integration Of Information Technology And Curriculum[J]. China's University Teaching, 2005, (5).
[4] Kedong Li. Digital Learning [J]. Electrochemical Education Research, 2001. (2).
[5] Kekang Li. The Main Body Of Teaching Structure [EB/OL].< http://www.etc.edu.cn/>
[6] Weihua Gao. The Integration Of Information Technology And Modern Education Technology [J]. Journal Of North China Coal Medical College, 2003, 5(4):527.
[7] Chunhong Gong,The Practice Of Information Technology And Curriculum Integration Research [J]. The China Science And Technology Information, 2006, (2):92–93.

Information, Computer and Application Engineering – Liu, Sung & Yao (Eds)
© 2015 Taylor & Francis Group, London, ISBN 978-1-138-02717-6

Optimization of fuzzy decision tree with software estimation based on genetic algorithm

Ying Gou & Xiao Zong Zheng
Department of Electronic Engineering, Chongqing College of Engineering, Chongqing , China

Ting Cai
Department of Computer College of Mobile Telecom, Chongqing University of Posts and Telecom,Chongqing, China

ABSTRACT: The genetic algorithm is a simulated biological search algorithm, and the decision tree method is an efficient method of data mining. This article aims at estimating the software process of uncertainty and inaccuracy; evaluation function is constructed based on the genetic algorithm, to solve the weights of decision trees of non-leaf nodes and then to combine more than decision trees, the optimal set of rules is formatted. Finally, the experimental results prove the feasibility and effectiveness of the method.

KEYWORDS: Genetic Algorithm; Decision Tree; Optimization; Software Estimation

1 INTRODUCTION

With the expansion of the scale of software system and the complex degree of increase, since the late 1960s, a large number of software project schedule delays, budget overruns, and quality defects have emerged as the typical features of software crisis, and these are still frequent. According to the Standish group, according to the CHAOS report published in 1995 in 8 000 projects from 350 organizations, only 16.2% are "successful" and can be made within the budget and within a time limit, 31.1% are "failed," which were either not completed or cancelled; The remaining 52.7% is called "challenged," although the complete budget overruns by an average of 89%. In 2004, in the group of statistics, the number increased by more than 50, 000; the results showed that the proportion of successful projects increased to 29%, and the challenged project was still 53%. Although some studies suggest that the CHAOS report about budget overruns 89% of the data that have been exaggerated, the actual situation should be in 30% ~ 40% on average.

Currently, in the aspect of software estimation, a lot of methods have been studied; for example, in software estimation model of mathematical approaches such as COCOMO, there is a need to use historical data for correction factor of parameter adjustment. Based on the analogy of the estimate for new projects and historical comparison, we need to find out the match history project and new project characteristics and then estimate the new project.

2 DECISION TREE

Model selection is always an important research problem; the decision support system of the existing model selection method can be roughly classified into two types: analytic method and the artificial intelligence method. Analytical method is the method based on the linear programming model; depending on the USES model, the historical information can be used to select the model [1, 2]. Artificial intelligence method can solve many of the traditional mathematical methods to solve the problem. Genetic algorithm (ga) is one of the artificial intelligence methods, and it simulates the genetic selection, in the process of solving constant evolution of species, to seek the optimal solution.

2.1 Weight calculation

Decision tree is a kind of simple structure, search efficiency of classifier. Currently, the genetic algorithm is used to construct the decision tree method and it is also used to solve the linear weighting of vector tree nodes while constructing the decision tree method[3]. This article uses the genetic algorithm of linear weighting of tree nodes and then according to the error rate of these tree nodes, to construct the decision tree.

The decision tree contains leaf nodes and non-leaf nodes, in which the non-leaf node contains the decision function and the left and right subtrees, and the leaf node is the class attribute. The decision-making process is as follows: The input values from the root nodes, according to the decision tree of non-leaf nodes of decision function, choose a child node, traverse to the leaf node, then the leaf node corresponding class as the class attribute of the input value [3].

Definition 1. Every node in the tree can use the only string {p, L, R}, p is the root node, L is the left node, and R is the right node. L and R can be a sub-string in the composition of nodes.

Definition 2. The training sample sets for $x_N = \{x_1, x_1, ... x_{N_i}\}$, size of the sample space for N_i.

The i attribute $A^{(i)}$ for the j attribute values $T_j^{(i)}$ fuzzy set of the potential for : $M_j(A^{(i)}) = \sum_{i=1}^{n} x_{N_i}$

Definition 3. Should be divided by the target class for $C = \{C_1, C_2, ..., C_h\}$, attribute values $T_j^{(i)}$ on the k classification C_k of correlation degree are defined as:

$$p_{jk}^{(i)} = \frac{M(T_j(A^i) \cap C_k)}{M(T_j(A^i))}$$

The number of sample set of attributes for s, each attribute for the A^1, A^2, ..., A^s. To solve the s attributes correlation degree, and according to the correlation degree from small to large order, the sorting result is A_1, A_2, ..., A_s. For the sorted attributes, different weights are given:

$$\begin{cases} w(A_i) = 1 + (i-1) * 0.5 \\ 1 \leq i \leq s \end{cases} \quad (1)$$

The probability $p(A_i)$ of each attribute is selected with the method of the roulette wheel; the probability of each attribute is selected as follows:

$$p(A_i) = \frac{w(A_i)}{\sum_{i=1}^{s} w(A_i)} \quad (2)$$

2.2 Construction of decision trees

From s attributes choose d attributes to build the decision tree; dt, in turn, build t: dt_1, dt_2, ..., dt_t. Building rules are as follows:

1 From all the attributes, select the most relevant attributes as the root node of extended attributes. Root node of sample space is all the fuzzy sets.
2 According to the root node, extended attributes of attribute values structure child nodes.

For a training sample x_i according to the j decision tree classification, x_i is the probability of the target class C_i for $p(x_i, C_i)$; the j decision tree for the weight of the target class C_i : $W_p(x_i, C_i)$. A training sample x_i after combination of the weighted value probability for the target class C_i is as follows:

$$\alpha_i = \sum_{j=1}^{t} p(x_i, C_i) \times W_p(x_i, C_i) \quad (3)$$

3 STRUCTURE OF EVALUATION FUNCTION

For an individual, F^k is composed of multiple genes, a decision tree for the target class weights is a gene, and the form of individual F^k is as follows: Among them, $W_p(x_i, C_i) \in [0,1]$

$$F^k = \begin{bmatrix} W_p(x_1, C_1) & ..... & W_p(x_1, C_i) & ..... & W_p(x_1, C_h) \\ \cdot & \cdot & \cdot & \cdot & \cdot \\ \cdot & \cdot & \cdot & \cdot & \cdot \\ \cdot & \cdot & \cdot & \cdot & \cdot \\ W_p(x_t, C_1) & ..... & W_p(x_t, C_i) & ..... & W_p(x_t, C_h) \end{bmatrix} \quad (4)$$

For the training sample x_i application of decision tree classification, each individual F^k and $p(x_i, C_i)$ are combined, to get a set of combination: $\alpha = \max\{\alpha_1, \alpha_2, ..., \alpha_{ht}\}$. Evaluation function $f(F^k)$ computation formula is as follows:

$$f(F^k) = \frac{\alpha}{\sum_{i=1}^{ht} \alpha_i} \quad (5)$$

With evaluation function as fitness function of genetic algorithm, the scale of the problem is solved and the specific algorithm is as follows.

Algorithm 1. Genetic algorithm to solve scale

1 Initialize the 0 generation of population scale: x_0.
2 The individual fitness is calculated by the formula (5).
3 Calculate selection operator:
 According to individual fitness, and using the roulette wheel method, choose the optimal parent individuals.
4 Calculate crossover operator: random cross to parent individuals.
5 Calculate mutation operator: the new individual genes to mutation probability P_i, among them, $p_i \in [0,1]$
6 According to the fitness update population.

4 EXPERIMENTAL RESULTS AND ANALYSIS

Training and test data are selected in this study from "the third international competition of knowledge discovery and data mining tools" test data [4]. The test data contain 23 target classifications; each sample datum contains 41 attributes, including 7 symbol

174

attributes and 34 continuous attributes. A random sample of 10000 data is taken from test data as the training sample. In the random selection of three groups of data, the number of each group of data is as follows: The first group is 9000, the second group is 9200, and the third group is 9400, as shown in table 1. Choose 10 attributes to build 10 decision trees. Crossover probability is 0.7, mutation probability is 0.1, and the evolution algebra is 500. For three groups, data are tested, and test results are shown in table 2.

Table 1. Attribute table.

Decision Tree		Attributes			
Tree.1	36	28	23	35	11
Tree.2	2	41	14	32	28
Tree.3	35	11	19	11	32
Tree.4	2	39	26	32	20
Tree.5	38	36	30	39	40
Tree.6	11	24	30	37	4
Tree.7	40	31	22	10	37
Tree.8	29	33	27	36	29
Tree.9	11	5	30	10	24
Tree.10	14	5	34	23	12

Decision Tree		Attributes			
Tree.1	15	40	29	32	24
Tree.2	30	10	5	21	41
Tree.3	23	38	13	40	24
Tree.4	32	17	8	4	3
Tree.5	28	14	30	18	23
Tree.6	26	25	34	40	28
Tree.7	22	15	35	31	30
Tree.8	17	32	37	21	31
Tree.9	16	11	36	11	5
Tree.10	22	41	31	26	40

Table 2. Result of the experimental table.

Experimental results before optimization

Group Name	Total Number	Error Number	Error Rate
G.1	9000	301	0.0334
G.2	9200	294	0.0320
G.3	9400	311	0.0331

Experimental results after optimization

Group Name	Total Number	Error number	Error rate
G.1	9000	274	0.0304
G.2	9200	218	0.0237
G.3	9400	235	0.0250

5 CONCLUSION

Analysis of the result of the experiment shows that using a combination of the genetic algorithm of the decision tree results in software estimation error rate being low; accuracy is improved than before optimization; and the accuracy of software estimation can be improved.

REFERENCES

[1] Klein G, Konsynsk B, Bec PO. A liner representation for model management in a DSS. Journal of Management Information System, 1985, (2):40–54.
[2] Klein G. Developing model strings for model management. Journal of Management Information System, 1986, (2):94–110.
[3] Chai Bingbing, HuangTong, ZhuangXinhua.Sklansky Jack. Piecewist liner classifiers using binary tree structure and genetic algorithm. Pattern Recognition. 1996, (11):1905–1917.
[4] KDDCUP99 data [DB/OL]. (1999). http://kdd.ics.uci .edu/databases/kddcup99/Fund project, (Project ID: 2013xzky02).

Information, Computer and Application Engineering – Liu, Sung & Yao (Eds)
© 2015 Taylor & Francis Group, London, ISBN 978-1-138-02717-6

Effective synchronization of data in distributed systems

P. Ocenasek

Brno University of Technology, Brno, Czech Republic

ABSTRACT: This paper proposes an approach for effective synchronization of data in distributed information systems. The first part of the paper deals with the synchronization. The second part presents the system that realizes the proposed approach. This approach supports its own protocol for exchanging data between nodes. The system is designed as platform independent.

KEYWORDS: Synchronization, data, distributed system, backup, snapshot

1 INTRODUCTION

Given the current increase in information technology and the gradual expansion of electronic documents and data, rises the need for sharing between different platforms and backup of the data. Data backup is a process which creates a copy of the data, in order to avoid the risk of loss. It is possible to return to backup data at any time. Synchronization and sharing of data is needed, where we work with different technologies, and even different platforms. Furthermore, it is needed when working in groups and in places where it is frequent to share data together. In both cases, you can automate these activities for the convenience of the user, which can also save much of the time.

This work aims to propose an approach (Martak, 2013) that should promote the general protocol for exchanging data, so that you can synchronize and backup data between different platforms. Nowadays, it is certainly the need to deal with the security of transmitted data, in order to prevent getting this data to unauthorized users. Therefore, when designing this tool, we put the emphasis on security and data retention. Another important factor for the design of this tool was the effectiveness. This would result in reduce demands on resources, such as bandwidth usage, network used and reduce the disk space required for backup data. (Tridgell, 2008; Martak, 2013)

2 SYNCHRONIZATION

Data synchronization should ideally bring the sharing of data between different places (Cohen, 2008). This process does not ensure data backup, since it does not provide the possibility of data recovery, but only duplication. The benefit of such sharing is that on all the places anyone can work with this data independently of each other.

When synchronizing with must ensure that out-of-date data was synchronized with the amended form of the same data from another location. This brings the issue of the conflict, so there will be the same data in other places. These conflicts can be solved in different ways and it just depends on the user and the tool, which is used for synchronization and what options for conflict resolution he offers.

These systems allow us to cooperate between the different participants and shall keep the status of the file system (most often in the form of a tree) as developed in time (Cougias et al., 2008). Users have the option at any time to return to previous states of the file system. An important feature is the file sharing solution and their conflicts between users (Pilato, 2008). The basis of such systems is a repository, which contains just mentioned various versions of files, and other information depending on the particular system.

One of the options, how can a user store different versions of his files is a local version control (Cougias et al., 2008). This form is applicable only to one user. Therefore, there is another form, called centralized version control system (Cougias et al., 2008; Pilato, 2008). This allows multiple users to work simultaneously. Centralization allows us to have the entire work in one place, and users download only the current version of the files. If we have to restore an older version of the file, we download the required version from the server. The disadvantage may be that the system is now in one place. In the case of failure, the user does not have the opportunity to work with the remote system.

Another type of management systems represent distributed version control systems (Cougias et al., 2008). These systems keep the contents of the entire repository on the user's storage place. The user does not need to face problems like with the centralized systems and can still operate with the repository. If a repository contains large number of files and versions of files, it can be difficult to store at local medium.

3 ARCHITECTURE

We decided for peer-to-peer design of the concept. (Martak, 2013) Each node holds the entire history, including the current version, each node can use this data even in the event that the another node on the network is not available. The distributed system also provides duplicate data in case of damage. Thanks to peer-to-peer architecture, it appears similar topology to the mesh topology. On Figure 1 (Martak, 2013) we can see the example of such a topology. Node A synchronizes with node B. Both nodes C and D synchronize through node B. Both the nodes synchronize indirectly with the node A.

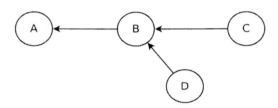

Figure 1. Possible synchronization topology.

For communication between two nodes, we need to determine their order. If the first node connects to another, we consider the first node as the client. This is due to that the BSD sockets are designed for a client - server architecture. During the whole communications, the nodes are equal. The system supports both IP versions due to the gradual expansion of IPv6. (McCabe, 2010; Martak, 2013)

4 COMMUNICATION PROTOCOL

For a description of the protocol we have used Protocol Buffers. It is ensured that the communication between tools supporting this protocol is platform independent. Anyone can start communication. Since both the peers (communication subjects) are equal, they can communicate with each other at once. (Martak, 2013)

Each message is created in the Protocol Buffers and is encapsulated in length-value structure like BER TLV 2 structure. Since the message types are contained in the structure of the Protocol Buffers, it is not necessary to indicate the type of message. We will create a simple length-value structure. The first four bytes of length-value structures represent the size of messages in the Protocol Buffers and are

followed by the contents of the serialized message. (Tridgell, 2008)

Communication is divided into the following types: notification, request and response. Each type of request or response message contains a transaction identifier. At the beginning of communication both devices exchanged a message about extensions of the Protocol and introduce its own identifier and name, followed by exchange of information about volumes. After the initial exchange of information, there starts the request-reply communication. One request may have more answers. The request may be followed by the next request, if this is specified in the previous request. If the tool detects the presence of any of the volumes, the remote user can start synchronization. First, other party is requested the identifier of the last known snapshot. If the remote user does not have any snapshot for any volume, there will be no identifier in the message. This synchronization will end, since there is nothing to synchronize. (Martak, 2013)

Finally, the tool receives the last snapshot, it checks its own database if that snapshot already contains. If the snapshot has been found, the database already contains all of the changes to the other party. If the snapshot has not been found, the tool continues with the request on the next snapshot. After the receipt of this request, it has to find out whether a snapshot of parents. If the parent exists, and does it in a local database, the tool must continue polling for so long, until it finds the last synchronized snapshot or until it finds one that does not already have parents.

After finding the last of the new snapshot, which is no longer in a local database, the tool can begin querying metadata objects and data themselves. If the tool downloads all the objects bound to this snapshot, it can save it to our database as the last snapshot with the same identifier and continues on the next new snapshot, which does not yet have. This continues until all remote snapshots are synchronized. After synchronization is complete, the tool sends notification that it is synchronized.

When saving the snapshot to the database, it is important to maintain the snapshot identifier. This is because the tool, which originally belonged to, could detect its own snapshots in the foreign database profiles and does not need to download them again. Otherwise this could cause a recurrence of downloading own data.

When posting a pull message, the party can specify whether the remote node has to send changes that gradually appear while working with the working directory of the volume. Another message can stop sending these changes at any time.

If there is an error in the determination of the response, the tool has an option to send an error message together with a description of the error.

For efficient data transfer between the nodes, we have used the rsync delta algorithm. This can be used in cases where one party requests data for a particular file, has also available an earlier version of the file. In other cases, it is necessary to always send the entire file. The use of this algorithm is not mandatory, even in cases where this would be appropriate. Some of the parties may not have this extension implemented. (Martak, 2013)

5 STATUS AND DATA RETENTION

Files and directories of the current working directory are synchronized according to the defined volume. Each volume has a fixed path on the system and an optional name. The version history of files may be kept according to the user settings.

Next, it is needed to store all the information for the current status, all metadata for files and the file data itself. Each node (peer), the object and the volume will be identified using the UUID3, which should guarantee that none of the listed entities will have the same identification number. Object is data, metadata, and snapshot (snapshot). The snapshot represent a new version of the changes and can contain one or more items of metadata. (Tridgell, 2008; Martak, 2013)

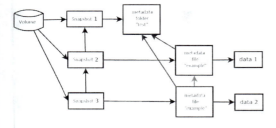

Figure 2. Object hierarchy.

We can see an example of the possible object hierarchy on the Figure 2 (Martak, 2013). There is a single volume named volume. In this volume, there were gradually made three changes, which represent the three snapshots. Snapshots are dependent on each other to make it possible to determine the sequence of changes. Snapshot 1 has no parents, so it is the first change. Snapshot 1 shows the object type metadata directory called test. The metadata of the test does not have any direct ancestor; this means that the directory was created. Since the directory itself cannot contain data, it cannot refer to a data object. The second snapshot indicates the type of metadata file named example. This metadata belongs to the directories test, this is expressed in the blue arrow. The metadata of the

type file can contain data, which in our example is the arrow on the object data 1. There is also a snapshot 3, which shows the metadata for a type of file that is named example. This metadata have an ancestor, which expresses the green arrow and also contains custom data, which expresses the arrow on the data object 2.

Another option is to identify objects by using the SHA-1 thumbprint as it makes Git. This could be difficult and inefficient on the platforms with limited resources. For simple detection of whether the data in the file has changed, it would be possible to use SHA-1 fingerprint file. This could simply detect whether file data has changed or not. If the data has not changed, new metadata object with the original data is created. (Tridgell, 2008; Martak, 2013)

If we want to save binary data in any of the databases, we have to count with their limits. All the considered database systems can save a maximum of 1 GB of binary data into one record. If files are larger than 1 GB, the data must be split into multiple records.

The stored data is always full. The tool will always make a full back up of the data. For the efficient storage of data, it would have been able to save only the changes between the two objects of data. It might be challenging, so the implementation will always save the data in the full form.

6 IMPLEMENTATION

The application is designed as a multi-threaded, as is the need to react asynchronously at a networking event and the events in the file system. Furthermore, it is necessary to use a blocking operation for user input. Overall, at least three threads together with main thread are needed. (Tridgell, 2008; Martak, 2013)

The main thread manages volumes and should contain the logic for the management of the entire application. The main loop in the main thread also ensures collecting requirements of class FSMonitor and larger clusters are added through the class VolumeController to the individual volumes. This ensures that if there will be more changes at once, all added with only one snapshot. If they were added gradually changes one at a time, for each such amendment would be needed to create a new snapshot.

After starting the tool, the initialization of the user comes, and the tool generates a certificate with the private key. If the user owns some volumes, their checking follows. Then, the three threads are created. One thread is for network communication and processing of remote requests, the other thread is for user input and the last thread is waiting on an event from the file system. For synchronization between threads, we have used mutexes and in some cases, a shared queues. The program has been used for the design on an object-oriented paradigm. (Tridgell, 2008; Martak, 2013).

7 SUMMARY

The aim of this work was the development of effective synchronization and backup system. The system was tested on a random generated sample data to test its functionality for backing up and synchronizing the data and to test its performance. The improvement of the transmitted data efficiency has been achieved using the rsync algorithm, delta. When using this algorithm, during the synchronization only the changed data is transmitted. This saves the synchronization time bandwidth.

ACKNOWLEDGMENTS

This project has been carried out with a financial support from the Czech Republic through the project no. MSM0021630528: Security-Oriented Research in Information Technology and by the project no. ED1.1.00/02.0070: The IT4Innovations Centre of Excellence; the part of the research has been also supported by the Brno University of Technology, Faculty of Information Technology through the specific research grant no. FIT-S-14-2299: Research and application of advanced methods in ICT.

REFERENCES

Cohen, M. 2009. *Take Control of Syncing Data in Snow Leopard*, 1st Edition. Take control, TidBITS Publishing, Incorporated. ISBN 9781615420032.

Cougias, D., Heiberger, E., Koop, K. 2003. *The Backup Book: Disaster Recovery from Desktop to Data Center. Network Frontiers Field Manual Series*, Schaser-Vartan Books. ISBN 9780972903905.

Dubuisson, O. 2000. ASN. 1: *Communication Between Heterogeneous Systems*. ISBN 0-12-6333361-0.

Georgiev, M., Iyengar, S., Jana, S. 2013. The most dangerous code in the world: validating SSL certificates in non-browser software. In *ACM Conference on Computer and Communications Security*. URL <https://crypto.stanford.edu/~dabo/pubs/abstracts/ssl-client-bugs.html>.

Martak, A. 2013. Synchronization and Backup of Data under OS Linux. FIT Brno University of Technology.

McCabe, J. 2010. *Network Analysis, Architecture, and Design. The Morgan Kaufmann Series in Networking*, Elsevier Science. ISBN 9780080548753.

Pilato, M., Collins-Sussman, B., Fitzpatrick, B. 2008. *Version Control with Subversion*. Holley Series, O'Reilly Media. ISBN 9780596510336.

Tridgell, A. 1999. *Efficient Algorithms for Sorting and Synchronization*, The Australian National University.

Analysis of temperature field for the spindle of machining center based on virtual prototyping

T.T. Lou, W.H. He & K. Lin

College of Quality and Safety Engineering, China Jiliang University, China

ABSTRACT: Thermal balance design for the spindle of machining center is important for its rotary precision and precision of retention. Thermal analysis for spindle-based Virtual Prototyping helps design thermal balance at a low cost. Pro/E software was applied to establish the 3-D Virtual Prototyping model. Bearing friction heat and coolant convective heat transfer were calculated. On this basis, the temperature field of the spindle system was built by finite element analysis and the maximum temperature rise was increased in the spindle journal. The regression model of temperature and coolant flow was established. The model can describe both the correlation and variation quantitatively. The results provide a theoretical basis for thermal balance optimization and experiments of the spindle.

KEYWORDS: spindle of machining center; Virtual Prototyping; finite element analysis; temperature field; flow; regression model

1 INTRODUCTION

Virtual prototyping technology is a numerical design method based on computer modeling and simulation. Since the early 1990s, virtual prototyping technology has been applied in many fields. Virtual prototyping technology avoids the build of physical prototypes, to shorten the period of development and to reduce the cost of design. However, it has the advantages of low cost, high speed, strong flexibility, reducing design iterations, and realizing concurrent design [1].

Machining center has played a more important role in industrial manufacturing, and the requirements of its machining precision is also increasingly high. Large number of experiments and surveys showed that machining errors caused by thermal deformation accounted for 40% ~70% in precision machining, whereas the thermal deformation of the spindle system is one of the main factors causing thermal deformation of machine errors. Therefore, in order to make the spindle maintain high accuracy at high speed, reducing temperature rise of the spindle is an effective method. In order to reduce the spindle temperature and improve cooling efficiency, coolant circulating cooling method is usually used for forced cooling of the spindle system [2]. This article uses finite element method to analyze the thermal characteristics of the spindle system and discusses the influence of coolant flow on the spindle temperature rise. The results provide a reference for optimizing the structure of machining center design and reducing the thermal error.

2 STRUCTURE OF THE SPINDLE SYSTEM

The research object is the spindle system of NBP1100 vertical machining center. The structure of the spindle system is mainly composed of spindle, bearings, sleeve, and end cap, as shown in figure 1. The spindle adopts FAG precision spindle bearings, which can bear radial force and axial force simultaneously. The spindle can automatically change the tools with an automatic oil cylinder. The spindle cooling system is water cooling. The oil cylinder is a double-acting hydraulic cylinder, with a signal ring for working position at the back of the cylinder rear spindle running signal output, to ensure the reliability while tightening or loosening the tool [3].

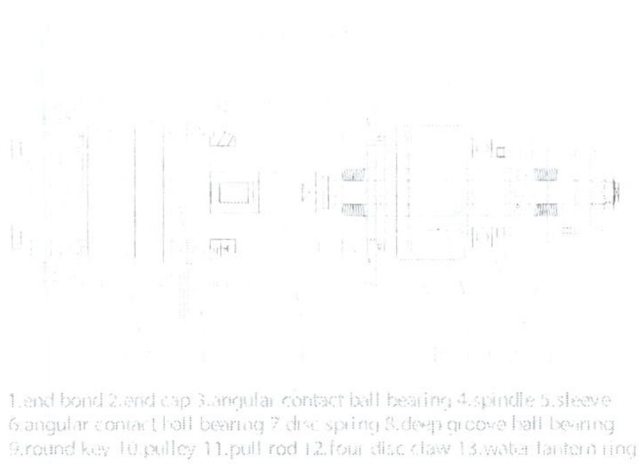

1.end bond 2.end cap 3.angular contact ball bearing 4.spindle 5.sleeve 6.angular contact ball bearing 7.disc spring 8.deep groove ball bearing 9.round key 10.pulley 11.pull rod 12.four disc claw 13.water lantern ring

Figure 1. Structure of the spindle system.

3 FINITE ELEMENT CALCULATION METHOD FOR THE TEMPERATURE FIELD OF THE SPINDLE SYSTEM

3.1 Spindle system modeling

The text is set in two columns of 9 cm (3.54") width, each with 7 mm (0.28") spacing between the columns. All text should be typed in Times New Roman, 12 pt on 13 pt line spacing except for the paper title (18 pt on 20 pt), author(s) (14 pt on 16 pt), and the small text in tables, captions, and references (10 pt on 11 pt). All line spacing is exact. Never add any space between lines or paragraphs. When a column has blank lines at the bottom of the page, add a space above and below headings (see opposite column).

The virtual prototype model of the spindle system is established by 3D software Pro/E strictly according to drawing dimension. The geometric size of the spindle system model is similar to the actual situation, so that the analysis and calculation can accurately reflect the thermal characteristics of the spindle system, as shown in figure 2.

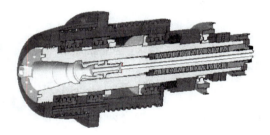

Figure 2.   Virtual prototype model of the spindle system.

In order to save computer space and improve the speed of finite element analysis, the model is simplified without affecting the analysis results. Fine structures that have no effect on the analysis results, such as threaded holes, chamfers, fillets, undercuts, and so on, are omitted. Considering the high spindle speed, the balls of bearing are simplified to be a circular ring with constant cross-section. Then, the simplified physical model can be imported into the ANSYS finite element analysis software. In order to make the various parts of the spindle system independent of each other and have a common boundary, using glue command, we ensure that all volumes and areas are attached to the boundary [4]. Considering the irregularity of the spindle system, the model adopts Solid7.0 unit for free meshing; the finite element model is shown in figure 3.

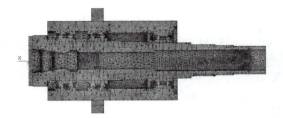

Figure 3.   Finite element model of the spindle system.

3.2 Boundary conditions for finite element analysis of the spindle system

3.2.1 Friction heat of bearing

With the function of every kind of heat source during the manufacturing process, machine tools have formed a non-uniform temperature field. The heat sources of spindle are complex, in which greater impacts on the machine tool are bearing friction heat. The heating intensity can be expressed as follows:

$$Q = \frac{M \cdot n}{9550} (KW) \qquad (1)$$

where M is the friction moment ($N \cdot m$), and n is the spindle speed (r / min).

The total friction moment of bearing is composed of two terms:

$$M = M_0 + M_1 \qquad (2)$$

M_0 is the moment related to bearing type, speed, and lubricant properties. Equation (3) and (4) are the functions applied for the calculation of M_0.

When $\upsilon n \geq 2000$, $M_0 = 10^{-7} f_0 (\upsilon n)^{\frac{2}{3}} d_m^3 (N \cdot mm)$ (3)

When $\upsilon n < 2000$, $M_0 = 160 \times 10^{-7} f_0 d_m^3 (N \cdot mm)$ (4)

where υ is kinematic viscosity (mm^2 / s), n is the spindle speed (r / min), f_0 is coefficient related to bearing type and lubrication, and d_m is mean bearing diameter (mm).

M_1 is the moment related to load on bearing. Equation (5) is the function applied for the calculation of M_1.

$$M_1 = f_1 P_1 d_m (N \cdot mm) \qquad (5)$$

where f_1 is the coefficient related to bearing type and load, P_1 is the equivalent bearing load, and d_m is mean bearing diameter (mm).

3.2.2 Convective heat-transfer coefficient

The form of heat transfer in the spindle sleeve is forced-convection heat transfer. Heat transfer coefficient is calculated by judging the Reynolds R_e in advance, then determining flow patterns, and finally choosing the appropriate empirical formula for calculation.

R_e about cooling fluid in the spindle sleeve can be obtained by Equation (6).

$$R_e = \frac{u \cdot D}{\upsilon} \qquad (6)$$

where D is the setting scale of geometric feature (m), u is fluid characteristic speed, (m/s), and υ is kinematic viscosity (mm^2/s).

The results of R_e in this article are greater than the critical value of the steady state, so the steady-state forced convection heat transfer criteria are used to calculate heat transfer coefficient α. The formula is used in Equations (7) and (8):

$$\alpha = \frac{N_u \cdot \lambda}{D} \qquad (7)$$

$$N_u = C R_e^n P_r^m \qquad (8)$$

where N_u is nusselt number, λ is fluid thermal conductivity, D is setting scale of geometric feature (m), and P_r is fluid prantdl number [5,6].

3.3 Temperature field of spindle system simulation example

Referring to the actual processing environment of the machine tool, analysis in temperature field of the spindle is conducted under the following conditions: (1) the ambient temperature is 20°C, (2) the viscosity of grease is 20 mm^2/s, (3) the spindle speed is $n = 4000r/\min$, (4) the coolant flow of cooling system is 2.4 $L/\min$, and the inlet temperature is 20°C. In this case, through the front of the formula, heat quantity of the front bearing is 583 W, the rear bearing is 416 W, and heat transfer coefficient is 4127 W/m^2K. The calculations are applied in the finite element model of the spindle system; then, the final temperature field of the spindle system can be obtained, as shown in figure 4.

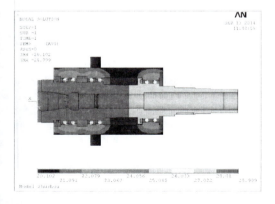

Figure 4. Temperature field of the spindle system.

Figure 4 shows that the maximum temperature of the spindle system is 28.999°C, located in the front bearing inner ring. The minimum temperature is 20.102°C, located in the sleeve surface. In this work environment, the maximum temperature rise of the entire spindle system is 8.999°C [7–9].

4 QUANTITATIVE MODELING BETWEEN COOLANT FLOW AND TEMPERATURE RISE OF THE SPINDLE

The main cooling measure in the spindle system is cooling by coolant. It has a decisive influence on rotary precision of the spindle. Because the rotary precision of spindle is mainly determined by the geometric precision at the journal, this article intends to set up a relation model of coolant flow and journal temperature. The model can describe both the correlation and variation quantitatively. Spindle temperature distribution is shown in figure 5. The maximum temperature appears in the front bearing end, because the front bearing adopts two pairs of angular contact ball bearings. By selecting 3 nodes in the journal, as shown in figure 5, the coolant flow ranges from 0.6 $L/\min$ to 2.4 $L/\min$, and other conditions of temperature field analysis are the same; the results are shown in table 1.

Figure 5. Temperature field of the spindle.

Table 1. Temperature of journal nodes.

No.	flow ($L/\min$)	No.1 temp (°C)	No.1 temp rise (°C)	No.2 temp (°C)	No.2 temp rise (°C)	No.3 temp (°C)	No.3 temp rise (°C)
1	2.4	28.989	8.989	28.742	8.742	27.602	7.602
2	2.2	29.056	9.056	28.796	8.796	27.651	7.651
3	2.0	29.133	9.133	28.859	8.859	27.708	7.708
4	1.8	29.225	9.225	28.933	8.933	27.777	7.777
5	1.6	29.337	9.337	29.023	9.023	27.859	7.859
6	1.4	29.476	9.476	29.135	9.135	27.962	7.962
7	1.2	29.653	9.653	29.278	9.278	28.094	8.094
8	1.0	29.889	9.889	29.467	9.467	28.270	8.270
9	0.8	30.224	10.224	29.735	9.735	28.519	8.519
10	0.6	30.742	10.742	30.151	10.151	28.909	8.909

The regression curves with coolant flow and temperature rise in journal nodes 1, 2, and 3 are shown in figure 6 (a), (b), and (c). S is the goodness of fit, accounting for the curve fitting degree. R-Sq is correlation index, expressed as a percentage for total error of the regression model error. It can be seen from the figure that temperature rise is significantly reduced with the increase of flow, and they are not linearly related but in a curve form. R-Sq is 99.9%, showing that the impact of flow on the temperature rise is significant, and also the regression model fits well. The regression equation of figure 6 (a) can be expressed by equation (9).

$$T_1 = 12.96 - 4.975Q + 2.307Q^2 - 0.3855Q^3 \quad (9)$$

where T_1 is the temperature rise in journal nodes1 (°C), and Q is coolant flow ($L/\min$).

The regression curve with coolant flow and temperature rise in journal node2 is shown in figure 6 (b). The temperature rise decreased compared with node1 at the same flow, but the decline trend is similar to node1. The regression equation of figure 6 (b) can be expressed by equation (10):

$$T_2 = 11.93 - 3.985Q + 1.846Q^2 - 0.3087Q^3 \quad (10)$$

where T_2 is the temperature rise in journal nodes2 (°C), and Q is coolant flow ($L/\min$). The regression equation of figure 6 (c) can be expressed by equation (11):

$$T_3 = 10.57 - 3.742Q + 1.741Q^2 - 0.2916Q^3 \quad (11)$$

where T_3 is the temperature rise in journal nodes3 (°C), and Q is coolant flow ($L/\min$).

In general, if R-Sq >75%, there is correlation between the variables, and the regression analysis can be used with caution. If R-Sq is more than 85%, the relationship is significant. In figure 6, R-Sq>85%, temperature rise is significantly reduced with the increase of flow. It shows that the coolant plays a good role in spindle cooling, and the change of temperature rise with coolant flow can be quantified by the regression equation to be predicted.

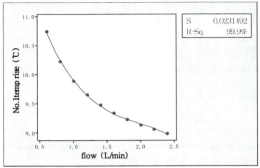

(a) Regression curve in node1

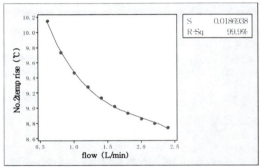

(b) Regression curve in node2

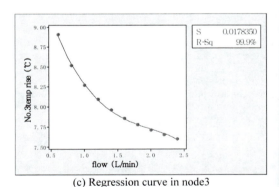

(c) Regression curve in node3

Figure 6.   Regression curves of temperature rise in different journal nodes.

5   CONCLUSIONS

This article adopts PROE, ANSYS software to build a virtual prototype of a spindle system. Boundary conditions for finite element analysis are determined by bearing heat and cooling heat dissipation theory. This method simplifies virtual prototype model reasonably, meshing by Solid7.0, improving computational efficiency, and it finds the highest temperature in spindle journal, which has a direct impact on the spindle rotary precision. This article also sets up a relation model of coolant flow and journal temperature. The model can quantitatively describe both the correlation and variation.

In this article, temperature field analysis of machining center spindle by virtual prototyping technology shows great superiority. The correlation analysis can also be used for the optimization design of spindle cooling. The results can improve factory precision and precision preservation of the spindle system.

REFERENCES

[1] Li R T, Fang M, Zhang W M, Peng L Z, Conception and application of virtual prototyping technology, Metal Mine. 7 (2000) 38–40.

[2] Qiu S H, Wu B C, Su C, Research on the situation analysis and improvement countermeasure for the reliability of domestic numerical control machine, Manufacture Information Engineering of China. 7 (2009) 1–4.

[3] Yu X, Research and analysis of the thermal deformation of the spindle system, Applied Energy Technology. 10 (2008) 47–50.

[4] Robinson D, Palaninathan R, Thermal analysis of piston casting using 3-D finite element method, Finite Elements in Analysis and Design. 37 (2001) 85–95.

[5] Nan J, Zhou X M, Sun Y, Thermal analysis on vertical machining center's motorized spindle, Equipment Manufacturing Technology. 7 (2009) 10–13.

[6] Li X S, Wang A L, Thermal structural coupling analysis of motorized spindle based on ANSYS, Group Technology & Production Modernization. 4 (2013) 5–8.

[7] Chen D, Feng M, Analysis in steady-state temperature field of high-speed motorized spindle, Modular Machine Tool & Automatic Manufacturing Technique. 5 (2014) 54–57.

[8] Chang C F, Chen J J, Thermal growth control techniques for motorized spindles, Mechatronics. 19 (2009) 1313–1320.

[9] Creighton E, Honegger A, Tulsian A, Analysis of thermal errors in a high-speed micro-milling spindle, Machine Tools & Manufacture. 50 (2010) 386–393.

Information, Computer and Application Engineering – Liu, Sung & Yao (Eds)
© 2015 Taylor & Francis Group, London, ISBN 978-1-138-02717-6

Disconnector mechanism operating force testing system

Wu Jin Li, Zhang Jun Ye, Jie Chen, Wei Heng Wang, Zheng Yu Yuan, Xuan Yao Wen & Hua Zhang
State Grid Sichuan Technical Training Center, Chengdu, Sichuan Province, China

Song Qi Li
Southwest Jiaotong University, Chengdu, Sichuan Province, China

ABSTRACT: Disconnector mechanism operating force test system is primarily intended to test the power system in each typical disconnector transmission component under various conditions of stress. Based on analysis of force characteristics parameter of the switching mechanism operating force, the disconnector fault detection rate would be, ultimately, enhanced and the maintenance time required for locating faults and examining a single device would be shortened.

KEYWORDS: Disconncetor; pull; test; characteristics parameters; fault detection

1 HARDWARE COMPOSITION PRINCIPLE

Disconnector mechanism operating force test system is primarily composed of torque sensors, signal transmitters, power modules, PC, and multiple cables. The block diagram of its hardware components is shown in Figure 1.1.

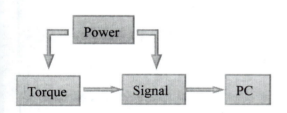

Figure 1.1. Block diagram of disconnector mechanism operating force test system.

The main working principle: The torque value that is detected by the torque sensor is converted to an electrical signal. The signal is then transmitted to the PC through USB interface after signal amplification transmitter and digital processor. Simultaneously, the data are manipulated by the supporting software to complete records, store, for operational analysis, generate reports, and so on.

1.1 *Power module*

Power module uses a lithium-ion polymer battery. Compared with the traditional liquid lithium-ion batteries, this kind of battery uses a solid electrolyte instead of a liquid electrolyte; therefore, it has a higher energy ratio; exhibits more safety and reliability, a longer life span, and other advantages.

1.2 *Torque sensors*

Static torque sensors used in this system adopt strain gauge technology. The extensimeter is pasted on the elastic axis to compose measuring a bridge. When the electrical resistance of the bridge changes, which is caused by slight changes in the elastic axis, the changes in the resistance of the strain bridge would be converted into the changes in electric signals in order to achieve torque measurement. The technology has high accuracy, fast response, good reliability, long life and other advantages.

1.3 *Signal transmitter*

The signal transmitter amplifies the sensor output signals precisely. And the inner circuit processes voltage stabilization, constant current for the bridge, the voltage-current converter, impedance adaptation, linear compensation, temperature compensation, and so on. This process converts mechanical quantity into a standard mechanical quantity current. The voltage signal output USB interface is connected with the host computer directly.

1.4 *Hardware connection*

The disconnector mechanism operating force test system consists of the following steps to complete wiring:

1 To connect the battery output port to power input port in the transmitter, the power supply is provided by the sensor transmitter;
2 To connect the signal cable to the transmitter of the sensor;
3 To connect the USB interface in transmitter signal output line directly to the host computer USB.

2 HOST COMPUTER SYSTEMS

After clicking the "Test" button, the system begins to collect data. And real-time data points are plotted on the right side of the chart. The system stops the collection after the "Stop" button is clicked.

After clicking the "Report" button, the selected data generate Excel reports, which includes data base information, data analysis, data curve, and the original data.

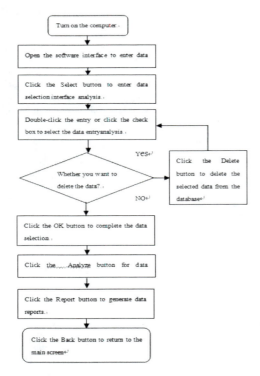

Figure 2.1.   Software system work flow chart.

3 SUMMARIES

Based on the comparison of test data curve and a normal curve apparatus, the following fault conditions can be deduced according to the steepness and the smoothness of the curve:

1 Low precision and big tolerance in processing bearing units without sealing would not guarantee an exact fit between the drive components, which results in an unstable operation and unreliable transmission For example, the bad mesh between the transmission racks and gears would cause wear in machine and non-self locking between the components. The rough processing part and inconsistent connection tightness would cause some problems in manual operation
2 Neither lubrication measures nor lubricating grease is used between drive links. With increasing growth of the running time and the number of operations, the lubricating grease would be dried or drained and bearings and axle pin would be corroded and worn, which results in clamping stagnation in rotating parts and changes in mass transport characteristics.
3 There is low strength and corrosion in axis pin. Rust is the greatest risk for causing malfunction. Rust would result in clamping stagnation when connecting rotary movement to rotation drive or the low strength of the mechanical components, which results in component deformation.

REFERENCES

[1] Zhang Yigang, & Peng Xiyuan (2006) 'Automatic testing', Harbin Institute of Technology Press, 2(3), pp 38–56.
[2] Lu Lu&Wang Boyong (2006) 'Software automatic testing technology', Tsinghua University Press, 5(4), pp 211–245.
[3] Zhang Jin & Du Chunhui (2008) 'Software automatic testing', China Machine Press, 3(2), pp 88–103.
[4] Chen Changling (2008) 'automatic testing and interface technology', China Machine Press, 7(1), pp 239–264
[5] Xie Shiliang (2004) 'Visual C#.NET 2003 Development and skills' Tsinghua University Press, 4(3), pp 57–71.
[6] Shi Jiaquan(2006) 'The database system course', Tsinghua University Press, 7(4), pp 358–379.

Information, Computer and Application Engineering – Liu, Sung & Yao (Eds)
© *2015 Taylor & Francis Group, London, ISBN 978-1-138-02717-6*

Prototype design of toilet reminders based on Arduino

Tie Cheng Zhang

Dalian Economic Technological Development Zone, Dalian University, Liaoning, China

ABSTRACT: This article demonstrates how to design the reminders used in toilets for people who have the bad habit of playing with the phone or reading newspapers in the bathroom for a long time. The prototypes were designed and realized based on the Arduino electronics platform. The Arduino electronics platform was the control center, and the infrared body sensor block was the sensor cell in the prototypes. 9V DC battery was the power supply; the functions of reminder and alarm were realized by the voice shield for Arduino connected with an external speaker. The appearance was designed by Creo, and the internal circuit was designed based on Arduino. The prototype realized the desired functionality after repeated testing and improved the bug.

KEYWORDS: Arduino; Prototype design; Reminding device

1 INTRODUCTION

More and more people have the habit of reading newspapers or playing games on mobile phones when going to the toilet, as they think it is an effective way to save time. According to the research from a published journal, it may trigger off diseases, especially habitual constipation, and even diseases such as anal congestion, hemorrhoids, and so on[1]. The bad habit of reading newspapers or using mobile phones in the toilet must be changed for one's personal health. It is necessary to design a new device to solve this problem. The new device was designed by the Arduino shield and sensor block, has the function that monitor whether someone before toilet, calculate the duration in toilet, reminders and alarm in a specify time. The goal behind developing the device was to inculcate good habits in young people in the toilet, to remind them to finish as soon as possible, and to leave the toilet as soon as possible.

2 OVERALL STRUCTURAL DESIGN

The prototype was designed by the Arduino electronics platform: The Arduino was the control center, the PIR sensor was the monitor, and the voice shield and speaker were the main parts. The main functions were switching the power, monitoring the status of people, timing the time spent in the toilet, supplying devices and the reminder to be left by the voices. The main principle of the reminders was that the predefined voice message embedded in the reminder will be played when people remain in the toilet beyond the specified time, and the alarm voices that are played exceed the maximum time. The overall structural system is shown in Fig. 1.

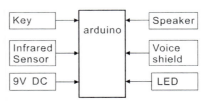

Figure 1. Overall structural system.

3 OVERALL WORK FLOW AND WORK PRINCIPLE

The device was installed in the family bathroom: The location may be directly in front of the toilet or on the left of the toilet, which was aligned with the people sitting position. The location is shown in Fig. 2.

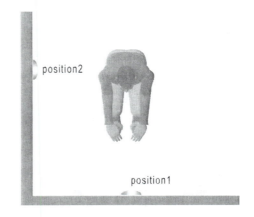

Figure 2. Installed location.

As depicted in the workflow given next, the PIR sensor block can sense the people status: When people are in the toilet, the timing function will be started automatically according to the duration when people are in the toilet for more than 1 minute. The reminders will play the predefined embed voices, if people go away later than the time assigned originating from the usual time of going to the toilet. The alarms sounds remind people to leave as soon as possible, if the time in the toilet more than the maximum time. The system will stop automatically when people leave, adopt the standby mode, which a low sense rates once every minute. The switch should be turned off to shut down the power when unused for long times or in an emergency state, such as when cleaning the bathroom. The workflow is shown in Fig. 3.

Figure 3. The workflow.

4 HARDWARE DESIGN

4.1 Controlled switch

The controlled switch is a simple button switch that is found in a series with the 9V DC power supply. It controls the circuit on and off and shuts down the reminders in emergency.

4.2 Controlled center

Arduino is the control center in the prototype; most functions are realized on the open electronics platform; the PIR sensor is the monitor used to sense the external environment whether people before the toilet, the voice shield and speaker were the mainly part to deliver the sound information, which stored in SD cards and called program written in Arduino. The power indicator and alarm states lights are a 5mm LED that is also connected and controlled by the Arduino.

Arduino is an open-source electronics platform based on easy-to-use hardware and software. It is intended for anyone making interactive projects. Arduino can sense the environment by receiving inputs from many sensors, and it affects its surroundings by controlling lights, motors, and other actuators. There are multiple versions of the development board, and the Arduino Uno developed in 2012 was the first choice in the prototype. It is a microcontroller board based on the ATmega328. It has 14 digital input/output pins (of which 6 can be used as PWM outputs), 6 analog inputs, a 16 MHz ceramic resonator, and a USB connection. It contains everything needed to support the microcontroller; it simply needs to be connected to a computer with a USB cable or powered with an AC-to-DC adapter or battery to get started[2].

4.3 Monitoring functions

The monitoring functions are realized by the Pyroelectric infrared sensor controlled by the open electronics platform; its composition is determined by sensitive unit, impedance transformer, and optical window. It has been widely used ever since it was developed as a new type of sensor in the late 1980s. Only the motion and temperature of the human body can be sensed by the sensor. The distance a common sensor can sense was inadequately 2meters. It can be enlarged and widely equipped with the Fresnel lens before the sensor[3]. The area of the bathroom was less than 6 m² in common families, and the sensor available can meet the design requirements.

4.4 Power supply

The Arduino is powered via the USB connection or with an external power supply. External power can come from either an AC-to-DC adapter or a battery. The board can operate on an external supply of 6 to 20 volts. If supplied with less than 7V, however, the 5V pin may supply less than five volts and the board may be unstable. If using more than 12V, the voltage regulator may overheat and damage the board. The recommended range is 7 to 12 volts[4]. So the 9V battery is the best option as the external power supply. The battery can be replaced easily when the devices prompt the low battery status with LED blink and dim.

4.5 Reminder and alarm function

The functions of reminder and alarm are realized by the external hardware. The Arduino is not built to be a synthesizer, but it can certainly produce sound through an output device such as a speaker. The sound produced from arduino does not meet its requirements for it is simple and boring. The SparkFun

VoiceBox Shield extends Arduino's capabilities. It can be mounted on top of Arduino board to give it access to all of the capabilities of the SpeakJet voice and sound synthesizer [5].

4.6 Indicator light

The 5mm LED lights used as the light emitting part remain alight when power is on, and they blank out when in the status of a voice reminder.

5 CIRCUIT DIAGRAMS

Circuit diagrams of the reminder drawn by "fritzing" constitute an open-source hardware initiative that makes electronics accessible as a creative material for anyone. Fritzing was installed with a parts library: In the library, parts frequently used were available and new parts were added in every version. It is very convenient to connect electronic components according to the desired goals; for instance, in the circuit, the 9V DC, button key series with the arduino's DC input pin. The Pyroelectric infrared sensor connected the Arduino with the pin of GND,5V and "13"; the LED was connected with Arduino in the pin "2" and GND pin; and the series used a 200Ωresistance to reduce the current. The speaker connected the SparkFun VoiceBox shield by the specified pin, and the SparkFun VoiceBox shield connected the Arduino in a designated pin with the wire. The circuit diagrams are shown in fig. 4.

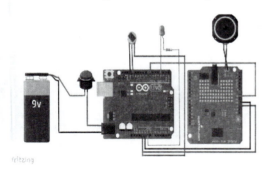

Figure 4. Circuit diagrams drawn by fritzing.

6 APPEARANCE DESIGN

The electronic parts were modeled accurately and the layout was simulated in Creo, which was a revolutionary 3D product design software issued by PTC company. Aesthetics and reasonable space were the first considered factor in appearance design. The appearance of the reminders is similar to a hemisphere: The Pyroelectric infrared sensor, button key, and LED were arranged on the spherical end; the Arduio shield, the SparkFun VoiceBox shield, and 9V DC were in the layout inside the hemisphere. The internal component layout in shown in fig. 5. The reminder was hanging by the screws, whose fixation was on the reserved hook on the wall. The appearance is shown in fig. 6.

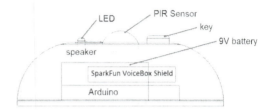

Figure 5. Internal component layout.

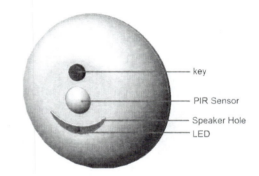

Figure 6. Appearance of reminder.

7 SUMMARY

The Arduino has the advantages of getting started quickly and having more parts available. It is very suitable for the prototype design for amateurs. The desired function can be realized quickly and exactly by the electronic building blocks with the help of a simple programming knowledge. The prototype of reminder was realized by the Arduino hardware and software. It realized the expected requirements after several debugging attempts.

REFERENCES

[1] Hyung Kyu Yang: Hemorrhoids (Springer Berlin Heidelberg, Berlin 2014).
[2] Massimo Banzi: Getting Started with Arduino (O'Reilly Media, Sebastopol 2011).
[3] Weiguo Gong, Ke Wen, Lifang He, Lihua Cheng, and Yong Li: submitted to Journal of Thermophysics (2012).
[4] Information on http://www.arduino.cc/
[5] Information on https://www.sparkfun.com

Information, Computer and Application Engineering – Liu, Sung & Yao (Eds)
© 2015 Taylor & Francis Group, London, ISBN 978-1-138-02717-6

Research on practical teaching reform of applied undergraduate computer major under the university—enterprise co-cultivation mode

Hang Chen

Department of English, Tianhe College of Guangdong Polytechnic Normal Uniersity, Guangzhou, China

ABSTRACT: With the popularity of colleges and universities as well as the reform of the educational system, applied undergraduate computer major must be reformed and innovated to adapt to talent training requirements under the university–enterprise cooperation mode. talents training under the university–enterprise cooperation mode must use new ideas and thoughts to break through traditional teaching modes, introduce international standards of computer industry technical knowledge, improve the computer talents training level in china, and realize positive and innovative talents training mode with resources to be shared. this article mainly discusses the applied undergraduate computer professional practice teaching reform measures under the university–enterprise co-cultivation mode.

KEYWORDS: University–enterprise co-cultivation, Applied, Computer major, Practical teaching, Reform

1 INTRODUCTION

According to different methods of talent cultivation, China's colleges and universities can be divided into three types: vocational technical colleges and universities, professional applied universities, and comprehensive research universities. Among them, professional applied universities are intended to cultivate applied talents and main components of China's institutions of higher education. This kind of university aims at cultivating adaptable talents, emphasizing cooperation and exchanges with enterprises, and realizing the goal of complementary advantages between universities and enterprises and resources sharing. Computer major in university is a subject with high applicability and practicality. It should actively conduct practical teaching reform and improve the teaching level under the university–enterprise cooperation mode.

2 INTRODUCTION OF UNIVERSITY–ENTERPRISE COOPERATION SCHOOL-RUNNING MODE

The cultivation of professional talents in colleges and universities lays stress on the combination of classroom teaching and practice teaching. "University–enterprise cooperation" refers to a kind of education mode in which school classroom learning, revenue plan, and practical work experience are combined together and this practical work experience should be associated with students' learning goals and professional goals. That is to say, schools and enterprises establish a new type of cooperative relationship and enterprises are involved with students along with their own advantages. University–enterprise cooperation is the key to implementing a combination of working and learning and talents training mode reform. University–enterprise cooperation is of great significance in the applied computer major. First, enterprises know markets and can provide talent training plans and a lot of market information. According to the information, universities optimize and integrate talent cultivation modes and make methods of talent training more adaptable to the market demand. University–enterprise cooperation can improve students' practical abilities, which are valuable resources for students majoring in computer. Students can experience the development trend and work environment of the computer industry in advance and study pertinently. University–enterprise cooperation can improve students' dedications to work responsibilities, professional skills, and professionalism.

3 PROBLEMS ENCOUNTERED IN PRACTICAL TEACHING REFORM OF APPLIED UNDERGRATUATE COMPUTER MAJOR UNDER THE UNIVERSITY–ENTERPRISE CO-CULTIVATION MODE

The cultivation of applied talents is different from those of higher vocational specialties ordinary undergraduate majors. Higher vocational colleges cultivate talents according to the requirements of professional posts and pay attention to the development of practical technology. Ordinary undergraduate colleges and universities mainly cultivate people with good scientific literacy and focus on the cultivation of computer theoretical knowledge and technology. Applied undergraduate colleges and universities lay emphasis on professional training of thinking and cultivate practical technical talents. The cultivation of applied undergraduate talents with university–enterprise cooperation has become the mainstream of the current personnel training, but there are still more problems in the university–enterprise cooperation training mode after years of practice.

3.1 Lack of good legal environment

After the implementation of education reform, the government heavily supports universities to develop university–enterprise cooperation. However, on the whole, China's university–enterprise cooperation policy is still imperfect. There is no protection of legal documents in some specific operation details and many defects exist in rights and obligations, which can cause the law to not effectively protect the interests of enterprises and universities and restrict the development of university–enterprise cooperation.

3.2 Applied undergraduate universities have insufficient abilities to meet the needs of enterprises

Applied undergraduate universities tend to implement talent training using research universities' talents training modes as references and pay attention to the teaching concept from theory to practice, often use traditional teaching ways in the process of teaching, consider theoretical knowledge and teachers as main bodies of teaching, and take exam grades as the main criteria for the teaching. This kind of teaching mode enables students' knowledge and abilities to be disjointed from enterprises' actual needs, and universities cannot effectively cultivate talents that enterprises really need.

3.3 Enterprises have inadequate impetus to participate in university–enterprise cooperation

The implementation of university–enterprise cooperation talent training mode can not bring about direct economic benefits for enterprises in a short time and may increase the cost of enterprise management and uniformity at the early stage. Therefore, some companies would not actively participate in university–enterprise cooperation and the embarrassing situation of universities' one-sided enthusiasm may appear, which can greatly influence the conduct of educational mode of university–enterprise cooperation.

3.4 Operation mechanism of enterprise cooperation does not conform to the requirements of talent cultivation reform

The key to university–enterprise cooperation is to introduce a batch of double-qualified teachers, but it is difficult for applied undergraduate universities to introduce double-qualified teachers due to restrictions in school-running conditions, title appraisal, scientific research funds, and evaluation systems. Traditional teaching modes of applied undergraduate universities have no reasonable evaluation system in the practical teaching link, and it is difficult for all aspects to adapt to the requirement of university–enterprise cooperation and restrict its smooth development.

4 PRACTICAL TEACHING REFORM MEASURES OF APPLIED UNDERGRUATE COMPUTER MAJOR UNDER THE UNIVERSITY–ENTERPRISE CO-CULTIVATION MODE

4.1 Construct applied undergraduate computer curriculum system with characteristics

The setting of curriculum should break through the traditional setting mode, combine curriculum setting and professional ability with training objectives around the cultivation of students' application abilities, screen out professional knowledge that is directly related to application ability training and has higher use frequency according to position requirements, and form a curriculum system that gives priority to comprehensive ability and breaks through the skilled application combining with the practical education. Meanwhile, it should also consider the market's demand for diversity and compound type, improve talents' abilities to adapt to society, and pay attention to training students' innovative and practical abilities. Moreover, it should build a professional curriculum system and set corresponding courses according to the requirements of position ability and work contents. Curriculum setting should highlight the necessary basic knowledge in engineering theory and professional knowledge training, lay emphasis on the mastering of practical work ability and basic skills in the field of computer engineering, and give prominence to practicality and professionalism in the teaching process.

4.2 Positively construct double-qualified teachers

The key to applied talents training is to combine the production with practice. Teachers are guides of talents. For talent cultivation, teachers should have strong abilities of application and abundant practical experience. Otherwise, they would be armchair strategists.

Therefore, applied undergraduate universities should possess a batch of double-qualified teachers. They can establish full-time and part-time teachers with external employment or internal training methods. Specific measures involve arranging a batch of young teachers to conduct professional practice in cooperative enterprises, enriching their practical experience, and improving their abilities to solve practical problems. Moreover, universities invite experts and engineering technicians with rich experience in cooperative enterprises as part-time teachers and connect the class with production.

4.3 Establish talent training platform based on collaborative innovation

Universities should actively cooperate with enterprises and promote the development of enterprise integration. For applied undergraduate universities with good students and strong comprehensive abilities of students, they should reform university–enterprise cooperation mode, actively introduce teaching resources of world famous enterprises, and positively carry out foreign technology service and development in aspects, such as learning, research and development competition, and skill certification. Universities should implement the combined teaching mode of production-studying research and establish a university–enterprise cooperation environment with complementary advantages, resource sharing, and mutual promotion. Second, they should well make international collaboration, actively draw lessons from foreign successful school-running experience of applied undergraduate universities in the process of teaching, introduce cutting-edge computer technology into teaching, and make students understand and grasp operation modes of top computer companies as well as expand their horizons. Meanwhile, four aspects of strength can be collaborated and connected to conduct in-depth cooperation, actively cultivate practice abilities and innovation abilities of application-oriented computer talents, and create a talent cultivation system support platform with era significance.

5 CONCLUSION

In a word, the university–enterprise system is an important approach that is used to cultivate applied undergraduate talents. For computer science students, the university–enterprise cooperation teaching mode should be carried out positively to perfect the cultivation method with a combination of disciplinary knowledge and practical ability; make students form learning habits of active learning, autonomous collaboration, exploration, and innovation; and improve students' abilities of practical application.

ACKNOWEDGMENT

This study was supported by Guangdong Higher University Teaching Quality and Teaching Reform Project "Professional Comprehensive Reform Pilot Project" (Y. J. G. H. [2013] No. 113).

REFERENCES

[1] Wu Yufeng, Ye Feiyue and Hu Jieqiong: Computer Professional Talent Training Mode under the Perspective of Collaborative Innovation, Journal of Education and Vocation, Vol. 24 (2014), pp. 121–122.
[2] Fan Xincan, Sun Yong, Xu Renfeng and Li Bin: On the Talent Training Model of Applied Bachelor Talents of Computer Specialty Based on Collaborative Innovation, Journal of Vocational and Technical Education, Vol. 34 No. 14 (2013), pp. 108–109.
[3] Hu Tianming, Li Hongzhi, Dang Yuexuan and Xue Lijuan: Exploration and Practice of University-enterprise Cooperation School-running Mode in the Teaching of Cultivating Applied Undergraduate Talents, Journal of Heilongjiang Science and Technology Information, Vol. 13 (2008), pp. 154–155.
[4] Gao Yan, Liu Yongjun and Chang Jinyi: Practice and Reflection on College-enterprise Cooperation Talent Cultivating Pattern for Application-Oriented Undergraduate Colleges, Journal of Changshu Institute of Technology, Vol. 12 (2012), pp. 145–146.

Information, Computer and Application Engineering – Liu, Sung & Yao (Eds)
© 2015 Taylor & Francis Group, London, ISBN 978-1-138-02717-6

Literary texts and aesthetic transformation under full media age

Xiao Dan Diao

Liaoning Jianzhu Vocational University, China

ABSTRACT: In recent years, with the advent of the all-media era, the production and consumption of literature occurred in the field of major change. Under the new development environment, based on the traditional aesthetics, the concept of literature has been unable to effectively guide the development of new literature, literary texts, and aesthetic transformation imperative. Based on the transformation of both the literary texts and literary aesthetics, the whole media age was analyzed.

KEYWORDS: All media; literary texts; aesthetic; transformation; analysis

With the continuous development of the new era of the Internet, television, and so on, people have entered the whole media age; literary texts have gradually realized the literary market, a market economy to achieve a commodity literature between consumers and producers of docking, thus gradually changing the literary text mode and route of transmission, and so on. In the all-media era, with new media emerging, for different levels of audiences, diverse media terminal information is carried out in literature with categories of consumption, to achieve a change in the means of literature, receiving mode, and means of communication, to provide literature with a vast space and bred Cai Jun, baby Anne, a large number of Internet writers, promote China's literary texts and aesthetic transformation, for the comprehensive development of our literature has injected fresh blood.

1 TRANSFORMATION OF LITERARY TEXTS IN FULL MEDIA AGE

With the development of the new era, there is an impact on all media, making literature go beyond the boundaries of the text; it broke free of political, moral, philosophical, and other ideological constraints, expanding the theme of literature.

In previous development, the paper text is the main form of literary texts, and stress illustrated, so the whole media environment under the new era of electronic text is the most respected and popular major literary text. We include animation, audio, and web site to form a variety of ways in combination with the arrangement, sound illustrations, thus facilitating multi-angle multi-channel audience to its browser. In addition to the traditional way of receiving newspapers, radio endures, adding a variety of media forms microblogging, WAP sites, and the context of

development of the whole media age; paying more attention to the development of literature in the field of screen reading audience aesthetic taste, with an emphasis on sound, light, electricity, a comprehensive show of diagrams; and deliberately chasing fashion, avant-garde, and fragmentation- related operations through the text. In any text, editing and pasting into a new time and space will no longer limit the audience reading of literary texts, such as under the circumstances, the audience has to have a lot of control over the text, in this ongoing process of development, and gradually promote the literary text to shift weight from heavy graphics technology [1].

First, the main aspects of literature, from the past into a keyboard to manipulate the author grip hand, the new era literature gradually network name of the author, in particular the creative process, the general framework of a good idea of the text, the real creative use a variety of computer software to achieve, writers identity often interchangeable, full two-way communication and interaction between the media and the readers of the identity of such a creative subject of transgression, showing the process of literary creation has been achieved to some extent, the whole Internet users operation, the formation of the popular development of creative subject. Development of the whole media era, combined with the actual situation, literary writers, and the use of computer technology is quite skilled, but it denied their language and writing skills in general. This is because the writers of literature are closely related to the duties and computers. The current course of development, the gradual emergence of a non-linear continuous text and other literary texts, such as the most representative of the "multimedia novel" and so on. With the development of the new era, the past emphasis on retouching phenomenon text language has changed, and then more attention is paid to the integration of diverse electronic text messages, and a way to change the past consumption, combined with the actual development, the development of complex consumer and multi-directional consumption, The audience needs to be adjusted according to literary texts, the gradual emergence of dynamic multi-literary texts, such as Flash poetry, books, and so on.

Second, with the development of literary media, the electronic substitution of this type, literary writing in literary texts should pay more attention to interactive media technology, interactive virtual reality presentation, and production methods. All these became media era literature, after the media published the text, through the hands of their use of multimedia technologies, in order to read the audience accepted contents of literary texts, removal of the traditional text cage reader with a mouse jumped from one space to another space, use the space fully interactive development literature, such as the screen name "I eat tomatoes," book "stars change", a short time has reached 36 million hits, which is affecting its broad, fans were writing after the end of its truly realized the roles were reversed and interaction all media under broke through to the creators of literary narrative-centric approach, the interaction changes the aesthetic tradition of literature to accept the mentality, to promote the development of a wide range of literature [2, 3].

Finally, in the acceptance of literature, there has also been a great change from reading the newspaper into becoming the screen reader; the media group has gradually become the main mode of existence of the audience, which is popular with the media that evolved form to promote the writers' grassroots, lowering the threshold of its creation, such as in Annie's "Farewell Wei An," Xing Yusen "online fair lady," and so on. At first, they were just avid readers; times have changed and they are media, let the self some mental demands put in writing for the web, the achievements of his literary road. Development of high-tech computer changes formed the basic structure of literature in the public domain, the mass media influences and guides the public domain, and they do not speak for the mainstream consciousness. Literature reflects no utilitarian aesthetic, which is the formation of the spirit of self-identity.

2 LITERARY AESTHETIC TRANSFORMATION UNDER FULL MEDIA AGE

In addition to features such as creativity and fiction, the literature focuses more on its aesthetic development. With the advent of the all-media age, literary texts mass public consumption, making aesthetic back to the folk literature, close to the people's production and life, to achieve the national development.

First, entertaining literary aesthetic transformation. The new era of development, relying on all media, literature, and literary aesthetic spirit of the 20th century, rejected the in-depth mode, thus achieving a pan-public nature. Nonlinear evolution and secular literature occupy a hallowed halls Wen Road, resulting in literary ethics landslides and value anomie situation, together. No matter how sublime the network technology is, it always cannot completely replace the literature itself. Aesthetic properties of the essential characteristics of literature is literature, has far-reaching significance for the development of literature, it must be showing a beautiful, sublime aesthetic style, but the new era of literary aesthetics tend to be entertainment, ending the aesthetic tendencies of personality, turn to move closer to the consumer, grandstanding, the pursuit of aesthetic perception of commonality. After all, media era literature on audio, pictures, and other hypertext style utilized for reconstruction of the spiritual world audience, subject, and object are created to meet the aesthetics of secularity, carnival freedom, to overturn the traditional, for the entertainment-oriented development [4].

Second, the technical and artistic unity. Development of the new era, "see the screen" has become the main way to accept the text, the specific practice, through the video text, images, animations, etc. use to stimulate the reader's senses, to pursue their visual experience, the whole media age aesthetic of literary techniques such as forward direction, together, show their perfect unity focusing its technology and art, aesthetic psychology of the audience, emphasis on intuition pursuit. Pastime has turned a simple swap of sensory stimulation, virtual and reality blend of literature, and literary aesthetics gradually making mediocre, however, although the content of literature is fictional, but relying on animation, video and other media, on the basis of their creative play promote the network literature and other aesthetic development, and promote the diversification of the current literature.

3 CONCLUSION

In conclusion, the development of literature in the new period is based on the terms of the whole era of communication media and the impact of the creation of space. It faced a heavier weight graphic technology transformation zone, and this also affected the literary aesthetic transformation. The author analyzes the specific content and direction of development of its restructuring, on which the irrational development was focused in order to be able to fully provide a useful reference literature media under development.

REFERENCES

[1] Zeqing Wang. Aesthetic ideal all media changing times [J] Northern Forum, 2011, 01:43–46.

[2] Mao Jianhua. Literary form and literary value of the whole media era [J] Chinese Modern Literature Research Series, 2011, 08:192–199.

[3] Jingxiu Li. Audience aesthetic psychology research all media Perspective [D]. Jilin University, 2013.

[4] Hongwei Zeng.full media treatment of literature: literary treatment of the new space and a new realm [J] Central Plains Journal, 2012, 06:187–191.

Information, Computer and Application Engineering – Liu, Sung & Yao (Eds)
© *2015 Taylor & Francis Group, London, ISBN 978-1-138-02717-6*

University library cultural construction under network environment

Xiang Shen Kong
Zaozhuang University, China

ABSTRACT: The University Library culture plays an important role in the building and development of students. In the network environment, university library should actively make use of network information technology, strengthen libraries' cultural development, and constantly enrich the library culture, to create a good library culture and promote the sound development of libraries and universities.

KEYWORDS: Network environment; Library culture; Universities

University Library is not only the cultural center of colleges and universities but also an important place for school quality education, which shoulders functions of guiding students, education, cohesion, motivation, and so on, and plays a very important role in the growth and development of students and campus culture construction [1]. With the development of information technology and advancement of modern society, under the Internet age, university libraries must adapt to the times and actively use modern technology to enhance the library culture construction, to better meet the diverse spiritual and cultural needs of teachers and students.

1 UNIVERSITY LIBRARY CULTURE

University Library is an important site for gathering knowledge and information and heritaging culture, which can continue to analyze, select, and absorb external culture. On this basis, it is derived from the emerging social undercurrents culture system, possesses advanced features[2], and can pre-boot sociocultural development. With social development and the popularity application of information networks, the lively, open, and promptness characteristics of the network are prominently manifested, so that the University Library becomes a variety of information tide surging positions, as well as a place where a variety of culture exchange penetrates, which largely contributed to the accumulation from various social cultures. College students have a strong thirst for knowledge, coupled with a multicultural university collection; only diverse library culture is able to meet the needs of university readers and provide better services.

Library culture can create guidance and influence its spiritual atmosphere and corresponding cultural

carrier, and, thus, unknowingly purify the minds of students, infect their emotions, cultivate their sentiments, causing their resonation, and thus produce a subtle impact on the development of the individual student, so that students grow up to become conscious, motivated, and possess excellent quality talents. Overall, the university library culture is teaching practice while servicing the university; to some extent, it complements the university teaching practice.

2 PATHWAYS OF STRENGTHENING UNIVERSITY LIBRARY CULTURAL CONSTRUCTION UNDER NETWORK ENVIRONMENT

2.1 *Enrich material culture using network resources*

Library material culture mainly includes library architectural culture and libraries collections. Libraries built environment can have an impact on the choice of students and teachers. Therefore, in material culture construction, there should be a reasonable distribution and layout. Use modern information technology and automatic card-style layout, or create a shared online space to combine the traditional information print carrier resources and information technology and provide a convenient and comfortable environment for teachers and students[3]. The equipment installation and configuration of university libraries should allow readers to use books; interior design should be user friendly, and it requires a well-lit, well-ventilated, clean, and tidy place; and the harmony layout should create a relaxed, cozy reading environment and give full play to the library to exercise the cultural influence function. Library collections are the material basis of the library[4], which is the construction of literature resources; the rapid development of the network provides advanced network information technology and rich information resources for the library. University libraries can establish databases through network and improve digital and network of libraries culture, establish the information network system and better transit, receive advanced culture, and carry forward the spirit of the library culture. As the network of information resources are rich, the construction of libraries collections should be strictly checked. Based on actual collections and the demand for teachers, choose literature benefit for the development of teachers, students, and library; enhance safety awareness; and strengthen the construction of a library management system to prevent the spread of harmful information in the library, to ensure the authenticity and reliability of library culture and promote their healthy development.

2.2 Use modern technological innovation to serve culture

Quality service is an important based capital of library, and servicing reader is the daily work of the library. The service quality directly affects the satisfaction and trust of readers. An important focus of library culture is to serve culture. In the network environment, build a library service culture. First, establish a correct sense of service and do humanity service around the "reader-centered," pay great importance to the readers' dominant position, and make adjustments according to readers' distribution. The demand level and nature change, in order to provide high-quality, efficient, and convenient reading service.

On the basis of consciousness transformation, transform service patterns, face audience, take advantages of network integration to change the way of providing resources, service approach, and services content, turning from document-based services into knowledge services. Apart from continuously accumulated library resources building also do a multi-level on the resources development, all-round development to form a new pattern setting audio and video data, text data, network resources, and digital resources to provide benefits of resource development and utilization, which has a positive meaning to enhancing the quality and efficiency of service and better use of knowledge services to work, study, and life, which also contributes to the development of a greater degree of potential reader groups and is senior transformation of knowledge. With University Library, as an important service supporting university research, service development should also turn toward research service. Information borrowing, using, or sharing are second, third, or even several times the development process. Along with research-based services, for the development path of the library itself, the research needs of readers, the data integration research, research-and-spoke service network, and so on are essential issues in the development process of the library and are also the inevitable road continuously structuring characteristics, providing personalized advantage in the future development, which is essential for the promotion of innovation in terms of libraries. In services innovation, advanced concept as a guide, continuously expand service areas, provide quality content, use database services, online services, and knowledge services to meet the needs of the majority of audiences, use a variety of techniques to integrate the knowledge resources as an organic whole and meet the diverse needs of the reader, develop potential audience to promote the diverse exploration and practice in library transformation development, accumulate development experience in practice, and burst innovation inspiration to better serve readers.

2.3 Leverage network technologies to shape spiritual culture

University library spiritual culture is the direction of the library practices, which also reflects the operational objectives of the library, the value content, and conduct codes of librarian and ethics, which is an invisible force for encouraging and constraining librarians' behavior and thinking. It is a library in long-term practice that is consciously formed to represent their style and image of spirit. Under the guidance of this spirit, the library staff can work hard toward a common goal and continue to accumulate and develop the library career. Spiritual culture of university library includes humanities content and innovation, which guide the majority of teachers and students to establish the correct values, worldview, and outlook on life, actively play into the creativity and intelligence of university students, thus enhancing their self-worth. In the spiritual and cultural construction, we should strengthen the training of library staff's ideological, moral, and cultural qualities, through a variety of training; change personnel work style and attitude to establish staff dedication, pioneering spirit, unity, cooperation, and the spirit of serving the readers. Simultaneously, we should actively use the Internet technology to enhance the library's spiritual and cultural dissemination, such as organizing network competition, using a network platform such as microblogging, and Wechat to recommend the latest reading resources for teachers and students, attract the teachers and students to the library, and prompt students and teachers to understand library culture to enhance the vitality of the University Library.

3 CONCLUSION

For the majority of teachers and students, the University Library faces a majority of teachers and students, and it owns the mission of popularizing scientific knowledge, spreading advanced culture, carrying forward the fine traditional culture, and integrating diverse ideologies. University Library, in the long-term social practice, formed its own unique culture. With the continuous development of network information technology, the rise of Internet culture and various IT applications in the library had a great impact on the library culture. In this context, the university library should actively use network technology to innovate library materials, services, systems, and spiritual culture, as well as continue to enrich and develop the cultural connotation of the library.

REFERENCES

[1] Zhou Jiufeng. *Analysis on University Library Culture Function*[J]. Library Theory and Practice, 2010, (12):87–89.

[2] Yu Zhongan. *Integrate Socialist Core Value System into the Whole Process of University Library Cultural Construction* [J] Gansu Science and Technology, 2012, 28 (14):94–95, 78.

[3] Luo Wenge. Thinking on University Library Culture Construction[J] Lantai World, 2013, (17):121–122.

[4] Liu Xia. *Analysis on Library Culture Construction in Network Environment* [J] Ningxia Normal University, 2012, 33 (4):154–156.

Information, Computer and Application Engineering – Liu, Sung & Yao (Eds)
© 2015 Taylor & Francis Group, London, ISBN 978-1-138-02717-6

The application 3D animation in electric transmission line construction

Chun He Li, Hong Ming Sha, Gang Liu, Ji Tao Zhu & Xiao Lin Wu
Liaoning Electric Power Supply Co. Ltd., Liaoning, Shenyang, China

ABSTRACT: With the continuous development of science and technology, the multimedia animation has been widely used in various fields, and can also realize the frequently-used, new and special construction technology and craft in the construction of transmission and transformation project. It shows the construction process with a vivid and visual animation, which is convenient for the vast engineers in infrastructure projects to learn and communicate, improves the efficiency and quality of construction, and is convenient to do direct planning, design, bidding, approval and management in large complex engineering projects. It is in favor of the designers and managing personnel to do aided design and to evaluate the design scheme for all types of planning and design schemes in order to avoid construction risk. Besides, it can do the applications of the construction technical disclosure, the management planning technical plan review, the accumulation and training of new technology and new craft, the construction process control, quality safety weakness analysis, and the accident repair.

KEYWORDS: 3D animation; Electric transmission lines; Construction

1 OVERVIEW

With the rapid development of the modernization of power system, its scale of electric power has been expanded, and there have been more and more jobs related to electric transmission line construction. As the computer technology has developed at full speed, 3D animation technology has already been mature in the application from the film to the game. Now, the personnel in electric power began to think about how to apply it to the electric power producing specification. Although in the power sector, the computer simulation technology has been applied in the distribution network and substation operation, the application of the computer virtual simulation technology in the domestic production and construction is much scarce due to the particularity of electric transmission line in electric power system. Therefore, we should draw lessons from the existing experience, and combine the reality of the electric power enterprise to make our own electric transmission line production 3D animation technology and bring the electric power production operation specification and operation risk control into a new field.

The current operation specification of the electric transmission line facility is based on the entity. Due to the various types of the line equipment and operating ways, working in high altitude and under high pressure may be dangerous, and it cannot record whether every step of the operators accord with the specifications or not. Besides, due to the increasingly outstanding problems of the single training mode and the long training time, the skill and quality of the new personnel have not been improved in the recent years, which eventually result in fewer and fewer frontline workers and cannot meet the requirements of safety in production. In order to reduce the cost of production, improve the business skills of the operating personnel, accelerate to cultivate the personnel, and reduce the labor intensity and operating risk of workers, it is an urgent to introduce a new 3D animation technology.

2 THE CHARACTERISTICS OF 3D ANIMATION

1 It can simulate the scene which cannot be completed and reproduced by real shot, and can also reduce the cost;
2 Its modifiability is stronger, and the quality requirements are more susceptible to be controlled;
3 It can simulate and complete the dangerous scene;
4 Its production is not affected by factors such as weather and season;
5 It has high technical requirements for producers;
6 Its production cycle is relatively long;

The threshold of 3D animation technology is low, but it needs to master and use proficiently, as well as unremitting efforts for many years.

3 THE CONTROL OF 3D ANIMATION THCHNOLOGY FOR THE RISK OF THE ELECTRIC TRANSMISSION LINES

The danger coefficient of the electric transmission line stringing construction is high, so it asks the production safety for high requirements. We can not only take the existing theoretical knowledge as guidance, but also should add more graphic and video data of 3D animation. Meanwhile, before producing, the staff should be given guidance so as to effectively control the construction risk.

3.1 The theoretical training

Before induction, the employees are given a kind of cognitive training using 3D animation, which can let the employees know the terms and content related to the profession, and establish a general outline of the profession the employees occupied in their mind.

3.2 Graph

It can clearly and easily demonstrate the gist, process and specification of the operation and the requirement of risk control of the key links and main risk in the operation and the usage of instruments using the forms of words and pictures. Its drawbacks are that it cannot demonstrate the details of the whole operation and the operation risk, the difficulty of training standard operating way is big and the time is long, and the expected effect is not obvious.

3.3 Video

It can shoot the whole operating process, and the gist and details of the key operating behavior can also be shown basically, but since the electric transmission line operation is usually did in high altitude, and the shoot is restricted by angle and space, so the safety performance and risk control measures in operation also need professional personnel and equipments to do post-processing, and the producing cost is high.

4 THE UTILIZATION OF 3D ANIMATION TECHNOLOGY

4.1 The utilization of 3D animation technology

The model material library of the transmission primary equipment, the standard safety facilities, the general instruments, and the necessary instruments in the process of production and operation is established using the 3D animation technology. After that, based on the thinking of risk control, normalization, and standardization, combining with the working instruction and the field operation and skill of experienced workers, organically falling together the three-dimensional model, characters, standard operating way, and risk control, the script will be written, which can realize to clearly, comprehensively and vividly show every operating action of the operators and the details of the operation process in the field environment. Then the phenomenon such as sound, light, and spark are added in order to stimulate the real transmission system operation scene. In the later period, 3d animation, subtitles, voice, and music will be connected by professional non-linear editing system. Through constant screening, cutting, editing, synthesis and production, the saved video shot can be formed an intact 3D animation film, which will be supplied for everybody to train and learn. The virtual world generated by computer will change the virtual environment into reality, and give the feedback about various stimulus for human's senses generated by the things in the virtual world to employees as realistically as possible, so as to make people produce the "immersion" feeling of going into the equal realistic environment. This technology can also use the 3D virtual stimulation to realize the process and effect which cannot be realized by the traditional training, operation and practice, and the accident playback, so it can provide abundant content for the skills training, evaluation and assessment, and risk control of the power industry and the skill and knowledge base of operation.

4.2 The application example

4.2.1 Establishing model

The application of 3D animation technology meet the clients' higher requirement for the design quality and efficiency, and meet customers' demand for the information management of the full life cycle of assets at the same time. The design of 3D animation platform needs to meet the following points: (1) to establish a unified and open system information model, and ensures the standardization and practicability of the data model through stratified and hierarchical data management, ensure the data; (2) to establish a broad and interactive aided design engine, and ensures the designers to process the most core decision problems with the intelligent and automated computer engine power; (3) to establish distributed and real-time communication architectures, and integrates the personnel, events and resources in the process of design through the collaborative design in the whole profession and process; (4) to establish a vivid and efficient 3D render engine, and realize humanized visualization of 3D design through the way of multiple view linkage.

4.2.2 Design thought

With the aid of the power of the computer artificial intelligence aided design, it can ensure the design quality and design efficiency to the maximization. The design efficiency, human-computer interaction, and the rendering effect can be shown by visualization.

5 OUTLOOK OF 3D ANIMATION TECHNOLOGY IN THE FUTURE ELECTRIC TRANSMISSION LINES

3D animation technology can make the process of the typical accident which happened reoccur vividly so as to better teach the staff. For the problems of the complex transmission line repair process, the difficult technology, the high safety risk, difficult solidification and almost lost unique operation skill, 3D animation course is produced. It realizes the details of each operation step by virtual stimulation way, and establishes and improves the 3D model of the transmission line equipment, instruments, and safety facilities. According to the characteristics of the electric transmission line standard operation, the real-time driver is developed in the environment. The scenario model is uploaded, and through the upload of the procedure engine, users can operate and observe the virtual equipment in 360° scope through moving in the 3D space, realize to operate the virtual equipment and even control the characters. It truly realizes the human-machine interaction, and integrates the electric power training system in the end.

6 CONCLUSION

To sum up, 3D animation is a valid try of the safety production training, drills, accident playback, risk control, and construction modeling for the electrical network enterprise in the application of virtual technology, and has more important practical significance. At the same time, 3D virtual technology is the most advanced and the most imaginative film and television performance means in China, and even in the world. The utilization of the means has brought vitality to the power production and the employee training. In the maintenance of the transmission line operation, the standard operation and risk control become more and more important. This paper realized the production safety and risk control in the power system transmission line. The 3D animation video demonstration course of operating with electric has already been applied to the employee training of Liaoning Power Supply Co., LTD, and gained a good result.

REFERENCES

[1] Ye Die, Zhong Liangwei, Luo Yun: Three-dimensional World's Simulation Roaming System Based on VegaMultiGen. Computer Engineering and Design, Vol. 2 (2005), pp. 362–364.
[2] Li Lingyun: The Research Review and Comparative Analysis of Interactive Multimedia Learning Environment. E-Education Research, Vol. 6 (2009), pp. 85–88.
[3] Liu Yijin, Wang Ruchuan, Zhang Ying: Virtual Reality Technology and Its Application in E-commerce. Microcomputer & Its Applications, Vol. 6 (2002), pp. 47–48.
[4] Lu Difei, Zou Wangong, Ye Xiuzi: Mesh Deformation Clone Based on Local Affine Transform. Journal of Computer-Aided Design & Computer Graphics, Vol. 5 (2007), pp. 595–599.
[5] Li Jiang, Wu Yanzhang, Hao Tengfei: Discussion the Graphic Theory of 3D Animation and Computer Graph. Management & Technology of SME, Vol. 23 (2008), pp. 224–225.

Information, Computer and Application Engineering – Liu, Sung & Yao (Eds)
© 2015 Taylor & Francis Group, London, ISBN 978-1-138-02717-6

Study of the application of multimedia technology in vocational English teaching reformation

Wei Li

ChiFeng Industry Vocational Technology College, Inner Mongolia Chifeng, China

ABSTRACT: Vocational English education is mainly used for strong application of the English language. The application of multimedia technology in teaching can enhance students' learning consciousness and learning efficiency, and this article will present discussions on this subject.

KEYWORDS: Vocational English Education; Multimedia; Teaching

1 APPLICATION OF MULTIMEDIA TECHNOLOGY IN VOCATIONAL ENGLISH EDUCATION CLASSROOM

1.1 Help create context

In the dialog class, creating contexts is often the key. Making the students come close to the actual context would inspire students' dialogue and learning desire. With multimedia, the original single context is able to become rich and audio, images, video, and so on are conducive for creating a context.

For example, in the context of simulating ordering, let students, respectively, act waiter, treat people as guest, and so on. When the waiter hands out menus, students need to ask the fellow "Would you like some ...?". As a treat person, he can order according to the picture that the teachers provide, so that students learn to use the sentences of "I'd like some ..." or "I prefer ... ," and the waiters can interact with them, indicating that the supply of certain dishes has been completed and they can recommend several popular dishes, and so on. Thus, students can create different dialogs according to the environment and their own preferences. In the course of practice, they have not only practice but also can learn table manners, names of dishes, and learn to deal with the embarrassment at the table. Such a contextual exercise can not only greatly enhance the fun but also prevent students from invariably imitating standard conversation of text, which is lively and interesting.

1.2 Help expand learning capacity and improve learning efficiency

In new textbooks, the second lesson of each unit focused on reading and requires teachers to complete it within a class. The traditional way of teaching finds it difficult to ensure the amount of information taught in a class. Using multimedia technology, it is possible to train students' oral and written ability; simultaneously, it is convenient to accelerate the classroom pace and ensure efficiency.

For example, you can play an audio tape for reading content. Let the students listen to it and have a general impression, and then release English subtitles a second time to enable students to examine the contents just heard; the third time, release the comparison of English and Chinese to allow students to test whether they can fully understand the meaning, which can exercise students' listening, reading, and English conversion thinking as much as possible for students to get more exercises in just a lesson.

1.3 Overcome difficulties and make anything easy

Based on the characteristics of vocational schools students' poor English basis and weak learning acceptance ability, teachers in the classroom must make everything easy and they should slowly and carefully teach difficult topics. Thus, students will find it easier to accept particularly boring grammar. Teachers can vividly demonstrate grammar by multimedia technology and give more examples.

For example, in teaching present progressive, teachers can play some movie clips for students through some GIF to try to describe the figures' action; at this time, teachers can use questions such as *what's he / she doing?* to prompt students to answer by using the present progressive. Students can describe a series of actions by progress *Subject + am / are / is + doing* according to the picture change while simultaneously reviewing the past tense *Subject + was / were + doing*, thus comparing the two grammar topics to allow students to more vividly distinguish between them.

2 REFLECTION ON THE APPLICATION OF MULTIMEDIA TECHNOLOGY IN VOCATIONAL HIGH SCHOOL ENGLISH TEACHING

2.1 To improve the quality of multimedia courseware

Multimedia teaching must be done combined with courseware, so the quality of the courseware will affect the quality of teaching. Teachers should make course-ware based on the lesson content, from the shallower to the deeper, combining with knowledge relevance, collecting wealthy material to improve the courseware quality. Specifically, courseware materials must be closely linked with teaching materials. Therefore, we must first study the textbook and focus on prominent teaching; for example, tenses teaching should high-light the characteristics of each tense and application and make it a chain. Second, we must collect material that produces courseware; thus, so-called good mate-rial must have forms of text, images, sounds, and so on. The transferred information should be clear and understandable. Finally, in order to facilitate the stu-dents to practice, we should set a practice session, which is mainly the creation of situations to make students practice what they have learned in class and fully consolidate and improve knowledge.

2.2 In the application of multimedia technology teaching, the leading role of teachers is crucial

In a multimedia technology classroom, teachers play a leading role and they have a grasp of rhythm, atmos-phere, and order. Most importantly, they depict how to avoid the boring traditional classroom and set a strong interactive teaching mode to encourage stu-dents to participate and stimulate students' desire to learn, which requires the mutual cooperation between the teachers and courseware to achieve the appropri-ate teaching purposes and effects.

2.3 Select the essence and discard the dross for the traditional teaching

Compared with traditional teaching, the application of multimedia technology in the English classroom can make the classroom rich, vivid, but traditional teaching has its own values, and we must not over-use multimedia technology for multimedia technology reformation. It should be used at the appropriate time or combined with the outstanding traditional teaching to achieve the finishing touch. The most important is the selection and application of the teaching mode around the teaching theme.

2.4 Deficiencies of the applications of current multimedia technology in the English classroom

Currently, many vocational school teachers do not really use live multimedia in lessons and just sym-bolically transfer blackboard writing to the course-ware without innovating or setting some interactive content according to each lesson to be taught. This classroom often becomes lifeless figure lessons or music lessons.

3 SUMMARY

As a teacher of a vocational college, the teaching methods and means must be distinguished from the teachers in colleges. Taking into account the situation of students and acceptance level, we must pay more attention to combining with new teaching methods and continuously updating their teaching ideas for building a good teaching environment.

REFERENCES

[1] Zhou Xiaoqun. *How to Implement Hierarchical Pedagogy in Vocational High School English Teaching* [J]. English for Middle School Students (Junior High School). 2012 (20).

[2] Chen Baguo. *The Application of Separate Class Teaching in Specialized Teaching* [J]. Occupation. 2012 (17).

[3] Sun Xiaoyu. *Professional English Teaching Strategies in Vocational Schools* [J]. Occupation. 2012 (17).

Information, Computer and Application Engineering – Liu, Sung & Yao (Eds)
© 2015 Taylor & Francis Group, London, ISBN 978-1-138-02717-6

Analysis of the realization of low carbon tourism with the use of internet of things

Ya Nan Li

Zhengzhou Tourism College, China

ABSTRACT: With the social development and national progress, the Internet of Things technology has a wide application prospect in the development of modern tourism and plays an important role in the development of tourism. The Internet of Things is primarily based on the Internet for the better implementation of information interaction between things to achieve the normal conduct of the market economic activities ultimately meet the needs of people's life and realize the low carbon life philosophy. How to promote the application of the Internet of Things in the tourism industry is the key. According to the feasibility of the Internet of Things in technology, policy, consumer demand and the actual application in the tourism industry, this paper discusses the concrete proposal that the Internet of Things technology is used to realize the low carbon tourism industry.

KEYWORDS: Internet of Things technology, Tourism development, Low carbon

1 INTRODUCTION

From the development of the Internet of Things and tourism industry, the Internet of Things technology has wide application prospects in the construction of tourism informationization technically, economically or on policy. The Internet of Things technology can be used to integrate physical information and resources in the tourist market and tourist hotels, tour transport and scenic areas that are related to tourism into a system and form another system with the integration of tourists, travel companies and government service through the information system. This can help to provide whole course services for tourists, change the operation ways of tourism enterprises and speed up the informationization construction and is beneficial to the realization of scientific tourism industry management. It can be seen that the application of advanced Internet of Things is of great importance in tourism development.

2 THE GENERATION OF THE INTERNET OF THINGS PLAYS AN IMPORTANT ROLE IN THE DEVELOPMENT OF TOURISM

At present, though the tourism market is huge, its service items are single, mainly for visiting places of interest. Part of the consumer demand can not be satisfied. Combined with the golden week, hot scenic spots can receive limited visitors. The Internet of Things has the characteristics of instantaneity and interactivity. With the rapid development of information and communication technology, its application has penetrated into all aspects of society. The Internet of Things almost covers all the network space, such as fields of informationization convergence, public management and science and technology. In terms of the application of the Internet of Things in tourism, Wang Hongkun has made a preliminary exploration, analyzed the applications of the Internet of Things in ticket, passenger flow and other aspects in Analysis of the Internet of Things Technology's Application in Scenic Spots.

The completion of Distribution Management System of Scenic Area Tourist Based on RFID written by Zhang Yongqiang, et al. provided technical support for the use of the Internet of Things in tourism industry. At present, the Internet of Things technology can be used in tourism management scientifically and reasonably. Each link in tourism activities has not only been realized, but also the intelligent tourism development has been achieved, which conforms to the concept of low carbon, safety, energy conservation and environmental protection, and can boost the efficiency of tourism services and management. The advantage of the Internet of Things technology should be effectively utilized to improve the level of intelligent management in scenic spots and provide personalized services for tourism consumers.

3 THE FEASIBILITY OF THE INTERNET OF THINGS IN TOURISM DEVELOPMENT

First of all, it is not hard to see from the concept of Internet of Things that the Internet of things ahs covered RFID technology, chip technology and mobile communication technology. With the rapid development of communication technology and information industry, there are no technical barriers for the application of the Internet of Things in the tourism information construction.

Second, Views on the Acceleration of the Development of Tourism Industry published by Chinese State Council presented clearly, "improve the efficiency of tourism services with informationization as the main channel. National Middle &Long-term Science and Technology Development Plan (2006–2020) also put forward to strengthen the modern information service industry information support technology and research and development of large-scale application software. Here it can be seen that policies have provided good policy environment for the Internet of Things technology in the development of tourism industry and established the status of tourism informationization construction.

Finally, with the economic development in China, most of the personalized needs of tourism consumption make the Internet of Things be used in tourism, which not only promotes the upgrade of tourism industry, but also improves tourist experience.

Combined with national research and investment on the Internet of Things technology, the Internet of Things technology will be spread to various industries with its great development potential.

4 PRACTICAL APPLICATION OF THE INTERNET OF THINGS TECHNOLOGY IN TOURISM INDUSTRY

Low carbon tourism has become a new direction of tourism development. It mainly focuses on energy conservation and emission reduction in hotels, green environmental protection enterprises and scenic spots. The promotion of energy conservation and emission reduction has greatly driven the application of the Internet of Things technology in tourism industry, mainly in green hotels, visitors' capacity control, intelligent transportation, hot scenic area security and other aspects.

4.1 The application of the Internet of Things technology can help master visitors' needs and provide services for tourists

To achieve the intelligent service and low carbon environmental protection, hotels first set up tourists' information for easy management, query and analysis of visitors' accommodation conditions, establish RFID identification system to connect terminal equipment in visitors' rooms with information management system for hotel guests and construct business cooperation with surrounding catering, entertainment

and other service facilities for providing service interfaces for tourists.

4.2 The application of the Internet of Things technology provides convenience for real-time understanding of tourists' distribution in scenic spots and tourist capacity control

RFID readers can be set up in key locations in the key scenic spot or a small area. The configured antenna should cover key locations. When tourists walk though the key locations of some scenic spot, their ID numbers can be identified and the system can determine visitor amount according to the information results.

4.3 The Internet of Things technology is also particularly important in tourist traffic

With the aid of this system, it can provide information collection processing release and other diversified services for traffic participants, which can make administrative staff know roads and vehicles' whereabouts clearly.

4.4 The application of the Internet of Things makes tour activities more humanized

Before a journey, people can complete the order of scenic spot ticket and reserve of other tourism links and enter the scenic spot using the mobile phone ticket. Tourists also can get advice and guidance of self-help travel in the scenic spot and the all-directional services can be realized, including scenic spot line guide and mobile tour guide. Tourists can be supplied with infinite convenience.

4.5 The application of the Internet of in scenic spots can inspect and prevent the occurrence of accidents and establish a security for tourists in scenic areas

For instance, in the scenic spot with complicated and changeable geomorphic environment, when tourists are lost or in danger, their specific locations can be found out through GPS positioning technology carried by tourists and the nearest rescuers equipped with GPS RFID handheld devices can be notified to rescue the first time.

5 THE PAN OF USING THE INTERNET OF THINGS TO REALIZE THE LOW CARBON TOURISM

Through the above analysis, the applications of the Internet of Things have made consumption pattern change and develop towards low carbon and provided consumers the convenience to a certain extent. Therefore, the Internet of Things plays an important role in development of tourism. To make tourism have more healthy development, the key is how to promote the applications of the Internet of Things in the tourism industry.

1 Related personnel should accelerate the process of core technology research and development of the Internet of Things and unify industry standards. Especially speeding up the technological breakthrough of core weak links is imminent. In addition to, relevant information security legislation system should be improved to promote the healthy development of the Internet of Things industry.

2 The government needs to support wisdom tourism pilot projects through various aspects and provide relevant financial preferential policies for wisdom tourism pilot projects, such as preferential loans, tax breaks and other preferential policies. The government can create a good environment for the application of the Internet of Things in tourism industry and encourage various market main bodies with strength to actively be involved in the project construction and accumulate experience for application and popularization of the Internet of Things to improve the quality of project construction and investment benefits.

3 The government can actively promote and encourage universities to cultivate professional talents. For instance, universities can set up majors related to the Internet of Things to solve the Internet of Things talent demand. As an intermediary, the government can actively promote exchanges and cooperation between companies of the Internet of Things, local scenic spots and higher vocational schools, carry out related cooperation projects with the tourism application of the Internet of Things and provide talent guarantee for tourism application fields of the Internet of Things.

The development of the Internet of Things industry is still in its infancy, but its status will rise with the improvement of tourism industry, rapid development of communication technology and changes of tourists' demands. Therefore, the application of Internet of Things technology has a broad development prospect in the tourism industry. In summary, the application of Internet of Things actively has promoted the tourism to develop more healthily. People should effectively grasp the opportunity, fully develop technological advantage of the Internet of Things, break factors that will hinder the development of the Internet of Things, raise the level of scenic spot intelligent management and realize the concept of low carbon tourism to make the Internet of Things provide personalized services for tourism consumers and become a driving force for the future development of tourism.

REFERENCES

[1] Liu Yu: Analysis of Tourism Innovation Evolution Pattern under the Background of the Internet of Things, Journal of Management, Vol. 5 (2012), pp. 23–24.
[2] Liu Hongjun: Analysis of the Application of the Internet of Things in the Development of Modern Tourism, Journal of Golden Years, Vol. 1 (2013), pp. 55–56.
[3] Zhan Taizhong: Analysis of China's Tourism Development under the Background of the Internet of Things, Journal of Management & Technology of SME, Vol. 2 (2014), pp. 88–89.
[4] Jia Huifeng and Chang Kun: Analysis of the Application of the Internet of Things in China's Tourism, Journal of San Jiang University, Vol. 6 (2013), pp. 100–101.
[5] Niu LIcheng: Analysis of Tourism Informationization Mode under the Internet of Things, Journal of Chinese Business and Trade, Vol. 19 (2014), pp. 77–78.
[6] Yao Guozhang and Han Linghua: Research on the Application of the Internet of Things in Wisdom Tourism, Journal of Chinese Business and Trade, Vol. 23 (2013), pp. 89–90.

Information, Computer and Application Engineering – Liu, Sung & Yao (Eds)
© 2015 Taylor & Francis Group, London, ISBN 978-1-138-02717-6

Research in vocational "web production" course task-driven teaching mode

Feng Yun Liu, Chun Hong Wu, Wen Zhong Xia & Yu Bing Qiao
Zhangjiakou Vocational College of Technology, China

ABSTRACT: Webpage making courses are compulsory courses in computer and e-commerce, which should not only focus on the study of knowledge but also focus on training the application of knowledge and skills. As in vocational education and training that are based on application-oriented talents for the foundation of society, therefore, we should strengthen students' daily teaching practice to make teaching points. In this article, task-driven teaching is used in order to enhance students' web institutional capacity.

KEYWORDS: Webpage making; task-driven teaching; Vocational

1 INTRODUCTION

For the purpose of personnel training colleges, we should actively seek reasonable teaching methods to promote vocational student web production capacity; currently, it is more important to enhance the teaching and research work. Although the teaching vocational web production environment is getting better, there are still a lot of teaching problems in this process. One curriculum software is a more independent course of study lacking a systematic teaching mechanism; whereas the other focuses students on Image Processing, page layout, animation, other low degrees of professional knowledge, and strong grasp [1]. In this regard, the reasonable arrangement of teaching content should facilitate students to acquire professional knowledge, with a need to actively explore the usual teaching work.

2 TASK-DRIVEN TEACHING MODE

2.1 Connotation of the task-driven teaching mode

Taking task-driven teaching about many problems in webpage making for vocational colleges teaching curriculum, by research in practice in a long time, get a better practice effect. Task-driven teaching model is based on constructivist theory-based teaching methods, through continuous exploration of issues to keep students actively exploring learning interest and enthusiasm. This teaching mode and teaching mode are exploratory consistent, independent innovation abilities for students, independent thinking, problem-solving abilities were cultured; its application in a long time, the formation of the main routes to the task of teaching, student teaching subjects the idea of teaching methods; teachers in teaching mainly intend to guide the development and the ability to help students learn the basics.

2.2 Significance of task-driven teaching mode

Many of our webpage making courses at vocational colleges teaching have launched a task-driven teaching mode, from which they have developed an interactive, project-based, and modular concrete method of the teaching mode; through specific empirical studies, it is shown how the task-driven teaching mode can mention the promotion of independent learning enthusiasm, initiative, while helping students strengthen the practice of operations and thinking innovation. In a particular practice, task-driven teaching methods through the integration of research on teaching specific content of the teaching task items, but more confusion inherent level, varying difficulty and complexity of the distribution of many courses, resulting in a messy student learning step in the direction of the loss of learning, so that it gradually decreased interest in learning and confidence. In the teaching process, the development of specific content and social reality does not match the skills, leading to students being unable to meet the needs of society and hindering its future development. In this regard, there is a need for a fundamental change for the vocational colleges web production courses to strengthen its hierarchical system of the teaching system. The teaching mode tries to achieve the basics of teaching and practice of teaching combined to promote two-way development of teaching skills and upgrade the knowledge of students.

3 TASK-DRIVEN TEACHING MODE PRACTICAL APPLICATION IN WEBPAGE MAKING COURSE

3.1 *Modular course structure optimization*

Vocational colleges using web authoring standards of vocational education curriculum materials for teaching, after China Software Industry Association CSIA identified. Its textbook knowledge is divided into two main areas; compulsory and elective, which, in turn, are divided into five areas of specific knowledge, with its modules corresponding to the appropriate selection of the five web production training content. For textbook content, by integrating it into 13 units, each unit has its own specific operation explain the content, by conducting specific tasks to carry out a specific explanation for their teaching content. Meanwhile, module for their projects is not related to the content, and it has a separate supplementary explanation. Carry out the project merit more modular curriculum structure, specifically including; 1, in the case of a combination of theory and practice with each other, a solid student learning goals, increasing the students' interest in learning. 2, to promote the realization of a dynamic combination of course modules and the implementation level teaching methods. 3, this may correspond to different levels of teaching students a reasonable potential for learning and development, for the learning ability of students to be allowed to complete multiple tasks at the same time, and for students with low learning ability, can be allowed in the regulations complete the task within the time to learn the basic efforts to achieve quality education individualized teaching.

3.2 *Carry out the project tasks of teaching practice*

Project teaching tasks need us to break the traditional curriculum teaching mode in the usual practice teaching work; the whole teaching process and everyday applications around the five sites were in relatively close contact, through the adoption of the theory and practice of teaching a combination of standardized teaching forms of conduct. The specific teaching process is divided into eight[2].

1 The creation of scenarios: The teacher's main task involves visiting the online course; students demonstrate previous works and good business works. To teach students how to learn, the contents were compared, the difference between their content was analyzed, and the process in this section points out the knowledge points. The students will need to listen carefully, based on a careful observation of the teacher, professor, to find out the specific problems in learning the key elements, whereas the corresponding problem is in accordance with the teachers' ideas and concrete solutions.

2 Task driven: In the teaching process, teachers teaching in this section should have clear objectives; teaching courses should be in conjunction with each other; d the actual instances of in-depth research and analysis should utilize the knowledge content; and teaching content should correspond to the specific task of teaching the students arranged to strengthen the mission design production standards, so that it can effectively promote the development of students' thinking ability to create. But it should explain to the students how to download network resources. Students need to learn to understand the specific task of teaching content, understand the task is completed during the required approach, and possible problems, for matters not understand questions to the teacher to be timely. Through application examples, vivid teaching content is so that students in the learning process easier to understand, promote self-learning ability of students to improve.

3 Creations: In the task process, teachers guide the work to be done, to keep abreast of the progress of the students' tasks and problems, to give students knowledge of coaching. Students conduct group discussions corresponding to task analysis, system implementation plan to be developed; do their task division of roles; actively collect task-related materials; and integrate complete the scheduled task team leader and team members in the division of labor and tasks summarize When finished, write a summary report. In the process, students learn through group discussions to enhance learning initiative and enthusiasm in communication with each other and also to strengthen the unity and cooperation among students' ability.

3.3 *Construction of the hierarchical structure of professional skills training*

Teaching for specific projects should carry out a hierarchical form of teaching, through the combination of practice and theory; enhance students' skills development; focus on the practical ability of students, with the help of students self-learning process; and build their professional accomplishment. Specific teaching level includes five aspects; one, focusing on the teaching of examples of web authoring application; 2, to enhance the process of teaching courses for teachers producing imitation application; 3, engage in more extracurricular activities of integrated training; 4, to promote student carry out independent creative practice; and 5, after the completion of correspondence courses, allowing students to complete web production alone. This form of teaching structure layer pushes help students establish a sound system of vocational skills and promote their career-oriented development.

4 CONCLUSION

Teaching and training for students in colleges, based on application-oriented social skills, are the basis for talent. Therefore, we should focus on the combination of theory and practice in teaching. Webpage making courses are compulsory courses in computers, and they have the ability to pay more attention to hands-on, in order to enhance students' vocational skills training, teaching process to take a task-driven teaching mode, and achieved good teaching results.

ACKNOWLEDGMENT

This study was supported by The Twelfth five-year Education Scientific Planning Project of Hebei "The Research of Vocational College <Web Design> Course Task-Based Teaching," No. 1414270.

REFERENCES

[1] Xiao Zhang. Vocational teaching reform of webpeg making [J]. Popular Science, 2009 (1), 22–23.
[2] Chun Liu. Task-driven teaching method in webpeg making course [J]. Health Vocational Education, 2008, (2), 55–56.

Application research of the campus library information management system based on 4G network

Ping Liu

Heibei University of Science and Technology, China

ABSTRACT: Nowadays, with the constant development of information technology industry, the mobile communication technology also develops vigorously. The advent of the 4G mobile communication technology promotes the communication technology to rise to a higher level, and it applies it to all industries to promote the development of industrial economic benefit. This article mainly analyzes the structure system of the 4G network technology and applies it to the campus library information management system to research and design its specific structure and establish a perfect network model construction; the NS2 simulation platform will conduct a comprehensive evaluation for the simulation of the network structure system. The results show that its overall network performance is strong, which can effectively promote the construction and development of the campus library information management system.

KEYWORDS: 4G network; Campus library; Information management system

1 INTRODUCTION

With the gradual perfection of the campus digital library construction, the number of the mobile terminal users is increasing; thus, the library carries out more functions pertaining to the mobile communication service. The continuous application of 3G mobile communication technology enables the mobile digital library information service to be more thorough in the process of being mature gradually. Users can use different network technology to search a library's collection resources via mobile terminal devices at anytime and anywhere, and they can simultaneously also manage their personal information [1].The arrival of 4G mobile network communication technology has also brought along two kinds of new network concepts—Broadband Access and Distributed Network. They mainly include broadband wireless local area network (LAN), broadband wireless fixed access, mobile broadband system, and broadcasting network operated by the Internet, all of which are applied and studied gradually [2].

2 BASIC CONNOTATION OF 4G MOBILE COMMUNICATION NETWORK TECHNOLOGY SYSTEM

The developmental history of mobile network communication technology is very long, and it has experienced three generations of network technology changes. The 1G mobile communication system was born in the 1980s, and its system mainly adopted the simulation technology and the technology of frequency division multiple access (FDMA), but due to the transmission capacity restriction of the bandwidth, there were lots of problems and defects, and the most serious problem is that it was unable to carry out long-distance roaming mobile communication. In the early 1990s, the 2G communication technology was born, and its technology mainly includes digital code division multiple access (CDMA) and time division multiple access (TDMA). The advantages were that its anti-interference and confidentiality were stronger, the call signal was stable, and the sound quality was clear. The disadvantage was that the data transfer rate was low, which was not matching with the continuously developing multimedia communication needs. As time goes by, the 3G mobile communication technology was born in succession. Its transmission rate was significantly faster than earlier; it could provide data, video, images, and other service functions and could access the Internet fast, so it is currently widely used in the market [3]. When 3G technology continuously developed into being mature, the emergence of 4G mobile communication network technology enables the mobile information technology to have a qualitative leap. The concepts are mainly Distributed Network and Broadband Access [4]; the data transfer rate is more than 2 MBPS, and the form is asymmetric; besides, it can provide digital imaging services with the speed up to 150 Mbps for the full speed mobile terminal users and can ensure the transport services of the three-dimensional image with a

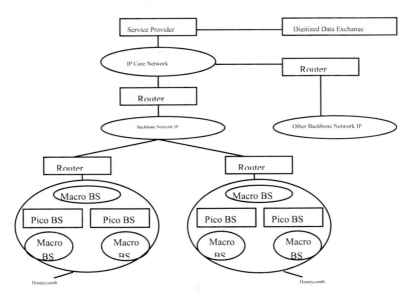

Figure 1. 4G network architecture.

high quality. The functions of 4G mobile communication network technology mainly include the complete fusion of various businesses, the perfect link among different systems in the high-speed mobile case, independent software platform, high intelligent network, and more convenient Internet connection. In 4G network system, the digitalized exchange is a much more important technology, and through this technology, the calls from different network platforms can be pretreated. Among them, the backbone network and IP core network mainly take the optical network technology combined with broadband IP technology to transmit the network information. The honeycombs in 4G networks can be mainly divided into Pico BS, Micro BS, and Macro BS according to the function.

3 CONSTRUCTION OF THE CAMPUS LIBRARY INFORMATION MANAGEMENT SYSTEM USING 4G MOBILE COMMUNICATION NETWORK TECHNOLOGY

3.1 Functions of 4G technology campus library information system

The application of 4G network technology in the campus library network can provide more diversified images, video, streaming media, and other related services for library users, and it is also the fundamental factor that it is different from the previous generation's mobile network communication technology. The functions of the campus library information

system aiming based on the 4G network technology mainly include the following: 1) It can expand and apply the broadband video on demand, the streaming media business, and other related services; it can provide users with more rich multimedia resources services, such as the colorful e-books, video lectures, multimedia database, and so on. Readers can enjoy the digital resources of the library at any time and at any place through mobile multimedia devices. 2) It can provide videophone service. Through this service, it can effectively extend the library's scope of service; no matter how far away, as long as through the terminal mobile multimedia devices, the reader's thoughts can be understood accurately and timely. 3) It can provide the mobile location service. The library can accurately locate the user's location and provide various services around the relevant positions, such as the domain name, the life information search, collection resources retrieval, and so on. 4) It can provide the multimedia distance education service. It can expand the remote education form through the 4G technology to provide users with infinite interactive video multimedia teaching and make the learning space be more personal and dynamic.

3.2 Campus library information system structure on 4G technology

The 4G network information management system aiming at the campus library is mainly based on the 4G network. In the work, the 4G network is accessible by the service of the base station by means of shunting and group, such as data, multimedia, and information instruction. It combines the transmission network

Table 1. Simulation parameters of each honeycomb subnet.

Specific parameters	Corresponding value	Specific parameters	Corresponding value
Routing protocol	AODV	MAC protocol	IEEE802.11
Size of scene	$1000 \times 1000m^2$	Type of channel	WirelessChannel
Type of queue	PriQueue	Type of antenna	OmniAnten
Simulation time	600s	Business	CBR
Numbers of subscriber	10.20.30.40.50	Queue length	50
station		Data rate & the maximum	100Mbps&3km/h
Mode of transmission	TwoRayGround	speed	20Mbps&60km/h
			2Mbps 8L 250km/h
			1Mbps&500km/h

with the bearer network and is a kind of versatile network system; therefore, it can carry out the bearing model of the whole business in the work. In the 4G network system, the remote server is its most direct visit access point of service, and it intensively processes the access, operation, and visit data flow of the video source. On the basis of buffering the users' requests and library resources database, the remote server, as the network management platform of the library system, uniformly manages the resource of the users and the library database. The remote server can not only process the video source to access the data flow, and promote the video visit to be accessed safely by verifying the authorization of the visitors and the access video source, but also buffer the video source data of the users' requests. Meanwhile, in the process, it can also arrange the queue model to meet the video access requirements that different users visit simultaneously. For the library information database system, we need to perform the optimal management constantly in the work; therefore, it is more convenient and efficient to promote the server to retrieve the database resources, and we should pay attention to the security and perform extensive construction of the 4G network campus library system.

3.3 Network model of the campus library information system based on 4G technology

In order to effectively verify the property of the campus library information management system based on 4G network, the network structure system is evaluated by applying the NS2 simulation platform; more specifically, it establishes a network model in NS2.32, and in which it sets five base stations. The number of users in each station increases from 10 to 50 gradually. The terminal nodes transmit data based on 4G, and its error rate of bit is zero. The topological data of the honeycomb network in these five base stations are generated in a square area, in which the base station is located in the center and directly covers 1/3 area of this district; there are about 2/3 subscriber stations that have 2–3 jump distance with the base station. There are 10 times of stimulation for each number,

and the resulting value takes 10 times as the criterion to obtain the relationship between the number of users and the three aspects of network message delivery ratio, average end-to-end delay, and routing overhead. Through research and analysis, all performance indexes of the network can meet the demand of practical application; with the constant increase of the honeycomb subnet users, the network message delivery ratio gradually decreased, which increased the average end-to-end delay and routing overhead. The main reason of this problem is due to the increasing number of the network users, which makes the volume of the business increase gradually. If the bandwidth resources are similar, they may cause more phenomena about packet loss and gradually accelerate the network performance deterioration.

4 CONCLUSION

From the study on 4G mobile network communication technology, we know its main performance characteristics and apply them to the campus library information management system. First of all, it is used to design the network architecture, then to build the network model, and, finally, to adopt the simulation platform to detect its specific performance. The results show that its performance is better and more feasible in practical application.

REFERENCES

[1] Sun Shujuan: Thinking about the Countermeasures of the Public Library Service Network Reading. Library Work and Study Vol. 12 (2008) pp. 97–98.
[2] Wang Wenjun, Yang Guofu, and Liang Zhenyi: Research on the University Library Information Management System Based on Network Technology, Journal of Modern Information, Vol. 4 (2006) pp. 77–83.
[3] Long Chaoyang and Wang Ling: First Exploration on the Library Information Service Model Based on 3G. Library Tribune, Vol. 3 (2008), pp. 8–11.
[4] Gu Zhisong, Huang Yunzhong, and Zhang Shiyong: The 4G Mobile Communication System Based on Internet. Computer Science, Vol. 5 (2003), pp.4–16.

Information, Computer and Application Engineering – Liu, Sung & Yao (Eds)
© 2015 Taylor & Francis Group, London, ISBN 978-1-138-02717-6

Construction of intelligent multi-agent model of virtual resources in cloud computing

Shu Ying Liu

The Institute of Information Engineering, XianYang Normal University, Xianyang, Shaanxi, China

ABSTRACT: With the development of science and technology and the popularization of network technology, Cloud computing begins attract more and more attention from the public. Cloud computing is a kind of important network service mode. It mainly uses network virtual technology to optimize and integrate information resources in different spaces to make resource management more convenient and realize efficient utilization of resources. This paper mainly discusses a kind of virtual resources processing method of intelligent multi-agent model in cloud computing to achieve the dynamic adjustment of resources and improve the efficiency and quality of network resource management.

KEYWORDS: Cloud computing, Virtual resources, Intelligent multi-agent model

1 INTRODUCTION

Virtualization technology can dynamically organize a variety of computing resources, realize transparent scalable machine damage system architecture and improve the use efficiency of network computing resources. In recent years, virtual technology has witnessed rapid growth and become a core technological form of cloud computing large-scale data center. In this virtual system, intelligent multi-agent model can meet the requirements of large cloud computing information services, large-scale operating data and high demand of calculation accuracy, realize the maneuverability and portability of data applications between cloud computing platforms and optimize data management, maintenance and configuration in cloud computing.

2 SUMMARY OF CLOUD COMPUTING AND VIRTUAL RESOURCES

Cloud computing is a kind of large-scale distributed computing model which is occurred based on economies of scale and developed along with the popularization of computer and network resources. In this mode, resource owners can transmit abstract, virtual, dynamic and scalable calculation, storage, applications and services in the resource pool to external users. Cloud computing has many characteristics. Firstly, it can centralize large amounts of resources and realize super-sized calculations. Secondly, the customized programs of cloud computing are very flexible and it has the ability to work fast and efficiently. Cloud computing can effectively improve the utilization of resources and reduce the scale of equipment. In addition, it also has dynamic extensibility and can realize energy conservation and emission reduction and low cost and stable resource consumption, which complies with the requirements of sustainable development strategy that modern people advocate.

Cloud computing virtualization technology is another major technological breakthrough compared with the information industry. The virtualization system of cloud computing is more perfect and diversified. Different from the network computing, the whole core of cloud computing is virtualized. It uses virtual resources to work and abstract material resource into unified virtual resources through virtualization technology. The advantages of virtual resources are higher availability, better security and good portability. In terms of the upgrading of investment equipment or programs' migrations to new environments, the configuration of virtual resources can move them to other platforms more easily.

3 VIRTUAL INTELLIGENT MULTI-AGENT

Virtual intelligent multi-agent is a kind of multi-agent model of social media services. It is established on the intelligent virtualization technology and is a kind of optimal allocation of resources. Virtual intelligent multi-agent is the core technology of cloud computing which only can manage a large number of information resources through it and readjust network resources through users' behaviors.

3.1 User agent

The user agent belongs to the data package type. One user's information includes hardware, software and personal use preferences. In cloud computing, the user agent is mainly used to receive mobile agents, provide information resources and identify the user's behavior and virtual information based on these information resources. There are two kinds of user agent service contents, analysis service and access service. It can understand the access type through data transmission and the user's specific location. According to the rules of network virtualization, in a distributed agent, when the user agent provides service ranges for the users involved, it also integrates the user's device information and connection information to analyze the user's access status and related service connection mode.

3.2 Analysis of distributed agents

Distributed agents in cloud computing are mainly used to allocate social resources. This process requires the correlation between users and services. In a distributed agent, intelligent virtualization management of intelligent virtualization management is realized through user identification and service demand state. The information of a distributed agent sends contextual information through the user agent and system resource manager. The contextual information contains judgment coefficient to determine what services it will provide users. In addition, distributed agents can provide material resources and replaceable systems and resources, expand the scopes of material resources by reordering system resources and improve the efficiency of resource utilization.

3.3 Collection agents

In cloud computing, collection agents are mainly used to collect information and data that cloud computing virtual system needs. The range of data collection is very narrow and generally limited to social web sites, such as SNS, news and some Weibo of bicycles and so on. Collection agents are applied to media tablets of search engines. When making a large amount of data parallel processing, there are many factors to consider and careful treatments are needed.

3.4 Analysis of agent managers

The role of the agent manager in virtual intelligent multi-agent is to manage the establishment, registration, deletion and other operations of each agent, provide relevant knowledge base resources for each manager in the system and monitor the group agent of the system. The monitoring ability of the agent manager is realized through failure data provided by the system administrator and event states between the agents recorded daily.

Information that the agent manager receives is relatively limited. Normally, it only gains some information resources from users' log information or contextual information in the system. This information can determine the type of information service and the form of agent structure. The agent manager manages distributed agent activities based on the system's online time, composes the control system through the agent manager's agent ID, related control information and control address, etc. The control signal is composed of user agent coefficient and Trap which includes the distributed agent's action status values and related events' status values received from the information system.

3.5 Analysis of virtual registration

In cloud computing, virtual registration is used for related social resources of virtual information inside the system and the reasonable utilization of system resources and resource status and submits the information to the administrator. Virtual registration can regularly analyze virtualized log information to make the system's resource management more efficient. Virtual registration management is different from the distributed agent in aspects, such as virtual creation and log data release. Generally speaking, log data of virtual registration include resource list, priority information, relevance rights and virtual resource number ID. In addition, to distinguish the weight of multi-agent, it is needed to update related weights of multi-agent. When reallocating resources, multi-agent can predict the virtual way in advance, redistribute resources to users and record the users' redistribution services in the historical data of the system service for users' management.

3.6 System resource manager

System resource manager mainly controls information recorded in the virtual registry. Based on the virtual registry, it manages resources provided by users. The system administrator can directly or indirectly control system resources through the distribution state of system resources. The structure of system resource manager is composed of virtual registration, resource utilization analysis, system context, analysis of system grade, system resource classification and information database of the system management. Meanwhile, it also controls and manages physical and logical resources in the cloud computing system. System context provides data analysis information for system registration and the context information formed in this way reallocates system resources based on the actual status of users. Analysis of system grade mainly provides users with the set grade of each resource, analyzes resource information in the whole system and obtains the analysis of system grade.

4 PERFORMANCE EVALUATION OF VIRTUAL INTELLIGENT MULTI-AGENT

Virtual intelligent multi-agent analyzed in this paper uses two forms of cloud computing network, system resources of cloud computing and virtual resources requested by users. Cloud computing network also includes cloud application monitor, workload manager, dynamic allocator of resources and reshuffle manager. Cloud computing can reasonably integrate servers with these network appliances, reduce the cost of information technology, increase the flexibility of information services, minimize downtime, improve the business increment, reduce the quantity of run servers and increase energy efficiency. Therefore, it has very high practical significance.

5 CONCLUSION

Cloud computing can have efficient virtual resource allocation strategies, convert the virtual allocation in cloud computing to the problem of path construction, optimize the allocation of information resources and rationally utilize virtual resources with the reasonable use of the server's overall resources. In short, virtual intelligent multi-agent can provide users with appropriate services and users can choose proper services to improve the efficient use of resources and simplify network resource management according to their own actual situations.

ACKNOWLEDGMENTS

Xianyang Normal University Research Fund (Grant No.13XSYK054).

REFERENCES

[1] Fang Jinming: Virtual Resources Scheduling Algorithm Based on NSGA II in Cloud Computing, Journal of Computer Engineering and Design, Vol. 33 No. 4 (2012), pp. 134–135.
[2] Wang Liuyang, Yu Yangxin and Zhou Huai: Design of Intelligent Multi-agent for Virtual Resources in Cloud Computing, Journal of Computer Applications, Vol. 32 (2012), pp. 147–148.
[3] Xu Li and Zeng Zhibin: Energy Efficiency Virtual Resource Allocation Strategy for Cloud Computing, Journal of Computer Systems and Applications, Vol. 20 (2011), pp. 158–159.
[4] Deng Dechuan: Non-cooperative Gaming and Bidding Model Based Resource Allocation in Virtual Machine Environment, Degree of Hangzhou Dianzi University, Vo. 12 (2012), pp. 1–44.

College English teaching reform based on multimedia network technology

Ya Fei Lv
North China University of Technology, China

ABSTRACT: The development of colleges in the new era, in order to meet the needs of college English teaching requirements for the Ministry of Education and foreign cultural exchanges, the current English teaching in colleges and universities deal with the many problems arise, and to respond positively to college English teaching multimedia network technology trends in the use of. In this article, the current status of college English teaching, as well as multimedia network technology in the reform of college English teaching mode has been actively explored.

KEYWORDS: multimedia network technology; English teaching; reform; application

1 INTRODUCTION

With the vigorous development of China's education, college admissions threshold is gradually adjusted so that more students have access to higher education to achieve a dream; a huge flood of students question the development of colleges and universities, which also presented greater challenges. The most obvious of these is reflected in the teaching of English, which is a required course, are involved in the most professional, but did not attract the attention of schools and students, resulting in a college English teaching quality is not high status. The development of multimedia network technology innovation for the reform of English teaching mode provides the opportunity to combine classroom teaching with computer networks, for students in all aspects of practical ability and comprehensive development of a very important practical significance.

2 CURRENT STATUS OF COLLEGE ENGLISH TEACHING

Amid the background of higher education development in the new era, the author of the English teaching situation analysis that was conducted concluded that their knowledge is still there re-light capability phenomenon, together, the majority of college English teaching model of a single, mostly teachers on the way to class Lord, small amount of language input, memorizing grammar and vocabulary knowledge, I heard that deficiencies in training, only to meet the primary objective of the examination, coupled with the impact of college enrollment continued, with the number of serious shortage of teachers,

teacher preparation classes and other struggling every day, students can study analyzed the characteristics and teaching needs, can not guarantee the quality of English teaching, long-term development process, students feel boring teaching atmosphere, gradually lose interest in learning, can not be achieved on the practical application of language ability. With the development of computer network technology, making advantage has been reflected in the college English teaching, more on adjunct to participate into the teaching of English in the implementation of the university, which has figurative, three-dimensional features in terms of expressions Construction for the development of English teaching in the wide-open spaces, and constantly optimize the teaching of English language environment, with its foundation, combined with the development of the campus network, will play an active role in the college English teaching and individual learning, for college English reform teaching model provides technical assurance [1,2].

3 COLLEGE-BASED ENGLISH TEACHING REFORM MULTIMEDIA NETWORK TECHNOLOGY FOR EXPLORATION

3.1 *Advantages of multimedia network teaching*

Development of university education in the new era should be combined with multimedia network technology, reform of teaching methods for students from colleges and universities to create a good learning environment; there are many advantages and features that are in line with the development trend of college education. First, the multimedia application process, combined to carry out the teaching of English, the use of sound, animation, text, and other technology to deepen the students' sensory stimulation, with its three-dimensional, vivid features captured the students' point of interest and attention for students to expand the amount of information and learning content. So a lot of the teaching content is not only difficult to understand on display through multimedia visualizing and deepening the understanding and memory of students, but it is also conducive to student learning related to teaching content. Second, through the use of multimedia network technology, personalized learning for students is possible; especially in college English teaching, the use of this technology provides students with a broader field of space time, creating an open-ended diversified learning environment, in particular the use of the process, by using a local area network or independent learning centers. According to their English proficiency and learning needs related network login, select appropriate learning content, and then implement targeted and personalized learning. Finally, this technology can be intended for students to create a more realistic locale, such as the appreciation of the excellent programs and movies in English, with the English text, or oral communication, and it can improve students' English level [3].

3.2 Teaching mode and teaching multimedia network technology combined

Leading to the secondary status of teachers and network, technologies were handled correctly. In college English teaching in the new era, a powerful multimedia network technology's capabilities is not enough to do everything their essence, it is merely a technical means to place undue reliance on them will cause the loss of classroom teachers in the dominant position, because it obviously not a substitute for full teaching mode, in order to be rational use, must strengthen instructional design, reasonable role of its secondary status [4].

3.3 Optimize the use of homemade courseware classroom teaching process

Based on the particularity of college English teaching, I found that in some of the traditional methods of teaching, some concepts to understand the content is difficult to break, but the multimedia teaching is different, the sound of their own culture and Mao and other advantages, can be carried out to improve the teaching content and rich, with a vivid, intuitive way possible to abstract concepts such as the contents of the demo came out, plus the amount of information

teaching, the teaching process optimization, and improve teaching efficiency on this basis. Courseware made by teachers encourages students' needs to learn to effectively integrate the content, use of multimedia network technology, creation of a vivid English network teaching environment, and promotion of the college students learning English.

3.4 Make use of the internet to expand the platform for learning and monitoring

Amid the background of the current network development, no teachers proceed step by step to teach students a lot of grammar and translation; this approach will obviously increase the students' English language teaching in conflict-related knowledge by teachers will be uploaded to the independent learning center, in order to basis, so that students can be based on their level of English and need to be targeted learning and reinforcement, select the appropriate content to practice self-self. To further promote student learning, students should be encouraged to actively use e-mail and other exchanges with teachers for the sake of discussion; through time and space constraints, a second class should be created for students. In addition, based on - English is the status of an actual application-oriented language training students should increase their listening and reading, for which you can use the appropriate educational software, increase training intensity listening and reading, teaching and constantly consolidating what they learn. In addition, in order to achieve effective supervision of student, teacher learning can be monitored through a teaching platform for students based on computer time. Oversee the integration of feedback information, a comprehensive understanding of the students' self-learning, in order to adjust the teaching schedule timely and continuously promote student learning corresponding to English.

4 CONCLUSION

In conclusion, for the problems existing in college English teaching, the author of the new era of multimedia network technology uses a starting point, which is a combination of classroom teaching, the traditional English teaching model for effective reform, strengthening student self-learning and e-learning, and constantly improving the learning of subsequent monitoring and evaluation. All these escort students to learn English in order to continue to promote the college students' English learning and then to train them for the construction of the motherland applicability talent.

ACKNOWLEDGMENT

This study was supported by Young Talents Plan of Beijing Comprehensive University, Program number: YETP 1434.

REFERENCES

[1] Yanli Liu, Rujia Xu, Hong Wang. Research of college english teaching situation in multimedia network environment [A] Chinese Institute of Electronics, Information Theory Branch [C] Chinese Institute of Electronics, Information Theory Branch:.., 2011:4.

[2] Hua Li. Innovative research college english teaching under multimedia network environment [A]. Hubei University of Technology, China.Proceedings of 2010 Third International Conference on Education Technology and Training (Volume 6) [C]. Hubei University of Technology, China:, 2010:4.

[3] Yangling Wang. A Trial Application of Multimedia Network Technology College English Courses reform of traditional teaching model [J] electronic test, 2014, 10:124–125.

[4] Haibo Mu. Research of teachers and students roles adjustment of traditional Chinese medicine colleges in the English Reformation and network technology environment [J] Chinese Medicine Modern Distance Education, 2014, 11:83–85.

Information, Computer and Application Engineering – Liu, Sung & Yao (Eds)
© 2015 Taylor & Francis Group, London, ISBN 978-1-138-02717-6

Risk assessment and early warning mechanism research of nature reserve ecological tourism development in China

Yan Mei
Chengdu University of Technology, China

ABSTRACT: Since the late 20th century, along with the rapid rise in the number of eco-tourists, a growing number of nature reserves develop eco-tourism. However, due to the growing number of tourists in nature reserves, we should make a lot of nature reserves ecological risk gradually due to tourism development. This article focuses on tourism ecological risk assessment and early warning mechanisms for the development of nature reserves for analysis and research.

KEYWORDS: Nature reserve, tourism, ecological risk, early warning mechanism

In recent years, with the annual growth rate of 4% of global tourism, eco-tourism and the annual growth rate of about 25 percent compared, and occupy a rapid upward trend in the world tourism market. Our EPA statistics show that the number of Chinese nature reserves in 2002 reached 1700, and its total land area is 14%. Since the late 20th century, with the rapid rise in the number of eco-tourists, more and more nature reserves began developing eco-tourism [1]. National Nature Reserve, in particular, has good resource conditions; most of them have developed eco-tourism and significant economic benefits. However, due to the growing number of tourists in nature reserves, makes a lot of nature reserves ecological risk gradually due to tourism development. This article focuses on tourism ecological risk assessment and early warning mechanisms for the development of nature reserves for analysis and research.

1 RISK FACTORS OF TOURISM DEVELOPMENTS IN NATURE RESERVES IN CHINA

1.1 *Interfere with wildlife, impact wildlife behavior*

In the development of nature reserves, protected areas will seriously affect the behavior of wild animals; wild animal behavior will change. Especially with the changes in domesticated animals, the influx of tourists, so many wild animals are often in anxiety and panic, increasing wildlife escape rate.

1.2 *Increase the risk of biological invasion*

Transport and tourism has become an important medium for the spread of foreign tourists in the tourist biological activities. Especially in recent years, the world's tourism industry gradually developed, making a lot of incoming alien invasive species to China. The Chinese customs have repeatedly checked out from exotic fruits containing dorsalis, the Mediterranean fruit fly, and other organisms. These biological invasions affect our ecosystem Nature Reserve to some degree.

2 NATURE RESERVE TOURISM ECOLOGICAL RISK ASSESSMENT

There are many risk types of tourism planning, and the role of the relationship between each risk and the extent of the impact of tourism planning often finds it difficult to precisely measure the existence of fuzzy characteristics. So, in some areas where precise mathematical description finds it difficult to deal with a complex problem system, the fuzzy comprehensive evaluation method has its own advantages.

2.1 *Set evaluation elements index system*

Setting the index mechanism is a prerequisite for fuzzy comprehensive evaluation risk; the choice of indicators to evaluate the reasonableness of the results of the reliability and accuracy of science has a direct impact. For different projects, the actual situation in the project area should be effectively combined to develop evaluation factors set: $U = \{U1, U2, ..., Un\}$, each sub-element set Ui ($i = 1,2, ..., m$) is : $Ui = \{Ui1, Ui2, ..., Uin\}$

2.2 Determine reviews collections

Risk assessment should be conducted in accordance with the needs of tourism development and should be determined for comment rating, record reviews; the collection is V = {V1, V2, ..., Vp}.

2.3 Determine the weight of a subset of the evaluation elements

In accordance with the actual situation of the project to determine the weight of evaluation factors, usually two weights are used. Marking each subset of a weight is as follows: mark each sub-centralization weight as $\tilde{A}_i = [a_{i1}, a_{i2}, ..., a_{ip}]$.

2.4 Establishment of the decision matrix of comprehensive evaluation and fuzzy evaluation

Build a fuzzy decision matrix, the first step followed by a group of experts to evaluate each judge index level in accordance with reviews, if there are twenty people judging panel evaluated members, the evaluation U1 evaluated by seven of them is V1, that is to say, in the Panel of Experts, the proportion of experts who agree to 35% (7/20), namely, the index of V1 corresponding evaluation index of 0.35. It follows that a subset Ui (i = 1,2, ..., m) evaluation and decision matrix, that is, Ri (i = 1,2, ..., n)

$$R_i = (r_{ijk})_{m \times p} = \begin{bmatrix} r_{111} & r_{112} & \cdots & r_{11p} \\ r_{121} & r_{122} & \cdots & r_{12p} \\ & & \cdots & \\ r_{1m1} & r_{1m2} & \cdots & r_{1mp} \end{bmatrix} \begin{bmatrix} i = 1,2,...,n \\ j = 1,2,...,m \\ k = 1,2,...,p \end{bmatrix}$$

Then, the calculation results according to the evaluation and decision matrix synthesis and weighting coefficient vector are $[b_{i1}, b_{i2}, b_{in}](i = 1,2,...n) = \tilde{B}_i = \tilde{A}_i \bullet R_i$.

According to the results of single factor fuzzy comprehensive evaluation, we should ultimately be able to arrive at a comprehensive evaluation of each subset of decision matrix in U:

$$R = \begin{bmatrix} \tilde{B}_1 \\ \tilde{B}_2 \\ \\ \tilde{B}_n \end{bmatrix} = \begin{bmatrix} b_{11} & b_{12} & \cdots & b_{1p} \\ b_{21} & b_{22} & \cdots & b_{2p} \\ . & . & . & \\ b_{n1} & b_{n2} & \cdots & b_{np} \end{bmatrix}$$

Finally, by U sub-centralization weight coefficient vector A and comprehensive evaluation matrix R, by merging operations, the final calculation of the Tourism Development fuzzy comprehensive risk assessment results is as follows:

$$[b_1, b_2, ..., b_p] = [\tilde{B}_1, \tilde{B}_2, ..., \tilde{B}_n][a_1, a_2, ..., a_n] = \tilde{A} \bullet R = \tilde{B}$$

3 TOURISM DEVELOPMENT OF ECOLOGICAL RISK WARNING MECHANISM IN NATURE RESERVES IN CHINA

For many tourism ecosystem risk factors, many scholars began studying the travel warning theory more and more, which is the practice of China's tourism having an important warning guidance.

3.1 Ecological risk warning mechanism

Tourism has a large nature reserve of an ecological risk warning system, a more complex structure, and related factors will be more co-ordinate their time, knowledge and runs three elements of the social sciences, tourism, system science, natural philosophy and computer science technology to be integrated, according to the evaluation of police intelligence, analyze police source, police ruled out the risk and determine the degree of police during the operation, etc., divided into specific systems analysis, data acquisition, implementation and design of the four stages of the study area ecological risk warning system, building a sound ecological warning mechanism [2].

3.2 Ecological risk warning regulatory mechanisms

a. Optimize the allocation of human resources, information systems, and services for tourism to be continuously improved. Strengthen institutions and domestic and international tourism cooperation to constantly improve the mechanism of a new form of tourism professionals and improve the level of China's coastal resort planning study tourism and marine tourism, for the construction and development of the international tourism island to provide intellectual support [3]. And our eco-tourism must also focus on risk management, travel risk, strengthen domestic eco-tourism risk management and evaluation of mobile platforms and the Internet platform, continuous improvement and improve the information and service management oversight function to ensure the accuracy and timeliness of information, will be national Tourism risk prevention and early-warning management role into full play.

b. Further strengthen the policies and regulations and the organization and management development. Strengthen efforts to promote tourism of new regulations, constantly regulate the tourism market order, and gradually improve the mechanism of law, tourism, while also strengthening the standardization of travel agencies, tourist-star hotels, shopping, tourist attractions, and so on.

4 CONCLUSION

In China, it is late forn Tourism Development and Ecological Risk relatively late , but also pay attention to explore the methods and theories, specifically including research tourist attractions, tourism, tourism products and tourism resources, tourism ecological risk, with more extensive research. However, these elements have not yet formed a complete system of method and theory; therefore, the establishment of national nature reserves develop an ecological risk warning mechanism on health, with sustainable development of eco-tourism guiding significance.

ACKNOWLEDGMENT

This study was supported by the Provincial key research project of Sichuan education administration (12ZA007).

REFERENCES

[1] Jianchao Xi, Ruiying Zhang, Meifeng Zhao. Travel safety risk assessment along the Qinghai-Tibet Railway[J] Mountain Science. 2012 (06):189–190.
[2] Yuping Tao. sports leisure and sports tourism risk management research [J] Chongqing University (Social Science Edition). 2012 (03):124–125.
[3] Yongfeng Zhao. Inner Mongolia tourism and environmental warning Evaluation Index System [J]. Geographic Environment Research. 2011 (03): 126–127.

Using 3D animation way to show the construction method of the transmission and transformation project

Hong Ming Sha, Ji Tao Zhu, Lei Wang, Jin Hui Liu & Jun Cheng Cui
Liaoning Electric Power Supply Co., LTD. Liaoning, Shenyang, China

ABSTRACT: At present, with the continuous development of economy, China has been paid more attention to the power transmission and transformation project construction, which asks for a higher requirements for the power transmission and transformation project construction. Meanwhile, due to the continuous improvement of the science and technology, 3D animation has already been widely put into engineering application, and its main purpose is to improve the accuracy of the engineering construction. Combining with the practice, this paper explores the application of 3D animation in the power transmission and transformation project construction for reference.

KEYWORDS: 3D animation; transmission and transformation projects; Construction; methods

1 INTRODUCTION

In order to further promote the application of "standard craft", State Grid Corporation of China (SGCC) organized and compiled the *Transmission and Transformation Engineering Technology Standard Library of State Grid Corporation of China* to guide the engineering project and carry out the planning work. In the concrete implementation process, although the technology content of the standard library is comprehensive, it lacks intuitionism and is not vivid enough in the process of learning, training, and applications.

2 THE NECESSITY OF USING 3D ANIMATION TO SHOW THE CONSTRUCTION METHODS OF THE TRANSMISSION AND TRANSFORMATION PROJECT

With the continuous development of science and technology, the multimedia animation has been widely used in various fields, and can also realize the frequently-used, new and special construction technology and craft in the construction of transmission and transformation project. It shows the construction process with a vivid and visual animation, which is convenient for the vast engineers in infrastructure projects to learn and communicate, improves the efficiency and quality of construction, and is convenient to do direct planning, design, bidding, approval and management in large complex engineering projects. It is in favor of the designers and managing personnel to do aided design and to evaluate the design

scheme for all types of planning and design schemes in order to avoid construction risk. Besides, it can do the applications of the construction technical disclosure, the management planning technical plan review, the accumulation and training of new technology and new craft, the construction process control, quality safety weakness analysis, and the accident repair.

The construction/engineering demo animation is a new 3D technology application generated with the development of computer hardware technology in recent years. Through the 3D and the software in the late period such as 3DMAX NUKE REALFLOW AE, it can do a comprehensive, stereoscopic and visual demonstration for the construction process, construction methods, materials application, engineering mechanical equipment, appearance after completion, and even the construction management and construction quality testing, "break the regular pattern that making a lot of bidding manual when encountering a bidding project", do full preparation for undertaking projects from the report form, and also win the project for the contractor or the contractor because of the novel report form.

In addition to playing a role in the bidding section, construction/engineering demo animation can also reflect comprehensively the possibility of the engineering plan the before project construction, which is the important guarantee to improve the quality of engineering. Making the construction demo animation in advance can avoid some mistakes in the construction process and modify and adjust in advance, and eventually brings safety and quality assurance to the construction. Besides, engineering/construction 3D demo animation can also make the construction personnel understand some complex construction technology in advance, and make them be safe in the process of construction through explaining the construction animation.

3 FEATURES OF 3D ANIMATION

The form of animation can show the construction technology, technical difficulties, etc. in different stages of the construction to the operators, experts and judges, and it is applicable to the guidance and training before and after the construction and the safety production technical disclosure. Showing the complex technology and security considerations during the construction to the workers can make the training and the technical disclosure become more intuitive, convenient, standardized, and economic, and the effect is remarkable.

3.1 Intuitive

It breaks the traditional educational form, simulates the whole construction process before construction in the vivid animation form, and emphasizes the technical difficulties in construction. Before construction, it can show reasonable construction process to the operators in the intuitive form. Combining the visual effect and audio effect may have a stronger appeal.

3.2 Cost saving

It needs a lot of manpower, material resources and financial resources in the engineering design phase of the construction. Stimulating the construction process at the engineering design stage using 3D technology can show the real material, which can effectively control the construction costs before the construction, save the construction cost, and improve the accuracy of the budget.

3.3 Concise

It can make operators more rapidly and accurately master the key points in the construction of the transmission and transformation project in order to promote and apply the new technology. Using this new technology has a forewarning function in improving staff's security risk awareness, improving the operators' safety protection awareness, and preventing accidents.

3.4 Wide range of application

It can be applied to all participation units in the power grid construction, such as the construction management unit, electric power company, design institute, transmission and transformation projects, power supply research and development institution, power supply companies and so on.

4 THE APPLICATION PROCESS OF 3D ANIMATION IN 3D DESIGN AND TRANSFER WORK IN THE PROJECT OF 660KV QINGDAO CONVERTOR STATION

4.1 Engineering background

Ningdong- Shandong ±660kV d. c. high-voltage transmission project is the first ±660kV direct current project with bipolar delivery of 4000 MW in the world, and the current key construction projects for State Gird Corporation. It is the first time for State Gird Corporation to organize the 3D design and transfer work. Our institution undertook parts of 3D design work of this project by the end of Qingdao convertor station, and mainly did the following work.

4.2 Platform comparison and selection

First of all, 3D software platform was compared and selected. In the PDMS platform where the power generation was widely used, it mainly aimed at the piping design of the power plant, but there was no special tools to create a large number of wires in substation project, and it can only use a curved steel pipe to replace the fitted curve such as sag which cannot meet the need of the practical engineering. SUBSTATION 3D design platform in Bentley Company has outstanding advantages in the 3D design of the substation, has special software to design the wire with radian and can fit the actual arc curve. Besides, there were special software to design and check the electric distance, and special professional software such as lighting protection and cable placing which were more suitable for designing the substation. Ultimately, Bentley was determine to be the 3D design and transfer software platform of the project.

4.3 3D modeling

After determining the platform, all professions firstly undertook 3D modeling work, and in order to keep the model have unified standard among different professions, the following rules were formulated specially:

4.3.1 All professional models should adopt a unified unit. "Millimeter (mm)" is recommended to be the unit of length, and degree (°) to be the unit of angle.

4.3.2 All professional sub models should adopt the unified coordinate system, and the Z axis is facing up.

4.3.3 When generating, all professional equipment and material models should be given the level and color which accord with the overall design regulation of the model.

4.3.4 When modeling, it is appropriate to use the basic element such as the cube and cylinder, and reduce to use the complex element such as rotating body and stretching body. Besides, it should avoid using the "trepanning" and "cutting" operation in the process of modeling. In the process of a single device modeling, it should try to reduce the number of 3D entity.

4.4 Assembly model

4.4.1 3D model can be divided into the overall assembly model, the regional assembly model (including all professions), the regional professional assembly model (single profession), and the device model. 3D model should be assembled according to the order of "professional assembly, regional assembly, and overall assembly".

4.4.2 The overall assembly model arranges all regional assembly models according to the regional layout and the location of the unified coordinate origin, and connects the interfaces among all regions.

4.5 Collision check

After completing the overall 3D assembly model, it needs to do the hard and soft collision check.

4.5.1 Hard collision: it needs to check whether there is location conflict between the electrical equipment and other professional 3D model, including the collision between the electrical equipment and the civil engineering, and the collision between the cable trench and equipment foundation, as well as the pipes such as water conservancy project.

4.5.2 Soft collision: it is the charged distance calibration. It is appropriate to adopt the method combining the direct 3D space measurement with the soft collision, and do the charged distance calibration according to different regions.

4.6 Creating and making use of the database information

4.6.1 The 3D electric model will be endowed certain properties, including detailed parameters, equipment manufacturers, installation location coding and equipment material coding.

4.6.2 The doors and windows library and the China material steel and concrete cross profile library were improved and supplied combining with the engineering perfect.

4.6.3 The statistic function related to the electric equipment and the civil construction was formulated preliminarily, and the equipment material can be designed by 3D model automatically. When it changes, the material list of the equipment can update synchronously.

4.7 Achieving the cut-through between 2D and 3D preliminarily

4.7.1 The association (Part Number) of 2D symbols with 3D model was established through Part Database, and the data cut-through between 2D intelligent main wiring and 3D model was achieved.

4.7.2 The 2D flat and cross-section diagram was cut by 3D

4.8 3D digital transfer

The examination and approval documents, design documents, 2D drawings, specifications about equipment and technology, and the file formed in the construction were tidied in the whole design process in order to realize 3D digital transfer in PDF format and i-model format if they were 3D files. The owners can access to the digital substation project model and data through Project Navigater software platform.

Through visiting the digital engineering model, users can browse and search the engineering data and the related design documents and information, and can open the design file and design data directly by links.

5 CONCLUSION

Today, with the rapid development of the animation, the influence of science and technology brought to the society is unprecedented, the new 3D special effects animation technology really makes the animation industry and other fields benefit a lot, and the 3D animation art has become an indispensable tool in modern visual media art design, especially in the field of power transmission and transformation station. It is closely related to people's life, and once making a mistake, it will bring a huge lost to the national economy. Therefore, how to master the 3D animation technology and make it play a proper role in the transmission station has become a problem to be solved urgently in the present in China.

REFERENCES

[1] Li Luo: The Guality and Schedule Control of the Substation Construction, Electrical Equipment Vol. 4 (2005), pp. 124–125.
[2] Luan Xidao, Ying Long, Xie Yuxiang: Advances in Study of 3D Modeling, Computer Science Vol.2 (2008), pp. 208–210.
[3] Li Yang, Guan Zhiwei, Dai Guozhong: Research on Development Tools for Pen-Based User Interfaces, Journal of Software Vol. 3 (2003), pp. 392–400.
[4] Men Yan, Xu Renping: The Theoretical Research Status of the Architectural Animation, Science & Technology Information Vol.31 (2009), p. 68.

Information, Computer and Application Engineering – Liu, Sung & Yao (Eds)
© 2015 Taylor & Francis Group, London, ISBN 978-1-138-02717-6

Innovation and change of marketing management in e-commerce environment

Hai Yu Wang
Zhengzhou Normal University, China

ABSTRACT: In recent years, electronic commerce has made vigorous development, which also has an important influence on the development of enterprise marketing management to some degree. Along with the advent of the information age, the network gradually plays an important role in people's life. Furthermore, the business marketing management needs higher requirements for the change. Based on the analysis of traditional enterprise marketing management, combined with the current development of enterprise, we actively discussed the innovation and change of modern marketing management.

KEYWORDS: Marketing management; e-commerce; enterprise; innovation

In the process of development of the information age, e-commerce has gradually appeared with a unique way of network marketing, and in the subsequent development it achieves rapid development. It also promotes the enterprise marketing management mode in the new period toward its close, making the enterprise in terms of their products publicity, not only on television advertising for the traditional marketing way, such as to use of network technology actively, to convey their products to people all over the world, not only increase the sales of products, and also enhance the visibility of their products. The development of this model will become the mainstream of the future, which will have far-reaching significance on the further development of the enterprise.

1 ENTERPRISE MARKETING MANAGEMENT IN THE TRADITIONAL ERA

1.1 *Enterprises' traditional marketing management*

In the process of enterprise development, the network did not achieve universal socialization in the past, general of the personnel management was done by document, but written files and other documents easily be tampered and lose, etc., and the trapezoidal distribution enterprise personnel management structure also need the corresponding pay in the same way, which make enterprises' senior management level employees do not know the situation of the employers clearly, cause of formation managers immoral behavior, such as bribery, influence the development of the enterprise.

Television ads, newspaper ads, telemarketing, direct mail, and so on are the traditional enterprise marketing modes. Advertising idiomatic stickers are everywhere in the streets. The cost of television advertising is generally high, so it blocks the TV advertising for the small and medium enterprises, and even if they invest in TV advertising, it will sharply cut the funds of product research and development, making enterprises to live in difficulty, and which will have a negative impact on the development of the enterprise.

1.2 *Consumers' consumption behavior in traditional markets*

In the process of development in the past, the main purpose of business was to sell, and this situation has led consumers in inaccessible areas to lose the opportunity to compare. Businessmen are able to hike, greatly hurt the interests of consumers; in addition, in the traditional market, too few types of goods also limits consumer choice, making a far cry from real goods and imagination and defeating the consumers' enthusiasm. The method of face-to-face trade does not fully respect the interests of consumers, but it affects the development of enterprises [1,2].

2 INNOVATION AND CHANGE OF MARKETING MANAGEMENT IN E-COMMERCE ENVIRONMENT

2.1 *Commodity marketing strategy*

In the development of traditional mode, in order to meet consumer demand, enterprises need a lot of inventory. This phenomenon is bound to increase its inventory costs, but inevitably causes the phenomenon of goods stored, which will bring some economic losses, had also led directly to a number of SMEs' collapse, and in the environment of e-commerce development, is involved in shrinking the company's inventory, reducing the cost of the development of SMEs, and prompting a chance to be bigger and stronger. On the other hand, the current marketing mode has changed the status quo that consumers and businesses could not communicate in the past. The network platform gives businesses a chance to know their customers: They can learn about the real needs of customers, nuanced communication for the feasibility of the transaction. Only in this way, enterprises can design and produce goods according to customer requirements in the subsequent development. In order to continue to accumulate customer trust, in current social development tide, whether enterprises are successful or not depends largely on consumer recognition; this factor constitutes the foundation of the business of marketing in gradual development. Under this mode, the enterprises can continue to promote the stability of their product marketing and social status, which has a great significance for the future development of enterprises [3].

2.2 Employees' management

In the development of the new era, e-commerce dramatically reformed the method of enterprise marketing management. we can take the number of workers as a distinct, different from the person responsible for every aspect of the design in the previous development, cut the number of workers greatly, this change also is closely related to the reduce of the network marketing, make the company's marketing more simple, the enterprise in e-commerce environment only need to train a number of professional talents, this training methods fully meet the needs of this entire vendor operations, this phenomenon is obvious in the land development of Taobao business currently, one person or a few people can improve operating a shop, this mode will continue to promote the development of e-commerce network marketing, which will become the mainstream mode.

2.3 E-commerce talent discovery and cultivation

In current e-commerce network marketing development, the most urgent need of human resources enterprise development is that for e-commerce professionals in the new era, and they should have appropriate practical skills. Obviously, there is not much talent; based on this, training electronic practical ability for business professionals to develop practical skills is very important for the future of enterprises. In general, professional e-commerce has a more extensive coverage, related to economy and finance, computer applications,

and related software development and other areas. It also requires professionals in the new period of higher demand. A wealth of theoretical knowledge not only is the foundation but also actively uses extensions and constantly enhances their own practical skills, the ability to actively promote their own website to customers, and to constantly improve their e-commerce development. On the other hand, a business manager should also be clear about the fact that personnel training is a long process where one needs to start from zero. Haste makes waste: In the specific operation, combined with their own developments, they take its actual needs as the starting point and goal, targeted training, in order to highlight the characteristics of e-commerce professionals, to recognize outstanding talent can reduce the cost of e-commerce enterprises, and promote their own, the more rational and efficient land development. In a specific operational process, we should actively do website promotion work, and constantly improve the site management background, increasing their sales. We must also enhance customer trust and business, so managers can easily insure their own shop, improve personnel management, and enterprises can develop faster. So, in general, in the current e-commerce environment, the most important part pertaining to the development of enterprises marketing is e-commerce professionals, who also constitute the basis of larger corporate profits. We only need to increase e-commerce talent and provide systematic and effective training, in order to enable enterprises to have a better foothold in popular e-commerce information society.

2.4 Network marketing function that the enterprise website should have

Under the environment of e-commerce development, the enterprises want to seek a wider development; it is very necessary to add some network marketing function, and it has a significant role in boosting the promotion of the development of enterprises.

First of all, one should have a certain ability to publish information, in continuous development of enterprises, in order to expand its business scope, attract financing, and so on. We need to be attractive to browse basic information, brand images that were published on the web; focus on product promotion information; and promptly get involved, such as the development of enterprises will bring the customers and profits. Second, we must establish the credibility of the site, the current complexity of the cyberspace, good and bad, in order to achieve good business development. Reputation is very important, and we should continue to accumulate the trust of customers; for instance, Taobao buyers are willing to shop in the crown of the consumer, because regardless of whether the quality is good or bad, at least you can make it have some psychological comfort. It is easy to see that the credibility of its business development will have a huge impact and, finally, it should develop an ongoing relationship with customers. In order to promote their own development, businesses need to micro-blog, blog, and so on to continue to enhance contact with customers and to promptly carry out some reciprocity policy. We should research customer needs in a timely manner, etc., to products the production and consumption which customer really needs, and constantly promote customer sales enthusiasm, in order to be bigger and stronger, and promote the continuous development .

3 CONCLUSION

In conclusion, in the current environment of e-commerce, if the enterprise wants to get bigger, enjoying more long-term development, we must actively use its advantages and characteristics, and constantly promote its marketing management methods, to create their own brand and credibility, adapt to the changing e-business environment, and improve their own way of operation.

REFERENCES

[1] Yunyun Su. Enterprise marketing management changes in the context of e-commerce [J] economic and technological cooperation, 2014, 18:106–107.
[2] Ke Zhu. E-commerce innovation networks and consumer behavior change based network [J] business, 2014, 24:72.
[3] Bing Li,Jianan Li.Innovation Research of business e-commerce marketing management [J] Market Modernization, 2013, 05:84–85.

Information, Computer and Application Engineering – Liu, Sung & Yao (Eds)
© 2015 Taylor & Francis Group, London, ISBN 978-1-138-02717-6

Data mining-based information extraction research of knowledge software

Pei Jun Wang
North China Institute Of Aerospace Engineering, LangFang, HeBei, China

Wei Guo
Dong Fang College, Beijing University of Chinese Medicine, LangFang, HeBei, China

ABSTRACT: This study analyzes and discusses the effective application of data mining in software repository information extraction in order to be able to provide useful data for software developers.

KEYWORDS: Software repository; data mining; information extraction

1 INTRODUCTION

In recent years, with more and more information to deal with, the size and complexity of software systems continue to expand, and the difficulty of its application and development has also increased. Our aim has been to extract valuable data in software engineering and ensure software researchers develop more high-quality software, in order to continuously enhance the stability of the software system.

2 DATA MINING

Data mining is intended to extract massive useful information from the data repository.

2.1 Data mining workflow

2.1.1 Data preprocessing

First, collect the data; regardless of the information being useful or not, try to collect and then clean the data, that is, delete or supplement incomplete data information to ensure a unified message format, and then pick out software information with a unified format, find out useful data information relevant to software and target, normalize the selected data, and ensure that the information data are in a certain range to make preparation for the next step [1].

2.1.2 Data mining

This is the foundation of information extraction workflow: Determine how to extract data; for the situation with the same target, the result will be a little different because of different algorithms, and

the selection of data mining algorithms typically is influenced by two factors, namely: a. Needs of users and goals;. b. Characteristic of information data [2]. For example, the software data are consistent; some software users need to predict the development trends of software based on data, but some users will need to collect summary results.

2.1.3 Model evaluation and knowledge representation

Obtain results from the earlier analysis and discussion; however, whether this result is consistent or not with the reality of the needs of the user can only be determined after the assessment results. "Mode" refers to a comprehensive assessment process for a certain degree of interest in knowledge-related software metrics identified. After the determination of the information results, the next step is how to effectively display information among users, which is a knowledge representation.

2.2 Data mining tasks

The main task of data mining is to build or identify a model that is consistent with the current targets, to ensure that the useful information can be found out in the numerous data. Common data repository data processing generally provides a summary description of the user. At this stage, there are a lot of data mining models, including correlation analysis, concept description, analysis and testing of heterogeneous, cluster analysis, classification and prediction, data reduction, and evolution analysis. In the specific operation, people not only apply a data mining but also make an integrated application of a variety of modes.

3 EXTRACT SOFTWARE REPOSITORY INFORMATION

In recent years, with the increasing of the complexity and scale of China's software, the single combat when developing software cannot meet demand, and in this situation, the Group of development emerged. However, the Group has developed the following problems: a. recover software status of a certain period; b. assess quality of software development; and c. only allow individual to modify or develop a program in a certain period of time [3]. To solve the earlier problems, you need to introduce the software engineering version control library. After generating software engineering version control repository, a lot of software developers gain favor. Its main advantages are as follows:

First, the software developer is divided into two roles, namely software administrators and ordinary programmers. Along with these two roles, entitling software administrator to freeze the program, the program has no right to freeze and the common software programmers are free to modify the program to freeze to fully control procedures that are freely modified. In software development, software administrator can modify the code section and clear program development progress, in order to ensure the coordination of software engineering [4].

Second, recover software to a certain time state at any time. This function can effectively compare the changes in software development and collaborate developers to accurately operate.

Third, evaluate software quality. Effectively analyzing modified time for a block of software, the principles have the authority to grasp the actual software development progress, to fully grasp the important and difficult points of the development of software.

Finally, full of changes. Allow only one person to complete a period of software development process; before completion, no one can modify the program library procedures to ensure the integrity and accuracy of the software system.

4 SOFTWARE REPOSITORY INFORMATION EXTRACTION DATA MINING APPLICATION

Typically, a software system treats software architecture as a highly abstract architecture; software and system function module description part is not much, but a detailed record of the relationship between the various subsystems. In understanding the architecture, the process is generally divided into three, namely suppose, compare, and investigate. These three processes are repeatedly executed throughout the entire system, and they finally complete the entire software system.

4.3 *Suppose*

With years of experience and expertise, software developers put forward an assumption of the relationship between software system architecture and software subsystems, and thus to speculate, based on these speculations to make a preliminary understanding of the software system architecture.

4.4 *Compare*

Before determining the system architecture, software developers will usually give a variety of assumptions and architecture. In terms of which is more consistent with the actual situation, we must compare these assumptions. It should be noted that, even if the software developers have the right assumption, the dependencies and differences of the software subsystem are still not clear [5].

4.5 *Investigate*

Software researchers use data mining to explore dependencies between subsystems, in order to save time, conduct proper guidance for development system framework through software reflection framework. Software reflection framework actual process is in Figure

On comparing the actual structure with the conceptual framework, software developers drew differences between the two programs, marked dependencies where inter-dependencies did not exist, so as to acknowledge the entire system more comprehensively, and thus modify the structure of the original concept. Difference between the actual architecture and concepts of software architecture is the key concern of the reflection framework law. Mutual dependence of subsystems can be divided into three kinds, namely lack of dependence, convergence dependence, and divergence dependence. We hope to get the results of which are dependent on convergence, and the dependence is lacking in real architecture. The factors leading to this phenomenon are mainly ill-considered software developers. Moreover, the divergence dependence is one kind of dependence that is not considered; the main factor for this case is that people modify or increase a subroutine BUG or function but do not join in the document in time.

5 CONCLUSION

In summary, through the discussion of the data mining technology, this study analyzes in detail the workflow and concept of data mining technology, and it discusses in detail the software repository information extraction. Data mining is actually a synthesis process that is used to draw from a number of data fields potentially useful and regular patterns, information, and other knowledge; to extract more useful software repository information, thus playing an important role in software development.

ACKNOWLEDGMENTS

This study was supported by KY-2014-26, application research of social exam data mining and analysis, and Youth Fund Project of North China Institute Of Aerospace Engineering LangFang HeBei.

REFERENCES

[1] Xu Zenghui. *Study on Data Mining Association Rules* [J] Science Technology and Engineering, 2012 (01): 148–149.
[2] Cao Luzhou. *Data Mining Technique and Its Application* [J] Science and Education (HEAD), 2012 (01): 120–121.
[3] Yu Shusi, Zhou Shuigeng, Guan Jihong. *Software Engineering Data Mining Progress* [J] Computer Science and Exploration. 2012 (01): 133–134.
[4] Yu Yonghong, Xiang Xiaojun, Gao Yang, Shang Lin, Yang Yubin. *Service-Oriented Cloud Data Mining Engine Research* [J]. Computer Science and Exploration, 2012 (01): 100–101.
[5] Chen Dongli. Genetic Algorithm-based Data Mining Technology Application in Library [J]. Computer and Modernization, 2012 (01): 129–130.

Analysis of modern information technology application in library management

Ying Hui Xu
Zaozhuang University, China

ABSTRACT: In recent years, with the rapid development of economy and the ever-changing nature of science and technology, currently, information technology has been widely used in various industries, especially in library management, bringing a positive impact on library management optimization. When studying the application of modern information technology in library management, this article first analyzes the importance of modern information technology and its application in library management; second, it analyzes the main problems faced by modern information technology in library management; and, finally, it discusses and summarizes specific measures for the application of modern information technology in library management.

KEYWORDS: Modern information technology; library management; application

1 INTRODUCTION

In the current 21st century, the rapid development of information technology has brought unprecedented opportunities for the current management of libraries, as well as how to use modern IT rational application in library management has always been a hot field in library industry research. Therefore, this article has some economic value and practical significance for the research and analysis of modern information technology application in library management.

2 MODERN INFORMATION TECHNOLOGY

2.1 *Definition of modern information technology*

The so-called modern information technology is primarily a technique that is used for information processing and management of science and technology; this information technology is mainly intended to use communication technology and computer science and technology, so as to realize development and design and installation. Modern information technology mainly includes sensor technology, communications technology and computer technology [1].

2.2 *Features of modern information technology*

In the actual application process, modern information technology tends to have certain basic characteristics; it not only has the digitization and networking features but also has the basic characteristics of multimedia, virtualization, and intelligence. And to some extent, an important technical feature of modern information technology is that it not only has a certain method of science but also has the advanced nature of the tools, equipment, skills, and a wealth of experience [2]. It also has a quick operation process and efficient function; the personality characteristic of information technology is the information, and in the course of services, its IT not only is mainly information but also has the basic features of a dynamic objectivity and dynamic nature.

3 IMPORTANCE OF THE APPLICATION OF MODERN INFORMATION TECHNOLOGY IN LIBRARY MANAGEMENT

With the rapid development of the economy and the progress of technology, currently, modern information technology is gradually being more widely used in various industries, and the current traditional library work in the actual work process, its main business processes and capture a Border collection, distribution and access to counseling and then to realize the work carried out in the course of business often requires the help of labor, so that staff has a relatively large work pressure and working time [3].

The application of modern information technology in library management, not only will the library automation business and its management of the staff to be realized, and the help of various management aspects of their work in the library, and promote the dissemination of information of the library, so as to raise the overall level of the library management.

Modern information technology in library management, by means of information technology to provide vast amounts of information to the user to help the reader have access to information, also saves labor. By means of a statistical information technology computer for museum work, we need to understand the functioning of timely and good management of the library of the business, so as to realize the form of network, distributed management, and digital resources management [4].

In summary, the application of modern information technology in the library often has a certain necessity, but we should strengthen efforts to publicize the application of modern information technology in library management.

4 PROBLEMS OF MODERN INFORMATION TECHNOLOGY APPLICATION IN LIBRARY MANAGEMENT

Some problems also exist in modern information technology in library management: Such problems are mainly copyright issues, occupation position of data, as well as network problems.

First, in terms of copyright issues, in the case of information sharing, there is a need to provide readers with a certain amount of electronic resources, and information technology that facilitates the realization was also attracted to a number of copyright disputes [5].

Second, the data occupy positions on the issue, the storage location is not clear, and it is difficult to ensure the protection of the fundamental operational information library management system, due to the lack of specialized agencies set up by some databases of the government.

Finally, there are network issues, but by means of modern information technology network technology in the same library management application, once the network is not clear, it will be difficult to achieve a smooth LEE Wing and share information.

5 REFLECTIONS ON MODERN INFORMATION TECHNOLOGY IN LIBRARY MANAGEMENT

In the process of modern information technology applications in the management of the library, we should combine with the actual situation of library management and rationally use modern information technology. Specific measures regarding the application of modern information technology in library management mainly include the following:

5.1 Establish a network reference service system

In the process of modern information technology applications in the management of the library, establish a certain network reference service system and aid in the form of e-mail and tables for online consultation and video consulting services, thereby providing an effective long-term service. Then, solve the problem that readers encountered when reading the information .

5.2 Carry out interlibrary loan service

Once the application of modern information technology in library management, we should vigorously encourage interlibrary loan; interlibrary loan and document information service system are perfect, and coordination and cooperation are used to reinforce regional libraries. Modern information technology in the library management application process, along with this relative openness of the service, not only attracts more readers but also enhances the reputation of the digital information library, thereby strengthening the overall formation of the current library and development.

5.3 Foster innovation management philosophy

In the process of modern information technology applications in the management of the library, according to the spirit of innovation management philosophy, break through traditional library management philosophy and management constraints, and in the context of modern information technology, constant innovation management philosophy, and a good cooperation and coordination of other library resource sharing platform to build on, and then the advantages of information technology library management fully play out, then maintain the stability and advancement of information.

6 CONCLUSION

With the rapid development of the economy and the deepening development of modern information technology, in the process of the current library management, we should seize the opportunity of the times, and with the help of modern information technology, conduct good management of the current library. According to the spirit of innovation angles, we should break through the traditional library basic management philosophy and fully mobilize the enthusiasm in all aspects of librarians; improve the information technology and professional ability of librarians; build a digital library information system; and improve the overall library quality and service reputation, thus comprehensively promoting the flourishing of the current library cause.

REFERENCES

[1] Zhang Chao. *Modern IT Advantages and Disadvantages of the Library Management Research* [J]. Office business, 2013, (7):134, 136.

[2] Wang Xiuping. *Modern Information Technology in Library Management* [J]. Youth Writers, 2011, (16): 184.

[3] Duan Lianjin, Yuan Xinyu. *Library Management Transition Under the Modern Information Technology-"Humane Management" Management Philosophy* [J] Jiamusi Education Institute College, 2014, (3): 173–173.

[4] Wang Chunlei. *Establish the Scientific Outlook on Development, Improve Library Management* [J]. Heilongjiang Science and Technology Information, 2011, (3):134.

[5] Meng Jie. *Effective Integration of IT and Library Management* [J]. Science and Technology Information, 2013, (25):245, 268.

Information, Computer and Application Engineering – Liu, Sung & Yao (Eds)
© 2015 Taylor & Francis Group, London, ISBN 978-1-138-02717-6

Innovate moral education in health vocational colleges by means of multimedia

Gui Qing Yang
Langfang Health Vocational College, China

ABSTRACT: With the innovation of education in China, "chalk and talk" mode of education depicted by a lot of vocational colleges is constantly being replaced with social progress, the improvement of the teaching quality, and students' initiative. Traditional education mode can better coordinate the role and exchange, hampering the cultivation of students' personality. Multimedia teaching has become the primary means of manners in many schools. To optimize the classroom, due to particularity of moral education in health vocational colleges, building a multimedia teaching platform in vocational colleges is very important. The multimedia teaching platform is consistent with the concept of moral education in health vocational colleges. This article focuses on the multimedia innovations in moral education in health vocational colleges.

KEYWORDS: Multimedia innovation; health vocational colleges; moral education

1 INTRODUCTION

Traditional teaching methods cannot keep pace with modern teaching methods, take advantage of the major institutions of multimedia teaching technology, expansion of new teaching ideas, changing a single teaching mode, and so on. So, the major theoretical university courses are being replaced with more and more multimedia teaching methods in teaching high school students to take advantage of multimedia technology on moral education in vocational colleges were to grasp every aspect of the use of the animated teaching content production into courseware, increasing the effectiveness and lively classroom, so that teaching students to learn more intuitive, comprehensive improve teaching efficiency and improve the teaching atmosphere, making health vocational colleges moral education teaching content richer, active and innovative.

2 IMPORTANCE OF MULTIMEDIA TEACHING IN ETHICS EDUCATION IN HEALTH VOCATIONAL COLLEGES

2.1 The sharing of resources between institutions

There are differences between institutions in educational philosophy, teaching style, and resources, so that each institution has its own vocational school characteristics. Traditional teaching management of resources has great limitations [1], so teaching can be considered a good resource to be more widely disseminated through multimedia technology, to enable students to better grasp the professional information through multimedia resources. Meanwhile, among the institutions that are also able to effectively transfer resources between each other to make teaching easier, more weak links have been strengthened to some extent. Courseware can enable different institutions to explore and learn; so complementing the weak point between the two can greatly enhance the sharing of resources.

2.2 Promote the friendly exchange between students and teachers and foster team spirit

Multimedia not only replaces traditional teaching, so that students have more learning space, more flexibility, and a ready nature but also enables teachers and students through multimedia for a more profound teaching on the moral teaching of great help action [2]. Students can enhance their own multimedia teaching weaknesses, but not by teachers to highlight the place, so that learning not only solves the problem but also deepens the impression, greatly improving the efficiency of learning, so that moral education is more firmly embedded in the hearts of students. Teachers can also clarify the doubts of students, highlight the classroom, which will not only solve the problems of students but also largely enhance exchanges between teachers and students, thus helping develop team spirit.

2.3 Reduce students' data access time

Moral education needs students to improve their own quality through a lot of reading and understanding, many students due course, the task is relatively heavy, not too much time for data query, appeared multimedia teaching students to solve this challenges students through multimedia teaching courseware making careful preparation and study, and often moral teaching learning knowledge, and lay a good foundation.

3 MULTIMEDIA TECHNOLOGY APPLICATION IN MORAL EDUCATION IN HEALTH VOCATIONAL COLLEGES

3.1 The use of multimedia largely explains the role of the curriculum, and the main content of the course also needs a variety of constituted factors

Multimedia teaching content for the implementation of programs is designed to provide reasonable; the main elements of the program mainly include content design, study guides, and teaching resources introduction.

3.2 Multimedia network teaching

Multimedia network teaching can use multimedia teaching to make teaching content scenarios by means of making courseware [3]. By producing content, including courseware, video, pictures, and text, students can get the information they want from the courseware in these different forms of content and save the more important courseware for a deeper study.

3.3 Assessment of learning situation

Multimedia teaching enables students to conduct network performance assessment, and teachers need to set up an evaluation system based on the characteristics of students and the teaching content [4]. If students need to grasp knowledge on their own, teachers should improve teaching content through the examination process and students' assessment results.

4 ISSUES OF CONCERN FOR MULTIMEDIA TEACHING IN THE MORAL EDUCATION OF HEALTH VOCATIONAL COLLEGES

4.1 Improve teachers' level of modern science and technology

Multimedia technology in university teaching has begun a comprehensive promotion, and no master teacher of modern technology presents a great challenge. Multimedia was introduced in our country, when the concept of multimedia teaching teachers apathy, lower operating levels, higher teacher arranged a job, and constantly improve their quality and ability of modern science and technology to master basic computer operations and operations related to moral education theory master courses teaching content production technology [5], optimize classroom teaching, improve teaching quality.

4.2 Combination methods of traditional teaching and modern teaching

Modern teaching methods consist of two parts: first, to enhance the teaching system in colleges and universities; the other is a multimedia teaching system. We should prove that multimedia teaching methods and traditional teaching methods have their advantages and disadvantages, and the two complement each other. Multimedia has developed into a modern educational model; the role of teaching is limited, its popularity is very broad, and inspite of all the characteristics of a part being applied, there are still limitations. Its many advantages not only compensate for deficiencies of traditional education but also depend on their own shortcomings of traditional education to compensate. We should objectively treat the role of media in physical education and combine the two organically and favorably.

4.3 Use of multimedia teaching should be appropriate

In the teaching process, it needs to be clear when using multimedia about what should be done with multimedia, to solve the problem, rather than blindly using multimedia teaching, that is, not to follow the trend, not only in order to make the classroom becomes more looking blindly use should not use the feature. Teachers should focus on teaching and difficulty in accurately grasping, so as to improve efficiency. The students can grasp the key points and difficult times, and multimedia applications can be considered as achieving their best teaching purposes.

5 CONCLUSION

With the rapid development of Chinese science and technology products, images, and sound recordings throughout the supermarkets, there are many occasions of using multimedia; the use of multimedia teaching has become a topic of people about teaching reform. Multimedia teaching methods have replaced the traditional ones, using modern technology to achieve the purpose of teaching. Multimedia teaching methods can enable a lot of students to organize learning in a manner independent of their own interest. Multimedia has a very powerful database that helps students solve difficult problems, in order to stimulate students' interest in learning, stimulate their potential, and enable them to better grasp computer applications. So, multimedia serves everyone and creates a bright future for everyone.

REFERENCES

[1] Li Fengju. *Of Moral Education Situation and Countermeasures* [J]. Weifang Higher Vocational Education, 2007, 04:23–26.
[2] Nie Chao, Zhang Lei, Zeng Qingqi. *PBL Teaching Philosophy Multimedia Network Teaching Platform to Build Health Vocational Colleges* [J]. China IT Education, 2014, 20:25–26.
[3] Gui Jinyu. *Innovation Mode of Moral Education in Higher Vocational Harmonious Campus* [J] Chinese educational technology and equipment, 2008, 06: 15–16.
[4] Liu Hui. *Vocational Colleges Use the Network of Moral Education Thinking* [J]. Teaching, 2009, 09:14–16.
[5] Xu Juan. *Vocational College Moral Education Research Network Sight* [J]. Changjiang Engineering Vocational College, 2012, 03:51–53.
[6] Qiu Xia, Yan Guorong, Li Ming, Wu Haiyan, Gao Guiling. *On Moral Education Mode of Higher Vocational Colleges* [J]. Chinese school education, 2013, 09:140.

Mobile devices-based personalized foreign language teaching system research

Ling Yang

School of Foreign Languages, Three Gorges University, China

ABSTRACT: Compared with general non-language subjects, mobile devices-based mobile learning is more suitable for foreign language teaching. This article discusses the application form of mobile learning on one hand as well as selection and design principles if foreign language learning materials on the other hand, puts forward a new form of foreign language teaching system based on mobile devices, and implements the primary attempt in teaching practice to obtain a certain effect.

KEYWORDS: Mobile devices; personalized foreign language teaching; system research

1   INTRODUCTION

The wide use of smart phones, tablet PCs, and other portable communication devices have made the mobile devices a new type of foreign language teaching form. Mobile devices teaching will not suffer the constraints of time and space and the learning form is flexible. Linguists believe that this form of learning combined with language learning has great assistance to language learning; the use practice in foreign countries confirms mobile devices with powerful foreign language teaching potential, especially reflected in the stories Solitaire and micro-fiction appreciation of foreign language writing, new words, and sentence as well as word usage tracking of words teaching. Video dictation, abstracts, reviews, and more of Listening teaching.

2   THE FEASIBILITY OF THE USE OF MOBILE DEVICES IN FOREIGN LANGUAGE TEACHING

2.1   *The outstanding advantages of mobile devices in supporting teaching*

First of all, take function phone as an example. First, smart phone has the proper size and high-resolution screen, supports the display of text, graphics, animation, and other contents, has many formats of audio file playback, and provides a good control interface. Second, the smart phone has network performance; you can go through the web throughout the WAP browser or a WWW browser. Third, the smart phone can also place all kinds of applications based on user requirements to perform the corresponding function.

Therefore, whether it is in the scenario building, or in the autonomous learning, or in terms of query data, smart phones can provide good service to the learner. In recent years, there was a new type of mobile device—I-pad. Since the appearance of I-pad on the market, people have been paying more and more attention to it, of course, people are also aware of the urgency and practicality of I-pad being used in teaching. Before, there have been schools at home and abroad introducing the I-pad to school teaching and get good results. For other mobile devices, I-pad has its own advantages, especially the advent of a new generation of I-pad, making its superiority as mobile learning devices become more pronounced.

Conclusions are drawn after the questionnaire by the relevant domestic researchers that the mobile device configuration of the students after 90s' in China is integral, the functions of mobile phones are powerful, and many students are willing to accept mobile interactive teaching tool.

2.2   *Mobile device is particularly suitable for foreign language learning activities*

Foreign language learning materials have the typical characteristics of the micro-learning materials, especially for mobile learning activities based on the mobile device environment.

2.2.1 Foreign language learning is not restricted by time and space, it is an autonomous learning, foreign language learning activities in the actual natural environment often allow students to rise emotional feelings because of the specific actual scenes or the atmosphere, being very beneficial for students to quickly and correctly understand foreign language learning contents.

2.2.2 Because mobile devices have characteristics of real-time, efficiency and so on, it can combine the foreign mobile language teaching activities with classroom instruction, form a foreign language training interactive multiple dimensions of time, space, body, etc., the interactive can be before-lesson preparation, class layout and after-school consolidation, of course, it can also be class, bedroom, and outdoor spoken language training, and so on. Such a form of interactive foreign language teaching can quickly enhance foreign language learning efficiency.

2.2.3 Foreign language learning materials can often be transformed into rendering mode featuring by simple content and highlights. Such data rendering model is not only adapted to the capabilities of the mobile device screen bearer learning content, but also protected the real needs of the most useful learning material placement in the limited space.

3 MOBILE DEVICES-BASED MOBILE LEARNING SYSTEM RESEARCH

There are three main forms of mobile learning based on mobile devices.

3.1 *Short message interactive-based mobile learning form*

Short message-based mobile learning form reflects the dimensional transfer relations between students and students, students and teachers, students and teaching servers, as well as teachers and teaching servers. Such dimensional transfer relation provides convenience for the contact between teachers and students, and is beneficial for the conduct of teaching activities, provides students with a lot of solutions to problem, being beneficial for the learners' autonomous learning. Specifically, students can ask their teachers through a short message, one student corresponds with many teachers, but the method to solve the problem of every teacher may not be the same, the student's question and teacher's answers are passed through the change and governance of teaching servers, then students can select the most appropriate solution to the problem through their own thinking and practice, which gives students a good independent learning opportunity. The relations among students, teachers, and teaching servers can be reflected in a model shown in Figure 1.

This form of learning is the easiest and most convenient form for teachers and students to complete two-way communication, through this mutual process, learners can enhance interests and confidence, form positive attitudes towards learning, at the same time it can strengthen communication skills

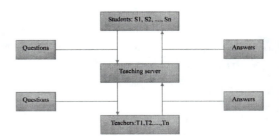

Figure 1. Multi-directional transfer model composed of students, teachers, and the servers.

of teachers and students, providing conditions for personalized learning of students.

3.2 *Video call communication-based mobile learning form*

The key objective of mobile learning is that it allows learners to study happily in the mobile shape. Because most importantly, mobile learning reflects non-fixed learning form anytime and anywhere, it can not only meet the demand of answers to ask of high school students in the traditional classroom, after the form of voice and video, the learners can implement synchronous two-way communication with their teachers and learning partners when they have difficulties in the learning process through the form of the sound pass and the screen display.

4 CONCLUSION

Through the construction of teaching theory, instructional design basic idea, and the reciprocal teaching method as the theoretical instruction, this paper proposes a new form of foreign language teaching based on mobile devices, provides a more real, vivid and effective learning environment for foreign language learners, making mobile learning more reliable. We have implemented a primary attempt in teaching practice and obtained certain results, gaining valuable teaching experience.

REFERENCES

[1] Zhang Jinhua. *Mobile Devices-Based Personalized English Learning System Research* [D]. Capital Normal University, 2013.
[2] Ma Kai. *Android Platform-Based Design and Implementation of Mobile Learning System* [D]. Beijing University of Technology, 2013.
[3] Wang Ai. *Mobile Learning-Based Literature Search Teaching System Study* [D]. Northeast Petroleum University, 2012.

Information, Computer and Application Engineering – Liu, Sung & Yao (Eds)
© 2015 Taylor & Francis Group, London, ISBN 978-1-138-02717-6

Design and implementation of equipment digital demo system based on B/S pattern

S.B. Yu, X.M. Li, D. Liu & L. Rao
Academy of Equipment, Beijing, China

ABSTRACT: This review is followed by an introduction based on the needs of data visualization technology that involved the current situation on equipment digital demo, combined with the designing equipment digital demo system. In the next section, a brief review of data visualization technology and characteristics with special regard to the advantage and the disadvantage of B/S pattern. Then, there is a short review of function module of equipment digital demo applied by the combination of B/S pattern architecture. Finally, an example demonstrates the specific implementation presents for equipment digital demo system. The system will provide an important platform for spreading equipment knowledge, providing decision support and analyzing equipment development trend.

KEYWORDS: B/S mode; equipment; digital demo system; design; implementation

1 INTRODUCTION

With the development of computer technology and Internet technology, and the popularity of the world wide web, the static web technology has matured application in many fields. With the appearance of page interactive, dynamic and scripting language, cookies, traditional description language has been unable to meet the requirements of users anymore. The B/S pattern as a kind of platform service gradually develops and grows based on Web technology, after the C/S pattern can't meet the new requirements of the current global network open, interconnection and information sharing. Compared with the C/S pattern and the B/S pattern simplifies the client, not need to install applications on the desktop, and users only need to install the browser for related business. For users, the B/S pattern simplifies the operation and with no need for study of related operation process and matters needing attention. Meanwhile, not only the complexity of system process and convenient maintenance personnel for maintenance are simplified, the reusability of data resources is improved, but also the pressure on the database is eased.

The data are the carrier of information, and information is a reflection of the characteristics of the objective world of things, in the reflection of objective things trace information after entering the brain, will form the knowledge[1]. According to IBM's research, obtained all the data in the whole human civilization, there are 90% form in the past two years. In 2020, the world of data scale will reach 44 times today. Based on the application of different types of data,

the birth of a variety of technology, there are used for data acquisition of data extraction techniques[2], used for data acquisition of data cleaning technology[3] and used for data storage, database, and data warehouse technology[4], etc.

The equipment is an abbreviation for weapons and equipment. There are two types of definitions of equipment in the 2011 edition Chinese people's liberation army (PLA) Military term, the one is as a noun which means used in operations and operations to ensure weapons and other military operations, electronic information systems and technology equipment, equipment, the other one is as a verb which is pointing to the troops or with weapons and other military equipment, equipment, fittings, and other activities, etc.

Data visualization is a higher level of visualization technology, which is following the scientific visualization and information visualization. Data visualization origins can be traced back to the early days of computer graphics in the 1950s, at that time, people use a computer to create the first graphic ICONS. Data visualization technology refers to the use of computer graphics and image processing technology, converts the data to graphics or image on a screen, and interacts with processing the theory, method, and technology. It involves computer graphics, image processing, computer-aided design, computer vision and human–computer interaction technology, and other fields. Data visualization is the science of science data visualization and abstraction of unstructured, the geometric data of information visualization, a visualization technology is gradually known by person.

Data visualization can be considered in the broadest sense of the scientific visualization, information visualization, and referred to in the field of data visualization[5], etc. Data visualization has the following characteristics[6]:

1 Interactive: The users can interactively conveniently data management and development.
2 Multidimensional: Can see multiple attributes of objects or events data or variables, and the data can be according to the value of each dimension, and its classification, sorting, combination, and display.
3 Visibility: The data can be used to image, curve, 2D graphics, 3D body and animation to display, and the patterns and the relationship between the visual analysis.

Data visualization not only provides a form of data display, but is also an important method of data analysis, and data visualization technology applied in the digital demo is still in the exploratory stage.

2 SYSTEM ANALYSIS AND DESIGN

The main purpose of this part is to build a logical model of equipment digital demo system, as the input for the development and implementation of the system. The following part will detail the instructions from the system architecture, function structure and development environment in three aspects.

2.1 System architecture

In order to achieve the demand of the operation is simple and fully functional, the digital demo system is divided into the foreground subsystem and the background data customization subsystem, and sharing the equipment database together. The overall module of equipment digital demo system is shown in Figure 1.

The foreground subsystem mainly includes: user register, user login, user logout, users' information feedback, and demo items list five modules. The demo module is the core part of the system, and is also equipped with digital system design of the main demonstration. The backstage subsystem mainly includes: system management, demo items management and user management. The design and maintenance personnel can through the backstage subsystem to improve and expand the equipment database, to add data visualization methods, at the same time, can constantly improve the system function, improve system performance, increase the demo module.

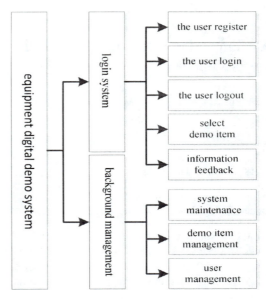

Figure 1. The overall module of equipment digital demo system.

2.2 System function design

Function modules indicate that the dependencies between the system and data processing capacity. According to the overall digital system module and system design objectives, the specific functions are as follows:

a) The main function of the foreground subsystem:

1 Homepage;
2 The user register& login & logout;
3 Introduced of digital demo system;
4 Digital demo lists;
5 The user information feedback.

b) The main function of the backstage subsystem:

1 Improving and extending the equipment database;
2 Adding data visualization ways;
3 Increasing the digital demo module;
4 Auditing & deleting users;
5 Maintaining the system daily operation.

2.3 System development and running environment

The equipment digital demo system based on Windows operating system, using the Java language and MyEclipse integrated development environment to develop. MyEclipse with a variety of good development tools, can effectively run the application design, development, debugging and deployment. Dreamweaver as a collection of web page creation and management site of a wysiwyg page editor, can create cross-platform, cross-browser web pages. After installing the Windows operating system, installed and configured the JDK (Java Development Kit) environment, set up prototype system Development environment based on MyEclipse and Dreamweaver, and the final will be deployed in Tomcat server developed prototype system. Prototype system software development environment includes:

Development languages: Java, HTML;

Development platform: MyEclipse 2015, Dreamweaver CS6;

Development and debugging environment: Tomcat v7.0, JDK VM 1.7.0 _u45 Windows;

The user's browser: Firefox 33, sogou browser 5.1 at a high speed.

The development environment in the laboratory, using 10 computer system platform, one server, and through laboratory internal hardware environment correlates of Ethernet connection.

3 IMPLEMENTATION OF EQUIPMENT DIGITAL DEMO SYSTEM

The development of science and technology promotes upgrading of equipment of speed, and applied to weapons and equipment more and more high-tech. Within the army, the masses of a line of soldiers, age generally small, lacking of equipment knowledge, cultural level, no one could use and dare to use of the equipment. For local people, some military sites for the purpose of attracting visitors, for the purpose of business profits, not objective, accurate to spread and popularize knowledge of some equipment. Therefore, it is necessary to stand in the perspective of a professional, under the premise of considering using the object, and design the equipment digital demo system.

In 2012, China's first aircraft carrier launching and equipped into the Chinese navy, China has become one of the few countries with an aircraft carrier, and the United States as the world's leading military power, the Nimitz-class aircraft carriers are almost half of the total aircraft carrier in the world. In recent years, Japan under the circumstances of the small land area, limited resources to illegal occupation of the Diaoyu islands in China, and the disturbance of purchasing an island is escalating. The situation of Diaoyu islands is not tested and verifying the rationality, feasibility, and applicability of equipment digital demo system optimistic. Chinese military since the research of military aircraft was once caused the press boom. To design scheme, this paper chose equipment and related data which are introduced above to display by

data visualization graphics. The figures shown below are system page screen-shots respectively.

As shown in Figure 2 for equipment digital demo system home page, also the Browser part of the B/S architecture, by the user input the correct user name and password, the page automatically jumps to the page. The page is made up of system notices, contact

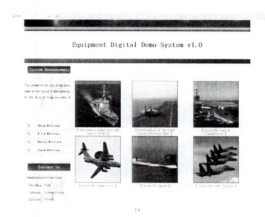

Figure 2. The homepage of digital demo system.

us and the main demonstration area. The photo in the top row from left to right consists of the Nimitz-class aircraft carrier of the United States, performance of an aircraft carrier (without the United States) and aircraft carrier competition. Below a row from left to right consists of the world aircraft competition, the military comparison of China and Japan and the analysis of China's military Aircraft. In the process of generating visual graphics, the visualization graph data are from equipment database, and the JSON data format is used for rule base. Finally, the visual graphics are generated (Figures 3 and 4).

As shown in Figure 3, the world countries with the carrier distribution diagram, this is the geographic information for a large innovation in the field of data

Figure 3. The distribution of countries with the carrier.

Figure 4.　The indent - hierarchy chart of the United States military establishment.

visualization technology. Here is the dark part of the aircraft carrier state distribution is useful. With concise and convenient visual graphics for the analysis of the world military pattern has a very important value. There are three commonly used kinds of geographic projection, equidistant projection with winker cylindrical projection, and sinusoidal projection.

As shown in Figure 4 for the United States military establishment system indent – hierarchy figure, through the levels of indentation figure can clear understanding of the American establishment. Here in the dark node said when click it can enter the next level.

There are some general pages such as user registration, user login, user information feedback, etc. Due to the limited length of the article could not provide screen-shots and further instructions, but the related work has been completed.

In the process of demo, equipment data from the database to the Web client to generate visual assessment in the process of the graphic, in data transmission using JSON (JavaScript Object Notation) data format. JSON is a lightweight data-interchange format, in many applications is to replace the XML (Extensible Markup Language), has the advantages of easy to parse, easy to read, easy to generate[7]. To meet the demand of equipment digital demo system database data call, the equipment data rule base is established. In this way, can through the equipment data rules database repository is mapped to the equipment, complete equipment data call, to meet user needs.

4 CONCLUSIONS

The paper embarks from the actual situation, in order to adapt the information age, and the maximum transmission equipment knowledge and contribution. On the way to disseminate and popularize equipment knowledge on the basis of the analysis of the present situation and demand, using the mature mode of B/S structure which is combined with the current popular data visualization technology, building a digital network equipment demonstration evaluation system based on Web. An equipment digital demo system based on B/S pattern has completed the equipment database building and data collecting, sorting and warehousing equipment, and has realized the module design of user register, login, logout, etc. According to the general design, the realization of the prototype system, visualization demo algorithm will be the main content of the work in the future. The implementation and operation of equipment digital demo system will play an important role to improve the efficiency of equipment knowledge popularization, and increase the acquaintance of equipment knowledge for non-professional person and the majority of grassroots officers and soldiers.

Finally, it is time to say thanks. X.M. LI & D. LIU, thank you for being mentors and friends, and the guidance and help to me. L. RAO, thank you for checking and modifying my paper. Of course, I have to say thanks to funding support from laboratory basic research project and Bu, Weiji fund project.

REFERENCES

[1] ZHONG, Yixin. 2007. Theory of "information - knowledge - intelligent transformation rule". Journal of Beijing University of Posts and Telecommunications 30(1):1–8.
[2] WANG, Jun. 2013. A brief introduction of Web information extraction technology. Henan science and technology 10:5–6.
[3] LI, Ming. 2009. In text mining the prospect of the data cleaning technology is introduced into the analysis. Information Research 5:103–104.
[4] ZHAO, Qiao. 2007. The comparison of the database and data warehouse. Journal of Liaoning Teachers College (Natural Science Edition) 9(4):62–64.
[5] TAO, Yijun. 2011. Data Visualization: making the statistics more "beauty". China statistics 4:13–14.
[6] REN, Yonggong & YU ge. 2004. Research and Development of the Data Visualization Techniques. Computer Science 31(12):92–93.
[7] LU, Xiaoyan. 2013. The Application of JSON data exchange language in Ajax technology. Journal of Henan Science and Technology 20:23–24.

Information, Computer and Application Engineering – Liu, Sung & Yao (Eds)
© *2015 Taylor & Francis Group, London, ISBN 978-1-138-02717-6*

Study on enterprise happiness management based on dual problem

Shao Zeng Dong & Xiao Ling Wang

Harbin University of Science and Technology, Weihai, Shandong province, China

ABSTRACT: With the proposal of the happiness index, happiness has become the common goal of enterprises and employees, and happiness management has also become the focus of enterprise management. On the premise that happiness can be fuzzy quantitative, this article analyzes how to understand and implement happiness management with O.R and puts forward corresponding guidance in order to enhance the common good of enterprises and employees.

KEYWORDS: Happiness; Happiness Management; Dual problem

With social progress and technological development, people do not pursue material wealth only but pay more attention to how to realize their happiness. As we all know, happiness is an eternal topic that we explore. How to enhance people's happiness is a social hotpot and the final target that management explores at present. Enterprises' final management target is to realize their happiness management. Based on this, this article starts from enterprises' and employees' happiness, thinks about changes in enterprises' management methods and explores new survival and development approaches of enterprises.

1 PROPOSAL OF HAPPY PEOPLE AND CONNOTATION OF HAPPINESS

Since Wanner Wilson's *Related Factors of Claimed Happiness* was published in 1967, the number of literatures about happiness has increased at an alarming rate over after 50 years' development [1]. Some behaviors of people can only indicate their specific pursuit and ways or processes in which they pursue happiness. We cannot make happiness equal to approaches, processes or conditions of happiness pursuit. A basic target and ultimate meaning of people's work and life is to pursue happiness maximization. From this perspective, it assumption about ultimate human nature in management should be 'happy people' [2].

Happiness is a target that people pursue and valued by people, so much research on happiness emerges. Happiness is mainly divided into two kinds, i.e., subjective happiness based on hedonism and mental happiness based on realization theory. Differences between the two views of happiness primarily depend on the situation that people have different comprehension about connotation of happiness. Although happiness is subjective and mental experience, it is

mental experience about whether needs and potential can be satisfied, which is objective and inevitable but will not be changed by our wills. Research based on this kind of happiness tends to integrate subjective happiness and mental happiness [3].

All behaviors in our daily work and life reflect our pursuit for happiness. Generally speaking, happiness is a kind of mental experience and acts as a kind of fact judgment on conditions of our work and life and value judgment about satisfaction with our demands. With social progress and technological development, material wealth is not the only factor that affects people's happiness. Indeed, mental wealth also has impacts on people's happiness to a large extent. The author deems that happiness is subjective reflection of satisfaction with our own demands under influence of our value orientation. It mainly contains the following two aspects. In the aspect of materials, happiness is experienced when our physiologic demands are satisfied [4], i.e., material happiness. Material happiness is closely related to people's materials survival and development environment and mainly reflects differences between individuals' demands and social material conditions are reduced. At the spirit level, happiness means people's social demands are satisfied, which reflects differences in value orientation of different individuals.

Traditional economics researches how to improve employees' happiness by increasing their wealth, while eudemonics studies how to maximize employees' happiness under the situation that material wealth is certain. Happiness index is a subjective index and number measuring the specific degree to which people feel happiness.

$$R = \sum w_i \frac{U}{Q}$$

R: employee happiness index, U: efficacy of the demand, Q: expected value of the demand and w_t- weight of the demand in total demands.

Thus, it is obvious that employee happiness index is weighted sum of realization about expected value of employees' demands. To improve employee happiness index, it is essential to satisfy expected value of employees' demands. As the urgent degree of employees' demands is different, its effect on employee happiness index is various. In the process in which employee happiness index is improved, we should analyze importance of employees' demands sufficiently and satisfy demands playing a decisive role in employee happiness index. The said employees' demands just refer to the happiness obtained after they have given a certain skill (it is supposed that happiness may be quantized), so arousing employees' certain skills is a key to realization of employees' happiness.

2 ENTERPRISE HAPPINESS MANAGEMENT

Happiness management is both a management method and value on management and serves as an operation mechanism that realizes happiness maximization of related organization parties by organizing and coordinating wisdom and advantages of enterprises' members.

2.1 Enterprises happiness target

A standard used to evaluate advantages and disadvantages of enterprise management level and enterprises' competitiveness is enterprises' ability to pursue profit maximization, i.e., wealth maximization. The wealth mentioned here mainly refers to material wealth. However, with social progress and economic development, enterprises not only focus on current interests but also pursue sustainable profits, i.e., total profit maximization. It is assumed that gross profits of several profit counting periods are certain. Then, how to allocate gross profits to each profit counting period becomes a target that decision-makers should pay attention to. If profits are not allocated well, enthusiasm for production will be affected. In case profits of a counting period are too high, enterprises will extend their scale blindly and social resources will be waster. If profits of a counting period are too low, enterprises will doubt prospect of economic development. According to this, it is found that enterprises not only pursue profit maximization but also need consider their subjective feeling. In addition to pursuing material wealth, enterprises also pursue spiritual wealth and happiness maximization. Under the operation

concept 'being larger, stronger and more sustainable', enterprises' happiness maximization is realized by minimization of information cost of organizational structure and employees' life cost and interaction and coupling between the two.

2.2 Employees' happiness is a premise for enterprises' happiness

Happiness is a kind of mental ability about status and with extensibility, which can be developed and improved and provide conditions for enterprises to obtain continuous human resources [5]. Employees are the most precious treasure. Enterprises can realize their happiness targets by replying on employees' happiness only. In the environment with fierce competition, the crisis of confidence between employees and enterprises become more and more obvious. When employees choose their work, they gradually consider their career planning to a larger extent and hope to master directions of their actions and future work. This leads to the situation that employees' mobility enlarges and damages stability of enterprise production. From the perspective of happiness, enterprise management aims at building and enhancing employees' sense of belong to enterprises and happiness, seeking a new approach of synergetic development between enterprises and employees and realizing maximum common happiness of enterprises and employees. Thus, we may find that the happiness target enterprises pursue turns to enhancement of employees' happiness to realize synergetic happiness maximization of enterprises and employees.

2.3 Constructing harmonious environment is a core of enterprise happiness management

Employees of enterprises have dual nature. On the one hand, they are members of enterprises; on the other hand, they are members of their families. Thus, balance between work and family is involved. In the current environment with fierce competition, conflicts among employees' different roles not only affect enterprises' happiness but also damage employees' family happiness. Thus, to realize enterprise happiness management, we must achieve the balance between work and family but also construct harmonious environment for work and family. When enterprises carry out career planning for employees, they should consider employees' family factors, make related measures with pertinence and realize harmonious and happy development of employees and families as well as synergic happiness maximization of enterprises and employees.

3 A DUAL PROBLEM ABOUT HAPPINESS MANAGEMENT

When happiness becomes a theme people pursue, it also is a problem that enterprises must face with. As people's brain becomes more developed and more enlightened, new discovery will appear and new truth will be disclosed. At the same time, with changes in environment, methods and suggestions will be changed and theory must develop and keep pace with times [6]. Enterprise happiness management in a process in which enterprises analyze employees' demands sufficiently and happiness maximization of employees and enterprises is realized by satisfying employees' demands, i.e., it is a process where enterprises pursue synergic happiness maximization of enterprises and employees. Happiness in enterprise happiness management contains two aspects, i.e., enterprise employees' happiness and enterprise organizations' happiness. Enterprise employees' happiness is employees' subjective judgment about the situation that enterprises satisfy their demands. The larger the degree to which demands are satisfied, the more the happiness of employees is, the higher their enthusiasm for work is and the more the wealth they will create for enterprises. Enterprise organizations' happiness is to realize profit maximization, i.e., the process in which minimum cost is used to satisfy employees' demand.

Amartya Sen (1993) deems that happiness is a function about all things he can do and his ability [7]. Ryff et al. (1995) thinks that happiness not only means people obtain happiness but also contains perfect experience people get by exerting their potential fully [8]. Employees' happiness derives from both increase in wealth and the situation that they give their skills. It is assumed that employees' happiness can be quantized and has linear correlation.

Let x_j, a_{ij}, c_j and b_i denote the happiness employees obtain when they take up the j job, the quantity of employees' i skills needed by doing the j job, unit price for the j job employees take up and the number of i skills owned by employees, respectively. Then, we may establish the following model about the maximum happiness that employees pursue.

$$\max f = \sum_{j=1}^{n} c_j x_j$$

$$s.t \begin{cases} \sum_{j=1}^{n} a_{ij} x_j \leq b_i \\ x_j \geq 0 (j = 1, 2, ..., n) \end{cases}$$

Enterprise happiness management is to allocate enterprises' human resources properly and its essence is to allocate and arouse employees' skills reasonably.

On the premise that limit is certain, the maximum profits enterprises pursue may be changed into minimum cost they seek, i.e., the minimum cost enterprises pay to obtain employees' skills. It is assumed that y_i stands for the cost paid by enterprises to obtain employees' i skills. We may establish a model about minimum cost of enterprise happiness management, as shown below.

$$\min z = \sum_{i=1}^{m} b_i y_i$$

$$s.t. \begin{cases} \sum_{i=1}^{m} a_{ij} y_i \geq c_j \\ y_i \geq 0, i = 1, 2, ..., m \end{cases}$$

Thus, it can be found that the employee happiness model and enterprise happiness model form a dual problem mutually. Solutions to the employee happiness model is to confirm optimal allocation schemes about employees' skills and solutions to its coupling problem the enterprise happiness model is to confirm appropriate evaluation on employees' skills. This valuation directly relates to the most effective utilization of employees' skills. It is impossible for enterprises to arouse or make full use of all skills of employees. Thus, how to arouse employees' specific skills is a key to happiness management. We may find some methods in shadow price. y_i is the valuation that enterprises make according to contribution of the skill i to enterprises, which is called shadow price [9]. When we use simplex method to solve linear programming problem, its checking number line is:

$$\sigma_j = c_j - C_B B^{-1} P_j = c_j - \sum_{i=1}^{m} a_{ij} y_i$$

Where c_j is unit price that employees give to their j job, i.e., the production value that the j job may bring to enterprises, and $\sum_{i=1}^{m} a_{ij} y_i$ represents sum of the shadow price of employees' all skills consumed by t the j job, i.e., implicit cost of work.

If production value is higher than implicit cost, it will mean the work is profitable, enterprises should pay attention to input into this job and x_j should be considered primarily in enterprise happiness management.

If production value is lower than implicit cost, it will indicate enterprises should apply employees' skills to other jobs and x_j may be a secondary target motivated by enterprise happiness management.

In the process in which a coupling problem is used to realize enterprise happiness management, we should pay attention to the situation that some errors may appear in happiness and skills judged by enterprises for employees' own level. This requires that measures should be taken effectively according to different people when employees' skills and happiness are motivated.

4 CONCLUSION

On the premise that employees pursue happiness increasingly, enterprise management should be transformed to happiness management gradually. It is no doubt that there will be some resistance in the process of transform. However, enterprise management can be realized and implemented better only when the relationship between enterprises' happiness and employees' happiness is understood correctly. As the enterprise happiness model and employee happiness model form a coupling problem, the happiness enterprises pursue does not conflict with that employees seek, which lays a solid foundation for enterprises to develop happiness management modes and promotes implementation of enterprises happiness management.

REFERENCES

[1] Ren Jun. Positive Psychology [M]. Shanghai: Shanghai Foreign Language Education Press, 2006.
[2] Pu Dexiang and Lin yeshu. Hypothesis about happy people and happiness management analysis of organizations [J]. Commercial Age, 2009(36):50.
[3] Wang Yan. Research overview about happiness [J]. Psychological Research, 2010(3):14–19.
[4] Pu Dexiang. Happiness management — a new concept about development of organizational behavior [J]. Technical Economy and Management Research, 2009(3):66.
[5] Feng Ji and Miao Yuanjiang. Happiness index and happiness management [J]. Enterprise Vitality, 2009(3):52–53.
[6] Keyes, C.L. M. Promoting and Protecting Mental Health as Flourishing: A Complementary Strategy for Improving National Mental Health [J]. American psychologist 2007(2):62.
[7] SEN A, NUSS—BAUMM. The Quality of Life[C]. Oxford University Press, 1993:30–53.
[8] RYFF. Psychological well being in adult life [J]. Current directory of psychological science, 1995 (4):99–104.
[9] Hu Yunquan. Foundation and application of operational research [M]. Harbin: Harbin Institute of Technology Press, 2009:49.

Information, Computer and Application Engineering – Liu, Sung & Yao (Eds)
© 2015 Taylor & Francis Group, London, ISBN 978-1-138-02717-6

Simulation research on a new type of marine steering gear control system

Lei Liu & Jin Yin Du
Tianjin Maritime College, Tianjin, China

Yong Yuan Wang
Tianjin Dredge Company, CCCC, Tianjin, China

Qing Guo Song
China Classification Society, Tianjin, China

ABSTRACT: The ship course automatic control is a complicated nonlinear control problem. Traditional marine steering gear control system is difficult to achieve the ideal control effect. This paper puts forward a kind based on t-s fuzzy model of dynamic fuzzy neural network algorithm. it can quickly generate an online dynamic neural network of high precision to streamline, which is used to implement self-organizing design and parameter identification of the fuzzy system. The simulation result shows that the designed control system has strong robustness and which can effectively improve the accuracy of heading control.

KEYWORDS: Marine steering gear; Fuzzy control; Simulation research; Mathematical model

1 INTRODUCTION

The steering gear is an important device to maintain or change the course and to ensure safe navigation of a ship. In the event of failure, the ship will be out of control, or even an accident would happen [1]. According to the International Convention on Safety Of Life At Sea (SOLAS convention) and Rules and Regulations for The Construction and Classification of Seagoing Steel Ships of CCS, steering gear must have a sufficient steering torque and steering speed under any conditions, and in the event of failure, it should be able to take emergency measures rapidly to ensure the steering capability of ship [2].

Classic PID control strategy is generally used for steering gear of modern ocean-going vessels [3], its control system is simple, stable, reliable, easy to regulate, and is suitable for modeling and simulation of marine hydraulic steering gear [4]. For there is a delay, nonlinear and other factors of the actual system, parameter setting of the control system is time-consuming and laborious. However, as the mathematical model is not dependent on the object, fuzzy control has good robustness and stability than PID control strategy. In combination with the advantages of fuzzy control and PID control, the author proposes the fuzzy control of steering gear system, realizing the adjustment of the parameters by fuzzy rules, and completes the steering gear control system of modular modeling and dynamic simulation.

2 ESTABLISHMENT OF MATHEMATICAL MODEL OF SHIP MOTION

The establishment of the mathematical model of ship steering gear and dynamic characteristic research is of great significance to improve the accuracy of ship course control. The ship motion control in water has many aspects of complexity, objectively, the ship has sailed with six degrees of freedom of movement, forward, drifting, ups and downs of three translational motions, besides three rotary motions like yawing, rolling, and pitching. At the same time, the external environment interference such as wind, wave force, and flow force makes the movement mechanism of the ship more complicated.

Coordinate system and coordinate transformation is the basis of mathematical analysis of ship motion. The coordinate system used in the ship motion analysis mainly includes Earth Centered Earth Fixed, North East Down, and BODY coordinate system. The relationship between the three kinds of reference coordinate system is shown in Figure 1.

The course control systems shall be equipped with some basic functions, and ship model modeling method is given to meet the test requirements of the IEC62065 standard. In this paper, in order to meet the test requirements, it should also establish the mathematical model of the ship in accordance with the standards. Therefore, the design of the ship motion mathematical model mainly includes

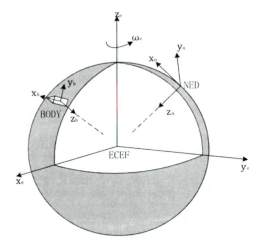

Figure 1. The reference coordinate system of ship motion.

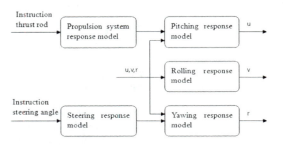

Figure 2. The basic components of the mathematical model of ship motion.

five basic modules, they are propulsion system response model, steering response model, rolling response model, pitching response model, and yawing response model.

3 ESTABLISHMENT OF MATHEMATICAL MODEL OF EXTERNAL INTERFERENCE

The interference factors in the external environment of the ship sailing mainly include wind, waves, currents, interference between adjacent vessels, quay wall effect, etc. In the research of the tracking control algorithm, the wind, waves, and the interference of non-uniform flow on ship belong to uncertain interference with the same properties. In the uncertain disturbance, therefore, this article only chooses the establishment of mathematical model of wave interference.

At present there are two main methods in the study of wave interference force: One is a first-order wave

disturbance force (also known as high-frequency interference wave power), the premise condition is that the oscillation of the ship caused by slightly waves is not very big, and there is a linear relationship between the wave forces that the ship's subjected to and the wave height, and the wave forces and wave have the same frequency. Another is the second order wave interference force (also called wave drift force). This wave force is proportional to the square of wave height. However, the first order wave interference force is the mainly cause of pitch and heave motion of the ship, the impact on the ship roll is relatively weak, so the impact of the drifting and rolling motion of the ship is even smaller. The second-order wave disturbance force with slow time-varying characteristics is non-linear, its effect will not only change the ship's course and track, but also will have a significant impact on ship movement of anchoring state and drilling platforms such as dynamic positioning system. Therefore, this paper uses a second-order wave interference force mathematical model, as shown below.

$$X^D = \rho L \cos \chi \sum_{i=1}^{n} C_{X\omega}^D \left(2\pi\omega_i^2/g\right) S_{\varsigma}\left(\omega_i\right)\Delta\omega$$

$$Y^D = \frac{1}{2}\rho L \cos \chi \sum_{i=1}^{n} C_{X\omega}^D \left(2\pi\omega_i^2/g\right) S_{\varsigma}\left(\omega_i\right)\Delta\omega$$

$$N^D = \frac{1}{2}\rho L \cos \chi \sum_{i=1}^{n} C_{X\omega}^D \left(2\pi\omega_i^2/g\right) S_{\varsigma}\left(\omega_i\right)\Delta\omega \qquad (1)$$

4 SIMULATION RESEARCH

The simulation test is started between the choice of using comprehensive methods or MATLAB command window. During the simulation runs if you want to view the results of a simulation, you can use scopes or other display module to realize the real-time monitoring of simulation. Its parameter can be changed and immediately see the results of the corresponding change, to obtain the influence extent of related parameters on the results, and the simulation results can be cast into the MATLAB workspace for visualization and the post-processing.

The above mathematical model should be deformed and the mathematical model of steering gear can be obtained. From the intuitive ship motion, simulation results can be directly obtained from the simulation software MATLAB, in order to research the performance of ship maneuvering, and don't have to study for further calculation. Among, the digital simulation results of steering gear position as shown in Figure 3.

When the interface display part is finished, adding button resources, the MATLAB simulation model of start-stop control and manipulation of the serial

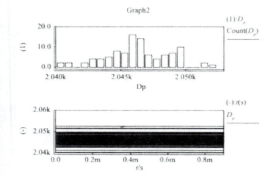

Figure 3.   Digital simulation results of steering position.

communication is realized. The drop-down list box is joined, and 5 simulated locations are added to the list box, then the ship location is optional in electronic chart. At this point, the simulation system operation interface is written completely.

5   CONCLUSION

This article is based on the ship motion model of the water surface, considering the influence of the waves and flow on ship motion and transforming it into the rudder angle disturbance, establishes the mathematical model of an automatic steering system of hydraulic control system. Through the establishment of the model and the simulation analysis, it is helpful to understand the static and dynamic characteristics of the new steering gear, it is a reference for the design of the steering gear, especially the design of the rudder angle position controller. The simulation result shows that the designed control system has strong robustness, if choosing reasonable coefficient, the heading control accuracy can be effectively improved.

REFERENCES

[1] Zhang Chunlai, Lin Yechun. *Marine Electrical and Automation* [M]. Marine Electrical. Dalian: Dalian Maritime University Press, Beijing: China Communications Press, P152 (Aug.,2012) (in Chinese).
[2] Fei Qian. *Marine auxiliary machinery* [M]. Dalian: Dalian Maritime University Press, (2004) (in Chinese).
[3] Yan Feifei, Chen Shengdong, Liu Yali. *Design for steering gear control system based on ADRC* [J].Ship Science and Technology. Vol.35,No.12. (Dec., 2013).
[4] Li Qi, Ren Longfei. *Modeling and Dynamic Simulation of Steering Gear System Based on Fussy PID* [J]. 《PLC&FA》 P83–85(Jan., 2013).

Information, Computer and Application Engineering – Liu, Sung & Yao (Eds)
© 2015 Taylor & Francis Group, London, ISBN 978-1-138-02717-6

A comparison of part-of-speech tag sets for transition-based dependency parsing

Xiao Yang Chen, Yan Ping Zhao, Jian Yun Shang & Xue Wen Shi
Beijing Institute of Technology, China

ABSTRACT: Recent years, transition-based dependency parsing has gained more and more attention since its simplicity and O(n) time complexity. Meanwhile, Part-of-speech tag features are essential to dependency parsing. But unlike English, Chinese part-of-speech tagging is more challenging. To find how the part-of-speech tagging impacts the accuracy of dependency parsing, we implemented a transition-based parser and compared with two well known transition-based dependency parsers. We also compared three different kinds of part-of-speech tag set in pipelined part-of-speech tagging and dependency parsing. Testing of the Chinese Penn Treebank data, our implementation of dependency parser with PFR tag set gave the best accuracy among other parsers and tag sets.

KEYWORDS: Dependency parser; Part-of-speech tagger

1 INTRODUCTION

Recent years, transition-based dependency parsing has gained more and more attention since its simplicity and O(n) time complexity. Meanwhile features with part-of-speech tag are more essential than other features to dependency parsing (Li et al., 2014).

But unlike most morphology-richer languages such as English, which have a high accuracy of part-of-speech tagging (97%). The accuracy of Chinese part-of-speech tagging (94%) is much worse than it in English. With error propagation, the accuracy of part-of-speech tagging has a large impact on the accuracy of dependency parsing (86.3% of gold-standard part-of-speech tag and 77.1% of the automatic tagged input data).

On the other hand, there is no systematic comparison of the different part-of-speech tag set for dependency parsing and determines which part-of-speech tag are most suitable for pipelined part-of-speech and dependency parsing task.

In this paper, we compared three different kinds of part-of-speech tag set: CTB, PFR and PFR coarse-grained. On the other hand, we also compared three different types of transition-based Chinese dependency parsers: MaltParser (Nivre et al., 2008), ZPar (Zhang et al., 2011) and our implementation of beam arc-eager dependency parser.

2 RELATED WORK

The transition-based arc-eager dependency parser is proposed by Nivre et al. (2008). Then Zhang et al. (2008) improves the arc-eager algorithm with beam search to do the global training and after three years Zhang et al. (2011) added rich, non-local features such as distance and valency to this model.

Che et al. (2012) compares three kinds of dependency parser as well as four kinds of constituent parsers, which used the training and test data provided by the Stanford tagger with the MEMM model in combination with 10-way jackknifing for the Chinese Penn Treebank data. Li et al. (2014) proposed a joint model for part-of-speech tagging and dependency parsing. From his paper we found that although the joint model gave the state-of-art accuracy, the accuracy of pipelined model is not much worse than the joint one and its training and predicting speed is much faster.

3 PART OF SPEECH TAGGER AND TAG SET

3.1 Part-of-speech tagger

We use Conditional Random Field (Lafferty et al., 2001) to predict the tag of each segmented sentence. Compared with the Hidden Markov Model, it can exploit much more features. For example, in part-of-speech tagging it can use not only the word itself, but also the head or tail characters in the word. Compared with other discriminating models like Maximum Entropy, it can take the context data into account. This model is widely used in the sequential tagging tasks in the area of natural language processing.

We used the wapiti CRF library (Lavergne et al., 2010) with the SGD-L1 optimization method. The features used in this part-of-speech tagger is

W_0, W_{-1}, W_1, $C_H L$, $C_T L$

W_0 indicates the current word, W_{-1} and W_1 indicates the previous word and the next word. C_H and C_T indicate the first and last character in the current word. L is the length of the word.

3.2 Sets of part-of-speech tag

We tested three kinds of different part-of-speech tag set: CTB, PFR and PFR Coarse-grained.

3.2.1 CTB

CTB part-of-speech tag set is the tag set used in the original Chinese Penn Treebank. There is totally 32 part-of-speech tags in CTB5 (excluded some rarely used tags).

3.2.2 PFR

PFR part-of-speech tag set from the PFR Chinese Corpus which contains the segmented and tagged data from the news materials of People Daily at the year of 1998. There are 54 different tags in this tag set (excluded some rarely used tags).

3.2.3 PFR Coarse-grained

Follow the part-of-speech tag set in LTP (Che et al., 2010). We tried the coarse-grained tag set of PFR Corpus. It merges the tags with the same coarse-grained type. For example, merges "wb", "wf", "we" with other tags started with "w" into the coarse-grained tag "w" since they are all punition.

The noun is special. Follow Che et al. (2010) we keep three fine-grained noun tags. "nr" for person name or organization name, "ns" for location name, "nt" for date and time, "nx" for foreign words and "nz" for other types of named entities.

4 DEPENDENCY PARSERS

There are two well-known algorithms for dependency parsing. One is graphed-based and the other is transition-based (Zhang et al., 2008). The graphed-based algorithms try to rank all the possible output trees to find the most probable one. It has an $O(n^3)$ time complexity of first-order and $O(n^4)$ for high-order decoding. Meanwhile, the transition-based algorithm divide the parsing procedure into a sequence of transition actions: shift, reduce, left-arc and right-arc. When starting parsing, puts all words from the sentence into the input queue and pushes the ROOT node into stack. Then, it uses a classifier to predict each transition, according to the current state. The shift transition gets a word from the input queue and pushes it into the stack. The reduce transition pops a word from stack. The left-arc transition gets a word from

the queue and makes the stack-top word as its head. And the right-arc transition gets a word from the input queue and makes it the head node of the word from stack-top, then push this word into stack. When there is only ROOT node in the stack and there is no word in the input queue, the parsing procedure finished. It takes 2n-1 steps to parse a sentence of length n, so the time complexity of transition-based parsing algorithm is just O(n) (Nivre et al., 2008).

Since the transition-based parsing algorithm is fast and simple, it is widely used in industrial projects. We use three types of transition-based dependency parsers for our work: MaltParser, ZPar and our implementation of beam arc-eager dependency parser.

4.1 MaltParser

MaltParser uses an SVM model to predict next transition and apply it step by step. So the decoding procedure just likes a greedy search algorithm. The feature used by MaltParser could be the word, part-of-speech tag or the dependency label on the stack top, or the head of input queue as well as their parents or children. The original transition-based parser (Nivre et al., 2008) uses a Support Vector Machine with quadratic kernel to combine each feature. But the later experiments are shown that using a linear classifier with joint features like STwtN0wt (the word and part-of-speech tag from the stack top and the word and part-of-speech tag from queue head) also work well.

The accuracy of this parser is not much worse than the first-order graph-based parser, but the speed is much faster than graph-based one (Che et al., 2012).

4.2 ZPar

ZPar uses an averaged-perceptron with beam search to find the best sequence of transitions. It can be regarded as the global training version of transition-based arc-eager dependency parser.

The parser follows the global linear model from Michael et al. (2004) to find the sequence of transitions with the highest score. The model of this parser could be written as:

$$Score(y) = \Phi(x) \cdot \vec{w}$$

When decoding, it puts a partial tree with only one ROOT node into the beam. Then, apply each transition to all items in beam, find the most possible generated partial tree, put them into the beam again. Repeat it until the stack only have ROOT node and the input queue is empty.

When training, it uses a strategy called "early update" that uses the above algorithm to decode, when there is no correct partial tree in the beam, stop decoding and update the model using the item with the highest score.

Zhang et al. (2011) shows that the beam arc-eager parser can use the non-local rich features to improve the performance of the parser and give the state-of-art accuracy of 86.0% in Chinese with the gold-standard part-of-speech tag.

4.3 Our implementation

We implemented another transition-based dependency parser with arc-eager algorithm and beam search. Like ZPar, we use the global linear model with average perceptron and "early update" strategy. The difference is that we also implemented the algorithm provided by Nivre et al. 2014 to constrain each output to be a tree. So besides the shift, reduce, left-arc and right-arc transitions, we also have an "unshift" transition. Furthermore, we use an "early-reduce" oracle to predict the sequence of correct transition that the stack-top word will be reduced as soon as it has the head node and has no child node in the input queue.

5 EXPERIMENTS

5.1 Part-of-speech tagging experiments

We evaluate the part-of-speech tagger using Penn Chinese Treebank (CTB) 5.0 and PFR Corpus for experimental data. For CTB data, we use 10-fold cross-validation to evaluate the accuracy and for PFR data, we use the Jan. 1998 data for train and Feb. 1998 data for the test.

The PFR fine-grained tags were converted into coarse-grained tags using a conversion table for the experiments of coarse-grained tags.

The test accuracies of part-of-speech tagging are shown in Table 1, where each column represents different sets of part-of-speech. PFR-C is the coarse-grained tag set of PFR Corpus and TA represent for tag accuracy. All of them use the CRF-based part-of-speech tagging model mentioned above.

Table 1.   Accuracy of part-of-speech tagging.

Tag set	CTB	PFR	PFR-C
TA	93.5%	95.6%	96.8%

The coarse-grained tag set gets the highest accuracy since it uses fewer tags than the fine-grained one and it avoids the "vn" and "v" problem since both "vn" and "v" tags are included in the coarse-grained tag "v". The accuracy of the CTB tag set is only 93.5%. There are two reasons. Firstly, the training corpus is smaller than PFR Corpus. And, also, for some contexts the tagger can't distinguish between "NN" and "VV" for verbs just like the "vn" and "v" problem in fine-grained PFR tag set.

5.2 Dependency parsing experiments

Our experiments for dependency parsing were performed using Penn Chinese Treebank (CTB) 5.0. We follow Zhang et al. (2008) to split the CTB5 into train development and test data as shown in Table 2.

Unlike Zhang et al., 2011, we use Stanford Parser to convert the bracketed sentences into dependency trees since Stanford Dependency provides rich dependency labels than penn2malt. And we use the

Table 2.   Training, development and test data from CTB5.

	Sections	Sentences	Words
Training	001–815	16, 118	437, 859
	1001–1136		
Dev	886–931	804	20, 453
	1148–1151		
Test	816–885	1, 915	50, 319
	1137–1147		

part-of-speech tagger mentions above to turn the training and test data into 8 files:

1.GTr: The training data with gold-standard CTB tag.

2.GTe: The test data with gold-standard CTB tag.

3.KTr: The training data with CTB tags by 10-fold jackknifing.

4.KTe: The test data with CTB tags by 10-fold jackknifing.

5.PTr: The training data with PFR tags by a CRF tagger trained with PFR Corpus.

6.PTe: The test data with PFR tags by a CRF tagger trained with PFR Corpus.

7.CTr: The training data with coarse-grained PFR tags by a CRF tagger trained with PFR Corpus.

8.CTe: The test data with coarse-grained PFR tags by a CRF tagger trained with PFR Corpus.

We test three dependency parsers mentioned above. The MaltParser uses the linear kernel (liblinear) and the features could be found in NivreEager. xml. ZPar uses the arceager configuration from its implementations. Meanwhile, we have implemented three types of transition-based parsers. The first one is the transition-based arc-eager parser (O-eager) use the same idea as Nivre et al., 2008, but the difference is that we use averaged-perceptron as the classifier instead of Support Vector Machine and the features we used are defined in Zhang et al., 2011. The second one is arc-eager parser with tree restrictions (Nivre et al. 2014) we call it O-eager-tr. The last one is arc-eager parser with beam search (O-eager-beam), we set beam size = 64 and use the "early-reduce" oracle to give the sequence of transitions for training.

The test accuracies are shown in Tables 3 and 4. Where each row represents a parsing model and each column represents training and test data. We measure the accuracies by Unlabeled Attachment Score (UAS).

Table 3. Unlabeled attachment score for gold-standard test or training data.

	GTr /GTe	GTr/ KTe	KTr/ GTe
MaltParser	0.828	0.728	0.805
ZPar	0.864	0.759	0.836
O-eager	0.822	0.728	0.794
O-eager-tr	0.823	0.728	0.794
O-eager-beam	0.861	0.758	0.838

Table 4. Unlabeled attachment score for automatic tagged training and test data.

	KTr/KTe	CTr/ CTe	PTr/ PTe
MaltParser	0.735	0.740	0.723
ZPar	0.771	0.771	0.768
O-eager	0.739	0.739	0.723
O-eager-tr	0.739	0.740	0.724
O-eager-beam	0.777	**0.781**	0.776

As we can see from the table, the accuracy of pipelined part-of-speech tagging and dependency parsing (GTr/KTe) is significantly decreased compared to the gold-standard part-of-speech tag one (GTr/GTe). This significant decrease is also mentioned by Li et al. (2014). We follow the work in Che et al. (2012), to train the parser with a CRF part-of-speech tagger combined with 10-way jackknifing (KTr/KTe), the accuracy is better than the pipelined parser with gold-standard training data (GTr/KTe).

When we use the 10-way jackknifing data for training and the gold-standard part-of-speech tag for the test, the accuracy is also better than other ones. This phenomenon reflects the importance of part-of-speech tag for dependency parsing.

The comparison of three different part-of-speech tag sets indicates that the PFR fine-grained tag set is the best for pipelined part-of-speech and dependency parsing among them. The accuracy of coarse-grained PFR tag set is worse than the accuracy of PFR tag set and CTB tag set, although its tag accuracy (TA) is the best among them (in Table 1). The reason is that the more different kinds of part-of-speech tag, the more messages could the part-of-speech tagger provide to the dependency parser. Our implementation of arc-eager parser with beam search (O-eager-beam) combined with PFR fine-grained

tag set gets the highest accuracy among the parsers and tag sets.

6 CONCLUSION

We implemented a CRF-based part-of-speech tagger and an arc-eager dependency parser with beam search. Then we compared three different transition-based parsers and three different part-of-speech tag sets for the pipelined part-of-speech tagging and dependency parsing.

We conclude that the arc-eager parser with beam search is better than the other, since its global training and having more search space. Meanwhile, the PFR tag set is the best for the pipelined part-of-speech tagging and dependency parsing.

REFERENCES

Li, Z., Zhang, M., Che, W., Liu, T., & Chen, W. (2014). Joint Optimization for Chinese POS Tagging and Dependency Parsing. IEEE/ACM Transactions on Audio, Speech and Language Processing (TASLP), 22(1), 274–286.

Nivre, J. (2008) Algorithms for Deterministic Incremental Dependency Parsing. Computational Linguistics 34(4), 513–553.

Zhang, Y., & Nivre, J. (2011, June). Transition-based dependency parsing with rich non-local features. In Proceedings of the 49th Annual Meeting of the Association for Computational Linguistics: Human Language Technologies: short papers-Volume 2 (pp. 188–193). Association for Computational Linguistics.

Che, W., Spitkovsky, V.I., & Liu, T. (2012, July). A comparison of chinese parsers for stanford dependencies. In Proceedings of the 50th Annual Meeting of the Association for Computational Linguistics: Short Papers-Volume 2 (pp. 11–16). Association for Computational Linguistics.

Lafferty, J., McCallum, A., & Pereira, F. C. (2001). Conditional random fields: Probabilistic models for segmenting and labeling sequence data.

Lavergne, T., Cappé, O., & Yvon, F. (2010, July). Practical very large scale CRFs. In Proceedings of the 48th Annual Meeting of the Association for Computational Linguistics (pp. 504–513). Association for Computational Linguistics.

Che, W., Li, Z., & Liu, T. (2010, August). Ltp: A chinese language technology platform. In Proceedings of the 23rd International Conference on Computational Linguistics: Demonstrations (pp. 13–16). Association for Computational Linguistics.

Michael Collins and Brian Roark. 2004. Incremental parsing with the perceptron algorithm. In Proceedings of ACL, pages 111–118, Barcelona, Spain, July.

Nivre, J., & Fernández-González, D. (2014). Arc-Eager Parsing with the Tree Constraint.

Information, Computer and Application Engineering – Liu, Sung & Yao (Eds)
© 2015 Taylor & Francis Group, London, ISBN 978-1-138-02717-6

Study on the reasons and countermeasures of college students' mobile phone media dependence

Xue Ting Liu

Harbin University of Science and Technology, Weihai, Shandong Province, China

Zhi Wei Song

NO.3 Middle School of Rongcheng, Weihai, Shandong Province, China

ABSTRACT: Nowadays, in the college campuses, "mobile phone media addicts" is everywhere who holds a mobile phone media or a tablet to search the Internet, play games or update microblog. The students depend too much on the mobile media, and this phenomenon has led to increasing problems of safety, health, and social life. So it is extremely urgent to find effective solutions to deal with this problem. In this paper, the author analyzes this phenomenon, and proposes some effective solutions to strengthen students' self-management, improve the quality of teaching, enrich the cultural life in campus, and create a favorable environment so as to guide the students to correct their bad habits and develop a healthy life style.

KEYWORDS: College students; mobile phone media; dependence

With the mobile phones being the audiovisual terminals and WAP is the platform, Mobile Media is a personalized information medium, which is called "the fifth media" following after newspapers, radio, television and the Internet. With the rapid development of the information age, the mobile phone media, featuring compact, portable, wireless and with functions of improved Web-surfing, videos and games, is increasingly becoming a vital communication and entertainment tool for students in their daily life. They are so attracted by such a small screen that they are called "mobile phone media addicts". In recent years, college students have become increasingly dependent on mobile phones and some of them have suffered from some mental illness like "mobile phone media addiction"—defined by the psychologists. This problem has affected the physical and mental health and students' school life, so it is a major issue to be addressed.

1 CAUSES ANALYSIS FOR "MOBILE PHONE DEPENDENT" PHENOMENON

1.1 *Individual causes*

As it is suggested in the Maslow's hierarchy of needs theory, there are five levels of human needs. They are physiological needs, security needs, social needs, esteem needs and self-actualization needs of low to high. According to the population characteristics presented by the dependent phenomenon, the individual causes can be summarized as lacking needs in the following aspects.

1.1.1 *Lack of security needs*

College students are in the golden stage of physical and mental development. They have a maturing mind while they also encounter many problems, such as indifferent relationships, poor academic performance and high employment pressure, etc. If they cannot handle these problems properly, they will be prone to be unconfident, feeling insecure, sensitive and vulnerable. Fiddling with their mobile phones in public places, the students can isolate themselves from the surrounding environment, thus creating a world of their own. Disguising themselves in the real world while expressing their real thoughts in the virtual world, the students show some certain escape and indifference to the reality through such kind of self-protection. This has become a common trend among younger people who lack of sense of security.

1.1.2 *Lack of social needs*

Everyone wants good social relations and hopes to take care of each other. Therefore, access to information has become essential. In the Internet era of information proliferation, digital terminal represented by smart phones has made it very convenient for people to get information. Refreshing the Internet, focusing on the latest news, finding a common topic to communicate with others, people can not only get all kinds of information, but also have great access to information.

1.1.3 *Lack of self-actualization needs*

As the top most level in Maslow's hierarchy of needs theory, usually the self-actualization need is the most difficult one to be gained in real life. However, some social networking services such as Wechat, microblog, and QQ space have made it easier to get "self-actualization". To express their personal opinions through the mobile phone software, to forward and comment on information, to show off photos, to focus on the recent lives of their friends or even strangers, people can get feedback from others without any cost so as to gain the greatest psychological satisfaction.

1.2 *School causes*

During the class if teachers just repeat what the book says and teach with a dry method, students will lose their interest in learning and become book-wearied and then they will gradually rely on the mobile media. At the same time, campus activities are miscellaneous and lack of innovation and diversity, and schools do not guide the students to have a proper career planning, as a result the students' college life become dull without a reasonable guide. Besides finishing the learning tasks, the students spend more time surfing the Internet, chatting, playing games, etc. As time passes, cell phones and computers become the best tool for the university students to look for some other needs.

1.3 *Social causes*

According to the social environment analysis, the fast-paced life has made people's private time fragmented. Young people have less integrated time, so that they can only seize the fragmentary time to look for leisure and entertainment via digital terminals. Digital terminal represented by smart phones provides abundant applications for people and brings convenience and entertainment for them. In recent years, the popularity of 3G and 4G network has added diversities to mobile phones and expanded popularity of smart phones.

2 SOLUTIONS

2.1 *Strengthen students' self-control and self-management abilities*

To solve phone-dependency problems, besides the teacher's supervision and guidance, the students need to have active coordination and self-discipline. First, students should be deeply aware of the damage

caused by the overuse of the smart phones, clarify the relationship between learning and entertainment, devote their main energy into learning professional knowledge and reinforcing self-accomplishment and learn to use the mobile phones rationally in their spare time. Second, college students should learn to correctly deal with the individual needs in all aspects. If lacking relevant needs, students should try to cope with such situation rationally. For example, if they lack social needs, they should actively learn to master interpersonal communication methods and techniques, try to improve their social relationships rather than indulge in phone networks and ignore the most sincere communications between people.

2.2 *Enhance the appeal and the quality of teaching*

During the class teachers should innovate the teaching methods, enrich teaching contents, and create interesting teaching situations so as to enhance the teaching effectiveness and quality through an entertaining way. Teachers should not see it as a great disaster that the students are addicted to mobile phones; instead, they should make the best use of the situation to distract students' attention from mobile phones. For example, teachers can encourage students to participate in the classroom activities through experiential and interactive teaching to achieve the "empathy" effect. Another example is that the teacher can use the search function of mobile phones, so that students can use a mobile phone to retrieve relevant information and to analyze, conclude, and summarize the information.

2.3 *Enhance the vitality and enrich cultural life in campus*

Nowadays, college students usually have a wide range of interests. They are curious and are willing to accept new things. Cultural activities on campus can not only enrich university life, but also fully exploit the students' interests and help them promote their qualities. Schools should extensively carry out cultural activities on ideology, science and technology, sports, entertainment and so on to guide the students to "leave their bedroom and network for the playground"; to guide the students to build friendships, improve skills, promote qualities and build self-confidence through professional learning and practical competitions. In such ways, the negative impact of mobile phones on college students will be reduced and gradually there will be fewer "mobile phone media addicts".

2.4 *Create a favorable environment in joint efforts*

Students live in groups—from bedrooms to classrooms, from classes to communities, so every moment they are dealing with others and affecting each other and creating a certain atmosphere. When individuals are affected by groups, they will doubt and change their views, judgments and behaviors towards the way in which most people do. This phenomenon is called the "bandwagon effect" in psychology, which is commonly known as "go with the flow". Supposing that the majority of the students in a bedroom or in a class is addicted to the mobile phone networks, other individuals will be susceptible and gradually change their thinking, cognition and behaviors, and then they will go consistent with the groups both in thoughts and behaviors. As for the "mobile dependent" phenomenon, schools should make full use of the positive effects of the psychology of conformity create a healthy and upward school culture, classroom environment and bedroom atmosphere through the thematic education activities, group guidance and slogans. For example, teachers can carry out "no cell phone" classes. Before class, teachers should guide students to hand in their phones into a unified bag and encourage them to supervise each other. Classes with outstanding performances can get appropriate incentives, and thus an excellent study style and a healthy concept will be gradually formed.

3 CONCLUSION

Over-reliance on mobile phones has led to increased risk of physical and mental health and social life. Schools, teachers and students should all pay high attention to this problem. Strengthening the students' self-management, improving the teaching quality, enriching the cultural life on campus and creating a favorable environment and atmosphere will help to guide the students to learn to correct their bad habits and establish healthy lifestyles.

REFERENCES

[1] Chengfang Xu. Psychological Causes of College Students' Over-Reliance on Mobile Phones and Countermeasures [J]. Theory, 2011, (32).
[2] Zihai Wang and Huijun Fu. Causes of "Mobile phone media addicts" in University Campus and Solutions [J]. Labor and Social Security in the World, 2013, (20).
[3] Qi Bao. Study on Education for College Students Based on Mobile Media [J]. Education and Teaching, 2014, (04).

Information, Computer and Application Engineering – Liu, Sung & Yao (Eds)
© 2015 Taylor & Francis Group, London, ISBN 978-1-138-02717-6

A product review score prediction model based on neural network

Jin Li, Guo Shi Wu & Jia Yin Qi
Beijing University of Posts and Telecommunications, Beijing, China

Yu Su
Communication and Technical Bureau, Xinhua News Agency, Beijing, China

ABSTRACT: In order to mine the useful information on consumer reviews and help product manufacturers obtaining feedback about products for improving product quality and services, this paper proposes a review score prediction method, the method uses the features and its sentiment information and scores in reviews to train a neural network based on back-propagation for score prediction. Experiments on Galaxy Note II reviews from real-world data are presented to demonstrate the results of the proposed techniques, experiment results demonstrate the effectiveness of this method.

KEYWORDS: score prediction; neural network; supervised learning; text mining

1 INTRODUCTION

Most of e-commerce websites such as Amazon, JD, etc., support scoring the features and functions of products and publishing short text comments after consumers buying goods. Mining the reviews can obtain useful information about the quality and services of products, while the review score is directly quantifiable data about product quality and user experience, it can be a composite score as a reference for user making purchasing decisions and be used for recommendations for merchants, etc., merchants can also track the users' feedback about product features and quality to improve the quality of their products and services. Some consumers aren't only commenting the products, but also give scores, while some others not give the evaluation scores. In this paper, we define complete user reviews as comments that contain both short text and comment scores, uncompleted user reviews as comments contain only brief short text.

This paper presents an uncompleted user review score prediction method, the method uses features and its sentiment information and scores in the complete user reviews to train a neural network model, the trained model can use for scores prediction.

The remainder of this paper is organized as follows: section 2 describes the related work, section 3 introduces the process of neural network model training, section 4 introduces the data experiment and evaluation, section 5 presents the conclusion.

2 RELATED WORK

Mining product reviews are mainly based on text mining techniques. In order to complete feature-based score prediction, generally we need to identify and extract product features and its sentiment orientation in product reviews.

2.1 Extracting product features

Recently, there are some research methods about extracting product features, for example, employ POS tagging (Hu & Liu et al. 2004) for review data, identify the nouns and noun phrases, and then extract product's features using the association rule mining method. A semi-automated manual intervention iterative method (Kobayashi et al. 2004) to extract the product features and user opinions. Bayesian classification is applied in the know-it-all web information extraction system (Popescu et al. 2005) for to extract the product features. As for sentiment orientation recognition, three kinds of machine learning methods (Naive Bayes, maximum entropy, SVM) (Bo Pang et al. 2002) is applied to mining the sentiment orientation. (Sista et al. 2004) Use a commendatory and derogatory term collection of seed vocabulary, and then use WordNet lexical to extend term collection automatically, last use the expanded collection as the effective polarity classification features.

In this paper, we don't focus on extracting product features and identifying features' sentiment orientation but the score predictions. We assume that product features associated with a product domain are given. We manually gather and tag product features and its sentiment orientation for the mobile phone category. Section 3 will describe in detail.

2.2 *Feature ranking and score prediction*

For feature-based ratings and rankings, there are some research methods, for example, calculate the feature words frequency (Lei Zhang 2010) and use the web link analysis algorithm HITS to calculate feature correlations, and obtain the importance of features for ranking. (Zhang et al. 2009) propose a model based on the PageRank algorithm. It builds a weighted directed graph model with sentiment orientation information and features comparative information, thus rank the products in various dimensions. While (Christopher Scaffidi et al. 2007) not uses the sentiment orientation of the comments, opinions, but only the review scores, he count binomial statistics on feature words frequency of the comments to rank product. (HaiYang Jiang et al. 2010) mine the history data of a user's reviews, learn the user's preferences, habits, thus predict the scores. While in this paper, we will use the supervised classification method to build a neural network model for score prediction.

3 MLP NETWORK-BASED ANALYSIS MODEL

In this paper, we propose a mechanism based on a multilayer perceptron network (a kind of neural network) training and modeling, the resulting MLP network is used for the review scores predicting, this is a supervised learning classification process. Figure 1 shows the whole process. The following will describe every step detail of the process.

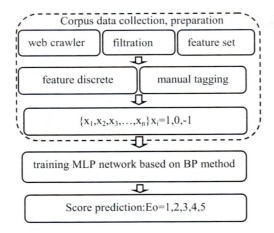

Figure 1. The process of MLP Networking model training.

3.1 *Data acquisition and pre-processing*

Use accurate information extraction tools downloading user reviews of the specific product category (such as mobile phones) from the e-commerce websites (such as JD, Amazon, etc.) and save it to a local database. Simple data preprocessing is required, such as regular expression matching, simple filtering rules (such as a simple repetition of words, etc.) to filter reviews, the reviews unrelated data and data containing some spam information is removed.

3.2 *Feature set and sentiment value labeling*

3.2.1 *Feature clustering*

Features are characteristics of the product itself, generally refer to the appearances of a product, quality, functionality and other attributes; different category of product is characterized by different. In this paper, we consider the mobile phone products. User reviews describe by natural language, each review contains only some portion of the features of the product. In order to predict the review score accurately and completely, we will first generate product feature set, refer as Set_F.

$$Set_F = \{F_1, F_2, ..., F_N\} \tag{1}$$

The feature set consists of two parts, one is the inherent and common characteristics or attributes of product, such as property in the product specifications, test reports, etc.; the other one is featured from mining the product review. Collect features of mobile phone manually to establish the feature set.

In user reviews, the similar product features may have different descriptions, for example, "Network", " Call quality", " Signal", etc., convey the same meaning. Considering the large number of product feature words, product features clustering is necessary for feature discretion processing and labeling of sentimental value and the realization of the system later. We will cluster the feature words to corresponding feature item in Table 1.

Table 1. Feature clustering.

Feature	Feature Item
endurance, battery, charging...	battery
color, line, profile...	appearance
Price, cost...	price
resolution, screen, clear...	screen

3.2.2 *Identifying feature sentiment value*

After establishing product features sets manually, discrete features in each review with this feature sets SetF, the set refer as to Sets after discretization in each review, this set Sets is a subset of Set_F.

Calculate the feature sentiment value base on its sentiment orientation, the method is that if the user give the attribute positive comments (positive word), the corresponding sentiment value is 1; in contrast, the sentimental value is -1 with a negative orientation; the review has no related feature turns out that user prefer to a neutral emotion, sentimental value is 0. In order to ensure the accuracy of the training data, in this paper, we label the sentimental value in each review manually, and each review score is retained. Finally, each review is as shown as follows.

$$S = \{a_1, a_2, ..., a_N, r\} a_i \in \{-1, 0, 1\}, r \in \{1...5\} \quad (2)$$

In order to get accurate results, sample reviews must be further filtered, involved: *Data De-noising*: remove the sample data whose sentiment values are all 0. *Data Correction*: correct the sample data which contains the same sentimental values to same score.

The resulting sample data are shown in Table 2, these sample data will be used for training the neural network.

Table 2. Input sample data.

#	Battery	Screen	Touch Feel	...	Score
1	0	0	1	...	5
2	0	-1	0	...	3
3	0	-1	0	...	1
4	0	-1	-1	...	5
5	0	1	1	...	5
...	...	...	...	...	...

3.2.3 Build MLP network model

In this paper, MLP neural network is composed of multiple layers, the first layer neuron receives input, here is the sample review data labeled with sentimental value; the last layer output the result, here is the corresponding review score. The MLP neural network also includes intermediate layers whose role as combining the input, they have also been called "Query layers" or "Hidden layers". Nodes in input layer are connected to the hidden layer nodes, nodes in hidden layers are connected to nodes in output layers (Toby Segaran 2007). In this paper, we use the only one middle layer. The network structure is shown Figure 2.

The model training algorithm, pseudo-code is shown in algorithm 1, η is the learning rate, the larger η, the larger correction of excessive weight matrix that may make instability in the learning process, and may skip optimal or even local optima that cannot converge; if η is too small, it will lead to too long time training, it is usually between 0.01 and 0.9 (Jianxin et al. 2013).

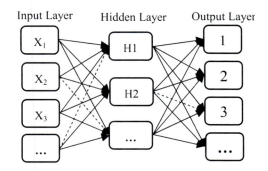

Figure 2. MLP Network Structure.

Algorithm 1 MLP Model Trainning

1: RawDataSet=Crawler(Cat); // Information Extraction

2: Set_F=Gather_Feature(Cat);

3: Define TrainDataSet, Matrix W_i, W_o,

$S(x)=\tanh(x)$, $S'(x)=\text{dtanh}(x)$, err;

4: **for** review in RawDataSet **do**

5: Set_S= Discrete (review, Set_F);

6: S=$\{a_1, a_2, ...a_N, r\}$=Label(review, Set_S);

7: TrainDataSet.append(Sample);

8: **end for**

9: **for** Sample in TrainDataSet **do**

10: H_i=Sample $[A_i]*W_i$; H_o=$S(H_i)$;

11: O_i= $H_o * W_o$; O_o= $S(O_i)$;

12: error=$1/2*($ Sample $[r]-O_o)^2$

13: **if** error > err **then**

14: deltaO=dtanh(O_o)* error;

15: deltaH= dtanh(H_o)*(deltaO*W_o);

16: $W_{o,N+1}$=$W_{o,N}$+η* H_o *deltaO+momtum($W_{o,N}$-$W_{o,N-1}$);

17: $W_{i,N+1}$=$W_{i,N}$+η* A_o *deltaH+momtum($W_{i,N}$-$W_{i,N-1}$)

18: **else**

19: **break**;

20: **end if**

21: **end for**

---

4 EXPERIMENTS RESULTS

Download GALAXY Note II phone reviews from JD and SUNING websites. We get a total of 4675 reviews, the review scores range from 1 to 5. After preprocessing, we use one of 3500 reviews as a training set to train the model, the remaining 1121 reviews as the evaluating set to evaluate the accuracy of the model. The feature set of GALAXY Note II phone generated manually is shown in Table 3.

Table 3.   Note II Feature Set.

Product	Feature Set
GALAXY Note II	CPU, Fitting, Battery, Price, Screen, Touch Fell, Appearance, After-sales, OS, Signal, Heat, Other

Use the sample review sentiment values and its review score of the training set to train the MLP Network according to section 3 above.

Correct Rate (P): Input the evaluating data to the MLP Network, if the output corresponding score is the same as the sample review score, the model classification is correct, that is to say the score prediction is correct, count the correct cases as $Set_{correct}$. *Correct Rate (P)* is defined as:

$$P = \frac{Set_{correct}}{Set_{Eval}} \qquad (3)$$

The evaluated results are shown in Table 4 and Figure 3 and 4 below.

Table 4   Evaluating result.

#	Training Times	Error rate	Correct rate
1	100	0.09493	62.21%
2	300	0.07537	71.63%
3	600	0.05728	74.35%
4	1000	0.04364	76.64%

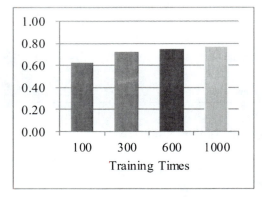

Figure 3.   Error rate result.

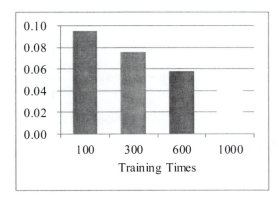

Figure 4.   Correct rate result.

The results show that with the increase of the number of training times, the error rate will be reduced, and the score prediction accuracy will be improved. When the iteration is more than 600 times, the prediction accuracy will reach to a certain value. Neural network training based on back-propagation would cause excessive fit and reduce the correct rate, so we should not make same samples too much of the training.

5   CONCLUSION

This paper proposes a method to establish a three layer MLP network evaluation model based on user review training, and with user reviews data experiment, the method can predict review scores and exhibit better performance. As the product features include product components, parts of property or function-level information, while the product feature set for model training in this paper is not processed to a finer level, the more valuable information (such as weighted user ratings, etc.), finer feelings grade and more network layers, etc. will be the direction of future research.

ACKNOWLEDGMENTS

This research is supported by the National Natural Science Foundation of China (Project No.: 71231002), and Research Fund for the Doctoral Program of Higher Education of China (Program No.: 20120005110015).

REFERENCES

C. Scaffidi, K. Bierhoff, E. Chang, M. Felker, and C. Jin, "Red Opal: product-feature scoring from reviews" 8th ACM Conf. on Electronic Commerce (EC), San Diego, California: 2007.

HY.Jiang. Score prediction collaborative filtering based on user preference learning reviews mining [J]. Application Research Compuers, 2010. 12, 27(12): 4430–4432.

K. Zhang, R. Narayanan, and A. Choudhary, Voice of the Customers: Mining Online Customer Reviews for Produc tFeature-based Rankings, Techni-cal Report, EECS department, Northwestern University, 2009.

Kobayashi, K. Inui, T. Fukushima. 2004. Collecting Evaluative xpressions for Opinion Extraction. InIJCNLP, pages 596–605.

L. Zhang, B. Liu, S. Lim. Extracting and Ranking Product Features in Opinion Documents. Coling 2010: Poster Volume, 1462–1470.

L. Lee, Pang B and S. Vaithyanathan. 2002. Thumbs up? sen-timent classification using machine learning tech-niques. In EMNLP, pages 79–86.

M. Hu and B. Liu, Mining and Summarizing CustomerReviews, Proceedings of the 10th ACM InternationalConference on Knowledge Discovery and Data Mining (SIGKDD-2004), 8 (2004), pp. 168–174.

O. Etzioni and A. Yates. Extracting Product Features and Opinions from Reviews. Proc. Joint Conf. on Hu-man Lang. Tech. / Conf. on Empirical Methods in Natural Lang. Processing, 2005, 339–346.

Popescu, A. Yates and O. Etzioni. 2004. Class extraction from the World Wide Web. In AAAI - 04 Workshop on Adap-tive Text Extraction and Mining, pages 68–73.

Sista, S., Srinivasan, S. Polarized Lexicon for Review Classification. Proc. Intl. Conf. on Machine Learning, Models, Technologies & Applications, 2004.

Toby Segaran Programming Collective Intelligence. Oreilly. 2007:97–98.

Ren Jianxin, Wu Guoshi, "A Research of Customer Value Hierarchy Model based on MapReduce and UGC", International Conference on Electrical Engineering and Computer Science (EECS), Hong Kong 2013, No 731, 1071–1077.

Information, Computer and Application Engineering – Liu, Sung & Yao (Eds)
© 2015 Taylor & Francis Group, London, ISBN 978-1-138-02717-6

LED street lamp energy-saving group control strategy based on wireless sensor networks

Gong Bo Zhou

School of Information and Electrical Engineering, China University of Mining & Technology, Xuzhou, Jiangsu, China

Peng Hui Wang& Zhi Xiong Cai

School of Mechanical and Electrical Engineering, China University of Mining & Technology, Xuzhou, Jiangsu, China

Shi Xiong Xia

School of Computer Science and Technology, China University of Mining & Technology, Xuzhou, Jiangsu, China

ABSTRACT: Street lamp is an important part of the city infrastructure. In pace with the gradual expansion of LED street lamp application scope, we need to decrease the junction temperature, improve energy efficiency and extend the lamp's lifetime through intelligent control under the premise of existing LED conversion efficiency. Based on wireless sensor networks, this paper designs an energy-saving dimming strategy that ensures the road lighting quality to achieve the goal of energy conservation. In this strategy, every node has three working modes. The node adjusts its working mode according to the working state of the two left adjacent nodes. Besides, this paper analyses the related performances about the designed dimming strategy in detail, the simulation results show that: the strategy can guarantee the road average illuminance and uniformity of illumi-nance, and compared with the LED street lamp lighting system with no dimming strategy, the dimming control strategy can save energy by 6.11%.

KEYWORDS: LED street lamp; Wireless sensor networks; Energy-saving

1 INTRODUCTION

Along with the accelerating of the city modernization, street lamp, as a kind of city infrastructure, plays an irreplaceable role in the urban construction. Recently, new-type LED began to enter the city public lighting fields, which is different from the traditional high pressure sodium lamps. It has many advantages of high luminous efficiency, long lifetime, high color-rendering index, fast-responding speed, small volume, no pollution, energy-saving and so on (Zhen, Zhang 2011; Wang, Wei 2011; Paddock, 2005). However, only 10–20 % power input of LED can be transformed into luminous energy, and most of the remaining energy is transformed into heat energy (Enke, Fan et al. 2011). Due to the large power density of the chip, PN junction and LED external environment thermal resistance causes LED junction temperature too high, thus LED chip rapidly deteriorates and its lifetime shortens. In addition, the luminous flux of LED decreases with the junction temperature rising, which has an impact on road lighting effect (Chen, Xuanping et al. 2011). In essence, even though to

solve the impact that the junction temperature has on lighting system needs further improvement in conversion efficiency of luminous energy and decrease in release of heat energy, in pace with the gradually extend of LED street lamp promotion and application, it is desiderate to decrease the junction temperature and extend the lamp's lifetime through intelligent control under the premise of existing LED conversion efficiency.

2 RELATED WORKS

Current research on urban lighting intelligent control, mostly made a single light dimming or on-off control through the microcontroller, or remote on-off control through GPRS, ZigBee-WSN technology, which cannot give full play to the advantages of LED. Single light control can control the junction temperature, but if a series of street lamps gain too high temperature at the same time, the road illumination cannot be guaranteed during the dimming process. Gao, Lei et al. (2011) has proposed a design of street lighting control system based on CAN Bus for achieving the improvement on the traditional street lighting system through intelligent control strategy, which can control a single light. (Wu, Guan et al. 2011) proposes a design of LED street lamp monitor system based on GPRS communication technology, the proposal is to construct a monitor platform, which can send the signals gathered from the sensor to the center through the monitor terminal installed on LED street lamps, thus can accomplish the purpose of monitoring. (Yu, Chunyu et al. 2011) proposes a design of LED street lamps intelligent control system based on RFID technology, by using the advantages of luminous efficiency of the illuminant and the advancement of RFID control technology. (Gong, Zhiwen et al. 2011) designs an LED street lamp remote monitor system based on ZigBee-WSN and GPRS/CDMA1x technology, achieving the best illumination control and attaining the purpose of energy-saving.

Above-mentioned documents are more concerned about the LED street lamp remote control and on-off control, these methods can improve the energy utilization rate and reduce the energy consumption, but they had not considered the junction temperature of LED chip. Therefore, this paper proposes an LED street lamp, energy-saving group control strategy based on wireless sensor networks, which can monitor the lamps' temperature in real time and control the power output through the wireless sensor nodes installed on street lamps, in order to achieve the control of LED junction temperature, thereby reduce its energy consumption and extend the lamp's lifetime.

3 PARAMETERS OF LED STREET LAMP

3.1 Structure of a street lamp

As shown in Figure 1, the street lamp node consists of sensor, wireless communication module, dimming module and LED light.

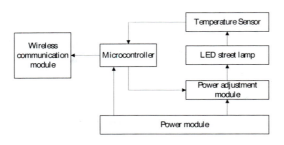

Figure 1. Schematic of the street lamp.

3.2 Deployment parameters

LED street lamp deployment parameters include the street lamps' overall arrangement on the two sides of the street, separated distance between street lamps and installation height of street lamps etc. The rational arrangement of street lamps play a key role in improving luminous effect, the common arrangements are single-side layout, both sides cross layout, both sides relative layout, crossroads layout and corner layout etc. The single-side layout should be adopted when the street lighting requirements are not high or the width of the street is less than 15m, the both sides relative layout or both sides cross layout should be adopted when the street lighting requirements are high or the width of the street is longer than 15m. In order to facilitate discussion and without loss of generality, the deployment parameters of LED street lamps are given in Table 1.

Table 1. Deployment parameters of LED street lamp.

Deployment parameters	Design formula	Design value
Installation height (h)	8~12 (m)	10 m
Separated distance (d)	$d \leq 3h$	30 m
Cantilevered length (l)	$l \leq 1/4h$	1 m
Lamp elevation angle (α)	5°~15°	Depend on lighting distribution curve

The deployment positions of street lamps are shown in Figure 2, for guaranteeing the street dimming in orders, each node has been allotted a unique ID number, the process of the dimming keeps in orders by ID number.

Figure 2. Sketch of street lamp deployment.

3.3 Electrical parameters

According to "City road lighting Design Standards" requirements, the luminous flux ϕ that the street needs is:

$$\phi = \frac{W_s \times d \times E_{av}}{q \times U_F \times K}$$

(1)

where W_s is effective width of street; d is separated distance between street lamps; E_{av} is the average illumination of road requirements; q is the arrangement of street lamps, U_F is the utilization factor of luminous flux; K is lamp maintenance factor.

In this paper, street lamps are both sides cross layout, the effective width of street equals the total width of street minus a cantilevered length, as $W_s = 14\,\text{m}$; $d = 30$ m; $q = 1$, illumination requirement is 15lx, for LED street lamps, $U_F = 0.95$, $K = 0.9$. Then, we have $\phi = \dfrac{15 \times 30 \times 14}{1 \times 0.95 \times 0.9} = 7368.42$ lm

According to "Road lighting with LED lights performance requirements"(GB/T24907-2010), this paper chooses LED street lamp whose power is 90W, in correspondence the luminous flux is $\phi_0 = 90 \times 85 = 7650$lm > 7368.42 lm, which meets demand. The parameters of LED street lamp are given in Table 2.

Table 2. Electrical parameters of LED street lamp.

Parameters	Design value
Power (W)	90
Luminous flux (lm)	7650
Input voltage (V)	AC220
Luminous effect (lm/W)	85

4 DIMMING OF LED STREET LAMP

Every node has three working modes, and it can communicate with the four adjacent nodes, the node adjusts its working mode according to the working state of the two left adjacent nodes. After each adjustment, the next adjustment won't begin until each street lamp reaches thermal equilibrium again. As shown in Figure 1, for example, the power input of street lamp 3 should adjust according to the working power of street lamps 1 and 2. Then, we define street lamp 3 as Dimming Node, street lamp 1 and street lamp 2 as Reference Node.

4.1 Working mode

As shown in Table 3, each node has three working modes. Set in the I mode, the power input of street lamp is P_i, th,unction temperature is T_{ji}, the lofia street lamps ϕ_i, the luminous intensity vertical to street lamp downward is I_{0i}.

4.2 Dimming strategy

M. Mn1 and Mn2 denote the working mode of Dimming Node, Reference Node 1 and Reference

Table 3. Parameters of street lamp in different modes.

Working mode	P_i(W)	T_{ji}(℃)	ϕ_i(lm)	I_{0i}(cd)
Mode 1	90	100	7650	1267.096
Mode 2	85	95.83	7315.95	1211.766
Mode 3	80	91.67	6972.16	1154.823

Node 2 separately. The pseudo code of dimming strategy is as follows.

```
if it is dimming time
then
    switch (M)
    case 1:
        {if((M_n2==1) || (M_n1 !=3 && M_n2 ==2))
        then (M=3)
        elseif((M_n1 ==3 && M_n2 ==2) || (M_n1 ==1&&
            M_n2 ==3))
        then (M=2)
        end}
    case 2:
        {if((M_n2==1) || (M_n1 !=3 && M_n2 ==2))
        then (M=3)
        elseif((M_n1 ==2&& M_n2 ==3))
        then (M=1)
        end}
    case 3:
        {if(|M_n2- M_n1 |==2) || (M_n1 ==2&& M_n2 ==2))
        then (M=2)
        elseif (M_n1 ==3)
        then (M=1)
        end}
End
```

5 PERFORMANCES

In this section, we applied Matlab to simulate and analyze the dimming algorithm. The simulation parameters are set forth as follows: node number is 20, the dimming period is 30 min, dim 24 times per-day.

5.1 Quality of lighting

The illumination quality on the road mainly depends on the road average illuminance, the uniformity of illuminance, and average power of the street lamps on the road. Figure 3 shows the changing process of the road lighting parameters during the 24 times of dimming.

From the Figure 3, during the dimming, the road average illuminance and uniformity of illuminance almost hold invariant, the average illuminance is $E_{av} = 13.74$,the uniformity of illuminance is $E_{jv} = 0.495$, through the analysis of the road

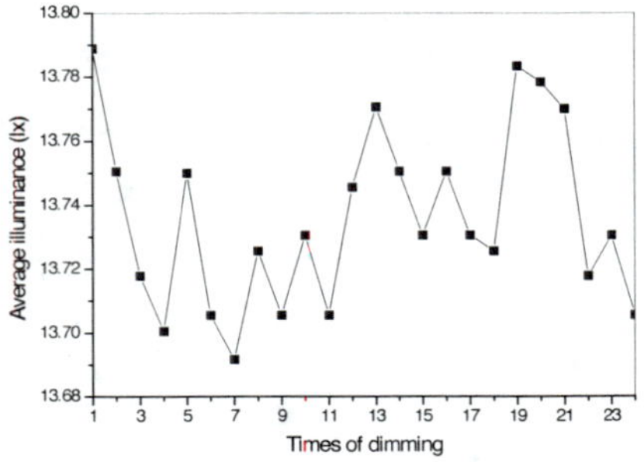

a) Average illuminance

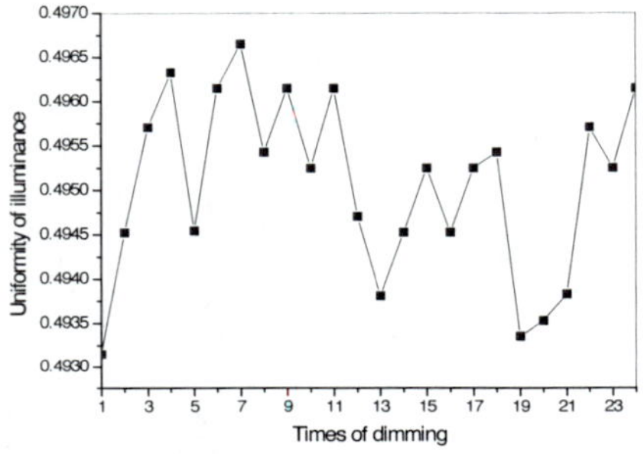

b) Uniformity of illuminance

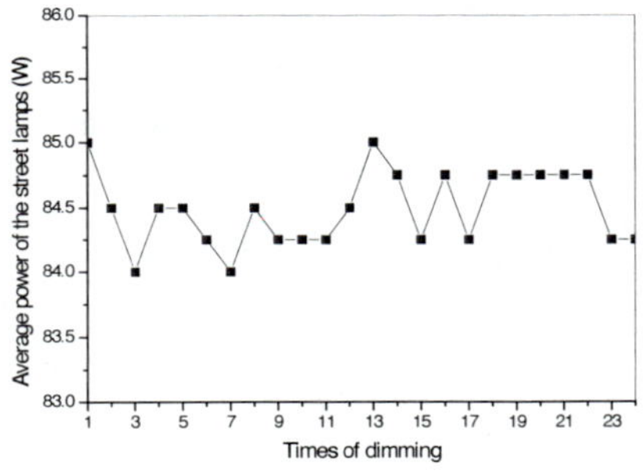

c) Average power

Figure 3. Quality of lighting.

average illuminance and uniformity of illuminance during 24 times dimming in one night, the variance of the road average illuminance is $\sigma^2_{Eav} = 7.84 \times 10^{-4}$, the variance of the uniformity of illuminance is $\sigma^2_{Ejy} = 1.017 \times 10^{-6}$, since randomness exists in one-time dimming, so we can estimate both road average illumination and illumination meet requirements of "City Lighting Design Standard" and "Lighting design manual", and the fluctuation of the road average illuminance and the uniformity of illuminance are very small during the dimming process.

5.2 *Energy efficiency*

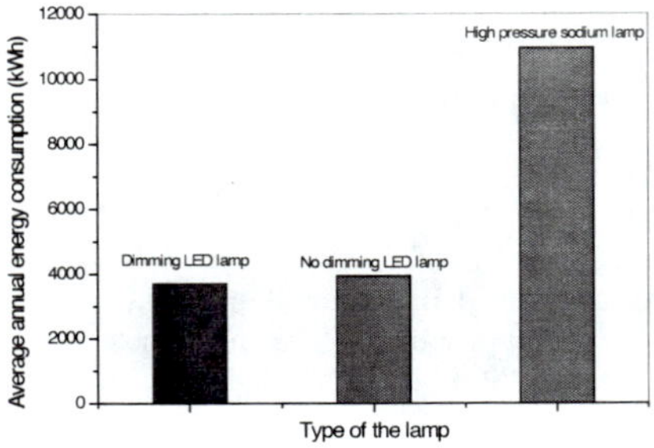

Figure 4. Average annual energy consumption per lamp.

Figure 4 shows the average annual energy consumption of the dimming LED street lamp, no dimming LED street lamp, and equivalent illumination High pressure sodium lamp. From the figure we know that the energy consumption of LED street lamp with dimming control strategy is only 33.84% of the high-pressure sodium lamp. Compared with no dimming LED street light, using dimming control strategy can still save energy by 6.11% under the premise of meeting the requirements of "urban lighting design standard". However, after that dimming light the road average illuminance falls from 14.72 to 13.74.

6 CONCLUSIONS

Based on wireless sensor networks, this paper designs an energy-saving dimming strategy that ensures the quality of road lighting quality to achieve the goal of energy conservation. Each lamp node has three work modes in the strategy, and according to the working condition of the left two adjacent nodes to adjust their work modes. The results of the simulation under the design strategy show that,

1 During the 24 times dimming, the variance of the road average illumination is $\sigma^2_{Eav} = 7.84 \times 10^{-4}$ and the variance of uniformity of illuminance is $\sigma^2_{Ejy} = 1.017 \times 10^{-6}$, which both can satisfy the "City lighting design standard" and "Lighting design manual".
2 Compared with the LED street lamp lighting system with no dimming control strategy, the dimming control strategy designed can save energy by

6.11%. After that, the road average illumination falls from 14.72 to 13.74.

ACKNOWLEDGMENTS

The material presented in this paper is based upon work supported by the National Natural Science Foundation of China (No.61104211), by the Research Fund for the Doctoral Program of Higher Education of China (No.20110095120005), by China Postdoctoral Science Foundation (No.20110491490 and No. 2014T70556), and by the Priority Academic Program Development of Jiangsu Higher Education Institutions.

REFERENCES

Zhen, Zhang. 2011. Street Lamp of LED and appliance. *Light & Lighting* 35(4):52–35.

Wang, Wei. 2011. The application prospect of lighting technology in the city lighting. *Electric Space* 17–18.

Paddock, A.N. 2005. LED Road lights shine in cradle-to-grave study. *Photonics Spectra* 44(5):40–42.

Enke, Fan & Zhang, Lihua. 2011. The design of heat radiation of high-power. *Journal of Baoji University of Arts and Sciences* 31(5):110–112.

Chen, Xuanping & Liu, Yunmei. 2011. The appliance and analysis of LED street lamp. *Light Sources and Illuminants* 2:8–10, 32

Gao, Lei Yu, Youling & Zhang, Zhiming. 2011. The design and implement of controlling system of street lighting based on CAN bus. *Technology of Measurement* 30(12):66–70.

Wu, Guan & Xiao, Jing. 2011. The design of monitoring system of street lamp based on the communication technology of GPRS. *Technology and Innovation* 38–40.

Yu, Chunyu Kong, Lingli He, Chunjiu & Zhang, Shendong. The design of intelligent controlling system of LED street lamp based on the technology of RFID. *China Lighting Electrical Equipment*, 2011, 35(1):33–35.

Gong, Zhiwen Gu, Yong Zhang, Changuo & Ge, Yu. 2011. The far-distance monitoring system of LED street lighting based on the technology of ZigBee-WSN and GPRS/CDMA1x. *Lighting* 30(14):36–39.

Information, Computer and Application Engineering – Liu, Sung & Yao (Eds)
© 2015 Taylor & Francis Group, London, ISBN 978-1-138-02717-6

Construction and practices of cultivating modes for students' innovative ability based on robotic technology

Peng Ju Zhao, Fang Qu, Wei Chen, Min Liu, Yi Li, Shi Hua Tong & Ming Hong She
School of Computer Science and Technology, Chongqing College of Electronic Engineering, Chongqing, China

ABSTRACT: Most Chinese students are deficient in innovation ability nowadays. Moreover, most of the majors and departments in colleges are running independently, which is not conducive to cultivating students' innovation ability. A cultivating mode for students' innovation ability based on robotic technology is proposed in this paper. The cultivating mode includes a series of measures, i.e., establishing a team of teachers, building a practice platform for students, exploring learning materials and resources and improving students assessing programs. Excellent achievements are obtained in improving students' innovation ability by taking the above measures.

KEYWORDS: Robotic Technology; Innovation Ability; Cultivating Practice

1   INTRODUCTION

Nowadays the change from Made in China to Created in China puts forward higher requirements for the new generation of Chinese students[1]. Innovation is the soul of national progress and the core of economic competition. Also the Chinese leaders put forward that it is the fundamental task of education to set a high moral standard in cultivating people and we should fully mobilize the innovative initiative and team-work spirit for students in higher vocational colleges. In addition, the Ministry of Education stated in the Twelfth Five-Year Plan of the national education development that the focus of higher vocational education is to cultivate developing, interdisciplinary and innovative technical talents for transferring, upgrading, and innovation of industry; thus we should enhance the cultivation of innovation awareness and ability; then we can carry out trials on ways to cultivate excellent students.

2   CURRENT PROBLEMS

Chinese students have a high examination ability, but weak practical ability, especially poor innovative ability. They are deficient in innovative thoughts and team-work spirit. With the purpose of cultivating skilled technical talents, although the higher vocational colleges pay a lot of attention to train the students' innovative ability, the innovative potential of students couldn't be fully explored because of the following two reasons. The first reason is that a strict division of majors is taken in the students' training process and different majors have their own different training programs. The second reason is too small majors

are divided and all the training programs are running independently, leading to narrow ranges of knowledge for the students, which cannot meet the requirements of the students to broaden their horizons in the knowledge structure. If the students are learning individually, without the basis of team-work participation, students' teamwork spirit and team-building awareness couldn't be cultivated effectively[2]. As a state-demonstrated higher vocational college, we proposed a cultivating mode for students' innovative ability based on robotic technology after seven years of exploration and practices. The three NOT CONDUCIVE problems are further systematically solved by the cultivating mode. The three NOT CONDUCIVE problems are: ① the fact that cultivation of professional talents is running independently is NOT CONDUCIVE to the exploration of innovative potential; ② the fact that the platform for innovation and practice is not perfect, is NOT CONDUCIVE to cultivation of innovative ability; ③ the fact that lacking of the evaluation and inspiration mechanism is NOT CONDUCIVE to the activation of the innovation motivation.

3   THE SCHEMES

The schemes for the cultivating mode for students' innovative ability is as follows. Firstly, an interdisciplinary, innovative team of teachers are formed by experts from the government, the enterprise and the college. They will teach the innovation awareness and innovation ability for the multi-disciplinary team of students[3]. Secondly, three-layer platform for innovative practice is established and robot clubs are formed under their supervision to provide effective support for students' innovation ability attracting more students to join in. Thirdly, corresponding inspiring measures are set by robot contest and innovative design of enterprise products to promote the growth of innovative talents. Fourthly, teaching and scientific achievement are obtained by summarizing the innovative design and practices, boosting a comprehensive upgrade of the quality of education and teaching.

Our purpose is to realize multi-professional integration and boost the exploration of innovation potential based on robotic technology. Robotic technology is a highly cross integration of multi-discipline and multi-profession and it involves embedded technology, pattern recognition (patrol track and positioning), drive control, information transmission, sensors and detection technology, interface circuit and driving technology, mechanical design, software design and so on. In the higher vocational college, it involves a series of majors, such as computer control technology, computer application technology, computer hardware and peripherals, embedded technology and applications, Mechatronics, mechanical design and manufacturing, applied electronic technology, microelectronics, automotive engineering and technology. All the related major clusters are integrated with the platform of robotic technology.

Robotic technology can guide students to explore new philosophy, and can guide students to explore a wide range of knowledge related with robotic technology. By robot education, students will learn to have overall thoughts, overall design and overall exploration in solving problems. Also, students will enjoy the joy of success through active investigation and practices related to the project theme. Because the robotic technology itself involves knowledge about computers, machinery, electronics and other majors, it can promote the students to develop good life habits and survival skills, such as communication and cooperation with people with different professional background. Thus, the robotic technology broadens their knowledge and visions and provides an essential guarantee for innovation.

4 IMPLEMENTATION OF THE SCHEME

4.1 Establishing an innovative team of teachers

An innovative interdisciplinary team of teachers is established by experts from the government, the enterprise and the college ensure the cultivation of innovation ability. In the joint-training process of talents, cooperation and the deep integration are strengthened among the government, enterprise and the vocational college explore the establishing mode of multi-disciplinary and innovative team of teachers. An innovative robot team in research and teaching is established based on the practical platform of high-intensive robotic technology education,

integrating all the teaching resources in related majors. Also, cooperative relationships are established with many companies in the Developmental League for Robots and Intelligent Equipments, and innovative, excellent teacher training is enhanced in a variety of cooperative forms such as survey in enterprises, internship practice and product research and development. Besides, ten CTOs and managing experts in the enterprises are brought in to serve as part-time teachers in the professional teachers team, in order to optimize the structure of the teachers team and to ensure the leading level and practicability of the teaching content. After seven years of exploration and practice, the innovative team of teachers of robots established by people from the government-enterprise-college provides a solid basis for the students' innovation ability.

4.2 Establishing practice platform for innovation

The three-layer evolutionary practice platform whose core is a robot training laboratory, robots innovation base and robot joint R&D center can provide a basis for cultivating innovation in hardware. The platform is established by making full use of the resources from the government, the enterprise and the college, such as the space, equipment, services and other tangible assets and personnel, technology, brand, industrial background and other intangible assets. At the robots training laboratory, comprehensive training curriculum and innovative production are completed. At the robots innovation base, the students' innovation ability is improved by holding all levels of competition, making invention, patent application. At the robot joint research and development center, the real robots R&D platform is established by making full use of the R&D ability at the college and the real working environment in the enterprise. The established platform can provide multi-level multi-professional innovative experiment and practice for students, can provide the latest development information in industry, can provide real business environment to enhance students experimental ability, thinking ability, activate creative thinking and develop integrated innovation, application innovation, technical innovation, process innovation ability, in a word, provide a solid and powerful support in hardware for students' innovation ability.

Robot training laboratory at college features combination of teachers-assigning and students-selecting on teaching-practice projects, and encouraging students to use their hands and brains to do comprehensive practice, designing practice, small manufacturing and small inventions. The innovation base established jointly by the college and enterprise features the purpose of innovation in science & technology, and also features the open managing mode of combing different majors, combing curricular and extracurricular, combing semester and holiday and combining college, enterprise, and the society. In the innovation base, complete arenas, equipment and facilities and the communicating convenience between the students and teachers are helpful to organize students to carry out multi-level innovative experiments and practice. The joint developing center features the guide of technology development in the enterprise and supervision of enterprise-sponsored projects, technical innovation, product pre-research and technology accumulation, combined with the business needs, and with enterprise technical staff, teachers and students as participants, and the industrial codes as the standard.

4.3 *Exploring teaching resources*

A module of teaching resources was established to ensure the effective implementation of the cultivation of innovative ability. The robot design and manufacture is the teaching content. The teaching resources module includes four parts, i.e., the design of the basic platform of robots, design of the robot sensor system, the design of the robot motion control system and design of the robots for competition. The purpose is to cultivate the basic innovation skills, innovation awareness, innovation spirit and application of innovation for students. Engineering objected teaching method is applied. Four-stage training method is applied in the teaching-learning process, i.e., demonstration of robots by teachers to complete certain tasks to stimulate students' interest, guiding students to do practical tasks to enhance their confidence, counseling students to complete practical tasks and students working independently to complete new tasks. According to the student's ability level, we choose teaching content with different difficulty levels to develop students' innovative ability, engineering practice ability, comprehensive professionalism. After teaching, we select outstanding members of the club to form a new training team, to participate in the domestic trials of the CCTV Asia-Pacific Robot Contest, the National Robots Competition for vocational colleges, Challenge Cup-Science & technology competition for college students, and to inspire and develop students' interest and innovation ability.

4.4 *Establishing a new evaluation mechanism*

Innovation evaluation and inspiration mechanisms are established to activate students' innovative motivation, to solve the current lack of evaluation and inspiration mechanism. The evaluation and inspiration mechanisms are composed of the process evaluation of campus, achievement evaluation of campus and inspiration mechanisms. The process evaluation is to evaluate the innovative awareness, innovative thoughts and innovative ability during the innovation practices by experts from the industry, enterprise and college. The achievement evaluation is to evaluate

students' achievements after the completion of innovative practice activities by experts from the industry and enterprise. Then students are rewarded according to the process evaluation and achievement evaluation as well as the inspiration mechanism. Students with outstanding performance, excellent innovation achievements will be rewarded timely and effectively, in ways of awards, credit for scholarships, credit for extra-curricular exploration and social practice, credit for professional award, exemptions of major courses, participating in academic writing, patent application and recommendation for province-level science and technology innovation award selection. Innovation evaluation and inspiration mechanism greatly cultivated students' interest in innovation, and fully mobilized and boosted students' innovative initiative, and stimulated the enthusiasm of the students to learn and study, and better stimulated students' motivation of independent innovation. After all, it makes innovative talents achieve more while do less, and solves the problem of students' deficiency in innovative motivation.

5 ACHIEVEMENTS

In the course of cultivating students' innovative ability, technical innovations are made through product R & D, patents application and other forms. Several national patents are obtained, such as a numerical-controlled electric cutting machine, an electric gripper robot and an integrated electronic assembly unit. In addition, outstanding achievements were made, i.e., two first-class rewards, five second-class rewards, five third-class rewards and two outstanding rewards in the National Skills Competition are achieved.

ACKNOWLEDGMENTS

This paper is sponsored by the key projects of teaching reform of higher education reform in the Chongqing construction practice of innovative talents cultivation based on robotic technology in vocational college (Project NO : 112101) and the projects of the city-level teaching team in Chongqing teaching team on embedded technology and robotic technology innovation.

REFERENCES

[1] Zhiguo Wang. Independent innovation: from made in China to Created in China. Studies on Mao Zedong and Deng Xiaoping Theories, 2010, (8):1–6.
[2] Jian Lin. Cultivation of innovation ability for excellent engineers. Research in Higher Education of Engineering, 2012, (5):1–17.
[3] Tao Li, Shizeng, Zong, Jiancheng Xu, Pengfei Li. Exploration and practices of establishing multi-displinar platformfor innovation practices. China University Teaching, 2013, (7):79–81.

Information, Computer and Application Engineering – Liu, Sung & Yao (Eds)
© 2015 Taylor & Francis Group, London, ISBN 978-1-138-02717-6

An advanced method for tag ordering and localization in mobile RFID systems

Qian Zhang
School of Software, Shanghai Jiao Tong University, Shanghai, China

Run Zhao
Computer Science Department, Shanghai Jiao Tong University, Shanghai, China

Dong Wang
School of Software, Shanghai Jiao Tong University, Shanghai, China

ABSTRACT: In many mobile RFID applications, not only tag detection and identification but also location and ordering of tagged items are attracting interest. In this paper, an advanced method for ordering and localization of tagged items on conveyor belts based on received signal characteristics is presented. We leverage the Doppler frequency shift to calculate the relative velocity between tags and RFID antennas, and then we build a relative motion model to get the order of the moving tags. Compared with the traditional approaches, we got a higher accuracy, the results show that this system achieves an accuracy of 97% even with a small neighboring tag space of 0.1m and a high conveyor speed of 4m/s.

KEYWORDS: Mobile RFID; relative velocity; Doppler frequency shift

1 INTRODUCTION

Radio Frequency IDentification (RFID) technology has gained significant momentum in the past few years. Recently RFID has been more and more frequently introduced in applications like retail and distribution, target monitoring and tracking, etc. In these applications, not only tag detection and identification but also location and ordering of tagged items are in need. A typical application is the items attached with RFID tags are placed on the conveyor belts for fast identification, these items pass through the RFID portal and tags are identified, the systems will do further operation like sorting using these identification information. For example, the express company attaches their packages with RFID tags, and in the distribution center, these packages are placed on the conveyor belts for identification and sorting according to their addresses. In order to ensure the correct sorting, the system not only needs the identification of the tags but also their relative orders when they are passing through the RFID portal on the conveyor belts. Similar applications also exist in warehouse sorting, airport logistics, garment sorting, as shown in Fig.1. In these applications, we need both the identification and the correct order of the tags on the conveyor belts, imagine if the system gives the wrong order of the luggage in airport sorting line, then the luggage may be placed in a wrong plane, which may result a serious consequence. In 2012, 1%

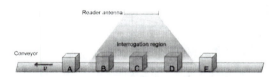

Figure 1. The scenario of tagged items placed on the conveyor belts.

of worldwide baggage was still mishandled, costing around USD 2.58 billion, therefore, the ordering and localization of items in mobile RFID systems are important to the development of RFID technology.

Items attached with tags pass through the RFID portal on the conveyor belts, the reader may not only read the tag which are just passing through the antenna but also read tags before or after this tag, as shown in Fig.1, this results that the system can't identify the correct order of the tags and further results wrong operation. In the traditional ways, we can enlarge the space between items to ensure only one tag can be read, but this will decrease the work efficiency. In addition, due to the probabilistic of tags identification and irregularity, the identification time can't be the judgment either. We need find other ways to obtain the order of the items

Recently some research focus on leveraging the received signal strength (RSSI) to localize the tag in mobile RFID systems, this can't work very well when the speed of the conveyor belts is high or the spacing between items is small due to the high sensitivity of RSSI to the fast changing environment. In this paper, we exploit the Doppler frequency shift data collected during tag movement along the conveyor belt to obtain the correct order and precise location of tags. We first use the Doppler frequency shift of the received signal to obtain the relative velocity between reader antennas and tags, and then we build a relative motion model based on the conveyor belts to obtain the correct order of tags passing through the RFID portal. Extensive experiments are conducted to validate the effectiveness of our method, the results show that our method achieves an accuracy of 0.05m even with a high conveyor speed of 4m/s.

2 RELATED WORKS

As in many applications, the ordering and location of RFID tags are needed, there are some research using the received signal characteristics of the tags to achieve these requirements.

It is expected that the RSSI of a passive tag is inversely proportional to its distance from the reader. Thus, when objects attached with RFID tags move through the reader on the conveyor belt, the RSSI of the tags will increase first and then decrease theoretically. And related works mostly use RSSI as a basis for tag ordering. Goller M presents an approach for tag ordering by computing the signal center of the RSSI pattern, which uses the RSSI value every time the tag is read as the weight to calculate the center time, and bases this time for ordering. Shangguan L et al propose OTrack, a practical protocol, to track the order of tags on conveyor belts. The authors define Response Reception Ratio (RRR), a measure of the tag's response reception, and use RSSI and RRR together to find the time that tag passing the original point on the belt.

Due to the high sensitivity of RSSI to its surroundings, there are some researches that rely on the received signal phase from tag to obtain accurate localization. P.V.Nikitin gives an overview of phase based localization, which includes Time Domain, Frequency Domain and Spatial Domain Phase Difference of Arrival, among them, the relative velocity can be obtained by the read time and received signal phase. It is shown that Time Domain technique was fairly robust to multipath. R.Miesen proposes a synthetic aperture radar (SAR) approach to localize a stationary tag, the localization method is based on phase values sampled from a synthetic aperture by a mobile RFID reader with a known relative trajectory. Furthermore, Andreas Parr proposes the inverse synthetic aperture radar (ISAR) approach to localize a moving tag with a known trajectory. The ISAR method enables precisely

and robust localization in mobile context but suffers substantial computation burden, thus can't satisfy real time order tracking in the case that both tags' density and moving speed are high.

3 RESEARCH METHOD AND SYSTEM DESIGN

In this section, we introduce major methodology used in this paper and the algorithm design. First we do a research on the way of ordering based on RSSI, and find that when the conveyor belt is in high speed, this way can't correctly obtain the order of the tags. Second we introduce the relative velocity between reader antennas and tags calculated by the received signal phase and the Doppler frequency shift. Finally we build a relative motion model based on the conveyor belts to obtain the correct order of tags passing through the RFID portal.

3.1 *Ordering based on RSSI*

In theory, the RSSI value is inversely proportional to the fourth power of its distance from the antenna, thus, some work use RSSI as a basis for tag ordering and localization. However, this can't work very well when the speed of the conveyor belts is high or the spacing between items is small due to the high sensitivity of RSSI to the fast changing environment. We choose 8 successive tags passing through the RFID portal on the conveyor belts with a high speed of 4m/s, and the results based on the time when RSSI is the maximum is shown in Fig.2, the TagID is numbered according to their relative orders passing through the RFID Portal. From the results we can see the order of the tags in high speed environments based on RSSI is quite wrong.

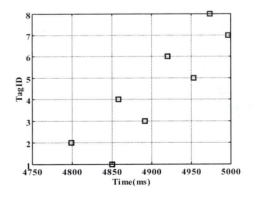

Figure 2.   The order based on RSSI.

3.2 Relative velocity

Most of today's commercial UHF readers can get the Doppler frequency shift from the received signal, Doppler frequency shift is the shift in frequency of the received signal at the reader due to relative motion between the reader and the tag. Let ΔT denote the time duration of a packet. Then, a Doppler frequency of f_d Hz introduces a phase rotation over this packet duration given by

$$\Delta\varphi = 2\pi f_d \Delta T \qquad (4)$$

The Doppler frequency experienced by the reader can thus be calculated by measuring the phase rotation across a packet and using the following expression:

$$f_d = \frac{\Delta\varphi}{2\pi \Delta T} \qquad (5)$$

The relative velocity can be calculated:

$$v_r = \frac{c\Delta\varphi}{4\pi \Delta T} = \frac{c}{2} f_d \qquad (6)$$

Estimating Doppler frequency over the duration of a single packet avoids many of the pitfalls (e.g. stochastic inventory protocol, antenna switching, channel hopping, and phase ambiguous) inherent to using the RF phase from two different packets, thus we will calculate the relative velocity based on the Doppler frequency shift of the received signal.

3.3 Velocity-based tag ordering

As shown in Fig.3, assume tagged items are moving along the conveyor belt with a speed of v, the angle between the tag moving direction and the direction from the tag to the center of the antenna is θ, and the relative velocity between the tag and the reader antenna $v_r = v\cos\theta$, the point that the tag is closest to the antenna center is zero point, and the time when the tag is at the zero point is zero point time, the distance between the tag and the center of the antenna when the tag is at the zero point is zero point distance. If we want to obtain the right order of the tags, we need find the zero time of each tag.

When the tag is moving towards the antenna, θ ill gradually increase to $\frac{\pi}{2}$, and then gradually decrease

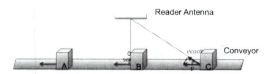

Figure 3.   Relative velocity on the conveyor belt.

when it is moving away, thus v_r will gradually decrease to 0, and then gradually increase, suppose the velocity when the tag is moving towards the antenna is positive, and when it is moving away the antenna is negative, then when the tag is passing through the RFID portal, its relative velocity is gradually decrease from positive to negative, as is shown in Fig.4. The data in Fig.4 is simulated, the speed of conveyor belt is 4m/s, and the zero point time is 0ms, and the zero point distance is 0.5m.

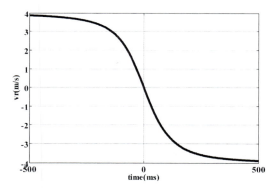

Figure 4.   The relative velocity of one tag.

First we calculate the relative speed using Doppler frequency shift, and then, theoretically we just need to find the last time that the relative velocity is positive and the first time that the relative velocity is negative, the time is t_+ and t_-, and then the zero point time is

$$t_0 = \frac{t_+ + t_-}{2} \qquad (7)$$

But in fact the relative velocity will be fluctuant and may not be monotonic decreasing, especially near the zero point. Therefore we use Algorithm1 below to find the zero point time.

Algorithm 1 Find the Zero Point Time

Input: the relative velocity of the ith tag identified v_i.
Output: the zero point time t_0.

1   $MaxpN=0, pN=0; MaxN=0;$
2   **For** i=1,2,... ,
3   **If** $v_i > 0$
4       $pN=pN+1$
5   **If** $MaxpN<pN$
6         $MaxpN=pN$
7         $MaxN=i$
8   **end if**
9   **else if** $v_i < 0$
10      $pN=pN-1$
11   **end for**
12   $t_0 = \dfrac{t_{MaxN} + t_{MaxN+1}}{2}$

13 **return** t_0

The system gives the order of tags based on their zero point time.

4   EXPERIMENT AND EVALUATION

We conduct extensive experiments to validate the effectiveness of our method. We use Impinj R420 RFID reader to implement our experiments, the tag is ALN-9740, the conveyor speed is 4m/s, which can satisfy most need in RFID applications, the experiment scene is shown in Fig.5.

Figure 5.   Experiment scene.

We choose 8 successive tags passing through the RFID portal with their neighboring space is 10cm, their relative velocity trend is shown in Fig.6, and we can see our method based on relative velocity can correctly give the order of tags.

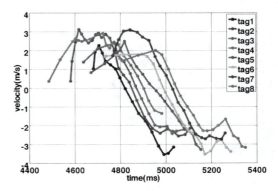

Figure 6.   Velocity based tag ordering.

We conductive more than one thousand times experiment, and our experiments results shown that the ordering method based on the relative velocity can achieves 97% accuracy even with a neighboring spacing of 10cm, and a speed of 4m/s.

5   CONCLUSION

In this paper, we present a method for RFID tag ordering in mobile environments, which is based on the relative velocity between the tag and the reader antenna, and we calculate the relative velocity using the Doppler frequency shift of the received signal. The experiments results show that our method can still work very well when the space between neighboring tags is small and the speed of the convey is high.

REFERENCES

M. Goller and M. Brandner, "Experimental evaluation of rfid gate concepts," in RFID (RFID), 2011 IEEE International Conference on. IEEE, 2011, pp. 26–31.L. Shangguan, Z. Li, Z. Yang, M. Li, and Y. Liu, "Otrack: Order tracking for luggage in mobile rfid systems," in INFOCOM, 2013 Proceedings IEEE. IEEE, 2013, pp. 3066–3074.

R. Miesen, F. Kirsch, and M. Vossiek, "Holographic local-ization of passive uhf rfid transponders," in RFID (RFID), 2011 IEEE International Conference on. IEEE, 2011, pp. 32–37.

A. Parr, R. Miesen, and M. Vossiek, "Inverse sar approach for local-ization of moving rfid tags," in RFID (RFID), 2013 IEEE International Conference on. IEEE, 2013, pp. 104–109.

P.V. Nikitin, R. Martinez, S. Ramamurthy, H. Leland, G. Spiess, and K. Rao, "Phase based spatial identifica-tion of uhf rfid tags," in RFID, 2010 IEEE International Conference on. IEEE, 2010, pp. 102–109.

Information, Computer and Application Engineering – Liu, Sung & Yao (Eds)
© 2015 Taylor & Francis Group, London, ISBN 978-1-138-02717-6

The audit risk and strategy of an enterprise in an accounting computerization environment

Li Li Liu

Qinhuangdao Institute of Technology, China

ABSTRACT: With the continuous development of China's market economy, the management environment of Chinese enterprises has undergone tremendous change. By analyzing the audit risk, the enterprise can effectively reduce its loss and promote the improvement of its management system to ensure its healthy developments. This paper mainly discusses the audit risk of an enterprise in an accounting computerization environment and puts forward strategies to prevent and cope with the audit risk.

KEYWORDS: Enterprise; Audit risk; Prevention and control; Accounting computerization

In recent years, with the rapid development of China's market economy, the enterprise's managerial and administrative level has also been improved. During the process of management control, the enterprise will be affected by many factors, such as economy, law, environment, etc. Therefore, it will have a certain audit risk for the enterprise.

1 REASONS FOR AUDIT RISK

The audit work, as one of the important driving force to promote the rapid development of the economy, can not only maintain the order of the market economy, but can improve the operational efficiency of the economy. At present, the research of audit risk in China is only limited to some superficial and symbolic research which won't have a timely control and constraint of the audit risk. If audit activities can be carried out orderly and rapidly in the market economy, legal restrictions and regulations are a must. Without the written legal norms, audit can not be carried out effectively.

1.1 Auditors have a lower ability and poor risk awareness

For an enterprise, the audit is a systematic work under the condition of the computerization, which involves a very wide range of knowledge. If auditors can't timely update their technical knowledge, it is difficult for them to audit quickly and scientifically. They should not only master the knowledge of auditing, accounting, and finance, but also grasp the knowledge of computerization. If the auditors do not have such knowledge, there will possibly be deviations of the accounting information in the process of audit which will eventually lead to the emergence of the audit risk. Some auditors are lack of risk awareness. Compared with traditional audit, audit in the accounting computerization environment has the feature of concealment and destruction. Auditors must pay attention to the audit risk and fully understand the importance of it. They should carry out an objective and fair evaluation of the authenticity and reliability of the financial revenue and expenditure of the unit under auditing. Auditors in Chinese enterprises often lack independence. The department responsible for auditing in enterprises should have an independent power. Audit work should not be interfered by other leadership. Only in this way can audit fully play the role of supervision and control.

1.2 The internal system of the audit unit is not perfect

The traditional control of audit work is usually controlled by handwork which specifies clear responsibilities. But in the accounting computerization environment, internal audit has had a revolutionary change and turned it into computer control. If the audit program is lack of scientificalness, it will reduce the computer control effect. Some lawbreakers may tamper with relevant programs, and in the computerization conditions the unit under auditing is difficult to modify and improve its internal control. The unit under auditing driven by its own interests may intentionally conceal some important information. Auditors should verify relevant information provided by the enterprise according to auditing standards and material. Therefore, audit work needs the cooperation of the enterprise. But in practice, some units may conceal or change some important information. Besides, personnel changes also affect the objectivity and fairness of the audit evidence.

1.3 The complexity of the enterprise's operating activities and the continuous expansion of the audit environment

The scale of operation of the enterprise is expanding and its operating activities are becoming more and more complex, especially when there are more and more trans-regional and transnational corporations. First of all, the audit content covers a wide range which not only includes the traditional audit work but also the evaluation of the accounting software. Audit trails are getting more and more hidden. Under accounting computerization conditions, the audit trails are stored in a magnetic medium, so they are invisible and untouchable. They can only be read by man-machine dialogue. Accounting information is stored in the magnetic medium, which reinforces its hiding degree and increases the audit risk. Finally, there are many types of accounting software. Because of many kinds of accounting software in the market, it is impossible for auditors to grasp all of them. Some accounting software will be upgraded which can affect data extraction. The software companies out of their security considerations may not easily reveal the software design programs to the auditors, which may also increase the audit risk to a certain extent.

1.4 Laggard auditing methods

Many Chinese enterprises are lack of due attention to the audit work and are unable to timely update new auditing methods and means, which increases their audit risk to a certain extent. The commonly used modern auditing method is sampling technique. Through the analysis of a certain amount of extracted samples the total feature can be inferred. Many of the Chinese enterprises develop relatively late and they have not yet set up a set of perfect enterprise management systems. They often use relatively simple methods, which may also exacerbate their audit risk.

2 FACTORS INFLUENCING AN ENTERPRISE'S AUDIT AND ITS RISK

The complexity of the modern enterprise's managerial and administrative environment has unceasingly intensified. Some enterprises, when dealing with their internal complex economic management activities, due to the lack of their own auditing means and methods, are often difficult to make scientific audit recommendations which can also strengthen the enterprise's risk of management. There are many factors affecting

audit risk including inherent risk, control risk and detection risk. Inherent risk refers to the risk having nothing to do with the enterprise's internal control programs and systems and is excluded from the enterprise's accounting and financial audit. The inherent risk of the enterprise is mainly caused by the complexity of its business which will make the enterprise have some illegal behaviors. Control risk: the lack of the enterprise system makes it unable to timely deal with some problems of operation and management. Modernization in most of Chinese enterprises is relatively slow and they haven't established a set of perfect operational management system, which also increases the enterprises' audit risk. Detection risk: some auditors' poor quality or their negligence may make it difficult for the enterprise to timely find out its operational problems, which also increases the enterprise's risk.

3 PREVENTION MEASURES OF AN ENTERPRISE'S AUDIT RISK

3.1 Countermeasures for the evaluation of audit risk

3.1.1 The application of risk analysis method
The method of risk analysis is to evaluate possible factors of audit risk so as to determine the enterprise's audit risk in its operation and management. This method generally consists of 5 steps: Firstly, having a systematic investigation on the sources of risk which is very important for the enterprise to reduce its audit risk. Secondly, identifying specific conditions for shifting risk. In business management, these conditions need to be controlled in order to reduce audit risk. Thirdly, to determine whether the enterprises have these conditions for shifting risk in their operation, and to control and correct some of their management problems timely. Fourthly, to evaluate the consequence if risk occurs according to the sources of risk. Finally, to have a systematic evaluation of the risk. Using this approach for risk evaluation is relatively simple, and decisions can be made timely. In the process of audit there may be sources of risk which may lead to a shortage of funds or the loss of customers. When investigating an enterprise's financial situation, auditors should fully learn whether the enterprise has some conditions for shifting the sources of risk, and carry out systematic investigations on some of specific situations. If the enterprise's cash flow has a sustained abnormal condition, it is necessary to prevent in advance and have a systematic evaluation of the enterprise's audit risk. It is of great necessity to evaluate the enterprise's audit risk. Although the method of factor evaluation may be subjective, it can prevent in advance on the inherent

risk and detection risk. Risk evaluation can be divided into different levels. Gather different risk factors and evaluating factors together and then evaluate them in accordance with "level one, level two, level three" or "high level, medium level and low level".

3.1.2 *The application of a comprehensive evaluation method of fuzzy quality*

The fuzzy evaluation method is formed on the basis of the theory of fuzzy mathematics. It can have a comprehensive check on all the factors and can distinguish the degree of importance of each factor by weight. The enterprise's audit risk is influenced by various factors existing in its operation. Fuzzy quality evaluation is carried out according to the enterprise's risk factors and the evaluation results should be graded, for example, reliability, safety, key, risk and other aspects can be divided into different grades. On-the-spot investigation should be carried out. Investigation results and evaluation results are mixed together to establish an evaluation system in the form of a matrix. Different risk may lead to different results and different factors have different effects. Hence it should be evaluated according to the effect of the risk, and corresponding weights need to be determined. The method of fuzzy mathematics will be applied in the evaluation results which need to be calculated.

3.1.3 *The application of the method of audit analysis*

The method of audit analysis refers to chartered accountants make a comparative analysis on important ratios or trends of audit units to adjust some abnormal variation trends or related information by analyzing relevant financial data. This is a comprehensive audit method covering a wide range of content and including not only accounting and finance knowledge, but also some complex statistical models of mathematics. If the validity and reliability of some data need to be verified, the method of comparison and analysis has to be used to analyze and verify relevant audit objectives.

3.2 *To establish a strict internal control mechanism*

Enterprises should establish a perfect internal control mechanism according to the modern enterprise management system, and they need to implement it by rules and regulations. The leaders of the enterprise should also play a leading role and comply with various rules and regulations. The enterprise must carry on the evaluation and prevention of some possible audit risk and make good preparations for its finance. It needs to control the audit risk in a tolerant range. For the uncontrollable audit risk, the enterprise should adopt a positive strategy and try to avoid it.

3.3 *To improve audit methods*

The enterprise should improve and innovate its audit methods to improve its audit efficiency according to its problems existing in the audit work and its own audit risk. It can adopt an advanced international risk-oriented audit method to have a timely systematic evaluation of various interior risk factors of the enterprise. The audit work of the enterprises can not simply rely on data. Other factors except data need to be audited comprehensively. The enterprise should continuously improve the comprehensive quality of auditors in order to improve their ability to cope with audit risk.

4 SUMMARY

With the development of market economy, the enterprise's problems of audit risk present new characteristics. The audit risk of the enterprise in an accounting computerization environment is an important problem for its development. The enterprise should take various measures to improve its audit efficiency and constantly reduce its audit risk.

REFERENCES

[1] Sun Xu. Risk and Countermeasures in the Accounting Computerization Environment [J]. Science and Technology Innovation Guide, 2012, 17 (16):32–33.
[2] Zhang Sufan. The Study of Tax Audit in the Accounting Computerization Environment [J]. Hunan Tax College, 2013, 35 (17):10–11.
[3] Wang Beilei. How to Carry Out Audit Work in the Accounting Computerization Environment Taking Xishan Coal and Electricity Group As an Example [J]. Youth Science (Teacher Edition), 2013, 34 (11):18–19.

Information, Computer and Application Engineering – Liu, Sung & Yao (Eds)
© 2015 Taylor & Francis Group, London, ISBN 978-1-138-02717-6

An iterated greedy algorithm for the integrated scheduling problem with setup times

Yu Jie Xiao
Jiangsu Key Laboratory of Modern Logistics, School of Marketing and Logistic Management, Nanjing University of Finance and Economics, Nanjing, Jiangsu Province, China

Yan Zheng
College of Automobile and Traffic Engineering, Nanjing Forestry University, Nanjing, Jiangsu Province, China

Qiong Jia
Business School, Hohai University, Nanjing, Jiangsu Province, China

ABSTRACT: This paper studies an integrated scheduling problem of machines and Automated Guided Vehicles (AGVs) in flexible manufacturing systems where sequence-dependent setup times are considered with the objective of minimizing the makespan. In most manufacturing systems, machines scheduling and AGVs scheduling are highly related and consideration of setup times is frequently encountered. That integrated problem considering setup times is much closer to the real-life industrial applications but much difficult to solve. Based on the structural property of this problem, an iterated greedy based algorithm is developed, within which a local search is also applied to diversify the solutions. Computational experiments are conducted to evaluate the proposed heuristic algorithm. The results show that our method is promising for dealing with the integrated problem.

KEYWORDS: Integrated scheduling problems; Sequence-dependent setup times; Iterated greedy; Makespan

1 INTRODUCTION

With the development of computer technologies, a fully automated and integrated manufacturing environment has become the current trend in the field of industrial applications. Flexible manufacturing system (FMS) is such an integrated manufacturing environment that responds rapidly to market changes. It consists of several multifunctional machines to produce various part types, is controlled by a distributed computer system and connected with an automated material handling system (MHS) (Groover 1987). In MHS, the wildly used material handling equipment is the automated guided vehicles (AGVs) due to its flexibility and compatibility (Vis 2006). This paper addresses a simultaneous scheduling problem in FMS where machines scheduling and AGVs scheduling as well as setup times are considered.

Machines and AGVs scheduling problems are two coordinated problems in the aspect of FMS scheduling. It is necessary to simultaneously make decisions on machines and AGVs scheduling as these two limited resources are concurrently involved in the production process (Zheng et al. 2014). Machines scheduling is concerned with the processing sequence of operations on machines, while AGVs scheduling is

focused on dispatching a limited number of AGVs to transfer tasks. The integrated scheduling problem of machines and AGVs arose in 1992 (Sabuncuoglu & Hommertzheim 1992) and has been studied by many researchers since then (Bilge & Ulusoy 1995, Abdelmaguid et al. 2004, Deroussi et al. 2008, Zheng et al. 2014). In their problems, setup times are either negligible or assumed to be part of job processing times. However, this simplification is theoretical for some real-life industrial settings and thus has adverse influence on the solution quality of the real world scheduling applications where explicit consideration for setup times is required. For all these reasons, a complex integrated machines and AGVs scheduling problem with setup times is studied in this paper.

Setup times can be roughly classified into two categories. One is sequence-independent setup times (SIST) which only depend on the current processing job on machines. The other is sequence-dependent setup times (SDST) that depend on both the current job being processed and its immediate predecessor on the same machine. Moreover, setup times can be either anticipatory or non-anticipatory. Anticipatory setup means that the setup can begin on machines before jobs arrive at the machines. Here, we focus on anticipatory SDST in our integrated scheduling problem.

This integrated problem has the same structure as that in the work of Bilge & Ulusoy (1995), but for SDST. It extends the classical job shop scheduling problems (JSP) by involving additional features: transportation times, AGVs dispatching and SDST. Since JSP with SDST has been proved to be NP-hard (Naderi et al. 2010), the proposed problem is more difficult to solve. For this reason, we develop an iterated greedy (IG) algorithm which has a simple principle and is easy to implement. IG algorithm has been successfully applied to difficult optimization problems such as the set covering problems and various scheduling problems (Jacobs & Brusco 1995, Ruiz & Stützle 2008, Bouamama et al. 2012), which motivates us to develop an IG based algorithm for our integrated scheduling problem.

2 PROBLEM DESCRIPTION

In an FMS, we are given N independent jobs which have to be processed on M machines. Each job p has a set of ordered n_p ($p = 1,2,...,N$) operations. The total number of operations for all N jobs is Q. Each operation i ($i = 1,...,Q$) has a dedicated processing machine with a known processing time p_i. Setup times s_{ijk} ($i,j = 1,2,...,Q; k = 1,2,...,M$) for operations i and j on machine k is sequence dependent and anticipatory. All of N jobs are transferred by V identical AGVs between machines. Jobs are loaded and unloaded at the predetermined pick-delivery (P/D) points. All AGVs begin at the load/unload (L/U) station and finally return to L/U station when they finish all transferring work. The objective is to concurrently find the processing sequence on each machine and trip of each AGV, such that the maximum completion time or makespan C_{max} is minimized.

To model our problem, the following assumptions are described. Each machine has sufficient buffer space for parking AGVs and storing jobs. For each machine, at most one job can be processed at a time and cannot be stopped until it is finished. Travel times between different machines are given. All AGVs are identical and move at the same speed. They carry a single unit load each time. Traffic congestion and conflict problems are not included in this study.

3 THE PROPOSED IG ALGORITHM

The IG algorithm is a simple and very efficient meta-heuristic algorithm which was first proposed by Jacobs & Brusco (1995). The original IG algorithm starts with an initial solution S_0 and iteratively searches better solutions through a main loop which includes two phases called destruction and construction. In the destruction phase, a partial solution S_p is created by

removing d elements from the current solution S, and then in the construction phase, a complete solution S_c is reconstructed by a greedy heuristic. Once S_c is completed, an acceptance criterion is used to determine whether the current solution S is replaced by S_c. Application of local search is optional, which is helpful to the diversification of the search. This loop proceeds until the termination condition is satisfied.

3.1 Solution representation and the initial solution

Each solution p is represented by a $2' Q$ array, which has a two fixed-length string in rows. The upper row, denoted as p_s, gives the processing sequence of operations on machines. The lower row, denoted as p_v, indicates the AGVs assignment to each operation. As the processing machine for each operation is known, the operation sequence on each machine and the trip of each AGV can be obtained by following the order of p_s and p_v.

The initial solution for IG is created using the initialization method introduced by Zheng et al. (2014). p_s is first initialized using topological sort and random priority assignment method to guarantee the feasibility of operation sequence. Based on the initial p_s, the nearest vehicle rule (Vis 2006) is employed to assign vehicles to each operation. Figure 1 shows a feasible solution for an example problem with three jobs including seven operations (operations 1, 2 and 3 for job 1; operations 4 and 5 for job 2; operations 6 and 7 for job 3) which are moved by two AGVs between machines.

π_s	1	4	2	6	3	5	7
π_v	v1	v2	v1	v2	v1	v2	v2

vi represents AGV_i

Figure 1. A feasible solution π for the example problem.

3.2 Procedure of the proposed IG

1 Step 1: Initialization. An initial solution $p_{initial}$ is created by applying the initialization method introduced by Zheng et al. (2014). Set the current solution $p = p_{initial}$ and the best solution $p_{best} = p_{initial}$.

2 Step 2: Destruction and Construction. Select d operations from p_s in p. A partial solution p_p is generated by deleting the selected d operations together with its assigned AGV in p_v from p. Then add the d columns to pR- a list of the deleted columns, in the order they are removed. Reinsert the removed columns in pR, one by one, to all possible and feasible positions of p_p. Each time one deleted column is reinserted into the partial solution, a set of candidate partial solutions is

generated. Only those satisfying the job routing are feasible, among which the one with minimum C_{max} is selected as the new partial solution p_p. After inserting all d deleted columns, a new complete solution p_{new} is then obtained. Check if $C_{max}(p_{new})$ £ $C_{max}(p)$, then $p = p_{new}$ and $p_{best} = p_{new}$; else if $C_{max}(p_{new}) > C_{max}(p)$ and r £ P, then $p = p_{new}$, where r is a generated random number between $[0,1]$, P is a parameter defined by users.

3 Step 3: Local search. Keep the operation sequence p_s of p unchanged and create a set of Q' $(V-1)$ candidate solutions by modifying AGV assignment of each operation one by one to any of other AGVs. Among the candidate solutions, the one p_{local} with the minimum C_{max} is selected. Check if $C_{max}(p_{local})$ £ $C_{max}(p)$, then $p = p_{local}$ and $p_{best} = p_{local}$; else if $C_{max}(p_{local}) > C_{max}(p)$ and r £ P, then $p = p_{local}$.

4 Step 4: IG algorithm iterates through Step 2 and Step 3 until the stopping criterion is met. When the counter of iteration n_{iter} reaches the maximum allowed number of iterations N_{iter}, IG algorithm stops.

4 COMPUTATIONAL EXPERIMENTS

To examine the proposed IG, two sets of experiments (E-1 and E-2) are conducted. E-1 includes eight instances (EX11-EX81) taken from the benchmark problem proposed by Bilge & Ulusoy (1995). E-2 is a set of sixteen problem instances generated by adding two levels of SDST to the problems instances in E-1. The ratios of SDST to the processing times are 30%, 70% of the maximum processing time of Bilge & Ulusoy's benchmark problems. Thus, setup times are generated from two uniform distributions over $U(1,6)$ and $U(1,14)$ for each problem instance. The proposed algorithm is coded in C language and all computational experiments are tested on a PC with a Dual-core 1.60 GHz and 1 GB RAM in the Microsoft Window XP environment. Each instance is run ten times and the best solution obtained together with its computation time is recorded.

4.1 Parameters settings

After conducting preliminary experiments, the parameters d, P and N_{iter} are set to be 4, 0.4 and 800Q, respectively.

4.2 Computational results for E-1

Our proposed methodology was applied to the benchmark problems of Bilge & Ulusoy (1995) by setting the setup times to be zero. Since the tabu search algorithm (TS) developed by Zheng et al. (2014)

performed better results than other previously published methods, we use TS algorithm as a comparison to our proposed method. Table 1 shows the comparison results, from which we can see that IG searches the same best solutions with those of TS.

Table 1. Computational results of TS and IG.

Instances	TS C_{max}	IG C_{max}	CPU time
EX11	96	96	1.98s
EX21	100	100	2.03s
EX31	99	99	2.08s
EX41	112	112	2.12s
EX51	87	87	2.14s
EX61	118	118	2.16s
EX71	111	111	2.61s
EX81	161	161	2.67s

4.3 Computational results for E-2

Two groups of sixteen problem instances are generated based on EX11-EX81 by adding two levels of setup times 30% and 70%, indicated as SDST_30 and SDST_70 in Table 2. All instances are solved by using the proposed IG algorithm. The convergence speed of IG algorithm is also examined by revealing the value of makespan changing with each iteration for problem instance EX81 with SDST_70 in Figure 2, which shows that IG algorithm reach convergence within 1000 iterations.

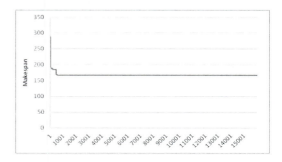

Figure 2. Convergence speed of IG algorithm for EX81 with SDST_70.

5 CONCLUSION

This study made the first attempt to study an integrated machines and AGVs scheduling problem with SDST which is closer to the realistic industrial situations, such as printing or textile industry, a label

303

Table 2. Computational results of problem instances with SDST obtained by the proposed IG algorithm.

Instances	SDST_30		SDST_70	
	C_{max}	CPU time	C_{max}	CPU time me
EX11	98	0.89s	103	1.00s
EX21	104	1.27s	108	1.24s
EX31	105	1.52s	112	1.52s
EX41	113	2.25s	117	2.25s
EX51	88	0.89s	89	0.89s
EX61	125	2.04s	130	2.01s
EX71	114	2.61s	118	2.59s
EX81	165	2.77s	167	2.64s

sticker company and other manufacturing systems, by considering complex scheduling constraints. The proposed integrated model could help reduce the total cost and the potential investment in the way of looking at the manufacturing process as a whole. An easy-to-implement and effective IG algorithm was developed to solve this integrated problem. Computational results show that our proposed algorithm is promising for solving the integrated scheduling problem. In the future research, the IG algorithm will be improved by incorporating sophisticated diversification and intensification strategies. The initialization method for AGV assignment will be modified via considering the workload balance of AGVs. Meanwhile, other heuristic algorithms will be designed to test out proposed IG algorithm and more experiments will be conducted to test our proposed method.

ACKNOWLEDGMENTS

The corresponding author is Dr. Yan Zheng.

This paper is supported by the Natural Science Foundation of Jiangsu Higher Education Institutions of China (Grant No.14KJD410001) and the National Social Science foundation of China (Grant No. 14BGL173).

REFERENCES

Abdelmaguid, T.F., Nassef, A.O., Kamal, B.A. & Hassan, M.F. 2004. A hybrid GA/heuristic approach to the simultaneous scheduling of machines and automated guided vehicles. *International Journal of Production Research* 42(2):0267–281.

Bilge, Ü. & Ulusoy, G. 1995. A time window approach to simultaneous scheduling of machines and material handling system in an FMS. *Operations Research* 43(6):1058–1070.

Bouamama, S., Blum, C. & Boukerram, A. 2012. A population-based iterated greedy algorithm for the minimum weight vertex cover problem. *Applied soft computing* 12:1632–1639.

Deroussi, L., Gourgand, M. & Tchernev, N. 2008. A simple metaheuristic approach to the simultaneous scheduling of machines and automated guided vehicles. *International Journal of Production Research* 46(8): 2143–2164.

Groover, M.P. 1987. Automation, production systems and computer integrated manufacturing. New York: Prentice-Hall.

Jacobs, L.W. & Brusco, M.J. 1995. A local search heuristic for large set covering problems. *Naval Research Logistics* Quarterly 42(7):1129–1140.

Naderi, B., Fatemi Ghomi, S.M.T. & Aminnayeri, M. 2010. A high performing metaheuristic for job shop scheduling with sequence-dependent setup times. *Applied soft computing* 10:703–710.

Ruiz, R. & Stützle, T. 2008. An iterated greedy algorithm for the sequence dependent setup times flow-shop problem with makespan and weighted tardiness objectives. *European Journal of Operational Research* 187:1143–1159.

Vis, I.F.A. 2006. Survey of Research in the Design and Control of Automated Guided Vehicle Systems. *European Journal of Operational Research* 170:677–709.

Zheng, Y., Xiao, Y. & Seo, Y. 2014. A tabu search for simultaneous machine/AGV scheduling problem. *International Journal of Production Research* 52(19):5748–5763.

Information, Computer and Application Engineering – Liu, Sung & Yao (Eds)
© 2015 Taylor & Francis Group, London, ISBN 978-1-138-02717-6

Exploration on literary form and value play under the age of multimedia

Xiao Dan Diao

Liaoning Jianzhu Vocational University, China

ABSTRACT: The development of literature is closely related to media technology development. In the process of literature writing and dissemination, mass media is an important medium and carrier. In today's multimedia age, new literary communication context is built and literary production, dissemination and consumption space are expanded. Literature gradually transfers from elite text to mass culture and literary form changes. This paper focuses on elaborating literary form under multimedia era and analyzing the value play in today's literature.

KEYWORDS: multimedia; literary form; value

Crystallization of human history and culture development is literature. Social form, media carrier changes lead to literary texts, literary form changes. Under the shock of multimedia mainly on the network, *Literary Extinction* has been raised once again. Literary value, style and existence profoundly change. In today's multimedia age, the way people accept literature and consume literary are diverse to provide space for the coexistence of various literary forms, so that the mass consumer culture and elite literature coexist. At the same time, the advent of digital existence way change the whole creation mechanism of literary. Network literature brings challenge for traditional literature aesthetic and value. In this context, make better use of the role of literature and be the responsibility of guiding values, to promote the prosperity and development of literature.

1 THE LITERARY FORM UNDER MULTIMEDIA AGE

1.1 *The elite literature is coexisting with popular literature*

Since entering the early industrial era, some communication scholars and literary scholars proposed that *Literary Extinction* and the existence and responded with a negative attitude to the development of literature. In the 21st century, the global economy developed rapidly, people's lives change and the entertainment and commercialization of era culture and literature. At the same time, under the shock of multimedia, such as TV, Internet, traditional literary marginalized. However, *Literary Extinction* remarks a bit sensational. Throughout the entire development process of literature, the enhancement of every human material transforms and impact traditional literature, but traditional literature still occupies an important role in our culture. In multimedia age, people watch drama, film and television work through television, network and read popular literature for the purpose of resolving survive the pressure and recreational, popular literature has some readers and consumer groups [2]. Meanwhile, the society and its level of civilization continue to develop and enhance, people's income gradually increase, higher education is universal, the middle class is expanding. In this case, more and more people pursuit high-quality cultural literature products. Literary elite readership expanded. For example, in today's multimedia age, some writers such as Tong Hua, Nan Pai San Shu, have a lot of readers through the network literature, whose novels are sought and was taken into the film and television works. Meanwhile, contemporary famous writers' novels, such as Mo Yan, Liu Zhenyun, Yu Hua, Wang Shuo, also have a large readership, popular literature and literary elite coexist.

1.2 *The combination of literature and film and television*

The modes, such as culture producing, accepting and communication are influenced by mass media. The mass media as a medium, carrier, play a crucial role in the literary communication, dissemination. As a form of culture, literature needs to use the media to express its full meaning, to achieve communication with readers, dialogue. Previously, literature spread mainly through paper media to realize their value. With the development of multimedia, literature gradually by means of imaging media demands and communication and literature morphological changes. Under the impact of multimedia, literature appeal by means of image medium and further highlights the visualization and figurative of literature. At the same time, when the film art meet multi-sensory pleasure of the audience, the status of traditional paper media declines. The writers expressed concern about the significance of literature, the development of literary forms film depth. Faced with this situation, the combination of literature and film is essential, which provides a new way to convey the literary value [3]. For example, some literature classic works were remade as a television drama, such as Yu Hua's *To Live,* Mo Yan's *Red Sorghum.* The public feels the emotional feelings of the characters, the author thought through television works and spread to make a wider audience aware of these classic literature works through the media and reread literature. Meanwhile, the interpretation of film and television media of literature is more specific and image, eliminating the haziness of literary image and meaning, so that the audience understand the implication behind the literature.

1.3 *Digital media rewriting literature*

Throughout the history of literature development, with the evolution of world situation and the passage of times literature, the form and content change. As for our literature, the poetry developed from ancient poetry in free verse, the article developed from classic to the vernacular and literary in every age express emotion and value in every age [4]. In multimedia era, subject to the impact of new media networks, digital literature is born and developed. Through the written words, the digital literary make language text evolves by new media arts symbols, such as sound, animation, text, pictures. Meanwhile, the openness, interactivity and convenience of the network provide a variety of literary transformation template for literary. Blog, dedicated literary site spring up in large numbers, which blurs the definition of writers and readers, reducing the threshold for publishing works, and develop a diversified, open. For example, through the network platform, people can make their own network of text editing, design, publish, and according to the text according to the network node integrated book, through the blog, paste it, the text published on the site.

2 THE LITERARY VALUE PLAY IN THE MULTIMEDIA ERA

2.1 *Highlight the idea spiritual values and reflect the spirit of China*

In the current various literary forms, the spread of literature are more or less with the aid of multimedia. Through multimedia communication, literary value changes occur literary value. For example, in network

literature, network convenience and openness make literary works diverse and digest traditional literary noble and serious dignity, and the status of the public raise and they have more right to speak, in the Hubbub, people writing interpret, create. Thus, people are free to express emotion through the network and explain the *writings are for conveying truth.* Feature of traditional literature. The network literature is full of banter, violence, catharsis, pornography and so on. The value of literature was threatened [5]. In this case, literary is needed to pay more attention to their own value sense and spread sense, highlight their ideological, spiritual values and need the writer to write and convey the right idea through media and network. Meanwhile, in literature reviews, readers should not do values valuation by the standard of readership, but by thought of spiritual, national culture embodied the spirit of life to participate in the national standard, and gradually change the writer and public understanding of literature, identity and highlights the idea spiritual values of literature.

2.2 *Spread Chinese culture and inherit literary tradition*

Literature is a culture, which is an important form of cultural transmission. Currently, literary forms and work type are rich but most of the works are public literature, literary tradition lacks of inherit-ion, Chinese cultural transmission is weak. In our literature, its greatest value is to spread Chinese culture and to raise awareness and spirit of Chinese culture, such as Lu Xun, Lin Yutang, Ba Jin, whose works documented Chinese phenomenon in that era and spread Chinese culture, resulting in the world effects. Therefore, in today's society, the writers should focus on theme selection, absorb and inherit Chinese fine literature traditions to form cultural support awareness [6]. Meanwhile, in the combination of literature and film, you should adapt the best literary works onto the silver screen to expand the scope of the audience, such as the film adaptation of Water Margin, Dream of Red Mansions, which spreads the Chinese culture and expands dissemination and influence of the work.

3 CONCLUSION

In multimedia era, classic literature does not end and it shapes and spreads by a greater variety of literary forms, such as film and television cooperation, digital rewritten and continue to be memorized. Meanwhile, in a variety of literary forms, contemporary writers need to assume responsibility for its literary value propagation and the improvement of cultural awareness to create more great work with Chinese culture, highlighting Chinese cultural values and improve the aesthetic of readers.

REFERENCES

[1] Feng Zengyu. *Literary Forms and Cultural Values Play in Multimedia Era* [J] Modern Media, 2013, 35 (4): 61–63, 67.

[2] Chu Xiaomeng. *Literary Transformation under New Media Era* [J] Henan Institute of Engineering (Social Sciences), 2013, 28 (4):78–81.

[3] Dong Xiwen. *Transmutation of Media Carrier, Text Form and Literature Idea*[J] Central Plains Journal, 2011, (2):230–234.

[4] Zhang Xiaofei. *Does Literature Really Dead or Suspended Animation? - Reread Hillis Miller's Is Literary Dead yet in Multimedia Context*[J]. Anhui literature (the second half), 2011, (2):116–117, 127.

[5] Yu Shenghua. *Exploration on Microblogging Literature Features*[J]. Prose parterre and Education Parterre, 2013, (11):12–13.

[6] Yao Feifei. *The New Media Environment and Moving Literature*[D]. Nanjing Normal University, 2011.

Information, Computer and Application Engineering – Liu, Sung & Yao (Eds)
© 2015 Taylor & Francis Group, London, ISBN 978-1-138-02717-6

Enhance vocational college counselors' information literacy in multimedia network background

Yong Hong Tao

Liaoning Jianzhu Vocational University, China

ABSTRACT: As China's economy continues to develop and technology continues to progress, network multimedia is used more and more widely in modern society. With the implementation of education reform, multimedia are affecting college teaching all the time, particularly, vocational college teaching has played a significant role in the promotion of the overall teaching environment. Faced with the rapid development of Internet and multimedia applications in schools have become a major teaching force after the new curriculum, so higher vocational counselors need to use this new teaching model to form good information literacy, how to improve the vocational counselors' information literacy is very important, but also an important factor to improve the modern teaching. This article focuses on the promotion of vocational counselors' information literacy in multimedia network backgrounds.

KEYWORDS: network multimedia; counselors' information literacy; enhance

1 INTRODUCTION

Since ancient times, college counselors are a major force in carrying out teaching content, with the ongoing reforms, counselors' information literacy has been largely unable to meet the development of modern education, we must promote information literacy, making the counselors' information literacy adapt to the development of modern education to continue to lead the students to continue to learn and develop. Counselors are college students' mentors, teaching and serving students, as a guide for students and ideological education pioneer within the university campus, their own actions demonstrate the importance of health knowledge dissemination and management. Having entered the network multimedia teaching era, knowledge increases, the entire learning environment makes greater demands and challenges for counselors, forcing them to enhance information literacy.

2 THE SUMMARY OF INFORMATION LITERACY

The information literacy concept was proposed in 1973, refers to the use of large amounts of information resources for effective problem solving skills. In 1989, the American Library Association for Information Literacy was further generalization [1], a country with high information literacy people, we must determine when to use information resources, and effectively query and use when utilizing it.

For higher vocational counselors, it would not be regarded as improving information literacy in terms of basic computer operations, it's just having the basic skills to get information, but also has the ability to query for information and use the query process useful information has become his thoughts. It should also be capable to identify information. Therefore, the core of information literacy is the ability to process and use information. Able to correctly evaluate the information and have a good grasp of the characteristics and application occasions of information. They should be able to draw inferences from the use of information resources.

3 THE IMPORTANCE OF ENHANCING VOCATIONAL COUNSELORS' INFORMATION LITERACY IN MULTIMEDIA NETWORK BACKGROUNDS

3.1 *Changes in education primary object relations by multimedia*

Network multimedia applications have brought great convenience and change to the school education, strongly impacting the traditional teaching model, in traditional education, education subjects and students have divided obvious status. In Vocational, the counselor symbolizes authority, students are educated. With the traditional teaching mode after being replaced by network multimedia, teaching lost its hierarchical nature [2]. With modern teaching high school students from passive to active learning through the sharing of resources so that counselors and students enjoy the same resources, students largely able to break through the traditional teaching model, to gain more knowledge in different ways. Access to information and allows instructors and students to use in an equal position. Faced with massive information network, counselor to be adequate learning, in order to enhance their own, if it does not, the birth of new information generated by the new things do not understand, it is easy during teaching in an awkward situation, a large is a challenge to the authority of the instructor degree. Therefore, the counselor must learn the network on the basis of a comprehensive understanding of network information and learning, in order to get more information resources and knowledge in today's society the latest developments. Changes in education primary object relations by multimedia provide good facilities to improve counselors' information literacy, which is significant.

3.2 New media to provide a new education path for counselors

Multimedia network has been popular in our country, communication tools are more and more on the network, there is a good platform for teaching in the performance, convey a message through different sensory stimuli to the human body, providing a very good path for the counselors to better teach, which updates the traditional propagation path.

3.3 New media enhance the timeliness and relevance of teaching

Continuous development of information technology makes every aspect of people's lives more convenient. This makes teaching more focused largely on the extent. Network multimedia teaching has great immediacy, counselors use of multimedia for students to answer questions effectively, so that work more effectively, and improve work efficiency. Whether it is various viewpoints expressed by people in the network or their feelings expressed in we-chat, as long as counselors are concerned about this, they are able to become effective information, to be concerned about the situation changes, and to respond effectively [3].

4 APPROACH TO ENHANCE VOCATIONAL COUNSELORS' INFORMATION LITERACY IN MULTIMEDIA NETWORK BACKGROUNDS

4.1 Reverse the concept, improve information awareness

Many vocational college counselors recognize the importance of information literacy, it is a very convenient channel to get through the network, network data and audio files to be accepted by more and more counselors. Therefore, network multimedia, the counselor by changing the traditional concept of the use of information technology networks and improve their information literacy is a priority. While awareness of the culture of information required to optimize the information environment, we should create network information conditions conducive to improve information awareness, making people have sensitive awareness of the information environment while using resources [5].

4.2 Strengthen the capacity of information

In the information age, the information itself is not important, it is important that how people use network technology to obtain information, and the information processing and evaluation of the information capacity of innovation is the ability of information, which is information literacy in a very important part, for example, counselors during the information search when they encounter problems, showing the kind of emotion that this unhappy mood reflects the information on the ability of weak, so the counselors need to improve the network of information sources and search capabilities to strengthen its information capacity. Colleges need to conduct necessary network information training for counselors, education authorities need to make plan and measures according to the specific circumstances, use different forms of training tools to improve information capabilities of counselors.

5 CONCLUSION

When enhancing information literacy, counselor hope all of our instructors can set up a suitable counselor information literacy research council, have a good evaluation criteria and assessment for enhancing information literacy. Establish a relatively complete information literacy system, this system contains all the upgrade process for counselors. I hope policymakers can make effective decisions, build virtualized information environments, at the same time the concepts of education in universities need to innovated. Conduct perfect planning for counselors' individual development, applicant interactive platform while increase training. Through the continuous improvement of the system and establishment of evaluation mechanism, carry out necessary optimization for the network environment better and enhance information literacy of counselors.

REFERENCES

[1] Liu Hong. *Thoughts on Vocational College Students under Network Culture Information Literacy Education* [J]. Education exploration, 2010, 05:26–27.

[2] Qiu Ping. *College Counselors to Enhance Information Literacy in the New Media Contexts* [J]. Nanchang College of Education, 2010, 03:79–80 + 82.

[3] Zhu Yicai. *Predicament Vocational Counselors Development and Countermeasures - Based on Individual Survival Status and Career Planning* [J]. Education Forum, 2010, 06:71–75.

[4] Feng Bo. *Information Literacy of College Counselors under New Media Environment Research* [D]. Northeast Normal University, 2013.

[5] Yu Aihua. *Enhance the Physical Education Teachers in Vocational Schools Information Literacy Strategy* [J]. China Educational Technology, 2014, 06:127–129 + 134.

Information, Computer and Application Engineering – Liu, Sung & Yao (Eds)
© 2015 Taylor & Francis Group, London, ISBN 978-1-138-02717-6

Wildfire smoke detection method with YUV-constraint and texture-constraint

W.C Zhang , J. Liu , P. Li & C.Q. Gao
Chongqing University of Posts and Telecommunications, Chongqing, China

ABSTRACT: The wildfire smoke detection plays an important role in early warning system of forest fire because smoke often appears in the surveillance video before flame. But the conventional motion detection method used in wildfire smoke detection is not efficient in following situation. The smoke is moving very slow or there has not a clear background. In order to solve these problems, a specific for the movement of the smoke detection with YUV-Constraint and Texture-Constraint is proposed. In this method, the contact between smoke brightness and intensity of illumination assumes as a linear estimation. We used SVM as a linear regression to obtain the approximate range of smoke brightness, and defined this as a YUV-Constraint. Based on detecting motion region, we can spread out the main part of smoke region by considering the combination of YUV-Constraint and Texture-Constraint. Then some features based region has been extracted for SVM classify. Experimental results show our method outperforms the conventional methods, and have more robustness.

KEYWORDS: Smoke detection; YUV-Constraint; Texture-Constraint; SVM

1 INTRODUCTION

Wildfire has been a long-term issue of public concern, and the method of effective monitoring of wildfire has been extensively studied. Smoke detection is the main method of monitoring of wildfire, because in wildfire situation smoke always appears on the early stage of a fire.

Recently, because of the development of video surveillance technology, the methods of vision-based smoke detection have been widely proposed. Yu Cui et a.l[1] use Gray Level Co-occurrence Matrices (GLCM) based methods to extract multi-scales texture features of smoke. But their study is only for a single image. Toreyin et al.[2] study the smoke features in the wavelet space, and Jiaqiu Chen et al.[3] improved their method by a sample RGB-constraint, they remove interference of solid objects. Hongda Tian et al.[4] detect the smoke by using Local Binary Patterns (LBP) in the gray space, Angelo Genoves et al.[5] fuse various features based region to detect smoke. B.C Ko et al.[6] extract the Histogram of Oriented Gradient (HOG) for detecting smoke.

In general, smoke detection is divided into two steps: (i) detect motion region. (ii) Identify whether motion region is smoke or not. For the first step, the general method is improved Kelman filter[2][3][5] or Gaussian Mixture Model (GMM)[4]. For the second step, the conventional method is extracting features of smoke to train a classifier, and using the classifier to distinguish the smoke region.

However, these methods may not work very well in the following situations: (i) Smoke is moving very slowly. (ii) The motion region, which is obtained through background modeling may be incomplete. (iii) There is already exist smoke in the background. Above methods based on a hypothesis that is the first frame don't exist smoke, because they want a clean background.

In order to cope with the possible above conditions, we propose an improved motion detection method with adaptive local smoothness model. This model is divided into two steps: (i) the statistical YUV-constraint is applied to obtain a certain size region. (ii) According to the texture feature of the obtained region, further spread out the entire candidate smoke area. The advantage of this method is that it can more precise segment the candidate smoke area compare with the simple morphological processing. For the each segmented candidate smoke regions. The texture, color, shape features are extracted from each region to support the analysis.

The remainder of this paper is organized as follows: The general smoke detection processes with the emphasis on the improvement of the proposed method are presented in Sect.2. Experimental results and discussions are shown in Sect.3. Finally, some conclusions and perspectives are described in Sect.4.

2 PROPOSED FRAMEWORK

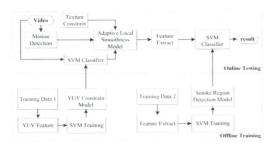

Figure 1.   Framework.

As shown in Fig. 1. This framework can be divided into two stages: the offline training stage and the online testing stage. In the training stage, we train two SVM models for candidate smoke region detection and smoke region detection respectively. In the testing stage, firstly, we adopt the conventional motion detection to generate the motion region, Secondly, we use an adaptive local smoothness model with YUV-Constraint and Texture-Constraint to spread out the main part of the candidate smoke region. Finally, we extract the features based the candidate region, and then use corresponding SVM classifiers to identify whether this region is smoke or not.

2.1   Detect motion regions

In this paper, considering the background may not be clean, we choose to be using the background estimation method developed by Collins et al. [7] instead GMM. But the essence of this method is a Kelman filter as previously described, This method can still be improved upon for adapting to the specific situation. Firstly, the foreground image is obtained with Eq. (1)

$$M_k(i,j) = \begin{cases} 1 \ \ if \ \left| I_k(i,j) - B_k(i,j) \right| > Thb \\ 0 \ otherwise \end{cases} \quad (1)$$

Where M_k denotes the pixel position of motion region, I_k denotes the current frame, B_k denotes the current background, Thb is a threshold which used to determine the pixel's location is a motion pixel and turn n-th frame into a binary image. The Thb is in the range of 20 to 40. In this paper, we set it to 30. When $M_k(i,j)$ is set to 1, it means this pixel belongs to motion regions, the value 0 represents the background. Then we can use Eq. (2) to update the background.

$$\begin{cases} B_{k+1}(i,j) = B_k(i,j) + g_k(I_k(i,j) - B_k(i,j)) \\ g_k = \beta(1 - M_k(i,j)) + \alpha M_k(i,j) \end{cases} \quad (2)$$

Where B_{k+1} denotes the updated background, g_k denotes the factor, β denotes the background factor, α denotes the motion factor. α and β are restricted to between 0 to 1. In this paper, we set $\alpha = 0.55$, $\beta = 0.1$.

2.2   Adaptive local smoothness model

2.2.1   YUV-constraint
According to the two fundamental atmospheric scattering models (the attenuation and airlight models) (Narasimhan and Nayar [8]), assuming that there are no specific point sources of light and the scattering coefficient of the smoke does not change appreciably within the visible wavelength, we can speculate the luminance information of smoke region is correlative to the luminance information of the whole image. Due to the input video is RGB color space, we need to convert the frame I_n from RGB to YUV color space by using the following formulas:

$$\begin{cases} Y = 0.299R + 0.567G + 0.114B \\ U = -0.147R - 0.289G + 0.436B \\ V = 0.615R - 0.515G - 0.100B \end{cases} \quad (3)$$

Then, let $f_k \in \Re^N$ mean luminance information of image at time k, it can be modeled as a linear blending of s_k and b_k, as follows:

$$f_k = \sigma_1 s_k + \sigma_2 b_k + n_k \quad (4)$$

Where $n_k \in \Re^N$ represents modeling noise, $b_k \in \Re^N$ represents the luminance information of background (or as if no smoke exists), and $s_k \in \Re^N$ is the luminance information of smoke region. σ_1 and σ_2 are fitting coefficient. So we can rewrite the formulas (5) as follow:

$$s_k = \frac{1}{\sigma_1} f_k + \frac{\sigma_2}{\sigma_1} b_k + \frac{1}{\sigma_1} n_k \quad (5)$$

We can use linear SVM to parameter fitting, and the $\left[\hat{s}_{k_{min}}, \hat{s}_{k_{max}} \right]$ will be obtained which means the ranges of luminance information of smoke in k frame. This range can be used as a constraint condition in adaptive local smoothness model.

2.2.2 Texture constraint

Generally speaking, the smoke is always gray, and its brightness has consistency and smoothness. Angelo Genovese et al. [4] think the luminance value of smoke regions should be higher than a threshold. In contrast, the chrominance values should be very low. But their constraint conditions depending on the threshold which is experimentally determined, this approach will bring more confidence risk. Through the observation we found: (i) In Y channel, there has not much difference between the adjacent two pixels in smoke region. (ii) The U channel and V channel almost does not contain any useful information,

By taking full consideration of the consistency and smoothness, we decide to use the texture feature of smoke YUV space as a constraint condition as follows:

$$
\begin{cases}
Y_{max} = u_\Omega + \varepsilon_1 v_\Omega \\
Y_{min} = u_\Omega - \varepsilon_2 v_\Omega \\
U < T_U \quad and \quad V < T_V
\end{cases} \tag{6}
$$

Where Y, U and V are the luminance and chrominance values of a particular pixel, respectively. Ω is the detected motion region of luminance space. u_Ω is the mean value of the pixel in Ω, v_Ω is the variance value of the pixel in Ω. ε_1 and ε_2 are constant coefficient, the range is $\varepsilon_1 \in [2,4]$ and $\varepsilon_2 \in [0,1]$. In this paper, we set $\varepsilon_1 = 3, \varepsilon_2 = 0.5$, and the thresholds T_U and T_V are experimentally determined.

2.2.3 Local smoothness model

Through the preliminary estimation of motion, we can get a motion region M. Carry out the following operations on the M, we can get a candidate smoke region $\bar{S}$ which is a binary image:

$$
\bar{S}(x,y) =
\begin{cases}
1 \ if \ s(x,y) \in [\hat{s}_{k_{min}}, \hat{s}_{k_{max}}] \ and \ M(x,y)=1 \\
0 \ otherwise
\end{cases}
$$

Then we use the structure element A to expand $\bar{S}$ as follows:

$$
\hat{S} = \bar{S} \oplus A = \left\{ s \,|\, (\hat{A}_s) \cap \bar{S} \neq \varnothing \right\} \tag{7}
$$

$$
where \ A = \begin{bmatrix} 0 & 1 & 0 \\ 1 & 1 & 1 \\ 0 & 1 & 0 \end{bmatrix}.
$$

This expand process is carried out until the number of iterations reaches a predefined value that we can obtain a certain size of an area which has a more clear texture and morphology features.

Then we do the same process in the texture Constraint condition as follows:

$$
S(x,y) =
\begin{cases}
1 \ if \ s(x,y) \in Texture - Constran \ and \ \hat{S}(x,y) = 1 \\
0 \ otherwise
\end{cases}
$$

$$
S_f = \bar{S} \oplus A = \left\{ s \,|\, (\hat{A}_s) \cap \tilde{S} \neq \varnothing \right\} \tag{8}
$$

The second expand process can spread into the main part of the candidate smoke region. It is carried out until u and v exceed a predefined value or the number of iterations reaches a predefined value.

2.3 Feature extraction

2.3.1 Smoke-color feature

According to early studies, we know the smoke is always grayish, which means in RGB space, the R, G and B values of smoke pixels are very close to each other. We choose the red-green(RG) and blue-yellow(BY) contrast which proposed by Walther et al.[9] used in our framework as shown in the following formula:

$$
RG = \frac{r-g}{\max(r,g,b)}, \quad BY = \frac{b - \min(r,g)}{\max(r,g,b)} \tag{9}
$$

Where RG and BY are the red-green and blue-yellow contrast respectively.

2.3.2 Permeter disorder feature

We choose the permeter disorder feature described in Angelo Genoves et al[5]. First, we can obtain a binary image S_f. And for each distinct 8-connected region of this binary image, the area and perimeter are computed. The perimeter disorder value for each region is computed as:

$$
R_i = \frac{P_i}{A_i} \quad \forall \ 1 < i < N \tag{10}
$$

Where R_i is the perimeter disorder value of the i-th blob (8-connected region), P_i and A_i are the perimeter and area of the i-th blob, and N is the number of the blobs.

2.3.3 *Texture features*

We choose three texture features as follows:

Entropy (ENT) can express the complexity of image:

$$ENT = -\sum_i \sum_j p(i,j) \log p(i,j) \qquad (11)$$

Contrast (CON), which is a measure of the image contrast or the amount of local variations present in an image, is given by:

$$CON = \sum_i \sum_j (i-j)^2 p(i,j) \qquad (12)$$

Smoothness (SMO) can express the relative smoothness brightness in the region:

$$SMO = 1 - \frac{1}{1+\sigma^2} \qquad (13)$$

where σ^2 is variance.

2.4 *Classification*

After region-based features are extracted, the feature vectors of positive and negative samples are used to train a classifier which will be further used to classify the new region. We selected Support Vector Machine (SVM) as the classifier to perform smoke detection.

3 EXPERIMENTAL RESULTS

The experiment settings are MATLAB R2013a. We use 7 video sequences to test the proposed method.

These video sequences can easily obtain in two open database which are Toreyin et al's[2](http://signal.ee.bilkent.edu.tr/VisiFire/) and B.C. Ko's.[6](http://cvpr.kmu.ac.kr).

The global evaluation measurements proposed by Jakovcevic et al.[10] are used to evaluate the proposed method. In this paper we choose true positive rate(TPR) and false positive rate (FPR) to evaluate our method is shown as follows:

$$TPR = \frac{TP}{TP+FN}, \quad FPR = \frac{FP}{FP+TN} \qquad (14)$$

Where TP (true positive) is the number of correct detections of smoke; FN (false negative) is the number of frames which have smoke but not recognized. FP (false positive) is the number of frames which does not have smoke but recognized as smoke; TN (true negative) is the correct detection of non-smoke. We also considerate the frame number of every test video and compare our method with others.

In experiments, 156 positive and 323 negative samples are selected to construct the training data 2 to train a Smoke Region Detection model. The result is shown in Table. 1. The '-' in Table 1 means author does not do experiment on the video.

Through the ATPR and AFPR of five methods we can see our method's performance slightly better than Hongda Tian's method, and equal to Jiaqiu Chen's and B.C.Ko's method. And the sum frame number of our method is 13000 that significantly more than the others. Specifically, our method has more effective than B.C.Ko's method in Video 1, because the TPR is 97.25% to 95.5%. The reason may be that the smoke is moving very slow in the Video 1. In addition, our method has more effective than Hongda Tian's method in Video 7,

Table 1. Comparison of experimental results of this method with other methods

Evaluation index	TPR(%)			FPR(%)			Frame		
	B.C.Ko's	Hongda Tian's	Ours	B.C.Ko's	Hongda Tian's	Ours	B.C.Ko's	Hongda Tian's	Ours
Video 1	95.5		98.3	0		0	781	-	2000
Video 2	100	Average: 97.69	99.13	0	Average: 1.165	0	475	-	1500
Video 3	-		99.65	-		0	-	-	2000
Video 4	100	98.71	99.7	-	-	0	520	2000	2000
Video 5	-	99.61	99.4	-	-	0		3000	3000
Video 6	100	94.59	99.77	5.2	-	0	735	2500	2500
Video 7	-	91.68	99.6	-	-	1.77		3000	3000
Average	98.875	96.92	98.9	1.73	1.165	0.25	628	2625	2286

because the TPR is 98.47% to 91.68%. This reason is that Hongda Tian's method needs 265 frames for background learning, instead we can perform on the first frame. Our method has 1.77% FPR in Video 7, this is because there has a white bright house in the video sequences. Our method may classify the house as smoke because of the drastic illumination variation in a few frames.

4 CONCLUSIONS AND PERSPECTIVES

In this paper, we introduce the Adaptive Local Smoothness model which can effectively detect the smoke in forest background. We can obtain a more complete smoke region to improve the detection method's robustness. Based on obtaining candidate region, we can get a good performance by using sample SVM classifier as shown in our experiments. But during we do the experiments, we find the test video sequences may slightly simple, we hope we can do experiments on more complex video. Further, we would improve our method to adapt the sky background.

ACKNOWLEDGEMENTS

This work is supported by the National Natural Science Foundation of China (No. 61102131, 61275099), the Natural Science Foundation of Chongqing Science and Technology Commission (No. cstc2014jcyjA40048), Cooperation of industry Education and Academy of Chongqing University of Posts and Telecommunications (No. WF201404), Undergraduate research training program (No. A2014-39).

REFERENCES

[1] Cui, Y., Dong, H., & Zhou, E. (2008, May). An early fire detection method based on smoke texture analysis and discrimination. In Image and Signal Processing, 2008. CISP'08. Congress on (Vol. 3, pp. 95–99). IEEE.

[2] Toreyin, B. U., Dedeoglu, Y., & Cetin, A. E. (2005, September). Wavelet based real-time smoke detection in video. In European Signal Processing Conference (pp. 4–8).

[3] Chen, J., Wang, Y., Tian, Y., & Huang, T. (2013, November). Wavelet based smoke detection method with RGB Contrast-image and shape constrain. InVisual Communications and Image Processing (VCIP), 2013 (pp. 1–6). IEEE.

[4] Tian, H., Li, W., Ogunbona, P., Nguyen, D. T., & Zhan, C. (2011, October). Smoke detection in videos using non-redundant local binary pattern-based features. In Multimedia Signal Processing (MMSP), 2011 IEEE 13th International Workshop on (pp. 1–4). IEEE.

[5] Genovese, A., Labati, R. D., Piuri, V., & Scotti, F. (2011, September). Wildfire smoke detection using computational intelligence techniques. In Computational Intelligence for Measurement Systems and Applications (CIMSA), 2011 IEEE International Conference on (pp. 1–6). IEEE.

[6] Ko, B., Kwak, J. Y., & Nam, J. Y. (2012). Wildfire smoke detection using temporospatial features and random forest classifiers. Optical Engineering, 51(1), 017208–1.

[7] Lipton, A., Kanade, T., Fujiyoshi, H., Duggins, D., Tsin, Y., Tolliver, D., ... & Wixson, L. (2000). A system for video surveillance and monitoring (Vol. 2). Pittsburg: Carnegie Mellon University, the Robotics Institute.

[8] Narasimhan, S. G., & Nayar, S. K. (2002). Vision and the atmosphere.International Journal of Computer Vision, 48(3), 233–254.

[9] Walther, D., & Koch, C. (2006). Modeling attention to salient proto-objects.Neural networks, 19(9), 1395–1407.

[10] Jakovčević, T., Šerić, L., Stipaničev, D., & Krstinić, D. (2010). Wildfire smoke-detection algorithms evaluation. In VI Int. Conf. Forest Fire Res (pp. 1–12).

Information, Computer and Application Engineering – Liu, Sung & Yao (Eds)
© 2015 Taylor & Francis Group, London, ISBN 978-1-138-02717-6

Design of the light information remote monitoring system for sugarcane fields based on Zigbee and Labview

Xiu Zeng Yang & Li Hui Li
Guangxi Normal University for Nationalities, Chongzuo, Guangxi, China

ABSTRACT: A set of light information remote monitoring system for sugarcane fields is designed for studying the photosynthetic characteristics of sugarcane fields, in which the technologies of both wireless sensor network and labview programming are adopted. The system is composed of four parts which are acquisition nodes, routing nodes, data sinking nodes and the PC. The sensor of BH1750 of collection nodes is used to collect light information in the sugarcane field, and the technology of labview programming is applied to program the monitoring routine. The test result shows that there are many advantages about the system, such as convenience and high measuring accuracy and so on.

KEYWORDS: Photosynthetic characteristics; CC2430; Protocol stack of ZStack2006; BH1750

1 INTRODUCTION

Sugarcane is one of an important sugar crops. According to statistics, sugar production accounts for about 60% of the world's total sugar and about 80% of China[1]. The dry substance and biological yield of sugarcane are mostly produced by photosynthesis of sugarcane leaves[2-3]. It is concluded that the photosynthetic efficiency is one of the important factors to affect sugarcane production. In order to facilitate the analysis of photosynthetic characteristics of sugarcane plants, a set of light information remote monitoring system for sugarcane farmlands is designed, in which the Zigbee wireless sensor network and Labview programming technology are both used. The system can real-time collect, display and store the light information of sugarcane fields, which provides the original data for further study on the photosynthetic characteristics of Sugarcane.

2 THE OVERALL DESIGN SCHEME OF THE SYSTEM

Figure 1 is the overall design diagram of the remote monitoring system of the light information collection of sugarcane. The system can be divided into light perception layer, transport layer and user layer on the basis of the system's function. The design scheme is composed of four parts of light information collecting node, routing nodes, sinking node and the PC. To get accurate illumination information in a sugarcane field, some light information collection nodes were distributed in the different points in the planting

regions of sugarcane, and the high precision digital light sensor BH1750 is installed in each light information collection node that can automatically collect light information in different points in the field. The light information collected by different collection nodes are automatically transmitted to their father nodes. The router node powered by the solar cell is active all time, and can relay light information to sinking node. The light information received by sinking node is packaged into a certain data frame and is transmitted to the PC. The PC stores, analyzes, and displays the light information.

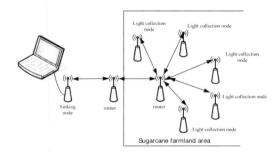

Figure 1. The overall design scheme of system.

3 HARDWARE DESIGN

3.1 *Hardware design of collection node*

Figure 2 is the hardware design diagram of light information collection nodes of the system. It is composed of a piece of solar battery board, charging management chip of LT1303, rechargeable battery pack, the light sensor of BH1750 and CC2430. In order to reduce maintenance cost of the system and improve the stability of the system, a solar radial cell with 10 watts is adopted to power the light information collection node. The chip of LT1303 is applied to improve the charging effect for radial solar cell, in which it can convert an instable input voltage to the fixed voltage suitable for solar energy storing. The CC2430 is applied in the system for reducing power consumption. The CC2430 is a true system-on-chip solution for a system in which low power consumption is required. The CC2430 combines the CC2420 RF transceiver with leading excellent performance with an industry-standard enhanced 8051 MCU, 32/64/128 KB flash memory, 8 KB RAM and many other powerful features. Combined with the industry leading ZigBee™ protocol stack (Z-Stack), the CC2430 provides the market's most competitive ZigBee™ solution.

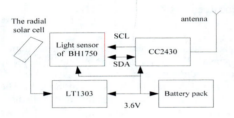

Figure 2. The block diagram of hardware design of collection node with solar energy.

3.2 The design of application circuit of BH1750

In order to improve the precision of light data acquisition, a high precision optical sensor named BH1750 is adopted in light collection node. BH1750 is an optical sensor with full digital high precision which can achieve functions of optical signal acquisition, optical signal amplification and an A/D conversion. The sensor has the similar spectral characteristics just as the human eyes. The highest resolution of the A/D converter of BH1750 can be set up to 16 bits, and it can be set to run on high, medium and low resolution models of instruction. In order to reduce the volume of the device, the BH1750 also uses I2C communication protocol to communicate with external CPU.

4 THE INTRODUCTION OF ZSTACK2006 PROTOCOL STACK AND TRANSPLANTATION

4.1 The introduction of ZStack2006 protocol stack

The Zigbee communication technology, which is based on the standard of IEEE802.15.4 is a duplex wireless communication technology. The Zigbee Alliance

released the first Zigbee protocol specification in 2005, which defined the physical layer, medium access layer, network layer and application layer in the protocol specification. ZStack2006 protocol stack must be transplanted into the system for facilitating different logical types to build a wireless sensor network. Before transplanting success, the principle of the protocol must be read and understood.

4.2 The process of starting and establishing a network

The sinking node of system is essentially a coordinator of the wireless network. The coordinator of wireless network is the first device in wireless sensor network, and can choose a channel and a network identifier (also called PAN ID) and then starts the network. Figure 3 is the flow chart that demonstrates the process of Coordinator how to start and establish the wireless network through the ZStack2006 protocol stack. When the coordinator is powered up, the ZStack2006 protocol stacks firstly carries out a series of initialization operation for the device, and then the wireless sensor networks are established according to the certain network parameters in the network layer. After the network is successfully established, the operating system

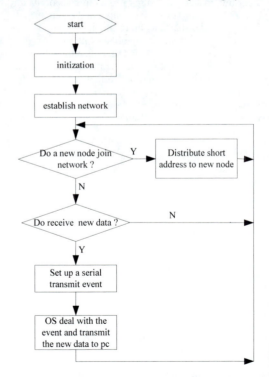

Figure 3. The flow chart of coordinator, starting and establishing the wireless network.

320

is running. The operating system is the multitask query type, operating system with different priorities, and its priority level is gradually decreased according to the order of a physical layer, medium access layer, network layer and application layer. If a new node wants to join the network, the network layer will set up a new node joining event. When the operating system queries the node joining event in the queue, it will run a certain program to assign a network address with 16 bits of the new node, and allow the node to join the network. If a new packet data is received by the coordinator, a serial transmit data event will be set up by protocol stack. After the serial transmit data event is handled by the operating system, the new packet data will be transmitted to a PC.

4.3 The process of sensor nodes joins the network

The sensor nodes distributed in the field collect the light information automatically. Figure 4 is the flow chart about the sensor nodes how to find and join the wireless network. When powered up, some devices of the sensor node are initiated by ZStack2006 protocol stack, and then the sensor node executes to seek network routine. When the sensor node finds a wireless network successfully, the sensor node executes routine to join the wireless network. After the sensor node entering the network is successful, the sensor node periodically sets up a light collection event, and

operating system deals with the light collection event to collect light information on different points.

5 THE DESIGN OF GRAPHICAL USER INTERFACE

Figure 5 is the graphical user interface on a PC which is divided into three child windows, namely node management child window, node status child window and user login child window. The user can enter into the system by a legal user name and a password in the user login child window. The node status displays child window status of nodes registered successfully in the system. Red node represents that the node is working while white node represents the node is wrong. The detail light information of node can be available, if we click one of the nodes in the node status child window. Figure 6 is the child window where the detail node's information is displayed. The legal user can click the button on the serial port parameter to set to modify serial port parameters, and click optional card on the top of the window to display light information about one of nodes in the field, such as historical data displaying, real-time data displaying and analyzing data displaying.

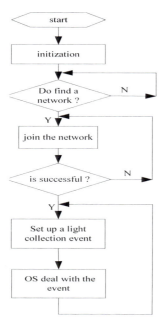

Figure 4. The flow chart of sensor nodes starting and joining the wireless network.

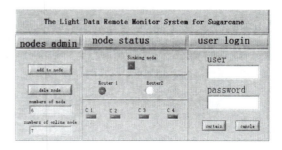

Figure 5. The graphical user monitor interface.

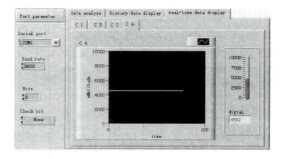

Figure 6. The graphical user monitor interface of detail information of node.

321

6 CONCLUSIONS

The test result shows that the system described above is applied in studying the photosynthetic characteristics of sugarcane fields.

ACKNOWLEDGMENTS

The corresponding author of this paper is Li Hui Li. This paper is supported by University scientific research project funding projects of the Guangxi Zhuang Autonomous Region (project number: 2013YB266).

REFERENCES

[1] Tang Shiyun, Lu Guoying, Han Shijian etc. Different Water Treatment on Influence of Photosynthesis for Sugarcane [J] Guangxi Sugar, 2001, 14 (2) pp:6–9.
[2] YANG Li-tao,CHEN Chao-jun,LI Yang-rui. Studies on photosynthetic characteristics in sugarcane varieties, Journal of Southwest Agriculture, 2005(3) pp:69–74.
[3] He Tieguang, Xu Jianyun, Wang Canqin. Cultural measures for increasing the photosynthetic efficiency of sugarcane. Guangxi Agricultural Sciences, 2005, 36 (6):pp:502–504.

Information, Computer and Application Engineering – Liu, Sung & Yao (Eds)
© 2015 Taylor & Francis Group, London, ISBN 978-1-138-02717-6

Research on computer modeling approach for children's toy products

Xi Yun Li

College of Art Design, Qilu University of Technology, Jinan, China

ABSTRACT: With the rapid growth of computer software technology, the application of computer aided design is wider and wider in product creativity performance, and the Rhino software, in particular, is favored by industrial designers with its specialty of curve modeling. From the perspective of fast and precise presentation of design creativity, this paper, taking children's toy products as an example, elaborates the computer modeling approach of Rhino software, and states its strong function of curve modeling and outstanding features of easy operation to make the product design personnel know the Rhino software well.

KEYWORDS: Computer-aided design; Rhino software; Children toy products; Modeling approach

With the rapid growth of computer software technology, the application of computer aided design is wider and wider in product creativity performance, and the Rhino software, in particular, is favored by industrial designers with its specialty of curve modeling. Rhino is a 3D modeling software based on NURBS launched by the US Robert McNeel Company in 1998. It is one of the most advanced professional NURBS modeling software, and obtains simple interface, powerful functions and low hardware configuration requirements [1]. From design sketching and an effective sketch to the actual product, the curve surface tools provided by the Rhino can accurately make all application models for rendering performance, animation, engineering drawing, analysis, evaluation and production. Children toy products obtain wide varieties, and different forms, which requires designers to maintain solid basic skills of modeling, and the powerful modeling function of Rhino software sets up a broad platform for designers to show their ideas and originality.

1 RHINO SOFTWARE FUNCTION INTRODUCTION IN MOULDING AND MODELING

First of all, the powerful curve modeling capacity of Rhino software is well-known, and it can cause fast and exact creation of the curved surface, and it can generate or edit the surface in reliance with tens of Rhino commands such as monorail sweep, double track sweep, grid face building and lofting. Secondly, in order to reveal the quality of the surface created in the first step, apply tools such as curvature analysis, zebra crossing, environment mapping, and reflection mapping, which can help intuitively observe the generated surface, and make the convenience for smooth modification in later stage. Thirdly, the cutting tool is a common tool used in Rhino modeling, and it can remove unnecessary modeling surface in the modeling process. Similarly, it can be used for entity model cutting and target surface keeping, making it easy to edit in the later processing. Fourthly, Boolean calculation is a common tool for entity subtract and combination in Rhino modeling, which shall execute partial subtraction from the completed mold, and perform partial structure addition to completed molding as well, where the final completion is still an entity component. Fifthly, the array is also a common tool used in the process of modeling, where rectangular array can be used to create objects with the same spacing, and ring array be used to create objects with the same angle permutation, which not only reduce the tediousness of the modeling process to a great extent, and at the same time also improve the accuracy of modeling and achieve an orderly and aesthetic effect [2]. Sixthly, in the course of complex shape modeling, some morphological structure may shelter the modeling sight, and obstruct the continuity of modeling work, and here the hide tool shall have the effect of eliminating the false and retaining the true, temporarily hiding the unnecessary structure for present modeling, and canceling component hiding when needed to avoid improper mistakes in the modeling work. Seventhly, in the final phase of molding and modeling, in order to achieve a certain need, different preview effects are available to show structure, scale, etc. of each part, where display mode shall play a crucial role, the commonly used modes of which are coloring mode, wireframe mode, translucent mode, rendering mode, X-ray mode, engineering drawing mode, artistic style mode, pen and other modes.

2 RHINO MODELING APPROACH AND MODELING IDEA

The rhino modeling method obtains three types: extrusion, rotation and square shaping. In Rhino software, the object extrusion and rotation are completed by extrusion tool and rotary tool. The rhino software extrusion tool is able to squeeze out along a straight line or a curve one, and also squeeze out to a point to form a cone. Rotate tool can rotate not only around rotation axis, but also along some path. The third kind of square shaping is similar to the curve surface thought below, which shall be created through the following four methods. One is sweep type, including monorail and double-track sweep; the second is free type; the third is grid type, forming a grid mesh surface by grid lines; the fourth is repair type, including the surface mixing, surface matching and merging.

Rhino modeling idea mainly includes: surface idea, entity idea and slice idea and other contents.

Surface idea obey the order to create the outer structure through the curve surface of the targeted object, during which, sweep tool, lofting tool and other tools shall be used. The surface construction method is very convenient, and shall rapidly present the body structure. However, over complicated surface shall fail to generate entity in the late skew-to-entity-forming process, or fail to conduct rapid chamfering due to huge calculation and lack of internal storage.

Entity idea is intended to divide the targeted structure to some smaller unit, and then conduct unit entity construction. After the completion of unit entity construction, apply Boolean Calculation for aggregation or subtraction to achieve final entity, where the disadvantage of this method lies in slow operation caused by huge calculation amount in the Boolean Calculation at a later stage. As for structure with thin wall, cap and others, after Boolean Calculation, chamfering shall lead to certain broken surface.

Slice idea is to cut the required structure form of surface or entity, which is the simplest, most direct and most effective method for product appearance pattern in modeling where any product pattern shall be achieved through cutting of several surfaces. In modeling, most people tend to choose the method of marking-off and sweeping as per the product appearance, which is convenient for simple body modeling, while it's not that simple for complicated surface structured body modeling. Slice idea is the simpler modeling method targeted to deal with the problems not solved by sweeping, through which entity structured object is achieved, and the problems are solved of inconvenient chamfering in surface idea and slow operation of late stage Boolean Calculation in entity idea.

3 APPLICATION OF RHINO SOFTWARE IN CHILDREN TOY DESIGN

3.1 *Import of toy sketch before modeling*

In children toy design, the critical design factor is the presentation of toy function, pattern, structure and its use mode. The original idea of children's toy before modeling is recorded by sketches, which is the most inspiring figure for designer and the great carrier to save the designer's creativity, whose success of creativity contributes to the pleasing result of later-stage modeling [3]. The purpose to offer better and more defined presentation of the sketch creativity shall be realized through 3D sketch modeling. The sketch import shall simplify and fasten the modeling process, achieve the original design concept of designers to the largest extent, and thus construct more reasonable 3D modeling, where the sketch can be drawn using 2D software (such as Photoshop, Coreldraw, and others), or in the papers by hand and then imported to Rhino through scanning or other methods, in order for modeling preparation.

3.2 *Process analysis before modeling*

Children toy shape modeling obtains slight difference in different shapes, but generally, it shall go through the design scheme scrutiny, structure analysis, modeling method selection, model construction, later-stage perfection, modeling construction completion and another process (see Figure 1). Structure analysis plays a crucial role in the modeling process, deciding the high efficiency of the modeling process and the smooth completion of modeling task. If fails to sort out the structure, there shall be piled difficulties in actual modeling work, and even more troubles in later-stage mold making.

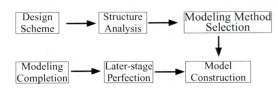

Figure 1. Modeling flow chart.

3.3 *Advantage highlight of curved surface modeling*

Rhino software obtains different types of modeling methods for the objects of the same pattern and structure. Child toy modeling shall consider its uniqueness, and fully present the design principles of enjoyment, entertainment, intelligence and safety [4] where the safety principle requires more camber shape.

It is a huge advantage for Rhino software of camber shape modeling, and it shall create more unexpected effects; in processing of modeling, add different view modes to view the model, and thus accurately combine the original design inspiration and creativity (see Figure2–5), and optimize the structure in the modeling process to realize uniformity and coordination among all components.

Figure 2.   Children's toy.

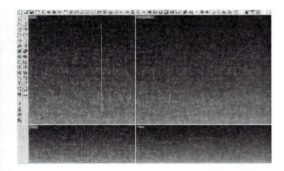

Figure 3.   Children's toy modeling.

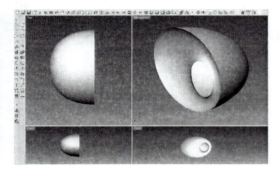

Figure 4.   Children's toy modeling.

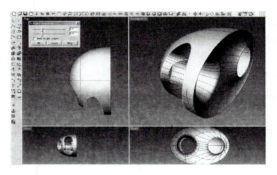

Figure 5.   Children's toy modeling.

4   CONCLUSION

Rhino software possesses incomparable curve surface modeling capacity, and its operating methods are duly added at the same time, thus offering different construction methods for the same structured objects. Since children's toys play a significant role in the children's world, which is irreplaceable by any other objects, more considerations shall be input in the modeling process, such as the more humanized way to present the enjoyment and intelligence of children's toy as well as its safeness which requires more careful chamfering shall be performed to later-stage modeling. Modeling learning is not a one-step thing, instead, it requires a large amount of practice and accumulation; learning multiple kinds of constructing methods is the premise for more unexpected effect creation as well as the chance for an interesting experience.

REFERENCES

[1] GE Liang, Xi ZHANG. Application of Rhino Software in Product Modeling Design [J]. China Packaging Industry, 2014. (18).
[2] CHEN Liang. Discussion on 3D Aided Software in Industrial Design Teaching [J]. Software Introduction, 2010(10).
[3] HUANG Mingjin. Modeling Comparison of Rhino and Pro/ENGINEER and Rhino Application in Industrial Design [J]. Enterprise Technology Development, 2012(4).
[4] WANG Zhenwei, Yongwei ZHAO. Toy Design Introduction [M]. Beijing: China Light Industry Press, 2013.

Information, Computer and Application Engineering – Liu, Sung & Yao (Eds)
© 2015 Taylor & Francis Group, London, ISBN 978-1-138-02717-6

Design and implementation of Cyrillic Mongolian syllable text corpus system

Quan Bao

Inner Mongolia Radio & TV University, China

ABSTRACT: Based on the building of Cyrillic Mongolian corpus, this paper makes statistics of relevant data on the word and syllable levels of Cyrillic Mongolian. At the word level, this paper respectively makes statistics of word length distribution with syllable and character as the unit, the distribution of high frequency words in the corpus as well as the distribution of syllable structure type in the word and the distribution of syllable type in the word. At the syllable level, this paper makes statistics of the distribution of high frequency words, vowel frequency distribution in the syllable and the distribution of syllable structure type.

KEYWORDS: Corpus; Cyrillic Mongolian; Syllable

1 INTRODUCTION

Mongolian belongs to Altaic family, and the main users are Mongol and Mongolian people in China. The recorded Mongolian language contains the traditional Mongolian and Cyrillic Mongolian. After 1941 the traditional Mongol script yielded to a modified Cyrillic alphabet in the Republic of Mongolia. Khalkha, or Mongol proper, is the most important Mongolian language. The official tongue of the Republic of Mongolia, it is native to more than 2 million people. (Columbia Electronic Encyclopedia, 6th Edition. Q1 2014, p1–1. 1p.)

Studies on Cyrillic Mongolian corpus are mainly conducted in Mongolia, Japan and China. Center for Research on Language Processing (CRLP) at the National University of Mongolia (NUM) completed 5 million words of Cyrillic Mongolian corpus with word-level labeling (Purev Jaimai, Odbayar Chimeddorj, 2009). In order to achieve the development of Japanese-Mongolian bilingual dictionary, Graduate School of Library, Information and Media Studies, University of Tsukuba used 1118 technical reports to build a phrase corpus containing 110458 phrase types (Badam-Osor Khaltar, Atsushi Fujii, Tetsuya Ishikawa, 2006). Mainly from the Research on Cyrillic and Mongolian Script's Morphology and Conversion System, China has built a small-size text corpus.

Mongolian syllable is a language unit naturally divided in pronunciation. It has a relative stable amount, and it is easy to listen and distinguish. Cyrillic Mongolian spelling is based on Khalkha dialect, having strict rules for syllable segmentation. Therefore, syllable is a good bridge to connect phonetics with text. Under this language condition, syllable becomes a good choice in studying and developing the language products.

2 BUILDING OF CYRILLIC MONGOLIAN TEXT CORPUS

2.1 Initial determination of corpus size

Mongolian and Uighur languages belong to Altaic family as well as the agglutinative language type in morphological structure, and they have the same language features in some aspects. Uighur language has 8 vowels and 24 consonants. There are 6 types of vowels and consonants matching in syllable and more than 2000 frequently used syllables (Mamateli Tursun, 2011). Cyrillic Mongolian has 13 vowels, 20 consonants. There are 8 types of vowels and consonants matching in syllables (Galsangpegseg, 2004). As compared, it can be preliminary judged that there are about 5000 Cyrillic Mongolian syllables. The ultimate purpose of this study is to build the Cyrillic Mongolian syllable text corpus, and thus make statistics of some language feature information for words and syllables. Kennedy (Huang Changning, Li Juanzi 2002: 33~34) believed that it was adequate to study the language rhythm with a corpus of 100,000 words. Therefore, in the corpus sampling strategy, this paper initially determines 100,000 words for the corpus size. The eventually built corpus size is as shown in Table 1:

Table 1. Corpus size.

Word Times of Corpus	Distinct Words	Syllable Type
273239	31567	5332

2.2 Encoding of corpus text

Before the popularity of Cyrillic Mongolian Unicode encoding, Mongolia has widely used non-Unicode encoding, such as Arial Mon, Times New Roman mon and other fonts. This study develops a Unicode encoding conversion tool, in order to ensure that all corpus texts eventually conform to the Unicode encoding.

2.3 Source of corpus

Under the premise of the initial determination of corpus size, this paper should further determine the corpus equilibrium structure, sampling principles and logical structure. This study collects the news content from the Mongolian Today News website in December 2013 and January 2014. Based on the content areas, this paper divides into 9 categories including society, politics, business, mining, nature, law, technology, leisure and international, a total of 882 papers. For specific paper inclusion in various fields, please refer to Table 2:

Table 2. Corpus structure.

Field	Society	Politics	Business
Proportion of Word Time	0.42	0.19	0.08
Field	Mining	Nature	Law
Proportion of Word Time	0.02	0.03	0.09
Field	Technology	Leisure	International
Proportion of Word Time	0.03	0.07	0.07

In addition, this paper also selects the included entries of *New Mongolian and Old Mongolian Dictionary* (Galsangpegseg, 2004) as the sample, and builds an entry corpus containing 24,575 words for the auxiliary research data.

3 DESIGN AND IMPLEMENTATION OF CORPUS MANAGEMENT SYSTEM

3.1 Features of corpus management system

Corpus management system is mainly responsible for the import, word segmentation, syllable segmentation and related query statistics for the collected raw corpus, which are specifically as follows: corpus file management including library import and recycling; separate storage of corpus text caption, main body and the author information; preprocessing of corpus text; conversion of the preprocessed text into word sequence; automatic segmentation of words into syllables; exporting of lexicon and syllable library, related retrieval and other features.

3.2 Working process of corpus management system

To implement the above features, the working process of corpus management system is shown in Figure 1.

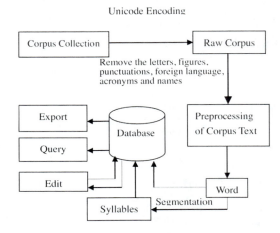

Figure 1. Working process of corpus management system.

3.3 Structure of corpus management system

This system mainly consists of four subsystems, including corpus file management, corpus preprocessing and word segmentation, syllable segmentation, query and statistics, as shown in Figure 2:

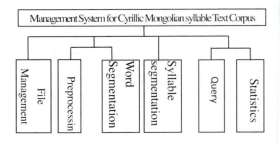

Figure 2. Corpus management system modules.

3.4 File management module

The file management module is used to import the raw corpus to the corpus management system, add the additional information of corpus source, areas and the author, and show the overall situation of collecting corpus.

3.5 Module of preprocessing and word segmentation

This module conducts the necessary preprocessing for raw corpus text and converts the processed corpus text into word sequence. According to the needs of building Cyrillic Mongolian corpus and syllable library, the preprocessing module is responsible for removing the characters, punctuation, figures, acronyms and names of non-Cyrillic Mongolian. Based on post-processing corpus text, this paper converts the Cyrillic Mongolian word sequence in the database for storage.

3.5.1 Syllable segmentation

Cyrillic Mongolian has 35 alphabets, and all of them have their own pronunciations in addition to two alphabets of "ь, ъ". The basic vowels are "а, э, и, о, у, ц, у", auxiliary vowels are "я, ё, е, ю, ы, й", and there are 13 vowels in total. There are 20 consonants, including "б, в, г д, ж, з, к, л, м, н, п, р, с, т, ф, х, ц, ч, ш, щ". Mongolian word can be divided into one or more syllables. The Mongolian vowel syllable structure consists of vowel combination and combination between vowel and consonants. If V stands for the vowels and C for consonants (long vowels and diphthongs are recorded by V), as a saying goes, modern Mongolian uses not only the common four forms of V, CV, VC and CVC but also another two forms of CVCC and CVCCC (Mongolian reference, 2011). As another saying goes, Mongolian words use the following syllable structures VCV, CVC, VVC and CVCC (Altangerel Ayush, et al, 2013). As still another saying goes, the Mongolian syllable structure consists of six types including V, VC, VCC, CV, CVC and CVCC (Huhe, 1999). In the process of actual syllable segmentation, this paper fully affirms that Cyrillic Mongolian has eight syllable structures of V, CV, VC, CVC, VCC, CVCC, VCCC and CVCCC, which was proposed by Galsangpegseg in *New Mongolian and Old Mongolian Dictionary*. Therefore, according to the type of syllable structure, this paper segments the rules-based word syllable. The study has also found that there are some common Cyrillic Mongolian transliterating borrowed words, and there are other types of syllable structure.

3.5.2 Query and statistics system

In this study, query, statistics, diagramming and other works use the R and EXCEL. Specific results are shown in "Conclusions" part of this paper.

3.6 Implementation of corpus management system

The corpus management system uses B/S structure design, applies Microsoft SQL Server 2008 as the background data server for storage corpus. Microsoft C # language is used for the development tool.

4 CONCLUSION

According to the common statistical methods in linguistics, this paper makes statistics at the word and syllable levels on the basis of corpus.

4.1 Words

4.1.1 Word length distribution - taking syllable for the unit

As corpus-based statistics found, Cyrillic Mongolian word consists of 1–6 syllables. This paper makes statistics on word length by taking syllable for the unit, and the word length distribution is shown in Figure 3:

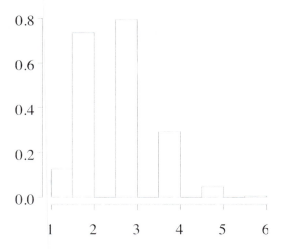

Figure 3. Word length distribution (syllable as the unit).

According to the diagram, dissyllabic and trisyllabic words account for the largest proportion in Cyrillic Mongolian, whereas monosyllable, quinquesyllabic and hexasyllabic words account for a small proportion.

4.2 Word length distribution - taking character for the unit

As corpus-based statistics found, the word length of Cyrillic Mongolian is 1 to 19 characters, and its distribution is shown in Figure 4:

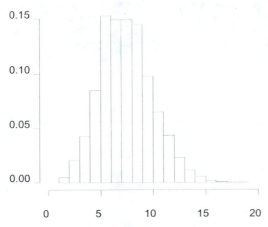

Figure 4.   Word length distribution (character as the unit).

According to the above diagram, the words with 5-8 characters are relatively common, whereas the words with over 14 characters are less.

4.3 Word frequency statistics

As corpus-based statistics found, among the top 20 high frequency words, there are 18 high frequency words in 50 high frequency words range proposed by J.Purev and Ch. Odbayar, and especially the top 5 high frequency words collocate. Top 5 high frequency words are shown in Table 3:

Table 3.   Top 5 high frequency words.

Word	Frequency (this study)	Rank (J.Purev)
нь	5703	1
байна	2740	3
энэ	2479	5
ч	1921	4
юм	1916	2

4.4 Syllable structure collocation of words

Mongolia transliterates the common borrowed words. The syllable structure type after transliteration has been beyond the type of modern Mongolian syllable structure. For example: плаза, syllable structure collocation is CCV/CV, but CCV does not belong to the syllable structure of Cyrillic Mongolian. As corpus-based statistics found, the borrowed words, including transliteration have 444 types of syllable structure collocation. After deleting the syllables that do not meet the type of Cyrillic Mongolian syllable structure, this paper sorts the collocation of high frequency syllable structure type. The collocation of top 20 syllable structure type is shown in Figure 5:

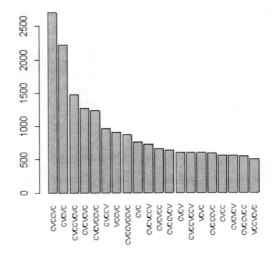

Figure 5.   High frequency syllable collocation type.

4.5 Syllable type distribution of words

This paper makes statistics on syllable types in each syllable position, and the results are shown in Figure 6:

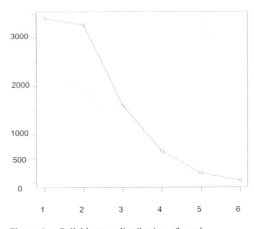

Figure 6.   Syllable type distribution of words.

330

According to the chart, from the first syllable position to the last syllable position, the syllable types generally show a declining trend, and the number of syllable type in the first and second syllable positions is relatively large.

4.6 Syllables

4.6.1 Statistics of high frequency syllables
In the corpus-based statistics, this paper sorts the syllable frequency. For top 20 high frequency syllables and frequencies, please see Table 4:

Table 4. High frequency syllables and frequencies.

Syllables	бай	га	а	сан	нэ
Frequency	0.013	0.010	0.009	0.009	0.008
Syllables	э	гуй	гэ	ла	гаа
Frequency	0.007	0.007	0.006	0.006	0.006
Syllables	гийн	нь	о	ны	на
Frequency	0.008	0.008	0.007	0.007	0.007
Syllables	сэн	ний	хэ	лийн	бо
Frequency	0.006	0.005	0.005	0.005	0.006

4.6.2 Vowel frequency distribution of syllables
In Cyrillic Mongolian, vowels can be divided into basic vowels, auxiliary vowels, diphthongs and long vowels. According to the statistical unit of syllable vowels, this paper makes statistics of its frequency, as shown in Table 5:

According to the above table, the syllables with basic vowels account for a large proportion, about 54% of all syllables.

4.6.3 Distribution of syllable structure type
In Cyrillic Mongolian, there are 8 types of syllable structure. According to the statistics, "CVC" and "CVCC" syllable structure type have a relatively higher frequency. The distribution of syllable structure type is shown in Figure 7:

5 PROSPECT

By collecting a Cyrillic Mongolian text corpus, this project builds a corpus with 0.2 million word times, and conducts the word and syllable segmentation over the collected corpus. Relevant linguistics parameters of Cyrillic Mongolian are obtained through query statistics. According to the analysis of acoustic and prosodic features, Khalha dialect and Chahar dialect are relatively closer (I Dawa, 2001). In the traditional Mongolian, there are significant differences between spoken Mongolian and written Mongolian

Table 5. Vowel frequency distribution of syllables.

Basic Vowels	а	э	и	о	у	ө	Y
Frequency	737	460	479	489	255	310	157
Auxiliary Vowels	я	ё	е	ю	ы	й	
Frequency	104	29	262	37	64	3	
Diphthong	ай	эй	ой	уй	Yй	яй	ёй
Frequency	162	31	72	36	57	0	0
Diphthong	юй	ей	ya	ay			
Frequency	3	34	13	21			
Long Vowels	аа	ээ	ий	оо	өө	уу	YY
Frequency	270	201	152	165	99	227	183
Long Vowels	яа	яу	ёу	еY	ёо	еэ	
Frequency	29	1	2	0	11	8	
Long Vowels	еө	юу	юY	иа	ио	иу	Yе
Frequency	0	11	2	59	42	39	5
Long Vowels	ео	ы					
Frequency	11	64					

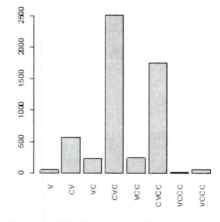

Figure 7. Distribution of syllable structure type.

(Chinggeltei, 1979). Therefore, the research findings are also valuable for the traditional Mongolian studies. In addition, the research findings have some values to the development of Mongolian speech synthesis, speech recognition, text proofreading and other linguistic related products.

ACKNOWLEDGMENTS

This study is supported by the scientific research project "Design and Construction of Cyrillic Mongolian Syllable Text Corpus System" of China's Inner Mongolia Autonomous Region Colleges and Universities (Key Project, Project Number NJZZ11135). The author would like to acknowledge all!

REFERENCES

J.Purev and Ch. Odbayar. *mongolian lexicon derived from corpus*. CRLP National University of Mongolia, http://www.panl10n.net/Presentations/Laos/RegionalConference/Lexicons/Mongolian_Lexicon.pdf.

Huhe Coyijungjab. *Analysis of Mongolian Speech Acoustics*. Hohhot: Inner Mongolia University Press, First Edition in October 1999, P14–P15.

Galsangpegseg. *New Mongolian and Old Mongolian Dictionary*. Inner Mongolia Education Press, December 2004, Hohhot City, P1108.

Altangerel Ayush, Bayanduuren Damdinsuren, *A design and implementation of HMM based Mongolian Speech Recognition System, Proceeding Title: Strategic Technology (IFOST)*, 2013 8th International Forum on Proceeding Time: 2013, School of Information and Communication Technology of MUST.

Badam-Osor Khaltar, Atsushi Fujii, Tetsuya Ishikawa. *Extracting loanwords from Mongolian corpora and producing a Japanese-Mongolian bilingual dictionary*. ACL 2006, Proceedings of the 21st International Conference on Computational Linguistics and 44th Annual Meeting of the Association for Computational Linguistics.

Mamateli Tursun. *Context dependent syllable based speech synthesis system for Uyghur*. Computer Engineering and Applications. 2011, 47(31).

Information, Computer and Application Engineering – Liu, Sung & Yao (Eds)
© 2015 Taylor & Francis Group, London, ISBN 978-1-138-02717-6

Depth image inpainting for Kinect using joint multi-lateral filter

Lei Ding, Yin He Du, Jing Lv & Chen Qiang Gao
Chongqing Key Laboratory of Signal and Information Processing (CQKLS&IP),
Chongqing University of Posts and Telecommunications, Chongqing, China

ABSTRACT: Since the Kinect sensor was released in 2010, depth images have been widely applied to many computer vision fields. However, inaccuracy and instability of Kinect depth data heavily influence the performance of its application. The Kinect depth images have the problems of depth holes and boundary being unmatched. In order to address these problems, we use the canny algorithm to detect the edges of the color image and the color image at the same time. Then the pixels between the edges of two images will be set to black holes. Finally, we propose a joint multilateral filter base on Anisotropic Diffusion. The filter is developed from bilateral filter, and utilize spatial correlation and pixel values similarity of guide image to fill in the black hole. As show in the experimental results, the proposed method can not only inpaint effective but also greatly improve the quality of depth image.

KEYWORDS: depth image inpainting; multilateral filter; black hole filling

1 INTRODUCTION

Kinect [1] has attracted the attention of many scholars since it was released in November 2010. It has been widely applied to many computer vision fields so far, especially in navigation, human computer interface, tracking and recognition.

The advantages of the Kinect are its low price and the possibility to acquire video and depth data with a good resolution (480x640). Compared with the traditional picture, the gray value of depth image represents the distance from the camera to the target object. So far there are two methods obtaining depth images: the first method is to use computer vision techniques to obtain the depth image, which calculates the geometric relationship between two-dimensional feature and three-dimensional image, such as multiview stereo[2]. However, the this method is costly in time complexity, with strict constraints and precise image correction, Another method is to capture three-dimensional scene through the time of flight (TOF camera)[3] or structured light (Kinect), compare with TOF camera, Kinect is cheaper.

However, the Kinect depth image is instable, and the image captured by Kinect has an ineligible quality to be used in computer vision. Thus, to inpaint the error becomes a challenging work. The reasons and types of errors include: (1) The most important error in the Kinect depth image is the absence of depth measurements in some image areas (shown in Fig.1(a)), which is called the black hole.

(2) The edges of the depth image don't match the corresponding edges of the color image (see Fig.1 (b)).

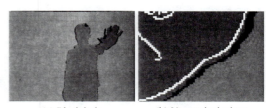

(a) Black hole   (b) Unmatched edges

Figure 1. Depth error.

The reasons of the first kind of error are in several respects: a) some areas are blocked by another object, b) The properties of the object surface also impact the measurement of point, smooth and shiny surfaces that appear overexposed in the infrared image impede the measurement of disparities. The second kind of error shown in Fig.1 (b), white pixels are the edges of the color image. It is obvious that edges of depth image don't match the corresponding edges of a color image (unmatched edges). Unmatched edges are caused by the optical distortion of infrared cameras[4].

2 RELATED WORK

Several methods have been proposed to post-process depth images. One of the most used is bilateral filter[5], such as Kopf et.al[6] proposed joint depth upsampling. The goal in these works was to reconstruct high resolution depth maps using the low resolution ones and their corresponding high resolution image. Matyunin[7], who proposed inter-frame motion compensation and median filtering of depth images to fill the black hole, however, also did not consider the boundary alignment and the situation where faced with a large area of the black hole. Junyi Liu [8] improves based on Telea [9] who proposed fast marching method (FFM) to reconstruct damaged portions of an image.

Considering all the special properties of Kinect depth image, we propose a new Kinect depth image inpainting approach. In our approach, we design a joint multilateral filter base on Anisotropic Diffusion, the filter is developed from bilateral filter, which utilize spatial correlation and pixel values similarity of guide image to filling the black hole. Besides, we introduce Anisotropic Diffusion equation as additional filter kernel.

Figure 2.   The framework of the proposed repair algorithm.

In the following section, Section 2 presents the relate works of repair depth map. Section 3 presents the algorithm framework and main steps. Section 4 shows the experiments of the proposed method and section 5 is the conclusion of the paper.

3   ALGORITHM FRAMEWORK

The algorithm framework of proposed inpainting algorithms for Kinect depth image is shown in Fig. 2.

Step 1 Cutting and aligning the Kinect texture image and the depth image.

Step 2 Remove the wrong pixels,

Step 3 Depth image processing with iterative joint multilateral filtering.

Step 4 Remove the noise of the depth image using Gaussian.

3.1   Remove the wrong pixel

It is clear that the wrong depth values locate between the edges of depth image and the corresponding edges of the color image. These wrong pixels are irregular,

so we can't move certain pixel to remove the wrong pixel.

In this paper, we propose a method to address this problem. First, we detect the edge of depth images and color images by Canny operation. The edge detection image in the depth image from the upper left to the bottom right pixel by pixel traverse to determine if the pixel coordinates of the point are greater than 0. If more than 0, this point is the edge of depth image. There must be the true contour in the vicinity of the object in the color image of the point, we set the point as the original point. Then we visit the areas close to the original point in color edges in four directions, If we find the pixel of color edges is 1, we set this point as an end point since the pixel is the color edges. Then the point between the original point and end point will be set to black hole (that's to say set the gray value to zero, see Fig.4).

We regard unmatched edge pixels as another kind of error pixels and remove their values. The next part shows how to estimate their exact values.

3.2   *Introduce Anisotropic diffusion equation*

Anisotropic diffusion also called Perona–Malik diffusion is a technique aiming at reducing image noise without removing significant parts of the image content[10]. It was improved from the traditional Gaussian filter, with a strong theoretical foundation.

(a) Color image   (b) Depth image

Figure 3.   the color and depth image captured by Kinect.

Figure 4.   Depth image after removing wrong pixel.

Anisotropic diffusion can be used to remove noise from digital image without blurring edges. This property makes it has been widely used in image smoothing, denoising, restoration, enhancement, etc.

In its original formulation, presented by Perona and Malik in 1987[10],

$$\begin{cases} \dfrac{\partial I}{\partial t} = div\left[c\left(\|\nabla I\|\right) \bullet \nabla I \right] \\ I(t=0) = I_0 \end{cases} \quad (1)$$

div is the divergence operator; ∇ denotes the gradient; $c(\|\nabla I\|)$ is diffusion equation. The basic idea of the anisotropic diffusion model is based on $\|\nabla I\|$ to achieve diffusion or not. The diffusion equation was proposed as:

$$c(\|\nabla I\|) = \exp[-(\|\nabla I\|/k)^2] \quad (2)$$

$\|\nabla I\|$ could be regarded as the edge detector. If $\|\nabla I\|$ far greater than 0, $\|\nabla I\|$ will be close to 0. This means diffusion is suppressed. On the other side, if $\|\nabla I\|$ far less than 0, $\|\nabla I\|$ will close to 1. This means diffusion is enhanced.

Previously we often regarded the image as a matrix. In this paper, image is regarded as a thermal field. Every pixel is regarded as heat flow. According to the relationship between the current pixel and the surrounding pixels, we determine whether to diffuse. If the surrounding pixels and the current pixel not similarity, this pixel will not be diffused, the surrounding pixels will not be taken into consideration to filling black hole. Otherwise, filling black hole will take surrounding pixel into consideration.

3.3 Black hole filling

As we presented above, We first briefly review bilateral filter[5]. Bilateral filter is a non-linear, edge-preserving and noise-reducing smoothing filter for images. Therefore the bilateral filter needs to create two templates. The first one is a domain template f, the weights depend on Euclidean distance of pixels. The second is range template g, which measures the photometric similarity between two pixels. Let us consider I_p the depth value of the pixel at position p and Ω its neighborhood, and the bilateral filter is defined as:

$$I_p^* = \frac{1}{W_p} \sum_{q \in \Omega} I_p \bullet f_s(p,q) \bullet g_s(\|I_p, I_q\|) \quad (3)$$

Where $W_p = f_s(p,q) \bullet g_s(I_p, I_q)$, which is a normalization factor. The function f and g are modeled as Gaussian.

In the previous section the most important problem that affects the accuracy of Kinect-generated depth images have been outlined: black hole, depth measurement fluctuations, and unmatched edge. In order to address these problems, this paper combines with the thought of iteration. This paper proposes an iterative joint multi-lateral filter under the consideration of spatial information, color information and diffusion coefficient. We need to introduce another template based on bilateral filters. The joint multi-lateral filter could be summarized as follows:

$$I_p^{n+1} = \frac{1}{W_p} \sum_{q \in \Omega} I_p^n \bullet f_s(p,q) \bullet g_s(I_p^n, I_q^n) \bullet g_I(I_p^n, I_q^n) \quad (4)$$

Where $W_p = f_s(p,q) \bullet g_s(I_p^n, I_q^n) \bullet g_I(I_p^n, I_q^n)$, W_p is normalization factor. I^n is a depth image after the n-th iteration (I^0 is the original input depth image).

The spatial filter $f_s(I_p, I_q)$ is as (3), which is modeled as a Gaussian. It determines their weight base on their distance from pixel q to the center pixel.

$$f_s(I_p, I_q) = \exp\left(-\frac{\|p-q\|^2}{2\sigma_s^2}\right) \quad (5)$$

The range filter g_s usually takes a gray value or intensity information for consideration. In this paper, we design a range filter g_s operate on the pixel value difference in color (RGB) instead of only in the intensity channel. The reason is that it is possible to have two regions with very similar intensity while their colors are quite different. The weights of filter g_s consist of three kernels, performing on the R, G and B channel, respectively, which the range filter can be summarized as follows:

$$g_s(\|I_p - I_q\|) = g_R(|R_p - R_q|) \bullet g_G(|R_p - R_q|) \bullet g_B(|R_p - R_q|)$$
$$= \exp\left(\frac{-|R_p - R_q|^2}{2 \bullet \sigma_R^2}\right) \bullet \exp\left(\frac{-|R_p - R_q|^2}{2 \bullet \sigma_G^2}\right) \bullet \exp\left(\frac{-|R_p - R_q|^2}{2 \bullet \sigma_B^2}\right) \quad (6)$$

Besides using color information, we further introduce an adaptive decay factor in the range filter in (6). According to the introduction of Ref[11], under the same amount of gray difference, the darker region and brighter region will exhibit smaller variation. Therefore, when the pixel is not in darker and brighter region, the weight should decay faster, so as to avoid over-smoothing the depth image. From Fig 5 we can observe that the smaller of the attenuation, the weight decay faster. In this paper, the attenuation factor in (6) is determined by color value and by the corresponding channel at position p. The relationship between of color value and σ is shown in Fig.6, we could not get the attenuation factor when the depth pixel value lower than 50, because when using

Kinect, the officially recommended distance is $0.8m$ to $4m$ [1]. Depth pixel value is 50 when the distance is $0.8m$, so Kinect can not capture the pixel value lower than 50.

In order to refine discontinuities depth at object edges. We introduce additional range filter $g_I(I_p, I_q)$, this filter is a diffusion coefficient:.

$$\begin{cases} g_I(\|I_p - I_q\|) = \exp[-(\|\nabla I\|/k)^2] & \|\nabla I\| < k \\ g_I(\|I_p - I_q\|) = 0 & \|\nabla I\| > k \end{cases} \quad (7)$$

where $\|\nabla I\| = \|I_p - I_q\|$, the constant k controls the sensitivity to the edges and is usually chosen experimentally or as a function of the noise in the image. Within the 5×5 window in depth image, if the diffusion coefficient of all three colors is enhanced, we determined that this neighborhood is homogenous in the color domain. The range filter will only process across such edge with same color on one side.

Figure 5.   The curve between weight and pixel difference.

Figure 6.   Adaptive range filter attenuation coefficient.

(a) Inpainting depth of Ref[6]

(b) Inpainting depth of Ref[11]

(c) Inpainting depth of proposed method

Figure 7.   Experiment result.

a) Original depth

(b) Method of Ref [6]

(c) Method of Ref[11]

(d) Our method

Figure 8.   Experiment result.

4   EXPERIMENTS

In order to evaluate the performance of the proposed algorithm, we capture more than 500 indoor scene color images and depth images from Kinect sensor under different light conditions base on Open-NI framework. The resolution of the color image and depth image are both 640×480, after cutting and aligning the resolution are 580×430.

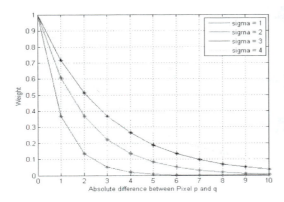

336

The proposed algorithm was implemented in MATLAB on PC with CPU 2.60GHz and 4 GB RAM. We can get the both edges of depth and color image by Canny Operator. The window size of wrong value pixels detection is 9×9, while another window size of the black hole filling and smoothing is 5×5. The smoothing parameter k of diffusion equation can affect the result, largely, thus we can set k to 15 through a large number of experiments. Compare with the result in Fig. 7, the proposed method clearly draws the contours of the person, especially in hand areas. Fig.8 is another scene, the red box presents the ability of refining edges. Blue box show that the proposed method can accurately fill black hole.

5 CONCLUSION

There are a lot of black holes and unmatched edges error in the depth image. In this paper, we proposed a method to repair depth image with the aid of a diffusion equation base on the characteristic of depth image. Firstly, we need to remove the wrong value pixels which locate between the edges of depth images and its corresponding edges of the color image. Then, we fill in the black hole with joint multi-lateral filter. Finally the repaired depth image is obtained by removing noises using the Gaussian filter. As shown in the experimental result, the proposed method can not only effectively fill in the black hole but also greatly refine the edges.

ACKNOWLEDGMENTS

This work is supported by the National Natural Science Foundation of China (No. 61102131), the Natural Science Foundation of Chongqing Science and Technology Commission (No. cstc2014jcyjA40 048), Cooperation of Industry, Education and Academy of Chongqing University of Posts and Telecommunications (No. WF201404).

REFERENCES

[1] "The xbox kinect," (http://www.xbox.com/kinect).
[2] Seitz S M, Curless B, Diebel J, et al. A comparison and evaluation of multi-view stereo reconstruction algorithms[C] Computer vision and pattern recognition, 2006 IEEE Computer Society Conference on. IEEE, 2006, 1:519–528.
[3] S. Foix et al., "Lock-in time-of-flight (tof) cameras: Asurvey," Sensors Journal, IEEE, , no. 99, pp. 1, 2011.
[4] Khoshelham K, Elberink S O. Accuracy and resolution of kinect depth data for indoor mapping applications[J]. Sensors, 2012, 12(2):1437–1454.
[5] Tomasi C, Manduchi R. Bilateral filtering for gray and color images[C] Computer Vision, 1998. Sixth International Conference on. IEEE, 1998:839–846.
[6] Kopf, J., Cohen, M. F., Lischinski, D., and Uyttendaele, M., "Joint bilateral upsampling," in [ACM SIGGRAPH 2007papers], SIGGRAPH '07, ACM, New York, NY, USA (2007).
[7] Malyunin S, Vatolin D, Berdnikov Y, el al. Temporal filtering for depth maps generated by kinect depth camera[C]. 3DTV Conference, 2011:1–4.
[8] Liu J, Gong X,Liu J. Guided inpainting and filtering for Kinect depth maps[C]. IEEH International Conference on Pattern Recognition, 2012:2055–2058.
[9] A. Telea. An image inpainting technique based on the fast marching method. journal of graphics, gpu, and game tools, 9(1):23–34, 2004.
[10] Perona P, Malik J. Scale-space and edge detection using anisotropic diffusion[J]. Pattern Analysis and Machine Intelligence, IEEE Transactions on, 1990, 12(7): 629–639.
[11] Lai, P., Tian, D., and Lopez, P., "Depth map processing with iterative joint multilateral filtering," in [Picture Coding Symposium (PCS), 2010], 9 –12 (dec. 2010).

Information, Computer and Application Engineering – Liu, Sung & Yao (Eds)
© *2015 Taylor & Francis Group, London, ISBN 978-1-138-02717-6*

Thoughts on development of Chinese digital campus

Bin Liu & Wei Liu
Huaiyin Institute of Technology, Huai'an, Jiangsu, China

ABSTRACT: Higher education information developed through the establishment of a campus network, the establishment of independent information systems, campus network information to improve several stages, the development of information systems as integrated digital campus concept. Colleges and universities solved the problem of financial, technical and management and established a school of their own "digital campus", the article also circumstantial evidence from the perspective of a software company developing the Digital Campus. Finally, the article summarizes the process of digital campus college and predicts the digital campus development prospects.

KEYWORDS: Chinese Higher Education; Digital Campus Development

1 INTRODUCTION

In the recent 10 years, significant results in higher education informatization Chinese is proposed and implementation of the concept of the digital campus. The concept of digital campus was used firstly by the Massachusetts Institute of Technology in the 1970s (Lei, Zhang & Yongzhong, Tang 2013). Digital campus construction and management. The Hexi University 28(2). In China, proposed the concept of digital campus has been well documented in 1999, digital campus early mainly refers to the network, school education informatization, unlike the concept in recent 10 years. With the development of information technology education, the "211 Project" led by schools such as Fudan University has presented "the construction of the public service system in digital library construction of digital campus, as the main content" (2, Shanghai Education Commission. 2002. http://www.shmec.gov.cn/html/article/200210/7224 .php.). Today, the digital campus has become a unified information portal, unified identity authentication, school data center core, including the ubiquitous campus network, campus infrastructure, learning resources, information technology, including the daily affairs of proper nouns. In December 2006, in accordance with *"Chongqing Municipal Education Commission on the selection of Chongqing University Digital Campus demonstration unit,"* Chongqing University, Southwest University and other six universities ranked to be the Chongqing University Digital Campus demonstration Units (3, Chongqing Municipal Education Commission. 2006. http://www .cqjw.gov.cn/site/html/cqjwportal/pypx/2007-09-10/ Detail_596.htm.). Education Department of Henan Province, the construction of digital campus as an important symbol of the modern university, from the beginning of June 2007, released 17 batches Digital Campus Demonstration Project Unit. Give rewards and subsidies for the inclusion of certain project demonstration project construction in Colleges and universities (4, Education Department of Henan Province. 2007. http://www.haedu.gov.cn/ 2010/4/21/633538090225512500.html.). The higher education tries to achieve this grand goal, or is in the process of implementation of digital campus.

2 THE DEVELOPMENT OF A DIGITAL CAMPUS

College education informatization development, there is a consensus that the university digital campus development can be divided into three stages (5, Tao, Liu. 2009. Digital Campus platform exploration and thinking. modern educational technology 2009 annual special issue, Vol. 19.). The first stage is the establishment of the campus network, the second phase of building campus network as the core of all kinds of business system; the third stage is the establishment of digital campus. Many colleges and universities recently, vendor proposed the concept of mobile campus, campus wisdom, etc., can be regarded as the continued development of the digital campus. As for the implementation, the course of the digital campus developed basically from a vague target to clear the concept of the digital campus of colleges and universities and achieve their goals. Tsinghua university in 2006, put forward the construction of the concept of "a new generation of digital campus" (6, Dongxing, Jiang. & Dayong, Guo.& Nianlong, Luo. & Qixin, Liu.Tsinghua University. 2007. A new generation of digital campus construction planning and practice. Xiamen University (Natural Science).), mainly including the people-oriented user environment, the associated application environment, the operation of the integrated data environment and high availability environment from several aspects, and put forward the concrete build information portal, application system integration, establish the school share the primary database and data exchange platform, based network data center and upgrade from several aspects. The digital campus building is very much in the budget and is relatively large. The university did not so much funds can be put into the information construction. Depending on the circumstances of each university, has taken the following ways to raise funds. Various types of construction funds and other key universities, Tsinghua first class, you can get national and relevant ministries. Such as Tsinghua University, Tsinghua University, won funding for basic research (JCqn2005042) and the Modern Distance Education Project central finance special (teaching skills Division 2006-86) funded. The second colleges with the operators take the market for funds, funds for the construction of digital campus or the direct cooperation of digital campus construction mode. This kind of colleges and universities is more, including the author's Huaiyin Institute of Technology.

The construction methods are also divided into several types. The First class universities are relatively easy to accessible to funds, so they can use some social source and technology solutions to build one aspect of digital campus, for example, the use of the unified identity authentication system in the Yale University building. The second type of cooperation with the manufacturers to take the mode of construction, these universities are generally the earliest start construction of digital campus , no domestic software companies have mature solutions and products, the need for software development companies assisted design goals school digital campus, specifically by the software company. Take Fudan University, for example, it is jointed with Kim Ji Education Information Co. Ltd. to explore and carry out the first digital campus developmenti domestic. The third kind of colleges and universities, they used an off-the-shelf software company with the solution to a certain amount of custom development basically. With the construction of digital campus, objectively has promoted the development of the concept of university informatization and institutional reforms, now many colleges and universities have to produce the special information management institution "informatization office", the management of the information system construction of independence from the network management. Basic mature and perfect, as the network construction of proportion in the construction of the informatization system in school will be more and higher, this trend will affect more and more colleges and universities.

The construction of digital campus in higher education school, first of all, the liberation of teachers and students from multiple system account, through the use of unified information entrance, solves the problem of information presentation; data standards and data transfer can solve the problem of isolated application system, and solves the problem of data sharing, but also facilitate the various departments through; data center data accumulation, which can be data mining, to better serve the students and teachers and school.

3 DIGITAL CAMPUS SOFTWARE VENDORS SITUATION

In the process of digital campus construction, we cannot leave a large number of higher education schools informationization service vendors, with them, the construction of digital campus can be carried out so smoothly, became a symbol of university network information development. Through the study of several most widely used software Services Company, also can see that in the development of the digital campus of colleges and universities in China. The author studied the tetragonal software co. LTD., Jiangsu Jin Zhi education information co., LTD., Guangzhou Lianyi information technology co., LTD., Chengdu concept information technology co., LTD., materials released by looking at four companies, can find some common ground. Firstly, the rapid expansion of enterprise scale increased the registered capital. Several companies have set up a number of branches or offices, provinces and cities all over the country. Some began to alter from the IT system integrators and Software Company to the information service provider. The golden wisdom education has successfully listed; other companies are in preparation for the listing. Secondly, complete product line, can provide solutions from a single information system to digital campus comprehensive business plan, covering the entire basic platform, teaching management, student management, administrative office, knowledge management, comprehensive services, mobile campus etc. Thirdly, the market performance is significant. According to the information disclosure is the most complete Affirmative software released information, Affirmative company currently has more than 1000 universities digital campus information platform for users, including users more than 100, the modern teaching management information system of more than 1000, more than 200 of the management system of student work, collaborative office system more than 100, more than 60 personnel management information system the scientific research management information system, more than 50 etc.. (7, ZhenFang Software Co. http: //www.zfsoft.com/type01/040000010201.html) Other company claims in the forefront of the market position, such as Jinzhi claims continued to maintain its market share in the domestic first, Lian Yi claimed that come out in front, Kangsai listed a number of cases, the Sichuan Chongqing area in safety.

4 THE EXISTING PROBLEMS IN THE DEVELOPMENT OF A DIGITAL CAMPUS

In the process of implementation of digital campus, there must be a variety of problems, summed up, including the following points: one is the project organizational difficulties, some schools lack of unified understanding, inadequate organization. The informatization of education must change some management mode, information flow and data processing mode, to cooperate fully with the requirements of construction of digital campus all up and down, to change the existing management methods and work process, if the organization is not enough, will cause regret. Two is the implementation of the project problems. Information management personnel in Colleges and universities less general, and the lack of professional experience in the implementation of the project, are learning by doing, the self-constraint of large programs depend on the software company, finally the effectiveness of the organization and implementation of large software company decision procedures, implementation of the project. In addition to the digital campus system involves many, many systems are by the relevant business department at higher level units designated for use, cannot be replaced. This gave the school coordination processing capacity of the test. Some business system integration unit for lack of motivation, and even has resentment; it also brought difficulty to work.

5 THE PROSPECT OF THE DEVELOPMENT OF A DIGITAL CAMPUS

Digital campus is significant results, will be popular in the education industry. From the company's website to check, they not only provide higher education school product digital campus, and also provides professional education, basic education plan. From the chart, we can see that on the Baidu search engine, the country since the end of 2011, the "digital campus" has maintained a very high level.

The construction of digital campus will continue to go deeper into the digital campus, the existing scheme for the establishment of a number of information systems and three foundation platforms, but the lack of mining large data using. Some schools have already started the construction of the digital campus in the two phases, the three phases of the project, such as South-Central University for Nationalities (8, South-Central University for Nationalities. http://news.scuec.edu .cn/xww/?view-6276.htm.). With the development of the times, some new applications are beginning to enter the campus, campus, such as the mobile cloud service, location service.

In the background of the global informationization development, with "Digital Campus" is the symbol of the domestic higher education informatization has made a good start. At the same time, we should have a clear understanding, the informatization level of domestic colleges and universities with the developed countries there is still a considerable gap, we must continue to emancipate the mind, change ideas, strengthen the input of capital and manpower, technology, the realization of the digital campus better for teachers and students, social service.

Information, Computer and Application Engineering – Liu, Sung & Yao (Eds)
© 2015 Taylor & Francis Group, London, ISBN 978-1-138-02717-6

On MOOC applied in English education for English major students

Feng Cun An & Yu Si Wu
Yanbian Univeristy, Yanji, China

ABSTRACT: The rapid development of MOOC brings large impacts to the traditional models of instruction, which influences English teaching and education obviously. On the basis of the resource characteristics of MOOC and the current issues of English teaching and education from English Major, the thesis discusses the application of MOOC.

KEYWORDS: MOOC; English Teaching; effects

1 INTRODUCTION

MOOC (Massive Open Online Course) is the most open network course model which is free for the public. With the help of internet, it provides the public with the resources of courses from worldwide famous universities, and the learners can select the course online according to their demands. The concept of MOOC was put forward by Bryan Alexander and Dave Cormier in 2008, and afterward, it made great progress in 2012, so the year was named the first year of MOOC era (Balfour 2013). MOOC is different from the former on-line study or online courses basing on the internet. It lays emphasis on the interaction between teaching and learning, on the basis of course resources issued by MOOC. The model of instruction with the unique micro course with quizzes and the forms of learning open in large-scale is very popular with the public, especially the young learners. MOOC symbolizes a new model of courses in the open educational field and reflects the trend, transferring from the simple resources to the resources of courses and instruction in the open educational field.

2 RESOURCE CHARACTERISTICS OF MOOC

MOOC makes it easy to acquire resources for teaching and learning with high quality. The whole process of education is systemic, transparency and unprecedentedly open, which brings the enormous impact on the traditional education. Course resources of MOOC own the following characteristics:

2.1 Open online in large scale

The supreme trait of MOOC resources is to online in large scale (Yin 2014). Compared with traditional classroom with dozens of students, MOOC can hold thousands of people, and even more, which is really large-scale. The learners with varying identities and background distributing around the world can register for MOOC. And the requirement to register is not high, and uses can register unrestrictedly. "Online", here, means that the form of learning is not restricted by the time and territory, and the users can study online with any mobile devices. MOOC can supply the learners with the complete learning experience just like the study online and a series of learning process with synchronous study, discussion, assignment, quiz and certificate. "Open" means that the learning resources of MOOC are completely free and open to the public, which broke the ideology that the higher only exists in the institutions of higher education and through which the learners can share the global teaching resources with high quality online. Meanwhile, the learning procedure is also open to the individual, and the learners can make a learning schedule of their own without any restriction like in the classroom study.

2.2 Multidimensional interaction environment offered

The platform of MOOC holds the learner-centered design philosophy from the beginning and constructs the modularized curriculum resources and the interactive teaching platform. MOOC manages to build up an interactive mode with the internet cooperation and interaction online. It tries to realize the interaction between learner and learner, creators of resources, and professors. The learners registering the same course can discuss the problems during online learning and share learning experiences.

2.3 Developing rapidly and high capacity

With the domestic and overseas active exploration on the teaching mode of MOOC, the quantity and quality of teaching resources to develop rapidly. Such development not only reflects on the rapid increase of the resources, but on gradual international trend. The course resources on the platform of MOOC have developed from several courses to hundreds of courses since the first establishment in 2012 and increase rapidly still. The covered disciplines also develop from inchoative simplification to diversification. At present, MOOC have been covered varied disciplines like foreign languages, arts, physics, economics, etc. In addition, the origins of course resources have improved from the several famous American universities to numerous colleges and universities spreading in more than 30 countries and regions. At present, some topmost institutions of higher learning have joined and pushed out their own distinctive courses onto the platform of MOOC.

3 THE CURRENT ISSUES ON TEACHING AND EDUCATION IN ENGLISH MAJOR

3.1 *Limitation of classroom teaching model*

The traditional teaching model with one book, teacher, and one class cannot satisfy the students' requirements. In classroom English teaching, the main content is the textbook. Before the teacher plan a course, the first task is to select a textbook as the teaching content in a semester. Generally speaking, the content of the textbook is hysteretic, and only one text book can't satisfy the students' requirements of knowledge. The whole process of the course is fulfilled by a fixed teacher in a fixed classroom. The limitation of teaching places will influence the students and only a few students in the classroom will benefit from the teaching. It's very hard to motivate the enthusiasm of the students in learning, and students with similar background in the same classroom can't have broad sense and scope of some certain questions. For the classroom teaching model, the teacher will meet limitations because of their own structure of knowledge and way of thinking. And the teachers have a different cognitive level and the abilities to analyze and resolve problems. The teacher will adopt the same methods to teach and influence the students without caring of individual differences, so it's very hard to train the students with distinct individuality and ability of creative thinking.

3.2 *Limitations of teachers' knowledge and methods*

With the development of the digitization and networking age, people come to be used to the convenience of new technology, especially their application in the educational field (Inge.de 2013). The teaching model with the modern technology advocates the student-oriented concept and lays emphasis on learner autonomy, which is being chased after by more and more learners. While the traditional classroom English teaching model can't reflect the dominant role of the students, and the English education is impelled by the concept of English instrument. The English teachers' teaching just stops at the level of language knowledge, and they put the emphasis on vocabulary explanation and reading skills. The teaching task is to explain the language knowledge, assess the acquisition of language knowledge and use the knowledge. This kind of teaching neglects the application of English in practice, overlooks the thought, emotion and social competence of learners. The former president of Tsinghua University in the period of the Republic of China, MEI Yiqi, said, "College gets its name of the great master, not for the building." The modern technology proposes the higher requirements for the teachers, but many teachers still come apart the advanced network technique and online courses resource.

3.3 *Pluralistic development of the students*

The evaluation criterion to talents in the modern society is more and more diversified. College students as a group of the modern time are influenced by the social and economic environment of the times to develop multiply. They have active thoughts, strong self-consciousness, and distinctive individuality. All of these cause them to have quite different learning goals and characteristics. The students in English major must have the common features of the modern college students, and have their own distinctive characteristics. Because they are long-term fed up with foreign culture, their ways of thinking, value orientation and moral outlook change profoundly into the tendency of personal independence of conduct. Their various learning motivations and goal must be accompanied with multiple learning paths and participation. Now the students in the English major have stronger individuation, autonomy and initiative in the ways of participation. They tend to choose the course materials, tools and the ways of participation autonomously which are fit for themselves according to their own motivation and goal.

3.4 *Learners' autonomy, lack of instructions and extracurricular resources*

Behavioral autonomy is the inevitable feature of human being's development, so the learner autonomy is the inevitable choice for the personality development of the students. To the students in English major, it is not enough to acquire the language only in the classroom, and the students can't give full play to their dominant role in the class, which cries for the avocation of the learner autonomy, awakening their potentials and promoting the autonomous development. Although, many teachers try to use the multimedia in their teaching, and encourage students to search the relevant learning materials and present their views and opinions with multimedia equipments, because the teachers can't instruct the learning style and approach to the students appropriately, and they can't construct the extracurricular learning resources well, the result of learner autonomy is not ideal.

4 APPLICATION OF MOOC IN ENGLISH MAJOR EDUCATION

4.1 Change of the teachers' roles

Teachers are the key elements to ensure the quality of school teaching, and the fundament of the connotation construction of the school. In the teaching environment of MOOC, some teachers will face the danger of being eliminated. At the age of MOOC, teachers are no longer sole teachers, and they will improve toward diversification, specialization and professionalization. Teachers not only undertake the task of teaching, but act the roles of inductor, analyst and researcher (Deng 2014). Teachers should strongly believe that students are capable of acquiring the knowledge by the way of learner autonomy, but not depending on the teachers' teaching. Teachers should study the successful experience of MOOC earnestly. In the learner-centered teaching system, teachers should take on the role of educators of information. Teachers are the instructors of teaching in the classroom and guiding on MOOC. And teachers should make full use of MOOC to improve their teaching on the platform of MOOC. Because the amount of the students on MOOC is very large and it's easy to analyze the big data, teachers can analyze the reasons of success and failure of the students' achievements with the database of MOOC and adjust the relevant contents and methods of the course. So teachers are also an analyzer. Furthermore, teachers should enhance the self digitization level and adapt the teaching model of digitization, combine the traditional classroom with the network.

4.2 Promoting the ability of learner autonomy and increasing sense of identity on MOOC

With the enhancement and amplification of the social functions of English, besides the universities, some institutions of different level continually push out many distinctive courses. Resources of English teaching from worldwide famous universities are integrated by MOOC. While these resources broad the learning approaches for English major students, they impact the traditional classroom teaching a lot. In the environment of MOOC, teachers are not the

only people who can help the students to resolve problems (Sandeen 2013). For example, if students want to improve their listening, they can ask MOOC for help; listen to the relevant resource of listening from other teachers. This awakens the students' subject consciousness in some degree. On the other hand, this can reduce the teacher's burden.

We investigated the 200 students in our English major about their cognition on MOOC through a questionnaire. We did the statistical analysis in six aspects: satisfactory, valuable, available, effective, necessary and not familiar with. The statistical results were as the chart below:

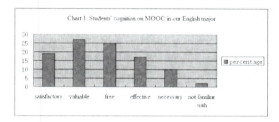
Chart 1 Students' cognition on MOOC in our English major

4.3 Teachers' devotion in the teaching design of MOOC courses

In traditional teaching, teachers can know of the students' progress of learning through the face-to-face way, but in MOOC teaching model with the students at the center, teachers transfer the initiative to students. And with the changes of teaching models, teachers should adjust the teaching design. The textbook is just the outline of the courses, and in the design of the course, teachers should think of many relevant reference materials, so the teachers should collect more materials about the teaching contents. It is found that man's attention can concentrate in 15 minutes, and MOOC teaching model breaks the defect of scattered attention by successional teaching in traditional class (Zhao 2014). Teachers can arrange the length of MOOC videos according to such characters. Knowledge module can be divided into many video fragments around 15 minutes, so students can arrange their time to learn different fragments according to their own individual characteristics. Meanwhile, the materials and resources of English teaching are relatively plentiful. In the teaching videos on MOOC, instructors can make full use of network resources and adjust the teaching contents.

4.4 Properly arranging the learning schedule on MOOC according to the teaching content

With the development of MOOC, the teaching sites become flexible, and students have more approaches to acquire knowledge. The teaching site is not limited in the classroom. Teachers can use any occasions, according to the teaching contents. Also, they can use the existing resource on MOOC, and they can make videos of teaching by themselves and send onto MOOC. Teachers can order students to finish studying relevant contents in the appointed time, which can promote the ability of learner autonomy and can build up a formation of resources sharing, the class extends in large scale, and more students with the similar interest can join English major students to discuss some questions. And teachers who teach the same course can form a teaching team to join efforts in developing courses. In addition, the teachers who teach the same course can communicate with each other or even form a teaching team before publishing resources in order to improve the quality of online courses. Online quiz is one of the characteristics of MOOC platform. The teaching team can design some online quizzes. After fixing answers to the objective questions in the system, the system can give students' scores automatically so that students can get feedbacks on time. Teachers can also give an assignment to the students who have passed the quizzes online. After a series of learning process, learners can enter the online forums to join in the discussion with peer academics, video publishers or course-related teachers. A specific process is shown in Chart 2:

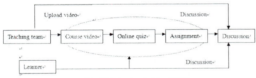

Chart 2: The learning schedule on MOOC.

4.5 Combination of MOOC and classroom

In English teaching, the effective combination of the two is necessary. That is to say we should not only give full play to the teacher's leading role of guiding, inspiring and monitoring in the teaching process, but also motivate the learners' enthusiasm and subjective activity in the learning process. A 'Blended Learning Model' combined the classroom teaching with MOOC learning process can help a lot. The combination of MOOC and classroom is shown in Chart 3:

Chart 3 Combination of MOOC and classroom.

5 CONCULSION

The emergence and development of MOOC is a unique product of the information-based society, as an educational innovation platform in the information age, it has a profound impact on the development of teaching in English Major Education. Teachers should exert their own experience and strengths to make international MOOC have local characteristics and try to improve the education level of the domestic teaching and service through using MOOC effectively.

REFERENCES

[1] Balfour S.P. 2013. Assessing Writing in MOOCs: Automated essay scoring and calibrated peer review [J]. *Research &Practice in Assessment*.
[2] Fang Luo&Changxing Yang& Weiguo Liu.2014.9. Studies on curriculum teaching design based on MOOC—take university computer class as an example. 13–9.
[3] Hedong Yin. 2014.9.MOOC's impact on higher education. *Journal of Chongqing Second Teachers College*. 5–27.
[4] Huijie Zhao.2014.6. Study on the reconstruction of higher education teaching system based on MOOC. *The Shandong higher education*. 6.
[5] Inge.de W. 2013.Analyzing the impact of Mobile Access on Learner Interactions in a MOOC [D]. *Athanasca Unibersity*.
[6] Manying Deng.2014.11. Curriculum resources construction of second language based on MOOC. *Journal of Changsha Railway University*. 15–4.
[7] Sandeen C. 2013.Assessment's place in the new MOOC world[J]. *Research &Practice in Assessment*.
[8] Yuqin Yang &Jianli Jiao. 2014.Design Framework of MOOC Learners' Individualized learning. *Audio-Visual Education Research*. 08–005.
[9] Zhansheng Mou &Bojie Dong. 2014.Research on Blended Learning based on MOOC—Coursera as an example. *Modern Education Technology*. 5.

Information, Computer and Application Engineering – Liu, Sung & Yao (Eds)
© 2015 Taylor & Francis Group, London, ISBN 978-1-138-02717-6

The application of all-optical network optimization algorithms based on grooming factor

Su Xia Cui & Yuan Yuan
Binzhou Polytechnic, Binzhou, Shandong, China

ABSTRACT: In all-optical networks, Add-Drop Multiplexer (ADM) is the basic electronic processing device using the point difference, which has the function of a separation or insert the desired wavelength to networks. With Wavelength Division Multiplexing (WDM) technology are increasingly improved, all-optical network is the most important thing is to reduce the cost of the network used in electronic processing equipment. The problem of minimizing the number of ADMs is a main research topic in recent studies. Traffic grooming problem is an important generalization of the ADM minimization problem, which is NP-hard. The goal is to try to minimize the network cost, such as the number of wavelengths required per fiber and the number of ADMs required in the networks by effectively packing lower rate traffic streams onto the available wavelengths. In this paper we pay attention to study the traffic grooming problem in bi-directional interconnected SONET/WDM ring networks. We apply the algorithm GROOMBYSC(k) to the bi-directional interconnected SONET/WDM rings, and we further improve the analysis of the algorithm, obtaining a similar result, and the approximation ratio is $2 \ln g + o(\ln g)$, where g is the grooming factor, and the grooming factor smaller than its approximation ratio of the approximation algorithm is also smaller, the better the performance of the algorithm can be seen, what's more we can obtain proved a general of the approximation algorithm.

KEYWORDS: Traffic grooming; ADM; SONET/WDM ring; Approximation ratio; Grooming factor

1 INTRODUCTION

With the rapid development of modern network technology, the demand for capacity and quality of service communication networks become the driving force of the growing network technology research. In order to overcome by the speed signal processing to achieve information exchange network, each node brings electronic bottleneck, it made the most promising technologies, the all-optical network. The so-called all-optical network is the network from the source node to the destination node of data transmission and information exchange between the whole processes is carried out in the optical domain, i.e., the end of the complete optical path, the intermediate electrical signal without any intervention, and the main medium of the transmission is an optical fiber.

WDM network is a cross- switching nodes are connected by a fiber link composed, A WDM wavelength channels carrying both low- speed flow is also able to carry traffic. In a communication system, most of the connections have a small bandwidth request, such as voice or text data, etc. Under no circumstances traffic grooming, each connection will occupy a separate wavelength channel. On WDM networks, the wavelength is a limited resource, so often there is not enough wavelengths to each connection provides

for an independent wavelength channel. Therefore, it is very necessary for low-bandwidth connections to allow multiple requests to share the same wavelength channel. Traffic grooming traffic is converged to a plurality of connection requests technical single wavelength channel, which greatly increases the bandwidth utilization rate of the wavelength channel, for example, if a wavelength channel bandwidth is OC-48, a bandwidth request is a linking group OC-12 (i.e., 0.622Gbit / s), then the connection may be to divert the four wavelength channels to an OC-48, wherein, to ease factor is 4.

In all-optical networks, basic electronic processing equipment using Add-Drop Multiplexer (ADM), whose functions is to separate the network or insert the desired wavelength, that is, the optical network temporarily unneeded wavelengths from the fiber separated, the wavelength of the network need to insert a channel in a data network for transmission of information.

In fact, the traffic grooming problem is converge the low-speed to high-capacity optical path of transmission traffic through effective multiplexing and switching processing technology, which is to improve the utilization of network resources. Using graph theory terms, traffic grooming problem can be seen as a path coloring problem, so most of which share an edge path g (g is the grooming factor). From the perspective of ADM, each optical path uses two ADMs, each endpoint requires one ADM. At a node case by the same edge with g optical paths having the same wavelength to arrive one node, and they may use the same ADM, thus it can save g-1 ADMs, its purpose is to minimize the use of the number of ADM [1].

With the maturity of the rapid development of modern communication networks and optical fiber communication technology, the mid -second century, the United States proposed a fiber -based network communication technology, synchronous optical network (SONET), which is a fiber-optic network for high-speed data communications transmission network, compared with the general network, SONET spread further example, the most important feature of the SONET network is support the ring topology structure network .

A SONET ring is composed of two or more nodes of a SONET/WDM closure ring, which is part of the fiber optic working ring, as part of the protection ring. Working ring is used for data transmission network need. Protection ring is a spare ring. The SONET / WDM ring has a self-test function, when the working ring failure in the data transmission process, SONET/ WDM ring will self-test and within a very short time the protection ring provides control information to the protection ring is in working condition. Therefore, a SONET / WDM ring is also known as self-healing ring, and it has the function of self-healing capabilities. The SONET / WDM ring is divided into two kinds of unidirectional and bidirectional ring. In unidirectional SONET / WDM ring, the direction of the working ring is the clockwise direction, i.e., in the data transmission is clockwise direction when carried out traffic transmission in SONET/WDM ring. In the case of bidirectional SONET / WDM ring, the direction of the working ring is bi-directional, i.e., both clockwise and counterclockwise directions opposite, when the two nodes adjacent is traffic transmission using two opposite clockwise and counterclockwise direction to data transmission.[2]

In electronic communications networks, SONET / WDM ring is widely used. Traditional SONET ring using a single wavelength, however, with the advent of WDM technology, in order to meet the growing traffic demands, these rings need to be upgraded to WDM, and multiple SONET ring cross- connected to provide a large area on the convergence.

In a cross-connected SONET / WDM ring, there are two kinds of node, that is, non-cross connections node and cross- connections node. A non-cross-connected node which with two interfaces to connect its two neighbors, as well as a local interface to implement to increase or decrease the flow request. In this article, we only relate to cross-connect nodes, and the nodes connected to the cross SONET / WDM ring,

which the two rings are interconnected between either a single node or multiple nodes.

As shown in Figure 1(a) and Figure 1(b):

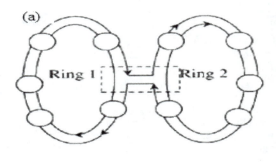

(a)

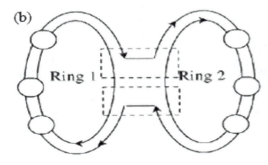

(b)

Figure 1.   (a) Single –point ring connected.

Figure 1.   (b) Multi-point ring connected.

If the traffic request goes to another ring by optical switch, which we call a single jump. If the traffic request from one ring to reach a digital cross-connect (DXC) firstly, then reach another ring, which we call a two-hop (or multiple jumps). This paper is based on the single-hop problem.

The literature has been given a ring network traffic grooming problem ($g > 1$) related concepts, and traffic grooming problem has proven to be NP-complete[3]. O. Gerstel, et al. who introduced the ring network, and minimized the number of the ADM in the case of g = 1, and proved this problem is NP- complete [4] [5]. In a ring network, for any fixed value of g, M. Flammini etc. who have proposed an approximation algorithm with approximate ratio $2 \ln g + o(\ln g)$[6]. In order to improve bandwidth utilization, B. Chen etc., Respectively, under different circumstances for the ring and star topology networks, traffic grooming technique using electronic processing equipment to minimize the problem of all-optical networks used in a profound study [7][8] .

M.Flammini, etc. proposed an approximation algorithm in a tree network with approximate ratio

$2\ln g + \ln \Delta + o(\ln(\Delta \cdot g))$, which Δ is the largest degree of the nodes in the tree network [9].

2 THE DEFINITION OF RELATED ISSUES

Traffic grooming problem can be abstracted as a graph instance, which can be modeled as a graph coloring problem. An example of triplet (G, P, g), Where G = (V, E) is a request diagram, P is a simple path s set of G, g is a positive integer, which called grooming factor.

Definition 2.1.1: Given a subset $Q \subseteq P$ and one edge e, $e \in E$, Q_e is the paths set using edge e of Q. $l_Q(e)$ is the number of these paths, or with network jargon, which is the load on edge e which exported from Q path. L_Q is the maximum load on any one edge of G which exported from path Q.

Definition 2.1.2: A coloring(Wavelength assignment) of(G,P)is a function $\omega : P \mapsto N^+ = \{1,2,...\}$. We further extend the definition of the function ω on any subset Q of P, $\omega(Q) = \bigcup_{p \in Q} \omega(p)$. For a coloring ω, one color λ and a subset $Q \subseteq P$, Q_λ^ω is colored λ by ω in a path from a subset of Q.

Definition 2.1.3: A right coloring (Wavelength assignment) of (G, P, g) is one coloring of P, for any one edge e, there are g paths with the same color coloring used edge e mostly. Formally, $\forall \lambda \in N^+, L_{P^\omega} \le g$.

Definition 2.1.4: If you use one color on the Q path coloring, then coloring ω is a 1-coloring of $Q \subseteq P$; if the presence of a right 1- coloring for path set Q, then Q is1-colorable; if and only if $L_Q \le g, Q \subseteq P$ is 1-colorable.

Definition 2.1.5: For one coloring ω of P and one node $v \in V$, Q_v is a subset which has an endpoint in the set of paths Q of v. v may be the source node or the destination node. Degree of a node v is the number of path which v is the source node or destination node. In form, $d(v) = |Q_v|$; The degrees of the set of paths Q is the sum of the degree of all nodes in Q. ADM_λ^ω is the number of ADM used in wavelength λ on network. ADM^ω is the total number of ADM used in the network, that is, the total number of ADM on different wavelengths of operation.

In form,

$$d(Q) = \sum_v d(v) = \sum_v |Q_v|; \tag{1}$$

$$ADM_\lambda^\omega(Q) = d(Q_\lambda^\omega); \tag{2}$$

$$ADM^\omega = \sum_\lambda ADM_\lambda^\omega(P) = \sum_\lambda d(P_\lambda^\omega) \tag{3}$$

Traffic grooming problem is an optimization problem that to find a right coloring ω of instance (G, P, g) such that the minimum number of ADM used in optical network.

3 GROOMBYSC (K) ALGORITHM

Known an instance (G, P, g) of traffic grooming problem, the algorithm has a factor g depends only on a parameter k of grooming factor g, and the value of k is determined in the algorithm analysis.

The algorithm of GROOMBYSC(k) to achieve in three stages. The first stage, calculation 1- colorable sets and their corresponding weights, considering the subset of paths P which size up to $k \cdot g$. When it finds a set of paths 1- colorable Q when put Q and their corresponding weights added to the relevant set, which makes the path set Q ADM required minimum number (when the path of Q colored with the same color). The second phase, it is to find a set cover of path P with the subset calculated in the first stage, which using a minimum weight set cover problem proposed GREEDYSC approximation algorithm[10] in this process. The third stage, by eliminating the set intersection, the set cover problem will be converted to a partitioning problem and then colored with the same color for each set of paths, each division are set to use a different color shading.

3.1 GROOMBYSC(k) algorithm is described as follows:

a) Phase 1- Prepare GREEDY input:

$S \leftarrow \phi$;

For each, $U \subseteq V$ such that $|U| \le k$ {

For every path set of every ring $Q \subseteq P_U$,

Such that $|Q| \le k \cdot g$ {

If Q is 1-colorabe , then{

$S \leftarrow S \cup \{Q\}$;

Weight[Q]=d(Q);

}

}

}

b) Phase 2-do GREEDYSC:

//Without loss of generality, set $SC = \{S_1, S_2, ..., S_W\}$

$SC \leftarrow GREEDYSC(S, weight)$.

c) Phase 3-Put set cover SC convert to a division PART:

$$PART \leftarrow \phi$$

$$\text{For i=1 to W } \{PART_i \leftarrow S_i\}$$

As long as there are two disjoint sets $PART_i, PART_j$ {

$$PART_i \leftarrow PART_i \setminus PART_j$$

$$\}$$

For $\lambda = 1$ to W $(\omega(Q) = W)$ {

$$PART \leftarrow PART \bigcup \{PART_\lambda\}$$

For each $p \in PART_\lambda$ { $\omega(p) = \lambda$ }
}

3.2 Algorithm analysis

3.2.1 Correctness and run time

Correctness and time complexity of the algorithm is analyzed and has been given a similar proof[5], Which has been demonstrated for any fixed value of k and the instance (G, P, g),the run time of the algorithm is polynomial time in $n = |P|$ and $m = |E|$. We can prove to be applied by a similar cross-connect ring GROOMBYSC (k) obtained by analyzing the algorithm, and its run time is polynomial time too in $n = |P|$ and $m = |E|$ [5].

3.2.2 Approximation ratio

Lemma: Given an integer t, there is a solution $\overline{SC}$ for the instance on the set covering problem which determined in the second phase of GROOMBYSC (k) algorithm, we have $k = 2t-1$, such that

$$weight(\overline{SC}) \le ADM^*(1 + \frac{2g}{t}).$$

Proof: Set $\omega^*(P) = \{1, 2, ..., W^*\}$ and $1 \le \lambda \le W^*$, considering the set V_λ^* of node v such that $ADM_\lambda^*(v) = 1$, that is on the node v has an operation wavelength λ in the ADM. Start from any node in the clockwise direction of the ring will be set V_λ^* into a plurality of at least t the set of nodes, and network nodes at most $k = 2t-1$ (as shown in Figure 2). For cross two SONET rings connected, each ring have taken this division, the optimization problem are independent of each ring, reach (or derived from) the other traffic on the ring comes from (or arrive) cross node. Set $V_{\lambda,j}(j = 1, ..., P_\lambda)$ is the subset of nodes in this way. Set q_λ is the number of subset with at least t nodes in V_λ^*, if there has present the subset is less than t nodes in V_λ^*, then set $r_\lambda \le t$ is the remaining number of nodes in the subset. In form, se $ADM_\lambda^* = |V_\lambda^*| = tq_\lambda + r_\lambda$, among them $r_\lambda = |V_\lambda^*| \bmod t$ and $0 \le r_\lambda \le t$.

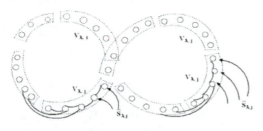

Figure 2. Set $V_{\lambda,j}$, $\overline{S_{\lambda,j}}$ and $t = 5$.

Note that if no fewer than t nodes in the collection division, then $p_\lambda = q_\lambda$ and $r_\lambda = 0$; Otherwise, $p_\lambda = q_\lambda + 1$. Clearly, $\forall 1 \le j \le q_\lambda$, $|V_{\lambda,j}| = t$ and in the case of $r_\lambda > 0$, we have $|V_{\lambda,q_\lambda+1}| < t$; In both cases, we have $|V_{\lambda,j}| \le t$, then $t \le k$, we have $|V_{\lambda,j}| \le k$ .So $V_{\lambda,j}$ is added to the set S.

For $V_{\lambda,j}$, we define $\overline{S_{\lambda,j}}$ is the path set of $P_\lambda^{\omega^*}$, and there is counterclockwise endpoint in $V_{\lambda,j}$ .Because $V_{\lambda,j}$ have t nodes at least, and each node may be a clockwise endpoint of at most g paths which is 1-colorable.Then, have $|\overline{S_{\lambda,j}}| \le g \cdot k$ .Therefore, $\overline{S_{\lambda,j}}$ is taken into account in the inner loop algorithm in the first stage. $\overline{S_{\lambda,j}}$ is a 1-colorabe,is added to set S, that is, $\overline{S_{\lambda,j}} \in S$.

Two endpoints of each path $p \in P_\lambda^{\omega^*}$ are in the set $V_{\lambda,j}$, specifically, for a certain value of j, it has its endpoint in a clockwise in $V_{\lambda,j}$ .So, it is an element of the set $\overline{S_{\lambda,j}}$, then we have $\overline{SC_\lambda} = \bigcup_j \{S_{\lambda,j}\}$ is a set cover of $P_\lambda^{\omega^*}$. We conclude, for $1 \le \lambda \le W^*$, we have $\overline{SC} = \bigcup_{\lambda=1}^{W^*} \overline{SC_\lambda}$ is a set cover of path set P. Let's prove that the weights have the following attributes:

We are known $ADM_\lambda^* = tq_\lambda + r_\lambda$, the sum of all possible values for λ, get $ADM^* = t\sum_\lambda q_\lambda + \sum_\lambda r_\lambda$, this means $\sum_\lambda q_\lambda \le \frac{ADM^*}{t}$. Two cases are considered:

1 When $\forall j \le q_\lambda$ $weight(\overline{S_{\lambda,j}}) = d(\overline{S_{\lambda,j}}) \le t + g$. Because:

 a) The source and destination nodes are the endpoints of the paths $\overline{S_{\lambda,j}}$ are also in $V_{\lambda,j}$, and $|V_{\lambda,j}| = t$.

 b) The source node and the destination node, only one is at most g paths in the path $V_{\lambda,j}$.

2 When $j = q_\lambda + 1$, $weight(\overline{S_{\lambda,j}}) \le g \cdot q_\lambda + r_\lambda$

Because:

 a) Only one endpoint is at most g paths in $V_{\lambda,q_\lambda+1}$.

 b) The source and destination nodes are the endpoints of the paths $\overline{S}_{\lambda,q_\lambda+1}$ are also in $V_{\lambda,q_\lambda+1}$, and $|V_{\lambda,q_\lambda+1}| = r_\lambda$.

Summing all $1 \le j \le q_\lambda + 1$, we have:

$$weight(\overline{SC_\lambda}) \le \sum_{j=1}^{q_\lambda}(t+g) + g \cdot q_\lambda + r_\lambda$$

Because $ADM^* = t\sum_\lambda q_\lambda + \sum_\lambda r_\lambda$, and

$$\sum_\lambda q_\lambda \le \frac{ADM^*}{t}.$$

Thus, we get the following results:

$$weight(\overline{SC}) = \sum_{\lambda} weight(\overline{SC_{\lambda}}) \leq ADM^* + 2g\sum_{\lambda} q_{\lambda} \quad (4)$$

$$\leq ADM^* + 2g \cdot \frac{ADM^*}{t} = ADM^*(1+\frac{2g}{t}) \quad (5)$$

(2)**Theorem:** For bidirectional SONET ring in the cross-connect traffic grooming problem, there is an approximation algorithms with approximate ratio $2\ln g + o(\ln g)$.

Proof: A greedy algorithm with minimize weight set covering problem is an H_f -approximation algorithm, among them, f is the largest potential of algorithm input set and $H_f = 1 + \frac{1}{2} + ... + \frac{1}{f}$ is the first number f harmonic number.

We know $ADM^{\omega} = H_f \cdot weight(\overline{SC})$ [5], and $weight(\overline{SC}) \leq ADM^*(1+\frac{2g}{t})$,
So, we have:

$$ADM^{\omega} \leq H_f \cdot ADM^*(1+\frac{2g}{t}) \quad (6)$$

Equivalent to

$$\rho \leq H_f \cdot (1+\frac{2g}{t}) \quad (7)$$

We were carried out on the value of t and f , take $t = g\ln g$, $f = (2t-1)\cdot g$ (Only consider bidirectional SONET ring which has only one cross -way intersection connected), then,

$$\rho \leq H_{(2t-1)\cdot g} \cdot (1+\frac{2g}{t}) \leq H_{(2t-t)\cdot g} \cdot (1+\frac{2g}{t})$$
$$= H_{t\cdot g} \cdot (1+\frac{2g}{t})$$
$$\leq (1+\ln(t \cdot g))(1+\frac{2g}{t}) = (1+\ln(g^2 \ln g))(1+\frac{2}{\ln g})$$
$$= (1+2\ln g + \ln\ln g)(1+\frac{2}{\ln g})$$
$$= 5 + 2\ln g + \ln\ln g + \frac{2}{\ln g} + \frac{2\ln\ln g}{\ln g}$$
$$= 2\ln g + o(\ln g) \quad (8)$$

Thus, the proof is completed.

4   CONCLUSION

This paper main highlights traffic grooming problem in cross-connect bidirectional SONET / WDM ring, and mainly studied the approximate ratio g of logarithmic approximation algorithm, and applied to solve the cross-connect bidirectional SONET / WDM ring traffic grooming problem, and through the analysis of the algorithm obtained approximate ratio $2\ln g + o(\ln g)$. By proven processes and results, the approximation ratio of the algorithm is independent of the number of nodes in cross-connect bidirectional SONET/WDM ring, only depends on the grooming factor g. Grooming factor smaller than its approximation ratio of the approximation algorithm is also smaller, the better the performance of the algorithm can be seen, which can greatly improve and optimize the network performance.

By the results of this verification in this article, GROOMBYSC (k) approximation algorithm can be applied to any topology, through more accurate analysis and prove similar results can be obtained.

This article will be bidirectional SONET ring GROOMBYSC (k) approximation algorithm is applied in the cross-connect, get better performance, proved a general of the approximation algorithm.

REFERENCES

[1] K.Zhu and B.Mukherjee. "A review of traffic grooming in wdm optical networks: Architecture and challenges." Optical Networks Magazine, 4(2):55–64, March-April 2003.
[2] O.Gerstel, R.Ramaswami, and G..Sasaki. "Cost effective traffic grooming in wdm rings." In INFOCOM'98, Seventeenth Annual Joint Conference of the IEEE Computer and Communications Societies, 1998.
[3] Walter Goralski. Optical Networking & WDM[M]. Post & Telecommunications Press. 2003, pages 118–122, 148–150.
[4] O. Gerstel, P. Lin, and G. Sasaki. "Wavelength assignment in a wdm ring to minimize cost of embedded sonet rings." In INFOCOM98, Seventeenth Annual Joint Conference of the IEEE Computer and Communications Societies, 1998.
[5] T. Eilam, S. Moran, and S. Zaks. "Lightpath arrangement in survivable rings to minimize the switching cost." IEEE Journal of Selected Area on Communications, 20(1):172 C 182, Jan 2002.
[6] M. Flammini, L. Moscardelli, M. Shalom, and S. Zaks. "Approximating the traffic grooming problem." In ISAAC, pages 915 C 924, 2005.

[7] B. Chen, G. N. Rouskas, and R. Dutta. "Traffic grooming in wdm ring networks with the min-max objective." In NETWORKING, pages 174 C 185, 2004.

[8] B. Chen, G. N. Rouskas, and R. Dutta. "Traffic grooming in star networks." In Broadnets, 2004.

[9] M. Flammini, G. Monaco, L. Moscardelli, M. Shalom, and S. Zaks, "Approximating the traffic grooming problem in tree and star networks," presented at the WG 2006 Workshop, Bergen, Norway, Mar. 2006.

[10] D. S. Johnson. "Approximation algorithms for combinatorial problems." J.Comput. System Sci., 9:256 C 278, 1974.

Information, Computer and Application Engineering – Liu, Sung & Yao (Eds)
© *2015 Taylor & Francis Group, London, ISBN 978-1-138-02717-6*

A secure data storage mechanism for SE-based ticket applications

Jian Chao Luo & Si Yu Zhan
School of Computer Science and Engineering, University of Electronic Science and Technology of China, Chengdu, China

ABSTRACT: Ticketing has become one of the major application fields of NFC technology. In order to secure the electronic tickets, secure elements are used for ticket data storage. However, as file system is not a part of the Java Card specification and is missing in the Java Card implementation, managing data storage of an SE-based ticket application is not a trivial task for developers. This paper presents a data storage mechanism for SE-based ticket applications, which provides simple interfaces for accessing and storing ticket data. In addition, security mechanism is provided for securely accessing the ticket information. As a result, development of a secure SE-based ticket application is significantly simplified. This paper describes the design and implementation of the data storage mechanism, and presents a sample application that shows its feasibility.

KEYWORDS: Electronic ticket; Secure element; Data storage; Near Field Communication

1 INTRODUCTION

Almost everyone has a mobile phone these days. Mobile ticketing has rapidly gained popularity, as it can reduce the need for infrastructure, because users already have the payment equipment in their mobile phones, and save time that users spend waiting in line. The mobile ticket application delivers tickets right to the user's phone and the user will not have to print them out. Usually, the ticket arrives as a text message with a barcode. When the user shows up at the event or the cinema, the barcode will be read by a barcode reader. However, as it is not safe to store the ticket information in the mobile phone, some ticketing systems use NFC (Near Field Communication) enabled mobile phones that operate in card emulation mode for exchanging ticket data and store the sensitive information of the tickets into the secure element (SE).

An NFC device working in card emulation mode emulates a contactless smartcard and contains an SE to store the information for the emulated card in a secure way. The SE provides a secure environment to prevent unauthorized access to sensitive data. GlobalPlatform issues a card specification for standardization and interoperability of application management within an SE (GlobalPlatform, 2011). However, as file system is not a part of the Java Card specification and is missing in the Java Card implementation, managing data storage of an SE-based ticket application is not a trivial task for application developers. They will have to allocate storage space and design interfaces for accessing and storing application data in the SE all by themselves.

This paper presents a data storage mechanism for SE-based ticket applications, which enables mobile ticket application developers to store and access ticket information in an easy and secure way. As a result, development and deployment of a secure SE-based ticket application are greatly simplified.

The rest of the paper is structured as follows. Section 2 reviews the related works. In Section 3, we describe the design and implementation of the proposed system in detail. Section 4 presents the sample application built using the proposed data storage mechanism that shows the feasibility of the approach. In Section 5, we conclude the paper.

2 RELATED WORKS

The use of electronic tickets has become more and more common. The Handy-Ticket system (Handy-Ticket, 2014) enables customers to purchase tickets with their mobile phone via SMS. The text message contains a ticket code. However, it is very inconvenient for the customers to manually input the ticket code into the verification device for ticket validation. Hu et al. (2010) present an integrated ticketing system based on barcode. The barcode can be read through scanning devices automatically so as to achieve automatic processing of information. But it is still not efficient enough, compared with the tap-and-go NFC technology. The NFC technology can solve these problems thanks to its ease of use.

Ticketing has become one of the major application fields of NFC technology. Many ticketing systems like Small Events NFC-Enabled Ticketing System (Chaumette et al. 2012) use NFC enabled mobile phones in peer-to-peer mode for exchanging data and validating tickets. However, the ticket information is stored in the mobile phone, which lacks secure operating environment. Card emulation mode is based on the use of SE, which is used to store sensitive data. Widmann et al. (2012) introduce a scenario for the integration of an electronic ticketing system into an existing public transport system based on NFC card emulation mode. But it is difficult for application developers to deploy mobile ticket applications on the SE, because application management and data storage within an SE are very complicated.

3 THE PROPOSED SYSTEM

As shown in Figure 1, the SE-based ticket applications are deployed on the NFC enabled mobile phones in card emulation mode. The NFC reader communicates with the mobile phone via the Application Protocol Data Unit (APDU) command. The NFC reader is connected to the ticketing system, where the server application is running for ticket validation.

The proposed system is essentially an SE applet, which is called Secure Storage Applet. The Secure Storage Applet is a Java Card application that runs on the SE and controls access to the secure storage space allocated for the electronic tickets.

Figure 1. Usage scenario of SE-based ticket applications.

To allocate data storage space for the electronic tickets, the Secure Storage Applet should be installed on the SE before the tickets are loaded into the SE.

Figure 2 demonstrates the structure of the storage space for the tickets, which contains fixed-size records. Tickets from different service providers are stored into the ticket records respectively. Each ticket record consists of three parts: ticket information, counter and key. The field "counter" is used for recording the number of the times that the ticket can be used, while the field "key" is used for controlling access to the ticket record.

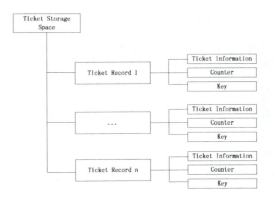

Figure 2. Structure of the ticket storage space.

In order to enable the NFC reader and the mobile ticket application to access the ticket information stored in the SE in a simple and secure manner, the proposed system provides a series of APDU commands used for adding, removing, reading and updating the ticket information. The off-card entity carries out these operations by sending the following APDU commands to the Secure Storage Applet.

- ADD TICKET
  The "ADD TICKET" command is used for adding a new electronic ticket record to the storage space. As described in Figure 2, the data field of the command should contain ticket information, counter value as well as record accessing key. However, in order to ensure that the command is sent by a trusted off-card entity, message authentication code (MAC) value of the command should be appended at the end of the command message. The MAC value is calculated by using the record accessing key. The initial vector used for the MAC calculation is a random number newly generated by executing the "GET CHALLENGE" command. Once receiving the "ADD TICKET" command, the Secure Storage Applet will verify the MAC value for identity authentication of the off-card entity. The content of the ticket information is defined by the ticketing service providers and hence different tickets (movie tickets, museum tickets, train tickets, bus tickets, etc.) have different contents.
- READ TICKET
  The "READ TICKET" command is used for getting information of one ticket. However, the key value of the ticket record cannot be read for security reason.
- UPDATE TICKET
  The "UPDATE TICKET" command is used for updating the ticket information. Before the

execution of the command, the command receiver verifies the MAC value of the command message to determine whether it is sent by a trusted sender.

- DELETE TICKET

The "DELETE TICKET" command is used for deleting a ticket. In the same way, the Secure Storage Applet will authenticate the identity of the command sender.

- READ COUNTER

The "READ COUNTER" command is used for getting the number of the times that the ticket can be used.

- INCREASE COUNTER

The "INCREASE COUNTER" command is used for increasing the times that the ticket can be used. To ensure secure operation, MAC value of the command has to be appended at the end of the command message.

- DECREASE COUNTER

The "DECREASE COUNTER" command is used for decreasing the times that the ticket can be used each time the customer uses the ticket. To avoid receiving counterfeit tickets, the off-card entity should verify the authenticity of the ticket by executing the "INTERNAL AUTHENTICATE" command before the "DECREASE COUNTER" command is executed. For security reason, MAC value of the command should be calculated and put into the data field of the command message.

- UPDATE KEY

The "UPDATE KEY" command is used for updating the ticket record key if needed. To ensure confidentiality of the new key, the key value is encrypted by the old key of the ticket record before its being written into the record. To ensure secure operation, MAC value of the command also needs to be appended at the end of the command message.

- GET CHALLENGE

The "GET CHALLENGE" command is used for generating a random number.

- INTERNAL AUTHENTICATE

The "INTERNAL AUTHENTICATE" command is used for verifying the authenticity of the ticket. The authentication process is shown in Figure 3:

1 The off-card entity sends the "GET CHALLENGE" command to the Secure Storage Applet for getting a newly generated random number.

2 The off-card entity sends to the Secure Storage Applet the "INTERNAL AUTHENTICATE" command, which contains the random number.

3 The Secure Storage Applet encrypts the received random number by using the ticket record key and returns the cipher text.

4 The off-card entity decrypts the received cipher text of the random number and gets its plaintext.

Authenticity of the ticket can be confirmed if the plaintext is the same as the generated random number.

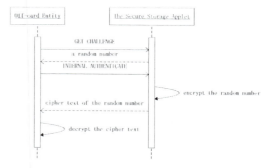

Figure 3. The process of verifying the authenticity of the ticket.

4 SAMPLE APPLICATION

For evaluating the feasibility of the proposed solution, one mobile ticket Android application for a swimming pool is developed. The ticket is stored into the SE of an NFC phone via the proposed system.

In order to add a new electronic ticket to the SE, the mobile ticket Android application developed for the swimming pool only has to send the "ADD TICKET" command to the Secure Storage Applet. Figure 4 shows the ticket information read by the mobile ticket application after its being added to the SE successfully.

Figure 4. The ticket information read by the mobile ticket Android application.

355

5 CONCLUSIONS

Java Card technology provides a runtime environment for applications that run on the SE. However, development of SE-based ticket applications is complicated, as file system is not a part of the Java Card specification and is missing in the Java Card implementation. The proposed secure storage mechanism provides easy-to-use interfaces for accessing electronic tickets securely, and therefore development of an SE-based ticket application is significantly simplified.

ACKNOWLEDGMENTS

This work is supported by the Mobile Payment Platform Research project (ID: 2013–265) of Important Science & Technology Projects of Chengdu Science and Technology Bureau and the National Natural Science Foundation of China under Grant No. 61202444.

REFERENCES

Chaumette, S., Dubernet, D., Ouoba, J., Siira, E. & Tuikka, T. 2012. Architecture and evaluation of a user-centric NFC-enabled ticketing system for small events. *Lecture Notes of the Institute for Computer Sciences, Social-Informatics and Telecommunications Engineering,* 95:137–151.

GlobalPlatform. 2011. GlobalPlatform Card Specification. <http://www.globalplatform.org/specificationscard.asp>

Handy-Ticket. 2014. <http://www.oebb.at/en/Tickets/Mobile_Tickets/SMS-Ticket/index.jsp>

Hu, L., Wang, Y., Li, D. & Li, J. 2010. A hybrid client/server and browser/server mode-based universal mobile ticketing system. In *Proceedings of 2010 2nd IEEE International Conference on Information Management and Engineering,* pp. 691–695.

Widmann, R., Grunberger, S., Stadlmann, B. & Langer, J. 2012. System Integration of NFC Ticketing into an Existing Public Transport Infrastructure. In *Proceedings of 4th International Workshop on Near Field Communication,* pp. 13–18.

Information, Computer and Application Engineering – Liu, Sung & Yao (Eds)
© 2015 Taylor & Francis Group, London, ISBN 978-1-138-02717-6

Evaluation method of selecting model for testing equipment for auto based on extension theory

Xiang Wen Dang & Rui Jun Liu

College of Automobile & Civil Engineering, Beihua University, Jilin, China

ABSTRACT: Automotive comprehensive performance testing equipment selection scheme affects economic and social problem, detection equipment selection evaluation criteria system based normative, economic, technical, service, energy saving environmental protection, human relations elements of the composition. Establish equipment selection decisions based on the feature set of extension theory, the use of expert experience to establish the desired selection program element model and matter-element model of alternative selection schemes, Extension associate degree through direct evaluation of the merits of the program. Examples show that automobile test equipment selection program evaluation based on extension theory is feasible.

KEYWORDS: Automobile; Testing equipments; Selecting model; Extension theory; Evaluation method

1 INTRODUCTION

Automobile comprehensive performance testing is a "regular testing, mandatory maintenance, repair, as appropriate," an important part of the vehicle maintenance system. Equipment used to detect the type of complex, different manufacturers of the same kind of testing equipment there is a big difference in the structure, quality, accuracy, price, life and reliability, service and other aspects, even in measuring principle, measuring method, measurement conditions, operating ease, intelligent and so also there is a big difference, resulting in detection equipment selection more difficult. Currently, equipment selection has a fuzzy comprehensive evaluation method and gray correlation method, but their deficiencies exist in the actual selection. The fuzzy evaluation method is a quantitative analysis method to establish the membership function has a lot of subjectivity, fuzzy information accurately process information will result in data loss. Gray evaluation method is an effective means to solve the problem of lack of evaluation information in the original data dimensionless treatment, different non-dimensional approach will directly affect the minimum and the maximum reference sequence and comparison of sequences. In addition, testing equipment selection is often conflicting and incompatible with the existence of circumstances, under certain conditions, cannot be directly addressed. In order to solve the problems and contradictions incompatibility issues, Tsai Wen proposed extension set theory, a combination of qualitative and quantitative methods to solve contradictions have been made in many areas a better application. Characteristics of automotive

testing equipment selection for multi-objective decision-making based on the detection equipment selection scheme proposed evaluation method extension theory to detect needs, to regulate the economy, technology, services, environmental protection, human relations system for the evaluation criteria to establish an alternative selection expertise program element model and matter-element model of the desired selection scheme, the introduction of correlation function, through the extension of the direct evaluation of the pros and cons associated with the selection program.

2 EQUIPMENT SELECTION CRITERIA

Automobile comprehensive performance is testing equipment selection evaluation criteria system by seven elements, namely security applicability, economic, technical, service, environmental protection, human relationships.

2.1 The applicability of the principle

The suitability of test equipment and inspection requirements is in line with the degree of indicators. Number, model structure and composition of the vehicle to detect, such as traffic conditions and the environment is to detect the composition of elements of demand, shall be determined in accordance with the inspection requirements and testing equipment to detect the number and detection models. Taking full account of the links, supporting reasonable at all stages of equipment, ensure mutual balance the device detection capacity, in order to ensure a reasonable convergence of all aspects of the entire inspection process among ensure efficient conduct detection.

2.2 Normative principles

The normative considerations device is not a regular product, there is no type of identification number, and there is no measurement equipment production license number. Detection methods provided are consistent with existing state and industry norms and standards, testing parameters, detection accuracy whether the requirements of the relevant norms and standards.

2.3 Economic principles

Economic considerations equipment purchase costs and use of the total cost of the whole process, in which the total cost of the entire process, including equipment before and after use, use and scrapping of the total costs. Early into the device in order to run the equipment carried, as reported loading hiring and training of relevant personnel, equipment, calibration, and certification, and so on. During normal operation of equipment, and those equipment-related costs in addition to working hours, fees, energy consumption occurs, such as equipment maintenance and minor repair costs. Admission to the sum of equipment installation and dismantling costs of removal and clearance fees. The residual value of the equipment scrapped after.

2.4 Technical principles

The detection efficiency of a direct impact on the completion of the comprehensive performance test line car inspection car number per unit of time. Despite the different test items, the time required for detection of the major categories of items with related, but the same item, different types of detection devices, the time it takes quite different. Reliability principles from two aspects to consider, first, the device can be a steady job, low failure rate. Second, the good performance of the equipment, high precision, fully guaranteed accurate test data. Manufacturing process to ensure that equipment is an important prerequisite for the use of reliability, which reflect and measure the manufacturer's technical and management level. In order to improve equipment utilization, equipment failure can require treatment at the scene, which requires equipment disassembly convenient, reliable supply of spare parts and interchangeability good, but should be designed for specific parts specializing in tools.

2.5 Service principle

Delivery or arrival period: in principle cycle as short as possible. Transport and requirements: different modes of transport without special requirements, the costs of this commitment, during transport risks are also different. Service: The service includes professional, technical and operational training, spare parts supply, technical testing and equipment maintenance and repair.

2.6 Energy saving and environmental protection principles

Equipment energy consumption refers to its primary energy or secondary energy consumption. Device units usually start time to represent energy consumption; the optional equipment must comply with the requirements of national standards, "Energy Conservation Law" provisions. Equipment selection should pay attention generated during its operation noise, vibration frequency and pest control emissions within the scope of national standards.

2.7 Principles of human-computer relationship

Operational safety: operations security refers to the degree to improve equipment safety guards. Operator comfort: comfortable operating adaptability means the operator, good operating condition mainly refers to machinery have comfortable seating space, better visibility and ventilation. Good conditions should be regarded as an important factor manipulation to improve efficiency and reduce accidents. This indicator also reflects the operability and automation equipment.

3 THEORETICAL BASIS EUTHENICS

3.1 Matter-element theory

The things that matter element, features and magnitude considered in a continuum, so that people have to consider the amount of dealing with problems, but also consider the quality. Meanwhile, the matter element of things is the internal structure of matter-element change and things change the internal structure of the three elements of matter element makes a difference, so that the matter-element variability describes things become basic tools.

$$R = (N, C, V) = \begin{bmatrix} R_1 \\ R_2 \\ \vdots \\ R_n \end{bmatrix} = \begin{bmatrix} N, & c_1, & v_1 \\ & c_2, & v_2 \\ & \vdots & \vdots \\ & c_n, & v_n \end{bmatrix} \quad (1)$$

Matter Element can make use of qualitative and quantitative description of matter-element model. Introduced multidimensional matter element can be formalized and comprehensive description of things, but also to build the equipment selection program element model provides a theoretical basis for the evaluation.

3.2 Extension sets

In order to quantify the problem-solving process, euthenics a mathematical tool which adapts its foundation is the extension set theory.

$$\tilde{A} = \{(x,y) \mid x \in U, y = K(x) \in (-\infty, +\infty)\} \quad (2)$$

$$K(x) = \frac{\rho(x, X_0)}{D(x, X_0, X)} \quad (3)$$

$$\rho(x, X_0) = \left| x - \frac{a+b}{2} \right| - \frac{b-a}{2}$$

$$D(x, X_0, X) = \begin{cases} \rho(x, X) - \rho(x, X_0) & (x \notin X_0) \\ -1 & (x \in X_0) \end{cases}$$

4 THE RIGHT TO BUILD AND INDEX THE INDEX SYSTEM OF DETERMINING THE WEIGHTS

For device operation status determination, we must first establish a suitable evaluation system. The traditional method for evaluating equipment running major concern is the economic and technological advancement, usually selected technical indicators and economic indicators as the device status evaluation, without considering the impact on the social and environmental equipment. However, with the development of society, the environment continues to ceteriorate, gradually scarce resources, environmental issues and resource issues more and more people's attention, which prompted companies to purchase equipment, the use of the device, the device must be considered when repairing equipment or modification of equipment environmental impacts and resources that meet the requirements of functionality and quality assurance, under the premise of reasonable cost, throughout the life cycle of a minimum negative impact on the environment, the maximum efficiency of resource use. In addition, building a harmonious society day, corporate social systems as part of their survival and development is closely related to social, to reflect the people's cultural ideology update equipment, will impact on occupational safety and health into the evaluation of various factors range is also a trend today.

This article from the four aspects to consider building equipment status evaluation system, analysis of advanced technology equipment, economy, environmental protection and occupational safety, comprehensive evaluation equipment running. For the CNC milling machine, through a comprehensive analysis of the above four evaluations subdivided into 14 secondary indicators, combined with the device factory parameter to determine the index value interval, determine the number of overall weight of each index using the analytic hierarchy process, get the CNC milling machine operating status evaluation system and the index weight distribution are shown in Table 2.

Table 1. Evaluation index system of carding machine operation state.

C_1	C_2	C_3	C_4	C_5
0.1262	0.1589	0.1196	0.0129	0.1062
<0,5>	<35,45>	<0,0.5>	<0,4>	<28,30>
<5,12>	<45,60>	<0.5,2.5>	<4,6.5>	<22,28>
<12,30>	<60,100>	<2.5,4>	<6.5,8>	<5,22>
<30,90>	<100,450>	<4,5>	<8,10>	<0.5,5>
14	52	2.4	6.3	28

$$\rho(x_i, x_{ji}) = \begin{bmatrix} 9 & 2 & -2 & 16 \\ 21 & -7 & 8 & 48 \\ 13.5 & -0.1 & 0.1 & 1.6 \\ 10 & -0.2 & 0.2 & 1.7 \\ 16 & 6 & 23 & 27.5 \\ 31 & 11 & -4 & 4 \\ -4 & 7 & -7 & 8 \\ 12.2 & 1.4 & 1.7 & 1.9 \\ 86 & 7 & 22 & 97 \\ 13.95 & 0.09 & -0.01 & 0.01 \\ 26 & -2 & 2 & 22 \\ 86 & 2 & 7 & 97 \\ -14 & -5 & 15 & 25 \\ 86 & 12 & 7 & 83 \end{bmatrix}$$

5 OPERATIONAL STATUS AND PRODUCTION TECHNICAL CHARACTERISTICS OF MACHINING EQUIPMENT

Production equipment is the basic condition, the state of the device determines the quality of the efficiency of the organization of production, products, costs, impact on corporate business results. The operational status of the device when the device is installed and put into operation to be scrapped from the level of technological, economic, environmental and other indicators of the various stages of the life cycle has. Generally divided into good condition - Maintenance, state general - of repair, poor state - overhaul, state difference - scrapped four stages. (1) Maintenance Phase: After the commissioning of new equipment for a long period, stable operation, only the fault occasional maintenance service, to be called the maintenance phase. Production technology features of this stage are the high production efficiency, low operating costs, high precision equipment, the impact on the environment is minimized. (2) Repair stages: With the use of time, the state is gradual deterioration, increased failure frequency; move on to the repair stage. At this stage, the shutdown station gradually increased, decreased productivity, maintenance costs are gradually increasing, and often by changing the process parameters to ensure product quality. (3) Overhaul stages: equipment after prolonged use, the state of deterioration, into the overhaul period. Technical characteristics of this phase of production performance for technical accuracy can not meet the product requirements, parameters pass rate below 85%, many defective products, and therefore must be overhauled in order to continue to use the recovery accuracy. (4) Decommissioning phases: If the device or after a major overhaul, with no economic value, or cannot be restored again after a few overhaul production requirements of technical accuracy, or cause greater adverse impact on the environment and should be scrapped.

$$K_j(x_i) = \begin{cases} \dfrac{\rho(x_i, x_{ji})}{\rho(x_i, x_{pi}) - \rho(x_i, x_{ji})} \\ -\dfrac{\rho(x_i, x_{ji})}{|x_{ji}|} \end{cases} \quad (4)$$

$$\left. \begin{array}{l} \rho(x_i, x_{ji}) = \left| x_i - \dfrac{a_{ji} + b_{ji}}{2} \right| - \dfrac{1}{2}(b_{ji} - a_{ji}) \\ \rho(x_i, x_{pi}) = \left| x_i - \dfrac{a_{pi} + b_{pi}}{2} \right| - \dfrac{1}{2}(b_{pi} - a_{pi}) \end{array} \right\} \quad (5)$$

Method development theory is an issue to resolve conflicts, the main elements and their transformation law Studies, is the system of science, thinking, science, mathematics, cross interdisciplinary, throughout the natural and social sciences and is widely used cross-sectional disciplines, can achieve better overall quality and quantity of things unity, in order to solve the problem of providing a comprehensive evaluation of new scientific methods.

Description of the CNC milling machine running a state has deviated from the good stage, but has not yet reached the stage in general, but somewhere in the middle of a good and general. With the use of time, its status will gradually deteriorate; move on to the repair stage. The performance of its status as a fault frequency increased gradually increased shutdown units, decreased productivity, maintenance costs have gradually increased, the need to change the process parameters to ensure product quality, but also a negative impact on the environment is gradually increased. Therefore, for the CNC milling machine in addition to normal maintenance, there is a need periodic repair items.

6 CONCLUSION

Equipment selection according to business principles, combined with auto comprehensive performance testing project requirements and manufacturers, the establishment of testing equipment selection evaluation criteria system model. Use extension theory to solve the contradiction of incompatible features, equipment selection proposed evaluation method based on extension of the force. With the extension theory is proved to evaluate the pros and cons of automotive testing equipment selection scheme is effective for the program provides a method and basis for decision-makers and realistic.

* Corresponding Author: Liu Ruijun

REFERENCES

Zhao Tao, Mao Hua, Liu Shu theory equipment management system and development trend [J] Industrial Engineering, 2001 (6):1–4.
H.Paul Barringer, PE and David P.Weber Life Cycle Cost Tutorial [C]. Fifth Intenational Conference on Process Plant Reliability, Gulf Publishing Company, Houston, TX, 1996:5–7.
Wei Jianzhong, Liu Rongying equipment updated technical and economic analysis [M] Shaanxi Science and Technology Press, 1995:15–18
Kenneth Lee, Wu Xinyu genetic algorithm with random switching control equipment used in solving the problem of updating [J] Nanjing Institute of Posts and Telecommunications, 1996,16 (2):26–28.
Zhang Xinmei equipment renewal plan based on genetic algorithm [J] Mechanical Science and Technology, 2000 (4):611–613, 616.
Yangyuan Liang equipment renewal directed graph model [J] Operations Research and Management, 1997 (4): 68–73.
Yangyuan Liang equipment renewal decision model of game theory [J] Forestry Machinery & Woodworking Equipment, 2001 (5):21–23.
T.Goldstein, SPLadany, A.Mehrez A dual machine replacement model:. A note on planning horizon procedures for machine replacements [J], Oper.Res.34 (1986):938–941.
Tsai Wen expand learning overview [J] Systems Engineering Theory and Practice, 1998 (1):76–84.
Yuan Tsai Wen object model and its application [M] Beijing Science and Technology Literature Press, 1994:161–202.

Information, Computer and Application Engineering – Liu, Sung & Yao (Eds)
© 2015 Taylor & Francis Group, London, ISBN 978-1-138-02717-6

Ideological and political education for college students in new media background

Hong Yu Yang
College of Medical Test, Beihua University, Jilin, China

ABSTRACT: In the context of new media, traditional college students' ideological and political education has been an unprecedented challenge to the current ideological and political education of college students to fully understand the importance of the use of new media, necessity, and through new media the student population is widely popular media to develop and strengthen ideological and political education of college students to continuously improve the effectiveness of ideological and political education, and improve the quality of personnel training. In this paper, the characteristics of the new media, the impact of new media and how to use ideological and political education of college students as well as to strengthen the ideological and political education of college students three elaborate.

KEYWORDS: New Media; Ideological and political education; College students

1 INTRODUCTION

The emergence of new media, the new technical support system of media forms, such as digital magazines, digital newspapers, digital broadcasting, SMS, mobile TV, network, desktop Windows, digital television, digital cinema, touch media. With respect to the media, press, outdoor, radio, television four traditional senses of the new media is aptly called the "fifth media." The new media have the following characters.

1.1 *Degradability*

The emergence of new media, traditional media boundaries between digestion, but also relieved the borders between countries and between communities, and boundary information between sender and receiver have been digested, the integration of characterization is also increasingly the more prominent.

1.2 *Interactivity*

The emergence of new media and the majority of the audience really established a relatively close contact, the spread of new media is not just a one-way transmission of information media to the audience, but for two-way interaction, not just a role of the media in the audience, the audience can also act on the media, including this interaction and interaction between man and machine and the machine, such as interpersonal machine.

1.3 *Immediacy*

Guo Wei, said: "The biggest difference between the new media and traditional media, is to change the propagation state: from a multipoint-to-multipoint changed." This multi-multipoint mode of transmission makes the timely dissemination of very fast, and information exchange between the subject and the object that is the point that pass, but also a great breakthrough in the region and even national boundaries.

2 THE POSITIVE IMPACT OF NEW MEDIA ON THE IDEOLOGICAL AND POLITICAL EDUCATION OF COLLEGE STUDENTS AND THE IMPACT OF NEW MEDIA ON COLLEGE STUDENTS

2.1 *New media to broaden the knowledge of college students*

As the new media have large amount of information, influence spread fast and wide range of other features, allow students to learn through books in addition to a large amount of network information. Students no longer rely on a single traditional teaching model to acquire knowledge, but in the traditional way of learning, the adoption of new media this convenient platform to learn their own interest, which continues to broaden their knowledge and learn the most cutting-edge knowledge. At the same time, help college students improve their own quality and students' awareness of innovation, but also conducive to the development of college students and personality formation.

2.2 The negative impact of new media on college students

New media is a double-edged sword, in a positive impact to the students, but also have a negative impact side, the main problems: the proliferation of network information and network information dissemination accept concealed cause weakening of the moral consciousness of college students and faith missing. Some students due to over indulge in the network, so that normal learning and life order is destroyed and lead to weariness dropouts, etc., but also harm the health of college students. In addition, long-term obsession and dependency may also lead to new media college students also may make students 'interpersonal, emotional light, and even lead to students' network autism suffer from "social networking disorder" and so on.

2.3 The impact of new media on the ideological and political education

The emergence of new media can make a high school student workers may have more channels are kept informed by a more realistic idea of dynamic students, targeted at college students ideological education management and other aspects will be further enhanced. At the same time, the ideological and political education is also made greater challenges and higher requirements.

3 WAYS AND MEANS OF STRENGTHENING THE NEW MEDIA BACKGROUND IDEOLOGICAL AND POLITICAL EDUCATION

3.1 Ideological and political education should give full play to the function of self-education students

In the new media environment, ideological and political education from the traditional fixed channels to implement new media education to the ideological and political education of the "free" paradigm shift, and new media ideological and political education is the most important feature of self-education. To give full play to the function of self-education of students, so that students in the new media platforms have broad freedom of choice, students learn more knowledge, independent thinking and other aspects of autonomy, so that students in the self-study and selection, and thinking interpretation of the process learn to grow.

3.2 Ideological and political education to achieve from the managed transition to a service-oriented

Due to the characteristics of the new media environment, the ideological and political education from the traditional paradigm shift in management based on student growth and success of service, to take advantage of the positive factors for the new media to college students, to give college students' growth help and guidance. In the new media environment, blindsided emphasis on education by way of education and do not want to instill students how to use new media like the platform that resonates with students, the effect of education is compromised.

3.3 Innovation in the education system and mechanism

To build a new interactive media is to ideological and political education, guidance and restraint. Through new media platforms, to carry loved, colorful and attractive college students' online activities, organized by the Internet and other civilized education campaign, and the focus of discussion of the hot social issues of current concern, the teacher can publish depth and persuasion net assessment force to the front guide online public opinion. At the same time, but also to build the new media environment and restraint mechanisms of ideological and political education, through the establishment of a related network management system, improve the network behavior of college students to promote self-discipline.

3.4 Innovation in terms of the content of education

With the traditional ideological and political education, science into the ideological and political education in the new media environment, ideological and political education should be combined with the characteristics of the new media and college students together with students willing to accept a form to attract college students. In the design of the content is the new media to strengthen moral education and quality education. New Media Students 'moral education important in the face of a lot of information in different ways of self-control and non-judgment, actively promote and strengthen the network of moral construction, improve college students' network moral self-discipline. At the same time, to reflect and pay attention to the quality of education, focusing on training and education to enhance awareness of the college students' network, effective use of sound science good new media to enhance their overall quality and capacity. Innovation in the form of means and education is to build a network to strengthen ideological and political education of the new platform through the new media. Through new media, college students discover the existence of ideological problems, conduct analyzes to guide and carry out targeted work-related education.

4 IDEOLOGICAL AND POLITICAL EDUCATION OF COLLEGE STUDENTS' PROBLEMS THAT EXIST UNDER THE NEW MEDIA

Since the reform and opening up, some Western countries tend to use their influence and ability to control the dissemination of information through the Internet, the new media means to transport their decadent ideology and culture of the socialist countries. This new channel of communication networks for some ulterior motives of Western countries, ideological and cultural infiltration facilitated to our country, they use the network technology advantage in foreign scientific and technical output, while output and decadent ideological and cultural values. Technology is relatively backward countries in the introduction of advanced technology; it is often affected by decadent ideology and culture. Because young students at a crucial stage of life, values, world view, which is often prone to hide behind the technology, ideological and cultural identity and values, leading to the blind worship of Western capitalist culture.

With the rapid development of computer popularity of the network, ideological and political education website becomes an important educational resource university. Many universities have established online school has its own characteristics, network Party School, a large auditorium and other characteristics of the network of ideological and political theme of the site, but in the process of building the application, there is still a lack of attractiveness, click on the site utilization rate is not high, lack of education and teaching resources, interactive network platform construction and poor management and other issues. Because of these problems makes it difficult to attract students to browse the ideological and political sites, thus losing the use of advanced network means the value of ideological and political education.

Ideological and political education of employees is the main ideological and political education is to optimize the ideological education of college students in the new media organizations reliable protection. They themselves often have a direct impact on the quality of the effect of ideological and political work. This new media on the internet ideological and political education practitioners of traditional ideas, tools and other proposed new challenge. Especially when they are faced with the network information analysis and processing, not only do ideological and

political work, more important is the need to master certain network technology, only a good combination of these two points, to handle under the new issues facing the new media age.

Thus, the network era of ideological and political education of employees proposed new requirements, they need to be with the times, constantly update their knowledge structure, learning new knowledge. But in reality the majority of employees are not ideological and political ideological and political education graduates, in the actual work needs continued exploration of ideological and political work law, but also a lack of expertise in the work system of learning opportunities for further study, and the other due to some schools for students ideological and political work is not enough investment, often resulting in the treatment of the ideological and political work of employees is relatively low, which leads to the ideological and political staff mobility is relatively large, the objective existence of these factors to carry out ideological and political work of the network played a certain impediment. Ideological and political education for employees' enthusiasm and initiative had not affected, some of these problems exist shows the current ideological and political ranks Universities insufficient. Network World has a massive information, which will inevitably there are some harmful information, in addition to network control laws and regulations is not perfect, who need access to self-awareness of their online behavior constraints, which often lead to college students in the face there will be more and more psychological conflicts network temptation.

Free and open network features, on the one hand to achieve a high degree of freedom of speech, it also makes pornographic, reactionary, and so some harmful impact on the dissemination of information and college students on the network in the next package of modern technology. This harmful information on the student's original ethics forms the impact, resulting in a dilution of some students' morality. In addition, college students and even easier to make laws contrary to moral behavior, such as the manufacture and dissemination of false information, such as the spread of the virus.

Thoughts on new media network ideological and political education of college students

Network provides a variety of information resources to the people, but these information resources in different ways, and some are even harmful to us instill negative information and ideas through a variety of forms, so during the ideological and political education work, we must adhere to the correct theoretical guidance, arm themselves with scientific theory. We should adhere to under the guidance of the important thought of Marxism-Leninism, Mao Zedong Thought, Deng Xiaoping Theory and the "Three Represents" and the scientific concept of development, from the reality of the students' actively socialist core values education. Through this education to help college students to establish a correct outlook on life and values, thus improving their ability to resist corruption.

First, the ideological and political education "Red Website" message content must be in accordance with the needs of living close to the students, close to the students thought the pursuit of growth closer to the principle of student needs to build, in order to attract students of resonance, thus creating favorable conditions for ideological and political education. Second, an interactive network platform between student teachers to create a good atmosphere for the network of ideological and political education work. We want to take full advantage of the interactive network to serve the ideological and political work of our colleges and universities. College To take full advantage of the interactive features of the network, to build an interactive platform for exchange between students and teachers, the use of micro-Bo, QQ and other carriers to facilitate communication and exchange between teachers and students.

Ideological and political workers should also take the initiative to participate in the discussion forum on campus, conscious appropriate guidance for students, for students to use the Internet to solve problems encountered in learning and life. Only fully use the network to provide services to students, in order to attract the majority of students come to the ideological and political education website. Third, we must do the work of public opinion analysis and opinion leaders. At present, many colleges and universities have attached great importance to the Internet public opinion analysis work, to collect and analyze information on student interest in various forms, for college students to carry out the work provided the basis for, and achieved good results.

Ideological and political education practitioners should strengthen political theory of Marxism-Leninism in order to continuously improve their theoretical knowledge. Second, the ideological and political education workers should strive to broaden their knowledge in order to adapt to online media under the ideological and political work. Third, the ideological and political education of employees must master certain network communication technology and network knowledge. Good skills and knowledge as a means of ideological and political education of employees work, is an important way to interact with college students. Ideological and political education of employees through Fetion, QQ chat, microblogging and other exchanges with students on an equal footing, to fully understand the students' interests, targeted work, so that the network becomes the assistant to carry out ideological education.

Colleges and universities to strengthen publicity and education on the one hand the network of laws and regulations, norms network behavior of college students, advocating civilized it has a good awareness of the law, efforts to improve college students the ability to identify the harmful information network. On the other hand, you want to enhance students' self-restraint to help them establish the correct network values and ethics, improve college students' self-awareness and self-control, to enable students to be responsible for their own network behavior, consciously resist the intrusion of bad information, do propriety civilized and law-abiding Internet users, and consciously safeguard the network order.

5 CONCLUSION

The emergence of new media, colleges and universities engaged in ideological and political education of the workers brought certain challenges. Therefore, colleges and universities to strengthen the building of student work. First, we must strengthen the ability to use new media culture, and to be able to use a variety of skilled operators and new media, to lay the necessary foundation for IT ideological and political education. The second is to improve the overall quality of student work teams. Under the requirements of the political workers the new media environment to be able to grasp the pulse of good times, make full use of the means of modern media to raise awareness of the hot focus and difficult problem analysis, judgment, and effective guide summarizes the student's ability, resulting in the new media environment has created the ideological and political education work situation.

REFERENCES

Komidori Lee, Chun "2005 Review of the new media, new media attention and hot in the front row[J]," People, 2006. 3.
TAN Jian-ping, Lin. Ideological and political education to build long-term mechanism [J]. Hunan First Normal University, 2009, (5).
Mao Zicheng blog: students ideological and political education of new carrier [J]. Tianjin Manager College, 2008, (1).
Wang Huancheng. New trends in the new media environment of ideological and political education of college students [J]. Contemporary Education Forum (Management), 2010 (8).
Lai Yong. Ideological and political education in the new media environment exploration [J]. Net wealth, 2010 (7).
Li Yan, who Velen, Haitao. Ideological and Political Education of College Students new carrier of the new media environment [J]. Chongqing University of Posts and Telecommunications: Social Sciences, 2010 (5).
Exposing the red and blue, Wangqi Si. On the use of new media platforms college counselor ideological and political education [J]. Technology Square, 2009 (12).

Information, Computer and Application Engineering – Liu, Sung & Yao (Eds)
© 2015 Taylor & Francis Group, London, ISBN 978-1-138-02717-6

The evolution of contemporary literature ecology based on new media environment

Qun Ying Gao

Teacher's College, Beihua University, Jilin City, China

ABSTRACT: Since the new century, the rise of new media and its culture represented by the Internet is having a profound influence on the society and culture as a whole. From the perspective of cultural development, this influence can also spread to the field of contemporary literature.

KEYWORDS: New media culture; Contemporary literature; Collision; Communication; Integration

1 INTRODUCTION

The term "new media" has long been familiar to the public, and new media related topics are from certain areas to the broader field furnace exhibition, theory and practice on the new media has also been deepening. However, the exact meaning of the new media, academia, but it is far from a consensus of experts and scholars on this also eyes of the beholder, the wise see the wisdom. I believe that, from the following three aspects "new media" concept to grasp. First, the new media is an evolving concept. The emergence of a new media concept of time is roughly the 1960s, as its growing wealth in the form of the same practice; people's understanding and awareness of this concept are constantly changing. In 1967, CBS television network technical director of the Institute of Radio Gold Mark (P.Goldmark) published a proposal on the development of electronic recording of goods, he plans to book in electronic video called "NewMedia" (New media), "new media" concept resulting.

Subsequently, the Chairman of the Special Committee on the spread of the US policy of President Rostow (E. Rostow) in a report to the then US President Richard Nixon submitted repeatedly mentioned "New Media" this concept, the term "new media" and then began in the United States popular and soon all over the world have been widely disseminated. At the annual meeting of the UN Committee in 1998, the "new media" is used to refer to the fourth media - the Internet, the Internet considered after newspapers, radio, television three traditional media, the fourth mass media. The rapid development of the Internet in all areas of social life has had a profound impact, which is no need to argue the facts, and from the current trend of the development of new media,

the new media have already gone beyond the contents include online media, to mobile phones forms of media, interactive television continue to penetrate.

Levinson (PaulLevinson) in the "new media", based on the concept of "new new media", the new media refers to the first generation of media on the Internet, which is characterized Once uploaded to the Internet, people can always get, rather than according to a schedule determined by the use of media, such as e-mail, iTunes players, chat rooms; and the new new media refers to the second generation of media on the Internet, and its salient features is the user to create content, there is no top-down control, everyone can become a publisher and producer, blog network, wiki networks, and now crested Masamori facebook, twitter, youtube is a typical representative of this area. Second, the new media of the "new" is relative to the terms of traditional media.

In the course of human history, the media update is the norm, there will be able to represent every era of new media forms that time period. US new media expert who? Crowe Spey (VinCrosbie) believes that the new media is able to provide personalized content while the public media mix is communicators and recipients who become peer exchange, and numerous mutual communicators room can be personalized communication media simultaneously. He pointed out the mode of transmission of new media – interpersonal media, "one" and the mass media, "one to many" mode of transmission, including "many to many" model specific to the level. "Connection" magazine New Media is defined as "the spread of all against all. "While this definition cannot be counted in the strict sense of the definition of the concept, but gave away the new media and traditional media, compared with the essential characteristics.

2 THE IMPACT OF NEW MEDIA CULTURE ON CONTEMPORARY LITERATURE

2.1 The impact of new media culture of literary aesthetic philosophy

The impact of new media on traditional literary culture, the first from the technical characteristics of new media, the impact of technical factors on literature than ever before are more clearly revealed. In the network literature, for example, the network literature technically realized interactivity, hypertext and multimedia, to subvert the traditional literary writing that relatively closed state. This makes the writer's thought to have a greater degree of freedom of divergence, thus greatly expanding the creativity and imagination, sound, pictures and video of the factors the authors not only by means of multimedia technology into the content of literary works to go, even more profound impact on the aesthetic style of literary works. New Media Age literature and art is not only "mechanical reproduction", it is everyone can participate and everyone is involved in the birth of an important condition for the works, which laid its digestion elite culture, tend to the basic position of grassroots culture. Being Digital status greatly inspired the creation of all aspects of human potential, in addition to significant progress in the field of artistic practice outside democratic consciousness raised more creative, and even the survival of the concept of artistic thinking and other aspects of innovation.

2.2 The impact of new media on the production and dissemination of cultural mechanisms of literature

In traditional literary institution before the birth of the Internet, a huge obstacle exists between the production and dissemination of literary works, and that is a strict publishing system. After the creation of the works only in accordance with the requirements of editors, publishers, and publication review bodies, in order to be published and distributed, content of the work in order to be exposed to the reader. This mechanism is essentially literary production and dissemination of literary discourse reflects the power monopoly. In this process, the controller of literary publications (editor, editor) and literary critic literary discourse on the right has a crucial influence. Control who can decide whether literary publication of literary works can come out into the literary circle holds the work of "tickets' embrace the ideal of literature in the end because "gatekeeper" is not recognized by the loss of the opportunity to enter the temple of literature this case in reality foreign literature is not uncommon; and literary critic holds the right to judge the works, literature dissemination "pathfinder", their comments and behavior research can explore the deep value of

literary works, in order to establish " classic ', but we can also kill a handful of works and an author.

2.3 The impact of new media culture of literary language

Heidegger said that language constructs the existence of the world, language is their home. Rich and expressive language for humans to build a diverse sense of the world, the world of literature is in this sense a member Cui dish, known as the language of art, ancient Chinese literature on the "meditation" "Sentimentality Profusion "for the aesthetic pursuit. Changes in the state of human existence is the norm, which requires carrying the meaning of language in the world of real-time updates of the state must maintain in order to adapt to the real needs. Recalling the history of Chinese literature can be found in the development, progression and change the language of literary history is closely linked. In the process of establishing modern Chinese literature, is one of the classical vernacular replace a crucial link that will". Words are the voice, Genbunitchi "from the theoretical level to advance to the level of practical operation, in order to adapt the Pat ran into modern society the modern Chinese ideographic needs, thus establishing the basic norms of modern literary language.

3 THE IMPACT OF NEW MEDIA IN CONTEMPORARY LITERATURE OF SELF-REFLECTION

3.1 Self-adjust and adapt to contemporary literature

The relationship between literature and society has been a hot topic in the field of cultural studies. Confucian culture dominant in the Chinese cultural context, literature has always been an important commitment by political education, who is entrusted with the literati Patriots political ideals, this tradition has been extended to the fate of the ups and downs of contemporary literature and this tradition has also inseparable link. Confucius from the Spring and Autumn Period, "Xing Guan," said Tang Dynasty Bai's "article together as sometimes the song Poems for things to do," and "Poetic Revolution" Liang made the modern, "fiction revolution ", and" 54 "New Culture movement literature once served as the vanguard of the revolution, literature is always in the center of political and social discourse, bear an important mission.

3.2 Contemporary literature on new media culture of acceptance and absorption

Network literature is a new thing after new media culture and literature collide birth, is a typical product of the new media culture, and thus contemporary literature on new media culture of acceptance and absorption change is reflected in the contemporary literature on the relationship between literature and the Internet. The initial network literature like rustic origins of "Wild Child", in a rebellion against the traditional gesture to show people, both with orthodox serious literature there is a clear difference between form and content. There are a lot of people are addicted to the nose, the network literature can only walk on the edge of literature. After ten years of development, the network literature has been very strong momentum of development had to re-examine its value and significance, mainstream literature network literature began to extend an olive branch, which opened up the network literature has been accepted by the mainstream literature course.

4 NEW MEDIA AND CONTEMPORARY CULTURAL SPACE WITH THE NEW FIELD OF LITERATURE DOMAIN

4.1 New media communication and literature

In the new media (Internet) era, writing has become a very convenient thing, no "gatekeeper" strict control, nor the power of discourse of oppression. Enter the text from the online publication, can be done in an instant, most people desire in writing forum, blog, microblogging and literary websites in cyberspace have been achieved, and this is becoming more and more common way of writing, have evolved into "writing addiction," the tendency of a large range, which opened the era of universal writing. Generated widespread alienation writing addiction, addiction and generalized writing and in turn strengthen aggravated alienation. From the invention of the printing press ago to make people better understand each other, but in the era of addiction writing, writing a book with the opposite meaning: Everyone is surrounded by its own terms, as being in between the heavy mirror wall, any external sound cannot penetrate in.

4.2 Vitality and dilemmas network literature

Network literature is the most mature level of development of new media literary style, vigor and literature bears witness to the plight of the current network of new media culture and literature can achieve the depth of communication. Network literature is booming in China jointly contributed by many factors. China's online literature did not bring too much "technical" color, but more a continuation of the popular Chinese folk culture colors. China eager to civil society the right to speak freely, making the emergence of the Internet to bring a more sociological sense and idealism, provides an ideal platform for those who are eager to express themselves, so popular in China capable of folk literature network tremendous vitality. In addition, China has the largest number of Internet users worldwide, as well as worldwide Chinese, Chinese reading has a broad readership, plus not perfect digital copyright protection system, jointly contributed to the rapid development of China's Internet literature citation.

Network World is a reality and the future vision of modern society, the rapid development of the human face, a metaphor, and people committed to the construction of the "imagined community." Specific to the Chinese context, due to the level of urbanization lags, significant urban-rural gap, only the city was better reflect modern social development goals, values and ideals of the Internet and its manifestation more to do with the formation of the corresponding contemporary urban life relationship. Network World is an extension of urban life, there is a network of resulting in a "city" and "people" double change. Spatial morphology changes occurred in the city, by a single physical space superimposed virtual space, constitute a "complex space, the network also enables writers habits, writing style, creative ideas and a way to experience the city is affected, not only to expand the writer the vision is to reconstruct the contemporary urban literature. Real significance lies not in the network literature literary achievements it has made, but in the literature for the city to provide a new living space, a new, and completely different from the books aesthetic experience of reading.

5 COMMUNICATION CONTEMPORARY WRITERS AND NEW MEDIA

Mo Yan, represented in this group have long active in the Chinese literary writers, including Jia Ping'ao Liu Zhen, Wang Anyi, Yu Hua and others, they have a strong appeal in the literary world, is a typical representative of contemporary writers. Study the behavior of these writers writing in the field of new media can be found, although they are willing to let their works in the field of new media to be more widely recognized, to develop their own literary space, but most people for online writing this behavior itself did not show a strong interest, also did a considerable effort betting online writing practice. Once upon a time, in the online blog, blogging has become a trend with the majority of writers' cultural figures, and numerous old and new writers have joined the ranks to experience the "net" fun.

But I blog and micro-blog writing case study in terms of some contemporary writers to blog and micro-blog is representative of the new media discourse places have not been effectively utilized, largely Cang blog writers to book advertising positions enthusiasm, not only failed to inspire writers to participate in public affairs, but also failed to become the ideal place for writers and readers to interact fully constructed literary discourse, and this is precisely the greatest positive significance of new media writers can bring.

Of course, the Internet as a representative of the new media space, there are a lot of boring, "noise", is not necessarily an ideal place to speak, but the sound is because of this, a writer, after serious thought was given more seem valuable, but also the society's most valuable and most scarce thing. Meanwhile, the literature is not a Utopian world, one can really have an impact on contemporary society "contemporary" writer must be able to skillfully use the "contemporary" the most influential social media tools to create the conditions for their literary activities.

6 CONCLUSION

We must admit that the new media culture's emergence in the literature to bring a new look, but also the kind of literary aesthetics inherent charm a threat, literature and reading in the new media era obtain the ideal of liberty, equality and openness after the space, but also inevitably caught up in some kind of shock theory. Cyberspace literature and reading activities are interactive and virtual sex. Interactivity makes the boundaries of identity between the reader and the author no longer exist, although the reader inspired creativity, but it also affects the independence and integrity of the literary works of literature tend flattened; virtual reality makes aesthetic has become an artificial environment automatic generation of process is in full accordance with the aspirations set up on reality "analog" instead of real feelings and emotions and nostalgia, a sense of mystery literature canceled. In this sense, the new media culture digestion of literature, "authority", and the unique charm of "authoritative literature" is the literature lies.

Hillis Miller in "literary authority," the article pointed out that literature is not "self-made authority,"

but "he was to authority" and "authoritative literature from performative use of language artistry (a performative use of languageartfidly), such use of language makes the reader when reading a work on it to create a virtual world creates a sense of trust "with the modern technology of virtual reality technology is different, literature always" hidden from some dark secret forever, but also a fundamental feature of "authoritative literature precisely because of this literature by means of language itself has aesthetic charm of the abstract symbol systems, creating a fictional world of aesthetic distance and imagination, the only making literature has a divine cult value, and was given the authority. In the new media to create a sweeping and transparent world, ancient and mystical literature encountered Kui gun with the dilemma has been self-evident. But does this mean that the new media culture and literature of the opposition will inevitably? The answer is not so pessimistic.

REFERENCES

Chen Ping. Modern Linguistics [M]. Chongqing: Chongqing Publishing House, 2003.

Cheng Gong. Generative grammar for Chinese "own" the word study [J]. Foreign linguistics, 2005, (1).

Gui Shichun, Ningchun rock. Linguistics Methodology [M], Beijing: Foreign Language Teaching and Research Press, 2004.

Hu Jianhua. Chinese long-distance reflexive pronoun of syntactic study [J]. Contemporary Linguistics, 2008, (3).

Hu Jianhua. Long-distance reflexive pronoun "their" function, semantics, pragmatics and discourse theory [Z]. To be published.

Huangzheng De Xu Debao. Contemporary Linguistic Theory Series • Preamble [A]. Beijing: China Social Sciences Press, 2005.

Song Guoming. Summary of syntactic theory [M]. Beijing: China Social Sciences Press, 2007.

Xu Liejiong. Semantics [M]. Beijing: Chinese Press, 2000.

Joos,M. (ed.) Readings in Linguistics:the Development of descriptive linguistics in America since 1925.American Council of Learned Societies, New York, 2008.

Lyons,J.etc.(ed.) New Horizons in Linguistics 2. PenguinBooks, 2007.

Information, Computer and Application Engineering – Liu, Sung & Yao (Eds)
© 2015 Taylor & Francis Group, London, ISBN 978-1-138-02717-6

Variation of the Chinese language in the context of network culture

Shu Wen Wang

Teacher's College, Beihua University, Jilin City, China

ABSTRACT: With the popularity of the Internet, and now China has entered the Internet age, people and businesses are now inseparable from the network, it is with people's lives, especially young people now, the network has become their indispensable part of life. Due to the timeliness of the network, more people have begun to get used to obtain information from the network. In addition to obtaining information from the Internet, it is also possible in the online dating, information sharing, and perform the other network activity. Increased network activity, network language began to flourish. In order to adapt to the special environment of the network, network language Chinese language in vocabulary, grammar, pronunciation and composition have changed a lot of variation. This article will generate network language network language to influence the development of Chinese language and produced for analysis.

KEYWORDS: New media culture; Contemporary literature; Chinese language; Network language

1 INTRODUCTION

According to statistics, the current number of authorities of Internet users has exceeded 400 million close to 500 million. The network is undergoing dramatic change in ten years of development, the popularity of the Internet is also affecting all aspects of real life. People's way of life is because of the popularity of the network in the event of a large number of application changes. The same language as the use of different occasion's scenarios, in order to be able to fully express the intention of the user, the language will change accordingly required according to the meaning of the expression, in a variety of forms.

2 CHINESE LANGUAGE AND VOCABULARY OF VARIATION

Open Baidu, a "network terms" search in the Baidu search engine, while search results found up to several hundred thousand. But in the end, which is a network of popular vocabulary words of the pioneers who would not be able to speak clearly. But one thing is beyond doubt, along with the development of the Internet, the network is always introducing new vocabulary, each period has netizens use their wisdom to create a new network vocabulary. Chinese language and vocabulary are in the network changes to reflect the change in language. As one scholar said, the special circumstances of modern life, so that people can not as a tool to communicate, they need a way to move people directly sign language or

a variety of traditional languages, so that they can be made in a short reaction time. Master's degree of freedom in the network is relatively high, the traditional language of expression, but also has limitations, which allows users on the network in order to overcome the limitations of traditional language, play to their creativity and imagination, publicize their personality, on the network continues to have on the network-specific vocabulary, the traditional Chinese language appeared variation.

The network environment, the use of words has broken the original way, no longer the main way of writing Chinese characters, appeared with the original Chinese different words, a large number of Chinese pinyin and Chinese homonym is widely used. With the popularity of the network, these words have been used by a large number of users.

2.1 Spelling of words

People on the network in order to be able to adapt to the rapid exchange of information and transfer abbreviation, in order to achieve the purpose of the exchange of briefs and effectiveness, a large number of Chinese pinyin is used, and in accordance with certain common rules fix its meaning, take the first letter as a Chinese spell words

As representatives of sister mm, gg representatives brother, dd representative brother, jj representatives sister, lm representative babes, lr on behalf of bad people, and so on, these words are reflected in the popularity of the Internet's impact on people. Pinyin Chinese characteristics have become shorthand for the network. The Internet has made it easier to contact people, closer to people's distance; they had no sense of the meaning of the letters become rich, give people a bit happy and relaxed.

2.2 Homonym of vocabulary

Network homophonic words may be because the pronunciation of the two words is very similar. For example: porridge representatives like jam on behalf of an award. There is also a more common type of digital homonym, for example: 9494 means that is, the 4242 mean is ah yes ah, 7456 meaning that mad at me, 748 means hell, meaning 88 byes, bye. Do not be angry mean 847, 987 is not meant to go, 5555 is an imitation of the sound of crying, weeping meaning is the meaning of meaning 1414, 3166 is the Japanese pronunciation bye is used in this meaning, 3q is thank you, 8147 is Do not get angry, is not it means 848, 886 is the meaning of worship myself. These figures will be subject to the application of the reason why the majority of Internet users are welcome, mainly because these digital inputs only need to use the keyboard to the right of the numeric keypad, while easy to use and fast, the quality difference between the characters is small, easy to understand, at the same time added a fun language.

2.3 Fold increase in the use of sound words

When people often use the Internet or at the time of the forum message reduplicated words to enhance the sense of the image of the text, with the spoken language instead of Chinese written language, the language will be adding a bit more intimate and playful, a greater emphasis on writing people than. Use this slightly tender tone vocabulary of children, the most representative than the "thing" called "stuff", and now the network terminology has been carried out into the real-life use, network language the impact is evident. There is also a representative of the "beautiful" as "splendidly", the bag called the "bag", the Apple called "fruit", so there are many, many examples. Adults frequently the child's use of language is often used on the web, it seems inconsistent with the law, contrary to common sense, in fact, this is not surprising, the real social life in high pressure and fast-paced let adults at heart there is a yearning innocent lives of children, longing for the time to be able to care for their parents and care back to childhood again.

On the network, in the virtual world, people can fully vent frustrations and stress, only here it is no competition and pressure. Where people are unfamiliar with each other, each other not know each other's true identity, and even each other's sex and age. This phenomenon is more common in young women's groups. They want the young at heart get to play freely in the free networks. Therefore, this is full of childlike fun vocabulary appeared on the network, the network a lot more fun.

2.4 The new generation of word

Rapid development of the network has spread to every corner of people's lives. Now more and more people are realizing the network not only on behalf of high-tech, and also a new culture and lifestyle. It has on people's lives has a huge and far-reaching impact. With the popularity of the network people's attitudes are changing. Generate new words is reflected in their youth psychology. The main changes in the new term, there are two, increase the prefix, and suffix increase. Add a new word prefix, such as: Zero: Zero compensation, zero income, zero complaints, zero defects, zero tolerance, zero growth in the high: high intelligence, high-quality, high-quality, high-return, high-quality, high visibility and the other is Add a new word suffix, such as: it: Disco, songs, lounge bar, some things off the Internet: blog, hackers, mob family: paycheck to paycheck, office workers, car owners, slaves: slaves, slave car, there is also a new word slave card production, it is relying on the original terms, users on the network by the pursuit of individual psychology, in modern Chinese language and vocabulary based on the original in the context of the network produced a new meaning. Because the network has a very good spread of freedom and widespread, many people have become used to this new vocabulary and have put these words into the daily life, has become part of their life, this occurs mainly word two cases.

Tamper with the concept of one kind of method, so that the original meaning of this new generation. These terms include: control means is a hobby person, Guards meant to make some counterintuitive thing is to get others to carry out surgery means spoof, anatomy mean someone parsing of Korean kamahi is call, meaning waste wood is useless person, meaning that flutter down the street on the street, which means diving behavior does not comment.

The second is the appearance of the association through words. For example idol mean people object you want to vomit, strong mean good, seductive mean looking, corruption mean eating and drinking, anti-corruption means that asked people to eat, dizzy mean read, Paizhuan means an opinion, because of who mean chasing girls, high mean particularly excited, wealthy single men are called diamond bachelor, not handsome girl is called dinosaurs. There are known to have the ability to feed, and so it goes in the vocabularies because of its specific meaning, has also been accepted by the majority of the young people, but these words because of its own inherent disadvantages, so it can not in real life is widely used.

3 GRAMMATICAL VARIATION

Characteristics of the network are simple language requirements quickly and efficiently. People do not particularly focus on the Internet when Chinese language syntax, but more attention to represent the meaning behind the words, so you can often see on the network does not meet the grammar and vocabulary of the statement appeared, mix and match the statement, the word before and after the word upside down, long after this situation occurs, it is slowly being accepted by the public, on the formation of a new language. These new language combination modes include the following situations. Omitting any case there are a lot of people in the use of networks for information exchange is carried out using computer keyboard input information, which makes people's thinking and text entry there is a certain time lag. Because typing speed is slower than speak, in order to compensate for this defect, fast-paced adaptation of the network, improving efficiency to save time, increase the amount of information in the input text appeared inconsistent with grammatical ellipsis.

3.1 Any new type of word formation and acronyms

Information network is very large, people can read a lot of information online, on hand and will have a lot of available information, coupled with the freedom of the people on the network is relatively high, some random combination of words have appeared. For example, love and so atypical. There are some exotic, some entirely by the Chinese language to express the English vocabulary, pronunciation of English in order to brief also be replaced, for example, in Chinese, "download" the English word "download", for simplicity, in the network language directly using the "when" instead. English-mixed phenomena abound, such as "My e-Home", "e happy life," and so on vocabulary, too numerous to mention.

3.2 Syntactic variation

On the network in order to show young people's personality, perhaps slips factors often have rear adverbial sentence appears. Which I think the most famous sentence is Stephen Chow's phrase classic lines, "I go first."

Extensive use of speech inflection of words caused a change in the network grammar, also contributed to the emergence of such terms. More common are: noun verb, you Baidu yet, in this sentence, "Baidu" is a noun, it is used as a verb; adjective verb use, there are more representative, your computer was hacked. Sentence in itself is an adjective in black here was as a verb; noun adjective, for example, you are too old fashioned bar. Which in itself is an antique term, where it is used as an adjective to describe someone thought obsolete; adjective adverb used on the network performance for exaggeration or enhance tone, for example: I love you, is a typical example.

4 PRAGMATIC LEVEL VARIATION

The network is relatively unfamiliar with each other, and in the initial exchange, when inevitably some self-protection awareness. For example, A, said:. "Who are you," B said:. "User" in these two conversation information exchange is zero, which is what we call bullshit. This is different from the traditional communication place. Network language variation is also reflected in the tone of exaggerated sentences can express a variety of tone, such as affirmative, questioning, appreciation, stressed, etc., compared with the traditional language, the language on the network even more exaggerated, as : "Guiqiu answer!", etc., on the network use very often. Use these language techniques, you can make a more vivid tone, occasionally, not a bad idea, but too much can only use the language of the user's state of mind impetuous performance and language skills of the poor, the traditional language does not have a significant impact. The direct cause of the network is the language of prosperity in 2010, "People's Daily" published an article entitled "Jiangsu to the power of" cultural province ",″ the article, the term of this network in the "People's Daily" is used, so that the vast majority of Internet users feels very cordial.

For a time "to force" has been used in the major media have. Network words can board the "People's Daily" phenomenon reflects, initially thought to not be humble online language has begun to gradually penetrate into daily life, and is a widely recognized and accepted. The vocabulary generation network is a reflection of social development and technological progress, but also the environment and their unique network inseparable. Out of the environment will be meaningless. Chinese strong inclusive while also promoting the development of the network vocabulary. Language is a living thing, with the change of society itself is bound to be updated; the new network will appear in the vocabulary. Vocabulary while the network is double-edged sword, the proper use of Internet language can become more complex ideas simple and straightforward in expressing. Once out of the network environment, many network words do not meet the requirements of Chinese language and grammar, it will give the development of Chinese language and heritage caused the greatest negative impact. Such adverse effects on young people a greater harm. Young people are at the learning stage of knowledge, self-knowledge system has not yet fully formed, together with youth itself is the subject of Chinese Internet users, access to the Internet for a long time, so it is susceptible to network culture is not conducive to the future of Chinese language in real life the specification and use.

Constructivist approach to evaluation of language learning, designed to promote and enhance the student with the text, and the students, and the interaction between society, the ability to construct a personal language in the dialogue, so that students' language learning process extension and development in the evaluation. Therefore, evaluation of the subject should be diversified, to break the monopoly of the evaluation of the current situation of teachers using student self-assessment, life and peer assessment, evaluation of teachers and multi-directional interaction combines commentary.

Student self-assessment is the process of learning to guide students, according to the objectives and requirements of self-monitoring to participate in learning activities. Self-evaluation is the students' positive self-awareness and self-assessment an important way of learning ability, is to promote self-learning students achieve true, to "teach is not to teach," the state and the use of an important tool. Life and peer assessment, the main method is to guide the students that say understanding and mutual inspiration in the exchange, the formation of an open learning environment to develop exchanges and cooperation ability to learn the spirit of mutual cooperation with students.

In actual teaching, we knew that the evaluation should be diversified, but often superficially, because the students carry out self-assessment, peer assessment life and time-consuming, the effect is unsatisfactory; it is because we do not have a method for the students. Learning is a purpose, not for show and performances, but should allow performers to promote students' reading classic literature. For each student a show, you can also shoot made into a video file, upload to the network, the exchange of learning, to encourage more students to participate. Tech Internet age, with the Chinese language and literature classics, there is no intersection between the two is not, in-depth study of one of the characteristics, weaknesses, will combine the two, the development of Chinese language and literature teaching in the Internet age, so that the network service to Chinese Language and Literature classics Reading inspire more students to actively read literary classics, outstanding cultural heritage.

5 CONCLUSION

The development of the Internet age, a lot of classical literature to bring a certain impact, however, networks and Chinese language and literature are not mutually exclusive, there is a definite link between its two. Use of network resources and avoid the habit of students formed in the network era in the teaching of Chinese language and literature, develop the network edge, blending classical literature, a different form of the show, to find a balance point of the network and literature, will combine the two in the network By age, continuing our tradition of excellent classical literature, to enhance students 'reading classic literature, to enhance students' literary quality.

REFERENCES

Chen Ping. Modern Linguistics [M]. Chongqing: Chongqing Publishing House, 2003.

Cheng Gong. Generative grammar for Chinese "own" the word study [J]. Foreign linguistics, 2005, (1).

Research Meijuan Cai. Chinese Language and Literature Students' Ability System [J]. Shandong University of Technology (Social Sciences), 2012 (04).

Kaolinite, Wang Ningning, Ma Ying research "Chinese Language and Literature Practical Teaching Research" report. [J]. Beijing Radio and Television University, 2008 (02).

Huang Yi. Chinese Language and Literature Teaching [J]. Times Education (Education Edition), 2009 (03).

Liu Chang. Teaching of Chinese Language and Literature [J]. Intellect, 2011 (22).

Gui Shichun, Ningchun rock. Linguistics Methodology [M], Beijing: Foreign Language Teaching and Research Press, 2004.

Hu Jianhua. Chinese long-distance reflexive pronoun of syntactic study [J]. Contemporary Linguistics, 2008, (3).

Hu Jianhua. Long-distance reflexive pronoun "their" function, semantics, pragmatics and discourse theory [Z]. To be published.

Huangzheng De Xu Debao. Contemporary Linguistic Theory Series • Preamble [A]. Beijing: China Social Sciences Press, 2005.

Song Guoming. Summary of syntactic theory [M]. Beijing: China Social Sciences Press, 2007.

Xu Liejiong. Semantics [M]. Beijing: Chinese Press, 2000.

Joos, M. (ed.) Readings in Linguistics:the Development of descriptive linguistics in America since 1925.American Council of Learned Societies, New York, 2008.

Lyons, J.etc.(ed.) New Horizons in Linguistics 2.PenguinBooks, 2007.

Information, Computer and Application Engineering – Liu, Sung & Yao (Eds)
© 2015 Taylor & Francis Group, London, ISBN 978-1-138-02717-6

Research on simulation for crash of complete auto based on forming results

Xiang Wen Dang & Rui Jun Liu
College of Automobile & Civil Engineering, Beihua University, Jilin, China

ABSTRACT: In order to improve auto crash simulation precision, the way of crash fine simulation which forming results of auto-body panel is as initial condition by mapping is proposed. The forming results of absorbing parts are as an initial parameter of crash model is loaded by mapping theory, and loading way is introduced. Simulation can apply to predict for results of complete car after comparing results between simulation and test.

KEYWORDS: Automobile; Forming results; Crash; Simulation

1 INTRODUCTION

The passive safety performance is more and more vehicle collision by the user's attention. Before the application of vehicle crash simulation software, automobile collision data are obtained through the real car collision mode. The real vehicle crash test disadvantages of long experimental period, high cost, single collision of a small amount of data etc. With the advent of computer hardware development and all kinds of simulation software, the use of a digital model of a vehicle engineering and technical personnel, simulation of collision can be carried out with the aid of the computer in the vehicle. At present, the digital model of the body of conventional vehicle collision simulation is of uniform thickness, zero residual stamping parts assembly force and zero strain and this body of digital model does not consider the forming results of stamping parts, stamping parts of the results include: deformation caused by the uneven thickness of residual parts, unloading stress generated and strain differences and the actual assembly body this body model. Through the analysis of the impact factors of known crash simulation accuracy, ignore the body model using stamping forming results is one of the factors leading to the distortion of the simulation results. A research team led by Professor Huh of South Korea's National Institute of science and technology, matched with the modern car company to carry out earlier work in this area. The domestic scholars have carried on the effects of forming factors on typical stamping parts collision results, but the results of these studies show the effects of the collision condition of forming factors of stamping parts of the vehicle directly.

In order to improve the precision of vehicle collision simulation results, proposed the stamping forming factors of vehicle collision simulation method based on fine, stamping assembly model of the method using the reserved the forming results of crash simulation. Forming calculation with KMAS/OneStep software of auto body stamping of independent research and development, through the method of physical quantity of mapping will form results (thickness distribution, residual stress and strain) into the crash simulation model, using elastic-plastic beam element instead of the joint of the body stamping parts for assembly. Collision contrast fine crash simulation, conventional collision simulation and vehicle practical results show that the fine crash simulation method can improve the accuracy of vehicle crash simulation.

2 SHEET FORMING AND MAPPING

2.1 *The one step inverse forming FEM theory*

To satisfy the incompressibility constraint in elastic-plastic large deformation process of sheet deformation assumption and the process is proportional loading. Departure, will a stamping C as surface deformation at the end of the work piece, by means of finite element method to determine the satisfy certain boundary conditions, location of the P0 node P work piece in the initial flat blank C0, strain by calculating the initial flat blank nodes in all the nodes and corresponding work piece position relationship can obtain work piece. The stress distribution and thickness. The obtained stamping results (strain, distribution of stress and thickness) method called one-step inverse forming FEM, as shown in figure 1.

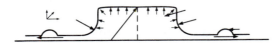

Figure 1. The one step inverse forming FEM schematic diagram.

2.2 *The physical volume mapping*

The collision simulation is used in the final parts of the finite element grid, so the grid and finite element in the one step inverse used in forming a different grid, and the former than the latter sparse grid. Therefore, in order to consider the effect of stamping results in crash simulation, need to be after press forming unit thickness and the equivalent strain, stress and other physical quantities from the one step inverse mapping of the finite element mesh nodes in the collision simulation. The mapping algorithm to transform this physical quantity process is called physical quantities, namely a fine mesh from a one-step inverse forming to the conversion between physical quantities of coarse grid in crash analysis method. Using the mapping method proposed in the literature can reasonably be one step inverse physical quantity of stamping forming method indicates the distribution of conversion to the finite element analysis model of collision.

2.2.1 *The finite element model*

In the finite element model of the body collision simulation as shown in Figure 2, using elastic-plastic beam element body, chassis and the seat body assembly and finite element model of vehicle collision, as shown in Figure 5, in which 2411241 nodes, each kind of unit 2847127.

Should first select the unit type in the mesh before, according to the C type frame shape, material and analysis requirements, selection of element type is SOLID45, the element type is mainly used for 3-D entity structure. The Unit is composed of 10 nodes, which are combined together, at the same time unit with the intermediate nodes, thus improving the analysis precision; each node has x, y, Z displacement direction of 3 degrees of freedom. The element has plasticity, creep, swelling, stress stiffening, characteristics of large deformation and large strain.

In determining the method, mesh selected cell types, but also to determine the accuracy of grid division. 3-D solid model, method of setting accuracy mainly include: SMRTSIZE and ESIZE. Where ESIZE is the size of a setting unit directly, while the SMRTSIZE is a precision grades set partition, this level do not directly reflect the unit size, this division method mainly depends on the shape of the complexity, namely shape regional division of more complex and more detailed, but the division of other region can be quite rough. It uses ESIZE command to set

accuracy, setting accuracy: ESIZE, 0.009, 0.009 indicates the length of the baseline for 9mm unit.

The main function of the rigid body displacement constraint is limiting the whole finite element model. If the finite element model in the presence of rigid body displacement, its last analysis results are necessarily wrong, even will lead to carries on the analysis of the whole model. At the same time, constrained by the other components are actually exist (or it can be ground) may occur in a finite element model generated by the motion of a rigid body displacement constraint, i.e. restricted model. These components may not be part of the analysis model, but always real, but only the actual contact place there will be a constraint function, that is a constraint on the loading finite element model, especially with particular attention to the rigid constraint, also should pay attention to is not a force can place as a constraint, although to reflect the restriction role constraints can also be used force, but the force and the constraint is not a concept. Contact place, there may be constraints force, is concerned, because it is the role of the displacement of the rigid body limit of finite element model, the limit function will usually produce force. Therefore the analysis of finite element model, there must be a clear

Figure 2. The finite element model of the body.

Figure 3. Chassis finite element model.

Figure 4. Seat finite element model.

Figure 5. Vehicle collision simulation finite element model.

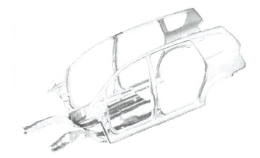

Figure 7. The main absorption fine crash simulation of components.

concept of the forces and constraints, which force to as constraint without a load in the finite element model, which is driven by the force model movement or deformation but not as constraints to loading.

3 THE VEHICLE FRONT COLLISION FINE SIMULATION PROCESS

3.1 *Determination of vehicle collision energy absorption is part*

Frontal collision process, energy transmission line as shown in Figure 6, the post collision energy transfer to the floor, from the front, side and other parts through the front longitudinal beam, and through the part deformation during collision kinetic energy generation absorption. Through the analysis of front collision energy transfer that the front side member and side absorb most of the energy generated in the collision. Select the front longitudinal beam and the side wall is main suction fine crash simulation components, as shown in figure 7. Application of one step inverse solver KMAS/OneStep to calculate its thickness, strain and stress distribution, the forming results into LS-DYNA collision analysis software, fine analysis of the vehicle frontal collision process of vehicle.

3.2 *The forming results loading*

Through the KMAS/OneStep after press forming the suction plate thickness, residual components of the stress and strain data according to the LS-DYNA Identification Format writes the specified key file, by setting the header file for the keyword *INCLUDE key file into LS-DYNA software model, complete stamping parts forming results of loading.

4 POSITIVE FINE CRASH SIMULATION RESULTS

Chinese automotive technology and research center on the basis of the thorough research and analysis of foreign NCAP combined with the automobile standards and regulations, China's road traffic and the actual situation of vehicle features, and to determine the C-NCAP of domestic and international technology exchange and the actual test widely (China-New Car Assessment Program) test and scoring rules. Frontal crash is which frontal impact requirements on a vehicle speed 50km/h and rigid, fixed barrier 100% overlapping rate.

Numerical simulation of the forming of frontal collisions and the real vehicle crash test in accordance with C-NCAP standard setting collision as the speed of 50km/h based on the comparison of real car collision and collision simulation results of the collision time deformation process results and vehicle body acceleration time results. A vehicle front collision time deformation process results, as shown in figure 8. Vehicle frontal impact time and acceleration curve, as shown in figure 9.

Through the observation of figure 8 can be seen between simulation result and real car collision results of time consistent deformation process.

It can be seen in Figure 10, the conventional simulated body time acceleration curve and crash test body time acceleration curve of the overall morphological

Figure 6. Frontal collision energy transfer route map.

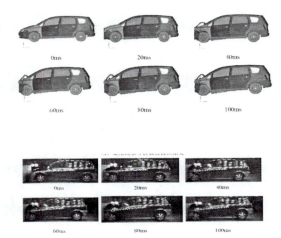

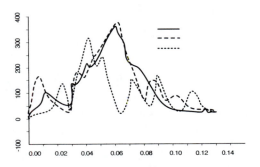

Figure 8. Time history results collision deformation.

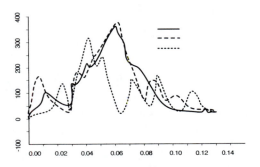

Figure 9. Collision body time acceleration curve.

differences exists, the peak acceleration and conventional simulation predates the fine simulation and collision test, which shows that the errors of conventional simulation results. Fine crash simulation, conventional collision simulation and collision test body time acceleration curve shape is basically the same, but the acceleration test collision in the maximum value should be smaller than the maximum fine simulation values of acceleration, which is the main vehicle collision test parts before experienced degusting, assembly and painting processes, equivalent to the press by the aging treatment, eliminate to the parts are part of the residual stress reinforcement stress, leading to the actual participation of stress is smaller than the collision forming fine simulation results stress loading.

Through the analysis of deformation time history results and body time acceleration curve that fine simulation result and crash test results fit well, fine simulation results can react to the automotive collision process is true.

5 CONCLUSION

The forming results of auto body stamping loading to the vehicle collision model through the mapping theory, vehicle frontal impact simulation of a new model in accordance with the C-NCAP standard, by comparing the conventional simulation, get the following conclusions based on the results of simulation and experiment of forming fine:

1 The car body stamping forming factors to affect vehicle collision simulation results.
2 The proposed fine crash simulation flow forming based on correct and feasible.
3 Fine crash simulation results forming is better than that of conventional collision based simulation, and the results can be used for automobile collision prediction results.

REFERENCES

Kim H S, Hong S K, Hong S G, Huh H. The evalua2tion of Crashworthiness of vehicles with forming effect, 2003, 4th European LS2DYNA Users Conference, 25–33.
Lee S H, Han C S, Oh S L, Wriggers P. Comparative crash simulations incorporating the result s of sheet metal analyses. Engineering Computations, 2001, 20(526): 744–758.
Lee S H, Han C S, Oh S L, Wriggers P. Comparative crash simulations incorporating the result s of sheet metal analyses. Engineering Computations, 2001, 20(526): 744–758.
Hu Ping, Bao Yidong, Robert Huth, Bo, et al. Car body crash simulation into the process factors analysis [J]. Journal of solid mechanics, 2006, 27(2):148–158.
Bao Yidong. Automobile body parts of one step inverse forming FEM and lean collision simulation of [D], PhD thesis, Jilin University, 2004.
Huh H , Kim K P, Kim, S H, Song J H , Kim H S , Hong S K. Crashworthiness assessment of front side members in an auto2body considering the fabrication histories. International Journal of Mechanical Sciences, 2003, 45:1645–1660.
Huh H, Kim K P, Kim, S H, Song J H , Kim H S, Hong S K. Crashworthiness assessment of front side members in an auto2body considering the fabrication histories. International Journal of Mechanical Sciences, 2003, 45:1645–1660.
For many years, Liu Li, Li Hongjian, et al. Study on [J]. automobile technology application of FFS method to the analysis of vehicle bumper crash simulation, 2008 (3):27–29.
He Wen, Zhang Weigang, Zhong Zhihua. Finite element simulation [J]. Chinese Journal of mechanical engineering, welding connection simulation of vehicle collision in 2005, 41(9):73–77.
Shi Yuliang, Zhu Ping, Shen Libing, et al. The relationship between the finite element simulation method for spot welding of car crash simulation based on [J]. Journal of mechanical engineering, 2007, 43(7):227–230.

Information, Computer and Application Engineering – Liu, Sung & Yao (Eds)
© 2015 Taylor & Francis Group, London, ISBN 978-1-138-02717-6

A filtering method for audio fingerprint based on multiple measurements

Bin Bin Han
Tianjin Tianshi College, Tianjin, P.R. China

Yong Hong Hou
Tianjin University, Tianjin, P.R. China

Lin Zhao & Hua Yu Shen
Tianjin Tianshi College, Tianjin, P.R. China

ABSTRACT: Tow fields in content-based music information retrieval are in rapid development. One is Query-by-Humming systems which firstly transcribe a sung or hummed query and secondly search for related musical themes in a database. The other is Query-by-Example systems which firstly record the music in complex environments, secondly extracts the robust features, thirdly search for the target songs, and finally return the name of the song. This paper presents a fusion of features and a novel searching method in Query-by-Example systems based on audio fingerprint. Firstly, we use the audio fingerprint composed of two kinds of peculiar features which are similar to that proposed by Shazam to search for candidates and then we compute the exact distance between the query and the segments in the music database. Finally, the candidate with the least distance to the entire query is returned to the user. Experiments show that our approach can achieve very high accuracy for Query-by-Example, which is much better than those systems proposed by Shazam.

KEYWORDS: Query-by-Example; Audio Fingerprint; Music Information Retrieval; Shazam

1 INTRODUCTION

The steep rise in music downloading over CD sales has created a major shift in the music manufacturers and multimedia content providers away from physical media formats and towards online products and services. Now more and more people download the music on the Internet. The collection of music has been approaching, millions of recordings and this has posed a major challenge for retrieving, searching and sorting those musics. Traditional music search engines are usually based on metadata, and the user must type in text information to get the intended song. Imagining when you have tried hundreds of query words in the web, but the song that you want never emerges, just because you forget the name of the song. This problem promoted the development of the query-by-humming (QBH) system, which allows the user to obtain the intended song via a few seconds of voice using a recording device.

The QBH task can be divided into two subproblems: 1) converting a query into a format which results in robust searching and 2) matching the query with the song in the database. However, the uncertainty and fluctuation of queries because of various music backgrounds raise a great challenge. Hence, a more robust retrieving method becomes vitally important. In the QBH system almost all the researchers use pitches as the stable features, and in the matching stage there are all kinds of different algorithms such as Dynamic Time Warping (DTW) [1], Earth Mover's Distance (EMD) [2], Recursive Alignment (RA) [3], Locality Sensitive Hashing (LSH) [4], and so on.

The QBH system can solve the problem when the user remembers the rhythm, but forgets the name of the song. Imagine the situation that you're listening to the radio and suddenly you hear a song which you are interested in but you missed the announcement and don't recognize the song. What should you do? You could call the radio station, but it is troublesome. By using the query-by-example (QBE) system, which makes an unknown audio clip as a query on a music fingerprinting database, the audio clip can be identified. Similar to the QBH, the QBE system also contains two parts: I) transcribing acoustic input into a sequence of fingerprinting segments ii) searching for music pieces that best match the provided transcription. In the system, we must consider both the efficiency and effectiveness. The core of the point is a highly robust fingerprint representation which is invariant even in a noisy environment and a very efficient fingerprint search strategy, which enables the computer to quickly retrieve from a large fingerprinting database.

In the development of the fingerprinting system, Philips Robust Hash (PRH) [5–9] and Shazam [10–14] are the best examples of this approach. The PRH algorithm uses the energies of the 33 non-overlapping logarithmically spaced subbands (e.g. Bark Scale) covering the frequency range from 300 Hz to 2000 Hz as fingerprints [5]. All the fingerprints computed are compared with the audio files stored in the database which are registered in a hash table with the fingerprints being treated as the keys. Shazam [10] utilizes the peaks of the spectrum as the feature which is robust to noise and takes advantage of the histogram to search the most-likely candidates.

In this paper, we use the features similar to Shazam to construct a system which is more efficient than Shazam. The peaks of the spectrum only contain frequency information, so we add the time information on the features using the peaks in the spectrum considering both the time abscissa and frequency ordinate. In the procedure of retrieval, in order to speed up searching, we make use of several peaks and the length of time between them to build up the music fingerprint. After finding the similar audio clips, we compute the exact distance between the query and the segments in the music database to get the best sorting.

The outline of this paper is organized as follows: Section 2 describes the Shazam algorithm, Section 3 deals with feature extraction, and Section 4 presents our method for searching survival candidates. Finally, Sections 5 and 6 give the corresponding experimental results and conclusion.

2 SHAZAM ALGORITHM

Shazam [10] is a deployed, flexible audio search engine. The algorithm is noise and distortion resistant, capable of quickly identifying a short segment of music captured through a cellphone microphone in the presence of dominant noise. The algorithm uses a hashed time-frequency constellation of the audio, which is composed of peaks of the spectrum, yielding unusual properties resistant of noise. The peaks are defined as frequency points with high energy. These peaks are found frame by frame for the entire length of the music. Thus, a complicated spectrogram may be reduced to a sparse set of coordinates called a constellation. We choose pairs of points from the constellations as fingerprints. Each pair of point is a point paired with the other in the fixed target region. For the audio files stored in the database, all the fingerprints computed are registered in a hash table with the fingerprints as the keys. In the stage of searching, the fingerprint of the query is compared to those in the database. The pattern of dots should be the same for matching segments of audio. If you slide the constellation map of a database song over the query constellation map, at some point a significant number of points will coincide when the proper time offset is located and the two audio clips are the same. Figure 1 shows an example of the matching audio [10].

3 FEATURE EXTRACTION

Most fingerprint extraction algorithms are based on the following approach. First the audio signal is segmented into frames. For each frame the features invariant to signal degradations are computed. Shazam uses spectrogram peaks that have the largest energies, due to their robustness in the presence of noise. In fact, the peaks are very invariant to the white noise, but not the colored noise. This paper uses two features, one of which was proposed by Shazam. And the other feature is robust to colored noise and is described as follows [15]: for every point in the spectrogram after short-time Fourier transform, we calculate the energy ratio:

$$R_k = \frac{\sum_{i=-m_1}^{m_1} \sum_{j=-n_1}^{n_1} E[k+i][k+j]}{\sum_{i=-m_2}^{m_2} \sum_{j=-n_2}^{n_2} E[k+i][k+j]} \quad (1)$$

In the formula above m_1 and n_1 are the length and width of inside rectangular, m_2 and n_2 are those of outside rectangular in the spectrogram. The rectangular box is depicted as figure 2.

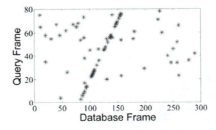

Figure 1.   An example of the matching audio. A significant number of matching fingerprints appear on a diagonal line.

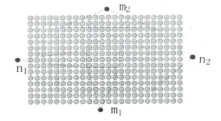

Figure 2.   The rectangle of the spectrogram.

Then, for every point we can compute the ratio of the sum of points' energy in the inside rectangular box and that in the outside rectangular box. The largest ratio represents the fastest changes in energy. The points having the largest increase in energy and longer time information are very robust to colored noise, so we choose the point which has the largest ratio of the energy as the second robust features.

4   2-LEVEL FILTERS

4.1   The back-end to features

In our system, we use two features to form the fingerprints. One is peaks in the spectrogram proposed by Shazam and the other is the newly proposed features as shown in Section 3. To form the fingerprints, if every feature obtained is treated as fingerprint, we will find lots of collisions in the hash function. Here, we employ the point paired with the neighboring ones as the fingerprint. In fact, we discover that the paired points should be neither too close nor too far. When the distance is small, much serious collision happens. If the distance is large, many spurious fingerprints are created. In our system, every point paired with several other points forms the fingerprints, and the best distance of paired points is obtained based on experiments.

4.2   The first-level filter

In order to locate the query clip in the music database, the Shazam retrieval algorithm can be used. First, we note the songs that have the same fingerprints from the query clip. Then we calculate the number of matching time-aligned fingerprints between the query clip and the music in the database as shown in figure 1. The song corresponding to the largest matching number is the one we want [10]. Since the music expected may not always rank the first, we preserve several tens of candidates.

4.3   The second-level filter

The first-level filter has eliminated most unlikely candidates, but the expected song may not emerge yet. We can process the results of the first-level filter to obtain a more accurate result. Since the exact position has been located for each candidate after the first level filter, we can calculate the Euclidean distance between every candidate clip and the query clip by using the original features as follows:

$$Dis_j = \sum_{i=0}^{n} (q_i - d_i)^2 \qquad (2)$$

Dis$_j$ is the fingerprint distance between the j-th candidate clip and the query clip, q$_i$ in the formula 2 is the feature of the query clip, and d$_i$ is the corresponding one of the candidate in the database. To calculate the distance, we only need to reserve the top N candidates in the first level filter, so this process is very fast and it can give us a more accurate ranking.

5   EXPERIMENTS

5.1   Experimental data

It is really difficult to compare different fingerprint systems, as the retrieval performance is influenced by many factors such as query samples, recording condition, music database, and so on. Still, we implement some existing approaches and attempt to make a comparison with identical test condition. The music database used in this study consists of 500 songs, including pop music at home and abroad. The query corpus is composed of 400 clips randomly selected from the database. To assess the robustness of the algorithm, the following distortions were applied to queries.

Set 1: Playing and recording in a noisy environment.
Set 2: Additive white noise with the SNR at 5 db.
Set 3: Additive white noise with the SNR at 0 db.
Set 4: Additive white noise with the SNR at -3 db.

The length of each clip is 10 seconds. All the songs in the database are 8kHz/8bit sampling. There are several parameters which should be determined from the experiments. In our system, the frame length is 30 ms, the frame shift is 30 ms, the small rectangular is (1*3), the large rectangular is (3*11), the interval points between paired points is 3 and 5, and the survival rate of the first-level filter is set to 4%.

In contrast, we compared our algorithm with Shazam on the same corpus. The performance is evaluated in four measurements: Mean Reciprocal Rank (MRR), Top-1 rate, Top-5 rate and Top-20 rate. MRR is calculated by the reciprocal of the rank of the first correctly identified cover for each query (1/rank). Top-1 rate is the percentage that the correct song is ranked in the first place of all returns. Top-5 rate is the percentage that the correct song is ranked within the top 5 of all returns. Top-20 rate is the percentage that the correct song is ranked within the top 20 of all returns.

5.2   Experiments for different intervals

Figure 3 reveals the performance of different fingerprints with different intervals. The last intervals in the figure are composed of intervals 3 and 5 which have the best performance. In the experiments, we test several fusions of intervals and the optimal composition is 3 and 5. In the following experiments, we all used these intervals.

5.3 *Experiments for different methods*

For each query set, three algorithms were applied, including Shazam method, one-filter method and two-filter method. It can be seen from these four figures, our method outperforms others, especially the percentage that the correct music ranks the first of all returns. By using these two features and the filtering algorithm, the accuracy improves greatly.

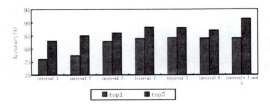

Figure 3.   The results of different intervals for set 1.

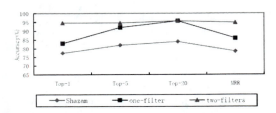

Figure 4.   The results of different algorithms for set 1.

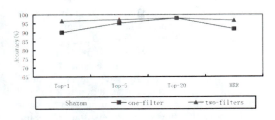

Figure 5.   The results of different algorithms for set 2.

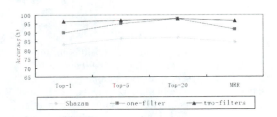

Figure 6.   The results of different algorithms for set 3.

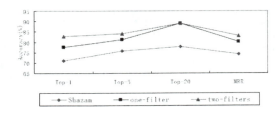

Figure 7.   The results of different algorithms for set 4.

6   CONCLUSIONS

In this paper, a new feature extraction algorithm and a two-level filtering, retrieval mechanism are adopted in the fingerprinting system to improve the retrieval accuracy. From the experiments we can see that our retrieval mechanism is better than using the original one level alone, for example Shazam, and the proposed method achieved satisfying result. The future work is to study the performance in the database consisting of tens of thousand songs and find more robust fingerprinting features.

REFERENCES

[1] Jyh-Shing Roger Jang and Hong-Ru Lee, "A general framework of progressive filtering and its application to query by singing/humming", IEEE Transactions on Audio, Speech, and Language Processing, vol. 16, no.2, pp. 350–358, 2008.
[2] Hongchen Jiang and Xu Bo, "Query by humming via multiscale transportation distance in random query occurrence context", in Proceedings of IEEE International Conference on Multimedia and Expo, pp. 1225–1228, 2008.
[3] Wu Xiao, Ming Li, Jian Liu, Jun Yang and Yonghong Yan. "A top-down approach to melody match in pitch contour for query by humming", in Proceedings of International Conference of Chinese Spoken Language Processing, 2006.
[4] Ryynanen Matti and Anssi Klapuri, "Query by humming of midi and audio using locality sensitive hashing." in Proceedings of IEEE International Conference on Acoustics, Speech and Signal QProcessing, pp. 2249–2252, 2008.
[5] Jaap Haitsma and Ton Kalker, "A highly robust audio fingerprinting system with an efficient search strategy", Journal of New Music Research, vol. 32, no. 2, pp. 211–221, 2003.
[6] Yu Liu, Hwan Sik Yun, Nam Soo Kim, "Audio fingerprinting based on multiple hashing in DCT domain", IEEE Signal Processing Letters, vol. 16, no. 6, pp. 525–528, 2009.
[7] Jin S. Seo, Minho Jin, Sunil Lee, Dalwon Jang, Seungjae Lee, Chang D. Yoo, "Audio fingerprinting based on normalized spectral subband moments", IEEE Signal Processing Letters, vol. 13, no. 4, pp. 209–212, 2006.

[8] Zhiyuan Guo, Qiang Wang, Gang Liu, Jun Guo, "A music retrieval system based on spoken lyric queries", International Journal of Advancements in Computing Technology, vol. 4, no. 8, pp. 173–180, 2012.

[9] Jixin Liu, Tingxian Zhang, "Wavelet-based audio fingerprinting algorithm robust to linear speed change", Computing and Intelligent Systems, vol. 234, pp. 360–368, 2011.

[10] Avery Wang, "An industrial strength audio search algorithm", In Proceedings of International Conference on Music Information Retrieval, 2003.

[11] Avery Wang, "The Shazam music recognition service", Communications of the ACM – Music information retrieval, vol. 49, no. 8, pp. 44–48, 2006.

[12] Peter Jan O. Doets and Reginald L. Lagendijk, "Distortion estimation in compressed music using only audio fingerprints", IEEE Trans. Audio, Speech, and Language Processing, vol. 16, no. 2, pp. 302–317, 2008.

[13] Pedro Cano, Eloi Batlle, Ton Kalker and Jaap Haitsmat, "A review of audio fingerprinting", Journal of VLSI Signal Processing, vol. 41, no. 3, pp. 271–284, 2005.

[14] Qiang Wang, Zhiyuan Guo, Gang Liu, Jun Guo, "Audio Fingerprinting Based on N-grams", International Journal of Digital Content Technology and its Applications, vol. 6, no. 10, pp. 361–368, 2012.

[15] Tang Jie, Liu Gang, Guo jun, "Improved algorithms of music information retrieval based on audio fingerprint", in Proceedings of IEEE Third International Symposium on Intelligent Information Technology Application Workshops, pp. 367–371, 2009.

Information, Computer and Application Engineering – Liu, Sung & Yao (Eds)
© 2015 Taylor & Francis Group, London, ISBN 978-1-138-02717-6

Study on distributed networked software testing

Y. Yao, Z.P. Ren, C.Y. Zheng & S. Huang
PLA University of Science and Technology, Nanjing, China

ABSTRACT: Networked software becomes more and more popular. Traditional software testing methods cannot be suitable for networked software testing. In this paper, concepts of distributed networked software testing and related category are mainly introduced. Distributed networked software testing technology is established based on the distributed testing environment, and model and architecture of distributed networked software testing technology are also given. Advantages of distributed networked software testing are analyzed.

KEYWORDS: Networked Software; Software Testing; Distributed environment

1 INTRODUCTION

Distributed Networked Software testing is a new testing technology based on distributed computing, distributed testing consists of both test and distributed computing, is mainly referred to software testing, software testing has its own test method, test method, test process. Due to the test environment and test management are built on distributed computing technology, through the distributed computing technology to realize the test process and method, therefore, means and methods in the test process, etc., distributed testing has some unique characteristics. Through on the server to deploy multiple virtual machines, distributed testing can provide remote testing service. Users no longer need to purchase and install the software locally test automation tool, can easily finish the test task on distributed testing platform, distributed service provider offers a variety of testing platform.

The advent of the era of distributed computing to provide a broad platform for software services, software test is one of them. We can with the help of the distributed computing platform infrastructure build test environment, by means of distributed computing platform tools for software testing, software testing, of course, also can use the platform for software testing service providers to provide professional services. Service providers offer a variety of hardware, software, testing tools, services and so on, users only need to make test plan, write good scripts, can accomplish on a distributed testing platform of software testing. With distributed computing technology matures, will greatly change the way people application information, delivery mode, research and development of software service enterprises and fundamental changes in the way of software testing.

Under the influence of the global economic crisis under cost pressure, in the market, more and more enterprises begin to consider the value of the distributed model can bring. Have some distributed testing enterprises on the market at present and application, but it is not standard and definition of distributed testing as a unified regulation, distributed testing is still in the steady development, with the rapid development of distributed computing, will certainly obtain further rapid development.

2 DISTRIBUTED NETWORKED SOFTWARE TESTING MODEL

Distributed Networked Software testing technology at present, relatively well-known companies in the development of their respective distributed testing products, distributed Networked Software testing model as shown in Figure 1.

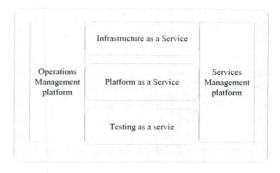

Figure 1. Distributed networked software testing model.

The model includes five parts:

1 IaaS (Infrastructure as a Service) is to set up test environment needed for some tools, computing resources, storage resources, network as a Service provided to users, according to the IT Infrastructure will take to be delivered to the user.
2 PaaS (Platform as a Service): the test software development, testing, and deployment environment as a Service provider to the user, and provide the database management, data storage, operating systems, testing development environment, etc.
3 TaaS (Testing as a Service): is the application of software Testing service mode, the user can through the network connection, using a variety of testing services.
4 Operations management platform: interface provides a unified and standard management, monitoring and management of all resources and activities, with recovery and reducing power.
5 Service management platform: distributed testing services platform, can provide the service request, consulting, summary, report, etc., can provide various services for the distributed testing.

3 THE CHARACTERISTICS OF DISTRIBUTED NETWORKED SOFTWARE TESTING

Because distributed testing based on distributed computing, distributed testing feature is closely related to distributed computing. Due to distributed computing network, computing, storage, operating systems, platforms and Web services basic services such as packaging, emphasis on the use of these resources, rather than the implementation details, so the use of distributed computing technology, operation process and the way of implementation is transparent, not visible for the user, the main supply in the form of all kinds of service users. So the distributed testing technology based on distributed computing is mainly provided a wide range of services, including infrastructure resources also became a kind of service, to provide a unified interface and presentation, just according to the regulation of distributed testing, access to these services in accordance with the requirements, you can easily complete the on-demand customized software testing work, rather than Guan Yun testing services for specific implementation details. For example, you want to test, one of your software you only need to mention

Hand in your software, submit the way might be the source code, executable, or have already deployed system, then you can access distributed testing services, execute the test directly, and get the test result. Secondly, through virtualization technology, the testing resources can be rapid deployment and dynamic allocation, meet the demands for multi-tenant, a

great convenience for the construction of the test environment and change, simply configure corresponding virtual parameter Settings, can be achieved provided test resources according to the needs of users, the dynamic extension.

Software system distributed computing environment, the traditional software, running on a broader system terminal, server migration, face a more extensive system with more vast amounts of data, more focus on the user experience, more abundant information provided. So the software quality model with the computing environment is paying attention to the original function under the premise of the weight of its ease of use and efficiency of software will be greatly enhanced, at the same time because of the characteristic of the elastic compute distributed computing system itself will pay more attention to the reliability and portability of the system.

4 CONTENT OF DISTRIBUTED NETWORKED SOFTWARE TESTING

According to the concept of distributed testing can be derived, need to use software tools and environment test can be carried out distributed testing, with the development of distributed computing, software testing, and application of distributed computing technology will be based on the rapid progress, more and more projects will be suitable for distributed testing, suitable for distributed testing items or content such as shown in Figure 2.

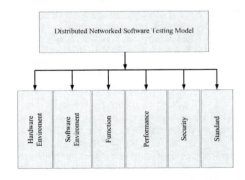

Figure 2. Content of distributed networked software Testing.

5 CATEGORY OF DISTRIBUTED NETWORKED SOFTWARE TESTING

Due to the current test field are mainly the traditional test mode, the lack of sufficient research on distributed testing problem, so based on the distributed computing, distributed testing theory and technology support is not perfect, the future distributed testing will be focused mainly on three aspects of software testing, that is, in a distributed environment for its distributed test, migrating existing tests to the distributed environment. The following were introduced:

1 Test software. Distributed test platform provides the appropriate test environment and resources for software testing, the user only needs to connect to the Internet access to the distributed testing services, will be able to write their own software professional, reliable and high quality testing, without concern for testing the use situation of environment and resources, tools, and distributed testing platform is responsible for scheduling, to the related resources optimization, modeling and other issues. The distributed testing platform can accelerate the software testing process, to ensure the quality of software testing, promote the efficiency of software development. Test software types, not only can test the traditional meaning of local software, and can also test deployment in Distributed environment software. Moreover, distributed testing platform can quickly configure the required test environment, a user can quickly and effectively get the testing environment and resources, with the rapid development of distributed computing, distributed testing technology will have a great impact on the traditional testing technology, change the traditional test way and method.

2 Distributed environment test. Test questions focus primarily on the distributed platform its architecture, environment, function, performance, to make the distributed computing itself to meet the requirements of various distributed technical indexes meet the performance rules, mainly the development of virtual platform performance consumption of distributed computing and virtualization platform of performance test, virtual link performance and virtual switch performance testing and storage performance test. In addition, in order to ensure the accuracy and safety testing, should also test a variety of distributed deployments in distributed computing environment software.

3 Migration test to the distributed environment. With the rapid development of distributed computing technology, its application is more and more, distributed computing is a key technology, will get the support and improvement of the corresponding technology, and provides the corresponding solutions, in order to facilitate the existing mature application and system migration to the distributed environment, at the same time, migration, traditional software testing to the distributed computing environment and it can reduce the hardware cost of acquisition, provide strong support for the development of enterprises and

organizations. The traditional testing migrates to the distributed environment in a test environment, to provide the required environment by using distributed testing technology, testing on the test software. Can also migrate will address itself to the distributed test environment, test for deployment in distributed networked software testing.

6 ADVANTAGE OF DISTRIBUTED NETWORKED SOFTWARE TESTING

6.1 Cost savings

After using the distributed testing technology, users do not need to buy a lot of computer, test equipment and various types of expensive test software or tools, and do not need to build the test environment is complex, the test cost is saved to achieve the maximization of profit. Just making a good test plan in advance, submitted to the distributed testing provider, you can get the corresponding service, according to the needs, to achieve on-demand pay. According to the test resource dynamic additions and deletions of test requirements, create, save maintenance and costs of approval test environment. No question need to consider the software upgrade costs generated in the networked software in a test environment, test the distributed service provider ready ahead of software upgrade and maintenance, user only needs to press the need to use on the line.

6.2 Improving the efficiency

Through virtualization technology, distributed testing platform can allocate resources, completed in the shortest time and least operation service configuration, service on-line and service activation and a series of operations, according to the needs of users in the shortest possible time to build test environment, you can save a lot of time, completely eliminates the need to purchase, the soft, hardware installation, management of time, environment and the management of the Dou You Yun test service providers are responsible for, the user need only as needed on the line. There are many test cases and test defect Library in distributed storage, convenient user access and reference, for the problem of software and hardware fault test generation process, all can obtain the distributed service provider support, embody the distributed testing services of professional level.

6.3 Build of environment

Distributed test environment, can be convenient, high-efficient environment system platform according to the need to build a variety of different, only need to submit the application test environment, distributed providers will for users to build well different test environment.

6.4 Immediate use

Distributed testing can according to the different needs of users to automatically build test environment need to meet different, testing personnel only need to log in through the network to the distributed testing environment, can be easily tested immediately, do not need to prepare the complex process, saving the time consumed a lot of environment deployment, testing cloud providers completely to provide the corresponding services. Through virtualization technology, testing the use of distributed testing standard environment, greatly reducing the test preparation time.

6.5 Convenient service

Distributed testing in the provision of services to users at the same time, when the user need to stop a service, distributed testing can automatically complete the service stops, service offline, delete the service configuration and resource recovery operation, according to the needs of the business or additions and deletions module, resource allocation. In the virtual technology support, the service is no longer needed; you can cancel the deployment to release the occupied resources. Can open the automatic rollback set service, by setting the restore point, will the service to restore to the specified state. The open state and the process monitoring service, state management, in each step of testing the whole monitoring implementation, are easy to find and solve problems. In addition, distributed testing platform can also provide all kinds of consulting services, including expert services, consulting, case consultation, case consultation and related services, to meet the different needs of testing personnel, to provide all the convenient service distributed testing.

6.6 Performance testing

Distributed testing can achieve more and more economic performance, scalability and pressure test. Distributed testing services can be time and cost to meet the traditional way is difficult to achieve the requirements. Almost do not have to spend much time, and then in these servers is installed on the test software to do the test. In general, a shorter testing time contributed to the development of more rapid. After the completion of the development of some new features that can be tested in different environments, then you can immediately get the effective reporting, and use in future development. Distributed computing can easily across different operating environment based on availability, enables developers to understand the

service user perspective. The services at a low price to provide some customers, in return, use the service to customer feedback and other defects such as suspected or found a bug. When considering whether to do the test in the distributed environment, also need to make a tradeoff between the internal test environment and test environment in the distributed networked software, the test environment is smaller than the production environment, and the distributed testing environment and production environment similar. Internal testing environment, safety, but quality will affect the test results. On the other hand, distributed testing environment can generate test results more real, but need to pay more attention to data security.

ACKNOWLEDGMENTS

This work is supported by the Natural Science Foundation of Jiangsu Province, China (Grant No. BK2012060, BK2012059) and the Prep-research Foundation of PLA General Armament Department (No. 9140A05040213JB25068). Resources of the PLA Software Test and Evaluation Centre for Military Training are used in this research.

REFERENCES

He, K.Q. Peng, R. Liu, J. He, F. Liang, P. & Li, B. 2006. Design methodology of networked software evolution growth based on software patterns. *Journal of Systems Science and Complexity* 19(2):157–181.

He, K.Q. Liang, P. Peng, R. Li, B. & Liu, J. 2007. Requirement emergence computation of networked software. *Frontiers of Computer Science in China* 1(3):322–328.

Li, H. Huang, B. & Lu, J. 2008. Dynamical evolution analysis of the object-oriented software systems. Proceedings of the 2008 IEEE Congress on Evolutionary Computation (IEEE CEC 2008), Hong Kong, June 1–6, 2008, 3035–3040.

Li, H. & Lu, J. 2009. A novel scale-free network model with accelerating growth. *Proceedings of the 2009 IEEE International Symposium on Circuits and System, Taipei, 24–27 May, 2009.*

Liu, J. He, K.Q. Ma, Y.T. & Peng, R. 2006. Scale free in software metrics. *Proceedings of 30th Annual International Computer Software and Applications, Chicago, USA, 18–21 September, 2006.*

Liu, J. Lu, J. He, K.Q. & Li, B. 2008. Characterizing the structural quality of general complex software networks, *International Journal of Bifurcation and Chaos* 18(2):605–613.

Wang, H.C. He, K.Q. Li, B. & Lu, J.H. 2010. On some recent advances in complex software networks: Modeling, analysis, evolution and applications *International Journal of Bifurcation and Chaos* 20(4).

Zhao, T. 2014. The exploration and research on software testing technology based on Cloud Computing. *Wuhan University.*

Information, Computer and Application Engineering – Liu, Sung & Yao (Eds)
© 2015 Taylor & Francis Group, London, ISBN 978-1-138-02717-6

Study on the lag of China's rural financial development and its coping strategies

Hai Yan Zhou
Department of Economics, Chongqing University of Science and Technology, Shapingba District, Chongqing, China

Yu Chen
The Center of Agricultural Information of Chongqing , North New District, Chongqing, China

ABSTRACT: China finance presents distinctive urban and rural dual economy structure. The lag of China rural financial development mainly appears in the urban and rural financial elements, fund flow, and the uneven progress of financial supply, demand and so on. The main reasons can be listed as follows, the defects of the rural financial system and the tent towards the interest of the financial institution. Establishing rural financial basic elements, strengthening rural financial risk control and expanding the direct financing channels for rural aspects can narrow the unbalance status between rural and urban to perfect the rural financial system.

KEYWORDS: Overall urban-rural development;rural finance; coordinated development

1   INTRODUCTION

The urbanization and modernization, generating in the process of dividing the urban and rural dual contributes to urban and rural dual economy structure. As the core of the national economy and the development of the booster - financial also present the characteristics of the dual structure between urban and rural areas, the development of the rural financial obviously lags behind the city, and to further restrict the development of rural economy.

2   THE PRESENT SITUATION OF THE LAG OF RURAL FINANCIAL DEVELOPMENT

Since China's reform and opening up, the degree of China's financial marketization has greatly improved; financial development has made great progress. However, as for the regional structure, urban and rural financial development uncoordinated remained significant, dual structure characteristic still being outstanding, mainly shown in the following respects.

2.1   *The urban and rural financial resource distribution*

2.1.1   *The urban and rural financial institution*
Development of rural financial institutions obviously lags behind the urban financial institutions. First of all, the type is monotonous. The basic rural financial institutions refer to the banking financial institutions. In the countryside, the insurance financial institutions

are rare, much less the securities and guarantee. On the contrary, the bank, insurance, securities, and guarantees take on an air of prosperity. Second, the layout of rural financial institutions is incompatible. Especially in the central and western regions, the distribution density of financial institutions is smaller. At present, our country owns three rural financial institutions: agricultural development bank, agricultural bank, and rural credit cooperatives. The branch credit business which set up in the villages is difficult to effectively meet the residents' basic financial services.

2.1.2   *Urban and rural financial market*
Financial markets are divided into the currency market and capital market. Currency markets, with agricultural bank and rural credit association as representative of rural financial institutions also took part in the financial activities, such as the acceptance discount, short-term loan, interbank leading, and discount. It increases the chances for the funds required for the rural financing channel, form and the urban financial institutions. But, in fact, in the interbank market, for example, the basic performance of rural credit association as the net lending, which makes the rural capital outflow. It will not benefit for the ultimate provider-rural capital agriculture operators. And it distorts the financing way of rural economic development. Capital markets, in addition to the bank financial institutions financing, some large agricultural enterprises (especially in the agricultural industry leading enterprises) also can be established

through the issuance of stocks, bonds or way of investment funds in the capital market financing. By the end of July 2013, the Shanghai stock exchange and ShenZhen stock exchange, forestry and fishing companies only 62, only 2.5% of all listed companies 2432, agriculture and agriculture in the national economy, the scale of direct financing is not in the position to match. In addition, rural secondary securities trading market shortage, so that the agriculture operators (especially the rural residents), despise the rest of the money in the bank, will have no other investment channel choice.

2.1.3 Urban and rural credit system

A sound credit system is the sign of the mature market economy. It is without good credit activities, much less perfect credit system. Relative urban finance, rural credit system also has many problems. First of all, the rural credit information collection and evaluation system is not perfect. Currently, farmers rely mainly on credit cooperative in the small amount of the carrier to collect credit. There is no complete management system, which leads to the lack of understanding to the basic situation of farmers timely, thus can not perform credit assessment and monitoring on time. Secondly, there are not too many laws and regulations to support the construction of credit system, contributing to the individual faithless with no punishment. Thirdly, farmers' credit consciousness is weak, for they are lack of knowledge. To evade the debt rural malicious behavior also exists, to some extent, the rural credit education. The cultivation of farmers' credit consciousness has not been seriously. They are short of the initiative and enthusiasm of credit system construction.

2.2 The urban and rural financial capital flows

Uncoordinated urban and rural finance also reflected in the flow of funds, serious outflow of rural financial resources, and financial net capital flow into cities. For a long time, rural financial institutions are saving more and borrowing less. Meanwhile, savings and loan is widening. In addition, due to the characteristics of for-profit commercial, financial institutions, most capital in commercial Banks' credit flows into the city, mainly the large enterprises, group, and urban development stable and profitable enterprises. This will drive the rapid development of urban economy, while it is lack of rural financial capital supply, thus exacerbating the financial structure imbalance between urban and rural areas.

2.3 The rural financial supply and demand

Compared with the urban financial supply and demand situation, the rural financing is still uncoordinated and imbalance.

2.3.1 Rural financial demand

Over these years, the country launched a series of policy which as a whole support urban and rural development and new rural construction, and promoted the rapid development of the rural economy, modern agriculture, the increase in the income of farmers, rural economic development and the increasing demand for credit. It diversified the rural financial demand for general use, quantity, scale. Financial demand gradually transforms the subject from the past single farming households planting to large households, individual businesses, professional cooperatives, agricultural leading enterprises and rural small and medium-sized enterprises and the main body; The rural financial capital needs are rising, farmers' average loan has amounted to ten thousand RMB. Rural residents' demand for financial products, tend to be diversified. Exceeded by farmers, small loans to expand the scope of production and operation, they also need to fund settlement, wholesale, consulting, insurance, securities, credit CARDS and other financial services to meet the needs of farmers in the consumer demand, education, housing, etc.

2.3.2 The rural financial supply situation

At present, the rural areas are mainly in the form of deposit financial institutions, the savings and loan business. Due to the level of rural financial service electronic informatization is still low, paper business, consumer durable loans, travel loans, credit loans, online banking, credit CARDS, ATM, securities and other financial business penetration rate are extremely low. It is difficult to meet rural residents' finance, insurance, consumer demand.

The uncoordinated phenomena showing in urban and rural financial development cannot exist in isolation. And there is a certain logic relationship. First of all, because of the rural credit evaluation system is not perfect, rural credit risk is high, which resulting in rural financial institutions' development lags behind the urban financial institutions. Since rural financial institutions is limited and unevenly distributed, urban and rural financial market development had serious imbalance and capital transferred from countryside to city. Great strides, on the other hand, in the rural economy, the demand for money constantly added up, the scale became larger. If without effective reform, rural financial supply, the contradiction between supply and demand of rural capital is difficult to solve the problem.

3 THE ANALYSIS OF THE CAUSES OF THE RURAL FINANCIAL DEVELOPMENT LAG

Since the reform and opening-up, the government has made a series of reform to the rural financial institutions, in order to better serving agriculture, rural areas and farmers, such as: restoring agricultural bank and rural credit cooperatives, expanding autonomy recovery of the rural credit cooperatives, and founding the agricultural development bank, and decoupling agricultural bank of rural credit cooperatives, which lead agricultural bank to the real commercial financial enterprises, outlaw, reform of rural credit cooperatives and rural cooperative foundation. After a series of reform, at present, the rural credit cooperatives take the system of rural financial institutions as the main body, agricultural bank and the auxiliary of agricultural development bank, the postal savings bank and the pattern of village Banks as a supplement. The overall layout is reasonable, but there are still some problems and deficiencies.

3.1 Agricultural policy bank

There is only agricultural policy bank, that is, the agricultural development bank of China, from the status quo, agricultural development bank financing source is not stable, and the use of the efficiency of funds needs to be improved, and its scope of business mainly in terms of its oil circulation, rarely involved in agricultural industrialization, agricultural information services, agricultural science and technology, and more areas. Therefore, agricultural development bank is still difficult to fully undertake the important task of China's agricultural policy finance.

3.2 Commercial rural financial institutions

In the main body status of the commercial, rural financial institutions mainly include the agricultural bank, state-owned or joint stock commercial Banks and other Banks, postal savings bank, etc. Profitability is one of their operating principles, profitability and becoming the main channel of rural capital loss. Agricultural bank is put forward: "committed to the construction of" three rural", urban and rural linkage oriented,, into the international first-class modern commercial Banks". Its for agriculture, rural areas and farmers to design some special product, but look from actual effect it absorbs deposit from the agricultural sector is more, with less loans for farmers and agriculture, leading to agricultural funds transfer to non-agricultural industries. Other state-owned (stock) the main business of commercial Banks and other commercial Banks are in the cities, but also can absorb a significant number of deposits from agriculture, rural areas and farmers, but very few loans for agriculture, rural areas and farmers. The postal savings bank serves as "agriculture, rural areas and farmers", taking small and medium-sized enterprises, a community service of commercial Banks as its orientation. Below 30% from county rural postal savings funds, but the basic are microfinance lending, quite a part of the People's Bank of China, the impact on the local rural economic development co., LTD.

3.4 Rural credit cooperatives

Rural credit cooperatives has been China's normal, the main force of rural formal financial cooperation the perspicuity of property right, management system and upgrade, and a series of reform, rural credit cooperatives of asset quality, operating conditions have obvious difference, but at the same time, also more and more show the tendency of commercialism, bank, make the ability of rural residents is not because of the development of rural credit cooperatives to improve and get better financial services. Leading to low income, the less money, and not easy to find a guarantor and rejected low-income farmers, and rural credit cooperative credit agreement.

4 THE COPING STRATEGIES OF IMPROVING THE SYSTEM OF RURAL FINANCIAL SUPPORT IN URBAN AND RURAL AREAS AS A WHOLE DEVELOPMENT

4.1 The establishment of rural credit reporting system

First of all, the linkage of financial institutions and government departments speed up the rural credit system construction. Formulating relevant laws and regulations, rural credit reporting systems and commercial financial institutions of the people's bank of China, the department of agriculture and forestry department, the environmental protection department and other rural credit reporting system, such as credit information collection, information sharing, information query, etc. Second, it strengthened rural credit publicity and the honesty education in the countryside. Through radio, telephone, newspapers and other media publicity credit, the concepts deeply rooted in the hearts of the people. Third, it constructed credit evaluation system as soon as possible. According to the actual circumstance of the local rural economy, it can build effective evaluation index system, determine the collection index, and evaluation standard, making the credit rating has quantitative indicators.

4.2 Strengthening the risk control and prevention mechanism

By sound policy-oriented agricultural insurance system, the government can strengthen rural financial risk prevention and control in the economic development. First, the government should make appropriate support to agriculture loan risk compensation, such as post political interest, policy of guiding financial institutions to the rural capital flow. Secondly, establishing a government-guarantee mechanism, led by the government and other social institutions, it can set up an agricultural loan guarantee fund to guide the capital flow to rural areas. Finally, we will accelerate the establishment of an agricultural reinsurance and catastrophe risk, disperse mechanism, establishing a fund of catastrophe risk. To large compensation, formed by the major natural disasters by the risk fund to a certain proportion of compensation, it can transfer gradually from agricultural catastrophe risk sharing mechanism. In order to increase the confidence of the financial institutions loans for agriculture, increase agricultural loans push.

4.3 Building diversified rural financial system to adapt to the urban and rural development as a whole

The focus of reform is to change rural formal financial institutions and standardize the development of rural informal financial institutions. In order to support urban and rural development as a whole, we need to develop the policy Banks to promote the role of rural economic development. In view of the countryside actual situation, it's necessary to provide relative preferential policies for supporting agriculture, carrying out targeted financial loans, meeting the needs of various funds on urban and rural harmonious development. Agricultural bank, China postal savings and other commercial, financial institutions should be operated in their commercialization, and keep the balance between the service "three rural". Under the commercial principle, as far as to "agriculture, rural areas and farmers" credit, it needs to increase rural service network. We will deepen reform of the rural credit cooperatives and the transformation of management idea, improving the innovation consciousness. At the same time, the government should give corresponding policy support, trying to build rural cooperative financial institutions and interactive cooperation mechanism of farmers' professional co-operatives, meanwhile, allowing conditional professional cooperative development of credit cooperation. Countries should formally recognize the rationality of the folk lending, letting them get the same development space. The formal financial authorities should allow the informal financial organizations to make registration, so that it will be effective, according to the requirements of the formal financial regulation.

4.4 Directly expanding rural financing channels

As income is growing, the farmers' financial surplus is increasing, inevitably which face problems such as the value and investment choices. The needs of using legal channels, that is, except bank farmers can be more easily get investment and fund. These are open to direct financing channels for rural areas. At the same time, along with the rural economic structure adjustment and agricultural industrialization, the development of rural industry and urbanization, the rural capital demand is growing. Especially, led by leading enterprises, and township enterprises, financing body is needed besides bank of indirect financing channels of direct financing. And the balance between urban and rural development in China requires to vigorously develop rural secondary and tertiary industries, economic development. The supply of public goods has a strong investment demand, in addition to the traditional way, such as finance, credit, open direct financing channels to mobilize the participation of private capital, which is necessary.

REFERENCES

[1] Lanhong, Zhanghua, & Liurong. Study on the Rural Financial Organization Innovation under Integrating Urban with Rural Development[J]. Southwest Finance, 2011(11).
[2] Chen Xiaodi. Research of Integrating Development of Urban and Rural under the Perspective of Financial Support[J]. Economic Research Guide, 2011(24).
[3] Li Yumei. Research on the Relationship Between Informal Finance and the Development of Rural Economy[J]. Pioneering with Science & Technology Monthly, 2012(5).
[4] Wang Bing, Su Lin. Research on the Contradiction Between Supply and Demand of Rural Finance and Its Solving Path[J]. Commercial Research, 2012(5).

Information, Computer and Application Engineering – Liu, Sung & Yao (Eds)
© 2015 Taylor & Francis Group, London, ISBN 978-1-138-02717-6

Analysis on the development of CDM projects in China

Ling Yu Huang & Xiu Ling Liu
International Business School, Dalian Nationalities University, P.R.China

ABSTRACT: With the fulfillment of CDM projects, advanced technologies and management philosophies are introduced to China. Based on the analysis of the current situation and problems of China's CDM projects, suggestions for promoting the development of clean energy are put forward in the paper, and the conclusion is drawn that CDM projects, as a "win-win" mechanism, bring a good opportunity for enterprises to raise their competition and make sustainable development in the global market.

KEYWORDS: Clean Development Mechanism (CDM); Certified Emission Reductions (CERs); Clean energy

1 INTRODUCTION

Clean Development Mechanism (CDM), as part of the UN Framework Convention on Climate Change (UNFCCC), is one of the three flexible mechanisms under the Kyoto Protocol. With the fulfillment of China's CDM projects, the Certified emission reductions (CERs) of China's CDM projects registered on the Executive Board (EB) of the United Nations have a rapid increase and will keep increasing. In view of the big consumption of energy and the low efficiency of energy utilization, there is a great potential in energy conservation and emission reduction in China. CDM projects, as a "win-win" mechanism, bring a good opportunity of sustainable development for China.

2 CURRENT SITUATION OF CHINA'S CDM PROJECTS

CDM projects came to China since 2002. The registered CDM projects in EB have got to 3804 by August 2014, and the expected average annual CERs of China's registered CDM projects in EB are 627,035,939 tCO$_2$e (see Table 1), which takes 50% of the global CERs. With 12 years development, China's registered CDM projects in EB and the expected average annual CERs have been increasing.

Table 1. China's CDM projects by August 2014.

CDM Projects	Numbers	CERs (tCO$_2$e)
Approved in China	5058	781,469,045
Registered in EB	3804	627,065,861
Issued in EB	1400	333,536,399

* http://www.cdm.ccchina.gov.cn.

The foreign investment of CDM projects in China has been growing in large scales since 2006. The U.K., taking 40% of the total international CDM projects in China, is the leading investment party (see Table 2). China and all the cooperative parties are mutual benefit from the CDM projects.

Table 2. Main international cooperative parties of CDM projects in China by August 2014.

Country	Approved Projects in China	Registered projects in EB	Issued projects in EB
The U.k.	1970	1485	491
Switzerland	534	427	94
Japan	466	339	197
Holland	339	262	148
Sweden	300	255	121
Germany	166	113	68
France	108	98	18
Austrilia	19	17	3

* http://www.cdm.ccchina.gov.cn.

New and renewable energy are the main scope of CDM projects, taking 83.3% of all CDM projects in China (see Figure 1), and among the new and renewable energy, wind power items take the largest percentage, some installed capacity of wind power has reached international advanced level.

2.1 Establishment of CDM supporting system

In order to administrate and support the development of CDM projects, China has set up functional organizations and institutions in administration

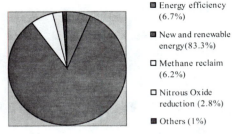

- Energy efficiency (6.7%)
- New and renewable energy(83.3%)
- Methane reclaim (6.2%)
- Nitrous Oxide reduction (2.8%)
- Others (1%)

Figure 1. Distribution of registered CDM projects by scope in China by August 2014.
* http://www.cdm.ccchina.gov.cn.

departments, such as Ministry of Energy, Ministry of Environmental Protection and Ministry of Land and Resources. A series of laws and rules which aim at encouraging the development of CDM projects have been issued, such as Energy Act, Atmospheric Pollution Prevention Law, Energy Conservation Law and Renewable Energy Act. The importance of CDM in national energy development policies is emphasized. Some policies and measurements have been carried out to support the development of CDM projects—subsidy policy for CDM industries, favorable tax revenue for use of wind power, solar power, heat utilization and new energy automobile, special funds for CDM technology innovation. The backward thermal power plants have been closed, instead clean energy has been developed in large scale.

2.2 Setting up of clean energy industrial chain and distribution

The development of clean energy industries has promoted the progress of relevant industries and industrial cluster in equipment manufacture, new materials and smart power grids. The relatively intact industrial chain covering clean energy equipment manufacturing, technology research and development, survey certification and supporting service has formed. In order to push the healthy and sustainable development of clean energy, China's Twelfth Five-Year Plan clearly stipulates the binding character, i.e. the use of non-fossil energy is to take 11.4% of primary energy by 2015. The overall development objective and distribution of clean energy have been determined in the specialized subject plan on wind power, solar power generation and biomass energy. So far clean energy industrial distribution has come into being.

2.3 Promotion of clean energy industrial competitiveness

With the increasing investment in development and utilization of clean energy in China, the key generic

technologies in the fields of clean energy, smart power grids and energy storage have been mastered. The successful research and utilization of the key technologies not only stimulate the development of clean energy, but also promote the industrial competitiveness of clean energy. Home made wind turbine generators have taken 90% of domestic market share, the production of photovoltaic cell takes two third of the total global output, the transfer efficiency has reached global advanced level with the strongest competitive advantages in technology and cost. Coal transformation and availability is on the way to modern coal chemical industry subject to oil substitute products. So far China's coal chemical technology, productivity and output have ranked the first in the world.

2.4 Social benefit in the development and utilization of clean energy

The development and utilization of clean energy is helpful to improve the ecological environment and the life quality, which lead to the social benefit. The continuous extension of clean energy utilization reduces the building energy consumption. Garbage incineration power generation changes the waste into treasure, and in the meanwhile reduces environmental pollution.

3 PROBLEMS OF CLEAN ENERGY DEVELOPMENT IN CHINA

Referring to the statistics of Roland Berg Strategy Consulting Company, on the basis of percentage of wind power and other clean technologies in GDP, China, among 38 countries all over the world, made the fastest growth in environmental protection technology industries with average annual growth of 77% and 44 billion Euro production value of clean technology, which takes 1.4% of GDP. China is changing herself from an import country to the main creating country of clean energy technology. China has set up a policy system in the field of energy, which promoted the development of clean energy industries of China to some extent, but the following problems still exist and influence the further development of clean energy in China.

3.1 Lack of financial investment in technology development of clean energy

Clean energy industry is a capital intensive industry. Large investment and high cost are the biggest obstacles restricting technology commercialization, popularization and application on the initial stage of industrial development, so it is of great importance to support the development of clean energy with national financial policy. The insufficient investment caused the lack of motivation in the development of clean energy. For example, the central government appropriation in clean energy technology research is limited, and the expense of operation and personnel management takes a large percentage of the funds, local governments haven't made a relevant budget of clean energy items, which lead to the lack of financial investment in the research and development of clean energy technology, and the failure to meet the need of developing clean energy.

3.2 Weakness of policy in supporting industry development

There is a long investment cycle for clean energy industry, its development will come true with a long industrial chain and the mutual links of all segments. Most of clean energy industries of China are in the initial period of development at a comparatively lower level. The imperfection of infrastructure and service system influences the development of clean energy industry. Although China's clean energy industry has made a rapid development, the proportion of clean energy out of energy industries in China is limited compared with developed countries, due to weak productivity and small scale. The development of China's clean energy industry too much relies on foreign partners, the master of the core technology is on the stage of introduction, digestion and joint design. The lack of proprietary intellectual property rights badly influences the development of China's clean energy.

3.3 Imperfection of CDM system

The development of clean energy involves many government departments and institutions, the lack of unified working mechanism in coordination, promotion and supervision causes problems in dispersal financial investment, limited resources share, repeated construction of platforms and project application. Though a series of policies were issued to support the development of clean energy, most of them are too macroscopic to execute.

3.4 Limitation of industrial entry and supervision policy

The new energy automobile is the important tool of transportation, and it is necessary to enhance its popularization, reliability and competition in the service of charging and battery change. The infrastructure construction of charging constraints the commercialization of new energy automobile, and there is not an overall plan and systematic design of charging and battery change in public areas and residential areas in cities. The industrial entry, supervision policy and technology standard of the new energy automobile are limited. The policy of supporting and subsidy is not systematic, which doesn't concentrate on enterprises, and particular emphasis on vehicle enterprises. There is a great demand and the potential prosperity for the distributed power in China's market, but the rights and liabilities in building integrated photovoltaic (BIPV) are vague, which is apt to cause ambiguity in profit distribution among project developers, building owners and energy users. The long term instability results in the difficulty in promoting the model of contract energy management.

3.5 Insufficient use of clean energy in urban

With the progress of China's urbanization, problems appear: backward facilities construction in sewage treatment and garbage disposal, lack of standardization in operation and declining air quality. Especially in garbage disposal there are many difficulties in site selection, collection, process supervision and reasonable benefit. In the process of urbanization, the development and utilization of clean energy should be taken into consideration to promote environmental protection.

4 SUGGESTIONS ON PROMOTING THE DEVELOPMENT OF CLEAN ENERGY IN CHINA

It is necessary to develop clean energy in China to promote the sustainable development of economy and to keep the promise made at the Copenhagen Climate Summit, "to increase the share of non-fossil fuels in primary energy consumption to around 15% by 2020, and per unit of GDP carbon emissions reaches 40–45% lower than that in 2005".

4.1 Establishing an industrial supporting mechanism of clean energy

International Energy Agency (IEA) predicted in 2009 in "Outlook of World Energy": most of carbon emission concerning energy is coming from Non-OECD countries, among which China's carbon emission is to increase by 6 billion tons of CO_2e, which press China to make an adjustment and optimization of industrial structure and energy structure. In order to increase the share of clean energy industry in the industrial system, it is necessary to establish an industrial supporting mechanism of clean energy. On the one hand, establishing industrial supporting system to promote the development of clean energy technology, including setting up quality criterion, perfecting test system and technology service system to increase the competition of China's clean energy industry; on the other hand, taking emission reduction measures to high energy-consuming enterprises, such as paper making, iron and steel manufacturing and cement producing, to relieve the pressure of environmental pollution and resources exhaustion.

4.2 Confirming the priority strategy of clean energy development

The share of fossil fuels in primary energy consumption is 87% in 2011, which aggravates global climate deterioration and endangers the sustainable development of global economic society. Nowadays, main countries and economies in the world are devoting themselves to clean energy development. Low carbon, high efficiency and clean energy have become the main trend of world energy development. Coal plays an important part in China's energy structure, the share of coal in primary energy consumption is 67%, the foreign trade dependence on oil reaches 57%. The problems of ecological, environmental protection and climate change coming from energy production and consumption have seriously restricted the development of economic society. Therefore, it is necessary to confirm the priority strategy of clean energy development and to improve the energy structure in order to protect the environment, master the market opportunity, get rid of the control of energy supply, ensure the energy security and realize the sustainable development of economic society.

4.3 Promoting the utilization of clean energy and environmental protection in urbanization

In the development of clean energy, complementary advantages are to be taken into consideration, all clean energies are to be in the process of harmonizing development. When developing wind power and photovoltaic power, it is necessary to speed up the facilities construction of sewage treatment and garbage disposal, to develop straw power and garbage power, to take advantage of methane in large scale, to encourage use of lithium ion battery especially in the field of stored energy and communication, to raise the entry threshold of lead battery industry, and to obsolete backward productivity, hence to improve the operation and management.

4.4 Cultivating human resources of clean energy enterprises

Human resources are the supporter of science and technology, and the talents team with initiation,

innovation and entrepreneurial spirit is important for clean energy enterprises make technology innovation. Nowadays the cultivation of clean energy human resources in China is beyond the need of the development of economic society in quantity, structure, quality and capability, the lack of human resources has been the bottleneck of developing clean energy industry. So the cultivation mechanism of human resources in the field of clean energy needs to be improved. Make wide connections with universities and institutions home and abroad in the way of academic research and technology training raising the level of operation, management and innovation. Set up the talents team in the field of low carbon by introducing and recruiting high qualified personnel home and abroad, and support and encourage them with relevant policies in scientific research, self employment and technology results transferring. Enhance the construction of the human resources, carrier by establishing a photovoltaic engineering research center, postdoctoral work station and personnel training base to promote the development of technology research and clean energy industry in the field of bio-fuel technology, new energy conservation technology and greenhouse gas emission reduction technology.

5 CONCLUSION

CDM brings China a great opportunity of investment in the field of energy, chemical industry, construction, manufacturing, transportation, waste disposal, forestry and agriculture. With the fulfillment of CDM projects, advanced technologies and management philosophies are introduced to China, which is helpful for enterprises to raise their competition and make sustainable development in the global market.

REFERENCES

Wang Bo. 2009. Exploring China's Climate Change Policy from Both International and Domestic Perspectives. *American Journal of Chinese Studies.* 16(2):87–105.

Wang Bo. 2010. Can CDM Bring Technology Transfer to China? *Energy Policy.* 38:2572–2585.

Huang Xiao & Wu Xin. 2009. States, challenges and strategies on carbon trade in China. *China Economy and Trade.* 16:1–3.

McCarthy J., Canziniani O. & Leary N. 2001. *Climate Change 2001: Impacts, adaptation and vulnerability,* Cambridge: Cambridge University.

Report on Low Carbon Economic Development in China 2011. Beijing: China Social Science Documents 96–111.

Blue white license plate location method based on openCV

Hong Bin Gao & Qing Ying Wan

School of Information Science & Engineering, Hebei University of Science and Technology, Shijiazhuang, China

ABSTRACT: This paper presents a method of vehicle license plate location based on Visual Studio 2010 platform and OpenCV 2.4.9. First transform the inputted original RGB color image into HSV color space, use the fixed background of the license plate region, construct the Binary image, process the Binary image by using morphological operation, detect the contour of the Binary image, Combine with the feature of the license plate geometry to tag the license plate region, locate the license plate. After verification, this method has a high accuracy rate.

KEYWORDS: OpenCV; License plate location; Color space; Contour detection

1 INTRODUCTION

The research on the automatic recognition technique of license plate has vital significance to the highway traffic management, which has important social value. Automatic license plate recognition system mainly consists of license plate location and character recognition. License plate location is the first step to realize the automatic recognition of license plate, the precision and speed of license plate location affects the whole performance of the vehicle license plate recognition system.

At present there are a lot of methods of license plate positioning, The most common location technology is mainly based on edge detection method, based on wavelet transform method, based on the genetic algorithm method, based on mathematical morphology and texture analysis of gray image. Although the above methods can achieve the division of license plate under a specific environment, but most of the license plate recognition system used in outdoor, affected by weather, background, and influenced by image tilt and license plate wear and other objective factors. These methods targeted relatively strong, have limitations in practical application, therefore the applicability of these methods is relatively small.

This paper focuses on the research of the blue and white plate of small vehicles, because of the particularity of the blue license plate color, consider using the HSV color space which was closer to human vision to locate license plate, make full use of the information provides by each component and the background color of the license plate region, constructed the Binary image. Morphological processing the binary image, the technology for contour detection combines the geometric characteristics of the license plate search out the license plate region.

2 CHARACTERISTICS OF TARGET REGIONAL OF LICENSE PLATE

The starting point of license plate location method is to determine the license plate using the characteristic of the license plate region, region the license plate from the vehicle image segmentation. The license plate itself has many inherent characteristics, these characteristics are different for different countries. From the point of view of human visual angle, the following characteristics can be used to locate the license plate in our country:

The license plate background generally has the biggest difference to the body color, color of the characters;

The license plate usually has a continuous or discontinuous frame due to wear;

There are multiple characters on the license plate, substantially horizontal arrangement, there are rich edges of the rectangular region of the plate, these characters have a texture feature with rules;

The interval of the characters in the license plate is uniform, the gray value of the characters and the background of license plate's existence large jump, the character itself and license plate bottom inside have a uniform gray;

In images of specificity, the size and location in different license are different, but the ratio of length to width range in a certain range, there are one maximum and one minimum.

The above several features are conceptual, various features alone seem not unique to the license plate, image, but combine them can uniquely identify the license plate. Among these characteristics, color, shape, location are most intuitive, easy to extract. The texture features are abstract, must go through a process converted to other features to obtain the corresponding available character index. Usually text feature requires at least through the character segmentation or recognition may later become features available, generally only used to judge the correctness of the license plate recognition.

3 LICENSE PLATE LOCATION

There are four types license plates in China: blue white, yellow background and black characters, white on black or scarlet, black bottom mispronounced character, namely there are blue, yellow, white and black four plate background colors. The blue background represents the small power vehicle license plate, yellow plate identifies high power automobile license plate, and white represents a military or a police vehicle license plate, black bottom plate on behalf of foreign or foreign institutions stationed in China license plate. Because the blue license plate accounted for all the largest proportion of bottom license plate, the focus of this paper is mainly aimed at the blue white small power automobile license plate.

3.1 *The choice of color space*

The RGB color model is the most common color model; all the colors in nature can be combined with the three primary colors of red, green, blue. When the three color components are 0 (weakest), mixed black light, when the three color components are 1 (strongest) mixed into white light. Any color F is a point of this cube coordinates, any adjustment of trichromatic coefficient r, g, b will change the coordinate values of F, i.e. change the color value of F. RGB color space was represented by physical three primary colors, and the physical meaning is clear, suitable for color picture tube work. However, this system does not adapt to the characteristics of human visual, and the brightness value of the three colors were more obvious affected by light. Even if two similar colors the values of R, G, B may be different. So it's very difficult to realize the location of the license plate in the RGB model. The RGB color model is shown in Figure 1.

HSV color space reflects the feel of human perception of color, H (Hue), S (Saturation), and V (Value) were the three parameters used to describe the color characteristics, H defines the wavelength of color, called tone; S defines shades of color, called saturation; V defines brightness, In the HSV color space, color H and saturation S contain the information of color, and the brightness V is nothing to do with the color information. This feature makes the color separation treatment in the HSV model can not be affected by the brightness, using the fixer threshold value can effectively be in a variety of colors separate all kinds of colors. This greatly simplifies the workload of image analysis and

processing, improve the processing efficiency, very suitable for image processing algorithm which is based on the human visual system to perceive color characteristics, so in this paper license plate location was implemented in the HSV color space.

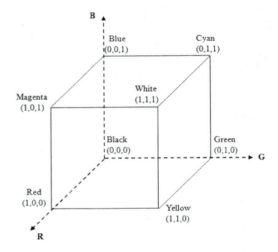

Figure 1. The model of RGB color space.

The model in HSV color space corresponds to a conical subset of cylindrical coordinates, the top surface of the cone corresponded to V=1. The color represented by the three surfaces of R=1, G=1, B=1 is brighter. Color H is given by a rotation around the V axis. Red corresponds to the angle 0°, green corresponds to the angle 120°, and blue corresponds to the angle 240°. In HSV color model, each kind of color differ 180° with its complementary. Saturation S values from 0 to 1, so the radius of cone top surface is 1. At the apex of the cone (i.e. the origin point), V=0, H and S without definition, represents black. The center part of the top surface S=0, cone V=1, H is not defined, represent white. From the point to the origin represents the brightness gradually darken gray, i.e. with different gray. For these points, S=0, the value of H is not defined. It can be said that the V axis in the HSV model corresponds to the main diagonal in RGB color space. The color of the circumference on the top surface of the cone, V=1, S=1, the color is pure. The model of HSV color space is shown in Figure 2.

In the HSV color space model, the range of the three components of HSV is 0<H<360, 0<S<1, 0<V<1. In the OpenCV, the data without sign character in 8 bits can't represent more than 255, so converse the three component values of HSV in OpenCV. The specific conversion formula is as follows: transform the three component values (h, s, v) in HSV to the value of the three components (h', s', v') in OpenCV. See for example, Equation 1 below:

$$h' = {}^h\!/_2$$
$$s' = s \times 255$$
$$v' = v \times 255 \qquad (1)$$

where $h \in [0,360]$, s, $v \in [0,1]$, $h' \in [0,180]$, s', $v' \in [0,255]$.

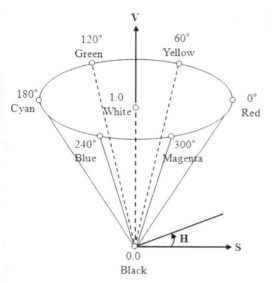

Figure 2. The model of HSV color space.

3.2 Convert image to a binary image

The binarization of image sets the grayscale value of the pixel of the image to 0 or 255; the whole image obvious shows the effect of black and white. In image processing, the binary image occupies a very important position. Binary image makes it easier for further processing of the image, the image becomes simple, and the quantity of data is reduced, can highlight the contours of target region of interest. The gray of the target area is set to 255; the gray of background region is set to 0, which can separate the object region and background region distinct, beneficial to extract target contours. In HSV color space, can use H and S components to recognize the blue area, the gray value of blue area in the image is set to 255, other color area is set to 0, constructed the Binary image, we can preliminary determine the approximate location of the blue license plate. Through extensive experiments, obtains the interval range defined by OpenCV blue is shown in table 1. The binary image of Figure 3 after blue filter is shown in Figure 4 in the experiment.

Table 1. The range of three component of blue HSV.

	Range
H	[95–123]
S	[89–255]
V	[76–255]

Figure 3. The original image of RGB.

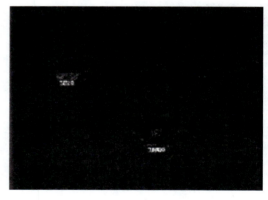

Figure 4. Blue filtered binary image.

3.3 Using morphological operations to process the binary image

From the perspective of image processing, the corrosion and expansion of Binary image are a small Binary image (i.e. structural elements, the size generally is 3*3) move and compare in a large Binary image point by point, make corresponding processing according to the results of the comparison. The expansion algorithm refers to scan each pixel of the Binary image by structural elements of 3*3. Use structural elements and the covered binary image progress "and" operation, if all are 0, the pixel of the structure image is 0, otherwise 1. The result is to make the Binary image enlarge. Corrosion algorithm using structural elements of 3*3 scans each pixel of binary image. Use structural elements and the covered Binary image progress "and" operation, if all are 1, the pixel of the structure image is 1, otherwise 0. Results the binary image reduced in a circle.

The binary image filtered by color generally has noise like snowflakes, so using morphological opening operation to denoising, which first make corrosion operation, and then make expansion operation. Due to the license plate area are easy to crack, so use morphological closing operation to crack closure, namely first make expansion operation and then make erosion operation. In the experiment, Figure 4 is shown in Figure 5 after morphological processing.

Figure 6.   The region of blue license plate.

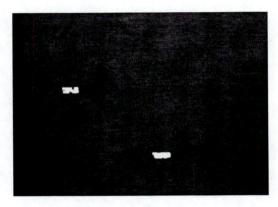

Figure 5.   The license plate image after morphological processing.

3.4   The selection of license plate region

Make contour detection of the Binary image after the morphological operation. Search possible connected region of the license plate. Acquire the outermost rectangular boundary of detecting contour, the rectangle is the possible license plate areas. According to the geometric characteristics of the license plate area, such as the area of the plate area, length and width ratio characteristics of the license plate region to exclude the rectangular region of non license plate area, draw the plate area with rectangular box, were screened out with rectangular box, the blue license plate area is shown in Figure 6.

4   THE RESULTS OF EXPERIMENT

In this paper, the 60 license plate images with the size of 2048X1536 have carried on the experiment. 123 blue license plates contained in the 60 images, there are 117 blue license plates positioned correctly, the effective localization rate reached 95%.

5   CONCLUSION

This paper presents a license plate location method based on OpenCV in the HSV color space model, conversion the RGB image to HSV color space, makes full use of the characteristics of HSV color space, filter out the blue area, and the image is processed by morphological operation, and use morphological operation process the image, search possible connected region of the license plate area. Use geometric characteristics of license plate accurately to locate the license plate region.

REFERENCES

[1] Chai Xiaorong, Liu Jingao. Algorithm of accurate license plate location based on texture analysis [J]. Application of computer system. 2010(02).
[2] Chen Zailiang, Liu Qing, Zou Beiji, Shen Hailan, Zhou Haoyu. The algorithm of extract interese area by combin visual attention and texture features [J]. micro computer system. 2012(05).
[3] Zhou Zehua, Pan Baochang, Zheng Shenglin, et al. License plate location method based on multi color model [J]. micro computer information. 2007(01).
[4] Amir Sedighi, Mansur Vafadust. A new and robust method for character segmentation and recognition in license plate images[J]. Expert Systems With Applications, 2011, 38(11).
[5] Mahmood Ashoori Lalimi, Sedigheh Ghofrani, Des McLernon.A vehicle license plate detection method using region and edge based methods [J]. Computers and Electrical Engineering, 2012.

Information, Computer and Application Engineering – Liu, Sung & Yao (Eds)
© *2015 Taylor & Francis Group, London, ISBN 978-1-138-02717-6*

Blind detection of JPEG composite image based on quantization noise analysis

Zhi Hong Wang

Shandong Yingcai University, Shandong, China

ABSTRACT: Quantization noise may be introduced to images after being compressed as JPEG format. For the synthesis or tampered images obtained in different ways, quantization noise in places tampered or not tampered will be different. Based on this feature, a new algorithm is presented for detecting image tampering. JPEG image will be recompressed after tampering. Using this algorithm, after contrasting the noise before and after compression, transforming correspondingly, processing binary, detects and locates the tampering position automatically.

KEYWORDS: quantization noise, recompression; binary

1 INTRODUCTION

With advances of many image editing software, people can easily carry out a variety of digital image tampering, which makes it difficult to distinguish and identify whether the images are true with the naked eye. If digital images-tampering technology is abused in forensic evidence, scientific discoveries, the official media, even people's life, it will cause incalculable harm to the stability of society [1]. Therefore, forensic studies have been pressing for a digital image tampering. The initiative requires prior evidence embedded in the image digital signature or digital watermarks in order to achieve the purpose of detecting image tampering. Passive forensics are inserted in the image does not require prior identification of data or information is added, to be detected by directly detecting a digital image work to determine whether the image to be detected after a work operation of a tamper artificial technique, and therefore more practical.

2 QUANTIFICATION ANALYSIS

The so-called quantization noise is the quantization process exist quantization error, and then reflected to produce the receiving end, the size of the display of the digital image and the original image differences. Quantization refers to the image pixel DCT coefficients after the discrete cosine transform, corresponding to the obtained quantization step size by dividing and rounding process, shown in the following formula:

$$P''_{ij} = round(P_{ij} / M_{ij})$$

JPEG image synthesis tampering generally lead to double compression, which will alter the characteristics of the original image of the original. By analyzing the characteristics can be achieved on JPEG image synthesis blind testing. JPEG is a lossy compression format, an image can be compressed into a small storage space, and from the visual point of view, the image quality is not significantly reduced. The JPEG compression process, the quantization is an irreversible operation [2]. Therefore, the image is decompressed, it will lose some information of the original image, which we miss information called image quantization noise. Quantization noise can be obtained from the difference between the original image and the image after compression.

Different from the original image after the JPEG compression to give an example of the average quantization noise. The two images of different formats have been recompressed different quality factors, respectively [3], to strike the original image and the difference image after compression, and then averaged to obtain the average quantization noise under different JPEG compression, the same quality factor of JPEG compression, especially the quality factor QF> 60, the lossless compression image (such as BMP, TIFF format) greater than the JPEG image quantization noise quantization noise. This is the first time since the JPEG image compression, many high-frequency information has been quantized to zero again when the image quality through higher compression factor, relatively little information may be lost.

If the background image is a JPEG image, the composite image again after a higher quality factor JPEG compression, its area and the non-tamper tampering regions having different quantization noise. This is because:

1 If the tamper region of the uncompressed image, such as BMP or TIFF format image, the image after tampering after JPEG compression, tampered area after only one JPEG compression, but the JPEG compression twice the background area;

2 If the tampered area of JPEG images, JPEG images are due to 8x8 block size by compressing, forger in order to obtain a better visual effect, often do not consider the case of the blocks aligned, resulting in the tamper block region and the background region location inconsistency. Therefore, the image is tampered with again after JPEG compression, compression characteristic loss before tampering area can be considered only after a JPEG compression, rather than tampering with the same area twice JPEG compression.

Therefore, after the composite image compression again, for the background area, because after two JPEG compression, its quantization noise and less widely distributed; for tampering area, since only through a JPEG compression, so the quantization noise distribution is more concentrated and more. Use PCA transforms, quantization noise can extract key information, and remove some less scattered some noise.

3 ALGORITHM DESCRIPTION

From the analysis shows that the quantization characteristics, the quantization matrix quantization loss of image compression and image selection process is closely related to, if the different areas are subjected to different image compression quantization matrix, a big difference in the occurrence of loss in weight after compression, utilizing the difference in loss as a basis for detecting the authenticity of an image. The existence of a composite image of the background area and tampering area quantization matrix inconsistencies, according to quantify the properties of the synthesized image if recompression, the composite image background area and distort history to quantify due to the different regions produce different degrees of damage. Thus, the composite image can be recompressed, extracted small areas of the quantization loss of image, the image small degree of loss distinction between the background region and tampering different regions. In order to achieve detection of the synthesized image.

When the quantization matrix selection recompression original background area of the quantization matrix[4], the minimum overall loss of

image. Meanwhile tamper area loss was significantly greater than the loss of the background area, accurately quantify the loss of each region extracted and examined treatment measured by measuring the loss of image, the algorithm can be very effective to achieve positioning tamper detection area. From the above analysis shows that effective detection of two core issues to be addressed: 1) estimate the original quantization matrix background area; 2) a measure to quantify the loss of the composite image.

4 DETECTION ALGORITHMS

1) Principal component analysis (PCA)

When processing the collected information, the information in the two variables sometimes there is a certain relationship between two variables such information to reflect this issue there is a certain degree of overlap[5], but in order to solve this overlapping relationship, directly reduce the variables the number is the simplest and most direct solution, but to reduce the number of variables means that the information be missing or incomplete information, so people will want to find a more effective way to resolve this contradiction, the new method should neither cause loss of information or incomplete overlap and reducing the number of variables, principal component analysis, one of analytical methods is to meet the above requirements and have been applied for.

$\xi_i (i = 1, 2...s)$ note of the original variable index is re-set combined into a set of smaller number of uncorrelated composite indicator is $\eta_j (j = 1, 2...t; t \leq s)$, then

$$\begin{cases} \eta_1 = \alpha_{11}\xi_1 + \alpha_{12}\xi_2 + \cdots + \alpha_{1p}\xi_s \\ \eta_2 = \alpha_{21}\xi_1 + \alpha_{22}\xi_2 + \cdots + \alpha_{2p}\xi_s \\ \cdots\cdots\cdots\cdots \\ \eta_t = \alpha_{t1}\xi_1 + \alpha_{t2}\xi_2 + \cdots + \alpha_{tp}\xi_s \end{cases}$$

α_{ij} principle factor in determining:

① η_i and η_j ($i \neq j$) are linearly independent;

② η_1 take on all linear combinations of $\xi_i (i = 1, 2...s)$ in which the variance of the largest; η_2 take the greatest variance in η_1 all linearly independent of all linear combinations of $\xi_i (i = 1, 2...s)$; η_t and $\eta_1, \eta_2...\eta_{t-1}$ are linearly independent is all linear combinations of $\xi_i (i = 1, 2...s)$ fetch maximum variance a.

2) Algorithm description

① quantization noise extraction: the difference between the original image and the image after the JPEG compression is the image of the quantization noise. Composite image after JPEG compression, its quantization noise tampered area is more obvious than the quantization noise of the background area.

② PCA conversion: the quantization noise is obtained from the previous step, not only the main information including the tampered area, background area also includes pieces of information, therefore, need to quantizing noise for further processing. Use PCA transform, quantization noise can extract key information for more focused, and remove some less scattered some noise. After the quantization noise after PCA transformation, respectively, to obtain the quantization noise of the first, second, and third principal components.

③ binarization process: After the PCA transform, the quantization noise component of the first main component in addition to containing the main information area outside tampering, as well as some background area information, and therefore also require some post-processing. Here the use of certain thresholds be binary processing. A threshold value of the normalized gray level histogram is determined according to the first after the principal components of the quantization noise. When a pixel value is greater than the set thresholds, it is considered the point of tampering point.

④ expansion and corrosion: Finally, morphological expansion and erosion operations on binary quantization noise after processing, to more accurately locate the tampered area.

5 SUMMARY

After the JPEG image compression introduces quantization noise. Based on the composite image tampering area with non-tampered area Quantization noise of inconsistency, we propose a blind JPEG image detection algorithm based on the synthesis of quantization noise analysis. Real Experience has shown that this method is simple and effective, and can pinpoint areas of the image tampering. Of course, there is an algorithm Given limitations. For example, if the background color of the color area and its surrounding area too tampering same region or background Contains a lot of texture information, the accuracy of the algorithm to detect decreased. For this type of JPEG image synthesis, We will further study.

ACKNOWLEDGMENTS

This work is supported by the National Natural Science Foundation of China (Grant No. 61402271) and the colonel issues of Shandong Yingcai University (Grant No. 14YCYBZR04) and Shandong Province technology innovation projects (201441908019).

REFERENCES

[1] Weihai Li, Yuan Yuan, Nenghai Yu. Detecting copy-paste forgery of JPEG image via block artifact grid extraction. In:2008 International workshop on local and non-local approximation in image processing, a satellite event of European signal processing confererence, 2008, 121–126.
[2] He JF, Lin Z C, Wang L F, Tang XO. Dectecting doctoredJPEG images via DCT coefficient analysis. Proceedings of 9th European Conference on Computer Vision Graz, Austria, 2006, 10423–435.
[3] J. Lukás and J. Fridrich, Estimation of primary quantization matrix in double compressed JPEG images[J]. in Proc. Digital Forensic Research Workshop, Cleveland, OH, Aug. 2003. 1–17.
[4] Dongdong Fu, Yun Q. Shi, Wei Su. A generalized Benford's law for JPEG coefficients and its applications in image forensics[J]. Proc. SPIE6505, Security, Steganography and Watermarking of Multimedia Contents IX, San Jose, CA, Jan. 2007:1–11.
[5] W. Chen, Y. Q. Shi and W. Su. Image splicing detection using 2-D phase congruency and statistical moments of characteristic function. Proc. of Security, Steganography and Watermarking of Multimedia Contents IX. San Jose: SPIE. 2007:65050R.

Information, Computer and Application Engineering – Liu, Sung & Yao (Eds)

© 2015 Taylor & Francis Group, London, ISBN 978-1-138-02717-6

The status and prospects of photovoltaic power generation

De Jun Xue & Xiao Lin Qiu
Nanchang Institute of Technology, Jiangxi Nanchang, China

ABSTRACT: This paper introduces the current situation of the development of photovoltaic power generation at home and abroad, describes the principle and key technology of photovoltaic power generation, and in-depth analysis of grid problems, and finally points out the two major development direction of photovoltaic power generation, namely the city scale photovoltaic power generation and desert of photovoltaic power station.

KEYWORDS: PV or photovoltaic; status quo. prospects

The basic principle of solar photovoltaic power generation is the use of solar panels of photovoltaic effect, i.e., When the object is affected by light, an effect of the charge distribution within the state of an object changes arising from electric potential and current. In the petrochemical energy shortage and environmental pressure increasing today, the characteristics of photovoltaic power generation with its non polluting, renewable, favored by the people, is a kind of renewable energy development prospect is broad.

1 RESEARCH ON THE DEVELOPMENT STATUS AT HOME AND ABROAD

Since the French scientist Becqurel in 1839 found that the photovoltaic effect so far, has a history of nearly two hundred years. During the period, the photovoltaic power generation from the laboratory into our real life. In 1954, America Baer made the first practical laboratory of monocrystalline silicon solar cell, solar energy is photovoltaic technology power was born. The 1990 World solar cell production reached 46.5MWp, while in 2005 has exceeded 2000MWp. Germany began implementing the "roof plan" from 1990, and built more than 10000 photovoltaic roofing system in 1997, the installation of PV modules as 33MW. Japan in 1992 to start the new sunshine plan, by 2003 the PV module production accounted for 50% of the world. In addition, Switzerland, Spain, Finland and other countries also have invested heavily in order to occupy a place in the field of photovoltaic power generation.

Photovoltaic industry in China in recent years the good momentum of development, in 2002 the implementation of the "bright project" and "send electricity to power sent to the village project Xiang" and 2006, use photovoltaic technology, to the end of 2007, photovoltaic power generation installed capacity of up to 100MW. Our country "1025" planning has installed photovoltaic power generation target to 10GW, according to the International Energy Agency forecast to 2050 12% power supply by the solar energy ratio, our country when solar energy generating capacity will reach 1.1 kwh trillion. Grid connected photovoltaic power generation is the general trend of photovoltaic power generation, grid connected photovoltaic power generation is mainly used in the desert of the photovoltaic power station and town combined with the construction of the urban scale photovoltaic power generation.

2 THE KEY TECHNOLOGY OF THE LARGE CAPACITY AND GRID CONNECTED PHOTOVOLTAIC POWER STATION

2.1 The principle and characteristics of large capacity of photovoltaic power station

Large capacity photovoltaic power station built in the sunny desert zone, the whole system consists of the capacity as the basic unit of the composition of 0.3 ~ 1.0MW, the core component of a solar cell array, the inverter and the solar tracking control system. Battery array is converted light energy as the core device of electric energy, the inverter converts DC power is converted directly to the alternating current, in order to facilitate the grid, and the tracking control system is designed to make power generation to achieve maximum efficiency. The photovoltaic power station system is composed of a plurality of mutual cooperation of subsystems, the subsystem between the central control center unified command.

Photovoltaic power station has the following characteristics:

① Due to the influence of the light intensity with the weather, photovoltaic power generation with stochastic volatility great;

403

② The access grid voltage, power distance, and power supply distance will inevitably bring about problems such as loss;

③ Photovoltaic power plant cell array area is large, the module performance due to manufacturers and models vary, so that can not reach the ideal array combination characteristics;

④ Since the inverter capacity, models and more generally used multiple low-voltage inverter combination form, by a transformer to 10kV/35kV and then incorporated into the power grid, resulting in circulation harmony wave discharge in increasing system redundancy and the phenomenon of the will produce impact to the power grid, voltage fluctuation, isolated island problem caused by;

⑤ From the literatures, the problems such as non ideal characteristics of temperature and multi peak characteristics, there are still a combination of photovoltaic array rise effect, hot spot effect and the inverter combination.

2.2 *The key technology of the large capacity grid connected photovoltaic power station*

Large capacity grid connected photovoltaic power plant mainly has the following key technologies:

① Photovoltaic array, monocrystalline silicon, polycrystalline silicon thin film is the main and battery technology;

② Transformation technology, mainly converter unified control, collaboration and inverter efficiency;

③ With the grid friendly interactive technology, mainly the rapid control of active power and reactive power, the maximum power point tracking technology, d ~ q axis current decoupling conveying control of active and reactive power, harmonic suppression etc.;

④ The photovoltaic power station standardization design, reliability evaluation and optimization of topology configuration.

3 CITY SCALE PHOTOVOLTAIC POWER GENERATION

3.1 *The principle and meaning of city scale photovoltaic power generation*

City scale photovoltaic power generation refers to the modern big city, the first is the construction of solar photovoltaic grid power generation facilities in the roof public welfare construction, make full use of the advantages of building roof light, an implementation of photovoltaic building.

It has very important significance for carrying out the city scale photovoltaic power generation: ① Can make the power structure has been optimized; ② To expand the domestic photovoltaic market, make its sustainable development; ③ To improve the city power supply safety; ④ Play the role of energy-saving emission reduction, Protect environment, conducive to building a resource-saving, environment-friendly city.

Now China has been part of the city and region to carry out pilot projects, such as Beijing, Shanghai, Guangdong, Shandong, 2010 one thousand national built a total installed capacity of 50000 kilowatts of rooftop photovoltaic power generation projects.

3.2 *The development of our country city scale photovoltaic power* generation conditions at present, our country city scale photovoltaic power generation is quite mature in the aspects of the basic conditions, industry conditions, engineering conditions and policy conditions, can develop

① Basis, the available building roof area in China reached 3 millions of square meters, each year to provide 31600000000 kwh electricity;

② Dozens of listing Corporation, industry annual output value of more than 300000000000 yuan. China has already mastered the key technology of polysilicon and solar cell manufacturing, industrial chain and constantly improve.

③ Engineering and technical conditions, in 2009 launched the "solar roofs plan" and "golden sun demonstration project", two of the installed capacity of up to 723MW, the same year Dunhuang large-scale grid connected photovoltaic power station began bidding, the following year, the second batch of start bidding capacity up to 280MW. By the end of 2010, China's photovoltaic power generation cumulative installed capacity of up to 800MW, the development momentum is good, has the engineering technical requirements of large-scale urban construction of PV grid connected generation;

4 PROBLEMS AND COUNTERMEASURES OF GRID CONNECTED PHOTOVOLTAIC POWER GENERATION

Research on the overall momentum of the development of photovoltaic power generation in our country is good, but there are many problems. ① The screening data and unified planning is not in place. ② The technology and the equipment aspect, our country still lags behind the advanced world level. ③ The high cost of power generation. ④ The standard system is not perfect. ⑤ Grid management and personnel training should be strengthened and perfected.

5 CONCLUSION

Solar photovoltaic power generation clean without pollution, is a green and renewable energy, while China as rich, broad development prospects of photovoltaic power generation. Academician Zhang Yaoming is analyzed and demonstrated considerable in solar photovoltaic generation prospect in China, put forward the idea of the Three Gorges of the sun. With the growing shortage of fossil energy and environmental pressures are increasing, the solar photovoltaic power generation will become the main force in the future of mankind's energy.

Information, Computer and Application Engineering – Liu, Sung & Yao (Eds)
© *2015 Taylor & Francis Group, London, ISBN 978-1-138-02717-6*

The development trend of foreign electric power market and enlightenment

Fang Yun Li & Wei Zhao
Institute of Technology, East China Jiaotong University, Jiangxi, Nanchang, China

ABSTRACT: From the view of the world, the electric power industry reform began in 1981 in Chile. But this reform developed into a world of the reform movement is started after the power industry reform in the late 80's Britain.

KEYWORDS: electric power industry reform; vertical monopoly; market competition

From the view of the world, the electric power industry reform began in 1981 in Chile. But this reform developed into a world of the reform movement is started after the power industry reform in the late 80's Britain.

1   REFORM THE MODE OF WORLD POWER

World differences in management mode of electric power industry are large, and is closely related to the domestic economy development level, economic development level, forms of ownership, etc. To sum up, about the following several models:

1.1   A single vertical monopoly mode

the national power generation, transmission, power distribution business is operated entirely by an electric power company, from state owned enterprises, by the national power company to implement the control of management, France, Italy, Portugal, Greece and other countries belong to this type.

1.2   Long monopoly mode

This mode has two kinds of situations, the first kind is the multi vertical monopoly, is composed of a plurality of power company division management, each regional electricity supply is relatively independent, power generation, transmission, distribution under vertical integration monopoly, the district power network is connected. Japan is a typical use of this model, a top 10 national power company, in addition to Okinawa Electric Power Co, the other 9 major power companies have networked with each other. Second is the number of power monopoly, distribution business with other companies, transmission network is the internet. At present, the electric power industry in California is using this model.

1.3   Single monopoly and multiple combination model

transmission business from a single power company, and the vertical management of power generation and distribution business, power generation and distribution also allows other firms to enter, but because the power transmission company monopoly advantage, to ensure fair competition, the state regulation of electric power company, the current power enterprises in Germany, USA this is the management pattern.

1.4   The open market competition mode

Full implementation of competition in the generation, placing the electric field, the transmission by a company's business, and government supervision. This mode is mostly formed nearly 10 years after the reform of electric power industry in the world, with the UK, Australia, Argentina, Chile, Norway, New Zealand, Singapore, Holland and other countries as the representative.

In the electric power system reform in the world and the establishment of the electricity market model, the basic trend of the global power industry reform is to break the monopoly, the formation mechanism of competition has become an inevitable trend. But the characteristics of power industry, transportation, and send the request for completion and to serve the public, to break the monopoly of the method is not the simple open, also is not the "laissez faire" market. The power industry restructuring should make shareholder value, service quality and price level to achieve the optimal balance, maximize the total value of the three parties.

2 CHARACTERISTICS OF FOREIGN ELECTRIC POWER REFORM

The electric power industry is the public utility has its particularity, therefore, even in the market economy system perfect, the organizational form and the operation mechanism of the electric power industry is to distinguish the particularity in the general competition in the industry and the long-term existence. Promote and makes the reform success there are two main reasons: one is the external pressure, i.e., the society to improve the efficiency of power industry requirements, because the monopoly is easy to make the enterprise management and market pressure disjunction, resulting in high cost and low efficiency, resulting in price rises, the society is to promote the reform of the external conditions on the to improve the efficiency of the electric power industry requirements. Two is the reform and material conditions, namely the development of the power grid security technology, the power in the hair, transmission, distribution and sale of separate and develop competitive conditions, still can maintain stable and safe production and supply.

The process of foreign electric power reform is generally introduced by the government reform, scheme design and set up an authoritative organization reform, and the drafting of the relevant policies and regulations, then the power companies of the integration of traditional decomposition, and set up a mechanism to supervise to the competition.

The power industry restructuring has no ready-made model to copy, a state of the industry pattern depends on the country's electricity industry characteristics and the political, economic and social factors and so on. The world is not a perfect reform mode of electric power industry, power industry reform should seek truth from facts, from the actual situation.

3 ENLIGHTENMENT OF EXPERIENCE OF FOREIGN ELECTRIC POWER REFORM TO CHINA

3.1 *Objective should have a clear understanding of the reform of electric power, to determine the reform steps and methods*

Objective to reform in different periods of different countries are different. The different reform objective, reform will bring the methods and steps of differentiation. The purpose of power reform in Britain, one is to bring benefits to users, the two is to improve efficiency, the three is to improve the social

benefit. The best way to realize this objective is the wide competition, promote the cost and the price to fall, give users more rights and better new services, users become the first beneficiaries, which is not only the guarantee of safety. To break the monopoly, encourage competition, improve efficiency, reduce the price, improving service, has become the most state power industry reform directions.

For the power industry reform in our country, is important to face the disadvantages of generation investment diversification and vertical integration management reality, through asset restructuring, the introduction of a competition mechanism in the field of power generation, fully reflect the fair, just and open principles, and improve national control on the power grid. This and other countries reform is not the same. Improvement will bring cost reduction, price reduction and service level, of course competition results.

3.2 *Establishment of the power market, the introduction of competition mechanism, not one size fits all, rushing headlong into mass action, and should emancipate the mind, seek truth from facts*

At present, China's resources are not fully effective use. Therefore, to optimize the allocation of resources, obtain the power balance in the greater region, established a solid network, is the implementation of the outstanding problems in the reform. The reform of the electricity market is advancing in the continuous improvement in the.

For the reform of the power market, we should start current generation:

① Net factory is separated by a separate operating mode as the starting point, with property right separation as the end point, no property right separation is hard to say fair competition.
② Hydropower plants and key power grid should be mastered, which plays an important role in the safe operation of the power grid in the competition.

3.3 *Electric power system reform should be government led, legislation*

The reform is to adapt to the need of the development of productive forces to adjust the relations of production, will inevitably involve the redistribution of power and interests, thus encountered some resistance, at the same time, the reform also needs cost. Therefore, reform must be carried out under the unified leadership of the government, under the coordination, can not go one's own way. That the British and Australian practice, the governments of the two countries play an active leading role by direct and indirect ways in the reform of electric power, a major decision before all pass system analysis and careful argumentation, and to fulfill the legal procedures, the two country's reform progress, continue to reduce costs, one of the decisive factors that effect more obvious.

3.4 *Must establish the matching market rules and regulatory system*

Foreign legislation first, after the experience of reform, in the present situation of our country is very difficult to do. But the legislative consciousness should be implemented in the whole process of electric power system reform, through the reform of certain things should be as soon as possible by laws or regulations or the form of fixed, timely amend the relevant laws or regulations, in order to ensure the smooth progress of reform. The rules of the market and supervision holds the extremely important status in the establishment of the electricity market competition, especially in China's legal system is not perfect, not matching conditions, a sound regulatory mechanism is particularly important.

In addition, we also deal with the supervision mode of in-depth study. It should be said that the reform model is different, the regulation on electricity market are not the same. The grid is owned by the state or the holding of the national, the power grid scheduling, power market and power network operation business management, to accept the strict supervision of the state, at the same time, the country still to transmission and distribution price regulation. For example, the British model is taken, the electricity market is strictly regulated by a government agency OFFER.

3.5 *Power market reform is a gradual, continuous development and improvement process*

From the overseas reform situation, the electricity market reform is a gradual process, involving the system, mechanism, policy, ideas and habits, especially with China's current reform of the entire economic system and political and economic stability is closely related to. Therefore, the reform of electric power should be timely, gradual reform in to protect the power investors, operators and users of the legitimate rights and interests of all aspects, to ensure the stability of.

In our country at present is concerned, the power generation sector is non-monopoly, should be strictly regulated. When the transformation and construction of urban and rural power grid reached a certain level, can consider to use separate electric business and distribution business. The distribution business is a monopoly, electric business should compete. Finally, transmission and distribution separation, the establishment of a unified national electricity market open, competitive and orderly.

3.6 *Reform must ensure the safety of the grid*

No true security issues are closely related to the economic efficiency of enterprises, users and social benefits. Therefore, the compensation problem seriously study the reactive power, voltage, AGC ancillary service system in the grid to the trading mode and operation rules, can be simple to complex, must establish the reactive power status and price. Because it is related to the safe and stable operation of the power grid, so all countries on the issue of compensation ancillary service. To adhere to the principle of unified scheduling in grid scheduling. The United Kingdom, Argentina, Chile is economic dispatch, and the British scheduling belongs to the power grid enterprises, further reform of the current power market or so. We insist on this point in the future.

Information, Computer and Application Engineering – Liu, Sung & Yao (Eds)
© *2015 Taylor & Francis Group, London, ISBN 978-1-138-02717-6*

On the production of multimedia courseware

Jin Sheng Song

Jiangxi Tourism and Commerce Vocational College, Jiangxi, Nanchang, China

ABSTRACT: multimedia courseware has become an important part of modern education, many teachers and their application in teaching practice, help students to better complete their studies. This article from the multimedia courseware type and manufacturing process two aspects elaborated the multimedia courseware.

KEYWORDS: multimedia courseware type production process

The computer aided teaching as a kind of advanced teaching methods, change the traditional education idea, teaching method and means, characteristic of the teaching mode and the amount of information, games, simulation examples and problem solving such vivid, easy operation, strongly interacting, more and more teachers and students of all ages. To strengthen the study and application of this technology, has the very vital significance to promote the reform of teaching content, teaching method and curriculum system. Therefore, good multimedia courseware can better help students master knowledge, but also can improve the teaching efficiency. This paper discusses multimedia courseware.

1 TYPES OF MULTIMEDIA COURSEWARE

Multimedia courseware according to the division of rules is different, can be classified into different types.

1.1 *According to the courseware structure, it can be divided into*

1 Linear make Courseware: the characteristics of simple structure, convenient demonstration, the courseware process like a run down the line. At present, the vast majority of teachers are using this type of courseware.
2 The branch type courseware: features as its structure is a tree structure, differences in the degree of change, according to the teaching content of students in the courseware process to selectively control the execution.
3 Modular Courseware: is a relatively perfect courseware structure, according to the teaching purpose of the teaching content will be a part of the production into a courseware module, the

teacher can choose the corresponding module teaching according to the teaching content.
4 Integrable Courseware: Based on the modular courseware as a blueprint, will be in the textbook on certain knowledge or teaching made small courseware independently, and then through the integrable ware system for these small courseware call, to the preparation of suitable teaching content of the courseware. Its advantage lies in its systematic, open and reusable.

1.2 *According to the way of courseware running, it can be divided into*

1 A stand-alone version of the multimedia courseware: its characteristics can only be run on a computer, corresponding design according to different computer configuration. The advantage is fast running speed, requirements for the technology is simple.
2 Network multimedia courseware: its characteristics through the network transmission, running in a user terminal. Advantage of resource sharing.

1.3 *According to the teachers' and students' participation ways, it can be divided into*

1 General demonstration courseware for classroom teaching, commonly used in the key point to solve some of the teaching contents, demo, added to a small amount of interaction.
2 The learning courseware: General for students' autonomous learning, teachers can be carried out through its individual tutoring to students, referred to as learningware. Should have a complete knowledge structure of this kind of courseware, can reflect the teaching process is explained, some answers, homework, detection, evaluation

and summary and so on, has a friendly interface, good interaction.

3 Practice courseware: generally used for training, consolidate, strengthen the students' knowledge and skills. Should have the function of a test at different levels; questions should be upgraded; can give students appropriate answers or hints; courseware should add fun and game note when making.

4 The simulation courseware: with the help of computer simulation technology, simulation test of the whole process,so that students can quickly grasp the methods and skills test, fast results.

2 THE BASIC PROCESS OF MULTIMEDIA COURSEWARE MAKING

2.1 Determine the form of teaching

According to the actual needs of teaching, design different teaching modes according to the different types of multimedia courseware, in order to achieve a good teaching effect.

2.2 Teaching design

Multimedia courseware is a computer program designed according to the teaching goal, as a kind of teaching media, it can according to the student's interaction, the presented control computer teaching information. Teaching courseware design division, the main teaching unit according to the syllabus and the student's choice of teaching mode, multimedia information appropriate selection, knowledge structure establishment and formation of design and practice of learning evaluation. The analysis should focus on the teaching objectives and teaching content, design of teaching activities when attention situation design, emphasizing the important role of context in the study, pay attention to the design of independent learning, pay attention to the design collaborative learning environment, focus on the design of multimedia teaching strategy based on. The following are the "center" and "learning centered" instructional design to teach for two:

1 Teaching design to teaching at the center

- To determine the teaching objective (we expect students through learning should achieve results);
- Analysis of the teaching goals and teaching content according to the results of the analysis (as required to master the knowledge unit achieves the teaching goal, teaching order (for the teaching of each knowledge unit sequence) analysis can be determined by the teaching goal, can also be determined by other methods;

- Analysis of the characteristics of learners (whether has the knowledge, need to learn the current content and with what cognitive characteristics and personality, etc.);
- Teaching starting point is determined according to the analysis of teaching content and learner characteristics (i.e.determine at what levels of difficulty and knowledge on the basis of the current study carries on teaching);
- According to the requirements of design and choice of teaching media, teaching objectives, teaching content and teaching object;
- According to the design of teaching strategies on teaching contents and learner characteristics;
- As a formative evaluation of teaching (to make sure students achieve the teaching objectives of the degree), according to the collected information on classroom teaching, teaching content and teaching strategies to modify or adjust, and make the appropriate feedback to students.

2 The design of teaching with learning at the center

- Analysis of teaching objectives
- The creation of context
- And to provide design information resources
- The design of autonomous learning strategies
- The design of collaborative learning environment
- The assessment of learning effect
- Strengthening exercises design

2.3 Selection of multimedia courseware making software

Can now be used to make multimedia courseware software many, every software has its own advantages and disadvantages of multimedia courseware, can choose according to their own needs making tools.

1 PowerPoint
Microsoft Corp export slide making software, the electronic manuscript made extensive application.
2 Authorware
The Macromedia Multimedia development tool company launched, with a strong creative ability, and simple user interface and good scalability, is the most widely used multimedia development tools.
3 Director
The full name of Macromedia Director Shockwave Studio, IsMacromediaThe company launched a heavyweight tool. Intuitive, easy to use user interface, and has a strong ability of programming (integrated Lingo language).
4 Flash
MacromediaProduced by the company on the Internet for dynamic, interactive shockwave. It has the merits of small volume, can be played while

downloading, users avoid complacent waiting. FLASH can do the animation can be added in the Webpage voice, and the generated file size is very small.

5 Founder author tool and other software

The software based on the courseware making, but in terms of function is single, operation characteristics, but in general the use of complex.

2.4 *Making courseware*

A collection of good material, multimedia production tools can use the selected for editing of various materials, according to the teaching process, teaching structure and the design idea for the script, the courseware is divided into modules to make, and then each module interact, links, finally set to become a complete multimedia courseware.

2.5 *Courseware debug*

After the integration of the need for careful testing: interactive function is realized, the effect how, how to control the multimedia program, whether it can meet the demand of the teaching and a series of testing.

2.6 *Courseware release*

After commissioning, package and package, and made into a multimedia CD-ROM. The whole production process is completed.

REFERENCES

[1] Xu Dinghua, Miao Liang, Chen Feng.Author ware multimedia courseware making practical tutorial [M]. Tsinghua University press, 2005, 3.

[2] Chen Bingmu, Zhang Jianping. The multimedia courseware design case tutorial [M]. Beijing Kehai Electronic Press, 2003, 6.

[3] Wu Jiang. Multimedia courseware design and making of [M]. of the posts and Telecommunications Press, 2002, 8.

Information, Computer and Application Engineering – Liu, Sung & Yao (Eds)
© 2015 Taylor & Francis Group, London, ISBN 978-1-138-02717-6

On the current situation of college computer teaching

Ren Fen Liu

Sifang College, Shi Jiazhuang Tiedao University, Hebei, Shijiazhuang, China

ABSTRACT: The goal of the computer-based teaching is to train non-computer science students to master certain basic computer knowledge, techniques and methods. On the status of key universities and local colleges and universities teaching research, analyzes the problem and its causes exist in teaching, concludes with a reform program for the four-point proposal.

KEYWORDS: Computer basics; teaching status quo; curriculum; network teaching platform; information literacy

1 BACKGROUND COLLEGE COMPUTER BASIC TEACHING REFORM

With the rapid development of communication technology, computer work and increasingly becomes an indispensable tool for learning. Our basic computer education began in 1978, when the university admissions resume shortly, only a few universities in the non-computer professional set up high-level language courses. From the early 1990s, China's universities gradually universal basic computer education. Countries in order to better supervise the development of computer-based education, the promulgation of implementation of the National University Computer Rank Examination in 1997. At the same time, some units in the recruitment of personnel, usually require graduates with some computer skills, which makes various colleges and universities attach great importance to computer-based teaching. But the employer of graduates in the computer skills feedback situation, a considerable part of the computer application ability of graduates is poor, a lot of theoretical knowledge and practical ability of graduates out of line, because this situation is a multifaceted specific analysis of the main points:

First, students' base varies widely. Although the Ministry of Education as early as 2006 had developed a primary and secondary school IT education programs and curricula, the goal is within 10 years of the 21st century, we should gradually spread IT education primary and secondary schools, so that more and more college freshmen computer basic level to get rid of "Beginners", but because of the huge difference in our region, urban and rural areas, there are still some students or "machine blind students", the gap between freshmen and more obvious, which is the current computer-based teaching first face issue.

Second, it is a unique way of teaching sufficient practice [1]. Computing is a strong operational discipline and application; it requires students to spend a lot of time practicing, to form the corresponding skills. Early computer equipment is more expensive, making students' practice time subject to greater restrictions. To improve most colleges' and universities' hardware conditions, the school usually invests a lot of money to buy computer equipment, without in the meantime increasing tuition and school, until the analysis of the ratio of people still in the state of the machine. Most of the students, in addition to the normal time in class, are rarely conducive to free time for practicing computer operation, even if a small number of students in their free time seized the computer. But also often chatting online or playing games on the computer for students cannot be guaranteed fundamentally based equipment will not be able to improve the quality of teaching. Third, the books' texts are not complete, the direct result of ineffective computer-based instruction [2]. The traditional disciplines of physics and chemistry in higher education have a long history; their curriculum for teaching, that is, the content has strict and clear rules and generally uses textbooks' texts, and thus has more authority. Teaching computer in the classroom Night, teaching computer at various colleges and universities are very different: the curriculum is reviewed by the universities, in general, despite the fact that there are many textbooks in society based on the computer. But with regard to a considerable part of textbooks of similar serious cases, many universities, in the selection of materials, make no analysis of their pupils situation of learning; theoretical knowledge is often not be overemphasized and practical ability to be ignored, so of course, it helps improve the skills of computer science students.

2 STATUS QUO COMPUTER EDUCATION IN UNIVERSITIES

2.1 *Curriculum*

In the computer foundation course setting, does not have what difference key university and colleges and universities, all are set up in accordance with the relevant documents of the Ministry of education. As a basic course in computer of Zhongshan University: "Computer Application Foundation (Wen Ke)", "science based" computer application, "C++ program design", "Visual of Basic language program design", "dynamic Webpage design and electronic commerce". Henan Normal University is: "University Computer Foundation", "basic" C program design, "VB program design", "basic" VFP program design "," Java program design basis. From the two typical examples can be seen, in fact, a basic course in computer of many colleges and universities are basically the same. In addition,according to each specialty in Colleges and universities is different, each have their own syllabus, rational engineering of Xi'an Jiao Tong University teaching syllabus, the medical humanities teaching syllabus, syllabus. According to the different needs of different majors in Peking University computer, divides the teaching system of computer basic course for computer majors (Class A), science for non computer majors (class B), Liberal Arts (class C) three class, the purpose is to teach students in accordance with their aptitude, as much as possible to increase the computer foundation teaching effect.

2.2 *Curriculum implementation*

The course mainly manifests in the teaching mode and teaching methods, in this "211" universities and local colleges have obvious differences. The Xi'an Jiao Tong University, Peking University, with Chongqing University as an example, we can see that the implementation of the present situation of our country's key university computer foundation curriculum: Xi'an Jiao Tong University to establish digital teaching platform, the platform of integration of curriculum website, examination system, multimedia digital resources management system. Through the teaching practice in recent sessions of the newborn, selected high quality multimedia courseware in numerous teachers produce courseware, the courseware will be selected after the professional artists, and strive to publish in the press. At the same time, to reform the classroom teaching mode, using large numbers of teachers from the original artificialteaching mode. Teaching mode adopts the digital network, at the same time, the realization of online real-time answering, submission and

correct homework, online self-help study and experiment, the Internet online self-assessment of computer basic course; Peking University will be posted on the Internet, has become the network curriculum; Chongqing University of computer network and multimedia technology, computer interface and the application of the 14 basic computer course into network courses for students' Autonomous learning. There are otherkey universities. They all have their own online courses, so it can promote teaching efficiency improvement.

While the other ordinary local colleges and universities, from the official Campus Online their point of view, there is no network course. Such as Henan Agricultural University, Guangxi University For Nationalities and so on. Visible, their teaching methods basically still adopt the traditional teaching methods, through the use of multimedia, computer and other methods organize teaching.

3 SUGGESTIONS AND COUNTERMEASURES

3.1 *Hierarchical classification, basic computer teaching planning and implementation*

Because the computer foundation teaching has its own unique rules, so many colleges and universities have carried out the reform of computer basis. New measures introduced to Renmin University of China. Computer basic teaching as an example, the National People's Congress implement hierarchical, classification of educational reform on basic computer education, reform of the "computer application foundation" course assessment methods: allow some better foundation, strong ability of self-study students in the final examination based on the completion of all work early on. The final exam questions and the difficulty and the final exam quite, if students advance examination score of 70 points or more, it may apply for an earlier node node class, after-school students can spend more time on more development content and advanced study. Some university in college students, the student first computer skills assessment, and then let them into different levels of learning, for those who are particularly good foundation of students allowed to exempt, the basis of poor students is divided into several grades, the implementation of hierarchical teaching.

In addition, the University also can according to the specialty of basic computer application ability requirements, professional teaching plan and syllabus. For example, some professional, in-depth to the computer specialized knowledge requirements, and some other majors, completely is to use the computer as a tool, therefore, should be based on the limited teaching hours, according to the different types of schools and professional, make a different teaching syllabus is very necessary. Basic computer teaching in Peking University as an example, according to the different professional, has a different teaching syllabus, the development of different materials, strive to make each student actively strive to learn, finally let all students can achieve the basic computer education the teaching goal.

3.2 To focus on building the foundation course

The development and construction of curriculum is the foundation of computer basic education, only a good course, and good teaching method plus, can have the computer teaching well, to improve the college students' level of computer,to cultivate qualified talents for the society. So do a good job in basic computer teaching, we must develop a good basic "University Computer Foundation" this course, each university should analyze their student learning, for students of different majors, some emphasis on teaching knowledge. As for the accounting specialty, should be in the database of knowledge to tilt, in order to achieve the students learn professional knowledge and computer knowledge integration, to lay a good foundation for the further study of students.

3.3 To train students in teaching the goals of informational literacy

The definition of information literacy is: can clearly recognize when information is needed, and can determine, evaluation, effective use of information and the ability to use various forms of information exchange. The essence of education is to cultivate the information literacy of the college students' ability of a rapid acquisition of knowledge. The rapid development of computer technology makes it like paper and pen, is a basic skill for all the necessary, it is also the ability of modern college students must have. In the aspect of teaching contents, much of the content of basic computer courses is related to information literacy, for example in the teaching of network knowledge, can further enhance the students' understanding of social informatization, improve the students' consciousness and ability of processing information, the use of information. In the aspect of teaching mode and teaching methods, through the design of teaching process of computer basic course, so that students in the learning process, and gradually realize what is the society of information learning mode and working mode, and the ability of education on how to get this knowledge training. In short, the education of basic computer education and information literacy is more close, basic computer teaching in universities should be more for the information literacy education goal.

REFERENCES

[1] Liu Liping. Reflections on computer teaching situation of [J]. science and technology information, 2010 (4): 236.
[2] Baojiejun. En la reforma de la enseñanza de informática de la universidad de [J]. La ciencia de China y tecnología de la información, 2005 (16): 264.

Information, Computer and Application Engineering – Liu, Sung & Yao (Eds)
© 2015 Taylor & Francis Group, London, ISBN 978-1-138-02717-6

An opinion on the development of secondary occupational education

Xin Min Xiong & Ling Zhang
Jiangxi School of General Engineering Technique, Jiangxi, Yongxiu, China

ABSTRACT: This paper focuses on the analysis of reasons for difficulties in the development of secondary vocational education from the angle of economics, and puts forward the corresponding countermeasures.

KEYWORDS: secondary vocational education; development; market; skills

Vigorously develop vocational education, not only can provide a large number of skilled workers for the economic development of our country, but also on the current employment pressure, to education and poverty alleviation, of the social comprehensive governance has a significant role to alleviate. But the current enrollment and vocational school graduates employment difficulty has directly affected the normal development of secondary vocational education.

1 MEDIUM OCCUPATION EDUCATION DEVELOPMENT DIFFICULT REASON

1.1 *There are its deep social and economic reasons of secondary occupational education development difficulties in China*

In the process of industrialization, our country is at the beginning of industrialization. Production technology and equipment are behind the stage. The enterprise and the mode of economic growth do not completely depend on the quality of human resources. Scientific and technological progress of the track influence the growth of a capitalist economy that is often greater than the skill of workers. The quality of economic growth influences education paper ring, causing the enterprises pay attention to the price of labor and ignore the quality of the labor force, and even enterprises in the downsizing process, in order to reduce expenditure, appear with ordinary minus two technical workers, reducing high wages for low wage. "After the first training employment" and "posts" shout for years, but to really implement and has its profound economic background. From the economic structure, China is the two Yuan economic structures exist typical advanced industrial sector and backward agricultural sector labor demand, labor supply relative to have unlimited tendency. Infinite labor supply is due to industrial departments of labor

and production died far higher than the agricultural sector, caused a large number of agriculture surplus labor transfer results. At the same time, overpopulation, laid off workers in state owned enterprises and institutions reform, reason of infinite labor supply. The long existing phenomenon of a pile up in excess of the requirement of the labor market faces a harsh external environment so that the development of secondary occupational education. The value orientation of view, "an official" under the influence of traditional culture, many families are often read as the main way children rise head and shoulders above others, and not to regard education as a kind of employment means. Secondary vocational education is the cultivation of a skill or technique type number of technology long laborers, unable to meet this reading out ahead of the psychological needs, this is even if many students don't high school rather than vocational and technical schools of the important reasons. The current social employment situation, some enterprises employment one-sided pursuit of high degree, making the university the college students do can do work some technical secondary school students. This "high low" phenomenon, not only caused a great waste of our limited education resources, but also have a negative impact on the labor market. To analyze the investment and earnings from the point of view, personal education cost is divided into two parts: one is the cost of education, two students is disposable income effect. The cost of education and includes two parts: one is the government appropriated funds, the other is the personal burden of tuition fees. The student valuation of their education after the earnings are divided into personal future and personal income higher future larger professional mobility and personal future bigger self - development space. Now the development of secondary vocational education is difficult, in fact the students' spend a greater cost, it is difficult to obtain the reason of high income.

1.2 The education system and the education structure reasonably have restricted the development of secondary occupational education

Since the reform and opening up, our country's education reform has made great achievements, but the education structure inherited from the planned economy period and have not been changed fundamentally. To some extent, education resources in our country the problem itself is not the main problem, the more important issue is education resources disposition. In 1997, according to age (the age in general students have graduated from high school) primary school graduates accounted for 28.6%, junior high school graduates accounted for 51.5%, namely 80.1% people did not receive vocational training in any form into jobs. In the structure of education environment in our country this kind of heavy degree education light skill education, once the entire education transition from the planned system to the market system, the development of secondary vocational education necessarily at a disadvantage. Under the condition of a market economy, the education of reasonable structure should be of higher education in the compulsory education on the basis of the emphasis on academic education, quality is more important than quantity; vocational education should focus on skill education, make it become the majority employment means. From the overall social development perspective, talent training should focus on improving the quality of most human resources, not just the limited funds completely to prepare for training senior academic talent.

1.3 In recent years, the internal structure of the secondary education unreasonable situation, blindly opened management, no restriction supplementary low level students, resulting in secondary education structure unreasonable phenomenon is more serious, resulting in a large number of invalid education, occupation education development team and make more difficult

The economic structure under the condition of a market economy, the medium occupation school enrollment number and professional structure should be with a certain level of economic development of the structure of the labor resource allocation decision of industrial structure and technological structure to adapt to the high level of talent resources, and not the more the better. A manifestation of the current "high low" phenomenon is the educational resource allocation. On the one hand, our country lacks the

skilled workers, enterprises after workers secondary vocational and technical training proportion is very low; and the emergence of the employment difficulty of vocational students is a common phenomenon on the other hand. A very important reason is that the secondary vocational education does not adapt to the need of market economic development, a lot of vocational education and training by the people with the skills and need not the social economic development skills. Education is indirect, hysteresis and long feature, now the professional, curriculum, admissions to adapt effective amount is three or four years after the market demand. But few schools or educational institutions analysis and forecast of market demand survey, with. Even more serious is that secondary vocational education management fragmentation, the lack of an authoritative coordination mechanism, which makes the professional settings between the schools overlapping, enrollment of vicious competition. Especially in recent years, driven by the interests, the schools rush to start economic courses, such as computer professional hot subjects, leading to the lower level talents cultivation of repetition, which restricts the development of secondary vocational education.

2 THE KEY SECONDARY OCCUPATION EDUCATION OUT OF ITS PREDICAMENT LIES IN THEIR CORRECT POSITION

Secondary vocational education should take employment as the goal, take the enterprise as the classroom. Take the market as the only standard to check the quality, according to the needs of society to organize teaching, construct the new system of secondary vocational education.

2.1 Secondary vocational education should be training base of unemployment and laid-off workers

From an economic point of view, unemployment will be a waste of resources, bring economic losses from the social point of view; education, unemployment will affect social stability. Secondary vocational education is to change the unemployed restricting economic development as a model of human resources. To solve the problem of unemployment is not only starting from creating jobs, more important is the use of existing educational resources, From the culture of the unemployed skills, the implementation of "open employment". Under the condition of a market economy, the structural unemployment (labor skills do not adapt to the economic structure, especially the transformation of industrial structure, technical structure and upgrading the generated unemployment) is inevitable, this is permanent secondary occupation education market. We should learn the experience of Germany. The unemployment relief and demobilized, transferred together, unemployed personnel must be trained, or can not give the unemployment benefits and subsidies.

Medium occupation education should be geared to the needs of the rural market. The key of rural modernization in China is to improve the quality of farmers, the modernization of farmers' training into hundreds of millions of peasants understand technology, understand the economic, which is the largest market for occupation technology education. Rural education vigorously popularizing compulsory education at the same time, should increase the proportion of occupation education, strengthening the construction of rural occupation technical school. The farmer agriculture school held the entrance exam, no age limit, no time limit, the laws of the market under the guidance of agricultural technology, to cultivate practical. Practical agricultural technology education has great external effects. Therefore, the government should put the agricultural vocational school as a public welfare undertakings to do.

2.2 The secondary occupation education content to specific practical skills based

The enterprise demand for skilled workers is the occupation education vitality. The German occupation education adopts the double track system, linking theory with practice, school and enterprise, which mainly operating technology enterprises. Students in the occupation technology education for three years, with the enterprise apprenticeship learning skills, in the occupation school or training center as a student learning technology theory, students generally four days a week in the enterprise to carry on the practical skills training, a day in the school learning theory course. In the enterprise to pay part of the apprenticeship period. The German experience is of great significance to our country's present secondary occupation education.

3 THE DEVELOPMENT OF SECONDARY OCCUPATION EDUCATION EXPOSED TO GOOD SOCIAL ECONOMIC ENVIRONMENT

3.1 According to foreign experience, China should formulate regulations on urban high school graduates before employment will be a mandatory 2–3 years of vocational training, without pre job training will not be allowed to employment

In this way, one can improve the quality of workers, on the other hand, the continuation of them into employment team time, alleviate the present employment problem is very serious.

3.2 The development of secondary vocational education should focus on quality, optimizing the structure

The school to carry on the analysis and forecast to the market in the detailed market research basis, determine the training direction, quantity, quality and time according to the market situation. The implementation of a vocational school admission system, determine the conditions of access from the condition of running a school, students, teaching conditions, market. Are not eligible for schools closed, stop, and, turn.

3.3 The local government should increase the investment in the secondary occupation education

Education has exterior sex is very strong, complete the education of the market, the role of education market will not play good configuration. Medium occupation education itself both individual and social benefits of the double, students to obtain personal income should pay certain fees, government as a social income beneficiary of occupational education, should also bear some of the cost of education. Therefore, occupation education is not compulsory education, but should be encouraged by the government.

Information, Computer and Application Engineering – Liu, Sung & Yao (Eds)
© 2015 Taylor & Francis Group, London, ISBN 978-1-138-02717-6

Design and management of intelligent transportation integrated command platform

Hong Ying Li
Automotive Engineering Institute, Jiangxi University of Technology, Nanchang, China

ABSTRACT: China's social development is in urbanization, motorization rapidly increasing stage, traffic safety, traffic environment, traffic congestion, transportation efficiency, transport services, and many other problems concentrated there, has become an important problem restricting social development, intelligent transportation systems and solutions to alleviate these problems have a direct role and significance, and intelligent traffic management system is the "Long-term Science and Technology Development Plan" (2006–2020) theme six priority development areas as transportation one. Integrated intelligent transportation technology has a strong crosscutting and multi-technology features, and integration innovation is an important mode of intelligent transportation technology development. In this paper, information systems, project management methods and project management knowledge, a large integrated application system is designed to build intelligent transportation field - Intelligent Transportation integrated command platform business model, and propose a feasible technology architecture to achieve various traffic infrastructure systems, deep integration of various traffic information to achieve in information services, improve management efficiency, improve management level of utilization.

KEYWORDS: Intelligent Transportation; motorized; platform

1 INTRODUCTION

Pass more and more incidents, road traffic safety situation is getting grim, integrated application alone has been unable to meet the needs of modern traffic management and urban development. Further use of modern science and technology management tools, the ability to "police training" to the technology to the police, scientific organizations transport, improve existing road capacity, coordination disposal emergencies, to ease traffic congestion and enhance traffic police rapid reaction capability and other aspects of based on research and development event, plan management, urban intelligent transportation Secret mission command platform integration is imminent.

Our country has started research to promote the use of intelligent transportation systems in the late 1990s. Over the past decade, China's intelligent transportation construction and development has achieved positive results. However, China's intelligent transportation is still in the development stage. The overall level of technology and application of the scale compared with developed countries, there are still a large gap in the field of cutting-edge technology is relatively backward. The core technology in the field of intelligent transportation many of the key technical issues are yet to be a breakthrough, but with the development of our society and economy, especially the whole society urbanization, accelerated the process and the rapid motorization of modern information technology, intelligent transportation technology application development prospects of China's vast potential.

2 THE MAIN CONTENT

Through domestic and intelligent transportation systems research and analysis, the whole idea of building intelligent transportation systems integrated command structure in terms of technology, business model, system implementation, the standard procedure using project management techniques in the construction of information systems management to complete the R & D intelligent transportation integrated command platform development.

1 To the modern intelligent traffic management system for analysis, practical application case studies proposed effective intelligent transportation integrated command platform viable business model.
2 Proposed a total solution integrated intelligent transportation command platform for business integration module and adaptability, transparency and sharing of information resources, the integrity and security of data, system application stability and scalability make unified planning and design.

3 In-depth study of standard project management techniques in the process of information systems, the use of relevant technologies for system implementation.

According to the current domestic problems of intelligent transportation systems and development trend analysis of ITS, we can see, the application of intelligent transportation systems through the traffic and road resources, equipment, facilities, management and control of traffic travel guide and command a reasonable person to achieve, which is the key to traffic management integrated command system. It is the urban traffic management department of traffic command center core systems, unified information platform for urban traffic control system services, information exchange and sharing, decision-making and rapid response unified command. Through the multi-source traffic flow data collection, analysis, fusion, and based on valid information for decision-making and ultimately obtain traffic dispatching, strengthen the capacity of road traffic macro-control and command scheduling emergencies quickly and efficiently respond to the formation of mechanism for timely processing of traffic incidents and traffic information through a variety of channels will release to the police detachment and other traffic participants, so as to effectively alleviate the traffic congestion situation and improve the capacity of urban roads.

This paper discusses the main system construction to achieve traffic information around management, traffic management-integrated dispatching, secret service management plans to expand and improve the ability to respond to rapid mobility traffic police as a whole, to achieve traffic police on information management effectiveness level, meet the urban transport management capacity building needs.

Specific system construction objectives are as follows:

1 Based on GIS dispatching function
Dispatching functions mainly realized through the construction of police positioning and wireless communication systems to achieve pavement visualization of police dispatch and other functions, and establish urban road traffic warning and rapid response mechanisms, effective scheduling of police, traffic police intelligence force disposition.

2 Multi-mode multi-level cooperative scheduling
Multi-level diversification collaborative dispatching technical support hierarchical control model, layered to achieve regional and brigade-level command center scheduling, dispatching network will cover the entire region in order to achieve data sharing and coordination of scheduling between the various brigade and detachment function.

3 Integration related infrastructure systems
Mainly refers to the integration of traffic control, video surveillance, incident detection, traffic guidance and other basic intelligent transportation systems to adaptive traffic signal coordination control, key sections of the road traffic incident detection and alarm, and display geographic information systems, and large screen full control of the main city traffic conditions in real time systems, key parts of the traffic flow of intelligent induction.

4 Support for rich plan
Though the plan is dispatching auxiliary functions, but it is also the key to its technical support. Depth study of the characteristics of the traffic business, plans to create a rich library that implements various types of traffic police dispatch events. On this basis, the business subsystem control, decision support equipment scheduling, improve the overall efficiency, adjustment mode and eventually reaches guarantee the purpose of the quality and level of service.

5 Provide comprehensive, accurate and timely decision-making data
Provide comprehensive, accurate and timely data through the acquisition and integration process of the traffic flow, as well as query applications for traffic control planning and decision making.

3 TECHNOLOGY ROADMAP

3.1 *Geographic information system technology*

Matured GIS geographic information technologies for a wide range of applications in the transport sector has created a foundation. Use of databases, computer graphics, multimedia and other latest technology, geographic information data processing, real-time and accurate capture, modify and update geospatial data and attribute information for decision-makers to provide visual support. GIS and traditional traffic information analysis and processing technology closely extend the geographic information system (GIS-T).

GIS-T through a variety of geographic information systems and traffic information analysis and processing technology integrated operating platform for providing transportation planning, traffic control, transportation infrastructure management, and real-time traffic. According to statistical analysis functions, dedicated map and flex-based interface display technology to analyze changes in traffic volume, to develop traffic plans. Traffic management department can use GIS_T on traffic conditions in real-time monitoring of the entire city.

3.2 *Transportation technology fusion technology*

System in accordance with the "data fusion, system integration, equipment integration" principle, to formulate exchange format based on the system interface specification mechanism and data messaging services, to include all existing public security and traffic management technology subsystem, business information systems integration to the next platform, "Technology subsystem, information and data, police equipment" as a resource platform can be called, usually, "monitoring, control, management," wartime "synergy command and Control." Platform can be at different times, different manufacturers, different hardware environment, under different software environments built subsystems according to the unified system interface specification, smooth access platform system, to facilitate seamless police traffic command center system upgrades, expansion.

More than 3.3 grade multi-dispatching technology are collaborated.

Multi-level diversification collaborative dispatching technical support hierarchical control model, layered to achieve regional and brigade-level command center scheduling, dispatching network will cover the entire region in order to achieve data sharing and coordination of scheduling between the various brigade and detachment function. Diversified scheduling of resources, on a virtual level, characterize the physical resources of various business systems services for dispatching users with available resources on business-level view. Through a combination of dynamic scheduling engine, the system involved in the combination of dynamic scheduling, sharing and collaboration on virtual organizations.

REFERENCES

[1] Meignan, David,.Simonin, Olivier, Koukam, Abderrafiaa. Adaptive traffic control with reinforcement learning. 4th Workshop on Agents in Traffic and Transportation (ATT"06). 2006.
[2] Ezzedine, Houcine, Bonte, Therese, Kolski, Christophe, Tahon, Christian. Integration of traffic management and traveler information systems: basic principles and case study in inter-modal transport system management. International Journal of Computers, Communications & Control. 2008.
[3] Zhou, J, Lam, W. H. K, Heydecker, B. G. The generalized Nash equilibrium model for oligopolistic transit market with elastic demand. Transportation Research. 2005.

Information, Computer and Application Engineering – Liu, Sung & Yao (Eds)
© *2015 Taylor & Francis Group, London, ISBN 978-1-138-02717-6*

Explore the application of computer simulation of patients in pediatric respiratory trainee teaching

Xue Jun Zuo

Langfang Health Vocational College, China

ABSTRACT: In recent years, with the rapid development of medical technology, medical education has also undergone tremendous changes. Many colleges are growing shortage of clinical teaching resources. It is difficult to meet the needs of modern clinical teaching. For this, optimizing the teaching process, implementing diversified teaching methods has become common in clinical practice initiatives. Currently, there is a great shortage of teaching resources in pediatric respiratory novitiate, making the search for new multimedia technology to simulate patients for clinical practice teaching opened up a new direction of development. This paper is mainly for pediatric patients with respiratory trainee teaching of computer simulation applications providing an important reference for clinical practice teaching.

KEYWORDS: pediatric respiratory system; probationary teaching; computer simulation; applications

With the continuous development of China's science and technology, the Internet has been widely applied to various fields within the range of social, online teaching has gradually played its convenient advantages, gradually applied to the education industry. However, as in the background of the current medical education, various colleges and universities often due to large-scale enrollment allows greatly increased the number of students, but also combined with clinical teaching are very scarce in typical cases, eventually leading to the development of clinical teaching can not meet the needs of medical education. The application of computer simulation patients, pediatric respiratory trainee teaching has brought new opportunities for development.

1   PEDIATRIC RESPIRATORY PROBLEMS IN TEACHING TRAINEE

Pediatrics is a highly specialized clinical medical discipline. However, in the actual teaching process, many medical students face pediatric respiratory patients are often at a loss, often appear during interrogation of unknown primary and secondary phenomena. How to comprehensively improve the clinical diagnosis of medical students the ability to become pediatric respiratory trainee teaching important issues. In pediatric respiratory trainee teaching, application of computer simulation patient manner and achieved satisfactory results. Investigate its reason, as a result of pediatric respiratory trainee clinical teaching is to master and consolidate theoretical

knowledge and clinical practice it is inextricably linked. Thus, pediatric respiratory clinical practice trainee teaching is an important part of pediatric respiratory trainee, the affection of teaching, good or bad, will directly affect the results of clinical practice. However, under the current circumstances, many problems still exist in pediatric respiratory trainee teaching. For example, pediatric respiratory trainee teaching resources are not adequate, clinical practice and disease pathogens as well as the relative lack of the traditional teaching model are still relatively simple. These have a direct impact on medical students' interest in learning, but also affect the quality of trainee teaching presence.

2   THE APPLICATION OF COMPUTER SIMULATION OF PATIENTS IN PEDIATRIC RESPIRATORY TRAINEE TEACHING

2.1   *Computer simulation of the patient's application*

Computer simulation of the patient is the use computer to reproduce, to achieve the purpose of probationary teaching. Typically, a computer simulation of patients includes two parts, one is a computer simulation of the patient's clinical scenario design. Select pediatric respiratory system, typical cases, while collecting the appropriate history, physical examination and laboratory examination data, initial diagnosis and treatment. In this process, use a computer, to establish the correct module in the corresponding area of the link. According to the analog auscultation, medical imaging data, images, and laboratory testing out the patient's condition changes, at the same time, design the appropriate questions, for example, by setting the multiple-choice questions to achieve the purpose of the man-machine dialogue, and on evolution of the cases can be achieved by way of the operation. Another part of the teaching process is mainly trainee. For each case of children, may be appropriate simply recording history and teaching given by the students will complete medical history to complete it, and then use the appropriate video, etc. to obtain information about the patient's examination, and ultimately achieve the purpose of issuing medical advice. In this case, the computer simulation system will change the condition in children after treatment displayed, so that you can provide students with the appropriate reference, in order to carry out further processing. If in the process, the students deal with a certain error, then simulating the system will directly show children sicker or directly to death. Finally, teachers deals with student trainee about pediatric respiratory and gives some evaluation process.

2.2 The characteristics of computer simulation patients

First, the computer simulation of patients mainly takes clinically typical cases as center. Computer simulations used in pediatric respiratory cases were common and frequently occurring. During the course of the application of computer simulation patients, students need to be able to use the pre-mastered comprehensive theoretical knowledge, combined with the appropriate checks and laboratory examinations were analyzed eventually complete the entire treatment process. And teachers can simulate the situation of medical records by the students to understand the student mastery of knowledge. Second, give full play to the status of the student body, but also play a guiding status of teachers. In pediatric patients using computer simulation respiratory trainee teaching process, but also for timely feedback based on the actual situation of the students, for the difficult problems encountered, teachers should actively organize group discussions for students to learn the characteristics of summarized history, to correctly identify the clinical signs, to make the correct clinical diagnosis. Finally, to take advantage of the computer, make sure to complete the whole process of computer simulation of the patient. Under normal circumstances, a computer simulation system operation is very simple, convenient and very easy for the students receive. At the same time, the use of computer simulation system operation without time and space constraints, effectively avoiding the lack of pediatric respiratory trainee teaching resources.

2.3 The advantages of the application of computer simulation patient in pediatric respiratory trainee teaching

3 SIGNIFICANCE OF COMPUTER SIMULATION IN PEDIATRIC PATIENTS WITH RESPIRATORY TRAINEE TEACHING

Computer simulation of the patient is an important part of medical trainee teaching, and clinical practice can encourage students to combine knowledge from medical theory and practice effectively, so as to nurture students' scientific and rational clinical thinking, improve their ability in clinical practice, the students into the future hospital laid a solid foundation. As traditional pediatric respiratory trainee teachers will be teaching mainly of typical cases selected, and then allow students to view the patient beds and bedside corresponding demonstration, and finally discuss the summary. This single probationary teaching model can not meet the needs of modern teaching and complete student progression in patients unable to observe the evolution of detail displayed. In this case, the computer simulation is applied to pediatric patients with respiratory trainee teaching, encourage students to put themselves in a more realistic clinical situations themselves, able to observe the evolution of the patient from admission to discharge the entire illness, accordingly design corresponding clinical thinking, so as to continuously cultivate independent thinking and problem solving skills. For example, respiratory syncytial virus pneumonia patient scenario simulation, you can take full advantage of playing video segments, etc., the symptoms of mental condition and breathing difficulties in children with a detailed dynamic form displayed, prompting students to observe full of children with the disease evolution. At the same time, by appropriate laboratory testing data, imaging studies, so that students to think independently, analyze the different problems faced by the clinical treatment process, the final by virtue of its ability to be resolved. In this process, each diagnostic or treatment decisions are made through the appropriate discussion, the students presented their different views, and ultimately draw the corresponding results. And teachers should give full play to the guiding role of teachers, the students made the appropriate decision-making timely guidance and analysis. This way, not only can effectively active in the classroom atmosphere, but also to encourage students to master the relevant medical knowledge, to further improve pediatric respiratory trainee quality of teaching.

4 CONCLUSION

In conclusion, the computer simulation is applied to the pediatric patient respiratory trainee teaching, to provide the free exercise of time and space for students to promote pediatric respiratory trainee teaching quality is improved. However, in patients with the use of computer simulation process, the need for more skilled computer and other related operations, so as to maximize the advantages of a computer simulation of the patient, effectively stimulate the enthusiasm of students, improve the quality of teaching trainee pediatric respiratory system.

REFERENCES

[1] Meihua Zhu,Ming Liang,Zhijian Wang.etc. the application of computer simulation of patients in pediatric respiratory trainee teaching [J] Chinese Journal of Medical Education, 2012, (5):522–524.

[2] Lianlian Li,Yanlian Hu,Yinjie Jiang.The application of simulation system in ophthalmology clinic clinical practice teaching [J] Yichun University, 2014, (3):140–141.

[3] Meihua Zhu,Ming Liang,Zhijian Wang.etc.The development of pediatric patient computer simulation system [J] Chinese Journal of Medical Education, 2014, (4):140–141.

Information, Computer and Application Engineering – Liu, Sung & Yao (Eds)
© 2015 Taylor & Francis Group, London, ISBN 978-1-138-02717-6

Intelligent property management model built in the era of IOT (Internet Of Things)

Da Jian Hu

Qingdao Vocational and Technical College of Hotel Management, China

ABSTRACT: With the rapid development of economic and science development and science and technology and the advent of Internet of things, all kinds of intelligent products are increasing, including a new building in the form of a fusion of modern building technology and network communication technology and the other - the intelligent building form into people's lives. The intelligent building has become a trend in the development of modern architecture. The need for the proper functioning of intelligent building intelligent property management. This article will establish a property management model for an intelligent networking era of intelligent property management, and model make this detail.

KEYWORDS: Internet of Things; intelligent; property management; Model

With the arrival of the era of things, more and more intelligent and smart buildings into commodities among people's lives, for example, we now use smart phones, smart TVs, etc. They provide a lot of convenience to our lives. Thus, modern intelligent building construction have been proposed, such as smart hospitals, schools and other intelligence. The intelligent architecture requires intelligent property management. Property management is a product of the same time intelligent information age, which combines a variety of disciplines, a variety of new technologies, is the direction of modern property development. We will build intelligent property management model analyzed below.

1 IOT OVERVIEW

IOT is an important component of modern information technology department, which uses a two-dimensional code recognition devices, radio frequency identification devices, infrared sensors, global positioning systems and laser scanners and other information sensing device, in accordance with certain provisions of the agreement, the Internet of Things and any five parts connected together to be able to transfer and exchange of information, to realize a modern information network of intelligent network system identification of goods, positioning, monitoring, and management operations. Under normal circumstances we call "connected to the Internet."

IOT generation is a huge market, rapidly developing, networking technology has now spread to all areas. With the rapid development of social economy

and information technology, people's living standards improve, people's way of life has also been greatly improved, China vigorously promotes the triple play technology, people's lives are moving in the pursuit of intelligence, personality oriented, healthy and comfortable, safe and convenient direction. Thus, in the era of things intelligent property management is welcomed by the people.

IOT is seen as modern to expand the Internet, the Internet of Things both characteristics, identification and communication features and intelligent features. Internet levy amount Angte performance networked objects need to achieve the interoperability of network systems; identifying communication characterized by being incorporated into the Internet of Things "things" do not have the automatic Haas and functional physical objects to communicate; intelligent performance characteristics automation and intelligent Network system.

2 MODERN URBAN RESIDENTIAL PROPERTY MANAGEMENT

With the acceleration of the urbanization process in China, the urban population is gradually increasing, along with the city's housing construction. There have been a number of columns in the Garden District residential and residential villas. In order to ensure the smooth progress of the work and events residential homes need to manage these residential quarters, which was set up and some property management companies. In this highly competitive market, the property company also faces enormous competitive pressure. Each residential area has its own property management companies, but these companies lack professional management staff, resulting in unreasonable of district management. In addition, the residential area is less than normal due to security personnel and security measures in place, to the district's security threat management market will appear the phenomenon of theft or damage to public facilities. And some residential quarters equipped with more security personnel, but the age of the security personnel is too large, poor wages, and it is difficult to retain people. If for Urban Residential property management involves a set of intelligent management system, simply hiring fewer people will be able to complete property management within the district, but also to achieve a real-time monitoring of the cell case, not only reduces the cell error the management of human input also improves the residential property management intelligence and security. Intelligent property management system not only to property companies brings great convenience, but also solves the property company to expand the scope of property and lack of manpower problems. Thus, in the era of things, for the residential property management with intelligent management model to bring greater convenience and efficiency.

3 BUILD INTELLIGENT PROPERTY MANAGEMENT MODEL

3.1 Make intelligent property management model

Throughout the modern property management there are still many shortcomings in the era of things, the popular forward the development of intelligent life, and therefore need to design a property management for intelligent management systems, to standardize on residential property Karma effort. Here, we analyze the current situation by residential property management, property management proposed intelligent model shown in Figure 1. This intelligent management model consists of a central control room, therefore, also need to establish a central control room to achieve the intelligent management of property.

This model is primarily established to identify the identity and residential property owners of vehicles, monitoring cell environment, security and surveillance as well as help the family can only terminal equipment for the foundation. This intelligent management model uses Wi-Fi, 2G / 3G networks to ensure the smooth transmission of system data.

3.2 Intelligent property management model subsystem

This intelligent property management model is composed of more than one subsystem, which is divided into four subsystems, namely the application layer, the platform server, network layer and the perception layer. First, the network system through the sensing layer face recognition device, the image, temperature and other data, and the network layer receives the read transfer to the server platform, the platform server to the data will be stored, and will need to be transmitted to the application device to the application layer of the layer. Here, we will analyze the various subsystems of this intelligent property management model.

3.2.1 Owners identification system model

This intelligent system is designed to identify the owners of residential security systems that are designed for a subsystem. The owners want to achieve recognition in the area of accessing ports need to install identification sensor to enter go to the owners of each facial image scanning, image scanning will be completed on the image information is transferred to the application platform, and with the information in the database for comparison, and relevant information is transmitted to the relevant staff, if other abnormalities in the cell is a cell can also access staff Recall the case of eleven investigation, to achieve cell theft.

3.2.2 Vehicle identification system model

This subsystem is mainly to facilitate cell vehicle management, and vehicle identification is achieved by the installation and the license plate image recognition sensor to be achieved by sweeping face recognition device and transmit information to the database for comparison, and then make a judgment on the vehicle and judge the results transmitted to the staff, it is easy to manage within a cell vehicles, but also save a lot of unnecessary disputes. In addition, through the vehicle identification to organize illegal vehicles into the area, to provide protection for the community's safety.

3.2.3 Community environmental monitoring recognition system models

Subsystem model has some difficulties in the establishment for this. This monitoring system needs to monitor objects, sound, temperature and quantity of data within a cell, so the sensor devices made high demands. This sensor device monitors the situation within the cell to the instrument and send timely relevant information to the intelligent terminal owners, the owners and businesses weather anomalies. Furthermore, in order to ensure the community recreation area sports equipment in good condition, also need to set up monitoring equipment It looks real-time monitoring.

4 CONCLUSION

With the introduction of smart buildings, having a huge impact of the new technology will have modern construction and development, networking technology applied to the construction of the security property management, environmental monitoring of residential property as well as the property management have a significant role. Property management model for the intelligent application of this technology in this paper has been established. By applying this model of intelligent management, and in improving the efficiency of the property company, while also reducing property company human input, increasing the economic benefits of the property of the company.

REFERENCES

[1] Janing Luo. The exploration of IOT technology and smart home remote control system Design[J]. Technological innovation and application, 2013 (03): 66.
[2] Jianjun Cheng.design of intelligent networking technology based property management system [J]. Silicon Valley, 2012 (06).
[3] Yanling Liu, the function of the intelligent building networking applications [J]. Power Information Technology, 2009, (10).

Information, Computer and Application Engineering – Liu, Sung & Yao (Eds)
© 2015 Taylor & Francis Group, London, ISBN 978-1-138-02717-6

Research on campus digital construction and application based on intelligently accessing teaching resources

Xiao Mao Hou, Ling Ma & Qun Hui Zhang

Hunan Institute of Information Technology, China

ABSTRACT: In recent years, with the rapid development of the economy, science and technology, the scale of the modern university gradually expanded, the construction of digital information has gradually become the primary developing direction of the colleges and universities. At the same time the construction of digital information on campus is main to realize the campus networkization, realize the office automation, intelligentization and integration and even realize the information publicity and then form an unimpeded information channel, and finally to achieve the teaching process intelligentized and integrated. But how to intelligently use the teaching resources is always the important focus point and a hard part in the current digital construction of colleges and universities. This article first analyses the necessity of intelligent accessing to the teaching resources based on the digital construction on campus, and then analyses the application of intelligent access to the teaching resources on the campus digital construction, lastly, analysis the relevant suggestions to strengthen the intelligent using of the teaching resources in the digital construction of campus when the author do this research.

KEYWORDS: intelligent; teaching resources; campus construction digital; application

1 INTRODUCTION

Today, in the twenty-first Century, along with the reform and opening-up policy, college is the important basic place to train the qualified personnel. According to the new requirements of diversified personnel training, now the digital construction is a kind a of systemic construction that integrates teaching, management, scientific research, life services and technical services. Through the construction of digital services platform to realize the integration of teaching management and teaching activities. And the intelligently using of the teaching resources is an important basic part of digital construction of the universities and colleges, so this paper has an important realistic meaning and economic value of the research on campus digital construction application based on the intelligently using of teaching resources.

2 THE NECESSITY OF CAMPUS DIGITAL CONSTRUCTION BASED ON THE INTELLIGENT USE OF THE TEACHING RESOURCES

With the rapid development of the information technology and the economy in this polynary era, along with the gradually expanding of the scale,the colleges and universities also pay more attention on the construction of informatization and intelligentization technology platform to realize the efficient and intelligent information management in colleges and universities. While the digital teaching resources as an important infrastructure in the current university must with the aid of an intelligent information platform to realize the complete new teaching.

The campus digital construction based on the intelligentization to use the teaching resources is not only conducive to the effective aggregation and sharing of the school resources, but also provides a good digital learning content for teaching, with the help of digital and intelligent information which done the personalized customization for the teaching resources to lay a solid foundation for all campuses to share the teaching resources.

In a word, the campus digital construction based on the intelligently accessing to the teaching resources not only protect the teaching knowledge property, at the same time have realized the modernize teaching resources sharing and organization in the process of integrating the teaching resources from different channels, it will also fully realize the complete sharing of teaching resources and knowledge, therefore the campus teaching resources digital construction based on the intelligently accessing to the teaching resource is a certain necessity.

3 APPLICATION OF INTELLIGENT PLATFORM TO ACCESS TEACHING RESOURCES IN THE CAMPUS DIGITAL CONSTRUCTION

During the intelligently accessing to teaching resources process in the campus digital construction, it should ensure that there is a reasonable teaching resource library, and provide necessary convenience for the classroom teaching, teachers electronically-preparing lessons and students extracurricular autonomously learning.

1 The basic principles of the intelligently accessing to teaching resources application in campus digital construction. During the intelligent using the teaching resources process of campus digital construction, first of all we must always adhere to the basic principles of education, do a reasonable and good design and configuration for all the text-books, and treat the students' learning motivation and improving the students' interest in learning as the basic starting point. At the same time also should adhere to the basic principles of teaching, with the aid of the network platform, comprehensively collecting relevant system resources, present the basic principle of teaching and zlearning, and hold on to the scientific principles, also adhere to the basic principles of the art, do a good job of teaching plot conception and flexible display screen.

The process of intelligently accessing to the teaching resources should always adhere to the basic principles of openness and rationality, and then make the teaching resource be rich and pertinent.

2 The basic content of the intelligently accessing to the teaching resources in the campus digital construction. During the process of intelligently accessing to the teaching resources in the campus digital construction. On the one hand,it is the public service platform for the teaching resources construction, and will realize the digitization, informatization and networking of campus.During the digital construction process in colleges and universities, through the combination of wireless campus network, to optimize the design of everyday teaching plan, and make the digital construction to keep abreast of the era trend of the information wireless technology development, ensure that the campus wireless network construction design and plan be reasonable. During the intelligently accessing to the teaching resources in the campus digital construction, should also ensure the service functions related to teaching resources can be taken via individual smart phones and tablet computers, and promote the convenience of

the intelligent using of the teaching resources, then construct the new standard of the teaching resources in universities and college.

In a word, during the process of using the teaching resources intelligently in the campus digital construction, not only should have an abundant material class teaching resource storehouse and excellent network course, but also should combine the digital library and related teaching website, provide substantive convenience for the intelligent tablet computers and mobile phones to access teaching resources.

3 The basic methods for intelligently using the teaching resources in the campus digital construction. As for the intelligently using of the teaching resources in the campus digital construction, the first thing is to use the online teaching resources reasonably, through the internet platform to download the teaching resources, and with the help of related software and plans to select the teaching resource, integrate the teaching resources. The second is to buy the good quality teaching resource on the network and some related application software, to improve the quality of teaching resources. Lastly, it will be necessary to do the research and development of the network teaching resources, and improve the application level of the school information technology and computer network technology, and construct the campus teaching resource using platform, ensure the service system of teaching resources to be more perfect.

4 RELEVANT THINKING ABOUT INTELLIGENTLY USING TEACHING RESOURCES IN THE CAMPUS DIGITAL CONSTRUCTION

During the process of intelligently using the teaching resources in the campus digital construction, should do the rational allocation of the teaching resources, and in the intelligent information technology development era, combined with the mobile network and the tablet computer to do the reasonable configuration of the teaching resources to realize the construction of campus digital teaching resource system.

First, we must ensure that there is a certain scientific, rationalization and standardization of the construction of digital campus network platform, and introduce the advanced information management methods, sharing teaching resources.About the teaching resources, should do good information management, performance management, curriculum management, teaching outline management, professional settings and teaching material management and so on, and finally realize the rational design and development of the current teaching resource library fundamentally.

Secondly, during the teaching resources downloading process, through the relevant information collecting and integrate processes, with the help of mobile network, the teachers and students can download the related software, and then select the most useful information among the acquired information. Do a good job of intelligently using teaching resources in the digital construction, and jointly safeguard the teaching resources service function system platform.

Lastly, do a good job of the teaching information resources integration of the intelligently using teaching resources in campus digital construction. Respect the related teaching resources property rights, construct the campus wireless network reasonably, optimize the everyday teaching design scheme, and pay much more attention on the foundation work of campus digital construction, combined with advanced wireless information technology, do a good job of the basic construction of the campus wireless to realize the new standards of the campus digital construction development.

5 CONCLUSION

In a word, with the rapid development of era economy, campus digital construction of intelligently access to the teaching resources should always adhere to the basic education, teaching and scientific principles, with the help of campus wireless network, optimize the everyday teaching design plan fundamentally, and with the help of the advanced modernize wireless information technology, do a rational planning for the construction of the feasible and economical campus wireless network to guarantee the service function of the current campus teaching resources has a certain degree of standardization and scientific, and then make it convenient for teachers and students to access to the teaching resource by personal smart phones and tablets.

ACKNOWLEDGMENT

Hunan Province Education Science and planning issues (Project No.XJK013CZY025).

REFERENCES

[1] Liu Peng:*Digital Construction of colleges and universities*[D]. Ocean University of China (2011).
[2] Huang Xin. *Analysis of the current situation of Hubei Province primary school digital application based on the comparative study* [D]. Huazhong Normal University(2014).
[3] Wu Xiaoyang. *Discussion on the teaching and service functions of the digital campus*[J]. Journal of education03, 63–65(2013).
[4] Gao Hongxia. *Find the existed problems from the utilization of natural resources respect in* campus digital construction [J]30, 7342–7343(2011).

Information, Computer and Application Engineering – Liu, Sung & Yao (Eds)
© *2015 Taylor & Francis Group, London, ISBN 978-1-138-02717-6*

The application of CAI in youth basketball skill training

Yan Long Li
Sports Department, Harbin Institute of Physical Education, Heilongjiang, China
College of Postgraduates, Wuhan Institute of Physical Education, Hubei, China

ABSTRACT: With the development and progress of computer technology and information technology, computer is used to teach in various fields and the youth basketball skill training also use the CAI technology. In this paper, the R & D process and its significances of CAI in Youth basketball skill training with computer technology conduct the comprehensive analysis. And in the end, put forward the principle of the CAI manufacture of youth basketball skill training.

KEYWORDS: Youth; Computer Assisted Instruction (CAI); basketball skill training

1 INTRODUCTION

In the 21th century, to develop education vigorously and to cultivate new-type talents with strong competitiveness, high quality has become the common concern all over the world. School is the base of cultivating talents, and its central work is teaching, therefore, teaching reform is the key to cultivating talent. With the help of modern learning theory, educational media theory, systematic science theory and modern teaching design theory, the mode of making multimedia CAI courseware for teaching technique in physical education courses was designed. And this paper mainly discusses the process of research and development of multimedia teaching courseware of juvenile basketball technical training and the role of CAI in juvenile basketball skill teaching and training. Hope that it can provide some recommendation and suggestion for the late researcher and teacher for the youth basketball teaching.

2 THE PROCESS OF RESEARCH AND DEVELOPMENT OF MULTIMEDIA TEACHING COURSEWARE OF JUVENILE BASKETBALL TECHNICAL TRAINING

Generally, the process of research and development of multimedia teaching courseware of juvenile basketball technical training is made up of Basketball technical training material preparation, the design of the CAI script of juvenile basketball skill training and the edition and process of CAI courseware of juvenile basketball training techniques. What's more, the diagram of the process of research and development of

multimedia teaching courseware of juvenile basketball technical training is shown in Fig1.

Basketball technical training material preparation. Youth basketball skill training includes footwork, shooting, dribbling, ball-passing, break, rebounding, defense, and other opponents. In the process of CAI courseware production, its basic job is acquiring, processing, screening, and sorting the basketball skills training material. And the material includes text, images, animations, movies, sounds, etc. With scanners, cameras, tape recorders, digital cameras and other tools, and combining with a lot of collecting material in basketball training practice, using the relevant software to process the raw material, will lay the foundation for post-production. The tools of collecting material in basketball training practice are shown in Fig. 2.

Design of CAI script of juvenile basketball skill training. Manuscript is directly basis of courseware design. Therefore, in the course of the courseware, teaching materials should not be copied to the courseware with no change. You must carefully analyze every technical training content, knowledge base of young people, knowledge difficulty and focus, how to inspire, etc. In addition, the design, organizational form of training content need the desired effect of making courseware with giving full play to imagination and creativity and descriptions in the form of graphics and text. For the characteristic of the teenagers' plasticity, intellectual curiosity, his the nervous system, form, functional development imperfection, we should select some little difficulty skill and less demanding action to practice to lay the foundation for future technical training, such as single-handed Shooting, road between underhand layup, hands chest pass the ball, etc.

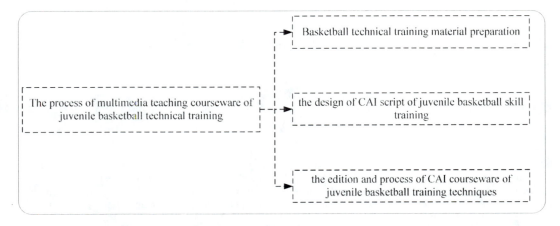

Figure 1.   The process of research and development of multimedia teaching courseware of juvenile basketball technical training.

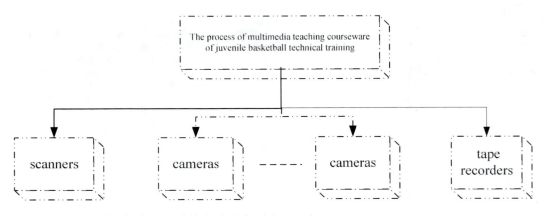

Figure 2.   The tools of collecting material in basketball training practice.

The edition and process of CAI courseware of juvenile basketball training techniques. According to the design of courseware manuscript, combine the material of basketball skill training in the order of pace action, ball-pass shoot, dribble, break, rebound and defend. Using FrontPage, Dreamweaver, Flash, PowerPoint software, etc. complete courseware page by page and ensure that multimedia courseware has the professional-quality. After the courseware manuscript of CAI of youth basketball, technical training is basically completed. Let the relevant professional teacher observe and conduct, and from the three aspects of training content, the training quality and software technology to evaluate and modify again, polish, error checking, error correction, and strive to

perfect courseware. The aspects of evaluating the CAI courseware are shown in Fig. 3.

3   THE ROLE OF CAI IN JUVENILE BASKETBALL SKILL TEACHING AND TRAINING

The role of CAI in juvenile basketball skill teaching and training can include three aspects: Help to stimulate students' learning enthusiasm and broaden their horizons, Help students establish the correct action concept and unify technical action norm and Help coach correct the wrong action by the method of contrast between right and wrong.

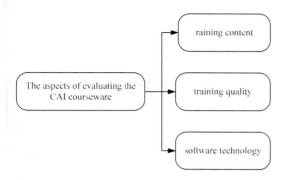

Figure 3. The aspects of evaluating the CAI courseware.

Help to stimulate students' learning enthusiasm and broaden their horizons. Teaching multimedia courseware combines video, audio, animation and other techniques, which can apply the multimedia technology to slowly demonstrate, action decomposition and action freeze for the difficult movements. And this is of great help of students' understanding. Because of their teaching methods by means lively and innovative and diverse, and overcoming the traditional basketball training mode, both in line with curiosity and novelty psychological characteristics of young people. What's more, create a better situation and emotional experience, cause and keep students' attention and interest, thereby can stimulate students' enthusiasm for learning, mobilize the enthusiasm of the students.

Help students establish the correct action concept and unify technical action norm. In basketball technical teaching, a lot of technical action is difficult to make the standard correctness, especially some difficulty and movement speed higher action. Students cannot be at the moment to accept and establish the correct action representation, so it is so difficult for students to grasp the correct action essentials from the explanations and demonstration to affect the training effect. However, CAI technology can help solve this contradiction; it can introduce advanced technology demonstration, specification of excellent basketball athletes of the world to students and through the slow and fixed the decomposition action to help students see action details, establish the movement imagery faster, more whole and more accurate which can shorten the process of teaching and improve the efficiency of teaching.

Help coach correct the wrong action by the method of contrast between right and wrong. In order to reduce and avoid the students the wrong actions in practice, at the same time in the demonstration of the correct action, basketball technology actions susceptible errors, causes and corrective methods should be noted. Let the students watch and think, avoid many common action error occurring. The students not only quickly master the action, and train the observation and analysis ability of the students.

4 CONCLUSIONS

With the help of modern learning theory, educational media theory, systematic science theory and modern teaching design theory, the mode of making multimedia CAI courseware for teaching technique in physical education courses was designed. In this paper, the research and development process and its significance of CAI in Youth basketball skill training with computer technology conduct the comprehensive analysis. All in all, through the research and analysis of the application of CAI in Youth basketball skill training, hope that there is more and more people design and research the CAI in Youth basketball skill training. And only in those ways, can our country, basketball is strong and develop and progress.

REFERENCES

[1] Wang Yongsheng. The Theory of the CAI. [J] Oxford: Peking University Press,. 1998 (1): 90–94.
[2] Qian Rulin, How to use CAI in Teaching [J]. Peking university, 1998(12): 32–33.
[3] Yang Wan. An Research of CAI on the gymnastics [J]. Peking Nomal University, 1999 (1): 62–69.
[4] Li Xuenong. The Design of Multimedia [M]. Guang Zhou University. 19(1–2), pp.67–74, 2008.

Information, Computer and Application Engineering – Liu, Sung & Yao (Eds)
© 2015 Taylor & Francis Group, London, ISBN 978-1-138-02717-6

The research of the distance education management system reform based on BPR theory

Zheng Ming Huang
Tianjin Open University, Tianjin, China

ABSTRACT: Distance education will improve the scientific and cultural quality of the people, the development of the economy and society, which has important theoretical and practical significance. This paper uses the method of theoretical analysis and empirical analysis, combining qualitative and quantitative analysis. It first introduces the characteristics of distance education, and the principle of doing some analysis, finally, puts forward several suggestions on the remote education according to the theory of BPR and.

KEYWORDS: BPR theory; distance education; management;

1 INTRODUCTION

The development of modern distance education management platform is a very important link in the process of modern distance education. It is the inevitable product of a certain scale of the modern distance education development, which is called the modern distance education management platform. It is the use of modern distance education processing knowledge and experience through the network. It makes the whole distance education domain knowledge and experience to get the communication, sharing and accessing, interoperability, maintainability and reusability of the data. The ultimate goal is to management platform for a breakthrough for the students to construct a good learning support service system.

2 THE CONNOTATION OF THE BPR THEORY[1-3]

Definition of BPR. According to the review, we can figure out that the connotation of BPR is a simple generalization, business process reengineering. It is the company's core business processes, which makes the fundamental thinking and completely rebuilt. The aim is to achieve a significant improvement in terms of cost, quality and speed, so that enterprises can maximize adaptation to "customer, the changes of competition" for the characteristics of the modern enterprise management environment.

BPR principle. In essence, the thought of BPR is with the materialist dialectical point of view, which is the establishment in the dialectic materialism thought on the system. The widespread contact stresses of things, the unity of opposites gives emphasis. So BPR

thought is a focus on long-term and overall situation, which highlight the development and cooperation reform philosophy. The principles are as follows:

1 The organizational structure should be based on the process at the center, not the task centric;
2 The information processing works into the actual work to generate the information;
3 The distributed resources are as a whole;
4 The parallel working together, not just to contact their output;
5 From the information source of one-time access to information.

3 THE CONNECTION BETWEEN BPR AND ORGANIZATIONAL CHANGE

3.1 *The necessity of business process reengineering of organization change*

The business process is contained in the organizational system. Therefore, even if the business process is with high efficiency, the whole organization, operation will not necessarily achieve high efficiency. Similarly, when the process in the operating cost is low, the whole operation of the organization efficiency does not necessarily improve.

The traditional organizational structure is built on the basis of the division of labor. The functional organization structure only focuses on the combination, but ignores the integration. The organization of internal door wall has the member seriously, which is accustomed to the use of other, narrow department goals instead of the whole organization operation target.

3.2 Promoting effect of BPR on organizational change

According to Adam Smith's theory of division of labor, we establish now most organizational structure. Therefore, according to the theory, we will firstly according to the functions, to distinguish between different independent departments. Then, in every major sector, it will again sub department until the final partition to individuals or groups. Each employee is responsible only for their own work and the scope of the duty is rather narrow. This form of organization in fact is the realization of the function of local maximum efficiency, rather than maximizing the overall flow of the organization efficiency.

Business process reengineering transformation of the organization is through the organization change, the process of reintegration. Through the optimization of the process, it will achieve the optimization of the organization's overall. Therefore, the effect of the business process re-engineering is not only manifested in the optimization of the organizational processes, just as the Fig1.

Feature	Before the reform of organization	After the reform of organization
Organization structure	Linear function	Linear function
Hierarchical status	Hierarchy more	The lower tier, Bain Pinghua
Evaluation	Objective to function as the judgment standard	In the process of target evaluation standard
Communicate	Vertical communication	Horizontal communication

Figure 1. Business process reengineering changes before and after the organization.

3.3 Pattern of modern distance education school

Modern distance education has a specific educational information transmission and communication means, which have to adapt distance education information resources. They have an educational management system and specific education management mechanism, etc. In our country, it is approved by the Ministry of education to carry out network education pilot institutions. In addition to the central RTVU, it is outside a distance education college. The other institutions are the traditional sense of the ordinary colleges and universities. In addition to technical knowledge network, multimedia, these colleges are for distance education, practical experience, which is not a lot. Using the mode, it will run into what kind of distance education. It is essential for these institutions at present. We already have seen the following five main clue running mode, just as the Fig. 2.

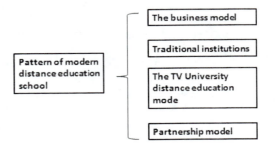

Figure 2. Pattern of modern distance education school.

3.4 The business model[2]

Taking Renmin University of China as a representative, University of China and Wantong Group cooperation will set up the East network company of distance education. In addition to the school policy, teaching plan, teaching is decided by the National People's Congress college enrollment, management network, and others are by the operation of the company. The advantage of this model is high efficiency, fast action, which occupies the market ability. It is conducive to innovation of distance education, especially conducive to the cultivation of individual learning mode. The disadvantage is a more commercial view of education.

3.5 Traditional institutions

To Beijing Normal University and other institutions, it is regarded as the representative. The school has the division of adult education, continuing education. The Department is in charge of education, director of the center for computer network courseware and network transmission. The advantage of this model is the staff familiar with the rules of education, experience, teaching implement. The problem is this mode of market consciousness, which is not strong, and the efficiency is not high, especially in the traditional mode of distance education institutions.

3.6 The TV University distance education mode

Association with the Central Radio and TV University, 44 provincial TV University is regarded as representing throughout the urban and rural areas. It is the source of students' ability and the teaching support service.

3.7 Partnership model

The combination of colleges and other educational institutions have mutual recognition of credits. Strong market in September, the establishment of the teacher education alliance aims to a partnership model, which has occupied the teacher education.

3.8 The implementation of modern distance education

The development of the modern distance education is supported by a complete set of network education platform. The set of platform will carry out all functions of modern distance education required, which can cover the distance education teaching, learning, management and public services in all aspects. The equivalent of a school will build on the network. Usually, the set of support platform is divided into the portal, learning platform, teachers' platform, and management platform. Management platform is in the center position of six on the platform, which is several other platforms to coordinate the operation foundation to realize the share of data among each other, data exchange guarantee[4].

3.9 Portal website

The service object includes the social public, teachers, students, managers and each substation management personnel. It is the Network Education Institute of foreign propaganda window, which is the bridge school contact the society. It is also the student teacher to enter the other platform portal.

3.10 Learning platform

The service object is mainly participated in distance education teaching and learning of the students and teachers. The learning platform is the support platform on the network environment of students' learning and teachers' teaching. It is the main tool to realize the network education.

3.11 The teachers' preparation platform

Its service object is the teacher and the resource providers, the assistant tool for teachers to prepare lessons. Its main functions include: graphic symbols required mathematical symbols. The keyboard is difficult to input the physical or chemical experiments, which provide some convenient conditions in the process of planning;

3.12 Management platform

The use of modern distance education field of technical means has digital processing knowledge and experience, and through the network makes the whole distance education domain knowledge and experience to get the communication, sharing and access.

4 CONCLUSION

This platform is a kind of through the network of quality management activities. It can not only collect, processing, producing process organization in Distance Education of all kinds of information, but also be able to dig out more knowledge from this information. In order for the teaching platform, it provides decision information, strengthen the service management platform. Summarizing the existing and looking to the future, the development direction of the platform will be smart.

REFERENCES

[1] Li Kang: To explore the several basic concepts in the theory of distance education, distance education Chinese, 2003.15.
[2] Liu Kai, Distance education call for service", distance education Chinese, 2003.16.
[3] Zhao Zhongjian, Gu Ji, School management system and IS09000 standard", East China Normal University press 2003.1, P18.
[4] Zhu Zhiting, Study of education network, audio-visual education research 2001.8.

Information, Computer and Application Engineering – Liu, Sung & Yao (Eds)
© *2015 Taylor & Francis Group, London, ISBN 978-1-138-02717-6*

Research on English test reliability and validity assisted by computer technology

Hai Ying Jin
Teachers' College, Beihua University, Jilin, Jilin, China

ABSTRACT: Computer technology has been increasingly used in language test. Communicative language teaching has been the principle and theoretical basis of language test assisted by the computer instead of Audio-lingual method. The passage aims at exploring training examiners before exams and motoring scoring quality on the basis of communicative language teaching to make sure of the reliability and validity assisted by computer technology.

KEYWORDS: Computer technology; Language test; Reliability; Validity

1 INTRODUCTION

With the development of computer technology, using computer technology in language test has become the trend of domestic and foreign language test. Many experts are working on many ways to make full use of computer multimedia technology to further improve the test's authenticity and types of diversity. They have developed a computerized test system that has the adaptive ability. TOEFL has been changed into a new type of exam based on Internet from the traditional paper exams since 2005. As one of the key tasks of English teaching reform in our universities, CET 4 and CET 6 will be gradually changed into computer-controlled exams from man-controlled exams. The reason for this is that the computer and Internet technology has its unique advantages in language test. Only with the help of modern technology, especially the computer and network technology is language test likely to break away from the traditional exams and really make language scientific and humane (Brown1997). Computerized language test also attracted many domestic scholars' attention. (Such as Yang Huizhong 2001; Zhang Baojun, 2003). In the pen based language testing and research of computer language equivalent test based on Li Qinghua in his article on the related research abroad is introduced; In the equivalent language test based on pen and paper and computer technology, Li Qinghua made some introductions to foreign relevant research. In the equivalent study of oral exams based on computer and oral exams based on interviewing, Cai Jigang made an experiment and demonstration. He found that 96.9% of the candidates obtained the same level or only half class between them in both exams. At the same time, through the

analysis and comparative study, Cai Jigang found that the computer test is more than CET4 and CET6 in both validity and reliability. The examination cost is also largely reduced. I will discuss the principles the theories of computer assistance in language testing and make some constructive suggestions in English test validity and reliability combined with the training of examiners and marking and monitoring.

2 THEORETICAL BASIS OF COMPUTER ASSISTANCE IN ENGLISH TEST

Theoretical basis of computer assistance in English test is to design the test system. Ideology and theoretical basis for computer aided English test in order to design computer assisted English testing system, improve its reliability and validity, we must have a basic understanding of theoretical basis of computer assistance in English test in order to improve its validity and reliability. Structural linguistics was dominating over the field of language testing in the last century 70's. Audio-lingual Method became the dominant language testing theory, which was popular in the fifties and sixties and has had a wide impact on world foreign teaching. This stage of the language testing stresses spoken language, but ignores the language situation. In order to examine the structure, test questions have obvious traces of artificial fabrication. When the world entered the last century 90's, the rapid development of language teaching and language testing theory and practice made the disadvantages of language testing gradually appear. Many experts of domestic and foreign universities and employers found that the non-English students who got high scores in the conventional English tests had a very low level in their speaking and writing communication skills. In view of this, American educational testing service at the end of the last century suggested that the English test system based on structural linguistics should be changed into a new type of test based on communicative competence. This idea is designed on the basis of Communicative Language Teaching. The aim is to test the candidates' practical communicative competence. Communicative Language Teaching was put forward by the British linguist in the late 60's and early 70's in the last century. This new teaching method focuses on communicative competence not on language structure, greatly improving the efficiency of language teaching and learning, which was widely praised. The main characteristics of communicative language testing is: focus on the meaning; context; acceptable purpose with language activities; meaningful language; the use of authentic language materials; text processing authenticity; unpredictability in reaction results; interaction as the basis; language ability in the real psychological state by the candidates ; scoring according to the actual communication results (Weir, 1993; Wu Zunmin, 2002).

We can say that grasping the spirit of the communicative language teaching is grasping the essence of modern English test assisted by computer technology. Because the communicative language teaching is the soul of the modern English testing system, using the Internet and the computer technology is the general trend in language test. Take the new TOEFL test based on the Internet environment t as an example. The general structure is the crossover design of listening ,speaking ,reading and writing, namely, speaking after listening, writing after listening, speaking after reading and writing after reading. This scientific scoring system owes to the Communicative Language Teaching.

3 CONSIDERATIONS OF IMPROVING THE ENGLISH TEST VALIDITY AND VALIABILITY BY COMPUTER TECHNOLOGY

The reliability of a test refers to the reliable degree of test results. The level of language teaching reliability mainly explains to what extent test results and reflects candidates' real language behaviors. The validity of a test is the extent to which it measures what it is supposed to measure. As we have seen, validity and reliability constitute the two chief criteria for evaluating language test. There are many factors affecting reliability and validity. They exist in every link of language test. Training examiners before exams is one of the most effective ways to ensure reliability and validity with the help of computers. Examination is a kind of educational measurement.

Not only does measurement have a standardized measuring instrument but also the examiners using this instrument are consistent in standards. For computer assistance in language testing, the examiners must have the same view of the measured language ability, have a common understanding of the evaluation standard, and must be consistent in score standards. So training examiners is very necessary. Training examiners includes two aspects: the first is about score criteria and the second is about choosing the typical sample questions for the test evaluation in order to improve the reliability of scoring. In the sample evaluation, we should refer to the expert evaluation, find out the key differences with the expert score (such as emphasis, scoring basis), and not just the score difference. At the same time, retrain examiners before every exam to keep score consistency between examiners to the maximum.

In the actual marking process, there are many factors affecting examiners, which include the ability to master correct marking standards, time tension, family and work pressure and so on. Artificial marking part of English test aided by computer, especially spoken English test score part, is usually separated from the candidates. They can't be face-to-face. Therefore, it is particularly necessary to strengthen supervision and scoring reliability monitoring in the marking process, especially the oral exam score part. In the marking process supervision, we can set the audio length of records in the machine scoring procedure. After marking, the man in charge checks audio length of records to ensure that each student has the same length of records. At the same time install cameras on the computers used by the examiners to make sure that the organizers can monitor the examiners' state at any time. The actual marking can be used in centralized marking or dispersed marking. The examiners won't leave any mark on the screen. The examinee scores are registered on another paper in order to do random sampling volumes that have been read. Sampling volumes should normally be reached 10% ~ 20%, which are reviewed by other examiners. If the examiner reading time can be ensured, in the scoring process we can also arrange two examiners to give the same candidate scores. Compare two examiners' scores at the end of the marking and if the difference between two scores is greater than a certain value, the third examiner gives the candidate scores. In the end the average scores given by these three examiners are the final marks of this candidate. The above several approach results can use the statistical method to examine the scoring reliability. Usually a correlation coefficient is requested to be greater than 0.80 and then it is qualified. In addition, establish archives, eliminate unqualified examiners, and gradually establish a team of qualified examiners.

4 CONCLUSION

In the face of domestic and foreign language testing computer trend, in order to design computer assisted English testing system, improve its reliability and validity, designers and managers need to understand and grasp the idea and theoretical basis of modern language testing, and also need to do more research and exploration in the training of examiners and computer aided English test papers quality monitoring. The problems and difficulties that happen in the unified examinations should be solved as soon as possible or given the constructive suggestions in order to further promote the College Specialized English test. Only ensuring the validity and reliability testing can reasonably reflects the true English level of students, and really promotes the development of English teaching.

REFERENCES

Tao Baiqiang, 2004, Development of world language testing theory [J] Basic English education, (9):29–30.

Cai Jigang, 2005, Study of reliability and validity in CET4 and CET6 T [J]. Foreign language world, (4):96–97.

Li Lin, 2008, Research on reliability and validity of university computer English oral test [J]. China modern education equipment, (2).80–81.

Michael Russell, Amic Goldberg, Kathleen O'connor, 2003, Computer-based Testing and Validity: a look back into the future [J]. Assessment in Education, (03)20–22.

Information, Computer and Application Engineering – Liu, Sung & Yao (Eds)
© 2015 Taylor & Francis Group, London, ISBN 978-1-138-02717-6

Research on the design of the art classroom teaching aided by computer technology

Chang Sheng Liu
Teachers' College, Beihua University, Jilin, Jilin, China

ABSTRACT: With the rapid development of computer technology, computer has been more and more widely applied in art teaching. With the coming of the twenty-first Century, computer technology has become the main means of teaching arts and transmitting teaching information. In the art classroom teaching, making full use of computer is an effective way to develop art teaching creativity, improve students' learning initiative and optimize the classroom teaching. It is also an innovation of classroom teaching methods, and it is favorable for teachers to update teachers' knowledge and improve teaching contents.

KEYWORDS: computer, optimize, art teaching, teaching ideas, harmonious development

1 UPDATING TEACHERS' AND STUDENTS' LEARNING IDEA

Art teaching is a modern information, communication activity of teaching and learning. Qualified teaching is a process of students' gaining experience, generating capacity and change under the effective guidance of teaching aim. It requires the teachers to arouse students' learning motivation, to tell students what to learn and to what extent and enable students to consciously participate in learning activities. Teachers' teaching ideas updating optimizes classroom teaching, which is as follows:

1.1 *Changing teaching methods*

The main content of the art curriculum reform is curriculum implementation, while the basic way of curriculum implementation is teaching. So the teaching reform is the proper meaning of the curriculum reform. The key to the teaching reform is the updating of teaching ideas. If the teaching ideas are not updated and teaching methods are not changed, curriculum reform will be a mere form, and even get half the results with double the effort.

The previous art teaching activities are teacher-centered, which is one-way information activities. Using the computer as a link between teachers and students, it forms a computer- teacher – student process, which is a two-way communication and exchange of knowledge and information. Teachers are no longer the inculcators of knowledge but become computer controllers and organizers. Teaching activities change the one-way transmission of information into a two-way information exchange.

The previous art teaching relies too much on the teachers' words and textbook information. The students' subjective initiative can't be fully reflected, and they lack awareness of innovation. Making full use of computer technology makes students more creative. All this, truly reflects the student-centered principle.

1.2 *Improving teaching contents*

In the traditional teaching concept system, "Curriculum" is understood as the standardized teaching the teachers have no right to change it, nor do they think about it. So the vitality and initiative of the teachers and students can't be fully developed. The use of computer teaching makes the art curriculum change from closed to open. The curriculum is no longer just a specific carrier of knowledge, but the process of the teachers and students exploring new knowledge together. The teachers and students have together become creators of the art curriculum contents.

In the art teaching, full use of the computer creates the environment for students appreciating. It provides the effective carrier for the students' accepting new knowledge, strengthening the inner experience, enriching the spiritual world, forming a positive learning attitude and setting up right values. Through the computer which is an effective medium, art teaching contents can be also improved again.

1.3 *Improving the teaching evaluation*

Computer technology provides text, graphic, sound, and video for teaching. The teachers encourage the students to turn what they sense and feel into their own words and express their views. At the same time, the teachers challenge students to debate, and have different opinions. The teachers should praise, but not criticize whether they are right or not, so that the students feel the joy of success and have more self-confidence.

In the art teaching evaluation, we should combine teachers' evaluation, students' self-evaluation and mutual evaluation organically, timely understand the needs of students in the development, find the potential of students, and timely give guidance. At the same time they give their judgments and evaluation of the students' language expression ability, participation consciousness, the spirit of cooperation and cognitive level, etc. All this creates favorable conditions for art education teachers establishing a scientific evaluation concept.

2 PROMOTING THE HARMONIOUS DEVELOPMENT BETWEEN TEACHERS AND STUDENTS

2.1 Creating learning environment

Art is visual. The computer is visual, too. They have the same characteristics. Creating a specific situation for the student with the help of digital media makes the contents more intuitive and vivid and attract students' a variety of sensory organs. The students better enjoy the art atmosphere and have the rich imagination and inspiration. The use of emotional background music, clear projection, color pictures and text, will display the contents intuitively and wonderfully. For example, in the appreciation of Qi Baishi's art works, because he has many representative works, if we teach in a traditional way, we can't have a full understanding of his works., nor does time permit. Instead, if we use media, that is to say, we put the good works of image text descriptions and diagram analysis into a courseware before class, we will have enough time both to fully appreciate and to focus on analysis. In this way, the students can easily appreciate the works, which also broaden their eyes and provide the students' opportunity to understand beauty, appreciate beauty. This kind of learning environment fully reflects the life of art teaching.

2.2 Developing individual learning quality

Previous learning was based on human's object and passivity. It ignores human initiative and independence. We need to change the passive way of learning, respect the individual way of learning, and let students become the master of learning, cultivate good individual psychological quality.

Using computer technology in art education and teaching provides advantages for students to form a good personality and psychological quality. It is easy for teachers to find the students' individuality, encourage students' imagination and enhance their self-confidence. The teachers let the students play the games in their opinions so that students have psychological excitement. They carry on the innovation according to the purpose and methods of their own. Gradually they form the positive emotion attitude and develop their own personality.

In art teaching, we use the computer to select students interesting image, well design the teaching methods, inspire students' active thinking, and develop their unique thinking. Through the form of "one question but different asking or one asking but different answers" and other forms, we train the students' creative thinking ability, enhance the flexibility and innovation of appreciation. This can fully express their self-confidence and positive attitude. With modern media as the carrier, students develop good personality and psychological quality in the art teaching.

2.3 Exploring creative spirit

The current social development needs to have innovation talents. Innovation ability training can be showed from the creative learning. Art teaching is one of the most effective disciplines to develop innovative spirit and creative ability. It can effectively cultivate students' ability of thinking in images and promote the creative activities.

Using computer technology in art teaching, the teachers guide students to observe, create a favorable learning situation, arouse students' learning enthusiasm, and arouse the students' positive thinking and creative desire, producing a kind of internal driving force in the excitement and fun. Through the computer, the teachers combine intuitive teaching principles and the teaching environment and let the students' eyes and brain develop coordinately. The teachers make the students lost in the individual knowledge system easily and encourage students to take part in creative thinking activities, and gradually master the thinking methods. In the teaching of "painting—morning glory", the teacher shows the students some paintings such as morning glory, calabash, wisteria, and Chinese trumpet creeper with the help of the computer and together with background music. Due to the ups and downs of each picture, the students' feelings are changing to the music. The students understand the beauty and feel the beauty of the fresh situation so that they make the active judgments and develop their creativity.

3 CONCLUSION

Art teaching should pay special attention to guide the form of creativity. Teachers should take the initiative to create a conductive learning atmosphere and inspire the students to express the original opinions and carry on the research learning in the art education activities. Art teaching can make people know the uniqueness of many things, the life value a lot of things. Art works have the original, distinctive personality, which is the essential difference between art works and others. The difference itself is also a kind of beauty of personality. Making full use of computer technology in the art teaching updates the teachers' teaching ideas, promotes the students' harmonious development and fully achieves the teaching aims of art teaching.

REFERENCES

Zhong Qiquan, 2001 August First Edition, the outline of basic education curriculum reform (Trial) interpretation, East China Normal University press.

Education ministry of the people's Republic of China, Full-time compulsory education art course standard (experiment draft), Beijing Normal University press, 2001 July 1.

Pi Liansheng, 2000, teaching design ——theory and technology of psychology, higher education press, 21–22.

Information, Computer and Application Engineering – Liu, Sung & Yao (Eds)
© 2015 Taylor & Francis Group, London, ISBN 978-1-138-02717-6

Lean production operator loop design

Guo Feng Yang

School of Automotive Studies, Tongji University, Shanghai, China

ABSTRACT: This paper discusses the design of production line operator loop using lean thinking principle. In comparison to traditional production line, a lean production line offers more flexibility in regard of cycle time, hourly output and operator usage. Consequently, design of a lean production line also requires a comprehensive concept including layout planning and operator loops for different usage level. This paper provides a concept to gradually generate well balanced operator loops to meet different output targets with maximized productivity efficiency.

KEYWORDS: Lean Production System, Operator Loop Design, Production Layout

1 INTRODUCTION

Lean manufacturing or lean production was firstly introduced in Jim Womack, Daniel Jones and Daniel Roos' book, The Machine that changed the world, in 1991. Lean production is the systematic elimination of waste. As the name implies, lean is focused at cutting "fat" from production activities. But more importantly, it is not only financially and physically leaner; it is emotionally much leaner than non-Lean facilities. People work with a greater confidence, with greater ease, and with greater peace than the typical chaotic, reactionary manufacturing facility.

In order to realize a lean production, it is essential to have a suitable production line concept. The lean production line concept usually has 3 characteristics: semi-automated machines with manual work content of loading/unloading; scalable output with variable operator input; balanced operator workload. Further, a concept is needed to gradually generate well balanced operator loops to meet different output targets with maximized productivity efficiency.

2 CHARACTERISTICS OF A LEAN PRODUCTION LINE

2.1 Semi-automated machines with manual work content of loading/unloading

In practice, a lean production line must not be highly automated, nor should all the processes and stations be fully integrated. It is very important that the workstations can be made in compact and independent units, which can be easily attached or removed. In this way, investment for high automation can be saved and in the meantime, maximal flexibility can be kept.

Therefore, semi-automated machines with manual load/unload process steps are often considered as a precondition for lean production line because of the best combination of medium cost, high process quality and high flexibility.

2.2 Scalable output with variable operator input

In a real customer pull production system, one basic requirement is to meet the different output targets day by day, even hour by hour. A dedicated production line with fixed tact time will not fulfill this requirement

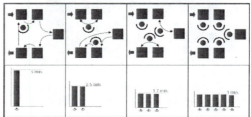

Figure 1. Scalable output with variable operator input.

efficiently. A lean production line concept must be able to offer several standard scenarios for low/medium/high output target with low/medium/high operator inputs.

2.3 Balanced operator workload

Operator's workload has to be well designed and balanced, in order to maintain the productivity at a high level. Mike Rother proposed a natural cycle time range between 20 to 120 sec., where most of the operators feel less tired and more concentrated during their 8 hours work. It is the idea target cycle time. When the cycle time is less than 10 sec., it is considered to be very unsuitable for manual work and need to be avoided.

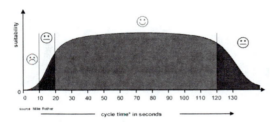

Figure 2.   Cycle time and the adaptability of operators.

3   LEAN PRODUCTION LINE OPERATOR LOOP DESIGN

The concept of lean production line operator loop design includes 3 steps.

Step 1: Design of single operator working loop

Single operator working loop is the simplest form of operator loop concept. It starts with the assumption, that only one operator works through the entire production line. He starts with the 1st. work process and carried the product to the next. The initial state of all workstations is "previous product finished, ready to be taken out". Meanwhile, all periodic works such as refill of raw material, sampling quality control etc. are to be left out of consideration at this stage.

All manual time the single operator spends on each work station to produce one product will be summed up and visualized on a stack diagram. The result is to be considered as the target cycle time for one operator loop. This accordingly, the operator hourly productivity can be calculated and set as benchmark for multiple operator working loops.

Step 2: Design of multiple operator working loops

Design of multiple operator working loops is based on the result of single operator working loop. Ideally, CTn (cycle time of multiple operators) is to be CTs (cycle time of single operator) divided by n (number

of the multiple operators). In this way, the operator hourly productivity can be kept at the same level. Bit in reality, CTn can't be simply set as CTs divided by n due to many constraints. One major technical constraint is the machine time of bottleneck process. It is the lower limit of production line's cycle time. Knowing the machine time of bottleneck process, the upper limit of n can be calculated. Above the the upper limit of n, additional operators will not be able to produce more output, more than n operators will be meaningless.

Once the upper limit of n is obtained, the next task is to find out the suitable combination of work content for multiple operators from 2 to n-1. Different from single operator concept, multiple operators concept must define the handover process of products between operators. The most common handover process is to identify a fixed handover point between workstations, where one operator always put his part down, and this part will be picked up by another defined operator afterwards. Handover process has to be very carefully designed and tested under all possible conditions, so that related quality risks can be minimized.

Step 3: Trials and confirmation of designed operator working loops

Complexity of the multiple operators working loops concept increases rapidly with increasing number of working stations and operators. It becomes difficult to design the complete operators working loops concept just based on theoretical calculation and estimation. Trials and simulations are necessary. Often, the physical production line has not been ready at this stage due to timing issues. The best practice to solve this conflict is to build a "mock-up" line for testing. Similar to product design, engineers need to ensure the production line design will meet the expectation, before it starts to be build. A real size 1:1 prototype of production line is needed to test and verify the ergonomics, line layouts and operator working loops. Trials should be conducted by both engineers and operators together, so that potential

Figure 3.   "Mock-up" line example.

improvements can be resolved right before start of production.

4 CONCLUSION

This paper has attempted to describe the characteristics of a lean production line and to explain the methodology of lean production operator loop design concept. A suitable lean production concept and operator loop design enhances and encourages the ownership and motivation of line operators. Hence, it sets the foot stone of a living lean production system.

REFERENCES

Womack, James P. & Jones, Daniel T. & Roos, Daniel, 1991. *The Machine That Changed The World*, New York: Harper Collins.
Womack, James P. & Jones, Daniel T., 2007. *Lean Solutions, How Companies and Customers Can Create Value and Wealth Together*, New York: Simon & Schuster.
Wilson, Lonnie, 2009, *How to Implement Lean Manufacturing*, New York: McGraw Hill Professional.
Rother, Mike & Harris, Rick, 2001, *Creating Continuous Flow*, Cambridge, MA: Lean Enterprise Institute.

Information, Computer and Application Engineering – Liu, Sung & Yao (Eds)
© *2015 Taylor & Francis Group, London, ISBN 978-1-138-02717-6*

Explorations on intelligent and interactive multiscreen integration design

Hao Li
Straits Institute, Minjiang University, Fuzhou, Fujian, China

Jie Yang
Information Engineering College, Wuhan University of Technology, WuHan, HuBei, China

Jun An Di
Department of Electronic Information Engineering, Minjiang University, Fuzhou, Fujian, China

ABSTRACT: This paper introduces the interactive technology and equipment management technology of the multiscreen business and builds a set of intelligent and interactive multiscreen solution. It uses the interactive multimedia document to set a unified orderly mode to enable the control order set by the users on the video collection equipment to be compatible with the OMA standards.

KEYWORDS: Interactive technology, intelligent application, mobile application

1 INTRODUCTION

There are indications that the business of the multiscreen (mobile phone screen, computer screen, IPTV screen) will be the next big business trend of the wired / wireless broadband. All members in the industrial chain, including the traditional SP (traditional carriers and full service carriers) and non-traditional SP (Google, Yahoo, Microsoft, Apple), device makers and content companies have already taken positive action to make product definition and planning to seize this opportunity. In North America, AT&T is working to promote the business of multiscreen. Multiscreen has been recognized by North American operators, as the business opportunity that can bring billions of dollars to the SP. As for a dazzling multiscreen experience based on a true connection, many new businesses can also be brought in to provide consumers with better service, and thus bring more business opportunities.

What needs to focus on is that the current multimedia business has a growing emphasis on interaction, and user participation. As for the multiscreen services which have crossed three kinds of heterogeneous networks (mobile communication network, wired Ethernet, Broadcast and TV Website/telecom IPTV network), and involved three kinds of terminals, interactivity has played a more prominent role in multiscreen business.

Therefore, we must give a set of programs that have crossed three kinds of heterogeneous networks to make interaction. In this way, users are enabled to enjoy the dazzling multiscreen business.

2 INTERACTIVE TECHNOLOGIES AND EQUIPMENT MANAGEMENT TECHNOLOGY IN OMA STANDARDS

As we all know, the Global System for Mobile communication is in its rapid development. China's mobile communication network has become No.1 in the world in its network size and the user scale. In order to improve the quality of mobile communication services, to meet people's needs, and to solve the interoperability problem between different systems, the urgent task is to establish uniform standards and specifications that can be used globally, safe and reliable. In addition, the standards and specifications should facilitate the operation of end-to-end mobile communications. A global standardization organization OMA is produced to meet the need of this new trend.

OMA was established on June 12th, 2002. Its purpose is to seek worldwide interoperability standards that are open and unrelated to the system, on which terminals of various applications and services can be realized. Since its establishment, the world's leading mobile operators, device manufacturers, and content providers have joined. Dozen standardization organizations have also joined the OMA or established a cooperative relationship with OMA. The output value of OMA members accounts for 90% of the output of the entire mobile communications. As members of the OMA unit cover the world's leading mobile operators, device and network providers, IT companies, application development agencies and content providers, all links in the entire industry value chain are able to cooperate together. In this way, seamless mobile services are ensured to be provided to the end-users around the world.

2.1 Interactive technologies in OMA standards

Interactive services provide interactive application information related to mobile TV programs. Users can acquire the background and introductory information, material and data that are relevant to the program content to participate in the voting, quizzes and activities through interactive services.

OMA interactive technology is a powerful interactive technology with better compatibility. It can support heterogeneous network (mobile communication network, wired Ethernet, Broadcast and TV Network/ Telecom IPTV network, etc.) and a variety of equipments. Therefore, it is especially suitable for bright multiscreen services across heterogeneous networks.

When the users are viewing the contents of the multimedia services, if the business provides interactive features and users want to engage in the interaction, users are required to acquire the interactivity contents (the prompted interface of users' participation in interactive services, for example, the draft voting list and the team symbol of quiz results of the basketball game) from the interactive application servers through such ways as the WAP, etc. Users can take part in the interaction according to the relevant interface message (click on the link and send text messages, etc.) to achieve the final interaction.

Interactivity Media Document has contained such levels as Media Object Group and Media Object Set levels. An Interactivity Media Document can contain a number of Media Object Groups. Corresponding to multiple interfaces of the interactive program content (for example, in terms of 3 groups of problems of

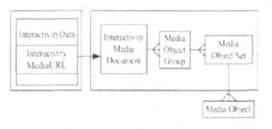

Figure 1. unit architecture of interactivity data regulated by OMA.

the quiz, there is an interface for each problem, and each interface information is stored in a Media Object Group) a Media Object Group contains a number of Media Object Sets that are corresponding to different

terminal selections (such as the terminal's ability etc.). The user terminal can obtain the URL, the interactivity content corresponding to the programs, (such as an interactive URL server address) from the Interactivity Data. When the users are watching the programs, they can click the interactive options and the terminal can get the interactive prompts and participation way via the WAP access.

2.2 Equipment management technology in OMA standards

OMA DM (Device Manage) is a set of management protocol specifically for mobile and wireless network, which is defined by the OMA organization. It is a kind of application of the OMA protocol. As OMADM agreement is clearly independent of the bearer network, so OMADM applications are built on top of various networks. In other words, OMADM commands can be transferred through a variety of network protocols.

We can see from Figure 2 that the underlying bearer network of OMA protocol has crossed GSM, CDMA, and WiMAX networks, which almost covers all of today's mobile transmission technologies.

The realization carriers of the OMA DM protocol contain the two types: TCP/IP and WAP (Wireless

TCP/IP		WAP	OSI
	OMA DM TREE		Application
	syncML		Presentation
OMA DM Protocol	HTTP	WSP	Session
	SSL	WTLS	Security
	TCP/UDP	WDP	Transport
	IP		Network
	2G 3G WiMAX Wi-Fi Bluetooth ...		Data Link
			Physical

Figure 2. OMA device management (DM) protocol framework picture.

Application Protocol). WAP acts not only as a realization carrier of OMADM agreement; more importantly, the original intent of this agreement is to establish a bridge between the Internet and wireless devices and define the communication way between the wireless mobile devices and the fixed network servers in the network. It establishes a network framework through the fixed Web server and wireless mobile devices for the OMADM agreement.

It can be seen from Figure 2 that the each device that supports OMADM must contain a Management Tree. The Tree organizes all management objects (MO) on the device to be a hierarchical tree structure.

Each node of the tree can use URI to be individually addressable. Figure 3 shows a management tree example:

The DM server can get the tree structure through Get command to access node. If the access is the internal node, the returned result is the name list of its child nodes. If the access is a leaf node, the value of the node is thus returned.

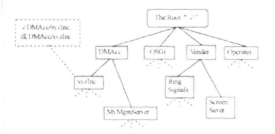

Figure 3. OMA device equipment tree organizational picture.

DM tree can be extended at runtime, for example, through the Add or Replace command, the new internal node or leaf node can be created. The device itself can be also extended to be the management tree (when the user inputs or attaches certain accessories to the device). In addition, the DM server will be responsible for maintaining the location information of nodes in the Management Tree.

Through the use of the OMA Device Management technical standards, the business of multiscreen TV (including set-top boxes), the PC, and the mobile phones can become a set of organic whole.

It is convenient for the operators to manage, and also gives users a better feeling.

3 INTELLIGENCE BASED ON OMA INTERACTION AND EQUIPMENT MANAGEMENT STANDARDS

Multiscreen interactive services figure 4 is a typical intelligent multiscreen interactive service network architecture diagram. Among it, multiscreen users (who both have mobile phones, PCs, set-top box and TV) can watch the real-time multimedia video shot by cameras, while they can also control the camera (including head) movement through interactive server. For example, control rights over the Panda camera for a period can be obtained through the way of auction. In addition, the Panda camera movement way can be determined collectively through the way of voting (including movement to and fro, and around

and also up and down, the camera tilt angle and focal length adjustment, etc.).

As is shown in Figure 5, we can use the OMA Interactivity Media Document to set up a unified command template. When the users who support OMA protocol have obtained a unified command format, they can edit OMA protocol-based interactive

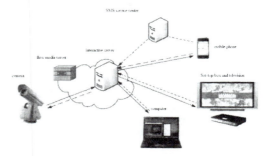

Figure 4. a typical intelligent multiscreen interactive business network framework picture.

command according to the unified command format.

The user terminal can directly send the SMS type of interactive commands based on the OMA protocols to the interactive server. It is unlike the sending way of the traditional short message command, which should be carried out by the short message server center and the short message gateway forwarding. Therefore, the transmission delay of the interactive command is reduced. In this way, it is convenient to support the OMA protocol servers to receive interactive commands in time and thus make corresponding treatment (this system also supports interactive SMS).

Because a unified command format has been adopted, the command edited by different user terminals is executed in the same order and with a unified unit. Therefore, it is conducive for interactive servers to quickly and accurately the interactive commands of the users. Through using the command format set by OMA Interactivity Media Document, users can easily and quickly build the interactive commands, and reduce the probability of the input errors generated by users, which will greatly improve the user's QoE (Quality of Experience).

In the example of Figure 4, we can allow users to combine the video capture device (video content source) and OMA standards and build a unified control mechanism for video capture devices, especially the control mechanism of the camera movement (including lenses and cloud units). Through the use of OMA interactive media document command template, users can easily and quickly build his/her control commands, which greatly reduces the probability of user input errors. Moreover, users can also obtain the command template before the start of the business. In this way, users can get familiar with the operation methods of the camera and the cloud units in advance, or they can even write the control commands in advance to avoid missing the best opportunity of the original video screen or control, due to the rush to control the video capture device before the start of the services. In addition, there is no need to build the circuit connected to the domain in advance. This will greatly improve the users' QoE and reduce the delay. In terms of the interactive servers, as the client terminal uses a unified command template, the server can make responses to users' control commands timely and accurately. In addition, there is interactive server that participates in the control process, who can provide a variety of businesses. For example, carry out the draft voting business facing multiscreen users; determine the camera motion in accordance with the election results or auction results and so on. Or even in the OMA Interactivity Media Document (see Figure 5), we can also provide the appropriate Web site links. Users can control the movement of the camera in the corresponding page with the mouse (including lenses and cloud units).

unlimited opportunities. What needs to pay attention is that the current multimedia business puts growing emphasis on interaction and user participation. In terms of multiscreen business, as it crosses the three kinds of heterogeneous networks (mobile communication network, wired Ethernet, wide power grid) and involves three kinds of terminals, interactive multiscreen business is more dramatic. Through the use of the OMA Interactivity Media Document, we set up a unified command template, making users' interaction combine with the OMA (Open Mobile Alliance) standards. Interaction mechanisms powerful functions and excellent compatibility can be built.

Through the use of the OMA Device Management technology standard, multiscreen TV (including set-top boxes), the PC and mobile phones can serve as an organic whole, which is convenient for operators to manage and also able to bring better feelings to the users. In short, with the globalization development of OMA, intelligent interactive multiscreen solutions based on the OMA standards will be able to enhance user participation. It brings unprecedented multiscreen experience while improving the customer loyalty of operators.

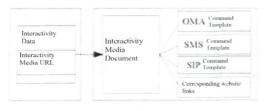

Figure 5. Intelligent multiscreen interactive media document.

4 CONCLUSIONS

Bright multiscreen business will be a new business trend for wired/wireless broadband. It brings

REFERENCES

YANG Xiangjuan, DONG Junhong. Study on the Human-Computer Interaction in Computer Assisted Language Learning. JOURNAL OF NORTHWESTERN POLYTECHNICAL UNIVERSITY (SOCIAL SCIENCES), 2013, 04.16.
QIN Yan-yan, TIAN Feng. An Interaction-Centered Hierarchical Post-WIMP User Interface Model. JOURNAL OF SOFTWARE, 2012, 10:21–31.
JI Lian-En, ZHANG Feng-Jun. 3D Interaction Techniques Based on Semantics in Virtual Environments. JOURNAL OF SOFTWARE, 2011, 08.02.
NI Yi-quan, CHENG Jing-yun. The progress and trends of human computer interaction. Computer Engineering, 2012, 04.12.

Information, Computer and Application Engineering – Liu, Sung & Yao (Eds)
© *2015 Taylor & Francis Group, London, ISBN 978-1-138-02717-6*

Assessment of the effect of team trust on GDISD team efficiency based on the AHP method

Ji Jiao Jiang & Yan Mei Zhang
School of Management, Northwestern Polytechnical University, Xi'an, Shaanxi Province, China

ABSTRACT: This paper conducted the research of the influence factor of team trust and team efficiency based on large quantities of data. By comparing various influence factors and utilizing the method of vague level evaluation, we assess the rate of various influence factors according to its weight thus getting the degree of the effect of various factors on team efficiency.

KEYWORDS: Trust; AHP; Influence factors; GDISD

1 INTRODUCTION

With the development of globalization, the multi-national and inter-organizational collaboration has become an important form of market operation. Geographically Distributed Information Systems Development (GDISD) refers to the process in which multiple teams in geographical different locations develop the same projects. Generally it is the cooperation of two or more branches or offices of the same company, mostly with no competitive contract relationship. While the mode of GDISD has not only effectively utilized both sides of the market but also increased time efficiency, communication is its biggest problem. One effective approach of increasing communicative efficiency and saving the cost of communication is to establish effective mechanism of trust among teams.

Analytic Hierarchy Process (AHP) is a multi-objective decision analytical method combining qualitative and quantitative analytical method. The main idea of this method is to divide complex problems into several levels and several factors, compare the degree of importance of two indexes, establish the judging matrix and finally get the weight of importance degree of different approaches by calculating the judging matrix's maximum eigenvalue and its corresponding eigen-vector, thus providing basis for selecting the best approach. The vague level evaluation, being based on AHP method in effect, is utilized to improve a series of problems in T.L Saaty's AHP method such as the differences of judging coincidence and matrix coincidence, the difficulties of coincidence testing and the lack of scientific. Using vague level evaluation to assess the importance of various factors that influence GDISD team trust and the state of trust can boost agility of the team thus setting efficient approach and avoiding risk resulted from lose of trust.

2 THE AHP METHOD

Analytical Hierarchy Process is initially proposed by an American operational research expert Saaty. Its basic idea is to see a complex multi-objective decision problem as a system, then construct a hierarchical structure model that can describe the system's characteristics, dividing one objective into several objectives or principles; ensure rating scales; compare elements on each level one by one and ensure its relative degree of importance by rating scales thus establishing judging matrix. By comparing weight relative to elements on the level immediately above which is calculated according to rating scale, we can calculate the synthetic weight of elements on different levels and rank them, thus providing decision maker the basis for scientific decision making.

The basic steps of AHP method is as follows:

Analyze elements of rating system, establish hierarchical level structure model;

Ensure rating scales. The rating scales usually adopted are shown in Table 1. The hierarchical level structure is comprised of objective level, principle level and index level.

Using hierarchical analytical model calculates weight of various elements. Beginning from elements on the top level, compare the n elements one by one on the lower level, thus establishing n order matrix, which is recorded as follow.

$A = (\omega_{ij})_{n}$

In which,

$\omega_{ii}=1$; $\omega_{ij}=1/\omega_{ij}$; $\omega_{ij}=\omega_{ik}/\omega_{jk}$.

Table 1. The hierarchical level structure rating scale.

Scale	Meaning
1	Two elements are equally important.
3	The former element is slightly more important than the latter element.
5	The former element is obviously more important than the latter element.
7	The former element is significantly more important than the latter element.
9	The former element is extremely more important than the latter element.
Reciprocal	If the importance ratio of element i to j is a, then the ratio of element j to i is a=1/a.

By using MATLAB software we can calculate the maximum eigenvalue λmax of matrix A. Through the input command: [V, D] =eig (A), we can get output D as the eigenvalue and V as the corresponding eigenvector, and the maximum eigenvalue is λmax.

The way of calculating uniformity index CI used for judging uniformity is as follows:

$CI=(\lambda(max)-n)/(n-1)$; (In which, λ max is matrix A's maximum eigenvalue while n is order of matrix A)

Because uniformity index RI (used for judging matrix uniformity) and matrix order has relationship listed in table 2, we can directly get the value of RI according to the value of n.

Table 2. Hierarchical analytical method uniformity index RI and its related value of n.

Matrix order	1	2	3	4	5	6	7	8
RI	0	0	0.52	0.89	1.12	1.26	1.36	1.41
Matrix order	9	10	11	12	13	14	15	
RI	1.46	1.49	1.52	1.54	1.56	1.58	1.59	

Calculation formula to judge uniformity ratio of matrix A:

CR=CI/RI;

If CR>0.1, then judging matrix A is mutually exclusive matrix and we need to ensure judging matrix again; if CR≤0.1, then judging matrix A has satisfying uniformity which is inclusive matrix.

When A is inclusive matrix, we can use MATLAB software to calculate maximum eigenvalue λmax's corresponding eigenvector:

$V''=(\omega'_1,\omega'_2, \ldots ,\omega'_n)$;

Then we normalize the eigenvector:

$$V''=(\omega'_1/\sum_{i=1}^{n}\omega'_i, \omega'_2/\sum_{i=1}^{n}\omega'_i, \ldots, \omega'_n/\sum_{i=1}^{n}\omega'_i)$$

V'' is weight of various elements in relation to various elements on the upper level.

When we have calculated weight of various elements in relation to elements above, we need to calculate various elements' synthetic weight to the general objective, rank various elements in precedence order, and get the needed conclusion. Section headings are in boldface capital and lowercase letters. Second level headings are typed as part of the succeeding paragraph (like the subsection heading of this paragraph).

3 ESTABLISHMENT OF RATING MODE

This paper focuses on the assessment of team trust on GDISD team efficiency. By referring a large amount of related articles, we sum up 9 main influence factors of team efficiency. Utilizing secondary data, we get the distributed frequency statistics information of influence factor. Using hierarchical analytical method in a comprehensive way, we establish rating model of team trust on GDISD team efficiency based on AHP method.

3.1 Establishment of hierarchical analytical model

Objective Z: team trust on GDISD team efficiency.

Principle P: GDISD team trust grade, namely trust grade 1, trust grade 2, trust grade 3, trust grade 4 and trust grade 5.

Objective D: elements of GDISD team efficiency are cost of system, mission understanding cost, cost of communication, cost of docking time, cost of mission follow-up efficiency, difference of mission importance, transition of staff, geographical factors and unpredictable event.

Establish multi-order structure analytical graph. Multi-order structure analytical graph is shown in scheme 1:

Scheme 1. Multi-order Structure Analytical Graph.

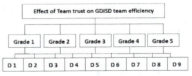

D1: system cost   D2: mission understanding cost

D3: communication cost   D4: docking cost

D5: mission follow-up cost   D6: difference of mission importance

D7: staff transition   D8: geographical factors

D9: unpredictable events

Establish and calculate judging matrix and weight of related importance of various elements

Construct the second level in relation to the first level judging matrix. Based on the hierarchical level structure rating scale, we give trust grade 1, 2, 3, 4, 5 as related scales as 1, 2, 3, 4, 5 as shown in table 3.

Table 3. Scale of trust grade.

Trust grade	Grade1	Grade2	Grade3	Grade4	Grade5
	1	2	3	4	5

The 5 trust grades when compared two by two can construct judging matrix P, 5 is highest while 1 is lowest. The matrix is as shown in table 4.

Table 4. Judging matrices P.

	Grade5	Grade4	Grade3	Grade2	Grade1	weight
Grade5	1	5/4	5/3	5/2	5	1/3
Grade4	4/5	1	4/3	2	4	4/15
Grade3	3/5	3/4	1	3/2	3	1/5
Grade2	2/5	1/2	2/3	1	2	2/15
Grade1	1/5	1/4	1/3	1/2	1	1/15

Based on data in table 4, we can use MATLAB to calculate and get:

λmax = 5.00, CI = 0.00, RI = 1.12

By calculation we can get CR=0.00<0.1, therefore the judging matrix P has satisfying uniformity, and it is inclusive matrix.

Construct the third level in relation to the second level judging matrix. By referring large quantities of data, we sum up 9 main influence factors that influence GDIST team efficiency. And by taking secondary data, acquiring different levels of trust, we get various influence factors' distribution frequency statistics figure, as shown in table 5:

We can similarly construct 5 judging matrix of three levels of various elements under different trust grades, namely judging matrix of trust grade 1, 2, 3, 4, 5.

Under the condition of trust grade 1, 9 different in fluency factors constructed judging matrix D1' by comparing one by one, as table 6 has shown:

Table 5. Distributive frequency statistics figure of various elements under different trust degree.

	Grade1	Grade2	Grade3	Grade4	Grade5
D1	55	843	219	39	27
D2	82	844	204	32	47
D3	82	852	202	31	42
D4	82	828	219	31	45
D5	74	825	223	31	46
D6	74	833	230	30	42
D7	73	838	213	37	48
D8	23	329	103	25	41
D9	56	822	227	28	54

Table 6. Judging matrices of various influence factors under trust grade 1.

	D1	D2	D3	D4	D5	D6	D7	D8	D9	Weight
D1	1	55/82	55/82	55/82	55/74	55/74	55/73	55/23	55/56	0.0915
D2	82/55	1	1	1	82/74	82/74	82/73	82/23	82/56	0.1364
D3	82/55	1	1	1	82/74	82/74	82/73	82/23	82/56	0.1364
D4	82/55	1	1	1	82/74	82/74	82/73	82/23	82/56	0.1364
D5	74/55	74/82	74/82	74/82	1	1	74/73	74/23	74/56	0.1231
D6	74/55	74/82	74/82	74/82	1	1	74/73	74/23	74/56	0.1231
D7	73/55	73/82	73/82	73/82	73/74	73/74	1	73/23	73/56	0.1215
D8	23/55	23/82	23/82	23/82	23/74	23/74	32/73	1	23/56	0.0383
D9	56/55	56/82	56/82	56/82	56/74	56/74	56/73	56/23	1	0.0932

Based on data in table 6, we can use MATLAB to calculate and get:

λmax = 9.00, CI = 0.00, RI = 1.46.

By calculation we can get CR=0.00<0.1, therefore the judging matrix D1' has satisfying uniformity and the matrix is inclusive.

Similarly, we can calculate different weights of 9 influence factors under other trust degree.

Calculation method is similar, use MATLAB to calculate and we can get:

Based on data in table 7, we can use MATLAB to calculate and get:

λmax = 9.00, CI = 0.00, RI = 1.46

By calculation we get: CR = 0.00 < 0.1, thus the judging matrix D2' has satisfying uniformity and is inclusive matrix.

Based on data in table 8, we can use MATLAB to calculate and get:

λmax = 9.00, CI = 0.00, RI = 1.46

Table 7. Judging matrix D2' of various influence factor under trust grade 2.

Factors	A	B	C	D	E	F	G	H	I
Weights	0.1202	0.1203	0.1215	0.1180	0.1176	0.1188	0.1195	0.0469	0.1172

Table 8. Judging matrix D3' of various influence factor under trust grade 3.

Factors	A	B	C	D	E	F	G	H	I
Weights	0.1190	0.1109	0.1098	0.1190	0.1212	0.1250	0.1158	0.056	0.1234

By calculation we get: CR = 0.00 < 0.1, thus the judging matrix D3' has satisfying uniformity and is inclusive matrix.

Table 9. Judging matrix D4' of various influence factor under trust grade 4.

Factors	A	B	C	D	E	F	G	H	I
Weights	0.1373	0.1127	0.1092	0.1192	0.1092	0.1056	0.1303	0.088	0.0986

Table 10. Judging matrix D5' of various influence factor under trust grade 5.

Factors	A	B	C	D	E	F	G	H	I
Weights	0.0689	0.1199	0.1071	0.1148	0.1173	0.1071	0.1224	0.1046	0.1378

By calculation we get: CR = 0.00 < 0.1, thus the judging matrix D4' has satisfying uniformity and is inclusive matrix.

Based on data in table 10, we can use MATLAB to calculate and get:

λmax = 9.00, CI = 0.00, RI = 1.46

By calculation we get: CR = 0.00 < 0.1, thus the judging matrix D5' has satisfying uniformity and is inclusive matrix.

Based on data in table 9, we can use MATLAB to calculate and get:

λmax = 9.00, CI = 0.00, RI = 1.46

Calculate levels and analyze general weights. After unitization of different weights figures of 9 influence factors under 5 trust degrees, we use the column vectors to construct 9×5 matrix. By timing it to matrix P under trust grade 1, 2, 3, 4, 5, we get general weights of 9 influence factors, as shown in table 11.

After calculating we get general uniformity rate CR=0<<0.1, which demonstrates that hierarchical analytical general weights calculation matrix of

Table 11. Hierarchical analytical general weights calculating matrix.

	Grade 1 (0.067)	Grade 2 (0.133)	Grade 3 (0.200)	Grade 4 (0.267)	Grade 5 (0.333)	Total weights
D1	0.0915	0.1202	0.1190	0.1373	0.0689	0.10552
D2	0.1364	0.1203	0.1109	0.1127	0.1199	0.11734
D3	0.1364	0.1215	0.1098	0.1092	0.1071	0.11208
D4	0.1364	0.1180	0.1190	0.1092	0.1148	0.11602
D5	0.1231	0.1176	0.1212	0.1092	0.1173	0.11635
D6	0.1215	0.1188	0.1250	0.1056	0.1071	0.11291
D7	0.1215	0.1195	0.1158	0.1303	0.1224	0.12274
D8	0.0383	0.0469	0.0560	0.0880	0.1046	0.11873
D9	0.0932	0.1172	0.1234	0.0986	0.1378	0.11873

hierarchical analytical general weights has very satisfying uniformity.

4 CONCLUSION

Based on large quantities of data, through research of team trust on team efficiency, and by utilizing vague level evaluation analysis on various influence factors to make weights rating, we get analytical ranking of various influence factors' effect on team efficiency as 1,2,3,4,5,6,7,8,9. This result has relatively high confidence level, and offered good guiding approaches to increasing teamwork efficiency and formulating related solutions.

ACKNOWLEDGMENT

This paper is supported by National Science Foundation of China: The research of geographically distributed information systems development team agility based on dynamic trust (G011201).

REFERENCES

[1] Nosek, J.T., The Case for Collaborative Programming, in Communications of the ACM. 1998. p. 105–108.

[2] Weinberg, G.M., The Psychology of Computer Programming Silver Anniversary Edition. 1998, New York: Dorset House Publishing.

[3] Dai Chunyan, Yang Yi, Trust Evaluation Model for Virtual Project Team Members, academic journal of Chongqing Technology and Business University.

[4] Youmin XI, Zhiqiang XU, Hongwen XIAO, The Trust Which Organization Gives to Team and The Analysis of Influencing Factors, Management Science, 2006, C936.

[5] Luhman, Niklas."Familiarity, Confidence, Trust: Problems and Alternatives.", in Trust: Making and Breaking Cooperative Relations, edited by Diego Gambetta. New York: Basil Blackwell, 1988, p94–107.

[6] M. Chiu, et al. "CIMSS - a case study in Web-based distributed project management", Systems and Information Engineering Design Symposium, IEEE, Washington D. C. , USA, 2003.

[7] L. Yadong, "Building trust in cross-cultural collaborations: Toward a contingency perspective", Journal of Management, vol. 28, issue. 5, 2002, pp. 669–694.

Information, Computer and Application Engineering – Liu, Sung & Yao (Eds)
© 2015 Taylor & Francis Group, London, ISBN 978-1-138-02717-6

Research of higher vocational network teaching platform based on B-learning idea

Lu Yan Lai
School of Business, Jiangxi Environmental Engineering Vocational College, Ganzhou, China

Fa Hui Gu
Department of Electronic Information Engineering, Jiangxi Applied Technology Vocational College, Ganzhou, China

ABSTRACT: Based on B-Learning idea, adopting technology of Struts+Spring+Hibernate, this paper establishes the higher vocational teaching platform through analyzing the current research of the higher vocation teaching platform, which is helpful to improve the quality of teaching and to promote the reform of higher vocational education.

KEYWORDS: blended learning; network course; network teaching platform

1 MOTIVATION

It is not difficult to find that most of the current network courses of higher vocational colleges have the following problems:

1 Most of the current network course designers have not design each teaching unit of teaching objectives, teaching content, teaching methods and so on, but for electronic notes, after-school exercise and others teaching resources in electronic information.
2 Most of the current network courses are lack of the guiding ideology and theoretical support or built by the guiding ideology of E-Learning.
3 There are phenomena of weak navigation and students' initiative in most of the current network courses.
4 Students' practical abilities cannot be emphasized on the cultivation in most of the current network courses, which are paid no attention to the characteristics of higher vocational education.

The existences of these problems greatly restrict the further development of higher vocational education. Therefore, it is important for the development of the higher vocational education to establish the network platform suitable for the characteristics of higher vocational education.

2 INTRODUCTION

All the world, there are many scholars researching the theory of the network course construction

idea of B-learning, such as Badurl Khan, Purnima Valiathan and Driscon etc. In China, Professor He Kekang of Beijing Normal University was firstly proposed B-Learning concept at the 17 session of the global Chinese Conference on computer application in 2003, and gives the new connotation named mixed type teaching, which emphasizes combining the combination of E-Learning and traditional teaching [6].

3 DESIGN OF HIGHER VOCATIONAL NETWORK TEACHING PLATFORM MODEL BASED ON B-LEARNING IDEA

3.1 Teaching design based on B-learning

The teaching process thought of B-learning should not only highlight the student's main body status, but also the leading role of teachers. Therefore, the teaching process based on B-Learning is designed as Figure 1:

1 Course introduction
   Course introduction is the beginning stage of B-learning teaching process, and it will form a consensus among teachers and students in the curriculum objectives, teaching content, teaching material, teaching plan, teaching activities, the organizational form of learning support method, examination methods and so on.
2 Learning
   In the process of student learning, the students can ask teachers or group internal or other group for academic or non academic help, when they encounter difficulties. Their purpose is to complete the task of learning.

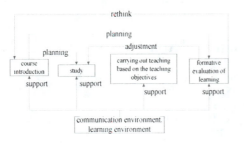

Figure 1. the teaching process based on B-Learning.

3 Carrying out teaching based on the teaching objectives

According to the classification of knowledge and teaching goal, there are a variety of activities in the form of the material organization of teaching process, such as collective learning based classroom teaching, team learning based collaborative learning, individual learning based autonomous learning. Through the organization of various teaching activities, it can increase the communication between teachers and students, improve the students' team spirit, and make students have enough chance to demonstrate their abilities.

4 Formative evaluation of learning

Formative evaluation of learning is a means to check and evaluate the expected effect of teaching learning between teachers and students, and it is the overall evaluation of the students, teaching design, the teachers' teaching activities and teaching process. Study of formative evaluation results can put forward reflection in the introduction to the course, adjust the learning support mode and teaching organization form to promote the teacher to change the teaching method, improve teaching quality, and put forward valuable suggestions and advice for students' further study.

3.2 Higher vocational network teaching platform model based on B-learning idea

According to the teaching design based on B-learning idea, higher vocational teaching network platform model can be designed as figure 2:

1 Classification and coding of basic information

The basic information about platform should be classification and coding in accordance with the principles of unified classification and coding, so as to ensure a unified platform for basic information sharing and exchange platform, promote the information resources to share and exchange.

2 Teaching resources

The main carrier platform is provided for the teachers as teaching resources to use, and for students as learning resources to learn.

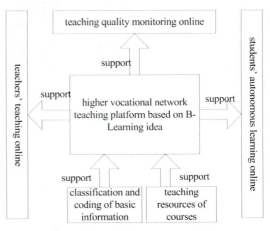

Figure 2. The higher vocational network teaching platform model based on B-Learning idea.

3 Teachers' teaching online

Through the relevant operation of the platform, such as publishing assignments, online interviews and other means of the leading teaching, teachers can promote students' Autonomous learning.

4 Students' autonomous learning online

Students can autonomous learn, according to their learning progress, learning interests, and selection of platform resources.

5 Teaching quality monitoring online

Whether the B-learning as a means of teaching aids, or an independent way of education, the teaching quality control of the B-learning is essential.

3.3 Higher vocational network teaching platform structure based on B-learning idea

The users of the platform include students, teachers and administrators. Each category of users is required to authenticate to gain corresponding operating authority, visitors can not question, collect and build their own space, but for browsing relevant curriculum information resource. The higher vocational network teaching platform structure based on B-Learning idea is designed as figure 3:

On the platform, the teacher to log on to the platform can be convenient for course management, professional organization for students to discuss and exchange information online. Students, to log on to the platform, can browse learning resources, and easily build their own learning space to improve the efficiency of learning. No matter what the user, if not through the identity authentication, can not do corresponding operations, such as the construction of their own space, publishing the relevant information resources, managing information resources.

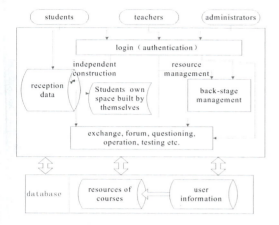

Figure 3. The higher vocational network teaching platform structure based on B-Learning idea.

4 IMPLEMENTATION OF PLATFORM

The implementation platform is designed as a typical J2EE three layer structure, which is divided into the presentation layer, business layer (middle layer) and the persistence layer (database service layer). The business rules, data access and verify the legitimacy of the whole system are set in the intermediate layer to the process. The client does not directly read and write the database, it read and write the database through connecting with the middle layer.

The presentation layer is implemented by the traditional JSP technology combined with the Struts tag library.

The business layer is designed thorough combination of Spring and Hibernate technology. In order to making the control layer and service layer separation, the business layer is divided into Web layer, Service layer, DAO layer, and PO.

5 CONCLUSIONS AND DISCUSSIONS

The network platform is implemented by adopting Struts, Spring, and Hibernate technology based on the J2EE. Platform can set columns independently. Under its help, an unprofessional teacher can built his net course easily as well as to promote his scientific research ability. On the other hand, the platform can help the students to change their way to learn and inspire their study interests. The thesis took "the computer network technology" which constructed a net course named "TEACH EQUAL TO LEARN " on the platform as its example. The course not only considered the universality of higher vocational education, but also highlighted its uniqueness.

The network platform that is based on B-Learning has been running at a vocational college. Many network courses have been constructed on it. The practice proved that the courses built on this platform combined both traditional teaching advantages and the characteristics of higher vocational education. It is helpful to improve the quality of teaching and to promote the reform of higher vocational education.

ACKNOWLEDGMENT

This work was financially supported by the Education Department of Jiangxi province science and technology research projects with Grand GJJ14807.

REFERENCES

[1] Zhong Yanhui, Liu Junhui, Chen Shaojun. Study on the construction of the network course in higher vocational education strategy [J]. Journal of Geography, 2008(5):110–111.[in Chinese].
[2] Jiang Xianzhong. Research on the construction of network course in Higher Vocational Colleges [J]. Journal of Wuxi Institue of Commerce, 2006(6). [in Chinese].
[3] Harvey Singh. Building Effective Blended Learning Programs[J]. Educational Technology, 2003, (6):51–53.
[4] Pumima Valiathan.Blended Learning Models[EB/OL]. http://www.learningcircuits.org/2002/aug2002/valiathan.html, 2002-08-10.
[5] Driscon M. Blended learning-How to Integrate Online and Traditional Learning[EB/OL]. http://www.koganpage.co.uk, 2003-16-20.
[6] He Kangke, the new development of educational technology theory from Blended learning[J]. China audio visual education, 2003(3):05–10.
[7] Hibernate Reference Document[M]. Jboss Professional Open Source, 2004.
[8] The Husted/Cedric Dumoulin. Struts in Action[M]. Manning Publictation, 2003.
[9] Jesse James Garrett. The Elements of User Experience[M]. New Riders Press, 2002.

Information, Computer and Application Engineering – Liu, Sung & Yao (Eds)
© *2015 Taylor & Francis Group, London, ISBN 978-1-138-02717-6*

Organic design in the digital age

Xin Xiong Liu, Lin Gan & Hai Ping Zhao
Huazhong University of Science and Technology, Wuhan, Hubei, China

ABSTRACT: This paper summarizes the design principle and development history of the organic design, analyzed the performance characteristics of the organic and the internal spirit. And put forward as time changes, the aesthetics characteristic of organic design in the digital age.

KEYWORDS: Organic design; Digital age; Optimize; Solid Thinking Inspire

1 INTRODUCTION

The organic design concept is a breakthrough of modernism design style. The term organic usually represents the natural growth of all beings, it is harmonious and unified, have vitality. Purely from the curve or imitate the nature level understand organic is not comprehensive.

2 ORGANIC DESIGN PRINCIPLES

Uniqueness. Just as no two leaves are similar in nature, the design of the organic there will not be copied or similar cases. Each form has its unique aesthetic features and functional requirements. And new material, new technology created wireless possibilities for these organic forms. Because plastics are renewable resources, easy to recycle, and with super plasticity, very easy to processing, can be a stamping forming. These characteristics make plastic products as organic modernism style representative of one of the material [1]. In combating modern design trends show the contrast between the culture, technology and aesthetics and the fusion.

Vitality. Organic design is vivid, and it can adapt to environmental changes and keep vitality. In the symbiotic relationship between man and nature, people have been always consciously or unconsciously imitate objects and phenomena in nature. From ancient Greek and Roman period to the Renaissance, building layout considerations of symmetry, the axis of the primary and secondary configuration, although not directly follow the creature, but more or less are constructed with reference to the organism's logic, that rooted in architect thought system [2].

Integrity. For the overall understanding of the theory of the organism is that function and form, harmony of environment–man–machine. The overall

requirements in the design of handling the relationship between internal and external relations and the correlation of the continuity of the movement relations. Such building structure will be reflected through the form, the internal space is caused by the design of the structure, the outline of the construction of the epidermis and form together with the surrounding environment harmony, building mutual echo, between the local and all keep in touch with the whole. Inside and outside the building space, form, structure, and the unity of the design, the relationship between buildings and the environment become the organic architecture characteristics, therefore the design principle of integrity is also a feature of the organic architecture [3].

3 THE DEVELOPMENT OF ORGANIC DESIGN

The New Art Movement (Art Nouveau) in the late 19th century early 20th century in continental Europe is popular, Indicates that the design of classicism tradition to modernity. And the natural forms of curve

 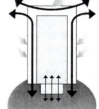

Built on the mountains   Built in the mountains

Figure 1. The difference between the organic architecture and ordinary buildings[4].

Figure 2.   Mila apartment (Gaudi).

Figure 3.   Falling water (Wright).

created a kind of abstract style, outstanding performance curve and the organic form in decoration. For the use of natural elements is Arts and Crafts Movement and New Art Movement of thinking about how to treat mechanical goods after the industrial revolution. As well as subsequent organic design provides reference and inheritance.

And at the same time the United States, architect Wright (Frank L. Wright, 1869–1959) proposed the theory of organic architecture. Wright absorbed and developed the Chicago School elite designers Sullivan (Louis H. Sullivan, 1856–1924) "Form Follows Function" thoughts, trying to form an organic whole of an architectural concept, namely, the function of the building, structure, proper decoration and construction environment be in harmony are an organic whole, form a kind of suitable for modern art. And emphasizes the integrity of the architectural art and make every small part of the building in harmony

with the whole [5]. "The building environment fusion, each building should be like out of the ground grow naturally" is the most vivid portrayal Wright's design style.

Spanish architect Gauti (Antonio Gauti, 1852–1926) said, "straight line belongs to the human, the curve belongs to god", most of his designs using curve with full of vigor and vitality to build organic types of objects to form a building. Fairy tale of the drama and plastic surface movement used by Gaudi in a romantic fantasy mixed together, create a dreamy, spiritual construction.

During the two world wars, the Scandinavian countries (Finland, Norway, Sweden, Denmark, Iceland) style of design become more common. Its rich humanistic appeal design and respect for the natural material, and is characterized in by the geometric diffusion and curve modeling. It broke the modernism hale and hearty, also known as the "Organic Modernism". Saarinen's (Eero Saarinen, 1910–1961) furniture design often reflects the "organic" free form, his furniture is overall more molding design.

4   THE DIGITAL AGE OF ORGANIC

With revolution brought by the advent of the computer technology, the organic also blooming in the era of digital technology a new vitality. Contemporary avant-garde designer spare no effort to explore the potential of new technologies, new processes, new materials, especially the dynamic simulation, the modeling of NURBS surface is generated by the form and CAM components such as digital technology made freely complex surface shape and space, so that the designer for the freedom of expression and rational logic is handy. So the contemporary design on organic possibilities open up farther than them [6].

Solidthinking Inspire is an optimization software, which integrates the mature finite element solver and optimization algorithms library, at the same time the software also has a wealth of secondary development interface. Solidthinking Inspire software at the beginning of the design is determining the nature of the design concept, and after topology optimization structure is amazing and organic design style is consistent. Again the nature is the most sophisticated designer.

In the past designer to ensure the stability and the materials optimized design of the structure is really hard. The solidthinking Inspire application Altair OptiStruct solver, the designer can be used in the early stage of the CAD design solidthinking Inspire quickly get optimal allocation structure, after in the CAD software design space defined in the product structure, to import the solidthinking inspire nodules after the best configuration for topology optimization,

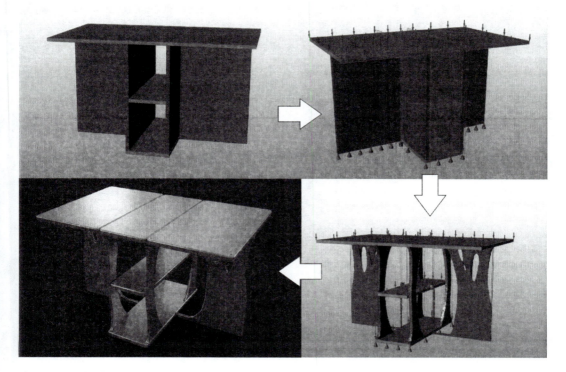

Figure 4.   Optimal process.

structure surface is drawn in CAD software, and cooperate with CAE software in structure strength test and verify [7].

It changed the thinking mode of stylist, injected in the form of simple engineering factor, let the designed results satisfy the need of designers to create beauty, and can guarantee the correctness of the design of the project, to achieve the balance of the two aspects of aesthetic design and engineering technology.

SolidThinking Inspire optimization process for folding table. First, define optimal design space. As shown in Figure, the red part is the folding table structure design optimization of space, is the desk-top support structure. Then, define material, load

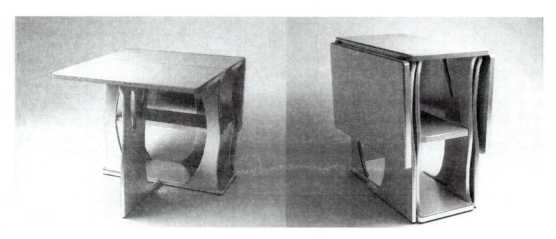

Figure 5.   Design sketch.

and define manufacturing constraints. Three pieces of each desktop are 80 n Applying the torque on the side panel. Finally, topology optimization results, and redesign the optimization results.

REFERENCES

[1] Lu Chen, Development of "theory that man is an integral part of nature" thought and organic style of interior design, (2008).

[2] Jiongde He, A new bionic artificial life of the era of new buildings, China building industry press (2009).

[3] Kan Feng-yan, Research on the interactive enlightenment between organic architectures and product design, (2011).

[4] James Wines, Green Architecture, published by TASHCEN, Italy, 2000.

[5] Renke He, A history of industrial design, Higher education press, (2004).

[6] Hua He, The foreign history of art and design, Daxiang press, (2013).

[7] Rui Fang, Bing Dong, From design to process design, from the perspective of contemporary design digital "organic", Art Education, p.158–159, (2013–08).

Information, Computer and Application Engineering – Liu, Sung & Yao (Eds)
© 2015 Taylor & Francis Group, London, ISBN 978-1-138-02717-6

Research of the employment-oriented practice teaching system for software engineering

Jian Meng Xie
Modern Education Technology Center, East China Jiaotong University, Nanchang, China

Xiao Fang Li
School of Software, East China Jiaotong University, Nanchang, China

ABSTRACT: Practice teaching plays a vital position of improving the ability of software engineering students' employment. According to the shortage of practice teaching in software engineering, this paper studies how to make the cultivation of employment ability embedded in professional practice teaching, and build practical teaching system based on real projects. This teaching model was applied in the Software engineering and good results were achieved.

KEYWORDS: software engineering; employ ability; practice teaching system; real project.

1 INTRODUCTION

Currently, there are two major Contradictory phenomena in the employment market of the software industry, one is that it is a sufficient demand of software industry, and Adequate supply of graduates Software engineering graduates; the other is that it is difficult for software organizations to recruit the right person and it is difficult for software engineering graduates to find jobs. The Contradictory "Booming and Dilemma" phenomenon, mainly due to the shortage of practice teaching in software engineering and software engineering graduates' practice capability, innovation capacity cannot meet the requirements of employers, which means that software engineering graduates lack experience in project development. Practice teaching is a bridge that leads students from theory to practice, which plays an irreplaceable special role to cultivate students' employ ability[1].

However, it is undeniable that there are some of the following problems during the practice teaching of software engineering. First is that related professional teachers lack the practice experience.[2] Second is that Professional Practice Curriculum System is unreasonable, There are no focuses and isolated between courses in The Professional Training Program. Moreover, the content of practice teaching has not closely related organizations with real project development, which Led to the students' practical competence away from the enterprise's expectations. The third is that Student performance evaluation system deviated from ability-oriented training. Most practical courses in schools still use test scores as the evaluation criteria, which led Most students to memorize experimental procedures and methods of operation instead of developing curriculum knowledge to solve practical problems. [3] For these questions, This paper is trying to focus on enhancing the employability of software graduates, combining with the requirements of enterprises to construct and apply the practice teaching system of software engineering, which is based on real project development, Practice has proved that it is good to improve the software's graduates' employment ability.

2 BUILD EMPLOYABILITY-ORIENTED PRACTICE TEACHING SYSTEM OF SOFTWARE ENGINEERING: A CASE STUDY OF SOFTWARE ENGINEERING OF EAST CHINA JIAOTONG UNIVERSITY SOFTWARE ENGINEERING

2.1 Build practical teaching system of software engineering based on real projects

During making software engineering professional practice program, some high-level managers of some information technology enterprises and some Outstanding graduates who work for famous information technology companies are invited to build practical teaching system of software engineering, which also based on questionnaires and interview survey in IT enterprises. First to analyze the ability of students majoring in software engineering practice, which consists of general practical ability, professional practical ability and Comprehensive practical ability; Followed by an analysis why the students of software engineering are so difficult employ is that the student lack of students 'practical abilities, and the shortage of practice ability mainly due to the shortage of project development experience; Finally Practical teaching system of software engineering based on real project is turned out, as the following table.

Practical teaching system of software engineering is based on real projects, In accordance with the order of "watching project, doing projects, leading the project, which is cultivated from the experimental curriculum, integrated curriculum designed to progressive arrangement for real projects.[4] Experiment and comprehensive course designed to train students to do a virtual experience on smaller projects, so that students become familiar with the software engineering project organization and basic procedures, principles and methods of engineering practice. Followed by Eight Weeks' Software Engineering Training students learn from discrete skill to a system of integrated skills. Eight Weeks' Software Engineering Training mainly adopts simulation training, which the school will bring in the establishment of a virtual reality of the business environment, the use of corporate real project case, creates a corporate real working pressure and take part in project planning, requirements analysis, design, coding, testing, and delivery process to make the students familiar with the whole process and method of Software project development.

Training instructors act as project director, to control the progress and quality of the project. The students play directly members of the project development role, including Project Manager (PM), Technical Manager (TM), Software Configuration Management (SCM), Database Administrator (DBA), Software Developer (Developer), Quality Testing Engineer (Tester), in order to understand the role of software development teams, processes, specifications and implementation methods, as well as the

ability to communicate the importance of cooperation in the team, to develop good professional practice. During the Training, Students' role consistent with the company's project team roles, commuting by using fingerprint, wearing badges, writing Work logs every day, summarizing weekly, Cultivating the students 'comprehensive ability, the model of learning by doing helps the student's accumulation of " "watching project' and "doing project", and lay the profound foundation of "leading project", Than Replacing virtual item for real projects, which relies on the tutorial projects. Generally, students qualified for project tasks and excellent students in Software Engineering Training can become members of the true backbone of the project to "lead project" and serve as team leader of the project to responsible for everything on the project team. Meanwhile, In order to ensure the correct implementation of the project, we adopt a project team mentoring and advisory system, to oversee projects and processes.

2.2 Constructing ability-oriented evaluation

Breaking through the shackles of traditional courses and test scores to evaluate students, and analyzing of students' employment ability structure and core employability, which includes practical ability, analytical ability, communication ability and problem-solving ability, team work and so on, we study how to guiding students to easily adapt to the quality of the labor market and maintain the power of learning and reform of practice teaching methods by the establishment of capacity-oriented evaluation mechanisms. We build a tri-level teaching and training mode, which consists of basic experiments, experimental, research and innovation-enhancing experiment[5]. Through the basic course standard experiment, increase experiments, research and innovative experiment to train students' analysis ability and imposed project management on the open

Table 1. Practical teaching system of software engineering based on real project.

Course Name	Nature	Term	Periods /weeks	Credits
Course experiment	compulsory	1–4	344hours	21.5
Basic software training	compulsory	1	1week	1
Comprehensive curriculum design	compulsory	2–6	10weeks	10
Software engineering training	compulsory	6	8weeks	8
Actual Project	compulsory	7	10weeks	5
Post practice	compulsory	8	8weeks	8
Graduation Project	compulsory	8	8weeks	8

experiment teaching[6], Through the establishment of forms for our team project, let the students as members of a team working together to complete the project to foster students' ability of communication and team corporation[7] Through extra-curricular, participation in ACM or related discipline events, participating in project design and other links to meet the market demand for engineering practice, which make students meet the employment market of software engineering. [8] Introducing the comprehensive assessment system of enterprise's personnel evaluation standards in Software Engineering Training, we can multi-affectedly evaluate from the perspective of students' ability of individuals.

Students' practice teaching achievements depart into "grades" and "final grades" two parts, by the 40% and 60% into a total score. Grades examine once a month by the team leader who responsible for the assessment. Final grade assessed by the tutor group and project team members according to student presentations and public answer. In order to give students access to comprehensive engineering training, we focus on students 'assessment from nine aspects of Ability to understand global project and every technical principles; Ability to learn the knowledge required for the project Ability to find out fault analyze capabilities and improve fault location; Practical ability of producing, processing, debugging, testing, programming; searching and sorting resources; Communicating skills and negotiation skills with mentors, team members, and project supervisor of internal communications, and with the crew, with partners, suppliers, and processors; Module design abilities; innovation ability to improve and optimize project design; Expression ability to discuss, explain and defense; Through real project combat experience, students gain real project development experience, feel the responsibility of the actual project by project developers comprehensive training, project risks necessary to withstand the pressure, the formation of systems engineering concept, and inspire a sense of accomplishment.

3 CONCLUSIONS

Our school in 2010 signed a cooperation agreement with the Jiangxi Microsoft Technology Center and Beijing Chuanxi Group to develop the "3 + 1"training model of school-enterprise cooperation, which means the students learn three years in school, then study in these enterprises in the fourth year to complete Graduation Project. Based on "3 + 1" training model of school-enterprise cooperation. The paper study the cultivation of employment ability embedded in professional practice teaching, and build practical teaching system based on real projects and ability-oriented evaluation mechanisms, This teaching mode was applied in the Software engineering and good results were achieved. The employment rate of software engineering graduates reached 100%, which reduces the cost of retraining, and have achieved good social effects. It is obvious that this teaching mode presents here are valuable. Of course, a lot of further research will need to be conducted in the future.

ACKNOWLEDGMENTS

The work was supported by East China Jiaotong University (No:14RJ07) and The Education Department of Jiangxi Province (No: JXJG-13-5-31).

REFERENCES

[1] Zhengzhong Zhou. Analysis on the Effectiveness of Practical Education of Higher Education system, Practice Education[J]. Adult Higher Education, 2010 (2):23–25.
[2] Changying Sun. Thoughts on improving employment ability of college students[J]. China's higher education, 2007, (11): 87–88.
[3] Xiaoming Zheng. Study on the Employability [J]. Journal of China Youth College for political sciences, 2002, (3): 91–92.
[4] Ruicui Liu. Outstanding practice teaching to promote students 'ability [J]. Adult Higher Education, 2008, (12):120–121.
[5] Liu BO. Exploration the practice teaching system Based on the real project.[J].Research and exploration in laboratory, 2013, 26(7):5–8.
[6] Jinjiu Zhu. Study and practice of experimental teaching of Digitalsignal processing course[J]. Research and exploration in laboratory, 2008, 27(5): 96–98.
[7] Changping Zhu, Bo Huang. fostering theelectronic specialized students ' practice ability by three levels of experiment[J]. Research and exploration in laboratory, 2007, 26(7):5–8.
[8] Zhen Li. 3 + 1 teaching model and cultivation of students 'innovative ability[J]. Experimental technique and management, 2007(1):128–130.

Information, Computer and Application Engineering – Liu, Sung & Yao (Eds)
© *2015 Taylor & Francis Group, London, ISBN 978-1-138-02717-6*

The heaviness of function and the lightness of cognition—commenting on a new discussion on P.E. education curriculum reform and what is supposed to be a good P.E. class

Zong Feng Huang & Ya Qing Wei
Institute of Physical Education, Hechi University Yizhou, China

ABSTRACT: This paper, under the background with the completion of Mao Zhenming's book, "New Theory of Physical Education Curriculum Reform: On what is good PE" will re-examine the "heavy" Chinese sports teaching functions, namely PE teaching should allow students a lifetime not only to participate in sports, but also to experience the fun and success. It is necessary for students to master motor skills, develop their security awareness and prevention capabilities. Doing this can enable students to understand the role of sport for physical shape, cultivate their will, morality and spirit of cooperation. What's more, it is essential to raise national awareness of the Physical Education Teaching. The phenomenon "light" in the Physical Education Teaching, in other words, is national contempt for physical education, national sports superficial and one-sided understanding. To sum up, national sport socialization has a slow pace in these three areas. And the Outstanding of good PE classes can be summarized as "knowing", "understanding", "enjoying", and "training".

KEYWORDS: P.E. teaching; Function; Cognition

1 INTRODUCTION

In social life, the contradictions of, "heaviness" and "lightness" exist everywhere. To put it simple, heaviness refers to responsibility, heed and conditions of bounding and implementing in due place. On the other hand, lightness assumes the absence of one's duty, abandonment of constrains, and contempt, etc. To refine the concept in P.E. teaching, the two words above have more specific and plentiful meanings. The book, *New Discussion on P.E. Education Curriculum Reform: And What Is Supposed to Be A Good P.E. Class,* has generated the consistent education perceptions the author, which highlights the discussion on the heaviness and lightness in P.E. Teaching. It not only reinforces readers' understanding of the meaning and value of spots, but also point out the concrete projecting routes for the reform in P.E. teaching, and enhance the integration process of education and society in our country.

2 THE BACKGROUND OF MAO ZHENMING'S BOOK, "NEW THEORY OF PHYSICAL EDUCATION CURRICULUM REFORM: ON WHAT IS GOOD PE"

The book, written on the background of new curriculum reform, is a systemic analysis of the author's research output of reform in P.E. Teaching for years. The author has carried on the research of P.E. teaching for a decade continuously, and he committed to integrating sports and health, thereby realizing integration of sports and society. He reframed the teaching philosophy of P.E. class, building a purposeful and effective teaching system. Mr. Mao Zhenming, as a prominent pioneer in China's educational discipline, has furthered his study on P.E. teaching in Japan on his early years and worked as a teacher for years. It is known to all that before World War II, the physique of Japanese could not equate with that of Chinese. However, after 1990s, the author, in his exchange to Japan, discovered that their physiques have witnessed an amazing improvement. Numerous surveys on physical conditions thereafter have shown that, the nutritional level, growth and development conditions, qualities such as strength and speed of Chinese teenagers lag behind their Japanese counterpart's. Moreover, the gap keeps on widening. The reason lies in the Japanese government's prompt concern on the importance of their civil physique, and their plan and implement of nutritional and exercising policies that improve the civil physique. Mr. Mao's years of experience in studying and teaching in Japan can render to us new teaching philosophy in P.E. teaching. We can learn from the excellent achievements of P.E. teaching of Japan, and optimize the deficiencies in our own country.

3 "HEAVY" FUNCTIONS OF THE SPORTS TEACHING

From the author's perspective, P.E. teaching is inevitably heavy, which demonstrates in its functions. In 2011, the latest version of *Sports and Health Curriculum Criterion* came out, whose principles are as follows: adherence to the principle of "health first", commitment to gearing up the healthy growth of students; stimulating students' interests in exercise, cultivating their spots consciousness and habits; focusing on their development, and helping them to learn sports and health curriculum; attention to individual and regional differences, guarantee the benefits of everyone. Having profoundly analyzed the new principles above, the author realized that sheer education of sports events is inadequate, and the addition of sports skills, sports knowledge and security consciousness of the former is just passable. And the core problem in P.E. curriculum reform is the transformation of exam-oriented education in a quality-oriented one. In term of P.E., it will be the transformation of exam-oriented P.E. curriculum to fitting exercise and life-long sports. In his opinion, high-quality P.E. education should serve in the interests of students throughout their life, which involves them in sports activities and brings on them entertainment and fulfillment; which makes them realize sports can shape their physiques, and cultivates their will, morality and team spirit.

In conclusion, P.E. is a comprehensive subject that integrates exercise, nourishing of life and morality cultivation. It mainly aims at cultivating teenagers with a all-round development physically and mentally. The youth, as the pillar, the future erector and successor of the nation, should be trained in this way to be equal to their great responsibility. However, the author also discovered that the P.E. education in China is undergoing unbearable lightness.

4 "LIGHT" NATIONAL AWARENESS OF THE PHYSICAL EDUCATION TEACHING

4.1 The people's contempt of P.E. education

In China, P.E., art, and music have been simply regarded as minor subjects separated from those major ones such as Chinese and maths. And there is a lack of a reasonable leveling standard. Parents and students concentrate unilaterally on entrance of higher-level schools, and they overlook sports and exercise, or even treat them as a waste of time.

4.2 The people look upon P.E. in a short-sighted and one-sided manner

Owing to schools' insufficient and inefficient implement of P.E. education, students usually think it is just some running and jumping deeds. But P.E. is actually a great subject rich in contents. Not only does it has diverse branches of sports, but it is closely bound up with people's health and personality cultivation. This kind of lightness of people's consciousness results in the heaviness in our real life. This bitter fruit usually demonstrates in four points: I. There is a general decent in the teenager's physiques, with which it is scarcely possible to realize the ambition of "strong teenagers usher in a strong country." II. Many students can hardly feel any interested in P.E. class, and then develop detest or fear towards the subject, which amounts to strangle their life-long exercise course. III. Though some students realize the positive effects exerted on their health through exercise, they can get irreparably injured for lack of correct sports skills and self-protection consciousness. IV. As for spiritual level, this lightness leads to the scarcity of team spirit and sense of collective honor, and the prominence of egoism.

4.3 People react clumsily to the socialization of sports

The heaviness of its functions and lightness of people's consciousness on the one hand demonstrate the severe current situation in our country, on the other hand, reflects the disjoint between P.E. Education plan and social demands. First, the single-target of P.E. education training can barely satisfy the development of social demands; second, the teaching content is dull, which is lacking in pertinence, and the students are not interested enough in the class; third, the teaching method is inflexible, which fails to give play to teacher's leading role and student's principal role; finally, the inadequate curriculum distribution has blocked the improvement of students' comprehensive quality. Currently, the function of sports has undergone diversified and enriched transform from single "fitting" and "medal grabbing" into "health first". Along with the economic growth in China, the socializing process of sports has been put forward. It has penetrated and spread in every sphere of the society, playing an increasingly important and influential role in society. It will not confine to a teaching and exercising means, but will be one that enriches people's cultural life and improved their living condition. Therefore, P.E. teaching should take the path to socialization. The teaching contents, teaching methods, and learning methods must adapt to the socializing demand of the curriculum. Schools should devise some innovative curriculum contents according to characteristics of the schools and demands of the society, committed to improving students' practice skills, creative skills and social adaptive capacity. These curricular should be adapted promptly to social demands, through which the students can achieve self-transcendence, cultivated temperament, and developed personality. In a word, the P.E. teaching of a school is ultimately intended to cultivate teenagers who can well adapt themselves to society and possess all-rounded development of mind and body. So the curriculum design should be flexible and purposeful. At the same time, the teaching content should match with students' quality and social demands. It should have a clear and definite direction for the schools to put on efforts. The curriculum should keep pace with the speedy economic growth in the society, and improve the general physiques of the teenagers in order to fit with the needs of social development.

5 SUMMARY

On the basis of years of practice experience, the author devoted a large section of his book to discussing the weighing standard of excellent P.E. class, and clearly stated his view through instance analysis. Citing the author's answer in an interview before, we can briefly summarize an excellent P.E. class in four words, i.e, understanding, ability, happiness, and practice."Understanding" emphasizes the students' systemic absorption of basic sports knowledge and security consciousness;"ability" requires students to learn sports skills, and participate in sports activities in person;"happiness" stresses that sports derived from games, it should bring along happiness and more than happiness; "practice" asks for students to

realize the indispensability of sports to health and fitness, and keeps them doing sports after P.E. Class. These four words, as a significant link between P.E. teaching and social demand are the highlight of my concern. My latest researching project—*Build a New Personnel-cultivation Mechanism of the 'Integration of Society and Class' in the P.E. Education Faculty of Newly-founded university in Guangxi* (Project code:2014JGA210), and the teaching reform project in Hechi University—*The Educational Reform and Practice of P.E. Teaching in Newly-upgraded University on the Background of New Curriculum Standard*(Project code:2012EA001)—will also center on these in order to improve the faculty's effectiveness.

REFERENCES

[1] Wang Jian. motor skills and physical education - primary and secondary school students in theoretical discussion and empirical analysis of the formation of motor skills [D]. Fujian Normal University, 2004.
[2] Shao Guihua. Teleology of Physical Education and evaluation of the comb - to build self-organizing sports teaching teleological [J]. Sports Science, 2005 (7).
[3] Li Qidi Zhou Yan. Different teaching methods in sports teaching [J]. Sports and Science, 2012 (11).
[4] Huo Jun. PE Teaching Methods Theory and Practice innovative educational concept study [D]. Beijing Sports University, 2012.

Information, Computer and Application Engineering – Liu, Sung & Yao (Eds)
© *2015 Taylor & Francis Group, London, ISBN 978-1-138-02717-6*

Study on the interaction of folk custom sports teaching practice and community sports activities

Yu Ting Yang, Yun Sheng Liu & Li Ming Feng
Wen Hua College, China

ABSTRACT: In this paper, on the basis of original subject of building foundation folk custom sports teaching practice and community sports interactive model, combined with the interactive practice method to explore the positive interaction mechanism of folk custom physical education and community sports activities are promoted for sports teaching reform and explored the relevant strategies and solutions. Finally, according to the combination method of comprehensive analysis and expert interviews, the interactive performance of folk sports teaching practice and community sports were analysed to provide the theory and method of reference for the college folk custom sports teaching practice.

KEYWORDS: folk custom sports teaching practice; community sports; interaction

Community sports activities and sports socialization process are based on the social action with the mutual conditions and results. Folk custom sports teaching practice is mainly about related courses teaching content, teaching method and organization, teaching objectives and evaluation way. Community sports team mainly refers to the members of the community team organization which takes the area of the natural environment and sports facilities as the material basis. The main task of provincial and Municipal Sports Bureau of social sports management center is: according to the legal provisions of the state and sports policy, unifing organize and guide the development and management of the owning social sports projects of the province, promoting the popularization of the project and improvement of sports technology level, industrialization development of sports projects. The private enterprise social sports organizations are gradually evolved into social sports industry pioneer, they take the opportunity of promotion and popularization of related projects. In this paper, through the analysis of interactive modes of organizational social sports private enterprise folk physical education and community sports organization, management center and agency found out the enlightenment of interactive activities to the folk custom sports teaching reform and the promotion of folk sports teaching.

1 FOLK CUSTOM SPORTS TEACHING PRACTICE AND COMMUNITY SPORTS INTERACTIVE MODE

1.1 *The mechanism for mutual assistance folk custom sports teaching and community sports organizations*

Folk custom sports teaching practice through the mutual aid with community sports team organization to re-examine the teaching content, teaching method and teaching evaluation reform measures. Folk custom sports teaching teachers and students can go deep into a community sports team organization involved in guiding and assistance services through the weekend and holiday, in order to promote the understanding and grasping of teachers and students to the community sports activities required for university sports professional training, and thus can provide a theoretical basis and reference to the current and future education reform. At the same time, students can also through this realizing social, professional status, understanding the course of study and enhance the learning significance for social understanding.

In the interactive process communityRoli ball tissue, undoubtedly will not only be benefiting groups, also has brought the project technology and theory practice and make up a missed lesson position for students of professional sports colleges of university folk softball sports curriculum. For college students, through many leisure time to participate in community soft ball service, will undoubtedly enhance the interpersonal attraction, through open communication, guidance and other forms to promote the healthy development of their personality and other psychological factors; taking soft ball technology movement as the main non-linguistic communication to promote the show and expression of body language , becoming the important basis for future cross-cultural interpersonal interaction smoothly.

1.2 Benefit mechanism of folk custom body teaching and service management center

Sports colleges and universities are shouldering the mission of university philosophy, teaching and training, scientific research and social service of the trinity. Among them, the social service work, scientific research and teaching practice are closely related to training students in the direction of future employment and social sports service and sports service people, this theory is a responsibility and refinement in the folk custom sports curriculum. As provincial and the Municipal Sports Bureau of social sports management center, their duty to serve society, serve the people and promote the project development responsibilities have similarities, the resemblance is no doubt the biased source of interactive cooperation in folk custom sports teaching process and the provincial and Municipal Sports Bureau of social sports management center.

On the one hand, provincial and Municipal Sports Bureau of social sports management center and related associations may formulate the relevant management system setting, the development situation and students overall situation to perfect the management system of combination of PE curriculum in College; secondly, provincial and Municipal Sports Bureau can through sharing of sports teachers of colleges to make up for their personnel, manpower, material resources deficiency in the management of numerous associations. On the other hand, the related teachers of professional sports colleges and universities projects can use their own advantages and students actively participate in the aspects of human resources of provincial and Municipal Sports Bureau projects related to the work of the association, with a view to promote the raise of discipline curriculum development, promote their own teaching, training, management, competition organization, promote the cycle base of university students, who practice and use what they have learned. Teaching teachers

to participate in the work of management of social sports organization itself is also a way to professional counterparts of social services, though, not the main forms and ways of promoting educational reform, but it is a meaningful thing that worth to affirm.

1.3 The reciprocity mechanism between folk custom sports education and related sports private enterprise

Although the related sports private enterprise organization for profit purpose, but the source of its take money in its business is based online idea. First of all, as the work of community sports for profit, sports private enterprise organization develop the residents sports participation, making efforts to promote the personal, social and community awareness and community sense of belonging to community sports activities, with the appeal of a new form of sports, interesting content to attract many residents to actively participate in community sports activities, to participate in the organization. Secondly, through scientific, healthy and civilized way of life helps to improve the quality of life of residents, maintaining social order and stability, to a certain extent, enrich the amateur cultural life of the residents, resist corrosion life content which is not healthy, and has played a positive role in improving the residents way of life. Thirdly, they play a positive role in strengthening social integration, enhance community cohesion. Finally, the related sports private enterprise organizations are taking more and more important role in the development of community culture. And the related sports private enterprise organization in the activities and organizations such plays more and more important role in the development of a direct or indirect influence on the community sports culture.

Professional sports colleges and universities may be appropriate to employ relevant private enterprise organization's elite team coach for the College Lectures, to make up for lack of theoretical level social sports organization of private enterprises, thus to better promote the two sides to serve the purposes of society. In addition, through the interaction between students and the organization of private enterprises of social sports, lead a correct understanding of students to the relevant enterprise, strengthen the relevant units to understand and grasp the situation of students, pour into early career planning of the employment problem for specialty students. In the process of guide teaching and amateur training of Wuhan Sports Institute folk custom sports softball project, often exchange employs, component technology, theoretical study with Interactive Development Co. Ltd. in Beijing Aobolong sports, to the conveying a successful move elite coaches is a successful case analysis of this idea.

2 THE INTERACTIVE PERFORMANCE OF FOLK CUSTOM SPORTS TEACHING PRACTICE AND COMMUNITY SPORTS

2.1 To explore the interaction of folk custom sports inspired practice teaching reform

Through the interaction of practice, deepen the scientific direction ideas of folk sports teaching reform. Innovation Research on the teaching reform of sports colleges and universities folk custom sports first focus on the judgement method of related projects, at the same time, pay attention to the study of standardization project of technical movements, in order to promote the science trend of the folk custom sports. The mainly object audience of folk custom sports is the community sports activities, because of the district of body condition the value orientation of community sports activities. To participate in the competition is not the ultimate activity point, so the provisions of non-standard, their technical movements reflect the routine exchanges and learning interaction in relative terms, it is more common to arbitrary change action specification. The interaction process of competitive exchange, showing a misinterpretation of the new projects which is related to their judgment law, even from ignorance.

Through the interactive practice, the reform of folk sports teaching in physical education institutes should pay attention to folk custom sports fitness-oriented increasing difficulty level and routine exercise levels, in order to promote the folk custom sports service for people. In a sense, the lack of rules, technical specifications, contest rules consciousness is the core problem existing in the scientific standard that folk sports in the community sports should be change scientifically, but their appearance is one of the objective reasons of esteem for the vast number of community sports activities. The reform is bound to focus on the measures of specific aptitude election, adjusting the match. The individualized adjustable game, namely because the majority of community sports activities of the main body of the age, body condition and adjust the format, game type, including competition award system adjusts the way. Inspired teaching reform content segmentation, hierarchical of sophisticated, in order to apply to a wider object audience change in demand. With the competitive market, between the different level routines at a reasonable level coefficient of distinction, in order to meet with the competitive market at different levels of teaching and learning group need.

2.2 The enlightenment to explore the interaction of folk custom sports social service

Through the interaction, make a profound understanding of the practical problems about community sports activities. For example, It's difficult to build a charismatic influence or organize a large-scale with profound influence community sports activities when the leadership lack of cohesion. Different organizations and different institutions act of their own free will restrict the physical resource sharing and optimization; the lack of sports facilities and funds which required for the activities the lack of relevant sports equipment and facilities' repair and maintenance; the commonness and individuality of problem between the sports organizations in different communities, the Interactive practice explicit the guidance service. The information service practice of lining publicity activities and service organization, etc. the different residents diversified and personalized demand problem; finally is the community sports exercise guidance personnel number appears insufficient, quality serious about showing relatively uneven situation.

Through the interaction, a profound comprehension of the community sports activities in the need urgent of "vulnerable" populations of physical exercise. They are usually more concentrated on the ideological understanding of vulnerable: the lack of correct understanding of fitness. Such as disease-free painless no fitness mistake idea perfections; disease with disaster causing Mai Bukai sports fitness, slim, Yangxin pace; although this is non mainstream phenomenon, but the lack of fitness for all important issues in the ideological understanding needs to be enhanced. These are the basis for community members to strengthen scientific consciousness of fitness guidance, strengthen the organization and management system, the construction of community sports to strengthen the government behavior of such measures.

At the same time, through the interaction of folk sports teaching practice and community sports exploration, deepening understanding of community sports service personnel shortage situation, and further deepen the understanding of the service system of community sports, sports college folk custom sports to the community sports service related to sports information service, the community sports organization service, community sports guide service to the community sports service range, combined with classroom teaching, can also provide the community sports facilities; through interaction research, clear the cultivation oriented innovative scientific research accomplishment of students to construct the security system of community sports talents, breaking the previous community sports guide practitioners and community sports management limited cognition practitioners.

ACKNOWLEDGEMENTS

The corresponding author of this paper is Liu Yunsheng.

This paper is supported by The research project of Humanities and social science in Hubei Province (2014G512).

REFERENCES

[1] Sui Ivan The mission of the University and its guardian[J] Educational Research, 2010, 372 (1):68–72.

[2] Wang Kaizhen Zhao Li Community Sports [M] Higher Education Press, 2004:59–60.

[3] Luo Xu Theoretical and Empirical Research on National Fitness Service System [J] Sports Science, 2008, 28 (08):81–94.

Information, Computer and Application Engineering – Liu, Sung & Yao (Eds)
© 2015 Taylor & Francis Group, London, ISBN 978-1-138-02717-6

Global asymptotic stability of a kind of third-order differential equations with two delays

Xiao Ping Song

Faculty of Science, Jiangsu University, Zhenjiang, Jiangsu, China

ABSTRACT: Application of the analogy method of constructing a Lyapunov function, discussing the global asymptotic stability of a class of third-order differential equations with two delays. Similar to methods of constructing Poincare-Bendixon ring outside the boundary, the positive orbit of the system is proved to be bounded. Thus, the global asymptotic stability of the trivial solution of the adequacy criterion is given, and by using the method of numerical simulation to verify the effectiveness of the theoretical results.

KEYWORDS: delay differential equation; global asymptotic stability; Lyapunov functional.

1 INTRODUCTION

The Lyapunov second method is very effective in the study of the stability of nonlinear systems, to make a Lyapunov function right is the key of studying nonlinear system stability, for nonlinear time-delay system is no exception. However, the structure of Lyapunov function has no general method. In the case of low order, it can use the analogy method. Using this method, the global stability of third-order nonlinear system, has made great achievements in [1]-[6]. But for third-order nonlinear system with time-delay, the delay phenomenon brings great difficulties to study the stability of the system, making it less documents. The literature [7], [8] respectively generalizes [3], [4] equation, the global asymptotic stability of the zero solution of the sufficient condition for a class of three order nonlinear delay systems are obtained. But it is a single delay case. Though the document [9] considers variable delay situation, it is also a single delay case., and only studies its asymptotic stability.

This paper studies the global asymptotic stability of the zero solution of third-order differential equations with two delays,

$$\dddot{x}(t) + g(\ddot{x}(t)) + f(\dot{x}(t-\tau_1))$$
$$+ h(x(t))\varphi(x(t-\tau_2)) = 0. \tag{1.1}$$

which $\tau_1, \tau_2 \geq 0$.

For the sake of convenience, we agree that function involved in this paper is continuous, and can ensure the uniqueness of the solution of the system. If necessary, it is assumed that they are continuously differentiable. At the same time, In order to prove, the theorem used in this paper will be given in the form of the lemma.

Considering the system

$$\frac{dX}{dt} = F(X), \ X(0) = 0. \tag{1.2}$$

Lemma1 [10] If there is a positive definite function, it makes

$$\left.\frac{dV}{dt}\right|_{(1.2)} = \nabla V(x)^T \frac{dX}{dt}(x)\Big|_{(1.2)} \leq 0,$$

and the set of $\dot{V} = 0$ except O and does not contain the positive orbit of the system (1.2), and all the positive orbit of the system (1.2) are bounded, then the zero solution of the system (1.2) is globally asymptotically stable.

2 MAIN RESULT

Consider the following differential equations with two delays

$$\dddot{x}(t) + g(\ddot{x}(t)) + f(\dot{x}(t-\tau_1)) + h(x(t))\varphi(x(t-\tau_2)) = 0,$$
$$g(0) = f(0) = h(0)\varphi(0) = 0. \tag{2.1}$$

(2.1) equivalent to the system

$$\begin{cases} \dot{x}(t) = y(t) \\ \dot{y}(t) = z(t) \\ \dot{z}(t) = -h(x(t))\varphi(x(t)) - f(y(t)) - g(z(t)) \\ \qquad + \int_{-\tau_1}^{0} f'_y(y(t+s))z(t+s)\,ds \\ \qquad + h(x(t))\int_{-\tau_2}^{0} \varphi'_x(x(t+s))y(t+s)\,ds \end{cases} \quad (2.2)$$

Theorem 1 If there exists a constant $a > 0, b > \dfrac{1}{2}$,
$L > 0, M > 0$, it makes $l \leq \sqrt{2}a \ (l = \max\{L, M\})$,
$a(\tau_1 + \tau_2 + a^2\tau_1) \leq \dfrac{1}{2}$, and it meets

I $\quad 0 < [h(x)\varphi(x)]' < a(b - \dfrac{1}{2})$

II $\quad f(y)\mathrm{sgn}\, y \geq (b + a\tau_1 + 2a\tau_2)|y|$

III $\quad |f'_y(y(u))| \leq L$

IV $\quad |h(x)\varphi'_x(x(u))| \leq M$

V $\quad \dfrac{(a^2+1) - \sqrt{\Delta_1}}{a} < \dfrac{g(z)}{z} < \dfrac{(a^2+1) + \sqrt{\Delta_1}}{a}$

$(z \neq 0, \Delta_1 = 1 - 2a(\tau_1 + \tau_2 + a^2\tau_1))$,

The zero solution of the system (2.2) is globally asymptotically stable, and the zero solution of (2.1) is globally asymptotically stable.

Proof: The corresponding linear system of the system (2.2)

$$\begin{cases} \dot{x} = y \\ \dot{y} = z \\ \dot{z} = -cx - by - az \end{cases} \quad (2.3)$$

The Lyapunov function of the system (2.3)

$$V(x,y,z) = \frac{1}{2}acx^2 + cxy + \frac{1}{2}(ay+z)^2 + \frac{1}{2}by^2$$

The Lyapunov function of the system (2.2)

$$V(x,y,z) = a\int_0^x h(x)\varphi(x)dx + h(x)\varphi(x)y$$

$$+\frac{1}{2}(ay+z)^2 + \int_0^y f(y)dy + \int_{-\tau_1}^{0}\int_{t+s}^{t} a^2z^2(u)\,du\,ds$$

$$+\int_{-\tau_2}^{0}\int_{t+s}^{t} a^2y^2(u)\,du\,ds$$

$$= a\int_0^x h(x)\varphi(x)dx - \frac{1}{2b}h^2(x)\varphi^2(x)$$

$$+\frac{b}{2}[y + \frac{h(x)\varphi(x)}{b}]^2 + \int_0^y [f(y) - by]dy$$

$$+\frac{1}{2}(ay+z)^2 + \int_{-\tau_1}^{0}\int_{t+s}^{t} a^2z^2(u)\,du\,ds$$

$$+\int_{-\tau_2}^{0}\int_{t+s}^{t} a^2y^2(u)\,du\,ds$$

(i) We first prove that $V(x,y,z)$ is positive definite.

$H(x) = a\int_0^x h(x)\varphi(x)dx - \dfrac{1}{2b}h^2(x)\varphi^2(x)$, it is easy to know $H(0) = 0$ by the notation of $H(x)$ above, $h(x)\varphi(x)$ increases monotonously by conditions (I), that is $\mathrm{sgn}\, h(x)\varphi(x) = \mathrm{sgn}\, x$, $H'(x) = \dfrac{1}{b}\{ab - [h(x)\varphi(x)]'\}h(x)\varphi(x)$, so when $x > 0$, $H(x)$ increases monotonously; $x < 0$, $H(x)$ decreases monotonously. Note that $H(0) = 0$, so at that time, $x \neq 0$, $H(x) > 0$.

The conditions (II) shows that when $y \geq 0$,
$f(y) \geq (b + a\tau_1 + 2a\tau_2)y$,
namely $f(y) - by \geq (a\tau_1 +; 2a\tau_2)y \geq 0$;

when $y < 0$, $-f(y) \geq -(b + a\tau_1 + 2a\tau_2)y$, namely $f(y) - by \leq (a\tau_1 + 2a\tau_2)y \leq 0$, so

$\int_0^y [f(y) - by]dy \geq 0$.

Again $\dfrac{1}{2}(ay+z)^2 \geq 0$, $\int_{-\tau_1}^{0}\int_{t+s}^{t} a^2z^2(u)\,du\,ds \geq 0$,
$\int_{-\tau_2}^{0}\int_{t+s}^{t} a^2y^2(u)\,du\,ds \geq 0$, so $V(x,y,z)$ nonnegative. When $V = 0$, there is $H(x) = 0$, there will be $x = 0$ by the above prove, generation into the system (2.2) can get $y = 0$, $z = 0$, so the $V(x,y,z)$ positive definite.

(ii) Prove $\left.\dfrac{dV}{dt}\right|_{(2.2)} \leq 0$.

From the above:
when $y \geq 0$, $f(y) \geq (b + a\tau_1 + 2a\tau_2)y$;
when $y < 0$, $-f(y) \geq -(b + a\tau_1 + 2a\tau_2)y$, we know that $f(y)y \geq (b + a\tau_1 + 2a\tau_2)y^2$.

Therefore

$$\left.\frac{dV}{dt}\right|_{(2.2)} = \{ah(x)\varphi(x) + y[h(x)\varphi(x)]'\}y$$

$$+\{h(x)\varphi(x) + a(ay+z) + f(y)$$

$$+\frac{1}{z}\int_{-\tau_2}^{0} a^2[y^2(t) - y^2(t+s)]ds\}z$$

$$+\{(ay+z) + \frac{1}{\dot{z}}\int_{-\tau_1}^{0} a^2[z^2(t) - z^2(t+s)]ds\}$$

$$[-h(x(t))\varphi(x(t)) - f(y(t)) - g(z(t))$$

$$+ \int_{-\tau_1}^{0} f_y'(y(t+s))z(t+s)\,ds$$

$$+ h(x(t))\int_{-\tau_2}^{0} \varphi_x'(x(t+s))y(t+s)\,ds]$$

$$= y^2[h(x)\varphi(x)]' + a^2 yz + az^2$$

$$+ \int_{-\tau_2}^{0} a^2[y^2(t) - y^2(t+s)]\,ds - ayf(y)$$

$$- ayg(z) - zg(z) + \int_{-\tau_1}^{0} a^2[z^2(t) - z^2(t+s)]\,ds$$

$$+ (ay+z)\int_{-\tau_1}^{0} f_y'(y(t+s))z(t+s)\,ds$$

$$+ (ay+z)h(x(t))\int_{-\tau_2}^{0} \varphi_x'(x(t+s))y(t+s)\,ds$$

$$= -\{ab + a^2\tau_1 + 2a^2\tau_2 - [h(x)\varphi(x)]'\}y^2$$

$$+ a^2 yz + az^2 - ayg(z) - zg(z) + a^2\tau_2 y^2$$

$$- \int_{-\tau_2}^{0} a^2 y^2(t+s)\,ds + a^2\tau_1 z^2 - \int_{-\tau_1}^{0} a^2 z^2(t+s)\,ds$$

$$+ \int_{-\tau_1}^{0} |ay(t) + z(t)| \, |f_y'(y(t+s))| \, |z(t+s)| \, ds$$

$$+ \int_{-\tau_2}^{0} |ay(t) + z(t)| \, |h(x(t))\varphi_x'(x(t+s))| \, |y(t+s)| \, ds$$

Because

$$\int_{-\tau_1}^{0} |ay(t) + z(t)| \, |f_y'(y(t+s))| \, |z(t+s)| \, ds$$

$$\leq \int_{-\tau_1}^{0} |ay(t) + z(t)| \, |Lz(t+s)| \, ds$$

$$\leq \int_{-\tau_1}^{0} \frac{1}{2}[(ay(t) + z(t))^2 + L^2 z^2(t+s)]\,ds$$

$$\leq \int_{-\tau_1}^{0} \frac{1}{2}(a^2 y^2 + 2|ayz| + z^2)\,ds$$

$$+ \int_{-\tau_1}^{0} \frac{1}{2}L^2 z^2(t+s)\,ds$$

$$\leq \int_{-\tau_1}^{0} \frac{1}{2}[a^2 y^2 + (a^2 y^2 + z^2) + z^2]\,ds$$

$$+ \int_{-\tau_1}^{0} \frac{1}{2}L^2 z^2(t+s)\,ds$$

$$\leq a^2 y^2 \tau_1 + z^2 \tau_1 + \int_{-\tau_1}^{0} \frac{1}{2}L^2 z^2(t+s)\,ds$$

Similarly, we can attain

$$\int_{-\tau_2}^{0} |ay(t) + z(t)| \, |h(x(t))\varphi_x'(x(t+s))| \, |y(t+s)| \, ds$$

$$\leq a^2 y^2 \tau_2 + z^2 \tau_2 + \int_{-\tau_2}^{0} \frac{1}{2}M^2 y^2(t+s)\,ds$$

Therefore

$$\left. \frac{dV}{dt} \right|_{(2.2)} \leq -\{ab + a^2\tau_1 + 2a^2\tau_2 - [h(x)\varphi(x)]'\}y^2$$

$$+ a^2 yz + az^2 - ayg(z) - zg(z) + a^2\tau_2 y^2$$

$$- \int_{-\tau_2}^{0} a^2 y^2(t+s)\,ds + a^2\tau_1 z^2 - \int_{-\tau_1}^{0} a^2 z^2(t+s)\,ds$$

$$+ a^2 y^2 \tau_1 + z^2 \tau_1 + \int_{-\tau_1}^{0} \frac{1}{2}L^2 z^2(t+s)\,ds + a^2 y^2 \tau_2$$

$$+ z^2 \tau_2 + \int_{-\tau_2}^{0} \frac{1}{2}M^2 y^2(t+s)\,ds$$

$$= -\{ab - [h(x)\varphi(x)]'\}y^2 + ay(az - g(z))$$

$$- zg(z) + (a + a^2\tau_1 + \tau_1 + \tau_2)z^2$$

$$- (a^2 - \frac{1}{2}L^2)\int_{-\tau_1}^{0} z^2(t+s)\,ds$$

$$- (a^2 - \frac{1}{2}M^2)\int_{-\tau_2}^{0} y^2(t+s)\,ds$$

$$\leq -\{ab - [h(x)\varphi(x)]'\}y^2$$

$$+ \frac{a}{2}[y^2 + (az - g(z))^2]$$

$$- zg(z) + (a + a^2\tau_1 + \tau_1 + \tau_2)z^2$$

$$- (a^2 - \frac{1}{2}L^2)\int_{-\tau_1}^{0} z^2(t+s)\,ds$$

$$- (a^2 - \frac{1}{2}M^2)\int_{-\tau_2}^{0} y^2(t+s)\,ds$$

$$= -\{a(b - \frac{1}{2}) - [h(x)\varphi(x)]'\}y^2 + \frac{a}{2}g^2(z)$$

$$- (a^2 + 1)zg(z) + (\frac{a^3}{2} + a + \tau_1 + \tau_2 + a^2\tau_1)z^2$$

$$- (a^2 - \frac{1}{2}L^2)\int_{-\tau_1}^{0} z^2(t+s)\,ds$$

$$- (a^2 - \frac{1}{2}M^2)\int_{-\tau_2}^{0} y^2(t+s)\,ds$$

By conditions (1) $-\{a(b - \frac{1}{2}) - [h(x)\varphi(x)]'\}y^2 \leq 0$ (if and only when $y = 0$, the equal sign is established); because $a^2 - \frac{1}{2}l^2 \geq 0$, then $a^2 - \frac{1}{2}L^2 \geq 0$, $a^2 - \frac{1}{2}M^2 \geq 0$, so $-(a^2 - \frac{1}{2}L^2)\int_{-\tau_1}^{0} a^2 z^2(t+s)\,ds \leq 0$, $-(a^2 - \frac{1}{2}M^2)\int_{-\tau_2}^{0} y^2(t+s)\,ds \leq 0$.

If $\frac{a}{2}g^2(z)-(a^2+1)zg(z)+(\frac{a^3}{2}+a+\tau_1+\tau_2+a^2\tau_1)$

$z^2 \triangleq H(z) \leq 0$ can be proved, $\left.\frac{dV}{dt}\right|_{(2.2)} \leq 0$ can be obtained.

Here is to prove that $H(z) \leq 0$. If $z=0$, then $H(z) \leq 0$; If $z \neq 0$, we can regard $H(z)$ as quadratic function about $g(z)$, as

$$\Delta = [1-2a(\tau_1+\tau_2+a^2\tau_1)]z^2 = \Delta_1 z^2 \geq 0,$$

so

$$H(z) = \frac{a}{2}[g(z)-\frac{(a^2+1)z-\sqrt{\Delta_1}\,|z|}{a}][g(z)$$

$$-\frac{(a^2+1)z+\sqrt{\Delta_1}\,|z|}{a}].$$

According to (V), when $z>0$,

$$\frac{(a^2+1)z-z\sqrt{\Delta_1}}{a} < g(z) < \frac{(a^2+1)z+z\sqrt{\Delta_1}}{a},$$

there is $H(z)<0$; when $z<0$, Similarly, we can get $H(z)<0$, so $H(z) \leq 0$. then $\left.\frac{dV}{dt}\right|_{(2.2)} \leq 0$.

(iii) Prove the set of $\{(x,y,z): \left.\frac{dV}{dt}\right|_{(2.2)} = 0\}$ except 0 and does not contain the whole positive orbit of system (2.2),and the positive orbit of system (2.2)is bounded.

a. Fiystly, prove the set of $\{(x,y,z): \left.\frac{dV}{dt}\right|_{(2.2)} = 0\}$ except 0 and does not contain the whole positive orbit of system (2.2).

let $K = \{(x,y,z): \left.\frac{dV}{dt}\right|_{(2.2)} = 0\}$, easily, we can prove that K does not contain the whole positive orbit of system (2.2) except for trivial solution.

In fact, $K \subset \{(x,y,z)\,|\,x,z \in R, y=0\}$, then when $y=0$, from (2.2) we get $x=0$, $z=0$.

b. Then prove the positive orbit of system (2.2) is bounded.

Taking such a large positive number L and N, which make any point $P_0(x_0, y_0, z_0)$ included into the area D. D is determined by

$$V(x,y,z)<L \text{ and } |y|<N.$$

Clearly, D is bounded region. We prove the orbit of system (2.2) which through point P_0 do not extend out D.

Factly, for $\left.\frac{dV}{dt}\right|_{(2.2)} \leq 0$, we know if the point of the orbit leave from the eare,it must go through the plane part of the region boundary, namely, there is such monment T, making $|y(T)|=N$.

Below prove this is contradictory.

For $V(x,y,z)<L$, there is $\frac{1}{2}(ay+z)^2 < L$,

Namely, $-ay-\sqrt{2L} < z < \sqrt{2L}-ay$ (2.4)

If $y=N$, from the right inequality of relation (2.4), there is $z < \sqrt{2L}-aN$;

If $y=-N$, from the left inequality of relation (2.4), there is $z > aN-\sqrt{2L}$.

This is to say, when N is large enough,

$$z \cdot \mathrm{sgn}\, y = \frac{dy}{dt} \cdot \mathrm{sgn}\, y < 0 \text{ holds. Therefore, when the}$$

orbit of system (2.2) intersect the plane part of the region D boundary, the orbit should go through from outside to inside ,so the positive orbit of system (2.2) is bounded.

From (i), (ii), (iii),we know zero solution of system (2.2) is global asymptotic stability

Then zero solution of system (2.1) is global asymptotic stability.

3 DISCUSSION

1 When $\tau_1 = 0$, system (2.1) is

$$\dddot{x}(t)+g(\ddot{x}(t))+f(\dot{x}(t))+h(x(t))\varphi(x(t-\tau_2))=0,$$

$$g(0) = f(0) = h(0)\varphi(0) = 0. \tag{3.1}$$

According to theorem 1, get

Deduction 1 If there is the constant $a>0, b>\frac{1}{2}, L>0, M>0$,and$l \leq \sqrt{2}a (l = \max\{L,M\})$,

$a\tau_2 \leq \frac{1}{2}$, holding $0 < [h(x)\varphi(x)]' < a(b-\frac{1}{2})$, $f(y)$

$\mathrm{sgn}\, y \geq (b+2a\tau_2)|y|$, $|f_y'(y(u))| \leq L$, $|h(x)\varphi_x'$

$(x(u))| \leq M$, $\frac{(a^2+1)-\sqrt{\Delta_1}}{a} < \frac{g(z)}{z} < \frac{(a^2+1)+\sqrt{\Delta_1}}{a}$

$(z \neq 0, \Delta_1 = 1-2a\tau_2)$, then

Then zero solution of system (3.1) is global asymptotic stability.

492

(2) When $\tau_2 = 0$, system (2.1) is

$$\ddot{x}(t) + g(\ddot{x}(t)) + f(\dot{x}(t - \tau_1)) + h(x(t))\varphi(x(t)) = 0,$$
$$g(0) = f(0) = h(0)\varphi(0) = 0. \qquad (3.2)$$

According to theorem 2, get

Deduction 2 If there is the constant $a > 0, b > \dfrac{1}{2}$,

$L > 0, M > 0$, and $\quad l \leq \sqrt{2}a \quad (l = \max\{L, M\})$,

$(a(\tau_1 + a^2\tau_1) \leq \dfrac{1}{2}$, holding $0 < [h(x)\varphi(x)]' < a(b - \dfrac{1}{2})$,

$f(y)\operatorname{sgn} y \geq (b + a\tau_1)|y|$, $|f_y'(y(u))| \leq L$, $|h(x)\varphi_x'|$,

$(x(u))| \leq M$, $\quad \dfrac{(a^2 + 1) - \sqrt{\Delta_1}}{a} < \dfrac{g(z)}{z} < \dfrac{(a^2 + 1) + \sqrt{\Delta_1}}{a}$

$(z \neq 0, \Delta_1 = 1 - 2a(\tau_1 + a^2\tau_1))$

Then zero solution of system (3.2) is global asymptotic stability.

(3) When $\tau_1 = \tau_2 = \tau$, system (2.1) is

$$\ddot{x}(t) + g(\ddot{x}(t)) + f(\dot{x}(t - \tau_1)) + h(x(t))\varphi(x(t - \tau_2)) = 0,$$
$$g(0) = f(0) = h(0)\varphi(0) = 0. \qquad (3.3)$$

According to theorem 1, get

Deduction 3 If there is the constant $a > 0$, $b > \dfrac{1}{2}$,

$L > 0, M > 0$, and $\quad l \leq \sqrt{2}a \quad (l = \max\{L, M\})$,

$a(a^2 + 2)\tau \leq \dfrac{1}{2}$, holding $0 < [h(x)\varphi(x)]' < a(b - \dfrac{1}{2})$,

$f(y)\operatorname{sgn} y \geq (b + 3a\tau)|y|$, $|f_y'(y(u))| \leq L$, $|h(x)\varphi_x'|$

$(x(u))| \leq M$, $\quad \dfrac{(a^2 + 1) - \sqrt{\Delta_1}}{a} < \dfrac{g(z)}{z} < \dfrac{(a^2 + 1) + \sqrt{\Delta_1}}{a}$

$(z \neq 0, \Delta_1 = 1 - 2a(a^2 + 2)\tau)$,

Then zero solution of system (3.3) is global asymptotic stability.

4 NUMERICAL EXAMPLES

In order to illustrate the application of the main conclusion, we are here to take just one example.

Example Considering the system

$$\ddot{x} + (2 + \frac{\sin \ddot{x}}{4})\ddot{x} + \frac{9}{4}(\dot{x}(t - \frac{1}{8})) + x(t - \frac{1}{16}) = 0, \qquad (4.1)$$

Then $g(z) = (2 + \dfrac{1}{4}\sin z)z$, $f(y) = \dfrac{9}{4}y$, $h(x) = 1$,

$\varphi(x) = x$, $\tau_1 = \dfrac{1}{8}$, $\tau_2 = \dfrac{1}{16}$.

Let $a = 1$, $b = 2$, $L = 1$, $M = 1$, we have $l = 1$. It is easy to conform the condition of theorem 1. So, The system (3.3) is global asymptotic stability (That is given in figure 1).

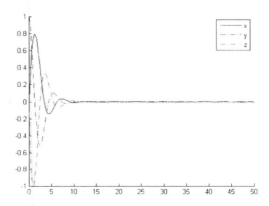

Figure 1. The solution of the delay differential equations.

5 CONCLUSION

In this paper, the global asymptotic stability of a kind of third-order differential equations with two delays is proved, and we obtain some deduction. But because here the delay is constant, the question is relatively easy to study. When the delay is variable, situation will be more complicated, it remains to be further research.

REFERENCES

[1] ZhANG Li-juan, Wu Yan. Global asymptotic stability of a class of third-order nonlinear system [J]. College Mathematics, 2007, 23(1):70–74.
[2] LIU Chang-dong. Global asymptotic stability of a class of third-order nonlinear PDE [J]. Journal of Zhanjiang Normal College(Natural Science Edition), 1999, 20(2): 16–18.
[3] KANG Hui-yan. On the deduction about the two theorems of global stability of a class of third-order nonlinear system [J]. Applied mathematics, 2001,14 (add):7–9.
[4] ZHANG Li-juan, SI Li-geng. Global asymptotic stability of a class of three order nonlinear system [J]. Acta Mathematicae Applicatae Sinica, 2007, 30(1):99–103.
[5] Qian C. On global stability of third-order nonlinear differential equations [J]. Nonlinear Anylysis, 2000, 42: 651–661.

[6] Omeike M O. Further results on global stability of third-order nonlinear differential equations [J]. KANG Hui-yan. Global asymptotic stability of a class of three order nonlinear system [J]. Journal of Jiangsu Polytechnic University,2004,19(3):41–42.

[7] ZHANG Li-juan, Kang Hui-yan. Global stability of a class of third-order nonlinear delay system [J]. Journal of Jiangsu Polytechnic University,2005,17(4):48–50.

[8] Cemil Tunc. New results about stability and boundedness of solutions of certain non-linear third-order delay differential equations [J]. The Arabian Journal for Science and Engineering, 2006, 31 (2A):186–196.

[9] WANG lian,Wang Mu-qiu. Qualitative analysis of nonlinear ordinary differential equation [M]. Harbin: Harbin Institute of Technology Press,1987.

494

Information, Computer and Application Engineering – Liu, Sung & Yao (Eds)
© 2015 Taylor & Francis Group, London, ISBN 978-1-138-02717-6

An advanced mobile indoor tracking system based on fingerprint

Yue Cheng Zhang
School of Software, Shanghai Jiao Tong University, Shanghai, China

Run Zhao
Computer Science Department, Shanghai Jiao Tong University, Shanghai, China

Dong Wang
School of Software, Shanghai Jiao Tong University, Shanghai, China

ABSTRACT: Radio Frequency (RF) fingerprinting, based on WiFi is a popular approach for indoor localization and it was attracting many research efforts in the past decades. However, most of the system is mainly focused on absolute positioning, but ignoring one important aspect that if the system is applied to a mobile phone, movement is constrained because of human native behavior. If we combine the method of continuous positioning with the fingerprint, the tracked location will be estimated more reasonable instead of being positioned in a strange place. In our system, we introduce an approach that makes use of fingerprint for navigation and we believe that such method is closer to the actual. Our system utilizes the sensors on mobile phones and is designed to run in a compact environment. With little human interaction except carrying mobile phones, the system reaches a satisfactory result. The main creation of our paper is (a) fingerprint-based navigation, (b) particle filtering to estimate location and (c) pattern training to adjust cumulative error.

KEYWORDS: Indoor localization; RSS Fingerprint; Mobile phone; Tracking

1 INTRODUCTION

With the rapid development of technology, mobile computing has caught much attention and became the most popular issue among researchers. In most of today's LBS applications such as hospital, smart mall and so on, location have been the most important issue in the context aware. Considering the huge business value in the pervasive computing, a lot of solutions have been proposed to solve the localization problem especially for the indoor localization.

A majority of past indoor localization approaches applies Received Signal Strength (RSS) to determine the location. As RSS is the easiest to obtain and a lot of models such as fingerprint and propagation could be applied, many systems use RSS to track mobile things based on WiFi or RFID signal strength. In these methods, training is essential in the first phrase. Researchers collect the RSS characteristic from one location to another and build the relationship between location and RSS in the database. When the system receives a location query, the estimated location will be calculated and sent back.

Besides, there is also a dead-reckon method introduced into indoor localization. This method makes use of a mass of speed sensors and adjustment algorithm to trace moving items. When user carries a

smart phone, these methods record the step by the accelerometer and orientation by compass.

However, the way of carrying phones has a big bottleneck. Compass need to be horizontal laying and this is impossible in reality. In our study, we use RSS fingerprints to navigate and reach the same accuracy as the traditional method with fewer sensors. In this paper, we will investigate the recent study on indoor localization technology in Section II. In section III, we'll describe the main architecture of the system into two phrases. In section IV, the method of fingerprint database construction will be specified into three key points. In section V, we will show you the algorithm in operating phrase.

2 RELATED WORKS

In the past two decades, many techniques have been proposed to solve indoor localization. Generally, they can be summarized into 3 categories: fingerprint-based, model-based and dead-reckoning-based method.

A. Fingerprint-based method

The main idea of the fingerprint-based method is to measure the signature fingerprint of every location in the areas and then build a fingerprint database.

The location is then estimated by mapping the measured fingerprints against the database. The first fingerprint-based system is Radar [2]. Radar approach makes use of existing WiFi infrastructure and establish the relationship between RF signals and position in database offline. Then the location is estimated by mapping fingerprints in database in online phrase. LANDMARC [1] use positive RFID tags to build fingerprint and works well. LiFS[3] investigate sensors integrated in android mobile phones and leverage user motions to construct the signal strength map of a floor plan, which is obtained only by site survey in previous studies.

B. Model-based method

Model-based schemes calculate locations based on geometrical models rather than search for best-fit signatures from per-labeled reference database. For example, the log-distance path lose (LDPL) model that is the most widely used in distance estimate, builds up a semi-statistical function between RSS values and RF propagation distances. EZ[6] is one well-known location system that calculate location by the physics of wireless propagation, but EZ assume many unknown signal transmitters.

C. Dead-reckoning-based method

Dead-reckoning is a simple idea that uses the accelerometer and the compass to track a mobile user. The main difficult points in dead-reckoning-based method is how to adjust cumulative error. Unloc[4] introduces organic landmarks and unsupervised clustering to adjust error. The decision tree is used for detecting SLMs. With pattern recognized, Unloc can calibrate the user location to landmarks. Zee [5] is another dead-reckoning system while the original position is not needed. The key idea of Zee is that when a user walk continuously in an indoor environment, through hallways and corners, the possibilities of the user's path and location shrink gradually.

3   SYSTEM ARCHITECTURE

In this section, we present the architecture of the system. Same to many other indoor localization system, there are two phases in our system, including training phase and operating phase (Figure1).

The training phase is based on uncertain fingerprints. During each step of the collectors, mobile phones record WiFi RF signal characteristics. With the step and orientation in the user's path, the location of a fingerprint can be measured. By mapping the path to the floor plan, the fingerprint is associated with the location and the fingerprint database is constructed.

Figure 1.   System architecture.

The next phrase is operating phrase that make use of the fingerprint database built by the training phrase. We introduce an RF-based method to measure orientation. With the particle filter approach, errors can be reduced and such method gives a good result. Details will be described in the following section.

4   FINGERPRINT SPACE CONSTRUCTION

In this section, we'll mainly describe the training phase, which include how to collect fingerprints and how to make use of these fingerprints.

A. Step Counting

Step Counting is important in both train phase and operating phase. In our system, we assume there are two states, idle and walking. When user is in idle status, it is expected that acceleration value is low. Thus the standard deviation of the acceleration would be a good indicator. However, is not enough. In order to figure out how acceleration will be when a user is walking, we use a method named repetitive nature. Acceleration shows a repeating pattern when walking. The reason is that the rhythm of nature walk is composite by a two-step sequence.

B. Fingerprint Collection & Space Construction

After a user setup is figured out, we use an electronic compass embedded in mobile phones. Suppose we have a map of one floor plan. A user walks around carrying mobile phones and leave his path as Figure 2 in our system, the fingerprint is also recorded in our

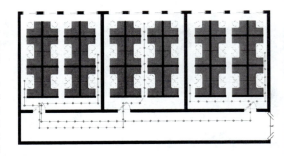

Figure 2.   Moving path while collecting.

system. After fingerprints are collected, we use a method called MDS to build the floor plan for mapping the path.

C. Landmark Recognition

As we all know, dead-reckon method may cause a huge error if we don't adjust the model, while adjusting to increase accuracy is a big challenge. In our system, we use multiple landmarks such as WiFi fingerprint landmarks and accelerometer landmarks.

Considering an area overheard a set of WiFi AP, each grid of locations can be corresponding to a set of fingerprints. We compute the standard deviation of each grid. If deviation is less than a certain value, we dim such point as a WiFi fingerprint landmark.

5 DEAD-RECKONING

In this section, the operating phrase will be introduced. Our method is to adopt probabilistic methods to compute the next position based on the fingerprints. Suppose we know the current position and movement of this time, we can measure the next position.

A. Fingerprint-based Orientation

As the RF propagation model shows, the power received in one place is interrelated with the distance between transmitters and receivers. In 2D axis,

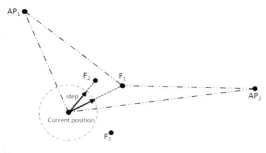

Figure 3.   Principle of next position generation for each AP & PF.

$$P_i = P_0 - 10\gamma \log d_i + X_\sigma \qquad (1)$$

P_0 means the power received from AP, while P_i represents the power in i^{th} location farther than the reference location. The distance between the reference location and the in i^{th} location is d_i.

Suppose a situation like Figure 3, there is one user which is at position p_0. Besides, there is two APs $< AP_1, AP_2 >$ and three fingerprints $< F_1, F_2, F_3 >$ around the current point. When a step is monitored, we consider the user will move a one-step distance.

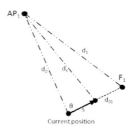

Figure 4.   Location relationship between AP, current position, fingerprint position and next position.

The orientation is possible according to the surrounding fingerprints. If the orientation is the same as that of fingerprint F_1, the theoretical value could be computed according to the RF propagation model. As Figure 4 shows, AP_1, $P_{current}$, F_1 three vertex constitute a triangle. With one step, the user will reach a next possible position which is one-step away from the origin. We name it as P_s and it must be on the line of $P_{current}$ and F_1. If we connect AP_1 with $P_{current}$, P_s and F_1 and choose three points on each line which is as far as to the AP_1, we can construct the propagation formula between those points.

Next, we use the Least Square Method to compute the best r_s. As systemic error exit, r_s and the observed signal strength R will not be same. Actually, there will be several APs and we can calculate each theoretical strength and observed value. The distance between the theoretical $< r_{1s}, r_{2s}, .., r_{ns} >$ and $<r_1, r_2, .., r_n >$ is measured be Euclidean distance. We use this distance as a factor of possibility to this point.

B. Particle Filter

Particle filters provide an effective way to generate samples from the distribution that does not need to assume the state distribution. Consider the location is a sample of state-space, we recur the next location by Bayesian equations and every fingerprint are observed samples. We assume the most possible location as average.

6 EXPERIMENT

The experiment is deployed in a club which has about 5 APs. The 4 APs are placed at the center of each border and one is placed at the center of the room in Fig 5. While, we choose Nexus 4 as a mobile phone which is equipped with accelerometer, compass and WiFi sensors. We compare the real locations to the estimated ones from one step to another and built the comparison diagram (Fig 6).

The estimated path quite close to the real one. The mean error is about 1.5m and assuming the 2m as

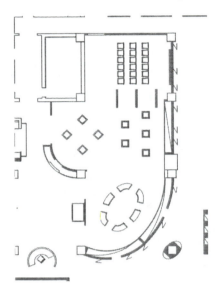

Figure 5.   Experiment environment.

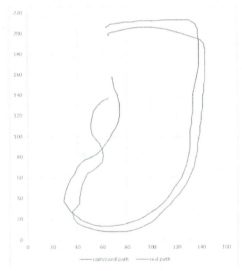

Figure 6.   Comparison between real and computed.

Table 1.   Estimated error.

Average	Max	Variance	Error rate
1.52m	2.98m	0.654	24.42%

error threshold, our system can reach the precision at 75% probability which is shown in Table I.

7 CONCLUSION

In our system, we choose a new thought that when a user moves from one fingerprint to another, the path will be restricted by the block of space such as walls, tables and etc.

On this basis, we design and implement the system based on accelerometer and WiFi sensors. Our system costs the same time as the other fingerprint-based system, but reach the same accuracy with few sensors.

ACKNOWLEDGMENTS

This research has been supported by the National High Technology Research and Development Program of China (863 Program) (2006AA04A105) and 2013 Shanghai Science and Technology Innovation Program (13dz1509800).

REFERENCES

L. Ni, Y. Liu, C. Yiu, and A. Patil, LANDMARC: Indoor Location Sensing Using Active RFID. In WINET, 2004.

P. Bahl and V. N. Padmanabhan. RADAR: an in-building RF-based user location and tracking system. In INFOCOM, 2000 Proceedings of IEEE, 775784.

Z. Yang, C. Wu, and Y. Liu. Locating in Fingerprint Space: Wireless Indoor Localization with Little Human Intervention. In MobiCom, 2012 Proceedings of ACM, 269–280.

H. Wang, S. Sen and A. Elgohary, No Need to War-drive: Unsupervised Indoor Localization. In MobiSys, 2012 Proceedings of ACM, 197–210.

A. Rai, K. K. Chintalapudi, and V. N. Padmanabhan, Zee: Zero-Effort Crowdsourcing for Indoor Localization. In MobiCom, 2012 Proceedings of ACM, 293–304.

K. Chintalapudi, A. Padmanabha Iyer, and V. N. Padmanabhan, Indoor Localization Without the Pain In MobiCom, 2010 Proceedings of ACM, 173–184.

Information, Computer and Application Engineering – Liu, Sung & Yao (Eds)
© *2015 Taylor & Francis Group, London, ISBN 978-1-138-02717-6*

Research on the storage virtualization of SVC in the enterprise data center

Chun Yu Li, Xiao Yuan Peng & Pei Ping Zhu
Department of Physical Science and Technology, Kunming University, Kunming, China

ABSTRACT: With the changes in business development and operations, The IT infrastructure of the enterprise has become particularly complicated due to the increased growth of new and unplanned applications. Enterprise users will face the constant increase of cost to store data. In addition, more and more complex environment makes it impossible to manage the storage data, This paper describes how "SAN islands" problem in the enterprise storage systems can be solved by using SVC storage virtualization to integrate enterprise data center, extending seamlessly effective solution to reducing management complexity and improving efficiency.

KEYWORDS: SVC; storage virtualization; enterprise; heterogeneous storage

1 INTRODUCTION

According to statistics, the global data storage is increasing rapidly. The Internet data is increasing by 50 percent annually. Gartner predicts that by 2020, the global amount of data will reach 35ZB, which equals to 8,000,000,000 4TB pieces of hard drive. Changes in the structure of the data will bring new challenges to the storage system [1]. Enterprise users will be faced with the two main problems. One is that the cost of storing data is constantly increasing, the other is that more and more complex environment makes it difficult to manage the data storage. So it has become an urgent issue for business users to classify and plan business data reasonably while reducing costs. In addition, it is important to achieve the integration of heterogeneous storage environments, to manage and solve "SAN islands" problem in enterprise storage systems, and to realize the seamless expansion. With IBM SVC storage virtualization technology to be used in the data center, those problems will be solved, reducing management complexity and improving efficiency.

2 OVERVIEW OF RELEVANT THEORIES

2.1 Storage virtualization

Storage virtualization technology means more than a storage medium module are centralized through virtualization technology. All the memory modules are managed uniformly in a storage pool. Thus multiple storage devices of homogeneous or heterogeneous can be centralized, providing users with a high-capacity and high data transfer bandwidth storage system. The separation of management of physical memory from the logical management will be achieved, making it transparent to access memory [2]. In fact, the storage system virtualization technology is essentially to virtualize the storage system [3]. Storage virtualization not only allows logical storage unit to be integrated within the range of WAN but also moves from one disk array to another without the need of downtime. In addition, by using the storage virtualization technology, the storage resources can be allocated based on actual usage of users and the need for additional physical storage hardware systems can also be reduced, resulting in the ultimate reduction of the overall energy consumption of the data center. For users, virtualized storage resource is like a huge "storage pool", in which users will not see a specific disk or tape. They do not need to care which path their own data will go through and what specific storage device they will go to.

2.2 The basic concept on SVC

The capacity of multiple disk systems can be integrated into a single "capacity pool" by IBM SVC (SAN Volume Controller). SVC can both help save space and energy and be merged to simplify the management of storage resources, which will greatly improve the utilization of existing storage as well as reduce the demand for additional storage [4]. SVC virtualizes storage by means of In-Band. SVC system is actually a cluster system, which is composed by nodes. An SVC system contains at least two nodes, and each consists of a two-node I / O Group, providing Host with I / O services (4).

In an SVC system, one or more memory cells in the storage subsystem are mapped into the internal storage unit SVC MDisk (Managed Disk). One or more Mdisk can be turned into a virtual storage pool (called MDG). MDG is a storage pool, which allocates the virtual storage units according to some allocation strategy. It is called VDisk. I / O Group provide LUN-Masking (also known as LUN-Mapping) service for Host in Vdisk units so that Host can have access to Vdisk through HBA.

3 INTEGRATING ENTERPRISE DATA CENTER BY MEANS OF SVC STORAGE VIRTUALIZATION

In order to solve "SAN islands" problem in the enterprise storage systems, the server farms can be effectively used and integrated. Thus the cost of IT budgets will be reduced and multi-platform interoperability of storage together with data sharing will be realized. SVC is added between the storage system and the host server with the help of IBM SVC storage virtualization technology. It does not affect the topology of existing SAN environment .As is shown in Figure 1:

Connect SVC to the fiber optic switch. The control functions will be centralized to the storage network through a certain configuration and the capacity of multiple disk systems will be integrated into a single storage pool, building a unified reasonably highly scalable architecture to manage the storage system in a unified way. The system service will be provided in the form of virtual volumes after virtualization

and has a dynamic creation and expansion of logical functions, which will greatly improve the efficiency of storage management. Although the administrator have to face a large number of heterogeneous storage devices, they do not need to consider which types of equipment the back-end physical storage belong to, as well as the disk space used by the server. SVC is able to detect the various subsystems of the memory cell and these storage units can be mapped into MDISKS, which masks the differences in various storage systems. System administrators can simply focus on the management of all storage operations

The application of SVC storage virtualization in enterprise data center results in the following aspects:

1 The flexible disk management functions

SAN Volume Controller in SVC has flexible disk management functions, which will greatly improve the efficiency of storage management, such as dynamically creating , extending a logical volume, and so on. Moreover, SVC will not only offer a unified data replication platform for a variety of storage devices, but also help administrators reallocate and expand the storage capacity without affecting application availability to customers. Besides, it will achieve replication across multiple storage systems. The enterprise unified disaster recovery strategy for data can be used even though each site supports different storage devices. Real-time disaster recovery

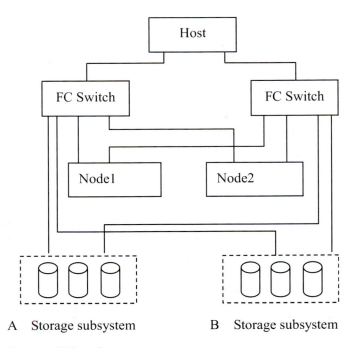

A   Storage subsystem   B   Storage subsystem

Figure 1.   SVC topology.

and data migration between storage systems from different vendors can be achieved, avoiding putting more costs at different time to ensure that data is not lost.

2 Transparent data migration.

When the SVC is added to an existing SAN environment, there is no need to do data migration. SVC will keep the existing disk configuration wholly intact. In this way the application on the server is completely transparent. When the SVC is fully configured, it can transparently migrate original volume and the data on the disk to other real virtual volume. All the migration process is transparent to the server, so there is no need to terminate the application

3 The conveniences of management

SAN Volume Controller of SVC has convenient graphical interface, which centralizes all the storage virtualization management through the GUI management interface [5]. It configures, manages, and serves the multiple storage systems from different vendors

4 Tiered Storage

An Volume Controller makes it easier to implement tiered storage because it can maintain a consistent management functionality between all of the storage layers. Furthermore, it will not cause disruption to the application while the data is moved between the layers. SVC has also cache so that it can provide a lower level of storage performance. It can be widely used in the data center, reducing costs.

4 CONCLUSION

The application of SVC in enterprise will fully integrate the existing different types and brands of storage resources. It will help achieve the storage capacity sharing, simplify the implementation of a tiered storage architecture optimized for storage portfolio, and increase its lower-level storage amounts. The reliability, credibility, scalability and manageability of the data center will be realized. The problem that the enterprise data center storage resource utilization is low and the storage system expansion capability is poor will be solved fundamentally.

REFERENCES

[1] The Global Data Storage Explosive Growth, http://www.d1net.com/storage/news/254753.html
[2] Tom clark, Storage virtualization [J]. Addison_wesley.2007.
[3] Shenglong Tan, Research of Storage Virtualization Technology [J], Microcomputer Applications, 2010 (001): 33–38.
[4] The Solution Proposal by IBM SVC heterogeneous storage Integration, http://www-03.ibm.com/systems/cn/resources/systems_cn_05_SVC_090709.pdf
[5] Yunying Li, The Application of Storage Virtualization Technology in Enterprise Data Centers[j]. Computer Knowledge and Technology, 2014.6

Information, Computer and Application Engineering – Liu, Sung & Yao (Eds)
© 2015 Taylor & Francis Group, London, ISBN 978-1-138-02717-6

Application of PSO-SVM model in logistics demand forecasting in Jiujiang

Feng Xiang Ren & Jing Tao
Jiangxi Vocational College of Finance and Economics, Jiujiang, China

ABSTRACT: Logistics demand forecast is an important basis for the development of modern logistics construction. This article from the status quo of logistics demand of Jiujiang, through the analysis of factors to measure indexes and influence its logistic demand scale, the scale of logistics demand to establish the comprehensive and systematic prediction index system, and for these specific is quantified, and then through the combination of PSO-SVM regression model to forecast and better prediction results, the results also indicate that Jiujiang logistics demand in the growth of the state, in line with the actual development trend of logistics in Jiujiang.

KEYWORDS: Support Vector Machine (SVM); Particle Swarm Optimization (PSO); Logistics Demand; Forecast.

1 INTRODUCTION

In recent years, our country towards the city of industrialization and economic globalization, the direction of rapid development, especially the countries introduced a number of supportive policies to promote the development of the logistics industry, logistics planning is put into the city strategy, has entered a rapid development period. To accurately predict the scale of logistics demand and the development trend, reasonable planning, the construction of logistics infrastructure, to the national and local making the development strategy, has the important meaning of sustained and stable development and economic.

2 INFLUENCING FACTORS OF LOGISTICS DEMAND SCALE AND QUANTITATIVE ANALYSIS

2.1 *Influencing factors of logistics demand scale*

The impact of regional logistics demand scale is very extensive and complex, but the logistics demand as a derived demand, is the product of economic development. It is closely related to regional economic level[2]. Therefore, this article mainly analyzes the influence of economic factors on the scale of logistics demand. Based on Beijing's economic and social development, there are several key factors affecting the demand scale as follows [3–4].

1 The level of economic development. At different stages of economic development, a lot of logistics demand in quantity and quality difference. The speed of the development of logistics and the development speed is proportional to the GDP.

2 The industrial structure. The difference of the industrial structure and the disequilibrium of decision of characteristics of logistics demand structure. The first industry occupies the dominant position in the gross national product, logistics demand structure for low value added product logistics demand of the dominant, the second industry accounted for the largest proportion in the gross national product, high value-added logistics demand is increasing rapidly, the third industry in the GDP share of more than 45%, high level logistics service demand for knowledge, technology characterized occupy an important position.

3 Regional total retail sales of consumer goods. Demand for regional total retail sales of consumer goods and logistics have a close relationship, the consumer decides production, regional consumption determines the flow of goods, consumer demand is the logistics demand, may also put forward higher requirements to the existing logistics service, create new logistics demand.

4 The regional foreign trade. To the port city or river city opening to the outside world, the foreign trade logistics in the total proportion of regional logistics scale cannot be ignored.

5 Total investment in fixed assets. The growth in fixed asset investment, can promote the development of modern service industry, the total investment in fixed asset level will directly affect the development of the logistics infrastructure equipment construction and logistics industry.

6 Level of Urbanization. The city is the commodity distribution and processing center, as the center of modern logistics, city level is an important condition for the development of the logistics industry,

at the same time, the development of the logistics industry to promote the city development.

7 The consumption level of residents. With the level of consumption increase, the customer's purchase ability increased. In the meantime, diversity and personalized consumer demand promote regional delivery volume increase quickly.

2.2 Analysis of logistics demand scale

There are two kinds of indexes, viz. quantity of work and service value, to evaluate the scale of logistics demand [5–6]. The number of index, physical indices is essentially a logistics demand. The actual volume of goods is essentially the quantitative index, such as the volume of freight transport, warehousing and distribution amount, etc. Value index is essentially the quality index of logistics demand. The actual volume of the value is essentially the quality index, For example, the total social logistics, logistics costs and logistics industry output value, etc. But because of the regional logistics is to meet the regional economic activities and the lives of the residents, the research object is all logistics activities within the region, involving the time and scope, its value is difficult to effectively measure, the state statistics are the lack of data in this area. From the physical perspective, transportation is the most important link in the logistics, through the logistics activities always freight volume determines the amount of work of other logistics activities. Although the transport demand is only a part of logistics demand, but the freight volume can reflect the change of logistics demand in a certain extent. Therefore, this paper chooses the freight volume to measure regional logistic demand scale.

3 ESTABLISHMENT OF PREDICTION INDEX SYSTEM

Based on the statistical data can be acquired and Jiujiang statistical data of the different selection, regional GDP (GDP) X_1 (100 million yuan), the first industry output value X_2 (100 million yuan), the output of the second industry X_3 (100 million yuan), the output of the third industry X_4 (100 million yuan), the total volume of retail sales of the X_5 area (100 million yuan), the regional foreign trade volume X_6 (100 million yuan), the total investment in fixed assets of X_7 (100 million yuan), city level X_8 (%), the consumption level of residents X_9 (yuan). As the economic index prediction of Jiujiang logistic demand scale. Select the volume of transport (Y tons) as Jiujiang logistics demand scale index, specific statistical data collected in table 1.

4 PREDICTION OF THE SCALE OF LOGISTICS DEMAND IN JIUJIANG

4.1 SVM regression model based on particle swarm optimization

In the application of the support vector machine method, two important problems need to be addressed.

1 Choice of kernel function. Although as long as the functions satisfying Mercer condition can be used as kernel function, but the return performance of different kernel functions have different.

$$K(x, x_i) = \exp\left[-\frac{\|x - x_i\|^2}{2\sigma^2}\right], \sigma > 0 \qquad (1)$$

Table 1. Jiujiang logistics demand forecasting index data size.

Year	X_1	X_2	X_3	X_4	X_5	X_6	X_7	X_8	X_9	Y
2003	306.8	51.6	147.3	107.8	85.6	2.4	119	35	2935	1220
2004	360.2	62.5	178.9	118.9	106.1	1.8	159.5	35.6	3414	1594
2005	428.9	72.1	214.8	142	121.3	2.4	213.5	38.5	3619	2540
2006	506.2	78.2	262.9	165.2	139.6	3.6	255.1	39	4307	3306
2007	592	82	315	195	163.8	3.7	358.3	40	5018	3705
2008	700.6	83.9	384.9	231.8	201.9	4.8	455	40.8	6035	3901
2009	931.4	92	441.1	298.3	240.5	7.1	659	43.1	6897	8992
2010	1032	98	579.7	354.3	284.1	18.1	877.5	44..4	7680	9746
2011	1256	107.7	735.9	412.8	329.7	37.9	1006.5	45	8925	10828
2012	1420	113	820	487	400	45.2	1300	46.1	10115	12541
2013	1570	117.5	924.9	525	430	47.4	1570	47.5	14356	13398

2 Optimization of the parameters. If you choose the RBF kernel function, the parameter to be optimized is the regularization parameter γ and the kernel parameter σ. Learning ability and the generalization ability of the two parameters of the model of the impact is very big, the general method of using artificial search to obtain. But the artificial search process is relatively trouble, time-consuming. And it often fails to get the optimal parameters. Particle swarm optimization (PSO) algorithm is a kind of excellent evolutionary algorithm of fast development in recent years. The basic idea of PSO is first initialized a bunch of random particle, and then through the iterative search for the optimal solution. In each iteration, the particles can be tracked through the two "extreme" to update their location. One is the optimal particle itself to find solutions, namely individual extremum pBest, another is the optimal overall population currently looking into a solution, the extreme called global extreme value gBest. Particles found above two extreme, to update the velocity and position themselves according to the following two functions.

$$v = wv + c_1 rand\ ()\ (p_{Best} - p) + c_2 rand\ ()\ (g_{Best} - p)$$
$$p = p + v \tag{3}$$

And v is the particle velocity, p is the current position of the particle, pBest and gBest are respectively the individual extremum and global extremum, $c1$, $c2$ are called the learning factor, usually between 0–2 value, rand() is a random function between 0 to 1. w is the weighted factor, as the iteration proceeds, w decreases from the maximum weighted factor w_{max} to the minimum weighted factor w_{min}.

$$w = w_{max} - \frac{i \times (w_{max} - w_{min})}{i_{max}} \tag{4}$$

And i is the current iteration times, i_{max} is the maximum number of iterations.

4.2 Parameter optimization of SVM based on PSO algorithm

If the formula is as follows:

$$\min \frac{1}{l} \sqrt{\sum_{i=1}^{l} (y_i - \overline{y_i})^2} \tag{5}$$

And y_i is the real value, $\overline{y_i}$ is a predictive value, l is the number of samples. The parameter optimization problem can be simplified to: To find the optimal parameters of (γ, σ), through SVM learning, makes the formula (5) the minimum value. This paper selects the formula (5) as the objective function of PSO algorithm, the choice of kernel function of radial basis

function, the parameters of the PSO algorithm using optimized (γ, σ). The specific algorithm as described in the following steps5:

Step 1. The sample set is divided into a training set and testing set.

Step 2. Initialization settings. Let $iter_{max}$ be the maximum number of iterations, w_{max} and w_{min} be respectively the largest weighting factor and the minimum weighted factor. Random generation k particles, the position of each particle is a two-dimensional vector, Representing the SVM parameters γ and σ.

Step 3. According to the current parameters γ and σ, we calculate the fitness value by formula (5).

Step 4. The corresponding memory individual and group the best fitness value position pBest and gBest.

Step 5. Position, velocity of particle is updated, in search of better γ and σ by formula (2) and (3).

Step 6. Repeat step 3 until it meets the maximum number of iterations.

Step 7. Predict the test sample by using the trained SVM.

4.3 Jiujiang logistics market demand forecasting method based on PSO-SVM model

Let the interval data sequence of the year be t, the data sequence is obtained as:

$$x(0), x(t), \cdots, x(it), \cdots, x((n-1)t) \tag{6}$$

The data sequence is also the input sequence predicted the future value. N years ago m values used to predict the n values can be expressed as a search for the following function relationship f.

$$x_{nt} = f(x_{t(n-1)}, x_{t(n-2)}, \cdots, x_{t(n-m)}) \tag{7}$$

Let $x(it)$ be simplified as x_i for convenience, then the formula can be simplified as:

$$x_{nt} = f(x_{t(n-1)}, x_{t(n-2)}, \cdots, x_{t(n-m)}) \tag{8}$$

The training sample is constructed as follows: Enter the corresponding to the output a time for xm+1. Enter the corresponding to the output of a year is xm+2. It can be drawn by the L training samples can build a training sample of l-m. By constructing the training samples to train the SVM model, the parameters of SVM are optimized by particle swarm optimization algorithm. After the completion of training the model, the future value of first step prediction is obtained with the following function:

$$x_{n+1}^{pre} = f(x_n, x_{n-1}, \cdots, x_{n-m+1}) \tag{9}$$

505

The second step forecast for:

$$x_{n+2}^{pre} = f(x_{n+1}^{pre}, x_n, x_{n-1}, \cdots, x_{n-m+1}) \quad (10)$$

Then we forecast after each step. xi is the i point of the real value, x_i^{pre} is the I point of the predicted value, i=1, 2,..., j.

4.4 Data sources and pretreatment

This paper selected the data from the Jiujiang Statistical Annals (2008) (see Table 1). According to the principle of selecting sample data, selecting 2004~2010 years as network training sample data, 2011~2013 data as the network of test samples, by setting a good network of logistics demand forecast 2014~2015 year. Because the data have different dimension, required to be normalized, the data processing for the interval between [1, 2] data. The normalized formula [1].

$$\bar{x}_i = \frac{x_i - x_{min}}{x_{max} - x_{min}} \quad (11)$$

This paper uses MATLAB to realize the normalization process.

4.5 The prediction results and analysis

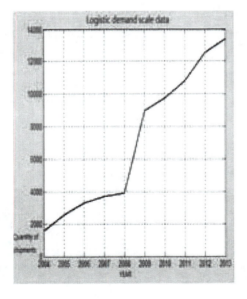

Figure 1. Logistic demand scale curve.

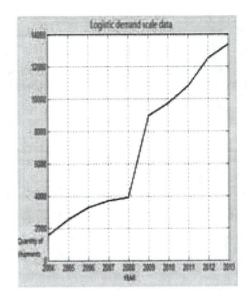

Figure 2. Logistics demand scale curve after normalization.

Best Cross Validation MSE = 0.01852 Best c = 8 Best g = 0.125

Mean squared error = 0.00277665 (regression)

Squared correlation coefficient = 0.980524 (regression)

From the Figures 4 and 5, we learn that the prediction method has better prediction accuracy and reliability of the support vector machine, its predictive value is close to the actual value, MSE = 0.00277665. The maximum relative error is 8.02%. Data in 2008 is rejected. There is a big error in the prediction of 2008 on account of the financial crisis in U.S 2008. Besides, the support vector machine prediction model to predict the effect is better, when the data appear larger fluctuation, can still have a strong sensitivity, and not because the data appear larger fluctuations in the predicted effect is particularly big impact. Therefore, the use of support vector machine to predict the freight volume of Jiujiang during 2004~2015 is very suitable.

According to the prediction model established, we forecast the scale of logistics demand in Jiujiang during 2014~2015, as shown in table 3.

5 CONCLUSION

This article is based on the present situation of logistics demand in Jiujiang. Through analysis of the factors that influence measure and its logistic demand

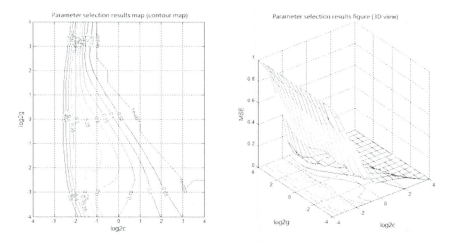

Figure 3. Optimize the parameters of SVM results based on the PSO.

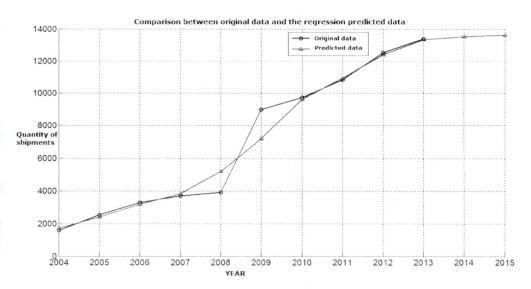

Figure 4. Comparison between the predicted effect of freight volume.

scale, the authors establish an index system of forecast the scale of logistics demand more comprehensive and systematic, and quantify these indicators. There is a good prediction result with the method of PSO-SVM. The results show that Jiujiang logistics demand in the growth of the state, in line with the actual development trend of logistics in Jiujiang, and provide the basis for the modernization of Jiujiang logistics industry development in the short term.

Due to the limited information available, there is some difference in logistics demand between the actual situation and predicted values. Hence, it is necessary to have a further study and discussion in order to have more accurate results.

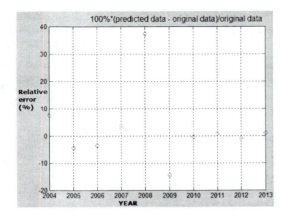

Figure 5.   Freight volume forecast relative error.

Table 3.   Prediction results of port cargo volume during 2014–2015.

Year	Logistics market freight volume (million tons)
2014	13 520
2015	13 650

REFERENCES

[1] CAI Xiaoli, CHEN Chouyong. The Prediction and Analyse of Regional Logistics Demand[J]. Logistics SCI TECH, 2004(09):15 (in Chinese).
[2] Zhang Lixue. Study on the Method of Urban Logistics Demand Forecasting[R]. Southeast University, Jiangsu, 2006.04. (in Chinese).
[3] Banerjee M, Chakraborty MK. A category for rough sets[J]. Foundations of Computing and Decision Sciences, 1993, 18(3–4):167–180.
[4] XU Yu. Factor Analysis of Logistics Demand in Hunan Province[R]. HuNan University, 2007.(in Chinese).
[5] Vapnik VN. The nature of statistical learning theory[M]. NewYork:Springer-Verlag, 1995.
[6] Xuegong Zhang. Introduction to statistical Learning theory and support vector machine[J]. Acta Automatica Sinica, 2000, (1):32–42.
[7] SVM based on improvement Particle Swarm Optimization Algorithm[J]. Computer Engineering and Applications, 2007, 43(15):44–49.
[8] Particle Swarm Optimization with Contracted Ranges of Both Search Space and Velocity[J]. Journal of Northeastern University(Natural Science), 2005, 26(5):488–491.

Information, Computer and Application Engineering – Liu, Sung & Yao (Eds)
© 2015 Taylor & Francis Group, London, ISBN 978-1-138-02717-6

A fast collision detection algorithm based on the hybrid bounding box

W. Zhao

School of Electronics and Information Technology, Zhejiang University of Media and Communications, Hangzhou, China

Z. Li

School of Information Technology, Jilin Agricultural University, Changchun, China

H. Y. Qu

School of Computer Science & Technology, Jilin University, Changchun, China

ABSTRACT: A kind of Sphere-OBB hybrid hierarchical bounding box algorithm aiming at the disadvantage of traditional hybrid hierarchical bounding box is put forward in this paper. The OBB bounding box is optimized by a new storage method, and a prediction of the OBB bounding box is added in earlier detection. The experimental results show that, the real-time, accurate of the collision detection and the testing efficiency were improved by the improved algorithm, the feasibility and superiority of the improved algorithm is fully proved in the detection efficiency.

KEYWORDS: OBB; Collision detection; Bounding box

1 INTRODUCTION

With the rapid development of the field of virtual reality, the requests to the reality of a virtual scene of the people become higher and higher, the collision detection algorithm has become an indispensable part, fast collision detection algorithm can meet the real-time and accuracy and now has become a hot social research today. The hybrid hierarchical bounding box algorithm can combine the merits between different bounding box, in order to improve the real-time of the collision detection, and thus to the extensive and in-depth research, and shows a more and more important trend. In this paper, through the study of temporal and spatial correlation of the motion of an object, give an improved in view of the defects of the algorithm, the paper puts forward a hybrid hierarchical bounding box algorithm based on temporal and spatial correlation. Through the contrast can be obtained, the improved method has the obvious advantage of multi model in the simulation scene.

2 IMPROVEMENT AND IMPLEMENTATION OF THE ALGORITHM

2.1 Arithmetic statement

Compared with the traditional hierarchical bounding box algorithm, improved hybrid hierarchical bounding box tree first used spatial segmentation to quickly eliminate the distance of the object, determine the adjacent objects. Then the adjacent objects using a method of hybrid hierarchical bounding box for exact collision detection. For the hybrid hierarchical bounding box continued use the three layers structure. Improved mixture bounding box adopted single bounding box form in addition to middle structure. Compared to the use of hybrid bounding box, it reduced the complexity of the intersection test, improved the speed and also reduced the time. The improved Hybrid hierarchical bounding box collision detection algorithm joining the predicted OBB bounding box collision detection before the hybrid hierarchical tree traversal. Use of the spatial of the object to exclude distant objects, So the root node only uses sphere bounding box. This reduces the burden of the update stage, to make it more simple, so as to achieve the purpose of reduce the bounding box intersection test time and improve the intersection test speed. The basic collision detection algorithm pseudo code description of the improved hybrid hierarchical bounding box tree is as follows:

```
{
    if (Wx, Wy is the root nodes)
    {
        if (B(Wx)∩B(Wy)=Null)
        {
            if(Wx, Wy is the leaf nodes)
```

{ Do Sphere-OBB bounding box intersection test;

   Do OBB intersection test between bounding box;

     if(Qx ∩ Qy=Null)

       {

          return (Ture)

       }

     else

       {

          Traversal the node of Wx and Wy

          Traverse(Gx ,Wy) ,Traverse(Wx,Gy);

       }

  }

    else

      {

         Traversal the node of Wx and Wy

         Traverse(Gx ,Wy) ,Traverse(Wx,Gy);

      }

  }

  Else

    {

      return (False)

    }

  }

  else

    {

      return (False)

    }

}

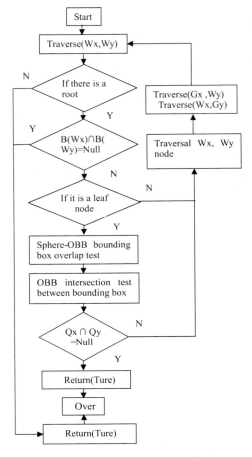

Figure 1. Flow chart of the improved hybrid hierarchical bounding box algorithm.

The improved hybrid hierarchical bounding box algorithm flow chart shown in figure 1.

Among them:

Wx,Wy: Representing the node detection of the two objects collision detection bounding box tree;

Gx,Gy: Are sub node set corresponding to node Wx and node Wy;

B(Wx), B(Wy): Represent the bounding box containing a collection of Wx and Wy;

Qx,Qy: Represent the basic elements of Wx and Wy;

2.2 Analysis of hybrid hierarchical bounding box

This hybrid hierarchical bounding box proposed in this paper using two kinds of bounding box, Sphere and OBB. The Sphere bounding sphere is selected as the root node bounding box, because Sphere bounding sphere has the simple characteristics, and easy to update after rotation and translation. The middle layer using Sphere, OBB hybrid bounding box is because of two kinds of bounding box can make objects relatively surrounded closely, and the intersect detection logarithmic of bounding box is minimum, because the 0BB bounding box is more suitable for the rigid body, so it is used as the underlying bounding box, and the bottom of all using a single OBB bounding box to improve the speed of detection, so, although the OBB bounding box structure is relatively complex, but do not affect the test results in the study of the primary stage of the collision detection process in this chapter, and can ensure the optimal speed of this phase collision detection, save time for the whole detection process. In the structure of the tree, using the classic two fork tree as the structural form of the tree. Although in a variety of tree structure, the two fork tree is the simplest, but it has fast calculation[5], select top-down method as constructing method of tree structure, mainly because of this mode in the process of collision detection using more, and the technology is more mature than bottom-up, and easier to implement and more robust. Finally, using the static structure method to structure the hierarchical bounding box tree to play the effect of accelerating algorithm. Improved hybrid bounding tree shown in Figure 2.

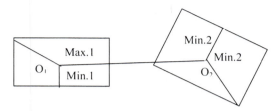

Figure 3. The position relationship diagram between two OBB bounding box.

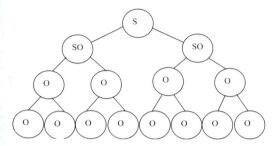

Figure 2. Improved structure diagram of hybrid bounding box tree.

2.3 Add prediction fast collision detection of OBB bounding box

For traditional bounding box algorithm, in the early stage of the collision detection, the use of bounding box can be quickly removed the objects impossible collision. However, if only plus bounding box on the object outermost layer, in the accurate collision detection, still do intersection test in all of the geometric elements of the two object. The accurately test time of the two collision objects to be tested still has no change. Aiming at the shortage, can add the fast collision detection of OBB bounding box before the traversal of hybrid hierarchical bounding box, predict their intersection by predicting the two bounding box relationship between the center distance and the furthest distance to the nearest distance of the surface.

Because of the high complexity intersection test of the OBB bounding box, so adopt separating axis algorithm intersection test. The basic idea is the first prediction before use the separating axis judge the collision between bounding boxes, use the intersection test method based on separating axis only in the situation of prediction results may intersect. The position relation of two OBB bounding box shown in Figure 3.

Among them:

OBB.1 and OBB.2: Represent the two tested OBB bounding box;

O_1 (x_1, y_1, z_1) and O_2 (x_2, y_2, z_2): represent the center of two bounding boxes;

Min .1 和 Max.1: Represent the minimum distance and maximum distance of the OBB.1 surface point to the center;

Min .2 和 Max.2: Represent the minimum distance and maximum distance of the OBB.2 surface point to the center;

L: Represent the center distance between OBB.1 and OBB.2;

$$L = \sqrt{(x_1 - x_2)^2 + (y_1 - y_2)^2 + (z_1 - z_2)^2} \qquad (1)$$

$$\text{Min.1} + \text{Min.2} < L < \text{Max.1} + \text{Max.2} \qquad (2)$$

The pseudo code of the prediction collision detection algorithm between two OBB bounding boxes as follows:

Pre.Collision B(OBB.1,OBB.2)
    {if (L>Max.1+Max.2)
     OBB.1∩OBB.2=Ø
     if (L<Min.1+Min.2)
       B1∩B2≠Ø
     else test OBB.1-OBB.2
    }

2.4 The optimization, storage mode of OBB bounding box tree

After the OBB hierarchical bounding box tree is established, do the two bounding box intersection test by traversing the two trees of the bounding box tree. The steps are as follows:

1 Take the root nodes of the bounding box tree X and bounding box tree Y, Box_X and Box_Y, If the pointer to Box_X or Box_Y is empty, return, on the other hand, step (2);

2 Judge the intersection of Box_X and Box_Y, if does not intersect, then return, on the other hand, step (3);

3 Judge whether Box_X and Box_Y is a leaf node, if they were all leaf nodes, then do the triangle intersection test, record the results, and make corresponding processing, on the other hand step (4);

4 Judge the contains the patch number of Box_X and Box_Y,if the number of patch Box_X contains less than Box_Y, then Box_Y take the root node of Y,Box_X take each child node of X, return, continue with step (1); if the number of patch Box_X contains larger than Box_Y, then Box_X take the root node of X, Box_Y take each child node of Y, then return continue with step (1);

Through the above steps,if put the information of original triangle leaf nodes in the parent node, skip the intersection tests between the bounding boxes of the leaf nodes,direct the basic geometrical elements intersection test, the number of the node needs to store is only N-1, it's basically removed half of the nodes, thus saved the storage space,The number of nodes needs traverse also decreased significantly when testing, the storage structure of the bounding box after optimization shown in Figure 4.

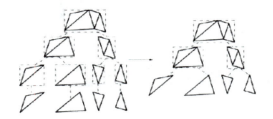

Figure 4. The optimization storage methods of bounding box tree.

3 EXPERIMENTAL RESULTS AND ANALYSIS

This experimental application in the PC machine of 2.80 GHz CPU and 2GB memory, and use the VC6.0 platform to realize. Since this paper study the collision of rigid body, so use the Rapid algorithm to compare and analyzed, detect efficiency of the collision. Table 1, Figure 5 respectively the contrast of the detection time and the average detection time between the improved algorithm and the Rapid algorithm.

Table 1. The contrast of the detection time between the improved algorithm and the Rapid algorithm.

The triangle number	Rapid/ms	The algorithm in this paper/ms
1468	4,5462	4,2283
3672	7,9825	7,4978
5983	12,1240	11,4318
7552	14,8913	12,9731
9867	19,1320	15,9854

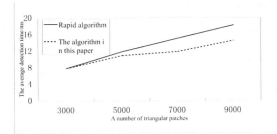

Figure 5. The contrast of the average detection time between the improved algorithm and the Rapid algorithm.

The experimental results show that, under the condition of the increasing triangle element number, although the improved algorithm rose relatively slowly in the detection time curve, but in detection time, it is obviously better than the Rapid algorithm. It can be proved that the improved algorithm in the collision detection process of the rigid body, shorten the collision detection time, speed up the object's collision response and improved the real-time of the collision detection.

4 CONCLUSION

For efficiency of the collision detection algorithm, after the characteristics study of various bounding box, this paper proposes a collision detection algorithm based on the hybrid hierarchy bounding box suitable for rigid body,in the early detection, join the predicted OBB bounding box, use temporal and spatial correlation of the motion objects speeds up the collision detection to achieve the effect of quickly removed the distant objects and determined the adjacent objects, then do exact collision detection for neighboring objects by using the method of improved hybrid hierarchical bounding box. Proved by experiments, the improved hybrid hierarchical bounding box is more simple and effective to improve the intersection test speed, and reduces the collision detection time with real-time in rigid body collision detection.

ACKNOWLEDGMENTS

This work was supported by the following: it was supported by the following Funds: Young and middle-aged talents and outstanding innovation team of science and technology plan projects of Jilin province (20121818), Natural Science Foundation of Jilin Province (20101521,201215182); Jilin Science and Technology

Development Project (201201095, 201201101). Changchun Science and Technology Support Program of Things major science and technology under item number 12XN14,13KG27. Education Department of Jilin Province "Twelve—five" key projects of science and technology research "Corn pest diagnosis and prevention system based on things and image reconstruction" (I ke Jiao he zi [2013] No. 58) Education Department of Jilin Province "Twelve—five" projects of science and technology research " key technology research of real-time simulation " (I ke Jiao he zi [2013] No. 73), key project of Natural Science Foundation of Jilin Province 20140101196JC.

REFERENCES

[1] H.B.LI,D.Y.ZHOU & Y.WU.Dec.2010.Collision Detection Algorithm Based On Mixed Bounding Box [J].Journal of Computer Applications.Vol.30, No.12.

[2] Z.WEI,W.H.LI.July 2009.Fast Collision Detection Algorithm for Spherical Blend Reconstruction[J]. Computer Science.Vol.36, No.7.

[3] J.H.GAN,Q.PENG,P.D.DAI.Oct.2011.Improved Collision Detection Algorithm Based on Oriented Bounding Box.Journal of System Simulation.Vol.23, No.10.

[4] T. NING,G.Chen,S.W.ZHANG.Oct.2011.Optimization of collision detection method using hybrid bounding box.Computer Engineering and Applications, Vol.47, No.1.

[5] W.ZHAO,R.P.TAN, Y.LI.Jan 2010.Real Time Collision Detection Algorithm in Complex Virtual Environment[J].Journal of System Simulation.Vol.22, No.1.

[6] Bay H,Ess A,Tuytelaars T.et al.2008.Speeded-Up Robust Features (SURF) [C]//CVIU.346–359.

[7] Christopher A. Otieno G N, Nyakoe C W W.2010. A Neural Fuzzy Based Maximum Power Point Tracker for a Photo voltaic System[EB/OL]. 11–21.

[8] S.Suri, P.M.Hubbard,J.F.Hughes.1999.Analyzing Bounding Box for Object Intersection[J]. Journal of the ACM,18(3): 257–277.

[9] Z.Qiu,L.Yang.2012.A new feature-preserving nonlinear anisotropic diffusion for denoising images containing blobs and ridges[J].Pattern Recognition Letters,33(3):319–330.

Information, Computer and Application Engineering – Liu, Sung & Yao (Eds)
© *2015 Taylor & Francis Group, London, ISBN 978-1-138-02717-6*

University library in the functional expansion of the digital age

Yan Qiu Wang

Jilin Teachers' Institute of Engineering & Technology, Changchun, Jilin, China

ABSTRACT: The function of university library has received more extensive attention in the ubiquitous information society. With the development of mobile Internet and intelligence, university libraries are facing many challenges. This paper analyzes the basic function of university library, and probes into the expansion of university functions from aspects such as improving the construction of digital resources, paying attention to and participating actively in open access etc.

KEYWORDS: Library; Digital age; Function

1 WAVE OF UBIQUITOUS INFORMATION SOCIETY SURGES

1.1 *Concept of ubiquitous information society*

Ubiquitous refers to interconnection between people and people, people and object, objects and objects, which can be achieved through wireless communication in any condition at any time. The goal is to "building a network environment of multi language, multimedia, mobility and semantics to promote the development and reform of information service".

Ubiquitous information society uses ubiquitous technique to perform data acquisition, processing and analysis based on intelligent ubiquitous network providing "ubiquitous" and "transparent" information service.

1.2 *Characteristics of ubiquitous information society*

Ubiquitous Object: Being no longer limited to objects between person and person, and various devices and terminals are also objects of ubiquitous society. All can be interconnected through network to realize transmission and sharing of global information.

Ubiquitous Network: Ubiquitous network refers to comprehensiveness of network interconnection, diversification of intelligent equipment and popularity of communication mode. It is an intelligent information environment in which any person can access to network and enjoy service at any time, anywhere.

Ubiquitous Information: Connect the information of matters itself and the related information properties, such as state, location, type etc. so as to allow all kinds of information to be more easily obtained and utilized by users.

Ubiquitous Application and Service: With the development of mobile Internet and intelligence, information service can be embedded into various application and operating systems for the convenience of timely access to needed resources and services by different information users.

1.3 *Ubiquitous information service*

Defining from the technical perspective, the ubiquitous information service refers to network developing service based on various wireless, mobile Internet, terminals of various cross screen, multi screen information service, cloud computing promoting intelligence of information service, data mining and the like technologies.

Defining from the content perspective, ubiquitous information service mainly refers to multimedia digital fusion, conducting big data mining and knowledge open access.

Defining from the service strategy perspective, ubiquitous information service refers to service based on location, embedded service, perceptibility of personalized service, benefit of coexistence of fee-based service and free service, socialization of instant service, multiple collaborative services.

2 INTERNATIONAL FEDERATION OF LIBRARY (FILA) TREND REPORT

2.1 *The concept of the "IFLA trend report"*

"IFLA trends report" is the report on future development trend in new information environment, which is obtained by many experts and related personnel who consult for twelve months. It defines five macro trends in the global information environment, involving information acquisition, education, privacy, citizen participation and technical transformation. Nowadays how to evolve to keep it not to be marginalized in the new information environment is one of the most urgent problems faced by libraries.

2.2 Content of "IFLA trend report"

Trend 1: New technology is a double-edged sword, and may enhance or restrict person's ability of access to information. In the constant expanding digital environment, information literacy becomes particularly important, such as basic reading and digital capability etc.. Person lacking information literacy will be heavily blocked when entering this highly-developing field. The characteristics of new online business model will influence to a great extent on whether person can successfully own, obtain, share and benefit from it.

Trend 2: Online education promotes global education to be fair, and its side effects can not be ignored. The rapid expansion of global online education resources makes one get more and cheaper learning opportunities conveniently. Lifelong learning will be accepted as universal values. Unofficial and informal learning experiences will get more and more recognition and affirmation.

Trend 3: Privacy boundary and data protection will be redefined. Big data owned by government and enterprises can be used to carry out individual analysis in detail. By means of method of complex detection and filtering data communications, tracking a will be easy and inexpensive. In network environment, personal privacy and integrity record is extremely easy to be leaked and cause serious consequences.

Trend 4: Hyperlink society will listen to new sound and built up new strength. In the highly interconnected society, the collective behaviors of some non party groups will appear more frequently and lead to the decline of traditional party politics. They speak for themselves, promoting the solution and development of certain social problems effectively. Information disclosure and open access to the government and public sectors make public services more transparent and Humanistic.

Trend 5: The new technology changes the global information economy. The fact of the constant increase of mobile device with highly interconnected function, Internet of things used in home appliances and infrastructure, and 3D print and language translation technology and so on will change the form of global information economy. The existing business model of many industries will be influenced and even overturned by new equipment generated in new technology environment. With the aid of these devices, people can develop economic activities anywhere without being limited by space.

3 TEN MAJOR CHALLENGES FACED BY A UNIVERSITY LIBRARY

Challenge 1: Resource access path and various resources providing platform appear to be complex and difficult for readers and librarians to understand. How to set access entrance to make resources more easily to be retrieved and utilized becomes an urgent problem.

Challenge 2: In addition to those provided by current COUNTER report, librarians need to know that content is used by which users in the school and how the content is used.

Challenge 3: To understand the using situation of their own and peer institutions, and to keep improving their own level. Compared with other institutions of the same level, whether the using situation is better or worse? What should be done to make it being used better?

Challenge 4: What is the significance and influence of the purchased resources on the output of institutions. That is, how does the library prove that it can effectively help to improve students, bring in more funds, or get new discovery, even obtain patent etc.. Showing these values is of great significance for obtaining financial support.

Challenge 5: Digital license model is complex. How to explain nuances well to ordinary users and senior management personnel respectively is also an attention-worth problem.

Challenge 6: How to embed service into the resources access process of researchers and students. To successfully achieve this, it is necessary to deeply understand user requirement and user behaviors during the interaction with library service. How to mining the related information and service really needed by user to influence and change people's daily life?

Challenge 7: Strengthen training for author and research personnel, especially for the beginners. Librarians give more and more training courses to the beginners as consultants of the school beyond basic retrieval skill. The trainings Include understanding of copyright, how to write project application, how to obtain the OA fund and put it into application and how to publish research achievement in the most influential magazine and so on.

Challenge 8: Make full use of data management tools to play new role of its own. Do the university library well play the role of supporting data monitoring and research management behavior of department and laboratory? If it is not the liability of library, then which department will do it?

Challenge 9: Improve the function of library staff and improve their ability. Support the combination of subscription services and open access, and meet the constant expanding demand of knowledge consultation in the school. It is required to reposition the role of librarian in the resources- tight era.

Challenge 10: How to reform the library policy to adapt to the mixed economy and the new reality? Do e-books and paper books need to be purchased at the same time? How much fund proportion should be distributed to DDA? Does textbook still needs to be bought? Do Libraries need to bear the cost of OA? All these new emerging problems are to be solved and lot of tests is going on.

4 PHOTOGRAPHS AND FIGURES

4.1 Basic functions of university library

The basic function of University Library is to serve personnel training, scientific research, social service and cultural inheritance in university. The main task is as follows: to construct literature information resource system adapting to the need of the school development, provide literature information service; to fully participate in the personnel training process of the school and campus culture construction ; to implement information quality education; to develop library science and academic studies of related fields; to organize and coordinate literature information work of the whole school; to participate in the national or regional literature guarantee system construction, implement resource co-construction, common knowledge and sharing; to play the advantages of information resources and professional advantages of librarian to provide the service for the society; to do well in resources evaluation and service evaluation, improve resources using benefit.

4.2 Improve digital resources construction

The change of user requirements and using habits, and the decline in paper documents service and growth in digital document service volume lead to the transition of literature resources construction from focusing on paper documents to focusing on digital literature. The content of digital resources construction mainly includes: buying or renting electronic journals, thesis, special topic full text library, e-books and so on; acquisition or integration of open access journals, open access books, network resources, institutional (thematic) website, social media; self built or co-built characteristic databases, institutional repository, research data etc..

4.3 Pay attention to and actively participate in open access

"Open access" becomes exchanging and sharing mode of new academic resource based on Internet, and then becomes a movement utilizing the Internet to promote free propagation of academic achievements It has developed from the initial simple coping with academic publishing crisis to reflection and process reconstruction of academic exchange and propagation mechanism. In the News briefing of global Research Council 2014 annual meeting held in Beijing in May this year, for the research papers gained public funds by China Academy of Sciences and the National Natural Science Foundation of China, the final validated version are required to be in stored in corresponding knowledge base and open access is to be implemented within 12 months after publication. This fully demonstrates the responsibility and hard work of promoting open access, benefiting society from knowledge, and driving development through innovation in the scientific circle in our country.

Open access involves two modes: one is institutional repository (Institutional Repository, referred to as IR), the other is OA Journal (Open Access Journal).

Institutional repository is convergence platform of university academic community. In this platform, knowledge producers can systematically display, and store academic achievement and research path, promote communication and exchange of achievements (faster, wider), improve the academic reputation and academic influence within field, and understand the peer progress and development trend of the field. The managers master the whole research situation and strength of institutions, understand the academic contribution of each member in the academic community, help to plan strategy of subject construction and establish developing strategy. Knowledge servers integrate explicit knowledge products of institutes, explore tacit knowledge and information, carry out knowledge service targeting personalization. What stores in Institutional Repository is scientific research achievements, therefore the main body of institutional repository construction is scientific research personnel in this institute, meanwhile, the construction of institutional repository needs close coordination cooperation of all aspects in the institute such as school management, college and department, network information center, library and so on.

University Libraries participate in domestic and foreign journals open publication. In April 17, 2012, Harvard University Library Advisory Committee expressed to the Harvard faculty in an open letter that now subscription fees of main academic journals keep rising, the subscription has been difficult to be maintained, and suggested teachers and students turn to open access to resources. According to the statistics of Open Access Journals (DOAJ) Directory of Lund University Library , as of the end of January,2014 , there are 9804 kinds of open access journals, of which, 5636 kinds can be searched at title level. There are also 1,573,847open access articles. Chinese science and technology journals open access platform (China Open Access Journals, COAJ) is the first open access, academic, nonprofit scientific and technical literature resource portal in China, and operated on line in October, 2010. The construction tasks of COAJ include: collect and store all domestic technology OA journal; conduct Chinese Open Access Journals authentication, data format, digital copyright, open style according to the international standard ; support OAI-PMH protocol, and gradually realize permanent storage of website data, and open data to library ; store data to related institutional repository, open data download and service ports; application of non- traditional measurement tool (alternative measurement) etc..

REFERENCES

[1] IFLA trend report introduced by the seventy-ninth IFLA General Conference (August 19th, 2013) http://www.360doc.com/content/14/0108/14/2011549_343589662.shtml.

[2] Xiaocong Qian. 2010, (8): 38-40) At the time of the ubiquitous network being truly ubiquitous [J].China telecom industry.

[3] University of Connecticut Libraries. Academic Liaisonprogram [EB/OL]. (January 18th,2010) http://www.lib.uconn.edu/using/services/liaison/programs.htm.

[4] University of Florida Health Science Center Libraries. Library Liaison program [EB/OL]. (January 18th, 2010) http://www.library.health.ufl.edu/servicea/liaisons.htm.

Information, Computer and Application Engineering – Liu, Sung & Yao (Eds)
© *2015 Taylor & Francis Group, London, ISBN 978-1-138-02717-6*

The exploration on teaching reform of fire investigation course based on CDIO mode

F. Qu, Y.Y. Cui & X. Wang
College of Safety Engineering, Shenyang Aerospace University, Shenyang, China

ABSTRACT: Aiming at the poor autonomic learning ability, weak engineering practice as well as the lack of team cooperation spirit and other issues, the CDIO engineering education mode is introduced into the teaching of Fire Investigation course system combination with the teaching status quo and reform objectives. The application of the mode of teaching reform can improve teaching quality and effectiveness step by step, and highlight the cultivation of various skills and professional quality for the undergraduates; all these lay favorable foundation for them to complete the fire investigation task in the future.

KEYWORDS: CDIO; Fire Investigation course; Teaching reform

1 INTRODUCTION

1.1 *Contents of CDIO*

In response to the situation of the industry development of economic globalization demands for innovation of engineering talents, the CDIO concept was originally conceived at the Massachusetts Institute of Technology (MIT) in the late 1990s. The CDIO engineering education mode stands for "Conceive", "Design", "Implement", "Operate" respectively, which represents a framework of "learning by doing", fortunately, it is an essential revolution to the lecture-based class mode. On one hand, which let the students actively practice, and take the organic connection approach between the course to learn and ponder over, as a result, the students' engineering ability are well cultivated. The comprehensive ability contains not only personal academic knowledge, also includes the students' lifelong learning ability, team communication ability and control of big system. In fact, the exercise just conforms to the training goal of CDIO, which is theoretical knowledge, and reasoning, personal and professional skills, teamwork and communication skills. So far, a few world famous universities to join the CDIO international organizations, with full CDIO engineering education philosophy and curriculum. In China, Shantou University first introduced CDIO mode on teaching reform in 2005, since then, Tsinghua University, Beijing Jiaotong University and some other domestic universities and colleges also gradually carried out research and practice of CDIO teaching mode (Bai et al. 2009, He et al. 2013).

1.2 *The necessity of the reform on fire investigation course*

Fire investigation is a course with emphasis on fire investigation theory and technology, whose aim is to train men for professional talents with the skills of on-site inquiry and interview, exploration of the fire scene, fire cause and responsibility identification theory, together with the mastery of the laws and regulations of fire investigation in our country. Theoretical and practical natures are both relatively strong. On one hand, in accordance with our college's demand in students' training target for safety engineering, fire engineering, insurance engineering and security engineering majors, that is, possessing themselves of the fire investigation and handling abilities; on the other hand, academic research and application of fire investigation at home and abroad has become increasingly active, the importance of the course is self-evident.

Along with the development of economy, the cause of fire is more and more complex, resulting in more new materials and new situation encountered in the process of fire investigation, thus, correspondingly, more new technology and new methods need to be employed thereinto. As a result, fire accident investigators are required to be able to keep pace with the times, at the same time, equipped with sustainable innovation consciousness. As an innovative fire investigator, he can not only quickly become aware of the problems existing in fire scenario, but also do his best to seek new solutions and detect more unusual details in seemingly mundane things, furthermore, to think deeply on all evidences. The current teaching method adopted in Fire Investigation course is mainly "multimedia & blackboard", which is centering on teachers, classrooms and teaching material, unfortunately, the students are objectum. Classroom teaching method still adopts priority to knowledge dissemination, and lecture are in excessive pursuit of Systematicness and integrity, What the teachers pay attention to is the quantity of instilling knowledge to students, to some extent, the teachers are more apt to use ready-made scientific knowledge to arm students' mind, correspondingly, the students would rather pursue the standard answers on the book or offered by the teacher, while seldom ask questions in class. So it's more difficult to form vivid and lively atmosphere in class, to a certain extent, against from the characteristics of the engineering talents' training mode. The cultivation content of the students is just limited to thinking practice; nevertheless, the ability of operation, practical ability and innovation ability have been greatly ignored (Zhao. 2011).

If the teaching method continues this way, that will obviously make the students lack of learning enthusiasm and thinking mode is restrained at the same time, likewise, it is unfavorable to the cultivation and improvement of innovation ability. In fact, as the social responsibility of engineering colleges and universities is to bring up interdisciplinary talents with comprehensive ability. Hence, the teaching mode reform of Fire Investigation course is extremely urgent, and the importance and necessity to perfect and build of curriculum system is beyond doubt as well.

2 THE TEACHING REFORM BASED ON CDIO MODE

The basic goal of the reform is to take CDIO mode as the background of Fire Investigation course teaching, to develop students with the survival and growth of knowledge, ability as well as quality in modern engineering environment. Moreover, it is advised to attempt to employ "project implementation" as the organization principle of the engineering practice, concentrating on developing students' self-learning ability, communication skills, team awareness and capacity, mainly through two ways, one is changing the relationship between teachers and students from the teacher's one-way impartment to the problem-solving interactive learning mode; the other is guiding the students to the active unknown field exploration instead of inheriting knowledge-based learning. For instance, to study the foreign advanced analysis and interpretation of fire scene evidence by literature search dependently.

To achieve the target, the project's basic content is: in accordance with CDIO educational concept, the interactive discussion-based teaching mode will be adopted. The students are ought to be the subject during the education process, firstly, the students actively express their viewpoints through communication, then learn from each other on the positive exchange, finally, enrich their knowledge and improve their ability step by step. All these need bold innovation on the existing teaching mode, the complex, random and probabilistic nature of fire means that it is often impossible to fully determine how the fire developed and spread. There will be times when an appropriate and necessary understanding of a particular incident requires a laboratory reconstruction and interpretation of laboratory data, as well the argument of testimony.

By means of interaction and guidance as well as self-study teaching mode between the teachers and students, such as simulated on-the-spot interview and interrogation, simulated exploration & prospection of fire scenario, simulated fire scene, the special case investigation ,etc. Consequently, the students can perceive, understand, comprehend and verify contents of teaching in the real or virtual fire scenario. I the autonomous concept, judgment and skill can be actively formed and independently mastered by the students in the acquisition of knowledge simultaneously. Specific contents are elaborated as follows:

2.1 The cultivation of self-study ability

Teachers are supposed to know well that the cultivation of the students' self-study ability can significantly improve their development potential. There are a lot of work to change the previous teachers as the main body in teaching situation, i.e., the students just passively received knowledge, for instance, first, the students should be required to collect and screen a large number of literature data, second, to find and study focus and difficult problem through preliminary analysis, next, the students integrate the preview and lecture in class, the other students can ask questions and make complements , last but not least, the above-mentioned contents would by summarized by teacher .This kind of mode can fully mobilize the students' learning enthusiasm, and obtain favorable teaching effect (Liu. 2012).

For example, the cause of a fire usually can be determined from a detailed inspection of the charred debris, combustibles, devices, and residues located at the point of (or within the area of) origin. Theoretically, the cause of a fire can be categorized into one of four classifications: 1.Natural: Act of God (e.g., lightning); 2.Accidental: Unintentional and explainable; 3.Undetermined: Cause unknown, unable to be identified; 4.Incendiary: Intentionally set (Redsicker & O'Connor. 1997). Focusing on these classifications, the students may open up discussion on details of some cases or definition to extend and deepen their preview.

2.2 The cultivation of innovative ability

Whether the fire scene takes minutes, hours, or days to investigate, the basic procedures are almost the same. Typically, fire scene investigations involve three broad areas: 1.witness interviews; 2. the physical examination; 3. forensic or engineering analysis. Since each fire is different and the circum- stances surrounding the fire are also different, the degree to which each component is involved varies from fire to fire (Daéid. 2004). Mainly depending on the complexity of the fire scene, but when facing the increasingly complex fire investigation work, innovative professional fire investigator resources are scarcer than ever before and of great significance. Fire investigation is a highly innovative work; to simply rely on the limited knowledge in the textbook or try to locate something by following up a clue rigidly is impossible to find out the fire cause. Each fire scene has its own characteristics, only taking specific analysis on the specific problem and adopt open mind can find the valuable clues. In the teaching process, it is necessary to pay attention to guide the students to take the initiative to discover the problems, and to tackle the problems through various ways like literature review and discussion. All these are helpful to foster their divergent thinking pattern.

2.3 The cultivation of the team cooperation consciousness and communication ability

The goal of a successful fire scene examination is to accurately identify the first fuel ignited and the ignition source that caused the fire. But many factors will enter into this process, and this is often not a simple task to accomplish, in another word, fire investigations can be complex endeavors that involve many different disciplines. As we know, witnesses must be interviewed, the scene examined, and analysis conducted. Therefore, no one factor or indicator can be used to determine the origin or cause of the fire solely. It is the analysis of all relevant data that leads the fire investigator to an accurate determination of the fire cause. Correspondingly, the actual fire investigation work generally can not be accomplished independently, which needs setting up the fire scene protection group, live interview group, on-site inspection group, experimental analysis group and other working group, it's important to communicate and discuss in the process of investigation timely, at last, the comprehensive information from various aspects will be used to determine the fire cause.

In order to strengthen the students' team cooperation consciousness, the comprehensive practice project would be a good choice to realize the aim by simulating fire scene or a specific case, such as an arson or a vehicle fire, different working groups should be divided to complete the whole fire investigation. This type of training can not only make students more realistically to practice fire investigation, but also give them chance to fully employ professional knowledge to find out the fire cause, on the other hand, the team cooperation consciousness could be greatly enhanced.

3 CONCLUSIONS

It is not hard to understand through comparison that the implementation of above-mentioned teaching trials can not only enliven classroom atmosphere, and perfect the teaching effect, but also make the students rack their brains trying to analyze and solve problems, in the meantime, teachers would be also inspired to novel things, thereby forming a win-win situation on teaching and learning. From the students' point of view, it is no longer the previous receiving passively, instead, they are eager to gather information, then actively conduct research by analysis and practice; what's more, the most important point is the integration of comprehensive knowledge, so that innovative ability, engineering practice ability, team cooperation ability and so on get overall promotion and exercise.

It is undoubtedly that when the CDIO engineering education mode stresses on strengthening professional fundamental education, at the same time, which is focusing on the wide range of engineering practice ability. As far as major's development is concerned, its introduction into Fire Investigation course teaching is operable, which can make the training objectives of safety engineering major more rational and feasible.

REFERENCES

Bai Mei et al. 2009. The exploration of teaching reform of architecture based on CDIO concept. *Journal of Chang chun University of Science and Technology (Higher Edueestlen Edition)* 4(4):152–154. (in Chinese)

He Guangxiang et al. 2013. Reform and Practice in Chemical Engineering Design Based on the Concept of CDIO Education. *Higher Education in Chemical Engineering* (2):42–44. (in Chinese)

Zhao Yanhong. 2011. Research and Practice on the Cultivation Mode of Talents of Innovative Fire Investigation Professional. *ESME 2011*:2165–2167. (in Chinese)

Liu Ling. 2013. Teaching method reform of fire investigation based on CDIO. *TEACHER* (32):98. (in Chinese)

David R. Redsicker & John J. O'Connor. 1997. Practical fire and arson investigation. In Elsevier (ed.), 1986. Florida: CRC Press, Inc.

Niamh Nic Daéid. 2004. Fire investigation. Florida: CRC Press, Inc.

Information, Computer and Application Engineering – Liu, Sung & Yao (Eds)
© 2015 Taylor & Francis Group, London, ISBN 978-1-138-02717-6

Research and application part-of-speech state transition in text categorization

Chong Wei Deng, Ya Nan Zhang, Zhao Peng Meng & Chao Xu
School of Software, Tianjin University, Tianjin, China

ABSTRACT: Large scale text categorization is a complex problem. A new method is proposed for text categorization based on the part-of-speech state transition. Through word segmentation and part-of-speech tagging, extracting the part-of-speech sequence of the text with specific part-of-speech. Calculate the probability of part-of-speech state transition, according to part-of-speech sequence. Text categorization can be finished fast and accurately using Bayes classifier characterized by the probability of part-of-speech state transition. Different from other current methods of text categorization which based on the feature of words, this method is based on the state transition of Markov model and does not require analysis and understanding of semantic. Experimental results show that, this method has broad applicability and remarkable effect with a small feature space and a small amount of calculation.

KEYWORDS: text categorization; part-of-speech sequence; feature extraction; Markov model; state transition; Bayes classifier

1 INTRODUCTION

With the rapid development of the Internet, the network text quantity rapid growth, to extract useful information from a large amount of text requires an effective way to organize and categorize text. How fast and reliable, large-scale text classification, is a research project with a large practical value. Text classification is mainly in two steps: First, the text feature extraction, text digitization makes it easy to identify and calculate; then choose a particular classifier for characterization of text classification

The current text classification feature extraction methods are mostly based on semantic analysis, such as term frequency-inverse text frequency (TF-IDF)[1], the information gain (IG)[2], mutual information (MI)[3], CHI statistical[4]. These methods are based on analysis of entry for features, such as TF-IDF method, statistical frequency of an entry appearing in the text of the article, as well as the total number of text entry occurs, to calculate the weight of the entries as features and then text classification; IG method is a statistical number of entries to appear or not appear to be classified in the article text. Characterized in terms of text classification method, you need a basic understanding of the text of semantic classification. First need to select entries feature item in accordance with specific rules, and then apply the appropriate classification for text classification. Because of the openness of Chinese language, as characterized by the number of entries is very large, that is, as text analysis feature

space is very large, the amount needed to calculate increases.

The very large number of entries, and continue to increase, and the part-of-speech is constant, therefore, we propose a text classification method based on part-of-speech transition characteristics. Based on part-of-speech feature extraction, classification corpus application in a foreign language, take good results[5-6]. Language is a random process, each part of speech seen as a random state, building part-of-speech sequence of text, part-of-speech sequences can be used to calculate sentence similarity[7], which was extended to the entire text, part-of-speech sequences of different texts also have different characteristics therefore calculate the state transition probabilities of part-of-speech as a feature of the text, text classification. In addition, the text length does not affect the construction of the transition probability matrix of speech, the build process detailed in the following sections. This method does not in terms of entry features, not semantic text analysis, features a small space, wide applicability, can quickly and accurately for large amounts of text classification.

2 PART-OF-SPEECH FEATURE

Part-of-speech is used to divide the parts of speech features. Modern Chinese words into two categories: content words and function words. Content words act as the main component of the sentence, function words to express all kinds of grammatical meaning or tone, emotion, content words and function words can be used as the characteristics of the text.

In this paper, Base on Institute of Computing Technology, Chinese Academy of Science (ICT) "ICTPOS3.0 Chinese POS tag set", tagging the part-of-speech, this tag set contains 98 part-of-speech, Including 22 first-class part-of-speech, 65 second-class part-of-speech, 11 third-class part-of-speech. Due to the low frequency of part-of-speech II and III appearance in the text, in order to reduce the dimension of feature vectors, so the obvious data sparse feature space phenomenon does not exist, make the part-of-speech II and III to its part of part-of-speech I, this final part-of-speech feature selection contains only 22 of a class of speech, reduce the dimension of feature vectors. This 22 part-of-speech label include content words, function words and punctuation, punctuation can reflect the rhythm of the language[8], therefore did not give up punctuation when extracting feature. Part-of-speech tagging names and symbols are as shown in Table 1.

Table 1. Part-of-speech tagging names and symbols.

noun	verb	adj	pron
n	v	a	r
num	quant	adv	prep
m	q	d	p
conj	aux	temp	loc
c	u	t	s
orient	dist	state	int
f	b	z	e
mood	onom	pref	suff
y	o	h	k
char	punct		
x	w		

3 PART-OF-SPEECH FEATURE EXTRACTION

Part-of-speech feature discussed in this article is based on the part-of-speech sequence calculated part-of-speech state transition probability matrix. First, the text segmentation and part-of-speech tagging, and then extract the part-of-speech sequences that can be regarded as Markov sequence, each part-of-speech as a state, obtained by calculating the part-of-speech state transition probability matrix.

3.1 Sequence of POS

We use ICTCLAS, a Word Segmentation System of the Chinese Academy of Sciences, for Chinese segmentation and POS tagging. 98 part-of-speeches are obtained after word tagging and speech extraction. By clustering similar words, A PA (Part-of-speech Array with 21 words) for each sample can be expressed in the following definition.

$$PA = <pa_1, pa_2, \ldots, pa_{i-1}, pa_i, pa_{i+1}, \ldots, pa_n>$$

Each feature is an element of SOP (Set of Part-of-speech) which is a set consists of {a, b, c, d, e, f, h, k, m, n, o, p, q, r, s, t, u, v, w, x, y, z}. $n \in Z$ and $n>0$, N is a positive number as there is no empty sample.

3.2 Transition probability matrix

The Russian mathematician Markov proposed a state transition matrix: The state of a system will change as time goes on. The Nth results are only affected by the N-1th time, namely the next moment states only related to the situation now, but has nothing to do with the past state. The so-called state refers to the state a system may appear or already exist[9].

The steps used for application of the state transition probability matrix and text extraction are listed following:

1 Extraction of part of speech sequence PA
2 Calculating the state transition probability

p_i stands for the speech of state pa_i and p_j stands for the speech of state pa_j. $N(p_ip_j)$ means the times speech p_i and p_j appear in WA. $N(p_i)$ means total times of speech in WA. $P(p_i|p_j)= N(p_ip_j)/ N(p_i)$ stands for the possibility of an appearance of p_i after p_j appeared.

3 Calculate State Transfer Matrix T

The transfer matrix elements in the $t_{ij}= P(p_i|p_j)$, $p_i, p_j \in SOP$.
State Transition Probability Matrix T can be obtained by calculating all the t_{ij}.

Part of the State Transition Probability Matrix is shown in figure 1.

The transition probability matrix is used for analyzing the sample's feature and text classification.

```
0        0.3333    0    0.3333   0.3333    0    0
1.0000   0         0    0        0         0    0
0        0         0    0        0         0    0
0        0         0    0        1.0000    0    0
0        0         0    0        0         0    0
0        0         0    0        0         0    0
0        0         0    0        0         0    0
0        0         0    0        0         0    0
0        0         0    0        0         0    0.0909
0        0         0    0        0         0    0
0        0         0    0        0         0    0
0        0.5000    0    0        0.1667    0    0
0        0         0    0        0         0    0
0        0         0    0        0         0    0
0        0         0    0        0         0    0
0        0         0    0        0         0    0
0        0         0    0        0         0    0
```

Figure 1. Part of the state transition probability matrix.

4 EXPERIMENT

4.1 Classification recall rate and precision

T We usually use the following indicators to evaluate the performance of text classifier.

1 Recall: After classifier completing the classification, the radio of the amount of text which has been divided corrected(N) into the total amount of the text in test set(M_1).

$$R_i = \frac{N}{M_1} \times 100\% \tag{1}$$

2 The average recall rate: the average of all kinds of text recall rate.

$$R = \frac{\sum_{i=1}^{n} R_i}{n} \tag{2}$$

3 Precision: After classifier completing the classification, the radio of the amount of text which has been divided corrected(N) into the total amount of text which are in the same kind (M_2).

$$P_i = \frac{N}{M_2} \times 100\% \tag{3}$$

4 The average precision rate: The average of all kinds of text precision rate.

$$P = \frac{\sum_{i=1}^{n} P_i}{n} \tag{4}$$

4.2 Experimental analysis

1 The constitution of the corpus and the selection of classifier

We choose five kinds of text in the Fudan corpus for multidisciplinary, cross experiment, and the corpus includes 200 texts in the domain of transportation, 200 texts in the domain of education, 200 texts in the domain of economy, 200 texts in the domain of the computer, 200 texts in the domain of the military. We selected 150 texts randomly for each type of text as the training sample, the remaining 50 as the test sample. The selected text length ranging from

200 to 20000 words. There is no further screening of the corpus. Text classification method based on the part of speech state transfer characteristic, which is described in this article, has good effect of extracting text feature, in the case of unified text. According to statistics, the entry appeared in a 200-word text, is just one over ten thousand of all entries, but the part of speech is more than half of all the parts of speech. This suggests that it is enough to extract the characteristics of the text. There are a many types of text classifier, such as Bayes classifier[10], the classifier based on the vector space model[11], the classifier based on the instance, the classifier using SVM[12] and so on. This paper uses Bayes classifier for experiments.

2 Classifier training

Experiments are conducted on binary classes, three classes and five classes. Naive Bayes models are trained on the train set and used to predict the results shown in Table 3 –Table 5.

Table 3. Binary class result.

	To Traffic	To Education
Traffic	145	5
Education	3	147

Table 4. Three class result.

	To Traffic	To Education	To Computer
Traffic	141	5	4
Education	3	143	4
Computer	4	6	140

Table 5. Five class result.

	To Traf	To Educ	To Comp	To Econ	To Mil
Traffic	131	5	4	6	4
Education	3	139	4	4	0
Computer	2	4	136	3	5
Economy	4	6	4	133	3
Military	5	6	6	7	126

In these tables, the column represents text categorization discriminated by classifier, row represents the fact belonging to corresponding text categorization. We can compute recall and accuracy rate of classification through the result of classification as shown in the table 6, table 7, and table 8.

Table 6. Recall and precision of binary classes(%).

Class	Recall	Recall
Traffic	96.67	97.97
Education	98.00	96.71
Average	97.33	97.34

Table 7. Recall and precision of three classes(%).

Class	Recall	Recall
Traffic	94.00	95.27
Education	95.33	92.86
Computer	93.33	94.59
Average	94.22	94.24

Table 8. Recall and precision of five classes(%).

Class	Recall	Recall
Traffic	87.33	90.34
Education	92.67	86.88
Computer	90.67	88.31
Economy	88.67	86.93
Military	84.00	91.30
Average	88.67	88.75

Through analyzing experiment results, we can know that with the increasing of text category recall and accuracy rate of classification will drop, which is the result of the interaction between various types of text. A traditional text categorization method based on entry, such as TF-IDF recall and accuracy rate is about 85%[13]. The recall and accuracy rate using information gain to extract features and adopting KNN classifier is about 86%[14]. While our method guarantees the accuracy rate, it also reduces the dimension of feature vector and increase the operation speed. The experiment result represents that the classifier using the state transition probability as characteristic training has a good classification performance.

3 Test of Classifier

In the test set, We select three kinds of text classifiers of test to get the result as shown in the table 9.

Table 9. Three classes result in the test set.

	To Traffic	To Education	To Computer
Traffic	46	2	2
Education	2	48	0
Computer	1	2	47

The results in Table 9 are graphically shown in Figure 2.

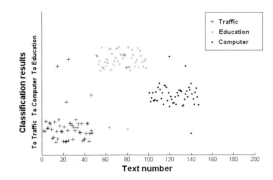

Figure 2. Three classes result in the test set.

In the Figure 2, abscissa represents the text number of the test set. In this test set, the number 1 to 50 is 50 traffic texts, number 51 to 100 is 50 educational texts and number 101 to 150 is 50 computer texts. Ordinate represents classification results. Each type of text distinguishes with different tag identification. The classification result has a good effect on cluster as shown in the Figure 2.

We use a column chart to display the result of table 9 as shown in the Figure 3.

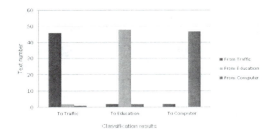

Figure 3. Column chart of three classes results.

In the Figure 3, abscissa represents the classification result, the vertical represents a number of texts. The classification result has a high recall and accuracy rate, according to contrast in the histogram. Through the result from Table 9, we can compute recall and accuracy rate of three type text in the test sample of the classifier.

The experimental results show that through analysis of POS state transition, text classification tasks can be completed quickly and accurately. This method is significantly different from text classification based on feature of the word. Performance in the following areas:

Table 10. Recall and precision of three classes result in test set (%).

Class	Recall	Recall
Traffic	92.00	93.88
Education	96.00	92.31
Computer	94.00	95.92
Average	94.00	94.04

1 Small feature space

Characterized of the POS(part-of-speech) state transition involves only few parts of speech. And Extracted the text feature by a simple calculation. The dimension of feature vectors is much smaller than text categorization method based on the analysis of the word.

2 Broad applicability

Text classification based on POS state transition, do not need to understand the semantics of the text, and extract the POS sequences directly. The experimental results show that this method has low requirements for the text standardization. It has a good classification results for long texts and short texts.

3 Versatility

Different language has different vocabulary, grammar and structure. But they can all be regarded as a Markov process. As long as the process is selected fom each state, the text feature can be extracted by calculating state transition probabilities. This method can be applied to other languages by Little change. It has good versatility.

5 CONCLUSION

Through deep analysis and research of the POS feature of text. Found that different types of text have different characteristics of POS state transition. Experimental results show that text categorization can be finished fast and accurately using Bayes classifier characterized by the probability of part-of-speech state transition. This method does not consider the semantics of the text. It is more concerned about the mathematical characteristics of the text. By analyzing these features, complete text classification tasks, and achieve a higher accuracy. The method also needs improvement. Such as when selecting POS features can apply statistics principle. Find the POS which has a large contribution to the text, and ignore the POS which has a large contribution. It can reduce the dimension of feature vectors, improve efficiency and increase accuracy.

REFERENCES

[1] ZHANG Yun-tao, GONG Ling, WANG Yong-cheng. An improved TF-IDF approach for text classification[J]. Journal of Zhejiang University. Science, 2005, 6A(1):49–55.
[2] Harun Uğuz. A two-stage feature selection method for text categorization by using information gain, principal component analysis and genetic algorithm[J]. Knowledge-Based Systems, 2011, 24(7):1024–1032.
[3] SAKAR C O, KURSUN O. A method for combining mutual information and canonical correlation analysis: predictive mutual information and its use in feature selection[J]. Expert Systems with Applications, 2012, 39(3):3333–3344.
[4] PEI Yingbo, LIU Xiaoxia. Study on improved CHI for feature selection in Chinese text categorization[J]. Computer Engineering and Applications, 2011, 47(4):128–130.
[5] Chua S. The role of parts-of-speech in feature selection[C]. Proceedings of the International Multi Conference of Engineers and Computer Scientists. Amsterdam: Reed Elsevier, 2008:124–132.
[6] Shi Kangsheng, Li Lerming. High performance genetic algorithm based text clustering using parts of speech and outlier elimination[J]. Applied Intelligence, 2012, 7(8):1–9.
[7] LAN Yan-ling, CHEN Jian-chao. Chinese Sentence Structures Similarity Computation Based on POS and POS Dependency [J]. Computer Engineering. 2011, 37(10):47–49.
[8] CHEN Fan, FENG Zhiyong. Research and application language nature rhythm in documents category[J]. Computer Engineering and Applications, 2012, 48(30):28–32.
[9] Li K, Chen G, Cheng J. Research on hidden Markov model-based text categorization process[J]. International Journal of Digital Content Technology and its Application, 2011, 5(6): 244–251.
[10] ZHOU Guoqiang, CUI Rongyi. Research on Korean Text Categorization Based on Naive Bayes Classifier [J]. Journal of Chinese Information Processing, 2011, 25(4):16–19.
[11] WEI Cheng, LIU Lu, ZHAI Ming. A Method for Web News-Text Classification with Four-dimensional Vector Space Model [J]. Microcomputer Applications, 2010 (3):58–62.
[12] Lee L H, Wan C H, Rajkumar R, et al. An enhanced support vector machine classification framework by using Euclidean distance function for text document categorization[J]. Applied Intelligence, 2012, 37(1):80–99.
[13] Zhang Baofu, Shi Huaji, Ma Suqin. An Improved Text Feature Weighting Algorithm Based on TFIDF [J]. Computer Applications and Software. 2011, 28(02):17–20.
[14] REN Yong-gong, YANG Rong-jie, YIN Ming-fei. Information-gain-based Text Feature Selection Method [J]. Computer Science,2012,39(11):127–130.

Information, Computer and Application Engineering – Liu, Sung & Yao (Eds)
© 2015 Taylor & Francis Group, London, ISBN 978-1-138-02717-6

A solution to attribute reduction of formal concept lattices on object set

En Sheng Zhang

College of Mathematics and Information Science, Anshan Normal University, Anshan, Liaoning, P.R.China

ABSTRACT: Formal concept analysis is a powerful tool for data analysis in machine learning, data mining, knowledge discovery and information retrieval. The concept lattices building is very important,and the attribute reduction of concept lattices can reduce the complexity of concept lattices. Currently researchers use the discernibility attribute matrix method to work out attribute reduction set, requiring all the formal concepts in concept lattices to be solved, which is also a difficult job. This paper proposes the structure theorem of reduction set and an algorithm to work out reduction attribute directly on object set, which significantly reduces the complexity and simplifies the calculation of concept lattice structuring.

KEYWORDS: Formal Concept; Concept Lattices; Attribute Reduction

1 INTRODUCTION

The concept of "formal concept analysis" was firstly proposed by German mathematician Professor Wille in the 1980s. [1] Later on it has been widely applied in knowledge engineering, data mining, information retrieval, and software engineering, and so on so forth. [2-5] The research on concept lattice is mostly focused on the theory and methods of solution, establishment, decomposition and composition. For deeper understanding of the structure of concept lattices, researchers begin to study on reduction theory and method.[6-9]

Reduction of concept lattices provides a means to research concept lattice by defining reductiveness and reductible object.[1] Reduction of concept lattice also makes tacit knowledge easily to be discovered and expressed, representing a new method to establish concept lattice, enriching the theory of concept lattice, which is significant to both research and application of the theory.[8] The methodology of attribute reduction is to find a minimum attribute subset that can exactly define the concept and structure of the original formal context. The concept lattice of formal context of reduced subset is isomorphic with that of the original, hence the reduced concept lattice can be used to work out the original lattice, which actually reduces the complexity of concept lattice of original formal context. Presently discerned attribute matrix is adopted to solve the reduced attribute set[9], as the solution to discerned matrix must be based on all the concepts, its time complexity is pretty great. This paper, by looking into reduced collective attribute characters, provides a reduced set solution based directly on the original formal context. This proposed

algorithm is highly efficient, with time complexity $O(mn)$, where m represents the number of attributes, n represents the number of objects.

2 DEFINITION AND PROPERTIES OF CONCEPT LATTICE AND ITS ATTRIBUTE REDUCTION

All definitions of concept in formal concept analysis are based on a formal context, which is defined as follows:

Definition 1.[1] We define (U,A,I) as a formal context, of which $U=\{x_1,...,x_n\}$ is the object set, every x_i ($i{\leq}n$) in this set is called an object; A= $\{a_1,...,a_m\}$ is the attribute set, every a_j($j \leq$m) in this set is called an attribute; I is the binary relation between U and A, $I{\subseteq}U{\times}A$. If (x,a) $\in$ I is true, then we define a as an attribute of x, expressed as xIa.

As for formal context (U,A,I) , we define operations on the object set $X{\subseteq}U$ and the attribute set $B{\subseteq}A$ respectively:

$X^* = \{a|\ a{\in}A,\ {\forall}x{\in}X,\ xIa\ )$;

$B^* = \{x|\ x{\in}U,\ {\forall}a{\in}B,\ xIa\ )$.

${\forall}x{\in}U$, $\{x\}^*$ is expressed as x^*, and ${\forall}a{\in}A$, $\{a\}^*$ expressed as a^*.${\forall}x{\in}U$, $x^*{\neq}{\varnothing}$,and ${\forall}a{\in}A$,$a^*{\neq}{\varnothing}$,then we define the formal context (U,A,I) canonical. All formal context discussed in this paper are canonical (before reduced).

Definition 2.[1] Assume (U,A,I) is a formal context, if there is a binary set (X,B) meets X^*= B, and B^*=X, then we define (X,B) a formal concept, or "concept" for short. X is called the extension of the concept, and B is called the intension of the concept.

We use L(U,A,I) to represent the set containing all the concepts of formal context (U,A,I), and express $(X_1,B_1) \leq (X_2,B_2) \Leftrightarrow X_1 \subseteq X_2$ (or $B_1 \supseteq B_2$),

then (L(U,A,I),$\leq$) is in partial ordering.

The definition of supremum and infimum are:

$(X_1,B_1) \wedge (X_2,B_2) = (X_1 \cap X_2, B_1 \cup B_2)^{**})$

$(X_1,B_1) \vee (X_2,B_2) = ((X_1 \cup X_2)^{**}, B_1 \cap B_2)$

then L(U,A,I) is called a lattice, and a complete lattice, also called a concept lattice.

For formal context, $(U,A,I), \forall X_1, X_2, X \subseteq U, \forall B_1, B_2, B \subseteq A$ has the following properties:

(1) $X_1 \subseteq X_2 \Rightarrow X_2^* \subseteq X_1^*, B_1 \subseteq B_2 \Rightarrow B_2^* \subseteq B_1^*$.
(2) $X \subseteq X^{**}, B \subseteq B^{**}$.
(3) $X^* = X^{***}, B^* = B^{***}$.
(4) $X \subseteq B^* \Leftrightarrow B \subseteq X^*$.
(5) $(X_1 \cup X_2)^* = X_1^* \cap X_2^*, (B_1 \cup B_2)^* = B_1^* \cap B_2^*$.
(6) $(X_1 \cap X_2)^* \supseteq X_1^* \cup X_2^*, (B_1 \cap B_2)^* \supseteq B_1^* \cup B_2^*$.
(7) Both (X^{**}, X^*) and (B^*, B^{**}) are concepts.

Definition 3.[8] For a concept lattice L(U,A,I), the set containing all its conceptual extension is expressed as:

Lu(U,A,I) = {X|(X,B)∈L(U,A,I)}

If for concept lattices L(U,A$_1$,I$_1$) and L(U,A$_2$,I$_2$),

Lu(U,A$_1$,I$_1$)=Lu(U,A$_2$,I$_2$) is true.

then we say the two concepts L(U,A$_1$,I$_1$) and L(U,A$_2$,I$_2$) are equal, expressed as

L(U,A$_1$,I$_1$) = uL(U,A$_2$,I$_2$)

and the two concept lattices are isomorphic, expressed as L(U,A$_1$,I$_1$) ≅ L (U ,A$_2$,I$_2$).

Definition 4.[8] For two concept lattices L(U,A$_1$,I$_1$) and L(U,A$_2$,I$_2$), if $\forall$(X,B)∈L(U,A$_2$,I$_2$),then there always exists a concept (X',B')∈L(U,A$_1$,I$_1$) which makes the object set of the concept X=X', then we say L(U,A$_1$,I$_1$) is thinner than L(U,A$_2$,I$_2$).

Definition 5.[8] On a formal context (U,A,I),if $\forall D \subseteq A$, expressed as I$_D$=I∩(U×D),then (U,D,I$_D$) forms another formal context, which is a child of the original context. The operation X*(X⊆U) on (U,A,I) is still represented by X*, but it should be represented X* by on (U,D,I$_D$).Here apparently we get X^{*A}=X*,X^{*D}=X^{*A}∩D=X*∩D,X^{*D}⊆X*.

Definition 6.[8] For formal context (U,A,I),if there is D⊆A making L(U,D,I$_D$) = uL(U,A,I),then we call D a consistent set of (U,A,I). And if $\forall$d∈D,L(U, D-{d},I$_{D-\{d\}}$)≠uL(U,A,I), then we call D a reduced set of (U,A,I). The intersection of all the reduced sets is called the core set of (U,A,I).

Definition 7.[8] On the formal context (U,A,I), all reduced sets are expressed as {D$_i$|D$_i$ is reduced set, i∈τ, τis an index set}. Attribute A can be divided into the following three types:

(1) absolutely required attribute (core attribute)
b: $b \in \bigcap_{i \in \tau} Di$.

(2) relatively required attribute c: $c \in \bigcup_{i \in \tau} Di - \bigcap_{i \in \tau} Di$.

(3) absolutely dispensable attribute d: $d \in A - \bigcap_{i \in \tau} Di$

Of which the non-core attributes are defined as dispensable.

e: $e \in A - \bigcap_{i \in \tau} Di$.

Theorem 1.[8] There is at least, but not limited to, one reduction for any formal context (U,A,I)

Theorem 2.[8] For formal context (U,A,I), if D⊂A, D≠Ø, E=A-D; then "D is a consistent set" $\Leftrightarrow \forall$e∈ E, $\exists$C⊂D, C≠Ø,such that C*=e*.

Theorem 3.[8] For formal context (U,A,I), if $\forall$a∈A, expressed as G$_a$={g|g∈A,g$^* \supset$a*}.Then the following conclusions are true:

(1) "a" is a core attribute $\Leftrightarrow$(a**-{a})$^* \neq$a* .
(2) "a" is an absolutely dispensable attribute $\Leftrightarrow$ (a**-{a})*=a*, and G$_a^*$=a* .
(3) "a" is a relatively required attribute $\Leftrightarrow$ (a**-{a})*=a*, and G$_a^* \neq$a*.

3   STRUCTURE OF THE REDUCED SET

Definition 8. For B⊂A, x∈U, if x contains every attributes in attribute set B, we say x contains attribute B, expressed as xIB.

Definition 9. For formal context (U,A,I), let a,b∈U, if a*=b*, then we call a, b equivalent to each other.

Theorem 4. For formal context (U,A,I) ,if a,b∈A, and a*=b*, then the following conclusions are true:

(1) For x∈U, xIa⇒xIb, xI{a,b};
(2) (a**-{a})* = (a**-{b})* = (b**-{b})* =(b**-{a})*.

Theorem 5. For formal context (U,A,I), if exists a,b∈A, makes a*=b*, then the following conclusions are true:

(1) If a is relatively required attribute, b is also relatively required.
(2) If a is absolutely dispensable, the same is b.

Theorem 6. For formal context (U,A,I), if a,b∈A, a is a relatively required attribute and a*=b*, then a,b cannot be contained in the same reduced set.

Theorem 7. For formal context (U,A,I),$\underline{D}$ is the equivalent set of its relatively required attributes and $\underline{D} \neq$Ø. If B is a reduced set of A, $\forall$[a]∈$\underline{D}$,B∩[a]≠Ø.

Theorem 8. For formal context (U,A,I), $\underline{D}$={[a$_1$],[a$_2$],....,[a$_t$]} is its equivalent set of relatively required attributes and $\underline{D} \neq$Ø. If K is the core set of A,$\forall$a$_i$'∈[a$_i$],i=1,2,....,t, we have K∪{a$_1$',a$_2$',....,a$_t$'} is a reduced set of A and all the other reduced set can be expressed in the same form.

4 DISCRIMINATION THEOREM AND SOLUTION FOR ATTRIBUTE CHARACTERS

4.1 *Discrimination theorem for attribute characters*

Now we present the discrimination theorem of attribute character.

Theorem 9. For formal context(U,A,I), $a\subseteq A$,if $a^{**}-\{a\}\neq\varnothing$, and $a^{**}-\{a\}=G_a\cup E_a$,among which $G_a=\{g|g\in A,g^*\supset a^*\}$, then E_a is the attribute set of all equivalents of a.

Theorem 10. (Discrimination for absolutely required attributes) For formal context (U,A,I),$a\subseteq A$, a is absolutely required attribute $\Leftrightarrow a^{**}=\{a\}$ or $\exists u\subseteq U$, u is free of a, $uI(a^{**}-\{a\})$.

Theorem 11. (Discrimination for relatively required and absolutely dispensable) For formal context (U,A,I), $a\subseteq A$, $(a^{**}-\{a\})^*=a^*$.

(1) If $G_a=\varnothing$,then a is relatively required.
(2) If $G_a\neq\varnothing$, and $\exists u\in U$, u does not contain a,and uIG_a, then a is relatively required, or else a is absolutely dispensable.

4.2 *Solution to attribute characters on object set*

Hereinafter presents the detailed method to solve core attribute, relatively required attribute and absolutely dispensable attribute, as well as the verification of the method.

The philosophy of this method is using n-tuple vector to represent the objects in the set of formal context (U,A,I). If $A=\{a_1,a_2,\ldots,a_m\}$,$|U|=n$, $\forall u_i\subseteq U$, $u_i=(u_{i1},u_{i2},\ldots,u_{im})$, of which $u_{ij}=1 \Leftrightarrow u_iIa_j$, or else $u_{ij}=0$. We use "attribute(a)" to represent the set of all attributes contained in u.

The following is algorithm description in C++ syntax.

Algorithm

Check$(U,a_j,\&M)$

Function: categorize a_j by core attribute, relatively required attribute, or absolutely dispensable attribute, and use M to represent the equivalent attribute set to a_j.

Input: Object set U, attribute a_j, a_j equivalent attribute set.

Output: If a_j is a core attribute, then returns 0; or else a_j is relatively required, returns 1; or else a_j is absolutely dispensable returns 2.

```
{
e=(1,1,1,...,1)
for(i=1; i<=n; i++) {   fetch u_i from U;
  if(u_ij==1)e=e&u_i;   } //herein attribute(e)==a_j**.
```

```
if(attribute(e)=={a_j}) return(0); //herein(a_j**={a_j}
else {   assign the No.j component of e to zero;
//the nonzero components in e represent attribute
set a_j**-{a_j}
  t=(0,0,...,0); flag=0;
  for(i=1; i<=n; i++)
  {   fetch u_i from U;
    if(u_ij==0) //check the object without attribute a_j
            {   if(e == e&u_i)return(0); //herein u_iI
(a_j**-{a_j})
        else if( t != (e&u_i)|t )
            {   t=(e&u_i)|t;
                if(t==t&u_i) flag=1; else  flag=0;
            }
        else  if(t==t&u_i) flag=1;
    }
  }
    M= attribute(e^t);
//M is the set of all attributes equivalent to aj
    if(flag==1) return(1);
    else  return(2)
}
}
```

5 CONCLUSIONS AND PROSPECT

In this paper we firstly discuss the characters of equivalent attributes and conclude that an equivalent to relatively required attribute is also a relatively required attribute (RRA); therefore the RRAs of a formal context can be divided into different equivalent groups. Then we discuss the relations between RRAs and reduced set, and we conclude that two equivalent RRA cannot sit in the same reduced set, if a formal context has at least one RRA, then each of the intersection of attribute reduced set with every RRA equivalent set is not null, that is Theorem 8, which describes the component and structure of a reduced set. Finally we present the theorem and algorithm on attribute discrimination by object set, tell how to know the type of an attribute and work out the equivalent set of RRA and absolutely dispensable attribute so to work out the reduced set of all the attributes. For future study, we can research on the quick solution to solve reduced concept lattice by reduced sets, and acquire the original concept lattice from an isomorphic reduced concept lattice.

REFERENCES

[1] Ganter B,wille R. Formal Concept Analysis :Matheatical Foundations, Berlin: Springer-verlag, 1999.

[2] Sergei O K, Sergei A O, Comparing performance of algorithms for generating concept lattices. The International Conference on Computational Science 01 Workshop on Concept Lattices-based KDD, Stanford, USA, 2001: pp.35–47.

[3] Sergei O.Kuaznetsov. Machine Learning and Formal Concept Analysis, ICFCA 2004, Second International Conference on Formal Concept Analysis, Berlin: Springer, 2004: pp. 288–311.

[4] Petko Valtchev, Rokia Missaoui, Robert Godin. Formal concept analysis for knowledge discovery and data mining. The New Challenges. In ICFCA 2004 ,Second International Conference on Formal Concept Analysis, Berlin: Springer, 2004: pp.353–371.

[5] Ben Martin. Formal concept analysis and semantic file systems. ICFCA 2004 ,Second International Conference on Formal Concept Analysis, Berlin: Springer, 2004: pp.88–95.

[6] Zhang Ensheng, Geng Xinqing, Lou Yabin. A quick solution to formal concept intensions, Journal of Guangxi Normal University, 2007 2007,25(4): pp.36–39.

[7] Hu Keyun, Lu Yuchang, Shi Chunyi. Concept lattice and its application progress. Journal of Tsinghua University (Science and Technology), 2000,40(9): pp.77–81.

[8] Zhang Wenxiu, Wei Ling, Qi Jianjun. Theory and method on attribute reduction of concept lattice. Science China (Compile E). 2005,35(6): pp.628–639.

[9] Xie Zhipeng, Liu Zongtian. A quick progressive construction algorithm of concept lattice. Chinese Journal of Computers, 2002,25(5): pp.489–496.

Information, Computer and Application Engineering – Liu, Sung & Yao (Eds)
© 2015 Taylor & Francis Group, London, ISBN 978-1-138-02717-6

The research on environment art design of computer-aided design technology based on expression technique

Ming Xi Yan

The Fine Arts Department, Hunan University of Humanities, Science and Technology, China

ABSTRACT: As the important the computer technology tools of the information age, computer have widely used in the field of environment art design and have rapidly impacted on the traditional expression technique. In this paper, through research the situation and prospects of our country environment art design of computer-aided design technology, obtaining the conclusions that currently with more and more attentions and applications of computer-aided design in the field of environment art design, people pay more and more attention to expression technique.

KEYWORDS: Expression technique, computer-aided design, environmental art, design

1 INTRODUCTION

Science and technology have greatly contributed to the progress and development of society and also have a huge impact on the field of art design. With the great development of computer hardware and software technology, the computer-aided design has obtained wide reorganization and application in the field of environmental art design. As the visual arts, computer-aided design in the creation of laws, rules and forms of traditional art methods and aesthetic design is the same. Faced with the wide applications of Computer-aided design in the environmental art design, we should not only recognize the nature of computer graphics software tools, and should also realize that it is not just a tool in common sense. Its appearance will change people's attitudes to environmental art design and cognitive performance in the more aspects of art design.

As an important tool for the information age, computer technology, have been widely used in the field of environmental art and design and have a rapid impact on the traditional expression technique. Development of computer technology has had a major role in promoting the development of social and artistic creation. Technology development and aesthetic advances make people demand more and more for difference and novelty. Therefore, only continuing to improve the visual designer's imagination, creativity and performance capabilities could meet the people on the growing demands of visual art. The relationship between technical performance and visual arts is better handled and computer technology is applied to the field of environmental art design.

So, our computer-aided environmental art design can have sustained development, but it also can be bound to get even more impressive social benefits.

2 THE STATUS OF OUR COUNTRY'S COMPUTER-AIDED ENVIRONMENTAL ART DESIGN

Computer-aided design came from computer graphics technology. And the concept of computer-aided design began in 1950. But our country computer-aided design originated in the 1970s. The same as the development trajectory, domestic research and application of computer-aided design is basically developed from various colleges and universities. The early 1990s, with the rapid development of China's modernization and further proliferation of computers in the field of environmental art design and creation, the value of computer technology, and gradually get people's attention.

As an important tool for the information age, computer technology have been widely used I the field of environmental art design. The so-called computer-aided environmental art design is that the designer use the computer technology perform design intent, and finally inform customers in the way of the image so that customers clearly understand the designer's design intent and creativity. It is a more direct and effective way to behave and often also is known as computer architectural renderings, divided into indoor architectural renderings, outdoor architectural renderings and garden renderings. And the classification of the computer architectural renderings can be seen in Fig 1.

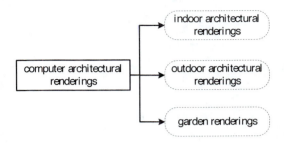

Figure 1. The classification of the computer architectural renderings.

With the development of the last decade of computer-aided environmental art design, the type of computer architecture effect performance already has a very detailed classification, which can be divided into: Computer architectural renderings, computer construction roaming animation and computer architectural renderings virtual reality. And the classification of the type of computer architecture effect performance can be seen in Fig. 2.

Computer architectural renderings mainly rely on 3DS MAX, Lightscape, Photoshop and other computer software to make the static renderings, which can be seen in fig3.

3 COMPARATIVE ANALYSIS OF THE DIFFERENT EXPRESSION TECHNIQUE OF ENVIRONMENT ART DESIGN

3.1 *The development trends of computer-aided environmental art design*

At present, with the rapid development of computer hardware and software technology, computer-aided design has received widespread attention and application in the field of environmental art design, such as a variety of programs reporting, bidding and investment can be seen everywhere. Therefore, there appeared a lot of drawing software, and pays more attention to the teaching course of computer graphics software.

People pay more and more attention to the computer technology, try various devices to master various drawing software and spend a lot of time in the model, material, light and various rendering techniques, while people ignore the art of the final renderings. Computer-aided design is an interdisciplinary science, which combines art and computer design and. Computer-aided design similar to the traditional art design in visual art creation rules, forms principle and aesthetic method.

Undeniably, the appreciation standards of people continue to improve. Due to the technology development and the technical, aesthetic progress, People's visual taste is more and more innovation and differentiation. The developer of the computer technology plays a major role in the society and art innovation. In the initial stage of the development of the computer-aided environmental art design, the designer's goal is to make the effect graph with a sense of reality, and simulate the real effect of a future scene with certain practicality. Based on the present development situation of China's computer-aided environmental art, the future computer architectural renderings should present the trend of the art, the human sentiment and the diversification.

3.2 *Using hand expressing space environment*

Before the 1990s, in domestic building indoor and outdoor space usually were showed by Hand sketching. And there are many painting techniques to draw architectural renderings, such as: watercolor, gouache, transparent color, airbrush, marker pens and so on.

3.3 *Using computer expressing space environment*

Since 1990s, computer-aided design is widely used in the field of environmental art design. Along with the progress of computer hardware and software,

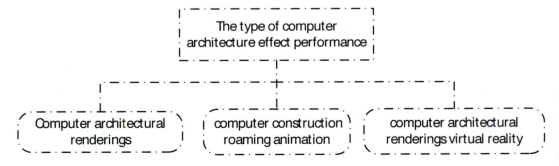

Figure 2. The type of computer architecture effect performance.

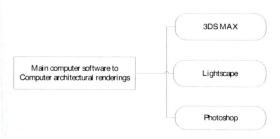

Figure 3. The main computer software to computer architectural renderings.

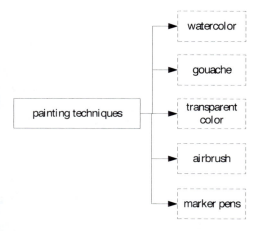

Figure 4. The painting techniques to draw architectural renderings.

computer-aided design has gradually become the mainstream of the construction effect chart performance, and has become one of the widely recognized art forms to interpret the designer's design concept and to achieve the best visual effect by simulating the real scene more realistic.

4 CONCLUSIONS

With the great development of computer hardware and software technology, the computer-aided design has obtained wide reorganization and application in the field of environmental art design. As the important computer technology tools of the information age, computer have widely used in the field of environment art design and have rapidly impacted on the traditional expression technique. Computer-aided environmental art design, develop and progress not only needing to master the technology of computer software, but also making work with the more artistic appeal. In this paper, through research the situation and prospects of our country environment art design of computer-aided design technology, obtaining the conclusions that currently with more and more attentions and applications of computer-aided design in the field of environment art design, people pay more and more attention to expression technique.

REFERENCES

[1] Journal of art&design education [M].Published for the National Society for Art Education by Carfax Pub, 1982.
[2] Schirato Tony. Understanding the visual [M].London Thousand Oaks, 2004.
[3] Richard Hickman. Critical studies in art&design education [M].2005.
[4] NarboniRoger.Lighting the landscape:art design technologies [M].2004.
[5] Clegg, Elizabeth.Art, design, and architecture in Central Europe[M].Yale University Press, 2006.

Information, Computer and Application Engineering – Liu, Sung & Yao (Eds)
© *2015 Taylor & Francis Group, London, ISBN 978-1-138-02717-6*

Design and implementation workflow management system based on JAVA technology

Wen Fu Zhang
Department of Electronic and Information Technology, Jiangmen Polytechnic, Jiangmen, China

ABSTRACT: The workflow management system is one of the most rapidly developed technologies in the computer application field in recent years. This paper firstly summarizes the related concept of workflow and workflow management system, which has conducted the thorough research in the current research on workflow technology and the existing deficiencies from the abstract level. Then it summarizes the characteristics of the workflow management system, the system structure and workflow model in workflow management system on the basis of careful analysis and study. Then, it puts forward the system structure of the workflow management platform based on JAVA according to the idea of MVC design pattern and module design.

KEYWORDS: JAVA technology; workflow management system; design; implementation

1 INTRODUCTION

The main characteristic of the workflow process is automation and the process contains the combination with people as the foundation activity, especially for those with the IT application process, tool interaction for automated processing. In the recent years, the rapid development of the technology for workflow management and its products are widely applied in office automation, computer integrated manufacturing system (CIMS), e-government and other fields. The workflow management system WFMS (Workflow Management System) is mainly based on computer networks, integration of enterprise, internal various resources, which achieved a clear definition in the enterprise business process.

The business process is automatic or semi automatically completed. At the same time, the workflow management system provides the tools, and the process of tracking and control, according to the need of business process reengineering. Thus, the workflow products are ready to develop their business, which will no longer need to start from the most primitive demand as long as the analysis of their production activities to concentrate.

2 INTRODUCTION OF THE WORKFLOW MANAGEMENT TECHNOLOGY

2.1 *Workflow Management Coalition (WFMC)*

Since the 90s time, the workflow management technology is developing very fast, which has been more

and more used in different business areas. Its most important characteristic is the combination of the human basis and the machine as the foundational activity.

Many software vendors have to provide those include workflow management technology of workflow products, but also more and more products to enter the market. In these products, suppliers must focus on a particular function, so it can adjust the product to meet the functional requirements. Numerous products brought technology innovation and promote, making workflow, workflow products between various system with other application system for interoperability. Standardization of workflow management is imminent.

2.2 *Basic concepts of workflow*

Workflow see from the literal meaning is "work flow" (Work Flow), namely between work and job transfer or transfer dynamic process. This concept is based on the production and office automation. It is aimed at the daily work with fixed program activities and brought out, because the activity is there are many personnel cooperate to complete. A portion of each participant is a complete activity, because it is fixed procedures. Therefore, it does not require manual selection can be in accordance with the fixed program, which will continue to promote the activities, go to the next processing personnel hands. People seize the characteristics of things done with the development of information technology, using computer automated processing to replace the original manual labor. The workflow technology will build into the enterprise business activities, improve enterprise efficiency, and reduces production cost.

2.3 The characteristics of the workflow management system

Because of the complexity of diversity workflow application field and the ways of implementation, so that different workflow tube has some physical systems differ in thousands of ways, but they also have many common characteristics. Look up from the higher level. Some workflow management system provides the following three kinds of function, just as the fig 1:

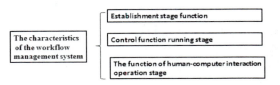

Figure 1. The characteristics of the workflow management system.

1 **Establishment stage function**: It mainly considers the definition and function of modeling workflow process and related activities. This stage is mainly the business models of enterprises in the process of workflow model can be recognized by computer.

2 **Control function running stage**: It is in the operation environment, the execution of the workflow process, and complete the activities of each process scheduling and scheduling function.

3 **The function of human-computer interaction operation stage**: The interaction is between a user and application tools to achieve a variety of activities in the process of execution.

3 DESIGN OF WORKFLOW MANAGEMENT SYSTEM BASED ON THE TECHNOLOGY

3.1 The system target

The goal of this paper is the use of JAVA language, to establish a relationship structure, extensible and reusable workflow management system based on. The system platform breaks through the limitation of the industry, provides the workflow management basic management functions.

3.2 The problems of workflow based on JAVA

Workflow technology has been in theoretical research and practical applications have made important

achievements after nearly 20 years of development. But the workflow in the following aspects technology still has some shortcomings, the need for more research work infrastructure is not perfect workflow operation, which must have a communication infrastructure underlying support, must the workflow management system. Following fig 2 is the problems of workflow based on JAVA.

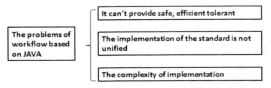

Figure 2. The problems of workflow based on JAVA.

3.3 It can't provide safety, efficiency and tolerancy

To build on top of the underlying communication based on certainty. Operation workflow products better require the underlying communication infrastructure to provide safe, efficient and fault tolerant, reliable distributed computing environment, but the current application is far at this point can't fully meet the user requirements.

3.4 The implementation of the standard is not unified

Different manufacturers provide different modeling, process definition and implementation of standards, each does things in his own way, so that heterogeneous system for cooperative work is very difficult, the application integration user to multiple products. Although the workflow management coalitions WFMC gives the recommended standards, but have not yet fully realize these standard products can be, most of workflow products is just accord with the small part of which has been. WFMC proposed standards to achieve as the relational database model.

3.5 The complexity of implementation

Workflow management system application in the enterprise not only need to complete the process definition, still need to do many other jobs, the complexity of the implementation process needs to complete tasks include: for external application systems packaging to workflow engine can activate it when necessary, establish distribution must workflow engine running computing environment, design the development of the corresponding user interface, also includes the corresponding management rules and user operation standard and so on. Current workflow for these business done to provide limited ability, workflow applications all need the workflow product suppliers and application developers for a long time cooperation can be completed by the end.

3.6 Workflow modeling technology

At present, there are many kinds of workflow modeling methods, mainly divided into the following several types. People have carried on some process language extensionsto help mechanism of workflow modeling and workflow control. Also scripting language directly uses people, such as process-oriented workflow language Camp, CO and APPLJA. It is the extension based on C and on the Ada language, MOBILE, which uses the composed of primitives specialized, text-based scripting language to describe the process of work.

3.7 The conversational model

It is the basis to development using the language of a behavioral theory on the language / action to establish the method of workflow model, the workflow is simplified as between the customer and the service of the circulation of the composition of the network session.

3.8 State transition diagram

Automata theory of Petri network and State based on Charts language is built foundation for this type of model is the most commonly used.

3.9 The mathematical models

The commonly used include mathematical methods to establish the process model: the temporal logic, CCS, CSP etc. Intuitive can be described with the executive ability and the process model of this kind of method is the issue to be improved.

3.10 ECA rules and active database

Event condition action (ECA) method in some projects has also been used for workflow modeling and execution. Active database in recent years has also been used for workflow management. Here, people use the ECA rule definition should be adopted in the specific conditions of the action and the use of trigger mechanism to perform. Some ECA rules have execution is the formation of a specific workflow.

4 CONCLUSION

In this paper, we give a simple method to realize the architecture of a workflow management system and its core. In this method, we by pure Java classes to implement. There are two reasons for doing so: in order to support all kinds of system structure based on Java. In our design and implementation, it embodies the design idea of a kind of layered. This also is we will workflow management system positioning to reflect an application development and integration platform through the continuous development of the application in this platform.

REFERENCES

[1] Li Feng. Research on collaborative modeling tool of workflow management system in WFLOW Computer application, 2001, Vol.21 (9):12–17.
[2] Li Weiping, Li Li. Research on implementation technology of workflow management system. Computer integrated manufacturing system CIMS, 2002, Vol.8 (3): 202–206.
[3] Li Weigang, Mo rong. The co-Flow workflow management system based on WebJournal of Computer Aided Design & computer graphics,2002, Vol.14 (8): 717–720.
[4] Shi Meilin a WEB [[J]. software of workflow management system based on Vol.10 journal, 1999, (11): 1148–1155.

Information, Computer and Application Engineering – Liu, Sung & Yao (Eds)
© *2015 Taylor & Francis Group, London, ISBN 978-1-138-02717-6*

A processing method of voice digital signal based on intelligent algorithm

Li Ya Wang
School of Mathematics and Information Science, Langfang Teachers University, Langfang,China

ABSTRACT: Voice digital signal is applied in many fields of human society due to its simple and fast. With the popularization of computer technology the popularity of the trend subjects of a number of computer-aided voice processing software for real-time signal processing which is based on an important development direction of general-purpose computer signal processing simulation system. Intelligent Algorithm such as genetic algorithm can be applied to improve computer assisted voice digital signal processing technology and proposed a set of related technologies to improve processing solutions.

KEYWORDS: Voice Digital Signal, Genetic Algorithm, RBF neural network, computer-aided, recognition,

1 INTRODUCTION

Genetic algorithms originated in the 20th century the sixties and seventies in the United States. It was evoluted in the nature of biological survival of the fittest and creatively used in computer simulation models. Previously it was used in nature rule, some students began to address some engineering optimization problems, and in which the proposed and other populations, selection and mutation related concepts. Genetic algorithms will soon be extended to actual application because its encoding technology and genetic manipulation method are simple and effective.

Genetic Algorithms (GA) belong to Evolutionary Computation (Evolution Algorithm, EA) that is an integral part Genetic algorithm. Genetic algorithms are a reference biological natural selection and the selection, crossover and mutation, a random search algorithm that genetic machine generated. Its object is to search the individual chromosomes in the form of a group of binary string. Exchange of information search strategy groups and groups of individuals between the two main features of the genetic algorithm. More representative of the traditional search methods are analytical method, brute-force method, and random method. Compared to these methods, the transmission method had no knowledge of the search space. Paralleling to climb peaks adaptability encoding method, the search did not depend on the gradient information, the continuity of the derivative function is defined and there is no other advantages.

2 THE MODEL OF SPEECH SIGNAL

In order to understand speech by applying the generation process, the analog voice signal was analyzed and processing through a reasonable mathematical model. Complete mathematical model of the speech signal was composed of incentive model, serial channel model and radiation model of the three sub-models.

$$H(z) = A \cdot U(z)V(z)R(z) \tag{1}$$

In where U (z) is excitation signal, V (z) is the channel transfer function, R (z) is the radiation impedance obtained by the channel. The two sources voiceless and voiced excitation were determined by the position of the switch. When the random noise was generated voiced excitation signal as an impulse sequence is represented by E (z) model.

$$U(z) = E(z)G(z) \tag{2}$$

The channel transfer function is based on the assumption that short-term steady analysis was derived. The speech signal in a short time within the shape of a stable pipeline was thinking as a quasi-commentary voice steady. V (z) is an all-pole function expressed by the formula as follows:

$$V(z) = \frac{1}{1 - \sum_{k-1}^{P} akz^{-k}} a_0 - 1 \tag{3}$$

In this formula, P represents the order in theory and P values are all-pole filter. When the low frequency band decrease by the end of the impact the radiation effect of the model is typically a high-pass filter. It is represented by the formula as follows:

$$R(z) = R_0(1 - z^{-1}) \qquad (4)$$

The waveform stored in the format sound files as WAV files. By reading the data analysis file, the voice data represented in digital form. WAV files are RIFF (Resource Imerchange File Format) as a standard. A WAV file includes three basic blocks: RIFF block, format blocks and data blocks. There are four bytes in the RIFF block in the front of the WAV file. Format blocks and data blocks are included in two sub-blocks of the RIFF. The structure was stored in WAV file and the information WAVE FORMATEX structure was contained in format blocks such as the waveform data sampling rate, number of channels, sample files, memory length and the like. The Data block stored voice data itself, and its end, which is the end of the RIFF block, namely the end of the WAV file.WAV file data storage structure are in accordance with all sorts of little endpin format type.

3 OPTIMIZATION TOOLS BASED ON RBF NEURAL NETWORK

A technique called RBF (Radial Basis Function) neural network not only can acquire better classification results, but also can cut down the time of training and recognition. Because of its simple structure, fast convergence, powerful characteristics, the RBF is widely used in many fields.

For improving its global optimization capability the process of genetic algorithm enters the training network of RBF. It can improve network performance pattern recognition that the network hidden layer can

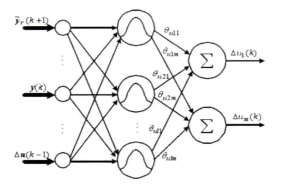

Figure 1. RBF neural network.

be trained by the genetic clustering algorithm and the output layer can be trained by the principle of the least squares gradient descent.

4 CONTRAST EXPERIMENT BETWEEN TRADITIONAL LBG ALGORITHM AND GA-LBG ALGORITHM

The sound was collected in this experiment by relating sound card and a mono microphone. The pronunciation of the word "LOVE" was collected for 30 times and was trained by 25 times. The remaining 5 times is as identification. MFCC audio library functions in the collected sound were used for storage with WAV format files in this experiment. The signal is sampled frequency is 5500Hz and SNR is 20dB. According to the order Voice input feature extraction includes pre-emphasis, windowing, subframe, endpoint detection. In this experiment the pre-emphasis filter is 0.95Hz and the frame length is 256 points. Since the results of MFCC coefficient ratio LPCC Costrel is better and more robust at runtime this performance characteristic parameter of MFCC represents the way of choosing a better voice on behalf of the respective personality traits.

Due to the state space of the objective function extremism points with the code of the length change, the traditional LBG algorithm easily regarded local optima. There are some limitations in the traditional RBF neural network clustering algorithm in this case under the same code. The improved GA-LBG algorithm and adaptive GA-RBF neural network succeeded in raising the codebook and the ability of network optimization effect is identified with great advantage in the experiment.

5 CONCLUSION

In this paper, the genetic algorithm applied to computer-aided speech processing was summarized and reviewed and improved by the basic ways of computer-aided speech processing. The effects of genetic algorithms can improve accuracy rates of recognition in recognition technology. In order to test the accuracy and simplicity, it should be available to choose a mode of measurement and analysis.

ACKNOWLEDGMENT

This work is supported by Langfang Administration of Science & Technology, Langfang Science and Technology Research and development Guidance Program 2014(2014023095).

REFERENCES

[1] M.Srinivas,L.M.Datnaik.Adaptive probability of cross-over and mutation in genetic algorithm.*IEEE Trans.On SMC*,2013.

[2] Lawrence R.Rabiner.Theory and Applications of Digital Speech Processing.*Springer Verlag*,2011.

[3] M A Potter,De Jong KA.A cooperative coevolutionary approach to function optimization.*The Parallel Problem Solving from Nature*,2010,5:249–257.

[4] Lin SC,Punch W,Goodman E.Coarse-Grain parallel genetic algorithms:Categorization and new Approach. *Proceedings of 6th IEEE symposium on Parallel and Distributed Processing*,Arlington,2011.

Information, Computer and Application Engineering – Liu, Sung & Yao (Eds)
© 2015 Taylor & Francis Group, London, ISBN 978-1-138-02717-6

Research on the exploration and practice of the design of flax fiber art product

Li Jie Qiu

High-tech Industrial Development Zone, University of Science and Technology of Liaoning, Anshan Liaoning Province, China

ABSTRACT: Since the last half of the twentieth Century, fiber art, experiencing the communion, inheritance and evolution from traditional culture to modern culture, shows the prosperity of the cultural interaction era in the cultural background of different countries and different regions. Many factors such as material comparisons constitute the basic characteristics of modern fiber art. To explore the function and value of modern flax, fiber art materials have become widely focus of fiber artists around the world. With the application of natural green flax fiber as materials, discuss and research the new flax fiber art design and product innovation.

KEYWORDS: flax fiber art design; product; innovation; development

1 INTRODUCTION

The flax fiber art product is the most potent green product of the functional fiber in twenty-first century. With the deepening of scientific and technical workers and designers to study the art process of flax fiber and to use flax fiber in fiber art design, people have a new understanding on the product design and development of functional fiber art product, so that flax fiber art product will have a much bigger development in the domestic and international market. In this paper, the author firstly focuses on the historical evolution of the flax fiber art development, respectively from the domestic and foreign research present situation of flax fiber art; secondly, discusses the flax fiber structure and product variety, flax fiber art product design and machining and flax fiber art product development; finally, outlook the application prospect of flax fiber makes designing in product innovation and art.

2 THE BACKGROUND OF THE RESEARCH ON FLAX FIBER ART PRODUCT

With the spreading influence of modernism literature trend, artists gradually discovered the various materials used in the fiber art can create new artistic forms and styles. By this means, essentially, breakthrough the traditional concept that materials in traditional art form are in the membership status. Flax fiber invisibly promoted to deepen the artists' understanding of the concept of traditional flax fiber art and at the same time also prompted the extensive exploration of contemporary fiber art materials unique language, developing and testing a bold design art fiber to fiber art form and showing the open and diversified style.

Flax fiber art can be flat and relief type wall hanging, three-dimension and device soft sculpture and display fiber art, which are shown in Fig. 1. Flax fiber as a material, will undoubtedly inject new vitality into this ancient art. All this marks that the modern flax fiber art language, pioneering creative thinking and free test characteristics and modern and post-modern art trend are synchronized and integrated, emphasizing the art forms of flax fiber diversity and multi-dimensional and comprehensive flax material and technology. With China's accession to the WTO accelerating, flax fiber art products will be more and more popular and the development of flax fiber art industry is very broad prospects.

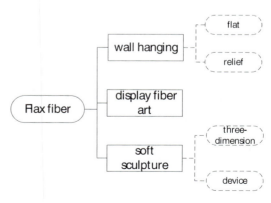

Figure 1.   The application of Flax fiber.

In recent years, although the research, development, design, production and application of China's linen fiber art have already had great progress, compared with the developed countries, there is still a big gap. Therefore, how to produce more varieties and high quality flax fiber products is the problem the research and design personnel urgently need to solve. With China's accession to the WTO accelerates, flax fiber art products will be more and more popular and it is very realistic significance of the development of the art industry of flax fiber.

3 THE SIGNIFICANCE OF THE FLAX FIBER ART PRODUCTS IN THE INDUSTRY

The flax fiber art product mainly is in the face of the international market. In recent years flax hot make our country flax fiber art industry all-round development from south to north. Therefore, flax fiber art industry chain are elongating increasingly, with obvious competitive advantage.

Our country has large Flax fiber output. Flax fiber has natural antiseptic, antibacterial and anti-ultraviolet function and flax product belongs to high-grade products, which is a symbol of fashion and identity. Because of its characteristics of natural green environmental protection people are in the favor of it thousands of years changeless. Therefore, according to foreign experts predict, the center of gravity of the next century world of flax fiber, production will shift to Chinese, China will become the world's flax fiber art production and science and technology center. With the continuous improvement of living standards and the flax textile technology improves, more and more people prefer the unique concept of flax fiber brings and flax fiber consumption will be more and more, after the consumption of flax fiber products will be in various kinds of natural fiber products of the first in 2020.

Therefore, the flax fiber products in the market, long-term good, potential and tremendous, playing an important role in the structure of fiber flax art industry, will promote the development of the regional economy.

4 RESEARCH ON THE EXPLORATION AND PRACTICE OF THE DESIGN OF FLAX FIBER ART PRODUCT

With the spreading influence of modernism literature trend, artists gradually discovered the various materials used in the fiber art can create new artistic forms and styles. The flax fiber art product mainly is in the face of the international market. In recent years flax hot make our country flax fiber art industry

all-round develop from south to north. Therefore, the research on the exploration and practice of the design of flax fiber art product is essential and cannot wait. Following, we introduce the research on the exploration and practice of the design of flax fiber art product from two aspects: the Flax fiber art product design and Flax fiber art creative thinking.

5 THE FLAX FIBER ART, PRODUCT DESIGN

In the design of flax fiber art product, we need to continue to use the essence of Chinese traditional culture, combine fashion content with modern humanistic art innovation in product design and giving priority to fashion style and trend, and make the flax fiber art product follow the time development trend. In the aspect of art design should also pay attention to the flax fiber China ancient cultural elements and flax fiber material together, make the Chinese traditional dragon and Phoenix, Tai Chi diagram, Zhu Lan Mei Ju into new elements of era, resulting in a number of both Chinese cultural characteristics and reflects the fashion of flax fiber textile products. At the same time, the design should be in line with the world advanced concepts. Absorbing and accepting the international fashion elements, the product has its own characteristics, be designed to balance the Chinese and Western, both new and old, ages and other characteristics. The Chinese traditional cultural characteristics are shown in Fig. 2.

Figure 2. The Chinese traditional cultural characteristics.

6 FLAX FIBER ART CREATIVE THINKING

In the creative process of fiber art, the selection and grasp of the creative expression way play an important role too. When we get creative material, would choose in what way to achieve the purpose of creation. Creativity is the soul of fiber art, which reflects the creative ideation depending on the breakthrough ideas, techniques, the influence of habit, with method of non-constancy to convey information also requires creativity: starting from the visual and psychological point of the visitors, from the performance of human, human entry, to find the best way of expression. In general, the theme can be divided into two categories: one category is the embodiment; the other is abstract, such as the Eleventh Lausanne Biennale's theme "fiber and space", The Twelfth theme "textile sculpture" and the thirteenth theme "return to the wall".

7 CONCLUSIONS

With the spreading influence of modernism literature trend, artists gradually discovered the various materials used in the fiber art can create new artistic forms and styles. With the deepening of scientific and technical workers and designers to study the art process of flax fiber and to use flax fiber in fiber art design, people have a new understanding on the product design and development of functional fiber art product, so that flax fiber art product will have a much bigger development in the domestic and international market. In this paper, we mainly discuss the background of the research on flax fiber art product, the significance of the Flax fiber art products in the industry and the research on the exploration and practice of the design of flax fiber art product from the aspects the Flax fiber art product design and flax fiber art creative thinking.

REFERENCES

[1] Physiology and Biochemistry, Volume 47, Issuel, January 2009.
[2] Wang Linlin, Zhang Hongjie. Science and technology development strategy of "light my flax industry [J]. Heilongjiang textile, 2002, third.
[3] Liu Hongman, GuoXiangyu and the market competitiveness of comparative advantage of Heilongjiang province linen analysis.[J]. Research of agricultural modernization, 2004, third period.
[4] XieLiren. Flax flagship helmsman – visit Heilongjiang province linen Group Chairman Wang Xibin [J]. textile information magazine, 2003, ninth. Fourteenth pages.

Information, Computer and Application Engineering – Liu, Sung & Yao (Eds)
© 2015 Taylor & Francis Group, London, ISBN 978-1-138-02717-6

The research on upper bounds of dynamic coloring in planar graphs

Yue Lin
Department of Mathematics, Qiongzhou University, Sanya, China

Zhe He Wang
School of Electronic Engineering, Qiongzhou University, Sanya, China

ABSTRACT: A proper vertex k-coloring of a graph G is dynamic if for every vertex v with degree at least 2, it that mean the neighbors of v receive at least two different colors. The smallest integer k such that G has a dynamic k-coloring is the dynamic number $\chi_d(G)$. In this paper, we will prove the following best possible upper bounds as an analog to five color theorem: If G is a planar graph, then $\chi_d(G) \leq 5$.

KEYWORDS: Planar graph, dynamic coloring, induced, polyhedral graph

1 INTRODUCTION

In this paper, all graphs are finite, simple and undirected. Any undefined notation follows you can find that in Bondy and Murty [1]. We denote the edge set and the vertex set of G by $E(G)$ and $V(G)$, respectively. For a graph G and $v \in V(G)$, $d_G(V)$ and $N_G(V)$ denote the degree of v in G and the set of vertices adjacent to v in G, respectively. Let $G_{i,j}$ denote the subgraph of G which induced by the vertices of colors i and j. Δ and d denote the largest degree and the smallest degree in G, respectively. The cycle of k vertices will be denoted by C_k, while $C(k)$ denoting a set of k colors. Let P_n be a path of order n. The plane graphs is a 3-connected graph in which every face is bounded by a polygon, such a graph is called a polyhedral graph.

A dynamic vertex k-coloring of a graph G is a map $c : V(G) \rightarrow C(k)$.
such that
(1) If $uv \in E(G)$, then $c(u) \neq c(v)$, and
(2) For each
vertex $v \in V(G)$, $|c(N_G(v))| \geq \min\{2, d_G(v)\}$.
The smallest integer k such that G has a dynamic k-coloring is the dynamic chromatic number $\chi_d(G)$.

By the notion of the above, we can say the dynamic coloring property holds that at v if $|c(N_G(v))| \geq 2$.

In other words, the neighbors of v receive at least two different colors respectively.

The dynamic chromatic number has been studied and initiated by B. Montgomery in 2001 in [3], it has the following upper bounds for any graph: If $\Delta \geq 4$, then $\chi_d(G) \leq \Delta + 1$.

In this paper, we investigate the coloring of dynamic chromatic for the plane planar, because that

there is little we can say about the chromatic number of an arbitrary planar graph, then we will prove the following theorem:

Theorem 1 If G is a planar graph, then $\chi_d(G) \leq 5$.

2 PROOF OF THEOREM

We will start with two Lemmas before we prove the Theorem of the proof.

Lemma 1 If G is a planar graph, switch the two colors on any component of $G_{i,j}$ yield another G, then $\chi_d(G) = \chi_d(G')$.

Proof. Without loss of generality, we can assume the graph to be connected for convenience. Since the $G_{i,j}$ is a subgraph of G, we are switching the two colors of $G_{i,j}$ also guarantee that $|c(N_G(v))| \geq \min\{2, d_G(v)\}$ for each vertex respectively, it that means the neighbors of v receive at least two different colors respectively, therefore $\chi_d(G) = \chi_d(G')$. This completes the proof of Lemma 1.

Lemma 2[4] Any continuous simple closed curve in the plane, separates the plane into two disjoint regions, the inside and the outside.

Proof of Theorem 1.

The proof is by induction on the number of vertices of G.

Basis step: $n(G) \leq 5$. All such graphs are 5-colorable.

Induction step: $n(G) > 5$. The edge bound implies that G has a vertex v of degree at most 5. By induction hypothesis, $G-v$ is 5-colorable. Let f: $V(G-v) \rightarrow [5]$ be a proper 5-coloring of $G-v$. If G is not 5-colorable, then f assigns each color to some neighbors of v, and hence $d(v) = 5$. Let $v_1, v_2, \cdots, v_5$

be the neighbors of v in clockwise order around v. Name the colors so that $f(v_i) = i$.

Let $G_{i,j}$ denote the subgraph of G-v induced by the vertices of colors i and j. We are switching the two colors on any component of $G_{i,j}$ yields another proper a-coloring of G-v. By the Lemma 1, dynamic number of G-v is the same as before. If the component of $G_{i,j}$ containing v_i does not contain v_j, then we can switch the colors on it to remove color i from $N(v)$. Now giving color i to v produces a proper 5-coloring of G. Thus G is 5-colorable unless, for each choice of i and j, the component of $G_{i,j}$ containing v_i also contains v_j. Let $P_{i,j}$ be a path in $G_{i,j}$ from v_i to v_j, illustrated below for $(i, j) = (1, 3)$.

Consider the cycle C completed with $P_{1,3}$ by v, this separates v_2 form v_4. By the Lemma 2, the path $P_{2,4}$ must cross C. Since G is planar, paths can cross only at shared vertices. The vertices of $P_{1,3}$ all have color 1 or 3, and the vertices of $P_{2,4}$ all have colors 2 or 4, so they have no common vertex.

By this contradiction, G is 5-colorable.

And this contradiction completes the proof of the Theorem.

The following example shows that the result is the best possible. Denote G_p as a polyhedral graph with 9 vertices (see Fig.1).

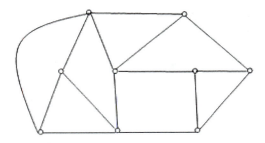

Figure 1.   polyhedral graph with 9 vertices of G_p.

It is easy to verify that $\chi(G_P) = 5$. We can easily construct graphs with arbitrarily high order of a planar G, hence $\chi_d(G) \le 5$.

3   CONCLUSION

In this paper, we investigate the dynamic coloring on planar graph first, used the similar analysis of five color theorem in the process of proof. And then find a better upper bound on it so far. Finally, we give a concrete example, showing that the result is the best possible it can reach.

ACKNOWLEDGMENT

This work is supported by Natural Science Foundation of Hainan Province of China (113008) and (2012YD29).

I would like to thank the referees for providing some very helpful suggestions for revising this paper. Also the authors are deeply grateful to the referees for their fruitful comments.

REFERENCES

[1] J. A. Bondy and U.S.R. Murty, "Graph Theory with Applications", American Elsevier, New York, 1976.
[2] H.-J. Lai, B. Montgomery, H. Poon, Upper bounds of dynamic chromatic number, Ars Combin. 68 (2003), p. 193–201.
[3] B. Montgomery, Dynamic coloring of graphs, Ph.D. Dissertation, West Virginia University, (2001).
[4] Berg, Gordon O. Julian, W. Mines, R. Richman, Fred (1975), "The constructive Jordan curve theorem", Rocky Mountain Journal of Mathematics 5 (2), p. 225–236.
[5] H.-J. Lai, J. Lin, B. Montgomery, T. Shuib, S. Fan, Conditional colorings of graphs, Discrete Math. 306 (2006), p. 1997–2004.
[6] A. Prowse, D.R. Woodall, Choosability of powers of circuits, Graphs Combine. 19 (2003), p. 137–144.
[7] V. G. Vizing, Vertex coloring with given colours, Metody Diskret. Analiz. 29 (1976), p. 3–10.
[8] V. G. Vizing, Some unsolved problems in graph theory, Uspekhi Mat. Nauk 23 (1968), p. 117–134.

Information, Computer and Application Engineering – Liu, Sung & Yao (Eds)
© 2015 Taylor & Francis Group, London, ISBN 978-1-138-02717-6

The research on the security of campus network based on core switch

Chang Jun Han
Eastern Liaoning University, Dandong, Liaoning, China

ABSTRACT: With the rapid development and popularization of the campus network, the campus network security has become the most concerned problem. The core switch with higher reliability and performance, not only ensures the backbone network high speed forwarding communication and performance optimization, but also can improve the security of the campus network through the anti-attack strategy. This paper introduces the specific application of core switch anti-attack strategy in campus network security.

KEYWORDS: campus network; core switch; anti-attack strategy.

1 INTRODUCTION

With the development of informatization construction at colleges and universities, campus network is becoming more and more popular, the campus network in teaching, scientific research, and campus administration has played a positive and significant role. Because of the particularity of college students and the campus network, campus network has been getting more and more attacks, including hacker attacks and unconscious attack. In the core switch with anti-attack strategy configuration, normal operation can achieve the protection of the campus network and can improve the security of campus networks.

2 THE CHARACTERISTICS OF THE CAMPUS NETWORK AND THE ATTACK TYPE OF DATA LINK LAYER

The own characteristics of the campus network also can lead to the occurrence of network security problems. Compared with the government or enterprise network, security management of campus network in colleges and universities is more complex and difficult.

2.1 *The characteristics of the campus network*

1 Limited funds

Network security is usually despised by the campus network constructors and managers. Since most campus network construction funds are insufficient, they put the limited funds on key equipment. For the network security construction, there are no systemic investments all the time, so that the

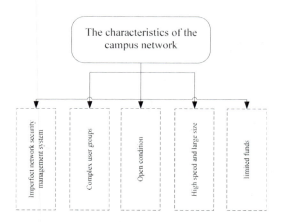

Figure 1. The characteristics of the campus network.

campus network is in an open condition, lacking of effective security warning methods and preventive measures, especially in the management and maintenance personnel.

2 High speed and large size

The campus network is the earliest broadband network. Because the Ethernet technology is widely used, the campus network original bandwidth is not less than 10Mbps. Currently, we have used wide spread 100 Mbps and even 10000 Mbps backbone interconnection. The user groups of campus network generally are large. Less is thousands people and more is ten thousand. Clearly, the Chinese college students generally get accommodation in school, so the user group is more intensive. Because of the features of high bandwidth and large number of users, the problem of network security is generally spread fast and the influence of network is more serious.

3   Open condition

Campus network must use the open network environment is determined by the teaching, scientific research and other characteristics. Students not only can query results by the network within the school or without school, but also can teach the student curriculum or answer questions online. Therefore, the backbone network of the campus network should not be limited too much, otherwise it is very difficult in using new technologies and applications in the campus network.

4   Complex user groups

In Colleges and universities, the complex user groups are mainly teachers and students, in which the students are the most active. Some students have a strong desire for knowledge and are curious to the network, trying bravely. Even some students use a variety of techniques learning from the Internet and books to attack, causing a certain impact and damage on the network.

5   Imperfect network security management system

The security awareness of campus network user is weak. Large amounts of abnormal access lead to the waste of cyber source, and also bring a great security risk to the campus network.

2.2   *The attack type of data link layer*

The attack in the data link layer will cause a great threat to the other layer of IOS model. By now, there are several attacks, including ARP attack, DHCP attack, VLAN attack, spanning tree jumping attack etc., which can be seen in Fig 2.

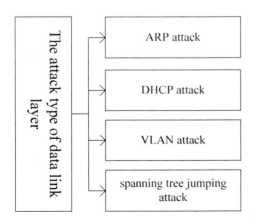

Figure 2.   The attack type of data link layer.

3   THE DEFENDING POLICIES AGAINST ATTACK IN THE CORE SWITCHES

From the above discussed, the attack in the data link layer brings great threat to the network security, while the security of the data link layer can be greatly improved through the defending policies against attack in the core switches. Following, the paper will introduce the attack and defense strategies based on the core switches.

4   ARP ATTACK-DEFENSE

DAI provides the binding of IP and MAC address in the core switch and dynamically build the binding relationship. DAI take the DHCP Snooping binding list as a basis. When it didn't use the DHCP server, it can be achieved by using statically add ARP Access-list. The DAI configuration for the same VLAN interface can open or close. In addition, we can control the ARP request message number of some port through DAI. After all, we can defense the "intermediator" attack.

The whole network using the IP address dynamic allocation mode. The access switch use Cisco or H3C products that support DIV. And the configuration is as follows.

Switch(config)#ipdhcpsnoopmvlan 10
Switch(confi},)#ipdhcpsnoopmvlan 20
Switch(config)#no ipdhcp snooping information option
Switch(config)#ipdhcp snooping
Switch(config)#iparp inspection vlan 10
Switch (config)#iparp inspection vlan 20 /* Define VLAN to detect ARP message*/
Switch(config)#iparp   inspection   log-buffer entries 1024
Switch(config)#iparp inspection log-buffer logs 1024 interval 10
Switch(confib if)#ipdhcp snooping trust
Switch(config-if)#iparp inspection trust /* The trustinterfacedefinition*/
Switch(config-if)#iparp inspection limit rate 20 (pps)// interface definition: The number of ARP message per second
Switch(config)#arp access-list static-arp
Switch (config)#permit ip host 192.168.1.1 mac host aaaa.aaaa.aaaa
Switch(config)#iparp inspection filter static-arpvlan 30

5   DOUBLE-LABELED VLAN JUMP ATTACKS DEFENSE

In order to defense double labeled 802.1Q VLAN jump attacks, what it needs to be determined is the

Native VLAN of the trunk link between the swithes, but not user VLAN. In general, the recommended approach is to assign VLAN without any users' port. The important configuration is as follows:

SWA(config)#interface range fastEthernet 0/1-fastEthernet 0/2

SWA(config-if)#interface fastEthernet 0/1
SWA(config-if)#switchpon mode access
SWA(config-if)#switchpon access vlan 10
SWA(config-if)#interface fastEthernet 0/2
SWA(config-if)#switchpon mode access
SWA(config-if)#switchpon access vlan 20
SWA(config-if)#interface fastEthernet 0/3
SWA(config-if)#switchport mode trunk
SWA (config -if)#switchport trunk native vlan 4094 //Native VLAN is VLAN 4094

SWB(config)#interface range fastEthernet 0/1-fastEthernet 0/2

SWB(config-if)#interface fastEthernet 0/1
SWB(config-if)#switchpon mode access
SWB(config-if)#switchpon access vlan 20
SWB(config-if)#interface fastEthernet 0/2
SWB(config-if)#switchpon mode access
SWB(config-if)#switchpon access vlan 10
SWB(config-if)#interface fastEthernet 0/3

SWB(config-if)#switchport mode trunk
SWB (config -if)#switchport trunk native vlan 4094 //Native VLAN is VLAN 4094

6 CONCLUSION

With the rapid development and popularization of campus network, the campus network security has become the most concerned problem. This paper introduces the specific application of core switch anti attack strategy in campus network security, especially the ARP attack-defense and double labeled VLAN jump attacks defense and their specific configuration.

REFERENCES

[1] Lu Kai, Zhang Kunli. The analysis and Countermeasure of the campus network security problem [J]. Journal of Kaifeng University, 2006-3:82–851.
[2] (US) Yusuf Bhaiji, CCIE#9305. Network security technologies and solutions [M]. Luo Jinwen et al. The posts and Telecommunications Press, 2009-3:177–186.
[3] Xia Haijing, Liu Guangwei. ARP attack prevention and solution analysis [J]. Fujian computer, 2011 (1):84–86.

Information, Computer and Application Engineering – Liu, Sung & Yao (Eds)
© 2015 Taylor & Francis Group, London, ISBN 978-1-138-02717-6

Research on the management mode of high schools based on network information technologies

Jian Bo Yang

Tianjin Radio and TV University, Nankai District, Tianjin, China

ABSTRACT: As a response to the era, Network Information Technologies have played an important role in changing people's life. In terms of the high-level education, China's high schools have applied these technologies in their management. This paper lists out the current situation and difficulties in high school management, probes the application of Network Information Technologies in high school management, finds the existing and potential problems of their application and finally proposes the solutions for the purpose to give insights on China's high school educators and management levels.

KEYWORDS: High school; High school management; network information technology; NIT; risks.

1 INTRODUCTION

With the advancement of network technologies and their rapid improvement, high school management is faced with new challenges and threats. The introduction of Network Information Technologies to high school is proven to be a great success. Thus, an emphasis on these technologies in high schools requires an attention on understanding and echoing on their influences. Network information technologies carry the characteristics of being open, cooperation, and sharing. For high school management, these technologies have a comprehensive influence. Therefore, a research on the influences and analysis is necessary.

This paper focuses on the application of Network Information Technologies in high schools. It gives an overall background introduction to the current high school situation, probes the application of Network Information Technologies in high school management, finds the existing and potential problems of their application and finally proposes the solutions for the purpose to give insights on China's high school educators and management levels.

2 PROBLEMS OF HIGH SCHOOL MANAGEMENT

Recently, with the expansion of high school enrollment, more students, either middle school graduates or other social sources, are able to accept the high school education. Then, the scale of high school is larger and larger. The management burden on high schools and the management level is growing as well. Insufficient management source, along with the gap between the desire for a better management system and the actual management source, has become the frustration for the upgrading of these schools.

On the other hand, a lot of evolving high schools which though have enrolled over one thousand students do not have a mature management system. As a result, the management efficiency is lower than expected or considerably fall behind the average level. Instead, they take the traditional management measure or sometimes a borrowing from the modern technical approaches. Then, the labor cost has been a large proportion of the total expenditure.

Meanwhile, online schools are set up. Then, it has a higher requirement for the management of high schools on how to utilize the online resources and integrates the online and offline schools. The main difficulties in high school management are listed in the below figure (Fig. 1).

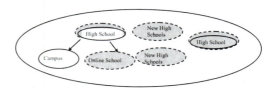

Figure 1. Main difficulties of high school management.

3 APPLICATION OF NETWORK INFORMATION TECHNOLOGIES IN HIGH SCHOOL MANAGEMENT

3.1 Application of network information technologies to manage teaching activities

Management of teaching activities via network information technologies refers to computer data management and information processing that is aimed to support teachers' teaching functions, assist to monitor, regulate, evaluate and guide the learning process of students. It also provides essential information on teaching decisions, thus, improves teaching effects and efficiency. Online teaching modes are diversified, some of which are teacher-oriented, some of which is student-oriented, some of which is teaching-oriented, some is based on discussion and others of which are for integrated purposes. In those modes, the discussion mode, probing mode and cooperation mode are better suitable for online teaching and among the online approaches students' enthusiasm about learning will largely be stimulated.

On the other hand, Multiple-media teaching material is the trend, to include both the texts, the pictures, sounds, movies, cartoons, etc. It is a development in succession to traditional teaching material. It is convinced to transfer the "written" books into "living" material. Moreover, other education sources, for example, the library, classroom, update of research activities, information on students and teachers, students' basic information, and even employment news, are achieved via online approaches. The application of network information technologies in teaching activities helps to save a lot of sources and investments. Thus, it, to the most extent, promotes the development and upgrade of high school management.

3.2 Application of network information technologies for resource management

Resource management is a key to application of Network Information Technologies for high school management, which collects and files the software database and database, including teaching reference data, historical material, multiple-media teaching material, classic teaching references, etc. This application allows students at different sites, with different education background, to choose what they need and what they would like to have via online approaches. Apart from teaching material, other information related to education, for example, the general information about teachers and students, enrollment information, registration information, education and research station, infrastructure information, balancing data-sheets etc., are integrated via this approach for the purpose to check, search, modify, do the

statistics and give timely feedback to related department. In this way, the administration departments are able to master the most updated information and to make the plan accordingly. Thus, the human resource and fund is possible to be allocated scientifically. [1]

Besides, the connection between high schools and the contact with research institutes are enhanced based on the possibilities to share both the software and hardware. More teachers are freed from the heavy work and then will be focused on research and communication.

3.3 Application of network information technologies for file management

With the popularity of office automation, file management in high school involves this technology as well. Modern file management system in high schools has a big storage size to restore students' files even 10 years ago. Daily file management is improved in this way to be secure and effective. Data input or transferred from other database can be shared among people who have the access. And the computer is assigned to do the statistics or generate the report automatically. From this aspect, the administrative staff does not need to spend most of their time on duplicate and careful data inputting work. Then, the resource of these departments is optimized.

4 ISSUES OF NETWORK INFORMATION TECHNOLOGIES' APPLICATION IN HIGH SCHOOL MANAGEMENT

Though network information technologies have such advantages to be introduced to high school management, it still threats the mental and physical health of high school students.

4.1 The potential moral threats brought out by the internet

Network information technologies are an outcome of the times, with which the internet has given more risks. It seems to give a key to an apartment, but not each room is purely beneficiary. For example, the pornographic websites, pictures, and videos are released to the public. Schools encourage students to find the useful data online, but students are put under the risk to touch the pornographic websites. If these students do not recognize the issues correctly, then they might be addicted to it. Later on, their mental and psychological problems will come up. [2]

Challenges for Students to Balance the Leisure and Study. It is true that more students are found to have an addiction to play online games full day, even at the cost of skipping classes. Then, all along, these students are lost in the fictitious world and no longer work on the necessary measures.

4.2 *The conflict between the equipments for network information technologies and online education administration management aim*

The achievement of Network Information Technologies requires a continuous financial support and human resource investment. As high schools are not commercial institutions, it is inevitable to have conflicts between the expenditure of equipments along with maintain and the actual aim.

4.3 *Finally, the destruction of virus risks on network information technologies' application in high schools*

Viruses are convinced to be the main threat to the modern computer environment, whose influence will be expanded in the whole system which includes thousands of computers. Considering the essential information is in this macro environment, then any single virus could destroy, modify, or remove the data and furthermore, has negatively influenced the regular high school management. Also, if unfortunately any confidential information is stolen and released to the public, then the loss to the high schools is unprecedented.

5 SOLUTIONS FOR ISSUES OF INTERNET TECHNOLOGIES' APPLICATION IN HIGH SCHOOL MANAGEMENT

To react to the potential negative influence on students, high schools are responsible for establishing a systematic education program on how to connect the internet right. By giving the lectures, sharing the technical lessons and having solid technical support, students will be clearer on what shall be done and what cannot be done. Certainly, a certain firework is the alternative to eliminate all the potential risky websites. [3] But mainly students shall be granted to the trustful environment and to judge on their own, because it is what they have to learn before stepping into the society. See the below figure (Fig. 2), a system is set up to eliminate such influences.

Technologies requires a continuous financial support and human resource investment. As high schools are not commercial institutions, it is inevitable to have conflicts between the expenditure of equipments along with maintenance and the actual aim.

To further improve the Network Information Platform in high schools, both from the hardware perspective and the software for sure. The understanding of the importance and the outcome of network

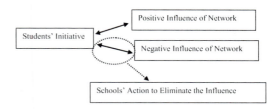

Figure 2. The system to react to the mental problems of students.

information technologies sets up the background of further improvement. Only with the qualified and upgraded hardware and software, the basic and regular high school management activities are able to be carried out. Moreover, the aim to apply the technologies and system can be achieved as expected. Most elementally, the anti-virus measurements are necessary to be taken.

Besides, an enhancement of staffs' network information technologies is the most effective support for the application of network information technologies' application in high schools. It not only guarantees the regular daily management work, but also assures the timely and effective reaction to any potential problems.

6 CONCLUSION

The current high school situation is complex and more schools have launched as well as the new teaching modes. Based on the change and update, China's high schools have applied network information technologies to set up a platform for sharing and organizing. As to the specific application, these technologies help to support the teaching activities, resource management and the file management. Though these technologies assist the high schools to better allocate the resource and improve the efficiency, they expose the whole system under a risky environment. To better react to the potential problems, the author has raised several solutions for the high school management level.

REFERENCES

[1] Ren Dong, Hu. Influence of Network Information Biotechnologies on High School Management and Their Solutions. Shandong Young Officers Newspaper. 2003 (01). (In Chinese)
[2] Jun. Yang. A Research on Normal High Schools Management Mode. Beijing Science and Technologies University Journal. 2001 (3). (In Chinese)
[3] Li Hua. Cai. Network Influence on Students' Mental Health. Jilin University. 2006. (In Chinese)

Information, Computer and Application Engineering – Liu, Sung & Yao (Eds)
© 2015 Taylor & Francis Group, London, ISBN 978-1-138-02717-6

The evaluation of ecological stoichiometry characteristics of different forest types

Cheng Yi Jiang

College of Tourism, Sichuan Agricultural University, Dujiangyan, China

ABSTRACT: The nitrogen and phosphorus characteristics of plants represent plant features and environmental factors. This paper tries to distinguish leaf and litter C: N: P stoichiometry characteristics, trees' nitrogen and phosphorus restoration, and the relationship between C, N content and temperature and precipitation. Based on the above content, this paper is divided into three parts. Part one discusses the study materials and methods, and showing the main species of trees and their life forms in tabular form. The second part is the most important section; this section introduces ecological stoichiometry characteristics of different forest types. The last part is the final summary of the conclusions of this paper.

KEYWORDS: Forest type, Nutrition limitation, Nutrition resorption

1 INTRODUCTION

This paper selects typical regional of temperate needle broad-leaved mixed forest, subtropical evergreen broad-leaved forest, tropical monsoon forest, subtropical plantation forest in China as the study sample. Through research on different type of natural mature forest, leaves of artificial near-mature forest and stoichiometries features of C: N: P of litter tries to answer the following problems:

1 The relationship between ecological stoichiometry features of litter and the stoichiometric ratio
2 The characteristics and stoichiometries features of C: N: P of different typical of living plant;
3 The relationship between temperature and precipitation of leaf C: N: P
4 The relationship between forest types and characteristics of N, P nutrient resorption.

2 MATERIALS AND METHODS

Plot overview. The resources of test materials come from the Chinese ecosystem research network: temperate needle broad-leaved mixed forest of the Changbai mountain station, subtropical evergreen broad-leaved forest of Dinghushan testing stations, tropical rain forest of Xishuangbanna testing stations and subtropical plantation forest of the Qianyanzhou testing station. The selected representatives of three types are shown in table 1. The element content in soil characteristics of 4 forest types is shown in table 2.

3 LEAVES AND LITTER C, N AND P CONTENT OF DIFFERENT REGION'S FORESTS

leaf C content of 4 types forest widely different (370.10-568.00mg·g-1), leaf C content of Cryptocarya concinna in dinghushan stations is the highest, leaf C content of Aporusa yunnanensis is the lowest; leaf N and P content in the leaves of 4 types forest widely different (n7.75-36.30mg·g-1, p-0.45-2.72mg·g-1), leaf N content of Gironniera subaequalis in Xishuangbanna station is the highest, leaf N content of Pinus elliottii in Qianyanzhou station is the lowest; leaf P content of Tilia anurensis in Changbai mountain station is the highest, leaf P content of Pinus elliottii in Qianyanzhou station is the lowest. The change range of C content of litter is (441.70- 545.90mg·g-1), litter C content of the evergreen monsoon forest of Xishuangbanna is the highest, litter C content of the Changbai mountain station is the lowest. The change range of litter N content is (8.60-18.80mg·g-1), litter N content of Pinus elliottii in Qianyanzhou station is the lowest. The change range of P content of litter is (0.23-1.19mg·g-1), litter P content of the Changbai mountain station is the highest, the lowest content of litter is Pinus elliottii in Qianyanzhou station. N and P content have dropped significantly compared with fresh leaf.

Table 1. Main species of trees and their life forms.

Station name	Geographic location	Elevation (m)	Forest type	Representative tree species	Life fom
Changbaishan Station	41°41′N, 127°42′E	736	Temperate needle broad-leaved mixed forest	Fraxinusmandschurica	DB
				Pinus koraiensis	EB
				Tilia amurensis	DB
				Castanea henryi	DB
Dinghushan Station	23°09′N, 112°30′E	300	Subtropical evergreen broad-leaved forest	Castanea henryi	DB
				Cryptocarya concinna	EB
				Aporusa yunnanensis	EB
Xishuangbanna Station	21°41′N, 101°25′E	570	Tropical monsoon forest	Gironniera subaequalis	EB
				Barringtonia fusicarpa	EB
				Garcinia cowa	EB
Qianyanzhou Station	26°44′N, 115°03′E	100	Subtropical plantation forest	Pinus elliottii	EB
				Pinusmassoniana	EB
				Cunninghamia lanceolata	EB
				Schima superba	EB

Table 2. Soil characteristics of four forest types (0–10 cm) (mean ± SD).

Forest type	pH	Organic carbon (mg·g–1)	Totalnitrogen (mg·g–1)	Total phosphorus (mg·g–1)
Temperate needle broad-leaved mixed forest	5.48 ± 0.14	164.32 ± 8.48	13.04 ± 0.84	1.56 ± 0.25
Tropical monsoon forest	4.66 ± 0.14	10.60 ± 1.68	1.13 ± 0.13	0.20 ± 0.04
Subtropical evergreen broad-leaved forest	4.08 ± 0.07	32.80 ± 3.67	2.50 ± 0.26	0.29 ± 0.02
Subtropical plantation forest	4.54 ± 0.21	11.83 ± 3.55	0.86 ± 0.2	0.11 ± 0.01

4 ECOLOGICAL STOICHIOMETRY CHARACTERISTICS OF DIFFERENT REGION'S FOREST

Ecological stoichiometry characteristics of different region's forest are shown in table 4. C:N:P (321:13:1) of temperate coniferous forest was significantly lower than that of artificial evergreen subtropical coniferous forests (728:18:1), subtropical evergreen broadleaf forest (561:22:1) is significantly higher than that of tropical monsoon forest (442:19:1); the lowest litter C:N:P are temperate coniferous forest (552:14:1), the highest is artificial subtropical evergreen conifer forest (1950:27:1), the litter C:N:P of subtropical

evergreen broadleaf forest is 1305:35:1; the litter C:N:P of tropical forests is 723:24:1.

Leaf C: N content differences in evergreen broad-leaved forest and evergreen coniferous forests were significantly (p<0.001, n=36). Leaf C: N content differences in the same forest type between the coniferous forests were also significantly; within the same forest communities, there are also significant differences between different species (p < 0.05, n = 12), leaf C: P content between different forest types of the same type or between the two communities are all significantly higher (p < 0.01, n = 12), the differences are similar to C: N. Litter N: P content between different forest types were not significantly different.

Comprehensive analysis, natural C: N and C: P was significantly lower than that of mature forest plantation the corresponding ratio, this study in 3 areas of mature forest C: N and C: P 25-26 and 321-561, respectively, C: N and C: P subtropical plantations were 42 and 728, show different areas of mature forest of C: N and C: P is characterized by consistent characteristics.

Table 3. Ecological stoichiometry characteristics of different forest types.

Forest type	C : N : P
Temperate needle broad-leaved mixed forest	321 : 13 : 1
Tropical monsoon forest	442 : 19 : 1
Subtropical evergreen broad-leaved forest	561 : 22 : 1
Subtropical plantation forest	728 : 18 : 1
Litter of temperate needle broad-leaved mixed forest	552 : 14 : 1
Litter of tropical monsoon forest	723 : 24 : 1
Litter of subtropical evergreen broad-leaved forest	1 305 : 35 : 1
Litter of subtropical plantation forest	1 950 : 27 : 1

5 N, P NUTRIENT RESORPTION CHARACTERISTICS OF VEGETATION

In four regional forest types, N resorption rate characteristic of tropical seasonal rain forest is similar to the subtropical evergreen broadleaf forest; N resorption rate characteristic of subtropical plantation forest is similar to temperate needle broad-leaved mixed forest. The N resorption rate of subtropical artificial evergreen coniferous forest has no significant difference compared to deciduous broad-leaved forests; temperate needle broad-leaved mixed forest has no significant difference compared to and subtropical evergreen broad-leaved forests ($p > 0.05$). But artificial mature forest with high N resorption rate compared to natural mature forest. In contrast, P resorption rates of temperate broadleaf forests overall lower than subtropical forest and subtropical evergreen broad-leaved forest, but all forests have a high P resorption rate.

6 THE RELATIONSHIPS BETWEEN C, N CONTENT AND TEMPERATURE AND PRECIPITATION

Elements content and stoichiometric in 4 species forest type, plant leaves C and P content with months temperature increasing, but only leaves C content has significantly relationship to months temperature ($p < 0.05$, n = 126), with months temperature increasing, leaves N content is decreasing, but has not significantly relationship ($p > 0.05$, n = 126); the relationship between plant leaves C, N, and P stoichiometric and temperatures is: leaves C: N and C: P content and monthly temperatures is negative related relationship, but not significant ($p > 0.05$, n = 126), plant leaves N: P and months temperature has significantly negative related relationship ($p < 0.01$, n = 126). The relationship between C content of 4 regional plant leaves and months precipitation is negative related relationship, The relationship between N and P content and months precipitation is positive relationship, but this 3 correlation are not significant ($p > 0.05$, n = 126); the relationship between plant leaves C, N, P and precipitation is: leaves C: N, C: P and N: P with monthly precipitation show a negative relationship.

7 CONCLUSION

Leaf C : N : P in subtropical plantation forests, subtropical evergreen broad-leaved, tropical rain and temperate needle and broad-leaved mixed were 728 : 18 : 1, 561 : 22 : 1, 442 : 19 : 1 and 321 : 13 : 1 respectively. Litter C : N : P of the four forest types were 1 950 : 27 : 1, 1 305 : 35 : 1, 723 : 24 : 1, and 552 : 14 : 1 respectively. The C: N of evergreen coniferous forest is higher than in evergreen broad-leaved and deciduous broad-leaved forests, but C: P has no relationship with forest type. Leaf N: P was highest in evergreen broad-leaved forest and lowest in deciduous broad-leaved forest.

REFERENCES

[1] Han WX, Fang JY, Guo DL, Zhang Y(2005). Leaf nitrogen and phosphorus stoichiometry across 753 terrestrial plant species in China. New Philologist, 168, 377–385.
[2] He JS, Fang JY, Wang ZH, Guo DL, Flynn DFB, Gentry Z (2006). Stoichiometry and large-scale patterns of leaf car-bon and nitrogen in the grassland biomes of China. Deco-logia, 149, 115–122.
[3] He JS, Wang L, Flynn DFB, Wang XP, Ma WH, Fang JY (2008). Leaf nitrogen: phosphorus stoichiometry across Chinese grassland biomes. Oecologia, 155, 301–310.
[4] Hogan EJ, Minnullina G, Smith RI, Crittenden PD (2010). Effects of nitrogen enrichment on phosphatase activity and nitrogen: phosphorus relationships in Cladonia porten-tosa. New Phytologist, 186, 911–925.

Information, Computer and Application Engineering – Liu, Sung & Yao (Eds)
© 2015 Taylor & Francis Group, London, ISBN 978-1-138-02717-6

The analysis and countermeasures of the investment and output performance in Maoming Binhai New Area

Qiang Wu
Maoming Polytechnic College, Maoming, Guangdong, China

ABSTRACT: Compared with other areas of Guangdong Province, Maoming industry foundation is relatively weak. The national economic departments did not achieve the balanced development, and the various industrial sectors are lack of close contact. In order to solve this problem, the establishment of Maoming Binhai New Area becomes necessary. And it's indeed necessary to do the analysis to the research of industry association of the Maoming Binhai New Area. This paper takes the current situation, the Maoming Binhai New Area Industry Association analysis as the foundation, find out the problems in the development of the Maoming Binhai New Area, and puts forward some suggestions and countermeasures for its development in the future.

KEYWORDS: input output table; induction coefficient; influence coefficient

In 1936, the famous USA economist Leontief was first put forward the input output model. This model is based on the general equilibrium theory of Walrus. It is thought that all economic phenomena can be manifested by some quantitative relation. And these quantities are interdependent, mutual influence and will be in equilibrium under certain conditions relations. Putting forward the input-output model provides a universal tool for the analysis of macro economy and micro economy. The writer takes Maoming Binhai New Area as the research object, using input-output model as the basis for performance analysis, and puts forward some suggestions for its future development.

1 ANALYSIS OF THE CURRENT SITUATION OF MAOMING BINHAI NEW AREA

Guangdong Province is the bridgehead of China's reform and openness. The speed of economic development has always been at the forefront of various provinces and cities in china. However, the economic development is a serious imbalance in its internal. Western area of Guangdong is far behind in Guangzhou as the center of the Pearl River Delta region. Therefore, the Guangdong provincial government was formally established the Maoming Binhai New Area in Author introduction: Wu Qiang (1960), male, native place: Huazhou, Guangdong Province Title: Associate Professor of mathematics, mainly engaged in the study of matrix theory.

Guangdong Province Project (GDJY-2014-B-b116) funded project. Maoming City Project (2013ZC08) funded project.

2012 April, in order to build a new economic growth pole in Guangdong, promote Guangdong economic development balance.

Compared with other areas of Guangdong Province, Maoming industry foundation is relatively weak. The national economic departments did not achieve the balanced development, and the various industrial sectors are lack of close contact. In order to solve this problem, the establishment of Maoming Binhai New Area becomes necessary. And it's indeed necessary to do the analysis to the research of industry association of the Maoming Binhai New Area.

2 THE ANALYSIS OF INPUT AND OUTPUT PERFORMANCE OF CURRENT SITUATION IN MAOMING BINHAI NEW AREA

2.1 Data sources

Data from the Maoming Binhai Economic Development Bureau home page(http://www.mgs.gov.cn/jjfzj/)

2.2 The research methods

For the system of national economic accounting, input, output table is one of the important component, which is usually presented in the chessboard balance sheet. Compared with other description method, the chessboard balance sheet is more vivid, concrete.

2.2.1 The calculation of some departments influences coefficient

The so-called influence coefficient is the national economy in the field of a certain sector production increases a unit, influence on the other sectors of the national economy. In this study, we use the formula as follows:

$$F_j = \frac{\frac{1}{n}\sum_{i=1}^{n}b_{ij}}{\frac{1}{n^2}\sum\sum b_{ij}} = \frac{\sum_{i=1}^{n}b_{ij}}{\frac{1}{n}\sum\sum b_{ij}} \quad (j = 1,2,\cdots,n) \tag{1}$$

In the formula (1), F_j that is the influence coefficient department j, Leon Leontief inverse matrix in the column data and b_{ij} is Department j of department i of the complete consumption coefficient, that n is the number of sectors of the national economy. The greater F_j and it means the greater divmore effectiveroduction more effects on other sectors. If F_j less than 1, it means the Department effects generated by the production of other departments must be lower than the average level of social influence.

2.2.2 The calculation of a department of the response coefficient

In the national economy, the sensitivity coefficient is reflected in all the departments of production will increase by one unit, which affects a department needs induction degree. Simply speaking, that is the department should provide the life quantity to cooperate with other departments of the production activities. We are using the calculation formula as follows:

$$E_i = \frac{\frac{1}{n}\sum_{j=1}^{n}k_{ij}}{\frac{1}{n^2}\sum\sum k_{ij}} = \frac{\sum_{j=1}^{n}k_{ij}}{\frac{1}{n}\sum\sum k_{ij}} \quad (i = 1,2,3\cdots,n) \tag{2}$$

In this formula, E_i that is department i induction coefficient, $\sum_{j=1}^{n}k_{ij}$ means the row data Leon Leontief inverse matrix. k_{ij} Refers to the total distribution coefficient sector i by sector j. n Is still the number of sectors of the national economy? In here, the larger E_i, and it means the larger that the production of this deanotherent is restricted by other department. Especially when E_i is smaller than 1,

it means the department of sensitivity is lower than the social average induction.

2.3 The analysis of the industries of Maoming Binhai New Area

According to the front calculation method, with the Maoming Binhai New Area in 2014 input-output table, we do the analysis of 16 departments in the area of influence coefficient and the sensitivity coefficient, as shown in table 1.

2.3.1 Influence analysis of the departments

We can see from the table 1, In 2014, there are 10 departments of influence coefficient more than 1 in Maoming Binhai New Area, and most of them are concentrated on the second industry, which make a great difference to the radiating and driving play of social economy. With the purpose of the balance of Maoming Binhai New Area to promote the economic development in Guangdong, this industry is advantageous to the Western Guangdong area industrial structure upgrade and optimization. Some high labor intensity, such as furniture manufacturing, construction, equipment manufacturing, is good to boost the employment. Oil, petrochemical industry, energy, information industry belong to the industry with highly concentrated on resources and capital, this kind of industrial products present the nature of the intermediate product. The coefficient of the third industry of shipping logistics, public management, and tourism resources is higher than the social average level, which is highly correlated with the production and life.

In addition, we also found that the influence coefficient of less than 1 the department has given priority to the first and third industry.

2.3.2 Analysis the department of induction degree.

From table 1, there are four class departments containing petrochemical industry, oil exploitation, information industry, energy, power, all of their sensitivity coefficient are more than 1. Among them, the petrochemical industry, oil exploitation, energy and power belong to the basis of industry. Petrochemical industry, oil exploitation of two sectors of the sensitivity coefficient is more than 11.43, which is five times of the social average value 2.05 more. This shows that the basic industry is more conducive to promoting the development of the national economy. The faster economic development, the greater pressure social demand of this department. It's the bottleneck to impede the economic development.

Table 1. statistical analysis of the influence coefficient and sensitivity coefficient in Maoming Binhai New Area various departments.

number	Sector type	Influence coefficient	sorting	Sensitivity coefficient	sorting
1	Agriculture	0.732	4	0.308	1
2	Petrochemical industry	1.273	5	11.445	2
3	Energy and power	1.309	1	2.334	3
4	Equipment manufacturing	1.313	3	0.936	4
5	Food manufacturing	0.746	2	0.243	5
6	Information industry	1.435	10	2.378	6
7	Furniture manufacturing industry	1.142	17	0.586	7
8	Real estate	0.598	7	0.576	8
9	Construction	1.253	16	0.207	9
10	Petroleum mining	1.272	12	11.435	11
11	Finance and insurance	0.589	15	0.623	12
12	Public management	1.030	14	0.201	13
13	Shipping logistics	1.014	13	0.683	14
14	Tourism resort	1.074	8	0.209	15
15	Entertainment	0.853	6	0.332	16
16	Education	0.836	9	0.241	17
17	Information transmission, computer service and software industry	0.876	11	0.371	10

3 SUGGESTIONS AND COUNTERMEASURES ON THE DEVELOPMENT OF MAOMING BINHAI NEW AREA

Combined with the previously made in the Maoming Binhai New Area of the national economy, industry association analysis and the regional strategic planning, we put forward the following suggestions on the future development of the Maoming Binhai New Area:

3.1 *Recognizing and promoting the main industry to drive to upgrade the continuous optimization of industrial structure*

3.2 *Continuously strengthen the basis of the industrial sector represented by energy, raw materials and the development of traditional industries*

From the data in Table 1, the sensitivity coefficient than the average level of social sector followed by the petrochemical industry, oil exploitation, information industry, energy and power. In view of this, energy is one of the factors restricting the economic development. Dueing to "set in the oil, promote for mine, prosperous due to the sea", there is a very

important relationship with the Maoming oil shale and the establishment of the Maoming Binhai New Area. Therefore, we should increase the relevant policy support, and constantly promote the rapid development of the energy industry, especially to increase investment in new energy investment and research efforts. In order to enable the development of various sectors of the national economy to obtain sufficient energy supply, we should insist on sustainable development energy strategy.

3.3 *Promote the steady development of the third industry, to provide the guarantee for the healthy and sustainable development of economic society*

From table 1, Entertainment, shipping, logistics, tourism and related industry has shown the leading role in Maoming Binhai New Area's economic development. Intensify investment in the related industry, with the help of a high and new technology to promote the rapid development, and promote the rapid development of the whole industry, has special strategic significance. Compared with entertainment, shipping and logistics, tourism industry, the induction coefficient is only 0.371 of information transmission, computer service and software in Maoming Binhai New Area. It's far less than the social average level, and has become a prominent conditionality factor. Therefore, in order to better promote the development of social economy, we should increase the high-tech introduction and product research and development efforts. Compared with the second industry, the third industry of less investment, low energy consumption, high output, it's more conducive to the ecological protection and the adjustment of industrial structure.

3.4 To layout reasonable industrial, optimize existing resources

By the feedback the information in table 1, if Maoming Binhai New Area wants to achieve the rapid development of the economy and society, we should pay attention to the development of the energy industry, the growing industry of new and high technology. Once we mention the energy industry, we can't do without oil, can't do without the petroleum chemical industry. However, as the troubled as the ethylene plant for Maoming Binhai New Area also faces the dilemma of the petrochemical industry. If it is limited to the western region, the petrochemical industry, environmental capacity is limited. It will make the whole area and image quality affected. If we lay out petrochemical industry, and organize a new petrochemical area in the east. Not only that, east of Binhai New Area is still in the upper direction of the city, which the layout of the petrochemical industry will seriously damage the ecological environment of the city. Based on the above consideration, it is suggested that the western region as a major petrochemical industry cluster, and eastern is more to consider the layout of high-tech industry. Specifically, it is the requirement of the western regions in the upper and middle reaches of the existing petrochemical industry as the foundation, utilizing mineral resources advantage and port advantage. According to the "base, scale, integrated, park" ideas, the government actively introduces large-scale petrochemical project, integrated development strategy. The government should focus on aromatic hydrocarbons and its post processing, Chlor alkali and its deep processing, polyurethane, synthetic ammonia and urea industry, fine chemical industry, and constantly improve the fine rate, and strive to create a high level of world-class petrochemical base. High tech industry in eastern area is constantly poly knowledge intensive, increasing marginal benefit, such as Zhongshan University, South China University of Technology to attract top domestic universities to establish science and technology incubation base here, to become the new engine of the economic development of the Maoming Binhai New area.

3.5 Grasp tightly the main function region strategy to realize the sustainable development

In recent years, under the leadership of the government, the effect was obvious in Maoming Binhai New Area of economic and social development. However, in the process of development, there have been various problems, such as the natural ecological conditions deteriorated sharply, further improve the quality of economic growth, developing power shortage, etc. If these problems cannot be corrected timely, Maoming Binhai New Area's planning and design can only be a mere scrap of paper, taking the construction of Binhai New Area to promote the development of the hinterland as an empty talk. Therefore, the relevant departments should take the "Guangdong Province City of Maoming Binhai New Area Planning (2012-2030)" as the basis, to integrate resources within the region and strengthen the industrial interaction. It is necessary to maintain rapid economic growth, but also should pay attention to the protection of the ecological environment, to truly become a model for the harmonious development of the economy and society in Western Guangdong.

REFERENCES

[1] Shuijin Zhuang. Development Strategy Considerations in Guangdong Maoming Binhai New Area[J]. Southern journal, 2012 (7):20–25.
[2] Jianwen Yuan Input and Output analysis of High Technology Industry in Guangdong Province. [J] Journal of Guangdong University of Business, 2005 (2):69–72.
[3] Chinese input output association research group Our country industry correlation analysis [J]. Statistical research, 2006 (1):3–8.
[4] Qunwei, Qiantao Mao, Yingtao Liu. Input output analysis and adjustment of industrial structure in Yunnan Province [J]. Science Educ, 2013 (29):200–202.
[5] Jianfeng Mao The Analysis of the Structure of Industry in Our Country — An Empirical Study Based on Input Output Method [J].Southern journal, 2005(6):15–17. [J]. Economic review, 2005 (6):15–17.

Information, Computer and Application Engineering – Liu, Sung & Yao (Eds)
© 2015 Taylor & Francis Group, London, ISBN 978-1-138-02717-6

The application of fuzzy estimate based on AHP

Wei Lv

Hangzhou Cigarette Factory, China Tobacco Zhejiang Industrial Co., Ltd, Zhejiang, Hangzhou, China

ABSTRACT: This paper describes the design principles of scene management evaluation on workshop, it includes systematic, feasibility, orientation, independence, flexibility and specificity at first, then we construct the structural model through six aspects of 6S. On this basis, through establishment the basic scores and weights for fuzzy estimating, to calculate composite scores. 6S is becoming a standardized management in workshop, using the theory and methods of fuzzy estimation to evaluate themselves, it redounds to find the weaknesses and problems of scene management for timely and accurately, proposed improvements, promoting the overall improvement of the level of scene management.

KEYWORDS: 6S; evaluation system; fuzzy estimate; membership degree

1 BACKGROUND

With increasing competition in the manufacturing sector, many companies take 6S method to improve the management level of the production scene. 6S is advance of 5S, it has gradually formed a complete theoretical system. It adopts effective management that they include personnel, equipment, material, system and environment of the scene through 6S, the purpose is to create a harmonious order, established a good condition, improved work efficiency, reduced failures, ensured product quality, ensured safe, shortened cycle, reduced waste, costed savings [1].

For now, while many companies pay attention to scene management, but the estimate research for scene management is rare. In the absence of a scientific evaluation index, and the lack of a systematic evaluation system, management's evaluation of the workshop will be a variational standard[2]. 6S is becoming a standardized method of scene management, only rely on management's estimate of the workshop is not enough. If the workshop appropriately selected estimate index and established a scientific system, used fuzzy evaluation method for scene management in the workshop, we would known the achieve results of 6S, help us to strengthen the confidence of implementation, discover the weaknesses and problems in scene management of the workshop, help up to nail down the thinking and measures of 6S, enhance the level and effectiveness of scene management. Visibly, the significance is great for establishing a scientific system of the estimate system.

2 DESIGN OF ESTIMATE SYSTEM

The estimate system means to achieve the purpose of estimate, it is according to the systems approach to build a system architecture that there is a series correlative index for objects. The estimate index should be taken out in these systems of 6S, fully reflect the characteristics and basic structure of scene management [3]. 6S is an integrated management of production line in the workshop, so the estimate index consists of several index groups hierarchically and orderly, in this group, the setting of indices related to estimating whether if comprehensiveness and accuracy.

2.1 Design principles

2.1.1 Systemic

6S is a systems engineering, the estimate should adopt the system principle, when we construct the system of estimate index, we should see the scene as a system, at first, made the factors to analyze, then, according to the intrinsic link of each aspect to construct a holistic system of covers index about scene management.

2.1.2 Practicability

The index, but also is able to express enough each aspect of the scene, but also to consider the index system whether if it has scalability and comparability. Thereby, we must determine the number of level and index for this system reasonably, the quantitative indicator made be calculated easily, and easy to analyze; the qualitative indicators made be unambiguous of meanings, and easy to estimate.

2.1.3 Oriented

Estimate the scene management of the workshop, the aim not only is graded and assessed, but also the more importantly is to motivate employees to continuously improve the level of self-management in the scene. Thereby, in the design of the system, we make sure the system has oriented effect activities for scene management [4].

2.1.4 Independent

The index should be independent of each other, to avoid overlapping.

2.1.5 Flexible

The system must have developmental concept. But it needs to change according to the management concepts and method to adjust properly, thus achieve flexible application [5].

2.1.6 Special

The estimation system not only to accord to "corporation scene management rule", but also it must incarnate particularity of different ranges and the scene in the workshop. 6S beyond area regional context, this system must incarnate the particularity of scene management.

2.2 Establish of the estimate system

According to 6S request in the "corporation scene management rule", we summarize and coordinate these rationalization proposals after extensively discuss the comments of related personnel of the workshop, we have designed a cover whole of the estimate system for scene management in the workshop, according to AHP, thereby, we can establish a hierarchical model from collectivity to monomer, and the model like a stair.

This model divided into three layers, the first layer is overall framework, it is the idiographic content of each estimate, the second layer is estimate frame, it is the main criteria, calculate method, notice to proceed of each estimate, the third layer is implementation framework, it is the content and request of each estimate. The choice of the second layer and under specific estimate, the content of main structure are seiri, seiton, seiso, seiketsu, shitsuke and security.

3 ESTIMATE MODEL

Seem from Table.1, the each estimate of 6S, not only there are quantitative indicators, but also there is qualitative indicators, this structure of the estimate system like ladder and multi-level. In the comprehensive

evaluation, we adopt fuzzy estimate, it is fit for dealing with complex matters of multi-factor and multi-layer under fuzzy environment, and it is fit for quantitative and qualitative evaluation of the combined estimate also. We estimate the scene management of the workshop through fuzzy estimate, the results are fuzzy vectors, this situation can objectively reflect the production of scene management. Based on this status, this paper adopt fuzzy methods to estimate the level of implementation and effects is viable.

3.1 Establish factors aggregate

The aggregate is factors of influence estimate results, this is represented by A, so the composing of A of scene management is: A = {a1, a2, a3, a4, a5, a6, a7}, the subset is: a1={a11, a12, a13, a14, a15}, a2 = {a21, a22, a23, a24}, a3 = {a31, a32, a33, a34}, a4 = {a41, a42, a43, a44}, a5 = {a51, a52, a53, a54, a55}, a6 = {a61, a62, a63, a64}, a7 = {a71, a72, a73, a74}.

3.2 Establish weight aggregates

In the fuzzy estimate, the set of weight is most important, this paper combine characteristic of scene management for the workshop, we adopt Analytic Hierarchy Process(Abbreviation: AHP) to confirm the weight of estimate. According to principium and program of AHP, at first, we compare twenty-two importance of each layer, then, we set basal values of each-layer elements based on this basic by workshop manager, these base values are composed judgment matrix, we use the method of weighted geometric mean, we made the matrix for a one-normal to obtain the estimate weight. We establish weight aggregatesW are: $W = \{w_1, w_2, ...w_j\}$ and $W_i = \{w_{i1}, w_{i2}, ...w_{ij}\}$, this Wi express weight of factors(ai), wij express the weight of factors(aij).

3.3 Establish remark aggregates

According to results of values to judge, the estimate objects carved up several grades, this grades use fuzzy language to describe commonly. At the time of estimate for scene management, setting five grade of remark for each estimate, given grade areas of remark: $V = \{v_1, v_2, v_3, v_4, v_5\}$, thereinto, v1 is best, v2 is better, v3 is common, v4 is more poor, v5 was most poor.

3.4 Single-factor estimate

We have structured grade-fuzzy subset, we use the method of membership function values(Abbreviation: MFV), we estimate each factor in factor aggregates, given the fuzzy relation matrix, shown exp.(1). Where the first element in row i and column j (rij) express membership from ai to vj of fuzzy subset. The acquit of level and effects in a factor(ai), it is described through the vector: $(R|a_i) = (r_{i1}, r_{i2}, ...r_{ij})$.

3.5 Fuzzy estimate

Using fuzzy synthesis algorithms, accounting fuzzy estimate() of the i-th factor, shown in exp.(2), and according to the principle of maximum membership degree evaluation.

$$B_i = W_i * R_i = (w_{i1}, w_{i2}, ...w_{im}) \begin{vmatrix} r_{11} & r_{12} & ... & r_{1n} \\ r_{21} & r_{22} & ... & r_{2n} \\ ... & ... & ... & ... \\ r_{m1} & r_{m2} & ... & r_{mn} \end{vmatrix} = (b_{i1}, b_{i2}, ...b_{in}) \quad (2)$$

Where i is the number of factors affecting, n is the number of i-th factor influencing factors, bij (j = 1, 2, ..., n) is the vector of fuzzy estimate. When there are multiple levels of factors, because of the high-level factors are determined by a number of lower level factors, so fuzzy estimate need to sort, at first, the lowest level should begin, then estimate a higher level until the highest level so far.

3.6 Calculated values and given the reviews

In order to compare between each scene, we transfer the vector of fuzzy estimate to comprehensive value. If the corresponding relationship between estimate rating and composite score is best, better, generally, poorer and poorer, it respectively corresponds scores range by [90, 100], [75, 90), [60, 75), [40, 60) and [0, 40), we calculate the vector of estimating value is D = (95, 82.5, 67.5, 50, 20), then the composite score is shown exp.(3).

$$A_i = B_i * (D)^T \quad (3)$$

We can calculate comprehensive value(Ai) to determine the grade of estimate, thereby, we can compare lengthways the management level for different time. We can analyze development trends for scene status through compare breadthwise of each scene, and we can analyze existent differences and cause of difference, making the measure of reparation, thereby, it promotes the continuous improvement of scene management.

4 APPLICATION EXAMPLE

Take Hangzhou Cigarette Factory power plant, for example, we use 6S method to comprehensive estimate. Firstly, we adopt the balance integral method to determine the degree of membership for each estimate. Where in the judgment results of each

estimate shown Table 2. We use the calculate model of fuzzy-comprehensive estimate:

$$B_i = [0.2, 0.2, 0.2, 0.2, 0.2] * \begin{pmatrix} 0.4 & 0.6 & 0 & 0 & 0 \\ 0.9 & 0.1 & 0 & 0 & 0 \\ 0.4 & 0.5 & 0.1 & 0 & 0 \\ 0.2 & 0.5 & 0.3 & 0 & 0 \\ 0.3 & 0.6 & 0.1 & 0 & 0 \end{pmatrix} = [0.440, 0.460, 0.100, 0, 0]$$

We can know in estimate vector, degree of beyond the better level is bigger, the value is 0.460. We according to the principle of maximum degree, we can get the criterion level(B1), namely, the holistic estimate is better of this workshop for 6S method. Similarly, we can calculate the results of other estimate results.

Table 2. The membership judgment result of each estimate.

Grade	a11	a12	a13	a14	a21	a22	a23	a24
Best	0.4	0.9	0.4	0.2	0.5	0.2	0.1	0.2
Better	0.6	0.1	0.5	0.5	0.4	0.5	0.4	0.6
General	0	0	0.1	0.3	0.1	0.3	0.5	0.2
Poorer	0	0	0	0	0	0	0	0
Poorest	0	0	0	0	0	0	0	0

B2 = [0.250, 0.175, 0.275, 0, 0],
Namely, the estimate of neaten is general;
B3 = [0.150, 0.200, 0.250, 0, 0],
Namely, the estimate of clean is general;
B4 = 0.050, 0.300, 0.270, 0, 0]
Namely, the estimate of depurate is general;
B5 = 0.680, 0.440, 0.480, 0, 0]
Namely, the estimate of safety is best;
B6 = 0.175, 0.325, 0.200, 0, 0]
Namely, the estimate of literacy is general.

We according to above results to build fuzzy-estimate matrix, shown exp.(4).

$$B_i = \begin{bmatrix} 0.440 & 0.460 & 0.100 & 0 & 0 \\ 0.250 & 0.175 & 0.275 & 0 & 0 \\ 0.150 & 0.200 & 0.250 & 0 & 0 \\ 0.050 & 0.300 & 0.270 & 0 & 0 \\ 0.680 & 0.440 & 0.480 & 0 & 0 \\ 0.175 & 0.325 & 0.200 & 0 & 0 \end{bmatrix} \quad (4)$$

We get the weight of estimate for 6S method is: W = [0.1128, 0.1118, 0.2028, 0.2428, 0.1106], and estimate result of target is: A = [0.1303, 0.4115, 0.2318, 0, 0].

According to the principle of maximum membership, the largest value is 0.4115 in five memberships, the corresponding level is better, namely, the effect of using 6S method is better. We can see that this workshop must regulate employee behavior further, improve literacy, continuing to optimize method of scene management, promote the holistic level more excellent of management.

5 CONCLUSION

This paper consults the relational standard of "enterprise scene management guidelines, adopt fuzzy estimate, to build estimate system of scene management. On this basis, we can actually estimate the scene of power workshop, from the effect of application to see. This method can be more comprehensive estimate of scene management. Basically, it can help the management of the workshop to understand directly what implementation case of each estimate, and it can help to find weak links of producing, than adopting measures. It is more important for improvement and enhancement scene management of the workshop. However, the scene management of workshop has a complexity and particularity job. There is a sill a gap that we give a very accurate and comprehensive estimate,

the estimate system will still continue to modify and improve, so there are many problems to be researched and explored.

REFERENCES

[1] Zhangmingque. Research of estimate system for 6S method[J]. Modern Property, 2011(3):69–71.
[2] Qiershi, Chengwenming. Construction and application of Toyota production method estimate[J]. Science And Technology Management, 2005(4):141–143.
[3] Zhaoshuanjun, Caohongmei. Problems and countermeasure research of scene management in workshop[J]. Shandong Textile Economy, 2011(6):37–40.
[4] Shanzhuangsheng. Construction and execution of executive power for scene management[J]. Modern Finance, 2007(3):77–81.
[5] Qiemingwei. The estiamte system research of scene management based analytic hierarchy process[D]. Chongqing University, 2010.
[6] W.Lv Research And Realization Of Pruning-branch To Adopt Data-Stream[J]. Management Information and Educational Engineering, 2014(4):1201–1205.

Information, Computer and Application Engineering – Liu, Sung & Yao (Eds)
© *2015 Taylor & Francis Group, London, ISBN 978-1-138-02717-6*

The exploration of university computer practical teaching reform and development

J.X. Wang, Y.L. Wang & W. Pang
Hebei Institute of Architecture and Civil Engineering, Zhang Jiakou, Hebei, China

ABSTRACT: With the rapid development of China's IT industry, computer teaching reform of engineering practice oriented has become the development trend of higher education curriculum reform. In this paper, from the reform of practice teaching of the Computer Information Engineering, College of Hebei Institute of Architecture and Civil Engineering, it constructs a model based on the project of the reform of teaching practice. It combines the successful reform of the teaching goal, teaching methods, teaching evaluation of teaching case. Practice shows that the engineering practice teaching mode can not only improve the teachers' scientific research ability, but also improve students' learning. It has a good interaction function for the subject construction and professional development.

KEYWORDS: Reform of teaching mode; Computer education; Engineering practice teaching; School-enterprise cooperation

1 THE NECESSITY OF COMPUTER TEACHING REFORM

The computer includes six basic professional: science and technology, software engineering, network engineering, information security, network engineering, digital media technology. It is characterized by: a professional cloth points, fast, professional expansion, the number of students and more strong practicality. In 2008, under the influence of the financial crisis, the IT industry suffered a great impact. But from the beginning of 2010, the computer industry began to recover, many software companies use many people. By 2012, 2013, 2014 year graduate student employment feedback, the number of recruitment and salary, generally increased year by year. But the ability of graduate students' practical requirements is higher than ever before.

The rapid pace of change of IT industry and so many computer professionals and students have brought opportunities and challenges to our teaching reform. With China's rapid computer professionals to society, different levels of personnel to undertake the different division of labor in society. So the training level and ability of students is decided the work division after graduation, even can find a job. Computer education as a group in undergraduate education, how to train themselves for computer professionals making accurate positioning? In this position, the student should have, what kind of ability? How to enable students to have this ability? These issues are worthy of serious consideration and research.

2 THE UNDERSTANDING AND TARGET LOCATION OF PRACTICAL TEACHING

The practical teaching is the core value of ability and talent cultivation, is the ability to analyze and solve problems, and is a basic theory of knowledge accumulation and use of ability training. Practical training is helpful for theoretical knowledge together and it is needed to complete all the teaching experiment platform final training. The present computer teaching culture is mainly for basic or applied science and technology, knowledge, innovation, technology innovation and integrated innovation talents. Although the doctor stage is often true of such talent cultivation period, the systematic education of undergraduate should prepare themselves for the follow-up study of high level and future talent showing itself, to lay the foundation. Through their undergraduate study, students should have the required talents, knowledge structure, fundamental ability and comprehensive quality. Graduates should have a more solid foundation in mathematics and the natural sciences, have a more systematic and thorough foundation in theoretical computer science and technology disciplines, expertise and practical skills. They have the ability to analyze and solve difficult problems, innovation capability, initially with an international vision. Further, our students should have loftier aspirations and ideals of life science, and have to achieve this ambition and ideal and stand on solid ground of the spirit of unremitting struggle, self-confidence and ability. We want to cultivate a number of potential developments, and able to lead the trend of computer science "masters" or even in the industry "industrial all-powerful person".

Innovation ability is high quality, research personnel of the soul, is also a measure of research talent standard. The cultivation of students' innovative ability and therefore has become the soul of undergraduate education. Establishment of the innovation ability of students usually goes through a relatively long training course from undergraduate to PhD. One of the main tasks of undergraduate education is to stimulate and catalyze the students' innovation. We should promote innovation germinates, and soil water, make the innovation of seedling is gradually growing up in a more natural environment. Therefore, we should train students to master a certain practical ability in the undergraduate stage. Let them test their ideas in the learning process to stimulate students' learning interest and inspiration.

3 THE SCHEME OF CURRICULUM REFORM

Curriculum reform must meet the training objectives of the current computer students. Based on many years of teaching and management experience, our institute took out a set of feasible reform scheme. Specific programs are as follows:

3.1 Programming training system

Program design, data structure and algorithm, software engineering, parallel computing and multi core programming courses constitute the basis of the student's ability of programming. We regard it as a whole, software department concentrates on their teaching goals and tasks. We specifically designed more than 60 teaching task, each task of teaching covers the specific knowledge points. Finally, it is through the completion of specific software to consolidate the knowledge points. Every six classes to complete a small task of teaching, it is by different teachers teaching small cross to complete each task. Students have the discontinuity results in the process of learning. Teachers in the teaching process must grasp the entire training system, which not only improves the students' enthusiasm for learning of knowledge, but also to avoid the teachers in the teaching process of fragmentation.

In the extracurricular practice, we formed a multi-level extracurricular practice chain. The specific method is the school organized a large number of computer software programming competition. This is not only a practical platform for students to consolidate the knowledge points, but also a playing field for the further selection of students to participate in high-level competition. Through the selection, we have the professional ability of students to form various groups to participate in the national and provincial some contest. At the same time programming contest base for support, development oriented program design capacity evaluation unit demand. The cooperation of enterprises has reached six. At the same time, the development of the network teaching support network, video courses, online Q & A, programming exercises and examination system in one of the online self-learning and self testing system.

3.2 System design training system

Digital logic, principle, system structure, interface technology, operating system, compiler system design and realization course principle, embedded system, constitute the basic training students' ability of the design system. We also regard it as a whole. Hardware department concentrates on their teaching goals and tasks. They overall plan experimental content. The experimental output leading curriculum is to follow-up courses experimental input. Mainly based on the hardware laboratory teachers, students participate in the research and development of certain computer system experimental platform, independent development of the working principle of CPU system, and the development of the school gate system and other practical hardware systems. Evaluation benchmarks for each student to participate in the hardware design before graduation. If not up to standard, in the graduation design finally must complete an evaluation of hardware design.

Another practice has expanded accordingly. From the second grade registration began to organize all kinds of hardware and interest groups, for a group of six, to complete their work within three years. By the standard testing, he will receive a computer expert certificate issued by the institute. This proves that he completed a computer design and production work.

3.3 Exploration of the training mode of research ability

We focus on the "three aspect of curriculum, project evaluation" to explore the training method of research talents with the combination of inside and outside class, combination of scientific research and teaching, methods. It is specific to the construction of "course" as the basis, to promote the reform of teaching method research. It is the senior seminar as a platform, through the research method for seepage subject leaders. It is projected as a carrier to provide a platform for the study of practical ability. We establish an open, comprehensive evaluation mechanism, to guide students to comprehensively improve the quality of scientific research.

Our institute in 2011 started the "course" project to promote the reform of teaching methods, research to "explore" and "criticism" as the core, realize the extent of the professional training to learn scientific research method. We stratified teaching method around the course explores the theme of pattern design: according to the characteristics of different levels of curriculum development mode of teaching methods combined with the features of the course. We with small class teaching and the teaching process as a means of teaching evaluation process intensification. And set the process appraisal, is worth 50%. There are more than 10 teachers has declared the course project. Each project, from the initial review, must go through three years of practice can be decided. So far there are three courses to obtain the title of "lapping course".

To set up an advanced seminar in discipline, academic leaders led by osmosis, the breadth and depth of teaching reform. By external professors and business excellence engineers to carry out the department of computer technology, introduces the new technology of computer science fields for undergraduate students, as well as the development of enterprise current situation and requirement. By subject, academic leaders and discuss research on their own strengths and students from a symposium. We obtain a high level of scientific research of students awarded credits, formed a research practice in traditional. The young researchers elective course closely tracking the forefront of academic progress.

We provide a practical platform for research ability training as the carrier for the project. We use the provincial college student innovation experiment program leader, college level research, research project for construction of scientific research projects, training system. We provide students opportunities to experience research methods project support, teachers' scientific research team intervention method. The students of grade three or four join in the group of teachers, second grade students in top-notch also participate in the team.

We build an open, comprehensive evaluation mechanism. We guide students to set up the lofty goal, the pursuit of excellence, all-round development, cultivating a healthy personality, and comprehensively improve the quality of scientific research. Selection for each session of top students, we designed the selection methods of "selection method programming + open written + free screenings of interview". A number of professional teachers and a political teacher interview with students and make a judgment, the students which pass the test can enter the tutor team. Establish the assessment of the quality of the diversified incentive system. We overcome the one-sided pursuit of achievement and abnormal development of the disadvantagement, encourage the development of comprehensive quality.

3.4 Invite enterprises to participate in part of the curriculum construction

Different from the curriculum design and teaching programs in internship programs, commercial projects usually adopt the most advanced technology, the project structure is complete, the feasible scheme. They are often used in practical areas, rather than just stay at the level of theoretical research and primary applications. In training, the students not only design system functions, ensure the normal operation of the system, but also in accordance with the complete process of writing commercial project development related project documentation. Therefore, commercialization project training greatly improves students' interest in participating in the project development, and strengthens the students' ability to analyze and solve problems, to broaden the students' knowledge. It helps the students from the multi angle understanding technology application situation of computer, lets the students in practice to master the practical development of technology.

4 THE ACHIEVEMENTS OF CURRICULUM REFORM AND DEVELOPMENT

We compared the employment situation of graduates with the reform of 2012, 2013 and 2014. In the employment rate, after the reform is 98.74% and before the reform is 81.62%. In the view of graduates are engaged in industry, the IT industry proportion of 76% to 95% after reform. To see from the wage, it is from 3500 to 5200. The reason is the quick development of IT industry and the wage increase, creating many jobs and higher wages. The second reason is we are strengthening practice teaching in recent years, especially since the practice teaching. Practical teaching can help students to achieve professional knowledge into professional skills. It can enable the students to practice more attuned to the needs of enterprises, improve the students' employment competitiveness. After the reform of enterprises employing certain graduates ability and performance, at the same time the college education level has further improved recognition. Some IT companies are satisfied with graduates in their jobs, they recruit graduates in our institute in recent years.

Since 2 years of teaching reform, great changes have taken place in teaching and scientific research work of teachers. In this period, it gets a provincial brand specialty; an electronic information, education innovation highlands; a college bilingual teaching team; Four college quality resources sharing class; a bilingual demonstration course; a ministry of education professional characteristics; a provincial personnel training innovation experimental area and Six practice bases.

5 IDEAS FOR FUTURE WORK

Probing into the reform of practical teaching aims to establish a new mode of talent training. It solves the present computer teaching divorced from theory and practice problems, to improve the college students' employment rate and employment quality, fundamentally and improve the innovation of teaching and scientific research teachers. It is a new path for the development and reform of computer education. In the future, we need to adjust the project is to increase the training time. According to the development needs of enterprises and technology dynamically adjusts the training plan. We should make the students grasp the new technology and strengthen practice teaching funds. We should make the use of funds reasonable and improve the software training base. We should continue to improve the standard practice teaching; support young teachers teaching reform and research and innovation from the economy. We need to grasp the direction in the future: better use of the world's elite open class; provide training atmosphere of innovation, consciousness through the valuable class; the role of teachers should be from the knowledge initiator into thinking guide; how to deal with the relationship between the content of the course and teaching content.

REFERENCES

Roberts FS Computer science and decision theory. Annalso/ Operations Research, 2008, 163:209–253RossiF, TsoukimsA(eds)p, vceedinga, ADT 2009, Venice, Italy, Oct. 2003.

Computer Vision Education Resource [OL]. 2013.

Bebis G, Egbert D, Shah M. Review of Computer Vision Education [J]. IEEE Transactions on Education, 2003, 46(1).

Zia LL. The NSF National Science, Technology, Engineering, and Mathematics Education Digital Library (NSDL) Program New Projects in Fiscal Year 2002 [J]. D-LibMagazine, 2002.

Wing JM. Computational Thinking [J]. Communications of ACM, 2006, 49(3):33–35.

Easley D, Kleinberg J Networks, Crowds, and Markets Cambridge University Press, 2010.

Session 2: Information, automotive and education engineering

Information, Computer and Application Engineering – Liu, Sung & Yao (Eds)
© 2015 Taylor & Francis Group, London, ISBN 978-1-138-02717-6

Neuroethics issues on smart car

Jian Feng Hu

Institute of Information Technology, Jiangxi University of Technology, Nanchang, China

ABSTRACT: In the future, a kind of smart car, which is controlled by driver's EEG signals, will appear. At the same time, it will bring about a series of ethical problems. This paper introduces the smart car, analysis, ethical problems, and put forward the suggestion.

KEYWORDS: Neuroethics, Smart car, EEG, Driver, Brain Computer Interface (BCI)

1 INTRODUCTION

The smart car is different from the conventional car, which refers to automatic driving of a car with a variety of sensors and smart highway technology. There may be a kind of smart car controlled by driver's EEG in the future. The driver wears a sensing device, which collect driver's EEG automatically and judge awareness or brain states through certain algorithm, and then control the smart car.

The special behavior of smart car includes the following aspects:

First, to control the movement of cars, such as forward, reverse, turn and stop. When the driver imagines to take the car to the front in his brain, his EEG signals are collected by acquisition sensor and analyzed by the signal analysis equipment, and instantly determine the driver's intention, then the car forward. When the driver imagines to brake in the brain, the devices judge the driver's intention, then car controls to brake.

Second, to control the internal environment of the car, such as adjusting the seat, air conditioning, music. When the driver in the car feels hot or cold, specific EEG signal will appear in the driver's EEG, the device capture this kind of changes in EEG waves, then open, air conditioning, adjust the temperature inside the car, until the driver feel good now.

Third, to test the driver's fatigue status. When the driver feels fatigue because of physical reasons or long time driving, this is very dangerous, the device can also automatically capture this kind of EEG signals, and remind drivers to pay attention to this situation, and will gradually down speed.

Fourth, to control the car to start, as the key of the car. When the driver gets in the car, he can start the car engine through a particular idea of himself. Other people can't produce the same EEG, it will be impossible to start the car. This function also serves as a cipher lock.

Of course, smart car has also other features.

2 HOW DOES SMART CAR WORK?

If this type of smart car above can be born, it will have very important significance. First, this type of smart car will be the upgrade edition of the current smart car, which just now is in the laboratory stage. Second, it will trigger "lazy time, the driver can control the car without the movement of hands or feet. Third, the potential of the human brain will be excavated in the depth. There may not be any secrets.

What is the secret of smart car? So magical. The core of this type of smart car described above is a technology, namely brain computer interface (BCI) technology that is a technology of communication between the brain and equipment. Generally speaking, it is through the brain to directly control the machine equipment, without needing hands or mouth. The core of BCI technology is to decipher the EEG signals. When the individual thinks or decides, his/her brain must have specific changes, expression of this change in the brain is a change of neuron activity, and expression of this change outside the brain is the EEG. EEG is a kind of electrical changes, as long as a man lives, there is EEG in the brain, but the EEG signal is changing all the time, also in the sleep time. Different dreams are also reflected in the EEG. EEG is more complex and have its temporal and spatial variation characteristics, which needs a certain algorithm to decode. There are more methods can be used for BCI to translate EEG.

3 NEUROETHICS ISSUES ON SMART CAR

1 Whether is the operation of smart car representative of the will of the driver?

The use of Smart Car is still rather controversial. One ethical problem is whether it is permissible to use Smart Car for the driver when their informed consent cannot be obtained.

The use of smart car is still rather controversial. One ethical problem is whether it is permissible to use smart cars for the driver when their informed consent cannot be obtained.

The smart car is in the control car by the automatic detection of EEG. There is a problem here, the correct recognition rate of BCI technology can not guarantee, if BCI judge wrong? Due to technical limitations and environmental impact, the BCI recognition rate is less than 100%. The limitations of technology include inaccuracy of samples and instability of the method. EEG of human changes all the time, just simple things are sufficient to influence and even change EEG. For example, when you imagine to stop, there may be sudden loud noise, then EEG was completely changed, BCI technology may not recognize that you want to stop.

2 Who should be held responsible for the accident of smart car?

If smart car had an accident, who should be responsible, driver of smart car? It is very difficult to distinguish. Because smart car is not controlled by the driver's hands, the driver should not responsibility. However, the car is operating according to the driver's imagination, it seems a little responsibility for the driver. However, the actual operation may not be driver's actual imagination, the driver should exemption. Of course, no method to determine whether the driver is not no imagination or how to imagine, it is difficult to distinguish the driver's responsibility.

3 Will there be a hacker invade smart car?

Hackers can invade the information system, including computer and mobile phone. Now smart car may be added. The smart car relies on information system to control the car, when the information system is hijacked by hackers, the car can only follow and execute the command of information systems faithfully. For example, the hacker remotely controls a smart car, he may make smart car to carry out terrorist activities, or random motion, or freedom of movement. This will make the driver feel very helpless, can not control the smart car, and worry about that he has been kidnapped by smart car. If the hijacking behavior result in serious consequences, it will be disastrous, for example, terrorists invade the smart car, control some amount of smart car to Dutch act attack. It cannot be imagined. Therefore, smart car should have adequate safety systems, including three aspects, one is the basic layer, preventing infection by virus; two is the program layer, preventing the execution of external programs; the three is the result layer, automatically determining whether some operation have resulted in adverse consequences, then refusing to implement. The three levels are executed automatically, and the driver cannot control, nor invasion. At the same time, the introduction of cloud control, cloud computing and cloud storage should be necessary. The cloud control is used to correct operation and behavior of the cloud car in

an emergency time. Cloud storage is used to story the driver's habits and ways of thinking, and determine whether the invasion is happening. Cloud computing is used to avoid the hijacker of information system of the car.

4 Whether do smart car enhance the function of the human body?

Human enhancement is an interesting topic, and will bring many ethical problems, such as fairness, addiction and equality, etc. Whether may smart car in turn have an impact on the human body? For example, controlling smart car frequently will result in the enhancement of some functions of the human body. We focus on the problem of addiction. When the driver operating a smart car may get pleasure, the pleasure will force the driver to keep "driving", in order to gain pleasure. This will constitute the addiction, somewhat similar to drugs or addiction. That is to say the "driving" addiction.

5 Whether let driver to operate a smart car free?

To allow or prohibit drivers operating smart car freely, there still exists a dispute. Operating a smart car is an inevitable thing, it is impossible trying to completely restrict its application to realize. For "enhanced or not enhanced?" There are two questions worth discussing:

Whether can operate? If the driver is disabled individuals, it is no doubt that he can operate a smart car not normal car. On the contrary, if you are normal, fun from operating a smart car is very attractive. In fact, everyone loves fashion. From this perspective, it is fashionable to operate smart car. As long as there are no strong side effects, it will not flush as prohibited smart car. But how to judge the side effects? Evolution of the car is similar to the growth of the iPhone. From the history of the iPhone, "intelligent mobile phone" development trajectory can be predicted, beginning with cautious, which considered expensive and fashion. However, it is a mainstream up to now, which improve performance. Smart car is also so, as long as all humans operating smart cars, the human will take smart car as normal car, then there are no inequality, it evolved into a way of life, become a part of earth culture.

6 Whether do smart car has disturbing effect?

Smart car may cause the formation of a kind of fluctuations in our daily life. After all, we operate the normal car by our hands and feet, but operate smart car by brain. In the process of driving the car, the motor function is so important, so that any small change will affect the main personal motor function, and change a person's memory. Memory is brilliant, can be difficult to forget all the pain of the past, difficult to extricate oneself. Why do people drink alcohol, marijuana and engage in other activities that they lost the feeling? Why psychiatric treatment to all want to lose the unhappy memories of patients drop a hint?

Why let the victim trauma, injury and other emotional events bear strong memories of the bitter?

4 RECOMMENDATIONS

Regarding smart car they should be fully available to those who need them. If smart car is the best option for a given driver, it should be a fully integrated option under that driver's health care provider. Unforeseen changes in the health care system aside, all realistic attempts to make it feasible and affordable should be taken. At the same time, regardless of whether the smart car is the best solution or not, the procedure should always be the driver's choice. The concept of informed consent should be strictly adhered to in all cases.

ACKNOWLEDGMENTS

This work was supported by Natural Science Foundation of Jiangxi Province [No 20142BAB207008], project of Science and Technology Department of Jiangxi Province [No 2013BBE50051].

REFERENCES

[1] Hu JF, Mu ZD. Smart Car Security System by Using Biometrics Based on EEG [J]. 2013 2nd International Conference on Frontiers of Energy and Environment Engineering, ICFEEE 2013: 1047–1049.

[2] Hu JF, Mao C.L. Neuroethics—a perfect intersections of humanities and neuroscience [J]. Journal of Jiangxi University of Technology, 2008, 2: 15–17.

[3] Hu JF. Review of Brain Computer Interface [J]. Journal of Jiangxi University of Technology, 2006' 1(2): 81–88.

[4] Santoni de Sio F, Faulmüller N, Vincent NA. How cognitive enhancement can change our duties [J]. Front Syst Neurosci. 2014 8: 131

[5] Douglas T. Enhancing Moral Conformity and Enhancing Moral Worth [J]. Neuroethics. 2014; 7: 75–91.

[6] Wolff W, Baumgarten F, Brand R. Reduced self-control leads to disregard of an unfamiliar behavioral option: an experimental approach to the study of neuroenhancement [J]. Subst Abuse Treat Prev Policy. 2013 8: 41.

[7] Maier LJ, Liechti ME, Herzig F, Schaub MP. To dope or not to dope: neuroenhancement with prescription drugs and drugs of abuse among Swiss university students [J]. PLoS One. 2013 8(11): e77967.

Information, Computer and Application Engineering – Liu, Sung & Yao (Eds)
© 2015 Taylor & Francis Group, London, ISBN 978-1-138-02717-6

Human capital and new generation migrant workers urbanization in China

Mi Zhou, Guang Sheng Zhang, Xiao Gang Li & Zhang Jing Ye
College of Economics and Management, Shenyang Agricultural University, China

ABSTRACT: Considering that the current Hukou system in China prohibits the settling down of potential immigrants in cities, the demand-identified biprobit model is used to estimate the degree of urbanization for New-Generation Migrant Workers (NGMW). We survey NGMW in Shenyang and Yuyao. The results suggest that the urbanization degree of Shenyang and Yuyao are 62% and 81% respectively. The following factors influence the settling down of NGMW in the cities they work in: the income level of the NGMW, whether they reside among the urban residents, the strength of their urban social network, the length of time they have already spent living in urban areas, and their occupational status.

KEYWORDS: New-generation migrant workers; Urbanization in China; Social capital; Human capital

1 INTRODUCTION

The new generation of rural Chinese migrant workers, who have migrated to the cities from the countryside, desire to merge into their new homes. Statistics show that China's current migrant worker population is 150 million, of which 100 million were born after 1980; they have been labelled 'new-generation migrant workers' (NGMW). Compared with their parents who are viewed as the first generation of migrant workers, most of the NGMW are more reluctant to return to their hometowns. When they visit home, they only stay for a brief period of time before heading back to the cities. Of the 346 migrant workers surveyed by the Social Survey Center of China Youth Daily (2010), 41.4% want to settle down in the cities they work in. However, whether they will be successful in realizing this dream depends on the degree of their assimilation in the cities and the factors that affect it.

This paper presents a new approach for thinking about and estimating the assimilation of NGMW, as we ascribe weightage to possible influencing factors through regression. Some existing studies exploited statistical methods, such as the geometric mean (Liu et al., 2006), to measure the degree of assimilation from the perspective of the labour supply behaviour of migrant workers. This framework has been troublesome, because it ignores the differences among various factors. Although a few papers did pay closer attention to this problem and made substantial progress in ascribing weights to influencing factors by using the Analytic Hierarchy Process (AHP) (Liu et al., 2009), they ignored the fact that the decisions of potential settlers to settle down in the cities were restrained under the Hukou system, i.e., the results were not really representative.

Notwithstanding the importance of potential settlers in China's Hukou system, relatively few empirical studies (e.g., Liu et al., 2008; Wang et al., 2008) have measured the assimilation degree of potential settlers in China. To our knowledge, this paper is the first to report the use of the identified-demand biprobit model for potential migrant settlers in China's cities.

2 MODELS AND METHODS

Whether migrant workers can assimilate smoothly as townspeople depends not only on the external environment, but also their own characteristics. This paper supposes that assuming constant external conditions, the determination of the transition from NGMW into townspeople is dependent on their need to become a citizen of that particular town, that is, the willingness of the migrant workers to reside in the city they work in. We use the term 'citizen needs' in this paper to refer to the need of the two particular categories of NGMW to become citizens of the cities they work in. On the other hand, the transition from NGMW to townspeople also depends on 'citizen supply', namely, their ability to settle down in the city. This paper investigates the citizen supply ability of NGMW from the viewpoint of their income only and supposes that a migrant worker has the ability to settle down in the city he/she works in only when his/her income level is higher than the average income level of the same city.

This paper estimates a simultaneous equations model of supply-demand equilibrium to measure the citizenship degree of NGMW and to reveal the factors influencing their decision to become citizens of the city. However, all the dependent variables are (0, 1) distributed variables, which means the classical simultaneous equations model cannot solve this kind of problem. Luckily, the demand recognizable biprobit model and its estimation method provide a way to solve this kind of problem. Moreover, this model surpasses the probit model and the partial observational biprobit model (Poirier, 1980; Evans et al., 1995; Huang et al., 2009). We use the probability of the NGMW to simultaneously possess citizen demand and citizen supply to express their citizenship degree.

3 DATA

The data used in this paper are collected from a survey of two regions. In July 2009, some teachers and graduate students of Shenyang Agricultural University conducted an interview survey of migrant workers in Yuyao Zhejiang province. The survey obtained 296 complete and valid questionnaires, including those from 114 NGMW, the latter accounting for 39% of the total sample. In November 2009, another such interview survey of migrant workers in Shenyang obtained 287 responses, including 154 from NGMW (54% of the total sample). Notably, the two investigations are comparable as the same questionnaire was used. We can thus conclude that NGMW comprise a sizeable proportion of migrants. The variables are listed as follows.

4 ANALYSIS OF ECONOMETRICS MODELS

4.1 Dependent variables

We establish three processes to identify citizen demand. The first step is to find permanent settlers. If the NGMW want to dwell in the city they work in permanently, we consider this as an indication of citizen demand. There are 95 such settlers, accounting for almost 35% of the total sample. However we cannot ignore the potential settlers, who choose not dwell permanently in the city they work in due to current policies, but could change their choice if the old regulations, such as the Hukou system, are abolished. Then, in the second step, we ask these nonsettlers question (2), in order to identify whether they prefer life in the city or that in the village, without considering the current policies affecting immigration. If the NGMW prefer life in the city to that in the village, we believe that they may exhibit a potential demand to become urban citizens. In the last step, we use question (3) to test whether the identified demand is a valid demand. If the NGMW think that their income level is the same as or higher than that of the average citizen in the city, then we believe that they have the ability to live in the city they work in. Thus, the potential settlers are those nonsettlers who enjoy the city life and believe that their income matches or exceeds that of the citizens of the city they work in. The results show that of the 268 NGMW, 122 (= 95 + 27) have citizen

needs. Thus, the dependent variable in the demand equation takes the value 1 for the 122 NGMW who exhibit citizen demand and 0 otherwise.

The dependent variable in the supply equation takes the value 1 if the real income of the worker exceeds the average personal deposited income (PDI) published by the Liaoning and Zhejiang Statistical Bureaus in 2009 and 0 otherwise.

Table 1. Regression results of the biprobit model.

Variable	Supply equation		
	Coefficient	Standard Error	P value
c1_1 = 1	0.0247	0.2834	0.9300
c1_1 = 2	−0.3039	0.2827	0.2820
c1_1 = 3	0.2167	0.2869	0.4500
xy	0.3863	0.1876	0.0390
hc_1	−0.3001	0.2351	0.2020
_lj8_2	−0.2609	0.2059	0.2050
_lj8_3	0.0022	0.2523	0.9930
j4	−0.4494	0.1103	0.0000
Marr	0.1584	0.2189	0.4690
a2	0.0195	0.1827	0.9150
Area	0.6150	0.2075	0.0030
b3	0.1033	0.0414	0.0130
Zy	0.0090	0.0048	0.0600
Constant	−0.5305	0.5880	0.3670
/athrho	0.7305	0.1383	0.0000
Rho	0.6234	0.0845	

Variable	Demand equation		
	Coefficient	Standard Error	P value
c1_1 = 1	0.3977	0.2844	0.1620
c1_1 = 2	0.1764	0.2790	0.5270
c1_1 = 3	0.5015	0.2918	0.0860
xy	0.1466	0.1775	0.4090
hc_1	−0.3989	0.2174	0.0670
_lj8_2	−0.9008	0.1981	0.0000
_lj8_3	−0.3707	0.2336	0.1130
j4	0.1501	0.0981	0.1260
Marr	0.2058	0.1998	0.3030
a2	−0.1270	0.1739	0.4650
Area	0.9138	0.1922	0.0000
jz	0.3845	0.1700	0.0240
Constant	−1.8582	0.5740	0.0010
maximum likelihood	−262.2672		
number of observations	260		

Note: There are six missing observations in the 'pay a new year's call net '.

582

4.2 *Estimation results and interpretation*

The results of the supply equation show that the higher social network height, the more likely they are to seek jobs by means of their network 'relationships', and thus, the more likely they are to obtain a higher and more stable income. Income satisfaction also has a remarkable positive effect on the citizen supply ability of NGMW. In addition, the longer the migrant worker has lived in the city, the deeper their integration with it. Such workers would have made closer and more numerous associations with the average city dweller, and so, they may be able to acquire the information required to change over to better-paying jobs or improve their occupational status. Moreover, migrant workers of certain professions often earn better and enjoy a higher social status. The inverse Mills ratio, considered as an extra regression element of the citizen supply equation, is remarkable. It indicates that the demand equation has influenced the supply equation.

The results of the demand equation provide several vital results. First, NGMW with a middle or high school education are more likely to return to their hometowns instead of staying in the city. The reasons for this are thought to be two-fold. On the one hand, it is not easy to convert the cultural knowledge mastered by such workers into requirements practical for a city life and/or into their social network relationships in a short time span. On the other hand, because of employment discrimination in urban enterprises, many migrant workers find it hard to settle down in the city. Secondly the more frequently they keep touch with their rural hometown, the more likely they are to return to their hometowns. Thirdly, NGMW who choose to live among the city dwellers show a stronger willingness to dwell in the city. Indeed, such living patterns can enhance communication between the migrant workers and the citizens of the city. Such experiences can help migrant workers break through the Hukou system.

5 CONCLUSION

Based on the survey data of the NGMW from Shenyang and Yuyao, this paper measured their citizenship degree and the various factors affecting the same from the viewpoint of citizen demand and citizen supply. Equation forecasts the urbanization degree of the NGMW of Shenyang and Yuyao as 62% and 81% respectively. NGMW with a high income, living among the city's citizens, possessing a good social network in the city, dwelling in the city for a long time, and having high occupational status are more likely to exhibit citizen demand willingness and citizen supply ability. Moreover, the living pattern of NGMW, namely whether they live among the city dwellers instead of confining themselves to living with other migrants from their hometown, is a key factor affecting citizen demand willingness, which in turn affects their citizen supply ability.

ACKNOWLEDGMENT

This research is funded by the National Natural Science Foundation of China (71203146), and supported by program for excellent talents in Liaoning province (WJQ2014016); and agriculture youth science and technology innovation talent training plan in Liaoning province.

REFERENCES

[1] Evans, W. N., & Schwab, R. M. (1995). Finishing high school and starting college: Do Catholic schools make a difference?. The Quarterly Journal of Economics, 110(4): 941–974.
[2] Huang Z.H., Liu X.Ch., and Cheng E.J. (2009) Explanations for the Low Participation Rate in the Formal Credit Market by Rural Households in the Poor Areas, Economic Research Journal, 4: 116–128.
[3] Huang Z.H., Qian W.R and Mao Y.C, (2004) The living stability and citizenship willingness of peasants in city. Chinese Journal of Population Science, 2: 68–73.
[4] Liaoning Bureau of Statistics, (2009) Appendix7-The basic situation of agriculture counties (2008), Liaoning Statistical Yearbook 2009.
[5] Liu Ch.J., Cheng J.L., (2008) Status Analysis and Process Measurement of the Second Generation Citizenship. Population Research, 5: 48–57.
[6] Poirier, D. J. (1980). Partial observability in bivariate probit models. Journal of Econometrics, 12(2), 209–217.
[7] Wang G.X., Shen J.F., and Liu J.B. (2008) Citizenization of Peasant Migrants during Urbanization in China, Population and Development, 1: 3–23.
[8] Zhejiang Bureau of Statistics, (2009) Chapter17–24 Main Indicators of National Economy by City and Country(2008), Zhejiang Statistical Yearbook 2009.

Information, Computer and Application Engineering – Liu, Sung & Yao (Eds)
© 2015 Taylor & Francis Group, London, ISBN 978-1-138-02717-6

Engine health monitoring based on remote diagnostics

Cun Jiang Xia

Civil Aviation Flight University of China, Guanghan, Sichuan, China

ABSTRACT: Engine health is very important for flight safety in civil aviation industry. There are many kinds of methods to monitor engine health, such as performance trends monitoring, borescope inspection, vibration monitoring, lubrication particle analysis, etc. In which engine performance trends monitoring is the most popular & effective method to speculate the engine health. Some years ago, the performance trends monitoring was done by airlines' own. As time passed, another method based on computer network called remote diagnostics is created, in which the performance trends analysis is executed by the engine's manufacturer. This article introduces the basic process of remote diagnostics, then gives the objective parameters of the engine to be monitored, after that the parameter correction and baseline concept are briefly introduced. It also gives the trends of parameters shift and reasonable explanation in the case of component failure on engine. At the end this paper gives one example to demonstrate how to troubleshoot the engine according to performance data trend plot.

KEYWORDS: Engine; Condition Monitoring; Parameter Trends; Shift

1 REMOTE DIAGNOSTICS INTRODUCTION

Engine Condition Monitoring (ECM) is a tool that allows engines to be maintained under "Condition-Based Maintenance" philosophy. Most regulatory agencies require operators to maintain an ECM system. Otherwise, the agency would not approve operator's operations.[1]

Traditionally, airlines were using an analyzing tool to monitor their own engines. Operators need to process operational data and review parameter trends to identify whether the engine condition is normal or not. If it's suspected abnormal, troubleshooting or inspection is required.

As the developing of network and computer technology, the manufacturer of the engine developed a new tool which is called out "Remote Diagnostics". It requires aircraft send operational data to the diagnostics server through ACARS, email or FTP method. A diagnostics system could process those data and generate parameter trends automatically. Additionally, this system could save the data & its trends, filter out the normal condition and generate alerts as required based on parameter trends. In recent years, most operators have already been using this advance tool.

Since this is a very useful tool to monitor engine's health and many operators are very interested, combined with Customers' experiences, this article gives the detail of Diagnostics knowledge.

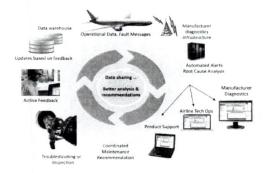

Figure 1. Collaboration Process of Remote Diagnostics.

2 PROCESS OF PARAMETERS

2.1 *Data used to analyze*

To monitor the engine health, these data must be gathered through ACARS system:

- Elevation,
- Total air temperature (TAT or T12 gotten by sensors installed on the engine)
- Mach number
- Bleeds configuration
- Operational data, e.g. N1, N2, EGT, Fuel Flow

In order to get better analysis, some optional data are recommended, e.g., Ps3, T25, T3, VSV & VBV position, HPTACC & LPTACC valve position, etc., all these data are very helpful for root cause analysis.

2.2 Parameters to be monitored

1 Performance data

- Exhaust gas temperature (EGT)
- Core speed (N2)
- Fuel flow (FF)
- Hot day EGT margin (EGTHDM)

2 Vibrations
- Fan & core vibrations (takeoff and cruise)
3 Oil
- Oil Pressure, Delta Oil Pressure and Oil Temperature

In this article, we will just focus on performance parameters and especially discuss the relationship between $\Delta N2$ trend and engine hardware.

2.3 Normalization of parameters and delta calculation

Because of the effect of altitude and climate on pressure and temperature, the parameters monitored must be normalized (corrected to standard day at sea level static condition).[2] The absolute valve is useless due to operational condition variation. To estimate engine's health, the trends of parameter's differential value should be tracked and reports of trends will be generated.

Generally, the parameters including Delta EGT, Delta Core Speed, EGTHDM and Delta Fuel Flow should be corrected.

Define θ_2 & θ_{25} as follows:

$$\theta_2 = \frac{T_{amb}(^\circ C) + 273.15}{T_0 (^\circ C) + 273.15}$$

$$\theta_{25} = \frac{T_{25}(^\circ C) + 273.15}{T_0 (^\circ C) + 273.15}$$

Where T_{amb}=Actual air temperature of the ambient, T_0=15°C (Air temperature at sea level standard), T_{25}=Compressor Inlet Temperature

The correction formula:

$$N_{1K} = C_{N1} \cdot N_{1meas} \cdot \frac{1}{\theta_2 f_1(N_{1K})}$$

$$EGT_K = C_{EGT} \cdot EGT_{meas} \cdot \frac{1}{\theta_2 f_2(N_1)}$$

$$N_{2K} = C_{N2} \cdot N_{2meas} \cdot \frac{1}{\theta_{25} f_3(N_2)}$$

$$W_{FFK} = C_{FF} \cdot W_{FFmeas} \cdot \frac{1}{\frac{P_T}{101325}} \cdot \frac{1}{\theta_{25} f_4(N_1)}$$

Where N_{1K}, EGT_K, N_{2K}, W_{FFK} are corrected parameter; N_{1meas}, EGT_{meas}, N_{2meas}, W_{FFmeas} are measured parameters, is the total pressure.

3 DELTA PARAMETER'S CALCULATION

3.1 Cruise "baseline" concept

Cruise "baselines" represents expected engine characteristics for operation at various flight conditions, for instance, expected EGT, Fuel Flow or Core Speed at some certain conditions.

The averaged parameter baseline curve for one engine model is calculated based on production acceptance test data of new engines. It takes into consideration of air temperature, mach number, altitude, power setting, pneumatic bleeds. It can be shown as:

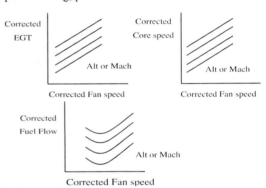

Figure 2. Baselines for cruise.

Baseline Model=f (TAT, N_{1K} Mach, Alt, Bleeds)

Individual engine baseline curves are different from the averaged parameter baseline curve because:

- engine part tolerances change within the build limits
- total time and cycles are different
- quality of engine refurbishment is different
- operating environments change

3.2 "deviation" or "delta" parameters calculation

The following cruise "deviation" or "delta" parameters are calculated:

Delta EGT - (°C)
Delta Core Speed - (%)
Delta Fuel Flow - (%)

These charts are representations of engine cruise performance characteristics. There is expected performance parameter variation with fan speed and flight condition. These characteristics derived from flight tests by OEM at various flight conditions.

To monitor engine health, "Delta" Parameters are typically trended at cruise. The parameters used to monitor cruise performance are usually differences between actual measurements and "baselines", such as Delta EGT (ΔEGT), Delta Core Speed (ΔN2) and Delta Fuel Flow(ΔFF):

The Formula for calculating deviations of cruise parameters is below: [3]

$$\Delta EGT = EGT_{Raw} - EGT_{Baseline}$$

$$\Delta N2\ (\%) = \frac{N2_{Raw} - N2_{Baseline}}{N2_{Baseline}} \times 100\%$$

$$\Delta FF\ (\%) = \frac{FF_{Raw} - FF_{Baseline}}{FF_{Baseline}} \times 100\%$$

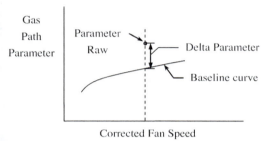

Figure 3. Deviation calculation.

3.3 *Performance parameter's shift*

If any component or system fails or deteriorates, raw value will shift accordingly, and then will lead to calculated cruise parameter change. This shift mode is called as "true shift".

As we know, baseline values vary by TAT, N1K, Mach, Alt and Bleeds, any false value or out of normal range or missing data in any of these parameters will impact the "baseline" value, consequently, the delta parameter such as delta core speed, Delta EGT, Delta Fuel Flow will behave shifts. This shift mode is called as "false shift", because the engine itself may have no any malfunctions, no inspection/troubleshooting is needed on the engine itself.

In this section, we will just cover the signatures that engine really has problem, and troubleshooting or maintenance action is required, e.g., HPT distress, air leaks, HPTACC Valve failure, etc. In these failure modes, the raw values of the parameters such as EGT, fuel flow and core speed will shift. Consequently, the Delta value will shift.

Most of the time, such signatures occur to just one engine on wing solely. Therefore, there will be shifts in the divergence of these parameters.

$$X\ Divergence_1 = X_1 - \frac{X_1 + X_2 + X_3 + ... + X_n}{n},$$

For the airplane which is just installed with dual engines, it will be:

$$X\ Divergence_1 = X_1 - (X_1 + X_2)/2 = (X_1 - X_2)/2$$

$$X\ Divergence_2 = X_2 - (X_1 + X_2)/2 = (X_2 - X_1)/2$$

It's obvious that any shift just in one engine parameter will cause divergence shift, e.g. EGT, Fuel Flow and Core Speed, which can be used for isolation of ambient condition impact or common parameters impact (TAT, ALT and Mach), it is to say, if the divergence has not shifted, which means the parameter shift might be caused by ambient condition impact or common parameters.

Table 1. On-wing performance diagnostics by statistics.

Failure	Parameter's shift		
	ΔEGT	ΔFF	ΔN2
Compression system Leakage	↑	↑	↑
Variable stator vane More close	↑	↑	↑
Variable stator vane More open	↓	→	↓
Core engine poor efficiency	↑	↑	↑
Fan &Booster poor efficiency	↑	↑	↑
LP turbine poor efficiency	↑	↑	↑

From table 1, it shows most of the anomalies cause up shift in Delta EGT and Delta Fuel Flow, the only difference is the shift direction of the delta core speed. Some of delta core speed goes up, some go down. The shift direction of delta core speed is a very important clue for troubleshooting. e.g., if core efficiency loss, it will cause shift-down in delta core speed, if air leaks or booster/LPT efficiency loss, it will cause shift-up in delta core speed. You will never take a wrong direction during troubleshooting/inspection if you know this phenomenon. But why will core efficiency loss cause core speed down? Why will air leaks or LPT efficiency loss cause core speed up? We will be talking about that right away.

4 EXPLANATION OF ΔN2 SHIFT TRENDS

1 Core efficiency

The core delivers energy to the LPT. As the core components deteriorate (efficiency gets worse, core temperature increases)

• Actual work required by HPC increases
• Actual work provided by HPT decreases

587

At a constant pressure ratio, EGT increases, Core slow down at constant fan speed

$$\eta_c = \frac{ideal\ work}{actual\ work} = \frac{\Delta h_i}{\Delta h_a} = \frac{h_{exit(ideal)} - h_{inlet}}{h_{exit} - h_{inlet}} \cong \frac{t_{exit(ideal)} - t_{inlet}}{t_{exit} - t_{inlet}}$$

$$\eta_T = \frac{actual\ work}{ideal\ work} = \frac{\Delta h_a}{\Delta h_i} = \frac{h_{exit} - h_{inlet}}{h_{exit(ideal)} - h_{inlet}} \cong \frac{t_{exit} - t_{inlet}}{t_{exit(ideal)} - t_{inlet}}$$

At a low HPT efficiency, HPT exit temperature rises, and inlet temperature of LPT increases, in order to keep the same N1 (constant energy to drive LPT), mass flow must decrease, then core speed slows down.

At an HPC efficiency loss, exit temperature of HPC rises, more actual work is needed to drive HPC rotor, then more energy is required, more fuel is needed, higher HPT exit temp is produced, and inlet temp of LPT increases, to keep the same N1(constant energy to drive LPT), mass flow must decrease, then core speed slows down .

From field, we experienced some failure modes related to core efficiency loss, such as HPC blade damage, HPT blade distress, HPT shroud burn through or drop, HPTACC failure and TBV failure, etc.

2 Air Leaks

When core leak occurs, mass flow through LPT module decreases, to keep the same N1 (constant energy to drive LPT), more mass flow is needed, then increasing core speed is required. To increase the core speed, more fuel flow is needed, then EGT increases. VBVs, customer bleed, parasitic bleed could be the air leakage sources.

3 LPT Efficiency

When loss of LPT efficiency, assuming there is the same enthalpy and same mass flow through LPT rotor compared to the conditions prior to LPT efficiency loss, then less energy will be converted to drive N1 rotor, then N1 will drop. However, in order to keep the same N1, more enthalpy at the inlet of LPT is required, therefore more fuel is burnt and more mass flow though LPT rotor, core speed goes up.

5 EXAMPLES BY REMOTE DIAGNOSTICS

The following plot was taken from remote diagnostic. Customer downloaded QAR data, it was confirmed that there are significant FF & EGT difference with the sister engine when N1 is greater than 75%. EEC BITE did find fault code recorded. BSI of HPT blades & shroud, and HPC stage 1-9, no abnormal found. Further analysis found ΔEGT &ΔFF shift up, but ΔN2 shifts down. According to table 1, it seems to be core efficiency problem. Customer Replaced HPTACC valve & TBV and performance parameter recovered[4].

The possible reason might be the TBV actuator lost its control of butterfly valve (would be mechanical broken), that's the reason why no EEC fault code and same TBV LVDT position between both engines

reading thru CDU input monitoring. TBV is always at open position and discharges HPC Stage 9 air to LPT Stg.1 Nozzle[5], which resulted in core efficiency loss.

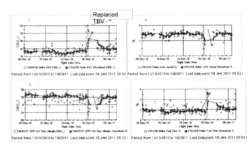

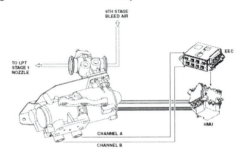

Figure 4. Parameter abnormal plot.

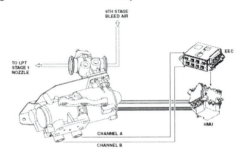

Figure 5. Transient bleed valve system.

6 CONCLUSION

Remote diagnostics is very helpful tool to monitor the engines' health, it uses the engine operational data to estimate whether the engine is in normal operation. During normal operation, it shows the engine's capability to get normal thrust without over-temperature and the deviation trends of parameters. If there is a large shift for parameters, the system generates some alert reports. This paper gives shift direction chart of parameters after component or system failure, which is like a fingerprint. Depending on the chart maintenance personnel can give the possible causes and some troubleshoot suggestion based on the understanding of the engine's operation.

REFERENCES

[1] GE, diagnostics training manual, Jan 2009, p1–3.
[2] Pratt & Whitney,engine condition monitoring using ECM, Sept 1994, p2–2.
[3] ZHONG Shi-sheng, CUI Zhi-quan & FU Xu-yun. Baseline mining method of RR's engine. Computer Integrated Manufacturing Systems. 2010, 16(10): 2265–2270. (In Chinese).
[4] Boeing, 737-600/700/800/900 aircraft maintenance manual, Oct 2008.
[5] CFM international,CFM56-7B Engine systems, Apr. 2007, p238.

Information, Computer and Application Engineering – Liu, Sung & Yao (Eds)
© 2015 Taylor & Francis Group, London, ISBN 978-1-138-02717-6

Analysis of location accuracy of low orbit dual-satellites passive location system based on differential method

Gang Li & Cheng Lin Cai
Institute of Information and Communication, Guilin University of Electronic Technology, China

Si Min Li
Institute of Electrical and Electronic Engineering, Guangxi University of Science and Technology, China

ABSTRACT: To improve the location accuracy of low orbit dual-satellites passive location system. An observation differential method based on differential station on the ground is advanced, and the positioning accuracy of dual-satellites passive location system can be greatly improved by using this approach in simulation. Meanwhile by utilizing this differential technique, this paper also analyses the impact of different orientations and baseline between the differential station and unknown radiation source on positioning accuracy. The results show that when the differential station and unknown radiation source at the same latitude with 3–5 degrees, the differential accuracy can easily achieve less than 2Km. And the differential accuracy of radiation source that is close to the sub-satellite point is not sensitive to the length of the baseline. Otherwise differential accuracy is sensitive to the baseline.

KEYWORDS: Low orbit satellite; Passive location; Differential station on the ground; Positioning accuracy

1 INTRODUCTION

Dual-satellites passive location system is to use double LEO satellites from the ground 800–1000Km to measure the time difference of arrival (TDOA) and the frequency difference of arrival (FDOA) of the unknown target on the ground to achieve the accuracy positioning of the target[1-3]. Compared with tri-satellites TDOA passive location system, it has the advantages of less receive platform, low cost, short cycle and reducing the complexity of achieving the system[4-5]. In addition, the system has such characteristics, long baseline and fast moving, therefore it's an ideal scheme to apply TDOA and FDOA measurements to locate the unknown target position. However, the current research concentrates on how to optimize the solution of TDOA and FDOA equation and achieve the Cramer Rao lower bound (CRLB) as far as possible. Fortunately the worse positioning accuracy can reach about 10Km by using these algorithms [6], but the positioning accuracy is still poor. [7–8] have shown that by introducing calibration emitter can improve the positioning accuracy in passive location. Thus the new positioning system should be explored to improve the positioning accuracy of dual-satellites passive location system.

In this paper, inspired by the ideas of differential GPS (DGPS), a method based on observations differential is presented and the positioning accuracy is also analysed before and after differential. Besides the influence of azimuth of differential station location and baseline between differential station and unknown target on positioning precision are analysed by using the proposed method.

2 THE PRINCIPLE OF DUAL-SATELLITES PASSIVE LOCATION SYSTEM BASED ON OBSERVATION DIFFERENTIAL

GPS differential technology includes positioning differential, Pseudo range differential, differential after carrier phase smoothing pseudo-range and carrier phase differential[9-10]. However, Pseudo range differential is often used. And its positioning accuracy is better than Positioning differential. Thus, in order to improve the positioning precision of dual-satellites passive location system, Pseudo-range(observations) differential technology is applied to correct TDOA/FDOA measurement in this paper. And the differential system model of dual-satellites is shown in figure 1.

Suppose that the location emitter on Earth, denoted by u (X, Y, Z), by measurements from two satellites, whose positions $S_i(X_i, Y_i, Z_i)$ and speeds $\dot{S}_i(\dot{X}_i, \dot{Y}_i, \dot{Z}_i)$ separately. The differential station position ref (X_r, Y_r, Z_r). The frequency of unknown radiation source is f_0. The signal propagation speed is c. $TDOA''$ and $FDOA''$ is TDOA and FDOA of the unknown target on the

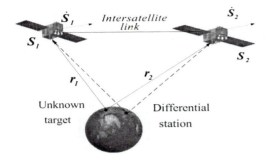

Figure 1. The differential system model of dual-satellites.

Earth separately. Therefore, as in figure 1, TDOA and FDOA equation of unknown target can be expressed as the followings.

$$TDOA^u = \frac{1}{c}(r_2 - r_1)$$
$$FDOA^u = \frac{f_0}{c}(\dot{r}_2 - \dot{r}_1) \qquad (1)$$

Where,

$r_i = \|S_i - u\|$, $\dot{r}_i = \frac{(S_i - u)^T \dot{S}_i}{r_i} = m_i(u).\dot{S}_i$, $\|*\|$ is the Euclidean norm.

2.1 Generation of correction numbers of low orbit dual-satellites passive location system

The observation differential method has two steps in low orbit dual-satellites passive location system. Firstly, using the differential station whose position is known to solve out the correction numbers, and then applying them to correct the measurements of unknown target. Assume that TDOA^r and FDOA^r that are measured by the two satellites represent TDOA and FDOA of differential station separately. Then the TDOA and FDOA of differential station in theory can be obtained by using equation (1).

$$\Delta t = \frac{1}{c}(\|S_2 - ref\| - \|S_1 - ref\|)$$
$$\Delta f = \frac{f_0}{c}\left(\frac{(S_2 - ref)^T \dot{S}_2}{\|S_2 - ref\|} - \frac{(S_1 - ref)^T \dot{S}_1}{\|S_1 - ref\|}\right) \qquad (2)$$

Where Δt, Δf is theoretical TDOA and FDOA of differential station, and Δt, Δf contain the ephemeris error. The differential correction numbers can be generated from equation (2).

$$\Delta_1 = TDOA^r - \Delta t \qquad (3)$$

$$\Delta_2 = FDOA^r - \Delta f \qquad (4)$$

Where Δ_1 and Δ_2 are the differential correction numbers of TDOA and FDOA accordingly. And using them to amend the measured TDOA and FDOA of unknown target, so equation (1) can be represented as

$$TDOA^u - \Delta_1 = \frac{1}{c}(r_2 - r_1) \qquad (5)$$
$$FDOA^u - \Delta_2 = \frac{f_0}{c}(\dot{r}_2 - \dot{r}_1)$$

In order to solve the unknown target position, the constrained condition that the unknown target is on the earth is needed to use. As for the earth model in this dissertation and taking into account the issue of positioning accuracy, the sphere earth model is rejected, and the WGS-84 ellipsoid earth model is given.

$$\frac{x^2 + y^2}{(N + H)^2} + \frac{z^2}{[N(1 - e^2) + H]^2} = 1 \qquad (6)$$

Where, $N = a / \sqrt{1 - e^2 \sin^2 B}$ is radius of curvature of local meridian circle. a is the Earth's semi-major axis. H and B are the altitude and latitude of unknown target separately. e is eccentricity of the ellipsoid earth model. Therefore, the problem of dual-satellites passive differential location is transformed into how to use (5) and (6) to solve the position of unknown target. However, it is uneasy to give the solution by solving the three equations, for they are nonlinear. Although the positions can be solved by Newton Iteration Algorithm through three equations, the algorithm require initial guesses closed to the real position and may would suffer from convergence problem. Thus, an analytical method is adopted in this paper.

2.2 Analytical solution of dual-satellites passive location system using differential method

The analytical method based on ellipsoid earth model has two stage approaches.

2.2.1 Coarse location
The sphere earth model donated as $u^T u = R$ is used in this stage, where R is the average radius of the earth. Combine the sphere earth model and the expression of r_i, equation (7) can be obtained

$$r_i^2 = R^2 + S_i S_i^T - 2S_i^T u \quad i = 1, 2 \quad (7)$$

Let $r_{21} = c(TDOA'' - \Delta_1)$

Squared the first part of equation (5), and then substitute it into equation (7). So equation (7) becomes

$$r_{21}^2 + 2r_{21}r_1 + S_2 S_2^T - S_1 S_1^T - 2(S_2 - S_1)^T u = 0 \quad (8)$$

(9) is obtained by taking time derivative of (8), giving

$$2(\dot{S}_2 - \dot{S}_1)^T u - 2S_2^T \dot{S}_2 + 2S_1^T \dot{S}_1$$
$$+2r_{21}\dot{r}_1 + 2\dot{r}_{21}r_1 + 2r_{21}\dot{r}_{21} = 0 \quad (9)$$

Where, $\dot{r}_{21} = \dfrac{c}{f_0}(FDOA'' - \Delta_2)$

r_1 and $\dot{r}_1$ can be regarded as known quantity, and apparently equation (7) (8) (9) can be written into matrix. $Gu = h$. The solution of matrix equation follows the approach in [2] to obtain an algebraic equation which is a 6th degree equation in one variable about r_1.

$$g_s^T r = 0 \quad (10)$$

Where, $r = [r_1^5, r_1^4, r_1^3, r_1^2, r_1, 1]$ and r_1 can be obtained after solving equation (10). So the coarse position of unknown target u'should be attained after eliminating ambiguous solutions using direction finding technology.

2.2.2 Precisely location

A sphere recursion algorithm is used to realize precisely location. The sphere recursion algorithm starts with the initial sphere whose radius is the distance between sub-satellite point and the centre of Earth and the coarse location u'. And each recursion consists of projection, structuring spherical equation and solving. Thus, after general 2–3 spherical iterations, accuracy position of target can be obtained under the WGS-84 ellipsoid earth model.

3 SIMULATION RESULTS

This section contains simulation results to demonstrate the theoretical development. In the simulation study, the receivers are LEO satellites having a distance 800km from the Earth surface. They are at $S_1(116.7, 26.4)$, $S_2(116.4, 26.1)$ respectively, and their velocity is $\dot{S}_1(-4057, -2016, 6471)$, $\dot{S}_2.(-4988, -2513, 5586)$, adding the ephemeris errors accordingly. And basing on the current level of error control, the measurement errors of TDOA is 100ns, the measurement errors of FDOA is 10Hz. The position of differential station is ref (112.5, 29.25). In the following analyses, the positioning accuracy using differential method is firstly analysed, and then we study the influence of baseline and orientation of differential station on positioning precision. "*"represent sub-satellite point, "o" represent the position of differential station.

3.1 The positioning accuracy after using differential method

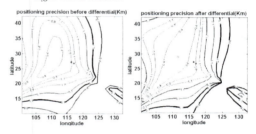

Figure 2. The positioning accuracy of radiation before and after differential.

As is known in figure 2, unknown target is closer to the differential station whose positioning error is less than 0.4Km area had increased significantly after differential compared with before differential. However, the farther the distance between unknown target and differential station is, the less distinct the position accuracy is after differential. (Such as the positioning accuracy of 5.1Km and 9.6Km of the regional, the improvement is not very difference). This is because no matter the distance between unknown target and differential station is short or long, measurement errors always can be eliminated by differential. But the ephemeris error is closely related to the distance between target and differential station. The nearest the target is to differential station, the more significant differential method can reduce ephemeris error. Thus, it's reasonable to conclude that positioning accuracy after using differential is related to distance.

3.2 Influence of baseline and orientation of differential station on positioning precision

In order to better study the influence of baseline and orientation of differential station on positioning precision. The satellite coverage area is divided into two parts, one is the sub-satellite point as the center within a circle of radius 900Km, and the other one is outside the 900 Km ring. Select a sufficient number of points in different regions as unknown radiation source.

Figure 3 (a) (b) (c) is the target map of differential station and unknown target with the same longitude and is separated 3 degree, 8 degree or 10 degree. Figure 3 (d) (e) (f) is the target map of differential station and unknown target with the same latitude and is separated 3 degree, 8 degree or 10 degree. The blue dot represent before differential and the red dot represent after differential.

As is known in figure (3), when keeping the same orientation between target and differential station, positioning accuracy decreased with the baseline distance from the radiation source to the reference station increase. When the baseline of the target and differential station remains the same, the positioning accuracy of the target and differential station located in the same latitude is much better than those in the same longitude. And also the differential station and unknown target at the same latitude with 3~5 degrees, the differential accuracy can easily achieve less than 2Km.

Figure 4 (a) (b) (c) is the target map of differential station and unknown target with the same latitude and is separated 1 degree, 3 degree or 8 degree. As is known in figure (4), differential positioning accuracy is much better under the condition of short baseline,

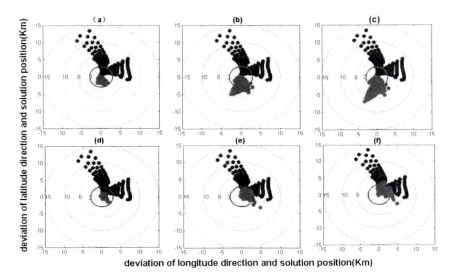

Figure 3.   The location deviation of radiation source close to the sub-satellite point.

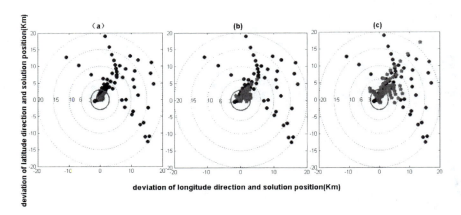

Figure 4.   The location deviation of radiation source far from the sub-satellite point.

but the differential positioning accuracy will become fairly bad, when the baseline becomes longer.

From the above analyses, the following conclusions can be drawn

1 Low orbit dual-satellites passive location system using differential technology not only can eliminate the system measurement errors, but can reduce the ephemeris error, so the difference technology can effectively improve the location accuracy.

2 The positioning accuracy of the target and differential station located in the same latitude is better than those in the same longitude.

3 The positioning accuracy is insensitive to the baseline between differential station and unknown target that is near to the sub-satellite point. However, when the target is far from the sub-satellite point, positioning accuracy would become worse with the baseline become longer.

4 In order to attain better performance in dual-satellites passive location system, we should prefer to choose the differential station as the same latitude with unknown target. And the baseline distance is as short as possible in theory, but if a suitable reference stations can not be found in the vicinity of the unknown radiation source, choosing a long baseline reference station is not very greatly impact on the positioning accuracy. However, when the target is far from the sub-satellite point, the longer baseline may have a distinctly impact on the differential positioning. Thus, the baseline between the reference stations and the target should be as short as possible under this condition.

4 CONCLUSION

In this paper, the principal of observation differential about dual-satellites passive location system is put forward. Simulation result shows the validity of this differential technology. In addition, the paper analyses the influence of the orientation and baseline of differential station and radiation source on differential positioning accuracy, so as to draw a few conclusions. The method and conclusions have developed positioning system of dual-satellites TDOA and FDOA, which has important value for improving its availability and the reliability of positioning result.

ACKNOWLEDGMENTS

This study was supported by the National Natural Science Foundation of China under Grant No. 60263028.

REFERENCES

[1] Patison T, Chou S T. Sensitivity Analysis of Dual-Satellite Geolocation. IEEE Transactions on Aerospace and Electronic System, NO.36, vol.1, 2000, pp.56–71.

[2] K.C.HO,Y.T.CHAN. Geolocation of a known altitude object from TDOA and FDOA measurements [J]. IEEE Transactions on Aerospace and Electronic System, 1997, 33(3): 770–783.

[3] Torrieri D J. Statical theory of passive location system. IEEE Trans on AES 1984, 20(3):183–198.

[4] GUO Fucheng, FAN Yun.A method of dual-satellites geolocation using TDOA and FDOA and its precision analysis [J]. Journal of Astronautics, 2008, 29(4):1381–1386.

[5] Wu Shilong, Luo Jingqing, Gong Liangliang. Joint FDOA and TDOA location algorithm and performance analysis of Dual-Satellite formations [J]. Signal Processing Systems, 2010.

[6] XUE Yangrong, LI Xiaohui, XU Longxia, etc. Research on position differential method dual-satellites TDOA and FDOA in passive location system [J]. Frequency Control Symposium, 2012.

[7] K.C.HO, Le YANG. On the Use of a Calibration Emitter for Source Localization in the Presence of Sensor Position Uncertainty [J]. IEEE Transactions on signal processing, 2008, 56(12): 5758–5772.

[8] Wen Zhong, Qu, etc., Algorithm of position calibrator for satellite interference location, Chinese Journal of Radio Science, October 2005, Vol.20, No.3, pp.342–346.

[9] Qiu Z H, Wang W Y. GPS principles and application. Beijing: Publishing House of Electronics Industry, 2002.

[10] Jay Farrell, Tony Givargis. Differential GPS Reference Station Algorithm-Design and Analysis [J]. IEEE Transactions on control systems technology, 2000, 8(3): 519–531.

Information, Computer and Application Engineering – Liu, Sung & Yao (Eds)
© 2015 Taylor & Francis Group, London, ISBN 978-1-138-02717-6

The design and realization of rotor vibration signal acquisition system based on STM32

Yu Li & Ping Liao
School of Mechanical and Electrical Engineering, Central South University,
Changsha, P.R. China

ABSTRACT: In order to study the performance degradation trend prediction and the fault evolution law of large rotors, a set of large rotor vibration signal acquisition system was designed based on STM32. And a practical circuit experiment was carried out. The sensor signals were processed by conversion, filter and some other signal processing. Then the external AD converter converted it to a digital signal. A digital filtering by the main control chip STM32f103x was used in the transformed signals, before it was outputted to PC for further processing through a serial port. The host user program was designed by LabVIEW. The two channel vibration signals were real-time displayed in the waveform diagram at the same time. And the data were fully saved in an Excel sheet. The fast Fourier transform was used in the historical data, which were read from the Excel sheet, to map out a frequency domain graph. The experiments show that the system can realize real-time vibration signal acquisition and storage. It also can analyze the frequency domain of the historical data. It has a great importance in the performance degradation trend prediction and the fault evolution law of large rotors.

KEYWORDS: STM32; data acquisition; condition monitoring; vibration signal

1 INTRODUCTION

Large rotating machinery is widely used in electric power, metallurgy, petroleum, ship and other major field. With the accelerating pace of the modernization of the society, the national demand for resources continue to increase. The operational reliability, stability and security issues of large hydropower, thermal power and other large rotating machinery have become increasingly highlighted. With the development of science and technology, the early artificial periodic detection method has been gradually taking the place of intelligent, automated monitoring system. The modern monitoring system can acquire, transmit and process the related machinery data in real time, and can make a state recognition, prediction of diagnosis. Our country is far behind foreign countries in the condition monitoring, fault diagnosis, and cannot meet the actual demand [1]. As the core component of the rotating machinery, the rotor's running condition is one of the important factors that affect the operation of large rotating machinery. Vibration signal processing, analysis and recognition are the foundation of equipment monitoring and diagnosis, and is one of the most common mechanical fault diagnosis methods. The vibration signals and its characteristics during the operation are main signals that reflect the actual changes in mechanical equipment and its running state [2].

Heavy duty gas turbines used in many fields have the advantages of large power, high efficiency and reliable operation, and have become a new generation of power plant [3]. But due to working for a long time in high temperature gradient, cyclic load and bad environmental conditions, the rotor system can be error prone, which affects the stable operation of the unit. For the study of heavy duty gas turbine combined rotor performance degradation mechanism and its forms in different periods, to reveal the cause of the potential failure may occur, this paper designed a set of rotor vibration signal acquisition system based on STM32. This design uses the STM32F103x series microcontroller as the main control chip, with an external ADC AD7606 of 8 channels and 16 bit high precision to transform the signal, to ensure the accuracy of the signal. The real-time acquisition system shows the rotor vibration signal of X and Y direction at the same time, the host user program can choose I/O port, set the baud rate and data bits, control the start and end. All data from start to finish are fully recorded in an Excel spreadsheet, for further data processing and analysis. The host user program also supports to read and do a fast Fourier transforms to the stored data. The time domain and frequency domain graph show in the same interface to regularly analyze the operation condition of the rotor, and maintain the rotor fault.

2 THE HARDWARE CIRCUIT DESIGN OF THE SYSTEM

This design uses a STM32F103x series microcontroller as the main control chip. STM32 with ARM Cortex-M3 inside is the smallest, the lowest energy consumption of existing 32-bit processor, which has advanced the kernel, outstanding power consumption control, maximum integration and outstanding and innovative peripherals [4]. The PC cannot directly read the output signal of the eddy current vibration sensor. And the test site work environment will produce certain effects of the sensor output signal, so a certain sensor signal processing is needed. The STM32 series microcontroller is integrated with a 12 bit ADC. But due to the high signal precision demand, the external ADC chip AD7606 of 8 channels and 16 bit high precision is needed, in order to satisfy the requirement acquisition system accuracy.

The hardware circuit design of the system mainly includes the main control chip module, power module, signal conditioning module, AD conversion module four parts. Sensor signal is first converted and filtered by the signal conditioning module to satisfy the demands of A/D input; After the AD converter, convert the input signal with high precision, it outputs the signal to the main control chip; Main control chip transmits data to PC through RS232 serial interface; Power supply module provides the working voltage for each module of the whole circuit. The hardware circuit design of the system is shown in Fig.1.

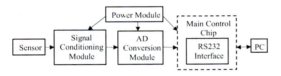

Figure 1. The hardware circuit design of the system.

2.1 Power module

Power module supply working voltage for the various modules of all components of the whole system. This design adopts the DC24V switching power supply. The K78XX - 1000 series high efficiency switching regulator convert the power separately to provide and the working voltage of + 12V, + 5V,3.3V for each element. The circuit diagram of power module converts 24V to + 12V is shown in Fig.2.

2.2 Signal conditioning module

Generally due to the electrical characteristics of the signals produce by sensors cannot direct input to the PC. In addition, in the environment of

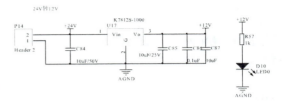

Figure 2. The circuit of power module.

laboratory and test site, the sensor signal is usually influenced by the electromagnetic and the sensor, amplifier circuit themselves. It contains a variety of frequency noise signal. Unhandled signals may be some error signal, and even lead to incorrect results. To obtain signal correctly, the signal must be handled. The structure of signal conditioning module is shown in Fig.3. The circuit diagram is shown in Fig.4.

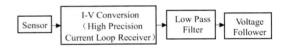

Figure 3. The structure of signal conditioning module.

In order to prevent the sensor output signal from attenuation with the transmitting distance, this design uses the current sensor. But the ADC's input is voltage, so we selected RCV420 high precision current loop receiver, a production of TI Company. It is capable to convert 4–20 mA current input signal into 0 to 5 V voltage output signal with high precision [5].

The introduction of the high frequency noise signal can cause serious interference to the acquisition of data. In order to further improve the accuracy of A/D input, this design uses the second-order voltage-controlled voltage source low-pass filter. It can effectively attenuate the high frequency noise of the signal, and has a good filtering effect [6]. The result of hardware circuit filtering is shown in Fig. 5. The upper figure curve is drawn by the signal without hardware filtering circuit (only convert by the I-V conversion as the input signal), the lower part of figure curve is drawn by the signal has already been filtered.

To make the circuit does not influence each other, a voltage follower is inserted in the circuit to realize the isolation effect. Voltage follower can reduce the output impedance of the pre-stage circuit effectively, improve the load capacity of the circuit, improve the input impedance, enhance the anti-interference ability of the circuit, and reduce the influence of load on the circuit at the same time.

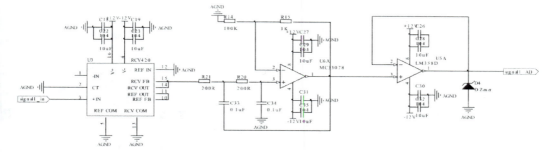

Figure 4.   The circuit of signal conditioning module.

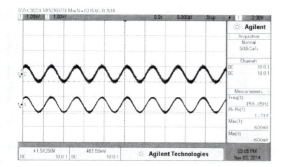

Figure 5.   The results of hardware circuit filtering.

2.3   *AD conversion module*

To meet the requirements of the design accuracy, this design chooses the AD7606 from ADI Company as a conversion chip in AD conversion module. The chip supports 8 channels synchronous sampling input. All channels have 16 bits analog-to-digital conversion accuracy. The sampling rate can be up to 200 KSPS. AD7606 uses 5V single power to supply. And it supports ±5V input signal, so the signal conditioning module has no need to enlarge the output signal. The circuit of the AD conversion module is shown in Fig.6.

3   THE SOFTWARE DESIGN AND REALIZATION OF THE SYSTEM

Software design of data acquisition system mainly includes two parts: the firmware program and the host user program. Overall system software flow chart is shown in Fig.7.

3.1   *The firmware program*

ARM's integrated development environment Keil4 is used as the development platform to design the firmware program (SCM). Modular design is adopted

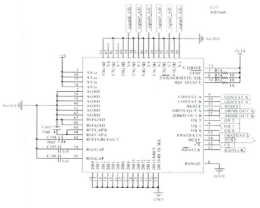

Figure 6.   The circuit of ADC.

to simplify the program design, debugging and maintenance.

When data acquisition and processing, the timer trigger an ADC conversion every sampling period. Low-pass filter circuit in the signal processing module can't effectively filter the sudden pulse interference signal. So after every AD conversion, there is an interrupt service routine to filter the collected data with the median filtering on average. Median filtering on average, also called the average pulse interference filter, combines the advantage of median filtering and arithmetic average filtering. It can effectively eliminate the pulse interference caused by sampling bias [7].

3.2   *The host user program*

LabVIEW is used to develop the host user program, which has the function of the development of communication, application interface and database access. It's mainly divided into two parts: real-time monitoring and the analysis of historical data. LabVIEW is based on the graphical programming method of the flow chart. It's NI - VISA

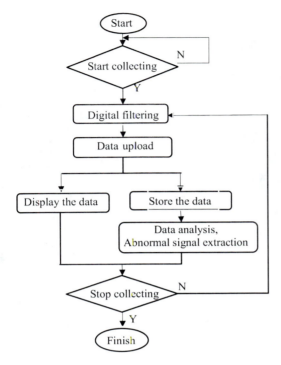

Figure 7. The software flow chart of the system.

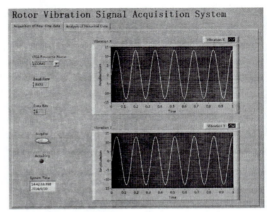

Figure 8. The interface of real-time acquisition.

can control all kinds of interface, and can invoke the corresponding driver, according to the type of instrumentation [8]. It has reduced the development difficulty greatly.

In real-time monitoring, real-time monitoring interface can display time the waveform graph of X, Y direction at the same, in order to monitor the working condition of rotor, make a preliminary judgment and discover the abnormal vibration of the rotor in time. The interface of real-time acquisition is shown in Fig.8. Every time at the beginning of monitoring, the host user program will remind you of data saving. After the monitoring, all the data will be recorded on an Excel sheet, used for subsequent data processing and analysis. The data stored in an Excel sheet after a certain monitoring is shown in Fig.9.

The host user program can read the data, which is stored while real-time monitoring, and can carry on the fast Fourier transform to convert the time domain signal to the frequency domain signals, which is relatively easier to analysis. The interface of historical data analysis shows the time domain and the frequency domain graph of rotor vibration signal in the same interface, which makes it more convenient to compare.

	A	B	C	D	E
1	No.	Date	Time	Vibration X/um	Vibration Y/um
2	1	2014/9/30	14:02:32	0.01	0.08
3	2	2014/9/30	14:02:33	0.15	0.11
4	3	2014/9/30	14:02:33	0.13	0.16
5	4	2014/9/30	14:02:34	0.23	0.28
6	5	2014/9/30	14:02:34	0.34	0.32
7	6	2014/9/30	14:02:35	0.35	0.36
8	7	2014/9/30	14:02:35	0.47	0.42
9	8	2014/9/30	14:02:36	0.48	0.48
10	9	2014/9/30	14:02:36	0.48	0.54
11	10	2014/9/30	14:02:37	0.61	0.62
12	11	2014/9/30	14:02:37	0.64	0.61
13	12	2014/9/30	14:02:38	0.65	0.73
14	13	2014/9/30	14:02:38	0.76	0.69
15	14	2014/9/30	14:02:39	0.78	0.8
16	15	2014/9/30	14:02:39	0.81	0.79
17	16	2014/9/30	14:02:40	0.84	0.87
18	17	2014/9/30	14:02:40	0.88	0.86
19	18	2014/9/30	14:02:41	0.93	0.93
20	19	2014/9/30	14:02:41	0.97	0.97
21	20	2014/9/30	14:02:42	0.94	0.99
22	21	2014/9/30	14:02:42	0.96	0.99
23	22	2014/9/30	14:02:43	1	1.01
24	23	2014/9/30	14:02:43	1.06	1.05
25	24	2014/9/30	14:02:44	1.02	1.03
26	25	2014/9/30	14:02:44	1.03	1.03
27	26	2014/9/30	14:02:45	1.03	1.06
28	27	2014/9/30	14:02:45	1.05	1.04
29	28	2014/9/30	14:02:46	1	1.02
30	29	2014/9/30	14:02:46	1.07	0.99
31	30	2014/9/30	14:02:47	1.05	1.01

Figure 9. The data stored in an Excel sheet.

4 THE EXPERIMENTAL RESULTS AND ANALYSIS

Extract a group of abnormal vibration historical data the system acquired to analyze. The sampling frequency of this group of data is 5 KHz, and the rotor's speed is 9600 r/min. The time domain and frequency domain graph the host user program drew from the historical data is shown in Fig.10.

The frequency domain signal diagram shown in Fig.10 is drawn from the original historical data after the fast Fourier transforms. It is shown in the diagram

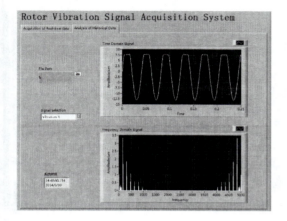

Figure 10. Abnormal vibration signal in time domain and frequency domain analysis diagram.

that there are dividing frequency at 0.2fx,0.3fx,0.5fx, etc and double frequency signal at 2fx,3fx,4fx,5fx, ect, except the vibration signal at the fx power frequency (160 Hz). It is caused by the rub-impact of the rotor.

In general, the fault features of rub-impact in the frequency spectrum is that there are vibration signals at fx, 2fx, 3fx and 4fx, and there is greater high-frequency signal at 5fx. The amplitudes at 2fx and 3fx are 0.2 to 0.5 of that at fx. The amplitudes at 4fx, 5fx and higher frequency are less than 0.01 of that at fx. The amplitudes in the dividing frequency area of 0–0.39fx, 0.4–0.49fx, 0.5fx and 0.51–0.59fx are less than 0.1 of that at fx [9-10].

In rotating machinery, when the amplitude of the rotor is greater than the clearance between rotor and stator, it will lead to rub-impact. The quality of the imbalance, bending parts or thermal expansion can also cause a rub - impact of the rotor. Rub-impact as one of the many faults, not only leads to a decline in the entire unit efficiency, but also causes the damage to the entire rotating machinery. Reading and analysis of rotor vibration signal data regularly, understanding the possible abnormal vibration and fault with the increase of working time and the prediction of the state of the rotor can avoid greater fault or the mechanical damage to other parts.

5 CONCLUSIONS

After debugging, data acquisition, transmission and analysis functions of the rotor vibration signal

acquisition system based on STM32 can meet the design requirements. Experiment results show that this system can realize the rotor real-time vibration data acquisition and storage, and be able to read the historical data and carry on the fast Fourier transform to analyze the rotor vibration signal in the frequency domain. The system has a great importance to the study of a rotor performance and degradation trend.

ACKNOWLEDGMENTS

The authors are grateful to the Natural Science Foundation of China (Contract No.: 51275535), and the Major State Basic Research Development Program of China (973 Program) (2013CB035706) for their support.

REFERENCES

[1] Yubin Liu. Dynamic Analysis for Rotor-Bearing System of Large General Machinery [D]. Dalian: Dalian University of Technology, 2009.6.
[2] Dongdi Chen. Research on Processing Method of Motor Rotor Vibration Signals Based on Wavelet Transform and Empirical Mode Decomposition [D]. Nanjing: Nanjing Normal University, 2013.5.
[3] Guoquan Feng, Duokui Wei. Investigation of Vibration Characteristics of a Typical Aero-Derivation Gas Turbine [J]. Aeroengine, 2002, 30(1):14–15.
[4] Zhiwei Huang, Bing Wang, Weihua Zhu. The Application Design and Practice of STM32F32 ARM Microcontroller [M].Beijing: Beihang University Press, 2012.8.
[5] Tongkui Tang. Several typical applications of RCV420I/V conversion circuit [J]. Process Automation Instrumentation, 1996, 17(8):37.
[6] Guie Luo, Jingqiu Zhang, Qun Luo. Analog Electronic Technology Foundation (Electricity) [M].Changsha: Central South University Press, 2005.2.
[7] Rong Hua. Signal Analysis and Processing [M]. Beijing: Higher Education Press, 2004.
[8] Weixuan Li, Jianxin Huang. The Driving Methods for Common DAQ card and Data Acquisition Based on LabVIEW [J]. Electronics Quality, 2005(07):13–16.
[9] BU Kaiqi, Ren Xingmin, Qin Weiyang. The System of Test and Analysis on Signal of Electric Generator of Gas Turbine [J]. Noise and Vibration Control, 2007(2): 40–42.
[10] Jun Zhang. The Research of Wavelet Transform Technique Application to Turbine Generator Fault Diagnosis [D].Beijing: North China Electric Power University, 2005.4.

Information, Computer and Application Engineering – Liu, Sung & Yao (Eds)
© *2015 Taylor & Francis Group, London, ISBN 978-1-138-02717-6*

A modeling for reconfigurable control I: Generic component

H.X. Hu
Agricultural and Animal Husbandry College, Tibet University, Tibet, P. R. China
Energy and Electrical Engineering College, Hohai University, Nanjing, P. R. China

Y.Q. Zheng & Y.F. Sheng
Energy and Electrical Engineering College, Hohai University, Nanjing, P. R. China

Y. Zhang
Computer and Information Engineering College, Hohai University, Nanjing, P. R. China

C. Shan
Business English Department, Shen-Yang Institute of Engineering, Shen-Yang, P. R. China

ABSTRACT: In this paper, we propose an approach based on generic component model to realize the model aggregation for reconfigurable control. This is a component – oriented method. Two elementary notions, the service and the operating mode, are introduced to construct a hierarchical system and to assure the coherences between specification and realization at each level and between levels.

KEYWORDS: Model aggregation; Generic component model; Service; Operating mode; Reconfigurable control

1 INTRODUCTION

Reconfigurable control considers the problem of automatically changing the control structure and the control law after a fault has occurred in the plant (Blanke, M. et al. 2003). The Generic Component Model (GCM) describes components from a point view of the users, who receive services and can use them differently in different operating modes (Staroswiecki, M. & Bayart, M. 1996). A formal analysis of the reconfigurability has been described by using GCM (Staroswiecki, M. & Gehin, A.L. 1998). Component interconnections are taken into account by considering higher level components, which result from the aggregation of lower level ones. Choukair and Bayart present an approach of modeling of a distributed architecture based on the Cartesian products of components (Choukair, C. & Bayart, M. 1999).

As mentioned in these researches above, the reconfigurable control based GCM relies on pre-designed alternative control structures. The main obstacle to perform these reconfigurable tasks is the combinatorial explosion of the products of numerous components, particularly for pre-designing alternative control structures in a large-scale system. The model

aggregation is a useful method by which a large-scale system is built as a hierarchical system composed of interacting subsystems (Aström, K. et al. 2001 & Hu, H.X. et al. 2006). Each subsystem or module involves only a few components and is therefore easy to design, analyze and maintain, but a global objective of the system must also be maintained so that the low modules can be coordinated to attain certain desired objectives.

In this paper, our goal is to realize model aggregation for reconfigurable control and to present a systematic procedure for the model aggregation of a hierarchical system. This procedure finds its justification from the coherence it achieves in a level and between levels, unifies the two elementary notions service and the operating mode during the model aggregation, and produces the important information about reconfigurable control.

The paper is organized as follows. The example used to illustrate the different notions is first presented in section 2. Next, we define the notions of generic component model and present the necessary notions, the services and the operating modes in section 3. In section 4 we point on the reconfigurability aspect. In section 5, we provide a brief summary of the main results of this paper.

2 EXAMPLE

The chosen example is a part of a level regulation process. It is composed of two identical connected tanks (see figure 1). Each tank is cylindrical of section A. The inflow Q_1 is provided by pump P_1, (controlled by the signal of level) filling tank T_1. The pump is continuous on a specific range. The flows Q_a, resp. Q_b between the two tanks are controlled by valves V_a, resp. V_b, connecting pipes are at level 0 and 30 cm. Valve V_o which is always opened is an outlet valve, located at the bottom of tank T_2. All the valves are on/off valves.

Tank T_1 is equipped with a continuous level sensor and tank T_2 with two discrete level sensors, indicating if the liquid is above or below the sensor level. Liquid is led into tank T_1 by pump P_1 and from tank T_1 to tank T_2 by valve V_a (V_b should always be closed in the nominal behavior). Some normal operating modes are considered: preparation, regulation, and emptying. During the regulation operation mode, the main objectives are to keep the liquid levels to 50 cm in tank T_1 and to 10 cm in tank T_2.

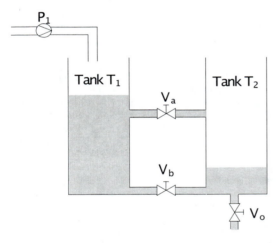

Figure 1. Hardware process description of two tanks system (TTS).

3 THE GENERIC COMPONENT MODEL

This section introduces the two main notions of the GCM: the services and the operating modes which consider firstly the components of the lowest level called elementary components (components that can not be decomposed into other components).

3.1 *Service*

From the user viewpoint, a system component provides one or several services. A service is defined as a procedure whose execution results in at least one modification of its output interface.

Inputs, outputs and procedures. A service is first described by the variables it consumes (Cons), the variables it produces (Prod), and a procedure (Proc) which transforms the former into the latter.

Requests. A service is run either on the reception of its specific request (for example, close, open for the valve) or permanently in time without any specific request presented to the component (for example: the storage service which is systematically provided by the tank, at all times and whatever the values of the inputs and outputs). A request (implicit or not) will be noted rqst.

Resources. The realization of a service rests on hardware/software resources (a tank with no leak for the storage service, a non faulty sensor for the measurement service ...). The service cannot be delivered when any of these hardware / software elements is not running properly; this is why they are called resources. Therefore, the model associates with each service the set of resources (res.) that are necessary for its normal running.

Set of services of a component. Formally summarizing the set of services provided by a component is defined as follow.

Definition 1: The set of services associated to a component is:

$S(k)=\{s_i(k), i \in I_s(k)\}$
$s_i(k)=<cons_i(k), prod_i(k), proc_i(k), rqst_i(k), res_i(k)>$

Where $S(k)$ is the set of services of component k, I_s the set of indices of the possible services, and the others are straightforward.

For the TTS example, the following services provided by the elementary components will be considered as in Table 1.

Table 1. Elementary services of TTS.

Elementary Components	Service
Tank 1 and 2	*Ti_store*
Valves a and b	*Vi_open, Vi_close*
Pump	*Deliver_Qi, Stop_Qi*

3.2 *Operating modes*

The notion of Operating Mode (OM) allows organizing the set of services into coherent subsets taking into account these two following requirements:

1 at a given time only a part of the services provided by a component are required to achieve the objectives linked to this component (for example, some objectives are regulation, initialization ...)

2 for safety reasons, incompatible services (for example, initialization and production services) must not be run simultaneously.

So, the notion of operating modes can be used to ensure that the aggregated model is effective in the sense of achieving specifications given either for the higher level model or for the lower level one.

Definition of operating mode. An operating mode is a subset of services of a component. The set of operating mode covers the set of services, i.e. each service belongs at least to one operating mode, and each operating mode contains at least one service.

Definition 2: An Operating Mode (OM) is defined by two elements:

1 one or several objectives to be achieved $O_j = \{o_j\}$.
2 a subset of the services of the component $S_j \subset S$ allowing the realization of the objectives.

The automaton of operating modes. Note that a component or a system is always in one and only one OM at the execution time. So the OM of a component can be described by a deterministic automaton given by definition 3.

Definition 3: The operating mode automaton associated to a component is defined by $A (M, T, m^\circ)$ where:

$M = \{m_i\}$ is the set of the OM, each of them being a vertex of the automaton,

$T = \{t_{ij}\}$ is the set of the transitions, each of them being defined by $t_{ij} = \{m_i, m_j, c_{ij}\}$ where m_i is the origin OM, m_j is the destination OM and c_{ij} is the firing condition defined from the requests associated to the services belonging to the destination OM.

$m^\circ \in M$ is the initial mode, i.e. the mode where the system stays at its initialization.

The conditions of transitions are very important, but are not shown here because they are not the central subject of this paper.

3.3 Services management

Nominal and degraded services are the key two notions. The execution of a service needs a set of resources, which includes hardware resources (sensors, memory etc.) and information resources (data with appropriate freshness status). These resources may belong to the intelligent vice or may be external. At a given instant, the service will run in a nominal way if the set of the resources it needs are able to perform normally. Unfortunately, this is not always the case, and it may happen that some of the required resources are faulty. The problem of tackling faulty resources first introduces the problem of evaluating the resource quality at each instant of time. Note that the notion of a quality index associated with a given resource has to take into account the fact that each

of the services whose execution needs that resource might have a different quality requirement. To introduce such a possibility, we distinguish two levels in a resources management procedure. At the first level, we assume that some index can be computed in order to characterize each resource state. At the second level, each service takes into account the set of the indexes that are associated with the resources it needs, in order to adapt its operation.

As soon as the resource can no longer be used by the service, it is labelled as faulty, and one has to analyze further. Note that a given resource might be non-faulty for some service and yet faulty for another one. In the case of a faulty resource, the service can no longer run in a nominal way, and two situations may occur. The first situation is that in which the intelligent instrument designer has implemented no replacement procedure for the service. In that case, the service becomes unavailable, and it should be taken away from the list of the services of the different OM in which it appears. The question now arises to decide whether such an OM keeps some sense in spite of the absence of this service. The second situation is a fault-tolerant one, in which the designer has implemented at least one replacement procedure to perform the service. When the resources that this replacement procedure needs are non-faulty, it can be run instead of the nominal one at each request. The list of the available services in the different OM remains unchanged, but the intelligent device operation is degraded.

So, introducing the notion of a degraded service increases the robustness and thus the availability of the intelligent instruments, since the faults that appear in some of the resources they need do not necessarily interrupt the services they render.

4 RECONFIGURATION ASPECT

The realization of a higher level service is reliant on the resources that allow the realization of the lower level ones it needs. The set of the hardware resources of a higher level service is the union of the sets of the resources of the lower level ones it needs. At the lowest analysis level, the hardware resource of a service provided by a component is the component itself. For example, the resource of the "open" service of a valve is the valve itself. Consequently there is only one version for a service of the lowest analysis level. Higher level services may have several versions due to the possibility to attribute the same functional interpretation to distinguish combinations of low level OMs. Each version is characterised by a different set of resources. In this case, the system reconfiguration may be possible in case of failure.

When a hardware resource is faulty, one or several services of the lower level become unavailable and it is possible that some others become active permanently in time. This is the case, for example of a blocked and closed valve. The opening service of this valve is unavailable and the closing service is permanent in time. Consequently to the loss of lower level services, some OMs are not reachable. For a given fault, several cases can be distinguished when one or several OMs disappear.

1 The unreachable OMs are not implied in the high level services allowing the realization of the system's current OM objectives. The system behaviour is not directly influenced by the fault.
2 The high level services allowing the realisation of the system's current OM objectives can not be provided under their nominal version
3 There are other versions allowing the realization of these services. The system reconfiguration is possible.
4 There is none version allowing the service realisation. The only possible reconfiguration is changing the system aims.

5 CONCLUSION

In this paper, two elementary notions, the service and the operating mode (OM) have been introduced to construct a hierarchical system and to assure the coherences between realization possibilities and given specifications not only at each level of the system decomposition but also between the levels of the decomposition. These two notions are the base of the model aggregation procedure for reconfigurable control. A service is perfectly determined as soon as its characteristics (consumed and produced variables and data processing) have been specified for each of its versions. Each OM provides a partition of the set

of the services into two classes. The services that do not belong to the OM cannot be executed and that belong to the USOM can be executed at any time. The list of services (the definition of their nominal and their degraded versions), as well as their logical organization (by means of the OM), constitute a generic model that allows one to describe any smart instrument, using a formal description language. The work about "what the aggregated system could do" and "what it should do" will be introduced in its sister paper "A Modeling for Reconfigurable Control II: Procedures of Aggregation".

ACKNOWLEDGMENTS

This work is supported by "The Natural Science Funds of Tibet", and "the Fundamental Research Funds for the Central Universities".

REFERENCES

Aström, K., Albertos, P., Blanke, M., Isodori, A., Schaufelberger, W. & Sandz, R. 2001. *Control of Complex Systems*. Springer.

Blanke M., Kinnaert M., Lunze J. & Staroswiecki M. 2003. *Diagnosis and Fault-Tolerant Control*. Springer.

Choukair C. & Bayart M. 1999. Application of External Model of intelligent Equipment to Distributed Architectures. *In ISAS'99, Orlando (U.S.A.), pp. 329–335, Juillet 1999*.

Hu, H. X., Gehin, A.L. & Bayart, M. 2006. Model Aggregation for Reconfigurable Control Based on Generic Component Model. *In ICSSSM'06, Troyes, France, 2006*.

Staroswiecki, M. & Bayart M. 1996. Models and Languages for the Interoperability of Smart Instruments. *Automatica* 32(6): 859–873.

Staroswiecki, M. & Gehin, A. L. 1998. Analysis of System Reconfigurability using Generic Component Models. *In Control'98, Swansea (UK), 1998*.

Information, Computer and Application Engineering – Liu, Sung & Yao (Eds)
© 2015 Taylor & Francis Group, London, ISBN 978-1-138-02717-6

Research on agglomerating trend and strategy framework for Guangxi industries

B. Ge
Guangxi University of Economics and Finance, Nanning, China
Huazhong Agriculture University, Wuhai, China

G.Z. Mo
Guangxi University of Economics and Finance, Nanning, China

ABSTRACT: The objective of this article is to provide a simple tool for industry policy analysis for local government. Based on the new structure theory, industrial cluster theory, using the data of Guangxi manufacture industry in the important time of Guangxi industrialization, we draw Guangxi industry agglomeration intensity profile with agglomeration degree and agglomeration speed coordinates, and find the developing trend and problem in the Guangxi manufacture industry. Then, the appropriate strategies of development are emerging.

KEYWORDS: industrialization, industrial agglomeration, new structuralism, strategy

1 INTRODUCTION AND REVIEW OF THE LITERATURE

Guangxi Zhuang Autonomous Region located in the southwest of China, whose level of new industrialization, has a big gap with the national overall level, is an undeveloped area with the unfinished process of industrialization, so that local people and the government eager to develop its' economic. Facing this historical period full of opportunities in the new century, Guangxi needs to achieve the upgrading of industrial structure and to establish competitive, modern industry so that it can obtain leapfrog development and build a comparatively well-off society in synchronization with the whole nation.

1.1 *From structuralism to new structuralism*

It is easy to find form observing the international experiences that along with the country's economic development and per capita GDP growth, industrial structure will change accordingly. Economic structure transformation is the most important topic of industrialization theory, thus it was called structuralisms school[1].

Structuralism asserted that because heavy industry in developing countries could not develop automatically, local government should draw up economic development plans and take mandatory measures to allocate resources, in order to achieve local conversion of economic structure[2]. But in practice, under the guidance of structuralism theory, developing countries appeared stagnant or even retrogression.

On the contrary, the new structuralism theory which emphasized to develop local comparative advantage succeeds in practices[3].

A new structuralism theory advocated that the structural characteristic of economies internal is determined by endowment structure and the market forces. Economic development is a spectral line of continuous change from an agricultural economy with low income to a postindustrial economy with high income, in which market forces efficiently allocate resources, promote independent innovation in the enterprise, and take shape in comparative advantages[4].

The new structuralism deconstruct the inner mechanism and the future road of economic growth, starting from resource endowment - comparative advantage - enterprise viability - industrial development strategy consistent with the comparative advantage- economic development[5]. On the basis of its resource endowment and comparative advantage, the undeveloped area should develop new and high quality factors of production and establish their modern industry. The government should solve the coordinated and external problem, support fundamental studies, and help to upgrade industrial structure [6].

1.2 *Industrial agglomeration and innovation*

Starting from the location theory raised by classical economists in the 18th century, scholars observed the phenomenon of industrial agglomeration. The industrial location theory suggested location of a factory was based on the minimum cost and maximum profit principle [7]. With further observation on the industrial agglomeration, scholars gradually realized that the relationship between industrial agglomeration and innovation. Schumpeter first pointed out the innovation which could change the production function appeared in groups in terms of time or space[8]. Potter's industrial cluster theory gave geographical characteristic of innovation a mature carrier[9]. Later studies researched on "innovation" and "cluster" together, for instance, they studied on the relationship between innovation and cluster, and innovation process of industry cluster. Moreover, the innovation theory system has absorbed some ideas of cluster's ideology and industrial cluster research has gradually focused on the industry cluster's innovation mechanism, and both of the two academic systems constantly integrate in the contemporary study[10].

The history of the industrialization is the historical changes of industrial structure, and also the history of industrial agglomeration. Structure transformation and upgrading are typically reflected by proportion of agriculture, industry, and heavy industry, light industry which is the depiction of the overall state of industrialization; and the change of industrial agglomeration can create a detailed picture of the origin and the trajectory of economic development, which reflects how the competitive industry starts and develops, providing important basis for tracing the industrial structure transformation, and can also reflect the direction and trend of industrial upgrading.

Thus, through the observation and analysis on industrial agglomeration characteristics of a region, it can be better known about the regional natural endowment and upgrading direction of the industrial structure.

2   JUDGEMENT OF NEW INDUSTRIALIZATION STAGE

According to Kuznets standard and Hoffman standard, the industrialization of Guangxi can be divided into four phases by the year1982, 1993 and 1999.

The first stage is the pre-industrialization stage before 1982, the period of high agricultural growth, which accounted for about half of the GDP and reached a peak of 48.7% in 1982. Besides, the Hoffman

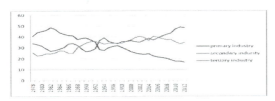

Figure 1.   The changing trend of the proportion of the first, second, and the third industry in Guangxi.

coefficient decreased from 10 to 1.5 with a low level of the second and the third industry; the second stage is the first early stage of industrialization stage, from 1982 to 1993, when the agricultural production continued to decline from the 48.7% to 28.2%, at the same time the proportion of industry and service industry kept rising. It was in 1993 that the industry accounted for 36.8% more than the proportion of agriculture for the first time, and since then it has not been surpassed with the Hoffman coefficient decreasing from 1.5 to 1; the third stage is from 1993 to 2009,the second early stage of industrialization. The agricultural proportion continued to decline while the proportion of the industry and service industry increased alternately ,and until 1999, agricultural proportion reached 18.8% ,with the Hoffman coefficient reducing from 1 to about 0.5, which means the Hoffman industrialization has completed; the fourth stage is from 2009 to now, the middle stage of Kuznets industrialization .

3   THE TRAITS OF INDUSTRIAL AGGLOMERATION IN DIFFERENT STAGES

The location entropy index[1] was used in this paper to calculate the industrial agglomeration degree in 1982, 1993, 2009, the three important time of the industrialization of Guangxi, also to calculate the aggregation situation in the near three years 2011, 2012, and 2013 to investigate its traits. All the dates are shown in Table1. (note:In this paper, agglomerating degree is examined through the location entropy (LQ). Although there are more advanced metrics such as the Beh Finn Dahl index (H index), spatial Gini data, and spatial aggregation index (EG), but in actual, because the acquisition cost of the advanced indicators was very high, and many studies estimated the important part in the index, so the location entropy is more intuitive and easier to operate.)

3.1   *Industrial development accompanied by the industrial agglomeration*

During the period from 1982 to 2009, the number of industries whose agglomeration index was above 1 increased a lot, it shows that in the process of industrialization in Guangxi, the agglomeration effect in advantaged industries is quite clear.

3.2   *Industrialization process accompanied by the industry diversification*

From 1982 to 1993, industries with agglomeration index above 1 substantially increase, the amount of them went up from 6 in 1982 to 15 in 1993, and 17 in 2009, which means more and more location industries accessed to faster growth than the national average level, and also implies that not only the most agglomerating industries contributed to development, but also the related industries did.

3.3 *Agglomeration of resource-intensive industries is most significant*

Observing the agglomeration degree in the top five industries in Table 1, Manufacture of food, forest and wood, paper and paper products, Production and Supply Electric Power in 1982 are all resource industries; in1993, not only the top five ones, but also the sixth to eleventh are resource-based industries; in 2009, it occurred a subtle change, manufacture of vehicle enter into the top five, but the other four industries in the top five are still resource-based. Then to year 2011 and year 2013, manufacture of vehicles fell out of the top five, ranked seventh. The Manufacture of Special Purpose Machinery which ranked thirteenth in 2009, is not in table1, showing its' agglomeration index lower than 1.

Table 1. Industrial aggregation rankings in different industrialization stages in Guangxi[2].

	1982	1993	2009年	2011	2013
1	Manufacture of Foods 1.96	Mining and Processing of Non-Ferrous Metal Ores 4.13	Mining and Processing of Non-Ferrous Metal Ores 3.07	Processing of Timber, Manufacture of Wood, Bamboo, Rattan, Palm and Straw Products3.01	Mining and Processing of Non-Ferrous Metal Ores 3.04
2	Manufacture of Forest and wood products 1.78	Manufacture of Foods 2.68	Processing of Timber, Manufacture of Wood,Bamboo, Rattan, Palm and Straw Products2.67	Mining and Processing of Non-Ferrous Metal Ores 3.04	Processing of Timber, Manufacture of Wood,Bamboo, Rattan, Palm and Straw Products 3.01
3	Manufacture of Paper and Paper Products 1.20	Mining of Other Ores 2.66	Processing of Food from Agricultural Products 2.52	Processing of Food from Agricultural Products 2.22	Processing of Food from Agricultural Products 2.22
4	Production and Supply of Electric Power 1.11	Manufacture of Rubber and Plastics Products 2.18	manufacture of vehicle 1.94	Non metal mining industry 1.97	Mining and Processing of Non-metal Ores 1.97
5	Manufacture of Articles for Culture, Education, Arts and Crafts 1.06	Manufacture of Foods 1.73	Production and Supply of El electric Power and Heat Power1.78	Comprehensive utilization Industry of waste resources 1.39	Manufacture of Liquor, Beverages and Refined Tea 1.70
6	Manufacture of Building materials 1.04	Mining and Processing of Ferrous Metal Ores1.72	Manufacture of Non-metallic Mineral Products 1.77	Manufacture of Liquor, Beverages and Refined Tea 1.70	Manufacture of Paper and Paper Products1.55
7		Manufacture of Paper and Paper Products1.44	Manufacture of Tobacco 1.70	Manufacture of vehicles 1.52	Manufacture of vehicles 1.52
8		Processing of Timber, Manufacture of Woods 1.32	Manufacture of Liquor, Beverages 1.61	Smelting and Pressing of Non-ferrous Metals1.45	Smelting and Pressing of Non-ferrous Metals 1.45
9		Manufacture of Tobacco 1.25	Smelting and Pressing of Ferrous Metals 1.53	Manufacture of Rubber and Plastics Products 1.50	Manufacture of Rubber and Plastics Products 1.50
10		Mining and Processing of Non-metal Ores 1.16	Smelting and Pressing of Non-ferrous Metals 1.48	Manufacture of Paper and Paper Products1.55	Utilization of Waste Resources 1.39

(note: Calculating by the data of "Main Indicators of Industrial Enterprises above Designated Size by Industrial Sector" in national and Guangxi Statistical Yearbook, here only the top ten above 1 in the list.)

All the phenomenon suggested that the industrialization road of Guangxi, which was significantly different. It is based on the original advantaged resources, especially natural resources.

3.4 Recycle economic industries start to agglomerate

In 2011, agglomeration degree of the Utilization of Waste Resources, jumped from 0.94 in 2009 to 1.97 in 2013, and above-scale enterprises from 10 in 2008 to 20 in 2013. In the past three years the output value increased from 1.04 billion to 7.85 billion at an annual compound growth rate of 96%, far more than Guangxi industrial average annual compound growth. Industries present agglomeration.

4 CHARACTERISTIC MATRIX

There are two important characteristics of Industrial agglomeration: Location entropy describes the agglomeration degree, higher agglomeration degree represents the more obvious the specialization advantage of the local industry; agglomeration speed is used to measure the dynamic changes of the industrial agglomeration, which is average growth of the agglomeration degree in the calculating years. The faster the agglomeration speed is, the more significant the agglomeration trend is.

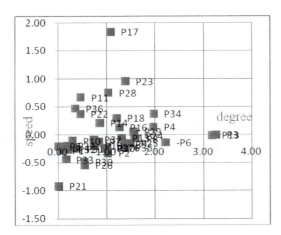

Figure 2. The agglomeration map of Guangxi manufacture clusters[3] (note: Because of consistent statistical caliber requirements, this table is calculated by the data from 2004–2013, industry P1–P36 is merged in the order of industry according to the statistical yearbook of the Nation and Guangxi, cording to " Main Indicators of Industrial Enterprises above D.)

With the above two characteristics as coordinates, agglomeration map of Guangxi manufacture clusters was drawn. And industrial agglomeration degree of 1 and a rate of 0 taken as the cutoff point, the agglomeration distribution map is divided into four regions.

According to the combination of agglomeration degree and agglomeration speed of every industry, it can be divided into four data areas of industries, we named them as competitive industry, potential industry, problems industry and mature industry.

In zone1, there are Mining and Processing of Non-Ferrous Metal Ores, Mining and Processing of Nonmetal Ores, Printing and Reproduction of Recording true of Articles for Culture, Education; General Purpose Machiry, Special Purpose Machinery. In zone2, there are Manufactures of Textile, Wearing Apparel and Accessories, Manufacture of Furniture, Production and Supply of Gas.

In zone3, there are Mining and Washing of Coal, Manufacture of Foods; Textile, Manufacture of Raw Chemical Materials and Chemical Products, Smelting and Pressing of Non-ferrous Metals; Metal Products, Manufacture of vehicles; Electrical Machinery and Apparatus, Manufacture of Computers, Communication and Other Electronic Equipment, Measuring Instruments and Machinery.

In zone4, there are Mining and Processing of Ferrous Metal Ores, Processing of Food from Agricultural Products, Manufacture of Liquor, Beverages and Refined Tea;Tobacco;Leather, Fur, Feather and Related Products and Footwear, Processing of Timber,; Wood Bamboo, Rattan, Palm and Straw Products; Paper and Paper Products; Medicines, Production and Supply of Electric Power and Heat Power, Production and Supply of Water; Manufacture of Chemical Fibres; Non-metallic Mineral Products, Smelting and Pressing of Ferrous Metal.

Table 3. Industrial Division Description.

type	character	degree	speed
Zone 1 Priority Industries		>1	>0
Zone2 Potential Industries		<1	>0
Zone3 Problem Industries		<1	<0
Zone4 Mature Industries		>1	<0

5 GUANGXI INDUSTRY AGGLORATION AND DEVELOPMENT STRATEGIES MATRIX

All manufacturing industries are distributed in the four regions of Figure 2, each region has its own strategic points as following.

5.1 Priority industries and the strategy

The industries in zone1 of Figure 2 which is accelerating gathering (speed >0) and of high degree of agglomeration (entropy index >1), are the Relative competitive advantage industries in Guangxi. They should be taken to promote for further agglomeration and development.

positive	2 Potential Industries **Strategy**: Observation, selection To cultivate or to give up	1 Priority Industries **Strategy** Planning and building industrial park Supporting policies To attract resources further agglomeration
Speed of aggloration	3 problem industries **Strategy**: Analysis Delays in processing ignore	4 Mature industies **Strategy**: Extend the industrial chain The development of circular economy Industrial and technology innovation Market innovation
negative		
	low Agglomeration degree high	

Figure 3.   Strategies matrix.

For example, Manufacture of Articles for Culture, Education, Arts and Crafts, Sport and Entertainment Activities in the zone has been developing rapidly, although its numerical value of output has just been ranked the seventeenth, not at the same numerical level of as this industry in Guangdong, Zhejiang and Shanghai, but its agglomeration speed ranked the fifth following Tianjin, Heilongjiang, Yunnan and Guizhou. The trend of agglomerating is obvious.

From the local government, the strategy of priority industry is to accelerate the agglomeration and development. Specifically, industrial park should be planned and constructed, supportive policies of land, finance, taxation, talent introduction and technical innovation should be formulated to support the development of the industries.

5.2 Potential industries and the strategy

The industries in zone2 of Figure 3, are also in a state of accelerating gathering (agglomeration speed > 0), but its agglomeration degree has not reached the national average level of agglomeration (the degree of agglomeration < 1).

The potential industries need further analysis and confirmation to find whether they can form the comparative advantage. The ones that got potential advantages, will be selected to be supported by industry policy.

5.3 Problem industries and the strategy

The industries in zone 3 of the Figure 3, of which agglomeration speed decreases (< 0) showed dispersed trend, and agglomeration degree is not high (< 1) suggested that its scale and advantage is not obvious.

Among these industries in Guangxi, Manufacture of Foods is the downstream industry of Processing of Food from Agricultural Products which the original Guangxi traditional superior industry; Smelting and Pressing of Non-ferrous Metals is the downstream of Mining and Processing of Non-Ferrous Metal Ores of Guangxi. Their upstream industry is all in zone 1. The fact is that the upstream industries which are comparably competitive industries in Guangxi did not grow and extend the existing advantages in the industrial chain, especially not generate senior manufacture industries from these existing advantaged industries. The local government should design more sophisticated policies to face this problem.

5.4 Mature industries and the strategy

In zone 4 of Figure 3, the agglomeration degree is more than 1, but the agglomeration speed is less than 0, the ones here showed that they were in the process of gradually disperse though they were highly agglomerating before.

The strategy for the development of mature industries is expanding the advantages, and to make full use of its comparative advantages. On the one hand, it can expand through the extension of industrial chain and circular economy; on the other hand, it can also promote by creating new products, adopting new technology, as well as searching for new market such as entering the international market etc.

In the continuous process of economic development, local government should act an important role but cannot replace the market. It formulates policies and guides the market, but not intervene too strongly or too actively. The foundation of guidance encompasses discovery, analysis and summary of the laws, trend, and reasons of the current economic operation, based on which the corresponding strategies can indeed have long-term and internal effects.

REFERENCES

[1] Zhang Shogun, Market leading with government induced —a review of the Lin Yifu's the new structural economics [J].Economics(quarterly), 2013(4):1079–1084.
[2] Lin You, structural economics [M].Beijing: Peking university press, 2008.
[3] BA Shoo-song ZHENG Jun ,Motivation and Direction of Chinese Industrial Transformation: Based on A Post

Structural Economics Perspective[J].Journal of central university of finance and economics .2012(12):45–52.

[4] Wei sen. Exploring the inherent mechanism of human social and economic growth and the future road— Comment on Professor Lin Yifu's new structure economics theoretical framework [J].Economics, 2013, 03:1051–1074.

[5] Ago Hongying, Review of the new structure of economic development theory [J]. Dynamic Economics,2011,02:111–116.

[6] Zheng Haoyu.The transformation of the economic development theory and the role of the state—in the 90's, Brazil as an example of structuralism and new structuralism[J]. Latin American Studies,2006,06:61–64.

[7] Joseph Alois Scumpeter. The Theory of Economic Development By Joseph Alois Scumpeter.English Translationg By Redvers Opie,Xian: Shaanxi normal university press,2007.

[8] Yang Liansheng, Zhe Yingming, Zhang Xin, Lv Huijun, Innovation Clusters: From Concepts to Practices——New Trends in International Researches on Innovation Clusters[J]. Journal of Nanjing University of Science and Technology,2013,01:77–85.

[9] Dong Weiwei , Li Beiwei. The economics analysis From innovation to the industry clusters[J]. Industrial technology economy ,2014,03:14–18.

[10] Li Beiwei, Dong Weiwei. Progress in innovation cluster research and future prospects [J]. technology economy and management study2013,07:36–41.

610

Information, Computer and Application Engineering – Liu, Sung & Yao (Eds)
© 2015 Taylor & Francis Group, London, ISBN 978-1-138-02717-6

An associative-model of learning resource for dynamic digital publishing

Lei Wu
Time Publishing and Media Co., Ltd., Hefei, Anhui, China
Wuhan University, Wuhan, Hubei, China

Qing Fang
Wuhan University, Wuhan, Hubei, China

ABSTRACT: Because of disorder and lack of semantic tags among various cross-media learning resources, it's difficult to dynamically publish these learning resources, or to retrieve associatively from them. Based on analyzing the relationships among knowledge units related to cross-media learning resources, a semantic associative-model and associative retrieval have been proposed. Experiment shows that the method in this paper performs better than the traditional method.

KEYWORDS: Learning resource; Dynamic digital publishing; Associative retrieval

1 INTRODUCTION

Dynamic digital publishing is a new emerging type of publishing. The core of dynamic digital publishing is associative retrieval technologies based on semantic relationships among learning resources. As most resources, lack of uniform semantic description criteria and sharing mechanism, these resources become many resource islands. Therefore, how to describe and model semantic relationships among learning resources is a fundamental problem in dynamic digital publishing.

Traditionally ontology is the most popular model used to describe semantic relationships among learning resources[1-3]. They emphasized on either relationships among ontology or similarities among keywords of learning resources. However, they have not considered relationships among learning resources deeply.

On the basis of analyzing association characters among knowledge, a new model based on improving semantic Web model has been proposed. With this model, associative retrieval of learning resources has realized.

2 SEMANTIC MODELLING FRAMEWORK OF LEARNING RESOURCES

Semantic models of learning resource is the key point of dynamic digital publishing. In fact, modeling learning resources is some kind of the packaging process. By describing the content and relationships of learning resources in a normative way, we can make learning resource shareable and reusable, and finally improve their interoperability[4].

Knowledge system of learning resources is generally stable and unchangeable. There are many complex relationships among knowledge units of a typical knowledge system, which is in fact a complicated relationship network. On the other hand, learning resources are mutable and changeable. So maintaining relationships among knowledge units is easier than maintaining relationships among learning resources. Therefore, the proposed semantic modeling framework of learning resources in this paper is hierarchical and is made up of three levels, i.e., learning resource level, knowledge object level and the top level, educational ontology level, as illustrated in figure 1. The framework is flexible and adaptive. The educational ontology level is unchangeable and the learning resource level is changeful. By separating mutable part (learning resources) from stable parts (knowledge object and educational ontology), we only need to maintain learning resources, which could greatly eliminate staff works and could reduce the system complication. If not so, frequent operations about learning resources such as adding, editing and deleting would always result in modifications to the educational ontology, which would make staff work's difficulty.

1 Learning source level

This level consists of various cross-media learning resources including photos, voices, videos, cartoons, texts, questions and etc. Basically, learning resource is a set of special explanation of some knowledge units and is a set of special manifestation of knowledge. Each learning resource at this level has a unique ID.

Figure 1. Semantic modeling framework of learning source.

Learning resource has different properties, each of which provides special information about the resource. Face is defined as a set of properties, which provide some higher and more abstract information. All properties related to learning resource could be divided into three parts, basic properties, knowledge properties and additional properties. Basic properties are the properties something like title, name, key word, language and arts; knowledge properties are used to tag semantic information, in fact, they are the labels of knowledge units included in learning resource; additional properties are added for additionally describing learning resources. Some basic properties and additional properties could be merged into faces.

(2) Knowledge object level

Knowledge object, which called a knowledge unit, is the abstract of semantic information included in isomeric cross-media learning resources. Each learning resource could be tagged by some knowledge objects, each of which is called a label of the learning resource. Then the learning resource is the multi-label learning resource. For some, learning resource sharing knowledge units, they would have same labels. That is to say, it's a "many-to-many" mapping between knowledge objects and learning resources. What's included in knowledge comes from knowledge ontology base.

(3) Educational ontology level

Educational ontology base describes all the concepts and relationships among them in the education field. Formally, these descriptions can be viewed as a directed graph. From the structure of the directed graph, we could find luxuriant semantic information hidden in the knowledge base. Similarly, there is "many-to-many" mapping between education

ontologies and knowledge objects. Many education ontologies could be used to tag a knowledge object while many knowledge objects could be tagged by an education ontology. Because knowledge in a field is generally stable and unchangeable, we could in advance set up the knowledge base, actually ontology base, in order to tag knowledge objects easily. In fact educational ontologies are the basis of semantically labeling learning resources.

3 KNOWLEDGE CHAIN MODEL OF LEARNING RESOURCES

3.1 Knowledge chain model

Semantic relationships among learning resources could be indirectly reflected in the relationships among knowledge objects. Considering the complexity between knowledge objects, a modified semantic description method called knowledge chain model is proposed.

Definition 1. Knowledge Chain Model is a directed graph, consisting of nodes, edges and the connections between nodes and edges. It is formally described as a triad S={V,E,R}, where V is node set, E edge set and R connection set.

Definition 2. Knowledge Chain Connection is a directed edge, standing for the semantic connections between nodes. It can be formally defined as E={sta, end, type}, where sta is the starting node, end the end node and type association type.

3.2 Modeling relationships

In the knowledge chain model, semantic relationships among learning resources can be extracted from semantic relationships among education ontology. So if we can correctly save the relationships, we can easily compute the semantic relationships among learning resources. Relationships among education ontology are also various[5]. Some main relationships are implication, hierarchy, sequence, reference and brotherhood. An example of knowledge chain is shown below.

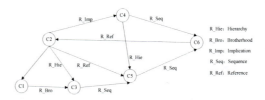

Figure 2. Illustration of knowledge chain model Caption of a typical figure.

4 DYNAMIC DIGITALPUBLISHING FRAMEWORK BASED ON SEMANTIC ASSOCIATION

4.1 Workflow of dynamic digital publishing

The core of dynamic digital publishing is to recommend proper learning resources based on user's associative retrieval.

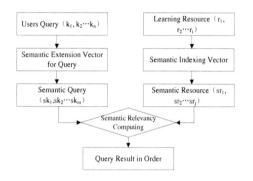

Figure 3. Dynamic digital publishing process based on semantic association.

There are 3 steps for dynamic digital publishing (Figure 3). First is to semantically expand users' query keywords to make semantic query vector. Second is to build semantic vectors of resources by combining related knowledge units and knowledge ontology based on the knowledge chain model. Third is to compute and sort the relevancy of semantic query vector and semantic resources vector to get Top-k appropriate learning resources. The method of semantic relevancy computing is recommended in the next part.

4.2 Compute semantic relevancy

Semantic relevancy is the relevance of extended semantic query vector and semantic label vector of learning resource. Based on knowledge chain model, semantic relevancy can be directly computed by comparing the differences between semantic query vector and semantic resource vector.

Based on mentioned above, the relationships in knowledge chain model contain hierarchy, sequence, reference and so on. Experts in education field evaluated the basic semantic relevancies of these relationships and defined the typical values.

Let $sk(sk_1,...,sk_m)$ be the extended semantic query vector, $sr(sr_1,...,sr_n)$ be the semantic label of learning resource. $Rel(sk_m,sr_j)$ is to quantitatively compute single semantic relevance between query keyword sk_m and semantic label of learning resource sr_j. The formula 1 is as follows:

$$rel(sk_m, sr_j) = \begin{cases} 1 & R_{SER}:\text{ Sequenced Relationship} \\ 0.8 & R_{Ref}:\text{ Reference Relationship} \\ 0.8 & R\_Hie:\text{ Hierarchy Relationship} \\ 0.5 & R_{Imp}:\text{ Implication Relationship} \\ 0.5 & R\_Bro:\text{ Brotherhood Relationship} \end{cases} \quad (1)$$

Semantic relevancy computation formula is as follows:

$$\text{SemanticR (sk, sr)} = \begin{cases} \frac{\sum(Max_{m=1;}^{m=m;}\,_{j=i}^{j=i}|rel(sk_m,sr_j))}{m} & \exists s, \exists E \\ rel(sk_m,sr_j) = 0 & \text{else} \end{cases} \quad (2)$$

We can quantitatively compute the relationship between learning resources and query keywords with formula 1 and 2.

5 PERFORMANCE

In order to prove the associative-model of learning resource proposed in this paper is efficient and practical, experiments are conducted in a dynamic digital publishing system for comparing the retrieval method and the traditional one based on keywords.

We analyze the performance of two methods with recall ratio and precision ratio. The recall ratio shows the coverage of search results and is calculated by the number of search results divided by the number of all of resources in one discipline. Precision ratio shows the accuracy of search results and is calculated by the number of effective resources retrieved divided by the number of the retrieved resources.

In this experiment, ten sets of keywords are chosen which is shown in Table 1.

Table 1. Keywords in experiments.

No.	Key Words
1	Linear function, Acute triangle, Sine function
2	Triangle ,The Pythagorean theorem
3	Positive number, Negative number ,Rational number
4	Number axis, Rational number, Irrational number
5	Rational number, Fraction
6	Quadratic equation with one unknown, Equilateral triangle
7	Coangle, Square function
8	Parallelogram, Sine function
9	Inequation,Irrational number, Polygon
10	Real number, Quadrilateral circumference

The experimental results are shown in Figure. 4. As a result, the performance of semantic search is better than that of the traditional method, no matter in recall ratio or precision ratio. Furthermore, we can find that the search results of the phrases with high association rate are much better than that of traditional methods, but the performance of lower association rate is not good enough. As for the reason, semantic search expands the concepts set with a knowledge chain model; on the other hand, it digs deep semantic association between keywords.

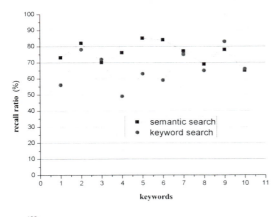

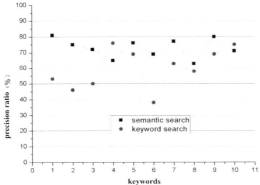

Figure 4. Recall ratio and precision ration between two searching methods.

6 CONCLUSIONS

We built a knowledge chain model used to model knowledge ontology, defined and analyzed the relationships between knowledge units. Then we built semantic models for resources by marking related knowledge units to find associative retrieval of learning resources. Experiment shows that an associative retrieval method based on this model outperforms the traditional method in recall ratio and precision ratio.

In future, research will focus on making more describable ways for modeling relationships and balancing depth search and breadth search to improve search performance in knowledge ontology base, so that the model can be directly applied in practice.

ACKNOWLEDGMENTS

This paper was supported by China Postdoctoral Science Foundation funded project (2014M562071) and National Key Technology Research and Development Program of the Ministry of Science and Technology of China (2012BAH89F01).

REFERENCES

[1] Huang W.,Eze E.,Webster D. Towards integrating semantics of multi-media resources and processes in e-Learning. Multimedia Systems, 2006, 11(3): 203–215.

[2] Zhong X., Fu H., She L. Geometry Knowledge Acquisition and Representation on Ontology. Chinese Journal of Computers, 2010, 33(1): 167–174.

[3] Zhang H.Y., Zhang M.Y., Li X. E-learning resource library model based on domain ontology. Journal of Computer Applications, 2012, 32(1): 191–195.

[4] GU F., ZHANG S.Y. Adjustable relevance search in Web-based parts library. Computer Integrated Manufacturing Systems, 2011, (04).

[5] Du L., Zheng G., You B., etal. Research of online education ontology model. 4th International Conference on Computational and Information Sciences, ICCIS 2012, Chongqing, China, 2012:780–783.

Information, Computer and Application Engineering – Liu, Sung & Yao (Eds)
© 2015 Taylor & Francis Group, London, ISBN 978-1-138-02717-6

Research on deformation by stress of main beam of tower cross arm based on Fiber Bragg Grating

J.Q Li
Faculty of Information Engineering and Automation, Kunming University of Science and Technology, Kunming, China
Postdoctoral Workstation of Yunnan Power Grid Corporation, Kunming, China

M. Cao & E. Wang
Energy Metering Key Laboratory, Yunnan Electric Power Test & Research Institute Group Co. Ltd., China Southern Power Grid Company, Kunming, China

Z.G. Zhao & C. Li
Faculty of Information Engineering and Automation, Kunming University of Science and Technology, Kunming, China

ABSTRACT: Tower structure healthy relationships of the entire power grid running. By ansys finite element analysis software, can know the that the maximum stress point of power tower is the crossarm strain. It Will be a crossarm can be converted into a variable center wavelength fiber Bragg grating strain sensors mounted changing tower crossarm, and this can achieve the detection of the tower structure. FBG sensor sensitivity experiments is 10–25pm / kg, repeatability error is 2.1%–4.3% FS.

KEYWORDS: Fiber Bragg Grating; Tower Crossarm; Strain; Wavelength Shift.

1 INTRODUCTION

Power tower is an important part of the grid system, tower crossarm are made of flat steel truss, and it's commonly fixed in the tower body. Crossarm main material damage mainly by snowfall, rainfall, folk dance and other causes[1]. Therefore, in the daily running of the electric power system, real-time monitoring of power tower is necessary. The traditional means of monitoring relies mainly on video surveillance, patrol officers and a helicopter patrol line technology, such monitoring ways are often used a lot of time and cannot keep abreast of changes in the status of the tower by force deformation [2-4]. Fiber bragg grating strain sensors are currently very wide range of applications. Due to its physical characterics, it is compared with electronic sensors have great advantages, fiber Bragg grating sensors with small size, light weight, high accuracy, corrosion resistance, long life and other characteristics, are very suitable for use in power transmission tower health testing.

This paper uses a fiber bragg grating strain sensors, power tower will be fixed on the surface of the crossarm. This can achieve real-time online monitoring of changes in stress power tower crossarm [5,6].

2 PRINCIPLES AND METHODS OF DETECTION POWER TOWER

2.1 Tower finite element analysis

The tower is a straight angle with bolt consisting of a plurality of bolts connecting angle so that the angle constraint increases. Using ANSYS finite element modeling, because the tower structure is more complex, so do not use solid modeling approach. In ANSYS modeling takes into account the selection BEAM188 directional angle, and transmission line tower closer to the real situation, the angle of the connection point is simplified to just node cells, the results of this modeling approach is relatively close to the actual situation, namely truss model. Rigid model is to consider its rigidity is large, the tower each rod are reduced to beam elements, selection BEAM4 space beam element in ANSYS, simplifying the connection point for the node angle unit is just a node; truss model is simply simplify each rod into the rod cells (general engineering calculations are treated as space truss tower model to deal with), use the space bar in ANSYS LINK8 unit, angle of connection points is reduced to a unit node hinge node. Linear finite element model of the tower in Figure 1. Tower of parameters to determine: the material is Q345 steel, Q345 yield strength of 345 MPa; Modulus of elasticity, E = 2.1e1 1 Pa, Poisson ratio of p = 0.3, the density ρ = 7800kg / m3.

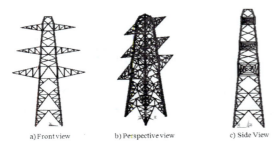

a) Front view b) Perspective view c) Side View

Figure 1.   Linear finite element model of the tower.

The gravity of the power tower and power lines hanging finite element analysis result in the Figure 2. The results show that power tower's variable cross-sarm should be much bigger than the other parts, so

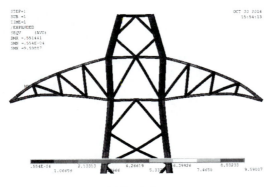

(a) Front view

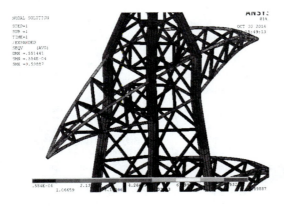

(b) Perspective view

Figure 2.   The result of tower finite element analysis.

it's necessary to reinforcement the crossarm at the tower, while detecting at the crossarm point is also very important. Therefore, this paper uses a small size, light weight, high accuracy, corrosion resistance, long life characteristics of fiber Bragg grating strain sensors, it will be fixed on the tower crossarm plane, achieve the purpose of the tower health detection.

2.2 *Principle of FBG strain sensing device*

Both ends of the FBG are fixed to the adjustment of the inner tube, the tube and the two adjusting screws fixed at one end by two desensitization tube, the other end of the reduction sensitizer is respectively fixed to the fastening pipe with a threaded pipe end, the tube passes between the two fastening a threaded connection Strain tube, the lower end of the two fastening pipe set a fixed fulcrum. Schematic strain sensor as Figure 3. Imposed load by hooking the force to the outer tube, the outer tube threaded through the tube force to the inner tube driven grating wavelength shift occurs. By measuring the wavelength shift amount, coupled with the appropriate mathematical model can be obtained in terms of the strain [7,8].

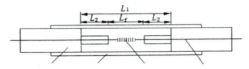

Threaded pipe   outer tube   FBG   Optical fiber

Figure 3.   Schematic strain sensor.

FBG strain response change when FBG is compressed or stretched, the Bragg wavelength shift caused by the fiber grating demodulator detection wavelength shift amount can be calculated with the Bragg wavelength corresponding to the strain.

3   TOWER CROSSARM FORCE LOADING TEST AND DATA ANALYSIS

When the sensor is arranged in the tower model crossarm as the standard, the corresponding position in the four cross arms arranged in four FBG, will be located in the cross arm left and right each 2 grating plane trusses were connected in series. In the tower cross arm load test trials, the test system consists of tower models, fiber, 3 dB fiber coupler, a broadband light source (ASE) and demodulation devices and other components. 4 wave peak using FBG optical fiber Fabry-Perot tunable filter is demodulated to determine the value of the wavelength of light reflected by the photo detector using the peak detection method.

Test by loading and unloading the weight of a load is applied to the tower the way the main material.

Before the experiment started, we must do pre-load test, mainly because of the elastic element to go through several loading and unloading cycles repeated, the deformation relationship can stabilized. Loaded from 0 kg Start, increase the weight increments 3 kg, until 30 kg; and then start unloading, in turn reducing the weight of each reduce 3 kg, 3 kg each increase or decrease in weight is recorded once FBG center wavelength. The value is calculated according to the test data of the sensor recorded a variety of static features.

Figure 4 shows Five times the amount of the relationship between the wavelength shift and strain between the experimental data, calculations show that the repeatability error of 3.2% FS.

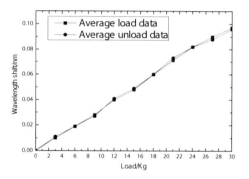

Figure 4. The wavelength shift amount relationship between loading and unloading.

Test results showed that: in the five times strain measurement experiments, the amount of strain measurement test load 0–30 kg, the wavelength variation 0. 092 nm, the maximum deviation from the least squares fit through value after 0.0025nm. Is calculated by nonlinear error seen. The paper developed

a sensor temperature test trial nonlinearity error of 3.2% FS. This one can meet the power tower crossarm strain accurate measurement.

4 CONCLUSION

In the use of fiber Bragg grating strain sensors detect experimental power tower structure, the wavelength shift of pressure-sensitized grating is modulated by the strain of metal elastic pipe caused by the structure pressure. This can have a real-time online monitoring of power tower crossarm stress. FBG sensor sensitivity experiments is 10–25pm / kg, repeatability error is 2.1%–4.3% FS.

REFERENCES

[1] Zou,L.,Liang,S.,Li,Q.S.,Zhao,L.,Ge,Y.Investigation of 3-d dynamic wind loads on lattice towers[J]. WindStruct.Int.11(4),323–340,2008.
[2] Kim,Y.-M.,You,K.-P.,Ko,N.-H. Across-wind responses of anaeroelastic tapered tall building[J].Wind Eng. Ind. Aerodyn.96(8–9), 1307–1319 ,2008.
[3] Zheng,C.Zhang,Y. Numerical investigation of wind-load reduction for a high-rise building by blowing control[J].Build.Struct.31(Suppl.2),S176–S181,2010.
[4] Irwin,P.A. Wind engineering challenges of the new generation of super-tall buildings[J].Wind Eng. Ind. Aerodyn. 97(7–8), 328–334,2009.
[5] Xinhua zhou, High-pressure structural tower structural strength analysis [D].Baoding: North China Electric Power University, 2002.
[6] He Zhu, Simplify the analysis of overhead transmission line tower member forces decomposition symmetrical structure stiffness method[J], Northwest Electric Power Technology, 2004(6):30–31.
[7] Yongqiang Chen,Wujun Bao, Transmission tower finite element analysis software design [J], Electrical machinery, 2003:55–57.
[8] Hongnan Li,Haifeng Bai, Status quo and development trend of high-voltage transmission line system disaster research [J], Electrical machinery ,2007(02).

Information, Computer and Application Engineering – Liu, Sung & Yao (Eds)
© 2015 Taylor & Francis Group, London, ISBN 978-1-138-02717-6

The innovation of teaching design on the university physical education

Ai Lin Yang

P.E. Department, Agriculture Science And Technology College, Jilin City, Jilin Province, China

ABSTRACT: Teaching design is not only a kind of teachers' innovation activities, but also thinking and practice activities. It reflects creative freedom which belongs to the teachers. Innovation contains the actual object innovation, as well as ideas and methods innovation. It is unreasonable that innovation is just a few people's talents. It is wrong to just consider great originals and inventions in science, technology, literature and art as innovation. Teachers' work is constant innovation. This kind of innovation is both predictable and unpredictable. Teachers put forward to imagine about teaching activities in advance, design activities and make plans, which is a predictable teaching activity.

KEYWORDS: university; physical education class; teaching design; innovation

1 TEACHING DESIGN IS A KIND OF TEACHERS' INNOVATION ACTIVTIES

What is innovation? Innovation is the procedure of reassembling the known materials to produce matters, ideas and methods which have new values. Three key points should be emphasized. The first is variable. Invariable can't be called innovation. Innovation is to create new things on the base of the existing materials and experiences. It is difficult to create new things and ideas from nothing. Combining or breaking down the existing materials, recombining the same materials, and analogizing according to relative materials can consist of new things, ideas or methods which are different from the present existing forms. The second point is novelty. Variety but not novelty can't ne called innovation. Not adhering to old habits, daring to destroy the old and establish the new and pursuing unknown results are innovating. The third one is practicability. Just novelty may be not innovation. Innovation should have values. The value of innovation is the new things which are produced should be fit for some desire and purpose. That is to say, the things which are formed by innovation should become valuable things.

Innovation has different standards and forms. It can be divided into three levels. One is the first creation in the human world, what others or forefathers didn't do; two is the first creation in a certain time or place; three is the first creation in personal practices. Teachers' innovation of teaching design has a variety of forms. Innovation of teaching design can be divided into three forms according to degrees of difficulties and complexities which are from solving problems. One is a simple combinational design. It refers to using the present teaching design to change the form slightly in order to use new aims. Or combine the whole functions of the present teaching design in order to have new compound functions. Two is new combinational design after decomposition. Analyzing or revolving the exiting teaching design, element functions which are departing from analyzing results make permutations directive which is different from before. Then, new functions are produced. Three is developing a design. It is the deepest level among teaching design. With the development of education and change of teaching conception, teachers should consider whether the original teaching pattern is reasonable or not. It also reflects the sensibility of teachers' creativity.

The long-term teaching practice proves that teaching design contains a procedure of studying. Teachers should have the sense of innovation. Teaching design should be looked on from the angle of studying. The art of PE teaching is teaching activities of teachers' creation. Innovation is its main soul. Teachers should develop and expand the spirit of innovation which dares to break and create and dares to be the world's first. They shouldn't be trapped in books, reference books, teaching methods and patterns. They should endow P.E. teaching with endless vitality of art and free beauty through certain Screening, selection and recombination.

2 HOW TO REALIZE THE INNOVATION OF TEACHING DESIGN

After the class, we often praise instructor for their teaching design: this class is so good, very enlightening; sometimes sigh: "why I didn't think of that!" It should be said that sigh is not a lack of knowledge of this condition, investigate its reason. One is not enough for teaching analysis to design, cannot grasp the essence of the nature. Two is bound by the teaching design of teaching box, can not jump out of the thinking mode of original design. Three is that the thinking is not active, lack of innovation, consciousness and innovation ability. Instructional design is a kind of innovation. Innovation is the soul of teaching design, so we must constantly improve the ability of innovation in teaching design work and study how to innovate the teaching design and solve the problems.

2.1 Breakthrough thinking trap and traditional ideas

In the teaching activity, our mind would collect a flood of information every moment, including all kinds of teaching scene, the ideas in the mind, the problems need to be solved and so on. When dealing with this information the mind includes the selection of information, analysis of problems to make a decision, do not need to each information sit down and think what to do, always consciously or unconsciously along the former thinking habits, familiar with the thinking direction and ways of thinking, and not the other path. Only in the face of new situations and new problems, when the need to use new methods and measures innovative thinking to come in handy. In general, the customary thing, old problems, without the need for innovation, thinking can solve. This kind of thinking to solve our problems sometimes beneficial, sometimes becomes a hindrance, plays a role in thinking bondage. The innovation from the point of view, we should weaken or to break the habit of thinking, continuously put forward new ideas, strategies and methods.

2.2 Development of innovative thinking perspective

From the perspective of creative thinking is to observe things by thinking perspective of an unusual, make things show unusual in nature. Learn to look at the same thing from the multi angle, as much as possible to increase the thinking angle of the mind, teachers will increase the teaching design more. Once the design has a bound, it is not easy to develop creativity. For example, the cup design, this design is very specific, but the glass may be the head of the designer, bound to this concept of the cup, again creative people also can only produce only in size, different materials have different, no decoration of the sub cup. If you change the perspective of thinking, to design from the creation of a more reasonable way of drinking water and tools, you can never appreciate water demand angle or the same person under different environmental conditions, time, occasion, to the drinking water demand angle analysis, seek truth from facts of science, also create an excellent work. Teaching design also has a similar situation.

The design of teaching creative thinking in the teachers should pay attention to two aspects of thinking, but they are unified. Thinking into a deeper level, but also open a new layer level of thinking. It also contains the breadth, depth development, but also the deep development can make the design more reasonable innovation of ours. So, thinking depth is an important manifestation of people's thinking ability, and it can ensure our innovation to achieve the best condition. Therefore, in the training of thinking, not only to the training thought breadth, but also pay attention to the cultivation of thinking depth.

2.3 Stimulate creative potential

An innovation must have faith, to continue to develop their innovation potential. Everyone has a treasure, and it is the potential for innovation. No matter what thing, always the first to do so after the innovation, also cannot do without the mind, cannot do without the mind thinking. Can the mind think, thinking to produce innovative? In the teaching design, to tap this treasure will have unexpected harvest. For the continuous development of thinking can combine their own teaching. For individuals, being good at thinking is the basis of innovation, no thinking of the change will not produce the innovation of action. So long as teachers have the consciousness of innovation, being good at stimulating their creative potential will have the harvest.

Teaching activities not only need to design in advance, and the need for timely design, they will be full of vitality. Both pre-design or timely design need to develop the potential for innovation, which requires teachers to have the courage to innovate, profound knowledge, the accumulation of experience and skilled teaching design skills.

2.4 Brave enough to practice

Practice is an important way to improve the ability of innovation or we can say that innovation ability needs to be improved in practice. Teaching activities of new innovative design is the goal of the teaching design and the most important way to achieve the goal is the practice. The key to innovate is new. If don't do or do it slowly, the new can become the old and will lose the value of the new. We have got some better innovation of teaching design after some efforts, but it doesn't mean to end. The more important problem is to use the teaching design. For the teachers, innovation practice is a development process and the teachers need to use the following main factors. One is the innovative spirit. It mainly means to explore the motivation actively, to have the initiative, curiosity, inquiry interest, have hundred-percent perseverance in innovation Two is the innovative thinking. Firstly, the teachers have high sensitivity to the problems, and soon will be able to pay attention to some problems in the teaching activities, and be aware of strange, unusual in the seemingly dull teaching activity. Secondly teachers' thinking fluency and diversity can solve the more ideas and schemes in the teaching and be good at choosing the best solution; Thirdly The flexibility of thinking can easily get rid of the original thinking, flexible to deal with the problem. Fourthly originality of thinking can put forward the innovative ideas and methods .Three is the innovation ability. The teachers should possess basic theoretical knowledge and practical ability of innovation. This is the comprehensive expression of the spirit of innovation, innovative thinking, the implementation of the innovative activities.

If the process of innovation in teaching design is regarded as a research activity, as a process of solving a problem, then this kind of innovation activities can be divided into 4 processes. One is to find the problem. Find out the problems usually neglected in teaching, collect data, analyze the problem, and make it into specific problems .This consciousness of discovery problem needs to accumulate some experience. Two is to design teaching plan. The teachers put forward innovation, several countermeasures, method, and select a kind to form of teaching plan. Three is to carry out the teaching plan. Put the teaching plan into the practice. Four is to verify. Teachers verify the implementation process and the degree of satisfaction of the results, reflect, put forward the improvement ideas and accumulate data etc.

3 THE INPLEMENTATION OF THE NEW CURRICULUM REFORM NEEDS TO PAY MORE ATTENTION TO THE TEACHING DESIGN

3.1 *Teaching research and design has the vital significance of the implementation of the new curriculum reform*

The experiment of the curriculum reform is a process of innovation, full of exploration, creation, and the combination of construction theory and practice. The purpose and task of the school sports curriculum reform experiment mainly include four aspects: to implement the spirit of the basic education curriculum reform, to construct a new teaching and evaluation system, to verify the rationality of the P.E. course standard, to try out and test the scientific validity and applicability of the new textbook, to develop curriculum products, to create teaching experience. To achieve these requirements, We need to study teaching theories and ideas, the content of experiment, teaching method, the method of learning; and how making their own learning. And we need to create and design related learning situation, to motivate students to learn actively in various ways, to guide the students to operate by themselves and acquire knowledge, to design teaching methods conformed to the basic education curriculum reform.

In the experiment of implementing curriculum standard, teachers should be brave enough to innovate and be good at teaching design. The way of the new curriculum reform experiment is not ready yet, and we must carry out creative work. How to implement the knowledge and skill, process and method, emotion, attitude and values of three-in-one teaching? How to organize the students to explore? How to select and organize the teaching activities is good for students' learning? How to make life, society, science and technology and the teaching link and so on, all needs to carry on a teaching design seriously. It means that the implementation of curriculum standards can't exist without teaching design.

3.2 *We should pay attention to innovate teaching design*

Every design must be innovative, new idea or concept, or new content and methods, or the new teaching method. The new design is not opposed to reference and it is necessary for it to draw on, but reference needs to be digested and absorbed. We must change the phenomenon of copying reference textbook and repeating the textbook.

The implementation of curriculum standards cannot be simply interpreted as the use of new textbook, but regard the new teaching material as a carrier. We carry out the reform of teaching method and learning styles to enable students to study creatively and lively, and thus we can truly achieve the goal of quality education.

3.3 *Teaching design should be flexible and pragmatic*

The teaching design is the real teaching work, convenient and applicable. In the past, sometimes we looked on our lesson plans as formalism. And it is not worthy of spending too much time to write the lesson plans and it is not effective to study the textbook. According to the idea, we can't achieve the goal of the experiment of the course reform.

The implementation of curriculum standards must give up these ideas and design the teaching carefully in the terms of the students, including education concept, the establishment and operation of teaching activities, the preparation and the use of the chart drawing, and the use of modern teaching methods. Thus, we can make each P.E. class prepared fully and achieve the goal of the basic education curriculum reform and P.E. curriculum standard.

We need to pre-plan teaching activities in some aspects, but in other aspects we need to design timely. We cultivate students' innovative spirit in the innovative teaching activity and use the methods or strategies to cultivate the innovative spirit. Or we cultivate the spirit of innovation in creative learning. Therefore, our first work should be that we study how to design creative learning effectively. People often say that teaching activities should be complex and indeterminate actionable. The teaching design is not to control the nonlinear, dynamic, complex and inde-terminate activity as much as possible, but to make the dynamic and complex activities promote students' development activities. The actual and uncertainty, we can find the source of creation and opportunity. To consider this teaching characteristic, grasp the opportunity and this source, combine pre design and timely design, and carry out flexible and diverse teaching activities, and we will get better results. Teaching content, organization and implementation of ideas: sports courses should be close links with academic courses, students form good exercise hab-its, developed independently for their own exercise prescription, sports and cultural qualities and has a high level view, and actively improve athletic skills, to participate in a challenge of outdoor activities and sports competitions, to choose a good sports environment, comprehensive development of physi-cal fitness, improve their ability to exercise science, trained in good health, in a challenging sports envi-ronment showed courage and indomitable will form a good behavior. In a nutshell, that is, apply the sports knowledge, technology in practice, the development of student ability, entertainment, and cultivate the awareness of students to participate in sports activi-ties and habits, to compete for students, collaborative spirit, will, and creativity training.

In short, the teaching design of physical education is the starting point of the teaching process, and it is also the premise of a good lesson. In order to achieve the best teaching effect, we must constantly innovate teaching design. This requires us PE teachers keep learning, keep pace with the times, pioneering and enterprising, and design more new and better class teaching for contemporary college students according to their own teaching project.

REFERENCES

[1] Tian Maijiu, Science of Sports Training [M] People's Publishing House, 2005.
[2] Xie Tietu, The Training Procedures and Methods of Basketball Skills Teaching[M]Beijing Sport University Publishing House, 2005.
[3] Wang PingChuan, Liu Shuxia, Knowledge econ-omy and higher education reform [J], Education Exploration, 2001.

Information, Computer and Application Engineering – Liu, Sung & Yao (Eds)
© 2015 Taylor & Francis Group, London, ISBN 978-1-138-02717-6

Regression-based execution time prediction in Hadoop environment

Xiao Dong Wu, Yu Zhu Zeng & C.T. Zhao
Faculty of Mathematics and Computer Science, Quanzhou Normal University, China
Fujian Provincial Key Laboratory of Data Intensive Computing, China
Key Laboratory of Intelligent Computing and Information Processing, Fujian Province University, China

ABSTRACT: Hadoop MapReduce has been proven an effective computing model to deal with big data for the last few years. However, one technical challenge facing this framework is how to predict the execution time of an individual job. In this paper, we propose a regression-based method to predict the execution times of Hadoop jobs. The proposed prediction method consists of three phases: learning data sampling, regression analysis and making prediction. Experimental results show that the predicted execution time obtained by our method is close to the actual execution time.

KEYWORDS: Cloud computing; Execution time prediction; Hadoop; Regression analysis

1   INTRODUCTION

We are now in a great new era of "big data", and the data is growing in an explosive way. Modern companies rely more and more on big data when making decisions. It was estimated that the total volume of digital data produced worldwide in 2011 was already about 1.8 zettabytes (one zettabyte is equal to one billion terabytes), while that produced in 2006 is 0.18 zettabytes (Gantz & Chute, 2008) [1].With the increasing scale of data, the traditional techniques are no longer able to satisfy the demand of data processing. Hence, Google introduced MapReduce (Dean & Ghemawat, 2008) [2] as a programming model for dealing with the large amount of data. In recent years, this programming model is widely used in big-data applications and has been proved to be an efficient way in processing big data.

Moreover, Apache hadoop (White, 2012) [3] is a popular open-source implementation of MapReduce. Hadoop is popular not only in the academic research area, but also in real industries. Many IT companies have already use hadoop to deploy their clusters to provide cloud services.

The cloud computing environment gives a shared resource infrastructure for various applications. The central scheduler decides how to allocate the resources to the applications. Generally, the scheduler does not know in advance how long the cluster will take to complete the execution of a job. Consequently, it usually can not make the optimized decision when scheduling jobs. Therefore, it is of importance for the cluster to have the ability to predict and estimate the execution time spans of newly submitted jobs.

There have been some research results concerning predicting the time span of the job execution. (Ge et al. 2013) [4] proposed a simple framework to predict the performance of hadoop jobs. (Bortnikov et al. 2012) [5] suggests a slowdown predictor to forecast how much slower a task will run on a given node. Using dynamic progress information and extrapolation, (Chtepen et al. 2012) [6] proposes an online method for execution time prediction. In (Ganapathi et al. 2010)[7], the authors use the statistical models to predict resource requirements for cloud computing applications. [8](Chtepen, 2012) using aspects of static analytical benchmarking and compiler to predict the execution time of jobs for grid scheduling. [9](van der Aalst et al. 2011) extends the process mining model to predict the completion time of running instances.

In this paper, we propose a regression-based method to predict the execution time for the hadoop jobs. The method consists of three steps: collecting and learning data samples, regression analysis and making prediction. The prediction is made according to a prediction model which is built based on the samples.

The remainder of this paper is organized as follows. In Section 2, we give a detailed description of the proposed regression-based prediction method. In order to evaluate the proposed method, the experimental results are provided in Section 3. Finally, Section 4 concludes the paper.

2 DESCRIPTION OF THE REGRESSION-BASED PREDICTION

In this section, we will give a detailed introduction of the proposed regression-based prediction method.

2.1 Overview of prediction method

The prediction method is composed of three steps, i.e. 1) collecting data and learning samples, 2) building the regression model and 3) making predictions. The model is built based on the machine learning samples that are obtained in the firs step. Then, the prediction is carried out based on the regression model.

Since the objective of the prediction is the execution time span hadoop jobs, the execution time is selected as the dependent variable of the regression model. We denote it by t in this paper.

In addition, as is known, there are many factors that could affect the execution time of hadoop jobs, such as the size of the input data that needed to be processed, the number of the processing nodes in the cluster, the CPU frequencies and memories of the processing nodes. Moreover, some configuration parameters with regard to the hadoop environment will also impact the execution span, e.g. the parameter dfs. block.size that specifies the block size of the HDFS system, the number of the mapper slots and reducer slots on each node, and so on. Suppose that there are n such variables. Those explanatory variables are regarded as regressors in our regression model, and denoted by $x_1, x_2, x_3, \ldots, x_n$. Thus the regression model can have the form:

$$t = \beta_1 \cdot x_1 + \beta_2 \cdot x_2 + \ldots + \beta_n \cdot x_n + \varepsilon \ (1 \leq i \leq n) \quad (1)$$

where ε is the error term, and $\beta_i (1 \leq i \leq n)$, which are the focuses of the prediction, are the n parameters of n regressors. Since there are more than one explanatory variable, model (1) is multiple linear regression.

After the establishment of the model, it can be used to make prediction.

2.2 Detailed description of the regression-based prediction

Once the n regressors are decided, the first step is carried out to collect the samples. The learning samples can be obtained by running jobs on the hadoop cluster in various environments, such as different hardware configurations, different parameters of the job and the hadoop environment as described by the regressors. The data that is needed to collect include the execution time of the job (i.e. the dependent variable of the regression model) and those selected parameters that could affect t (i.e. regressors x_i $(1 \leq i \leq n)$).

Then, the samples obtained in the first step are used as the input data set to construct the prediction model. There are many kinds of software that can be used to do the statistics analysis (including regression analysis) as well as build prediction models. Among them, R (R website) [10] is an open-source software that gives users a great deal of techniques, and has been widely used for computing with data. R provides comprehensive support for multiple regression. Therefore, we use R software for the regression analyzing in this work. In R, the models can be conveniently defined by syntaxes $Y \sim md$ where Y is the objective dependent variable that is be predicted, and md is the formula for the mathematical model. For example, syntax $Y \sim A$ defines the straight-line model $Y = \beta_1 \cdot A + \varepsilon$, $Y \sim A + B$ defines the first-order model in A and B without interaction terms: $Y = \beta_1 \cdot A + \beta_2 \cdot B + \varepsilon$, and $Y \sim A + I(A^2)$ uses function I() to define the polynomial model $Y = \beta_1 \cdot A + \beta_2 \cdot A^2 + \varepsilon$. After the model is defined, the regression analysis can be made by the command lm(). Note that the purpose of the regression analysis in this step is to develop a model that can be utilized to predict the execution time of jobs.

Finally, the predictions can be performed by using the results of the regression model. The predict() command can be used to do the prediction given the new values for which predictions are desired. The new values should be corresponded with the regressors and they are organized as a data frame as parameter of predict() in R. Consequently, as long as these parameters of a job are given, the time needed to complete this job can be predicted.

It should be noted that in our regression-based prediction method, the regressors in the model are not fixed. Consequently, the model can be flexibly established by chosen regressors according to the specific conditions of different hardware and software environment.

3 EXPERIMENTAL RESULTS

In this section, we report the experimental results of our proposed prediction method.

In order to evaluate the performance of the proposed regression-based prediction method, we set up a hadoop cluster environment. The machines that we used are 9 Lenovo personal computers, which are connected with each other by 100M switching exchange. Thus, the cluster can be deployed with up to 9 nodes, namely, 1 namenode plus up to 8 datanodes. All of those computers used in our experiment are equipped with the same hardware configuration, that is, a Pentium 4 dual-core processor (3.0G), 1G RAM and 80G disk space.

The benchmark used in the experiment is terasort. In order to obtain the learning data, we submit terasort jobs to the hadoop cluster, and collect the execution time as well as some relevant configuration parameters of each job after it is completed. Then, based on these samples, the regression model is built. We use R software (R website) to analyze the sample data. Finally, the regression analysis and the prediction are performed in R environment.

We simply select three variables to build the model, namely, the execution time the cluster spent to complete a job, and two parameters that impact the execution time as the regressors, i.e. the input data size of the job and the number of the datanodes in the cluster. Moreover, the input data size that used for terasort jobs is in {0.1G, 0.3G, 0.5G, 0.7G, 0.9G, 1G, 3G, 5G, 7G, 9G}, and the number of datanodes of the hadoop cluster is{1, 2, 4, 6, 8}. The parameter dfs. block.size, i.e. the block size of the HDFS, is set to 32M. We use the programme teragen, which is in the same package as terasort, to generate the input data for terasort with different size. Further, in our experiment, for each case (with the same input data size and datanode number), six executions are carried out so as to collect more samples for the regression analysis.

Figure 1 shows the relationships between the three variables. Note that all the sample points with different parameters are drawn in the figure. It can be found

that there is an obvious linear function relationship between the execution time and the input data size. However, the linear relationship between the number of datanodes and the execution time is less noticeable.

Figure 2 illustrates the linear regression analysis result between the execution time and the input data size. In this case, only the data with 8 datanodes are used as the samples to conduct the linear regression analysis. It can be seen from this figure that most of the sample points are very close to the linear regression line. This also demonstrates the linear relationship between execution time and data size.

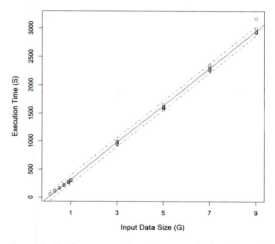

Figure 2. The linear regression between execution time and input data size, the number of datanode is 8.

In order to get more precise model, we divide the sample data into two groups according to the scale of input data size (i.e. less or not less than 1G). Figure 3 describes the average value of the six execution times for each case.

Based on the aforementioned sample partition, the two groups of data are respectively analyzed. And two regression models are built. Then, the two models are used to predict the execution time of newly submitted job. The input data size of the job (less or not less than 1G) decides which model is used to make prediction. We evaluate the performance of the proposed prediction method through comparing the predicted time with the corresponding actual execution time. In addition, we run each test 3 times and take the mean execution time as the actual execution times in this experiment.

Figure 4 depicts the result of the comparison of the predicted execution time and the actual execution time at different input data size. In figure 4(a), the input data size is {0.2G, 0.4G, 0.6G, 0.8G}, while the input data size in figure 4(b) is {2G, 4G, 6G, 8G}. As

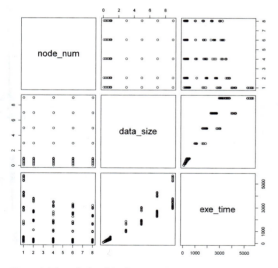

Figure 1. The relationships between datanode number, input data size and the execution time.

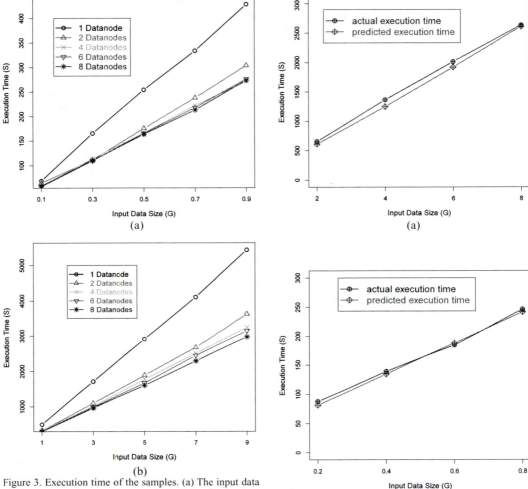

Figure 3. Execution time of the samples. (a) The input data size is less than 1G. (b) The input data size is not less than 1G.

Figure 4. The comparison of the predicted execution time and the actual execution time in different cases. (a) The input data size of the test jobs are in small scales {0.2G, 0.4G, 0.6G, 0.8G}, and the number of the datanodes is 4. (b) The input data size of the jobs are in larger scales {2G, 4G, 6G, 8G}, and the number of the datanodes is 8.

is shown in the figures, the prediction values match well with the actual ones.

4 CONCLUSION

In this paper, we present a regression-based method to predict the execution time of jobs in hadoop cluster. The prediction method consists of three phases, i.e. collecting and sampling learning data, regression analysis and making prediction. The prediction is based on regression models, which are established by analyzing the learning samples. Experiment results show the accuracy and efficiency of the proposed prediction method.

ACKNOWLEDGMENTS

This work is supported in part by the National Natural Science Foundation of China under grant number 61173045, 61472150, the Education Department of Fujian Province grant number FJJKCG13-170, and by Quanzhou Normal University Scientific Research Initiative Foundation.

REFERENCES

Bortnikov E, Frank A, Hillel E, Rao S. 2012. Predicting execution bottlenecks in map-reduce clusters. In: Proceedings of the 4th USENIX conference on Hot Topics in Cloud Computing. USENIX Association, Boston, MA, p 18.

Chtepen M, Claeys FH, Dhoedt B, Turck F, Fostier J, Demeester P, Vanrolleghem PA. 2012. Online execution time prediction for computationally intensive applications with periodic progress updates. J. Supercomput. 62: 768–786.

Dean J, Ghemawat S. 2008. MapReduce: simplified data processing on large clusters. Commun. ACM 51:107–113.

Ganapathi A, Yanpei C, Fox A, Katz R, Patterson D. 2010. Statistics-driven workload modeling for the Cloud. In: Data Engineering Workshops (ICDEW), 2010 IEEE 26th International Conference on, Long Beach, CA, pp 87–92.

Gantz JF, Chute C, Manfrediz A, Minton S, Reinsel D, Schlichting W, Toncheva A. 2008. The Diverse and Exploding Digital Universe: An updated forecast of worldwide information growth through 2011. IDC White Paper-sponsored by EMC, Framingham, MA, USA.

Ge S, Zide M, Huet F, Magoules F, Lei Y, Xuelian L. 2013. A Hadoop MapReduce Performance Prediction Method. In: High Performance Computing and Communications & 2013 IEEE International Conference on Embedded and Ubiquitous Computing (HPCC_EUC), 2013 IEEE 10th International Conference on, Zhangjiajie, pp 820–825.

Teng F, Magoulès F, Yu L, Li T. 2014. A novel real-time scheduling algorithm and performance analysis of a MapReduce-based cloud. The Journal of Supercomputing 69:739–765.

R website. http://www.r-project.org/

van der Aalst WMP, Schonenberg MH, Song M. 2011. Time prediction based on process mining. Inf. Syst. 36:450–475.

White T. Hadoop: the definitive guide. 2012. O'Reilly.

ENDNOTES

[1] Gantz JF, Chute C, Manfrediz A, Minton S, Reinsel D, Schlichting W, Toncheva A (2008) The Diverse and Exploding Digital Universe: An updated forecast of worldwide information growth through 2011. IDC White Paper-sponsored by EMC, Framingham, MA, USA.

[2] Dean J, Ghemawat S. MapReduce: simplified data processing on large clusters [J]. Communications of the ACM, 2008, 51(1) : 107–113.

[3] White T. Hadoop: the definitive guide[M]. O'Reilly, 2012.

[4] Ge Song, Zide Meng, Fabrice Huet, Frederic Magoules, Lei Yu, et al. A Hadoop MapReduce Performance Prediction Method. HPCC 2013, Nov 2013, Zhangjiajie, China. pp.820–825.

[5] Predicting Execution Bottlenecks in Map-Reduce Clusters.

[6] Online execution time prediction for computationally intensive applications with periodic progress updates. Maria Chtepen, Filip H. A. Claeys, Bart Dhoedt, Filip De Turck, Jan Fostier, Piet Demeester, Peter A. Vanrolleghem.

[7] Statistics-Driven Workload Modeling for the Cloud. Archana Ganapathi, Yanpei Chen, Armando Fox, Randy Katz, David Patterson.

[8] Execution Time Prediction of Imperative Paradigm Tasks for Grid Scheduling Optimization. Maleeha Kiran 1,2, Aisha-Hassan A. Hashim1, Lim Mei Kuan2, Yap Yee Jiun2.

[9] Time Prediction Based on Process Mining. W.M.P. van der Aalsta, M.H. Schonenberga, M. Song.

[10] R website. http://www.r-project.org/.

Information, Computer and Application Engineering – Liu, Sung & Yao (Eds)
© 2015 Taylor & Francis Group, London, ISBN 978-1-138-02717-6

Establishing the quality assurance system and improving the quality of teaching

Xue Mei Zhang

Yanan University School of Historical Culture and Tourism, Yanan, Shaanxi, China

ABSTRACT: Development process from scale to connotation in the university must establish a quality guarantee system of science, the highly effective teaching management system is the basis and effective teaching feedback rating supervision system is the promotion price, high quality teaching service support system is the guarantee, and will play an important role in the teaching and research. With only the four coordinate systems to comprehensively improve the quality of teaching.

KEYWORDS: Teaching quality; guarantee system; improve

The fundamental task of university is to train senior specialized talents with innovative spirit and practical ability, and the training of personnel is mainly through the teaching activities to achieve, the teaching work is the central work of colleges and universities, teaching quality directly determines the quality of talent training, which determines the survival and development of the high school, quality is the lifeline of higher education, to improve the teaching quality is the eternal theme of higher education. Therefore, the university must establish a scientific, standardized, operable, effective teaching quality guarantee system, to guarantee and improve the education quality.

The teaching quality guarantee system is to comprehensively improve the quality of teaching the work system and operation mechanism. The concrete is to improve the quality of teaching as the core, with the objective of cultivating high-quality talent, each link of teaching process, the Department's activities and functions are organized to form a reasonable, tasks, responsibilities, authority is clear, can organic whole mutual coordination, mutual promotion.

1 TEACHING MANAGEMENT SYSTEM IS THE BASIS TO ENSURE THE QUALITY OF TEACHING

Teaching management system is refers in charge under the leadership of the school leadership, teaching management system to the Academic Affairs Office dominated the formation of flexible operation, on issued, authoritative, high efficiency. The perfect teaching management system should include the command system, consulting system, operation system. Command system is sound teaching middle school leadership system, the preparation of Teaching Guidance Committee, the target system

overall consensus formed in the school leadership, follow the overall goal of building school, engaged in the teaching work, composed of experienced teachers and understand the teaching work, management expertise to teaching management, study and solve major problems of teaching management in the work. Operation system refers to the dean's office and teaching the basic unit, office of academic affairs is the operation system center. The registry working condition directly reflects the overall state of teaching in a school. Therefore, we must strengthen the office of academic affairs management functions, improve the structure of the Registry Department, equipped with high quality and relatively stable management cadres. In the implementation of operating system, college and department level teaching management work is also very important, the work efficiency and quality will have a considerable effect on the teaching work, should be equipped with teaching, management staff. Thus forming a group of excellent teaching management from the headmaster to specific staff. The operation model of the teaching management system can use the "ring structure". The so-called Circular refers to the teaching work from the leadership of the school director of teaching to the teaching of coordination and management department, to the teaching of basic units, until the teachers and students, and then from the teachers and students through a certain channel and the link back to the teaching management departments and the competent leadership of the school, to form a closed ring. As the implementation of the operation center in the coordination of the implementation of the educational function at the same time, we should know about the teacher's teaching effect, the students' learning quality information, constantly study and solve the teaching problems in the work, and to provide the command system, in order to put forward the new management target, make the teaching work and management always running and in the development of a new starting point and a higher level.

2 THE FEEDBACK OF TEACHING INFORMATION, SUPERVISION AND TO GUARANTEE THE QUALITY OF TEACHING, EVALUATION AND IMPROVEMENT OF TEACHING IS THE TEACHING FEEDBACK EVALUATION SYSTEM'S MAIN FUNCTION

The learning quality of students of teachers in the teaching running in the process of teaching effect, and information about the problem, should be able to pass the speed is quick, sensitive, accurate and reliable multi channel information feedback. First, adhere to the school, the hospital leadership, uninterrupted random lectures system teaching management cadres term (not just teaching in the mid-term examination), dynamic grasp first-hand materials, teaching work for classroom teaching in the most direct way, timely teaching problems in the work of research and improvement. Second, to develop students' evaluation of teaching activities. Student evaluation of teaching operation and teaching quality is the most direct, the most convincing, and truly embody the "teachers teach students how to say", therefore, each semester at least once about the teaching status of students forum or questionnaires, including teaching methods, teaching contents, teaching effect, teachers' image and so on each aspect, each scoring excellent, good, poor, four grades. The implementation of computerized management, questionnaire in unregistered machine reading card form, in order to ensure the fairness and efficiency of the questionnaire. The establishment of evaluation database teaching quality of teachers, the results will be evaluated the quality of teaching and the evaluation of professional titles, post allowance and worry that the hook award evaluation, evaluation of teaching quality is not up to the good above, hire will not be through the comments the title in the next year, the post allowance level down, cancel all kinds of awards appraisal qualification. Third, the establishment of various incentive mechanism to improve the quality of teaching, such as "elite teachers" award. Each semester each class selected a "elite teachers", for three consecutive terms was rated "excellent teachers" teachers, directly into the Department of "excellent teacher", all schools of various teaching awards such as the award for outstanding teaching quality, resulting from the "excellent teacher" directly. Fourth, the establishment of teaching supervision mechanism, by the old teachers with high academic level, a strong sense of responsibility, rich experience in teaching≈mainly composed of full-time and part-time teaching team. Its members should always go deep into the classroom, using classes, investigation, discussion, visiting various forms, the order and the quality inspection and supervision of each link of teaching. In the supervisory process, should pay attention to play the old teachers "guiding" role, namely please them in the process of "Doc" found in the foundation of analyzing the above problems, put forward a proposal and feasible measures of rectification. The maximum elimination of weak links. Fifth, feedback the students' exam is teaching situation. The organization course examination, the establishment of the system of examination paper analysis, based on the test results and analysis of student achievement distribution, check students' learning effect, summarizes the achievements and shortcomings in the teaching work, improve, improve the quality of. The school's teaching work should be often in supervision and evaluation, to discover problems, sum up experience, continuous improvement. The cause of higher education development, education reform continuously thorough, the problem or issue new emerging, the school should be regularly or irregularly on their teaching work to self evaluate, establish self teaching quality monitoring systems, and make it become a regular and systematic work. Evaluation and supervision as a whole, forming a unified evaluation of supervision mechanism; teaching evaluation mainly is the basic construction of evaluation, the whole school teaching work discipline and curriculum evaluation, teaching evaluation, teachers' teaching quality and students' learning quality evaluation. Carrying out teaching evaluation to clear goals, to take the target and content of teaching evaluation as the main contents of daily teaching construction and management, to realize the combination of the teaching work evaluation and daily teaching management. Teaching evaluation is the principle of promoting construction by evaluation, focusing on construction. The level of quality and problems by assessing the current master of school teaching work, analyze and solve the influence of school teaching quality factors, for the establishment of school puts forward valuable constructive opinions on quality standard, and give full play to the role of teaching evaluation work of the staff.

3 PLAYING AN IMPORTANT ROLE IN IMPROVING THE TEACHING QUALITY OF UNIVERSITY TEACHING AND RESEARCH OF THE GUARANTEE

China's higher education is experiencing a profound transformation of the hitherto unknown, in the process of deepening the reform of higher education, will continue to put forward the theoretical and practical issues related to the reform and development of higher education, the urgent need to answer and for guiding the reform practice of correct theory through the research. With the development of society, the progress of science and technology, many in the past that is correct, appropriate concepts, contents, methods, means, today has become the restriction even factors to reduce the quality of teaching, must be one by one to analyze, study and reform, abandoning the old, backward, content and method of the addition of new, advanced. At the same time at university to complete the daily teaching management, teaching guarantee normal operation, indicating that it has a good working basis and teaching environment. But if it wants to develop, to continuously improve the quality of teaching, it is necessary to carry out teaching research. Teaching research system can be embodied in two forms: one is to mobilize and organize the teaching first-line teachers and management personnel to conduct teaching research combined with their own work, two is the institutions and researchers specializing in the setting of teaching, teaching and research work, the formation of a combination of part-time teaching and research team. At the same time pay attention to teaching and research in theoretical research, more closely combined with the teaching work of the research has practical significance, further development should along with the progress of the society, the national economic construction and system reform. The research on talent training of new situations and new problems in the process, improve teaching, it can meet the requirements of social development and economic construction of the school.

4 HIGH QUALITY TEACHING SERVICE SUPPORT SYSTEM IS THE BASIS OF IMPROVING THE TEACHING QUALITY ASSURANCE

Renewing teaching content, improving teaching methods, try the new teaching methods to improve teaching quality must have two conditions, one is the passion and enthusiasm of teachers in education teaching reform, the two is harmonious, good service environment and the necessary material support. Logistics service department and staff of the school should further enhance the service consciousness and service level, should be fully aware of the work and school teaching quality is closely linked, to provide high quality logistics service for teaching, be warm and thoughtful, simplify procedures, convenient for teachers and students, teachers and students in solving life speaks for the gifted. Schools should make every attempt to provide material support to the teaching and reform. The information age, the teachers should prompt understanding domestic and foreign related disciplines and cutting-edge developments and related knowledge, need the support of the

Internet and related technology; today only depend on the blackboard and chalk teaching method is not enough, the audio-visual teaching means, teaching apparatus manufacture high-tech, need modern equipment and facilities, etc. A reform scheme, only the enthusiasm of teachers, the lack of the necessary material conditions, will not be completed. The work of teaching the past scattered in various departments, complex procedures, efficiency is not high, resulting in a waste of time and energy. Envisaged by the reform of inner management system reform of school and social logistics, establish a coordination mechanism, belonging to the competent leadership, unified, coordinated management relates to the teaching work of the service department, material supply department, printing department, the Department, maintenance department audio-visual. To ensure that the teaching service support system to the normal operation of teaching and teaching reform to provide convenient, fast, high quality service and material.

Teaching management system, teaching feedback framework supervisory evaluation system, teaching and research system and teaching support services system composed of quality guarantee system of teaching in Colleges and universities. The four system is an organic whole, need mutual coordination and cooperation. Only four of high efficiency and high quality operation system. In order to ensure the high quality of teaching in the university.

REFERENCES

[1] Ma Weiqiang on the construction of university teaching quality guarantee system of education and occupation[J]. 2009 (35).

[2] Chen Juan. Construction of Jiangsu higher education teaching quality guarantee system in Colleges and universities[J]. 2010 (03).

[3] Shen Yushun of college teaching quality guarantee system of organization strategy [J]. Fudan Education Forum. 2010 (04).

[4] Wang Yan, Jiang Wusheng. Application Research on constructing the quality assurance system of Undergraduate Teaching—Taking Tianjin Normal University as an example [J]. Heilongjiang higher education research, 2010 (07).

[5] Wu Guanghua, Wang Hui, Li Yan, Deng Hongzheng, Li Yanping. The teaching quality assurance system of institutions of higher learning of [J]. education academic monthly. 2009 (09).

[6] Gao Guizhen. Teaching quality guarantee system of internal universities undergraduate education research and experimental study on construction of [J]. 2009 (S1).

[7] Zhu Yongjiang. Contemporary education science construction of college internal teaching quality guarantee system based on total quality management [J]. 2010 (11).

Information, Computer and Application Engineering – Liu, Sung & Yao (Eds)
© *2015 Taylor & Francis Group, London, ISBN 978-1-138-02717-6*

The predictive validity of distance-to-default on financial distress: Evidence from Chinese-listed companies

Y.L. Cai & CH. X. Qian
School of Business Management, South China University of Technology, Guangzhou, Guangdong, China

ABSTRACT: Using the financial and stock exchange data of Chinese listed companies in later years, this paper examines the predictive validity of Distance-to-Default (DD) based on the Merton model on corporate financial distress by the use of Logistic regression and ROC curve analysis. We find that DD does have some but limited power to identify distressed and normal firms. When using DD as a predictor for financial distress, it is necessary to consider some accounting information. Firm's profitability, sales activity, capacity of cash to repay debt and capital structure are important for forecasting distress. Especially, the profitability can significantly improve the effectiveness of DD. However, DD is not irreplaceable. Appropriate combination of accounting variables can also bring about pretty good prediction results, DD just makes it even better. To some degree, DD behaves no better than accounting ratios. This means there is no obvious superiority of the Merton model in Chinese stock market.

KEYWORDS: Financial distress; Distance-to-Default; Predictive validity

1 INTRODUCTION

Financial distress refers to an enterprise in operating cash flow is insufficient to cover the existing maturing debt (such as commercial credit or interest), but was forced to take action to correct immediately. It can lead to defaults on the contract, and may relate to debt restructuring among enterprises, creditors and shareholders. What's more, it can even lead to bankruptcy. So there are four similar terms that often used interactively in domestic and foreign literatures. They are corporate failure, insolvency, default and bankruptcy. We will treat them equally. Since the seminal work of Beaver (1966), the study of predicting corporate financial distress has become a widespread concern. Nowadays, predicting and managing corporate financial distress have become increasingly important parts of doing business and making investing and lending decisions (Beaver et al. 2011). Shareholders, creditors and employees all have given high priority to the risk of financial distress and imminent bankruptcy.

Distance-to-default (DD) is the key indicator of the Merton model, and is widely used in measuring and modeling default risk or corporate failure. Moody's KMV first measures the DD as the number of standard deviations the asset value is away from the default, which highly summarize and integrate the following three important factors affecting defaults: the value of firm's assets and its volatility, and the leverage (Crosbie & Bohn 2003). Due to a strong theoretical foundation and overcoming some

shortcomings of accounting-based models, DD shows certain superiority for measuring defaults. Vassalou & Xing (2004) find that the default probability given by DD can indeed predict actual defaults and that the volatility in DD provides important default-related information. Hillegeist et al. (2004) demonstrate that the Merton model provides more information about bankruptcy than alternative accounting-based models. They recommend that researchers use it in their studies, because it is a powerful proxy for bankruptcy probability. Reisz & Perlich (2007) argue that the Merton model is more reliable for long-term prediction than accounting-based measures.

However, Beauer & Agarwal (2014) point out that the implementation of a DD framework for bankruptcy prediction is far from straight forward. First, it requires a number of assumptions such as the normality of stock returns. Second, Avramou et al. (2010) argue that distressed firms are prone to suffer from market microstructure problems such as thin trading or limitations to short-selling which might result in prices deviating from fair values for the long term. Third, some key variables (e.g. The value of a firm's assets and its volatility) for DD are unobservable and need to be estimated indirectly. Therefore, some researchers acknowledge that the Merton model is not effective as its theoretical description. Campbell et al. (2008) indicates that in the absence of market leverage and asset volatility, DD based on the Merton model almost provide no additional information. Castagnolo & Ferro (2014) use the data of listed companies in the UK during the year 1991-2010 to compare the performance of accounting-based and market-based models and find that DD is not a sufficient predictor of defaults, which is in line with Bharath & Shumway (2008), and that the default probability is inferior to the O-Score in the trade-off between type I and II errors.

In china, we have found that related researches are still focused mainly on the adaptability test and extension of the Merton model. Most researchers acknowledge the predictive ability of DD, few focus on how well its actual power. Some studies concern this point find that DD is not the best measure of defaults, only relying on DD can't forecast the financial distress of Chinese listed companies accurately (Liu et al. 2005), and there is no substantial improvement in the predicting power when adding DD into accounting-based models (Tan 2005; Pan & Ling 2012). Some scholars in Tanwai (e.g. Chen et al. 2004; Su & Lin 2006) also find that DD performs inferior to Z-Score or O-Score. Kong & Li (2012) show the accuracy of the Merton model for default predicting is only 31.26%.

Due to the asset value and its volatility cannot be observed directly, there are different algorithms in calculating DD. The most commonly used and direct method that our domestic scholars generally use is solving simultaneous equations. However, Crosbie & Bohn (2003) point out that in practice the market leverage moves around far too much for the relationship between asset and equity volatility to provide reasonable results, and worse yet, the model biases the probabilities in precisely the wrong direction, so they recommend an iterative procedure to derive the asset value and its volatility. While, the iterative procedure is too complex. Bharath & Shumway (2008) demonstrate that the solution of DD is not important for forecasting default, but its functional form is of vital importance. They propose a naïve alternative which is relatively simple in the calculation. We have tested the applicability of naïve DD in previous studies. Given these, we will use this simple algorithm to test the validity of DD on predicting financial distress for Chinese listed companies to further reveal the applicability of the Merton model for Chinese stock market.

We view companies that are specially treated (ST) as distressed firms by convention. Our final samples consist of 47 ST and 108 non-ST firms. After a series of quantitative analysis, we find that DD does have some ability to distinguish ST companies from non-ST ones, but the area under the ROC curve they can achieve is only about 0.7, which is in line with Shi & Ren (2005) and Zeng & Wang (2013). This means the predictive power of DD is limited, it is necessary to consider some additional information when using DD as a predictor for financial distress.

In a subsequent analysis, we find that firm's profitability, sales activity, solvency and leverage are important for predicting corporate financial distress. Especially that the profitability has a significant improvement on the effect of DD. DD is not irreplaceable. Some proper combinations of accounting ratios can also perform quite well, such as the combination of ROA (or ROE) and sales activity. In accordance with Kong & Li (2012), this means DD, the key indicator of Merton model, has no inevitable superiority in our Chinese stock market.

The rest of the paper is organized as follows. Section 2 presents the estimation of DD based on Bharath & Shumway (2008). Section 3 discusses our date and methodology. Sections 4 and 5 report the empirical results on logistic regression and ROC curve analysis with considering accounting ratios or not. Section 6 draws some conclusions from the analysis.

2 ESTIMATION OF DD

Crosbie & Bohn (2003) have specified the derivation of DD in detail. Here we give the results directly, as shown in Equation (1).

$$DD = \frac{\ln(V/F) + (\mu - \frac{1}{2}\sigma_V^2)T}{\sigma_V \sqrt{T}} \qquad (1)$$

Where V is the market value of firm's assets; σ_V asset volatility of assets value; F is the face val; of the firm's debt; μ is the expected continuously compounded return on V; T is time-to-maturity.

According to Bharath & Sshumway (2008), the following step is necessary for estimating DD. Firstly, approximating the market value of each firm's debt with the face value of its debt, namely:

$$D = F \qquad (2)$$

Thus, V is equal to the market value of firm's equities, E, plus the face value of its debt, F. That is:

$$V = E + F \qquad (3)$$

Given the existence of non-tradable shares in our stock market, we will use a weighted method to calculate the market value of equity, that is, E= the closing price × the number of shares outstanding + the net worth per share × the number of non-tradable share. For F, following the practice of Moody's KMV, we set F equal to current liabilities plus 50% percent of long-term liabilities.

Secondly, calculating σ_E, the volatilities of equities, from firm's historical stock returns (for details, see Tang et al. (2013)), then the volatility of each firm's debt, σ_D, can be approximated as:

$$\sigma_D = 0.05 + 0.25\sigma_E \qquad (4)$$

Where the five percentage points in Equation (4) represent term structure, volatility, and 25% times equity volatility allows for volatility associated with default risk.

Thirdly, using the following weighted algorithm to estimate σ_V:

$$\sigma_V = \frac{E}{V}\sigma_E + \frac{F}{V}\sigma_D$$
$$= \frac{E}{V}\sigma_E + \frac{F}{V}(0.05 + 0.25\sigma_E) \qquad (5)$$

Fourthly, setting μ equal to the firm's stock return over the previous year:

$$\mu = r_{it-1} \qquad (6)$$

Where, we use the return on stocks without considering the cash dividend reinvestment in the CSMAR.

Finally, substituting the inputs V, σ_V, μ into the Equation(1) with time-to-maturity equal to one year, we can obtain each firm's DD.

3 DATA AND METHODOLOGY

3.1 Sample and data

By convention, we view companies that are ST for financial abnormity as distressed samples. Our initial distressed samples are selected from 89 A share firms during the period January 1, 2012 to June 30, 2014 in the JuYuan database. We then perform the following three processes.

First, deleting companies that: (1) issue B shares and H shares simultaneously; (2) are specially treated for other non-financial reasons, including operating impaired, illegal guarantee, violations of information disclosure, etc.; (3) recover after being an ST for several months. As a result, there are 26 firms are removed.

Second, seeking normal firms (NST) to match each distressed company by the standard that fiscal year and industry sector are strictly equal, but difference of their asset size does not exceed 10%. According to the research by Shi et al. (2012), the best matching ratio is 1:3. We follow this rule as far as possible. If there is not sufficient normal firms, then reducing the ratio to 1:2 or 1:1. In short, we will make sure each distressed firm matched with at least one normal firm. Ultimately, we find that there are 13 ST companies that can't be paired.

Third, defining the previous year of financial distress occurred as T-1 year, the last two years as T-2 year, and the last three years as T-3 year, we collect their financial data and stock trading data, and find that 3 firms have no complete data or are in abnormal.

Therefore, the final samples consist of 47 ST and 108 non-ST companies, where 15 ST companies are in 2012, 13 ones are in 2013 and 19 ones are in 2014. All data are derived from CSMAR and Ju Yuan database.

3.2 Empirical methodology

We find that most of our researchers just use the simple hypothesis testing to assess the forecasting ability of DD, few of them using the analysis of ROC curves does not report the AUC (the area under the ROC curves), while the AUC reported by Zeng & Wang (2013) is only 0.68. Given this, we will adopt Logistic regression and ROC curve analysis to prove our expectations.

Since Ohlson (1980) pioneers the use of logistic regression to predict business failure, Logit model has become one of the methods commonly used in international studies. As we know, the most basic tool for understanding the performance of a default prediction model is the "percentage right" which is derived from a contingency table or confusion matrix. However, Stein (2005, 2007)argue that using contingency tables (or indices derived from them) to evaluate models can be challenging due to the relatively arbitrary nature of cut-off definition, while ROC curves generalize contingency table analysis by providing information on the performance of a model at any cut-off. They recommend using ROC curves to analyze the prediction results. Nowadays, ROC curves are widely used in the comparison of different models for forecasting. Sobehart & Keenan (2001) argue that the AUC is the decisive indicator of a model's predictive ability. In this paper, we will plot the ROC curve of each model by the SPSS soft to compare the AUC, and evaluate the performance of DD along with other indicators (goodness of fit) produced by our models.

When using the logistic regression, we set companies that are whether ST or not as binary dependent variable (ST=1, NST=0). We first perform three simple regressions only with DD in each year as a predictor. Then plot the ROC curve by means of the predicted probability produced by each model to compare the AUC. Based on this preliminary conclusion, in order to further validate the effect of DD on predicting financial distress, we select some accounting ratios to enter into the models to observe the performance of DD included or not.

4 EMPRICAL RESULTS

4.1 Summary description

Table 1 presents the summary statistics of DD for the year T-3 through T-1. It shows that, except for the standard deviation, all the statistics of distressed firms is less than normal companies during the three years. Figure 1 visually depicts the changes of average DD for the three years. We can see that DD of distressed firms has been always below normal firms. This indicates that the situation of normal firms is obviously better than distressed firms, and there is significant difference between the two groups. What's more, as ST approaches, DD of distressed firms reduce gradually, and the gap between distressed and normal firms increase slowly. It is clear that the smaller the DD, the greater the probability of occurring financial distress. DD can effectively identify firm's financial distress.

Table 1. Descriptive statistics.

		Mean	Std.dev.	Min	Max
T-3	ST	3.335	2.115	−0.950	7.101
	NST	4.015	2.117	−0.575	9.249
T-2	ST	2.116	1.664	−0.497	6.577
	NST	3.336	1.882	−0.058	8.792
T-1	ST	1.858	1.669	−1.251	6.270
	NST	3.356	1.785	0.222	8.988

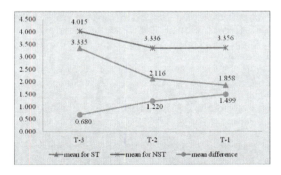

Figure 1. Change of average DD during the past three years.

4.2 Logistic regression

Table 2 gives the results of logistic regression with DD alone as predictor over the previous three years of ST. It shows that, the coefficients of DD are significantly negative, which indicating the larger

the DD, the less the likelihood of occurring financial distress. This is in line with our expectations. As ST draws near, the absolute value of the DD's coefficient becomes increasing, and the Wald statistic is also growing. This indicates that DD highlights its significance more and more. In total, the log likelihood of models shows a declining trend, and the chi-square value increase gradually, which signify the significance of the models themselves. In a word, all these indicate that DD does have a significant predictive power for corporate financial distress.

However, table 3 demonstrates that the AUC of DD in T-3 year is only 0.581, and its upper limit of 95% confidence interval is less than 0.7. In the year T-2 and T-1, although the AUC is significantly larger than 0.5, the maximum is only 0.732. If we translate it into the accuracy rate (AR) according to Engelmanne et al. (2003), then the largest AR is just 0.464. It forebodes that the predictive power of DD is not outstanding. This means relying solely on DD to predict financial distress is not enough, in order to further reveal the occurrence of a company's financial distress, it is necessary to consider some other predictors.

Table 2. The results of simple logistic regression.

Model		B	SE	-2LL
T-3	DD	−0.154*	0.085	186.819
	C	−0.267	0.349	
T-2	DD	−0.395***	0.111	175.419
	C	0.232	0.329	
T-1	DD	−0.542***	0.126	166.303
	C	0.554	0.345	

This table 2 reports the results of regressions with DD as a predictor. B is coefficient of DD. SE represents standard errors, and -2LL is log likelihood of model (*** significant at the 1% level, * significant at the 10% level).

Table 3. The area under the ROC curve.

	AUC	SE.	Sig.	95% confidence interval	
				Lower limit	Upper limit
T-3	0.581	0.051	0.108	0.482	0.681
T-2	0.688	0.046	0.000	0.598	0.778
T-1	0.732	0.044	0.000	0.645	0.819

This table 3 presents the AUC of each model. Note that the Sig. is progressive. The test is under the non-parametric assumption, and the hill hypothesis is that the real area equals to 0.5.

4.3 *Comparing with accounting ratios*

Given the above results, we select several financial indicators into the model to illustrate: (1) which financial indicator has a supplementary role on DD; (2) whether there is some superiority for DD compared with accounting ratios. In the selection of financial variables, since unable to distinguish which variable is necessary, our scholars usually pick a number of ratios representing firm's profitability, liquidity, solvency, sales activity and growth, ability to conduct factor analysis, and then extract common factors to enter into the model to use all available information. However, there are less variables, but that still can illustrate issues, in international studies (e.g. Shumway 2001; Beaver et al.

2005; Campbell et al. 2008; Tinoco & Wilson 2013). We integrate and compare the financial indicators in these studies with considering the finding of Liu et al. (2005), and at last select 9 accounting ratios as following: the ratio of EBIT to total assets(ETA), the return on total assets (ROA), the return on net assets (ROE), the current ratio (CA), the working capital ratio (WCTA), the ratio of EBITDA to total liabilities (ETL), the ratio of liabilities to total assets (LTA), the total assets turnover (ITA), the growth rate of total income (GR). They represent corporate profitability, liquidity, solvency, leverage, sales activity and growth ability, where EBIT equals to total profit plus finance expense and EBITDA is the sum of EBIT and depreciation/ amortization.

Table 4. Regressions with each accounting ratio.

	M1	M2	M3	M4	M5	M6	M7	M8	M9
DD	−0.395***	−0.63***		−0.46*		−0.39*		-0.25	
		(0.23)		(0.25)		(0.23)		(0.20)	
ETA		−70.13***	−65.56***						
		(12.70)	(11.33)						
ROA				−89.77***	−88.61***				
				(17.02)	(16.14)				
ROE						−36.36***	−36.26***		
						(6.95)	(6.90)		
ETL								−34.69***	−35.79***
								(6.24)	(6.30)
-2LL	175.419	70.93	81.60	57.17	61.35	56.52	59.66	79.17	80.68
AUC	0.688	0.969	0.957	0.979	0.977	0.978	0.977	0.951	0.946

Table 5. Regressions with combinations of 2 or 3 accounting ratios.

	M10	M11	M12	M13	M14	M15	M16	M17	M18	M19
DD	−0.36		−0.48*	−0.40*			−0.26		−0.21	
	(0.27)-		(0.28)	(0.27)			(0.21)		(0.30)	
ETA	72.36***	−70.10***							−75.93***	−75.23***
	(13.54)	(13.00)							(15.60)	(15.29)
ROA			−99.40***	−99.675***						
			(20.30)	(19.64)						
ROE			−40.32***	−40.68***						
			(8.30)	(8.41)						
ETL							−34.79***	-36.14***		
							(6.47)	(6.50)		
LTA	4.194*	5.97***							6.55**	7.61***
	(−1.28)	(1.90)							(2.64)	(2.25)
ITA			−2.52**	−2.57**	−2.44**	−2.48**	−1.18**	−1.21*	−2.31**	−2.44**
			(1.15)	(1.51)	(1.16)	(1.11)	(0.69)	(0.67)	(1.02)	(0.99)
-2LL	67.28	69.25	49.56	52.92	49.74	52.17	74.78	76.36	60.01	60.53
AUC	0.974	0.972	0.982	0.980	0.980	0.977	0.956	0.954	0.977	0.975

The table 4 reports the results of regressions with each accounting ratio. We omit the constant term for limited space. Standard errors are in parentheses. (*** significant at 1% level; **significant at 5% level; * significant at 10% level).

The table 5 reports the results of regressions with combinations of 2 or 3 accounting ratios. We omit the constant term for limited space. Standard errors are in parentheses. (*** significant at 1% level; **significant at 5% level; * significant at 10% level).

Following the study of Shi et al. (2012) and Tang et al. (2013), we use the data of the T-2 year for empirical analysis. We first combine the 9 ratios with DD into the logistic regression with attempting alternative forward and backward entrance, and find that the best model is composed of only three accounting ratios. Thus, we perform different combinations with not more than 3 indicators to test our issues, and aggregate all models whose AUC is about 0.95 and relevant variables are significant in table 4 and 5.

We find that when this nine indicators separately enter into the model with DD, although all ratios are significant at the level of 1% and coefficients confirm to our expectation, most accounting ratios have no substantial improvement for DD except the three profitability measures and ETL. Especially that, in the subsequent combinations, we find CA and WCTA has little predictive power, while the combination of GR with any variable can not make the AUC achieve 0.9. As a consequence, we remove these three variables.

However, all the profitability measures have the largest absolute value of the parameter, especially the absolute value of ROA's coefficient is up to about 90, and are significant at the 1% level. Comparing with model 1,3,5,7, the combination of any profitability measure with DD (model 2,4,6) can perform better than their alone. Visibility, profitability indicators play a significant role in upgrading the effect of DD for predicting financial distress. Therefore, when measuring corporate financial distress by DD, including some financial information provided by corporate profitability is very necessary.

Note that, DD is no longer insignificant when ETL included (model 8 and 16), but it is powerful (model 9), especially combining with ITA (model 17). Furthermore, LTA has little effect when entering into the model separately and make a DD insignificant when combined with other ratios (model 10 and 18). While removing DD, model 11 and 19 has not changed much. This should attribute to DD itself already contains a leverage measure.

In addition, we find that ITA is another important predictor of forecasting financial distress. Model 12 though 19 show that any combination of ITA and the profitability measures can perform quite well. The optimal prediction model (model 12) in this study is composed of DD, ROA and ITA. With their integrative action, the AUC can up to 0.982 and the log likelihood value reduce the least. However, there is almost no difference between model 12 and 13, which means DD may be instability in the significance. The comparison among all models also shows that, in fact, DD does not an indispensable predictor for predicting corporate financial distress. As shown in model 13, 15, and 19, the proper portfolio of accounting variables without considering DD can also produce a very good performance.

5 CONCLUSIONS

This paper examines the validity of DD based on the Merton model for forecasting, corporate financial distress using the financial and stock exchange data of Chinese listed companies in later years. We find that DD does have some ability to identity distressed and normal firms, which supports the test of applicability of KMV model by our scholars. However, this ability is limited. When using DD as an indicator for predicting financial distress, it is necessary to consider some information provided by financial statements. Our empirical results show that firm's profitability (EBIT / total assets, ROA, ROE), sales activity (total income / total assets), capacity of cash to repay debt (EBITDA / total liabilities), and capital structure (total liabilities / total assets) are important for predicting corporate financial distress. Especially, the profitability can significantly improve the validity of DD. The best predictive model is composed by DD, ROA and sales activity. This may mean the mix of accounting and market information can make optimal predicting effective. However, DD is not irreplaceable. Appropriate combination of accounting variables can also bring about pretty good prediction results, DD just makes it even better. We can see, DD based on the Merton model is inferior to accounting indicators in some way. In other words, the elegant and concise model of Merton has no superiority in our Chinese stock market. There is some particularity for Chinese companies vs. overseas enterprises, we should treat them separately.

REFERENCES

Avramov, D. et al. 2013. Anomalies and financial distress. *Journal of Financial Economics* 108(1): 139–159.

Bauer, J. & Agarwal, V. 2014. Are Hazard Models Superior to Traditional Bankruptcy Prediction Approaches? A Comprehensive test. *Journal of Banking & Finance* 40: 432–442.

Beaver,W.H. et.al. 2011. Financial statement analysis and the prediction of financial distress. *Now Publishers Inc.*

Beaver,W.H. et.al. 2005. Have financial statements become less informative? Evidence from the ability of financial ratios to predict bankruptcy. *Review of Accounting Studies* 10(1): 93–122.

Beaver,W.H. 1966. Financial ratios as predictors of failure. *Journal of accounting research*, 71–111.

Bharath, S.T. & Shumway, T. 2008. Forecasting Default With the Merton Distance to Default Model. *Review of Financial Studies* 21(3): 1339–1369.

Campbell, J.Y. et al. 2008. In Search of Distress Risk. *The Journal of Finance* 63(6): 2899–2939.

Castagnolo, F & Ferro, G. 2014. Models for predicting default: towards efficient forecasts. *The Journal of Risk Finance* 15(1): 52–70.

Chen, Y.L. et al.2004. Which Method is More Powerful in Predicting Financial Distress in Taiwan? Credit Scoring vs. Option Pring. *Journal of Risk Management* 6(2):155–179.

Crosbie, P & Bohn, J. 2003. Modeling Default Risk. *White Paper*, Moody's KMV, Revised December 18.

Engelmann, B. et al. 2003. Testing rating accuracy. *Risk* 16(1): 82–86.

Hillegeist, S.A. et al. 2004. Assessing the probability of bankruptcy. *Review of Accounting Studies* 9(1): 5–34.

Kong, D.L. & Li, X.F. 2014. Default Prediction: Market-Based Versus Accounting-Based Models. *Investment Research* 31(9):127–139.

Liu, G.G. et al. 2005. A Study on Logistic Model Taking into Accout Distance to Default. *Journal of Finance and Economics* 31(11):59–117.

Pan, B. & Ling, F. 2012. The Application of Financial Crisis Alert System for Listed Companies by Introducing Breach Distance. *Systems Engineering* 30(3):45–51.

Reisz, A.S. & Perlich, C. 2007. A Market-based framework for bankruptcy prediction. *Journal of Financial Stability* 3(2):85–131.

Saunders, A & Allen, L. 1998. Credit Risk Measurement: New Approaches to Value at Risk and Other Paradigms, *John Wiley and Sons*. New York (March, 2002).

Shi, X.J & Ren, R.E. 2005. Empirical Tests of Consistency Between Market-based and Accounting-based Credit Models: Evidences from China. *System Engineering Theory and Practice* (10):11–20.

Shumway, T. 2001. Forecasting bankruptcy more accurately: A simple hazard model*. *The Journal of Business* 74(1): 101–124.

Sobehart, J. & Keenan, S. 2001. Measuring default accurately. *Risk* 14(3): 31–33.

Stein, R.M. 2005. The Relationship Between Default Prediction and Lending Profits: Integrating ROC Analysis and Loan Pricing. *Journal of Banking & Finance* 29(5):1213–1236.

Stein, R.M. 2007. Benchmarking Default Prediction models: Pitfalls and Remedies in Model Validation. *Journal of Risk Model Validation* 1(1):77–113.

Su, M.X. & Lin, X.W. 2006. A Research for Limitation in Use Merton Model. *Review of Financial Risk Management* 2(3):65–87.

Tan, J.J. Forecasting Model of Financial Distress for Listed Companies Based on Financial Index and Distance-to-Default. *Systems Engineering* 23(9):111–117.

Tang, Sh.X. et al. 2013. Empirical Test of Credit Default Risk of Listed Companies in China. *Finance* 3:31–49.

Tinoco, M.H. & Wilson, N. 2013. Financial Distress and Bankruptcy Prediction among Listed Companies Using Accounting,Market and Macroeconomic Variables. *International Review of Financial Analysis* 30:394–419.

Vassalou, M. & Xing, Y.H. 2004. Default Risk in Equity Returns. *The Journal of Finance* 59(2):831–868.

Zeng, Sh. H. & Xu, Ch. The Risk Management of Companies Credit Risk in Emerging Industries based on the KMV. *Scientific Management Research* 32(1):63–66.

Information, Computer and Application Engineering – Liu, Sung & Yao (Eds)
© 2015 Taylor & Francis Group, London, ISBN 978-1-138-02717-6

Simple analysis of reactive power compensation technology in electrical automation

Kai Xu
Liaoning Jianzhu Vocational University, Liaoyang City, Liaoning Province, China

Zhe Wang
Liaoyang Electric Power Supply Company, Liaoyang City, Liaoning Province, China

ABSTRACT: With the rapid development of electrical automation in China, unidirectional power load in electrical automation equipment has the changeable nature, so the nonlinear factors will also increase, which makes people in electrical automation prospects must make full use of reactive power compensation technology. This article mainly discusses and analyses the influence of reactive power compensation technology to electrical automation.

KEYWORDS: Electric automation; Reactive power compensation technology; Application.

1 INTRODUCTION

In the near future in the development of society, China has gradually increased the amount of storage grid, the corresponding reactive power requirements of the standard are also gradually increased. Reactive power and active power functions are the same, is the part which cannot be an integral part to protect the power system power quality, voltage quality, reduce network losses and the security and stability. In the power system to make the reactive power balance, if not doing so would make the system voltage drop, to a certain extent will cause equipment damage, system islanding. In addition to this, if power factor and voltage network are reduced, then the electrical equipment will not be fully utilized, resulting in decline of capability of network transmission and loss increase. Therefore, solve the network compensation is a key problem to reduce the energy loss process in the network.

2 APPLICATION OF THE SIGNIFICANCE OF REACTIVE POWER COMPENSATION IN ELECTRICAL AUTOMATION

Electrical automation applications develop quickly follow the industry's pace forward rapid development, the domestic power transmission network in increasingly complex, the smart grid in this case is also arises in the historic moment. The maximum risk exists in the power system is harmonics, its presence on the safety and stability of the power system is a threat. This is because the changeable load, the linear change of nature that have no rules can be judged. Harmonic maximum damage on the power supply system performance is line losses, especially in the country's power transmission network, this situation is very obvious damage. In addition, damage to the supply and distribution system is also very significant: voltage and current cannot be stabilized, will destroy equipment, severe cases even connected to the whole system crashes, and cause huge economic losses. From these advantages and disadvantages can be seen, reactive power compensation technology is extremely important for growth stage of electrical automation, to decide its running smoothly or not, this means the necessary to intensify efforts to research the project, put forward a reasonable solution is imminent.

3 CHARACTERISTICS OF REACTIVE POWER COMPENSATION

Reactive power compensation is to installed in the local reactive power in the user substation or electrical equipment, to change the power system reactive power flow, this can improve the voltage of power system, reduce network losses and the cost of distribution lines and save electrical energy. The aim is to protect the safety and stability of the power grid operation and economy. With the same name is a reactive power support service or reactive voltage control service, the power injection or sucked out between generator and grid electricity, is a standard of system normal operation node voltage fluctuation, within the limits of can, and can be in electric power system fault later can still have enough reactive power support services to support the system voltage damage. Power system various electric equipment to absorb reactive power is associated with the voltage. In the vicinity of the rated voltage range, the reactive power and voltage is positive proportional changes. The system with or without power, and in the situation of the reactive power supply can't provide enough reactive power, the load voltage will cause the decency link everywhere reduced, absorb reactive power from the system, this will make the reactive power balance. So the reactive power compensation is the only method.

The increase of reactive power and voltage is adjusted method belongs to the auxiliary service class, is the security of the power system, is one of the stable operation of the elements, but also guarantee the necessary conditions for energy trading smoothly implemented. This can prove that it is effective measures to improve power quality, reduce grid power loss, and the method to improve the economic benefits of the quality of power supply and power grid operation. The characteristics of the reactive power service are: ① Analysis of complexity, the cost of the different cost sources of active and reactive power, one is because operating costs less but more investment costs, it is not easy to analysis, so it is difficult to point out the cost of the source. Reactive power production without fuel combustion link, and unit is working together and together, so put on reactive power costs in the cost of using the general calculation summed up will be very time-consuming and laborious. In the grid, all relevant parts of the work bears the entire operation of the grid voltage, power plants and users will need to use in their own can withstand limited voltage and power factor, and reactive power adjustment feature is regional and diverse resistance, resulting in the presence of reactive advantage pricing problems than active price factors to consider much more. ② the way the diversity of active power from the generator to produce, and reactive power generation is by generator, load, condenser and static reactive power compensator, sometimes with transmission lines. ③ decentralized control, similar to the active balance control frequency, voltage control will also have reactive power balance. The frequency of the whole network is the same, according to the active power balance of the whole network, all nodes in the same voltage is not possible, thanks to a control voltage corresponding to this node. ④ powered geographical features, during the long-distance transmission, transport conditions are necessary to make both electricity generation and acceptance among a large voltage difference, and the accompanying loss of active power, so it is not economical alternative view, therefore, reactive power is not suitable for long-distance transmission, only suitable for local geographical balance characteristics.

4 THE POWER FACTOR OF POWER LOAD

4.1 Reactive power

Based on the electromagnetic induction principle to design the application of grid equipment. Electric transducer changes the voltage in the magnetic field to transfer out of energy, the motor can turn the leading machine load. The magnetic energy of the magnetic field is brought by the power supply, motor and electric transducer can make the energy produces alternating magnetic field in the transformation, in this cycle of absorbing and releasing power is equal, this is the inductive reactive power.

4.2 Natural power factor of load

The power factor of the power supply device itself has its own power factor, is without the use of any compensation case, is the natural power factor. In the absence of the use of reactive power compensation settings. The ratio of active power equip it with the existing power is the natural power factor, the comprehensive load natural power factor, which is not compensated is approximately 1.6–0.9.

Characteristics of electric load are a comprehensive electrical load, which include various types of electric equipment, equipment characteristics and their load rate is a factor to determine the level of natural power, as shown in the following table:

Table 1. Power factor distribution electric transducer power factor.

loading power	0	25%	50%	75%	100%
cosΦ	<0.15	0.67	0.73	0.75	0.76

Table 2. The power factor of the induction motor under various loads.

loading power	0	25%	50%	75%	100%
cosΦ	0.2	0.5–0.55	0.7–0.75	0.8–0.85	0.85–0.9

4.3 Find a way to enhance the power factor

On the reactive power or reactive power compensation equipment with electrical capacity, which is the conventional way to improve the power factor, after doing so, the above-mentioned equipment needed reactive power supply can be obtained from and connected electrical capacity.

5 OPTIMIZATION OF REACTIVE POWER COMPENSATION CAPACITY AND THE INSTALLATION LOCATION

Reactive power compensation capacity is from the network loss and annual operation cost of both the smallest as an optimization method, this is the method to to calculate the optimum compensation capacity and makeup location under the condition of the best compensation capacity and considering the load distribution. These common algorithms have consistency is: Calculation of the value of the expression and function extreme value, then compensation capacity and location of the expression are obtained. In the field of mathematical point of view, this is an old algorithm, so it can become a classic optimization method. Until today, this method is still being used to determine the network capacity and location of the compensation practice.

5.1 Computer monitoring of distribution network compensation vessel scheme

There are two basic methods to control the compensation of a capacitor, a time control mode, a current mode, and both have certain advantages. In electric power system, SCADA function mainly provides the control method of computer monitoring and control of compensation capacitors. The collected data, the actual export current time from SCADA system can be back to the computer, and then let electrical switch control. The use of this method can be more accurate and scientific, and adjust the way is very flexible and reliable, and can observe the operating status in monitoring the capacitor.

5.2 System hardware configuration

The central control system is a computer equipped with a wireless communications link monitoring system and SCADA host can monitor the switching device distribution online compensation point. Compensation capacitor switching is a digital radio receiver that each compensation point will have, and link the self-locking type electric appliance, the equipment also is equipped with a counter to record the number of operations. SCADA system of computer audio oscillators and transmitters can help monitor capacitor issue instructions.

5.3 New requirements for reactive power compensation

The power to carry out a number of assessment and incentive has been charging the old reactive approach, if not reach the required level of power factor value of the standard would be financial penalties. This way does not only give an appropriate compensation method, but also does not provide an economic method, which can make the reactive power price of reactive power demand with the beat, and then change. It does not conducive to the mobilization the mobility of all aspects of the system voltage adjustment. After the electric power industry regulation was loosened, the voltage and reactive power support have become members of ancillary service. Because of low running costs, so under normal circumstances, the price is including reactive capacity and electricity prices. The establishment of capacity price is for the purpose of recovering the investment of reactive power compensation to declare and Reactive power equipment value, electricity price is to compensate for reactive power devices total operating costs as the standard. In the electricity market in foreign countries, difference on reactive power auxiliary service boundary, compensation, incentive and compensation for target cost is huge. In the power market USA, independent operators usually put the reactive power support generated by the generator active power loss and cause the cost attributed to compensate reactive power within the boundaries. In Australia, the independent system operators put the generator and capacitor placed within the boundaries of reactive power compensation, so the generator loss can compensate the cost, on the capacitor is only authorized to pay and charge fee. Power Grid Corporation is the only reactive consumers, it is necessary to pay for reactive power plant costs, in order to get the full compensation due from the reactive capacity of both electricity and reactive paid in.

5.4 System software configuration

Periodic data adjusting SCADA system software and the design of SCADA system transfer into a temporary storage area is computer control system the main function of the capacitor. Set sequence database is substation, feeder, distribution network capacitor. Each group of database compensation capacitor will install the two seasons of winter and summer and the load current events relative capacitors of relational tables, and will reserve each compensation point maximum, minimum load current at the same time. When the SCADA system has the data input, the capacitor monitoring system will send data and the compensation capacitor switching contact phase contrast to judge if the capacitor is needed. To avoid the SCADA system appears faulty to influence the normal operation of the capacitor, can control method as the protection measures of alternative choices.

5.5 The distribution of dispersion compensation capacitor

In order to let the main line transfer a small amount of reactive load as far as possible, we should reduce the drainage density of the main trunk. And it can reduce the maximum loss. It can also prevent the voltage is too high, so that capacitor high fever. The conclusion is that in the first half of a distribution network, backbone is not suitable for installation of capacitor. Low voltage will affect the efficiency of capacitor, In order to reduce this phenomenon; the last paragraph of the distribution circuit should not install capacitors. The most appropriate method is based on the actual line voltage testing conditions scattered installed at each suitable branch point. The installation points are no more than the minimum voltage standard voltage of 1.05 times. Under normal circumstances in the branch line is a third - 2/3 the length of the middle part of the best.

6 CONCLUSION

Following the growth of technology during the power electronics, in a large hotel, the hotel has many installed in the air conditioning system, water supply, gas supply system, etc. And in this kind of equipment, motor usually may be used. The model of the motor must be large capacity and large quantity, and a fluorescent lamp also can be used for lighting. In these parts, natural power factor is relatively low, and now the part of electricity load is not lower than the electricity load in the large and medium-sized enterprises. In future power period, capacity will gradually get the promotion. Using active filter for harmonic suppression, the technology of moderate AC transmission system can compensate the reactive power. These will be the direction of development in the electrical automation system. So, Great efforts to the publicity and promote the application of reactive power compensation on the spot is the power to promote all sectors of economic growth, and it also has great significance for reducing the pressure of power supply.

REFERENCES

[1] Chen xuelin .220KV Single-phase Traction Systems And Analysis Of The Negative Sequence Current Calculation Method [J] .Journal of Jiangxi Vocational & Technical College. 2006(20).
[2] YAO Jin-xiong, ZHANG Tao, LIN Rong, LUO Di. Impacts of Negative Sequence Current and Harmonics in Traction Power Supply System for Electrified Railway on Power System and Compensation Measures [J]. Power System Technology. 2008(9).
[3] Zhang weilin, Song xiuchen. Discussion On The Methods Of Carrying Reliability Testing Of Electrical Automation Control [J].Management & Technology. 2009(7).
[4] jiang qiusheng, Chen huifeng, Luyan. Application of the Concentrated Static Var Compensation Technology of Dynamic Stateto.[J].Colliery Mechanical & Electrical Technology. 2009(6).
[5] Wangchao. Electric Automation Technical Analysis Of Reactive Power Compensation.[J]. Guangxi Journal Of Light Industry. 2008(5).

Information, Computer and Application Engineering – Liu, Sung & Yao (Eds)
© *2015 Taylor & Francis Group, London, ISBN 978-1-138-02717-6*

Research on factors related to the use of ICT in Chinese higher education

P. Xu
Department of Sociology, Wuhan University, Wuhan, China

ABSTRACT: Based on Changing Academic Profession (CAP) survey data, this paper empirically analyzes how demographic, individual cognition and teaching environment factors influence the use of ICT in Chinese higher education. Results show that the use of ICT for teaching will be more likely to be found among those teachers who are female and older, prefer to spend more time on teaching, have positive evaluation on technology for teaching and emphasize the teaching content diversity. Findings also indicate that some Chinese teachers may feel worried that using ICT can detract the students' attention and reduce the teaching quality. In order to let more teachers realize the advantages of using ICT, Chinese higher education institutes can provide various demonstration lessons that employ ICT for teaching in different academic departments.

KEYWORDS: Information and communication technology; Higher education; China; Influential factors.

1 INTRODUCTION

With the astounding development of Information and Communication Technology (ICT), the human society has entered an unprecedented information era. ICT not only significantly influenced our style of life, but also had a major impact on the traditional teaching methods in higher education. As Polanyi (1944) clearly showed that transitioning countries faced the "great transformation" in the mid of last century, I tend to point out that the higher education is also facing the "great transformation" in contemporary information era. Undoubtedly, an increasing number of teachers become accustomed to the more efficient ICT-based teaching, which seems to gradually take the place of the traditional chalk talk model of instruction.

Empirical studies indicated that compared with developed countries, developing countries experienced more challenges in terms of the spreading and application of ICT for teaching (Bhuasiri et al. 2012). Grönlund & Islam (2010) reported that many developing countries were actively promoting a variety of ICT programs (e.g. virtual interactive classroom in Bangladesh). However, insufficient infrastructure construction, inadequate institutional support and traditional culture dependency restricted the application of ICT for education in developing nations (Raab et al. 2001, Brinkerhoff 2006, Shraim & Khlaif 2010). China is the largest developing country in the world, but limited work has been done to explore the status quo of the use of ICT in Chinese tertiary education. This research will help to fill the gap in the literature. In particular, I will focus on a number of factors that may affect the use of ICT for teaching in Chinese higher education institutes by using the Changing Academic Profession (CAP) survey data. I categorize these factors into three sets for expositional convenience.

1.1 Demographic factors

Previous studies showed that demographic factors affected the adoption of ICT. For instance, Norris (2001) found that there was "digital divide" between the younger digital natives and the older digital immigrants, and younger people seemed to have a natural affinity for information technology. Coffin & MacIntyre (1999) argued that compared with women, men were more proactive to use ICT and were not likely to be anxious when faced with technical problems. In this study, I will analyze five demographic factors, namely age, gender, marital status, income and education level.

1.2 Individual cognition factors

Bhuasiri et al. (2012) emphasized the importance of teachers' traits of character for the success of ICT-based tertiary education. More specifically, the traits of character could be represented by the following aspects: how much effort they are willing to spend for teaching, how they understand the teaching activity and how they evaluate the ICT for teaching. I treat the above three aspects as individual cognition factors, and will describe the operationalization of these factors in next section.

1.3 Teaching environment factors

Relevant research demonstrated that the heterogeneous conditions of teaching environment imposed influence upon the use of ICT for teaching (Watson et al. 2011).

In China, there are two types of higher education institutes: universities and colleges. Generally, universities have better working environment than colleges. In addition, teaching environment also varies in different academic departments. In this paper, I will analyze the impacts of institute type and academic department factors upon the use of ICT for teaching.

2 METHODS

This empirical study was based on the data from the 2007 wave of Changing Academic Profession (CAP) survey in China. This survey was part of an international cooperative survey in 21 countries across the globe. The questionnaire covered six aspects of information: career and professional situation, general work situation and activities, teaching, research, management, personal background and professional preparation. There were 3612 Chinese respondents from 68 different academic institutes located in 11 provinces or municipalities of mainland China (i.e. Beijing, Shanghai, Hubei, Guangdong, Shanxi, Shandong, Liaoning, Heilongjiang, Hebei, Jiangsu, Sichuan). All of the respondents were teachers in higher education institutes (Shen & Xiong 2014). After deleting the samples containing missing value, 2265 samples were selected for the following statistical analysis.

In this study, the dependent variable is the use of ICT for teaching. In CAP questionnaire, there is a relevant question: "during the current (or previous) academic year, have you been involved in the following teaching activity: ICT-based learning/computer-assisted learning?" The answers are Yes or No. This dichotomized variable (recoded as Yes = 1 No = 0) is used as the dependent variable for the analysis.

As noted above, the factors used as independent variables were divided into three sets for expositional convenience. The first set consists of demographic factors, i.e. (1) Interval level variables: age, age2 × 100, annual income (logarithmic); (2) Dummy variables: gender (male = 1 and female = 0), marital status (married = 1 and not married = 0), education level (have doctoral degree=1 and not have doctoral degree = 0).

The second set consists of individual cognition factors that represent teachers' traits of character. To indicate how much effort teachers are willing to spend for teaching, I used the following question in the questionnaire: hours per week spent on teaching (preparation of instructional materials and lesson plans, classroom instruction, advising students, reading and evaluating student work). Besides, I also set the variable "teaching hours2 × 100" to explore the non-linear relationship between teaching time and the use of ICT. Additionally, to measure how teachers understand the teaching activity, I used a set of Likert scale in the questionnaire (Scale of answer 1 = strongly

disagree to answer 5 = strongly agree). I employed factor analysis method to extract two common factors: F1 (Emphasis on the teaching content diversity) and F2 (Emphasis on teaching quality). Table 1 presents the factor analysis results. Furthermore, to know about how teachers evaluate the ICT for teaching, I used this question: "At this institution, how would you evaluate the following facilities you need to support your work: Technology for teaching; Telecommunications (Internet, networks, and telephones)?" Scale of answer ranges from Poor = 1 to Excellent = 5. I viewed the above two kinds of evaluation as two independent variables.

The third set of factors considered the non-uniform conditions of teaching environment in different institutes and academic departments. I constructed two independent variables to indicate the influence of teaching environment on the use of ICT: Institute type (college = 1 and university = 0); Academic department (0–1 dummy variables: science and engineering departments, teacher training and education departments, business and economics departments; Reference group: humanities and social sciences departments).

Table 1. Factor analysis on teachers' perception of teaching.

Description	F1	F2
You emphasize international perspectives or content	0.783	0.105
Practical knowledge and skills are emphasized	0.735	−0.056
You incorporate discussions of values and ethics	0.686	0.160
You inform students of the implications of cheating or plagiarism in your courses	0.622	0.185
Your research activities reinforce your teaching	0.606	0.074
Grades in your courses strictly reflect levels of student achievement	0.047	0.823
At your institution there are adequate training courses for enhancing teaching quality	0.130	0.647
Eigenvalues	2.535	1.037
KMO	0.775	
Bartlett's Test of Sphericity	Sig. = 0.000	

* Extraction method: Principal component analysis.
** Rotation method: Varimax with Kaiser normalization.

Descriptive analysis was used to demonstrate the sample characteristics, including mean and standard deviation. Binary logistic regression model was then employed for exploring factors related to the use of ICT in Chinese higher education. If y is dependent variable, then y = 1 means that teachers use the ICT for teaching, and y = 0 means that teachers do not use the ICT for teaching. "$x_1, x_2...x_n$" are independent variables (n = 15). If pi means the likelihood of teachers using ICT for teaching, then 1- pi means the likelihood of teachers not using ICT for teaching:

$$p_i = F(y) = F(\beta_0 + \sum_{j=1}^{n} \beta_j x_{ij})$$
$$= 1 / \{1 + exp[(-\beta_0 + \sum_{j=1}^{n} \beta_j x_{ij})]\} \quad (1)$$

Then use the logarithmic transformation for $p_i/(1 - p_i)$. Binary logistic regression model takes the following form:

$$Ln\left(\frac{p_i}{1 - p_i}\right) = \beta_0 + \sum_{j=1}^{n} \beta_j x_{ij} \quad (2)$$

where β_0 is constant term; β_j is regression coefficient, indicating the direction and magnitude of the independent variables' influence. In the following statistical analysis, I will put aforementioned independent variables into the logistic regression model to find out the factors that affect the use of ICT for teaching in China.

3 RESULTS

As shown in Table 2, the variables' mean and standard deviation were illustrated to present the sample's characteristics.

Table 2. Descriptive analysis of the sample characteristics.

Variables	Mean	Standard Deviation
Use of ICT for teaching (yes = 1, no = 0)	0.3329	0.47135
Gender(male = 1, female = 0)	0.6512	0.47669
Age	39.0358	8.26654
Marital status (married = 1, not married = 0)	0.9046	0.29378
Doctoral degree (have = 1, not have = 0)	0.3373	0.47289
Annual income (RMB)	45727.81	25007.388
Business and economics departments	0.1249	0.33073
Teacher training and education departments	0.0737	0.26139
Humanities and social sciences departments	0.2181	0.41305
Science and engineering departments	0.5832	0.49313
Institute type (college = 1, university = 0)	0.1408	0.34793
Teaching time per week	19.90	13.112
Evaluation on technology for teaching	3.5265	0.94059
Evaluation on telecom	3.1369	1.17471

Table 3 described the results of logistic regression analysis in order to explore factors associated with the use of ICT for teaching in Chinese higher education. As seen in Table 3, demographic variables have some significant correlation with the use of ICT. Females were more likely to use ICT for teaching than male counterparts. The odds of using ICT increased with age, which indicated that older teachers were more likely to use ICT. Teachers with doctoral degree had 18.9% lower odds to use ICT for teaching than those without doctoral degree. However, the income factor had no significant effect on the use of ICT. In addition, individual cognition factors were also related to the use of ICT for teaching. More specifically, those who were more willing to spend time on teaching had higher probability to use ICT. The odds of using ICT climbed by 16.1% for each level increase in teachers' evaluation on technology for teaching. Teachers who emphasized teaching content diversity were more likely to use ICT. By contrast, those who emphasized on teaching quality were less likely to use ICT. Furthermore, college teachers had 29.1% higher probability to use

Table 3. Logistic regression results.

Variables	B	Wald	Exp(B)
Gender	−0.242*	5.134	0.785
Age	.097+	3.787	1.102
Age² × 100	0.000*	4.154	1.000
Doctoral degree	−0.209+	3.269	0.811
Annual income (logarithmic)	0.108	0.907	1.114
Teaching time	0.042***	14.558	1.043
Teaching time² × 100	0.000**	7.035	1.000
Evaluation on technology for teaching	0.149**	6.727	1.161
Evaluation on telecom	0.063	1.960	1.066
Emphasis on content diversity	0.264***	26.582	1.302
Emphasis on teaching quality	−0.084+	2.891	0.920
Colleges	0.255+	3.056	1.291
Teacher training and education departments	0.417*	4.014	1.517
Science and engineering departments	0.255*	3.876	1.290
Business and economics departments	0.249	2.018	1.282

* Dependent variable: Use of ICT for teaching (yes = 1, no = 0)
** + p<0.1; * p<0.05; ** p<0.01; *** p<0.001

647

ICT than university teachers. Teachers who worked in training and education departments or science and engineering departments were more likely to use ICT than those who worked in humanities and social sciences departments.

4 CONCLUSIONS

This paper used Changing Academic Profession (CAP) survey data to analyze how demographic, individual cognition and teaching environment factors influence the use of ICT in Chinese higher education. Several findings have emerged from statistical analysis. Female and older tertiary education teachers have higher probability of using ICT, while those who have doctoral degree are less likely to use ICT for teaching. In addition, choosing to use ICT will be more likely to be found among those teachers who are willing to spend more time on teaching, have positive evaluation on technology for teaching and emphasize the teaching content diversity. One unexpected finding is that if teachers put more emphasis on teaching quality, the odds of using ICT would be lower. One possible explanation is that some teachers may feel worried that using ICT can detract the students' attention and reduce the teaching quality. Moreover, results also indicate that Chinese teachers who work in universities (rather than colleges) and humanities and social sciences departments have lower likelihood to use ICT for teaching.

Above findings provide some practical implications which may help to enhance the use of ICT in Chinese higher education. Firstly, male and young teachers need to be encouraged to use ICT in their classrooms. Secondly, stimulating teachers' teaching enthusiasm (e.g. make them spend more time on teaching and pay more attention on content diversity) will help to promote the use of ICT. Lastly, to let more teachers realize the advantages of using ICT, higher education institutes can hold various demonstration lessons that employ ICT for teaching in different academic departments.

ACKNOWLEDGEMENTS

This research was supported by National Social Science Foundation in China (Project No.: 12AZD016 & 13@ZH024) and Graduate School of Wuhan University.

REFERENCES

Bhuasiri, W., Xaymoungkhoun, O., Zo, H., Rho, J.J. & Ciganek, A.P. 2012. Critical success factors for e-learning in developing countries: A comparative analysis between ICT experts and faculty. *Computers & Education* 58(2): 843–855.

Brinkerhoff, J. 2006. Effects of a long-duration, professional development academy on technology skills, computer self-efficacy, and technology integration beliefs and practices. *Journal of Research on Technology in Education* 39(1): 22–43.

Coffin, R.J. & MacIntyre, P.D. 1999. Motivational influences on computer-related affective states. *Computer in Human Behavior* 15(5):549–569.

Grönlund, Å. & Islam, Y. M. 2010. A mobile e-learning environment for developing countries: The Bangladesh virtual interactive classroom. *Information Technology for Development* 16(4): 244–259.

Norris, P. 2001. *Digital divide: Civic engagement, information poverty, and the Internet worldwide*. Cambridge: Cambridge University Press.

Polanyi, K. 1944. *The great transformation: The political and economic origins of our time*. Boston: Beacon Press.

Raab, R. T., Ellis, W.W. & Abdon, B. R. 2001. Multisectoral partnerships in e-learning: A potential force for improved human capital development in the Asia Pacific. *Internet and Higher Education* 4(3–4): 217–229.

Shen, H. & Xiong, J. 2014. Sex segregation in academic profession and the gender difference of faculty income (in Chinese). *Journal of Higher Education* 35(3): 25–33.

Shraim, K. & Khlaif, Z. 2010. An e-learning approach to secondary education in Palestine: Opportunities and challenges. *Information Technology for Development* 16(3): 159–173.

Watson, W.R., Mong, C.J. & Harris, C.A. 2011. A case study of the in-class use of a video game for teaching high school history. *Computers & Education* 56(2): 466–474.

Information, Computer and Application Engineering – Liu, Sung & Yao (Eds)
© 2015 Taylor & Francis Group, London, ISBN 978-1-138-02717-6

Application of GIS in analysis of geological disasters in space: Taking Fugu county as an example

Xiang Zhi Huo & Xin Hu Li

Xi'an University of Science and Technology, College of Geology and Environment, Xi'an, China

ABSTRACT: Geological disasters occurred frequently in recent years, people for the geological disaster prevention and control of channel is becoming more and more widely, the terrain analysis based spatial analysis is more and more attention from people. In spatial analysis, the terrain slope, slope direction and the ups and downs are an indispensable factor. GIS technology to deal with the problem of terrain space analysis has great advantages. Through the introduction of GIS technology in the valley of the Shaanxi provincial new area for the application of spatial analysis, geological hazard assessment, can be more comprehensive understanding of the GIS in spatial analysis of geological hazard assessment.

KEYWORDS: GIS; Geological disaster; Spatial analysis; Slope; Aspect.

1 THE GENERAL SITUATION OF THE STUDY AREA

Fugu county location between $38^0 42'$~$39^0 35'$ N, and $110^0 22'$~$111^0 14'$ E, which located in the northern of Shaanxi province reduces the Yulin city in the administrative divisions, and among the Jin, Shaan and Meng provinces bordering areas, it is also the center of the Shenfu-Dongsheng coal field. The study area overall terrain from northwest to southeast tilt, and it is elevation between 780 meters to 1426.5 meters.

The main development of geological disasters in the study area are landslide, collapse, debris flow and surface collapse, the distribution of geological hazards in the area strictly the restriction of the natural geological conditions and human factors, the geological disasters are relatively concentrated on the space and the distribution regularity of zonal distribution.

2 MODEL AND FACTOR EXTRACTION

2.1 Constructing the grid DEM

Digital Elevation Model is referred to as DEM, It is in the form of a numerical array is a set of orderly ground elevation of a physical model of the ground and it is a branch of the Digital Terrain Model, and other kinds of terrain characteristic value can be derived there. Normally, DTM is described with gradient, slope direction, slope gradient as the main content of various landscape factors on the spatial distribution of the linear and nonlinear combination, and the DEM is zero order pure single digital terrain model, and including slope gradient, slope direction and rate of change of geomorphic features is derived on the basis of the DEM. DEM as the core of the geographical terrain analysis and the most important of the discrete element method which in the extraction of terrain ups and downs, slope and slope to the terrain parameters is also the basis of quantitative analysis of the landform and geological disaster evaluation.

Covering the DEM divisions into rules with same size and shape are the grid DEM grid, with a corresponding number of matrix element to realize the grid point two-dimensional geographical spatial orientation, the third dimension can be elevation is one of the characteristics of attribute values.

Here is based on MapGIS software to generate the grid DEM, on the basis of the work platform using the powerful DTM analysis module of the software, the discrete data grid method to generate the GRD terrain data, mainly uses the inverse distance power function, weighted grid discretization method, and by a large number of experiments to prove that the grid spacing selection around 10 m as the best spacing.

2.2 Extraction of topographic factors

Topographical factors mentioned here mainly refers to the aspect and slope of the study area, which aspect refers to the range 0 to 360^0 GRID face in toward each pixel, Where 0^0 represents North, 180^0 on behalf of the south. GRID is the rate of change of gradient of the pixel value of the calculation result of the elevation in degrees, as a decimal or a percentage of the pixels stored in the attribute, represented by 0 to 90, with different colors for each interval of a certain degree. So far, slope and aspect of the calculation method the average slope of the main promising Solutions of four laws designed, space vector analysis, fitting plane method. Another major demand for the solutions of the maximum slope of the ground surface is fitting method and direct solution. In the analysis of the slope of the ground model, when the slope, you need to compare sparse raw data is encrypted before it can be calculated, is worth mentioning that in order to follow the mapping or analysis needs to be easy to use the results to ". GRD" save or ".BMP" format of the specified file.

3 APPLICATION OF GIS TECHNOLOGY

Geological hazard assessment, also known as geological disaster, assessment refers to the activity level of geological disasters and the destruction of human losses resulting work estimated to be assessed.

3.1 Elevation analysis

Figure 1.   Elevation analysis of the study area.

An elevation analysis chart that is a different elevation range of the terrain filled with different colors, as shown by the case of a different color values undulating terrain reflect, through the ". GRD" format of the data generated by the analysis of discrete data, contours 3.1 understand the topography of the area and thus provide the fundamental basis for the land user.

3.2 Three-dimensional topographic map

The three-dimensional topographic map is a three-dimensional digital map database is based on a certain percentage of the three-dimensional real world or part of one or more aspects of the abstract description. Figure 3.2 with different colors represent different

elevation values can be clearly seen visually by a three-dimensional topographic map of the region's topography different circumstances.

Figure 2.   Three-dimensional topographic map of the study area.

3.3 Aspect analysis

Aspect has an important impact on the ecological and precipitation mountain. It reflects the direction of the slope face, but also through a digital elevation model calculated. As shown in Figure 3, the study area can be divided into eight slope range.

3.4 Slope analysis

Is the slope steepness of the degree of surface unit, usually vertical height and horizontal slope than the slope is called the distance, is an important indicator when geographic characteristic description, which indirectly shows the structure and morphology characteristic undulating terrain and can reflect the degree of tilt topography slopes. The key is to get the slope

Figure 3.3   Aspect analysis of the study area.

of the study area is calculated by DEM. They analyze and earth science related monitoring soil erosion, soil erosion analysis, the establishment of hydrological model to simulate the topography and other fields has an extremely wide range of applications. As shown in Figure 3.4, gray on behalf of rivers, namely the Yellow River, the slope is 0 degrees.

Figure 3.4   Slope analysis of the study area.

4   GEOLOGICAL DISASTERS TYPE ANALYSIS AND COUNTERMEASURES

4.1   Collapse

Study area with hard and soft mudstone inter-bedded sandstone and white as the main rock formations, plus the external impact of human engineering activities, so collapse as the region's most developed type of geological disasters. During the construction of residential land and to pay attention to small and medium preschools do the following, the upper part of the slope collapse block thereof; in those areas may have only a small collapse in the rain, you can get in on the toe or the construction of banpo walls, slope area prone to weathering and flaking, you need to build retaining walls, while the gentle slope of cement reinforced.

4.2   Landslide

Landslides in the region after the collapse, mainly in the western slope of the mountain and north- west slope, size range, formation conditions, consistent mechanism.

Counties during the construction of the slide you want to be effective in preventing the "early detection, prevention, identify the situation in a comprehensive way, strive to cure, leaving troubles" principle binding factor slope instability and the formation of internal and external conditions landslide.

4.3   Mudslides

Debris flow gully region mainly in the north-west. To avoid unnecessary losses caused mudslides in the construction process should pay attention: Do not put houses built in the channel, not the gullies garbage discharge course.

Considering the variety of geological disasters that may occur and in a timely manner after land use control measures in the study area may be as so in the area outside the new loop to the main multi-storey buildings, may be appropriate to build some low-level; in the case of some complex geology district peripheral hillside area, low-rise residential buildings should be based; in the heart of the new district, geology can arise and in good condition mainly high-rise building.

5   CONCLUSION

In this paper, the application of GIS technology in spatial analysis of geological hazard assessment, combined with the geological conditions of the study area was roundly terrain analysis, through a variety of terrain analysis chart provides a terrain analysis of the data, to broaden the application of GIS technology channels, meanwhile the assessment and prevention of geological disasters in the region has played a significant role in promoting.

REFERENCES

[1] New Fugu distribution and risk assessment of geological disasters. Xi'an University of Science and Technology 2009 Geological Engineering Master Thesis.
[2] ZHOU Qi-ming, LIU Xue-jun. Digital Terrain Analysis [M].Beijing: Science Press, 2006:52–75.
[3] TANG Chun, LUO Ji-xiang. GIS and counties Division of Geological Disaster Risk Progress [M]. Hebei Agricultural Sciences, 2009,13(5):144–145.
[4] CHEN Fen. Based on GIS terrain analysisFujian [M]. Donghua University of Technology, 2009.6(32),185–187.
[5] ZHOU Song-lin,TAN Shu-cheng,JIANG Shun-de, etc. Spatial distribution of geological disasters in Yunnan Longyang District, Baoshan City and prone zoning and Evaluation[J].Earth and Environment, 2010,38 (3).

Information, Computer and Application Engineering – Liu, Sung & Yao (Eds)
© 2015 Taylor & Francis Group, London, ISBN 978-1-138-02717-6

Research on the application of "double-qualified teachers" teachers team building

Hai Xiu Zhang, Cheng Dong Wang & Kun Yao Wang
Harbin University, China

ABSTRACT: University "double Teachers" team building, affecting the application-oriented Universities in terms of quality improvement in the sustainable development of application-oriented colleges and universities have an important influence. This paper describes the importance of the university "Double" and "Double Teachers" ranks the problems put forward specific solutions.

KEYWORDS: applied; double-qualified teachers; team building

1 INTRODUCTION

Towards education in our power in the process, "Double-qualified Teachers" capacity building as well as certification in coaching, showing a crucial role. Before people generally think "Double-qualified Teachers" Vocational education mainly in specialized requirements of professional teachers and teaching has nothing to do with regular undergraduate institutions, so in college "double-qualified teachers" build theoretical research is not a lot. With the rapid development of China's higher education, teachers in the teaching level of certification, has important significance.

2 IMPORTANCE OF THE "DOUBLE-QUALIFIED TEACHERS" TEACHERS TEAM BUILDING

2.1 *The only way for university development*

At present, China's higher education, the widespread emphasis on knowledge of teaching, training contempt and lack of knowledge of the problem into the ability of making theoretical teaching and practical teaching out of line phenomenon. Caused a lack of competitiveness as well as personnel training colleges lack of features, and finally lead to the ability to apply knowledge of graduates made the problem worse. However, in order to adapt to the development of China's higher education popularization of education, "Double-qualified Teachers" team building should focus on the actual importance of the highlights. Meet the demand for application-oriented talents, is a source of non-priority groups in other countries and the development of colleges and universities to survive. So, from their actual College, according to school sources in accordance with the main features,

cultivate a high capacity to adapt to the needs of society dedicated application-oriented talents, to achieve local economic and social development in colleges and universities to meet the application-oriented talent cultivation area.

2.2 *The objective need of training applied talents to the community*

When "Double-qualified Teachers" teach the students, they mainly teach "dual" teaching mode from the "unitary" model.

Type conversion, that is, the use of combined classroom and practical teaching methods, students ability in knowledge applications. "Double" quality teachers need to have a good ideological and moral cultivation, but also education and teaching ability. Lard focus on building private colleges have "Double" quality team, we can achieve important applied talents training objectives. Theory and practice teaching achieving seamless student learning. Also note that the students good industry professional quality and a solid level of expertise inspired culture. So that when students have a professional application capability, both in the professional teaching theory, they can engage in the practice of teaching students the guidance skills. Therefore, the construction of "Double-qualified teachers" Teachers team consistent with current applied talents training in teaching mode is mainly inherent requirement.

3 "DOUBLE-QUALIFIED TEACHERS" TEAM BUILDING PROBLEMS

3.1 *Systematic training is not strong*

Currently, part of the university teachers no systematic plan in culture, blindness more common. Targeted training is not strong.

On the one hand, the training content and training needs of teachers do not match. Makes a considerable part of the teacher training college has planned and actual operation, use the "mismatch" phenomenon. Although some schools also planning to develop, but because targeted, scientific, and operability is not strong enough, there has been no implementation of the program, according to the situation, making the actual situation and the teaching does not comply. On the other hand, the school's faculty of management, at the time of the school faculty status quo analysis, yet personalized research and analysis, it did not make a specific training program for different teachers. Leading to some requirements of training opportunities for teachers no training.

3.2 Inadequate investment in culture, the training effect is not obvious

Inadequate investment performance in terms of time and money. Because application-oriented Universities in terms of the cost itself is high, making the financial investment is relatively inadequate, so many schools in the "Double-qualified Teachers" Training in financial strength is not enough. In addition, because of the constraints of some objective factors, such as occurs in the overall number of faculty shortage problem, making some professional teacher shortages, and even seriously inadequate. There is no way to arrange a special time to conduct teacher training. In order to adapt to the personal training, quality assessment system, reflecting significantly solve the "Double-qualified Teachers" weaknesses in teaching, and finally making the quality of personal training is generally not high, we want to carry out application-oriented university "Double-qualified Teachers" training does not attach importance to the issue and urgently.

3.3 System is not perfect, the enthusiasm of training is poor

3.4 The pertinence of training measures is not strong

The main way is to train "Double-qualified Teachers" teachers in-depth practical learning. But the positive is that teachers selected face is not wide enough, and because the training is too long, the training of teachers are not enthusiastic. Delayed the teacher holiday. So it is not very popular with the teachers. Teachers go to business practices, mainly large enterprises, need to get business support. Because production uncertainty in [1]. The formation of an unstable situation training effective.

4 "DOUBLE-QUALIFIED TEACHERS" TEAM BUILDING WAYS

4.1 Expand the channels, the implementation of teachers appointed special combination

There are two ways to "Double-qualified Teachers" main building, on the one hand, teachers can go to the relevant companies and industry sectors employing technology so that practical experience and technical ability of the operator to the school teachers to participate in teaching over. Professional and technical personnel and can appoint a certain degree, as a part-time teacher for the school's teaching brings new technology industry, narrowing the distance between the school and the community, so that the actual teaching of theory and practice, so that the students theoretical knowledge both strengthen and increase the ability to train students in practical operation, ease of student employment. On the other hand, the school must continue to introduce the rich experience of professionals.

4.2 Complete job classification system and incentive system

Teachers only in escalating job classification, will improve the training of teachers to participate in the initiative. Application-oriented colleges and universities should actively job classification system needs to deepen diversification, making the existing theoretical job classification system has been upgraded, the practical ability of the assessment to join the teaching job classification in which the whole system, in practice grasp "Double" teachers inclined direction [2]. Improve teacher incentive mechanism in teaching, while at the appraised awards as well as financial support, etc., to increase the "Double Teachers" tendencies, the establishment of a special allowance for those who served both theoretical teaching while working as a teaching practice incentive for teachers.

4.3 Increase the current training, colleges and enterprises training mode

5 CONCLUSION

"Double-qualified Teachers" team building construction should pay attention to two aspects, one enough to fully understand the importance of "Double-qualified Teachers" culture, the development trend of teachers, more teachers to participate in "Double-qualified Teachers" initiative. On the other hand, should be taken in application-oriented universities and effective measures to correct guidance of teachers to carry out exchanges and demand-building "double" Teachers accelerate speed "Double Teachers" building. The only way to continue to meet the community applied talents, promote prosperity and development of future college and achieve the implementation of the meaning of "Double-qualified Teachers" teaching.

REFERENCES

[1] Kaiqin Li.Exploration and Reflection "Double-qualifeid Teachers" Team Construction [J] Chinese university teaching, 2010, (13).
[2] Daowen Zhang.Local Colleges "Double-qualififed Teachers" Teachers team building exploration [J]. Higher Education Research, 2010.

Information, Computer and Application Engineering – Liu, Sung & Yao (Eds)
© *2015 Taylor & Francis Group, London, ISBN 978-1-138-02717-6*

Application of multimedia technology in the teaching of basketball

Li Li Zhang & Da Zhi Wang
Langfang Teachers University, China
Langfang Vocational and Technical College, China

ABSTRACT: Now in the colleges and universities the specialized theory of the sports education exists this mode "cramming education", with the progress of society and the constant reforming of teaching, students' subjective initiative is more and more strong, the traditional mode of education has seriously hindered the students personality cultivation. The emergence of the multimedia technology in the basketball teaching breaks the traditional mode of teaching, it improves the teaching quality, cultivates the students' interest in learning. Here the author just takes the application of the multimedia technology in the theory course of college physical education major as an example to do some preliminary discussion and research.

KEYWORDS: Basketball teaching; multimedia technology; application

1 INTRODUCTION

Nowadays, the traditional basketball teaching methods can not keep up with modern teaching requirements, the colleges and universities are making full use of multimedia teaching technology, and extending the new teaching idea, have changed the former monotone teaching mode. Therefore, more and more theory course in the colleges and universities have been replaced by the multimedia teaching mode. The students can be able to have a better understanding and mastering of each basketball moves and skills in the use of multimedia technology, the use of animated way to make the teaching content into courseware, increased the effectiveness and vividness of the classroom, so that students' learning can be more intuitive, also have improved the teaching efficiency in all directions, and the new teaching mode can be a very good improvement for the basketball teaching atmosphere, makes the colleges and universities sports theory teaching content can be more rich, active and creative, and greatly improve the students' interest in learning.

1.1 Application of multimedia technology in the teaching of basketball

1 The software selection of basketball teaching courseware making. The students can from the basketball teaching courseware which use the multimedia technology to learn the basketball theory intuitively, the selection of software for courseware making mainly should follow the following principles: (1) efficient principle. The

advantage of using the courseware teaching is that the teaching content can be shown in animated form, and the same action can be repeated plays, so that the students can watch the section which they do not understand and then master their blind spot through the repeatedly watching. The courseware screen and pictures should be very clear, and also with coherent action, also should have very strong learning efficiency, this software had better has very rich effect settings and be easy to use, this in the very great degree reflects the efficient using of teaching courseware; (2) sharing principle. Because of the teaching content changes very fast, so the courseware also must update with the renewal of the teaching content, the development of software and multimedia teaching format must be compatible with each other, such as when the teachers explain some basketball tactics, they will use the network video or any other teaching video, then this courseware making has the great network sharing properties (3) economic principles. Teaching courseware making should base on the science to reduce the cost of software to the max.

Generally the users always use the PowerPoint and Flash software to make the courseware. Using the PowerPoint software to do the courseware design is relatively simple, it is convenient for students to through the limited time to design excellent courseware. The process of using the PowerPoint to make the courseware is divided into many steps, and each step is explained in great detail, including the pictures, video and text, etc. The courseware content is lively and interesting, and orderly movement, very suitable for the teaching which need the step-by-step explaining, and the software also has an auxiliary teaching function for the difficult and doubtful teaching content. The courseware produced by Flash software also has many advantages, one of the most outstanding advantage is a dynamic and interactive courseware. This interactive courseware mainly uses the introduction and controlling of the content, such as videos, vector and audio and so on. The effect of courseware made by Flash software is very obvious, it can control the rhythm of skipping very conveniently, and the volume of this kind courseware is very small, and can be fast transmitted. Both of these two software have the best effect, the PowerPoint and Flash are highly complementary, complementary using of them will be better.

2 The production of multimedia courseware. (1) The content of courseware design. The production of basketball teaching courseware is mainly according to the teaching curriculum, teaching objectives and teaching nature to do the specific production. The teaching contents include a basketball theory summary and basketball basic skills and basketball contest rules and so on.

(2) The functional design. When using the multimedia technology to do the functional design always should base on the teaching requirements to complete the design of the courseware function, according to the requirement to do the niche targeting auxiliary teaching. The functional design of courseware through using multimedia technology mainly includes pictures, video, text and other specific content. Around each point of the teaching knowledge to do some pointed design.

(3) Production of the teaching text. The text in the teaching mainly uses the this following two form-static text and animated text, and this two kind of text form is the classic form to make a teaching text. The static text and animated text producing mode usually use the two software Flash and MS PowerPoint to complete the specific matters plan and typesetting. When making the text, it should according to the different teaching content and the different teaching subject by different making form to do a unified layout.

(4) Image production. At the courseware making price, the picture production and text production have a certain difference, a lot of the pictures pixel fabricated by a scanner, and then use Photoshop software to carry on some specific processing for the picture effect. Also the users can capture some pictures which are suitable for the teaching from the related video and save them, and then through the use of stored to facilitate courseware. In the courseware, it also is necessary to show the specific route of basketball, draw the route, all these routes need to be directly produced by the flash software.

(5) The audio production. Most times the courseware making also need to insert a sound to coordinate with the other content, audio production is mainly using the courseware texts to explain the sound or the explaining voice captured from the related video to explain the courseware texts. Audio production must be clear, synchronization and accuracy and meet other requirements.

1.2   *The problems which should be paid attention to in the multimedia technology basketball teaching*

1 The user should pay attention to the appropriate blank. When the teachers use the multimedia courseware to explain the educational theory in the class, the using of the teaching courseware should be appropriated. The maximum purpose of the courseware is to reach the efficient teaching, only from the perfect using of the multimedia courseware can it gain good teaching effect. In the practical courseware teaching, many teachers just in the pursuit of technology of the multimedia, they make the teaching courseware be very complicated, and the useful information is not enough or have been covered by the riotous colors, so that students cannot obtain the useful information from the courseware, the teaching effect is greatly reduced. So making a simple courseware in the teaching is very important, it is good for the students to catch the useful information in the courseware, and also set aside the practice time for students to in-depth study each action and do some exercise, this can greatly improve the teaching effect.

2 Grasp the teaching focal point. Compared with the traditional teaching mode, the multimedia technology teaching has a lot understanding space, it can make the students understand the content of teaching more easily and deeply. Teachers should grasp the focus points of the teaching, reflected the emphasis and difficulty in everyday teaching through the courseware, and this enables the student to study bolt to the bran and to revise the lessons clearly. Because the courseware producing time is very long, so it greatly shortens the time for the teachers to prepare the lessons, so the courseware must show the emphasis and difficulty of the knowledge, and mention once about the knowledge which can be easily mastered.

3 The interaction between teacher and student. When in the use of multimedia courseware, the teachers often focus on the multimedia controlling, this behavior to some extent just neglected the interaction with students, because of the teachers too

many times operations of the multimedia then reduce the emotional communication with the students, so it is not conducive to improve teaching effect and learning efficiency. Because the lack of communication between teachers and students, makes the teachers can not grasp the learning situation of students, and leads to students did not have the opportunity to question the difficulty problem, and the difficult part just been mentioned once, and finally it is not conducive to improve the learning efficiency. The whole class just appear the "cramming education" phenomenon. So the teachers should do the knowledge exchange with the students in the class when they in the teaching process, to let students to cooperate with teacher actively, in a certain degree, it can enhance the enthusiasm of the students, fully mobilize the classroom atmosphere, it can not only improve the classroom effect and it can also make the students to grasp the difficulties efficiently.

2 CONCLUSION

With the rapid development of China's science and technology. Video, recording are all over the store, more and more occasions are using the multimedia, the using of multimedia has become the topic of teaching reforming people. Instead of the traditional teaching approach, the use of modern technology to achieve the goal of teaching. The multimedia education method makes the students can independently learn the knowledge according to their own interested way. Multimedia has a very powerful database, provide the solution for difficult problems of students, solve problems for the students to stimulate students' interest in learning, inspire students' potential, can better master computer application. Therefore, the multimedia is in service for everyone, for everyone to have a good future and create a good effect.

REFERENCES

[1] Song heping: *The research and application of multimedia technology in the teaching of basketball* [J]. Journal of weinan teachers university2, 75–76(2009).

[2] Chen Xuesong:*the multimedia technology application research in the basketball teaching* [J]. science and technology information, 32 ,192(2013).

[3] Lu Zhiwei: *Discussion about the application of multimedia in the college and university* [J] natural science edition 02, 109–111(2010).

Information, Computer and Application Engineering – Liu, Sung & Yao (Eds)
© 2015 Taylor & Francis Group, London, ISBN 978-1-138-02717-6

The application of mobile internet technology in artwork online transaction

C. Wang
Department of Common Required Courses, Hubei Institute of Fine Arts, Wuhan Hubei, China

L. Li
School of Computer Science and Technology, Wuhan University of Science and Technology, China

ABSTRACT: With the rapid development of network speed and the popularization of smart phone with large screen and high performance, the mobile internet technology has been gradually mature. Mobile e-commerce can effectively save transaction cost and improve transaction efficiency. Therefore, it is a general trend to develop online transaction for artwork using all kinds of mobile internet application technologies. The applications and characteristics of Wechat and mobile applications (Apps) in the online trade of art works were analyzed. The research suggests that the online transaction of artwork can be effectively promoted by combining the advantages of Wechat and mobile Apps.

KEYWORDS: mobile internet; online transaction of artwork; Wechat; mobile Apps;

1 INTRODUCTION

With the improvement of national wealth and people's living level, the investment market for artwork develops continuously. Besides, owing to the high return rate, more people join in the investment market for artwork. There is increasingly intensive demand for investment market of folk art. Presently, as the internet technology develops, the online transaction for artwork has been an important approach for trading artwork. The intervention of internet technology completely changes the traditional transaction mode, breaks the time-space limit of traditional trade, and enables transactions to be carried out anywhere at anytime. Moreover, as the mobile internet technology grows mature gradually, the above advantage of online transaction becomes more significant. The recent 33rd statistical report on the development situation of China internet network published by China Internet Network Information Center (CNNIC) showed that up to the end of December of 2013, the users shopping online using mobile phone network reached 0.144 billion, with a growth rate of 160.2% annually and a utilization rate of 28.9%. Customers have developed an internet consumption habit using mobile internet [1].

2 DEVELOPMENT SITUATION OF ARTWORK E-COMMERCE

The major artwork portal websites, including Arttrade, SSSC, Artxun, Artron, and Artintern, have constructed e-commerce platform. Comprehensive e-commerce enterprises such as Taobao and EBay also have program for artwork. It has become a common trend for selling artwork online based on e-commerce. However, owing to the limited purchasing power of the primary online consumers and the risks such as consumers can not see real goods, logistics and online payment, low prices artwork are mainly traded online [2]. The consumers are generally young people with certain artistic culture, frequently communicate and share information with friends using communication software including Wechat, and shop online using mobile Apps. Therefore, the development of e-commerce of artwork using mobile internet technologies such as Wechat and mobile Apps conforms to the shopping habit of the kind of consumer group. The paper analyzed the application and characteristics of Wechat and mobile Apps in artwork e-commerce.

3 APPLICATION OF WECHAT IN ONLINE TRANSACTION FOR ARTWORK

Wechat, as a subordinate communication tool of Tencent, has been used more and more owing to the high emotion viscosity and perceptual consciousness of QQ user resource. For user group, the built relational network shows incomparable stability than other social networks in China. The decentralization of Wechat assists the interaction and communication of users in the group and improves their viscosity.

Owing to the excellent popularization function of Wechat, art gallery, Auction Company, and art organization begin to promote their goods using the additional tools of Wechat including public account, friend circle, Wechat group and Wechat shop.

3.1 Online transaction for artwork using Wechat

The first problem for artworke-commerce is to find customer resource. On one hand, by sharing relevant articles of art collection, investment, auction, etc. using public account or friend circle of Wechat, users that interested in artwork can be attracted to focus on the account. Meanwhile, similar promotion can be transmitted using other popular instant communication tools and famous art forums. The method shows low requirements for the work experience of practitioners, who can carry out online transaction of artwork with basic art investment knowledge, while it attracts low customers. On the other hand, according to six degrees of separation theory, that is, one is able to know anyone he wants to know through six people [3]; Wechat can extend the contacts in reality. In other words, Wechat establishes relationships among relevant persons in art circle of art media, auction company, art gallery, and enterprise senior by the introduction among friends and conducting delightful activities such as rush for red packet. The method requires the practitioners having years of relevant experience of art investment and certain contacts in art circle, to obtain top customer resource. As top customers learn more about art investment and have adequate money, they present higher close rate in online transaction of artwork.

The online transaction of artwork using Wechat is based on public account, shop, and group of Wechat. Wechat auction in Wechat group is popular presently. It refers that after bidders uploading information and pictures of auction goods in friend circle or directly releasing them in Wechat group, all the users in the group can auction and bid for the goods by means of sending messages in the group.

3.2 Characteristics of Wechat auction

Wechat auction shows characteristics of brief, ordinary, and fast, as it handles businesses without completely using the regular procedures of auction company, but applies oral contracts. The trust is based on the mutual knowledge of Wechat users in the group. Meanwhile, the commission is low, and there is generally no default, debt and non-payment for a long time. Additionally, the other characteristic of Wechat auction is that the members in the group are updated continuously. By getting the inactive users out and inviting potential buyers, the users in the group are updated and the activeness is guaranteed.

There are many people in Wecaht friend circle working in art circle. They are mainly prospect customers that know art or having artwork purchase experience. They are interested in Wechat auction because: first, there is no margin of Wechat auction; second, the communication mode of Wechat group enables users in the group to communicate art based on the auction goods; and third, the goods are cheap and the transaction cost is low. Owing to the above advantages, Wechat auction constructs customer group easily. Besides, by displaying artwork using friend circle of Wechat, the pictures and buying links spread in viruses-like mode, which contributes direct sales.

The great success of Wechat auction is mainly because of users' interaction and word-of-mouth marketing, which spreads consumption culture and builds trust relation by paying lots attention on users' active evaluation, discussion and share of their consumption feelings [4]. Owing to most of the users in a Wechat group know each other, the communication about evaluation and bidding of artwork makes the auction more real and interesting, which can not be realized by taobao trade, live auction, and auction of ordinary e-commerce enterprises.

3.3 Problems in Wechat auction

Although Wechat auction shows many advantages, there are lots problems in its development. First of all, the former messages in Wechat group are likely to be crowded out by the later ones, and therefore some users don't know the information of current auction goods. Secondly, auctioneers have to keep an eye on the progress of the auction all the time, and this artificial operation possibly misses information. Meanwhile, Wechat auction fails to track, accumulate and analyze the trade data, and doesn't supply the entire service of auction trade. Thirdly, though synchronous group can be established or the inactive users can be gotten out, the amount of users in Wechat group is limited and is likely some potential customers are likely to be lost. Fourthly, Wechat auction is easily copied without specificity and can be realized as long as there are seller and buyer. Fifthly, though Wechat auction presents a large proportion of mid-low price original artwork, there is ignorable demand for consumer artwork and art derivative commodity of lower price, as people's material culture develops. The auction mode is obviously not suitable for the trade of this kind of artwork.

The above problems of Wechat auction is because Wechat is not designed for the artwork e-commerce specially. Therefore, to better develop artwork e-commerce and solve these problems fundamentally, more specified online trading platforms for artwork have to be constructed. According to the current e-commerce development trend, it is a favorable approach to develop special mobile Apps.

4 APPLICATION OF MOBILE PHONE APP IN ARTWORK E-COMMERCE

With the improvement of mobile network speed and the performance of smartphone, and the perfection of mobile payment technology, the advantage that mobile e-commerce performs trades anywhere at anytime is more distinct. Therefore, mobile phone shopping is accepted by more and more people in varied age groups, particularly young people, who are the major group of electronic shopping. Based on this, many large e-commerce websites, such as tao-bao, JD, Shanghai YiisaAgel Ecommerce Ltd, etc., popularize their specific mobile Apps. By integrating all the functions of online shopping and optimizing user interface and goods presentation, these mobile Apps enable users to purchase happily and rapidly in mobile internet network. E-commerce websites for artwork, including Zhaoonline, Artron, and SSSC also made attempts from the aspect and developed distinctive website Apps.

4.1 Advantages of mobile Apps

The advantages of mobile Apps in developing online trade of artwork are mainly displayed in the following aspects: first, owing to the mobility and portability of mobile phone, mobile Apps can be used as long as there is mobile network. Therefore, the user amount of mobile network is greatly larger than that of computer, which is also the reason why online shopping using mobile phone develops so fast. Second is the security of mobile phone payment. Because of the uniqueness of phone number, by binding it with payment platforms such as Alipay, Wechat payment, and mobile wallet, the security of online payment is effectively improved. Third, mobile Apps are highly customizable. Based on each e-commerce demand, developers develop mobile e-commerce Apps with corresponding functions. Compared with the secondary development of microblog and Wechat, mobile Apps show stronger functions and more flexible design of user interface. This is also the major reason why mobile Apps are more suitable for developing artwork e-commerce.

As mobile Apps are customizable in functions, they make up the limitations of Wechat in art auction from the following aspects: in auction preview module, apart from using pictures to display auction goods, videos can also be introduced to fully present the goods and enable buyers to learn the

goods three-dimensionally. By adding functions such as praise for goods and leaving message, the interaction and communication among the users are intensified and users' viscosity is improved. In online auction module, the whole auction process is performed by software automatically without auctioneers. All the auction data are recorded in backend database, and by analyzing these data, lots valuable information can be obtained. Additionally, with the customizable e-commerce Apps for artwork, new commerce modes can be developed to meet customer's individual demands for original artwork. For example, by imitating Didi taxi App, different App client ends are developed for customer and artist. Customer releases the requirement information including subject, content, and size of artwork customized; while artist obtains the information through the client end and offers prices. Afterwards, customer makes decision according to the prices and the quality of previous artwork of the artist. In the commerce mode, the goods customized can be ordinary artwork such as paintings, ceramic decorations, personalized sculptures, etc., also can be art service, including paintings on walls of houses, indoor decoration design, etc.. In this way, the commodity range of e-commerce of artwork is extremely broadened.

4.2 Disadvantages of mobile Apps

Although there are many advantages of mobile Apps in developing online transaction of artwork, mobile Apps developed by most portal websites for artwork show low download and usage amount in all large application markets. This is because the inadequate market promotion on one hand. Many art websites don't put in enough manpower and fund to promote their mobile Apps. On the other hand, their function lacks of innovation. The mobile Apps of many art websites are merely copies of portal websites and don't utilize their advantages to develop innovative functions.

5 CONCLUSIONS

Wechat auction accumulates excellent customer resource in art circle and realizes word-of-mouth marketing effect easily by group communication and friend circle functions. The high close rate and low default rate enable sellers to find more and better auction goods; through the virtuous circle, artwork e-commerce develops in favorable environment. However, as Wechat is not specifically developed for online transaction of artwork, there are many technological problems to be solved.

Owing to the mobility and portability, mobile internet has more end-users. The convenient and reliable mobile payment guarantees shopping using mobile phone; mobile Apps are more flexible in function and interface design due to high customizability. Therefore, it is a nice choice for the construction of online transaction platform of artwork based on mobile e-commerce by developing specific mobile Apps for art websites. But the problem of how to promote mobile Apps with limited manpower and fund has to be solved.

In summary, an excellent e-commerce website for artwork has to be built based on favorable customer resource, professional transaction platform, and artwork that complied with market demand. Therefore, Wechat auction and mobile Apps have to be combined to build more prefect online transaction system for artwork by promoting mobile Apps using the favorable customer resource of Wechat auction and improving customer experience using the specific function and service of mobile Apps.

ACKNOWLEDGMENTS

The research work was supported by Wuhan University of Science and Technology Youth Training Plan (2014xz017). The corresponding author is Li Lin.

REFERENCES

[1] China Internet Network Information Center (CNNIC), 33rd statistical report on the development situation of China internet network, January, 2014.
[2] Jiang Zhe and Wu Yanjie, China artwork market in network times, Beauty & Times, December 2010.
[3] Ding Xin, Li Rao, Li Ye, Victory of SNS—product design, operation and openness platform, and social marketing, Beijing: Posts and Telecom Press, 2009, (5).
[4] Jin Liyin, The effects of online WOM information on costumer purchase decision: an experimental study, Economic Management, 2007, (22).

Information, Computer and Application Engineering – Liu, Sung & Yao (Eds)
© 2015 Taylor & Francis Group, London, ISBN 978-1-138-02717-6

Instruments of tax regulation in a modern tax policy

T.Y. Tkacheva, L.V. Afanasjeva, S.N. Belousova & M.L. Moshkevich
South West State University, Kursk, Russia

ABSTRACT: Instruments of tax regulation and a mechanism of their effect on economic processes are considered. Tax benefits are tax regulation tools for certain categories of taxpayers to solve a number of social and economic problems. The proposed technique of performance evaluation of tax benefits allows us to make a complex analysis of the effectiveness of benefits and goals of their initiation to optimize the list of benefits by avoiding inefficient opportunities for tax minimization, as well as to send shortfall to the budget in the form of benefits to support sectors of the national economy.

Article is executed within the state task of Southwest state university, a project code: 2090.

KEYWORDS: tax regulations, tax benefits, shortfalls in budget revenues, the effectiveness of tax benefits, compensation payments.

1 INTRODUCTION

Origination and improvement of interbudgetary relations is closely linked with the development of public finances (emergence and development of the state budget) and the development of administrative-territorial structure of the states (creation of several levels of governance, delegation of responsibilities of the center to the regions and local authorities) (Tkacheva, 2012).

Currently, federal authorities in the Russian Federation regulate the level of territorial budgets' income by establishing standards for deductions from federal taxes, providing subsidies, subventions, grants and other financial assistance, because tax revenues of the territorial budgets of the Russian Federation generated by only fixed taxes cannot cover a large part of expenditure commitments of the budgets.

The research is of current importance because the need to improve financial security of territorial budgets of the Russian Federation depends on monitoring tax benefits and assessing effectiveness, by building up territorial budgets' own tax potential.

Moreover, presence of own stable revenue sources at the local level will provide a better balance in the distribution of inter-budgetary funds, optimize financial flows through the levels of the budget system, as well as create conditions for fiscal sustainability in the medium and long term (Belousova, 2012).

2 TECHNIQUE AND RESULTS

Creating a stable society with a market economy, the state should have an effective tax system that would meet public interests, form favourable economical treatment, ensure national security and development, and that would reflect interests of particular social groups.

The process of tax regulation is a specially-oriented state influence on the behaviour of economic agents through using different techniques and tools of tax policy in order to achieve the desired socio-economic results (Mishina, 2010).

Tax regulation is used by applying special tools and techniques of tax policy. The tools of tax policy include a set of tax legislation norms, which influence the economic activity of a tax payer through his economical interests to achieve a desired economical, social or any other profitable result (Mayburov, 2010).

Tax benefits take a special place among the methods of tax regulation in the system of fiscal relations, as their provision is aimed at solving specific social and distribution tasks (support socially vulnerable groups of the population), the stimulation of certain types of economic activities (investment in fixed assets, agricultural production, etc.) and economic growth, including individual sectors.

Tax benefits at different levels of management lead to falling budget revenues. As a result there is a necessity to assess the effectiveness of tax benefits for extending or cancelling the benefits and there is also a necessity to introduce new ones.

As our analysis shows, the amount of losses from the incentives established by federal law, is equal to half the amount provided the subjects of the Russian Federation with grants from the federal budget for leveling its budgetary security, and these grants exceed the subsidy in twenty regions (in the

Komi Republic, Kaliningrad, Sakhalin and Yaroslavl regions they exceed in 3–6 times (Table 1).

Table 1. Assessment of losses of consolidated budget revenues of the Russian Federation in connection with the provision of tax benefits for 2009–2011.

Activities	2009 г.	2010 г.	2011 г.
GDP, billion rubles.	39100,7	45172,7	54585,6
Tax revenues of the consolidated budget of the Russian Federation			
- the amount, bln.	10629,6	11886,8	12420,3
- % GDP	27,1	26,5	22,7
Expenditures of the consolidated budget of the Russian Federation - the amount, bln.	16048	17617	18320,0
- % GDP	41,0	39,2	33,6
Tax benefits - the amount, bln.	2014,3	2230,7	2460,5
- % GDP	5,1	4,9	4,5
The share of tax benefits in the total tax revenues of the consolidated budget of the Russian Federation, %	18,9	18,7	19,8
The share of tax benefits in the total expenditures of the consolidated budget of the Russian Federation, %	12,5	12,6	13,4

* calculated by the authors according to the Federal State Statistics Service, the official website http://www.gks.ru/bgd/regl/b12_14p/Main.htm.

Since the revenue base of the local budgets is insufficient to cover costs, then the territorial budgets depend on centrally allocated funds from the federal budget.

Due to the fact that the existing diversity of tax benefits increases shortfall in budget revenues and decrease of expenditure powers performed, there is a necessity to assess the effectiveness of tax benefits as a separate direction of tax regulation.

Government bodies' activity and the overall objectives of the regional budget policy are directed to reach a balance and stability of budget-fiscal relations in the region so that it could help to solve the main social and economic tasks of the region (Tkacheva, 2013).

Nowadays the model of main directions to increase tax revenues of regional budgets in Russia should be aimed at regulating the system of tax benefits.

In our opinion, the preferred direction of increasing tax revenue is a complex assessment of the tax benefits effectiveness offered to legal and physical persons.

3 CONCLUSIONS

Replacing the tax benefits for compensatory payments to preferential categories of taxpayers will promote the growth of tax revenues. It should be considered in order to further strengthening the financial autonomy of territorial budgets and to optimize the amount of tax revenues to the budget system.

REFERENCES

[1] Belousova, Sv.N., 2012. Estimation of efficiency of tax privileges under local taxes. Izvestiya of the Southwestern state university, 2: 191–195.

[2] Mayburov, I. A.,2010. Tax policy. Theory and practice: The textbook for undergraduates. YuNITI-DANA, pp: 518.

[3] Mishina, S.V., 2010. Tax regulation of the income of local budgets. Accounting in the budgetary and non-profit organizations, 4: 17–22.

[4] Tkacheva, T.Yu., 2012. Development of the budgetary and tax relations: theory and practice: monograph/ Southwestern state university, pp:182.

[5] Tkacheva, T.Yu., L.V. Afanasjeva and Sv. N. Belousova, 2013. Risk Management in Providing the Region for Fiscal Security. World Applied Sciences Journal, 28(4): 485–489.

[6] Federal State Statistics Service. Date Views 12.03.2014 http://www.gks.ru/bgd/regl/b12_14p/Main.htm

Information, Computer and Application Engineering – Liu, Sung & Yao (Eds)
© 2015 Taylor & Francis Group, London, ISBN 978-1-138-02717-6

Online path planning for detecting threats of UAVs

Wei Zhou, Hong Tao Hou & Bo Hua Li
College of Information Systems and Management, National University of Defense Technology, Changsha, China

ABSTRACT: The path planning of Unmanned Aerial Vehicles (UAV), mainly focuses on finding out the lowest-price flight trace to meet the goals of the tasks, and avoiding the kinds of threatens in the battlefield, under the condition that the vehicles could finish all the movements in the flight plan. In fact, three dimensional online planning is based on the flight of offline planning. The online planning correct parts of the offline flight to avoid new battle-filed threatens. This paper presents the modified Sparse A-star algorithm to meet the requirements of online planning. The experiments show that the modified algorithm is efficient to avoid new threatens. Comparing with offline flight, the online flight reduces 36.73% price of threatens. And it costs only 0.535 seconds to generate new flight on average.

KEYWORDS: unmanned aerial vehicles; three dimensional online path planning; sparse A-star

1 INTRODUCTION

Path planning of UAV, which bases the missions of objectives and comprehensively considers the constraints of maneuverability, performance, fuel consumption, environment and threat factors, and then programs the optimal or nearly optimal secure flight route from the origin point to the target (Fu & Zhou & Ding 2010). Path planning is the effective means of improving operational efficiency of an aircraft and implementing accurate long-distance operational tasks. The analysis of influence factors that include constraint condition of the UAV maneuverability and external environment, such as threat modeling and the aircraft restrains, is the basis of path planning. This paper discusses the UAV online planning is based on the referenced trace of offline planning, and then we will use Sparse A-star (SAS) (Shi & Du & Bi 2006, Szczerba 2000, Zhou & Chen & Qin 2005) algorithm to solve the problem for UAV online planning. The trace of offline planning is yielded by ant colony optimization (ACO).

2 THE INTEGRATED COST MODEL OF UAV

The indexes of UAV path planning contain the cost of fuel consumption and flight threats. The goal of online path planning is to get the overall minimal cost. Given the integrated cost of a UAV trace (Hu 2011):

$$\min W = \min \int_0^L [\delta_O w_O(s) + \delta_R w_R(s) + \delta_M w_M(s)$$

$$+ \delta_A w_A(s) + \delta_C w_C(s) + \delta_T w_T(s)]ds \quad (1)$$

Where, $w_O(s), w_R(s), w_M(s), w_A(s), w_C(s), w_T(s)$ are respectively denoted fuel consumption, radar, missile, anti-aircraft gun, atmosphere and terrain; the weight of $\delta_O, \delta_R, \delta_M, \delta_A, \delta_C, \delta_T$ is respectively denoted fuel consumption, radar, missile, anti-aircraft gun, atmosphere and terrains.

Assuming UAV velocity is a constant, and the trace has n nodes, given the cost of fuel consumption,

$$w_O = c_1 \cdot \sum_{i=1}^{n-1} l_i \quad (2)$$

Figure 2.1 shows that a trace is divided into six parts, and radar cost is the average of these five samples.

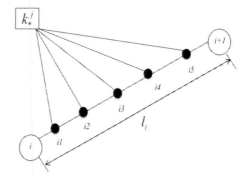

Figure 1. Computation model of threat cost.

Given the radar cost of trace l_i from radar k_R^j,

$$\omega_{Ri,k_R^j} = \frac{1}{5}\sum_{t=1}^{5} P_R \cdot d_{k_R^j,it}$$

Where, threat of radar j is denoted k_R^j, Euclidean distance between radar k_R^j and it is denoted $d_{k_R^j,it}$. Let the programming area having n_R radar threats, given the cost of radar along trace l_i:

$$w_{i,i+1}^R = \sum_{j=1}^{n_R} w_{Ri,k_R^j}$$

And the radar cost of the whole trace is given:

$$w_R = \sum_{i=1}^{n-1} w_{i,i+1}^R \qquad (3)$$

Other cost models are same with radar's so will not be described here.

3 PROBLEM STATEMENT OF THREE DIMENSIONAL ONLINE PLANNING

In order to efficiently settle the research, hence we make an important hypothesis (Scherer & Ferguson & Singh 2009):

Hypothesis: The offline referenced trace is optimal, when the threats remain changeless.

There exist several facts about this hypothesis: (1) the offline referenced trace, which is produced under circumstances of synthetically considering maneuverability, threats and fuel consumption, is more feasible. (2) The algorithm may not discover the more optimal long-distance trace, when time is limited and resource restricted. Given the formalization description of the hypothesis:

$$T_{i,j}(t_0) \equiv T_{i,j}(0) \quad \Rightarrow \quad o_{i,j}^S \leq o_{i,j}^D \qquad (4)$$

Where, $T_{i,j}(t)$ is the integrated threat condition between node i and j at $t = t_0$; $o_{i,j}^S$ is the general integrated cost of an offline section trace between node i and j; $o_{i,j}^D$ is the general integrated cost for an online section trace between node i and j; finally $o_{i,j}^S \leq o_{i,j}^D$ is statistical rule about offline planning and online. Above all, offline trace is the optimally choice, when the condition between node i and j is unchanged. Also, we present two principles for the online path planning:

Principle 1: Online planning occurs when the restrains is changed in the airspace under planning.

Principle 2: Online planning correct parts of the offline flight.

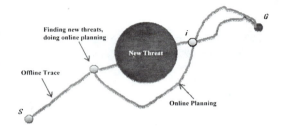

Figure 2. The situation not following principle 2.

By hypothesis 1, Figure 2 shows that the cost of UAV path planning will largely increase when going against Principle 2.

$$o_{i,G}^S \leq o_{i,G}^D$$

In Figure 2, the cost of online planning traces between node i and G will increase. So the UAV should choose the offline trace, when offline trace intersects the online and the threat is unchanged in the later airspace.

4 THREE DIMENSIONAL ONLINE PLANNING BASED ON SAS

4.1 Grid of three dimensional

Let the programming area be a $p \times q \times l$ cuboid, Figure 3 shows the area has a set including 27 nodes, Figure 4 shows ants can choose next node from not more than 15 feasible nodes, while not 26 nodes because of the restrains.

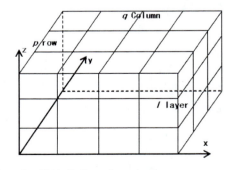

Figure 3. Grid of a three dimensional space.

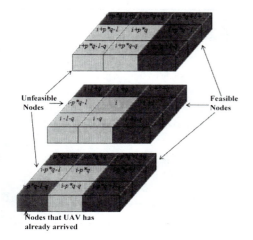

Figure 4. The set of feasible nodes in the three dimensional space.

4.2 The cost function

Given the cost function of SAS algorithm:

$$f(n) = g(n) + h(n) \tag{5}$$

Where, $g(n)$ is a general integrated cost between the starting point to the current node n that is waiting for planning, the computation model is described in Chapter 2; $h(n)$ is an estimated cost from the current node n to the target G, the computation method is same with $g(n)$.

4.3 The method for dealing with the constraint

The traditional A*algorithm has two questions; 1) the generated trajectory does not meet the UAV's maneuverability; 2) it takes too much time to produce trace, and then does not meet the requirements of real-time online planning.

So the nodes that we choose not only meet the condition of traditional A*algorithm, but also the constraints of UAV's maneuverability. We choose nodes according to flight-orientation, self-adopt, and then we can decrease the complexity of A*algorithm.

4.4 The termination of online planning

Figure 5 shows that the target of the termination of SAS is not the ending point, while the target of the missions is also G. Firstly, we give the definition about effective offline trace and disable invalid offline trace:

1 Effective offline traces: the trace does not run into threats, when new threats yield;
2 Invalid offline trace: the trace runs into threats, when new threats yield.

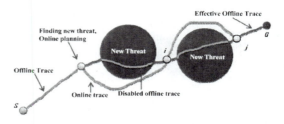

Figure 5. The termination of SAS.

Figure 5 shows that node i is the first point of intersection between effective offline trace and online planning, while the trace that includes node i is not the last effective flight path, so node i is not the target of the termination, and the node j is the point that meets the condition of online planning.

4.5 Flow chart of SAS

Figure 6 shows the steps of SAS algorithm.

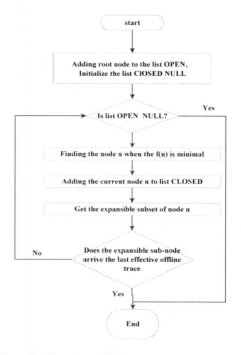

Figure 6. Flow chart of SAS.

Table 1. Parameters of three dimensional offline planning.

Symbol	Value	Description	Symbol	Value	Description
S	(1,1,3)	Starting point	Mh	5	Height of missile attack
G	(10,10,4)	The target	θ_A	$\pi/8$	Angle of anti-aircraft gun
$TR1$	(3,1,1)	Radar 1	Ah	3	Height of anti-aircraft gun
$TR2$	(6,8,2)	Radar 2	ω_{max}	90	Maximal turning angle
$TR3$	(9,4,1)	Radar 3	σ_{max}	90	Maximal pitching angle(°)
$TM1$	(1,7,2)	Missile 1	m	50	Numbers of artificial ants
$TM2$	(5,6,1)	Missile 2	α	1	Degree of IBH
$TA1$	(7,5,3)	Anti-aircraft gun 1	β	0.9	Degree of heuristic factor
$TB1$	(7,6,5)	Terrain threat 1	ρ	0.8	Remaining degree of IBH
$TB2$	(4,3,2)	Terrain threat 2	e	2	Increasing coefficient of IBH of optimal trace
θ_R	$\pi/6$	The angle of radar detection	λ	0.9	Local weight of Multiple -heuristic factor
Rh	4	Height of radar detection	NC_max	80	Maximal numbers of its iteration
θ_M	$\pi/8$	The angle of missile attack			

5 SIMULATION FOR THREE DIMENSIONAL ONLINE PLANNING

The computer CPU is Intel Core2 Duo T6600 @2.20Hz, and operating system is Windows7, We implement the method that puts forward by the paper used Matlab 2011B. Table 1 shows the parameters of the simulation.

Figue 7a and b shows a UAV detects an aircraft gun bastion when it arrives at node (4, 4, 4) along the offline trace.

Then, we make a simulation for online planning based on SAS, and Figure 8 shows the planning result. The method based on SAS can effectively avoid the new detected threat and decrease the fuel cost; and the convergence of online planning is offline trace. Table 2 shows five different online simulations and their result. Also, Table 3 shows that computation speed is comparatively quick; the mean time is 0.535s;

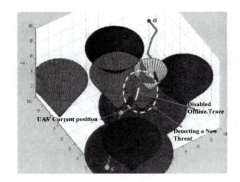

Figure 7b. UAV detects a new threat.

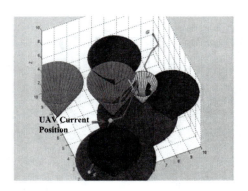

Figure 7a. UAV doesn't detect new threat.

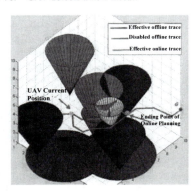

Figure 8. The simulation result for the supposed scene.

Table 3. Five simulations for online planning and their results.

UAV position	Type of the detected threats	Online costs time(s)
(8, 6, 5)	TM(9.5, 7, 2)	0.476
(7, 6, 5)	TA(8, 8, 2.5)	0.647
(7, 6, 5)	TR(9, 7.5, 1.5)	0.448
(3, 3, 4)	TM(4, 4, 1)	0.523
(5, 4, 4)	TR(8, 6, 1.5)	0.578
Mean time of Online planning		0.535

towards UAV, its speed is more or less 300km/h, on the period of online planning, the distance of UAV flying is only 44.583m.Comparing with the airspace, the distance can be ignored.

6 CONCLUSION

According to the simulation, we proposed that the capability, the feasibility and real-time performance of SAS algorithm is good; and then the algorithm can improve ability that UAV dodges the new detected threats. Comparing with offline planning trace, general integrated cost of online planning can decrease 36.73% and evidently improve UAV's security. Also, the time that online planning costs is just 0.535s basically meets requirements of real-time performance. In future, but the speed of UAV is more and more quick, so the next research is focused on how to improve the computation speed of SAS or use new algorithm to solve this question.

REFERENCES

Yangguang Fu,Chengping Zhou,Mingyue Ding. 3-D Route Planning based on Hybrid Quantum-Behaved Particle Swarm Optimization [J]. Journal of Astronautics.2010,31(2): 2657–2664.

Zongpeng Shi,Ping Du,Yiming Bi. A Study of real-time route planning approach based on A* algorithm[J].Journal of Naval University of Engineering. 2006, 18(5):79–82.

Szczerba RJ. Robust algorithm for real-time route planning [J].IEEE Transactions on Aerospace and Electronic Systems. 2000, 36(3) :869–878.

Chengping Zhou,Qianyang Chen,Xiao Qin.Parallel algorithm of 3D route planning based on the sparse A* algorithm[J]. Journal of Huazhong University of Science and Technology,2005,33(5):42–45.

Zhonghua Hu. The Key Technology of UAV Route Planning Based on intelligent Optimization Algorithm [D].Wuhan:Huazhong University of Science and Technology, 2011.

Scherer S, Ferguson D, Singh S. Efficient C-Space and cost function updates in 3D for unmanned aerial vehicles[C]2009 IEEE International Conference on Robotics and, 2009: 2049–2054.

Information, Computer and Application Engineering – Liu, Sung & Yao (Eds)
© 2015 Taylor & Francis Group, London, ISBN 978-1-138-02717-6

Innovation of training mode in private undergraduate university

Chen Lei Mao

School of Management, Jiangxi University of Technology, China

ABSTRACT: With the constant increase of private undergraduate universities, the private higher education in China has entered a new historical stage. If private undergraduate university wants to be healthy, coordinated and sustainable development, training mode must have the innovation. This article discusses the innovative ideas of training mode for the professional setting, training plan, teaching content and teaching evaluation, and expounds the measures from discipline construction, management innovation, teacher and vocational guidance education.

KEYWORDS: Private Undergraduate University, Training mode, Innovation, university-enterprise cooperation

1   INTRODUCTION

Along with the continuous deepening of the reform of the management system of higher education in China, private undergraduate education has become a new phenomenon with the development of China's contemporary education career. At present, private education in China has begun to take shape. If private undergraduate university wants to be healthy, coordinated and sustainable development, training mode must have the innovation.

Private universities start relatively late and running time is short. In the course of development, although mostly having a more perfect management system and development characteristics, but private undergraduate universities expose some problems. First, teaching management idea and management mode need innovation. Most of the private universities are exploring the process of cultivating applied talents and of can learn no template. By its constraints, it is difficult to get rid of the influence of traditional management ideas and management mode, so that with the background of the rapid development of market economy, popularization of higher education, and the society urgent need for applying talents, private universities need to use the ideas of development and innovation to think about and solve. Second, private universities need a rational thinking for target of personnel training specification. The cultivation mode of some private universities is still pursuing to train academic talents, pay attention to the basic theory, cultivate students' academic ability, far from the actual needs of society for talents. Third, teachers need to be improved, the structure of teachers needs to be improved. The resource for teachers in private universities is a shortage, young teachers are many,

school pay more attention to use teacher and neglect to cultivate, which is not conducive to the promotion of teachers and school quality improvement.

Education and training systems are under increasing pressure to respond to the new skills demands generated by a rapidly changing and globalized labor market. The acceleration of scientific and technological innovations requires new strategic responses to train the workforce and make an effective use of existing skills. This presents challenges at all levels of schooling, but especially at the tertiary level. Universities need to respond to the cycle of innovation and adapt their organization and pedagogies to serve increasingly heterogeneous student profiles and improve the teaching and learning of a variety of skills for innovation. Across countries and in multiple occupational fields, the private higher education sector is a prominent actor in introducing innovative approaches to the development of 21st century skills.

2   THE POSITION OF PRIVATE UNDERGRADUATE EDUCATION

Because of the level of both students and teachers, private university's personnel training target must be firmly fixed in the cultivation of applied talents. Only separating from the Research University, Teaching and Research University in the training objective level, private university must show its "application type" characteristics. At the same time, private undergraduate university should distinguish with higher vocational education. In the future career, vocational graduates are gray collar, then applied undergraduate talents are between the gray color and the white-color, the latter shall rise space and have a stronger ability to a higher level.

3 FOUR INNOVATIONS OF TRAINING MODE IN PRIVATE UNIVERSITIES

1 Innovation of professional setting. To cultivate applied talents as the goal should be reflected in a professional setting, according to social needs. Professional setting must have innovative thinking, and continuously explore a unique, new cross professional need for society urgently.
2 Innovation of training program. The training program is the templates of cultivating work. Training program in private universities is not a simple change of the past three year specialist training program, also cannot copy the training program of public universities. To highlight the characteristics of application type, according to the specific circumstances of the students, a training program should be created for students carefully. Three groups of knowledge, ability and quality should be constructed, and the implementation of the "modular teaching", students with the spirit of innovation and entrepreneurship awareness, and strong competitive strength and the actual operation ability are cultivated. Private undergraduate universities should pay more attention to students for cultivation of comprehensive quality and practical ability, and let students "learn to be, learn to do, learn to learn, learn to innovate".
3 Innovation of teaching content. The cultivation of high quality application type students is not empty talk, it should be embodied in the teaching contents. The design of teaching content should not only pay attention to play professional foundation, but also should emphasize the training of application ability. In the innovation of teaching content, it should be paid attention to three aspects: firstly, it should be clear what the student will need to work with application ability; secondly, in the teaching process, it should make students get more empirical knowledge and knowledge work processes, the production practice should be in the course; thirdly, operation of teaching content is important.
4 Innovation of teaching evaluation. Based on the principles of the evaluation scheme of national ordinary higher school, private undergraduate education should set up application type and strong, flexible operating mechanism of teaching quality guarantee of evaluation standard. Considering the particularity of the application type of undergraduate course teaching, it should be paid attention to "individualized assessment". The evaluation system should be diversified, that is to say,

students have the right to speak, the employing companies have the right to speak, experts and professors have the right to speak, training teacher should also have the right to speak.

4 MEASURES OF SAFEGUARDING

The strategic choice of private undergraduate universities, cannot follow and copy the public universities. In addition to draw the other valuable experience of similar institutions for construction of strategy, but also private university should avoid "what you do, I do what". In the reform of training mode, it should be paid attention to flexible and innovative, according to its own reality.

1 Take discipline construction as the key point. Because of the shortage of the congenital resource, private universities cannot construct all disciplines equally, and should concentrate on looking for a breakthrough, according to their own existing internal resources and external social resources. When some subjects form advantages, it not only contribute to the school level and the social impact of ascension school, but also conducive to the transformation of scientific research strength of private universities.
2 Take management innovation as the key. With the development of private universities, its management is gradually changing from "experience management" to "scientific management", so the innovation of management of private universities is a long-term strategic task. No grade difference of the actual meaning exists in the private universities, only the bold authorizing and encouraging creative work, in order to realize the modernization of management on real significance, flattening management improve the management execution. Because of its short history, Private universities should cultivate the university spirit, update ideas, and develop its own characteristics of the university spirit.
3 Take teachers as the key. The competition among university is in fact competition of teachers. Especially in private undergraduate universities, a shortage problem of talent teachers is always a difficult problem. The teacher is a major bottleneck restricting the development of private universities. In the school, the construction of teacher team should be in the priority position. Application of teaching requires teachers not only to master the professional knowledge and skills, but also to be familiar with the production practice, and practical application of knowledge and skills. The effect of applied talent training in China is not ideal,

one of the important reasons is that the construction of teachers team cannot keep up. In order to strengthen the training of applied talents, we must strengthen the construction of team teachers, realize effective communication between teaching and social practice.

4 Take the education of occupation guide as the key. This will be one of the characteristics of private universities. Students of private universities are poor cultural basis, in order to improve their employability and competitiveness, the school should cultivate emphatically to enhance students' entrepreneurial spirit, strengthen the training of occupation moral quality, strengthen the cultivation of students' ability of coordination, organization and management. The occupation guidance, education improve and encourage students to consciously change future employment pressure into motivation for learning during the school.

In the market system environment, the university should establish long-term, sustainable cooperation between school and enterprise, the key lies in the formation of a benign operation mechanism. First, to establish the dynamic mechanism of university-enterprise common development. The basis of cooperation is mostly carried out in the resources on the combination of both university and enterprise. University and enterprise all are willing to cooperate, and driving force should be enough. Only the needs and vision of government, universities and enterprises are stimulated, the power will be the formation of cooperation, cooperation will have greater vitality. Second,

to establish the driving mechanism based on mutual win-win interests. Third, to establish the security mechanism of university-enterprise cooperation. Fourth, to establish the sharing mechanism based on the complementary.

A private university is a bellwether of private higher education, and also hope of private higher education in the future, the quality and speed of development are directly related to the whole social reputation of private higher education. Only accurating position casting characteristics, managing innovation, in the future, private universities can come out on top, set up "the sign pole" for the private higher education, but also contribute to the promotion the whole social image of private higher education.

ACKNOWLEDGMENT

This work was supported by Jiangxi Province Office of Education of Humanities and Social Science Research [No. GL 1409].

REFERENCES

[1] Xu CD, Qu HM, Wang S. Research on innovation of application talents training in private university. Technological Pioneers, 2014(4):218–219.
[2] Li YP. Research on training mode of innovation talents in private university. Economic Research Guide, 2013(9):286–287.

Information, Computer and Application Engineering – Liu, Sung & Yao (Eds)
© 2015 Taylor & Francis Group, London, ISBN 978-1-138-02717-6

Research on the innovation of university student management

Xin Guang Ren

Harbin University of Science and Technology, Harbin, China

ABSTRACT: With the rapid development of society, the demands for high-quality talent are gradually increasing. At the same time, new requirements on the cultivation of high-quality talents are put forward. Facing the complex and volatile situation, combining the constantly changeable management environment and management object, it is becoming more and more important to further improve the quality of management of college students.

KEYWORDS: University; University student; Management; Innovation

1 THE PROBLEMS IN THE MANAGEMENT OF UNIVERSITY STUDENTS

College students management work is an important part of college management, with the training objectives of cultivating the college students with innovative spirit and practical ability. Therefore, to improve the quality management of college students, college students, ensure scientific standardized management level, is an important guarantee to achieve the training goal of college students. With the continuous development of society, there are many problems in the students' management in colleges and universities. It is difficult for the traditional management concepts, models and means to adapt to the need of social development. In the new situation of increasingly fierce competition for talent, we have to reform the work of student management, innovative management ideas, in order to meet the social demand for innovative talents. Therefore, the innovation research to further strengthen the management of college students work in colleges and universities is particularly important.

1.1 *Relatively backward of the traditional management concept*

With the continuous development of our society, the university student supervisory work is more and more complex and increasing difficulty. The traditional management idea and the system has not been able to meet the needs of the new social situation which has restricted the quality of university students management work to improve. The traditional mode of student work management focuses o the routine management, lack of students' service consciousness, students cannot participate in the management, which is not correspond with the

"people-oriented" concept of student management. But, at present, many colleges and universities in student management are still using the traditional students management mode. The students are not regarded as the main body by the administrators, they only give passive obedience to the administrators in the management process. In management, managers often used the method of disciplining, punishment, instill management mode, while neglecting the students' individual development, lack of equal exchange, and the students' active communication, which is not conducive to the cultivation of students' comprehensive quality and improve the innovation ability, there is a certain lag.

1.2 *Single management restrict the management work of ascension*

The society is in constant development, while the college management environment and management objects are promoted to the constantly changing, according to the current situation, with the analysis of the college students' characteristics, it is difficult for the past single management to achieve good management effect. Further optimizing the university student management means, has become the important problems of college student management work thinking. For a long time, the management of college students mainly through the implementation of the regulations to achieve the purpose of the management, the management means is too single. Along with steadily intensified of the trend of social diversification, the scale of the college students is growing steadily, and their values has been quite different from before. The traditional remedies have reduced the quality of the college students management which is not conducive to the personal growth of the students, and already can not adapt to the need of social development to raise high-quality personnel training.

1.3 *Lack of professional management team*

Along with our country higher education continuously popular trend, the number of college students increases unceasingly, while the corresponding student management team can not be added, in the complex of the management of students, management personnel shortage becomes more and more outstanding, become the major challenges facing the college students management work. At present, the management of college students in China is constantly refined, and management difficulty is getting heavier, the requirements on the college students management work are also gradually improved. And the team of student management work in colleges and universities is not stable, the quality is uneven, managers are lack of occupation identity, student management work experience is also insufficient. at the same time, the lack of professional training and education system, the lack of opportunities for improving their professional level, it is difficult to adapt to the reform and development of higher education in the management work. The existence of these problems will directly lead to the college management level of students can not be improved for long, which directly affects the cultivation of comprehensive quality of college students.

1.4 *Lack of perfect incentive mechanism*

At present, many college students management work is through the establishment of various student organizations to improve the college students self management ability, but in the specific implementation process of management, student organizations have become an important tool for the management of college students', and often can not play a good effect of self constraint. In the process of student participation, management operators will improve their ability as the reason to encourage the students. The student organizations lack of long-term effective incentive mechanism, has certain influence on the students' personal development. In addition, the evaluation system for college students is not perfect, which makes the student rewards and punishment means difficult to play a positive role. The appearance of these new situations and new problems, makes the traditional student management unable to meet the demand of social development, and also affected the effect of student management work. The innovation of college students management work and exploring a new mode of management have become an urgent need of the management of college students.

2 THE NECESSITY OF COLLEGE STUDENTS MANAGEMENT INNOVATION

With the change and development of social economy and situation in our country, college student management is facing new challenges. Only by constantly changing the management mode and methods of traditional management, continuous innovation, it can cultivate students to become all-round development of high-quality personnel.

2.1 *Innovation of college students management work is the need of higher education popularization*

With the continuous development of China's education, all colleges and universities continue enrollment expansion, the expanding of the enrollment scale, the number of college students increases unceasingly, and gradually to change the mode of higher education popularization, followed by college students overall quality level continuously, having the huge change, put forward higher request to the student management work in colleges and universities. The previous management mode is difficult to meet the development requirement of contemporary college students, college students management work only has to explore new ideas, active and innovative management, which will be able to adapt to the requirements of the development of higher education popularization.

2.2 *Innovation of college students management work is a "need to be people-oriented" services for students*

The current society is in the transition period, social life style gradually diversified, college students' concept of value and life style is undergoing great changes. With the rapid development of network technology, contemporary college students can receive and learn the new knowledge and new technology brought by the network faster, but also affected by the network deeply, which caused great impact to the traditional management mode and method. The college students management is the focus of the the university work. The realization of "people-oriented" service concept of training college students is an important goal of all-round development of high-quality personnel. Therefore, to train high-quality talents needed by the society, it must carry on the innovation of the university student supervisory work, which is not only the need to strengthen the management of the students, but also to improve the quality of education needs.

2.3 Innovation of college students management work is the need of training high-quality talents

With the rapid development of science and technology, new technologies are emerging, the demand of society for talents is higher and higher, which cause the college to continue promoting the innovation of education, not only making innovations in education concept and education system, but also exploring a new path on the students talent cultivation mode and the management work actively, in order to improve the college students' quality and ability. In the process of innovation of college students' management work, it must be continuously strengthened compared with the external environment, and constantly adapt to the trend of social development, only by this, it can not be eliminated by the times.

2.4 Innovation of college students management work is the need to promote the process of rule of law of the university

With the establishment of market economy system, the changing of the social situation, accelerate the social legal process, but also promotes the students gradually strengthen democracy and legal awareness, and make them know how to safeguard their own legitimate rights and interests. The development of the new situation, the requirements of college student management workers must go according to the provisions of relevant laws and regulations, according to standardized management, increase the intensity of student management reform and innovation, the legal process to promote the management of college students.

3 THE EFFECTIVE WAY TO STRENGTHEN COLLEGE STUDENTS MANAGEMENT WORK IN COLLEGES AND UNIVERSITIES

3.1 Update the scientific management concept

Adhere to the people-oriented management philosophy, to carry out the management of students, management operators should fully consider students' subjectivity and individuality development in specific work, to avoid or reduce the mandatory requirements. To put students in the main body of work status, to care for students' physical and mental health, reasonable service to students, make the students overall development. For poor students, psychological problems of the students focus on the healthy growth of the students, guide, help students to solve practical difficulties encountered in life, learning, and actively create a suitable environment for the growth of students, help students to establish a correct outlook on life. In the college students management work, and actively carry out democratic management idea, encouraging students to personally participate in the school management, help students to establish a sense of ownership. Management of workers in the process of democratic management, a comprehensive study of the various aspects of morality, innovation and practice ability of students, to ensure the subject status of students at the same time, promote the development of students' personality. In daily work, ensure the system transparent, pay attention to extensively listen to the students' opinions and suggestions, let the students really into the whole school, let the students self education, self management, self service and self excitation function can get full efforts to create a favorable environment for the development of the higher education management of College students.

3.2 Method of innovation of college students' management in colleges and universities

With the continuous development of social education, facing the continuous increase in the number of university students, the management mode of college student only has to be continuous innovation, in order to meet the social demand for students' comprehensive development. Therefore, we have to innovate the management of university students constantly, mobilize the enthusiasm of the students, then the effective combination of the higher education management can improve the ability of students' personal.

3.2.1 The establishment of fully utilizing the methods of network information management mode

With the rapid development of information technology, the Internet has become an important way for college students to expand their knowledge, to obtain the useful information. But the network information providing convenience to the management, also make it faces a severe test. Therefore, colleges should cultivate and establish a scientific, specialized student management team, student management workers should first make full use of modern network information technology means, to establish the network communication mechanism of administrators and students, make full use of the network to carry on the management of college students. To help students develop the correct network use habit, form a healthy network psychology, resist the bad influence that information on it. With the network information to implement information monitoring, timely detection of problems, solve problems, develop the student management work.

3.2.2 To strengthen the effective way of combination system of management and the management of campus culture

In the student management in colleges and universities, we should not only strengthen the system management, but also should learn how to use the educational function of the campus culture, combining the management system and the management of campus culture effectively. Perfect the management system of student evaluation system, in order to better services for the healthy growth of students, the students management work will be more scientific. In the process of college students management system evaluation system innovation, to keep open and scientific management system of the evaluation system, at the same time of the system of evaluation system of learning in students examine, pay more attention to the ability to exercise and quality cultivation, give students a free space for the development of the personality cultivation. Colleges and universities should combine the management system, and continuously carry out the rich campus culture activities, to provide a communication and show their talent, enhance the self platform for college students, but also for the student management provides further understanding of students of the channel, can help managers to constantly improve the methods of management, improve the management level of. College student management workers through the various student organizations management, strengthen students' standardized guidance, enhance the effect of College Students' self management and self education and management, make the students in the good campus cultural atmosphere under the influence of continuous self perfection.

3.2.3 To strengthen the ranks of college management major construction

The good management team of college students is beneficial to protect the student management work effectively, is to help the students to grow up, plays an irreplaceable role in creating a civilized, harmonious campus. To do well the work of college students management, occupation must strengthen management and professional construction team. Occupation training to strengthen management, improve the management team of the occupation level; through the pre job training, professional training and so on many kinds of ways to strengthen the management of the business of team learning effect, improve the professional quality, provide the chance for further study. Colleges and universities should constantly improve the management of team learning, working and

living environment, actively mobilize the enthusiasm of managers, improve the students' management work achievement. Colleges and universities should according to the actual situation of student management, establish a scientific appraisal system, benign competition to realize the management team, so as to promote the healthy development of the management team, ensure that students management work orderly, stable, healthy development.

3.3 Improving the incentive mechanism of college students management work

To improve college students' management incentive mechanism has a positive effect on the students management work, appropriate incentives to an important role in improving the students' management work efficiency. To perfect the incentive mechanism of college student management, stick to the combination of spiritual and material incentives of managers in principle, to give students a spiritual incentive at the same time, also want to through material incentives to mobilize the enthusiasm of the students. Ensure that students management work is carried out smoothly, adhere to the theory education and emotional education to students, adhere to the timely and appropriate, the principles of fairness and justice, to mobilize the enthusiasm of students to participate in school management, to enable students to trust and obey the management management. In the college students management work, establish and correct use of incentives are very important, management through the analysis of the students' emotional changes of understanding, cultivate rational consciousness of the students, guide the students toward the correct direction of development. Colleges and universities should take the example motivation, reward and punishment incentives, target incentive incentives, arousing students' learning motivation and positive, help students set goals, improve the students to realize self management, self restriction ability, promotes the university student supervisory work effect.

In short, with the continuous development of the social environment, the traditional management mode of university students has been difficult to adapt to the current needs of social development, there is a lag of student management work. Under the new situation, the innovation study of strengthening students' management work, strengthen students' management activity in innovation, and to improve the overall level of management of college students management, play an important role in training personnel, which has great theoretical significance and practical significance.

REFERENCES

[1] Tongfeng Cao.Research on college students' management work under the new situation[J]. The Journal of Shandong Agricultural Engineering College, 2014.

[2] Weihong Fan. Research on the innovation of college student management[J]. Hebei Normal University, 2013.

[3] Linfang Tang. On the innovation of college students management work in the new period[J]. The Examination Weekly, 2014.

[4] Yangbo He. Research on management innovation of college students[J]. Henan Science and Technology, 2014.

[5] Jianshen Song. New exploration of College Students' management work in universities under the new situation[J]. Journal of Heilongjiang College of Education, 2013.

[6] Wei Feng. A brief analysis of college students management in the new period[J]. Modern Economic Information, 2010.

[7] Yao Hao. Research on "post 90s" college students characteristics analysis and management countermeasures[J]. Economic Research Guide, 2011.

Information, Computer and Application Engineering – Liu, Sung & Yao (Eds)
© *2015 Taylor & Francis Group, London, ISBN 978-1-138-02717-6*

Rapid integration solution for online charging management in colleges

Hui Lin Zhang, Su Ping Xie, Yan Xia Li & Qiu Li Tong
Information Technology Center, Tsinghua University, Beijing, China

ABSTRACT: In order to simplify the integration work between charging business systems and payment system, proposes the construction of common online charging platform, introduces the working mechanism of the platform, analyses the key attributes of charging projects and items, expounds the application mode of the common charging platform including users and functions, service patterns, charging processes and financial reconciliation mode, finally provides a rapid integration solution to supporting online charging business in colleges.

KEYWORDS: Online Charging Management; Rapid System Integration; Charging Project Management

1 INTRODUCTION

Charging business is routine financial work in colleges and online charging is becoming a trend. In order to support unified management and monitoring on the charging businesses on campus for the financial department, it is an effective solution to build a unified payment system to integrate with third-party payment platforms and support the campus charging businesses. The third party payment platform refers to a mature platform on the market integrated with E-Bank platforms which is in charge of the safety and verifying procedures of the online transaction. It helps saving a lot of cost on negotiations with banks.

But on the other hand, since the online payment regulations have high requirements on security and integrity of the transactions, the integration of charging business systems which generates the charging data and payment system needs some development and debugging work. When more charging business systems put forward integration requirements, the integration work is becoming continuous and repetitive, which is not an effective construction mode from the development point of view. So in order to adopt a rapid and effective online charging solution, proposes the construction of common online charging platform, which packages the online transaction and safety control processes and provide simplified interfaces to charging business systems, so as to promote online payment better and more efficiently.

2 THE WORKING MECHANISM OF THE COMMON CHARGING PLATFORM

A prime difference between payment platform and the common charging platform is that the former one carries out real-time interactions with charging business systems in procedures of online payment, refunding and financial reconciliation while the latter one completes the whole transaction independently with charging data imported in advance and charging results transferred to the charging business systems after the transaction. The advantage of integrating directly with payment platform is that the business system obtains the status of the transaction instantly but on the other hand it need some more efforts in the integrating and troubleshooting procedures.

The common charging platform packages the online transactions to saves the above costs. It provides data interfaces to charging business systems to preset charging data and charging business systems don't have to pay attention to the interaction processes of online transactions. For some charging business without system supporting , the common charging platform supply a merchant working bench for the administrator to manage the charging projects and items, presetting charging data, tracking orders and checking account. It is a fast integration solution and also costs less during the maintenance period.

In order to meet the personalized needs of merchants and adapt to the different characteristics of charging projects and items, the common charging platform supports flexible attribute definitions. The following is differences of key attributes of charging projects and items.

3 ANALYSIS OF CHARGING PROJECTS AND ITEMS KEY ATTRIBUTES

With the extension and diversification of the services provided on campus, the volume of the charging work is growing rapidly. Charging project is the basic unit for charging management and different charging business needs to define separate charging projects. The establishing for charging projects need to be approved by the college financial department. To define a charging project needs to confirm the charging fee type, charging standards, administrative department, and charging period and so on. The financial reconciliation, bank transfers and statistics analysis are all based on charging projects

There may be single or multiple items in one charging project and attributes of Items are distinctive. Key attributes of charging projects and items determines the design of the order pages and processing procedures. The openness of projects, the fee standard types, the total amounts limit, and the single buyer amount limit and real-name purchase are key attributes that influence the design of the platform.

• Openness of the Charging Projects

Some of the charging projects are supplying services and commodities to all the public, so every user can buy the items under a total account limit, such like activities registration and online books. This charging services are not limited to specific users.

And Some other charging projects have preset charging list like temporary house rent expenses which are generated because a user is about to rent the house. So the payers and the payment amounts are settled in a list. The payer's identity need to be verified before the charging data are extracted.

• Fee Standard Types of Items

There are several fee standard types for items in different projects.

➢ Unified fee standard
Some items are charged with a unified amount for every user like books.

➢ Personalized fee account
Some items are charged different depending on the services content. For example, the house rent fee is related with the house conditions and rent period, so it's calculated according to the rent contract and not allowed modified.

➢ User-defined fee account
Some amount of the items is defined by users like internet fee pre-payment. Users decide the charging amount according to their own demand.

• The Total Amounts Limit

Some items are under total amount limit like activity registration or physical commodities and they are not available if the limit is reached. While some other items are totally unlimited like CET 4 or CET 6 registrations.

• The Single Buyer Amount Limit

Some items are in short supply like limited edition T-shirts or tickets for some popular shows and the organizers hope to let as much as possible people to have the opportunity, so the single buyer amount limit is set.

• Real-name Purchase

Some items have to be bought and bound with specific people at the same time like activity registrations. When a buyer pay for several persons, everyone's personal information has to be registered and only the registered user can take the charged service. Generally physically commodities don't need to collect real-name information.

The following is some examples to explain the different key attributes of charging projects and items.

4 THE APPLICATION MODE OF THE COMMON CHARGING PLATFORM

4.1 Users and functions

The main user groups of the platform include administrators, merchants, buyers and tourists. Platform administrators are in charge of charging project and items auditing, user management and orders management. Usually staff in financial department in university take the administrator job. Merchants are administrative departments responsible for charging businesses on campus and they establish charging

Table 1. Key attributes of charging project examples.

Charging Projects	Openness	Charging Standard	Total Amounts Limit	Single Buyer Amount Limit	Real-name Purchase
Limited edition T-shirt	Yes	Unified fee standard	Yes	Yes	No
CET 4 registration	Yes	Unified fee standard	No	Yes	Yes
Books online	Yes	Unified fee standard	Yes	No	No
Internet fee pre-payment	Yes	User-defined fee account	No	No	Yes
House rent fee	Yes	Personalized fee account	No	Yes	Yes

projects, distribute items, and handle orders and complete reconciliation every day. Buyers are campus and external user who pay for services or physical commodities supplied on campus and they can review items, submit orders, complete payment and query history orders on the platform, receiving the arrearage reminders at the same time. Tourists can review public items available but need registration if they want to submit an order.

4.2 Service modes

Common charging platform can work as an independent system to support a complete process of online payment service and can also serves as a background system to the charging business system supporting the online transactions.

Working as an online charging service hall, common charging platform supply a unified online charging services for all the users. For each logged-in user, they can query and pay all the personal bills and won't miss any charging reminders. The platform can even support paying for other peoples' bills. The payer only needs to obtain the other peoples' charging auto code and input it in the system to verify and extract charging data and complete the online payment which is convenient and safe. And at the same time, by distributing some public items, the platform is working like a simplified e-commerce platform.

When users log in the charging business system to start the online payment, they will skip to the common charging platform to complete the final online payment operation. The charging projects and items data have been preset in the charging platform in advance, so the charging business systems only need to maintain the links and won't need put too much efforts in development.

4.3 Charging processing

When the user are authenticated, the platform extracts the charging project list including public projects and non-public ones. Users choose a charging project and the system shows all the available items of the project. Then the buyer choose items and fill in necessary information according to the attributes of the items like purchase quantity or User-defined account, invoice information and so on, then finally submit the order. When the order is successfully generated, it will skip to the payment page. The buyer chooses a bank and it links to the bank's online platform to

verify the bank account and finish the deduction operation. When the charging platform receives the online payment result, it will update the order status and inform the charging business system later.

4.4 Financial reconciliation

The common charging platform supplies the merchant with a working bench to query the accounting data and provide daily and monthly reports which help the merchants to complete routine reconciliation. At the same time, the charging platform can obtain the amount of the daily bank transfer through third-party payment platform and check the account data in system with the remittance automatically, further reducing the daily workload of merchants. The automatic reconciliation tasks improve the efficiency and accuracy of financial work and support the whole procedure management for campus charging business.

5 CONCLUSION

Due to the rise of the internet applications, online payment has become the current widely used means of payment. For both the payees and the payers, online payment is more convenient, fast and safe. In comparison, the traditional payment way by cash, remittance and check spend more use and management costs. Colleges and universities need do more exploration in efficient financial management means to provide better support to teaching, scientific research and training on campus and common charging platform is one of the beneficial explorations.

REFERENCES

[1] Huilin Zhang, Qiuli Tong, Yanlong Wang, Mingyuan Li. Research on Planning and Design of Unified Payment Platform in Colleges and Universities[C]. Applied Mechanics and Materials, 2014, vols 631–632, 1160–1166.
[2] LvYiwen.The Designing and Realization for Sun Yat-Sen University Fee Management Platform Base on .NET [D]. Guangzhou, Sun Yat-Sen University, 2013.
[3] Yi Meichao, Qin Yuhua. Research on the Model of the Bank-College Payment System in Colleges [J]. Computer Knowledge and Technology, 2011, 7(1):158–160,171.
[4] Zhu Jing. The Research of Universities Net Payment Platform Based on Net Bank [J]. Journal of Guangdong Industry Technical College, 2012, 11(2):11–14.

Information, Computer and Application Engineering – Liu, Sung & Yao (Eds)
© 2015 Taylor & Francis Group, London, ISBN 978-1-138-02717-6

Perturbation bounds for eigenspace of generalized extended matrix under additive noise

Qiang Wu

Maoming Polytechnic College, Maoming, Guangdong, China

ABSTRACT: Under the situation that Matrix $\mathbf{A}$ and its perturbation matrix $\tilde{\mathbf{A}}$ have the same partitioning of spectral decomposition, the separating degree of spectral decomposition is used on the generalized extended matrix $\mathbf{F}_k(\mathbf{A})$ and its perturbation matrix $\tilde{\mathbf{F}}_k(\tilde{\mathbf{A}})$, both of which have $\mathbf{A}$ as original matrix, to pursue the separation of the eigenspace of the generalized extended matrix, and to study the perturbation bounds of the eigenspace of the extended matrix $\mathbf{F}_k(\mathbf{A})$ and its perturbation matrix $\tilde{\mathbf{F}}_k(\tilde{\mathbf{A}})$ under Frobenius Norm.

KEYWORDS: eigenspace, Frobenius Norm, generalized extended matrix, additive noise

1 INTRODUCTION

For the dimension of the matrix is very large, the use of the singular value decomposition of computation and storage is very big. Extension matrix is composed of a matrix (we called the mother matrix) by a larger matrix obtained after a series of transformation. Document [1–2] studied the extension matrix singular value decomposition. This paper used Spectral decomposition method and mother matrix perturbation bounds to study extension feature space matrix.

Definition 1[1] Assume A and the perturbation matrices $\tilde{\mathbf{A}} = \mathbf{A} + \Delta \mathbf{A}$ are n Hermite matrices of order, and the following is the spectral decomposition

$$\mathbf{H} = (\ \mathbf{U}_1 \quad \mathbf{U}_2\ ) \begin{pmatrix} \wedge_1 & 0 \\ 0 & \wedge_2 \end{pmatrix} \begin{pmatrix} \mathbf{U}_1^* \\ \mathbf{U}_2^* \end{pmatrix} \quad \text{and}$$

$$\tilde{\mathbf{H}} = (\ \tilde{\mathbf{U}}_1 \quad \tilde{\mathbf{U}}_2\ ) \begin{pmatrix} \tilde{\wedge}_1 & 0 \\ 0 & \tilde{\wedge}_2 \end{pmatrix} \begin{pmatrix} \tilde{\mathbf{U}}_1^* \\ \tilde{\mathbf{U}}_2^* \end{pmatrix} \quad \text{Among them}$$

$\mathbf{U} = (\mathbf{U}_1\ \mathbf{U}_2)$ and $\tilde{\mathbf{U}} = (\tilde{\mathbf{U}}_1\ \tilde{\mathbf{U}}_2)$, are a unitary matrix $\wedge_1 = diag(\lambda_1, \lambda_2, \cdots, \lambda_k)$ $\wedge_2 = diag(\lambda_{k+1}, \lambda_{k+2}, \cdots, \lambda_n)$ $\tilde{\wedge}_1 = diag(\tilde{\lambda}_1, \tilde{\lambda}_2, \cdots, \tilde{\lambda}_k)$ $\tilde{\wedge}_2 = diag(\tilde{\lambda}_{k+1}, \tilde{\lambda}_{k+2}, \cdots, \tilde{\lambda}_n)$ Definition 2 Assume $\mathbf{A} \in C^{n \times n}$ is the matrix, then the matrix of A

$$\mathbf{F}_k(\mathbf{A}) = \begin{pmatrix} \mathbf{A} & \mathbf{A} & \cdots & \mathbf{A} \\ \mathbf{A} & \mathbf{A} & \cdots & \mathbf{A} \\ \cdots & \cdots & \cdots & \cdots \\ \mathbf{A} & \mathbf{A} & \cdots & \mathbf{A} \end{pmatrix} \in C^{kn \times kn} \quad (1)$$

is known as the first class of k generalized extension matrix, which is called the mother matrix.

Lemma 1[4] The first class of k generalized extended matrix, $\mathbf{A} \in C^{n \times n}$ is Hermite matrix, A Spectral decomposition as shown in (1.1)

$$\mathbf{F}_k(\mathbf{A}) = (\mathbf{P}_1, \mathbf{P}_2) \begin{pmatrix} \Sigma_1 & 0 \\ 0 & \Sigma_2 \end{pmatrix} \begin{pmatrix} \mathbf{P}_1^H \\ \mathbf{P}_2^H \end{pmatrix} \quad (2)$$

and (1) $\sigma_i = k\lambda_i, i = 1, 2, \cdots, n$;

(2) $\mathbf{P}_1 = \mathbf{R}_k \left(\dfrac{1}{\sqrt{k}} \mathbf{U}_1 \right)$, $\mathbf{P}_2 = \mathbf{R}_k \left(\dfrac{1}{\sqrt{k}} \mathbf{U}_2 \right)$ Among them

$\mathbf{P} = (\mathbf{P}_1, \mathbf{P}_2)$,

$\Sigma_1 = diag(\sigma_1, \sigma_2, \cdots, \sigma_k)$,

$\Sigma_2 = diag(\sigma_{k+1}, \sigma_{k+2}, \cdots, \sigma_n)$.

Lemma 2[5] Assume M and N are Hermite matrix, and S is a complex matrix with proper dimension, assume the existence of two distance interval does not intersect δ, all the characteristics of an interval matrix which contains M values, another interval contains all characteristic value of matrix N. If $\delta > 0$, then the matrix equation $MX - XN = S$ exists only answer is X, and to any unitarily invariant norm $\| \bullet \|$

$$\|X\| \le \|S\| / \delta \quad (3)$$

Among them $\delta = \min\limits_{\tilde{u} \in \sigma(\mathbf{M}), u \in \sigma(\mathbf{N})} |\tilde{u} - u|$ Lemma 3[6] Assume $\Omega \in C^{s \times s}$ and $\Gamma \in C^{t \times t}$ are Hermite matrix. $\mathbf{E} \in C^{s \times t}$, if $\lambda(\Omega) \cap \lambda(\Gamma) = \varnothing$, then the matrix equation $\Omega \mathbf{X} - \mathbf{X}\Gamma = \Omega \mathbf{E}$ has a unique solution $\mathbf{X} \in C^{s \times t}$, and there is

$$\|\mathbf{X}\| \le \|\mathbf{E}\| / \rho \quad (4)$$

in $\rho = \min\limits_{\lambda \in \lambda(\Omega), \tilde{\lambda} \in \lambda(\Gamma)} \dfrac{|\lambda - \tilde{\lambda}|}{|\lambda|}$

2 THEOREM AND PROOF

Lemma 1 Assume A and the perturbation matrices $\tilde{A} = A + E$ are n Hermite matrices of order, and the following is the spectral decomposition (1.1) $F_k(A) \in \mathbf{C}^{kn \times kn}$ taking A as the mother matrix of the generalized extension matrix, including $F_k(A)$ such as (1.2) decomposition, so $\left\| \sin \Theta(P_1, \tilde{P}_1) \right\|_F \leq \dfrac{1}{2} \dfrac{\|E\|_F}{\chi(\delta_1, \delta_2)}$

Among them $\chi(\delta_1, \delta_2) = \dfrac{\delta_1 \delta_2}{\sqrt{\delta_1^2 + \delta_2^2}}$ proof:

$F_k(\tilde{A}) - F_K(A) = F_k(A+E) - F_K(A) = F_k(E)$ Form (4) Both $\tilde{P}^H$ and P left by right multiplication, is

$$\begin{pmatrix} \tilde{\Sigma}_1 & 0 \\ 0 & \tilde{\Sigma}_2 \end{pmatrix} \tilde{P}^H P - \tilde{P}^H P \begin{pmatrix} \Sigma_1 & 0 \\ 0 & \Sigma_2 \end{pmatrix} = \tilde{P}^H F_k(E) P$$

$$= R_k \left(\frac{1}{\sqrt{k}} \tilde{U} \right)^H \begin{pmatrix} E & E & \cdots & E \\ E & E & \cdots & E \\ \cdots & \cdots & \cdots & \cdots \\ E & E & \cdots & E \end{pmatrix} R_k \left(\frac{1}{\sqrt{k}} U \right)$$

$$= k \tilde{U}^H E U$$

By mean

$$\begin{pmatrix} \tilde{\lambda}_1 \tilde{P}_1^H P_1 & \tilde{\lambda}_1 \tilde{P}_1^H P_2 \\ \tilde{\lambda}_2 \tilde{P}_2^H P_1 & \tilde{\lambda}_2 \tilde{P}_2^H P_2 \end{pmatrix} - \begin{pmatrix} \tilde{P}_1^H P_1 \wedge_1 & \tilde{P}_1^H P_2 \wedge_2 \\ \tilde{P}_2^H P_1 \wedge_1 & \tilde{P}_2^H P_2 \wedge_2 \end{pmatrix}$$

$$= \tilde{U}^H E U = \begin{pmatrix} \tilde{U}_1^H E U_1 & \tilde{U}_1^H E U_2 \\ \tilde{U}_2^H E U_1 & \tilde{U}_2^H E U_2 \end{pmatrix} \text{ is}$$

$$\tilde{\lambda}_1 \tilde{P}_1^H P_2 - \tilde{P}_1^H P_2 \wedge_2 = \tilde{U}_1^H E U_2 \tag{5}$$

$$\tilde{\lambda}_2 \tilde{P}_2^H P_1 - \tilde{P}_2^H P_1 \wedge_1 = \tilde{U}_2^H E U_1 \tag{6}$$

According to (5) and (6) separately apply lemma 2, it can get $\left\| \tilde{P}_1^H P_2 \right\|_F \leq \dfrac{1}{\delta_1} \left\| \tilde{U}_1^H E U_2 \right\|_F$ $\left\| \tilde{P}_2^H P_1 \right\|_F \leq \dfrac{1}{\delta_2} \left\| \tilde{U}_2^H E^H U_1 \right\|_F$ According to (3), (5)and (6) it can get $2 \left\| \sin \Theta(P_1, \tilde{P}_1) \right\|_F = \left\| \tilde{P}_2^H P_1 \right\|_F + \left\| \tilde{P}_1^H P_2 \right\|_F$

$$\leq \frac{1}{\delta_1} \left\| \tilde{U}_1^H E U_2 \right\|_F + \frac{1}{\delta_2} \left\| \tilde{U}_2^H E^H U_1 \right\|_F$$

$$\leq \frac{\sqrt{\delta_1^2 + \delta_2^2}}{\delta_1 \delta_2} \left(\left\| \tilde{U}_1^H E U_2 \right\|_F + \left\| \tilde{U}_2^H E^H U_1 \right\|_F \right)$$

$$\leq \frac{\sqrt{\delta_1^2 + \delta_2^2}}{\delta_1 \delta_2} \|E\|_F$$

$$\therefore \left\| \sin \Theta(P_1, \tilde{P}_1) \right\|_F \leq \frac{1}{2} \frac{\|E\|_F}{\chi(\delta_1, \delta_2)}$$

Lemma 2 Assume A and the perturbation matrices $\tilde{A} = A + E$ are n Hermite matrices of order,

and the following is the spectral decomposition (1.1)moreover, A is a non singular matrix, $F_k(A)$ and the perturbation matrices $F_k(\tilde{A})$ are based on A and $\tilde{A}$ respectively for the parent matrix of generalized extended matrix , and There are such as (1.2) decomposition. Assume $\rho_i > 0, i = 1, 2$, and then $\left\| \sin \Theta(P_1, \tilde{P}_1) \right\|_F \leq \dfrac{1}{2} \dfrac{\left\| E A^{-1} \right\|_F}{\chi(\rho_1, \rho_2)}$ Among them

$$\rho_1 = \min_{\lambda \in \lambda(\wedge_1), \tilde{\lambda} \in \lambda(\tilde{\wedge}_2)} \frac{|\lambda - \tilde{\lambda}|}{|\lambda|}, \quad \rho_2 = \min_{\lambda \in \lambda(\wedge_2), \tilde{\lambda} \in \lambda(\tilde{\wedge}_1)} \frac{|\lambda - \tilde{\lambda}|}{|\lambda|},$$

$\chi(\rho_1, \rho_2) = \dfrac{\rho_1 \rho_2}{\sqrt{\rho_1^2 + \rho_2^2}}$ Shown by equation $\tilde{A} - A = E$ right Times A^{-1}, is $\tilde{A} A^{-1} - I = E A^{-1}$

$$F_k(\tilde{A} A^{-1}) - F_k(I) = F_k(E A^{-1}) \tag{7}$$

Form (7) both side $\tilde{P}^H$ and P left by right multiplication, is

$$\begin{pmatrix} \tilde{\Sigma}_1 & 0 \\ 0 & \tilde{\Sigma}_2 \end{pmatrix} \tilde{P}^H P - \tilde{P}^H P \begin{pmatrix} \Sigma_1 & 0 \\ 0 & \Sigma_2 \end{pmatrix} = \tilde{P}^H F_k(E A^{-1}) P \begin{pmatrix} \Sigma_1 & 0 \\ 0 & \Sigma_2 \end{pmatrix}$$

$$= R_k \left(\frac{1}{\sqrt{k}} \tilde{U} \right)^H \begin{pmatrix} E A^{-1} & E A^{-1} & \cdots & E A^{-1} \\ E A^{-1} & E A^{-1} & \cdots & E A^{-1} \\ \cdots & \cdots & \cdots & \cdots \\ E A^{-1} & E A^{-1} & \cdots & E A^{-1} \end{pmatrix} R_k \left(\frac{1}{\sqrt{k}} U \right) \begin{pmatrix} \Sigma_1 & 0 \\ 0 & \Sigma_2 \end{pmatrix}$$

$$= k \tilde{U}^H E A^{-1} U \begin{pmatrix} \Sigma_1 & 0 \\ 0 & \Sigma_2 \end{pmatrix} \tag{8}$$

$$\begin{pmatrix} \tilde{\wedge}_1 \tilde{P}_1^H P_1 & \tilde{\wedge}_1 \tilde{P}_1^H P_2 \\ \tilde{\wedge}_2 \tilde{P}_2^H P_1 & \tilde{\wedge}_2 \tilde{P}_2^H P_2 \end{pmatrix} - \begin{pmatrix} \tilde{P}_1^H P_1 \wedge_1 & \tilde{P}_1^H P_2 \wedge_2 \\ \tilde{P}_2^H P_1 \wedge_1 & \tilde{P}_2^H P_2 \wedge_2 \end{pmatrix}$$

Form (8) is written in the form of block

$$= \begin{pmatrix} \tilde{U}_1^H E A^{-1} U_1 \wedge_1 & \tilde{U}_1^H E A^{-1} U_2 \wedge_2 \\ \tilde{U}_2^H E A^{-1} U_1 \wedge_1 & \tilde{U}_2^H E A^{-1} U_2 \wedge_2 \end{pmatrix} \tag{9}$$

By form (9), and get

$$\tilde{\wedge}_1 \tilde{P}_1^H P_2 - \tilde{P}_1^H P_2 \wedge_2 = \tilde{U}_1^H E A^{-1} U_2 \wedge_2 \tag{10}$$

$$\tilde{\wedge}_2 \tilde{P}_2^H P_1 - \tilde{P}_2^H P_1 \wedge_1 = \tilde{U}_2^H E A^{-1} U_1 \wedge_1 \tag{11}$$

According to the topic design $\rho_i > 0, i = 1, 2$ knows $\lambda(\tilde{\wedge}_1) \cap \lambda(\wedge_2) = \lambda(\tilde{\wedge}_2) \cap \lambda(\tilde{\wedge}_1) = \varnothing$ .then by (10), (11) lemma3 can

$$\left\| \tilde{P}_1^H P_2 \right\|_F \leq \frac{1}{\rho_1} \left\| \tilde{U}_1^H E A^{-1} U_2 \right\|_F \tag{12}$$

$$\left\| \tilde{P}_2^H P_1 \right\|_F \leq \frac{1}{\rho_2} \left\| \tilde{U}_2^H E A^{-1} U_1 \right\|_F \tag{13}$$

by (3), (12) and (13), gets

$$2\left\|\sin\Theta(\mathbf{P}_1,\tilde{\mathbf{P}}_1)\right\|_F \le \frac{1}{\rho_1}\left\|\tilde{\mathbf{U}}_1^H \mathbf{EA}^{-1}\mathbf{U}_2\right\|_F + \frac{1}{\rho_2}\left\|\tilde{\mathbf{U}}_2^H \mathbf{EA}^{-1}\mathbf{U}_1\right\|_F$$

$$\le \sqrt{\frac{1}{\rho_1^2}+\frac{1}{\rho_2^2}}\sqrt{\left\|\tilde{\mathbf{U}}_1^H \mathbf{EA}^{-1}\mathbf{U}_2\right\|_F^2 + \left\|\tilde{\mathbf{U}}_2^H \mathbf{EA}^{-1}\mathbf{U}_1\right\|_F^2}$$

$$\le \sqrt{\frac{1}{\rho_1^2}+\frac{1}{\rho_2^2}}\left\|\mathbf{EA}^{-1}\right\|_F$$

means $\left\|\sin\Theta(\mathbf{P}_1,\tilde{\mathbf{P}}_1)\right\|_F \le \frac{1}{2}\dfrac{\left\|\mathbf{EA}^{-1}\right\|_F}{\chi(\rho_1,\rho_2)}$ Lemma 3 assume A and the perturbation matrices $\tilde{\mathbf{A}}=\mathbf{A}+\mathbf{E}$ are n non singular Hermite matrices of order, and the following is the spectral decomposition (1.1) moreover, $\mathbf{F}_k(\mathbf{A})$ and the perturbation matrices $\mathbf{F}_k(\tilde{\mathbf{A}})$ are based on A and $\tilde{\mathbf{A}}$ respectively for the parent matrix of generalized extended matrix , and There are such as (1.2) decomposition. Assume $\rho_i>0, i=1,2$, and then

$$\left\|\sin\Theta(\mathbf{P}_1,\tilde{\mathbf{P}}_1)\right\|_F \le \frac{k}{2\chi(\eta_1,\eta_2)}\left\|\mathbf{A}^{-\frac{1}{2}}\mathbf{E}\tilde{\mathbf{A}}^{-\frac{1}{2}}\right\|_F$$ among

$$\eta_1 = \min_{\lambda\in\lambda(\wedge_1),\tilde{\lambda}\in\lambda(\tilde{\wedge}_2)}\frac{\left|\lambda-\tilde{\lambda}\right|}{\sqrt{\tilde{\lambda}\lambda}}, \quad \eta_2 = \min_{\lambda\in\lambda(\wedge_2),\tilde{\lambda}\in\lambda(\tilde{\wedge}_1)}\frac{\left|\lambda-\tilde{\lambda}\right|}{\sqrt{\tilde{\lambda}\lambda}},$$

$$\chi(\eta_1,\eta_2)=\frac{\eta_1\eta_2}{\sqrt{\eta_1^2+\eta_2^2}}$$

Shown by equation $\tilde{\mathbf{A}}-\mathbf{A}=\mathbf{E}$ left times $\mathbf{A}^{-\frac{1}{2}}$ and right times $\tilde{\mathbf{A}}^{-\frac{1}{2}}$, is

$$\mathbf{A}^{-\frac{1}{2}}\tilde{\mathbf{A}}^{\frac{1}{2}}-\mathbf{A}^{\frac{1}{2}}\tilde{\mathbf{A}}^{-\frac{1}{2}} = \mathbf{A}^{-\frac{1}{2}}\mathbf{E}\tilde{\mathbf{A}}^{-\frac{1}{2}} \quad (14)$$

Form (1) and (14) and both sides take k time for extended matrix, is

$$\mathbf{F}_k(\mathbf{A}^{-\frac{1}{2}}\tilde{\mathbf{A}}^{\frac{1}{2}})-\mathbf{F}_k(\mathbf{A}^{\frac{1}{2}}\tilde{\mathbf{A}}^{-\frac{1}{2}}) = \mathbf{F}_k(\mathbf{A}^{-\frac{1}{2}}\mathbf{E}\tilde{\mathbf{A}}^{-\frac{1}{2}}) \quad (15)$$

Put (15) written block form, is

$$\begin{pmatrix}\Sigma_1^{-\frac{1}{2}} & 0 \\ 0 & \Sigma_2^{-\frac{1}{2}}\end{pmatrix}\mathbf{P}^H\tilde{\mathbf{P}}\begin{pmatrix}\tilde{\Sigma}_1^{\frac{1}{2}} & 0 \\ 0 & \tilde{\Sigma}_2^{\frac{1}{2}}\end{pmatrix}-\begin{pmatrix}\Sigma_1^{\frac{1}{2}} & 0 \\ 0 & \Sigma_2^{\frac{1}{2}}\end{pmatrix}\mathbf{P}^H\tilde{\mathbf{P}}\begin{pmatrix}\tilde{\Sigma}_1^{-\frac{1}{2}} & 0 \\ 0 & \tilde{\Sigma}_2^{-\frac{1}{2}}\end{pmatrix}$$

$$= \mathbf{P}^H\mathbf{F}_k(\mathbf{A}^{-\frac{1}{2}}\mathbf{E}\tilde{\mathbf{A}}^{-\frac{1}{2}})\tilde{\mathbf{P}}$$

$$= R_k(\frac{1}{\sqrt{k}}\mathbf{U})^H\begin{pmatrix}\mathbf{A}^{-\frac{1}{2}}\mathbf{EA}^{\frac{1}{2}} & \mathbf{A}^{-\frac{1}{2}}\mathbf{EA}^{\frac{1}{2}} & \cdots & \mathbf{A}^{-\frac{1}{2}}\mathbf{EA}^{\frac{1}{2}} \\ \mathbf{A}^{-\frac{1}{2}}\mathbf{EA}^{\frac{1}{2}} & \mathbf{A}^{-\frac{1}{2}}\mathbf{EA}^{\frac{1}{2}} & \cdots & \mathbf{A}^{-\frac{1}{2}}\mathbf{EA}^{\frac{1}{2}} \\ \cdots & \cdots & \cdots & \cdots \\ \mathbf{A}^{-\frac{1}{2}}\mathbf{EA}^{\frac{1}{2}} & \mathbf{A}^{-\frac{1}{2}}\mathbf{EA}^{\frac{1}{2}} & \cdots & \mathbf{A}^{-\frac{1}{2}}\mathbf{EA}^{\frac{1}{2}}\end{pmatrix}R_k(\frac{1}{\sqrt{k}}\tilde{\mathbf{U}})=k\mathbf{U}^H\mathbf{A}^{-\frac{1}{2}}\mathbf{E}\tilde{\mathbf{A}}^{-\frac{1}{2}}\tilde{\mathbf{U}}$$

means

$$\begin{pmatrix}\Sigma_1^{-\frac{1}{2}}\tilde{\mathbf{P}}_1^H\mathbf{P}_1\tilde{\Sigma}_1^{\frac{1}{2}} & \Sigma_1^{-\frac{1}{2}}\tilde{\mathbf{P}}_1^H\mathbf{P}_2\tilde{\Sigma}_2^{\frac{1}{2}} \\ \Sigma_2^{-\frac{1}{2}}\tilde{\mathbf{P}}_2^H\mathbf{P}_1\tilde{\Sigma}_1^{\frac{1}{2}} & \Sigma_2^{-\frac{1}{2}}\tilde{\mathbf{P}}_2^H\mathbf{P}_2\tilde{\Sigma}_2^{\frac{1}{2}}\end{pmatrix}-\begin{pmatrix}\Sigma_1^{\frac{1}{2}}\tilde{\mathbf{P}}_1^H\mathbf{P}_1\tilde{\Sigma}_1^{-\frac{1}{2}} & \Sigma_1^{\frac{1}{2}}\tilde{\mathbf{P}}_1^H\mathbf{P}_2\tilde{\Sigma}_2^{-\frac{1}{2}} \\ \Sigma_2^{\frac{1}{2}}\tilde{\mathbf{P}}_2^H\mathbf{P}_1\tilde{\Sigma}_1^{-\frac{1}{2}} & \Sigma_2^{\frac{1}{2}}\tilde{\mathbf{P}}_2^H\mathbf{P}_2\tilde{\Sigma}_2^{-\frac{1}{2}}\end{pmatrix}$$

$$= k\begin{pmatrix}\mathbf{U}_1^H\mathbf{A}^{-\frac{1}{2}}\mathbf{E}\tilde{\mathbf{A}}^{-\frac{1}{2}}\tilde{\mathbf{U}}_1 & \mathbf{U}_1^H\mathbf{A}^{-\frac{1}{2}}\mathbf{E}\tilde{\mathbf{A}}^{-\frac{1}{2}}\tilde{\mathbf{U}}_2 \\ \mathbf{U}_2^H\mathbf{A}^{-\frac{1}{2}}\mathbf{E}\tilde{\mathbf{A}}^{-\frac{1}{2}}\tilde{\mathbf{U}}_1 & \mathbf{U}_2^H\mathbf{A}^{-\frac{1}{2}}\mathbf{E}\tilde{\mathbf{A}}^{-\frac{1}{2}}\tilde{\mathbf{U}}_2\end{pmatrix} \quad (16)$$

from (16)it can get

$$\Sigma_1^{-\frac{1}{2}}\tilde{\mathbf{P}}_1^H\mathbf{P}_2\tilde{\Sigma}_2^{\frac{1}{2}}-\Sigma_1^{\frac{1}{2}}\tilde{\mathbf{P}}_1^H\mathbf{P}_2\tilde{\Sigma}_2^{-\frac{1}{2}}=k\mathbf{U}_1^H\mathbf{A}^{-\frac{1}{2}}\mathbf{E}\tilde{\mathbf{A}}^{-\frac{1}{2}}\tilde{\mathbf{U}}_2 \quad (17)$$

and $\Sigma_2^{-\frac{1}{2}}\tilde{\mathbf{P}}_2^H\mathbf{P}_1\tilde{\Sigma}_1^{\frac{1}{2}}-\Sigma_2^{\frac{1}{2}}\tilde{\mathbf{P}}_2^H\mathbf{P}_1\tilde{\Sigma}_1^{-\frac{1}{2}}=\mathbf{U}_2^H\mathbf{A}^{-\frac{1}{2}}\mathbf{E}\tilde{\mathbf{A}}^{-\frac{1}{2}}\tilde{\mathbf{U}}_1$ (18)

Put the form(17) 式(i,j) to write down, is

$$\sigma_i^{-\frac{1}{2}}(\tilde{\mathbf{P}}_1^H\mathbf{P}_2)_{ij}\tilde{\sigma}_j^{\frac{1}{2}}-\sigma_i^{\frac{1}{2}}(\tilde{\mathbf{P}}_1^H\mathbf{P}_2)_{ij}\tilde{\sigma}_j^{-\frac{1}{2}}=(k\mathbf{U}_1^H\mathbf{A}^{-\frac{1}{2}}\mathbf{E}\tilde{\mathbf{A}}^{-\frac{1}{2}}\tilde{\mathbf{U}}_2)_{ij}$$

$\eta_1>0$, so $\lambda(\wedge_1)\bigcap\lambda(\tilde{\wedge}_2)=\varnothing$, so $\lambda(\Sigma_1)\bigcap\lambda(\tilde{\Sigma}_2)=\varnothing$.

Therefore, we can have

$$\left|(\mathbf{P}_1^H\tilde{\mathbf{P}}_2)_{ij}\right|\left|\sigma_i^{-1}(\mathbf{P}_1^H\tilde{\mathbf{P}}_2)_{ij}\tilde{\sigma}_j^{-1}-\sigma_i^1(\mathbf{P}_1^H\tilde{\mathbf{P}}_2)_{ij}\tilde{\sigma}_j^{-1}\right|=\left|(k\mathbf{U}_1^H\mathbf{A}^{\frac{1}{2}}\mathbf{E}\tilde{\mathbf{A}}^{-\frac{1}{2}}\tilde{\mathbf{U}}_2)_{ij}\right|\left|\frac{\sigma_i-\tilde{\sigma}_j}{\sqrt{\sigma_i\tilde{\sigma}_j}}\right|$$

$$\le \frac{1}{\eta_1}\left|(k\mathbf{U}_1^H\mathbf{A}^{-\frac{1}{2}}\mathbf{E}\tilde{\mathbf{A}}^{-\frac{1}{2}}\tilde{\mathbf{U}}_2)_{ij}\right|$$ then we get

$$\left\|\mathbf{P}_1^H\tilde{\mathbf{P}}_2\right\|_F \le \frac{1}{\eta_1}\left\|k\mathbf{U}_1^H\mathbf{A}^{-\frac{1}{2}}\mathbf{E}\tilde{\mathbf{A}}^{-\frac{1}{2}}\tilde{\mathbf{U}}_2\right\|_F \quad (19)$$

Similarly, by form (18) available

$$\left\|\mathbf{P}_2^H\tilde{\mathbf{P}}_1\right\|_F \le \frac{1}{\eta_2}\left\|k\mathbf{U}_2^H\mathbf{A}^{-\frac{1}{2}}\mathbf{E}\tilde{\mathbf{A}}^{-\frac{1}{2}}\tilde{\mathbf{U}}_1\right\|_F \quad (20)$$

By (19)and(20)available

$$2\left\|\sin\Theta(\mathbf{P}_1,\tilde{\mathbf{P}}_1)\right\|_F = \left\|\mathbf{P}_1^H\tilde{\mathbf{P}}_2\right\|_F + \left\|\mathbf{P}_2^H\tilde{\mathbf{P}}_1\right\|_F$$

$$\le \frac{1}{\eta_1}\left\|k\mathbf{U}_1^H\mathbf{A}^{-\frac{1}{2}}\mathbf{E}\tilde{\mathbf{A}}^{-\frac{1}{2}}\tilde{\mathbf{U}}_2\right\|_F + \frac{1}{\eta_2}\left\|k\mathbf{U}_2^H\mathbf{A}^{-\frac{1}{2}}\mathbf{E}\tilde{\mathbf{A}}^{-\frac{1}{2}}\tilde{\mathbf{U}}_1\right\|_F$$

$$\le \sqrt{\frac{1}{\eta_1^2}+\frac{1}{\eta_2^2}}\sqrt{\left\|k\mathbf{U}_2^H\mathbf{A}^{-\frac{1}{2}}\mathbf{E}\tilde{\mathbf{A}}^{-\frac{1}{2}}\tilde{\mathbf{U}}_1\right\|_F^2 + \left\|k\mathbf{U}_1^H\mathbf{A}^{-\frac{1}{2}}\mathbf{E}\tilde{\mathbf{A}}^{-\frac{1}{2}}\tilde{\mathbf{U}}_2\right\|_F^2}$$

$$\le \frac{k}{\chi(\eta_1,\eta_2)}\left\|\mathbf{A}^{-\frac{1}{2}}\mathbf{E}\tilde{\mathbf{A}}^{-\frac{1}{2}}\right\|_F$$

So $\left\|\sin\Theta(\mathbf{P}_1,\tilde{\mathbf{P}}_1)\right\|_F \le \frac{k}{2\chi(\eta_1,\eta_2)}\left\|\mathbf{A}^{-\frac{1}{2}}\mathbf{E}\tilde{\mathbf{A}}^{-\frac{1}{2}}\right\|_F$

Inference Assume A and the perturbation matrices $\tilde{\mathbf{A}}=\mathbf{A}+\mathbf{E}$ are n Hermite matrices of order, and the following is the spectral decomposition(1.1). $\mathbf{F}_k(\mathbf{A})$ and the perturbation matrices $\mathbf{F}_k(\tilde{\mathbf{A}})$ are based on A and $\tilde{\mathbf{A}}$ respectively for the parent matrix of generalized extended matrix , and There are such as (1.2) decomposition. If $\rho_i>0,i=1,2$, If

$$\mu_2 = \left\|\mathbf{A}^{-\frac{1}{2}}\mathbf{E}\tilde{\mathbf{A}}^{-\frac{1}{2}}\right\|_2 < 1$$

Then $\left\| \sin\Theta(\mathbf{P}_1, \tilde{\mathbf{P}}_1) \right\|_F \leq \dfrac{\sqrt{2}k\mu_F}{2\sqrt{1-\mu_2}\min(\eta_1, \eta_2)}$

proof by $\left\| \mathbf{A}^{-\frac{1}{2}}\mathbf{E}\tilde{\mathbf{A}}^{-\frac{1}{2}} \right\| = \left\| \mathbf{A}^{-\frac{1}{2}}\mathbf{E}\mathbf{A}^{-\frac{1}{2}}\mathbf{A}^{\frac{1}{2}}(\mathbf{A}+\mathbf{E})^{-\frac{1}{2}} \right\|_F$

$$\leq \left\| \mathbf{A}^{-\frac{1}{2}}\mathbf{E}\mathbf{A}^{-\frac{1}{2}} \right\|_F \left\| \mathbf{A}^{\frac{1}{2}}(\mathbf{A}+\mathbf{E})^{-\frac{1}{2}} \right\|_2 \qquad (21)$$

by (21) and $\mu_2 = \left\| \mathbf{A}^{-\frac{1}{2}}\mathbf{E}\tilde{\mathbf{A}}^{-\frac{1}{2}} \right\|_2 < 1$

is $\left\| \mathbf{A}^{\frac{1}{2}}(\mathbf{A}+\mathbf{E})^{-\frac{1}{2}} \right\|_2^2 = \left\| \mathbf{A}^{\frac{1}{2}}(\mathbf{A}+\mathbf{E})^{-1}\mathbf{A}^{\frac{1}{2}} \right\|_2$

$$= \left\| (\mathbf{I}+\mathbf{A}^{-\frac{1}{2}}\mathbf{E}^{-1}\mathbf{A}^{-\frac{1}{2}})^{-1} \right\|_2 \leq \dfrac{1}{1-\left\| \mathbf{A}^{-\frac{1}{2}}\mathbf{E}^{-1}\mathbf{A}^{-\frac{1}{2}} \right\|_2}$$

So

$$2\left\| \sin\Theta(\mathbf{P}_1, \tilde{\mathbf{P}}_1) \right\|_F \leq \sqrt{\frac{1}{\eta_1^2}+\frac{1}{\eta_2^2}} \sqrt{\left\| k\mathbf{U}_2^H\mathbf{A}^{-\frac{1}{2}}\mathbf{E}\tilde{\mathbf{A}}^{-\frac{1}{2}}\tilde{\mathbf{U}}_1 \right\|_F^2 + \left\| k\mathbf{U}_1^H\mathbf{A}^{-\frac{1}{2}}\mathbf{E}\tilde{\mathbf{A}}^{-\frac{1}{2}}\tilde{\mathbf{U}}_2 \right\|_F^2}$$

$$\leq \dfrac{\sqrt{2}k}{\min(\eta_1, \eta_2)}\left\| \mathbf{A}^{-\frac{1}{2}}\mathbf{E}\tilde{\mathbf{A}}^{-\frac{1}{2}} \right\|_F \leq \dfrac{\sqrt{2}k}{\min(\eta_1, \eta_2)}\dfrac{\mu_F}{\sqrt{1-\mu_2}}$$

So $\left\| \sin\Theta(\mathbf{P}_1, \tilde{\mathbf{P}}_1) \right\|_F \leq \dfrac{\sqrt{2}k\mu_F}{2\sqrt{1-\mu_2}\min(\eta_1, \eta_2)}$

3   A NUMERICAL EXAMPLE

Assume $\mathbf{A} = \begin{pmatrix} 3 & 2 \\ 1 & 2 \end{pmatrix}$, $\mathbf{R}_2(\mathbf{A}) = \begin{pmatrix} \mathbf{A} & \mathbf{A} \\ \mathbf{A} & \mathbf{A} \end{pmatrix}$ Matrix

$\mathbf{A}$ is Hermite matrix, spectral decomposition is

$\mathbf{A} = \begin{pmatrix} 1 & 2 \\ -1 & 1 \end{pmatrix}\begin{pmatrix} 1 & 0 \\ 0 & 4 \end{pmatrix}\begin{pmatrix} \frac{1}{3} & -\frac{2}{3} \\ \frac{1}{3} & \frac{1}{3} \end{pmatrix}$ 则 $\lambda_1 = 1$,

$\lambda_2 = 4$; Make $\mathbf{E} = \begin{pmatrix} -0.2 & 0 \\ 0.1 & 0.1 \end{pmatrix}$, then $\tilde{\mathbf{A}} = \mathbf{A}+\mathbf{E}$

character root is $\tilde{\lambda}_1 = 0.92$, $\tilde{\lambda}_2 = 3.9805$;

(1)   As

$$\delta_1 = \min_{\lambda\in\lambda(\wedge_1), \tilde{\lambda}\in\lambda(\wedge_2)}\dfrac{|\lambda-\tilde{\lambda}|}{\sqrt{|\tilde{\lambda}|^2+|\lambda|^2}} = \dfrac{|1-3.9805|}{\sqrt{1^2+3.9805^2}} \approx 0.7262$$

$$\delta_2 = \min_{\lambda\in\lambda(\wedge_2), \tilde{\lambda}\in\lambda(\wedge_1)}\dfrac{|\lambda-\tilde{\lambda}|}{\sqrt{|\tilde{\lambda}|^2+|\lambda|^2}} = \dfrac{|4-0.92|}{\sqrt{4^2+0.92^2}} \approx 0.7504$$

$$\chi(\delta_1, \delta_2) = \dfrac{\delta_1\delta_2}{\sqrt{\delta_1^2+\delta_2^2}} \approx \dfrac{0.7262\times 0.7504}{\sqrt{0.7262^2+0.7504^2}} \approx 0.5219$$

Based on theorem 1 $\left\| \sin\Theta(\mathbf{P}_1, \tilde{\mathbf{P}}_1) \right\|_F \leq 0.2347$

(2)   As $\mathbf{A}^{-1} = \begin{pmatrix} \frac{1}{3} & -\frac{2}{3} \\ \frac{1}{3} & \frac{1}{3} \end{pmatrix}$, then

$$\rho_1 = \min_{\lambda\in\lambda(\wedge_1), \tilde{\lambda}\in\lambda(\wedge_2)}\dfrac{|\lambda-\tilde{\lambda}|}{|\lambda|} = \dfrac{|1-3.9805|}{1} = 2.9805$$

$$\rho_2 = \min_{\lambda\in\lambda(\wedge_2), \tilde{\lambda}\in\lambda(\wedge_1)}\dfrac{|\lambda-\tilde{\lambda}|}{|\lambda|} = \dfrac{|4-0.92|}{4} = 0.77$$

$$\chi(\rho_1, \rho_2) = \dfrac{\rho_1\rho_2}{\sqrt{\rho_1^2+\rho_2^2}} = \dfrac{2.9805\times 0.77}{\sqrt{2.9805^2+0.77^2}} = 0.7455$$

Based on theorem 2 $\left\| \sin\Theta(\mathbf{P}_1, \tilde{\mathbf{P}}_1) \right\|_F \leq 0.1118$

4   CONCLUSION

Matrix $\mathbf{F}_k(\mathbf{A})$ is taking $\mathbf{A}$ as base matrix k generalized extended matrix, If the dimensions of the matrix are large, or for large matrices: When it has the generalized extended feature extension matrix, it can first find mother matrix $\mathbf{A}$ and the number of times k, then according to the properties of the parent matrix $\mathbf{A}$ launch of generalized extended matrix $\mathbf{F}_k(\mathbf{A})$ related conclusion. This paper discusses the relative perturbation bounds of generalized k time delay characteristics of spatial extension of matrix $\mathbf{F}_k(\mathbf{A})$ and its perturbation matrix $\mathbf{F}_k(\tilde{\mathbf{A}})$ in additive perturbations.

This paper only discusses the relative perturbation bounds of generalized k time delay characteristics of spatial extension of matrix $\mathbf{F}_k(\mathbf{A})$ and its perturbation matrix $\mathbf{F}_k(\tilde{\mathbf{A}})$ in additive perturbations. Using a similar way, it can also do more special extended matrices (such as weighted extended matrix and unitary extension of matrix) for further study.

REFERENCES

[1] Zou H-X, Wang D-J, Dai Q-H et al. Singular value decomposition of extended matrix [J]. Acta Electronica Sinica, 2001(3): 289–292.

[2] Zou H-X, Wang D-J, Dai Q-H et al. Singular value decomposition of extended matrix [J]. Chinese Science Bulletin, 2000(7): 1560–1562.

[3] Lin X-L, Wang Z, and Jiang Y-L. Singular value decomposition and Moore-penrose inverse for unitary extended matrix [J]. Pure and Applied Mathematics, 2008,24(1): 49–53.

[4] Sun J-G. Perturbation Analysis of Matrix [M]. Beijing: Science Press, 2001.

[5] Li R-C.Relative perturbation theory:I.Eigenvalue and singular value variations[J], Natrix Anal Appl,1998(19):956–982.

[6] Li R-C.Relative perturbation theory:II. Eigenvalue and singular subspace variations[J], Matrix Anal Appl,1999(20):471–492.

Information, Computer and Application Engineering – Liu, Sung & Yao (Eds)
© 2015 Taylor & Francis Group, London, ISBN 978-1-138-02717-6

"Province administrating county" financial reform: Based on the analysis of Shaanxi Province

Fan Gu & Miao Miao Li

Xi'an University of Architecture & Technology, China

ABSTRACT: After implementing the Province administrating county finance reform for seven years, Shaanxi Province had made significant progress.using the field survey method, induction and deduction method,mathematical statistics and other research tools, concluded that what the Shaanxi Province Straight county finance reform further improve is to standardize the relations between the government governance,financial powers and property rights,and to promote the urbanization process, straighten out the financial allocation of intergovernmental relations ,accelerate the implementation of the direct control of the county administrative reform step-by-step, and establish an effective monitoring mechanism. The county government also should highlight the change of perfecting government functions.

KEYWORDS: Province administrating county; Financial reform; Urbanization Process; Government Functions.

1 INTRODUCTION

"Provincial governing county" fiscal reform that is "fiscal provincial governing county" system reform, administrative reform is a Chinese major institutional innovation. Mainly refers to the inter-governmental fiscal adjustment and configuration management responsibilities in revenue, expenditure responsibilities, budgets pipeline, financial settlement, transfer payments, income reported solution, capital allocation, regulatory and other aspects of debt, the provincial finance and city and county finance A financial management system in direct contact[1]. Shaanxi province government departments, respectively, in 2007, 2009, 2012 issued implementation file about direct governing county. This paper describes the concrete implementation about the fiscal reform content in detail, problems appeared in the process of implementation and the what measures should be taken.

which have been included in the pilot counties have five counties; Guanzhong area containing 35 counties (cities), which included the pilot counties have 13 counties; Southern Shaanxi containing Ankang, Hanzhong, Shangluo three cities, 25 counties in nine counties included in the scope of the pilot counties. Specific pilot counties in all regions of the proportion shown in Figure 1.

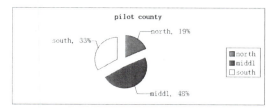

Figure 1. Regions included in the pilot counties in Shaanxi Province: the proportion diagram range.

2 THE IMPLEMENTATION PROCESS AND RESULTS OF SHAANXI PROVINCE'S FINANCIAL REFORM

2.1 The pilot program

Shaanxi Province have 83 counties (cities). According to geographical points, the Shaanxi Province is divided into northern Shaanxi, Guanzhong, Southern three regions, Yan'an and Yulin City, a total of 23 counties,

2.2 Initial results of "province administrating county" fiscal reform

From the perspective of zoning analysis, Shaanxi Province is divided into northern Shaanxi, Guanzhong, Southern three regions, that from northern Shaanxi, Guanzhong, Southern comparative analysis of these three regions were the direct control of the effectiveness of the implementation stage of the county finance reform.

In addition, according to the survey Shaanxi Province in 2007 to 2013 totals ranked counties, JIAXIAN rankings in economic reforms forward after 10ranking, from 2007 year's 83 to 2013 year's into the 73; Dingbian by 2007 year ranked 34, to the 2013year ranking in 15 overall forward close to the 19 to do so. Specific data shown in Table 1. Yijun County, Guanzhong area 56 from 2007year to 2013 year carry 50. For this reason advanced six. Town, County Southern region by 2007 year of 58 year carry a 15 ranking, the 2013year ranking of 43, also from 2007 Ningshan County carry 79 to 2013 year 68, carry a 11 ranking. Specific data shown in Table1.

Table 1. Shaanxi "Province governing county" reform pilot counties ranking table.

Region	County Name	GDP (1)	GDP (2)	Into ranking
Northern	Dingbian	34	15	19
Northern	Jiaxian	83	73	10
Guanzhong	Yijun	56	50	6
Southern	Zhenan	58	43	15
Southern	Ningshan	79	68	11

From the above data, the pilot regions made more obvious results since Shaanxi fiscal reforms implemented . Mainly in two ways. On the one hand is on the county's economic development. The pilot county finance system reform effectively alleviates the financial difficulties of counties and townships, and the major economic indicators improved significantly. On the other hand, from the direct control of the county finance reform institutional innovation perspective, fiscal reform to reduce the direct control of the county government level, the government realized the flat organizational structure.

3 THE MAIN ISSUES OF "DIRECT CONTROL OF THE COUNTY" FINANCIAL REFORM PROCESS

3.1 *The uneven development of Shaanxi regional economy and society*

Shaanxi land form is long and narrow, natural and geographical condition is hugely different, The province is divided into three distinct areas of different geography, history, culture and language etc. from north to south. Moreover, northern Shaanxi located on the plateau, the natural environment, there is more severe and the region is ranked less developed areas [2].

3.2 *The backward county economic development*

From the view of development of three major industries, our province economy facing following problems:In agriculture, arable land decreases, agricultural production base is weak, poor ability to withstand natural disasters, agricultural production base is weak, poor ability to withstand natural disasters, which restricts the steady development of the rural economy; In industry, goods prices, resulting in difficult in industry operating, industrial enterprise facing financing difficulties and the pressure of saving energy and emission reduction;In services, the consumer has little role in boosting the economy, service level is low.

3.3 *Defective of finance reform oversight system in the direct control of the county*

Shannxi Province implements "the direct control of the county" in a relatively short time. Little relevant experience to reference. Related polices and measures are not in place, being short of an appropriate institutional guarantee and legal basis, and relevant functional departments and oversight mechanisms have not been established.

3.4 *Lack of personnel resource becoming increasingly prominent*

Generally, the backward county economy is less attractive to talent which makes the outflow of personnel more serious and lead to the extremely scarce of talent.The survey found that in our province many grassroots financial staffs lacking professional knowledge and their management level is low.

4 SUGGESTIONS

4.1 *Promoting the urbanization process vigorously, and developing county economy positively*

Research to advance of the Shaanxi "direct control of the county," the pilot reform discovery that direct control of the county has a significant role in promoting the development of urbanization. Did the county the fiscal decentralization reform has important significance for urban economic development and the formation of urban circle [3].

4.2 Straighten out the financial allocation of intergovernmental relations

The current important factors influencing the fiscal system reform are the division between all levels of government powers is unclear [4]. Therefore, it is very important to improve the next provincial transfer payment system. To fully mobilize the enthusiasm of the provincial, city, county and other financial departments at all levels, and establish mechanisms to protect the interests of its municipal expenditure responsibilities linked. Fine-tuning of the provinces, cities and county's income in the case of basic maintenance of vested interests.

4.3 Accelerate the implementation of the direct control of the county administrative reform, and establish an effective monitoring mechanism

Flat reform of the administrative system is the strategic choice of Chinese current administrative system reform, the reform mainly triggered by social changes in China's political mobilization method, system transformation, service administration and information technology. The next strategic tasks and goals of China's local administrative division system and accompanying students further reform government management system is to adjust scale, reduce levels, enhance vitality.

4.4 The county government should highlight the change of perfecting government functions

An important prerequisite to achieve a flat zone management is rationalize government functions, a clear management responsibility at all levels of government and at all levels is an important basis of the flat area of reform. The basic protection of achieving a flat management is to improve the grass-roots social management, foster grass-roots social organizations, increase community capacity for self-management. Comprehensive administrative reform is an important guarantee for regional reform. In addition, there should have a good institutional environment.

REFERENCES

[1] Kexue Min. "Provincial County" Related Concepts [J], Administration and Law, 2014.
[2] Zhaoxiao E, Zhang Jun giant, Ren Menge research reform difficult issues, "provincial County" - in Shaanxi Province [J], Shaanxi Institute of Socialism, 2014 (1): 43–47.
[3] Tan Ling, Pan Xinhua. Shaanxi Province to support research finance "direct control of the county" under the umbrella of urbanization development [J], enterprise technology development, 2014,33 (3): 136–138.
[4] Wu Yongyi, Zhang Qian.Deepen Provincial Fiscal Reform - Reflections on Shaanxi Province "County" perspective of the reform [J], northwest of accounting, 2013:4–7.

Information, Computer and Application Engineering – Liu, Sung & Yao (Eds)
© 2015 Taylor & Francis Group, London, ISBN 978-1-138-02717-6

A study on management of foreign exchange and facilitation of trade investment

Wei Liu & Fang Fang Chen
Guizhou University of Finance & Economics, Guiyang, Guizhou, China

ABSTRACT: Analysis of current development of China's foreign exchange management, provided a detailed analysis and summary of some of the existing problems, suggested relevant solutions with suitable measures in order to make greater contributions to the development of the related fields in China.

KEYWORDS: foreign exchange management; trade and investment facilitation; effects

1 INTRODUCTION

With the rapid and healthy development of China's current economy, in the aspect of foreign exchange management, the main starting point and end point refer to the development of the service economy. In order to more effectively deal with the economic situation in the volatile international economy, China should continue to adjust the foreign exchange management measures and promotion, make the market subject meet the relevant environment to the maximum. At the same time, it also requires relevant staff to think and explore management concepts, innovative methods. By doing so, it ensures a normal, healthy operation and development of a more effective foreign exchange management, thus more effectively promote China's comprehensive national strength.

2 THE MEANING AND THE STATUS QUO OF THE TRADE AND INVESTMENT FACILITATION IN CHINA

2.1 Definitions of trade and investment facilitation in China

We can summarize trade and investment facilitation as a simplification of the trade and investment program, also suitable for co-ordination and other related laws and regulations. It is designed to respect the legal system and so be open, fair and impartial, this way, they can effectively provide relatively simplified procedures and steps for trade and investment to effectively implement the relevant elements of cross-border flows.

2.2 The purpose of trade and investment facilitation in China

In China's current foreign exchange management, its main goal in promoting trade and investment facilitation is the integration and unification of enterprise fund management and the management objectives of the foreign exchange bureau. In the present stage of China's rapid economic development, foreign exchange management mainly serves the purpose of promoting the international balance of payments; in order to maintain the relationship of mutual restriction between nations, effectively avoid some unnecessary disputes.

Enterprises have relevant requirements in terms of foreign exchange management, it mainly includes the following aspects:

First of all, it refers to the reduction of examination and approval link. The implementation of it can effectively reduce the cost of management to realize the reasonable distribution of resources.

Secondly, it refers to the opening of a convenient method of financing. In this way, they can have a more flexible allocate one of funds, and thus speed up the realization of simplified and convenient operation steps in trade investment.

Finally, it refers to the effective reduction of exchange rate risk. In this process, enterprises can analyze and decide on the types of currency they are

Holding, according to their own independent views in accordance with the current changes in the exchange rate. In addition, within the controlled risk range, the most important starting point and the foothold is to meet the demands of related enterprises to the maximum extent.

2.3 The relationship between capital account convertibility and trade investment facilitation

At present, in terms of trade and investment facilitation, China has been persistent in carrying out secure and advancing programs during the process in which capital projects are promoted. However, in this process, capital account convertibility in this regard, a lot of sectors are involved, which need not only to advance our reform of the exchange rate and interest rates, but also progress accordingly.

3 ACTIVELY CATER THE DIVERSE REQUIREMENTS IN CHINA'S TRADE AND INVESTMENT FACILITATION

Under the premise of China's current rapid economic development, facilitation of trade and investment has gradually involved in many aspects, including: policy support, management, transmission and other aspects of the information. Among them, the foreign exchange bureau as an important foreign currency administration, its development and service conditions should advance accordingly, at the same time, it also needs constant innovation and progress.

3.1 Continue to extend the range of foreign exchange services

With the development of China's current urbanization process, and the effective results from the reform and opening up on the current economy, China's foreign trade and economic development have been rapid, over time, the scale of China's foreign trade development has been constantly expanding, it became one of the indispensable components under China's rapid economic development. It is the rapid development of China's foreign trade economy that demanded more requirements from the current foreign exchange services.

3.2 Constantly adjusting the foreign exchange management services approach

In recent years, China's pace of development and advancement of State Administration of Foreign Exchange, has followed the pace of speculative trade facilitation, in the process, it developed a number of effective regulations and policies, which played a positive promoting effect on China's current foreign exchange sector. At the same time, China specially formulated "personal foreign exchange management approach" in accordance with our actual current development situation. The promulgation and implementation of these regulations and measures, not only effectively changed the current foreign

exchange management methods and means, but also to some extent, impacted the traditional measures in foreign exchange services.

3.3 The organic integration of foreign exchange services resource

Combined with our current situation, China is still at an important stage of transitioning to a market economy, it has not yet established a very regulated market subject behavior, this way, it indirectly led to China's presence in terms of foreign exchange links seriously out of touch with disharmony.

4 MORE EMPHASIS ON LONG TERM WORKING MECHANISM OF FOREIGN EXCHANGE SERVICES

To do a better job of foreign exchange services, it must have a long-term preparation. This is because foreign exchange service is not to rely solely on innovation and improving working methods, it must establish a sustainable, efficient foreign exchange service. In order to better promote the development of related undertakings, we need relevant staff to give more concern and attention to foreign exchange service in the future, meanwhile, we also need them to continually improve and enhance foreign exchange services.

In recent years, our society and communities do not have a full and correct understanding of the functions and responsibilities of the foreign exchange bureau, to a certain extent, this eased the management of related business, which results in inefficient services. At the same time, it is because foreign exchange bureau's relatively low management's ability, it indirectly caused the service to be inefficient, this will lead to awkward situations like passive services. Therefore, the relevant staff in the foreign exchange bureau, should strengthen the publicity and guidance of foreign exchange policy in their daily work, ensure relevant information can be known to the public in time. Meanwhile, the foreign exchange bureau should continue to strengthen communication and exchanges between itself and the people, ensure it can solve existing problems in the shortest possible time and in the most effective way.

5 SUMMARY

In summary, our country is currently in the case of rapid economic development, in the trend of economic globalization, foreign exchange management departments should appropriately play their role and exert their significance. Achieve the facilitation of trade and investment through foreign exchange management; this has a very important significance to our current economic situation. As a result, the staff of foreign exchange management faces more new requirements, it asks relevant staff to make certain contributions the development of related business in our country.

REFERENCES

Qiang SHAN. Foreign exchange management promotes Trade and investment facilitation research. Beijing Financial Review. 2013 (01).

Shangpu FANG. Strengthen foreign exchange managment system to actively promote trade and investment facilitation. Financial Development Review. 2013(09).

Juanjuan NIU. Foreign exchange management reform: Aimed at trade and investment facilitation objectives . Financial Times. 2012 (02).

Information, Computer and Application Engineering – Liu, Sung & Yao (Eds)
© *2015 Taylor & Francis Group, London, ISBN 978-1-138-02717-6*

Research on higher vocational education of Linyi city and harmonious regional economic development

Min Zheng

Business School of Linyi University, Linyi, Shandong, P.R.China

ABSTRACT: It points out the disjoint phenomenon between higher vocational education and regional economic development of Linyi city by analysis of the current situation. It can be improved by perfecting the professional structure with the regional supporting industry at the center, linking the main regional industry to make the applied and developmental research and constantly expanding the regional service.

KEYWORDS: Higher Vocational Education; Regional Economic; Harmonious Development

1 INTRODUCTION

Coordinated Development of Higher Vocational Education and Regional Economic directly affects the development prospects of local vocational colleges and socioeconomic transformation and upgrading. But generally speaking, whether national or Linyi City, the development of higher vocational education centers, which are training bases of higher technical staff, is still the weak point, there is a series of difficulties and problems in the development process, which does not meet the needs of local economic and social development. Strengthening the research on Higher Vocational Education and local economic and social development in harmony, will not only solve practical problems in the development of vocational education, but also promote the coordinated vocational education and local economic and social development.

2 STATUS QUO OF HIGHER VOCATIONAL EDUCATION OF LINYI CITY

Linyi city is located in the southeast of Shandong Province, it has a population of about 10 million. In recent years, higher vocational colleges provide high school graduates opportunities to accept college education and promote the development of regional economy of the city. With the help of local government, the higher vocational colleges actively take part in the joint school with the local enterprises and colleges, which not only expand the source of funding, but also cultivate hundreds of creative and practical high- quality talents for the enterprises. However, there are still many serious problems to be resolved.

2.1 *Shortage of source of funding*

Development of higher vocational education needs a corresponding increase in economic investment as a precondition. In recent years, China has been keeping fast in economic development, but input of education accounts for only 4% of the total economy. Limited education investment has been put into the construction of the famous universities. Very little source of funding is invested towards higher vocational colleges. Taking Linyi Vocational College as an example, scarce investment is one of the most serious problems they face. Because of the shortage of source of funding, the college education funds cannot meet the demand for school expansion, renewal of experimental equipments and introduction of teachers, which make it hard to meet the needs of local economic and social development.

2.2 *Aging teaching mode*

Currently higher vocational schools in Linyi mostly focus on employment, emphasize on career orientation, and cultivate skills-based talents. However, from the perspective of social demands, more attention should be paid to cultivate a general occupational ability and interdisciplinary talents not just the aim of employment. This requires vocational colleges to change the traditional educational pattern and fixed long-term system. Measures should be taken to change the old tendency of focusing just on the diploma, improve the hysteretic professional teaching facilities and rigid teaching methods. In a word, a new teaching mode should be established to satisfy the diverse vocational education needs which vary with the development of society.

2.3 Inappropriate major settings from the regional needs

On the whole, higher vocational education of Linyi cannot be connected to the leading industry, which results in the lack of high-quality applied talents and supply-demand contradiction. From the major setting of the four vocational colleges in Linyi, one common serious problem is disconnected combination of local economic development and convergence of major setting, which makes the leading industry hard to find the talents they need. The basic function of higher vocational colleges is to serve for the regional economy and social development. In order to better serve local social and economic development, higher vocational colleges need to know the local culture, history, regional characteristics and talents it needs to develop regional economy. As an agricultural city, the agricultural industry is the primary industry in Linyi. However, the lack of high skilled talents in this area seriously affects the development of modern agriculture and agricultural science promotion. According to the statistics, there is only one higher vocational college designing agriculture-related major. Most of the majors are so-called hot majors or already saturated majors such as accounting, marketing, computer science, etc. Such major setting, due to the inability to be connected to the leading industry, makes it difficult to cultivate talents who will adapt to the regional economic development of Linyi. Moreover, lack of regional characteristics directly impacts the recruitment and employment of vocational colleges, which affect the realization of long-term development.

2.4 More professional teachers should be concluded in the faculty

At present, teachers in Linyi higher vocational school are more, but the "expert" teachers are not so many. Scarce professional knowledge, unskilled working skills and weak practical abilities have a certain impact on the development of Vocational education. Four Linyi higher vocational college teachers generally exist unbalanced structure, weak, teaching enthusiasm not higher question, specifically in the part of the teacher's education idea lag, knowledge structure, professional structure and the title structure is unreasonable, the lack of stamina to improve teaching ability, especially the "double teacher" teacher shortage etc. This is in appropriate to the positioning, cultivating targets and developmental directions of higher vocational colleges. In addition, due to lack of funds, the introduction of high level teachers becomes more difficult. The existing teachers cannot meet the old specialty reform, new course teaching needs, high skill "double teacher" teacher ratio is not high, which affects the promotion the quality of teaching.

3 HIGHER VOCATIONAL EDUCATION IS NOT IN LINE WITH THE SOCIAL AND ECONOMIC DEVELOPMENT OF LINYI

Higher vocational education in our country is facing a crucial stage of reform and development, play system, regional economic consciousness is not strong, the market mechanism is ineffective and so on restricted the regional economic function of higher vocational education to a great extent. Linyi higher vocational education also exists the problems mentioned above, the specific performance in the following aspects:.

3.1 Levels of higher vocational education is not high, branded characteristics are not formed

While in Linyi regional economic positioning drive, higher vocational education level of the vocational colleges within the region have obvious development, the higher vocational education level is gradually improving, but on the whole, the higher vocational education level has to be improved. At present, Linyi higher vocational education is mainly simple skill based training, such as auto repair, hairdressing, computer repair services and so on, these training programs cannot meet the demand of modern service industry in Linyi city. At the same time, higher vocational education in Linyi City, the characteristics of the project is still relatively small, still needs to be improved in a number of brands, in the depth and breadth are pending further excavation.

3.2 Lack of management and financial investment of government and inaccurate positioning of higher vocational colleges

The government is insufficient investment in higher vocational colleges, should further increase investment in vocational education, especially in the government allowance training project, should increase investment, to expand all kinds of occupation training propaganda range. At the same time, the city of Linyi has occupation colleges and various training institutions, the different system, different attribution, they basically do not communicate with each other. On its location many vocational colleges is not clear enough, in a professional setting and curriculum development it still stays in the original follow level, often only see the immediate interests, while ignoring the influence of higher education to the entire regional economy's long-term development, also not be combined with the urban area of Linyi region economy orientation to form the characteristic of Higher Vocational Education of its own, this will directly affect the sustainable development of Higher Vocational Education in Linyi city.

3.3 Higher vocational colleges and enterprises are not closely linked

Higher vocational education is different from general education, compared to the general education, it has direction and goals - clear that adapt to the demand of regional economic development, the cultivation of applied talents for regional. In other words, occupation colleges should transport all kinds of talents for enterprises within the region. Enterprises should not only the subject carrying out higher vocational education and training, but also the biggest beneficiaries of higher vocational education and training. The school must rely on industrial enterprises to make the characteristic and provide the support for the regional human resource. At present Linyi city occupation colleges is not completely get rid of the entrance education and good project to follow the trend of the pattern, relationship and enterprise is also relatively loose, not consciously play the advantages of industry, enterprises in resources, technology, information, professional set up schools not according to the actual needs of enterprises, not only affects the occupation college students the employment rate, but also directly related to the sustainable development of regional economy.

3.4 Lack of social service ability

One of the most important functions of local higher vocational college is to serve for the economic and social development of district. The actual situation of four Higher Vocational Colleges from Linyi City, the breadth and depth of social services is very limited, as few involved in Yimeng Mountain Area *New Farmers* training, not to mention the promotion to participate in the new rural construction and the modern agricultural technology; rarely participate in the local laid-off workers training; docking local leading research, application develops industrial area by offering support services and more inadequate.

4 PATH CHOICE OF HIGHER VOCATIONAL EDUCATION AND HARMONIOUS REGIONAL ECONOMIC DEVELOPMENT

4.1 Accurate self-positioning

From the view of vocational education characteristic, their training mode should orient and have characteristic.. Higher vocational education must adhere to the service for the purpose, the employment as the orientation, face the society and the market direction. Three aspects should be focused on: 1) Positioning of training objectives. The development goal of occupational education, is to cultivate high-quality, high-skilled works. They should build strong competitiveness of technical talents facing the market competition, cultivate the students' innovation ability, promotes students' employment rate. 2) Positioning of specialty. Each school should aim at market professional, highlighting the characteristics of specialty, the cultivation of students' occupational skill in a prominent position education, to the positive transformation of the old professional, development of new specialty, build quality professional. 3) Positioning of educational levels. Higher vocational education should adapt to the market demand for multi-level of technical personnel, do a good job of positioning the school level, to adapt to the requirement of professional personnel training.

4.2 Deepen the reform of the personnel system, strengthen team building

Teachers in local higher vocational colleges should have the responsibility of cultivating high-quality applied talents for the regional economy and society. To strengthen the construction of teaching staff is the most important guarantee for strengthening the basis for capacity building of Local Higher Vocational colleges. One is to stabilize the contingent of teachers, to the cause of cohesion of the people, improve treatment from humane care, and improve working conditions and other aspects of stable teacher team. Next is to pay attention to the innovation of teachers. The characteristics of higher vocational colleges require the teachers should not only have the higher moral level and the knowledge structure system, but also must have a wealth of practical experience and the actual operation ability. Last is the innovative way of the teachers' training. Through training, we should enhance the teachers' teaching level and practice ability, to adapt to the requirements for teachers of local economy and the development of leading industry.

4.3 Connection of higher vocational education and regional economic development

4.3.1 Focus on the regional pillar industry, unceasingly optimize the specialty structure

Local higher vocational colleges should effectively connect major specialty and leading industries in this area, have a clear idea of group construction as a breakthrough. In accordance with the pillar industry of the effective docking place construction of key specialty based on task, as the link, in order to focus on the professional leading, by one or several similar related professional and the professional direction of the common component of the construction of idea of professional group. The construction of specialty group work carefully planned school. In addition, special attention should be paid to the professional setting must be based on the local leading industry development and the pace of upgrading Pay full attention to the leading industry in the development process of new technologies, new processes and new materials, production, timely opening of new specialty; also to offer a professional update and supplement the relevant knowledge and skill. According to the 12th Five-year plan of Linyi city, professional construction and adjust the direction of Higher Vocational Colleges in Linyi city are: constructing the electromechanical integration technology, modern agriculture, culture industry (especially the red cultural industry) and professional logistics and building professional as the leader of the five characteristics of professional group, adding agriculture, horticulture gardening techniques, technology professionals, electromechanical integration technology specialty has been set should focus on the direction of agricultural machinery manufacturing, building professional courses added to the construction of new rural areas.

4.3.2 Connection of local leading industry, launch developmental and applied research

Linyi city is a big agricultural city, modern agriculture is the leading industry of the city, the largest scale cultivation of agricultural products, agricultural products especially in the main planting garlic. In recent years, from the export of garlic consumption situation, garlic and its products in Linyi city has been a great impact, its profit space is gradually shrinking. In order to improve the added value of garlic, the Millennium Yimeng Mountain garlic has a qualitative change, is a major issue facing modern agriculture in Linyi city. Linyi Career Academy School of this problem as a local leading industry to carry out docking main subject of application, research and development, by reference to the relevant information of food rich in selenium. In 2008, it invested RMB1700,000 funds for scientific research, research and from the Agricultural Research Institute of Linyi city invited experts engaged inactive organic selenium garlic. Through unremitting efforts, in 2009 the development of Se enriched garlic planting in Shandong Province Medical Food Research Institute research organs, etc. after testing, the indicators have reached the normal value, the content of selenium is 6650 times the ordinary garlic, potential value is inestimable. This achievement made Linyi garlic industry completed optimization and upgrading.

4.3.3 Expanding the ability of serving regional development

Local higher vocational college should actively use various kinds of resources, establish training platform type which is rich and diverse. In the case of Linyi, the local higher vocational colleges should actively participate in the new rural construction, train new farmers, train more agricultural technology promotion personnel for the development of modern agriculture

Linyi city's economic development is in a transition stage, the adjustment of local industry and upgrading continuously, especially the differentiation of jobs in the technical level, the post of technical level, thus to the practitioners of technical requirements significantly improved, have strong demand for local higher vocational education development. Therefore, in the background of regional economy under the existing local higher vocational colleges should be the first step to change the concept, to meet the different levels of market demand, to win a new round of development opportunities.

4.3.4 Government should assume more responsibilities and expand financial investment

From 2008 to 2013, Linyi city general budget revenue at an average annual growth rate of 27.1%, while investment expenditures on education in Linyi city general budget revenue at an average annual growth rate of 27.1%, while investment expenditures on education in Linyi City, the Treasury has remained at about 2.8%, especially the investment of vocational education less investment the proportion of funds is obviously insufficient, has become one of the major factors restricting the development of Linyi City Vocational Education activities. Therefore, the local government in particular, to increase public financial investment in higher vocational education, the gradual implementation of the system of financial allocation per student, and in the discount interest loans, project declaration, laboratory construction and so on to give more support and tilt, to ensure sound and rapid development of Higher Vocational colleges.

REFERENCES

[1] X.F.Sun, Vocational education servicing regional economic the research of development strategy, *Hunan Agricultural University Journal*,vol.115, pp.85–90,2011.
[2] Y.L.Yan, Study on the relationship between Shandong Province higher occupation education and the regional economic coordinated development, *Engineering Master Papers of Shandong University*, pp.41–45, 2013.
[3] C.L.Hou, Higher vocational education reform and regional economic development, *Science Press*, 2011.
[4] W.Li and J.Liu, Planning and Development of Higher Vocational Education Institutions in China: Problems and New Policies[J/OL]. [2013-10-1].
[5] M.Coles and T.Leney, The regional perspective of vocational education and training, *International Handbook of Education for Changing World*, pp.411–425, 2009.

Information, Computer and Application Engineering – Liu, Sung & Yao (Eds)
© 2015 Taylor & Francis Group, London, ISBN 978-1-138-02717-6

The research on enhancing self-efficacy in college English learning

Shi Fang Wen
School of Foreign Languages, Shandong University of Traditional Chinese Medicine, Shandong, China

Wen Shuang Bao
School of Accountancy, Shandong Management University, Shandong, China

ABSTRACT: Low learning efficiency, lack of self-efficacy always lead to anxiety, impulsive, even resentment in English learning for college students. Lack of self-motivation methods and low self-efficacy are the main reason. You can't solve this problem if you focus on the perspective of students only. For the college English instructors, best teaching methods and reasonable teaching goals are also of great significance for the learners. More interactive sessions and emotional integration can highly improve the learners' interest in English learning, thereby enhancing their self-learning efficacy.

KEYWORDS: English Learning; learning efficiency; self-efficacy.

1 INTRODUCTION

College English Teaching Curriculum Requirement issued by Higher Education Bureau of the Ministry of Education of China In January 2004 made the personalized, autonomous internet teaching mode become goal and trend of college English teaching reform, which leads to a significant educational shift in English teaching from teacher-centered to student-centered, and it's now no longer the pattern of learning English that you do what the teacher ask you to without thinking much which is the so-called exam-oriented education.

Since the far-reaching impact of social attention on English learning and university CET exam, the freshman shows a great enthusiasm in learning English after they come to the college. But what they don't expect is that college English learning is completely different from that in high school, which makes them down. They do recite the words, read the text as they did in high school, but they get nothing from that. Some of them can't pass the CET-4. All the reason is that college English learning is no longer exam-oriented, it comes with a shift from teacher-centered to student-centered.

Deviation from expectations, the pressure and frustration of learning English lead to their anxiety and other allergic reactions as well as frustration, depression, impulsiveness and any other negative emotions, which reflects the fact that they are low in self-efficacy and lack of motivation to learn. So the most urgent task in college English teaching is to help students to get rid of exam-oriented learning,

rebuild their confidence and temper their will thus to strengthen the sense of self-efficacy.

2 SELF-EFFICACY

2.1 *The definition of self-efficacy*

According to Bandura, self-efficacy is a major determinant of human motivation, which is affected by outcome expectancies and efficacy expectancies. An outcome expectancy is defined as a person's estimate that a given behavior will lead to certain outcomes. An efficacy expectation is the conviction that one can successfully execute the behavior required to produce the outcomes. Individuals can believe that a particular course of action will produce certain outcomes, but if they entertain serious doubts about whether they can perform the necessary activities, such information does not influence their behavior. Briefly, people behave in accordance with what they believe, rather than in accordance with their actual capabilities; it is individuals' beliefs about their capabilities, rather than their actual capabilities, that accurately predict performance attainments.

Self-efficacy beliefs, which lie at the very core of social cognitive theory, determines"whether coping behavior will be initiated, how much effort will be expended, and how long effort will be sustained in the face of obstacles and aversive experiences". The higher the perceived efficacy, the greater is the sustained involvement in the activities and subsequent achievement.

In Bandura's self-efficacy theory, several other terms such as efficacy beliefs, self-efficacy beliefs, efficacy expectations, the perceived efficacy or Perceived self-efficacy, sense of efficacy and perceptions of efficacy mean the same as self-efficacy. They are interchangeably used. Moreover, self-efficacy is a domain-specific assessment of competence to perform certain tasks. Therefore, there are different specific feelings of self-efficacy. For example, we have mathematics self-efficacy, writing self-efficacy, sport self-efficacy, etc. In this presentation, we are concerned with self-efficacy in English learning, which only refers to students' beliefs in their capabilities to learn English.

2.2 The sources of self-efficacy

Generally, people's beliefs about efficacy could be developed in four major sources: mastery experience, vicarious experience, verbal persuasion and physiological and emotional states.

The most influential source is the interpreted result of one's previous performance, or mastery experience. Outcomes interpreted as successful raise self-efficacy; those interpreted as failures lower it. After strong efficacy expectations are developed through repeated success, the negative impact of occasional failures is likely to be reduced.

The second source of self-efficacy information is the vicarious experience of the effects produced by the actions of others. Seeing or visualizing other similar people perform successfully can raise self-percepts of efficacy in observers that they too possess the capabilities to master comparable activities. By the same token, observing others perceived to be similarly competent failing despite high effort lowers observers' judgments of their own capabilities and undermines their efforts.

Verbal persuasion is the third source of self-efficacy beliefs which is widely used to try to talk people into believing that they possess the skills needed for success at a given task. Positive persuasions may work to encourage and empower; negative persuasion can work to defeat and weaken self-efficacy beliefs.

Physiological states such as stress, tension and anxiety also provide information about self-efficacy beliefs. Mood also affects people's judgments of their personal efficacy. Positive mood enhances perceived self-efficacy; despondent mood diminishes it.

2.3 The influences of self-efficacy

Self-efficacy beliefs as determinants of behavior have great influences on human accomplishment and personal well-beings in several ways. They influence the choices people make and the courses of action

they pursue. In general, people are more comfortable attempting tasks and activities at which they believe themselves capable and tend to avoid those situations they believe exceed their capabilities. Self-efficacy beliefs also help determine how much effort people will expend, how long they will persist in the face of obstacles or aversive experiences and how resilient they will be when confronting setbacks and obstacles. The higher the sense of efficacy, the greater the effort, persistence, and resilience. Additionally, people's judgments of their capabilities influence their thought patterns and emotional reactions. Self-efficacy beliefs influence the amount of stress and anxiety people experience as they engage in an activity. Inefficacious individuals dwell on their personal deficiencies and imagine potential difficulties as more formidable than they really are. Confident students engage in tasks with serenity; those who lack confidence can experience great apprehension. Consequently, self-efficacy beliefs powerfully influence the level of accomplishment students ultimately realize.

3 ENHANCING SELF-EFFICACY IN COLLEGE ENGLISH LEARNING

The results of the present study manifest that students' self-efficacy in English learning is significantly correlated with their academic outcomes and their autonomy in English learning. Therefore, students' self-efficacy is worth stressing and full attention should be given to the enhancement of efficacy precepts. English teachers are recommended to make every effort to help students develop students' self-efficacy in English learning.

3.1 Providing students with opportunities to experience success

Among the four sources that develop efficacy beliefs, the interpreted result of one's purposive performance or mastery experience, which refers to an individual's experience with success or failure in the past situations, is the most influential. In other words, previous success is an important source of self-efficacy. According to the self-efficacy theory, successes raise efficacy appraisals while repeated failures lower them, therefore the teacher, as an educator, should always try to increase the chance of every student's success. Generally speaking, if students have more chances to experience success in English learning or they succeed in finishing a learning task that they themselves think a little difficult, they will feel more confident in their ability to learn English, i.e. their English self-efficacy will be raised. However, past failures will lower perceived self-efficacy, so the effects of past academic failures in English learning need to be minimized. How to make the students own more successful experience is an issue that teachers are supposed to care about. For example, teachers can assign students learning tasks according to their English proficiency: successful English learners can be assigned to perform more challenging tasks, whereas intermediate learners can be asked to take tasks that are moderately difficult with unsuccessful learners given the easiest tasks. Hence, every student can find their progress and experience the pleasure of success, leading them to gain confidence and enhance their English self-efficacy.

3.2 Setting students appropriate peer models

Modeling is a form of social comparison and much human behavior is developed through modeling. Social comparative information would enhance skillful performance and percepts of efficacy. Social comparative information constitutes a vicarious source of efficacy information. Vicarious experience is another important source of self-efficacy. Individuals compare themselves with peers who they think are similar in ability and intelligence to themselves. Peer models may be especially helpful to students who hold self-doubts about their capabilities for learning or performing well. We believe that students acquire information about their own capabilities through knowledge of how peers perform. Observing similar peers successfully perform a task can raise self-efficacy in students because they may believe that if the peers can learn, they can also improve their skills. In classroom activities, teachers can provide learners with social comparative information. Moreover, teachers can set different levels of peer models for students with different learning levels so that every student is encouraged by his own model.

3.3 Encouraging students to use self-talk positively

Physical states such as tension and anxiety exert influence on efficacy beliefs. Many learners, especially low achievers, have been strongly affected by years of negative self-talk such as "I'll never get this", "I'm always making mistakes", and so forth. They can be taught to tell themselves "I did that well", "I can learn this" or "I can do better next time" in order to reinforce their beliefs about their ability to learn.

3.4 Instructing students to attribute

Attribution theory states that what we see as the causes of our past successes or failures will affect our expectations and, through them, our performance.

According to Weiner's dimensional taxonomies of the causes of success and failure, ability and effort are internal factors, whereas task difficulty and luck are external factors; task difficulty and ability are stable factors while effort and luck change in different situations. Students will experience success or failure in the course of learning. Self-efficacy beliefs and their attribution style are interrelated: high efficacious students attribute their success to internal factors (ability and effort) and failure to controllable factors (lack of effort), which leads to the enhancement of self-efficacy. In contrast, inefficacious students attribute success to external factors and failure to uncontrollable factors (lack of ability), which may decrease their precepts of self-efficacy. It is advisable to teach students to link their success in English learning to effort and ability and failure to lack of effort rather than lack of ability.

3.5 Guiding students to set appropriate goals

This study demonstrated that goal setting is one of the predictors of self-efficacy in English learning. We hold that students' goals should be realistic- challenging but attainable. With realistic goals, students can monitor progress and decide on a different task approach if their present one is ineffective. Self-efficacy is increased as students note progress, attain goals, and set new challenges. Goals set too high or too low do not enhance achievement beliefs. Students perceive little progress if their goals are too high, which is likely to lower their self-efficacy and lead them to give up readily. And easy goals do not produce high self-efficacy because they do not inform students about what they are capable of doing. Therefore, it is recommended to guide students to set specific and proximal goals to increase their self-efficacy in English learning.

4 SUMMARY

As we all know the Chinese saying goes,'Give a man a fish and you feed him for a day. Teach a man to fish and you feed him for a lifetime.'College English Learning is a matter related both students and teachers, good teaching method and reasonable teaching goal make you a good teacher; a high sense of self-efficacy and learning project make you a better learner. In the present study, a simple strategy of English-learning was presented to both instructors and learners, we truly hope that the instructors can make a better teaching-program and at the same time the learners can enjoy the process of learning.

REFERENCES

Arnold, J. Affect in Language Learning. Beijing: *Foreign Language Teaching and Research Press*, 2000.

Benson, P. Teaching and Researching Autonomy in Language Learning. Beijing: *Foreign Language Teaching and Research Press*, 2005.

Bouffard-Bouchard, T. Parent, S. & Larivee, S. Influence of self-efficacy on self-regulation and performance among junior and senior highschool aged students. *International Journal of Behavioral Development*, 1991, 14:153–164.

Brown, H.D. Principles of Language Learning and Teaching. *Longman Press*, 2000.

Chan, V. Readiness for learner autonomy: What do our learners tell us? *Teaching in High Education*, 2001, 6(4):505–518.

Cotterall, S. Promoting learner autonomy through the curriculum: principles for designing language courses. *ELT Journal*, 2000, 54(2):109–117.

Davies, P. & Pearse, E. Success in English Teaching. Shanghai: *Shanghai Foreign Language Education Press*, 2002.

Gardner, D. & Miller, L. Establishing Self-access: from Theory to Practice. Shanghai: *Shanghai Foreign Language Education Press*, 2002.

Pajares, F. Gender and perceived self-efficacy in self-regulated earning. *Theory into Practice*, 2002, 41(2):116–125.

Shih, Shushen. Children's self-efficacy beliefs, goal-setting behaviors, and self-regulated learning. *Journal of National Taipei Teachers College*, 2002, (1):263–282.

Information, Computer and Application Engineering – Liu, Sung & Yao (Eds)
© *2015 Taylor & Francis Group, London, ISBN 978-1-138-02717-6*

Curriculum development of higher vocational college based on employment-orientation

R.J. Wu

Henan Information and Statistics Vocational College, Zhengzhou, Henan, China

ABSTRACT: A model associated with employment-oriented of curriculum development is disclosed. The model, which relates to the roles of the occupation, includes three steps, namely Role Analysis, the Evaluation Model, the UCSD Model. With the model, the development of curriculums can be more standardized and quantified.

KEYWORDS: Curriculum Development; Employment Oriented; Role Analysis; AHP.

1 INTRODUCTION

The curriculum development of Higher Vocational College should emphasize the knowledge and skills. With the deeper research on occupation, there are many methods of curriculum development have been found, such as MES (Modules of Employable Skills) and CBE (Competence-based Education).

For the curriculums in unstable fields where the responsibilities not clear, the author adopted RAT (Role-Analysis-Technique) to find the requirements of knowledge and competence for the professional role of the students, and developing curriculums.

2 OCCUPATIONAL ROLE ANALYSIS

Occupational role is the intermediary of the interaction between the individual and the social environment. The structure of occupational roles includes the rights, obligations, ideas, abilities, behaviors and ways of thinking. The rights and obligations of the role can't be changed via education, so the analysis of occupational role can proceed from the ideas, abilities, behaviors and ways of thinking. Role attributes and the characteristics of typical activities can be obtained through interviews combined with literature review, the author then further defined the role and role ability using the Delphi method.

An example for role analysis of administration is shown in table 1.

3 ESTABLISHMENT OF THE EVALUATION MODEL

According to the role analysis of administration in table 1, the author established an indicator system for

Table 1. Role analysis of administration.

Role	Typical Activities
Abilities	Communication
	Teamwork
	Business ability
	Innovation
	Training
Behaviors	Expression
	Planning
	Organization
	Leadership
	Controlling
Ideas	Economy
	Law
	Administration
	Operation management
	Marketing
	Human resource
	Financial management
Ways of thinking	Code of conduct
	Ideology
	Value orientation
	Mode of thinking

business administration specialty, and then adopted AHP (Analytic Hierarchy Process) to build the evaluation model.

The evaluation mode is shown in Equation 1.

$$M = \begin{bmatrix} m_1 \\ m_2 \\ ... \\ m_j \end{bmatrix} \tag{1}$$

Table 1. Indicator system.

A	B	C	D
Training target (A_1)	Employment oriented (B_1)	Abilities (C_1)	D_1 D_2 D_3 D_4 D_5
		Behaviors (C_2)	D_6 D_7 D_8 D_9 D_{10}
	Knowledge oriented (B_2)	Ideas (C_3)	D_{11} D_{12} D_{13} D_{14} D_{15} D_{16} D_{17}
		Ways of thinking(C_4)	D_{18} D_{19} D_{20} D_{21}

Table 2. Ranking scale for criteria and alternatives.

Intensity of importance	Definition
1	Equal importance
3	Somewhat more important
5	Much more important
7	Very much more important
9	Absolutely more important
2,4,6,8	Intermediate values

Table 3. Pairwise comparison matrix for A_1.

A_1	B_1	B_2	$\left(\prod_{j=1}^{2} S_{ij}\right)^{\frac{1}{2}}$	Priority vector
B_1	1	1	1	0.5
B_2	1	1	1	0.5

CI = 0, CR = 0 < 0.1 OK.

Table 4. Pairwise comparison matrix for B_1.

B_1	C_1	C_2	$\left(\prod_{j=1}^{2} S_{ij}\right)^{\frac{1}{2}}$	Priority vector
C_1	1	2	1.41	0.67
C_2	1/2	1	0.71	0.33

CI = 0, CR = 0 < 0.1 OK.

Table 5. Pairwise comparison matrix for B_2.

B_1	C_{31}	C_{42}	$\left(\prod_{j=1}^{2} S_{ij}\right)^{\frac{1}{2}}$	Priority vector
C_{31}	1	2	1.41	0.67
C_{42}	1/2	1	0.71	0.33

CI = 0, CR = 0 < 0.1 OK.

Table 6. Pairwise comparison matrix for C_1.

C_1	D_1	D_2	D_3	D_4	D_5	$\left(\prod_{j=1}^{2} S_{ij}\right)^{\frac{1}{2}}$	Priority vector
D_1	1	2	4	5	7	3.09	0.44
D_2	1/2	1	3	5	7	2.21	0.32
D_3	1/4	1/3	1	3	2	0.87	0.12
D_4	1/5	1/5	1/3	1	2	0.48	0.07
D_5	1/7	1/7	1/2	1/2	1	0.35	0.05

CI = 0.0384, CR = 0.0343 < 0.1 OK.

Table 7. Pairwise comparison matrix for C_2.

C_2	D_6	D_7	D_8	D_9	D_{10}	$\left(\prod_{j=1}^{2} S_{ij}\right)^{\frac{1}{2}}$	Priority vector
D_6	1	1/3	1/7	1/9	1/3	0.28	0.04
D_7	3	1	1/5	1/7	1/3	0.49	0.06
D_8	7	5	1	1/3	2	1.88	0.24
D_9	9	7	3	1	6	4.08	0.53
D_{10}	3	3	1/2	1/6	1	0.94	0.12

CI = 0.0550, CR = 0.0491 < 0.1 OK.

Table 8. Pairwise comparison matrix for C_3.

C_3	D_{11}	D_{12}	D_{13}	D_{14}	D_{15}	D_{16}	D_{17}	$\left(\prod_{j=1}^{2} S_{ij}\right)^{\frac{1}{2}}$	Priority vector
D_{11}	1	2	4	2	1/2	4	0.5	1.49	0.17
D_{12}	0.5	1	3	1	1/3	3	1/3	0.91	0.10
D_{13}	1/4	1/3	1	1/3	1/5	1	1/5	0.38	0.04
D_{14}	1/2	1	3	1	1/3	3	1/3	0.91	0.10
D_{15}	2	3	5	3	1	5	1	2.39	0.27
D_{16}	1/4	1/3	1	1/3	1/5	1	1/5	0.38	0.04
D_{17}	2	3	5	3	1	5	1	2.39	0.27

CI = 0.0159, CR = 0.0142 < 0.1 OK.

Table 9. Pairwise comparison matrix for C_4.

C_4	D_{18}	D_{19}	D_{20}	D_{21}	$\left(\prod_{j=1}^{2} S_{ij}\right)^{\frac{1}{2}}$	Priority vector
D_{18}	1	1/4	5	1/3	0.80	0.14
D_{19}	4	1	8	2	2.83	0.51
D_{20}	1/5	1/8	1	1/5	0.27	0.05
D_{21}	3	0.5	5	1	1.65	0.30

CI = 0.0492, CR = 0.0439 < 0.1 OK.

Table 10. Priority vector for each criteria.

	A_1	C_1 0.33	C_2 0.17	C_3 0.33	C_4 0.17
D_1		0.44	0.15		
D_2		0.32	0.11		
D_3		0.12	0.04		
D_4		0.07	0.02		
D_5		0.05	0.02		
D_6		0.04		0.01	
D_7		0.06		0.01	
D_8		0.24		0.04	
D_9		0.53		0.09	
D_{10}		0.12		0.02	
D_{11}		0.17		0.06	
D_{12}		0.10		0.03	
D_{13}		0.04		0.01	
D_{14}		0.10		0.03	
D_{15}		0.27		0.09	
D_{16}		0.04		0.01	
D_{17}		0.27		0.09	
D_{18}		0.14			0.02
D_{19}		0.51			0.08
D_{20}		0.05			0.01
D_{21}		0.30			0.05

4 THE UCSD MODEL

In order to make each course corresponding to the abilities, the author established the UCSD (Unitary Course Synthetically Distribution) model, in which each course will get 10 points to be distributed to the criteria in the evaluation model. The distribution of the points should according to the teacher's experience. Considering different courses has different significance, the author introduced the credit of the course to endow differ weight for the points. The model is shown in Equation 2.

$$S = k_i A \times M = \begin{bmatrix} k_1 a_{11} & k_1 a_{12} & \cdots & k_1 a_{1j} \\ k_2 a_{21} & k_2 a_{22} & \cdots & k_2 a_{2j} \\ \cdots & \cdots & \cdots & \cdots \\ k_i a_{i1} & \cdots & \cdots & k_i a_{ij} \end{bmatrix}$$

$$\times \begin{bmatrix} m_1 \\ m_2 \\ \cdots \\ m_j \end{bmatrix} = \begin{bmatrix} S_1 \\ S_2 \\ \cdots \\ S_j \end{bmatrix}$$

k_i: Credit of the course
A: Points of the criteria
M: Priority vector for the criteria

5 SUMMARY

Based on employment-oriented, the author focused on the role analysis, determining the key activities and abilities. Then building evaluation model with AHP, and the structure of the abilities is further clear in the model. With the UCSE model, the development of curriculums can be more standardized and quantified.

REFERENCES

Guo, J. & Zhu, Z.T. 2011. Study on the Method of Occupation Education Curriculum Development Based on Role Analysis. Chinese Vocational and Technical Education 3.

Lai, J.L. et al. 2006. Analysis of Higher Vocational Mechanical Engineering Professional Occupation Ability and Training Mode. Vocational and Technical Education 22.

Norton, Robert. 2004. The DACUM curriculum development process [C]. IVETA Conference: 14th, 2004.

W. Kouwenhoven. 2007. Design and Development of a competence based curriculum for a new Faculty of Education. Journal of vocational education and training Vol.87, No.6.

Information, Computer and Application Engineering – Liu, Sung & Yao (Eds)
© 2015 Taylor & Francis Group, London, ISBN 978-1-138-02717-6

Speeding up the development strategy research on modern vocational education

Jun Hua Yan

Handan Polytechnic College, Hebei, Handan, China

ABSTRACT: In recent years, vocational education in China developed rapidly, pushed forward system construction steadily, cultivated many senior skilled talents, which is important to improve the quality of the laborers, promote economic and social development and made important contributions to promote employment. However, the current modern vocational education can't adapt to the needs of economic and social development completely, the structure unreasonable, quality needed improvement, managerial condition was weak, poor systems and mechanisms, couldn't keep up with the pace of social and economic development. Speeding up the development of modern vocational education, is important to further implement the strategy of innovation driven development, speed up the transformation of the economic development patterns and industrial structure upgrade, to satisfy the economic and social development of high-quality laborers and technical skills talents which is an urgent need to diversity and the masses of the people to accept vocational education.

KEYWORDS: Modern vocational education; Problem; Strategy.

1 THE DEVELOPMENT OF MODERN VOCATIONAL EDUCATION

1.1 *The guiding ideology*

In the spirit of the third plenary session of the eighteenth, as guide, adhere to moral and talents as the basis, adhere to service development the purpose to promote the employment as the guidance, as a whole of government and market play a good role, as a whole the ordinary education, vocational education and continue education harmonious development, to speed up the construction of modern vocational education system, deepen the teaching fusion, university-enterprise cooperation, efforts to meet the needs of the people to accept diversified professional education, for our country's economic and social development to cultivate many high-quality laborers and technical skills talents.

1.2 *The basic principle*

Adhere to the service requirements, integration development; Adhere to the optimization of structure, development as a whole; Adhere to the system of leading, innovation and development; Stick to sorted guidance, the character development.

1.3 *Modern vocational education features*

1.3.1 *Modernity*
The modern vocational education system must adapt to the national modernization construction is the essential characteristics of the demand for all kinds of professional talents, must keep up with the rapid development of modern social economy. In the next few decades, China's demand for talent trend, including: the demand for skilled talents continue to rise, and may become the main body of the future labor; Rising demand of senior technical personnel, and asked them to deeper and broader knowledge and skills, need more diverse interdisciplinary talents; All kinds of technical personnel move further higher education levels, vocational and technical training will be multiplied. Modern vocational education system to suit the characteristics of the new era, in order to develop more suitable for the modern industry system development needs of talents, promotes the development of a modern economy.

1.3.2 *The type features of vocational education*
Modern vocational education to reflect the particularity of vocational education personnel training types completely, not attached to ordinary education

system, and cannot be administered system. On the function of education embodies the unity of the social public welfare and economic marketability; Education emphasizes on the mode of running schools and enterprises to participate in the form of cooperation; Pay attention to school education and job training in education form compatible with each other; On the content of the education emphasizes the skills training and teaching and moral education; In learning form pays attention to the campus learning and social practice to learn a variety of ways. In the education teaching all aspects of highlighting the type characteristics of vocational education, to do is different from the normal education system.

1.3.3 Systemic

Modern vocational education should have a complete set of system structure, pay attention to the interaction between the vocational education of various elements and harmonious development. On the one hand, embodied in the system of internal development of vocational education, vocational education internal education, management, research and system coordination and system planning; On the other hand, reflecting the development of vocational education in combination with systemic. The vocational education system must keep clear of the internal and external environment of communication, maintain and relevant government departments, enterprises, trade and related institutions of cooperation and communication, fully activate the elements in the vocational education system, strengthen the communication with the social and economic system, enhance the adaptability.

2 THE PROBLEMS IN DEVELOPMENT OF MODERN VOCATIONAL EDUCATION

2.1 Penetration and communication difficulties in modern vocational education to the general education

General education in China emphasizes the subject standard, contact practice and professional life is less, in the stage of compulsory education teaching little infiltration professional enlightenment, education, students of social, industrial structure, technology development and career change without basic understanding, very conducive to future development of the students. Some students choose to receive vocational education, as a result of the compulsory education stage no professional enlightenment, education, lacks basic judgment of their own career interests, most of the students according to the wishes of parents choose professional led to many of the students in vocational schools, lack of interest in the specialty, appear even

disgusted, dropping out of school. In addition, the communication between the vocational education and general education is one way. General education graduates can accept higher vocational education, and vocational education graduates receive a higher normal education opportunity is very small. Although relevant documents prescribed in China, the school of technical secondary school, ability, high school graduates can enter oneself for an examination the ordinary institutions of higher learning, but due to the teaching content, the college entrance examination system, the influence of the university entrance exam content and so on, these graduates basically do not have ability to ordinary colleges and universities, and the proportion of secondary vocational school graduates enter the ordinary university is strictly limited, the number of colleges and universities to choose is very few to make secondary vocational education "termination" education. This closeness has seriously affected the enthusiasm of people to receive vocational education, which limits the development of vocational education.

2.2 Unclose cohesion, in the higher vocational education

Preliminarily established in our country's higher vocational education system, just solved the cohesion, in the form of the internal still exist many problems, the core problem is the unclear orientation of higher vocational education, leading to a series of problems in higher vocational cohesion. The first is in terms of students, the higher vocational education requires students not only has to accept the common cultural level of higher education, and have an intermediate level of professional knowledge and professional skills, students only have these two aspects of quality, higher vocational education to cultivate qualified technical talents. Secondary vocational graduates as the main students reflects the higher vocational education "professional" one of the most important aspects of the characteristics. However, at present, the students of higher vocational education are given priority to with ordinary high school graduates, and graduates of this part belong to common the university entrance exam score low, and do not have professional skills and knowledge, is bound to lead to higher vocational education personnel training "professional". Second is in terms of talent training, secondary and higher vocational education belong to the same type of two different levels. In both depth and breadth of professional theory knowledge, practice ability, the nature and scope of higher vocational should build on the basis of the secondary, and higher than the level. However, this is not the case. Higher vocational talents training design based on ordinary high school graduates, therefore, higher vocational, professional theory courses and practical training are difficult to high school, and even the curriculum and teaching content and repeated serious technical secondary school, vocational education resources waste. The main reason for the unclear orientation of higher vocational education is vocational education, lack of hierarchy in the management of internal unit cohesion. Higher vocational education of the department of higher education management, school management of vocational education department, this according to the hierarchical management system for medium and higher vocational education of cohesion, is not conducive to the construction of a modern vocational education system.

2.3 "Double type" teachers are still in short supply

Teachers of vocational colleges is the main source of teachers colleges and universities from their although some degree is high, but not into the factory, mostly from book to book, lack of practical skills and work experience, not to teach real skills to students; Because of university-enterprise cooperation mechanism is not perfect, make the vocational colleges steadily for a long time very hard to enterprise technology, management personnel for part-time teachers, teacher carries on the practice learning opportunity to the enterprise is also very limited. Title promotion at the same time, teachers only pay attention to education and academic level, the contempt practice ability, blocks the construction of "double type" teachers.

3 MODERN VOCATIONAL EDUCATION DEVELOPMENT STRATEGY

3.1 Reform of vocational education and ordinary education infiltration and integration

Vocational education and ordinary education are the two most important part of national education system in our country, the development of national education, both be short of one cannot. To build new modern vocational education system, the priority is to actively promote vocational education and ordinary education mutual penetration and mutual accommodation. It is in the regular primary and secondary schools to carry out vocational preparatory education, will have certain professional orientation course in primary and secondary school education teaching plan, carries on the preliminary vocational education for primary and middle school students. Open labor technology course in primary school, for example, by labor technology course to cultivate the students' labor concept, emotion and labor attitude, undertake to the student labor values education; In junior middle school stage, courses in career development and employment guidance, guides the student to understand

the social status quo, economic structure, industrial structure, the professional and technical development trend, to guide students in a career choice and career orientation of suitable; And open vocational education courses, such as agriculture, industry, business foundation courses, etc.) for the students, to encourage students to establish the agriculture, industry, business knowledge concept, understanding of the field of social, economic, cultural and environmental relationship between, establish professional consciousness, awareness of employment and work; In addition, for the part no longer continue to provide basic education graduates employment training for the knowledge and technical skills. Second, the vocational education training base, open to the ordinary high school curriculum and teachers, take many forms, for graduate of average high school students are not bound to provide vocational education. Conditional according to the need to appropriately increase of average high school vocational education contents, cultivating middle school students' professional ability and innovation ability. Ordinary high school and secondary vocational school's exchange platform for the high school student status management, a high school education and secondary vocational education curriculum mutual authentication, credit mutual recognition and achievement transformation, for the ordinary high school students to receive vocational education and secondary vocational students in ordinary high school education. Three reflects the characteristics of vocational education in secondary vocational school at the same time, should further attention and deepen the knowledge of humanities and culture courses teaching reform, increasing the necessary cultural courses (such as Chinese, math, foreign language), the curriculum standard shall be basic, and the corresponding curriculum standard of ordinary high school to study for part of the spare capacity and study interest of students, makes some students according to their own learning ability and learning interest, choose the ordinary high school curriculum and continue to learn. Only in this way can we truly promote vocational education and ordinary education mutual penetration and mutual accommodation to realize the overall development of the national education and balanced progress.

3.2 Building cohesion system of higher vocational education the

Study and establish our country "the modern vocational education system construction plan (2015–2020). Construction of secondary vocational education, higher vocational education, vocational education graduate student cohesion and financing system of interchanges. To speed up the proportion of different categories of higher vocational education, to establish technical skills talents growth pattern selection mechanism. Focus on exploring and perfecting "knowledge + skill", a separate admission and skill, talent of an exemption of admissions examination way, gradually expand the rights of autonomous enrollment for vocational colleges and learner autonomy option. Improve junior high school starting point in the higher vocational college enrollment and training system, gradually expand the secondary vocational and general undergraduate section of "3 + 4" pilot coverage scope and specialty, explore vocational and general undergraduate section of "3 + 2" pilot. We will accelerate reform of the professional degree, postgraduate entrance examination system, enlarge recruit have certain work experience and practical experience in proportion to the Frontline workers. Moderately improve the specialty of higher vocational colleges and universities to recruit the proportion of secondary vocational school graduates, the proportion of recruits graduates of vocational college undergraduate institutions of higher learning. Strengthen vocational college credit system reform, the establishment of credit bank, credit accumulation and transformation system, promote mutual recognition of cohesion study results.

3.3 To strengthen the construction of "double type" teachers

In ordinary high schools and colleges and universities, according to the characteristics of vocational education, vocational schools of check and ratify, according to the reasonable establishment standard equipped with teachers, constantly optimize the structure of teachers. Expand vocational school autonomy of choice and employ persons, allows schools to recruit teachers themselves within the establishment, and 15% of the document according to the relevant provisions of the specific number of part-time teachers, fiscal appropriations standard according to the organization of personnel expenses for funds. Multi-channel hire enterprise engineering and technical personnel, skilled personnel, to undertake a professional course or practice to guide the teaching tasks. To improve vocational college teachers' access system, if have appropriate professional qualifications (qualification) the university graduates, recruitment has priority enterprise working experience and corresponding professional qualification (qualification) or juicer Vieira series of professional technical position. Perfect vocational college's teachers' position (title) appraisal method, classification review. Actively promote the reform of secondary vocational school teachers' title system, improve the teachers' professional development, evaluation mechanism, set up are senior teachers in a secondary vocational school position (title). Vocational college professional teachers, can be in accordance with the relevant professional technical position qualifications, apply for the second professional title evaluation. Vocational colleges to hire with technicians, engineers, senior technician) professionally qualified personnel as full-time teachers, can enjoy the treatment of related jobs advertised. Strengthen the "double division type" teacher training improving teachers' hands-on, practical application ability as the core, strengthen vocational college professional skills training for teachers. A build system of teachers to the enterprise practice, professional teachers must have two months every two years to the enterprise or the production practice.

REFERENCES

[1] Ma Shuchao. A number of policy thinking on modern vocational education system [J]. Journal of education development research, 2011.
[2] Xiao-ling sun. The basic characteristics of modern vocational education system in our country and the construction strategy [J]. Journal of education and profession. 2014 (05).
[3] Zhang club. Modern vocational education system to construct obstacle factor analysis [J]. vocational education BBS.2014 (01).

Information, Computer and Application Engineering – Liu, Sung & Yao (Eds)
© 2015 Taylor & Francis Group, London, ISBN 978-1-138-02717-6

Functional movement screen test and analysis in Chinese national level men's gymnastic athletes

Yong Sheng Sun

Capital University of Physical Education and Sports, Beijing, China

ABSTRACT: The men's gymnastics team 10 athletes tested for FMS. The test results show that 10 athletes in the highest score is 14 points. The lowest score is 7 points. The overall situation is relatively low. In their daily training injury risk is bigger. In the shoulder activity, accounting for 60% of the athletes showed pain and dysfunction; trunk stability in pain screening athlete's performance 70% out of low back pain like; rotational stability in pain screening athlete's performance 80% out of low back pain - shaped; deep squat ankle, knee or waist pain appeared in the test are 30%. 10 men gymnastics athletes in front of the body and body muscle strength asymmetry and after compensatory action and thoracic spine, hip and ankle joint flexibility is the cause of lower value.

KEYWORDS: Functional movement screen; Gymnastic athletes

1   SUBJECTS AND METHODS

1.1   *Subjects*

Ten male athletes in the Chinese national gymnastics team were enrolled as the research subject.

1.2   *Methods of study*

The diagnostics was gotten with FMS when the subjects were preparing the 2016 Olympic Game.

1.3   *Mathematical statistics*

Excel software was used to analysis the FMS test in the 10 Chinese national level gymnastic athletes. This analysis is not only useful for illustrating diagnostic indicators, but also providing a scientific basis for quantitative research.

2   RESULTS AND ANALYSIS

2.1   *Basic situation of athletes*

Table 1 shows the basic situation of the ten males gymnasts according to their main projects. Except four athletes have two main projects, including the vaulting horse, floor exercise and Individual all-around, the other six athletes are specialized in one main project such as floor exercise, vaulting horse and so on. Since different projects cause different sports injury, the athletes have their own injury characteristics. Therefore, the basic situation in the ten

male gymnastic athletes will be useful to analysis the cause of injury.

Table 1.   Athletes age and basic situation.

Subject's number	Age	The main project
1	24	Horizontal bar
2	24	Vaulting horse, Floor exercise
3	18	Individual all-around
4	20	Vaulting horse
5	19	Floor exercise
6	21	Individual all-around
7	25	Vaulting horse
8	23	Floor exercise
9	25	Rings
10	19	Individual all-around

2.2   *Injury location of the subjects*

As shown in Table 2, the Whole body injuries of the 10 athletes are located in the shoulder, low back, knee and ankle, the site of injury is concentrated in the shoulder, waist, knee and ankle, the specific situation is the number of the waist and ankle injury most, each for 5 people, accounted for 50%; knee injury of 4 athletes, accounting for 40%; the injury of shoulder for 3 people, accounted for 30%. There is no wrist and elbow injury at all. Thus, our data demonstrated that the injuries of the lower back and ankle are more common than wrist and elbow injury in the Chinese gymnastic athletes.

Table 2. Male gymnastic athletes injuries distribution.

Injury location	Person-time	Percentage(%)
Wrist	0	0
Elbow	0	0
Shoulder	3	30
Low back	5	50
Knee	4	40
Ankle	5	50

2.3 Analysis of gymnastics athletes FMS situation

FMS is a functional evaluation method designed by Gray Cook. FMS score was standardized in 4 levels. 3 points means that the action performs in line with operation mode. 2 points demonstrates that the action can be completed, but with compensatory. 1 point elucidates that the action cannot be completed. 0 points presents that there is pain when the action is performed. Usually, if athletes get 1 or 0 points in FMS, their risk of injury could be increase in training or competition. In this study, we use four FMS actions to evaluate the Whole body function and injury.

As shown in Table 3, score 10 athletes tested in the highest 14 points, the lowest score is 7 points, the overall situation is relatively low, the athletes in the daily training of high injury risk.

Table 3. The men's gymnastics athletes in individual FMS test scores.

Subject's number	Deep squat	Hurdle step	In line lunge	Shoulder mobility	Active straight leg raise	Trunk stability push up	Rotary stability	The total score
1	3	3	2	0	3	0	0	11
2	0	2	2	0	3	0	2	9
3	3	2	2	2	3	0	2	14
4	3	2	2	1	3	3	0	14
5	3	2	2	1	3	0	0	11
6	2	2	2	0	3	3	0	12
7	0	2	2	3	3	0	0	10
8	0	3	2	3	3	0	0	11
9	2	2	2	3	3	1	0	13
10	2	2	1	0	2	0	0	7

From Table 4, we can see that, in the shoulder mobility, accounting for 60% of the athletes showed symptoms of pain (40%) and poor function (20%); and the trunk stability pushes up pain screening in athletes performed 70% out of low back pain; rotary stability in pain screening athletes performance 80% out of low back pain; in the squat test, 30% of the pain comes from the ankle, knee and low back. From the test situation, 4 athletes in the hurdle step, linear lunge squats, shoulder flexibility three actions appear around the asymmetry. Our study finds that the main reason of the above phenomenon is due to the athletes muscle strength asymmetry, compensatory action and shoulder, hip, ankle flexibility caused by poor. Therefore, athletes need further training improving the shoulder joint, hip joint and ankle flexibility, strengthen the athlete body balance ability training,

Table 4. FMS score in the 10 gymnastic athletes.

	3points		2points		1point		0point	
	n	%	n	%	n	%	n	%
Deep squat	4	40	3	30			3	30
Hurdle step	2	20	8	80				
In line lunge			9	90	1	10		
Shoulder mobility	3	30	1	10	2	20	4	40
Active straight leg raise	9	90	1	10				
Trunk stability push up	2	20			1	10	7	70
Rotary stability			2	20			8	80

relevant error correction exercises to improve the movement of each joint pattern, make the pain symptom gradually relieved.

In addition, From table 4 we can see 10 athletes supine straight leg raises this action, 9 athletes get full marks (3points), the test results are very good. We suggest that there are two reasons account for the problems. One is the gymnastics athlete's lower extremity flexibility of demanding, so coach attaches great importance to athletes lower extremity flexibility exercises; The other is that the strength of lower extremity muscles and the gluteus muscles is not strong enough in the 10 athletes. Therefore, although the test results of the athletes performed very well, but there are also problems.

3 CONCLUSIONS AND SUGGESTIONS

Our study finds that the athletes muscle strength asymmetry, compensatory action and shoulder, hip, ankle flexibility are poor. Ten gymnastics athletes in FMS test, the highest score is 14 points, the lowest is 7 points, the average score is 11.2 points, the overall test score relatively low. Therefore, in their daily training the athletes injury risk higher.

We suggest that in the training and rehabilitation of athletes, attention should be paid to the players to relax and activate the muscle fascia, strengthen the easy damage stability force joint muscle and joint

flexibility, improve the athlete's body pillar strength and lower extremity muscle and gluteus muscle strength. These are effective methods to reduce men's gymnastics athletes incidence of injury.

ACKNOWLEDGMENT

The Young Talent Program of Beijing Municipal Education Commission (YETP1712).

REFERENCES

Gray Cook, Lee Burton,Kyle Kiesel, et al.Move-meet:Functional Movement Systems-Screening, Assessment, Corrective Strategies. Human Kinetics, 2011.

Francis G. O'Connor[1], Patricia A. Deuster, Jennifer Davis[1], Chris G. Pappas[2], and Joseph J. Knapik[3], Functional Movement Screening: Predicting Injuries in Officer Candidates. Medicine & Science in Sports & Exercise, Publish Ahead of Print.

Elizabeth Parenteau-Ga, Nathaly Gaudreaulta,*,1, Stéphane Chambersa, Caroline Boisverta, Alexandre Greniera, Geneviève Gagnéa, Frédéric Balgb. Functional movement screen test: A reliable screening test for young elite ice hockey players .Physical Therapy in Sport 15 (2014) 169–175.

Cao Lichun. Analysis of national gymnastics athlete trauma characteristics. Journal of Capital Institute of Physical Education: 2007 July, the nineteenth volume fourth issue, p70.

Information, Computer and Application Engineering – Liu, Sung & Yao (Eds)
© 2015 Taylor & Francis Group, London, ISBN 978-1-138-02717-6

Enlightenment from Dewey's moral education philosophy

Ping Zhang & Shao Guang Du
Harbin University of Science and Technology, China

ABSTRACT: Dewey's "education is growth" thought holds that education process has no purpose except for itself. Education is a process of constant restructuring and transforming. This thought is different from internal inspiration and external shape, and the ideal of growth boils down to that education is the constant restructuring and transforming experience. The so-called moral problems are in fact, behavioral problems. Therefore, it is particular important to persist in the people-oriented moral education method and strengthen the social education method of moral education.

KEYWORDS: Moral education; Dewey; People-oriented; Sociality

1 MORAL EDUCATION`S POSSIBILITY

Human beings are social animals, and thus, an absolutely isolated person is nonexistent. It is morality makes a society possible, which is also a fundamental condition of the social order. Morality makes human beings themselves. Moral problems have drawn great attention in modern society; so many people want to solve them by education. Therefore, a prerequisite problem that still needs to be discussed emerges: is morality teachable or not? Even education`s possibility needs discussing. There is a serious debate about whether education can undertake this mission in academia. Some people strongly support the viewpoint that "education is useless".[1]Of course, Either academically or in practice, education`s role has proved that human beings are teachable and education is possible. It is education that makes the continuation of human society. This is determined by human nature.[2]Dewey expounded the necessity of education convincingly. "In the broadest sense, education is the tool of social life continuation". "The members of a group die and give birth, and this unavoidable fact determines the necessity of education".[3]That is to say, the continuation of human beings needs education which transfers experience and lessons. Through education, "the older generations' behaviors, thoughts, and feelings are transferred to the new generation. If the members who are leaving the group do not transfer their ideas, hopes, expectations, standards and opinions to those who just come into the group, social life cannot go on".[4]"Due to the lack of survival instinct of human infants and their helplessness, they must rely on education if they want to survive and develop. Education is a career which helps children to get the essence of survival and development".[5]A new social member needs not only to adapt to the natural environment and get food and a mate, but to know the game rules of participating in social life, and most rules belong to the moral area. A wolf child raised in a wolf pack only has the natural attributes of humans, but he or she cannot understand the survival rules of human society about how to communicate, cooperate and exchange with other social members. Sociality is the essence of human beings, and man is the sum total of social relationship.

2 DEWEY'S MORAL EDUCATION THOUGHT

Dewey believed that "education is growth", and the ability of people's growth relies on both others` help and their own plasticity.[6]The thought of education is growth holds that education process has no purpose except for itself. Education is a process of constant restructuring and transforming.[7]This thought is different from internal inspiration and external shape, and the ideal of growth boils down to education is experience constantly restructuring and transforming.[8]It emphasizes experience and practice. In practice, especially in the specific social practice, it cannot only constantly stimulate the potential of people and arouse the creativity of people but shape people. People render some relative experiences to habits and memory. A person who knows bike balance principles may not ride a bicycle, while a cyclist may not be able to express a balance principle, but he or she may know how to balance it.

Dewey's moral education thought generally includes the following two aspects: one is the socialite of moral education; the other is the psychological aspect of moral education. These two aspects are inseparable. Dewey believed that moral education does not transmit concepts about moral to children, but transmits moral senses through teachers' moral character, teaching methods, the atmosphere and ideal of the school and teaching materials, etc. Of course, this kind of teaching is mainly realized through participating in various activities organized by the teacher and farming in practice.[9]He holds that the simplest moral knowledge teaching is incorrect. People should pay more attention to the actual effect of moral education. "The duty of educators is to make children and juveniles acquire the greatest possible number of ideas actively and make these thoughts their active concepts and the motivation to conduct behaviors".[10] This theory conforms to Dewey's pragmatism philosophy and is consistent with Dewey's moral education thought. On the basis of this thought, Dewey proposed that the purpose of moral education was to cultivate children to adapt to the society, make them quickly integrate into society after leaving school and make contributions to the society. If school moral education breaks away from the society, there is no morality. Without a social life, there is no moral purpose for school. Dewey's moral purpose serves society, especially for the capitalist society. He also thought that the purpose of moral education was to make children adapt to capitalism society requirements.[11] "The child is a unified whole including intelligence, sociality, morality, as well as sports. Thus we must regard children as a member of the society and ask schools to take necessary measures to enable children to wisely recognize all his social relations, and does his part to maintain the relationship".

The so-called moral problem actually is behavioral problem. Dewey's conscious education can be said to maintain "internal continuity" of knowledge and action. Education is to accomplish this task and achieve the ultimate purpose—the formation of personality. Dewey held that people should express moral purpose in all teaching process, and no matter in what subjects it should be in a common and ruling position.[12]From his democratic idea, Dewey discussed the social and moral purpose of education. He pointed out that education has three main points: children's life, school and teaching materials and society. Among the three elements, children are the basis and starting point of education, and the adult society is the purpose of school education. School is a special social institution, and it must enforce its special function to maintain and enhance social welfare and serve in society. School is a bridge linking children and society, and the purpose of education is to make children pass the bridge and become a useful person in an adult society. Education should cultivate "good citizens for society in a democratic country".[13]

3 MORAL EDUCATION METHOD

The educational thought of growth is different from internal inspiration and external shape. It is, to some extent, a combination of the advantages of the two schools. Thus, moral education method conducted by this thought is different from the ways advocated by internal inspiration and external shape. This kind of moral education thought and purpose determines that we must persist in the people-oriented concept and strengthen moral education's sociality.

3.1 Persisting in people-oriented moral education method

the traditional concept of moral education attaches great importance to the society rather than people. However, an individual is not a subject of the society, but the social management object, an object. This binds that the moral education method is forced, didactic and inculcating, and also makes it impossible for man's all-round development. This kind of education is "a submitted and objective education mode. Moral education sometimes becomes tools that limit and restrain people. It brings up good and abjectly obedient citizens rather than people with a sense of independence, initiative and creativity".[14] Dewey emphasized that moral education should be people-oriented and pay more attention to people's initiative and creativity in the development of society and history. People should not only passively adapt to the social environment, but also change and create an environment. People who want to adapt to the environment must have their own motivation and needs. Society is not abstract. It is reflected by specific persons and specific human practice activities, so the social development depends on the contributions of each member, and social progress and personal growth are unified. The society's progress must fully arouse individuals' action potential and allow them to fully realize their growth. To hit this target, society should have institution like schools to cultivate positive growth person. It is in this sense that Dewey proposed that education is growth. Without growth, an individual could not live and a society would not develop. Therefore, moral education must pay attention to the needs of people, human actions, and the development of people. Only when social, moral values and people's development of various aspects are linked can moral education's purpose be achieved.

Thus, moral education must undertake the function of cultivating students' personality.[15]Adhering to the people-oriented moral education method is to make humans human. "Educatee's inner moral demands should be the starting point. According to an educatee's thinking habits, cognitive style, personality tendency and values, an educator, through his initiative and creativity, cultivates and inspires an educatee's subjective consciousness and personality".[16] The school should cultivate students with moral sense proposed by Dewey rather than blindly impart moral ideas.

3.2 Strengthen the social education methods of moral education

Since the cultivation of students' moral concepts is more important than the concepts about morality, specific social practice plays an irreplaceable role in moral education. Moral education socialization is even the only mode of moral education. Other moral education methods and the design of teaching materials are around socialization and students' adaptation to the society. John Dewey believed that value came from experience, so the value is relative. Traditional value cannot be accepted because of its fixed and absolute value standard, which is the only criterion to measure people's behavior. From Dewey's viewpoint, morality does not have an absolute standard, that is to say, there is no absolute moral precepts that everyone should abide by. All the moral standards will change in the development process of the human culture. He thought that morality generates from the adaptation to the environment, a certain morality suited to a certain environment but did not suit to another.[17]Then environment is the basis of the product of moral beliefs. How can students adapt to the environment and cultivate moral beliefs become a vital part of moral education. The school moral education problem is transformed into how to make students adapt to social problems. Then Dewey's proposition can make sense, "The school is a small society". School, is fundamentally an institution with special effect—maintaining a social life and improving social welfare. It is an institution built by the society. Dewey's social outlook depends on biology and regards society as the totality of the organic relationship rather than the system and the structure of the material.[18]School is in society, and it is not an organization which is free from the society, but an important part of it. It is even an organization that transfers valuable experience to human continuity. "Though the purpose of education is the harmonious development of a person's all abilities, if these abilities break away from the social life, then any terms such as ability, development and harmony will be pointless and abstract".[19]

Moral education ideas and methods of Dewey did give us inspiration and useful reference. However, education is a social process, and there are various societies in the world, and the criticism of education and the construction of education standards contain a particular kind of social ideals.[20] For to comply, with education in our country, we should comply with people-oriented ideas, strengthen the social education methods of moral education, and accumulate management experiences in the premise of the construction of socialism.

REFERENCES

[1] Daniel Cottom. *Why education is useless* [M]. Jiangsu People's Publishing House 2005: 1–2, quoted from Liang Jinxia, Huang Zuhui. *The global view of moral education* [M]. South China University of Technology Press. 2007: 24.

[2] Liang Jinxia, Huang Zuhui. *The global view of moral education* [M]. South China University of Technology Press. 2007: 28.

[3] John Dewey. *Dewey's educational masterpieces* [M] Compiled and Translated by Zhao Xianglin & Wang Chengxu. Education and Science Press. 2006: 113.

[4] John Dewey. *Dewey's educational masterpieces* [M] Compiled and Translated by Zhao Xianglin & Wang Chengxu. Education and Science Press. 2006: 114.

[5] Liang Jinxia, Huang Zuhui. *The global view of moral education* [M]. South China University of Technology Press. 2007: 30.

[6] John Dewey. *Dewey's educational masterpieces* [M] Compiled and Translated by Zhao Xianglin & Wang Chengxu. Education and Science Press. 2006: 123–126.

[7] John Dewey. *Dewey's educational masterpieces* [M] Compiled and Translated by Zhao Xianglin & Wang Chengxu. Education and Science Press. 2006: 123.

[8] John Dewey. *Dewey's educational masterpieces* [M] Compiled and Translated by Zhao Xianglin & Wang Chengxu. Education and Science Press. 2006: 127.

[9] John Dewey. *Dewey's educational masterpieces* [M] Compiled and Translated by Zhao Xianglin & Wang Chengxu. Education and Science Press. 2006: 83.

[10] John Dewey. *Dewey's educational masterpieces* [M] Compiled and Translated by Zhao Xianglin & Wang Chengxu. Education and Science Press. 2006: 82.

[11] John Dewey. *Dewey's educational masterpieces* [M] Compiled and Translated by Zhao Xianglin & Wang Chengxu. Education and Science Press. 2006: 84.

[12] John Dewey. *Dewey's educational masterpieces* [M] Compiled and Translated by Zhao Xianglin & Wang Chengxu. Education and Science Press. 2006: 84.

[13] Liu Changhai, Du Shizhong *Comments on the spatiality of moral education: Dewey's viewpoint and its value* [J]. *Education Research and Experiments*. 2004 (3)

[14] Hu Jianbin. *The school moral education modernization of society transformation period* [M]. Central Compilation & Translation Press. 2006: 64.

[15] Zhou Taopin. *Discussion on the methodology of Dewey's thoughts of moral education* [J]. *Jiangxi Education Scientific Research*. 1992 (5).

[16] Hu Jianbin. *The school moral education modernization of society transformation period* [M]. Central Compilation & Translation Press. 2006: 64.

[17] Ge Xianping. *On moral education thoughts of Dewey* [J]. *Journal of Anqing Teachers College*, Social Science Edition. 2002 (5).

[18] Zhong Qiquan, Huang Zhicheng *Western moral education principle* [M]. Shanxi People's Education Press. 1998: 71.

[19] Zhong Qiquan, Huang Zhicheng *Western moral education principle* [M]. Shanxi People's Education Press. 1998: 72.

[20] John Dewey. *Dewey's educational masterpieces* [M] Compiled and Translated by Zhao Xianglin & Wang Chengxu. Education and Science Press. 2006: 84.

Information, Computer and Application Engineering – Liu, Sung & Yao (Eds)
© 2015 Taylor & Francis Group, London, ISBN 978-1-138-02717-6

Stratified teaching of water conservancy talents training

Ai Min Gong, Hai Yan Huang, Yong Qiu & Tian Wen Song
Yunnan Agricultural University, Kunming, China

ABSTRACT: The methods and application of water conservancy talents training in the stratified teaching are discussed in this paper. The training object is meeting the demand of the diversity and particularity of water conservancy talents in the process of regional economic development in Yunnan province. Over six year's research and practice of teaching reform, "Professional recruitment, Training categories, Shunt in the middle period, Stratified teaching" is advised to culture the students. The training achievements show that the train mode can develop the strong adaptability, the individuation, and the diversity of students.

KEYWORDS: Stratified teaching, Water conservancy talents, Talents training, Training categories

1 INTRODUCTION

Water conservancy is the important infrastructure of national economy; it also is the important material condition of supporting economic and social sustainable development. There has not only many mountains, but also many rivers and lakes in Yunnan province. The region of Yunnan province spans seven climate types, and it obviously has the characteristic of three-dimensional climate, and there are significant changes in water resources during the year. There have unique regional characteristics and the characteristics of the typical topography needs in Yunnan Province to train hydraulics and technical students who have the ability of adaptable, personalized, and diversified, work independently under specific environmental. The traditional teaching mode apparently have been unable to adapt to the requirements, therefore, education teaching reform is the necessity of social development.

Quality is the root of the survival and development of a school. Thinking from the frontier facts, diversity and specificity of the regional economic development process in Yunnan province for water conservancy talents, a new training plan of "Professional recruitment, Training categories, Shunt in the middle period, Stratified teaching" is advised to culture the students and train the teaching faculty. The practice of the teaching reform plan starts at 2008; and the shunt in the middle period is completed in the February 2011; and then the stratified teaching is in the way.

Stratified teaching is for the teaching strategies of all students, focusing on individual differences, teaching students in accordance with their aptitude as a starting point, combining each student's own objective reality, implementing stratified lesson planning, teaching, practice and evaluation, coordinating the relationship between the teaching objectives and teaching requirements, targeting students at different levels to choose the ways and means of education, promoting each level students' learning ability and teaching requirements to better adapt to each other, so that each level of the students get a good education.

2 THE CHARACTERISTICS OF THE STRATIFIED TEACHING

2.1 *The stratified teaching is the necessity of the development of higher education*

The unique hydraulic characteristics and the regional economic development demand of water conservancy technology talented person in Yunnan province present the diversification and more stratified trend. It not only needs technology applied talents who can apply professional knowledge to practical, also needs excellence engineers type talents who have a certain engineering ability, innovation ability and comprehensive management ability. At the same time, it also needs academic research talents who focus on theoretical research and high-tech research and development. The three different types of talent are different to the requirement of students' professional quality. It has the characteristics of Stratified. Therefore, for the sake of cultivating professional talents of different types and different levels of water conservancy, making every student who can get the best development, water conservancy professional education must keep up with the pace with the times, changing the traditional teaching mode, trying to cultivate to adapt to the social needs of high-quality water professionals, the stratified teaching mode has to do.

2.2 Stratified teaching is a requirement of improving students' comprehensive quality

The traditional teaching mode is not conducive to mobilize the enthusiasm of students learning, explore students' potential, teach students in accordance with their aptitude, and limit the student individuality developers of the shortcomings gradually appear, so it is unable to adapt to modern social development of the higher education schools.

The stratified teaching starts from level and cognitive ability of the students who had got some knowledge, respecting the students' personality characteristics. Stratified teaching fully arouses student's enthusiasm, initiative, making students study, and loving study. It emphasizes the students taking the initiative to "learning" and teachers' positive exchanges and dialogues in the process of "teaching". It allows students in the learning platform to form correct and stable professional ideas, and it is a positive role in promoting that cultivating students' interest, mastering professional skills for life, and autonomous learning, self exercise, independent thinking and cultivating creative ability. Students choose professional competition after stratification, it can stimulate students' learning enthusiasm, strengthening students' self-confidence and effective construction of the students' learning, motivation mechanism, setting up the correct learning consciousness, forming a good learning atmosphere. It will be conducive to the students thought quality, style of study construction. At the same time, the students in the stratified learning process not only understand their strengths, but also know their own shortcomings. So the students can foster strengths and circumvent weaknesses, comprehensive development.

2.3 The stratified teaching comprehensively improves the level of school construction and education quality of teaching

The implementation of stratified teaching will make students to choose their teachers according to their own ability, requirements. So it will be bound to inspire college specialty construction and the enthusiasm of teachers' teaching ability. Then it can effectively promote the professional construction of the university motivation mechanism and the teachers' teaching motivation mechanism. It is advantageous to the integration of college or school teaching resources, and to improve the professional construction level and education quality of teaching in school. At the same time, in the process of stratified teaching, the students' information feedback can make teachers find their advantages and disadvantages. It can be targeted to improve teaching, reversing to promote teachers' quality, teaching quality. In addition,

the teacher may improve the teaching conditions in various professional levels, who fully understand to the relative position of their courses, getting a higher goal. The teacher who has high evaluation given by his students can play an exemplary role in the teaching process, encourage the other teachers who have a poor effect to improve his teaching method.

3 STRATIFIED TEACHING

3.1 Training mode of stratified teaching

According to "Professional recruitment, Training categories, Shunt in the middle period, Stratified teaching", the students are divided into three levels: technology talents, engineer talents and research talents. The students in the three kinds accept different training: basic quality training, engineering quality training and study quality training. After the training, the students have different kinds of abilities: project application capabilities, engineering design capability and science and innovation capability. The training model is shown in Figure 1.

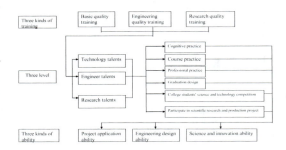

Figure 1.   Training mode of stratified teaching.

In the "Three level", the means of the technology talents are as follows: according to both the training plan, the students are strengthened their capacity and industry software application ability, their general education and practical skills, and cultivated them as the qualified designers and engineers. The means of the engineer talents are as follows: according to the general standard and industry standard for cultivating engineering talent, the students are asked to take deeply part in the practical activity in the industrial enterprises. The students are strengthened to cultivate their engineering ability and innovative ability in the classroom and company through solving the practical problems, and strengthened their comprehensive management ability. They are cultivated as the excellent engineer. The means of the research talents are as follows: for the excellent students, their minds are activated by adopting heuristic, exploratory, discussion-based, participatory teaching method. We focus on stimulating their research and innovation ability through directly taking part in the real projects.

In the "Three kinds of training", the means of the technology talents are as follows: the basic quality training is in relation to the training of the students the basic professional quality; the engineering quality training is in combination with engineering practice, training students' ability to solve practical problems; the research quality training is the combination of scientific research projects, training students' teamwork or independent scientific research ability.

In the "Three kinds of ability", the means of the technology talents are as follows: the project application ability is how to apply theory to solve the problems of the general ability; the engineering design ability is how to use the independent ability to solve practical engineering problems; the science and innovation ability is the teamwork or independent scientific research ability.

3.2 Practice of stratified teaching

The new stratified teaching mode "Professional recruitment, Training categories, Shunt in the middle period, Stratified teaching" is advised to culture the students and train the teaching faculty. The practice of the teaching reform plan starts at 2008; and the shunt in the middle period is completed in the February 2011. This mode vividly it is "1.5+1.5+X" mode.

The first "1.5" is the study 1.5 years that the students accept the categories of training after the entrance.

The second "1.5" refers to the follow-up a year and a half after the shunt in the middle period that the students accept the professional training. During this period, the students study in the professional way, according to 6 professional directions of the college that the teaching course has a different specialized fundamental course and specialized course, such as water resources and hydropower engineering direction for hydraulic steel structure and hydrology and water planning, hydraulic structures, such as curriculum, civil engineering direction load and structure design, mechanics of elasticity, masonry structure. In the process of studying, the students can also be based on their interest or need to study the application of fluid mechanics, environmental protection, mathematical model, material engineering, and according to the characteristics of the schools themselves the various types of courses.

The last "X" refers to the 10th and the 11th semesters' stratified culture. "The water resources and hydropower engineering class" and "The civil engineering class" is a branch of the "X" mode. The two classes are organized by the school of outstanding teacher resources, which have more and better software and hardware condition, such as new computer room, more modern laboratory, etc. We focus on how to strengthen the foundation of mathematics and mechanics, increasing the depth of the core curriculum, and encouraging the postgraduate exam, cultivating strong mechanics foundation and ability of analysis and calculation of structural engineers and high-level research talent. For other students, in the 10th term they still study professional courses according to professional direction. And in the 11th term, according to the market demand, we give them a variety of studies of chooses or vocational training, and give them full play to the college resources advantage. The training mode is similar to the order form training. The students take part in different types of projects through the experiment, practice, design, construction, research, and research activities in different company.

In order to guarantee the implementation of the stratified teaching, according to the first four semester students' credit grade point ranking, the students are shunted to six major. After the majors' adjustment, the number in the majority of the civil engineering increases from 75 to 137. The number in the majority of the soil and water conservation and desertification decreases from 46 to 20. The number in the majority of the resources and environment and urban and rural planning and management decreases from 63 to 32. The number in the majority of the agricultural water conservancy engineering decreases from 62 to 30. The number in the majority of the agricultural construction environment and energy engineering decreases from 57 to 26.

3.3 Achievements of stratified teaching

After 6 years of stratified teaching reform practice, the following results are obtained:

1 The team has been developed as the provincial "Comprehensive reform of water resources and hydropower engineering project", the provincial "Water conservancy and hydropower projects the professional excellence engineers education training plan construction project", the provincial "Agricultural soil and water conservation engineering teaching team" in 2012.

2 In the national construction engineering cost member training center, the rate of students who get a certificate rate is more than 70%. In Yunnan province water conservancy engineering training center of our school, the rate of students getting the certificate more than 95% who take part in the training of water conservancy project budget, the quality inspectors, and the inspector' student.

Authorized by provincial ministries of water resources, Yunnan provincial water resources bureau, our school will train county (city) bureau chief in Yunnan province for on-the-job training four periods. The students' number is more than 1780 people.

3 The professional core course of the students in the grade 2009 and 2010 increase about 8%~26% compared with the course with the grade 2008.

3.4 Experience of stratified teaching

The stratified teaching is not different from the levels' teaching. We must make students distinguish the difference between the stratified teaching and the different levels of competency teaching. We must prevent students from grade discretion misconceptions. Students should understand the significance of each level specific connotation, and carefully evaluate themselves objectively, their position, and then make an appropriate choice. The foundation of implementing of the stratified teaching is how accurately divide different level's students according to their individual differences. According to the students' choice, based on the principle of seeking truth from facts, we adjust some students' choice.

In the teaching process, we should pay more attention to the teachers' guidance, and timely pay attention to the students' motion at different levels.

The stratified teaching reform gives the student another chance to choose a new specialty according to their own ability development. It can effectively arouse the enthusiasm of the students' learning.

At the same time, for all students, the counselors, teachers and undergraduate tutor have to cultivate their different ability.

4 SUMMARY

For the regional economic development in Yunnan province, which the demand of the water conservancy talents has the characters of diversity and particularity, the College of water resources and hydropower and architecture of the Yunnan agricultural University advise and practice the stratified teaching reform. After six years' practice, "Professional recruitment, Training categories, Shunt in the middle period, Stratified teaching" is advised to culture the students. The training achievements show that the train mode can develop the strong adaptability, the individuation, and the diversification of students.

REFERENCES

[1] Huang Haiyan. Practice and exploration of basic mechanics experimental teaching reform[J]. Experimental Technology and Management. 2008,25 (1): 119–122.
[2] Huang Haiyan. The construction and implementation of digital and networked mechanics laboratory[J]. Laboratory research and exploration. 2008, 27(2):140–143.
[3] Huang Haiyan. The design of the engineering mechanics experiment test system [J]. Vocational education research. 2010,7:155–157.
[4] Sun Haiyan. Discussion of specialized course teaching employment under the guidance[J]. Journal of Jilin College of education. 2012,8(6):17–18.
[5] Sun Haiyan, Gong Aimin, Zhang Ling. A discussion on the teaching method of Building Material [J].Science Journal. 2012,18:155–156.

Information, Computer and Application Engineering – Liu, Sung & Yao (Eds)
© 2015 Taylor & Francis Group, London, ISBN 978-1-138-02717-6

Construction of a new university-based financial accounting system financial system

Jian Zhong Tian, Hui Jie Zhu, Chun Bo Wei & Lan Luan
Shenyang JianZhu University, Shenyang, China

ABSTRACT: Through the analysis of the new accounting system and the new college financial system, we proposed a different old system and new system of innovation in the reform, to build a new system based on the university financial accounting system, elaborated on the accounting treatment of accounting colleges the specific means and methods, and under this system, the proposed construction of the university financial accounting system. Design of specific functional modules and implementation requirements. Through the system design and construction, is conducive to enhance the school's management, financial work more scientific and after i joke, to achieve data centralized, consistent and scientific management of financial integration.

KEYWORDS: Finance; system; system; system; building

1 INTRODUCTION

With the new accounting system and the new "high school financial system," the issue, university financial accounting methods had a great change, making the management and allocation of capital projects related to the change, and promote the economic system and universities and other educational activities were subsequently changed. In today's rapid development of information technology, and efficient operation of financial accounting system and financial accounting system determines the overall level of university financial management level. Diversified funds, is no longer just the financial allocation, research into models, etc. occupies an increasingly important component. Financial management is gradually core funding for the school management body core management transition from finances. College financial management and the ability to work, the increasing requirements of financial change has become a major driving factor. Therefore, the accounting system and propose a software system for the current situation of development has become an important task to be performed imminent.

various colleges and universities have a certain level of computerization management software. Including relatively comprehensive GRP and other software, along with continuous development and financial software design and development, financial control and accounting for the level reaches a certain level and so, basically realized the function of financial accounting and statistical information and other functions. Financial management requirements have been increased from the simple to the strategic development of the accounting hierarchy levels, including decision theory school development planning, lesson plans and other aspects of the introduction of talent support. A handful of domestic development and even better universities have relatively comprehensive digital office concept, realized the online reimbursement, information auditing and financial support, according to measurements. However, requirements and design are carried out for the use of the existing accounting system and the old "college financial system", under the new system also cannot fully meet the specific requirements of financial reform. It needs to build a new university financial accounting system and software to meet the needs of the new system reform.

2 THE UNIVERSITY FINANCIAL ACCOUNTING SYSTEM AND SYSTEM STATUS QUO

Our level of education continues to progress, and promote the continuous improvement of financial management, in particular the development of information technology in general today,

3 ACCOUNTING SYSTEM CHANGES THREE NEW ACCOUNTING SYSTEMS AND THE NEW "HIGH SCHOOL FINANCIAL SYSTEM"

With the enactment of the new accounting system and the new "high school financial system", a number of new accounting requirements have been proposed, the existing accounting system does not fully comply with the requirements of the new system, the new system of specific accounting changes are as follows:

3.1 The objective of the new financial accounting shift

The existing accounting system, more focused on the balance of payments reflects the unit's budget, the lack of further explanation of the financial situation, the new system is required for each user can provide a similar accounting information administrative units as well as the financial situation of the implementation of the budget relevant accounting information. Tilt to the entire financial accounting objectives more clearly reflect the financial condition;

3.2 The new system requirements for financial reform more responsive

The new system increases the government procurement and accounting-related content, such as a centralized treasury payment, more complete implementation of the fiscal reform and fiscal policies as well as a reasonable convergence corresponding accounting method to ensure the full implementation of the specific requirements of financial reform;

3.3 The new system some of the subject accounting method changes

In the new financial and accounting system, the double points recorded in the accounting method used more, budget execution and accounting methods to reflect the current financial status information for, on the one hand to meet the specific requirements of budgeting, cash basis, on the other can truly reflect the true financial information, the use of accrual accounting for revenues, costs and expenses and other operating conditions. In addition to the requirements of the new system of fixed assets subject to the implementation of the double entry accounting method of construction in progress, intangible assets and prepaid accounts payable and eight subjects;

3.4 A more accurate method for the calculation of assets

A former accounting system for accounting and classification of assets, relatively loose, the new system for the provision of public services in asset management is very strict, subject to payment redrawing temporarily, so that the two proposed accounting content material reserves and public infrastructure. It also increases the intangible assets and construction in progress, the previously neglected intangible full inclusion into the new accounting system, to determine the appropriate accounting methods in construction.

3.5 Long-term assets are amortized

Because the use of cash basis accounting method, the original system of fixed assets are not depreciated, nor accounting and amortization of intangible assets. The new system of fixed assets and intangible assets was redefined to increase its depreciation and amortization related provisions. This satisfies management and financial management needs of both parties to the administrative budget, while ensuring accurate budget, but also to reflect the true value of the assets. To better reflect the financial position and prepare accrual-based consolidated financial statements provide a means.

3.6 regulate accounting liabilities

The suspense of the original payment accounting system has great range, and all other liabilities except capital transactions between sibling organizations can be attributed to a temporary account in payment, thus making accounting information can not be true and accurate reflection of the specific the causes and the nature of the liabilities. In the new accounting subjects, depending on the nature and causes of debt and other objects were divided, the newly added "tax payable", "other payables" and "accounts payable" and other details of the six subjects, thus ensuring accurate accounting information, and elements of the full record. But also for the long-term payables and accounts payable liability account to follow these two double entry accounting.

3.7 The new system requires capital accounts in months incorporated into school accounting records

The old system, the implementation of the university is the basic construction investment accounting system, thus leading to the capital within the accounting records of the accounts not in school, belonging to additional accounts. In the new system, the universities of infrastructure investment, in accordance with the relevant provisions of the accounting infrastructure investment accounting system, and the regular monthly accounts and other related infrastructure to be incorporated into the school's accounting records. This will improve the integrity of the university accounting information from a comprehensive point of view, so that the university can be more comprehensive and in-depth grasp of their own debt management level, improve the ability to prevent financial risks.

3.8 Accounting method on the net assets of more standardized and refined

The old system, the accounting methods of the net assets is quite simple, and the balance is divided into fixed capital. In the new financial system, which was divided into carry over balances and financial allocations and carry-over balances and other funds. Implement a more comprehensive fiscal reform requires funding and meticulous management model, especially for the management of budget funds and project funds become more standardized, more practical significance. And in asset accounting, increased capital assets and net assets to be sinking two subjects. Double points were recorded in accordance with accounting tools, able to independently reflect the percentage of non-monetary assets and net assets of the university because of write-downs amounting payable, long-term payables Amounts subjects generated, and the balance can be carried forward distinguish.

3.9 The balance of payments accounting management colleges under the new system becomes more standardized

The old system of financial appropriation revenue account in the new system, to be refined into basic expenditure allocations and project expenditure allocations, and funding basic expenditure has been refined as personnel expenses as well as public funding subjects, and to regulate the classification of expenditure. Expenditures and other capital expenditures to the old system in accordance with the nature of capital expenditure for the financial allocations refinement, and were on the financial allocation expenditures and other capital expenditures are subdivided into basic expenses and project expenditures, expenditures in the basic specification classification to project expenditures are classified according to the corresponding project makes financial balance system for college has become more standardized.

3.10 Reporting information under the new system more comprehensive and accurate

The new university financial system in the financial reporting system added a corresponding financial allocations of income and expenditure table, mainly reflecting the implementation of the financial allocation for the funding of appropriate budgetary arrangements, as well as financial allocations budget instructions. Moreover, the balance sheet and income and expenditure tables are generated changes to improve the original information content, making the university accounting information is reflected in the year to be more comprehensive and accurate.

4 UNIVERSITY FINANCIAL ACCOUNTING SYSTEM UNDER THE NEW SYSTEM

In line with the principle of canonical construction comply with the new financial accounting system, platform-based finance, integrated financial, health-based finance, financial decision-making, university financial accounting system under the new system integrates a budget, finance, planning, statistics, management and regulations and other information, focused on documents and accounting data and other information, financial data to achieve centralized storage, centralized financial management settings, process system centralized management, centralized information resource sharing, to enhance the level of automation of financial management. To enhance the college financial management, real-time analysis and risk budgeting scientific and standardized management foundation vent.

4.1 Basic application platform

Basic application platform UAP use technology to build, providing management software platform operating environment, integration interfaces to external systems, education cloud service access, data security management, mobile applications, multi-dimensional data warehouse management education, scalable secondary development interface and other senior system application.

4.2 Comprehensive online information services platform

Provides a unified interface for office staff, provide a comprehensive business management service window, providing their login information to the user, to-do and so on. Provide single sign-on capabilities for query and sharing of information, is a specific office platform.

4.3 Budget management

Budget management is budgeting system management application for the proposed university. Establish budget project database, each unit of work through the system to declare the budget by the Treasury plans to confirm and feedback, the formation of the school budgeting system, to achieve normalization of budgeting.

4.4 Budget indicators management

According to financial regulation, financial analysis, financial statistics requirements, access to budget management, budget data and consistent information in accordance with the relevant budget, index information and accounting systems.

4.5 Payroll

Payroll university consists of three parts: the financial system of wages, school salaries, other income, the system aims to co-ordinate the construction of university salary information platform planning, implementing complete, comprehensive coverage, fully shared data, interrelated processes, management harmonization resource university payroll management system.

4.6 Grants and subsidy management

Grants and subsidy management includes two aspects. One is for the university to send Wu personnel payroll needs, on the other hand is a reward for the subsidies paid to the needs of the students, specially designed management system.

4.7 The student fee management

Student fee management to achieve the student billing records, standardize the process of fees, the student tracking smart security of payment, student payment situation inquiry efficient, students pay statistics convenient, print automation requirements.

4.8 Bill management

Standardize inventory management college fee bills, storage bills to achieve coverage, recipients, use, check the fare, financial processing, for examination, notes lifecycle information management and other aspects of the destruction, colleges and universities to meet the bill for the use of other means of integrated management objectives.

4.9 Online claims management

Online reimbursement is based on a reimbursement basis on the Internet or campus network financial processes. Teachers and students at any time, any place to submit reimbursement claims, leading to approve the use of digital signature technology, finance department audit bill is correct, generate vouchers, payment through online banking.

4.10 No cash reimbursement management

Cashless funny college cash settlement change management work to bring inconvenience, helping cashier staff easy work to provide efficiency.

Net realized silver credited inquiry and various banking-related business.

4.11 Accounting treatment

The financial accounting treatment system is the basis of the core functions of financial accounting. With accounting, economic classification subjects, subjects such as multi-dimensional financial accounting function mode. On the basis of the information system through the accounts, balance of payments accounts, project accounting, a number of current accounts and other auxiliary projects for management accounting. Ensure seamless connection system with other business systems, basic data consistent. Unified interface standards, to ensure the accuracy and timeliness of the data.

4.12 Treasury management

Treasury management to complete the cash, bank deposits, treasury management accounting work. Use of this system can improve the efficiency of the cashier and reduces errors occur. With a bank teller cash management functions, and highly integrated accounting system that can automatically generate income payments monomeric automatically generated in accordance with the relevant accounting documents of different types of business, and passed to the general ledger system.

4.13 The electronic report management capabilities

The electronic report management function is one of the core system of the financial meticulous management platform. Electronic reporting and accounting treatment of the system used in conjunction with automated access by user-defined formats, automatically generate reports needed to complete the preparation of the financial statements. Including units daily financial reports, summary reports, tables and other financial analysis.

4.14 The system of accounting records

Accounting records management system developed and designed according to the latest industry standards promulgated by the National Archives, records management departments at all levels in order to collect information on the incorporation of electronic files for managing objects in a disk-array storage media, use of relevant technology of computer networks, information technology and digital encryption and compression technology for retrieving stored information, and information management file, and can more easily query profile information.

5 SUMMARY

Based on the accounting system based on university financial accounting system under the new analysis and research to build a university financial accounting system framework and function realization. The new accounting system through university research, will help improve the financial management of universities and scientific and refinement, but also conducive to achieving double the budget and financial management, contributing to the whole process of regulatory capital, to avoid a certain degree of financial risk, and makes data sharing within small centralized, establish a unified platform released the corresponding information is also beneficial to the school funds to achieve integrated management of the budget, accounting and accounts.

ACKNOWLEDGMENTS

Fund Project:
Fund of Department of Education in Liaoning Provincial (L2013224);
China Accounting Society research project (2014).

REFERENCES

[1] The new university financial performance evaluation system explicitly implement expenditure. "China Accounting newspaper" .2013-1-11
[2] Li Yao Ming Department of Education Finance Division Head of the Financial interpret the new system. "China Education" .2013-2-4
[3] People's Republic of China Ministry of Finance and universities accounting system. Lixin Accounting Press [S] 2013.
[4] Kao Hui Feng, Tian Jianzhong, high capital construction will Wei College Accounting Problems and Solutions [J] Shenyang Jianzhu University;. 2013,15 (1); 81–83.
[5] People's Republic of China Ministry of Finance. Institutions Accounting Standards Economic Science Press [S] 2012.
[6] Huangmin new Variance analysis and suggestions convergence of old and new accounting system of colleges and universities. [J]. Education ACCOUNTING.2014.25 Compare.
[7] Wang Guosheng. Old "high school accounting system" of. [J]. Finance and Accounting Monthly .2014 4 (a) 77–79.

Information, Computer and Application Engineering – Liu, Sung & Yao (Eds)
© 2015 Taylor & Francis Group, London, ISBN 978-1-138-02717-6

The study enthusiasm of the students' learning

Ying Song & Jing Sun
Harbin University of Science and Technology, China

ABSTRACT: The reason of influence of contemporary college students' learning enthusiasm and countermeasures.

KEYWORDS: College students, learning enthusiasm, countermeasures

1 INTRODUCTION

Higher education system reform and problems brought by social transformation have a great effect on the learning behavior of college students. College students' learning enthusiasm is not high; rate of class attendance is low; and they fail the exams. Those phenomena are widespread in college. In fact, there is a mutual restriction relation between learning, motivation and effect learning. So, activating students' learning, enthusiasm can improve the quality of higher education

2 EXTERNAL FACTORS

2.1 *Impact of information network*

According to a survey of more than 4300 college students in Beijing, 66.3% of them spend more than 3 to 4hours a day on the Internet. What's more, most of the time is spent doing things that have indirect contact with learning activities, such as playing games, chatting and watching movies. Thus, most students do not handle the relationship between information network and real life, and make themselves addicted to playing games, chatting online and watching movies. Hence, some students fail exams, repeat a year, leave school, and cannot graduate or find a job.

2.2 *Incomplete education system*

At present, there are some factors of an education system that affect college students' student enthusiasm, such as offering new majors blindly, which cannot adapt to the need to market and lead to bad employment situation. Besides, practice course and quality development courses are not included in the compulsory courses, and it leads to a fact that students lack practical ability. Moreover, college students enrollment expansion leads to shortage of college resources and reduces the quality of teaching.

2.3 *Imperfect employment system*

Each year, college graduates will encounter some unreasonable and unfair competition phenomena during the process of job-searching. It reflects a fact that the construction of the national employment system is not sound. Some college students even have a point that optimal points and great abilities have no effect of applying a job. So it frustrates their learning enthusiasm, and distorts their values.

2.4 *Inappropriate ideological education*

Nowadays, college students generally think that there are tiny differences in salary between them and migrant workers, then it causes weariness. However, current ideological education in colleges is limited to education of elective psychological courses, compulsory political courses, youth league and party organization activities. That is to say, ideological education forms in college are too single and formal to achieve the goal of guiding and teaching.

3 INTERNAL CAUSE

Lacking learning interest is one of the major subjective reasons why college students learning motivation is insufficient, but It is the lack of correct values that contributed to this phenomenon. First of all, the students are overburdened by entering into a higher school from primary school to senior high school, as long as their aim is achieved, they become much more proud than before, and at the same time, they do not build a higher learning motive timely, causing the phenomenon that they lack learning interest because "motivation drops". Secondly, knowledge values deviation of the college students is the deeper reason why they are lacking learning motivation. Some students understand the "learning is put into use" lopsidedly, and some students regard material over spirit, reality over the future, as long as they think it's no use they are not willing to learn. The dynamic function of the emotion has forced or force reduction effect on individual activity. The college students are a group of people who are very sensitive to emotion. As long as they are in a stable emotion situation, learning motivation can continue to play in a role. After they enter the college, with the physiology and psychology become mature than before, their emotional world has changed greatly: the yearning for family, the longing for friendship, the confusing for love, all affect their learning greatly. The improper handling of emotional problems is one of the major factors that affect college students learning enthusiasm. College Students' self-cognition and cannot correct attribution for it is another subjective reason that they are lacking learning motivation. The self-cognition refers self- efficacy perception, this is a term means subjective evaluation and confidence of a person's ability to engage in a job.

4 STRATEGY TO IMPROVE COLLEGE STUDENTS' STUDY ENTHUSIASM

4.1 Strengthen college education in thinking and social responsibility

Deng said the lad was the mistake in revolution and open up is education, which is the education of thinking. So to improve the enthusiasm of study in college students, firstly, the College should improve sinking education, Guide students look directly into the bad things in society, and seek a balance between network and Life. secondary left of college students know the condition of our country let them know the responsibility to study for our country development. For the more the forms of education could be more various, which should be combined with study practice and the life of college students.

4.2 Established the emotional bridge between teachers and students

Form study group, and the sign teachers as tutors of each group, who will notice the Academic condition of group members. After same time tutors and the members take party activities parties Macs together, improving relationship and promote the enthusiasm. Bring more communication between students and teachers, which will benefit both students' learning ability and teachers teaching methods.

4.3 Strengthen teaching management

First, we should arrange reasonable curricula to meet the needs of the market. Second, we should regulate students with strict attendance records,

including date participants of teacher and mentor. Students who play too much truant would be excluded. We should make the regulations. Third, they should take the computer level examination and a college English test into management. Forth, credit revolution. Students whole get enough credits can graduate in advance. Fifth, make a link between students synthetic test score and job enterprises which are better can get better students.Sixth Open up the job market for the graduates, let the better students get better Jobs.

4.4 Develop students better value in both need and target

1 Have students to build long aspiration. Teachers should combine students personal ideal with the nation's development. Encourage them to learn skills to serve our nation.
2 Develop their study interest through more reading more seeing, more practice and more participation.
3 Promote study enthusiasm ethical methods. There are always three kinds of atmosphere self-study competition and cooperation. In corporation classes, the basis of evaluation is the performance of the group. The members depend on each other, share the success and failure so it can motivate a mutual effort, the motivation of the honor of the collective.

5 CONCLUSION

College students study enthusiasm promotion is necessary for the quality of students. So we should apply the promotion into every chain of the education revolution Bring motivation into this strategy of Science and technology.

REFERENCES

[1] Ruizhen Shao, Liansheng Pi. Psychology Of Learning And Teaching [M]. Shanghai: East China Normal University Press, 1990: 97.
[2] Junhong Jiang, Xiaoju Li. Psychology Tutorial [M]. Beijing: People's education press, 1996: 123.
[3] Qian Fu. College students' study enthusiasm of study [J], Science and technology information, 2009 (32): 59–60.

Information, Computer and Application Engineering – Liu, Sung & Yao (Eds)
© 2015 Taylor & Francis Group, London, ISBN 978-1-138-02717-6

Discussion on students' education management work in multi-campus universities—taking Harbin University of Science and Technology as an example

Xiao Fei Zhou
Harbin University of Science and Technology, Weihai, Shandong province, China

ABSTRACT: In recent years, with the expansion of university enrollment, the upsurge of new university campuses appears in our country. However, there are a series of problems in the development of university new campuses. This paper takes Harbin University of Science and Technology as an example, analyzes the common problems, and expounds the development and innovation in students' management of multi-campus universities.

KEYWORDS: Multi-campus universities; Students' management; Working mode

In recent years, with the rapid development of China's higher education cause, the number of college students has increased year after year. In our country, there appears the upsurge of building new university campus. Harbin University of Science and Technology is an engineering institution of higher learning with tremendous strength, comprehensive subjects and a sizeable scale. It has four campuses. The East Campus, the West Campus, and the South Campus are located in Harbin, while the Rongcheng campus is located in Rongcheng, Shandong Province. The establishment of the multi-campus mode expands the university school space, increases the university scale, makes up for the deficiency of the education resources, and improves the university's competitiveness. It builds a platform for further improvement of schools' running level and rapid development. But it also brings with it plenty of difficulties and challenges to the managers. This paper takes Harbin University of Science and Technology as an example, through on-the-spot investigation and interviews, trying to discuss students' education management of multi-campus in our country's colleges and universities from the empirical point of view.

1 PROBLEMS IN STUDENTS' MANAGEMENT WORK

1.1 *Students' management work lacks innovation and still takes the "vertical and horizontal relationship mode, and primarily vertical" approach*

Students' management mechanism of Harbin University of Science and Technology takes the approach of "vertical and horizontal relationship mode, and primarily vertical" and persists in "one school" management belief. Power is relatively concentrated. Therefore, the straight line type "vertical" operating system will inevitably result in problems and conflicts between schools' unity requirements and diversity requirements of each campus. Except the separated location, the four campuses of Harbin University of Science and Technology have a consolidated requirement in education management. But there are some specialties for each campus student's management institution which leads to the differences in the cultivation goal of student's management. In addition, Harbin University of Science and Technology is different from single campus school in the decision-making process and communication between the superior and the subordinate under the circumstance of multi-campus. Problems such as poor communication and long-time decision-making seriously impede the student's management targets of speediness, efficiency and diversity, especially in solving new problems in each campus student's management work.

1.2 *Imbalances in each campus resources allocation*

An imbalance in each campus resource allocation is another main factor that influences multi-campus management. A large number of students enrollment expansion leads to serious problems in resources allocation. Hardware facilities are not balanced, and it is difficult to achieve effective integration between the new campuses and the old ones. Generally speaking, hardware facilities in new campus are superior to the old ones. Fundamental hardware resources such as new educational and experimental facilities, library, and students' dormitory can provide a more solid material condition for the development of the school. But other education resources allocation in different campuses goes against the equal opportunity of individual development. Especially in the new campus, because of the progressive process constrain of construction and management, it is difficult to enjoy the same rich resources as the main campus for teachers and students in new campus. The new campus is generally young, lacking university culture atmosphere and deep-going cultural traits, while the main campus possesses the study atmosphere and academic environment which is absent in new campus.

1.3 Unsound system of the students' management team

The most basic problem of multi-campus school is decentralized campuses. The increasing number of students makes it convenient to expand enrollment scale. While it increases the difficulty in students' management work. The work of student management needs more people, finance and energy and the pressure of students' management staff increases continuously which makes students' management staff are unable to understand students deeply and meticulously. When solving problems, main problems will be dealt with emergently, while minor problems will be postponed and ordinary problems are neglected. The management effect of the main campus cannot be accomplished and the original refined management mode changes into an extensive one.

1.4 Campus functional orientation does not adapt to the students' developed requirements

There are different functional orientations and assignments between each campus from the viewpoint of current university multi-campus practice. In some universities and colleges, freshmen enter the new campus, while the seniors and post-graduates stay in the old campus. Some are divided according to specialties. Some universities and colleges adopt the cross division standard, and Harbin University of Science and Technology is a typical one. No matter how to divide, the situation of each campus' specialties cannot be avoided. Therefore, each campus forms its specialties in teaching process, students' activities and university culture. This also impedes mutual communication. On the basis of this situation, if students' management approaches are not to be adjusted, while students, to some extent, will be constrained in the development of knowledge learning and skills training. It also generates a certain contradiction with the comprehensive developed requirements. At the same time, there will be some negative influences on the cultivation of student cadres and the comprehensive development of them.

1.5 Teachers and students lack communication between each campus which goes against students' management work

Communication between teachers and students is an important way of interactive teaching which promotes students to learn to behave, to do and to learn things. Lack of communication between teachers and students is a problem that needs to be solved in Chinese contemporary higher education. In the case of multi-campus school-running, this problem is more prominent. Because of scattered campuses, teachers live relatively concentrated on a campus, or the teachers and students live on different campuses. Some teachers take the school bus to their dormitories after class directly which reduces the opportunity of after-class communication except for limited contact with students in the class. Lack of communication results in poor interactivity between teachers and students, and students' psychological and ideological issues are ignored. It also increases the diaphragm between teachers and students, even resulting in trust crisis. This situation will influence the operation of students' management work and makes it hard to carry out further work and accomplish pertinence, timeliness and effectiveness.

2 COUNTERMEASURE ANALYSIS ON MULTI-CAMPUS UNIVERSITY STUDENTS' MANAGEMENT WORK

2.1 Establishing new operation mechanism

Mufti-campus are dispersed geographically. Additionally, there are differences in functional orientation and specific managed target between each campus. If universities persist in the straight line type "vertical" operating system of single campus, problems and conflicts will emerge inevitably. Therefore, students' management work of the multi-campus should focus on each management unit—campus. The student affairs office of the headquarters should do well in macro decision-making, correctly grasp the direction of the all-round development of students and create a service platform for the coordination of various campus student management work. In general, "vertical and horizontal relationships union, by block primarily" approach is undoubtedly an effective choice for multi-campus university. Building the integrated operation system of school—management center—grassroots managers is not only beneficial to help the school make long-term plan and reduce management level and range, but improves the work efficiency and the degree of specialization on the premise of meeting diverse needs of students.

2.2 Improving management approach according to specificity of different campus

Due to different historical background and development goals, students' management work on different campuses has its own specificity. Management approach should be adjusted timely according to this specificity. For application-oriented campus, students' management work should base on the theory and practice activities should be encouraged to cultivate students' ability in analyzing and solving problems. For research campus, students' management work should focus on the construction of academic atmosphere and academic forums should be held to cultivate students' academic accomplishment.

2.3 Constructing dynamic internet information system to achieve communication across campus

Multi-campus university should pay attention to the construction of Internet platform and strive to achieve the smooth communication project across the campus to improve the management efficiency of multi-campus. On the one hand, through building students' management information system and students work Internet platform, the university can provide the information of school and society duly through smooth information dissemination channels which makes students in different campuses get the latest information of school teaching, students' management, scientific research and social service with the fastest speed in a relatively closed environment. On the other hand, office information of students' management work should be achieved. In this way, all kinds of files and spirit of the meeting can be conveyed timely without the negative influence of area and space. It also solves the problems of distance-span and time-consuming and strenuous information delivery. Teleconference and network video conference can also solve the problems of emergent meetings and emergencies.

2.4 Improving the professional level of student management workers

Under the multi-campus university mode, the "human" factor is always at the core position in management. The precondition of a sound students' management work of multi-campus university is to build a management team with creative ability,

dedication, working ability and responsibility. We should promote the development of student affairs management discipline and the professional ability of students' management staff should be improved through systematical theory study and practice. Relative training system should be established to train the students' management workers regularly. In daily work, students' management staff should reinforce the communication with the students and understand the students' developments timely. They should also go deep into the learning and living place of the students to do research and solve various emergencies. In addition, relevant system should be established strictly.

Above all, in the process of construction and development of multi-campus university, especially during the fusion stage, there must be different kinds of complicated situations. Influenced by the overall development, students' management work of multi-campus university confronts plenty of obstacles and challenges. This requires us to proceed from the overall development and the practice of the university, continuously explore, improve and research on students' management modes, approaches and contents. At the same time, we should base on the school situation, persist in student-oriented policy, advance with the times and explore students' management rules in the case of multi-campus universities, so as to make scientific the students' management work in multi-campus universities.

REFERENCES

[1] Ou Zhou. Problems and countermeasures of student management in multi-campus university [D]. Hunan Normal University. 2006.

[2] Yu Lei. Studies on new campus management of multi-campus university [D]. Tianjin University. 2009.

[3] Ren Zhihong, Zhao Ping. Reference and enlightenment from American colleges and universities students' management work [J]. Beijing Education Publishing House. 2010 (5).

[4] Lv Xuefeng. Studies on students' management problems of multi-campus university in China [J]. Career Horizon, Comprehensive Edition. 2007.

[5] Huang Zhirong. Countermeasure analysis of student management problems based on multi-campus university [J]. Journal of Jinggangshan Medical College. 2008.

[6] Wu Ximei. How to do a good job in multi-campus university student management work [J]. Career Horizon. 2008.

Information, Computer and Application Engineering – Liu, Sung & Yao (Eds)
© *2015 Taylor & Francis Group, London, ISBN 978-1-138-02717-6*

Green concept application in low-carbon tourism economy

Zhi Fei Han
Hebei Institute of Foreign Languages, China

Hai Xia Qi
Northeast Petroleum University, Qinhuangdao, China

ABSTRACT: As a more integrated economic form, tourism is also known as the sunrise industry, involving transportation, shopping, accommodation, environmental protection and other aspects, based on the background of low-carbon economy and green concept development, low carbon tourism has become a new approach to development, whose great model role will promote the rapid development of the tourism industry. This article starts from the necessity and advantages of the development of low-carbon tourism economy, introduces effective measures to implement a low-carbon tourism economy.

KEYWORDS: green concept; low-carbon tourism economy; application

1 INTRODUCTION

With the rapid development of high-tech, people's pursuits for material and spiritual aspects are increasing. Serious environmental pollution problems also emerged as a result of socioeconomic development. How to protect the environment has become a focus of the entire humanity. With the rapid development of the tourism industry, the tourism industry is not only concerned about the economic effects, but pays more attention to change tourism ways, adjust the structure of tourism and social values. Low-carbon travel is a new breakthrough in response to global warming and in the implementation of the sustainable development concept, low-carbon tourism economy is to abandon the traditional growth model, use low-carbon economy model and low-carbon lifestyle an important strategy. Low-carbon economy is to reduce energy consumption and pollution to develop a green economy [1]. This text takes low-carbon tourism economy as research perspective, introducing a series of measures to implement low-carbon tourism economy.

2 A BRIEF ACCOUNT OF LOW-CARBON TOURISM ECONOMY

2.1 *Low-carbon tourism*

As a new term, low-carbon tourism is to use the concept of a low-carbon economy, with low power consumption, low pollution as the basis for green tourism. When the low-carbon tourism development of tourism resources for the new requirements proposed to carry out all aspects of the energy savings, reduce pollution, the whole action is a harmonious society, building

society savings. Low-carbon travel is to protect the wildlife and other resources of tourist attractions by way of reducing carbon dioxide, respect for culture and lifestyle in the region, making a positive contribution to the natural environment in the tourist areas.

2.2 *Necessity of developing low-carbon tourism economy*

With the continuous development of China's tourism industry, tourism is not just concerned about short-term economic effects, the analysis must begin with the long-term development of the tourism economic development goals, change the existing short-term values. A low-carbon economy based on economic and social development as the premise, through technological innovation, changing attitudes, the maximum reduction of greenhouse gas emissions, to achieve the purpose to reduce energy and resources. Development of low-carbon tourism economy is an important option to address global warming. Environment for the development of its tourism industry capacity and higher requirements. Low-carbon tourism industry not only to meet their development needs, but also to ensure the healthy development of the national economy. Tourism must follow a low-carbon ideas, to minimize their own development in emerging environmental pollution and energy consumption issues [2]. Compared with other industries, tourism facilitates rapid development of low-carbon industries. Low-carbon not only open up new markets for the development tourism, but also reasonably inhibits operating costs. Therefore, the development of low-carbon industries helps enhance the competitiveness of enterprises, but also the inevitable direction of tourism enterprise development. Tourism plays an important role in the development of tourism field, but also the content to adjust the socioeconomic structure. Tourism helps the traditional extensive economy to change towards the intensive economic, playing an important driving force to adjust the economic structure of our society.

3 MEASURES TO IMPLEMENT LOW-CARBON TOURISM ECONOMY

Tourists and tourism enterprises are the main body in tourism activities, and also an important force in the development of low-carbon tourism. In order to develop low-carbon green tourism economy, we must let tourists, government and tourism-related departments to develop low-carbon measures, and work to make a great contribution to the low-carbon business.

3.1 The reasonable control of ecological capacity

Ecological capacity refers to the number of eco-tourism activities in a certain period can be accepted tourist destination. Factors that determine the ecological capacity of the natural environment to absorb pollutants and purification capacity. Tourism resources, balance between tourists and local residents, is a basic requirement for building a low-carbon tourism. Therefore, a low-carbon economy must be the concept of governance, from the successful experience of international low-carbon economy, the local conditions of the ecological environment and resource characteristics destinations Expand rational planning, create green tourism resource recycling system, to minimize the ecological investment cost [3]. When the protection of the natural environment, the use of energy-saving technology to reduce wastage of resources, vigorously research efforts using renewable resources, the impact on the ecological environment of the tourism industry to the lowest level, prompting the development direction toward green, low-carbon of the tourism industry.

3.2 Government strengthens management and policy support

Government departments should introduce corresponding low-carbon travel services and product management system, regulate the consumption and business activities of the tourism industry, and strengthen cooperation with transportation, resources, taxation and other departments, to ensure the implementation of low-carbon green tourism concept. The Government should also be reasonable to use market-based instruments, to develop a series of preferential policies to compensate for the low-carbon tourism economy, further affecting the tourism enterprise management tools [4]. For high pollution, high energy consumption of the industry must be given harsh treatment shut down, giving a low-carbon green energy

industry the appropriate incentives. Government and the community must vigorously promote low-carbon tourism joint media, green concept development, contribute to cognitive change jobs, to guide the public towards the direction of low-carbon environment [5]. Meanwhile, induce the low-carbon environmental philosophy into the national education system, educating more young people to foster excellent environmental awareness and behavior from childhood.

3.3 Develop low-carbon tourism economy, according to the regional characteristics

Tourism economy has regional characteristics, therefore, the choice to meet the region's tourism economic development model is particularly important. Analysis must be expanded from the actual situation of the region, set reasonable planning objectives to ensure the optimization of low-carbon tourism investment and receipts. Tourism resources of each region have its own advantages and disadvantages, people must carefully analyze the geographical environment, economic, market and other local conditions, develop low-carbon tourism economy, according to the very region, in order to ensure that regional economic development towards diversification and high-level direction.

4 CONCLUSION

As an important part of the development of low-carbon economy, low-carbon tourism is a relatively large and complex projects, in order to promote the tourism industry to develop toward the direction of low-carbon and green, tourism and related sectors need to continue to explore. This paper takes the concept of the development of low-carbon green tourism economy as the research perspective, introducing measures to implement low-carbon tourism economy.

REFERENCES

[1] Yin Qifeng. *Study on China's Low-Carbon Tourism Economy Development Model* [J]. Productivity Research, 2011, (9): 73–74.
[2] Zhao Qiong. *Research on Green Concept Application of Low-Carbon Tourism Economy* [J]. Farmhouse Technology (HEAD), 2012, (11): 32.
[3] Xu Xue. *"Low-Carbon Tourism"-based Henan Tourism Industry Cluster Building* [J]. Tribune, 2011, (12): 88–90.
[4] Huang Ying, Long Fuliang. *Initial Construction of Low-Carbon Tourism Product Life Cycle Assessment* [J]. Commercial Times, 2011, (16): 113–115.
[5] Wu Qian. *Low-Carbon Tourism Economy Green Concept Application Research* [J]. CPC Fujian Provincial Committee Party School, 2012, (7): 97–101.
[6] Wu Xiong. Low-carbon tourism: *Strategic Choice of Low-Carbon Economy Context of Sustainable Development of Tourism* [J]. Chinese e-commerce, 2011, (8): 50, 52.

Information, Computer and Application Engineering – Liu, Sung & Yao (Eds)
© *2015 Taylor & Francis Group, London, ISBN 978-1-138-02717-6*

Ways to explore vocational schools building a learning organization

Geng Xin Liu, Ling Tong, Dong Jie Wang & Run Guo Chai
Langfang Health Vocational College, China

ABSTRACT: This article first elaborated the basic theory of vocational school learning organization, defines the type of vocational school learning organization; combined with targeted analysis theory vocational school status quo and current management issues and the corresponding proposed building specific strategies and measures vocational school learning organization, and hope that through this study, our vocational schools able to provide better reference materials when building a learning organization.

KEYWORDS: Occupation school: learning organization: construction exploration

1 INTRODUCTION

In the culture of continuous development of scientific knowledge, the learning organization is constantly promoting the application; after a successful business practice, and their use continues to expand, for its application in vocational schools in both the development trend of the times, but also the organization building needs. Moreover, by building a learning organization, but also a good solution to the development of vocational schools more difficult to solve many problems (building, teaching issues, etc.). In the current economic and cultural competition more intense, how to achieve good vocational school building a learning organization, we need to recognize specific vocational schools for specific learning organization significance; Research Learning Institutions promoting professional organizations, is our country for construction of vocational education needs for basic vocational schools is to train technical personnel for the purpose of teaching, which can be very good to promote China's social development and progress; research to promote vocational school learning organization can help vocational schools to strengthen their overall competitiveness; at the same time, it also has to promote the comprehensive development of their school teachers, as well as enhance the overall quality of students' learning abilities and capabilities.

2 THE THEORETICAL BASIS FOR THE CONSTRUCTION OF A VOCATIONAL SCHOOL OF LEARNING ORGANIZATION

Learning organization management philosophy began from the 1990s, with the constant process of development, a new theory through practice and continuous combination of theory formation. The theoretical basis include; organizational learning, development theory, behavioral science theory, system dynamics theory, human resource management theory, organizational behavior science theory. Scholars believe Peter Senge; building a learning organization is to strengthen the transformation of the organization building it into thinking the content contains intrinsic and extrinsic behavior. By constantly changing, to discover the true self, in order to achieve self-value creation. Meanwhile, Peter Senge also thinks you want to achieve true learning organization theory assumes that the system must be thinking, self-transcendence, the establishment of a common value target, improve mental and strengthen the construction of team learning. As a new management organization model, learning organization in the daily work, primarily through the construction of well organized learning environment, to promote the work of innovative thinking construction workers, making it more user-friendly organization, flat, flexible, thus to promote the continuous development of its work. In the process of building a learning organization, and also contains a knowledge management theory, teacher professional development theory, and Lifelong Education. Where knowledge management is the management of knowledge and people, again in this dynamic among the fundamental human management, the static knowledge as managed objects. For basic functions of knowledge management include; internalization, externalization, intermediaries, and cognitive four areas. For teacher professional development theory, the main thought of as a professional teacher, his professional stage of development should be systematic, hierarchical, range, improving their professional development through school-based research degree, training and other means. Teacher professional development also includes; professional skills, knowledge, and emotional aspects. According to the different level of the teaching profession, will be divided into different stages of their professional development; preparation phase, survival phase, the consolidation phase, the update phase, mature phase. Teachers work by strengthening the system construction, to carry out the work of teacher education and teaching diversity, promote the teachers' ability to constantly update knowledge to make teaching a more perfect harmony. Through various activities, to enhance students' lifelong education thought and teaching activities to be student-centered, focusing on its ability to enhance the quality; while also focusing on the continuity of teaching activities, integration, and related principles may develop resistance.

3 RESOURCE MANAGEMENT ISSUES AND SCHOOL ORGANIZATION BUILDING STRATEGY

3.1 Traditional vocational school organization building management issues

1 A lifelong learning environment lacking: it mainly in three aspects; the first is the authority of principals and teachers is too strong, resulting in a lower intensity of implementing quality education, education for schools, students should pay attention to their own development, and resolutely resist coercion style teaching method of education as a means to control. Second; classification of organizational structure; this structure results in poor flow of information on the extent of the teaching process, that is not conducive to the professional development of teachers, is not conducive to the promotion activities, which also led the students in the teaching work always in a state of passive acceptance, and finally to the requirements of individual and organizational behavior from each other. Third; vocational schools inside the human environment and organizational culture is missing. For the school's organizational culture is the organization to integrate its own modules culture, but also its cultural characteristics of the whole school. In the actual teaching process, the traditional teaching thinking too strong, resulting in a deeper degree of individualism. There is now mostly confined to pursue vocational school facilities in a modern architectural forms, cannot reflect its own cultural characteristics, feel the characteristics of the human environment.

2 Students and teachers have no desire for common development: For the traditional school organization, its overemphasis on external require the internal members, erroneous understanding of individuals, ignoring the people's self-improvement and development of motivation.

3 The vocational school communication platform is not perfect: Prolonged individual consciousness teaching methods, resulting in poor awareness of teaching resource sharing, resulting in a serious waste of resources in the school culture. Due to the expansion of vocational education enrollment, student quality landslides, causing many teachers appear to meet the class phenomenon, rarely between plus and students to communicate, resulting in the relationship between students and teachers appear opposed.

4 VOCATIONAL SCHOOL LEARNING ORGANIZATION BUILDING STRATEGY

By analyzing the traditional organization for vocational school management issues, so that we learn the importance of building a learning organization and urgency. Meanwhile, in the continuous research and analysis of the problem found in a concrete building strategy;

1 Actively create a lifelong learning environment: it is the construction of the vocational school basic content of the learning organization; only create a good environment for lifelong learning in schools in order to make a better learning organization activity carried out.

2 Establish and improve the overall development of the school common desire: to improve through the construction of this content, and promote vocational school learning organization, enthusiasm for building atmosphere, under a unified vision to promote and accelerate the speed of learning organization vocational schools.

3 Improve the exchange of learning campus network system platform construction: the building of a learning organization in the process, through the improvement of the campus network, to expand its communication platform, in a wide range of interaction, and promote exchanges between students or teachers more closely in the close exchanges during the continuous improvement of its teaching management.

4 Give full play to the principal role in promoting the construction of organizational learning: For schools, the principal character of its central management is to achieve a learning organization and management focus on vocational schools. In teaching activities, one should always carry out educational activities of the organization of

work in improving the environment of the campus culture, based on the introduction of a variety of policy measures to promote the strengthening of the campus learning organization, engaging teaching building activities.

5 CONCLUSION

Traditional vocational school organization and management problems of many drawbacks, no longer apply to the current developments in education and teaching, so that students cannot get the skills and knowledge to effectively train. In this regard, to carry out a learning organization has become the focus of the current work in vocational schools, students normalization through lifelong passion for learning and positive mental attitude, to enhance the quality of its overall capacity levels and promote its ever comprehensive exhibition.

ACKNOWLEDGEMENTS

The eleventh five-year plan key subject of education in HeBei province science, No10030042.

REFERENCES

[1] Jinyu Gu. Vocational school organization construction strategy, [D]. University of Electronic Science and Technology. 2007.
[2] Jianzhong Zhao.Learning organization-building elements of inquiry. [J]. Reading and writing (Education Journal), 2009, (12), 108–109.
[3] Hua Wang. Learning organization to build knowledge management system [J] Commercial Times, 2012 (27); 92–93.

Information, Computer and Application Engineering – Liu, Sung & Yao (Eds)
© 2015 Taylor & Francis Group, London, ISBN 978-1-138-02717-6

Analysis and design of college students' ideological and political guidance management system

Dong Mei Sun
Public Infrastructure, Yunnan Open University, China

ABSTRACT: With the rapid development of the society, the ideological and political guidance for college students needs to be fully updated before they can follow the footsteps of the times. In terms of problems which restrict college students' ideological and political guidance task, the premise and basis of the good management of ideological and political guidance is to strengthen the effectiveness, which requires analysis of the management system of ideological and political guidance, implements exploration from updating concept, working mechanism, system construction and organization building, and many other ways.

KEYWORDS: college students; ideological and political guidelines; management system

1 INTRODUCTION

To strengthen the effectiveness is the main problem which needs the focus in college students' ideological and political tasks. For the current new situations and new problems of college students, universities need to strengthen ideological and political tasks of college students, playing the real political power.

2 EFFECTIVENESS OF COLLEGE STUDENTS' IDEOLOGICAL AND POLITICAL GUIDANCE MANAGEMENT SYSTEM

2.1 Effectiveness of ideological and political guidance management content

Universities red ideological and political education is not made out without evidence, it needs to be established on the basis of Marxism-Leninism and Mao Zedong Thought theoretical principles, has its own systematic principles and theoretical basis. Now, our country is at an important stage in the reform progress, based on this reality, according to the actual needs of building a socialist harmonious society, the basic contents of ideological and political education of college students need to be realistic and show effectiveness in addition to the traditional ideal and belief education, outlook on the world, life, values, etc., such as advanced education, Honor and Disgrace education, pay attention to learning, politics, righteousness education. Family virtues, social ethics education, dedication, and many other spiritual education.

2.2 Effectiveness of ideological and political guidance function

The performance of ideological and political work is based on the essence and objectives of ideological and political education. There are two most important properties, the first is the ideology performance, and the second is the humanities cognitive performance. Performance of ideological and political education of college students, in essence, should be based on social as standard, so it has a political sovereignty, showing an obvious political performance. As the basis for the performance of college students ideological and political tasks, the key to such political performance is to spread advanced political thought and theory, is the only manifestation of social progress in performance.

2.3 Effectiveness of ideological and political guidance management objectives

The immediate target of college students ideological and political education is to solve a wide range of specific ideological and political issues among students, the fundamental objective is to through the solve of specific problems, gradually increase awareness of the value of life of young college students, and proactively struggle to achieve their life values, and therefore make corresponding contributions to the survival of mankind as well as the progress of society. The main purpose of college students ideological and political task is to "cultivate outstanding college students". Its main purpose is to cultivate actual practitioners of the important thought of "Three Represents" and the concept of sustainable development. Its main task is to educate students, according to the main ideas of Marxism-Leninism, Mao Zedong Thought, Deng Xiaoping Theory and "Three Represents" and the scientific outlook on development, cultivate socialist successors who have ideals, morality, culture, and discipline.

3 THE MAIN PROBLEMS LIMITING THE MANAGEMENT OF UNIVERSITY STUDENTS' IDEOLOGICAL AND POLITICAL GUIDANCE

3.1 Education concept fall behind

With the continuous progress of the socialist market with funds, diversification trend appeared in ideology and culture of young people. The concepts of ideological and political tasks of many college students lag behind, and there are backwards in practice, resulting in a lot of problems. The first is that political ideals and concepts slowly fade. An investigation found that only eight percent of college students identify their life directions on the realization of communism and the common ideals in the current period, but more than sixty percent of college students take the realization of their own value as their own positions. The second is that the ideological and political concepts are quite different. The study found that, in the question "What do you think of the socialist system", about thirty percent of the students choose the "both socialist and capitalist systems have their own advantages and disadvantages". However, for college students, about a third of the students have not read Marx and Lenin. There are some students expressing hesitation for the advantages of socialism. The third is the weak sense of social responsibility. Investigation shows that about half percent of college students have wanted to be a rich man of high status, etc., this phenomenon requires everyone's attention and thinking. The fourth is the obvious trend of the diversification of values. With the development and changes in social, economic, political and cultural structures, etc., historical change and social change and others are reflected in the values of college students. So when recognizing the mainstream values of university students, diversity needs to be reflected.

3.2 Single form of ideological and political guidance management education

The first is the relatively old teaching method. Before, many university teachers did not implement teaching with the flexible use of multimedia, network and other modern teaching methods in accordance with the demand of teaching work, teaching object and the background, many of the traditional ways of teaching cannot be well adapted to the timed pace, can not even inspire college students learning interest

and enthusiasm. The second is the single information delivery. One-way transmission teaching methods such as "I say you listen" and "I teach you do" are still prevalent and very popular in colleges and universities, but these popular methods which are concerned with students such as student-centered, equitable and democratic two-ways communication, are put aside, or merely in theory, lacking necessary practice teaching.

4 DESIGN OF IDEOLOGICAL AND POLITICAL GUIDANCE MANAGEMENT SYSTEM

We need to establish a people-oriented ethics. Scientific outlook on development needs to be dominant. Further strengthen college students' ideological and political management tasks to adapt to the new developments, new situations and new problems of the current college enrollment and market funding. To update education concept is to be people-oriented, highlighting the students' dominance. Such emphasis on people-oriented thinking in education among college students is the main achievements of college ideological and political theory and the current practice of the innovative education, is also the actual needs of the scientific outlook on development, which points the direction of the current university students ideological and political guidance management.

5 CONCLUSION

In short, college students' ideological and political task is full of challenging and innovative, with a long way to go. Progress forward with times, use the socialist core value system and the scientific outlook on development to decorate our ideas, adhere to depart from the reality of university reform and development, innovate mission concepts, and take timely new program, so as to energize college students ideological and political management systems and opening up new roads.

REFERENCES

[1] Ou Yang, He Mingchang. *Analysis and Design of College Students' Quality Management System* [A]. *The Third Teaching Management and Curriculum Construction Conference Proceedings, School of Law, Hunan University of Technology* [C]. Hunan University of Technology, School of Law, 2012: 5.
[2] Dai Xiaolan. *Analysis and Design of Private Colleges Career Guidance Management Information System* [D]. Yunnan University, 2013.
[3] Xie Jintao. *Analysis and Design of College Laboratory Teaching Management System* [D]. Yunnan University, 2014.

Information, Computer and Application Engineering – Liu, Sung & Yao (Eds)
© *2015 Taylor & Francis Group, London, ISBN 978-1-138-02717-6*

A study on higher vocational college libraries' serving for new rural vocational education

Lu Sun
Qinhuangdao Institute of Technology, China

ABSTRACT: In the new rural construction, the development of new rural vocational education is weak. As service institutions which possess abundant library resources and advantages of personnel and technological equipment, higher vocational Colleges can utilize their advantages to serve for the development of rural vocational education, accelerate and promote the new countryside construction. Starting from the analysis of the present situation of new rural vocational education, this paper expounds the advantages and ways of higher vocational college libraries' serving for new rural vocational education, hoping to provide a reference for the development of rural vocational education.

KEYWORDS: Library, New countryside, Vocational education

1 INTRODUCTION

The new rural construction is an important historical task in the current development of China and an important component of establishing a harmonious society. The new rural construction culture puts forward many requirements and is facing the effects brought by land circulation, urbanization and many other problems. In the process of new rural construction, how to coordinate the problem that rural labor forces are of low quality and comprehensively enhance peasants' education level have become one of the most popular research issues. As the important cultural resource bases, libraries of higher vocational Colleges should take advantage of their own advantages to serve for new rural vocational education, fully realize the improvement of social and economic benefits. They should fully excavate their potential service capacities from multiple aspects to promote the cultural construction in new rural areas. Combining with the rural vocational education situation, this paper will discuss how higher vocational college libraries make use of their advantages to serve for the new rural construction and development.

2 PRESENT SITUATION OF NEW RURAL VOCATIONAL EDUCATION

In China, the rural vocational education starts late with low starting point and level. It has congenital deficiencies on development and lacks high-quality educational resources. With the current active promotion of new rural construction, rural culture conditions in many areas have been improved to some extent and peasants' quality has also been improved. However, taking a panoramic view of the situation, congenital deficiencies and poor acquired development still exist in the rural vocational education, which has caused the overall quality of rural labors are low. In the process of new rural construction, the quality of rural vocational education and the quality of labor forces have become an important bottleneck to restrict the development. The present situation of rural vocational education can be summarized as low cultural quality of rural labors, low quality of science and technology, low quality of management and inability to better develop and utilize rural resources advantages to serve for economic development and social progress. Five goals and requirements of the new rural construction are enhanced productive forces, higher living standards, civilized living style, an orderly and clean environment and democratic administration, which are the important development goals of harmonious socialism new countryside with the purpose of cultivating a large number of well-educated and high-quality talents with ability of technology and management to serve for rural development. Therefore, goals of rural vocational education can simply come down to high quality and understanding both techniques and business operations, which also are ones of current new rural vocational education development that can not reach [1].

3 ADVANTAGES OF HIGHER VOCATIONAL COLLEGE LIBRARIES' SERVING FOR NEW RURAL VOCATIONAL EDUCATION

Higher vocational college libraries have many advantages in serving for new rural vocational education, such as the advantage of technical talents, the advantage of collection resources, the advantages of equipment and advantages of scientific and technological achievements. They can effectively support the development of new rural vocational education, expand the teaching contents and improve the teaching level.

The advantage of collection resources is the most typical and critical one of a library. Higher vocational college libraries have enormous literature information resources, excellent professional system settings and abundant corresponding category resources which are also synchronous with the scientific research and advancing with the times. The library collections are richly structured. Especially resources involving disciplines, such as agriculture, forestry, animal husbandry are plentiful and have outstanding advantages in supporting rural professional development. There are not only a large number of domestic and foreign literatures, but also physical resources with a variety of carriers, such as CDs, cassettes and video tapes within libraries. Libraries also are linked with multiple databases, such as CNKI, Wangfang, Superstar and CqVip, and share documents and data resources with many domestic universities. This extraordinary advantage not only meets teachers' and students' demand of application services, but also is a big boost for vocational education resources in rural areas, the key supporting power to make the vocational education better be carried out and has important significance to the promotion of the quality of rural labors.

In terms of talents, higher vocational college libraries enjoy a cluster of highly qualified professionals who have outstanding professional knowledge, high scientific and cultural qualities, superior service consciousness and excellent professional responsibility in the library management. They know demands of readers and users, are familiar with all kinds of scientific research literatures and can well serve for the development of rural vocational education. These professionals are highly qualified and proficient in the applications of digital libraries and digital library management. They can rapidly and accurately find out all types of literature service resources in massive amounts of scientific research resources and have a positive role in enriching the contents of rural vocational education and reforming the teaching mode.

With respect to the technical equipment, current higher vocational college libraries have been committed to vigorously promote the construction of digital resources and enhance the level of library information. Literature storing, transferring, integrating and sharing have formed new situations. Applications of diversified audio-visual equipment, duplicating machines, miniature machines computers and wireless network systems provide convenient technical supports for resource utilization and development [2].

In the aspect of scientific research achievements, higher vocational colleges have all kinds of experts and scholars and bear many provincial or national scientific research tasks. They have plentiful relevant professional scientific and technological achievements and a large number of high-level scientific research literatures which can be an advantageous auxiliary team for rural vocational education development, have great practical value for the exploitation of rural resources and technological development and have a role in promoting the situation change of rural vocational education development.

4 HOW HIGHER VOCATIONAL COLLEGE LIBRARIES SERVE FOR NEW RURAL VOCATIONAL EDUCATION

If higher vocational college libraries utilize their own advantages to serve for rural vocational education development, they should combine theory with practice, effectively use the advantage that information resources serve for vocational education to promote the new rural cultural construction, comprehensively improve the quality of rural labors and propel the progress and development of vocational education.

Higher vocational college libraries should be based on requirements of the current new rural vocational education to complete literature information collection and carry out all kinds of business work around the present situation and development needs of rural vocational education to feasibly offer knowledge resources services for the quality improvement of rural labors. In practice, the library staff should go deep into the countryside to carry out field surveys and interviews, know the present situation of peasants' knowledge demands in new rural areas, learn about the local development and make full investigations around the literature information demand, such as local policy information demand, economic information and scientific information demand and information technical demand for wealth, and provide corresponding services of literature resources and educational resources on the basis of full understanding. In addition, considering rural vocational level and cultural quality, they should also pay attention to the requirements of carrier types of document information resources, such as books, newspapers and magazines or audio-visual material and electronic publications, select the most appropriate way to spread information resources, formulate specific repositories for the development of local vocational education and supply high quality comprehensive information services for new rural areas.

After investigating and understanding needs and carrier means of rural vocational education information resources, library staff also should survey the literature circulation, actively expand service objects and provide various kinds of resource services according to the local education service situation. They can draw lessons from the public library service mode, lower libraries' thresholds from service concept, means of service and other aspects, provide personalized services based on peasants' needs and make them enjoy many convenience brought about by the information resources. In terms of service contents, they can achieve service goals through a variety of ways, such as regularly holding the activity of library resources to the countryside and using local rural activity rooms to establish branch libraries or mobile libraries. They can provide a wide range of technical knowledge and literatures, such as regularly organizing training courses and reading festival activities. Higher vocational college teachers and students can commit themselves to rural vocational education and teach peasants the latest agricultural policies, science and technology and technologies of being rich, such as regular arranging students to go to the countryside. Especially in summer and winter vacations, students can combine local peasants to conduct social practice activities. Students can serve for local rural development and the implementation of vocational education activities in the process of self-training [3].

Remote educational platforms can be established with the use of higher vocational college library equipment and technological advantage to provide diversified online information services. The establishment of remote education platforms can be not restricted by time and space. Combining with rural vocational education, various types of learning activities can be carried out and face to face teaching between teachers and students also can be realized. Teachers can timely answers students' questions. Carrying out all kinds of learning discussion activities is a big support for rural education work. The effective use of internal teaching resources in higher vocational colleges can change close teaching into open one. Educational activities can be closely and really combined with practice to enhance the level of rural labor force quality and improve the current status of rural vocational education.

5 CONCLUSION

To sum up, as the current important development goal of China and in the face of low rural labor force quality and weak vocational education, the new rural construction can use advantages of information resources of higher vocational college libraries, talents and science and technology to serve for the development of rural education activities, achieve the goal that the current education situation and the quality of rural labors are improved and promote that the new rural construction can be better carried out.

REFERENCES

[1] Cui Yi: An Analysis of Libraries of Agricultural Colleges and Universities and Rural Informationization Construction, Journal of Information of Agricultural Science and Technology, Vol. 16 (2012).
[2] Wang Furong: Present Situation and Development Countermeasure of Libraries in Newly Built Universities-Taking Libraries of Newly Built Universities in Jiangsu as Examples, Journal of Library Science, Vol. 10 (2013).
[3] Yu Xiangqian: Exploration on "Three-dimensional Rural Issues" Service in Agricultural University Library under the Network Environment, Journal of Agriculture Network Information, Vol. 6 (2013).

Information, Computer and Application Engineering – Liu, Sung & Yao (Eds)
© 2015 Taylor & Francis Group, London, ISBN 978-1-138-02717-6

Transportation development, spatial relation, economic growth: A literature review

Yi Tang

Chang'an University, Xi'an, Shaanxi, China
Chang Sha University, Chang Sha, Hunan, China

ABSTRACT: This paper briefly introduces the literature of the theoretical studies of the relationship between transportation development, spatial relation and economic growth. The literatures suggest that the development of transportation is conducive to expanding the scope of the market, reducing regional gaps and promoting the long-term economic growth. In the meanwhile, the literatures also suggest that the speed of economic growth and regional economic characteristics determines the overall scale and transportation structure.

KEYWORDS: transportation development; spatial relation; economic growth

1 INTRODUCTION

With the rapid economic development, the government further increases the investment for transportation infrastructure and accelerates the speed of construction, which has achieved rapid overall development of transportation infrastructure and promoted the economic growth powerfully. Meanwhile, the regional and structural issues of comprehensive transportation system are prominent along with this process, which have attracted many researchers' attention. This paper aims to survey the related theoretical literatures about the relationship between transportation development, spatial relation and economic growth at home and abroad.

2 TRANSPORTATION DEVELOPMENT AND ECONOMIC GROWTH

In classical economic theory, Smith (1776) pointed out that division of labor is one of the most important driving forces of economic growth, however, it is restricted with the range of market. Transportation development enlarges market, therefore, he pointed out "traffic improvement is the most efficient one in all reform". However, in neoclassical economic model, represented by Debreu's general equilibrium model (1991), in order to deal with economic issues in a mathematical way, the researcher constructed a utility maximization model with no spatial dimension through redefining goods. Taking its base into no consideration makes neoclassical economics lose the

extensive explanation for modern economic phenomenon. The new trade theory and institutional theory have improved the flaws of neoclassical economics. Through the study how people make choices between different forms of increasing returns and different types of transport costs, Krugman (1991) suggested that transport costs determine the size and spatial distribution of economic activities by affecting centripetal force and centrifugal force of the accumulation. New institutional economists, represented by Coase make the transaction cost as tool to analyze. During the analysis, North (1981) explained the effects of institutional change on economic development by applying the change of transaction cost and ownership structure. The circulation of goods or products between production and consumption includes the "transaction" of ownership and "transport" in spatial place. Most of institutional economists and new trade economists regard transportation as a skill, which excludes transportation cost from transaction cost and economic growth depends on the exogenous technical progress of transport. However, the paradigm of new institutional economics provides a reference for transportation included into economic system. It is the breakthroughs mentioned above that enhance the theoretical level of study about the relationship between transport and economic growth.

Generally speaking, there is positive correlation between transport and economic growth in modern economics. Most economists think that transportation development is one of the main sources of the regional economic disparity. Technological advance in transportation and the reduction in transport costs can better explain the economic concentration and the change of land use pattern. Based on the price theory under the perfect competitive market mechanism, western classical location theory studies the land use features in ideal condition, with transport as one of the main factors affecting the locational advantage. Scholars studying urban structure stress that transportation is the base of formation of land use patterns. Burgess and R. E. Park (1925) proposed that the final urban land use forms concentric circle pattern because of economic decline curve due to different traffic conditions. Homer Hoyt (1939) thought that the transportation accessibility would promote urban developing into fan-shaped mode. Harris and Ullman (1945) provided multi-core mode for urban land structure, in which traffic location advantage determines the formation of sub-center. Bid rent theory studies the relationship between transport and land use with the method of economic equilibrium theory. In Haig's view (1926), "rent and transportation cost are closely linked by the relation with special friction. Transportation is a kind of means to reduce such friction at the expense of time and money. " Meanwhile, he proposed that land market reaches balance when the sum of rent, transport and negative effects is minimum. Alonso (1964) supposed that residents' consumption include land, commuting and any other general consumption. He constructed the biding curve of residents and manufacturers to analyze the balance of land supply and demand. Also he elaborated effect of transport on urban location from the perspective of land use. The above theories discussed the issue of regional economic growth with the assumption of transport development as an exogenous factor. However, empirical studies show that transport development is not an exogenous factor independent of regional economic system, while it is affected by different factors, such as, regional economic development level, the level of urbanization and characteristics of regional economy. Therefore, it is an endogenous variable determined by the regional economic system.

Transportation is the cause for the classical location theory, the new trade theory, the new institutional theory and the bid rent theory in economics, namely, transportation development can promote economic growth. Some economists argue that transportation is the result, namely, economic growth brought about the transportation development. In mainstream economics, transportation development is a black box, and we don't know exactly what factors affect the progress of transportation, thus further affecting economic growth. Generation after generation of economists try to open the black box of transportation development. Aggregated model and disaggregated model forecast traffic transportation demand by analyzing the relationship between land use and transportation. Through applying the typical engineering methods and collecting cross section data, Lowry (1964) and

Wilson (1974) analyzed the relation between traffic volume and population location to explore the spatial distribution of human travel. The aggregated model represented by Lowry model and Wilson model, also known as the four-step models, suggests that the traffic demand is a decision based on the four stages of travel generation, travel distribution, mode split and traffic assignment. With the development of economy, traffic demand is showing a trend of diversification. Because the traditional aggregated model neglects the influence of social, psychological and other factors on the transport demand, it is difficult to accurately grasp the changing characteristics of transportation demand from a macro perspective. Based on probability theory, behavior science, economics and other relevant theories, the disaggregated model, paying more attention to internal cause analysis of the generation of transportation demand is used more widely. Multi-logit model and multi-probit model are two kinds of common disaggregated model. McFadden (1974) established Logit model based on random utility theory and explained its characteristics, basically forming disaggregated model theory system. Lerman and Manski (1977) applied Probit model into transport demand forecast. Disaggregated model, which pays more attention to the influence of individual behaviors on transport demand, is suitable for the perfect transport stage when integrated transportation system is basically formed. At such stage, socioeconomic requirements for transport have transformed from quantitative into qualitative change to meet the multifaceted, multi-level and full range of transportation demand. However, disaggregated model assumes that the transport supply is given. And if the conclusion predicted by transport demand under the same influence factors for transportation is applied in different places, it is only applicable to short-term and specific transport market analysis. The factors of transport demand embrace specific features because of different stages of economic development and different levels of technical development. Therefore, it becomes an important subject for experts studying transportation economic theory to explain the economic system endogenously determining the developmental speed and structure of the transportation system.

Some transportation economists believe that technological progress is a decisive factor to influence the speed of the long-term transport development. Button (1993) elaborates the rules of transformation of transportation structure in the perspectives of the scale economy, scope economy and density economy and analyzes the impact of transport demand elastic changes in different types and modes of transport on transport demand. With the analysis of defects of existing transport demand theories, Zhao Jian (2005) raises the way to adopt transportation income elasticity and the elastic demand of

Table 1. The difference between the aggregate model and the disaggregate model.

	the aggregate model	the disaggregate model
The independent variable	The date of traffic area	Individual (or family) data
Common methods	Fratar method, Furness method, The maximum entropy method, Traffic network equilibrium theory etc.	Multivariate Probit model, multivariate Logit model etc.
The characteristics of the application domain	The perfect stage of social economy and infrastructure	Traffic demand diversification, The stage of economic development
The basic process of traffic demand forecast	1, The occurrence and attraction of travel	1, The travel frequency
	2, Trip distribution	2, The choice of destination
	3, the division of traffic mode	3, The choice of modes of transport
	4, the assignment of the path	4, The choice of the path

the traffic volume in order to analyze the long-term changing trend of the transportation demand. The theories above are to discuss the long-term developing problems in the exogenous premise of economic growth and technological progress. However, many theoretical and empirical studies show that transport and technology progress and economic growth are not independent of the transport system. There is an interaction between the change of land use pattern caused by economic growth and the transport infrastructure supply, and this interaction mechanism is a process of bidirectional feedback, and transportation demand is exactly changing in the process. Wu Qunqi (2000) focuses on the correlation between transportation system and market environment, advocates the mainstream of the value movement to study economic theory of the transportation systems, and introduces the consumer surplus theory into the economic analysis of transportation. Yang Xiaokai (2003) studies how the market—this "invisible hand" makes trade-offs between economy of specialization and transaction cost with transportation cost included. The new classical economics, founded by Yang Xiaokai, explore the relationship between endogenous transaction cost, the evolution of division of labor and factories, and further provides the proper analytical framework and theoretical system for transport economics to melt into the mainstream economics.

infrastructure for people to overcome space obstacles and implement spatial association. In the study of the literature of transport and economic relations, it is a very important problem to set the forms of shipping space link. Thünen (1826) investigates the relationship between the spatial allocation of agricultural land use and the distance from markets from the perspectives of the proportion between freight and the distance and weight, and different transport rates due to different agricultural products. Weber (1909), adopting weight and distance as the determinants of freight, studies the relations between freight and the index of raw material weight (i.e., raw material weight and manufactured goods weight ratio), and hence uses transport location rules to determine industrial location. W. Christaller (1933) assumes the traffic cost to be proportional to distance and deduces the city grade system. The new classical economists, such as the representative Debreu (1991), defined different locations of the same item as two different commodities in order to remove the space dimensions in economic models. Krugman (1991), trying "to avoid a single transport modeling", introduces the "iceberg" transportation technology. The "iceberg" transportation technology is that if the price of some products in the production r is P_r, then the FOB or CIF P_{rs} of the products in the consumption s is:

$$p_{rs} = p_r T_{rs} \tag{1}$$

3 TRANSPORTATION DEVELOPMENT, SPATIAL RELATIONS AND ECONOMIC GROWTH

Spatial dependence and heterogeneity is the main factor for the regional economic growth and the formation of the difference, and transportation is the main

Here, $T_{rs} \geq 1$. Formula (1) shows that If the 1 unit of products from the production r is delivered to the consumption s, only a few of them $1/T_{rs}$ can reach, and the rest has been lost in the course of transit.

Different from the above model about the monotonic relationship between transportation cost and distance, some literature states other important

753

factors to influence the differences in transportation development space. Zhang Wenchang (1994) believes that the regional freight formation rules show different laws in different economic development stages and the development levels because of different transport objects. Button (2006) "isolates the industry which is insensitive to transport costs by observing the relative importance of transportation cost in the production cost". Wu Qunqi (2000) does not think that the transportation price is the main factor to decide the relation between supply and demand of transportation, but he holds that transportation cost, in addition to the potential losse, reflected from the price and the full cost, should include the loss of transportation time and space value and should raise the quality of transport in aspects of the quality and quantity service and comprehensively promote the development transportation system. In empirical research, Fogel (1979) adopts counterfactual argument quantitative methods to calculate the impact of the railway on American economic growth at the time, and contends that we should use historical description and the traditional method of economics to explain the role railway plays in economic growth. Based on the analysis of the historical data of the changes of community transportation structure in German, Zhao Yiping and Hu Anzhou (1993) discusses the influence of the factors of economic evolution stage on transportation structure change. They believe that transportation structure in a certain period of time should be suitable to the stage of economic development and the technical level.

4 CONCLUSION

Based on the introduction and evaluation of the theories about the transportation development and economic growth, thie paper shows that, since the early 1990's, the studies of the relationship between transportation development and economic growth are an important direction of transport economics and the theories of economic growth with a great number of papers, and the understanding of the relationship between transportation and economic growth has been greatly enhanced. This paper briefly reviews the literature about the relationship between transportation development and spatial relation, economic growth. Much literature show that, transportation development is conducive to expanding the scope of the market, reduce the regional economic disparities and promote long-term economic growth, and in the meanwhile, the speed of economic growth and regional economic characteristics determine the overall scale of transportation development and the structure of transportation. However, in order to possess a more comprehensive and profound understanding of the interaction between the economic growth and the transportation development, it is still necessary to carry out further researches in many aspects.

REFERENCES

Adam Smith: An Inquiry into the Nature and Causes of the Wealth of Nations. University of Chicago Press (1776).
Debreu G: The Mathematization of Economics Theory. American Economic Review Vol. 81 (1991), p.1–7.
Paul Krugman: Economics of Geography. Journal of Political Economy Vol. 99 (1991), p.483–502.
Douglas North, ed: Structure and Change in Economic History. US: Norton (1981).
Harold Carter: The Study of Urban Geography(Fourth Edition). US: The Hodder Arnold Publication (1995).
Haig R. M: Toward an Understanding of Metropolis: II. The Assignment of Activities to Areas in Urban Regions. Quarterly Journal of Economics Vol. 40(2) (1926).
Alonso: Location and Land Use: towards a General Theory of Land Rent. US: Harvard University Press (1964).
Lowry I. S: A Model of Metropolis. Rand Corp (1964).
Wilson A G: Urban and Regional Models in Geography and Planning. US: Wiley (1974).
McFadden D: Conditional Logit Analysis of Qualitative Choice Behavior. US: Academic Press (1974).
GUANG H Z: Disaggregate Model—A Tool of Traffic Behavior Analysis. Beijing: China Communications Press (2004).
RONG Z H: On Transportation. Beijing: Chinese Social Science Press (1993).
Kenneth Button: Transport Econimics (the third edition). Beijing: China machine Press (2013).
ZHAO J, CHENG Y: The Theory of Transportation Demand and Transportation Demand Growth Trend. Comprehensive Transportation Vol. 11 (2005), p. 11–16.
WU Q Q: Study of the Value Analysis on the Transportation System. Shangxi: Chang'an University (2000).
Yang X: Economics: New Classical Versus Neoclassical Frameworks. Beijing: Social Sciences Academic Press (2003).
Debreu G: The Mathematization of Economic Theory. American Economic Review Vol. 81 (1991), p. 1–7.
ZHANG W C, JING F J: The Formation and Growth Regularities of Spatial Transport Linkage. Acta Geographica Sinica Vol. 49(5) (1994), p. 440–448.
Robert W. Fogel: Notes on the Social Saving Controversy. The Journal of Economic History Vol. 39(1) (1979), p. 1–54.
ZHAO Y, HU A. A: Theoretical Study on Evolution of Transport Structure. Journal of Chongqing Jiaotong University Vol. 12(2) (1993), p. 1–8.

Information, Computer and Application Engineering – Liu, Sung & Yao (Eds)
© 2015 Taylor & Francis Group, London, ISBN 978-1-138-02717-6

The path exploration on party organization construction of health professional colleges and universities with an aim of excelling in performances

Dong Jie Wang

Langfang Health Vocational College, China

ABSTRACT: Higher professional colleges and universities are important strategic bases of scientific studies and talents cultivation and the important field of Party organization construction. Conducting Party organization construction with an aim of excelling in performances in colleges and universities is a new task of strengthening the Party's primary construction and the requirement of developing advanced Party members. Health professional colleges and universities should actively carry out the Party organization construction, combine education activities with serving students, improving the quality of school education and efficiency, further strengthen the school Party member activities and improve the influence of Party organization construction. This paper will mainly discuss the path of colleges and universities' Party organization construction.

KEYWORDS: Excelling in performances, Health professional colleges and universities, Party organization construction, Path

1 INTRODUCTION

China's educational enterprise shoulders the important responsibility and mission of cultivating socialist builders and national successors, an important place of training Party activists and a momentous way of establishing Party organization construction with an aim of excelling in performances and strengthening the cohesion and creativity of primary Party organizations. In building Party organization construction with an aim of excelling in Performances, people should profoundly grasp the important connotation of excelling in performances, positively make innovation on contents and forms of campus Party organization construction activities, insist considering the excelling in performances as the premise and condition of school Party organization development and improve the comprehensive quality of campus primary Party members.

2 ROLE OF EXCELLING IN PERFORMANCES IN PARTY ORGANIZATION CONSTRUCTION

2.1 *Excelling in performances is the common goal of Party organization construction*

Colleges and universities are important bases for cultivating talents. The basis of Party construction in colleges and universities is to insist the socialism path and adhere to the leadership of the Communist Party of China. All Party members are required to keep the combat effectiveness and cohesion of Party organization construction under a common goal, strengthen the organization's purpose and consciousness and improve the enthusiasm of Party members and realize the comprehensive, compatible and sustainable development of school Party members' overall business and individuals. The overall business of colleges and universities is to cultivate high-quality talents. Party members in colleges and universities should put themselves in a business platform and make joint efforts to enhance self-worth.

2.2 *Excelling in performances can inspire strong learning desire and advanced studying concept*

Party organizations of colleges and universities require building learning-type ones of Party members who must establish the concept of lifelong learning, constantly improve their work ability and knowledge level, play the positive role of Party organizations and improve the cohesion and creativity of Party organizations in the process of learning. That is to say that excelling in performances requires Party members should adhere to the unity of work and study, actively solve practical problems in the process of learning, achieve the unity of knowledge and action, be bold in innovation and practice and stimulate learning desires to promote the overall development of the team, encourage general Party members to make new contents and forms in activities and maintain their progressiveness and initiatives.

2.3 Improve party members' team consciousness

Party organizations of colleges and universities take the excelling in performances as the goal, which is a reflection of creating an advanced Party branch and striving for being excellent Party members and also a kind of expression of paying attention to team consciousness and Party members' subject consciousness. Through the integral collocation and overall activities grasping, Party organizations of colleges and universities positively make innovations on activity forms, excavate the potential of each Party member, improve the cohesion of Party member construction, boost Party members' ability to learn, organize and manage, promote the overall development of Party members and better achieve Party members' overall goal.

3 PROBLEMS EXIST IN PARTY ORGANIZATION CONSTRUCTION OF HEALTH PROFESSIONAL COLLEGES AND UNIVERSITIES WITH AN AIM OF EXCELLING IN PERFORMANCES

3.1 Ideological building is weak

Now ideological and political work launched in many universities has not high quality, often becomes a mere formality and the daily management is not standard. Activities are antiquated, single and less innovative and it is difficult to have charisma and persuasion. When organizing Party member activities, they would read a newspaper, attend a lecture or tour the scenic spots in the slogan of going out to learn, which has seriously affected the seriousness and instructiveness of Party organization construction. In addition, higher professional colleges have short lengths of schooling and greater mobility. Therefore, it is difficult to effectively to carry out the Party members' education management. Some party members display sloppy thinking, weak consciousness of responsibility and utilitarian motives of joining the Party.

3.2 The party organization construction lacks vitality

Health professional colleges and universities generally set up organizations According to the constructing requirements of ordinary colleges and universities and do not form relatively complete mechanism of party construction in the work function, which manifests as weak primary Party organization construction. Workers who are Party members inside

universities are rare, so they have heavy tasks. Lack of unified standards in specific work makes staff's work enthusiasm not high. When developing student party members, they often only pay attention to the quantity and ignore the quality. The lack of effective training of party members leads to not high development purity of Party members, incongruous relation between the party and the government, not good implementation of Party organization activities and low work efficiency.

3.3 Working styles have problems

When developing Party members, colleges and universities adopt conservative and traditional educational methods of primary Party organizations. Students' interest in learning can not be aroused and educational methods are difficult to adapt to the development requirements of modern society, which would affect the smooth development of Party organization work. There are many reasons which lead to the lack of regularity on activities of primary Party branches and the lack of flexibility on all kinds of systems of the Party organization construction.

4 THE PATH EXPLORATION ON PARTY ORGANIZATION CONSTRUCTION OF HEALTH PROFESSIONAL COLLEGES AND UNIVERSITIES WITH AN AIM OF EXCELLING IN PERFORMANCES

Organization and implementation ways and paths of university Party construction work are the main problems that the Party construction is facing in the new period. Colleges and universities should to make arduous efforts to explore, actively carry out activities on excelling in performances, provide good internal environment for campus Party building work and make excelling in performances become important measures for teachers and students in colleges and universities to improve the cohesion and combat effectiveness.

4.1 Actively carry out party member activities which conform to the actual condition of colleges and universities

Carrying out Party member activities should attach great importance to the methods and ways. First, it should abandon traditional ways, pay attention to the role of propaganda and guidance and make efforts to create a dense atmosphere of excelling in performances. Students can be encouraged to actively participate in the activities through opening activity columns of excelling in performances on the campus website and timely reporting methods and strategies of school activities on excelling in performances. Meanwhile, external publicity can be expanded to report advanced characters and deeds that have sprung up in campus activities on excelling in performances with the city's media and create a good atmosphere for campaigns.

4.2 *actively promote construction of teacher's ethics and style of study*

In the Party construction work, colleges and universities should consider the positive promotion of teacher's ethics and style of study as the important juncture of Party-masses' excelling in performances and strength the construction systems of Party conduct, teachers' morality and style, style of study and school spirit. Carrying out activities on excelling in performances in colleges and universities should pay attention to the construction activities of the teacher's ethics and study style first, consider teachers' and students' requirements and suggestions as the starting point, integrate the construction of Party conduct into the one of teacher's ethics and study style and promote the development of teacher's ethics and style of study with the construction of Party conduct. Party members and students work together to create the campus Party organization construction of excelling in performances. The study of campus Party organization theories should be done well. In addition, colleges and universities should attach great importance to Party members' ideological style construction, strengthen the construction of teacher's ethics and lay emphasis on the positive guiding function of publicity and education in the campaigns. Party members and cadres should serve as role models to others, actively create a learning atmosphere of respecting teachers and valuing education and set up good images of party members for teachers and students. Moreover, the campus Party organization system construction and supervision and administration should also be strengthened to restrain and promote activities of university teachers and students, establish and improve various university Party organization management systems. Organizational disciplines of Party members and cadres must be actively implemented to improve the cohesion of Party organization construction.

5 CONCLUSION

There are many problems that Party organization construction of health professional colleges and universities may confront with an aim of excelling in performances. People have to think seriously, make thorough investigations, have innovative ideas and respond positively. People should also attach great importance to the implementation of the party construction of colleges and universities, take the premise of excelling in performances and do well in every link of the party construction and cultivate excellent party member successor with high comprehensive quality for the national Party organizations.

ACKNOWLEDGMENTS

The twelfth five-year plan project of education in HeBei province science, No11050102.

REFERENCES

[1] Zhang Xiaomeng: Problems and Countermeasures Which Higher Professional College Students' Party Construction Confront, Journal of Health Vocational Education, Vol. 30 No. 20 (2013), pp. 192–193.
[2] Zhang Daiyu: Role Playing and Path Exploration of Excelling in Performances in University Learning Party Organization Construction, Journal of Da Guan Weekly, Vol. 4 (2012), pp. 108–109.
[3] Wang Dongjie, Liu Bisong, Wang Lijuan and Wang Yan: Implementation Strategies of Party Organizations with Excelling in Performances in Health Professional Colleges and Universities, Journal of Software (Education Modernization) (Electronic Version), Vol. 1 (2013), pp. 105–106.

Information, Computer and Application Engineering – Liu, Sung & Yao (Eds)
© 2015 Taylor & Francis Group, London, ISBN 978-1-138-02717-6

Analysis of competitive sport talents' training mode

Wei Wang & Chun Ping Dong
Military Physical Education Department, Jilin Agricultural University, China

ABSTRACT: In the 21st century, the tendency of competitive sports professionalization has intensified competitive sports competitions between countries. To achieve the cultivation of competitive sports reserve talents, a variety of competitive sports talent training modes have appeared gradually in recent years. Each of them has its own characteristics and has laid a solid foundation for China's competitive sports reserve talents. Starting from the concept of competitive sport talents, this paper analyzes various existing training modes of competitive sports talents.

KEYWORDS: Competitive sports, Talent, Training mode

1 INTRODUCTION

As the core of economic sustainable sports development, the development of competitive sports talents cultivation in the new era plays an important role in the development of a country's sports undertakings. With the prosperous development of competitive sports in recent years, economic sports training mode has become the most important factor that affects athletes training. The previous simple reverse talents cultivation depending on the state can not meet the needs of development. it requires to composite powers from various social organizations, such as colleges and universities, to realize the cultivation of reserve sports talents and form a diversified personnel training mode of competitive sports, which is of important practical significance for the development of competitive sports in China.

2 THE CONCEPT OF COMPETITIVE SPORTS TALENTS

According to the author, "talent" in Ci Hai refers to knowledgeable people with excellent morals and academic skills who can carry out creative work and make contributions to material, spiritual and political civilization construction in social development. The most important thing is embodied in the knowledge and ability. Competitive sports talent means talented people who specialize in sports trainings and competitions, including coaches, athletes, foreign affairs staff and support crew. They together constitute competitive sports. This paper mainly discusses the training mode of elite athletes and aims to provide the beneficial reference for China's cultivation of competitive sports reserve talents [1].

3 CHINA'S COMPETITIVE SPORTS TALENT TRAINING MODE AND ANALYSIS

3.1 *University training model*

As a talent training base, the university plays an important role in the aspect of sports talents cultivation. Student athletes have showed up prominently in competitive sports competitions. The cultivation of competitive sports talents has gradually inclined to combined cultivation of sports and education from national cultivation and the distinctive training pattern of colleges and universities has been gradually formed. Compared to the individual training mode, this channel has alleviated the problem of contradiction between leaning and training and realized the cultivation of outstanding talents of competitive sports from lower ages. Through systematic training and education in elementary and middle schools and professional cultivation colleges and universities, excellent sports talented person's potential and gifts can be best displayed and excavated with the most systematic training. Their first identities are students, followed by high-level athletes. The effective combination of knowledge and learning can promote the comprehensive development of competitive sports talents and lay a solid foundation for their subsequent development.

In the training model of university, the university bears most of the cost of athletes and athletes just need to receive professional training at specific times for achieving the goal of talents cultivation. On the selection of student athletes, the university should choose appropriate talents to develop combined with its own faculty, sports characteristics, key sports in its area and other various factors. This training mode reflects that China conforms to the development philosophy of times in the cultivation of competitive sports talents, effectively combines education with sports and preliminarily opens channels for universities to cultivate elite athletes. It has cultivated excellent athletes with both ability and political integrity for the national competitive sports development, but it also has exposed some problems in the development. First, athletes who have achieved excellent results in world competitions are still professional athletes in essence, because they enter universities on the basis of enjoying national welfare and policies and they are not student athletes cultivated entirely by universities. Second, in terms of training facilities, only specific universities have good systematic training facilities and most universities' training facilities can not keep up with needs and weight against the development of competitive sports talents. Finally, high level and experienced coaches are insufficient. Athletes' training is constrained [2].

3.2 State training mode

This is the most common and primary talents training mode of competitive sports, also known as the whole-nation system mode. It pays attention to layers of selection of athletes who can enter the national team for training only they achieve the corresponding standards. For the selection of athletes, the country has independent choice and is fully responsible for all costs, such as training, and competition entry. Athletes' first priorities are to improve their levels of competitive sports and win honors for the country when participating in large and small events. Adequate financial and material support can make athletes achieve excellent results in a short period [3].

State training mode has made significant achievements in recent years and this idea was more obvious at the 2008 Beijing Olympics. Before 2000, China knew nothing about trampoline. After many aspects of exploration and research and intensified training on related talented athletes, China made a proud achievement at the Beijing Olympics and champion figures emerged, such as He Wenna. But together, it still has some problems. For instance, lots of resources are concentrated on dominant events and the development of inferior events can not be supported effectively. The government's monopolization of sports leads to No Distinction of Management and Conduct". The operation of institutions was too dependent on administrative means, which violates the principle and law of market economy. The athletes' training relies too much on administrative input, which has demoralized the enthusiasm of sports socialization and gone against the cultivation of competitive sports talents.

3.3 Individual training mode

In the development progress in the new period, individual training mode of competitive sports talents is emerging. It extricates from the training mode from sports schools to provincial team and then to national team. Competitive sports talents are cultivated through family investment. The most representative athlete is snooker champion Ding Junhui whose success totally came from the cultivation of his family. He cultivated himself through self-study and overseas study, which also made this kind of training mode of competitive sports come into the public view gradually. Under this cultivation mode, athletes do not have to consider "Program of Striving for Olympic Glory" or achievements in a short period. In addition, benefits obtained after competitions can be fully owned by individuals. In the process of athletes' cultivation, families provide training and entry fees and bear risks entirely by themselves.

As an emerging training mode of competitive sports talents, it has a large operational development momentum to encourage families to pursue the benefit maximization, which has significantly enhanced the investment enthusiasm and autonomy of competitive sports. It integrates organization, training and competition in one, possesses high flexibility and creativity and opens up a new path for the development of competitive sports talents training. High-risk family investment may bring about high economic benefits, which is the motive of family training mode. Taken together, golf players and tennis players in China are also cultivated with this kind of training mode which has cultivated a large number of excellent athletes for China. There are still two problems to be solved in this mode. The first one is the high risk of training subjects. The second one is that athletes can not have both studying and training. They can only abandon their studies to train and take part in matches. After the retirement, many problems would appear and they should actively seek solutions [4].

3.4 Club training mode

Driven by the professional league system in the new era, club training mode emerges as the times require. As a new economic sports talent cultivation mode, it takes the club as the operating carrier, pays more attention to the athlete's initial interest when he comes into contact with certain sport and raises the possibility of its development based on this. Compared with the above mentioned training modes of competitive sports talents, club training mode has a few advantages. First of all, this is invested by social groups and needs no national funds. The improvement of athletes' economic levels more embodies the ability of the club and increases its popularity. Second, the club can reserve more outstanding sports talents. Based on the club's popularity ascension, more teenagers are attracted to the club by its reputation and the club reserves a large number of educable talent resources [5, 6].

4 CONCLUSION

In the club training mode, the most iconic sports event is football. For instance, Barcelona Football Club and Guangzhou Evergrande Taobao Football Club have reserved a large number of football talent resources and become cradles to foster and provide excellent athletes. Clubs give most teenagers training bases and opportunities to be professional athletes and have cultivated numerous excellent talents for the development of competitive sports.

To sum up, as main channels and ways of cultivating competitive sports talents in the new period, all kinds of training modes own different characteristics. This paper has analyzed the university, state, individual and club training modes and dissected processes and costs of various training modes of competitive sports talents and analyzed existing disadvantages in some training modes to provide beneficial references for China's competitive sports talent cultivation and promote the continuous development of China's competitive sports course.

REFERENCES

[1] Sun Qilin, Zhang Jianxin and Mao Lijuan: Analysis of the Combined "School-Sports-Enterprises" Cultivation Mode for High Quality Competitive Sports Talents, Journal of Shanghai University of Sport, Vol. 4 (2010), pp.65–68.
[2] Wang Hongjiang, Wang Hongkun and Wuwen: Survey and Improvement Countermeasures on the Current Situation of Athletic Sports Talents Cultivation System in Sichuan Province, Journal of Chengdu Sport University, Vol. 3 (2011), pp. 11–15.
[3] Yang Xuemei: Research on the Training Mode of Competitive Sports Talents in China, Journal of Chinese Talents, Vol. 14 (2011), pp. 177–178.
[4] Fan Chengwen and Liu Yayun: Research on Training Modes of Reserve Talents in Athletic Sports, Journal of Hunan University of Technology (Social Science Edition), Vol. 6 (2011), pp. 133–136.
[5] Kong Qingbo and Ge Yushan: Research on the Training Mode of Competitive Sports Talents in China, Journal of Nanjing Institute of Physical Education (Social Sciences), Vol. 4 (2014), pp. 48–52.
[6] Bai Xilin, Zuo Wei and Li Suo: Exploration on the Training Mode of Competitive Sports Talents in China in the New Period, Journal of Sports Culture Guide, Vol. 10 (2013), pp. 15–18.

Information, Computer and Application Engineering – Liu, Sung & Yao (Eds)
© *2015 Taylor & Francis Group, London, ISBN 978-1-138-02717-6*

Analysis of higher vocational college students' cultivation of entrepreneurial ability

Jun Jie Xu & Min Fei Shan
Hebei Engineering and Technical College, China

ABSTRACT: The Report to 17th National Congress of the CPC clearly pointed out that "entrepreneurship promotes employment". Therefore, strengthening the cultivation of higher vocational college students' entrepreneurial abilities will not only alleviate the national employment pressure, but also fundamentally solve the difficult problem of university students' employment to make their dreams come true and promote economic development. Thus it can be seen that cultivating college students' entrepreneurial abilities is of great significance. Combined with the problems appeared in the cultivation of higher vocational college students' entrepreneurial abilities, this paper will analyze, study and put forward corresponding measures for the problems.

KEYWORDS: Higher vocational college students; Entrepreneurial ability; Important significance; Exploration

1 DISCUSS THE PRACTICAL SIGNIFICANCE OF HIGHER VOCATIONAL COLLEGE STUDENTS' CULTIVATION OF ENTREPRENEURIAL ABILITIES

First, it can solve the difficult problem of university students' employment in China. It can fundamentally solve the difficult problem of university students' employment and alleviate the national employment pressure. In addition, the vitality of entrepreneurship also can form the multiplication scale effect on creating more employment opportunities.

Second, it is conducive to the creation of social value. To start their own business, college students will choose industries that they know well, bring their expertise and talents into full play and skillfully unify the self-value and social value to create value for society and ultimately achieve their goals in life.

Third, it is advantageous to the development of modern teaching modes. The Report to 17th National Congress of the CPC clearly pointed out, "As the senior form of employment, entrepreneurship can effectively solve problems of entrepreneurs' survival and development." Therefore, improving the level of higher vocational college students' entrepreneurship education is the need of social and economic development and conducive to scientific development of modern teaching modes.

Finally, it benefits the construction of powerful nation of human resources. Through the cultivation of higher vocational college students' entrepreneurial abilities, college students choose to be self-employed, which will make potential talents become talents in reality and make China move towards a powerful nation of human resources from a populous nation and then advance towards a modern powerful country from a powerful nation of human resources.

2 ANALYZE PROBLEMS THAT APPEARED IN THE PROCESS OF HIGHER VOCATIONAL COLLEGE STUDENTS' CULTIVATION OF ENTREPRENEURIAL ABILITIES

To make their entrepreneurial dreams come true and show the value of life, higher vocational college students should cultivate their entrepreneurial abilities, which is critical. However, at present, the cultivation development of China's higher vocational college students' entrepreneurship abilities is still in its infancy and potential problems exist in many aspects. Subjective problems come from college students' individual consciousness and quality. Objectively speaking, problems are from the attachment to their families, entrepreneurship education cultivation systems in colleges and universities and national policies. These factors mutually restrict and influence each other.

2.1 *Higher vocational college students have weak personal consciousness of entrepreneurship*

Many students are still armchair strategists and engaged in idle theorizing. They have weak personal consciousness of entrepreneurship and also once dreamed to show themselves in the business and realize the value of life. However, due to the lack of self-confidence, management experience, funds, human factors and other factors, they shrink back.

2.2 Parents lack of awareness of higher vocational college students' entrepreneurial ideas

Whether the entrepreneurship can be successful ultimately needs tests for a long time and investment of time, money, human and material resources. There are some difficulties and risks for general families. Therefore, many parents hold attitudes of indifference and rejection. They do not support their children to start their own business with risks and pressures and they would rather their children find reliable jobs with relatively stable income, which can directly affect the college students' entrepreneurial intention and the cultivation of entrepreneurial abilities.

2.3 Colleges do not pay enough attention to the combination of theory and practice on students' entrepreneurship education

Many colleges and universities only focus on theoretical education. For instance, they open some selective courses of entrepreneurship education and hold some lectures on entrepreneurship. The contents of textbooks are not innovative and practical and completely do not conform to social needs. Moreover, some colleges and universities lack teaching staff in entrepreneurship education and most teachers are temporary substitute ones which have no venture investment experience and entrepreneurial practical experience. The entrepreneurship education also lacks necessary practices. In this educational background, college students' entrepreneurship and entrepreneurial abilities can not be given full play.

2.4 Colleges and universities are short of consciousness to create entrepreneurial campus cultural atmosphere

Though a lot of colleges have conducted entrepreneurship education activities, many colleges' ways of entrepreneurship education are formalistic. They completely do not realize the importance of developing entrepreneurship education courses and do not really make every student become the protagonist of entrepreneurship education. The atmosphere of entrepreneurship education in colleges is also insufficient, which can affect students' enthusiasm of entrepreneurship and the full display of their innovation abilities.

3 ANALYZE HIGHER VOCATIONAL COLLEGE STUDENTS' CULTIVATION MEASURES OF ENTREPRENEURIAL ABILITIES

Cultivating abilities of higher vocational college students' self-employment and strengthening college students' entrepreneurship consciousness not only need the active participation of colleges and students, but also need support and help of the government and other enterprises and institutions from all walks of life. Now through the analysis of problems appeared in the cultivation of higher vocational college students' entrepreneurial abilities, the following measures will be taken for discussions.

3.1 Higher vocational college students should possess entrepreneurial ideas and consciousness and improve their own quality comprehensively

Higher vocational college students plan their careers as early as possible, change the traditional employment view, set up correct outlooks on career and entrepreneurial ideas, enhance the consciousness of entrepreneurship, cultivate entrepreneurial spirit, actively seek for business opportunities and have the courage to challenge when addressing opportunities. In addition, besides students master some professional knowledge and skills, they also should strive to learn some knowledge and skills related to entrepreneurship, such as sales skills and negotiation skills needed in enterprise property systems and in the process of entrepreneurship. They should exercise their abilities of communication and work, accumulate experience and improve their own quality comprehensively.

3.2 The concept of colleges' management should be undated and entrepreneurship education should be standardized

Colleges should not only provide students with solid education of theoretical knowledge, but also offer them opportunities of practices and actual operations. In addition, colleges need to strengthen responsibilities of entrepreneurship teachers and make students possess necessary awareness and abilities of entrepreneurship. Moreover, colleges can hold unscheduled lectures and invite entrepreneurs and successful people from all walks of life to give lectures. Teachers can be part-time advisors to tutor students with entrepreneurial dreams one-to-one. In this way, the level of entrepreneurship education can be improved and students also can be encouraged and helped to start their businesses. Students' entrepreneurship enthusiasm can be boosted. They not only possess necessary theoretical knowledge, but also are primary advanced technical applied talents with strong practical abilities.

3.3 Cultivate higher vocational college students' entrepreneurial quality and abilities from different fields and pay attention to the combination of theory and practice

Besides basic knowledge of entrepreneurship, colleges should more effectively combine extracurricular activities with knowledge learned. For instance, they can hold science and technology contests, establish entrepreneur clubs, provide students with opportunities of entrepreneurship and practice and cultivate their innovative abilities and consciousness of entrepreneurship. Thus, students not only master basic business regulations and relevant knowledge of entrepreneurial culture, but also recognize the necessary entrepreneurial quality and responsibilities and improve their business interest.

3.4 Colleges can cooperate with the government, enterprises and pubic institutions, establish entrepreneurship education training bases with internal and external efforts and optimize the entrepreneurial environment

Colleges can provide students free office space, policy advisory and other practical places to make them learn more about practical skills that entrepreneurship needs and lay foundation for higher vocational college students' future self-employments.

In addition, a good public opinion atmosphere should be created to make people treat entrepreneurship more rationally. The national support for college students' entrepreneurship can be promoted through television, network and other means of propaganda. Colleges can provide the most preferential policies for college students to start businesses and offer to provide higher vocational college student entrepreneurs

help and support from venture financing, project management and other aspects to create a better market environment.

4 CONCLUSION

Higher vocational college students are groups with the greatest innovation and growth potential. The cultivation of entrepreneurship abilities has a very important impact on college students' entrepreneurship. In the process of entrepreneurial ability cultivation, the awareness of entrepreneurship is very vital for starting a business, followed by the influence of entrepreneurial quality. Entrepreneurs should have the courage to make bold changes and innovations, defy frustrations and dare to take risks. They need a variety of comprehensive qualities. Therefore, they not only need the guidance of school teachers, but also need the mutual help of the government, society and themselves to comprehensively conduct the cultivation education of entrepreneurial capabilities. Through the campus entrepreneurial experience and the cultivation and practice of entrepreneurship, colleges can escort students' self-employments and improve their entrepreneurial capabilities until students finally and really enter the society.

REFERENCES

[1] Mao Hongwei: Research on the Cultivation of Vocational College Students' Entrepreneurial Ability, Journal of The Science Education Article Collects, Vol. 2 (2014), pp. 22–23.
[2] Huang Qunying: The Study on the Cultivation Situation and Countermeasures of Higher Vocational College Students' Entrepreneurial Ability, Journal of Knowledge Economy, Vol. 5 (2011), pp. 45–46.
[3] Hu Peng: The Study on the Cultivation Situation and Countermeasures of Higher Vocational College Students' Entrepreneurial Ability, Journal of Urban Construction Theory Research (electronic Edition), Vol. 36 (2013), pp. 10–11.

Information, Computer and Application Engineering – Liu, Sung & Yao (Eds)
© 2015 Taylor & Francis Group, London, ISBN 978-1-138-02717-6

Analysis of artistic gymnastics athletes' sense of apparatus training methods in sports institutions

Xiao Ping Xu
Faculty of Physical Education, Ningbo University, China

Xue Song Du
Faculty of Physical Education, Hengshui College, China

Jia Jin Liu
Faculty of Physical Education, Ningbo University, China

ABSTRACT: With the development of sports undertakings, China's sports art colleges and universities have been developing rapidly and become main training bases of artistic gymnastics athletes. The technical training of gymnasts is greatly influenced by sports apparatus. The training of sense of apparatus is a major means of athletes training techniques. The formation of an athletes' sense of apparatus is a long-term adherence process. To improve skill levels of athletes, the researcher should start from the apparatus, understand features of various sports equipment, make sports and apparatus from a good tacit agreement to better play its technology. This paper mainly discusses the training methods of athletes' sense of apparatus.

KEYWORDS: Sports institutions, Artistic gymnastics athletes, Sense of apparatus, Training methods

1 INTRODUCTION

Artistic gymnastics has quite a high technical difficulty. It is a kind of sports with the combination of sports and art, focuses on the reflection of athletes' excellent body movements and superb equipment skills and has very high appreciation value. The close combination of complex equipment actions and body movements is the remarkable characteristic of rhythmic gymnastics. Therefore, artistic gymnastics athletes must master sense of apparatus which is a perceptual ability that an artistic gymnastics athlete forms through long-term special training. Sense of apparatus has direct influence on athletes' technology improvement and plays an important role in the training.

2 SENSE OF APPARATUS'S SIGNIFICANCE ON ARTISTIC GYMNASTICS ATHLETES

2.1 Sense of apparatus is the key factor for athletes to master techniques

Different from other sports, rhythmic gymnastics requires the perfect performance of sports and art and lays more emphasis on artistry based on the demonstration of body movement. In the process of training, athletes should master technical motions faster and have a better sense of apparatus to accurately judge the characteristics of the apparatus. If athletes own better sense of apparatus, they can make vision, sense of touch, auditory sense, motion perception and other various body nerve sensations participate in the movement and constitute a perfect combination of athletes, apparatus, music, space, time and movement through harmonious activities. Only athletes have a good sense of apparatus, they can accurately grasp the speed, height, trajectory and time in accordance with the requirements and precisely complete a series of difficult movements to make the whole process be like a flowing picture and give people aesthetic feelings.

2.2 Sense of apparatus is the psychological characteristic that athletes must have for cultivating enormous skills

Sense of apparatus is important for athletes and directly affects the athletes' skills. In the long-term training, as long as athletes have formed acute sense of apparatus, they can potentially improve their self-confidence and strength and can make apparatus actions more flexible in the gymnastics competitions and effectively integrate body movements and music with apparatus actions. Athletes can be free from apparatus and pay more attention to action, music, scene and expressive force. If athletes have poor sense of apparatus, their apparatus actions would seem to be very stiff and they can not smoothly perform the whole movement. Thus, a good artistic effect cannot be achieved. Therefore, sense of apparatus is the psychological characteristic that highly skilled athletes must have.

3 IMPORTANT FACTORS THAT AFFECT ATHLETES' MASTERY SENSE OF APPARATUS

Sense of apparatus is important for artistic gymnastics athletes. If athletes want to have superb skills, they must have a good sense of apparatus. Characteristics of China's sense of apparatus of artistic gymnastics include diversity, complexity, continuity, instability, performance and risk. Artistic gymnastics athletes must master sense of apparatus skills according to the features of the apparatus and better display gymnastics skills. However, there are many factors that can affect athletes' mastery of sense of apparatus.

3.1 Accuracy of hand grip strength

Athletes' accuracy of hand grip strength is an important factor to affect the athletes' sense of apparatus in the use of apparatus. It can judge athletes' hand muscle strength and self-feeling accuracy. There are two main judgments indicates: sensitivity and muscle fine feeling which are performed by the human sensory system. Athletes could estimate apparatus movements according to the feeling of hand grip strength. Through such training, an athlete can form a conditional reflex perception as time passes. If an athlete has good sensitivity and strength, his sense of the apparatus would be good.

3.2 Depth perception

Depth perception, also known as stereoscopic perception and distance perception, is one of the main factors that affect the athletes' sense of apparatus. Depth perception is the reflection that the individual has on concave and convex of the same object or different objects' brocades. Through the determination of depth perception, athletes' distance perceptions can be accurately judged. Faults appeared in athletes' performances are generally caused by their wrong distance perception judgments. Therefore, the distance perception is an important factor affecting an athlete's sense of apparatus. Athletes should accurately grasp the distance perception in training, cultivate a sense of distance, accurately grasp the distance perception of apparatus and enhance the accuracy of movements.

3.3 The stability of hand muscle force

Test on the stability of an athlete's hand muscle force can accurately understand his sense of apparatus. It mainly reflects an athlete's control of his own state and motion in rhythmic gymnastics training and whether he can flexibly control his own body and motion. Good stability of hand muscle force shows that the athlete's perception of apparatus is relatively good and the action fluency in gymnastic practice will be better.

4 MAIN WAYS OF GYMNASTS' SENSE OF APPARATUS TRAINING

The cultivation of an athlete's sense of apparatus requires repeated training and groping. First, the athlete's training should be strengthened and the athlete should be providing more opportunities to contact apparatus. Especially athletes who have a poor sense of apparatus should more contact apparatus. Athletes should attach great importance to the basic training and adhere to long-term unremitting training. In this way, athletes' sense of apparatus can be certainly improved.

4.1 Strengthen the unity training of apparatus and body movement

Athletes' physical training should be combined with apparatus. In gymnastics training, athletes' body movements cannot be separated from apparatus movements. Otherwise, it is difficult to form a good sense of apparatus. Some outstanding athletes have perfect actions in individual practice activity. However, once actions are combined with apparatus, their performances are good enough. Superfluous movements or fewer movements would occur. Only strengthening the unity training of apparatus and body movement, they can overcome such faults. Body movements can not exist without apparatus movements and apparatus movements also can not be independent of body movements. In gymnastics training, whether ball, stick, ring and other apparatus or throwing, coiling, catching, rolling and other apparatus technology must have clear requirements on athletes' body gestures and movements, which is the basic premise to improve an athlete's sense of apparatus.

4.2 Enhance various body parts' application abilities of apparatus

Hands are the main body parts for athletes in gymnastics and all kinds of instruments are shown perfectly with hand movements. Besides hands, the application of other body parts of athletes is the development trend of future gymnastics. The use of other body parts can show athletes' apparatus skills. In the usual training, athletes should increase the use of legs, feet, backs, waists and other parts. The ability of the apparatus can improve the ability of the various parts' apparatus user. Thus, athletes can show more complicated and superior skills.

4.3 Customized training

Athletes should have customized gymnastics training. They should make some overall analyses on some difficult movements in technical training and practice repeatedly in view of incoherent movements. In training, different athletes should be trained in different ways. For instance, some athletes have better space feeling in ball gymnastics, but have a poor hand touch. It is difficult for them to control the apparatus's moving trajectory, time and height after the apparatus is tossed out or the apparatus cannot roll smoothly in hand. They should increase practices of hand touch to overcome this difficulty and improve the sense of apparatus.

4.4 Pay attention to natural and smooth connection process between apparatus

In the process of training, the naturalness and fluency between various apparatus that athletes use should be paid attention to. First, unnecessary warm-ups can be reduced and the speed of apparatus alternation should be improved to make it a natural cohesion process. Second, athletes cannot be simply trained with single apparatus. They must connect other apparatus movements when practicing some apparatus movement. Thus, movement fragment and incoherence can be avoided. If athletes pay attention to these problems in the process of training at ordinary times, they can produce good movements and form natural, smooth and coherent apparatus techniques in competitions.

5 CONCLUSION

In short, the training and cultivation of artistic gymnastics sense of apparatus cannot be in a hurry to succeed. It is an imperceptible process. Only planned, diverse, intense and practical training can lead to effective cultivation in usual training. The main channels of sense of apparatus training include better training and guiding ideology and certain means of technical training for in-depth training. In this way, athletes' best sense of apparatus can be cultivated to lay a solid foundation for the gymnastics training.

REFERENCES

[1] Zhao Xiang and Xu Ming: Sense of Apparatus for Rhythmic Gymnastics, Journal of Sports University, Vol. 40 No. (2014), pp. 55–56.
[2] Liu Huiling and Diao Zaijian: Characteristics and Cultivation Approaches of Artistic Gymnastics Athletes' Sense of Apparatus, Journal of Data of Culture and Education, Vol. 15 (2006), pp. 114–115.
[3] Fan Ming: Research on the Apparatus technique level of Chinese Rhythmic Gymnastics Gymnasts and the Influence Factors, Degree of Beijing Sport University, Vol. 12 (2009), pp. 1–44.
[4] Chen Zhen and Zheng Chunmei: The Research of Rhythmical Gymnastics, Journal of Boji Wushu Science, Vol. 6 No. 3 (2009), pp. 192–193.
[5] Wang Jing: Present Situation Analysis of Basic Training of Rhythmic Gymnastics in Higher Sport Universities, Journal of Contemporary Sports Technology, Vol. 4 No. 1 (2014), pp. 124–125.
[6] Pu Yunfei: Exploration of the Meaning and Methods of Building Apparatus Sense during Youth Rhythmic Gymnastics Training, Journal of Athletics Forum, Vol. 11 (2012), pp. 178–179.

Information, Computer and Application Engineering – Liu, Sung & Yao (Eds)
© *2015 Taylor & Francis Group, London, ISBN 978-1-138-02717-6*

The application of the green concept in low carbon tourism economy

Nian Ping Zhang
Guilin Institute of Tourism, China

ABSTRACT: Based on the background of global warming, developing low-carbon economy has become the developing direction of the world economy, and the development direction of economy also guides the tourism to develop towards the low carbon travel. This paper introduces the low carbon economy and the advantages of developing low-carbon tourism economy in China, and puts forward the strategy of developing low carbon tourism economy with the green concept.

KEYWORDS: Green concept; Low carbon tourism; Application

1 INTRODUCTION

Low carbon tourism is an important measure to ensure the sustainable development of tourism industry, takes low carbon as its aim, and plans and manages the tourist area [1].For the existing tourism resources, perfect low-carbon facilities have to be created to effectively curb the high carbon emissions. Low carbon tourism attaches great importance to cultivate the low carbon technology and the low carbon consumption patterns, advocates to change the changes tourism consumption mode by reforming the low carbon technology to achieve the aim of developing the sustainable tourism. A series of negative influence brought by the tourism in China, especially the negative influence on the ecological environment, is much more apparent. The blind development of the tourism resources caused the excessive deforestation and soil erosion become increasingly serious, and the serious environmental pollution caused in tourism management pattern leads to the public lack the environmental consciousness, and the carbon emissions increase accordingly. Therefore, the carbon dioxide emissions must be controlled rationally to ensure the tourism economy, environment, and social benefit to develop together. This paper will introduce the strategy of developing low carbon tourism economy in the green concept from the perspective of developing the low carbon tourism economy.

2 THE BRIEF DESCRIPTION ABOUT THE MEANING OF LOW CARBON ECONOMY

Low carbon economy takes the economic and social development as the premise, and reduces the greenhouse gas emissions to the maximum limit through technical innovation and concept shift in order to achieve the goal of reducing energy and resources [2]. Low carbon economy lies in the high-effect utilization efficiency and the clean energy, its core is the energy technology innovation, system innovation and the change of the concept about human survival. With the continuous development of practice, the connotation of low carbon economy also expands gradually, and it featured by low energy consumption, low pollution, and low emissions, and obtaining a new economic development model by using the least amount of greenhouse gas emissions.

3 THE ADVANTAGES OF DEVELOPING LOW-CARBON TOURISM ECONOMY IN CHINA

There is a unique advantage for China to develop low-carbon tourism. From the perspective of governing the principle politics, the development of low carbon tourism can not only optimize the industrial structure, promote economic, ecological and social benefit to develop harmoniously, and also match with the concept of the scientific outlook on development implemented in China. From the perspective of integrating with the international, the continuous development of the industries related to low carbon also becomes the worldwide political responsibility that the governments of all countries must bear. A series of related opinions of speeding up the development of tourism were issued by the state council, and the policies about developing low carbon economy were also put forward one after another by the provinces, municipalities and autonomous regions of all countries in order to provide important guarantee for speeding up the development of low carbon tourism. In the aspect of research on low carbon technology, the capture and purification using the carbon dioxide super gravity method and other related low carbon technologies have been developed, and if applying these technologies in various industries, the carbon dioxide emission reduction can be at least 40 million tons every year [3]. So, there are no technical obstacles in China when developing the low carbon tourism.

Low carbon tourism means that applying the energy conservation and environmental protection consciousness in the tourism activities when tourists are in the experience of low carbon tourism in order to make visitors take the consciousness of energy conservation and environmental protection into the tourism activity when they are felling low carbon tourism, and meet their spiritual demand, which also promotes the industry related to the tourism to develop towards the green and low carbon direction. The generalized tourism takes the nature protection, the ecological balance maintenance, and low-carbon life as the foundation, and promotes the city's development to transform to be a sustainable development model of tourism industry [4].

4 THE STRATEGY OF DEVELOPING LOW CARBON TOURISM ECONOMY UNDER THE GREEN CONCEPT

Tourism is one of the biggest and strongest industries in the global economy. Based on the background of global warming, developing low carbon tourism with green concept helps tourism develop towards the fine direction.

a. Taking the green and environmental protection as the concept, and promoting low carbon of the tourism products

The related supply chain serving the whole tourism products has to be provided in the food, accommodation, travel, transportation, shopping and other aspects. First, the most basic requirement of creating low carbon catering is to use ecological food, and the important aspect of establishing the ecological restaurant is advocating to eat organic and green food, and to consciously reject to use and sell some poisonous and fake food. The traffic condition is also an important support for the tourism industry, and has a great energy saving space. At present, the energy utilization rate of car in transportation is not very high, so some low-carbon or carbon-free means of transportation such as the public transportation, the electric vehicles, and automobile can be strongly advocated. The accommodation facilities provided by ecological tourism must also the circular economy must also follow the principle of the recycling economy

reduction and cyclic utilization, must effectively save the water resources and comprehensively utilize the water resources, rationally utilize rainwater and use the diversion system of rainwater and sewage, reasonably set up the path that the rainwater on the surface and the roof flowing through and reduce the surface runoff to the largest level, and use different penetrating measures to increase the osmotic quantity of the rain. The natural light and emitting light should be applied fully to light the basement and corridor; pollution should be reduced to the maximum limit, and the clean and low-carbon hotel accommodation should be promoted to ensure to provide visitors with a green and comfortable living condition.

b. Implement the most reasonable low-carbon tourism operation model

Low carbon tourism includes not only the establishment of low-carbon tourist attractions, low carbon serving products, and low carbon equipments, but also the low carbon concept, low carbon tourism management, low carbon technology and other soft environment. The establishment of the low-carbon tourist attractions is the key to attract tourists to participate in low carbon tourism, and it can affect the tourists, and let them quickly integrate into and immerse the scene and relax physically and mentally from the low carbon education [5]. The education of ecological protection must be strengthened to improve the public's in-depth understanding about the ecological consciousness, and to make tourists, tourism management departments and local residents establish the ecological protection concept. In those cultural tourist areas, the historical and folk culture should be attached importance to protect in order to avoid the excessive impact of foreign culture, and to prevent the local unique culture from developing towards the commercial and vulgar direction. At the same time, more ecological tourist attractions should be opened up, and different types of ecological tourist activities should be carried out to ensure the ecological tourism become a hot spot of tourism industry.

c. Completing the green growth of low carbon tourism

With the increasing fierce competition in tourism market, tourism service is an important performance of the competition among tourism industries. Here, the tourism services not only refer to the tourism services throughout the tourism activities, but also include the previous sales and after-sales services [6]. According to the principles of comparability and maneuverability, the examination and evaluation index system of the ecological construction should be made, the green audit mechanism should be introduced, the situation of the low carbon tourism region should be monitored regularly or irregularly, and the behavior of the service staff and visitors in the tourist area should be investigated. According to the related requirement of the scientific outlook on development, it should ensure to reduce the destructiveness of the ecological environment to the lowest level, and ensure the tourism to develop towards the direction of low energy consumption, low environmental protection, and low carbon emissions. Based on the characteristics of the tourism resources in each region, it should allocate the tourism resources accordingly, build tourism resources recycling system, and reduce to occupy the resources such as cultivated land, mountains and rivers. In scenic spots, it is best to use new energy public transportation, on foot, by bicycle walking, and other ways. It should save resources within the maximum limit, reduce waste of resources, and do not make too much rubbish in order to reduce emissions for low carbon tourism.

d. Continuously optimizing the tourism management
Low carbon tourism development is mainly the government's management system for low carbon tourism development and the administrative system for government. It should draw lessons from the administration mode and experience of low carbon economy which was carried out by foreign countries, introduce the technology of energy conservation and emissions reduction comprehensively, and set up low carbon system. First of all, it should integrate the internal resources continuously, highlight the special features and management level of low carbon tourism, and enhance the joint development of the low-carbon tourism products and the traditional tourism products, in order to achieve the win-win effect between the low carbon tourism products and the traditional tourism products. At the same time, it should strengthen the standard management for low carbon tourism and grasp the degree of management reasonably.

5 CONCLUSION

The development of low carbon tourism is a relatively large and complex project, which needs tourism and other related departments to explore continuously in order to complete. This paper introduces a series of strategies of developing low carbon tourism economy within the green concept from the perspective of low carbon tourism economy.

REFERENCES

[1] Zhao Qiong: Research on the Application of the Green Concept in the Low Carbon Tourism Economy. NONGJIA KEJI Vol. 11 (2012), p.32.
[2] Xu Xue: The Construction of Henan Tourist Industry Cluster Based on the "Low-carbon Tourism". Economic Herald Vol. 12 (2011), pp. 88–90.
[3] Wu Qian: Research on the Practice of Green Concept in Low Carbon Tourism. 2011 Collected Papers on the Forum of China's Sustainable Development Vol. (2011), pp. 90–92.
[4] Wu Xiong: Low Carbon Tourism: the Strategic Choice for the Sustainable Development of Tourism Industry Under the Background of Low Carbon Economy. Discovering Value Vol. 8 (2011) pp. 50–52.
[5] Bian Xia, Yu Shasha, Yang Yang et al..: The Feasibility Analysis of Combining "Zero Carbon" Concept with Green Rural Tourism. ZHONGGUO KEJI BOLAN Vol. 16 (2011), pp. 232–233.
[6] Fu Jingbao: The Development of Low Carbon Tourism Lies in the Change of Civil Consciousness. Journal of Southwest University for Nationalities (Humanities and Social Science) Vol. 7 (2011), pp. 121–124.

Information, Computer and Application Engineering – Liu, Sung & Yao (Eds)
© 2015 Taylor & Francis Group, London, ISBN 978-1-138-02717-6

The enlightenments of MOOCs on the teaching of the applied undergraduate English majors in social network environment

Mei Fang Zhang

Department of English, Tianhe College of Guangdong Polytechnic Normal University, Guangzhou, China

ABSTRACT: There are advantages and disadvantages for the application of MOOCs in the university applied undergraduate English teaching in a social network environment, so it should try to be school-based and localization, give full play to the advantages of its service, create the high-quality curriculum resources, and realize the objectives of training students' practical ability. This paper analyzes the enlightenment of MOOCs on the undergraduate English teaching, explores English teaching reform thought in MOOCs, and wishes to provide reference for English teaching.

KEYWORDS: MOOCs; Applied English; Teaching; Enlightenment

1 INTRODUCTION

MOOCs means the open and online courses with a massive scale. It is a new way and a new means to learn under the current Internet background, and plays an important role in supporting the teaching work. MOOCs itself has many applied advantages such as strong interactivity, strong openness, strong autonomy, famous teacher's lecture, teaching free and so on, which is helpful for the implementation of the online course and the long-distance course teaching, and is favored to accelerate to promote the current new curriculum reform and let students use their fragmented time to complete the efficient study [1]. In the following, the enlightenment of applying MOOCs for teaching will be explored combining with the situation of the applied undergraduate English majors.

2 THE ENLIGHTENMENT OF MOOCS ON THE UNDERGRADUATE ENGLISH TEACHING

The major characteristics of MOOCs itself determined that its impact on the undergraduate English teaching mode mainly concentrated on two aspects, which were the enlightenment on the teaching elements relationship and the enlightenment on the construction of teaching model respectively.

In the aspect of the teaching elements relationship, MOOCs is an important medium for teachers and students to communicate. The emergence of the media means that there are more possibilities for teachers to let students to actively participate in the teaching process, to transform their roles and outstand the position of guide and leader, and let students to give full play their own learning ability and subjective initiative to participate in solving the problems in the process of participation, which is in favor to promote the transition of the teaching mode, and highlight the position of student-centered [2]. Teachers, teaching content, teaching media and so on play their role around students' learning. The impact of MOOCs on the teaching elements relationship can be seen. The application of MOOCs in teaching would promote the teaching to continuously enhance its knowledge and skill level, which made students' have more interest in furthering their study and better master its application skill in the process of participating in learning MOOCs course, and made teachers yield twice the result with half the effort in teaching to be more skillful and pointed to lead students to participate in the learning process, and it has a very good effect to stimulate and consolidate students' learning interest undoubtedly, and changes the former passive learning into active knowledge construction. It is the serve and to be served relationship between MOOCs and the teaching content, and teachers in MOOCs course can actively seek localization and school-based, and give their advantages full play to support students' independent learning [3].

In the aspect of the teaching mode, in view of the characteristics of MOOCs, all classroom elements in university applied in English teaching will be deconstructed again. A new teaching mode will be reconstructed according to the current actual course demand and students' ability, will be reformed from the teaching organization, teaching objects, teaching environments and other aspects, and will complete teachers' work in face-to-face teaching and answering questions and train students' listening, speaking, reading, writing and translation ability in MOOCs course. Under the new teaching mode, teachers must make full preparations before class, and provide two regional teaching methods that face-to-face teaching + MOOCs mode, and raising a question- MOOCs- self answering, because different regional teaching modes have different influence on the key of students training and exercise [4].The first face-to-face teaching + MOOCs model is similar to the teaching mode of retroflex classroom. Under this model, teachers play the role to raise questions and lead students, and students need to learn how to apply MOOCs and use their ability to independently complete to solve the problems, which is to train the students' ability to use auxiliary means to answer questions, and in the process, students' autonomy and independence are expressed to the maximum limit. The second raising a question- MOOCs- self answering mode is a new teaching mode formed through organically combining the traditional classroom teaching with the MOOCs course. This kind of teaching mode is universally applied. It is convenient for teachers to give classroom tests, and it can timely understand the students' ability through the evaluation function of MOOCs. Under these two kinds of teaching models, students' five abilities of listening, speaking, reading, writing and translation have been improved. The sustained and in-depth application of MOOCs is to further deepen and inspect these abilities. In the whole process, the teachers' role has been transformed, and the students' center position has been prominent, which undoubtedly better accords with the teaching requirements of the current new curriculum reform and the concept of quality education. It can comprehensively cultivate and train students' ability to learn many kinds of language, and finally enhance their language application level, which undoubtedly provide powerful supports for applying language teaching.

3 EXPLORATION OF THE UNDERGRADUATE ENGLISH TEACHING REFORM UNDER THE APPLICATON OF MOOCS

The application of MOOCs in the social networking environment has become inevitable, but when confronting the impact of MOOCs on the current applied English classroom teaching in colleges and universities, how to ease the impact and how to use the advantage MOOCs are the key of the current teaching exploration. Under the application of MOOCs, the undergraduate applied English teaching reform needs to start from the following points to improve students' comprehensive language application ability.

Continuously promote the information process. The application of MOOCs is closely related to the improvement of the information level in teaching. The language teaching in colleges and universities needs to do from the hardware and software facilities construction, environment construction, perfect rules and regulations, and other ways to improve the information level in the classroom teaching. These auxiliary equipments can be used to make high-quality and excellent MOOCs course curriculum in order to serve the teaching work and give full play to the advantages of teaching resources. Through reasonable allocation and utilization, the teaching can be improved and reformed, and students' learning effect and teaching quality can be improved. Based on the technical preparation, it can provide powerful supports for cultivating the high-quality applied language talents.

Strengthening the construction and sharing of resources, and making high-quality online courses. There are prominent features in applied undergraduate English major teaching industry, which is mainly to cultivate professional English talents for the development of different industries. In teaching, it should pay attention to cultivating towards the important application-oriented direction, pay attention to the practical and comprehensive application of knowledge, pay attention to tending to train and cultivate the students' practical ability, and strengthen to train and cultivate their pioneering and innovative ability. Especially in the current critical period of the teaching reform transformation in colleges and universities, it should reasonably determine the training target, stand out the role of English application and training from the teaching content, teaching mode and other aspects, use school-school cooperation and school-industry cooperation to make high-quality curriculum resources, and strengthen the resource sharing [5]. For the important goal to cultivate applied students, it should pay attention to reasonably meet the talent training demand reasonable in the constructing process of the classroom system. For example, a domestic university actively reformed its teaching model using the applied module teaching method in colleges and universities which they learned from Germany. On the basis of a thorough investigation into its reality, this university designed a series of reform schemes, carried out the reform slowly from the basic courses of the discipline, the basic course of technology, the general practical courses and other multiple aspects, and gradually established an applied talents cultivation system. After the basic system construction, this university widely listened to experts' opinion, enterprises' opinion and social opinion in the subsequent preparation work, searched for jointing the enterprises' demand according to different training directions of applied talents, analyzed the specific requirements of talents knowledge, ability and quality in different groups, had important points in the course resource construction, created high-quality curriculum resources with characteristics, broke the subject barriers, formed the characteristic module teaching system, and completed the real docking between the applied talents cultivation and the position requirements.

The integration of MOOCs in the applied curriculum system construction in colleges and universities is beneficial to the construction of the trinity practical teaching system, can be used to establish a high-quality practice platform, complete to find the weakness of the traditional teaching with the support of a lot of high-quality teaching resources, and let students lay solid theoretical foundation in the MOOCs environment. Meanwhile, using the advantages of cooperating with experts, other schools and industries, and the MOOCs resources, it has made some progress in practice, completed the trinity of cognition, observation and attempt, got exercise and grow in the aspect of its own ability in the practical teaching platform [6].

4 CONCLUSION

Above all, the application of MOOCs under the social networking background, as a great opportunity in the university applied English curriculum reform, needs to complete the school-based and indigenization reform from the perspectives such as the teaching elements, teaching content, teaching mode, create high quality resources, construct the trinity practical teaching system, and complete the cultivation and training of students' language application ability.

REFERENCES

[1] Wang Zhonghua, Li Xiaofeng: Research on Informationized College Teaching and Its Development, The Chinese Journal of Ict in Education, Vol. 23 (2012).

[2] Zhan Lijuan: The Application of Mobile Learning in Vocational Colleges, Journal of Wuhan Technical College of Communications, Vol.3 (2013).

[3] Wan Li: The Current Situation and Improvement Measures of English Teaching in the Garment Major in the Vocational Colleges. Light Industry Science and Technology Vol. 12 (2013).

[4] Wang Yu: The Enlightenment of "Real Voice" on the College English Writing Classroom Activity Design, English Square (Academic Research), Vol. 12 (2012).

[5] Wu Hong: The Construction of the Regional Network Teaching Platform and the Thought about the Application Strategy, Modern Society, Vol. 11 (2013).

[6] Jiao Jianli: The Influence and Enlightenment of MOOCs on the Basic Education, The Information Technology Education in the Primary and Secondary Schools, Vol. 2 (2014).

Information, Computer and Application Engineering – Liu, Sung & Yao (Eds)
© *2015 Taylor & Francis Group, London, ISBN 978-1-138-02717-6*

Research on the system based on B/S structure of corporate human resource management

Liang Zhong, Hong Qi Chen & Lei Yun Zhang
Jiangxi Science & Technology Normal University, China

ABSTRACT: In recent years, with the increase of the number of corporate officers and staff mobility, the company's human resource management has become more and more complex and diversified, many companies have introduced a human resource management system. Friendly interface, provides powerful reporting tools, analysis tools, and information sharing capabilities, reducing the workload of the human resource managers, the most important is to improve the management efficiency. This article places the characteristics of human resource management system as a starting point, based on the proposed design and implementation of B / S structure of corporate human resource management system based as much as possible to meet the required human resource management.

KEYWORDS: enterprise; human resource management; systems; B / S architecture

Talent is the lifeblood of business survival, companies only have the talent to push it forward. Human Resource Management is to match the modern tools development and application of computer and other business development needs of human resource management systems, and thus enhance their market competitiveness. The main achievement is usually corporate human resource management system for recruitment, personnel files, training modules, and connect businesses to some extent, wage management ERP financial modules. According to business development needs based on B / S structure of corporate human resource management systems, resource data sharing in the true sense.

1 B / S STRUCTURE UNDER THE HUMAN RESOURCE MANAGEMENT SYSTEM OVERVIEW

1.1 *Enterprise human resource management system*

Human resource management system that will be applied to the human resources best value for money, especially the advent of the knowledge economy, human capital concepts gradually, so that companies recognize the importance of human capital no less equipment, land, factories, capital, and even will play a more economic benefit than the content. Who is a carrier of knowledge, for knowledge play a greater utility, you need proper use of human resource management. Traditional human resource systems using

C / S structure, although greatly reformed the way human resource management, but with the growing demand and system management capabilities, current corporate human resource management requires processing capability, update fast, flexible high and other systems. In response, many companies in recent years, the introduction of B / S architecture has been widely used.

1.2 *B / S architecture features*

2 DESIGN BASED ON B / S STRUCTURE OF CORPORATE HUMAN RESOURCE MANAGEMENT SYSTEM

2.1 *System design*

Taking a small-scale enterprise, for example, although the establishment of a separate internal personnel department, but the specific functions of the division the more obscure mold, managers and employees are not subordinate relationship is formed. Department managers understand the characteristics and needs of corporate officers when you want more than the personnel departments, and checks the recruitment, training, performance appraisal, compensation management and other core businesses. Coordination of various departments to develop business development strategy, business requires the use of human resource management system quickly achieve scientific and standardized management of human resources in order to meet the required business development.

2.1.1 System roles identified

First, determine the system role, because different users use the system objects have different permission level differences play a role in supporting the entire database system, a great lot of communication exists between the system and the database, it can be used as the database and create the role as super the role of users, system administrators, business leaders, personnel manager, department heads, the general staff, and database systems.

2.1.2 System function modules

Human resource management system can be divided according to the specific needs of multiple functional modules, each functional module consists of sub-modules, specifically shown in Figure 2. Figure 2 system function module, personnel management module includes basic information mainly on the management staff, changes in all kinds of experiences, positions, contract management, personnel inquiries, leave management, human resource planning and management of their daily work. Job management module provides recruitment needs for the system to collect and manage vacancies and help recruit managers develop a specific recruitment detailed plan, and even the subsequent recruitment interviews to hire management. The training includes the development and management of business-related training programs. The attendance Management module includes a summary attendance, overtime management, leave management, attendance management rules, the flexibility to deal with attendance business. The performance management module includes workers handling individual assessment, performance, processing and statistical analysis. Payroll includes a salary system to handle multiple business functions to help companies achieve diversification and flexible pay tax accounting methods. The system management module includes data maintenance, user management, such as providing accounts, passwords, automatic backup and restore capabilities for all types of users.

2.1.3 Database design

All the information stored in the database management system of human resources employees, can be said that the basic conditions for operation support systems, database design occupies a very important part in database application development process. In order to better understand the business needs of the design only after the module division of the system after the system requirements for the design of 28 data table structure

3 B / S ARCHITECTURE OF CORPORATE HUMAN RESOURCE MANAGEMENT SYSTEM IMPLEMENTATION

3.1 Login module

The Human Resource Management System is the first system login using the functions, mainly to complete the user identity authentication system function. Users in the system login page, just enter the correct category, user name, the password can be encrypted transmission and by comparing the original personnel information database table records the information, in order to identify the user's identity, authority. Users according to their permission to use the system-related functions primarily to protect the safety performance of the software, if multiple systems are continuously input error, is forced to exit the system, to further strengthen the security and confidentiality of user information and system security. The following is a partial realization of the code:

```
{if (textName .Tex t ! ="" " & tex tPass .Text! ="" ")
bool ifcom =temDR .Read();
else
......
{MessageBox .show ("User name or password is incorrect!", "Prompt"
,MessageBoxBut tons .OK , MessageBox lcon .Info rmation);
...... }
else
{... ... }}}
```

3.2 Business logic layer implementation

Business logic layer and business logic of the system are closely related, it is primarily responsible for business-related logic processing, such as processing, data retrieval legality and update the table, the convergence of the page layer and data access layer, but does not directly operate the business layer itself table, it is in the processing business logic layer functions can be mobilized. Each class will correspond Yewuluojiceng a page of their own business logic layer will be written to the corresponding page in the logical processing, convenient to modify and manage pages. However, data management is associated with a table, corresponds to a model set of classes in multiple classes.

Business Logic layer implementation the class reference:

```
using WS.Empl.Lib.Modal;
using WS.Empl.Lib.IDal;
using WS.Empl.Lib.DataAccess
```

4 CONCLUSION

In conclusion, the talent is the lifeblood of business survival, companies only have the talent to push it forward, corporate human resource management system that will be applied to the human resources best value for money, B / S architecture system greatly reduces the processing pressure client data to improve user access to the system, convenience and flexibility of operations, business development and talent pool to provide a good platform.

REFERENCES

[1] Zhimin Zhang.based SME Human Resource Management System Design and Implementation of BS structure [D]. University of Electronic Science and Technology, 2014.

[2] Yajie Wang,Shuqiang Ma,Suying Gao.Britain,human resource management system of enterprises in Beijing, Tianjin and Hebei [J] Hebei University of Technology, 2011,40 (2): 77–81.

[3] Xiaoqing Li. design and implementation of IT for small and medium enterprise human resource management system [D]. University of Electronic Science and Technology, 2014.

Information, Computer and Application Engineering – Liu, Sung & Yao (Eds)
© 2015 Taylor & Francis Group, London, ISBN 978-1-138-02717-6

The research on enterprise education diversified model architecture for college students

L.P. Yang & W. Wang
Computer Science and Technical Institute, Changchun University, Changchun, China

D.W. Xu
School of Electronic and Information Engineering, Changchun University, Changchun, China

ABSTRACT: Based on the full study on the major problems in China's college students entrepreneurship education, from multiple levels, multiple angles and multiple levels to determine the necessary elements of college students' entrepreneurship education. The full study internal diversified characteristics each element entrepreneurship education and the relationship between various elements, and the classification and stratification, forming a complete set of enterprise education diversified models architecture, and provide the guarantee for entrepreneurship and employment of college students.

KEYWORDS: entrepreneurship education; diversified model; employment

1 INTRODUCTION

The main goal of this paper is to study a feasible diversified mode of entrepreneurship education, entrepreneurship education from the participation of personnel to entrepreneurial team, from the method of entrepreneurship education to entrepreneurship education content, from enterprise education fund to the source to promote entrepreneurship education achievement, forming a complete set of diversified architecture of entrepreneurship education, to solve our universities entrepreneurship education problems, provide a theoretical basis and implementation plan for college students' entrepreneurship education. This can change the university students of employment attitude, knowledge choosing and behavior target, so that more students to entrepreneurship and self employment as the occupation choice, enhance the students' employment ability in the enterprise, stimulate students' creativity and creative potential.

2 MAIN PROBLEMS OF OUR COUNTRY UNIVERSITY ENTREPRENEURSHIP EDUCATION

Survey activities of entrepreneurship education in colleges and universities in China, it is not difficult to find some problems to be solved and improved problem [1].

2.1 *Textbook construction and curriculum system of rigid*

At present, the current entrepreneurship education textbooks in many colleges and universities introduced from Europe and the United States, the localization of entrepreneurial experience rarely infiltrates, more the lack of teaching institutions of its own characteristics, has not yet formed a relatively mature curriculum system in Colleges and universities, most attention is the small number of people "entrepreneurship", rather than the majority "of entrepreneurship education". Teaching, pay attention to classroom teaching, the lack of theory and practice, lack of diversity, system and hierarchy, affect the actual effect of entrepreneurship education. Planning, lack of effective, the lack of multi-level. Rigid, featureless teaching mode is difficult to arouse the students' interest in learning, not to mention to develop innovative spirit and pioneering the concept of talent innovation, resulting in business school students mostly empty talk, the lack of execution, the entrepreneurship of college students overall participation rate is not high[2].

2.2 *The shortage of qualified teachers and lack of professional ability*

Because of China's business education started late, lack of professionals, our country university entrepreneurship education teacher for more part-time teachers, mostly daily affairs busy instructors, the teacher in charge of part-time teaching, have not received systematic entrepreneurship education and training, their own lack of entrepreneurial experience, more a lack of entrepreneurial experience entrepreneurs, workers, commercial bank workers and the entrepreneurial success of alumni involvement. In the entrepreneurship education for students, pure knowledge, teaching more than actual combat in the high perspicacity, a simple echo what the books say, cannot make the students grasp the soul and the essence of entrepreneurship, often in the "empty talk" embarrassing position, the cultivation of students' awareness and ability of starting an undertaking is adverse, the teaching quality is guaranteed [3]. Therefore, the establishment of professional teachers is the biggest difficulties and problems existing in Entrepreneurship Education in Colleges and universities.

2.3 Entrepreneurship education has not been integrated into the personnel training system

The current university entrepreneurship education tends to focus only on course teaching, not the entrepreneurship education into the personnel training system, entrepreneurship education has not penetrated into every link of teaching, not penetrate to all aspects of the students' training. Professional education of entrepreneurship education and discipline has not formed the organic connection, often just outside professional education plan, additional activities or projects in extracurricular and spare time, entrepreneurship education lost discipline this one of the most powerful on entrepreneurial passion, resulting in students are over, but the internal strength is insufficient.

2.4 The development of enterprise education mode is single

University entrepreneurship education way is generally combined with the course of study and business plan competitions or a combination of lectures and business plan competitions, simply with courses or seminars based university also exist. While the students most want to participate in the entrepreneurship education is the practical form of entrepreneurial activity and practice in the enterprises of this kind can accumulate the experience of actual combat activities, and business plan competition and virtual entrepreneurial activity is not the students' ideal of entrepreneurship education form, course of study is the final selection.

2.5 The social environment support entrepreneurship education is not enough

Entrepreneurship education in Colleges and universities of the social environment is the external support

system includes not only the local government policy support, Students Pioneer Park building, also includes the whole society to support and boost to the entrepreneurship of college students, including all aspects of entrepreneurship of college students venture financing support, technical guidance, the whole society to the concept of change etc. Due to the lack of publicity and understanding, resulting in some colleges and university entrepreneurship education sometimes hot and sometimes cold, some even only stay on the stage shouting slogans, most school systems, training objectives, evaluation system, campus culture is also not around entrepreneurship education tenet forming system, a sound incentive oriented mechanism, restricting the development of entrepreneurship education in Colleges and universities [4].

3 COLLEGE STUDENTS ENTREPRENEURSHIP EDUCATION DIVERSIFIED MODE

3.1 College students entrepreneurship education component elements

Aiming at a series of problems in the process of development of entrepreneurship education in universities, the paper from many aspects of teaching, the school, the society and the government to start, students, teachers, and corporate leaders such as multiple point of view, the component elements of entrepreneurship education include the teaching staff, course system, teaching methods, evaluation system and guarantee system, its structure is shown in Figure 1.

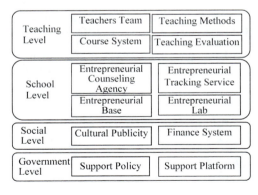

Figure 1. The component elements of entrepreneurship education structure diagram.

3.2 Architecture diagram of college students entrepreneurship education diversified mode

This paper takes the undergraduate college university as the research object, combined with the current status and existing problems of entrepreneurship education in Chinese colleges and universities, in consideration of the employment needs of students and people with talent demand unit, in order to realize the task of higher entrepreneurship education as the target, fully study the internal diversified characteristics of each elements and the relationship between various elements, and the component elements of the classification and stratification, formed an architecture diagram of entrepreneurship education diversified mode, that is shown in Figure 2.

All students are required to complete basic training plan learning. On this basis, according to the students' interests and occupation planning respectively choose skill intensive and sub module of professional expansion. The skill intensive type students, mainly to strengthen the ability of a programming language using and programming to solve problems, to strengthen the ability of communication, they go to the designated enterprises for development practice of the actual project, directly into a corporate job after graduation. The choice of professional expansion module of students, their practice is mainly guided by professional teachers. According to the professional module, their own interests and basic knowledge level, let the students to assist the supervisor to complete research projects, training their scientific research ability. They may go on to earn a master's and doctoral degree after graduation, or choose to enterprises and institutions work [5].

All the above training process is completed by the variety of factors, such as the constrain of the basic requirements of higher education, the control of the diversified curriculum system and service and guarantee system,

and the auxiliary of the entrepreneurship education diversified network platform.

4 CONSTRUCTION OF ENTREPRENEURSHIP EDUCATION DIVERSIFIED MODE ARCHITECTURE FOR UNIVERSITY STUDENTS

The specific construction process of diversified models entrepreneurship education architecture is as follows:

- Strengthen the construction of the teaching team to improve effectiveness of entrepreneurship education. Colleges and universities should guide the enterprise education course teachers to carry out the innovation of theoretical research on entrepreneurship education, support teachers to enterprise hangs duty to take exercise, encourage teachers to participate in the practice of innovation and entrepreneurship. Positive from all circles of society to hire entrepreneurs, entrepreneurs, experts and scholars as part-time teachers, build a professional, part-time combination of high-quality, innovative entrepreneurship education teachers, too inclined to give from the teaching assessment, assessment of professional titles, training, funding support. Also regularly organize teacher training, training and exchanges, and constantly improve the teaching, research and guidance of students' innovation and entrepreneurship practice level.
- Set diversified course system. Because of the diversity of "entrepreneurship", the course setting and plan arrangements of entrepreneurship education must be decided according to the difference of individual classification and business objectives. Teaching content increased to broaden our horizons, stimulating the entrepreneurial

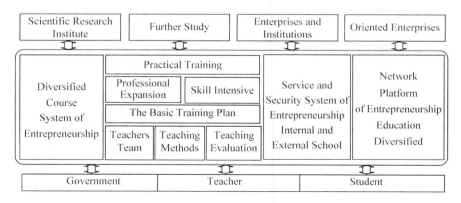

Figure 2. The architecture diagram of entrepreneurship education diversified mode.

awareness, improve the necessary entrepreneurial skills course. Course content covers the critical knowledge, such as the entrepreneurial cognition, entrepreneurship, entrepreneurial skills, entrepreneurship simulation of entrepreneurship education, to solve the "edge" problem of entrepreneurship curriculum.

- Exploration the entrepreneurship education teaching mode. In the way of teaching, should be based on different teaching goal, select appropriate teaching methods, vigorously carry out the heuristic teaching, discussion teaching, experiential teaching, case teaching, the successful entrepreneurial spirit, method, process and rules of visually, vividly show students, achieve the purpose of the inspire the students' entrepreneurial ideas, broaden the students' view of innovation, cultivating entrepreneurial ability and quality, the basic quality.
- Diversified evaluation system. Entrepreneurship education evaluation content is the value judgment for the goal of entrepreneurship education, the task of implementation and completion of the degree, level, state. First of all, the evaluation of college students, emphasis should be placed on the course setting and teaching evaluation, entrepreneurship education direction of the University can be guided through self-assessment, peer assessment, evaluation and school overall way, and ultimately achieve the purpose of entrepreneurship education. Secondly, the graduates' evaluation should focus on feedback and communication. At the same time, the school and society should establish the system of social assessment that feedback entrepreneurship quality situation of graduates, that is, to establish a long-term cooperation feedback mode with the government and enterprise, the real situation objectively reflect the students' Entrepreneurship and entrepreneurial activities, so that the school can recognize their own entrepreneurial education and the social demand gap, thereby better adjustment of educational ideas, to adapt to the needs of society, to further enhance the level of entrepreneurial education.
- Powerful guarantee system. The implementation of entrepreneurship education also needs some related guarantee besides education field. These elements include the three parts of funds, policy regulations and social culture. In the financial support, the government should consciously to entrepreneurship, provide financial support to the school site for settling in the project, and the development of education, to promote the business smoothly. In the national policy, our government should increase in the entrepreneurial support at the same time, strengthen coordination among various departments, do a good job in the implementation of entrepreneurship policy. In the social cultural aspects, using the radio, television, the Internet and other means, should widely publicize the entrepreneurship policy, successful case of entrepreneurship, necessary for entrepreneurship education to create a good social atmosphere and psychological atmosphere for entrepreneurship, promote the construction of entrepreneurship culture and social norms, and guide people through the efforts of entrepreneurship, achieve the purpose of continuous innovation.

- Construction of entrepreneurship education diversified network platform. The multi-level, full range, network entrepreneurship service, can make the students entrepreneurs find, identify, until the development of entrepreneurial opportunity, can greatly reduce the cost of enterprise establishment and growth, the whole entrepreneurial services network system should be covered each level from high technology to low technology.

5 CONCLUSION

The paper is mainly aimed at the situation of entrepreneurship education in college students, in order to improve China's College students' entrepreneurship and entrepreneurship education resources and project management. This paper is based on the network of our existing enterprise education, the diversified model introduced in entrepreneurship education, can provide a new theoretical research model for entrepreneurship education of our country university, put forward a new model of entrepreneurship education, help and encourage college students' entrepreneurial opportunity recognition of entrepreneurial intention, implementation of entrepreneurial action.

REFERENCES

[1] Z.L.Fu & H.X. Xie & X.X.Lan. 2009.The college students entrepreneurship and success education tutorial. Hefei:*University of Science & Technology China press.*
[2] G.C.Zhang. 2011. Discussion on several problems of entrepreneurship education for College students. *Education and Science*27:58–64.
[3] J.S.Lei. 2011. The status of University entrepreneurship education in our country and should make adjustments. *Youth Exploration*:15–18.
[4] Z.Y.Fan. 2009. Study on entrepreneurship education in College students oriented by employment. *Polytechnic Higher Education Research*3:57–59.
[5] H.Liu. 2010. Research of College students entrepreneurship education in China. *Southwest Jiao Tong University.*

Information, Computer and Application Engineering – Liu, Sung & Yao (Eds)
© 2015 Taylor & Francis Group, London, ISBN 978-1-138-02717-6

Computer controlling technology reformed in undergraduate teaching by research and practice of the application-oriented education

Dan Ping Zhang

University of Science and Technology, Liaoning, Anshan, China

ABSTRACT: Our University set up the course of Computer Controlling Technology according to the demand of higher education of training applied engineering and technical person, this is one of professional course of automation specialty which need theory and practice especially. This Article discusses the reason to set up this course, the matter in the teaching method and the purpose of reform and content on the basis of Computer Controlling Technology, introducing a concrete measure with good effects in the in the practice teaching method, it is an attempt and exploration to deepen higher education reform.

KEYWORDS: computer control, Education reform, Teaching practice, Chinese standard document classification:G64

Computer Controlling Technology is a very important required course to train engineers and technicians who work on automatic control and automation instrument and other professional fields, it has requires lots in theory and practice. The major study is integrated computer technology, communication technology and automatic control theory into productive process, work out optimal Computer control system with the demand of engineering design, lay a certain theoretical basis for the Students who work in the computer control system design, modification, maintenance work in the future, to the students of automation technology, it is a basic skill to master Basic knowledge of computer control technology, Construction of computer control system Use and analysis method.

1 The existing problems in the curriculum teaching Student only know well the basic content of Computer Control System but they cannot design a control system base on the teaching result these years. There are following:

1 Relative of Course. Computer Controlling Technology is a course with certain theoretical depth relating classical control theory and Modern control theory, but it also is closely related to the development of kinds of intelligent chip. With development of Information technology, the teaching of Computer control technology course base on the intelligent chip is changing accordingly. That demand that students should have the ability to circuit and software design together with good control theory, they can bring together all the individual parts of the course. Students report to the problem are class hour is too little to achieve

the teaching requirement of this course and more time is used to study theory and methods. Such complicated course is hard to be learned good during one term, that affect the study of Computer Controlling Technology course and restrict their ability to solve actual engineering problem with speculative knowledge.

2 The divorce between theory and Practice. This course helps student to lay a more solid theoretical, but it is hard to student to make digital control theory and computer and detection technology relationship to other courses in fact. How to combine the theory and Practice is still one of the teaching points. Thus, keeping the teaching content to fit in it that will develop pertinence and realism of teaching and mobilize the enthusiasm of learning of students. So it is a real problem of reform in education to update teaching materials and improve teaching quality. We will practice and exploration of course reform from the theory of teaching method of computer control system.

3 The lack of scientific experiment. In many Universities, Computer Controlling Technology is class taught, but lack of experimental class, it is only base on design of computer software (MCGS configuration software), comprehensive design experiment is not good enough (DC motor closed-loop speed control experiment, temperature control experiment and the water level control experiment, e.g, the divorce between hardware and software, limit to active hand design ability training of the student, theory teaching and experiment teaching cannot combine.

2 The concrete measures of teaching reform

In Computer Controlling Technology teaching, to the characteristics of Technology University, it needs to insist on following Teaching philosophies: First, integrate" quality education, engineering education, creative education" into a group, strengthen basic training, engineering concept, to stimulate the spirit of innovation. Second, theoretic Teaching, experimental teaching, computer assisted instruction are combined, improve student to comprehensive analysis and ability to solve problems. Third. Stick with the related curriculum to form in the series and Completely to improve the efficiency of teaching a course. Fourth. In - class study and after - class study is closely connected with teaching inspiration, guide and practical training, to create better conditions and environment which is good for the student to play an engineer. In Computer Controlling Technology course reform, to do the following practice.

1 THE REFORM OF TEACHING CONTENT

This Subject is less on teaching hour, but it relates many fields, including the hardware of computer control system and software and control algorithm, etc. So Teaching Materials should be concise according to teaching goals of knowledge study and ability training, According to the actual class hour, trimming Teaching Materials that make student accept new knowledge and practice learnt knowledge. The classroom teaching content organization can be unfolded around typical control system (i.e., Water level control system), describes the whole structure, sense and the background of the system, then lead to elements and systems, hardware and software, algorithm and program, overall system design and optimization of the theme finally. In the teaching process, it should make student clear different leading to a computer control system, highlight the relationship between Discrete system and Continuous system. The analysis part of the system, point out emphatically that signal inner Computer Control System is discrete signal, Judging the stability of the system, Steady state error analysis, the dynamic process of the system theory is different from Continuous System of analog signal processing. In the system design part, it should make student to understand accommodation and the advantage of the indirect design method and direct design procedure, pay more attention to the minimum time control and improved digital PID control algorithm which are most popular in use. For the foundations of mathematics of the Computer Controlling System, as z transforms, difference equation, etc., which are repeating knowledge, it can be explained student according actual situation.

2 USING A VARIETY OF TEACHING METHODS

1 Mainly to explain the main multimedia courseware demo, theory part explained assisting with blackboard writing focusing on emphasis and difficulty. The lecturer can give closer analysis to student to understand and use knowledge effectively by blackboard writing. Using multimedia teaching showing plentiful in graphics and animation, it Active student thinking, to stimulate students' learning initiative and creativity, improving the ability of analysis and solve problems.

2 Some project cases introducing in teaching should involve both in traditional principle and new technology, which helps student form a correct view of Engineering and the master solution through project cases, if student study hard and lecturer is good at lead up to synthesis and application of studying knowledge. Experimental session provided is more important except for rich and colorful content and teaching method.

3 THE EXPERIMENTAL AND PRACTICAL TEACHING

1 Computer Controlling Technology as a course of very practical, the experiment content configuration will affect teaching effect, for improve student to relate theory with practice, manipulative ability, reducing content validation experiments, increasing integrated and designed experimental content of teaching, for example, furnace temperature control system, loop DC motor control system debug, water level control system, etc. It demands student to set up hardware system on experimental platform and according to the function and the performance index of the system to meet the general design. Installed MCGS configuration software, let student to finish system circuits design, process for the preparation, emulation and debugging of the whole progress, improving the practical ability of usage of hardware and software.

2 Teaching of Practical training systematic and comprehensive engineering are essential quality of student of application technology. To set 4 weeks engineering Practical training is the purpose to train a student's ability to solve actual engineering problems. In Practical training, let students to group by themselves, that request student combines engineering practice with controlled object, control objectives, design scheme, aspects of software and hardware, finishing the whole Computer Controlling systems analysis, design, installation and adjustment. In this creative work, In the way to help student to enter a role of engineering, know

how of the incoming situation of work, and by this training, providing them a room of freedom to imaginative and audacious, arouse the curiosity of the students. To break through close type teaching method before, help those students of Automation Engineering know well of relative knowledge with practice, foster their ability to innovation, imagination and science techniques.

4 IMPROVING THE ABILITY OF THE LECTURER

In today's society, cause the development of Computer Technology, in the reform of teaching, Lecturer should keep track of technological trends and academic development both at home and abroad, studying the latest achievements, concluding experiences of teaching, and introducing new theories and technology to teaching materials, to make student understanding development trend of this field. It is also let student have an overall understanding the subject to catch academic developments both at home and abroad and know the technical gap between domestic and foreign. By training and serving temporary position practicing, joint cooperation of the University and Enterprise to build a contingent of Lecturer, cultivate a teaching staff with professional and sideline, make sure reform of curriculum system to implement projects smoothly. On the other hand, it can employ some senior engineers with rich practical experience from Enterprise to teach and arrangement of the experiment, set up a popular experimental platform, combine University and Enterprise. Improving professional quality and perfecting the team of Lecturer of Computer Controlling Technology course.

5 CONCLUSION

As the purpose of our University is trained student of an application of skilled talent, so the reform of course is concerned the quality of the students as superficial and training high skilled talents to meet the demand of Enterprise and Society for the deep administrative levels, to make contributions to the country and the enterprise. Form reform and practice of Computer Controlling Technology course, continue to advance steadily of " uniting practice and study, occupation guide" to improve teaching quality and students' learning results and efficiency, train ability of practice and analysis of problem solving independent. As developing of computer technology and teaching method, this course will be modified and improved constantly.

REFERENCES

[1] Yu Haisheng Microcomputer Controlling Technique (Second edition) Tsinghua University Publishing House Beijing 2009.
[2] Shi Yan, Weng Yifang, Hai Liqun 《Computer Simeulation for Control System》 Teaching reform and Practice[J]. China Power Education 2008, (5): 74–75.
[3] Yu Hongtao,Zhu Shangzhen, Lei Yanhua <<Computer Controling Technology>>Application in Teaching [J]. Scientific and Technical Information (25): 182–183.

Kinematic analysis of the Fr. giant. sw. bwd. dbl salto t. to up. arm hang of elite gymnast Cheng Ran on parallel bars

Min Jiang & Wei Jiang

Chengdu Textile College, Chengdu, China

ABSTRACT: The parallel bars is one of men's gymnastics projects, it has been listed as the Olympic Games event since 1896. In 2012 World Gymnastics Championships in Shanghai, Cheng Ran has successfully completed the "Fr. giant. sw. bwd., dbl salto t. to up. Arm hang" action, this action is currently one of the highest difficulty movement of the parallel bar project, thus make movement analysis over the movement in qualification trial, and obtained cinematic references, which provides theoretical and technical basis for the development and improvement of Chinese athletes

KEYWORDS: Kinematic, Analysis, Elite Gymnast, Parallel Bars

1 STUDY METHODS

The whole process of the competition was recorded by two GC-PX10 video cameras (JVC, Japan) at 50 Hz from different angles (the included angle of the principal optic axes of the two cameras was about 60°). After the match, we put three-dimensional scale with 24 control points in filming area. We used 3-D Signal TEC V1.0C software, and built the coordinate system. In order to meet the research needs, we added the parallel bars as the 22 test points in Songjing Model (16 links, 21 joints).The original data was smoothed by a low-pass filter with a cut off frequency of 8 Hz.

2 STUDY RESULTS

In order to better analyze the "Fr. giant. sw. bwd., dbl salto t. to up. arm hang" action on parallel bars, according to its operating characteristics we divided the movement into three stages:

The swing stage in giant swings: From the handstand moment, the giant hem stage, the giant swing stage into the hands off the bar moment.

Somersault stage: two hands pushing off the bar moment to the hands re grasp the bar moment

Stomach to support stage: the hands re grasp the bar moment to put the body into stomach support moment.

3 ANALYSIS OF THE SWING STAGE IN GIANT SWINGS

3.1 *Giant hem stage*

The stage is inverted moment movement to the body center of gravity moves to the lower center of gravity

Table 1. The kinematic parameters of the giant hem stage of Cheng Ran.

Shoulder angle in the handstand moment(°)	Hip angle in the handstand moment(°)	Time needs from the highest to the lowest of the center of gravity(s)	Speed of center of gravity(m/s) (m/s)
169.6	173.2	1.14	5.71

moment. This process is the human body potential energy into kinetic energy of rotation, so as to obtain the backflip energy.

As we can see in Table 1, Shoulder angle and hip angle at the handstand moment are respectively 169.6° and 173.2°, which show that Cheng Ran has successfully completed the action, time needs from the highest to the lowest of the center of gravity is 1.14s. This may be due to the shoulder joint used force, resulting in the maximum speed of center of gravity(5.71m/s).

3.2 *Giant swing phase*

Giant swing phase refers to the body from under the lowest center of gravity moment after the bar surface before two hands pushed away from the moving process. When the body is put to the bar sagging face, legs should be rapid upward around the legs, the shoulders sink, which can make the bar maximum bending deformation, bar elastic potential energy for the body to provide more energy movement.

Table 2. The kinematic parameters of the giant swing phase.

Time(s)	The maximum speed of the ankle joint (m/s)	shoulder Angle in Sagging face moment (°)	The distance of Center of gravity to bar in the hands off the bar moment(m)	The speed in the hands off the bar moment (m/s)	Shoulder angle in the hands off the bar moment(°)	Hip angle in the hands off the bar moment(°)	Trunk angle in the hands off the bar moment(°)
0.38	11.33	175.6	0.09	3.45	113.5	108.0	48.2

(The maximum speed of the ankle joint refers to the maximum speed of the ankle joint from the body center lowest moment to the hands off the bar moment, the trunk angle refers to The angle between the body vertical plane and horizontal plane)

The total time of Cheng Ran from the center of gravity lowest moment to the hands off the bar moment is 0.38s, working time is long, shoulder angle in sagging face moment is 175.6°, which shows the heavy torque shoulder obtained. In the hands off the bar moment, the distance of Centre of gravity to bar in the hands of the bar moment is 0.09m, the speed of the center of gravity is 3.45m/s, the hip angle is 113.5°, shows when Cheng left bar, he has a high center of gravity, large velocity, hip flexion fully, thus to increase flight height and flip angle velocity, the trunk angle is 48.2°, a little bigger. The maximum speed of the ankle joint is 11.33 m/s, which shows Cheng completed the action with a high quality.

4 SOMERSAULT STAGE

Somersault stage is from hands pushing off the bar moment after flight to the holding time at the end of the motion bar again. Push hands off the bar into the somersault stage, this time the body into the non support state, by the influence of gravity, the height of flight and flight time is limited.

As we can see in Table 3, the total time from the hands off the bar moment to the crotch hold moment is 0.20s, total time from the hands off the bar moment to the arm hang moment is 0.66s, Flight height is 0.63m, the average shoulder angle in the crotch hold moment is 27.9°, the hip angle in the crotch hold moment is 89.4°.

The arm hang position of Cheng Ran is not so good, the larger velocity of the center of gravity (3.17 m/s), hip Angle is small (153.2°), demonstrating that he extend his body late, center of gravity position in general, the buffer time is short, the impact is slightly bigger. So the overall quality he completed the arm hang action is not high.

5 STOMACH TO SUPPORT STAGE

Prone after arm hang with inertia is commonly used in the parallel bars, this technology should be sinking shoulders, quickly to the above pocket leg in order to overcome the gravity, increase the speed of the objects.

From the kinematic parameters in the arm hang forward swing to the center of gravity lowest moment (Table 5), we can see that, Total time from the arm hang moment to the center of gravity lowest moment is 0.28s, the speed in the center of gravity lowest moment is 1.30m/s, Shoulder angle in the center of gravity lowest moment is 70°, hip angle in the center of gravity lowest moment is 142.2°.

The sink shoulder and pocket leg control is the most important part of the arm hang forward swing stage. The sink shoulder is mainly to see the shoulder angle in the lowest point of the center of gravity, while the shoulder angle of Cheng Ran is 70°, a little smaller. Pocket leg mainly depends on the hip angle in the lower center of gravity, and the hip angle of Cheng Ran is 142.2°, it's also a little smaller.

Table 3. The kinematic parameters in the somersault stage.

Total time from the hands off the bar moment to the crotch hold moment (s)	Total time from the hands off the bar moment to the arm hang moment(s)	Flight height (m)	The distance from the centre of gravity to the bars in the crotch hold moment (m)	Shoulder angle in the crotch hold moment (°)	Hip angle in the crotch hold moment (°)
0.20	0.66	0.63	0.56	27.9	89.4

(The flight height refers to the distance from the highest centre of gravity to the bars)

Table 4. The kinematic parameters in the arm hang stage.

The distance from the centre of the gravity to the bars (m)	The speed of centre of the gravity (m/s)	Hip angle (°)
0.22	3.17	153.2

Table 5. The kinematic parameters in the arm hang forward swing to the centre of gravity lowest moment.

Total time from the arm hang moment to the centre of gravity lowest moment (s)	The speed in the centre of gravity lowest moment (m/s)	Shoulder angle in the centre of gravity lowest moment (°)	Hip angle in the centre of gravity lowest moment (°)
0.28	1.30	70.0	142.2

A stretched action needs to show in a smaller time from the arm hang forward swing to the stomach to support stage, a bigger distance from the center of gravity to the bars in the stomach to support stage, and a bigger shoulder angle and hip angle in the stomach to support stage. As we can see in Table 6, total time from the arm hang forward swing to the Stomach to support stage of Cheng Ran is 0.88s, the distance from the center of gravity to the bars in the Stomach to support stage is 0.40m, shoulder angle and hip in the stomach to support stage are 68.0° and 100.1°, so the action is not fully stretched.

Table 6. The kinematic parameters from the arm hang forward swing to the Stomach to support stage.

Total time from the arm hang forward swing to the Stomach to support stage (s)	The distance from the centre of gravity to the bars in the stomach to support stage (m)	Shoulder angle in the stomach to support stage (°)	Hip angle in the stomach to support stage (°)
0.88	0.40	68.0	100.1

6 STUDY CONCLUSIONS

According to the kinematics parameters from the three-dimension picture analysis, we can make conclusions as follows:

1 In the swing stage in giant swings, Shoulder angle and hip angle in the handstand moment are respectively 169.6° and 173.2°, the maximum speed of center of gravity is 5.71m/s. In the hands off the bar moment, The distance of Centre of gravity to the bar in the hands of the bar moment is 0.09m, the speed of the center of gravity is 3.45m/s, Shoulder angle , hip angle and trunk angle are respectively 113.5°, 108.0°, 48.2°.

2 In the Somersault stage, the total time from the hands of the bar moment to the crotch hold moment is 0.20s, total time from the hands of the bar moment to the arm hangs moment is 0.66s, Flight height is 0.63m, the average shoulder angle in the crotch hold moment is 27.9°, the hip angle in the crotch hold moment is 89.4°. In the arm hang stage, the distance from the center of the gravity to the bars is 0.22m, the speed of center of the gravity is 3.17m/s, the average hip angle is 153.2°.

3 The total time from the arm hang forward swing to the Stomach to support stage of Cheng Ran is 0.88s, the distance from the center of gravity to the bars in the Stomach to support stage is 0.40m, shoulder angle and hip in the stomach to support stage are 68.0° and 100.1°.

REFERENCES

[1] Wang Guoqing, Qian Jingguang, A research for Huang Xu's techniques of backward somersault with 2 saltos piked on parallel bars[J], Journal of Nanjing Institute of Physical Education, 2001, 2(15):7–9.
[2] Shi Guojun, Qian Jingguang, Two weeks on the parallel bars tuck somersault pike two weeks hanging and hanging comparative study of technology[J], Journal of Nanjing Institute of Physical Education, 2009,8(4):1–3.
[3] Wang Jun, Yao Xiawen, Wang Xiangdong, Kinematics analysis of Huang Xu's parallel bars upright support to stoop backflip 2 turns into hanging position[J], Journal of Shandong Institute of Physical Education and Sports, 2005, 21(6):96–99.
[4] Li Wei, Yao Xiawen, Wang Xiangdong, Research on the big returning to wreath, bending, and backward two-round flip and hanging the arms of the parallel bars, [J],Journal of Beijing Sport University, 2006, 29(6):856–858.
[5] Kinematics analysis of upright support to stoop backflip 2 turns into hanging position on parallel bars.

Information, Computer and Application Engineering – Liu, Sung & Yao (Eds)
© 2015 Taylor & Francis Group, London, ISBN 978-1-138-02717-6

Comparative analysis of knee isokinetic test results between national master athletes and first-grade athletes of male Judo

Ji He Zhou & Shuai Wang
Chengdu Sport University, Sichuan, Chengdu, China

Yun Fei Jiang
Chengdu Physical Education Institute, Sichuan, Chengdu, China

ABSTRACT: In this paper, the centripetal knee force of national master athletes (top rank athletes in China) and first-grade athletes (second top rank athletes in China) of male Judoka was measured by the ISOMED2000 Isokinetic system. Judo-related specific knee strength indices, such as peak torque in relation to body weight, flexor torque in relation to the extensor torque and average power of knee, were analyzed and compared between the two groups of athletes using Excel 2003 and SPSS 13.0 statistical software. The results showed that the first-grade athletes did not significantly differ from the national master athletes in terms of relative power of the knee flexors, but significant difference existed in the absolute power and fast power of the knee extensors between the two levels of athletes.

KEYWORDS: Judo, knee joint, isokinetic centripetal test, comparative analysis

1 INTRODUCTION

Compared with other sports events, Judo requires much higher lower limb strength, especially event-related specific strength of the knee joint, which is a significant factor in the competitive ability of an athlete. The isokinetic centripetal strength test of the knee joint shows high accuracy and repeatability. This study used the isokinetic test system to evaluate the isokinetic centripetal strength of two levels of male Judo athletes, namely, national master athletes (top rank athletes in China) and first-grade athletes (second top rank athletes in China). The knee joint strengths of the two groups were then analyzed and compared. The purpose of this study was to provide scientific evidence for the power training of Judo athletes.

2 OBJECTS AND METHODS

2.1 Subjects

Eight male national master Judo athletes (height: 167.5cm ±6.2cm, weight: 72kg ±11.6kg) and nine male first-grade Judo athletes (height: 165.3cm ±5.4cm, weight: 71kg ±12.3kg) volunteered to participate in this study.

2.2 Methods

The ISOMED2000 Isokinetic system (D&R Company, Germany) was used to measure the isokinetic centripetal strength of the knee joints of the participants. The subjects were instructed to do warm-up exercises before the measurement, which consisted of knee flexion and extension for 10min. According to the specifications of ISOMED2000, the subjects were positioned on the system, such that the power head center of the system was pointed to the center of the knee. During the measurement, the subjects initially performed knee flexion and extension 10 times to familiarize themselves with the system. Afterward, the subjects were asked to perform the maximum knee flexion and extension 5 times at an angular velocity of 60°/s and then 5 times at an angular velocity of 240°/s, respectively.

Four specific strength indices were calculated: 1) peak torque of the knee flexor in relation to body weight (PT/BW), 2) peak torque of the knee extensor in relation to body weight (PT/BW), 3) flexor torque in relation to the extensor torque (F/E), and 4) average power of the flexor and extensor. Each index was compared between the two groups of athletes. Excel 2003 and SPSS 16.0 software were used for the calculation and statistical analysis of data. T-test was performed to compare the two groups. The statistical significance was set at $p < 0.05$.

3 RESULTS

3.1 Basic strength of the knee

The results listed in Table 1 showed that as the angular velocity increased, the PT/BW of the knee flexors and extensors decreased, whereas F/E increased. This result supported the findings in the literature. Generally, the angular velocity of 60°/s is considered low speed for knee flexion and extension, which reflects the relative power of the muscle. The angular velocity of 240°/s is considered high speed for knee flexion and extension, which reflects the fast power of the muscle. In this study, the PT/BW of the extensors reached 2.63N·m/kg ±0.53N·m/kg at 60°/s, indicating that the relative power of the quadriceps femoris was in a relatively normal range. No significant difference existed in the F/E between the two groups of athletes, and the normal range was from 60% to 69% at an angular velocity of 60°/s. This strength characteristic is beneficial in the facilitation of joint stability and prevention of injury.

3.2 Comparative analysis of PT/BW

The results listed in Table 2 showed that when the angular velocity reached 60°/s, the PT/BW of the flexors of the national master athletes (1.73 N·m/kg±0.38N·m/kg) did not exhibit a significant difference from that of the first-grade athletes (p>0.05). Thus, evident difference did not exist between the first-grade athletes and the national master athletes in the relative power of flexors. However, the first-grade athletes (2.36N·m/kg ±0.57N·m/kg) and the national master athletes (2.80N·m/kg ±0.43N·m/kg) showed a significant difference (p<0.05) in the knee extensors. This result indicated a remarkable distinction between the first-grade athletes and the national master athletes in terms of the relative power of their knee extensors. When the angular velocity reached 240°/s, the two groups of athletes showed differences in the PT/BW of the knee flexors (p<0.01) and extensors (p<0.05). This result indicated that significant difference in the quick power of the flexors and extensors existed between the two groups. The national master athletes had higher quick power than the first-grade athletes in both flexors and extensors of the knee.

3.3 Comparative analysis of F/E

The results of F/E of male Judo athletes are presented in Table 3. The index of F/E reflect the athletic ability to coordinate the contraction of the whole muscle or the muscle group. This characteristic has an important role in preventing injury and facilitating joint stability. Based on documented data, when the angular velocity reaches 60°/s, F/E is from 60% to 69%; when the angular velocity reaches 240°/s, F/E is from 80% to 89%. In this study, the F/E of the national master athletes fell under the 60% to 69% criteria at low speed, whereas the F/E of the first-grade athletes was outside the range. No significant difference existed between the two groups of athletes. As the angular velocity reached 240°/s, the findings in the two groups of athletes showed some differences from the normal data provided in the literature. This result could be attributed to the specific skills and techniques of Judo. Moreover, the differences between the national master athletes and the first-grade athletes were significant.

3.4 Comparative analysis of the average power

The results of the average power of the flexor and extensor of male Judo athletes are presented

Table 1. Data sheet of isokinetic centripetal test of knee ($\overline{x} \pm SD$).

	PT/BW of the flexors [N·m/kg]	PT/BW of the extensors [N·m/kg]	F/E	average power of the flexor [w]	average power of the extensor[w]
60°/s	1.76±0.35	2.63±0.53	0.70±0.23	79.0±24.5	115.1±35.9
240°/s	1.25±0.36	1.74±0.39	0.74±0.18	113.6±42.4	163.6±55.3

Table 2. Data sheet of PT/BW of the knee flexors and extensors ($\overline{x} + SD$).

	The national master athletes		The first-grade athletes	
	PT/BW of the flexors[N·m/kg]	PT/BW of the extensors[N·m/kg]	PT/BW of the flexors[N·m/kg]	PT/BW of the extensors[N·m/kg]
60°/s	1.73±0.38	2.80±0.43*	1.79±0.33	1.84±0.38*
240°/s	1.31±0.28**	2.36±0.57*	1.22±0.40**	1.57±0.33*

Table 3. Data sheet of F/E ($\bar{x} \pm$ SD).

	The national master athletes		The first-grade athletes	
	60°/s	240°/s	60°/s	240°/s
F/E	0.66±0.13	0.74±0.13**	0.74±0.28	0.70±0.21**

Note: "**" means P<0.01.

in Table 4. The average power is calculated as the work done by the muscle or the muscle group in a unit of time, which reflects the work efficiency of the muscle or the muscle group. The literature shows that the average power of the muscle within a certain range increases as the movement speed increases. However, when the speed of the muscular movement reaches a critical value, the average power decreases. In this study, the test results corresponded with the documented data. However, when the angular velocity reached 240°/s, the average power did not reach the critical value. Significant difference did not exist in the average power when the flexors of the first-grade athletes were under low-speed condition. By contrast, when the flexors were under high-speed condition, a significant difference existed in the average power. The national master athletes and the first-grade athletes demonstrated significant difference when the extensors were under low-speed or high-speed conditions.

Table 4. Data sheet of average power of the flexor and extensor ($\bar{x} \pm$ SD).

	The national master athletes		The first-grade athletes	
	average power of the flexor[w]	average power of the extensor [w]	average power of the flexor[w]	average power of the extensor [w]
60°/s	81.4±26.6	118.0±31.4**	77.3±22.9	113.0±41.4**
240°/s	122.1±29.9**	170.1±45.2**	107.3±48.7**	158.8±61.3**

Note: "**" means P<0.01.

4 CONCLUSIONS

1 The first-grade athletes did not exhibit a remarkable difference in the absolute power of the knee joint flexors compared with the national master athletes, but showed a remarkable difference in the quick power. This finding suggests that attention should be directed on the quick power training of first-grade Judo athletes in the future.

2 The absolute and quick power of the knee joint extensor of the first-grade athletes was evidently weaker than those of the national master athletes.

3 The F/E of the national master athletes fell under the 60% to 69% criteria at low speed, whereas the F/E of the first-grade athletes was outside the range. No significant difference existed between the two groups of athletes.

4 When the flexors were under high-speed condition, a significant difference existed in the average power. The national master athletes and the first-grade athletes demonstrated significant difference when the extensors were under low-speed or high-speed conditions.

REFERENCES

[1] T. Yang, Z.J. Li, Application of Isokinetic Test in Evaluating Muscle Force, J. SSR, 28(3)(2007) 68–71.

[2] Abe T, Kawakami Y, Ikegawa S, et al. Isometric and isokinetic knee joint performance in Japanese alpine ski racers, JSM, 31 (1992) 353–357.

[3] J.H. Qu, Research on the Features of Isokinetic Contract Strength of Elite Water Polo Athletes' Knee Muscles, J. Journal of Beijing Sport University, 30 (5) (2007) 643–644.

[4] D.M. Lu, X.D. Wang, The muscle research of six joint in Young people, BSUP, Beijing, 2004, pp16–38.

[5] P.J. Abernethy, A. Howard, B.M. Quigley, Isokinetic torque and instantaneous power date do not necessarily mirror one another. JSCR, (10) (1996) 220–223.

Information, Computer and Application Engineering – Liu, Sung & Yao (Eds)
© *2015 Taylor & Francis Group, London, ISBN 978-1-138-02717-6*

Exploration and analysis of legalization approach of college student management

Hao Yu Huang
Harbin University of Science and Technology, Weihai, Shandong province, China

ABSTRACT: This paper puts forward detailed approaches to legalize student management from legislation, law enforcement, judicature and legal concept. School governance mode based on administrative means in China cannot adapt higher education needs under globalization tendency. It is required to change the traditional education concept and establish the idea of ruling by law in college student management. In other words, it is required to regard rights as the growing point and foothold of talent personnel. Through the efforts for more than 20 years, legal construction for student management in China has had certain foundation, but there are still some problems. For example, conflicts exist among legal systems; there are loopholes during law-making; there is blank in legislation; legal operability is weak. Thus, it is necessary to propose countermeasures in combination of practical conditions.

KEYWORDS: Student management; Ruling by law; Students' right; Judicial supervision

To achieve the legalization of college student management, it is required to not only make clear the legal relationship between college students and colleges, but also take action, answer and solve various practical problems with the idea and spirit of ruling by law, reforming management system, perfect legislation, abide by laws, enhance legal supervision and judicial remedy, standardize college management order, respect and protect students' right.

1 TO REFORM THE MANAGEMENT SYSTEM

In the 21st century, higher education system reform will be faced with greater opportunities and more severe challenges. One of the most important problems is system construction and innovation based on higher education system transition. The transition from the old system to a new system will inevitably experience the process of constructing and perfecting a new system. The basis of systematic innovation is to set up a new system involving government's macro-management, proper market regulation, wider social participation, autonomous school running and students' direct participation in management.

Firstly, government's macro-management. The government changes to indirect macro-management from direct public administration, which involves regulation of responsibility, right and interest and will trigger reform of government management means and ways. Enhancement of government's macro-management is reflected in the relationship between government and macro-management, i.e. the government turns to macro-management according to laws from direct public administration to ensure school's autonomy in running schools. Besides, it is also reflected in the relationship between government and market, i.e. the government formulates and implements market access and operation norms, standardizes market operations and gives play to the proper regulation function of market on education. To carry out macro-management with a market mechanism, the government must stick to the principles of fairness and efficiency, survival of the fittest, efficiency first and benefit maximization.

Secondly, proper market regulation. The function of markets in education and mutual relations between government and market Are basic problems of education system innovation. On the one hand, moderate regulation means market exerts regulation functions of education within a certain scope to some extent. On the other hand, such regulation function must be realized under government's macro-management and abidance by education rules. The key point of systematic innovation is to solve the field and mode of market regulation. For the relationship between factor market and education operation, effects of labor market, capital market and technology market in higher education are more direct. Sufficient and proper application of market regulation mechanism will promote reform and promotion of higher education system.

Thirdly, wider social participation. The educational system which achieves government's macro-management and proper market regulation provides a wide space for social participation and also increases practical needs. Social participation becomes an indispensable constituent part of the new education system and becomes a bond and bridge between government, market and school.

Fourthly, autonomous school running. In government's macro-management, under the precondition of proper market regulation and wide social participation, the school becomes the real education entity which runs a school autonomously according to laws. Meanwhile, whether autonomous school running can become true will depend on whether the internal school operation can establish a mechanism which accords with education rules, adapt and promote economic and social development and changes. Such mechanism needs a full set of organization, system, operating system and program. This is the major content of innovation of the autonomous school running system.

Fifthly, pay attention to students' self-management. The situation of students' participation in management is an outstanding mark which measures college management level. The resolution about teenagers' right passed by the United Nations Assembly in 1989 confirms the students' participation right: teenagers have the right to talk about and convey their experience and express their opinions in any related activity. This should contain students' participation in discussions about higher education problems, participation in evaluation and reform of course and teaching method and participation in policy making and school management work within the existing system.

2 TO ESTABLISH THE CONCEPT OF RULING BY LAW

The supremacy of law intensively reflects the concept of ruling by law. Meanwhile, the supremacy of law means the law should be the value standard of other social systems, become the controller of right and be the first channel to solve social conflicts. The faith of the supremacy of law is not the water without a source or a tree without roots, but the perfect unification of right concept, legal faith and law-abiding spirit. At present, cultivation of the concept of ruling by law should start from eliminating wrong ideological understanding and establishing scientific educational idea. In particular, it is required to enhance the right consciousness of education subject, cultivate legal faith and enhance the law-abiding spirit. Firstly, eliminate wrong ideological understanding; secondly, enhance subjects' right consciousness; thirdly, cultivate subjects' legal faith and ruling by law; fourthly, strengthen subjects' law-abiding spirit.

To get rid of the traditional legal concept and its negative effects on the legal construction cannot be completed in one day. Thus, Chinese education legalization is a progressive process and cannot be achieved at one stroke. Currently, cultivating people's concept of ruling by law must popularize law education. Besides, it is especially strengthening peoples' practical experience in actual legal practice.

Popularizing law education is achieved through popularization of comprehensive modern legal knowledge, which contributes to stimulating citizens' sense of identity and dependency on laws, forming general faith in modern laws and establishing the concept of ruling by law. In addition, it will generate very profound effects on forming scientific, legal thinking mode, emotional experience and behavior mode as well as implementing Chinese socialist laws in actual life in an efficient and smooth manner. The practice has proven that popularizing law education is a major means to cultivate people's legal consciousness and also a characteristic of legal construction in China. Workers popularizing law education should update their ideas, and change previous roles as administrators and educators and organization of law study and use through administrative means, but make the best use of the circumstances, actively do service and guarantee work well and guide citizens master their own legal weapons. Moreover, schools should cultivate managers' legal consciousness (especially democratic thought, the concept of equality, the spirit of justice, right consciousness and idea of ruling by law) through holding lectures about legal institution and especially education law and urging managers for self-study in accordance with special provisions of the education act. In this way, managers can standardize their words and deeds through laws, fairly treat each student in management work, respect students' rights and create the best conditions for students' integrated development.

Generation of the concept of ruling by law stems from people's approval and respect of ruling by law and depends on experience in practicing ruling by law. Relatively, people's practical experience in actual legal practice has great functions on cultivating the Chinese idea of ruling by law. This is because for concept certitude and belief persistence, what's most important is not language preach, but behavioral inspiration. It is not general advocacy, but specific demonstration. Such inspiration and demonstration to a large extent depend on the degree of government's (including schools) law abidance. It is required to provide people with actual legal experience and make them perceive specific laws. The most effective and direct method is that behavioral demonstration and inspiration of the state, government and the employees must comply with legal provisions. They should become the models of abiding by laws and enforcing laws.

3 TO PERFECT THE LEGISLATION SYSTEM

Seeing from the whole process and composition elements of legal construction, legislation is both the starting link of legal construction and basic precondition of legal construction. To legalize college legal construction, firstly, it is required to establish a legal system for student management with consistent contents, complex and uniform form and orderly hierarchy sorting which accords with students' benefits. There must be laws for people to follow. Student management legislation involves many contents. It is necessary to establish a set of complete system which includes the relevant provisions of the constitution, basic laws, laws, administrative laws and local regulations.

Firstly, further transform legislation thought. We should sufficiently recognize the function of student management legislation in national democratic legal construction from the perspective of ruling by law, establish legislation acceleration awareness ideologically and seize student management legislation in a down-to-earth manner. It is required to pay attention to correcting the tendency of attempting and accomplishing nothing in the student management legislation and focus on correcting negative practice of slowing down legislation to speed with the excuse of insufficient experience and immature conditions to make sure each department involving student management has laws to abide by.

Secondly, improve quality of student management legislation. Student management legislation should continuously perfect with situation development. It is required to clear existing laws and regulations regularly, modify, supplement or abolish them. It is required to not only prevent the tendency of getting something done once and forever in problems, but also focuses on maintaining unification of the national legal system and handle the relationship among legal establishment, modification and abolishment.

Thirdly, the perfect overall frame structure of student management legislation, make student management have laws to abide by and construct a complete legal frame system for student management. Currently, existing main legal systems about student management only include Teacher's Law and Law on the Protection of Minors. Other subjects' rights and obligations are mingled in relevant laws or comprehensive laws, such as Education Act and Higher Education Law. There is still lack of legal basis on duties and mutual relations between schools and administrative departments of education at each level, among administrative departments of education at each level and between social intermediary organizations and the education department. Macroscopically, there should be Organization

Law for Student Management which standardizes rights and obligations among student management subjects and specifies mutual duties and right boundary.

Fourthly, formulate the basic laws of college student management. This is because in the national legal system, student management law is just a part. If legal construction of student management is only focused on in the student management field and the research on student management law is not based on development situation of socialist legal system, student management law will lose the foundation of survival. Besides, the due functions cannot be exerted. Finally, the direction and vitality will be lost. Therefore, it is required to actively study national legal construction during studying legal system of student management, understand national laws macroscopically, make legal construction of student management comply with social development direction and then explore the rules and direction of national legal system development, enlighten legal construction of student management, make them own superior consciousness, make development of student management law basically synchronize national legal development, make it become a part of socialist legal system and finally occupy a position in Chinese legal system.

4 TO ACT BY LAWS

Legislation is the foundation and law enforcement is the key. Sound framework just solves the problem that there are laws to abide by. More importantly, it is required to ensure that the laws are strictly observed. Executive law enforcement is a key link of administration by law and also a weak link needing enhancement and improvement. Currently, in the student management field, relative to legislation, law enforcement problem is more prominent. So, strict enforcement of law and enhancement of students' level of management by law are current cardinal tasks.

Firstly, improve the ability and level of school administrators in the aspect of administration by law. The quality and working level of student administrators will have direct effects on student management level and effects, students' interests and school image. Therefore, it is required to vigorously strengthen the student management, team building, practically improve student management level, really reach the management of law and people first and make student management level continuously improve. Student management personnel at each level should carefully learn laws, understand laws, abide by laws, handle affairs in strict accordance with laws, regulations and policies and practices change working style.

Secondly, strengthen legality of management and law enforcement. Take disciplinary sanction for example. "Legality" should include the following aspects: 1) subjects and their limits of authority should be legal; 2) the contents should be legal; 3) the purpose should be legal; 4) there should be procedure guarantee.

Thirdly, intensify routinization of school management behavior. From the perspective of jurisprudence, procedure refers to a process of making legal decisions in accordance with certain sequence, ways and steps. Due procedure is a basic principle of ruling by law and also a basic requirement of ruling by law. A basic characteristic of modern ruling by law is that it increasingly pays attention to due procedure. The college management procedure is a very important means to guarantee correct and effective execution of school administration authority and restrain peremptoriness of the college management. On the one hand, the college management procedure can restrict randomness of college management behaviors and reduce the dangerousness of school management right violating individual legal interest; on the other hand, it should reserve certain freedom of choice to ensure vigor of school management right. Without school management procedure, a college can select the opportunity, mode, method and step of college management behavior at will, or cancel rights and interests given to students by the laws through blindly setting procedure barriers or delaying law enforcement. Meanwhile, students' obligations may be aggravated through selecting right execution method lacking scientificity and legitimacy.

Fourthly, vigorously make administrative affairs of a school public. There are at least three advantages. 1) It contributes to making all know school affairs and those involving vital interests of teaching staffs, reducing intermediate links of administration, avoiding faults of intermediate links and making government decrees smooth. 2) It can promote transmitting an order from above and conveying conditions at the lower level to the high level, contribute to bothway information between leaders and the masses and reduce conflicts. 3) It can facilitate one to hear the truths and good ideas, promote democratic decision-making and scientific decision-making and make us better exercise our rights.

Fifthly, carry out hearing system. Hearing refers to "listening to others' opinions". The origin of legal principle is "rule of natural justice" and American "due legal procedure" in English Common Law. Since hearing doesn't just require making decision-making basis, information and decisions openness, fairness and justice, but also allows and encourages each party participating in the hearing, cross-examination and argument to make a statement on decision-making matters. So, it integrates scientificity, democracy and legality of modern decision-making and owns multiple functions such as information, consultation, participation, quality control, supervision and feedback. It is one of the most effective system arrangements to drive and ensure scientification, democratization and legislation of public decision-making. In the face of modern reform of internal management, colleges should listen to statement, defense and opinions of the parties concerned in great fields involving students' legal interests such as punishment of students. This is the most important guarantee of the parties' right to know and right to supervise and also a basic requirement of ensuring scientificity, democracy and fairness of management behaviors.

5 TO ENHANCE JUDICIAL SUPERVISION

The practice has proven that to achieve the legalization of student management, legislation, law enforcement and judicature are far from enough. To make sure legalization of student management goes well, it is required to perfect related democratic supervision system and make student management legally receive supervision from authority, administrative body, judicial office and society as well as the masses, sufficiently guarantee a college students' right to receive education and related rights. Meanwhile, it is required to specify corresponding legal responsibilities, legally investigate illegal behaviors destroying law enforcement through judicial review and practically guarantee the legitimate rights and interests of college students.

Judicial review is the final guardian angel of justice. The judicial review system of administrative acts is a significant legal system established by modern democratic states. Besides, it is a legal system which investigates legality and appropriateness of corresponding administrative acts, repeal illegal administrative acts and repair damages caused by illegal acts at the request of public administration counterpart suffering unlawful infringement or adverse effects of administrative acts. Colleges must have judicial safeguard during managing the college by law. For illegal college management behaviors, college management counterparts can apply judicial review to a people's court to investigate legal responsibilities of college management subjects breaking the law. Such approach can more effectively implement legal supervision of college management behaviors than any other approach.

1) Gradually establish and perfect special functional organizations of education and justice; 2) try to set up and perfect arbitral institution for education; 3) establishes and perfect students' appealing mechanism.

REFERENCES

[1] Cheng Yanlei, Necessity of valuing and enhancing China's education legislation, Educational Research, 200(2).

[2] Peng Shuanlian, Legalization of college student management, Journal of Ideological & Theoretical Education, 2002(10).

[3] Shen Shuping, Analysis of legal status of British and American colleges, Comparative Education Study, 2002(5).

[4] Zheng Ruolin, Relationship between higher education and society, Modern University Education, 2003(2).

[5] Li Rong, On rights of main market subjects of higher education services, Petroleum Education, 2000(4).

[6] Cheng Jinkuan, Analysis of judicial system of American education, Studies In Foreign Education, 2002(1).

[7] Jiang Shaorong, Discussion on legal status of Chinese schools, Theory and Practice of Education, 1999(8).

[8] Liu Dongmei, On legal status of colleges, Educational Review, 1998(1).

[9] Zhong Hua, On modernization of educational legal system, Hunan Social Sciences, 2002(2).

[10] Chen Binli, Exploration of basic contents of college students' rights, Comparative Education Study, 2003(1).

[11] Chen Xiaobin, Educational Administration, Beijing Normal University Press, 1999.

[12] Lao Kaisheng, Introduction to higher education regulations, Beijing Normal University Press, 1999.

[13] Hao Weiqian, Li Lianing, Comparative study on educational legal system in each country, People's Education Press, 1999.

[14] Chen Yuanqing, Precaution and treatment of students' injury accidents, Jilin Photography Publishing House, 2002.

Information, Computer and Application Engineering – Liu, Sung & Yao (Eds)
© 2015 Taylor & Francis Group, London, ISBN 978-1-138-02717-6

Discussion on the principles of Marx and Engels' moral education

Zhuang Zhuang Zhang

Harbin University of Science and Technology, Weihai, Shandong province, China

ABSTRACT: Marx and Engels were not professional educationists and did not write special monographs on moral education. However, in the long-term revolutionary and practice and theory work, they paid great attention to the moral education work and combined it closely with the foundation of historical materialism and proletariat liberalism theory. In their works, many profound and important discussions about morality and education reflected their abundant connotation of moral education. Marx and Engels explored the principles of moral education. They proposed the principles, including: against the insane moralization; the fundamental way of moral education was the practice; insisting on persuading people by scientific theory; insisting on the combination of education and self-education; moral education is a long process.

KEYWORDS: Moral education; Practice; Criticism

The critics of the revolution are a key characteristic of Marxism. From Umrisse zu einer Kritik der Nationalkonomie to Critique of Political Economy and to Das Kapital, and from Zur Kritik der Hegelschen Rechtsphilosophie to The Holy Family to The German Ideology and to Critique of the Gotha Program, the criticism was in the whole life of Marx and Engels. Just in this process of criticism, Marx and Engels gradually established a series of principles for moral education.

1 AGAINST INSANE MORALIZATION

Marx and Engels were against insane moralization in moral education. They really mentioned more than once that they were against the moralization in their works. Their most obvious statement was found in The German Ideology when they opposed against Stirner, a Young Hegelian. Combing with Marx and Engels' works and corresponding background, seen from the critics of the revolution, the understanding about Marx and Engels' opposition against the moralization can be summarized as the following three aspects:

Firstly, Marx and Engels were against the behavior that the solution of social issues simply relied on moralization. The behavior mainly refers to that in front of increasingly obvious social issues in the development of capitalism, such as economic crisis, increasing gap between the rich and the poor, mass worker unemployment, and shortage of houses, some bourgeois economists or scholars believed that these issues could be solved by moral criticism or preach.

These bourgeois economists did not realize the social and economic root before these issues. Namely, these issues were inevitable once with the existence of the capitalist system. Marx and Engels were firmly against the behavior that solved the social issues simply from moral angle.

Secondly, Marx and Engels were against the bourgeoisie's moralism to the proletariat. Since the birth of the capitalist society, the bourgeoisie publicized the moral slogans, "Freedom, Equality, and Charity", while these slogans played a positive role in fighting against the feudal autocracy. However, along with the foundation and development of the capitalist system, "Freedom, Equality, and Charity" were more obviously the benefit of the bourgeoisie. The proletariat could never enjoy the same freedom and equity with the bourgeoisie. The bourgeoisie was also worried that the workers contacted with and were influenced by views, including freedom and equality and even believed workers were dangerous if they were educated. Thus, the education provided by them to the working class was more about the obscurantism education combining with the religious principles, which publicized moral slogans including "infinite loyal and obedient", "penitent", "self-denial", and "timid". Or, it could be said that the religious moralization was conducted. Such a mode had been already the key means and weapon of the bourgeoisie to control the working class and keep their ruling position. Marx and Engels were greatly against the bourgeoisie's moralism to the proletariat.

Thirdly, Marx and Engels were against the abstract moralization away from the reality, which referred to the abstract, insane, unique, and even permanent moralization in other moral theories. Marx and Engels' criticism against Stirner mentioned before was an example. Stirner regarded the egoism as the only moral principle and determined it as the nature of self-awareness, emphasizing the absolute freedom of human beings. In addition, just as the Marx and Engels' criticism, Stirner greatly publicized his egoism. The sanctified and abstract understanding of the single moral principle was also refused by Marx and Engels who insisted on historical materialism.

When Marx and Engels criticized the above moralizations, they also expressed their alternative principles and proposals on moral education via negative aspect in this critical process, including: the fundamental way of moral education was the practice; insisting on persuading people by scientific theory; insisting on the combination of education and self-education; moral education is a long process.

2 THE FUNDAMENTAL WAY OF MORAL EDUCATION WAS THE PRACTICE

Marx and Engels highly emphasized the role of the practice in the moral education process. They pointed out that the environment should be transformed by the practice, the human should be transformed by the environment, and the transformation of the environment and the transformation of human should be combined together. They suggested combining the personal moral education with the social transformation. The transform of environment was the same with the transformation of human activity and human, which could only be regarded and explained as the practice of revolution.

Insisting on the principle of practice shall firstly emphasize the role of benefit in the generation and development process of morality. They expounded that the life was not created by the principle, but the principle was created by the life. Marx and Engels also proposed that the benefit had an obvious class nature. The objective of all revolutions in history was for the benefit of different classes. On the Situation of The English Working Class, Engels praised the behavior that the union members, Chartism followers, and socialists constructed the schools and reading rooms at their cost, and believed this was a good method to give workers moral education. The bourgeois schools gave a training education. Of course proletariats were unwilling to go to those schools. They preferred to read in the reading room and discuss issues directly related to their practical benefit. The proletariat started the communist revolution for the benefit. Just based on the theory of communism revolution and practice activities, they abandoned the private system and realized the liberation of themselves and all mankind.

Insisting on the principle of practice relied on the moral education in practice. The practice not only included living practice and production practice, but also included the revolution practice for the liberation of the proletariat. According to the historical materialism, the morality was the product of social and economic relations. After all, people obtained their moral view from the economic relationship of their production and exchange. Thus, the moral education away from people's practice could hardly achieve the objective of moral education. Just because of the living activity and production activity of the working class, workers were given the best education and formed the moral view different from the bourgeoisie. Besides, they had the advantage of morality. For example, compared to the bourgeoisie, workers were kinder and were more ready to help others. They also began to know their benefit and the national benefit. Certainly, Marx and Engels also suggested that the working class had their disadvantage in morality, which was also determined by their living situation under the capitalism. Therefore, the moral education away from the reality given by the bourgeoisie to the proletariat could not improve the morality of the workers.

3 INSISTING ON PERSUADING PEOPLE WITH A SCIENTIFIC THEORY

Marx and Engels believed that the fundamental way of moral education was practiced, but they did not deny the influence of scientific theory on the working class. Marx and Engels' moral education was featured by that they explained the issue scientifically from the historical materialism instead of insane moral criticism, though they were full of moral indignation in the critical process against the real social issues. Marx and Engels never denied the function of the morality as the social consciousness in the social development, and never clearly denied the key role of moral education. Marx and Engels' criticism, different from the bourgeoisie economists, suggested that the criticism should be complete and should be from the root of social issues. If the fundamental function of the capitalist private system in all social issues could not be known, the moral criticism was really the pure moralization. The moral criticism was based on the scientific analysis. Namely, people should be persuaded by scientific theory in the moral education.

In order to give the guidance of scientific theory to the working class, Marx and Engels constantly denied themselves through constant explorations and created and developed the historic materialism by theory innovation. By defending the scientific nature of Marxism, they founded the scientific theory basis for the moral education of the proletariat. Firstly, they insisted on that the working class should accept the education of scientific theory. In 1847, Marx and Engels proposed the first condition that the League of the Just must accept their scientific theory after they received the invitation from the League.

The theory was scientific, but the population must accept the propaganda. Vast people could only improve their morality under the premise of accepting the theory. Therefore, when Marx and Engels created the historic materialism and communism, they emphasized the publicity. Marx and Engels carried out the publicity about moral education to the working population by various means. They improved the class consciousness of the proletariat and cultivated the class emotion and revolutionary spirit of the proletariat by many methods, including: printing the booklets, founding the press, constructing the publicity institutions, making speeches in meeting and assembly, and had communications with activists and friends of labor movements in various countries.

4 INSISTING ON THE COMBINATION OF EDUCATION AND SELF-EDUCATION

The formation of the moral view was the result of internalization of individual social moral standards. In order to change the social standards into individual believe, the moral formation process applied by the educatee for combination of self-discipline and heteronomy was needed besides the influence from external education. Therefore, the moral education activities were not only the activities which educator gave moral influence on the educatee intentionally, but also the activities which the educate carried out for education and self-education through the initiative and motility process.

Marx and Engels always suggested that the proletariat should seek for a kind of self-liberation. The moral education of the proletariat was a key component of the proletariat's self-liberation. Therefore, the moral education of the proletariat was not only the implantation, but expressing the initiative and motility of the proletariat. Thus, the proletariat transformed the objective world when transformed the subjective world.

Marx and Engels believed that the educator must have certain education to carry out the moral education. To give the working class the moral education, the educator must firstly have a high theoretical level and a high moral quality. Therefore, besides constantly improving the theoretical quality through study and research, Marx and Engels emphasized the selection and cultivation of other educators in socialist movements in various countries.

5 MORAL EDUCATION IS A LONG PROCESS

Moral education is a long historical process. Every era had its specific moral education task, which promoted the development process of moral education generations by generations. The education of the proletariat was not a thing to be completed in a day, but should be regarded as a development process promoted forward constantly. Seen from the development of human nature, the initiative activities of people were restricted by the objective material conditions. People could not choose the productivity and productive relations freely. All human activities were in certain historic conditions. Since every society had different social and economic relations, every morality had the social content and characteristic of that era and was corresponding to the social, economic, and political relations. Being an activity with the object of morality, moral education was also restricted by certain social and economic relations similar to morality. The moral requirement beyond the era and the social and economic relations could never be proposed. Any moral education should be the moral education in certain social and historical conditions. Specific era determined the subjective and objective bodies, object, content, object and modes for the moral education. With the era development and change, all elements of moral education activities also changed. Therefore, the moral education could never be completed in a time, but developed together with the era development. Firstly, seen from the object of moral education, the working class was constantly updated. The persons newly in the working class should accept the education of excellent morality, and the persons with poor morality should accept the enhanced education. Along with era development, the definition of the working class also changed. Correspondingly, the content of moral education should change. Secondly, seen from the content of moral education, the social existence determined the social consciousness. Social consciousness must develop and change together with social existence. The development of human civilization and the progress of the society added new factors in the moral education and proposed the new requirements. The moral education should be updated and improved constantly according to these new factors and requirements. For example, the era of Marx and Engels was the era that the proletariat did not grasp the regime in the revolution. However, when Lenin constructed the socialism, the proletariat already founded the national regime. Now, China is different from the era of Lenin. Therefore, the content of moral education shall also develop and change with the era development. Thirdly, seen from the mode of moral education, along with the diversifying human interaction modes and creative development of scientific technology, current moral education mode shall not be the kept. Instead, some more scientific and effective moral education modes shall be created based on the initiative, motility, and creativity. Fourthly, seen from the subjective bodies of moral education, people engaging in the moral education work are not born sages and men of virtue, but shall also accept constant new moral education and improve their moral quality.

REFERENCES

[1] Karl Marx and Frederick Engels, Vol. 1, Vol. 2, Vol. 3, Vol. 10, Vol. 11, Vol. 21, Vol. 25, Vol. 30, Vol. 31, Vol. 44, Vol. 45, Vol. 46, Vol. 47. People's Publishing House, Published in 2004.
[2] Capital Vol. 1. People's Publishing House, Published in 1975.
[3] Written by David McClellan, translated by Wang Zhen: Karl Marx, China Renmin University Press, Published in 2005.
[4] Li Yixian et al.: Discussion on Marx and Engels' Education Theory, Social Science Press, Published in 1986.
[5] Chen Guisheng: China's Moral Education Issue. Fujian Education Press, Published in 2007.
[6] Tan Chuanbao: University Moral Education Principle (Amended), Education Science Press, Published in 2003.
[7] Yuan Guilin: Contemporary Western Moral Education Theory, Fujian Education Press, Published in 2005.
[8] Written by Du Wei et al., translated by Wang Chenxu: Moral Education Princple, Zhejiang Education Press, Published in 2003.
[9] Li Kangping: Contemporary Marxism Moral Education Thought Study in China, Social Sciences Academic Press, Published in 2009.
[10] Li Zheng: Marx and Engels' Ideological and Political Education Theory and Practice Study, Peking University Press, Published in 2011.

Information, Computer and Application Engineering – Liu, Sung & Yao (Eds)
© 2015 Taylor & Francis Group, London, ISBN 978-1-138-02717-6

Project-based training programs for exhibition industry students

Hong Li

School of Economy and Management, Tianjin Sino-German Vocational Technical College, China

ABSTRACT: Exhibition industry has strict requirements to the skills and competencies of the practitioners. One of the challenges that the colleges face is how to train their students in practices so that they could solve real world problems. Not only are most of the companies in exhibition industry small businesses, but also because of the unconventionality of exhibition industry, colleges and enterprises should find an innovative way to work together. As a result, both of them will be better off if they could design a cooperative education program together. This article discusses the purposes, requirements, and methods of the cooperative education program.

KEYWORDS: Exhibition industry; project-based training; cooperative program

1 INTRODUCTION

Along with the development of convention and exhibition economy, more and more colleges are offering courses in exhibition industry. Despite the fact that the exhibition industry is a high practical job, students have to be professional in order to meet company requirements [1]. Since the exhibition industry has to focus on the development of practical teaching, colleges and enterprises need to work together to provide more opportunities for students. Although colleges are seeking ways to cooperate, enterprises are lack of interests because of their conflict of interest [2]. Even if companies agree to cooperate, most of them are using students as a cheap labor force. Students are assigned to trivial jobs, which will not meet the purpose of occupational skills training [3]. Exhibition business is more difficult than other businesses to accept large quantity of student to work as coop student at the same time because most companies in the area are small and undeveloped as well as the unconventional of exhibition industry projects. In fact, some colleges do offer cooperation opportunities in their commercial events, but since these projects are diversity, time consumed, seasonal as well as imbalanced personnel needs, these events can only offer reception jobs to qualified students instead of practical jobs for every student[4]. Moreover, these job offers usually conflict with school schedules. Colleges and enterprises should have developed a cooperative education program (hereinafter referred to as coop program) to change the current situation. What is a cooperative education program? According to a document of the Ministry of Education and experts' opinions, coop program can be defined as a teaching method that is adopted by post secondary schools cooperating with government or companies to develop projects meet market needs and provide practical training at the same time. Unlike commercial projects, coop program mainly focuses on experiencing working environment and process, developing students' professional quality and skills [5]. This new model can benefit both college and enterprise.

2 SIGNIFICANCES OF COOPERATIVE PROGRAMS

2.1 *Improve practical conditions of the exhibition industry*

Practical training has always been a problem in management and economics courses. Because the exhibition industry is an emerging course, only a few schools are equipped with training room or software [6]. Most of the schools are training students using scenario simulation and case analysis. The development of coop program can remit the problem of loss of equipment and lack of capital. By projects taking down, position designing, and task assigning, students can participate in project operation. Students can improve their professions and abilities by working in real situations. To be more specific, they can gain work experience; understand work ethics; build up skills; find and make up their problems. As a result, students can work and learn at the same time. Also, by taking part in projects, students will discover their potential abilities and interests to prepare for the future.

2.2 Fulfill training method based on working process

One of the most popular education models internationally is to train students based on working processes; however, it is difficult for the exhibition industry to adopt this teaching method in the current situation. The development of the cooperative education program enables students to experience a whole working process. Professors can redesign course curriculum and content based on coop program needs. Isolated topics can be reunited through a practical training program by applying knowledge and skills learned in actual jobs. Furthermore, students can improve communication skills, teamwork and problem solving abilities, critical analysis skills, and emergency handling; thus improve their professional behaviors and overall executive abilities [7].

2.3 Develop new cooperating method to benefit both schools and enterprises

As discussed above, colleges can benefit from the program by completing teaching objectives, improving students' professional abilities, and teaching standards. At the same time, enterprises can achieve the following benefits.

Gain cheap, but fine intellectuals and clients. Planning, exhibitor recruiting, and field service stages are built up with teachers' and students' intellectuals, such as research reports, proposal, apps, AV materials, and exhibitor and audience databases. These intellectuals can be turned into company productivity as knowledge resources, especially exhibitor and audience databases.

Improve awareness. Along with the marketing of coop program, enterprises can create a presence in media, event promotion, and news release; therefore, improve their awareness, popularity, and reputation. Besides, coop program can add more methods of marketing corporate image.

Provide high potential talents. Enterprises can reduce investments in human resources by discovering and training the right people in cooperative program.

3 CRITERIA OF COOPERATIVE PROGRAM

Cooperative program selection has to meet the following requirements to be successful:

3.1 Marketization

Program selection has to consider market needs. Although the program is not profit driven, marketing results directly affect its viability and sustainability. It is really important for the program to meet the market. Consequently, market research based on the target market and program content is necessary.

3.2 Feasibility

Feasibility study needs to be done after completion of program proposals. The feasibility study should include the following contents: market environment feasibility, program viability, human capital and capital needed, site and equipment required.

3.3 Difficulty

Although cooperative program adopts general market rules, it is not exactly same as a normal commercial exhibition program. Coop program should consider students' adaptability along with learning objectives, course setting, and teaching schedule. The program should meet the level of most students after special training. The program should be small or medium sized instead of large ones. When the number of students exceeds a certain amount or the length of schooling will be three years or longer, colleges should develop multiple programs so that students can be assigned to jobs according to their grades. Teaching design and position separation will be better completed when one or more projects with the complete working process are selected. The analysis should base on learning objectives and project difficulties.

4 METHODS OF COOPERATIVE PROGRAM

4.1 On campus exhibitions and events targeting students and teachers

Undergraduates and teachers form a large consumer base. Some comprehensive universities and education parks contains more than tens or even hundreds of thousands of students. Besides, most education parks in China are built in the suburbs. Therefore, creates more exhibition and exhibition industry opportunities. Since students are familiar with the market, it is easier to propose, recruit, and manage. Campus exhibitions and events can be held on campus or in residential areas with a low capital requirement. According to the above advantages, campus projects have a low degree of difficulty.

4.2 Company conferences and celebrations

Companies have high volumes of meetings and events in their daily operations, such as annual meetings, incentive conferences and tours, company anniversaries and festival celebrations. These events used to be held by the company; however, as market division emerging, companies are outsourcing these events, which create cooperative opportunities for students. These conferences and celebrations usually have high standards, students (preferably senior students) need to work under professor's guidance. For example, Constellation culture expo hosted by Shanghai Institute of Tourism, Fengxian District Tourism Bureau, and the Gulf tourism area has been operating for five years with a great success. Also, successful in annual meeting of Institute of Exhibition of Zhejiang Province hosted by Zhejiang Economic and Trade Polytechnic indicates the feasibility of these programs.

4.3 Non-profit social events

Along with changes of Chinese government functions and developments of social organizations, more exhibition and exhibition industry opportunities are provided. Due to the non-profit characteristic of the events, most events enterprises are not willing to participate. However, these events will easily gain government support and funding. Tianjin Sino-German Vocational Technical College cooperate with Ministry of Education as a co-organizer to hold National Vocational Skills Competition entry race to provide opportunities for exhibition industry students.

4.4 Existing exhibition and exhibition industry projects

Events enterprises also can offer opportunities in their commercial exhibition projects as organizer, co organizer or exhibition hall provider. By participating in these cooperative programs, colleges can meet their teaching objectives while enterprises can hire cheap but high quality labors. For instance, Tianjin Sino-German Vocational Technical College has been cooperating with Meijiang Exhibition Center and has achieved phased objects by participating in Northern Bicycle Show, Tianjin International Automobile Exhibition, Tianjin Fair, and etc. The main challenge is to reduce the conflict of concentrated hiring and teaching schedule. Thus, these programs will be more suitable in students' final internships.

5 CONCLUSION

Cooperative programs suit the main principle of vocational education and create a better situation for students. Yet, the development of the exhibition industry cooperative program is still at the start-up phase with lots of challenges, such as enterprise participation, initiative, government funding, program running and student managing.

ACKNOWLEDGMENT

The paper work was supported by the project of cooperative training program for exhibition industry students (No. zdkt2013–016) at Tianjin Sino-German Vocational Technical College.

REFERENCES

[1] Dan Zheng, "Research on Practice Study for Advertisement and Exhibition Industry Professionals." Education and Profession. vol. 8, pp. 106–107, 2014.
[2] L. Gupta, P. Arora, R. Sharma and M. Chitkara, "Project Based Learning: An Enhanced Approach for Learning in Engineering," Proceedings of the International Conference on Transformations in Engineering Education, p.581, Sept. 2014.
[3] V. Bartsch, M. Ebers and I. Maurer, "Learning in project-based organizations: The role of project teams' social capital for overcoming barriers to learning", International Journal of Project Management, vol. 31(2), pp. 239–251, 2013.
[4] L. Helle, P. Tynjälä and E. Olkinuora, "Project-Based Learning in Post-Secondary Education – Theory, Practice and Rubber Sling Shots," Higher Education, vol. 51(2) , pp 287–314, 2006.
[5] X. Jin, K. Weber and T. Bauer, "Relationship quality between exhibitors and organizers: A perspective from Mainland China's exhibition," International Journal of Hospitality Management, vol. 31(4), pp. 1222–1234, 2012.
[6] B. K. Dawson, L. Young, C. Tu and F. Chongyi, "Co-innovation in networks of resources - A case study in the Chinese exhibition industry," Industrial Marketing Management, vol. 43(3), pp. 496–503, 2014.
[7] S. Ahmad, M. Y. Abbas, W. Z. M. Yusof, M. Z. M. Taib, "Museum Learning: Using Research as Best Practice in Creating Future Museum Exhibitio," Procedia - Social and Behavioral Sciences, vol.105(3), pp 370–382, 2013.
[8] M. Young, The Technical Writer's Handbook. Mill Valley, CA: University Science, 1989.

Information, Computer and Application Engineering – Liu, Sung & Yao (Eds)
© *2015 Taylor & Francis Group, London, ISBN 978-1-138-02717-6*

Strategies of liquidity management for China Construction Bank

N. Ye

Department of Finance and Economics, Xinyang College of Agriculture and Forestry, Xinyang, China

ABSTRACT: The 1998 Asian financial crisis was a painful lesson, which brought the global economy into the recession with anxiety. Although China was not dragged down from this financial crisis, it still decided to maintain large foreign exchange reserves to guard against the next financial crisis. Therefore, it must construct the overall and a sufficient banking system for building the basis for future growth. Liquidity management will be paid more and more attention. This text analyzed the strategies of liquidity management of China Construction Bank, which sets an example for other financial institutions. It pointed out four important aspects of liquidity management to deal with some common liquidity problems.

KEYWORDS: Liquidity management; commercial banks; China Construction Bank

1 INTRODUCTION

Commercial Banking plays a significant role in the financial industry, through the implementation of the related bank's policy; countries can control and adjust the pace of economic development very effectively. With the globalization and economic diversification, commercial banks greatly satisfied the needs of all kinds' enterprises, financial centers and clients. Therefore, Commercial Banks have already become the major component element in economic development of a country.

In China, there are five largest nationalized commercial banks, which have floated shares in the market: Industrial and Commercial Bank of China (ICBC), Bank of China (BOC), China Construction Bank (CCB), Agricultural Bank of China (ABC) and Bank of Communications (BOCOM). Comparing with other four banks, China Construction Bank (CCB) has the best quality of assets and the radical business strategies, so that there will be an excellent growth in the future. Therefore, China Construction Bank might be the most representative commercial banks in China. The whole contents will focus on solutions and strategies to deal with the liquidity management of China Construction Bank since 2000.

2 LIQUIDITY MANAGEMENT

Liquidity management is the core content of cash management. Under the condition of rational payment capacity, effective liquidity management can involve in managing cash assets and reserves to maintain sustainability by the ability of maximum benefits at minimum cost. Banks control and manage effectively the resource of cash, improve cash flow to enhance the profitability. In addition, banks need to satisfy the requirement of the clients about withdrawing, payment and normal loans. In a word, effective liquidity management shows the high quality and strong capability of cash management with meeting financial crisis.

"What about the present economic situations in the world?" It is a question often asked after people have had the worst years of the great economic depression. Actually, it's a fact that people always worry about it even though nobody can make sure what will happen next. At all events, the economy has been one of the most important things for countries to pay attention.

The Asian Financial Crisis was a period of financial crisis that is happening in July 1997. The financial crisis began in Thailand because the greater impact on economic benefits and policy in the United States and the economy of scale in Asia.

During the Asian Financial Crisis, Indonesia, South Korea and Thailand were the countries got a lot of affect. Hong Kong, Malaysia, Laos and the Philippines were also hurt by the crisis. Word Bank (2000) reported that many countries experienced large scale of exchange rate depreciation and stock market crashed, which brought into the economic downturn lasting for a long time.

Radelet, S. & Sachs, J. (1998) pointed out the reasons of Asian Crisis was poor management in the banking system and lack of transparency in corporate governance.

Even though there had low level of economic development in China at that moment, the Asian Financial Crisis provided a great useful notice for China, which some crucial problems of many non-performing loans and relied on the relation of the exchange rate in the United States. However, there was another aspect needed to worthily consider: the weak awareness and shortage of liquidity risk management.

In China, the liquidity management existed these problems. Commercial banks cannot focus on liquidity management for a long time. On the one side, the banks had little experience and operation principle. On the other side, it had a high savings ratio, so there was no incentive for banks to improve the liquidity. Meanwhile, high saving ratio could support the liquidity, but it also showed the low quality of short-term loans. If bank loan loss incurred, the bank could increase cost of debt to aim at increasing liquidity. So the next financial crisis would come.

Since China accessed to the WTO, because of high lending and high investment followed by high saving ratio, it caused a high quantity of exports with the low consumption in domestic country, so banks had to meet the excess liquidity problem in China. Liquidity management would be paying more and more attention.

Under the status quo of economy in China, China Construction Bank (CCB) took some advantage measures to manage the effective liquidity. These actions and approaches were also can be used for reference and to learn.

2.1 Balance management of assets and liabilities

According to the CCB annual report from 2006 to 2013, it said that the liquidity ratio of CCB approximately remained between 1.05~1.2. It showed that liquidity ratio was very stable, and each ratio was over 1, which explained the supply of short-term liquidity was sufficient and short-term debt paying ability was strong; it also had the area of capital widening (Fig. 1). Therefore, the disadvantage was the majority of long-term loans during the short term liabilities. Meanwhile, the low ability of bearing interest in liquid assets leaded to low margin income from liquidity assets and liabilities. It meant the liquidity liabilities were greater than liquidity assets. Irrational assets-liabilities structure would incur the heavy liquidity risks. Once an unexpected incident had recurred, all of depositors speedily gathered on the bank to withdraw money, but banks were unable to timely satisfy the depositors' requirements, the liquidity risk would be produced.

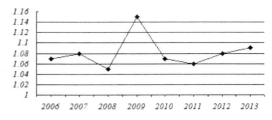

Figure 1. Liquidity ratio (%) of CCB during 2006 and 2013*.
* All the data from China Construction Bank's annual report 2006 to 2013.

Under these circumstances, CCB maintained the principle of balance management of assets and liabilities to work out the best liquidity management about relying mainly on managing liabilities while making asset management subsidiary.

In China, social credit structure and social credit environment should be improved. At the same time, CCB focused on making a careful analysis of its marketing environment and assets-liabilities structure, comparing the strength and weakness of its credit asset operation. For example, as the end of December in 2012, the non-performing loan ratio of CCB was 0.99%, which obviously reflected to decrease 0.1% points from the previous years. And the non-performing loan ratio of CCB kept on a steady value at 2013, which reflected the basis of non-performing loan was solid (Table 1).

Table 1. Non-performing loan ratio (%) of CCB during 2007 and 2013.

Year	2007	2008	2009	2010	2011	2012	2013
non-performing loan ratio	2.60	2.21	1.5	1.14	1.09	0.99	0.99

* All the data from China Construction Bank's annual report 2007 to 2013.

From Figure 2, it indicated the security and liquidity of funds used by five commercial banks within a certain period. Four largest commercial banks controlled their non-performing loan ratios within the range of 0~1%, except Agricultural Bank of China. It meant that commercial banks did non-performing loan management with prudence and kept the banks' balance sheets were quite healthy.

Moreover, CCB tried to keep the relationship of distribution between liquidity requirement

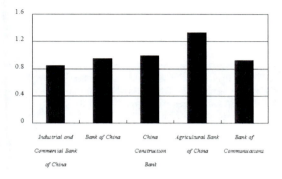

Figure 2. Non-performing loan ratio (%) of five largest China-based commercial banks in 2012*.
* All the data calculated from each nationalized bank's annual report in 2012.

and liquidity sources, so it attained the dynamic balance of the speed of assets expansion and capital adequacy ratio.

2.2 Strengthening the real estate credit management

With the rapid improvement of economic and enhancement of consumption level in China, excessive growth of credit availability in real estate caused some great potential crises about liquidity risks. Real estate lending had already become the highest quality assets in commercial banks. In order to attain the most market shares, it led to the fierce competition between each commercial bank. Some commercial banks took the low standard of lending with decreasing investigation steps, thus it seriously impacted bank asset safety. If banks could not sufficiently manage the real estate credit risk, it produced the real estate bubbles or expanded bubbles. After the bubble burst, it would be a horrible threat to bank asset safety at first.

According to this situation, with the purpose to improve the safe development of real estate by strictly controlling the credit growth, CCB decided to issue some new policy recommendations to avoid the liquidity worsening results for adopting vigorous measures to support affordable housing and personal housing loans, strictly control the lending of villas, high-grade housing and hotels, office buildings and shopping malls.

For the personal housing loans, CCB highly focused on the improvement of living level, it realized the detail information about ordinary commercial housing and affordable housing in time to speed up the credit rating and project valuation for developers. At the end of 2012, CCB's balance of personal housing loans already had exceeded RMB 1.54 trillion in

China, and the balance of personal housing loans had a growth of 16.7%. Thus, CCB ranked first among its peers in both loan balance and new balance.

For the high-grade housing and hotels, CCB must claim developers to have the highest percentage of equity funds, and to reduce the proportion of lending in the total project. So that it cut down lending risks for the high-grade housing, generally, CCB uses the mortgage lending to control the liquidity risks.

2.3 Improvement of the lending project to cut the excess liquidity

According to Diamond, D. W. & Rajan, R. G. (2001), it said that commercial banks not only made loans for illiquid borrowers, but also supported current capital for depositors' demands. In addition, Diamond, D. W. & Dybvig, P. H. (1983) told that banks could transform illiquid assets into more liquid demand deposits. All of these situations showed that it might be caused lots of excess liquidity problems.

During the period between 2008 and 2013, the amount of bank savings deposit rapidly raised, especially in 2010. In 2010, the growth of savings was more than 85% (Fig. 3). Meanwhile, deposits by governments and enterprises would go up quite substantially as well. Under the influence of liquidity short-term accumulation, CCB and other commercial banks had the sufficient capital, so banks had more opportunities to make loans.

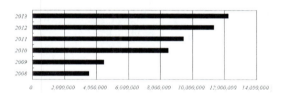

Figure 3. Savings deposit (million) of CCB during 2008 and 2013*.
* All the data from China Construction Bank's annual report 2008 to 2013.

In fact, excess liquidity problems might make the bank achieve the sufficient capital, attracted deposits easily and had lower capital cost. But it still had potential and serious consequence for a long period. CCB must look for the proper lending without non-performing loan.

In order to decrease excess liquidity problems, CCB adopted to increase the fiscal expenditure in the public. Not only purchased many treasury bonds, corporate bonds or financial bonds, but also carried on a variety of businesses, such as trust and investment. It was necessary to invest in social security, health care, education and environmental protection for expanding domestic demands, so that improved future expenditure to reduce the current high savings rate. For instance, CCB strictly enforced a "single-vote veto system" to give loan support the environmental protection. As at 31 December 2012, the loans of green credit project reached RMB 1.27 trillion, exhibiting a rise of 21.43%. Meanwhile, the balance of small business loans had already reached RMB 0.75 trillion from the beginning of the year.

2.4 Developing the secondary loan market

From term structure, the loans of fixed asset investment, project and personal consumption belonged to long term loans, this kind of loans needed to a quite long time to return. The proportion of long term loans to all loans was increasing rapidly, on the one hand, it showed that lending range had extended and the scope of profit increased to ensure the rapid development of commercial Banks. On the other hand, because of the large increase in long-term loans, it reduced the liquidity to decrease the ability of funds controlling. A large amount of cash assets existence resulted in restricting funds to adequately use and weakening the banks' profitability, at last it was the outcome of reducing the bank's business efficiency.

CCB had the advantage of high quality assets, which has better quality of credit assets, so it had great opportunities to develop the secondary loan market. Developing the secondary loan market was able to deal with this problem above. Bray, M. (1984) supported that it's broken down traditional barriers in developing the secondary loan market, and also could develop new products to meet the need for more liquid assets. Meanwhile, it transferred the liquidity risks to reduce the ratio of non-performing assets. It was also benefitted for CCB to use its capital, which increased the fund efficiency. Secondary loan market maintained the balance relationship between liquidity risk and profitability. CCB only made sure the enough funds to liquidity, the remaining funds could provide for lending to increase the profit level. Once the financial environment changed that CCB was difficult to pay, the loans were transferred into money to make sure the payment requirement, thus improving the asset liquidity management.

3 CONCLUSION

Nowadays, with the improvement of open economy and deepening of economic system, commercial banks must meet the various risks; the most risks are like liquidity risk, credit risk and interest rate risk. China Construction Bank (CCB) gradually establishes and improves risk management, especially to pay more attention on liquidity management. Effective liquidity can involve in managing cash assets to maintain sustainability, so the bank has the ability to achieve maximum benefits at minimum cost to improve the profitability. The most important is liquidity management can help banks to fight with the financial crisis.

For China Construction Bank (CCB), there are four best strategies concerning liquidity management. Fist of all, the strategy of balance management of assets and liabilities can transfer the situation of the liquidity liabilities are greater than liquidity assets, to preserve the liquidity stability. Then, enhancing the real estate credit management might be helpful for CCB to sufficiently supervise the process of the real estate credit, make sure to make the full use of the funds and avoid the liquidity worsening results. Next, to a certain extent, improved the lending project can cut the excess liquidity problem, so that improve future expenditure to reduce the current high savings rate and CCB achieve more profits. Finally, developing the secondary loan market not only can deal with non-performing loans, but also can financing through the loan transaction. Meanwhile, the loans can be easily transferred into money to improve the asset liquidity management, so it is able to avoid the liquidity risks problems.

REFERENCES

Bray, M. 1984. Developing a secondary market in loan assets, International Financial Law Review (October 1984): 22–25.
Diamond, D. W. & Dybvig, P. H. 1983. Bank Runs, Deposit Insurance, and Liquidity, J.P.E. 91 (June 1983): 401–19.
Diamond, D. W. & Rajan, R. G. 2001. Liquidity risk, liquidity creation and financial fragility: A theory of banking, Journal of Political Economy 109(2): 287–327.
Radelet, S. & Sachs, J. 1998. The Onset of the East Asian Financial Crisis, Harvard University, February 1998.
World Bank, 2000. Global Economic Prospects and the Developing Countries, World Bank, Washington, DC, USA.

Joint BF-CFS and AdaBoost-REPTree for traffic classification

Yan Shi, Min Wang & Bing Yao Cao
Key Laboratory of Specialty Fiber Optics and Optical Access Networks, Shanghai University, China

ABSTRACT: Network traffic classification is an important means for network administrators to supervise the traffic flows in order to manage the network efficiently. Therefore, accurate classification of flow is of great significance. The key of flow classification lies in the selecting proper features and classifier's efficiency and accuracy, this paper proposed a practical traffic classification solution, which is the joint of BF-CFS and AdaBoost-REPTree. In this solution, in order to determine the optimal flow feature subset, we used Best-First algorithm as the search strategy and evaluate with correlation-based feature selection (BF-CFS), and finally determine 11 feature subset out of 248 feature set space as the optimal feature subset successfully. Considering classifier's efficiency and accuracy, we proposed a new classifier which will take REPTree as the base classifier and boost the REPTree with AdaBoost boosting algorithm, to produce a strong classifier (AdaBoost-REPTree). Experiments indicate that this solution is highly accurate and efficient for the classification of our traffic flow samples, thus, it is quite practical and usable.

KEYWORDS: traffic classification; CFS; best-first; BF-CFS; REPTree; AdaBoost; AdaBoost-REPTree

1   INTRODUCTION

Traffic classification is the basis of traffic statistics, traffic monitoring, intrusion detection and quality of service (QoS). Therefore, classifying network traffic accurately is significant.

The main approaches of traffic classification include: port-based techniques, payload-based techniques and flow-based techniques. The port-based techniques are efficient and easy to implement since they assume the application uses the well-known port numbers assigned by IANA. However more and more applications use dynamic port numbers, some applications may even share the same port. Thus, port-based techniques will not work. Payload-based techniques classify traffic by checking the application's unique signatures extracted from packet payload. They can overcome the problems of port-based techniques with high classification accuracy, but they are not efficient since they need to analyze the packet payload. And they cannot deal with encrypted applications. What's worse, they involve users' privacy. In order to address the problems of port-based and payload-based techniques, flow-based techniques were proposed by researchers. With the assumption that flows generated by applications have unique statistics characteristics, flow-based techniques can discriminate the flows. They are efficient and practical since they analyze the flow statistical features instead of the packet payload signatures. Therefore, flow-based techniques become the research direction of traffic classification.

Machine Learning (ML) algorithms are widely used for flow-based classification.

Moore et al. in [1] presented an approach based naïve Bayes classifier to solve the classification problem of TCP traffic. Yang in [2] presented an approach based on C4.5.

The challenges of flow-based classification are selection of flow statistical features, selection of classification models and so on. These challenges determine the accuracy and efficiency of the classification, which is important for a usable and practical traffic classification system. In this paper, we proposed a new and practical flow-based traffic classification with BF-CFS and AdaBoost-REPTree solution. This solution adopts CFS[3] evaluation function for Best-first[4] search algorithm (BF-CFS) as the feature selection solution, and uses REPTree as AdaBoost's base classifier to classify the traffic flows which were filtered by BF-CFS.

2   OUR APPROACH

The architecture of our classification system is shown in figure 1. As is shown in the figure, the system includes three stages: feature selection stage, training stage and classification stage. The feature selection stage will select the best feature subset of the traffic flows, and the feature subset is used as the input to train the AdaBoost-REPTree classifier, which is the training stage. Then, we will apply the trained AdaBoost-REPTree classifier to unknown application flows, and classify these flows correctly.

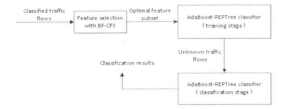

Figure 1. Architecture of the classification system.

2.1 Feature selection with BF-CFS

Traffic feature selection is the key of flow-based traffic classification since it directly determines the performance of classifier's efficiency and accuracy. In this paper, training data used for traffic classification is from Moore [5] et al., who collected and classified these traffic flow manually. These training data are very popular, widely used for research purposes, and called "MooreSet" generally. The MooreSet includes traffic flows of ten applications [1], which are BULK, DATABASE, INTERACTIVE, MAIL, SERBICES, WWW, P2P, ATTACK, GAMES and MULTAIMEDIA. Each sub dataset of MooreSet has more than 20000 traffic flows, and each flow contains 248 feature and a label attributes (which marked the application type of the flow). Among these features, more than 100 features are obtained by Fourier transformation [6] so some of the features will be redundant and irrelevant. Thus, we need to select a proper and optimal feature subset and remove the redundant and irrelevant features, to decrease the computational complexity and increase the accuracy of the classifier, which is the challenge and key of flow-based classification. This paper adopts BF-CFS as the feature selection solution, to select the optimal features subset and reduce dimension of the dataset. In order not to rely on the port number, the port features (the 1, 2 features in MooreSet) will be neglected.

2.1.1 The impact of feature numbers on classifier's efficiency

We did some experiment to study the impact of feature numbers on the classifier's efficiency in this paper. We used different number of traffic flow features to test the build time of AdaBoost-REPTree classifier. The results are shown in Figure 2:

As is shown in Figure 2, with the increasing of feature numbers, the classifier's build time also increase rapidly. It concludes that the computational complexity of the AdaBoost-REPTree classifier will be impacted by feature numbers significantly, which will make the classifier inefficient and unusable. Therefore, the feature selection is critical to the efficiency of the traffic classification system.

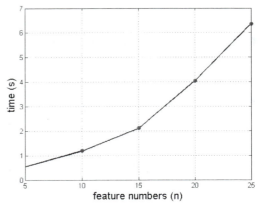

Figure 2. The build time of classifier vs the feature numbers.

2.1.2 BF-CFS algorithm

This paper proposed below BF-CFS framework (as shown in Figure 3) for feature selection, which is proved to perform well for traffic classification. We firstly used Best-First to generate the subset of the traffic features, and then evaluate the feature subset with correlation for the feature selection (Correlation-based feature selection, or CFS).

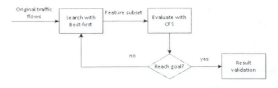

Figure 3. BF-CFS framework for feature selection.

The flow of BF-CFS algorithm is as below:

Step1: calculate CFS value of each single feature and add best N features (with higher CFS evaluation) into an initial priority queue

Step2: Select the best feature subset from the priority queue, if the CFS value fulfills target goal, return the subset

Step3: Expand the subset with a new single feature and re-evaluate CFS value

Step3.1: If the new value is no better than previous, back trace and go to Step3.

Step3.2: Otherwise, adjust the priority queue, go to Step2

The CFS value is calculated with formula (1):

$$G_s = k\overline{r_{ci}} / \sqrt{k + k(k-1)\overline{r_{ii}}} \tag{1}$$

In the formula, k is the number of features in the subset, $\bar{r}_{ci}$ is the mean feature correlation with the class, r_{ii} is the average feature inter-correlation, and G_x is the final evaluation result. So if the features in the subsets are highly correlated with the class and lowly correlated with each other, then this feature subset is good.

Using BF-CFS algorithm, we finally select 11 optimal features out of 248 features. The number of the features in MooreSet are: 60,66,79,83,84,95,96, 98,101,133,137.

2.2 AdaBoost-REPTree classifier

AdaBoost is an adaptive boosting algorithm first posed by Freund et al [7]. It was originally used to address binary classification problem, and then many varieties were raised to solve multi-classification problems, such as AdaBoost OC, AdaBoost ECC, AdaBoost M1, AdaBoost M2 and AdaBoost MH [8, 9]. The core idea of AdaBoost is training different weak base classifiers with the same training set, and then combines these weak classifiers to construct a stronger classifier with corresponding weights. Theoretically, as long as each weak classifier outperforms random guess, the error rate of the combined strong classifier will be close to 0 if the number of weak classifiers tends to infinity [7]. The advantages of AdaBoost are: it is a classifier with high accuracy; it provides a framework and different base classifiers can be used to improve the accuracy; it has a low possibility of over-fitting.

In supervised machine learning, decision tree algorithm is very simple, smart and easy to implement. It is widely used for classification due to its fast construction and high efficiency. For complex decision tree, we need to prune it, to avoid over-fitting for the traffic flow samples (an over-fitting decision tree has a higher error rate of traffic classification than simplified decision tree after pruning). Thus, the pruning of decision tree is the foundation to optimize the computational efficiency and classifying accuracy. This paper implements the REPTree decision tree algorithm, which has the post-pruning functionality. The principle to delete a tree node in REPTree is that, the tree after pruning will always outperforms the original tree. REPTree is proved to perform well from the aspects of both classifying accuracy and efficiency, on our traffic flow samples.

We proposed AdaBoost-REPTree classifier for traffic flow classification, considering the advantages of REPTree and AdaBoost. The implementation of AdaBoost-REPTree is as below.

We use x to represent the traffic flow samples with the features selected by BF-CFS, from MooreSet. And y is a set of the corresponding traffic flow application labels. Thus, $S = \{(x_i, y_i) | i = 1, 2, .., n\}$ is our training set for AdaBoost-REPTree, in which $x_i \in x$, $y_i \in y$, $y = \{1, 2, 3, ..., k\}$, k is the numbers of application labels, which is 10 for MooreSet.

1 $D_1(x_i, y_i) \leftarrow 1/n$
2 For $t \in [1,T]$ do (T is the number of iterations)
$\quad S_t \leftarrow$ Sample(S) // *create temporary training set S_t with distribution D_t on traffic flow set S*
$\quad h_t : x \rightarrow y$ // *classify S_t with REPTree base classifier, thus get the base REPTree classifier h_t*

$$\varepsilon_t \leftarrow \sum_{h_t(x_i) \neq y_i} D_t(x_i, y_i) \text{ // calculate the error rate}$$
of h_t

$$\alpha_t \leftarrow (1/2) \ln[1 - \varepsilon_t / \varepsilon_t] \text{ // calculate the weight of}$$
h_t in $H(x)$

$$D_{t+1}(x_i, y_i) \leftarrow \frac{D_t(x_i, y_i)}{Z_t} \times \begin{cases} e^{-\alpha_t}, & h_t(x_i) = y_i \\ e^{\alpha_t}, & h_t(x_i) \neq y_i \end{cases}$$

// *Z_t is the normalization factor, which is used to ensure $\sum_i D_{t+1}(x_i, y_i) = 1$*

3 $H(x) \leftarrow sign(\sum_{t=1}^{T} \alpha_t h_t)$ // *$H(x)$ is the final classifier*

3 IMPLEMENTATION RESULTS

3.1 Evaluation metrics

This paper evaluates our joint of BF-CFS and AdaBoost-REPTree traffic classification solution with accuracy and efficiency. The efficiency is measured by the build time of the classifier. To evaluate the classifier's accuracy, we introduce precision and recall metrics, as defined in (2) and (3) respectively. TP (True Positive) is the number of flows correctly classified as class c, FN (False Negative) is the number of flows of class c incorrectly classified as other classes and FP (False Positive) is the number of flows of other classes predicated as class c incorrectly.

$$Precision = TP / (TP + FP) \tag{2}$$

$$Recall = TP / (TP + FN) \tag{3}$$

3.2 Experiment results and analysis

This paper experiments with 5 datasets filtered by BF-CFS from the first 5 original datasets in MooreSet. To ensure the accuracy of the testing algorithm, we adopt the 10-folder cross validation methods. It divides the dataset to 10 portions, then, takes 9 portions as the training data for the traffic classifier, and 1 portion used as the testing data for validation, in turns. Finally, calculate the mean of all the 10 accuracy values, and use it as the evaluation of the classifier's accuracy. The results are shown as Table 1.

Table 1. The traffic classification performance of BF-CFS and AdaBoost-REPTree.

Dataset	Precision (%)	Recall (%)	Build Time (s)
Set01	98.9	98.9	0.79
Set02	99.3	99.3	1.54
Set03	99.2	99.3	1.42
Set04	99.1	99.1	1.46
Set05	99.2	99.2	0.92
Average	99.14%	99.16%	1.226

As is shown in Table 1, the joint of BF-CFS and AdaBoost-REPTree traffic classification has a very high accuracy in terms of both precision and recall, and is also efficient in terms of build time (average is about 1.226s, training on over 20000 flows). Additionally, the classifier performs stably on different datasets. Therefore it can be well applied to the real time traffic classification.

In order to illustrate the boosting of REPTree by AdaBoost, we did an experiment on Set05 using REPTree classifier and AdaBoost-REPTree classifier. The results are shown in Table 2:

Table 2. The traffic classification using AdaBoost-REPTree vs REPTree.

Classifier	Precision (%)	Recall (%)	Build Time (s)
REPTree	97.9	98.1	0.23
AdaBoost-REPTree	99.2	99.2	0.92

From Table 2, the traffic classification accuracy has improved from 97.9% to 99.2% after boosting REPTree by AdaBoost. It indicates that AdaBoost can improve REPTree classifier.

Compared with paper [1] and paper [2], our classification system has higher accuracy. And compared with paper [10], it uses GA_CFS to select 17 flow features but our BF-CFS select only 11 optimal features and achieves even higher accuracy. From the aspect of build time, it costs about 32 second, but we only take about 1 second. Therefore our joint of BF-CFS and AdaBoost-REPTree classification system outperforms and is more practical.

For further validating the performance of AdaBoost-REPTree classifier, we apply our AdaBoost-REPTree classifier to the 17 features selected by paper [10] on the same machine the efficiency comparison is shown in figure 4 (the horizontal axis represent the dataset, the vertical axis represent the build time of classifier). As is shown in the figure, the build time of our AdaBoost-REPTree classifier is about 1/3 of the AdaBoost+C4.5 classifier used by paper [10], which means our efficiency is 3 times better. And compared with the results of table 1, which used the flow features selected by BF-CFS, we can conclude that BF-CFS has selected optimal features and it helps to reduce the build time of AdaBoost-REPTree classifier.

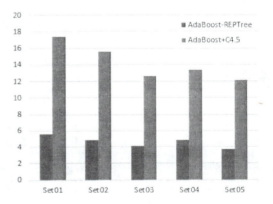

Figure 4. The comparison of efficiency for AdaBoost-REPTree vs AdaBoost+C4.5.

4 CONCLUSION

Selection of flow statistical features and selection of classification models, which will determine the accuracy and efficiency of the classification, are the two important challenges for flow-based classification. Thus, this paper proposed a new and practical classification solution with BF-CFS as feature selection module and AdaBoost-REPTree as classifier module. In the BF-CFS module, Best-first is used to search the minimal and optimal feature subset from a set of traffic flow features. And CFS is used to evaluate the feature subset selected by Best-first, finally the optimal feature subset is generated with only 11 features out of 248 flow features. Then the optimal feature subset is used as training set to train the AdaBoost-REPTree classifier. AdaBoost-REPTree is built on REPTree with the boosting of AdaBoost to improve the classification accuracy.

A good balance of accuracy and efficiency is important for practical traffic classification system.

Experiments indicate that our joint of BF-CFS and AdaBoost-REPTree solution can reach higher than 99% accuracy, and at the same time, has a high efficiency with the build time of about 1 second on over 20000 traffic flow samples. Also, the results are quite stable across different traffic flow datasets. Therefore the joint of BF-CFS and AdaBoost-REPTree traffic classification solution is practical with high accuracy, efficiency and stability.

REFERENCES

[1] Moore A, Zuev D. "Internet traffic classification using bayesian analysis techniques." ACM SIGMETRICS Performance Evaluation Review. Vol. 33. No. 1. ACM, 2005.

[2] Yang B, Hou G, Ruan L, Xue Y, and. Li J. "SMILER: Towards Practical Online Traffic Classification," inProceedings of the 7th Architectures for Networking and Communications Systems, 2011.

[3] Hall MA, Smith LA. Practical feature subset selection for machine learning[J]. 1998.

[4] Lu B, Liu Z. Prolog with best first search[C]//Control and Decision Conference (CCDC), 2013 25th Chinese. IEEE, 2013:917–922.

[5] Moore A, Zuev D, Crogan M. Discriminators for use in flow-based classification[M]. Queen Mary and Westfield College, Department of Computer Science, 2005.

[6] Moore A, Zuev D, Crogan M. Discriminators for use in flow-based classification[M]. Queen Mary and Westfield College, Department of Computer Science, 2005.

[7] Souza D, Erico N, Matwin S, Fernandes S. "Network traffic classification using AdaBoost Dynamic." Communications Workshops (ICC), 2013 IEEE International Conference on. IEEE, 2013.

[8] Schapire R, Singer Y. Improved boosting algorithms using confidence-rated predictions[J]. Machine learning, 1999, 37(3):297–336.

[9] Friedman J, Hastie T, Tibshirani R. Additive logistic regression: a statistical view of boosting (with discussion and a rejoinder by the authors)[J]. The annals of statistics, 2000, 28(2):337–407.

[10] La T.T, Shi J. Network traffic classification based on GA-CFS and AdaBoost algorithm[J]. Jisuanji Yingyong Yanjiu, 2012, 29(9):3411–3414.

Information, Computer and Application Engineering – Liu, Sung & Yao (Eds)
© 2015 Taylor & Francis Group, London, ISBN 978-1-138-02717-6

The ways of enhancing applied competence of engineering students in newly founded colleges

Jing Sun & Ying Song
Harbin University of Science and Technology, Weihai, Shandong province, China

ABSTRACT: This paper summarizes the status of newly built undergraduate course colleges and universities graduate students in engineering, and puts forward some ways to strengthen its cultivating applied ability.

KEYWORDS: Newly built undergraduate course colleges and universities; Engineering Students; Application ability

Engineering is to adapt to the development of new undergraduate colleges an important way of regional economic and social development, but also the urgent requirement of local government proposed the University in the region. Newly built undergraduate course colleges and universities the school time is short, the training of engineering students cannot be carried out according to the traditional ordinary university or research university model, we must proceed from the school of history and reality, and regional economic and social development closely, pay attention to the cultivation of strengthening the ability of application type.

1 THE PRESENT CONDITION OF ENGINEERING OF NEWLY-BUILT UNDERGRADUATE COLLEGE STUDENTS TRAINING

1.1 *The pressure of learning, the complexity of learning subjects and learning tasks*

The professional course knowledge points, a high degree of difficulty, the homework task, before and after the knowledge of cohesion is compact, students need to learn more on weekdays on input energy, not only to pay attention in class, but after class but also by doing lots of exercises to consolidate. In addition, in addition to the theoretical teaching, experiment course of certain hours is necessary.

1.2 *The subjective desire to communicate with people and the lack of favorable objective environment*

The engineering students of the newly-built undergraduate college were at the intermediate level of learning in the high schools. In the university, they are more eager to improve their various aspects ability. The new undergraduate colleges 90% engineering students from rural areas, lack of education and guidance of parents to their contacts with the people in the personal growth of family education, cause they want to contact the society, but the lack of effective methods and way. Many engineering students ineloquent, communication range is small.

1.3 *The chance of practice will be higher, pay attention to practice ability, and be willing to participate in various forms of practice, because practice is an important way to improve the ability*

Newly established undergraduate college engineering students have many opportunities of participating in professional practice, but the involvement outside of the professional social practice is less. Generally believe that the personal, practical ability is not strong. Many students participate in social practice is limited in the school, and the form is single, the lack of rich and colorful social practice form.

1.4 *Do not pay enough attention to the planning of their own*

Newly established undergraduate colleges of Engineering aware there must be planning to develop, but with the popularization of employment guidance course, the majority of students is recognized to have their own life planning to achieve their desired objectives, but because self-control is not strong wait for a reason, he will be not in accordance with the planning and implementation of the plan into action.

1.5 Addicted to the virtual world of the more prominent problems

With the popularization of computer and network in Colleges and universities, more and more college students addicted to online games, and caused many problems. Engineering students more easily lead to a network game addiction phenomenon due to environmental, psychological and personality differences and even some because of indulging in online games and the emergence of absence, waiting up phenomenon.

2 WAYS TO STRENGTHEN THE TRAINING OF NEW UNDERGRADUATE COLLEGES ENGINEERING STUDENTS' APPLIED ABILITY

2.1 Strengthen the construction of style of study, the theoretical basis of tamping the specialty

Strengthen students' entrance education, to consolidate students' professional and ideological education to promote the style of study. Student education stage is a key period for the construction of style of study. A lot of engineering students didn't know much about the profession which he studied after the admission. The professional thought is uniform. Therefore, students enrolled in education, we should pay attention to professional ideas and the best students and students to share professional learning experience, invite occupation guidance teachers to students to develop occupation career planning education, and organize the students to visit the laboratory and experimental practice bases and other ways, let the students have a general understanding of the professional school, the consolidation of the professional thought, grasp learning method science, make university career planning, clear learning objectives of University, strengthen the construction of style of study.

Actively carry out professional competitions, taking subject competition promote the style of study. We should organize students participate in various academic competitions, develop and enhance students' innovative spirit and practical ability, in the process of the competition to make students like a professional, promote students to take the initiative to explore the professional knowledge. First fully mobilize all grades outstanding students to join and guide them to effectively participate in various scientific and technological innovation activities; The second is to expand the competition of resources, strengthen cooperation with specialized laboratories, extending laboratory development time, to encourage students to actively participate in laboratory research, the research team to go into teaching, so that students in extra curricular science and technology activities carried out at a higher level.

Advocated the building of good class atmosphere, promote the style of study in class. Class is the basic unit of study style in Colleges and universities, the class construction is the cornerstone of the construction of style of study. To create a good learning environment, we must start from the class style of study, starting grabbed from the daily work. First, to establish supervised learning, instructors regularly check student absenteeism. Then, leave early and so on, to strengthen the communication with the teacher, understand the students in class and after class performance. Secondly, according to the characteristics of students' learning consciousness is poor, grade level organizations to carry out the reading in the morning, night activities, and to the class as a unit, called on all the students to actively participate. Finally, to support the students to engage in a colorful, healthy and beneficial class activities, we should create hard-working, good class atmosphere of solidarity. At the end of each school, according to each class in the test scores, violations of the law and discipline, attendance, awards, etc., conduct supervision and appraisal to each class construction of style of study, comprehensive performance in all aspects of the selection of advanced classes and advanced League branch, the granting of recognition awards at grade level.

"Grasp both ends, to promote the middle", Management education promotes the style of study. To highlight the power of example, play an exemplary student party members, party activists and students cadres in the lead role, strengthen guidance and education, let the students do the example cadre style construct. Meanwhile, in order to highlight the importance of learning, student achievement can be appraised with various awards, linked to the development of party members; Secondly, to strengthen the guidance for students with learning difficulties, mental health education students in difficulty; To make full use of various media, vigorously promote the construction of study style, to create a good learning atmosphere, such as the selection of the top ten per semester learning star or excellent learning model, and through publicity column, briefing, micro-blog, Micro message and other ways, actively promote the advanced deeds, let the outstanding student struggle deeds infection surrounding classmates.

To promote integrity in learning, serious examination discipline, in order to promote the study style of examination. Aiming at working copy homework, examination cheating phenomena existing in engineering students, instructors should not to mind taking the trouble to faith education of students.

Strengthening the cooperation between colleges and enterprises, to provide for the improvement of students' practical ability cultivation of applied type ability good social environment is a project cooperation, from teachers to practice site need to cooperate with enterprise, society, build a platform for college students' social practice, production practice and innovation practice activity, make the process of learning from the classroom to the workplace, to learn how to learn, use of knowledge is seen as an important part of students' learning, encourage students to explore the unknown, the courage to practice, meet the challenge, training to improve their practice, and coordination and cooperation more its personality perfect. The government, education department should make relevant policies to encourage enterprises, social forces and school communication and cooperation, the establishment of the school enterprise cooperation more and practice base construction practice platform, the lifting type application ability of college students become the responsibility of the whole society; social practice is an important platform to improve the students ability of application type of social practice, but also often faced with no social support and other problems, the need for students the social practice activities for more social care and support, to improve the students encounter in the social practice of the employing units do not receive, not allowed to wait for a problem, with the care and help provide practice opportunities and platform for the College Students, to improve their ability of social practice.

2.2 *The transformation of the teaching method of education, to provide a good learning environment for the improvement of students' practical ability*

Constructing cultivation scheme and engineering requirements of the teaching program of three-dimensional, diversified skills based on, through the integration of "vocational skills competition" and "applied type ability big lecture room", "non professional quality contest", "social practice" activities such as resources, guides the student to set up the professional clear development direction. In the aspect of curriculum, we should prepare students to enhance the ability of "application type" professional textbooks, break through the traditional education as the leading to the phenomenon of teachers, students centered, outstanding student's main body status, to the real task oriented curriculum system design based on working process; In the aspect of teaching, changing the traditional focus only on scores way of education, to the "ability standard" as the guiding ideology, deepen the teaching reform, with applied ability development for the purpose of constructing the system of practice teaching ability; Management students, the construction of evaluation system of students' application type ability, teachers guide students open,

fair evaluation, implementation of "science and technology innovation mode of tutor + students", by a tutor on a certain innovative experiment, led several students to carry out in-depth study lasting learning, improve student learning and innovation ability; In campus culture, integration and use of campus resources, creating a workplace environment, promoting the students' ability of general application form and development, for students to provide a better atmosphere, exerting influence of recessive courses to students applied ability of leading role; In the aspect of practice, aiming at the phenomenon of the past practice and theory from the practical teaching should make full use of the professional engineering conditions, social practice should make full use of local colleges in the regional advantage and industrial advantage, make the practice teaching and theory teaching in the formation of organic links. Teaching methods through series of education, enhance student team consciousness, subject consciousness, cooperation consciousness to enhance and strengthen personal psychological quality in study, work and life, things in new ways of thinking have understanding, enhance the confidence to overcome difficulties and the courage.

2.3 *Actively improve their practice ability*

Create opportunities for improving their practical ability, college students should take the initiative to out of school to learn the courses at the same time, participate in the social investigation, make the combination of theory and practice, find the problems, think and solve the problems in the social practice, and in practical activities in the timely feedback, to exercise their own ability and deal with the problem of communication with people ; Pay attention to the social practical activity of college students, and actively participate in the college students' quality development training, entrepreneurship training camp, the ability to enhance the construction of the application oriented training camp; To take the initiative to create an active atmosphere for innovation, based on reality, carefully plan their career, timely find their career anchor, detours, fully prepared to prepare professional challenges as soon as possible in the idea, attitude, knowledge, ability and quality etc; At the same time in the practice of college students also reflect on their own shortcomings, and continue to learn and accept new knowledge, learn from each other, seriously sum up experience, gradually cultivate the learning and developing their own; Students need to cultivate the scientific attitude of learning habits, habits of thought and science, also must continue to accumulate basic knowledge to lay a solid foundation, and is good at considering the problem of the use of reverse thinking to improve the ability of innovation.

REFERENCES

[1] Qiuxiang Tao, Manlan Deng, Students of engineering characteristics and management research of [J]. Science Wenhui, 2007, (6): 95
[2] Jinjun Fu etc. "Student activity theory under the background of quality education of university extracurricular education", Science Press, 2008.
[3] Changxi Huang, Lei Ye, Xinjie Han. "Engineering students' practical ability and the cultivation of innovation spirit and the whole process of the enterprise employment practice", "China Adult Education", 2012 fourteenth.
[4] Yan Sun. The construction of college students of science and engineering construction of style of study of the new mode of thinking [J]. Journal of North China Institute of water conservancy and hydroelectric power (SOCIAL SCIENCE EDITION), 2012, 8 (4): 160–162.

Information, Computer and Application Engineering – Liu, Sung & Yao (Eds)
© 2015 Taylor & Francis Group, London, ISBN 978-1-138-02717-6

Switching cost, perceived value and repurchasing intention—detecting the experiences of Taobao buyers' online-buying

B. Ge
Guangxi University of Economics and Finance, Nanning, China
Huazhong Agriculture University, Wuhai, China

G.Z. Mo & Q. Fang
Guanxi University of Economics and Finance, Nanning, China

W.B. Zhou
Guangxi Industrial and Commercial School, Nanning, China

ABSTRACT: The objective of this study was to explain why most of Taobao buyers who had the failure experience would still repurchase in Taobao. Participants, who were all Taobao users, completed questionnaires to assess their perceived value, switching costs, customer satisfaction and repurchasing intentions. After data analysis was conducted by SPSS 17 and AMOS17.0, the results obtained revealed that switching cost is the key factor. The results suggest that for the leading enterprise, there was a good switching cost strategy of constructing a business ecosystem.

KEYWORDS: Customer perceived value; Switching costs; Customer satisfaction; Repurchasing intentions; Online shopping

1 INTRODUCTION

The stock of Alibaba Group was successfully published in America shortly before. Alibaba Group subsidiary—Taobao.com is currently the largest online shopping platform in China. Since Taobao established in 2003, its registered users are constantly growing. Taobao statistics showed that in 2013, Taobao has nearly 500 million registered members and more than 60 million regular visitors in a day. According to the Scoreboard announced by an internet laboratory on the e-commerce site, Taobao ranked first as early as 2004, far more than the eBay and other shopping sites. Although the buyers shopping on Taobao have mostly bought some inferior goods or been suffered by bad service, their enthusiasm in online shopping has not been diminished. In the past two years in Taobao's "Double eleven" and 'double twelve" sales promotion period, Taobao's transaction volume is still showing impressive growth, and more and more buyers buying frequently on Taobao.com.

Why buyers were willing to visit and buy on it again and again, though goods and services provided by Taobao were not good enough? Referring to some related research, it may be the customers' switching cost was a rather important factor affecting customer satisfaction and repurchasing intentions.

2 LITERATURE REVIEW

2.1 Customer perceived value

Since 1990s, customer perceived value has become widely used in the field of corporate marketing, Zeithaml (1990) [1] and Monroe (1991) [2] proposed customer perceived value was what customer evaluated after using the product about its perceived loss and profits.Butz (1996) [3], Woodruff (1997) [4] and Litian Liang (2012) [5] all consider the customer perceived value concerned how customer perceived and assessed about the attributes (product quality, service quality, etc.)of product to meet his preference after purchasing. Customer perceived value which is subjective, is a kind of evaluation made by the customer based on the products or services. In the research, customer perceived value is becoming a catch-all construct that can explain all the reasons for buying. In order to reveal the real and specific reasons for buying, it was defined as the customer's cognitive appraisal and evaluation on the price and the attributes of a product or service which could satisfy their preferences in this reach. Besides, product property referred to product quality, service quality, information authenticity and safety, customer perceived value refers to the evaluation made by customers about these four product attributes.

2.2 Switching cost

"Switching costs" was first proposed in 1980 by Michael Porter, referring to the one-time cost when the buyer gived up his past seller. There were many discussions on swiching costs classification in the academic circles. It contained the meaning of actual cost (such as transportation costs, information costs), and the cost of the buyer's time and vigor, as well as the psychological process of adapting to the new products or services. Relatively representative views are more concentrated in the foreign scholars as the following table.

Synthesizing the views, switch costs, mainly include transaction costs, customer habits, communication convenience, etc.

Table 1.   Classification for switching costs.

Classification	Scholars
1.Search costs 2.Learning costs 3. Transaction costs 4.Cognitive effort 5. Customer Loyalty Discounts 6. Financial, social and psychological risks 7. Emotional costs 8. Customer habits	Fornell(1992)[6]
1.Opportunity cost 2.Risk costs 3.Behavioral and cognitive costs before conversion 4.Behavioral and cognitive costs after switching 5. Set-up costs 6. Sunk costs	Jones et al.(2002)[7]
1.Program conversion costs 2.Financial switching costs 3.Relationship switching costs	Burnham et al. (2003)[8]

2.3 Customer satisfaction

Westbook (1980)[9] pointed out that customer satisfaction was the overall satisfaction of the consumption experience accumulated. Earlier studies in academia generally agreed with that customer perceived value leading to customer satisfaction, and customer satisfaction and repurchase intention or customer loyalty have a high correlation.(Anderson & Sullivan, 1990) [10]

2.4 Repurchasing intention

The customer's repurchasing intention is a reliable indicator for measuring customer repurchasing behavior. Customer's repurchasing intention means customer formed intention of using the same brand product or buying from the same sailor or buying on the same website after his previous shopping. (Jones, 2002) [11].

Both customer loyalty and repurchasing intention would promote customers repurchasing behavior; there was no difference in the actual measurement. Customer loyalty emphasized the customer emotional commitment. Thus the theory of customer loyalty makes a distinction between the "behavioral loyalty" and the "emotional loyalty".However repurchasing intention had nothing to do with emotion, it can objectively describe the phenomenon. Thus construct of repurchasing intention will make the model more concise.

2.5 The relationship of switching costs, customer satisfaction and repurchasing intention

The existing researches rarely focused on the direct effect between the switching cost and customer satisfaction. Most of them agreed customer perceived value was the decisive factor of customer satisfaction, and switching cost was a mediator between customer satisfaction and repurchasing intention. As Jones' (2002) [12] study confirmed, customer satisfaction had a weak effect on customer loyalty when switching cost was higher; at the same time, customer satisfaction had a strong effect on customer loyalty when switching costs were low.

Jones' model was the occupied idea that was consistent with the logical thinking of people: cost could not make people satisfied. Could it be possible that the satisfaction degree of buyer's had been enhanced when the buyers had no other choice or the switching cost was too high? In the real world, it can be observed that most of people are willing to compromise with the reality rather than resistance; they tend to keep mind consistent with their behavior to keep their spiritual health. If the buyers repurchase or keep their loyalty to specific product or seller or website just because people are so "lazy" to change, they will be satisfied. In that case, switching cost is likely to influence customer satisfaction directly.

3   THE HYPOTHESIS

According to the literature review, customer perceived value has a positive effect on customer satisfaction, customer satisfaction also had a positive influence on repurchasing intention. Despite the fact that TAOBAO buyers had experience in shopping failure (low customer perceived value) but still repurchased suggested that there might be other factors which caused repurchasing. Such other factors can probably be the relatively low transaction costs of searching, learning, negotiation, purchasing executive, returning goods, etc. Besides, online-living habit of customers once formed, it was not easy to change. Moreover, TAOBAO provided a great platform of communication – Ali Wangwang with which TAOBAO sellers could rapidly respond the buyers' needs. Therefore, switching costs and perceived value both affected customer satisfaction and customer repurchase intention.

In our model, customer perceived value and switching costs serve are independent variables, and the repurchasing intentional acts as the dependent variable, together with customer satisfaction acting as mediator. Therefore, we proposed the following hypotheses in the context:

Hypothesis 1: Perceived value will be positively relate to customer satisfaction. Hypothesis 2: Customer satisfaction will positively relate to repurchasing intention. Hypothesis 3: Switching costs will positively relate to customer satisfaction. Hypothesis 4: Switching costs will positively relate to repurchasing intentions of customers. Hypothesis 5: Switching costs' impact on customer satisfaction will be much higher than perceived value. Hypothesis 5: Switching cost has a direct impact on customer repurchase intention.

4 METHODS

4.1 Procedure

Before the formal survey, 10 experienced Taobao buyers were asked to respond the questionnaire to ensure that the words and sentences can be correctly understood and answered. Then a survey was conducted on 'Questionnaire Star Site' from Feb.15 to Mar.15 in 2014.

A total of 298 questionnaires was collected in the survey and 266 of them were valid. Of these subjects, 71.8% were female, and 28.2% were male, their ages ranged from 18 to 46 years old; for the occupation, 63.2% of them are employees and 36.8% are university students; they all had 1-5 years online-shopping experience, 52% of the respondents shopping online more than once a week, 73.7% of them had failure online shopping experience.

4.2 Measure

We synthesized maturity scale of existing research from the Literature, modified the statements at the scale according to the characteristics in this context, and the questionnaire was pretested and the items which could not be correctly understood and answered were revised. The questionnaire was used Likert scale, containing 28 alternative response items, ranging from 1 (strongly disagree) to 5 (strongly agree), focusing on four areas of evaluation: perceived value (10items); 2) customs satisfaction (3 items); 3) switching cost (12 items); 4) repurchasing intention (3 items).

4.3 Method

The research focused on the buyers on www.Taobao .com, measured the buyer's perceived value, switching cost, customer satisfaction and customer repurchase intention; surveyed the relations among the variables. Data analysis was conducted by SPSS 17 and AMOS17.0 which includes descriptive analysis, reliability and validity analysis, variance analysis and model fitting analysis.

5 RESULTS

5.1 Reliability and validity analysis and structural equation model fitting

After the liability analysis of 4 dimensions of customer perceived quality, 3 dimensions of switching cost, customer satisfaction and customer repurchase intention, the results are as follows: the total Cronbach's alpha coefficient of 28 items in the questionnaire is 0.911. Cronbach's alpha coefficients of perceived value, switching cost, customer satisfaction, repurchasing intention are 0.758, 0.874, 0.758, 0.857. Alpha coefficients are all above 0.7, which means that the questionnaire has high reliability and internal consistency. The principal component method has been selected for factor analysis and the KMO value turns out to be 0.868, which pass the spherical Badett's test ($p<0.00$).

Then it has been analyzed by structural equation and the best model was chosen from the competing models, and then the result shows that the model has a good fit. (the following criteria equations):

Table 2. The final model fit index.

CHI-SQUARE	CMIN/DF	CFI	NFI	TLI	RMA	RMSEA	AIC	CAIC
230.03	2.52	0.946	0.914	0.928	0.036	0.076	240	471

5.2 Analysis of the relationship about the variables of AMOS model

In the available 266 questionnaires, there are 196 people who had experience of failure in online shopping which the percentage was 74%; the average of perceived value of who had experience was lower than who had not. By independent sample SPSS test of variance, the difference of perceived value between customers who had a bad experience and who had not bad experience is significant.

5.3 The influence path and model modification

When calculating by AMOS, the path from perceived value to repurchasing intention was not valid, so this part was removed. Then, with further analysis on items of higher ML value, original 28 items in the model were simplified to 20 items according to its necessity in the theory, and the model finally reached a better fit as shown in Table 2.

5.4 The maximum likelihood estimation

After the maximum likelihood estimation calculation, the results are as follows:

Table 3.   Path coefficient and path coefficient test.

	Standardized path coefficient	S.E.	C.R.	P
Customer satisfaction ←Customer perceived value	0.385	.039	3.101	.002
Repurchasing intention ←Customer satisfaction	0.238	.051	3.354	***
Customer satisfaction ←Switching cost	0.56	.690	2.479	.013
Repurchasing intention ←Switching cost	0.147	.168	1.911	.050

Note: the value of P * * * represents in 1 per thousand level path coefficient is zero

The standardized coefficients in table 3 show that the four paths are all positive and the path coefficients are all significant at the 99.5% level, which confirm the hypothesis 1 and hypothesis 2.

The path coefficient of switching cost to customer satisfaction is bigger than that of perceived value

to customer satisfaction (0.56 > 0.385) means that switching cost is the more important effective factor of customer satisfaction than perceived value. Therefore, even if the customer perceived value is low, customers will still be satisfied when switching cost is high, which confirms the research hypothesis 3 and 5. Furthermore, perceived value only has indirectly effect on repurchasing intention through the influence of customer satisfaction. The path coefficient of indirect effects is 0.385 x 0.238 = 0.09; at the same time, switching cost indirectly affects repurchasing intention directly and indirectly, the total effect is: 0.56 x 0.238+0.147 = 0.280. Comparing with 0.280 and 0.09, we deduce hypothesis 4 and hypothesis 6– when switching cost is high, even if customer perceived value is low, customers will keep their custom and to repurchase.

6 CONCLUSIONS AND SUGGESTIONS

6.1 conclusions

With the questionnaire, static data and structure model, this research analyzes the influence factors of repurchasing of Taobao's buyers, confirms the hypothesis raised before and gives the main conclusions are as follows:

6.1.1 Bad experience will reduce perceived value

After a comparison between Taobao buyers with a bad experience and with no bad experience, it can be found that the perceived value of them shows statistically significant differences. Perceived value of Taobao buyers with bad experience is apparently lower than that of who with no bad experience.

6.1.2 Perceived value positively influences customer satisfaction and repurchasing intention

The path coefficient is positive indicates when perceived value increases, customer satisfaction improves then customer repurchasing intentionally improves.

6.1.3 The higher switching cost, improves the degree of customer satisfaction and repurchasing

Switching cost of Taobao buyers had a positive impact to customer satisfaction. On one hand, switching cost directly and positively affects on customer satisfaction, through which it indirectly and positively influence customer repurchasing intention; on the other hand, switching cost directly and positively influence customer repurchasing.

6.1.4 Switching cost is the most important factor influencing Taobao buyers' satisfaction and repurchasing intention

When the perceived value was low, but customers' satisfaction and repurchasing were not low because of the switching cost was high as Taobao buyers believed Taobao was the largest and best online business platform which provides tools of safe and fast payment, instant communication, trustworthy commentary, convenient transportation and insurance, etc. The business ecosystem was valuable for the buyers. Buyers can give up the bad businesses and commodities, but can not give up Taobao.

6.2 Suggestions

Although the models developed in this study are simple, it can draw some meaningful conclusions and give Taobao and other online shopping site useful suggestions.

6.2.1 Leading enterprises create a better business ecosystem

Taobao.com is the largest e-commerce site, When accustomed to use it, buyers are not willing to give up or turn to other dealers because of the high switching costs. Therefore, to build a more complete system platform with better functionality and perfect service which have a great brand and high switching cost, have become the most important development strategy of leading enterprise.

6.2.2 Following enterprises innovate products and services

In the business ecosystem, small and medium-sized following enterprises whose switching costs are low should adopt different strategies. They have to innovate their products or service to make a difference from other enterprises in order to increase customer perceived value to improve customer satisfaction.

6.2.3 Further discussion on switching costs

The strategy on improving switching cost to keep customers is not strange for us. In an industry controlled by a monopoly, the customers have no other substitute to choose and the switching cost is highest in this situation. The monopoly does not need to pay attention to products and services' improvement while only need to focus on maintaining its monopoly position.

The welfare of consumers have been transformed into monopoly profits, which is undoubtedly the worst method to increase switching cost.

In the emerging internet industry with little government controlling, Taobao.com has become the leader of e-commerce industry by continued constructing the infrastructure, building its system platform, enriching and improving its business ecosystem etc., in which the process is similar to nature competition and selection. Its rapid growth in transactions is the best proof of recognition from consumers. Above this, it can also perform platform strategy which the core is to improve its switching cost to satisfy the users' needs, attract potential users and binding the original users which are all good strategies of switching costs.

REFERENCES

[1] Zeithaml, V.A.Consumer Perceptions of Price, Quality and Value: A Means-end Model and Synthesis of Evidence [J]. Journal of Marketing, 1990, 52(3):2–22.

[2] Monroe K.B. Pricing-making Profitable Decisions[M]. New York:McGraw-Hill,1991:59–65.

[3] Butz Howard E. Jr. Leonard D. Goodstein. Measuring customer value: gaining the strategic advantage[J]. Organizational Dynamics, 1996: 63–77.

[4] Woodruff. Customer value: The next source for competitive advantage [J]. Journal of the Academy of Marketing Science, 1997(2): 139–153.

[5] Tianlang, Li. Research on the influence of perceived value on usage intention for Mobile phone game users[D]. South China Technology University.2012.

[6] Fornell C.A National Customer Satisfaction Barolneter: The Swedish Ex Perience [J]. Journal of Marketing,1992,56(1):6–21.

[7] Jones M.A. Why customer stay: measuring the under lying dimensions of services switching costs and managing their differential strategic outcomes[J].Journal of Business Research.,2002,55(6):441–450.

[8] Burnham T A, Frels J K, Mahajan V. Consumer switching costs: a typology, antecedents, and consequences [J]. Journal of the Academy of Marketing Science, 2003,31(2):109–126.

[9] Westbrook, R A.and M D. Reilly. Value- Percept Disparity: an Alternative to the Disconfirmation of Expectations Theory of Consumer Satisfaction. Advances in Consumer Research, 1983, (01):256–261.

[10] Anderson, E.W&Sullivan, M.W. The Antecedents and Consequences of Customer Satisfaction for Firms [J]. Marketing Science, 1993, 12:125–143.

[11] Jones M.A. Why customer stay:measuring the under lying dimensions of services switching costs and managing their differential strategic outcomes[J].Journal of Business Research.,2002,55(6):441–450.

Information, Computer and Application Engineering – Liu, Sung & Yao (Eds)
© *2015 Taylor & Francis Group, London, ISBN 978-1-138-02717-6*

An exploration on the intermittent search approach for global optimization

Q. Jia
Hohai University, Nanjing, China

C. Guo
Shantou University, Shantou, China

Y. Zheng
Nanjing Forestry University, Nanjing, China

Y.J. Xiao
Nanjing University of Finance & Economics, Nanjing, China

ABSTRACT: This work explores the Intermittent Search (IS) strategy of biology and physics in association with evolutionary programming for application to global optimization. Four modes (static, ballistic, diffusive and ISPSO) of the IS strategy in two phases are designed to efficiently tackle optimization problems. In the computation, the benchmark unimodal and multimodal functions are tested and the outcomes show the effectiveness and potential of the IS approach as a useful search method, which is attributed to the fact that the approach attains desired local search ability in the first phase while the second phase offers benignly larger potential search space and accelerating characteristics.

KEYWORDS: Intermittent search; static mode; diffusive mode; ballistic mode; particle swarm optimization; global optimization

1 INTRODUCTION

With the continuing generation of more complex problems in computational researches and practical applications in engineering, advanced optimization techniques are persistently essential. A variety of optimization techniques have been proposed and prevalently used in the area of optimization, such as genetic algorithm (GA), simulated annealing (SA), tabu search (TS), evolution algorithm (EA), particle swarm optimization (PSO), and so on. These approaches are mimicked after the foraging behaviors of animals (e.g., PSO), the patterns of genes at the microscopic level (e.g., GA), or the evolution behaviors of animals (e.g., EA) (Riche & Haftka 2012). Of interest, intimating the foraging behaviors of animals and DNA motion at the microscopic level, intermittent search (IS) strategies are derived from the search patterns commonly existing in the nature, which combine two phases of motion, the detection phase of slow motion that allows the searcher to detect the target and the relocation phase of fast motion during which the searcher only moves fast without detecting the

target (Bénichou et al. 2011). The strategies are widely observed at various scales, for example, in the case that animals hunt for food at the macroscopic scale and the intermittent transport patterns happening in a reaction pathway of DNA binding proteins as well as in the intracellular transport on the level of cells. Bénichou et al. conduct a systemic review of the IS research and apply generic stochastic models in biology and physics, reaching the conclusions that the IS strategies permit the minimization of the search time suggesting the intrinsic efficiency of the IS strategies regularly encountered in the nature (Bénichou et al. 2011). It is noted that IS strategies can be used in a broader context to design and accelerate search processes, which directly gives motivation of this study focusing on IS-based optimization techniques. The IS approach as proposed in this work is designed to solve global optimization problems, formulated as a D-dimensional minimization problem as follows: *Min* $f(x)$, $x = [x_1, x_2, \ldots, x_D]$, where D is the number of the parameters to be optimized. The paper is organized as follows. Section 2 presents the description of IS followed by the four modes of the IS method in Section 3. The computational experiments and results are described in Section 4, with the conclusions drawn in Section 5.

2 IS STRATEGY

As shown in Figure 1, a generic IS modeling is illustrated based on the general features of IS, including four different modes for the detection and relocation phases in one, two and three dimensions. It is assumed that the searcher evolves in a D-dimensional spherical domain which has reflective boundaries and one centered immobile target of the radius r. Unknowing the target's initial location, the searcher starts the walk from a random point of the D-dimensional sphere and averages the mean target detection time over the initial position. The four patterns of the IS phases are described in more details in the following. (a) In the "static mode", the searcher is static and detects the target with certain probability per unit time if it is located at a distance less than r. (b) In the "diffusive mode", the searcher moves in an uninterrupted diffusive way with a diffusive coefficient, and finds the target instantly if it is positioned at a distance less than r. (c) In the "ballistic mode", the searcher moves ballistically in a random direction and reacts immediately with the target if it is found at a distance less than r. (d) In the "ISPSO mode", the searcher changes its velocity and position via the given PSO recipe, which traces the corresponding global best value and local best value to access the optimal value.

3 IS APPROACH FOR GLOBAL OPTIMIZATION

This study proposes IS schemes with the four modes for the global optimization, as shown in Figure 2.

3.1 *Representation and initialization of the searchers*

In the IS process, it is assumed that there are a swarm of searchers in the D-dimensional search space. Every searcher i ($i = 1, 2, ..., N$) has a position $P_i(t) = \{p_{i1}(t), p_{i2}(t), ..., p_{id}(t), ..., p_{iD}(t)\}$ and a velocity $V_i(t) = \{v_{i1}(t), v_{i2}(t), ..., v_{id}(t), ..., v_{iD}(t)\}$ during the tth iteration ($d = 1, 2, ..., D$; $t = 1, 2, ..., Max\_iteration$). The searchers move following a prescribed mode in the first detection phase, for which the velocity v_{id} and the positions p_{id} are calculated using the update function.

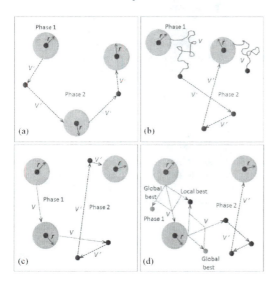

Figure 1. Four modes of the IS strategy in two-dimension.

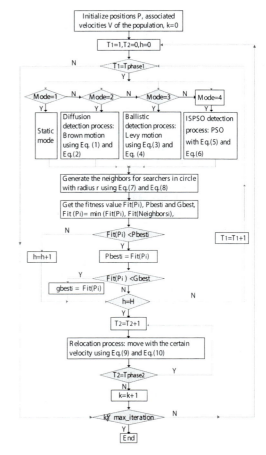

Figure 2. IS strategy for global optimization.

834

The fitness value of the solution is calculated through substituting the value of position $P_i(t)$ into the tested function and the *Gbest* value is the optimal fitness value among all the solutions. In the computation, the position and velocity values of the searchers are generated randomly in the range of [R_min, R_max], which has been set for every test functions.

3.2 Four different modes of the IS process

3.2.1 Static mode

In the static mode, the first mode shown in Figure 2, the searcher is designed to be static and finds the target lying at a distance less than the radius r.

3.2.2 Diffusive mode

The search process is assessed by a formula for the average distance moved in a given time during the Brownian motion as follows:

$$(\Delta x)^2 = <(x(t))^2> - <(x(t-1))^2> = 2\mu \cdot \Delta t \qquad (1)$$

$$p_{id}(t) = p_{id}(t-1) + \alpha \cdot \Delta x \qquad (2)$$

In the equation, $<x^2>$ is the average value for the square of the distance; μ is the diffusion coefficient assumed to be a reduced diffusion coefficient which obeys the Gauss distribution at the step t, $\mu \sim N(0, \sigma^2 t)$, and $\sigma = 0.5$; Δt is the time which may be assumed to be one time unit for one step.

3.2.3 Ballistic mode

This mode is equivalent to the model of a Lévy flight search, revealing advantages in intermittent search (Lomholt et al. 2008). The times between successive reorientations are Lévy distributed and the corresponding Lévy flight patterns are characterized by a distribution function,

$$Prob(l_j^{-v}) \sim l_j \qquad (3)$$

$$p_{id}(t) = p_{id}(t-1) + l_j \qquad (4)$$

with $1 < v \leq 3$, where l_j is the flight length. The power parameter v is set to 2, which is widely observed in both wild and fenced areas and can become optimal in the search process in any dimension as proved by numerical simulations (Viswanathan et al. 1999).

3.2.4 ISPSO mode

In the ISPSO mode of Figure 2, the searcher is viewed as a particle and pursues the PSO update pattern in the detection using equations below.

$$V_i(t) = w \cdot V_i(t-1) + c_1 \cdot rand \cdot [Pbest_i - P_i(t-1)] + c_2 \cdot rand \cdot [Gbest - P_i(t-1)] \qquad (5)$$

$$P_i(t) = P_i(t-1) + V_i(t) \qquad (6)$$

In Eq. (5), w is the inertial weight controlling the exploration of the search space, usually limited to [0, 1], and c_1 (resp. c_2) is a self-learning (resp. social-learning) factor that measures how much one particle will believe in $Pbest_i$ (resp. *Gbest*), the memory of a particle's own best position in the history of $Pbest_i$ for particle i (resp. the best solution of the total population).

3.3 Neighborhood detection

In the detection phase, the searchers scan the circled areas centering with the searcher's position, with the radius r. Then two neighbors are generated and compared whether they are close to the target, and the best solution among the searcher and its neighbors is chosen as the updated searcher. The neighbors are generated by using the equations

$$P\_neighbors_i = P_i \pm r, \qquad (7)$$

$$V\_neighbors_i = V_i \pm r. \qquad (8)$$

3.4 Switch strategy and relocation phase

In the detection phase, the searchers scan the circled areas centering with the searcher's position,

A parameter H is used to initiate the second phase. As Figure 2 shows, the second phase will not start until the H steps of the stagnant fitness value. The number of H steps of the second phase is determined by the parameter experiment. The relocation phase is initiated by the transition strategy as addressed above. The update of the velocity and the position follows the equations

$$v_{id}'(t) = \alpha v_{id}'(t-1) \cdot rand_2, \ rand_2 \sim N(0, \sigma^2 t) \text{ with } \sigma = 0.5 \qquad (9)$$

$$p_{id}'(t) = p_{id}'(t-1) + v_{id}'(t). \qquad (10)$$

4 EXPERIMENTS

4.1 Test functions, parameter settings and measurements

The benchmark functions which simulate interesting properties of model problems can be referred to

(Chen et al. 2011). They can be divided into unimodal (U) with only single optimum value and multimodal (M) functions with two or more distinct local optima near the global solution. For the four different modes of the IS approach, the special parameters of a selected mode is set based on the experience or characteristics of the algorithm in several preliminary tests. The population size N is set to 20 and the dimension of the problem is 30 for all the functions, with the radius r is set to 1/10 of the maximum value. In the ballistic mode, the natural parameter v is set to 2. In addition, the popular parameters settings for PSO applications are used in this study, with the learning factors $c_1 = c_2 = 2.0$ and the inertia weight $w = 0.4$ (Shi & Eberhart 1999). The algorithm is coded in the Matlab language and executed on an Intel Core Duo Processor T2300, 1.66G HZ personal computer with 512 MB RAM. As shown in Table 1, the best settings of the detection phase (T_1)and the relocation phase (T_2) change with the modes, namely, T_1=50 and T_2=25 for both static mode and ballistic mode, T_1=50 and T_2=10 for the diffusive mode, and T_1=100 and T_2=10 for the ISPSO mode. Accordingly, the parameters are concentrated on the lower parts of the T_1/T_2 rate, as marked in italics in the table, which implies the delicate balance between the search capacity of the first and second phases.

Table 1. The testing average fitness results of the four IS approach for the Rastrigin function with different settings of T1/T2.

T1/T2	50/25	50/10	100/10	100/5	150/5
Static	0.29	0.348	1.3	1.62	5.36
Diffusive	1.38E-5	3.12E-6	5.06E-6	9.86E-6	3.67E-6
Ballistic	4.93E-6	7.68E-6	8.64E-6	6.28E-6	2.71E-5
ISPSO	16.1	16	14.7	37.1	21.9

4.2 Experimental results and discussion

The proposed algorithms are tested over 20 independent trials for the seven benchmark functions as 30-D problems in the limit of the 120000 maximum fitness values. The $Avg.$ and $Min.$ values reached by the algorithms and the computation time (T) used by the computing process is recorded in Table 2. For comparison, representative results of the parameter $Avg.$ from typical non-IS approaches are also tabulated. The non-IS methods concerned include the conventional PSO without IS studied in this work. The non-IS methods concerned include the conventional PSO without IS studied in this work, including PSO with a bilateral objective function

(PSO+BOF) and random walking PSO with a bilateral objective function (RW-PSO+BOF) proposed recently (Chen et al. 2011), Gaussian PSO with jumps (G-PSO) (Krohling 2005), the mutation combined PSO with an acceleration coefficients algorithm (M-PSO) (Ratnaweera 2004), the comprehensive learning PSO (CL-PSO), the fast evolutionary programming algorithm (Fast-EP) (X. Yao, Liu, Y., Lin, G. 1999) and the self-adaptive differential evolution algorithm (SAD-EA) (Brest 2006). In the table, the figures in italics denote the best or the optimal zero solution searched by the corresponding IS approaches or non-IS algorithms, respectively.

Evaluated by the value of the $Avg.$ parameter, as given in the table, the diffusive mode clearly gains the best performance for the Sphere function, the Step function and the Schwefel function, indicating the efficiency of this algorithm for both unimodal and multimodal optimization problems. Similarly, the ballistic mode shows the best performance for the Step function, the Rosenbrock function and the Rastrigin function. The observation hints that these two modes have better capacity to avoid stagnation and trapping in local minima. Though the diffusive itself is slow and random, it can become faster and have the capacity to visit a broader area with the assistance of the intermittent search strategy. In contrast, the ballistic mode in this study makes use of the Lévy flight motion, which has been proved beneficial as a consequence of larger jumps (X. Yao et al. 2004). Comparatively, The ISPSO mode is moderately able to attain high-quality solutions for the Quadric function, the Rastrigin function and the Schwefel function. Nevertheless, the static mode is fast in convergence but easily to be trapped in a local minimum for the most of the seven test functions at the early stage.

5 CONCLUSIONS

We have combined the intermittent search strategy and the evolutionary method for application to solve the optimization problems. The IS methods of the four modes, including the static mode without the detection phase, the diffusive mode with Brownian motion random walking, the ballistic mode with the Lévy flight random walk and the ISPSO mode with the PSO motion pattern, are proposed and tested. The experiments show that the IS strategy with the ballistic and diffusive modes have superior performance followed by the ISPSO mode, while IS with the static mode is the most time-saving method.

The comparison with other non-IS algorithms evidently gives the convincing competitivity of the IS approach as proposed. The four modes of the IS strategy may offer a comprehensive framework for the application of the intermittent search strategy in solving other optimization problems. Future research for the improvement of the proposed design may coordinate IS with other evolutionary tactics and transformation methods of optimization. Experiments on much larger benchmark functions will be made and concluded. The parameters can be designed to be adaptive. In addition, the transition stratagem between the first and second phases may be further overhauled and exploited, with potential application of such methods to specific engineering problems.

Table 2. Results of global optimization derived from the test functions based on the different modes of the proposed IS approach, in comparison to that of typical non-IS methods.

	Sphere (U)	Step (U)	Rosenbrock (U)	Quadric (U)	Elliptic (U)	Rastrigin (M)	Schwefel(M)
IS							
Static							
Avg.	1.56	1.5	4.21	236	74.1	7.01	73
min.	4.57E-3	0	1.05E-2	0	5.03E-2	2.81E-02	1.21
T.	4.77	4.65	4.66	5.68	4.92	4.65	5.12
Diffu.							
Avg.	7.13E-6	0	7.42E-5	2.91E-3	9.27E-4	4.66E-6	3.92E-4
min.	2.75E-8	0	1.81E-8	5.48E-6	2.10E-5	2.55E-9	3.82E-4
T.	29.8	27.6	28.4	88.4	44.1	29.4	32.3
Ballis.							
Avg.	1.56E-5	0	4.06E-5	2.55E-3	1.15E-3	2.51E-6	3.24E-2
min.	1.57E-7	0	7.26E-10	2.79E-6	2.78E-7	1.98E-11	3.84E-4
T.	25.7	28.2	25.4	91.1	43.6	30.1	29.6
ISPSO							
Avg.	21.9	26.6	29.70	2.33E-7	35800	24.1	270
min.	4.01E-2	0	5.96	0	148	8.91E-4	4.25E-4
T.	69.9	73.8	70.9	134	86.3	71.9	75.4
Non-IS (Avg.)							
PSO	33.8	37.7	49.30	1.43E-06	37800	145	1470
PSO+BOF	5.98E-98	23.67	13.59	306.18	543992.02	44.32	4637.25
G-PSO	2.49E-8	0	24.99	5.93	601.37	16.31	1293.46
M-PSO	4.73E-39	0	18.65	9.16E-06	7.17E-29	12.09	1999.74
CL-PSO	7.29E-15	0	21.28	1596.29	1.58E-10	8.19E-5	2.82E-13
RW-PSO-BOF	1.93E-301	0	1.47	7.76E-04	0	0	5.52E-11
Fast EA	1.88E-5	0	29.45	1079.01	135.32	8.10	1462.27

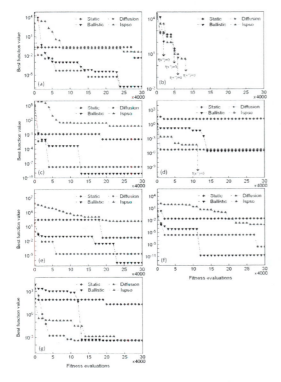

Figure 3. The convergence characteristics of 30-D test functions. (a) Sphere, (b) Step, (c) Rosenbrock, (d) Quadric, (e) Elliptic, (f) Rastrigin and (g) Schwefel.

ACKNOWLEDGMENTS

This work was financially supported by the Fundamental Research Funds for the Central Universities (Grant No.2014B01414), Social Science Research Program of Shantou University (Grant No. 07404861) and the National Science Foundation of Jiangsu Higher Education Institutions of China (Grant No.14KJD410001).

REFERENCES

Bénichou, O., et al. 2011, Intermittent search strategies, Reviews of Modern Physics, 83 (1).

Brest, J., Greiner, S., Boskovic, B., Mernik, M., Zumer, V. 2006, Self-adaptive control parameters in differential evolution: A comparative study on numerical benchmark problems, Ieee Transactions on Evolutionary Computation, 10 (6), 646–57.

Chen, Y.C., Chang, K.C., & Ho, S.H 2011, Improved framework for particle swarm optimization: Swarm intelligence with diversity-guided random walking, Expert Syst. Appl., 38 (10), 12214–20.

Krohling, R. A. 2005, Gaussian particle swarm optimization with jumps, Proceedings of the IEEE congress on evolutionary computation (21), 1226–31.

Lomholt, M. A., et al. 2008, Levy strategies in intermittent search processes are advantageous, Proceedings of the National Academy of Sciences of the United States of America, 105 (32), 11055–59.

Ratnaweera, A., Halgamuge, S., Watson, H. 2004, Self-organizing hierarchical particle swarm optimizer with time-varying acceleration coefficients, Ieee Transactions on Evolutionary Computation, 8 (3), 240–55.

Riche, R. L. & Haftka, R. T. 2012, On global optimization articles in SMO, Struct Multidisc Optim, 3.

Shi, Y. & Eberhart, R.C. 1999, Empirical study of particle swarm optimization, Proceedings of the 1999 Congress on Evolutionary Computation (Washington: IEEE), 1945–50.

Viswanathan, G.M., et al. 1999, Optimizing the success of random searches, Nature (London), 401, 911–14.

Yao, X. , Liu, Y. , & Lin, G. 2004, Evolutionary programming using mutations based on Levi probability distribution, Ieee Transactions on Evolutionary Computation, 8 (1), 1–24.

Yao, X., Liu, Y., Lin, G. 1999, Evolutionary programming made faster. IEEE Transactions on Evolutionary Computation, 3 (2), 82–102.

Information, Computer and Application Engineering – Liu, Sung & Yao (Eds)
© 2015 Taylor & Francis Group, London, ISBN 978-1-138-02717-6

Influence of culture industries on Xi'an's economic growth

Qing Mei Jin & Li Zhang

Xi'an University of Architecture & Technology, China

ABSTRACT: With the globalization of the economy and the progress of science and technology, cultural industry, as an emerging industry, plays an important role in promoting economic development and social progress. Its impact on economy improvement includes direct effect and indirect effect. The direct effect lies in culture industry output value of economic growth, and the indirect effect is that the development of the culture industry boosts industrial structure optimization. Through the structure optimization and adjustment, the development of the economy can be greatly promoted as a whole. This article mainly analyzed the direct contribution of the culture industry to economic growth. By applying the production function, it established the culture industry contribution on economic growth model. Using Xi'an's cultural industry development as the study case, the research studied the influence of the culture industry on the economic growth in order to promote the industrialization of xi 'an culture industry development, which is of great practical significance.

KEYWORDS: The culture industry; Economic growth; Industry development

1 INTRODUCTION

With regard to the impact of culture industry to economic growth, and scholars from different aspects of a more in-depth study. There are also some scholars began to try to take advantage of time-series data or panel data validation role in promoting a country or a region of the culture industry to economic growth. The economic role of the domestic culture industry less research, studies on the economic role of the cultural industry just to stay in the general discussion of the culture industry to economic growth impact, lack of systematic study of the economic role of the culture industry, basically does not involve the promotion of economic and culture industry underlying growth mechanism and the role of law and other issues.

2 CULTURE INDUSTRY TO BUILD MATHEMATICAL MODELS ECONOMIC GROWTH

Culture industry for economic growth in the contribution rate of the mathematical model can be used to construct the production function. Assume that G is the total output I_1 for the primary industry, s, ondary industry is I_2, I_3 for the tertiary industry, I_4 for cultural industries. Therefore, the total output function is as follows:

G=F (I_1, I_2, I_3, I_4)

The formula for the time t derivative, we have:

$$\frac{dG}{dt} = \frac{\partial F}{\partial I_1} \bullet \frac{\partial I_1}{\partial t} + \frac{\partial F}{\partial I_2} \bullet \frac{\partial I_2}{\partial t} + \frac{\partial F}{\partial I_3} \bullet \frac{\partial I_3}{\partial t}$$
$$+ \frac{\partial F}{\partial I_4} \bullet \frac{\partial I_4}{\partial t}$$

Dividing both sides with G, we have:

$$\frac{dG}{dt} / G = \frac{\partial F}{\partial I_1}$$
$$\bullet \frac{I_1}{G} \bullet \frac{\partial I_1}{\partial t} / I_1 + \frac{\partial F}{\partial I_2} \bullet \frac{I_2}{G} \bullet \frac{\partial I_2}{\partial t} / I_2 + \frac{\partial F}{\partial I_3} \bullet \frac{I_3}{G} \bullet \frac{\partial I_3}{\partial t} / I_3$$
$$+ \frac{\partial F}{\partial I_4} \bullet \frac{I_4}{G} \bullet \frac{\partial I_4}{\partial t} / I_4$$

Order

$$\beta_1 = \frac{\partial F}{\partial I_1} \bullet \frac{I_1}{G} ; \beta_2 = \frac{\partial F}{\partial I_2} \bullet \frac{I_2}{G} ; \beta_3 = \frac{\partial F}{\partial I_3} \bullet \frac{I_3}{G} ;$$

$$\beta_4 = \frac{\partial F}{\partial I_4} \bullet \frac{I_4}{G} ;$$

β_1 represents the first industry output elasticity, ie marginal contribution to the economic growth rate of the first industries;

β_2 represents the output elasticity of the secondary industry that the marginal contribution to the economic growth rate of the secondary industry;

β_3 represents the output elasticity of tertiary industry, the marginal contribution to the economic growth rate that the tertiary industry;

β_4 represents the output elasticity of cultural industries, namely the marginal contribution to the economic growth rate of cultural industries;
The original equation simplifies to:

$$\frac{dG}{dt}/G = \beta_1 \cdot \frac{\partial I_1}{\partial t}/I_1 + \beta_2 \cdot \frac{\partial I_2}{\partial t}/I_2 + \beta_3 \cdot \frac{\partial I_3}{\partial t}/I_3 + \beta_4 \cdot \frac{\partial I_4}{\partial t}/I_4$$

Without considering the time factor t of the case, the above equation simplifies to:

$$\frac{dG}{G} = \beta_1 \cdot \frac{\partial I_1}{I_1} + \beta_2 \cdot \frac{\partial I_2}{I_2} + \beta_3 \cdot \frac{\partial I_3}{I_3} + \beta_4 \cdot \frac{\partial I_4}{I_4}$$

$$\frac{\Delta G}{G} = \beta_1 \cdot \frac{\partial I_1}{I_1} + \beta_2 \cdot \frac{\partial I_2}{I_2} + \beta_3 \cdot \frac{\partial I_3}{I_3} + \beta_4 \cdot \frac{\partial I_4}{I4}$$

At present, China has made breakthroughs in the cultural system, continuous integration and cultural formats, in Xi'an City, in order to study the need, Xi'an Bureau of Statistics data on the "core", "outliers", "relevant layer" and the statistical study of culture industries to promote economic development or have some guidance. Among them, in 2011 and increase the value of units in the Xi'an cultural industry in 2012 has been in a dominant position, increase the value of individual cultures is also rising, the growth rate slightly higher than the increase in the value of the unit. 2012 increase the "core", "outliers", "relevant layer" values are higher than in 2011, the total added value "core" is always higher than that of "outliers" and related layers. Peripheral layer growth is 36.5% higher than the "core" and related layers. Description of cultural industries to promote economic growth, more and more significant role in the table below.

As an emerging industry, the development of cultural industries will change the configuration of the structure of resources, thus affecting the development of the whole industry, and promote the adjustment of industrial structure, promote economic growth. To do this, we must consciously planned way to explore the value of cultural resources.

3 SUGGESTIONS

3.1 Capital investment

Capital efficient allocation of resources can make a lot of resources to achieve greater efficiency has been

Table 1. Increase the value of cultural industries in Xi'an Shaanxi Province.

Value added (million)	2011	2012	Growth rate(%)
Gross value added	256.26	334.68	30.6%
Unit added value of cultural industries	209.00	277.34	30.2%
Individual added value of cultural industries	43.21	57.34	32.7%
"Core"	105.03	134.44	28.0%
"Peripheral Layer"	85.93	117.30	36.5%
"Relevant layer"	65.36	82.94	26.9%

able to grow the enterprise is good at capital operation in a short period. The accumulation of capital through cultural resources, community and cultural organizations to attract investment and financing resources, and by means of access to capital, and thus accelerate the development of cultural industry. For example, in 2013 Qujiang New District Administrative Committee of Shenzhen Fair leveraging this broad platform, through a comprehensive and international brands, international teams and international capital, the depth of cooperation, investment reached 87 billion yuan, the project involves up to 32. Laid a solid foundation for the Xi'an Qujiang realization of economic, social, cultural sustained by leaps and bounds, but also the Qujiang truly become a powerful engine of economic and social development of Xi'an.

3.2 Market development

Market-oriented operation is the basic law of the development of cultural industries and cultural resources in the market before they can only become a cultural capital and cultural products. For example, the Big Wild Goose Pagoda, Ancient City Wall, Terracotta Warriors, the Daming Palace Ruins Park belongs to the historical and cultural resources, before entering the market just a few historical sites, not market value, but once the market entry, the role of the market by the original ecology of these things exhumed for the market, these cultural resources would have a market value and commercial value. Xi'an cultural market has great potential, the key is to actively foster cultural adaptation and cultural needs of the Xi'an market, relying on the strength of the culture to encourage businesses to market-oriented, capital-based, active play to the basic role of the market in allocating resources to promote cultural the healthy development of the market.

3.3 Integration of resources

In the culture industry development, the level of industrial development and regional economic development of cultural industry is closely related to their degree of development is also greatly affect the region's economic development level and degree. Integration of cultural industry refers to cultural resources originally dispersed, fragmented by a certain way, it be centralized, optimization, so that it can form a market outlook, the market value of an effective form of cultural industries. Integration of cultural industries is cultural industry development and utilization of an important part of the culture industry in a variety of forms are usually distributed in our daily life, some scattered distributed not by the people's attention. For example Guanzhong area Tang Ling eighteen sites, which is distributed in six counties, it is the Tang dynasty nearly 300-year history of the essence concentrated to a unique position to become a prestigious cultural resource. The government, through the full integration of resources to break the boundaries of administrative areas, to achieve the coordination and cooperation across the region. Through market-oriented operation, to achieve the optimal allocation of resources and effective cultural integration.

3.4 Derivatives industry chain

Industry chain is a concept in economics, as early as 1958, Hirschman in the "economic development strategy" on the industry from the perspective of contact before and after the entry link discusses the concept of the industrial chain. Establish chain to maximize their access to social and economic benefits to meet the needs of different consumers, development and growth of the culture industry has become a top priority. By way of the industry chain, can further accelerate the industrialization of cultural resources. For example, Xi'an Qujiang Film Investment (Group) Co., Ltd. is committed to building the whole film industry chain, the company's film and TV investment, production, distribution and cinema-based investment management industry, investment brokerage artists, film and television production base, the new media business and other for the direction our country is now a well-known film and television industry chain operators. By way of the industry chain, can maximize the benefits of cultural resources, culture industry to bring new value.

3.5 Innovative system

Cultural resource industry is the core of cultural innovation. Development of cultural and creative industries, Qujiang text brigade a useful exploration. Qujiang text brigade by Chinese traditional cultural resources mining history, heritage, conservation, integration and innovation, to achieve the full depth of the butt traditional cultural resources, modern technology and creative ideas and cultural tourism market, produced a fantastic poetry and music, such as large-scale ballet "Dreaming Datang" large technology water dance "Monkey", a large Tang welcome ceremony for the "Royal horse dance" and many loved by the majority of tourists and has a good social and economic benefits of the arts and cultural tourism products. Xi'an City, there are other advantages unmatched industrial resources, cultural resources in Xi'an did not translate into a large part of promoting economic development, Xi'an need a more powerful sound policy and financial support for cultural industries on the cultural history of Xi'an integration and creativity, to make good use of its advantages of resources in order to maximize the development of the enormous potential value of the culture industry.

REFERENCES

—Greco. A. The Impact of Horizontal Mergers and Acquisitions on Corporate Concentration in the U.S.Book Publishing Industry:1989–1994[J], Journal of Media Economics, 1999.
—Hjorth-Andersen.C.AModel of the Danish Book Market[J], Journal of Cultural Economics. 2000, 4(1):27–43.
—Li Wuwei. "Creative change in China" [M]. Xinhua Publishing House, 2009.
—Huang Xiaojun& Zhang Renshou. "From the input-output analysis of the impact of cultural industries to economic growth in Guangdong, for example—." Guangzhou University (Social Science Edition) [J], 2011,10 (7).

Information, Computer and Application Engineering – Liu, Sung & Yao (Eds)
© 2015 Taylor & Francis Group, London, ISBN 978-1-138-02717-6

Learning social influence probabilities from consuming behaviors

Ling Geng, Run Zhao & Dong Wang
Shanghai Jiao Tong University, Shanghai, China

ABSTRACT: Social influence maximization problem aims to find the top influential people in a social network that maximize the spread of influence. Almost all the literature studying this topic relies exclusively on online social network services to look for user relationship and study influence probabilities. This logic has the following disadvantages: 1. It excludes those people who do not use an online social network service or who rarely log in or participate on that; 2. It can't provide a solution for on-site marketing.

In this paper, we establish an influence maximization model based on user consuming behaviors, captured from querying event from EPC network or "check-in" behaviors on business review site mobile apps, such as Yelp and Dianping.com. We propose and provide a shop-perspective solution for viral marketing companies. Instead of finding a set of seed users, we propose the algorithm to find a set of entity stores which help companies execute viral marketing in a more effective and cost-saving method.

KEYWORDS: influence maximization; viral marketing; IoT

1 INTRODUCTION

Recent years, the study of social influence and the phenomenon of influence-driven propagations in social networks is becoming a popular topic. This is motivated by researches showing people trust the information obtained from their close social circle far more than information from general advertisement channels such as TV, newspaper and online advertisement[1]. Thus, many companies are turning to believe viral marketing or word-of-mouth marketing strategies are the most effective marketing strategy.

Influence maximization is presented as a direct graph G where the weights of the edges capture the degrees of influence between two users. E.g., in Figure 1, the weight of edge (B,E) is 0.5 meaning that user B influences user E with probability 50%.

Almost all the literature studying this topic relies exclusively on online social network service, such as Facebook, Myspace and Twitter, to look for user relationship and study influence probabilities. However, this logic excludes people who do not use an online social network service or who rarely log on or participate in that. In other words, it only considers the virtual connections other than the physical connections between people.

When companies decided to influence seed nodes, they have multiple choices, such as: online advertising, email advertising, face-to-face marketing. Considering the difficulty of getting user's contacts and the fact that 40% of prospects converted to new customers via face-to-face meetings, and 28% of

current business would be lost without face-to-face meetings, according to report from Meeting Professionals International (MPI), the method of on-site face-to-face marketing is probably the most appropriate one especially for those possessing many entity stores[2].

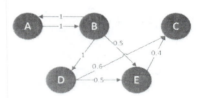

Figure 1: Influence maximization graph.

Our research reaches the following contributions:

• We first learn users' consuming habits and discover "user group" from the users' consuming behaviors. Users in the same "user group" are supposed to have the same consuming habit (e.g. always shop together). Thus, they tend to have higher probabilities to influence each other than those who are not in the same "user group". The consuming behaviors are captured from querying event from the EPCglobal network or "check-in" and reviewing behaviors on business review site mobile apps, such as Yelp and Dianping.com;

• We then develop a shop-perspective influence maximization algorithm for companies who wish to market their products through on-site face-to-face marketing. Our algorithm helps companies to find the best shops to implement their on-site marketing where they regard this shop's customer as initial users in the network to use it (by offering trial products or gifts). These customers are supposed to start influencing their friends to use and their friends will start to influence their friends and so on. With the help of the word-of-mouth effect, a large population would adopt their products;

- We conduct experiments on large real-word check-in datasets. We discover "user group" from that and compare the result to the original user network. The results show that our approach is accurate and effective. We also demonstrate the scalability of our algorithm by showing the performance on very large real-world datasets, which is much faster than the classical approaches.

2 BACKGROUND

Social influence maximization was first proposed by Domingos and Richardson[3]. David Kempe et al[4] formulated this problem: In a network, for a given number k, find a k-sized "seed" node of users, so that by activating them, we can maximize the expected spread of propagation (that is, the final intended end nodes is activated) under a certain propagation model.

Propagation model is to simulate the spread of influence through the network. David Kempe et al mainly focused on two propagation models: Independent Cascade (IC) model and Linear Threshold (LT) model. In both models, each node is either active or inactive in every time step. An active node will never return to inactive as long as it is activated. In IC model, each active node has one shot change to influence its neighbor with the influence probability. While in LT model, each node chooses a threshold T. Whenever the total weight from the active neighbors of it is at least T, it is activated.

Almost all the literature studying this topic relies exclusively on online social network service, such as Facebook, Myspace and Twitter, to look for user relationship and study influence probabilities. Many have proved their method to be effective based on real online social networks service[5] and get input action log data from that.

3 DISCOVER USER RELATIONSHIP FROM CONSUMING BEHAVIORS

With smartphone technologies and usage advances, more and more people habituate themselves to carrying their smartphones whenever looking around the shopping malls, enjoying dinners at restaurants or watching a new film at the cinema. In the above-mentioned scenarios, one usually performs consuming behaviors with one or more companions who could be one's friend, work-mate or lover. Nowadays, popular business review sites have developed smartphone apps customized for consumer devices which are "location aware" and set up with GPS system. Moreover, the "check-in" feature on the apps provides consumers to exclaim their feelings for their current visiting shops.

EPCglobal network was created by EPCglobal. It is used to store and share product data between trading partners, especially the supply chain data. It includes data about the movement of a project throughout its life cycle. This network is commonly used in anti-counterfeiting. The behaviors of querying whether a product is fake performed by customers can be regarded as Query Event (UserId, Time, Latitude, Longitude).

Social scientists have raised a theory: it appears that products and brands that individuals select can be influenced by their reference groups[6]. This theory inspires an idea that people who usually behave consuming behaviors together have strong influence probabilities with each other. Our approach to discover the user relationship is divided into 2 steps: cluster shopping places and discover User Group.

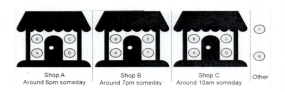

Figure 2. Query location.

3.1 Location cluster

As shown in Figure 2, the location where Query Event takes place could be really a shop or somewhere else. If some users always execute Query Event in the same shop at a very close time, they are likely to be a user group. But there would be some dirty data to concern: how do we filter Query Event happening outside a shop? Although EPCglobal network can trace most transportation nodes of the supply chain, many final retail stores cannot be tracked and stored because of budget. So in the first step, we cluster and discover shopping places.

The location data, latitude and longitude in Query Event is usually obtained by smart phones' GPS systems. Thus, location of Query Event taking place in one shopping place could be in a range due to shops' sizes and GPS devices' estimation error. We first introduce the notion of location cluster.

DEFINITION 1 (Location Cluster). For a set of locations L(x, y), set C is a location cluster, if

1. $c \subseteq L$;
2. For $\forall li \in c$, dist(li, Cen(c)) $\leq \delta r$. Where Cen(c) denotes the physical center of set C, δr is the radius threshold of set C. Intuitively, the distance between centers and the farthest is less than a threshold.

As we require location cluster with the threshold of radius instead of the number of clusters. Therefore, the clustering method we use is hierarchical clustering method, also known as bottom-up approach; the specific process is as follows:

1. Treat each point in L as one cluster;
2. Merge each cluster with its nearest cluster;
3. Run step 2 iteratively until no more clusters can be combined.

This approach requires a very big calculating space and inefficient time which can be unacceptable when the size of L is big enough. Therefore, an improvement is mandatory. It is worth considering that in the cluster merging process of the original algorithm, we need to store and traverse all the current temporary cluster status to decide which to merge. For any temporary cluster c, however, the potential merging cluster targets are those whose x coordinate ranges from [Cen_x(c) - 2δr + Rc , Cen_x(c) + 2δr - Rc]. For clusters outside the scope, they are totally negligible. Using this idea, we can greatly reduce the cost to maintain and compare with temporary clusters. The cluster shopping places can be computed in Algorithm 1.

Algorithm 1 Compute location clusters

INPUT: All Query Location L(Longitude, Latitude), δr
OUTPUT: Location cluster set C
1:     Sort L by Longitude ascendingly
2:     Set C to empty
3:     **for** l_i in L do
4:         **for** c_j in C Longitude in [Longi - 2δr, Longi]
5:             Output c_j
6:             Remove c_j from C
7:     **End for**
8:     Find subset C' in C allowing l_i to merge
9:     **if** C' is not empty **then**
10:       Find c_k in C' nearest to l_i
11:       Merge l_i into c_k
12:       Update C
13:     **else**
14:       Create new cluster c_N containing l_i
15:       Add c_N to C
16: **End for**
17: Output the rest cluster in C

3.2 Query cluster

At this time, we have got the location clusters which help us filter dirty data and find user group more easily. Based on location clusters and its algorithm, we define and find out query cluster:

DEFINITION 2(Query Cluster). In a set of Query Event Q(x, y, t, u), query event QA, QB is in the same query cluster where QA$\in$Q, QB$\in$Q, if

1. L(QA) = L(QB) where L donates the corresponding location cluster
2. |T(QA) - T(QB)| $\leq \delta t$ where T denotes when query event takes place
3. U(QA) $\neq$ U(QB) where U denotes which user performs the event

An intuitive understanding of query cluster is that if QA, QB is in the same query cluster, then UA, UB is regarded to perform a consuming behavior together. Although the query event itself is a three-dimensional data, with the results of location cluster in the last step, we only need cluster the query event on the basis of time variable. The idea is analogous to algorithm 1.

The cluster operation of Query Cluster is based on each location cluster, which means query clusters in the same place will not be intersected in the next steps. This approach applies to the situation where users usually only visit one place once. In the scenario of visiting the same shop many times, slight changes are made: traverse data in a location cluster of each day instead of the whole each location cluster. This would allow intersection between queries in the same shop at different days and solve the problem.

3.3 Discover user groups

Next step, we discover user groups on the basis of the previous result. We call 2 or more users in the same user group if they perform query event in the same query cluster more than a certain threshold time. To discover user groups, our work is to make intersection operation in the query clusters outputted in the last step. Detailed steps are shown in algorithm 2.

4 SHOP-PERSPECTIVE INFLUENCE MAXIMIZATION

Previous literature devoted themselves to finding a set of seeds without considering user profiles. This would be hard for companies who wish to market their products through on-site face-to-face marketing. To meet the needs of them, we develop a shop-perspective influence maximization algorithm to help companies to find the best shop to implement their on-site marketing where they regard their shop customers as initial users in the network to use it (by offering trial products or gifts).

Algorithm 2 Discover user group

INPUT: Query Event cluster EC(U, Time, ECID), δq
OUTPUT: User Group set AC
1:   Set AliveCluster to empty
2:   **for** EC(l$_i$) in EC do
3:      Intersect EC(l$_i$) to AliveCluster by CurrentCluster
4:      **for** c$_j$ in CurrentCluster
5:        **if** c$_j$'s size is less than 2 *then break*;
6:        **if** AC$_j$ = c$_j$ then frequency(AC$_j$) ++;
7:        **else** //c$_j$ contained in AC$_j$
8:          frequency(c$_j$) = frequency(AC$_j$) + 1;
9:          Add c$_j$ to AliveCluster
10:     **End for**
11:     Add ec$_k$ in EC(l$_i$) into AliveCluster with frequency(ec$_k$)=1 if ec$_k$ not in
12:     Output ac$_x$ in AC where frequency(ac) ≥ δq
13:       Remove ac$_x$
14: **End for**

Our idea comes from MIA model[7]. Let the probability that u activates v through path P is pp(P). Also MIP$_G$(u,v) (maximum influence path) defined as:

$$MIP_G(u,v) = \text{argmax}_p\{pp(P) \mid P \in P(G, u, v)\} \quad (1)$$

Where *P*(G, u, v) denotes the set of all paths from u to v.

DEFINITION 3 (maximum influence in-arborescence)

$$MIIA(v, \theta) = \bigcup\nolimits_{u \in V, pp(MIP_G)(u,v) \geq \theta} MIP_G(u,v) \quad (2)$$

To calculate the influence spread under IC model by giving the seed node set S, we assume the influence from S to v is only propagated through edges in MIIA(v, θ). Let the activation probability of any node u in MIIA(v, θ)), denoted as ap(u, S, MIIA(v, θ)), be the probability that u is activated.

To calculate influence maximization from a shop perspective, it is important for us to discover the relationship between users and shops. This probability can come from Query Event.

$$P_{shop}(u, sp) = \sum\nolimits_{shop=sp} Q(u, shop) / \sum Q(u, shop) \quad (3)$$

The activation probability of each node can then be computed recursively as given in Algorithm 3.

Algorithm 3. Activation probability from shop perspective

INPUT: *A User visiting shop probability* P$_{shop}$(u, sp)
1: **if** u ∈ S *then*
2:    ap(u) = P$_{shop}$(u, sp)
3: **else if** Nin(u) = ∅ *then*
4:    ap(u) = 0
5: **else**
6:    ap(u) = 1 − ∏w ∈ Nin (u)(1 − ap(w)·pp(w, u))
7: **end if**

In our model, we assume that seeds in S only influence node v through MIIA(v, θ). Thus, we have the influence spread of S in our model:

$$\sigma_M(S) = \sum\nolimits_{v \in V} ap(v, S, MIIA(v,\theta)) \quad (4)$$

5   EXPERIMENTAL EVALUATION

We conduct experiments on Gowalla check-in datasets. Gowalla is a location-based social networking website where users share their location by checking-in. Check-ins are from Feb. 2009 to Oct. 2010.

Table 1.   Gowalla dataset statistics.

Nodes	196591
Edges	950327
Check-ins	6,442,890

We discover "user group" from that and compare the result to the original user network. On discovering location clusters, we select threshold δr in 3 values (0.0005°, 0.001°, 0.0015°) to perform 3 individual experiments. Based on that, we calculate the user groups.

Table 2.   Result of user group.

δr	0.0005°	0.001°	0.0015°
Count of user group	21081	42651	74323

To evaluate the quality of user groups, we compare the user group from the real user relationship (edges) in datasets and calculate the proportion.

The results show that our approach is accurate and effective.

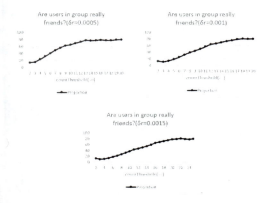

Figure 3: Quality of user group.

REFERENCES

[1] J. Nail. May 2004. The consumer advertising backlash. Forrester Research and Intelliseek Market Research Report.

[2] Meeting Professional international, ROI on the Rise, http://www.mpiweb.org/Magazine/Archive/US/January2010/ROIOnTheRise.

[3] M. Richardson & P. Domingos. 2002. Mining knowledge-sharing sites for viral marketing. In Proceedings of the 8th ACM SIGKDD Conference on Knowledge Discovery and Data Mining, pages 61–70, 2002.

[4] David Kempe , Jon Kleinberg & Éva Tardos. 2003. Maximizing the spread of influence through a social network, Proceedings of the ninth ACM SIGKDD international conference on Knowledge discovery and data mining, August 24-27, 2003, Washington, D.C.

[5] Amit Goyal & Francesco Bonchi, Laks V.S. Lakshmanan. 2010. Learning influence probabilities in social networks, Proceedings of the third ACM international conference on Web search and data mining, February 04–06, 2010, New York, USA.

[6] WO Bearden & MJ Etzel. 1982. Reference group influence on product and brand purchase decisions, Journal of Consumer research, 1982.

[7] Wei Chen , Chi Wang & Yajun Wang. 2010. Scalable influence maximization for prevalent viral marketing in large-scale social networks, Proceedings of the 16th ACM SIGKDD international conference on Knowledge discovery and data mining, July 25–28, 2010, Washington, DC, USA.

Information, Computer and Application Engineering – Liu, Sung & Yao (Eds)
© 2015 Taylor & Francis Group, London, ISBN 978-1-138-02717-6

Basic concepts and techniques of residential landscape design

Jin Ying Liu, Li Xia Song, Xiao Hua Wu & Yuan Ya Guo
Langfang Polytechnic Institute, China

ABSTRACT: With the development of our country's economy and social progress, people's standards of living and quality of life have been continuously improved. At the same time, people attach great importance to the construction of eco-friendly green homes. Thus it is crucial to conduct residential landscape design. With the development of garden industry in our country, remarkable achievement has been made in the level of design, industry and technology team. Many construction companies and real estate developers are also paying more and more attention to the residential landscape techniques. Residential landscape has become an important part of people's living environment. From the basic concepts of residential landscape design, this paper describes the principle of residential landscape design, and proposes specific techniques of residential landscape so as to promote the development of residential landscape.

KEYWORDS: Residential Landscape; Basic Concepts; Design; Techniques

Nowadays, with the rapid progress of our society and improvement of people's quality of life, people's requirement for living environment is becoming higher and higher. Traditional residence has been unable to satisfy people's requirement for residential environment. Therefore, we should innovative current residential design concept to improve the residential environment and construct residential gardens greening in order to satisfy the requirements for green living environment. Meanwhile, emphasis should be put on the improvement of design, techniques and technology team of landscape industry. Along with the residential construction, landscape design and construction are to build more beautiful and green living environment for people. This paper discusses the basic concepts of residential landscape design, analyzes the basic principles of the residential landscape design, and puts forward specific design methods, thus contributing to the promotion and implementation of landscaping, and promoting the development of the garden industry.

1 THE BASIC CONCEPTS OF RESIDENTIAL LANDSCAPE DESIGN

1.1 *The concept of residential landscape design*

Residential landscape design is the process of greening design in residential areas with the requirement of being able to combine the function of a residential layout, site conditions, and the construction environment to build a green dominated environment landscape, which has a certain ornamental value. Green residential areas is a place which serves for resident and the public green residential areas provide resident with sites for outdoor recreation. Green residential areas mainly consist of the green area between buildings, green area of public service facilities, and green area of residential roads.

1.2 *The basic concepts of residential landscape design*

First, residential landscape needs a definite theme. The rest of the green design and implementation should be carried out around the green theme. To make the green image idealized and artful, designers need to fully understand the local culture and environment, and have certain artistic creativity and imagination. In this way, the residential landscape will be a combination of imaginary artistic creation and the natural environment. The green theme gives a new expression and artistic charm to the garden, and can enhance the communication between man and nature by drawing closer the distance of man and nature. Second, residential landscape has rhythm and rhyme. Landscape is inseparable with rhythm and rhyme, almost all the beauty of art is connected with rhythm and rhyme, whereas the rhythm of landscape mainly includes strength, length, height, and a series of rigid relativities. Rhythm is similar to the beat, which is a way of undulating melody. When the lines, shapes, colors alternately repeat, it forms a unique sense of rhythm. For rhyme, it is more of a homonym. With well-arranged patchwork of green gardens, a rhyme comes into being, subtle in its design and revealing the rhythm of life. Third, green garden should have form beauty. Form and content cannot exist alone. Beauty in form is an important condition for the development and survival of art, while art aesthetics is embedded in form and should be started exploring with form. To make the garden designed with aesthetic form, attention should be focused on innovation and exploration. Through the accumulation of life as well as subtle observation and discovery, imagination and creativity will be stimulated to complete the goal of form beauty. This form beauty in the landscape is not only a natural beauty, but also a recreation of beauty by human hand. Fourth, landscape design should be able to use abstract ways. Two main techniques are used, simple abstraction and geometric abstraction. For simple abstraction, it is a rational induction through which natural images are sorted out and unnecessary specific images are abandoned, and then cross the boundaries of specific image in order to achieve the abstract realm. As for geometric abstraction, it is the patchwork of basic geometric shapes like point, line, and surface, in accordance with the laws of beauty, which go through a series of changes, together with the corresponding shift, rotation, dislocation and amplification to create a special landscape image.

2 BASIC PRINCIPLES OF RESIDENTIAL LANDSCAPE DESIGN

First, the landscape design should follow the basic principles of the unity of function and form. Landscape design should be based on the characteristics of the regional environment and climate, for example, the hot areas in summer should be covered with plants that throw shades during hot days; attention should be paid to windshield for those easy to wind areas during winter; in residential roads, stops, and places to rest, trees are to be planted to give shades. Second, landscape design should agree with local conditions. Landscape design should be fully integrated to local conditions, including humanities, social and economic development, topography etc. The landscape design should be economical and practical, adapting to the economic development of the region. Besides, make full use of existing local landscape features, and reduce earthworks; Plant trees properly, which could be good local native species so that they could adapt to local climate easily. Their survival rate will be high and the cost is relatively low.

Third, landscape design should be able to beautify the environment. Landscape design is a complex multidimensional space design. It is an environmental design based on natural beauty. It is to have a unified form, and then subtle changes of each part are to be carried out. A lot of design techniques such as contrast, rhythm and rhyme are used to optimize the whole design.

3 TECHNIQUES FOR RESIDENTIAL LANDSCAPE DESIGN

3.1 Landscape design should adhere to people-oriented policy

Landscape design should adhere to people-oriented policy. It emphasizes the affinity feature of the garden and its contribution to people' outdoor activities. Green area is the extension and expansion of people's lives, which is an important place for people to relax and entertain. People go into green space for the purpose of entertainment, sports or communication with others. So landscape design should be able to take full account of the needs of people, creating an atmosphere that is helpful for human activities. Thus for the basic principles of landscape design, it is necessary to start from the elementary needs of people, to fully study people's daily lives, and to fulfill the purpose of satisfying people's need and be user-friendly. For example, places for rest and stay when people are tired could be added up, exercise equipment for adults to keep fit, and sports equipment for children to play are to be increased.

The arrangement of plants in landscape process should take the season and weather into consideration. Improving the quality of green plants and planting variety of species in residential area can not only reveal the diversity of species, but also make the scenery change with the seasons. Those different varieties of plants have certain cooling effect during summer, and the coexistence of different plants can meet different people' needs. To a certain extent, it could enhance people's appreciation interest and aesthetic appeal. In addition, plants' health functions cannot be ignored. The introduction of some plants that can emit fragrance could make people refreshed and eyes bright.

3.2 Emphasis on landscape ecology and improvement of landscape quality

Residential landscape is not simply planting trees, grass, but a sensible choice and allocation of different plants and buildings from the perspective of the food chain and biological material recycling to make sure that plants and their living areas could survive and live harmoniously. Plants are an important element for achieving landscape beauty, and a key factor in achieving the lively ecological landscape. Therefore, in a residential area, the plants are to be able to adapt to climate and natural growing conditions. To ensure the ecological and landscape value of garden areas, great efforts need to be made to take advantage of the diversity of species and arrange trees and shrubs, slow growth and fast-growing plants, evergreen and deciduous plants scientifically and reasonably. In addition, the rational distribution of landscaping not only needs to consider environmental factors, but also need to increase the humane elements. Cultural and sports facilities could be introduced appropriately. And the introduction of a wide range of colors and styles could make the landscape layout more diverse and promote integration of the natural environment and human environment, and enhance the communication among diverse cultures, thus effectively improve the quality and the value of the garden landscape.

3.3 Landscape design should be adapted to local conditions

Residential landscape design should follow the local conditions, which include the following two points: first, use the existing earth scientifically and rationally, change the traditional construction methods, protect the original topography as much as possible, take advantage of favorable local topography, reduce the use of earth, in this way the cost of inputs can be reduced, the original ecology can be protected in its maximum, so the balance of the ecosystem could be maintained. Second, plant trees appropriately. Local types of trees should be planted, and trees with high appreciation value and trees of easily adaptable species are to be added, which could fully integrate with the local climatic conditions, the natural growth foundation, and the ecological environment. As a result, the diversity of the landscape are fulfilled, the richness of species of the gardens are maintained, and the garden structures are improved. So while under landscape design, attention should be paid to the coordination among plants and that of plants and the buildings. The resident buildings and landscape are to be melted into a whole organic.

3.4 Landscape design should organize space scientifically and rationally

Residential landscape are inseparable with people's lives, therefore, landscape should have strong function and usability. For residential landscape design, rational organization of space mainly has methods of division and infiltration. First, it is the division of space. Green space division is to be able to meet the specific requirements of people's feelings and activities at the green areas. Through division of space, enough space scales, rich visual perspective are created, multi-level spatial depth vision, like medium shot, and close-range are reached. Second, it is the infiltration of space. In terms of space infiltration, it is closely connected with space division. Only a simple partition of space will make people feel depressed and cramped, while by space infiltration this phenomenon can be effectively improved. An extension and expansion of space to nearby areas could achieve seamless connectivity, and generate hierarchical changes, which strengthen the links between the divided space, and avoid a sense of oppression and cramped.

3.5 Landscape design needs innovation and exploring new green garden space

Designers should be able to find beauty from life. Through good observation and reflection, designers could make the landscape aesthetically, and can be accepted by the public. So it is necessary to continue to pursue innovative design approaches, and actively expand the space for landscape. First, open up the vertical line of greening, forming a three-dimensional green space. Residential landscape will help to improve the ecological environment of residential areas, with the most notable feature of reducing the indoor temperature during summer. Therefore wall greening program can be implemented. In the small gardens close to the wall, some type of climbing plants such as ivy, and other plants could be planted in order to achieve the green wall effect. Second, construct the roof garden, open up new green space. Green roofs not only allow more plants to grow, but also have the function of effective insulation. But roofs greening needs to fully study the load-bearing capacity of the roof and deal with the roof leak properly. Third, carry out shelf and balcony greening. With the improvement of people's living standards, people have higher indoor environment requirements, balcony greening can meet people's appreciation needs, but also improve the indoor environment. There are also shelf greening, which gives shade and beautify the environment, improving the ecological environment of residential areas.

4 CONCLUSION

With the improvement of people's quality of life, we have put forward higher requirements for the living environment. Therefore, attention should be paid to the residential landscape in order to improve people's ecological environment. This paper mainly describes the basic concepts and principles of residential landscape design, and puts forward some specific measures for residential landscape design, so as to promote the landscape construction, enhance the value of the landscape.

REFERENCES

[1] Jia Changsong. A brief analysis of basic concepts and techniques of landscape design in residential districts [J]. Heilongjiang Science and Technology Information, 2009, (23) 32–33
[2] Wang Yingwen & Shi Ju. Basic concepts and techniques of landscape design in residential districts [J]. Real Estate Biweekly, 2014, (30) 40–41
[3] Wu Min. Talking about the landscape design of residential districts [J]. Shanxi Science and Technology, 2013, 28 (3)88–89

Information, Computer and Application Engineering – Liu, Sung & Yao (Eds)
© 2015 Taylor & Francis Group, London, ISBN 978-1-138-02717-6

Study of enterprise's integrated financial management mode

Ling Chang
Weifang University of Science and Technology, Shou Guang, China

ABSTRACT: Research the integrated financial management mode of enterprise. Firstly studies the basic theory of group financial management mode, then the study compared the pros and cons of the financial management modes, including the centralized type management mode, distributed management mode and incompatible type management mode. This paper studies the new financial management mode: Integrated financial management mode. Explorer from its three aspects: meaning, advantages and effects.

KEYWORDS: Group corporation; Financial management mode; Integration.

1 THE BASIC THEORY OF GROUP FINANCIAL MANAGEMENT MODE

1.1 *The connotation of the group company*

Group Co., Ltd is a large parent company has control as the core, to the subsidiary, taking a stake in the company and other companies or organization controlled by the parent company of the company union. Group corporation of capital output as the major means of subsidiaries and subsidiaries. Group members of the company maintained an independent legal person status, are equal in law, but on the management status of inequality, core enterprise plays a leading role to realize the harmony of the group. To ensure the flexibility and creativity of member companies, which are beneficial to exert the enthusiasm of each member of the company, ensure the consistency and coordination of each member company action, is advantageous for the company to achieve its overall development strategy.

1.2 *The characteristics of the group company*

1 Legal characteristics of the group company
As a corporate group, the group company is composed of multiple has the legal personality of the company form a consortium, including as a core group company's parent company, and has the legal person status under the core parent, subsidiaries or other companies dominated. Group Company's parent company, subsidiaries and other members of the company with legal person qualification, are a legal person enterprise. The relationship between shareholders and company, parent and subsidiary or parent company is a subsidiary of shareholders. Subsidiaries are independent of the company legal person. Parent-subsidiary has special regulations on ownership and obligation: a man holding each other; independent responsible for liabilities. There is no common debt; unless special circumstances, the debts of the subsidiary to the parent company promise the guarantee.

2 The characteristics of the group company
The characteristics of the modern corporation are mainly combined tightly coupling type and the network connection type mixed economic organizations, which is based on group inside the core and the main company, after a central organization coordination and management, in the group internal information, personnel, market, technology, capital, etc., the coordination, thus realizes the close connection and network connection of the combination of mixed economic organizations.

3 Operating characteristics of the group company
A group company of relatively large scale of operation, and constantly expands the scope of business, to develop in the direction of diversification, integrated, many group companies cross-regional, cross-grade and cross functional or even transnational operation.

4 Financing characteristics of the group company
Group company organizational form and the size determine the group company has a close relationship with financial institutions, which directly lead to the group company has a strong financial ability.

2 THE TYPE AND AMOUNT OF GROUP FINANCIAL MANAGEMENT MODE OF COMPARISON

2.1 *The pros and cons of centralized financial management mode*

2.1.1 *The meaning of the centralized financial management mode*

Centralized financial management pattern refers to the group's subsidiary to the parent company financial, investment, profit distribution and other financial matters have absolute decision-making, subsidiary of financial data and unified set, business accounting, the parent company control unit in the form of direct management of operating activities, each subsidiary of the financial sector itself no autonomy.

2.1.2 The advantages and disadvantages of centralized financial management mode

First of all, the company can focus on money to complete group's strategic goals. Second, the group company has with its high quality and good reputation, for effective financing decisions. Finally, group company on tax, execute unified accounting and tax, pay the income tax, each subsidiary company doesn't have to pay for themselves.

Group decision-making information is ineffective, easy to cause low efficiency; Decision flexibility is poorer, difficult to cope with the complex environment; Because decision-making concentration, lower efficiency, easy to delay business opportunities; Group company restricting the enthusiasm of the subsidiary company financial management, management autonomy and creativity, lead to group, lack of energy; Is not conducive to the establishment of the modern corporation system, cannot regulate the behavior of property management; Group company performance evaluation system not perfect, it is difficult to subsidiaries reasonable performance evaluation.

2.2 The pros and cons of distributed financial management mode

2.2.1 The meaning of distributed financial management mode

Distributed refers to the fiscal management mode; The power-sharing deal between the parent and subsidiary, significant financial decision-making power to the parent company, according to the principle of importance to group holding company with subsidiaries of financial control, management, decision-making properly classified.

2.2.2 The advantages and disadvantages of distributed financial management mode

Strengthen subsidiary reaction speed to market changes, enhance subsidiary flexibility; Subsidiary corporation to financing, to cultivate a subsidiary of financial management ability and risk awareness, make it more careful use funds, attaches great importance to the money; Subsidiary play to subjective initiative fully, enhance the flexibility of decision-making, can focus on the market, seize the opportunity, can create more profits.

Group company's financial power is affected by the autonomy in operation of the unit, and weakened the

group company capital optimization and allocation of resources; Distributed will surely lead to the separation of excessive, make the group's production and operation of contradictions and not harmonious, lead to repetition and waste of resources, weakened the group company's competitiveness and centripetal force, against the development of corporation, the realization of strategic goals; Give the subsidiary company of when enough power, subsidiary tend to be less, financial management activities from the initial target group company, not regulate the use of funds.

2.3 The pros and cons of harmony type financial management mode

2.3.1 The meaning of harmony type financial management mode

Centralized financial management mode and distributed mode the shortcomings and the insufficiency of financial management to a new financial management mode: Harmony type financial management mode. The financial management model is a model by combining centralization and decentralization, emphasized the advantages of the two modes at the same time, and do our utmost to overcome the shortage of the two modes at the same time.

2.3.2 The advantages and disadvantages of harmony type financial management mode

The financial management pattern of harmony is centralized financial management mode and distributed mode benefits of combining the financial management model. By unified command, unified arrangement, unified target, reduce the administrative cost, is advantageous to the group company internal all our subjective initiative, reduce the risk of group company collectively, reduce the group company cost of capital, increasing the service efficiency of funds. Enhance the enthusiasm of the subsidiary within the group, group internal centripetal force and cohesion, anti-risk ability, makes the group decision making is more reasonable.

Nominally, the centralization and decentralization combination in essence or centralized financial management, therefore, is not conducive to subsidiary zeal, initiative and creativity into full play. Improper system and strategy, easy to make the centralized financial management and distributed the combination of fiscal management system in name only.

2.4 Three kinds of financial management mode comparative analysis of pros and cons

Group internal relations and management features, determine the distributed financial management mode must be used. But in order to ensure that the company's scale, strengthens risk prevention awareness enough, asked we must attach much importance to the centralized financial management mode. Grasp the characteristics of a group company, to make the right decisions and choices, each group is the company's financial management problems. Suitable choice of the mode of financial management, but also the priority of each company decision.

3 NEW GROUP FINANCIAL MANAGEMENT PATTERN: INTEGRATION

3.1 The meaning of integrated financial management mode

Defined as the integrated financial management mode; Through the computer network technology, continuous real-time financial management information collection and processing, financial management and group company, on the basis of the organic combination of specific business in such aspects as capital, capital circulation, information exchange, strengthen management, to help the group, improve production efficiency, operational efficiency and economic benefit. The integrated financial management model is to apply the concept of integrated management in group company's financial management work. It emphasizes the whole optimization of financial systems and enterprise resource optimization configuration.

3.2 The advantage of integrated financial management mode

1 Information sharing
  During group company's fiscal management, financial management department need to each link to achieve effective information collection, information transmission and information processing. By management information system, the group of financial information processing, in order to achieve the purpose of the integrated management of group financial activities.
2 Resources concentrated configuration
  Integrated management of funds; The integrated management personnel; The cost of integration management
3 Centralized control of power
  Power is centralized control in order to make the group company financial management system of different links between separate departments, to achieve coordination and cooperation, further improve the system order degree.

3.3 The effect of integrated financial management mode

1 Economic effect of scope
  Integration of financial management, apt for this kind of advanced manufacturing technology as the core of the manufacturing system, is the basic requirement of economies of scope. Scope of pecuniary success, cannot leave all kinds of advanced manufacturing technology and management information system, all of them are also inseparable from the integrated financial management support.
2 Agglomeration economy effects
  Integrated effect is a kind of agglomeration economy effect, because it is a kind of integrated representation. Group company, through the integration of monetary management to produce the agglomeration economy effect.
3 Speed economic effect
  Because the current changes in the field of economy, the company's competitive landscape changing, more and more fierce competition. Speed economic effect refers to the company on the running speed has more advantages than its competitors and wins a kind of economic interest. Integration of financial management, for the company's financial management ability, the company's economic benefit, has a huge role.

4 SINOPEC BEFORE THE IMPLEMENTATION OF INTEGRATED FINANCIAL MANAGEMENT PROBLEMS IN FINANCIAL MANAGEMENT

4.1 Group's lack of effective budget control

Sinopec group to analyze budget beforehand and afterwards, but to effective supervision and control of not having it as a direct result of: unit budget and performance is not comparable in advance; Later analysis occurs at the end of the year, not accompanied by corresponding accounting information; After analysis the validity of the results don't have enough conviction. This makes the control of the parent company financial management efforts to weaken and subsidiary company financial management supervision strength is not enough.

4.2 The group internal control system is not perfect

Because of small size, small company requirements for the internal control system is generally not high; But because the large group company, the internal control system is very demanding. Reason: the group company level 3 and level 4 acceptable external supervision strength is small, and within the group complex. So, it is necessary to strengthen the internal control of the group company. China petrochemical group company internal control system is not perfect. Because of a large group company, at the same time, to take into account the financial management of distributed model and centralized financial management pattern, this period appeared many problems: the degree of centralization too deep; Degree of separation of powers is not fine; The scope of planning are not clear. Until 2006, the Sinopec group company preliminary establish an internal control system, this system still is not very perfect, needs to be strengthened.

4.3 Group financial information transmission speed is slow

Because it is a group company, the headquarters statistics all the financial information depends on the hundreds of the financial statements of the subsidiaries, information transmission speed is slow, information is easy to appear the phenomenon such as distortion. Different levels of the subsidiary company of Sinopec financial information. Because there is no strict internal control system as the basis, the subsidiary financial department's responsibility is not clear, in the process of statistics of the financial statements, ideas are not clear, vulnerable to outside interference or superior leadership department. Eventually lead to the production department and finance department coordination.

4.4 Group financial information software

The subsidiary company of Sinopec, the accounting computerization start earlier, but did not use uniform accounting computerization software. In the future for a long time, the subsidiary of financial information reported to headquarters is varied and headquarters makes it difficult to put all the financial information about organic together. After 2000, all subsidiaries unified use an accounting software, to make things get better.

4.5 Group information distortion is serious

Due to poor levels of each subsidiary part of the financial personnel, individual staff to cater to the company leadership department, falsification of financial information. Not strictly comply with the accounting ethics and accounting laws and regulation system, the financial and accounting information distortion is serious. Later, gradually establish the internal control

system, cancellation within group subordinate level 3 and level 4 company accounting independently, this kind of situation was improving. But because of the complex internal situation are too large, information distortion will still happen.

4.6 Consolidated is not authoritative

Subsidiary company financial management department power is larger, reported to the group's consolidated distortion, as a direct result of the group consolidated financial statistics out of groups authoritative [13]. Subsidiary in financial reporting, take the higher-yielding do low, low income, highway, human intervention in the financial statements to declare. Boom in oil prices, which had no significant effect to the whole group; But in the year of a drop in oil prices, a false prosperity, easy to make the whole group affect the group decision making.

5 CONCLUSION

Discusses issues related to group financial management models, and around the main problems in the practice, through the analysis of domestic and foreign research dynamic, investigates the variation law of the research object, the financial management mode of modern corporation made beneficial exploration. Group company financial management model research is a theoretical problem, and a robust operational practice, the content involved a lot, in many of the issues remains to be further studied and discussed.

REFERENCES

[1] Jiasheng Liao. Enterprise group finance companies risk prevention research [D]. Southwestern university of finance and economics, 2011. 17–28.
[2] Songsen Li, Yue Wang. Theory of Enterprise group and its formation way [J]. Journal of Northeast university of finance and economics, 1999, 05:34–38.
[3] Xuemei Wen. Group company financial control problems Research [D]. Southwestern university of finance and economics, 2003. 125–128.
[4] Quanzhong Guo. Corporate governance and the management system of research [D]. Renmin University of China, 2004. 326–329.
[5] Xingwen Zhang. Group company's financial risk management research [D]. Guangxi University, 2007. 32–35.
[6] Ning Hui. The regional economic effect of industry cluster research [D]. Northwestern University, 2006. 41–44.

Information, Computer and Application Engineering – Liu, Sung & Yao (Eds)
© 2015 Taylor & Francis Group, London, ISBN 978-1-138-02717-6

Research on the book binding design based on the concept of low carbon ecology

Hao Kun Ma
North China Electric Power University, Baoding, Hebei, China

ABSTRACT: Based on the concept of low carbon and it related to the essence of book binding design, combined with the present situation of Chinese book design, book binding design to explore the ecological. Ecological design and humanity design as the basic principle, discusses the background and significance of book binding and layout of ecological design, points out the main problems in the graphic design of a book to hinder the ecology, on the basis of this, put forward on strengthening the ecological consciousness a few points in the graphic design of a book. Hope this study can provide references for our country's book binding and layout design, and enhances the reader's appreciation ability and artistic accomplishment, establish the correct value, to promote the construction of harmonious society, and the development of low carbon economy and a green economy.

KEYWORDS: low carbon concept; ecological consciousness; book binding; problems and thinking

Prospective industry research institute forecast, to 2015 the total print run of book publication will be increased to nearly 8000000000 copies. In the range of books, book design more attention. With the change of the ideas of culture communication, more and more people begin to realize the important value of the book of environmental protection and the practical significance, whether it is the books shape material, or internal layout and structure, all of the readers' aesthetic experience and aesthetic habits formed a profound influence.

In the deteriorating ecological environment background, people began to re-examine their own behavior, how to scientific consciousness and correct ideas to create a new form of books, with reasonable structure make books really make the best use of everything, based on meet the readers' reading requirements, reduce damage to the environment, is a subject worthy of further study.

1 THE MAIN PROBLEM IN THE GRAPHIC DESIGN OF A BOOK THAT HINDERS THE ECOLOGY

The last century since the ninety's, along with the progress of the society, and the mature market economy, book binding design has been extensively, publishing field, become a symbol of the publication to improve the level of. In addition, the book binding design also brings about a series of ecological problems, if not promptly resolved, will probably be the destruction of ecological environment of the ills of book design. The current status of domestic development, the main problems in the graphic design of a book to hinder the ecosystem mainly includes the following five aspects:

1.1 *The ecological problem of excessive packaging*

In 2010, the central government formally promulgated the "notice" on the governance of excessive packaging work in 2013, the CPC Central Committee General Secretary Xi Jinping has made "enforce the important instructions of thrift, oppose extravagance and waste", which is an important measure for effective government protection and reasonable exploitation of ecological resources. In recent years social extravagance and waste problem is increasingly serious, books as cultural products have not been spared, excessive packaging, luxury and waste phenomenon is widespread, precious materials using a wide variety in the outer packaging design in some ordinary books and new materials. If the literature edition or classics, or personalized gifts of books, the precious, luxurious binding materials inverted youqingkeyuan, the problem is in the very ordinary books also uses complex binding mode and binding materials, such as some fast reading of children's reading picture books and there is no need to adopt special paper hardcover the [1]. Binding material weight many children's class gift book is far more than the basic and the most simple books weight, excessive packaging not only causes a waste of resources and ecological damage, and disrupt the market order books, caused great damage to the interests of consumers, contributing to the bad atmosphere of extravagance and waste, do not accord.

1.2 The ecological problems of abuse of process

With the development of modern printing technology continues to mature with each passing day of new materials and new technology, provides a wealth of material resources to the book binding design, create a good technological condition, the extensive application of new technology, greatly enriched the vocabulary designer design, effectively expanding the book binding design space for development. Books market increasingly intense market competition, leading to press too much emphasis on the external form of binding books, each press hope the first time to seize the reader eyeball in graphic design, often on the complex process as the standard of design and aesthetic requirements. The abuse of special layout process, such as the use of many hot stamping, process of silver anodized aluminum, anodized aluminum for the bottom membrane after transfer is difficult to be degraded processing, unrestrained abuse will inevitably increase the emissions of pollutants [2] anodized aluminum process. Since the end of last century the extensive use of plastic film mulching binding technology, coated with a layer of transparent plastic film on the covers of books, although can effectively extend the life of the picture books, more bright beautiful, but not harm problem considering the predeposited film itself, and after the coating process of books in two times in the use process often due to recovery the problem of high cost was forced to give up, thus exacerbating the problem of ecological destruction.

1.3 Ecological problems of paper applications

Book paper, monochrome printing basically adopts double gummed paper, writing paper, coated paper is basically used four-color printing, in order to enhance the whiteness of the paper, the wanton adds fluorescent whitening agent, whitening agent exceed the standard cause harm to human health. In recent years, along with the books, the increasingly fierce competition in the market, the characteristics of the paper is applied in many book binding and layout design. Traditional paper is mainly from wood or straw as raw materials, paper industry wastewater emissions of waste due to large, always is the most serious one industry for China's pollution. Special paper is to add the production of chemical raw materials in a variety of traditional paper basis and into, thus exacerbating the pollution of the environment, bring a series of ecological problems.

1.4 Ecological problems of binding method

The present book binding, popular lard in two ways, namely, lock line binding and wireless glue binding, these two ways to the use of the adhesive, and adhesive containing lead benzene two carbamate harmful substances, harmful to human body health. Many ancient books, hardcover books and classics are generally to choose in hardback form, while the hardcover book cover need to paste to a minimum of 2.5 mm thickness of cardboard, the cardboard and are generally made of grey board and pressed, and grey board paper mainly come from papermaking waste parts processing cannot papermaking in the slurry, belongs to the environmental protection material, grey board after several times of adhesive pressing became grey board, and many harmful substances contained in the adhesive is released slowly in the use of the books or the preservation process, cause the harm of to human body health and life.

Ecological problems of four aspects above, the current graphic design in the book binding design process is very easy to ignore, these ecological problems on ecological and human health has brought the huge negative influence, smooth hinder the construction of low carbon economy and harmonious color will.

2 REFLECTIONS ON STRENGTHENING THE ECOLOGICAL CONSCIOUSNESS IN THE GRAPHIC DESIGN OF A BOOK

Book binding actual need to stand on the aesthetic aspect of overall consideration and design, from conception to production link, must fully consider the binding design integrity and ecotype, height to achieve the combination of form and content, to focus only on the intrinsic value of books, advocating ecological environmental protection, low carbon green modern design idea, use the most suitable for the binding material and technology, achieve the content and form of books harmony, fully manifest the theme object aesthetic taste and cultural character, to give the book the aesthetic appeal, so that readers have the pleasure in the process of reading the emotional resonance and aesthetic experience.

2.1 Strengthen the consciousness of the whole book binding design, precise orientation of the audiences, the reasonable design

Two dimensional design emphasizes traditional books cover and format, modern book emphasizes the overall design of the full three-dimensional, involving more widely, including every link of the format of books, the cover, back cover, title page, waistband, correspondence, color, layout, font design, illustration, graphic material and process selection, printing and binding style, is the books by two-dimensional plane to six three-dimensional surface of a process, which belongs to the modern three-dimensional design. Style books audiences to determine the books, so it is very important to the audience positioning precision. Design to the intrinsic value of tightly around the book launch, height to achieve the combination of form and content, the harmonious unification.

2.2 To strengthen the research on personalized binding function design

At present, fragmentation reading times have fully come, book designers to strengthen design portability strong pocket book products, with low weight paper, reduce the volume and weight of books, makes it easy for readers to carry and reading. Many books inside pages for beauty, blindly pursue the paper thickness and stiffness, such as child class color books often use a higher weight coated paper to print pictures, children easily in the read process is coated paper sharp edges of the sheet to scratch, the reason behind the problem, mainly because of graphic design in the process failed to fully of humanities research. A good book designer must design and Study on the binding of standing in the angle of human nature, strengthen human color, books only so, can design the outstanding works to make the reader love books.

2.3 Using the low carbon environmental protection ecological materials

Printing paper, process materials are the realization of green production and ecological products, reduce the negative impact of product harm and lead to personal health and ecological environment in the process of production and use, resulting from the books in updating books, rubbish is easier to realize the natural degradation and timely conversion, and can basically realize the two use, this is the future development trend of book binding material. The latest eco-friendly inks and other related binding products, different level land to ecological standards above. Because there are many used in book binding product fails to meet the ecological standards, such as partial leather chemical processing of preparation of ecological problems, ecological problems such as high-temperature sterilization, so the conception of book binding and layout design of the part, the book binding materials the selection, selection of some special paper and special process, should give priority to the choice of ecotype material without harmful substances, no radioactive material, low carbon emission.

2.4 Strengthen the books shape design technology research and development

Make full use of the limited book binding material resources, improve the reading of humanized design and comfort design, fully meet the needs of each audience demand personalized reading. Especially in the development of the text paper color style multiple success, makes the text printing can not only realize the black text on a white background, but also with other different color words or graphs, in low carbon under the guidance of the concept, in the protection of readers vision principle, combined with the different connotation of book to create a different atmosphere for reading, open the reader door of the mind through the rich and varied forms of books. The use of modern technical means, adhere to the aesthetic principles, to create artificial ecological beauty, artistic strong so that it can enrich the connotation of graphic design, and can obtain the good ecological effect. Book binding design objective under the low-carbon concept is a harmonious unity of form and content in order to realize the books, for the two time in lifting the book function, scientific and rational use of modern technical means, to the overall planning and reasonable design of the shape of book, closely around the books the intrinsic value, create a broad ecological reading space design techniques most simple, strengthen the concept of low carbon design, strengthening ecological design consciousness.

3 CONCLUSION

Really good books, in addition to the content of outstanding, also includes the book binding design excellence. Books of ecological design should adhere to the reduction principle, the realization of low carbon environmental protection, not only embodies the ecological, environmental protection concept, but also to highlight the human spirit, this is the trend of the times and the development trend of the future of book binding design. Book binding design should stand in the perspective of the people, not only to create a good ecological reading space for the reader, but also to try to achieve a low carbon environmental protection, promote the harmonious development of man and nature. Stick to the road of sustainable development, based on the realization of economic development, but also to strengthen ecological consciousness, construction of low carbon environmental protection, harmonious and friendly type society. Therefore, strengthening ecological book binding design based on the low carbon concept is very necessary and has important practical value to guide.

In emphasizing the concept of low carbon environmental protection, seeking social cultural development at the same time, strengthen the protection of the ecological environment, to realize the harmonious unity of economic development and environmental protection. The book of ecological design as a special subject of systematic research, is very important, it will promote the sustainable development of China's book publishing industry. The low carbon concept in the book binding design, provides a new perspective for our book design practitioners, effectively expanding the creative space. As a graphic designer, even more need to strengthen the technical system and aesthetic thought in learning and absorbing. Advocate book binding ecological design, provide harmonious with nature and cultural products for people to create sustainable ecological culture space, will become the future mainstream book binding design based on ecological concept of low carbon.

REFERENCES

[1] Li Yameng. The regeneration design research on Ecological Design under the principle of books based on [J]. Association for science and Technology Forum (second half). 2013 (11).
[2] Wu Sword. Materials and process in book binding design form [J]. Journal of Hubei Normal University (PHILOSOPHY AND SOCIAL SCIENCE EDITION). 2013 (06).
[3] Zhang Jie. Book design art harmonious beauty [J]. arts. 2010 (05).
[4] Liu Yu, Zhou Yaqin. On the consideration of binding books emotional design elements of the [J]. packaging engineering. 2011 (02).
[5] Li Yuhui. On the negative space of book binding and layout design in the performance of the [J]. China Publishing. 2011 (02).

Information, Computer and Application Engineering – Liu, Sung & Yao (Eds)
© 2015 Taylor & Francis Group, London, ISBN 978-1-138-02717-6

Study on diversification strategy based on comparative analysis of Haier group and Giant group

Xiao Fang Li
School of Software, East China Jiaotong University, China

ABSTRACT: Diversification is to expand the scope of the organization that provides products or services. Many organizations have adopted a diversification strategy, but the results are very different. The paper presents some basic measures to implement diversification strategy more effective, which are based on comparative analyzing the successful experiences of Haier Group's application of diversification strategy and the lessons from Giant Group's failure of application of a diversification strategy, and on combining the diversification strategy theory basics.

KEYWORDS: Diversification strategy; Comparative analysis; Measures.

1 INTRODUCTION

Diversification is a corporate strategy to enter into a new market or industry that the business is not in vogue currently, whilst also creating a new product for that new market. This is most risky section of the Ansoff Matrix, as the business has no experience in the new market and does not know if the product is going to be successful.

Diversification is part of the four main growth strategy defined by Ansoff's Product/Market matrix: [1]

Products

Figure Ansoff's Product/Market matrix.

Ansoff pointed out that a diversification strategy stands apart from the other three strategies. The other three strategies are usually pursued with the same technical, financial, and merchandising resources used for the original product line, whereas diversification usually requires a company to acquire new skills, new techniques and new facilities. So diversification strategies are used to expand firms' operations by adding markets, products, services, or stages of production to the existing business. The purpose of diversification is to allow the company to enter lines of business that are different from current operations. Diversification is a form of growth strategy, which can use fully the excess resources to realize technical synergy, management synergy, marketing synergy. Rumelt[2] classifies diversification strategies into concentric diversification, core business-based diversification, related diversification and unrelated diversification. Haier Group and Giant Group have adopted the same diversification strategy, however Haier succeed and Giant failed, the results are very different.

2 COMPARATIVE ANALYSIS OF HAIER GROUP AND GIANT GROUP'S APPLICATION OF DIVERSIFICATION STRATEGY

2.1 Haier group's application of diversification strategy

Haier Group is an integrated large enterprise which has developed on the basis of the Qingdao Refrigerator Factory. Haier Group has taken mainly related diversification, which is orderly from related diversification into the highly relevant, moderate and low correlation of related industries, and to carry out non-related diversification carefully and gradually. It was during the 1990s that state policies encouraged business mergers and acquisitions. Haier Group over time acquired altogether eighteen domestic businesses, ushering in a broader development dimension in terms of diversified operation and expansion in scale, and gained a sustainable power development. In 2013, Haier's global revenue and profit reached

RMB180.3 billion (USD 29.5 billion) and RMB10.8 billion (USD 1.76 billion) respectively. Haier Group has transformed itself from an insolvent collectively-owned factory on the brink of bankruptcy into the number one global home appliance brand [3].

2.2 Giant group's application of diversification strategy

Giant Group relied on its original M - 6140 desktop publishing system has created the miracle that it had been developing at the speed of 500% for 3 years. In 1993, Giant Group has fallen into the bottom with the western computer companies entering into the Chinese computer market, so Giant Group started to adopt a diversification strategy and blindly entered an unrelated diversification, market, which invested 1.2 billion to build a giant building to enter the real estate industry. Meanwhile, it spent 5 million on health care products industry, and launched more than 100 varieties in less than one year. Because of the lack of the necessary management experience of real estate industry, Giant Group gained nothing in real estate industry, health care products market has achieved partial success, but it was not enough to support the development of the overall business [4].

2.3 Comparative analysis of the reasons of Haier's success and Giant's failure for their application of diversification strategy

Haier Group and Giant Group have adopted the same diversification strategy, however the results are very different, Haier Group succeeded and Giant Group failed, and the reasons are different. The following were comparative analysis from four aspects, which are core business and core competencies, synergistic effects, opportunities, approaches.

Firstly, the prime condition of diversification is to cultivate the company's core business and to develop fully core competencies [5]. Core competencies determine the depth and breadth of diversified business and core business is the foundation and security of diversification strategy. So organizations must build tightly its core business and core competencies when it plans to adopt diversification strategy. The Haier Group takes home appliance industry as the core business, and constantly explores their main business, upgrade product and expands horizontal integration of businesses. Meanwhile, Haier Group was the first to launch the "Star Service" system. While other home appliance manufacturers were engaged in a price war Haier Group had been already well positioned to win with its differentiated services and business. On the contrary, Giant Group applied diversification strategy without cultivating its core

business and core competencies, it didn't consider what can do and what suits to do, it chooses to enter a new business by a profit-oriented.

Secondly, whether organizations can establish a strategic match and the form synergy effect is the key to a successful application of diversification strategy. [6]Organization should prior choose closely matched to its core business, and easy to obtain the synergies areas as diverse objectives. Haier's successful diversification has two main experience: one is that its diversified businesses are based on its professional businesses; the other is that it supports related products and make new products related to original technology, marketing, management, which forms an effective synergy and builds its own irreplaceable competitiveness. Giant Group blindly pursued unrelated diversification strategy, while ignoring the synergy effect from strategic match, which is one of the main causes of the failure of its diversification strategy. It entered into unrelated businesses such as software, pharmaceuticals and real estate, which could not form a synergy in R & D, manufacturing, marketing and management, so Giants' diversification strategy doomed to lose.

Thirdly, Diversification opportunity is also important, which needs to consider two factors: the proposed industry stage and organization's strategic targets.[7] According to the theory of product life cycle, organizations should choose the imported products or growing products, and should be the first to choose those closely related to its core business, and easy to obtain the synergy effect. Moreover, whether organizations can enter the diversification industry also should make its strategic target match the industry opportunities. If the industry opportunity does not match its strategic target, organizations must abandon diversification; if the industry opportunity matches its strategic target, organizations must also consider the industrial technology, resources and competition. The Haier Group considered the proposed industry stage and organization's strategic targets fully when it entered a new industry, Giant Group did not consider them when it entered a new industry.

Fourthly, Diversification has two broad approaches to practice, one is internal entrepreneurship, and the other is External acquisition. [8] Internal entrepreneurship often faces to long cycle, high costs and great risks. External acquisition is easy for organizations to a new industry in the short term, and easily to get a variety of assets. So, the best choice to implement diversification should be acquisitions. Haier Group entered into air conditioners, freezers, washing machines, microwave ovens and other industries by external acquisitions, which it got efficient access to relevant assets. Combined with Haier's own high management ability, brand value and good sales service network, Haier made the success of a merger. When Giant Group entered into Real estate and medical by internal entrepreneurship, its cash flow began to decline and finally ran out.

3 CONCLUSIONS

Based on the above analyses, this paper finally presents three measures for organizations to implement diversification strategy more effective. The first is that diversification must build on its core business and core competencies. Secondly, diversification business must produce a synergy effect. The third is that Diversification must seize the right opportunities and choose the appropriate approaches. It is obvious that the measures present here are valuable. Of course, a lot of further research will need to be conducted in the future.

REFERENCES

Ansoff,Strategies for Diversification, Harvard Business Review, Vol. 35 Issue 5, Sep-Oct 1957, pp. 113–124

RichardP.Rumelt, Diversity and Profitabilit, Stra-tegic Management Journal, 3 (4) (1982), pp. 24–36.

http://www.haier.net/en/about_haier/

Bixia cai,T hought Unrelated Diversification based on Giant Group,[J].Commercial Economy, 2011, (4): pp.59 (in chinese).

Jing li, Diversification Strategy, [M]. Shang-hai,Fudan University Publisher, 2002(11): pp.126-128.(in chinese).

Lang,L.H.P.corporatediversificationandfirmperformance[J]. JournalofPoliticalEconomy,1994,(102):1248～1280

Richard A. Bettis,Performance Differences in Related and Unrelated Diversified Firms,Strategic Management Journal, 2 (1981), pp. 379–393.

Chakrabarti, Abhirup, Kulwant Singh,Ishtiaq Mahmood,Diversification and Performance: Evidence From East Asian Firms,Strategic Management Journal, 28 (2007), pp. 101–120.

Information, Computer and Application Engineering – Liu, Sung & Yao (Eds)
© *2015 Taylor & Francis Group, London, ISBN 978-1-138-02717-6*

Research on the application of interaction design in household appliances

Xi Yun Li

College of Art Design, Qilu University of Technology, Jinan, China

ABSTRACT: With the development of new technology and rising product features, the demand for interaction design also becomes more and more urgent. In this paper, based on household appliances, gives an introduction of the concept of interaction design and the necessity to emerge interaction design of products, and elaborate the application principle of interaction design in household appliances. It is aimed at designing a more humanized product by the designer; let users to clearly and accurately understand the product and effectively use the product.

KEYWORDS: Interaction design; Household appliance product; Human nature; User experience

With the improvement of people's living standard, the demand for household appliances is no longer met by the simplification functions, but pursues life taste enhancement and living environment improvement, and chases more intelligent and humanized products. Household appliances, along with the rapid development of information era, have ushered in a greater development opportunities and intelligentization is the inevitable trend for the future development of household appliances. Integration of interaction design in the household appliances design can achieve more convenient and comfortable, intelligent products with easy operation, and a pleasant function of "dialogue" with human.

1 CONCEPT OF INTERACTION DESIGN

Interaction design, also known as interactive design, is a kind of design field for the behavior of defining and designing the man-made system. Interaction design, as a new subject focused on interactive experience, was coined by Bill Moggridge, an IDEO founder in the 1980s. Interaction design mainly considers user experience, emphasizing the design of "humanization", and the "people-oriented". Interaction design combines research results and research methods in ergonomics, psychology, control technology, information technology and other subject, and creates the most harmonious interpersonal relationship [1] in the product and the user interaction process. The main research methods of interaction design are focused on analysis of the people and products interaction experience process, and make overall consideration of the user background, user experience, use feelings and other factors, in order to design the most suitable products.

2 NECESSITIES FOR THE APPLICATION OF INTERACTION DESIGN IN HOUSEHOLD APPLIANCES

From the view of the market, a large part of household appliances are still dominated by technology, product designers, instead of design from the demand of consumers, tend to put some technology together in reliance with existing technology, which eventually leads to products lack of humanization.

Facing with the increasingly complex products and user experience, function addition, the constant appearance of new artificial objects, the user's cognitive friction duly caused is also increasing, thus people's demand for interaction design is also becoming more and more urgent. Base on the present situation, how to apply interaction design to household appliances, designers shall firstly clear-out the basic process of the interaction design in product design (see figure 1). The basic purpose of interaction design application is to obtain user-satisfied products by designers.

At present, the application of interaction design in household appliances has been extended to many fields, and intelligentization will be the inevitable trend of household appliances design in the future, which shall be more and more favored by consumer, and brings convenience to people's life. The application of interaction design in household appliance control causes more harmonious human-computer interaction, and contributes to make simpler and more convenient products use [2].

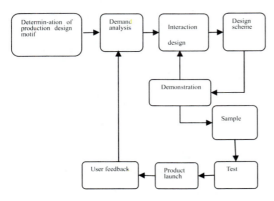

Figure 1. Flow chart for interaction design in product design.

Human-computer interaction design shall, in nature, be understood as "intercommunication" between human and machine. Designers shall, in the design process, involve themselves into "machine", namely product, and recreate the product function setting through the analysis of user experience (see figure 2); the realization of interaction design while product use appears to be the relationship between user and product, but in fact it is the communication between users and designers. For instance, apply LED technology in small household appliances such as water heaters and water dispensers for the product information presented to users in a timely manner, such as temperature, power, speed, capacity, and other info, in order to achieve the effect of man-machine communication, and build a more harmonious man-machine relationship.

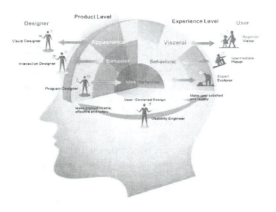

Figure 2. User experience and interaction design.

The lack of interaction design in the traditional household appliances, shall often cause behavior conflicts between users and machine. Take washing machine as an example, after the user starts washing machine, what shall he do if he suddenly changes his mind? How shall users operate the washing machine to satisfy his own requirements? And are these requirements realized immediately, or be achieved next time when he uses the washing machine? These problems shall all be considered by designers in interaction design, that is to say, designers shall conduct an analysis and summary of the user experience information, modify the existing washing machine design, and add midway clothes put-on, appointed laundry, power-off upon completion, laundry weighing and other functions on the control panel, and emerge the application of interaction design through function improvement household appliances.

3 APPLICATION PRINCIPLES OF INTERACTION DESIGN IN HOUSEHOLD APPLIANCES

Human-computer interaction design often appears problems in the practical application process, and the reason lies in the lack of common ground between designer and user, which makes it hard for many conversational interaction, so the designer, in the process of interaction design, shall follow certain basic principles, it is anything but the designer's "imagination", or "random arrangement". The basic principles are mainly as follows:

3.1 Simplification principle

Namely, make operation as easy as possible. For users, complicated thing is difficult product, which requires the designer to keep the information clear in design, and obtain concise, clear, and understandable interactive language. Take washing machine as an example, different kinds of washing mode shall be simple in design, and provides one-click service to the most possible extent, and the main design process is to simplify the tedious operation, and make it convenient for consumers to accept.

3.2 Dominance principle

That is to fully reveal the dominance status of product customers. Take washing machine use as an example, when consumers issue an instruction for washing machine, it shall give corresponding "respond"; and well respond shall bring a sense of pleasure to consumers, thus consumers will be in a dominant position to complete the better interaction process.

3.3 *Ease principle*

That is to say, the product shall always make customers at ease. Ease is a very typical use need, a kind of emotional needs. Designer, in ignorance of this problem, shall cause a feeling of anxiety and tension to the user, which will affect the acceptance ability of new products of the user.

3.4 *Intelligence principle*

In recent years, in high-end intelligent household appliances design, a key factor of interaction design is the intelligent automation, and its system shall conjecture or default instruction of the next step through the data analysis deconstruction. Environmental intelligence is a relatively new research project, and its goal is to create a kind of intelligent environment, which shall, in a caring, comfortable, active way, respond to man's existence and activities to provide the services required or may be needed for the people living in this environment. Many daily necessities and household appliances is embedded with internal computers, such as kitchen appliances, TV, heating system, etc., with the refrigerator, freezer knowing what's inside itself and, kitchen appliances knowing what time to do what and so on. Designers achieves real-time status tracking and analysis through data analysis as well as the application of modern technology (infrared induction, cameras, wireless radio frequency, etc.), enhances the intelligent security to complete the whole process of interaction design [3].

3.5 *Comprehensiveness principle*

Namely, the interaction interface shall provide rich, comprehensive, natural information. Designers, in the interaction design shall provide users with comprehensive information as far as possible, in order to meet the various needs of users. Richness and fullness are the basic requirements, while it is not only the requirements for designer to offer natural info that is easy to remember, but also the basic principle designers shall follow.

3.6 *University principle*

In the design of household appliances, the use of inter-action design also obtains a lot of standardized aspects.

Button design, for example, adopts uniform circular button design as a start button in many household appliances. Therefore, in the design process, while the function improvement and interaction perfection are pursued, it shall also be stuck to of information accuracy, comprehensiveness and generality for the basic design.

In addition to these basic principles, designers, in the product design process, may encounter many practical problems, such as precise semantic expression, good conceptual model, understandability, and so on, which shall all be practically considered in the household appliances design.

4 CONCLUSION

Future household appliances product design should accurately grasp the man-machine intercommunity, and accurately manage the interrelationship between machine and user, where interaction design is an indispensable important process. Interaction design in the future, should embody a more invisible form of communication in interaction pursuits, which is a way of undisturbed notification; invisible interaction is very interesting, and very helpful for intelligent system design, where no language, and no force are needed for information control and transition [4]. Therefore, intelligent automation will possess superior application value for household appliance in the future. It will be the important tasks faced by designers in the household appliances design to find a way to apply intelligent tools into the household appliance product, to improve the additional features of household appliance, to further develop the function of interaction design, and to bring more comfortable user experience to users.

REFERENCES

[1] LI Shiguo. *Product Interaction Design* [M], Nanking: Jiangsu Fine Arts Press, 2008.
[2] YANG Minglang, Hong WANG. *Perceptual Analysis in Man-machine Interaction Interface Design. Packaging Engineering* [J]. 2007, (11).
[3] JIA Hongwei, Weiqing TANG. *Research on Collaborative Design Human-Computer Interaction Resource Model. CIMS* [J]. 2007, 13 (12).
[4] Norman, Songtao Liu (translator). Future Product Design [M]. Beijing: Publishing House of Electronics Industry, 2009.

Information, Computer and Application Engineering – Liu, Sung & Yao (Eds)
© 2015 Taylor & Francis Group, London, ISBN 978-1-138-02717-6

Interdimensional association rules in drug influence research

Long Zhao, Xiao Bing Yang & Kang Jian Wang
College of Information Engineering, China Jiliang University, Hangzhou, China

ABSTRACT: With the development the information age, large-scale database system is popularized in our daily life. Data mining is a process of fetching valuable and important information. Association rule mining has become increasingly important in many areas. This article describes the use of interdimensional association rules on the influence of drugs. To discover the relationships between the drug influences and some physical examination items, a lot of experiments have been done. The hidden intrinsic relationships were discovered by mining the experiments data. And the drug influence discovered can be used to guide practical application which has an important significance.

KEYWORDS: Data mining; interdimensional association rules; drug influence.

1 INTRODUCTION

Data mining has gradually become one of the most important research directions. Data mining has made considerable achievements after so many years of efforts. For now, the study of data mining abroad focused on theory, technology and applications. The latest trend of data mining is the use of more in-depth study of algorithms in KDD.

The start-up and the development of the domestic data mining research in many ways behind foreign, and therefore there is no mature theory or technique. Research and Application of data mining is still in the development stage.

Data mining is rapidly evolving areas of research that are at the intersection of several disciplines, including statistics, databases, AI, visualization, and high-performance and parallel computing [1]. People in business, science, medicine, academia, and government collect such data sets, and several commercial packages now offer general-purpose Data Mining tools.

Data mining adopted its techniques from many research areas including machine learning, association Rules, neural networks, and so on.

Association rule is a powerful data mining technique used to search through an entire data set of rules, revealing the nature and frequency of relationships or associations between data entities. The resulting associations can be used to filter the information for human analysis and possibly to define a prediction model based on observed behavior [2]. How to find, understand and use association rules mining effectively is an important means to complete the data mining tasks, and thus the study of association rules has important theoretical and practical significance.

Today, the mining association rules have made remarkable results, the main research work is as follows.

After R. Agrawal proposed association rule mining, a number of efficient algorithms for mining association rules have been considerable development in the past few years [3]. The main research directions of a data mining algorithm contain multi-cycle mode, incremental updating algorithm, distributed and parallel, multilevel and multi-valued association rule, association rule mining algorithms based on concept lattice and so on.

The core idea of multi-cycle mode mining algorithms is hierarchical algorithm, the entire mining process is divided into several levels, after each level mining completed, and then combined each level into a final result. Such algorithms include Apriori, AIS, Apriori Tid and Apriori Hybrid [4]. DHP proposed by Park [5]. FP-growth, DIC and sampling algorithm were proposed by Toivonen [6]. A priori and FP-growth algorithm are the most effective and influential algorithms.

Incremental updating mining algorithm contains two cases [7]. 1) The update when database records changes. D.W. Cheng proposed update algorithm FUP, and the FUP2 algorithm was proposed based on FUP. The FUP2 can not only handle the increased transactions, but can also handle the deleted or modified transactions. 2) The update when the measure of association rules changes. Feng Yucai, etc. studied such case and proposed IUA and PIUA [8]. Feldman proposed association rules update technology called Border Algorithm [9]. The algorithm simply examines all true subset are frequent item sets under the lowest constant support for user-specified, while it is not frequent item sets. However, the algorithm still needs to store the results related to the frequent item sets for reducing the cost of updating the association rules.

Distributed and parallel association rule mining algorithm is generated under the background of the data to be processed is very large and geographically distributed. Distributed algorithms for mining association rules proposed by the majority of the literature are based on distributed processors in parallel mode. The main representative algorithms have PDM, FPM, and HPA and so on [10]. These algorithms can be seen as a parallel version of Apriori algorithm [11].

Multilevel association rule mining algorithm is that the minimum support threshold is defied based on each abstract layer on concept layer, and this algorithm is unlike the approach based support-confidence framework, it uses a variety of strategies and mining multilevel association rules [12]. Many mining multilevel association rule algorithms have been proposed so far. ML_T2L1 and its variants ML_T1LA, ML_TML1, ML_T2LA proposed by Han[13]; Cumulate, stratify and its variants Etimate, EstMerge proposed by R. Srikant[14].

Multi-valued association rule mining algorithm is different from the Boolean association rules. Now, the proposed multi-valued attribute association rule mining algorithms mostly translate multi-valued attribute association rule mining problem into a Boolean association rule mining. Multi-valued attribute's value is divided into a plurality of sections, each section as an attribute, each category of the attribute category as a property. G. Michael proposed a multi-valued attribute association rules and the form is $x = qx \Rightarrow y = qy$.

An association rule mining algorithm based on concept lattice is the most widely used and fruitful areas in data mining [15]. Domestic and foreign scholars were in-depth study in this area. Godin proposed the concept lattice model to extract implication rules, but the rules are deterministic, so that the method does not have the ability to describe the approximate rules. Hu Yun, etc. proposed a more efficient shopping basket analysis association rules algorithm based on the concept lattice structure, and the algorithm realized the association rule mining visualization.

Association rules that involve two or more dimensions or predicates can be referred to as multidimensional association rules. Database attributes can be nominal or quantified. The values of nominal attributes are "names of things". Quantitative attributes are numeric. Techniques for mining multidimensional association rules can be categorized into two basic approaches regarding the treatment of quantitative attributes. In the first approach, quantitative attributes are discretized using predefined concept hierarchies. In the second approach, quantitative attributes are discretized or clustered into "bins" based on the data distribution. Because this strategy treats the numeric

attribute values as quantities rather than as predefined ranges or categories, association rules mined from this approach are also referred to as quantitative association rules. We search for frequent predicates rather than searching for frequent item sets in multidimensional association rule mining. A k-predicate set is a set containing k conjunctive predicates. For instance, the set of predicates {age, occupation, buy} is a 3-predicate set. An example of a rule according to this definition would be:

$$
\text{sex } (X, \text{``Female''}) \wedge \text{age}(X, \text{``20~29''})
$$
$$
\Rightarrow \text{wage}(X, \text{``\$500''})[\text{confidence} = 85\%] \tag{1}
$$

Given this definition, provides an algorithm which approximately finds all rules by employing a discretization technique [16].

Now, the database association rule mining has made remarkable achievements, but also has a lot of challenging work. The trends of a data mining algorithm contains more efficient data mining algorithm, Mining, Visualization, Unstructured Data Mining, Parallel Data Mining Association Rules, More reasonable measure of evaluation criteria, Association rule mining technology under the network environment.

2 INTERDIMENSIONAL ASSOCIATION RULE

Association rule support and confidence are two measures of rule interestingness. They respectively reflect the usefulness and the certainty of discovering rules. If association rules satisfy both a minimum support threshold (min_sup) and a minimum confidence threshold (min_conf), then they are considered interesting.

$$
\text{support } (A \Rightarrow B) = P(A \cup B) \tag{2}
$$

$$
\text{confidence } (A \Rightarrow B) = P(B \mid A) \tag{3}
$$

$$
\text{confidence } (A \Rightarrow B) = P(B \mid A)
$$
$$
= \frac{\text{support } (A \cup B)}{\text{support } (A)} = \frac{\text{support\_count } (A \cup B)}{\text{support\_count } (A)} \tag{4}
$$

Rules that satisfy both min_sup and min_conf are called strong. Once the support counts of A, B, and $A \cup B$ are found, it is easy to derive the corresponding association rules $A \Rightarrow B$ and $B \Rightarrow A$ and check whether they are strong. If the relative support of an itemset I satisfies a prespecified min_sup, then I is a frequent itemset. Thus, the problem of mining association rules can be reduced to that of mining frequent itemsets.

There is a two-step process in association rule mining.

(1) Find all frequent itemsets: All these item set will occur at least as frequently as a predefined minimum support count.

(2) Generate strong association rules from the frequent itemsets: These rules must satisfy both minimum support and minimum confidence.

Because the second step is much easier than the first, the overall performance of association rules mining is determined by the first step.

A priori algorithm, proposed by Agrawal and R. Srikant in 1994, uses prior knowledge of frequent itemset properties. A priori property: All nonempty subsets of a frequent item set must also be frequent. And the a priori property is used to reduce the search space.

Apriori algorithm contains two important steps: connecting step and pruning step. But there are some differences when using in the application of drug effect.

(1) Connecting step: In order to find L_k a k-itemsets (C_k) candidate is generated by connecting L_{k-1} with itself. And the members of L_{k-1} are connectable if their first (k-2) items are in common. But the predicates can occur only once in an itemset.

(2) Pruning step: C_k is a superset of L_k whose members may or may not be frequent, but all the frequent k-itemsets are included in C_k. frequent, but all the frequent sets are included in C_k. While scanning database, the count of each candidate in C_k can be determined, also L_k (frequent candidates whose count is not less than the minimum support count). However, C_k may be large, its calculation amount also be lots. For compression of C_k, the Apriori may be used: any non-frequent (k-1) items can be not subsets of frequent k-items. Therefore, if (k-1) items of a candidate k-items is not in L_k, then the candidate cannot be frequent, which can be deleted from C_k.

Algorithm: Find frequent item sets using an iterative level-wise approach based on candidate generation.
Input:
D, a database of transactions;
min_sup, the minimum support count threshold.
Output:
L, frequent itemsets in D
L_1 = find_frquent_1-itemsets(D);
for (k = 2;L_{k-1}≠∅;k++) {
C_k = apriori_gen(L_{k-1});
for each transaction t∈D {//scan D for counts
C_t = subset(C_k,t);//get the subsets of t that are candidates
for each candidate c∈C_t
c.count++;
}
L_k ={c∈C_k| c.count ≥ min_sup}
}
return = U_kL_k ;

The a priori_gen procedure performs two kinds of actions, join and prune, as described. In the join component, Lk-1 is joined with Lk-1 to generate potential candidates. The prune component employs the Apriori property to remove candidates that have a subset that is not frequent. Then we can gain some itemsets about drug effect by the algorithm.

3 APPLICATION IN PHARMACEUTICAL

Medical information resources provide a good foundation for mining research work. Association rule mining techniques are good at finding meaningful patterns and knowledge from vast amounts of data, and it provides a strong technical support for medical knowledge discovery. Therefore, we have reason to believe that the medical research will gradually be promoted by the development of databases, artificial intelligence techniques, association rules, data mining technology.

Using the multi-dimensional association rule to a mining intrinsic relationship based on the patient's age and gender in the study of a drug to cure the disease. The rule contains many predicates (age, sex, drug, item), each of which occurs only once in the the rule. Besides, this rule must draw a result. First, considering whether gender alone would affect the results, if gender is just an unrelated property, and it has no effect on the drug curing disease, then just need to check the effect of age. Dividing the age range into groups rather the specific age, so that it will be more convenient. For example: age (X, "60~69"). There will be a test on the patients and the test is used to observe the patient's recovery condition after the patients are treated by the cure. Such as item (X, "albumin"). What medicine has been used on patients also need to be represented. If the potassium chloride injection is used on patients, then we can describe it like this: drug (X, "Potassium chloride injection"). There will be an effect can be obtained in an ideal case. For example: influence (X, "high before", "normal using", "normal after"), effect (X, "high before", "high using", "normal after"). Then we can get a conclusion like this:

Sex (X, "M") ∧ age (X, "60 ~ 69") ∧ drug

(X, "Potassium chloride injection") ∧ item

(X, "albumin") ⇒ influence (X, "low before",

"normal using", "normal after").   (5)

We have done a lot of experiments and got the result that contains 228700 records. And the objects of our experiments are for the elderly. Besides, we found that the sex and the age have little effect on effects of drugs. For example, 265 records are used to describe the effect of Furacilin I Solution on albumin. 135 records comply with the rule which is like effect

(X, "normal before", "low using", "low after"). The support of the rule is 0.12% and the confidence is 50.94% by calculating. We can get a result like (5) to describe the effect.

$$\text{drug}\,(X, \text{"Furacilin I Solution"}) \wedge \text{item}\,(X, \text{"albumin"})$$
$$\Rightarrow \text{influence}\,(X, \text{"normal before"}, \text{"low using"}, \quad (6)$$
$$\text{"low after"})[\text{support} = 0.12\%, \text{confidence} = 50.94\%]$$

4 CONCLUSION

Interdimensional association role plays an important part in drug application. And it can discover the effect of single drug on some physical items. Not only does it discover the effect of the drugs, but provides an important way to cure the patients.

ACKNOWLEDGMENT

The work is supported by the National Natural Science Foundation of China (Grant No. 61100160), Research on intelligent visualization of complex flow fields.

REFERENCES

[1] Xindong Wu. Data Mining: artificial intelligence in data analysis. *Proceedings of IEEE/WIC/ACM International Conference on Intelligent Agent Technology.* p7, 2004.

[2] Qi Luo. Advancing Knowledge Discovery and Data Mining. *2008 Workshop on Knowledge Discovery and Data Mining.* p3–5, 2008.

[3] Zaiane OR, EI-Hajj M, Lu P. Fast Parallel Association Rule Mining Without Candidacy Generation[C]. In: Proceedings of the 2001 IEEE International Conference on Data Mining. San Jose, California, USA, p665–668, (ICDM 2001).

[4] R. Agrawal, R. Srikant. Fast algorithms for mining association rules. Very Large Data Bases, International Conference Proceedings. p487–499, 1994.

[5] Park. Jong Soo, Chen, Ming-Syan; Yu, Philip S. Using a hash-based method with transaction trimming for mining association rules. IEEE Transactions on Knowledge and Data Engineering. p813–825, 1997.

[6] Hannu Toivonen. Sampling large databases for association rules. Proceedings of the international conference on very large database. p134–145, 1996.

[7] Zi-guo Huai, Ming-he Huang. A weighted frequent itemsets Incremental UpdatingAlgorithm base on hash table. 2011 IEEE 3rd International Conference on Communication Software and Networks. p201–204. (ICCSN 2011).

[8] Feng. Yucai, Feng. Jianlin. Incremental updating algorithms for mining association rules. Ruan Jian Xue Bao/Journal of Software. p301–306, 1998.

[9] Feldmann, Uwe; Schultz, Reinhart. TITAN: An Universal Circuit Simulator with Event Control for Latency Exploitation. Solid-State Circuits Conference. p183–185, 1988.

[10] Park. Jong Soo, et al. Efficient parallel data mining for association rules. International Conference on Information and Knowledge Management. p31–36, 1995.

[11] Garg, R. Mishra, P.K. Some Observations of Sequential, Parallel and Distributed Association Rule Mining Algorithms. 2009 ICCAE. International Conference on Computer and Automation Engineering. p336–342, (ICCAE 2009).

[12] Rajkumar, N. Karthik, M.R. Sivanandam, S.N. Fast algorithm for mining multilevel associationrules. Conference on Convergent Technologies for the Asia-Pacific Region. p688–692, (TENCON 2003).

[13] Han. Jiawei, Fu. Yongjian. Mining multiple-level association rules in large databases. IEEE Transactions on Knowledge and Data Engineeri. p798–805, 1999.

[14] R. Srikant, R. Agrawal. Mining generalized association rules. Future Generation Computer Systems. p161–180, 1997.

[15] Qing Xia, Sujing Wang, Zhen Chen. ARCA: An Algorithm for Mining Association RulesBased Concept Lattice. International Conference on Wireless Communications, Networking and Mobile Computing. p1–5, (WICOM 2008).

[16] Li. Yuefeng. Multi-tier granule mining for representations of multidimensional association rules. 2006 International Conference on Data Mining. p953–958, (ICDM 2006).

Information, Computer and Application Engineering – Liu, Sung & Yao (Eds)
© 2015 Taylor & Francis Group, London, ISBN 978-1-138-02717-6

The importance of the native language education and learning

Y. J. Cui, Y.S. Wu & F. C. An
Yanbian University, Yanji, Jilin, China

ABSTRACT: Nowadays people have not realized the value and position of the native language. Because of some one-side understanding of the language, people usually ignore the importance of the native language education and learning. In fact, language ability is one of the most foundational and core qualities of the people. Strengthening the native language learning will improve the development of thinking ability and also the study of the other subjects

KEYWORDS: Language; thinking; literal arts and science.

1 INTRODUCTION

Enthusiasm in learning Chinese has become a global phenomenon. More than two thousand universities open the course of Chinese all over the world. Chinese learning is popular abroad, but few people show much interest in it in china. Most people think to learn Chinese is to prepare for the college entrance examination.

People usually ignore the importance of the language.

Firstly, language is the tool of the communication.

We usually follow the direction of the language. There is a famous example of this, when Jill found an apple on the tree, she told Jack "I want to eat an apple." Jack got her thinking, and picked the apple for her. It's the function of the communication.

Secondly, language is a knowledge store.

After the appearance of the written language, we can read the text across the time and space. The accumulation of knowledge becomes possible; the civilization also speeds up greatly.

Thirdly, language is the medium of the thinking.

Language is an important medium of the thought, people usually do the judgment and inference by the means of language. The process of the listening, speaking, reading and writing is the process of the imagination, analysis, comprehension and judgment. All these procedures are realized by the thinking activities. Language and thinking are inseparable.

So the process of language learning is also the thinking exercise.

2 THE TEST DATA ANALYSIS

2.1 *What do the data tell us?*

In the senior high school in china, the students no matter majored in science or the liberal arts, should learn Chinese, mathematics, and English. As the basic and required class, the students should pay high attention to them. We found a lot of the parents and students pay great on the private math and English tuition, but nobody pays attention to Chinese. Chinese has not received the attention it needs. This paper chooses the final exam score of Chinese, math and English from a senior high school grade one in Yanji city. The final exam in this city is organized in the entire city with the unified examination paper and exam time. The samples are representative.

Table 1. 2013–2014 2nd term final exam sample (Literal arts).

Subject	Item	
Chinese	Average	109.8
	Good rates	49.3%
	Pass rate	90%
Math	Average	90.31
	Excellent rates	15.1%
	Pass rate	60.6
English	Average	115.6
	Excellent rates	65.8%
	Pass rate	96.1%

Table 2. 2013–2014 2nd term final exam sample (Science).

Subject	item	
Chinese	Average	108.4
	Good rates	43.2%
	Pass rate	90.2%
Math	Average	94.3
	Good rates	21.6%
	Pass rate	56.5%
English	Average	119.6
	Good rates	75.5%
	Pass rate	92.5%

From Tables 1 and 2, we can see the differences in Chinese are smaller than in math and English. And science majors did better in both math and English. It is different from our common understanding. We usually think literal majors should do better in both Chinese and English. We can not just judge the science majors are slower than the literal majors simply. But why is there small difference in Chinese, but great difference between math and English? How to solve the problem? How to narrow the gap? What's the secret in it?

People think and infer by the means of language, our mother language. But we just ignore the importance of it. People think Chinese learning is nothing but knowing some characters and sentences or reciting some knowledge. There is no need to do long-time learning or the further research. Actually, there is great relation between the Chinese and math, and also Chinese and English.

Table 3 is the correlations among Chinese and math, English of the literal arts majors.

Table 3. Correlations (Literal arts).

	C	M	E
C Pearson Correlations Sig. (2 tailed) N	1 225	.475** .000 225	.541** .000 225
M Pearson Correlations Sig. (2 tailed) N	.475** .000 225	1 22	.529** .000 225
E Pearson Correlations Sig. (2 tailed) N	.541** .000 225	.529** .000 225	1 225

**Correlation is significant at the 0.01 level.

Table 4 is the correlations among Chinese and math, English of the science majors.

Table 4. Correlations (Science).

	C	M	E
C Pearson Correlations Sig. (2 tailed) N	1 801	.383** .000 801	.444** .000 801
M Pearson Correlations Sig. (2 tailed) N	.383** .000 801	1 801	.444** .000 801
E Pearson Correlations Sig. (2 tailed) N	.444** .000 801	.444** .000 801	1 801

**Correlation is significant at the 0.01 level.

From the tables, we found the correlation between Chinese and meth, English is significant no matter among the literal arts majors or the science majors; Chinese influence math and English learning, so we can imagine if we enhance Chinese education, the other subjects can be promoted, because Chinese learning actually is a way of the mental exercise.

2.2 The relations between the native language learning and the second langue acquisition

Because of the similarities of the thinking mode among the human being, there must be same nature among the different languages. If the learner can make full use of their native language in the second language acquisition, and find the way to incorporate the recognition of language universals got into the native language learning into the foreign language learning, the comprehension and learning ability can be improved very quickly.

The native language, the first language, is got by osmosis and learning in the unified language environments. It plays the important role in the communication and thinking in the daily life. The second language acquisition is different from it. Under the premise of the mastery of the native language, the second language is acquired after that.

There is a common view about the first language transfer; there exists the transference feature in the language learning. Every learner will use this feature. At the beginning of the second language learning, the learners lack the overall understanding of the new language. In order to improve the learning efficiency, they will try to enhance the second language learning with the help of the language similarities. For this reason, we think that the native language plays a positive role in the second language learning.

In the process of long-term learning and use of the native language, the adult learners have formed a set of time, space and place concepts; they have the mature understanding and recognition. In the second language acquisition, they can make full use of the language similarities to promote the positive transfer. In the bilingual learning, the influence between native language and second language is mutual, native language background is the cognitive basis of the second language learning. Objectively, the level of people's native language has a direct influence on second language acquisition. During the period of comparative analytical theory, people thought there was a lot negative native transferred. But now more and more people believe the use of the native language is a process of recognition and regulation in the second language acquisition. In the learning, the learners don't focus on the overcoming the interference, but use it as a tool to master the new language and skills. Many scholars as Cook, 1993; Flynn, 1996; Gregg, 1996; Schachter, 1996; White, 1985, 1996 also believed that the native language plays a considerable role in the second language learning.

In our country, there are some common questions in the second language acquisition, especially the English learning. First, in the reading, learners always find they almost know all the words in the sentence, but don't know the meaning of it at last; second, it is difficult to get the implied meaning in the article; third, learners can not write a composition conforming to the requirements of the title; fourth, what the learners use and speak out is Chinese English. In fact, it is because the learners can not use Chinese properly to reconstruct the knowledge the new language, and this leads to the unsuccessful acquisition of the new language.

The learners with high levels of Chinese can perceive the sentiment color and artistic conception from the English originals, and can engage in the situations and resonate inside it. This ability can also help the learners master the English vocabulary. While learners with low levels of Chinese find it is difficult to feel the atmosphere and the meaning of the article. Two kinds of the situations led to the difference between the students in learning interest and the degree of persistence and finally influence their academic records. Although there is a difference between English and Chinese in the way of expression, the learner with high level of Chinese actually can understand and present English sentences sufficiently.

The learners with rich Chinese vocabulary and wide reading always feel easy to find the context of the original speech in the translation. Take Chinese-English translation as an example, good Chinese learners can always meet the requirement of the English expression. Good translation relates to the exact choice of the words and conveyance of the original spirit. It also relates to not only the foreign language knowledge, but also the level of the native language.

In the English writing, the good Chinese learners tend to use the words correctly and organizing passages properly according to the requirement. The better the learners' Chinese level, the easier for them to find the difference between the two languages. The composition is the convergence of the Chinese influence on the English acquisition. The positive influence of the native language on the mastery of the contents, structure and integrative view in the writing benefits from the accurate understanding of the difference of the two languages, the wide reading and also the accumulation of experience in writing. Learners with the rich textual knowledge and reading will have a better and quicker understanding of the style, topic and the hidden meaning of the article. If English and Chinese reading is all limited, the writing will be difficult.

The influence of Chinese on the English permeates all aspects of language learning. And it is continuous and bidirectional. In order to maximize the positive influence of Chinese, we should pay attention to the Chinese education, and also the learning and inheritance of the Chinese language and culture.

2.3 The relations between the native language and math

As everyone knows, math is ubiquitous in real life, and the contribution of the math is almost everywhere. If we want to solve the math questions, we must depend on the Chinese reading comprehension. We can see that Chinese is the basis of all the other subjects.

Actually, math is a kind of language. In the past, people thought math is the language and the tool of the natural science. And now it is thought to be the tool and the language of all the subjects, as social science, management and so on. It is little different from the everyday language. The math language is quite deliberate, intentional and well designed. So the famous American psychologist Bloomfield once said math language is the highest level the language can reach.

Math textbook is compiled by the words, math symbols and graphs. Sometimes math questions are only given in the form of written language. To solve this kind of questions and other the math questions, the topic examination is very important. If the Chinese reading and understanding ability is weak, the learners will find it is difficult to understand the question, or sometimes they will misunderstand the questions, finally, cannot work out the question or get the wrong solution. As the common reading, math reading is a complete psychological activity. Because of the high abstraction, math reading needs the strong logical thinking. The readers should know every terms, symbols; analysis its logic relations correctly, then work out the truth of the problem; and finally form the structure of knowledge by themselves. In this process, a lot of logical thinking involved. The math reading should also be very careful. The reader needs to analysis every sentence and graph, and figure out the content and the meaning. For the new definitions and theorems, learners should read them carefully again and again, and analysis it until they can understand and use them. And the semantic exchange is frequent in the math language; it requires the flexible thinking. The math language is the integration of the usual language, math symbol and the graphic language. The math reading emphasizes the understanding. One of the ways to achieve the goal is language transformation. Change the math language into the general accepted language; change the formal definitions and theorems into your own words. If the learners do not have the strong basic skill in Chinese, they cannot understand the question completely such as the word problem, probability and permutations and combinations.

In a word, if we want to solve the math problem, we must make full use of Chinese comprehension ability to examine the questions and the math learning. Lack of the Chinese knowledge and ability may cause the difficulty in math problem solving directly.

3 DISCUSSION

From the research, we can see the influence of Chinese to other subjects. We can also infer that language plays an important role in promoting the development of other subjects' learning. In the 1960s, an American professor found that the left and the right brain is completely different way to think. The learning of science, mainly depends on the left brain, while the perceptual activity depends on the right brain. There is the fiber connecting the two brains. In that way people can learn literal arts and science well at the same time. In the other word, if we only focus on one aspect of the learning, one part of the brain can be developed, but the other part will degenerate, and the thinking will become less and less effective. Language is a part of the intelligence; disabled language function will cause the mental problem. The ignorance of the Chinese learning will certainly influence the learning of other subjects.

In fact, Chinese learning not only influences the students' scores, but also the chance of the study and employment. People do not realize the importance of it. A lot of science majors do not like Chinese and also the politics and history. It is because they have a kind of mindset. The science like math is very logic; there are relations between the each chapter they learn. They do not need to memorize the knowledge, what they should do is to understand. For the good students, they can use the simple formula to derive any complicated formula and solve the question. So they like to challenge difficult questions. By the flexible use of the formula and the ability of deduction, they can get the highest mark in the exam. As for the Chinese learning, students need to remember a lot of knowledge, but the science majors' logic doesn't fit for this learning style. The knowledge, like poem makes them more stressful. For them, they can take Chinese as the relaxation; take the change of the learning style as a way of relaxation. In this way, they can devote themselves in the Chinese learning. For the students interested in the Chinese, they can improve the Chinese level by improving the learning method.

In people's life, there are two important periods in the language acquisition. The first one is from three to six years old; it is the formative stage of the native language. The second one is the high school; it is the change period of the mother language usages. In this period, people's sights widen, and their thinking comes to maturity. The language they use will change. It is also the time of the mental development and the formation of the language ability. Chinese education should take full use of this opportunity to accelerate the process.

4 CONCLUSION

There is a great correlation between Chinese and other subjects. Language plays an important role in mental development and also the people's real life. So Chinese teaching and learning is an important aspect which cannot be ignored in the high school education.

People tend to take the language just as a kind of the tool, and neglect the other functions of it. It is just because of the ignorance of its value and position; people do not pay much attention to the language training. It goes against the personal ability perfection. So the improvement of the language mastery and ability is the new task for all of us.

REFERENCES

Ye M. L. & Ye L. H. 2006. An experimental study on the stereotype of high School students Expecting to Major in Liberal Arts or Science. *Psychological Science* 29: 991–993.

Lei H. Z. 2011. The direction of the intelligence and the choice of science or liberal arts in high school. *Education Research Monthly* 1:46–48.

Dong H. N. 2012. The positive influences of the native language transfer on the second language acquisition. Journal of Chongqing University of Technology (Social Science) 26:125–127.

Liu Y. 2012. Relationship between language and thought. *Foreign Language and Literature* 28:89–92.

Yang G. S. 2014. The importance of the Chinese in the high school. *Education Teaching Forum* 25:276.

Chen F. 2013. The importance of the math reading in the math teaching. *Education and Teaching Research* 28:151.

Zhao S. J. 2013. Role and functions of individual's linguistic ability in the perspective of linguistic functions. *Journal of Yunnan Normal University* 45:37–42.

Chen H. 2014. The positive effects of the native language-strategy on the second language acquisition. Journal Kaifeng University of Education 34:51–52.

Information, Computer and Application Engineering – Liu, Sung & Yao (Eds)
© 2015 Taylor & Francis Group, London, ISBN 978-1-138-02717-6

Research on open and distance education model in English teaching

Guo Fen Wang
Langfang Television and Broadcasting University, China

ABSTRACT: Applying open and distance education model in English teaching can produce a good effect during the teaching process. It's a kind of new teaching methods to learn language. By means of Multimedia teaching aids and research on the English Teaching method to improve the quality of teaching, students' interest and ability in learning English. This paper focuses on open and distance education model research in English teaching.

KEYWORDS: Open and distance education, English teaching model.

1 INTRODUCTION

Open and distance education is a new mode of education after correspondence education, though the dissemination of computer information expanding the space of learning. In order to develop the students' ability to learn by themselves, teachers can though Internet to have lessons with their students. In the teaching process, in line with the philosophy that students are the main body, teachers are main leader. Applying open and distance education model in English teaching can achieve the goal of teaching science and make the students understand and master knowledge better.

2 GETTING STARTED

2.1 Teaching model of open and distance education

Teaching model is a kind of specific teaching operation to achieve certain teaching goal. Therefore, teaching model is the actual act of teaching theory, still a lot of teaching philosophy and experience summary. The general method of English teaching is to build a network of idea, strong purpose and instructional design emphasis on English teaching. It can be seen, the actual teaching of English has four main focuses: theoretical knowledge, media application, teaching content and process. Emphasis on theoretical knowledge is a kind of theoretical knowledge used in English teaching to make it more standardized and scientific. In the day of science and technology development, multimedia and evolving network information also penetrated in the teaching. Some scholars believe that the new English teaching should pay attention to the application of multimedia teaching, the use of network

information resources spread knowledge of English. The emphasis on the process of teaching content and teaching methods of English teaching goal is to identify and understand the impact of cultural origins and cultures English-speaking countries English language learning students can cultivate students' language proficiency. There existing differences not only where subject of language focus on, but also in teaching methods, and we cannot evaluation arbitrarily which is the best, however, in the application of open and distance education, teaching English need to adapt to adult learning, the teaching of English to be updated and improved.

2.2 Open and distance English teaching applying student-learners model

Only a theory cannot explain what the problem is real, this article through the student-learners, English teaching model to have experimental study.

(1) Focus on the ability of the students' self-learning
Applying open and distance education in English teaching for the sake of improving the students' self-learning and comprehensive abilities to accomplish the goal of teaching English. Although learning from them, it is not able to boot from the correct teacher, students need the proper guidance to learn accurate knowledge of English, or it will cause uncertainly, especially English language such disciplines, pronunciation, spelling deviation are not, or it will become another meaning. Self-learning mode consists mainly of learning programs for students to develop their own learning situation, and their choice of study time and place, to learn the scope and amount to develop their own learning, self-learning and self-evaluation record. The independent study of the opposite sense, independent learning require that students need to learn by their actual requirement

as their learning basics, by learning method as a countermeasure, by guidance as teaching goal, and the correct guidance of teachers and appropriate teaching plan suiting for students, student-centered learning autonomy will be measured.

Teachers correct guide to show the change of the teachers' role in the traditional sense, the previous teaching mode is the teacher as the main body, the "spoon-feeding" teaching to the student, especially in English this language in this teaching will only become more boring, so, after the application of open and distance education teaching way, change the old way of teaching, the students as the principal part of teaching and teachers has become a study guide and evaluators. Teachers need to be before the teaching to the teaching task, goal and content, learning mode and evaluation method for making, and to communicate with students in my study plan, lets the student can also on the learning plan. Teachers can use the teaching guidance, E-mail and other ways to communicate with students, and help students to guide students' autonomous learning plan, then supervise the students' autonomous learning, promote the development of autonomous learning and the improvement of evaluation.

(2) Teachers to scientific design, face-to-face teaching attaches great importance to the classroom atmosphere

According to the build socialism teaching idea, the student is the main body of learning, and students' autonomous learning ability and cognitive consciousness itself can affect the English learning process. But also don't tell the teacher's guiding role will have limitations; in contrast, requires teachers in teaching English knowledge when face to face can be teaching guide and the designer. In the application of open and distance education in English teaching, must reflect the status of the main body of students and teachers to guide in coordination.

Face to face, a professor at the design of the learning process for students' general situation, system plan scientifically, and to attach importance to students' and teachers' teaching status, realize the classroom interaction between teachers and students, for students to do a good examples of autonomous learning. Professor's face to face, mainly include teaching goals, teaching of listening, speaking, reading and writing and so on four important part of learning English, but also to solve the problem of the students' oral English teaching. The preparation of teachers to students in class, when teaching in class to use writing blackboard writing and multimedia teaching methods, to ensure that the students can not deviate from the study of key problem, and can organize students to discuss a question, help students to practice the spoken language power of expression, not only improve the learning ability of

students, to students provides a good English learning environment. In learning English, the focus of the chapter to teacher inspires and introductory explanation as the main way of teaching, teachers can adopt the method of interactive questions in class to let the students to participate in the study, make students understand knowledge in memory. That distinguish the key way of teaching can improve students' English application abilities, to a certain extent, improve the teaching quality.

(3) Combining teaching auxiliary resources and network resources

Teaching auxiliary resources, mainly include the students in the learning of printed materials, teaching materials needed in counseling information, practice examination paper, word CARDS, etc. And the main network resource in teaching is the teaching resources of the network, especially in open and distance education teaching model. In the teaching, the teacher needs to be both organic, common knowledge help learning how to learn English.

Today's students when they study English all have a common problem, is listening to don't understand, and also said not export. And now less than the number of English teachers, which makes it impossible to realize the teacher can undertake tutoring for each student, but with the open and distance education can make use of information network platform, make teachers to communicate with adult students, students can learn according to their own situation to teacher ask how to improve my English, and teachers can inspire and guide students in English listening, speaking, reading and writing practice, and students can also through the network information to real people imitate to study English, could hear the application of accurate pronunciation, to practice listening and speaking, to solve the problem of listening and speaking. The second is the innovative network information class. Online education resources will provide students with some innovative, full of joy of learning materials, such as audio equipment, animation and English film, etc., enrich the students' knowledge of English. Finally is the open and distance education network classroom resources requires teachers to reasonably to use of equipment resources, effective management students in the network autonomous learning curriculum in the classroom.

(4) Teaching guidance of class teaching

Such remedial teaching work mainly auxiliary teaching of after-school learning teaching face to face. The extracurricular teaching method has a more flexible, more vivid teaching characteristics. Remedial teaching can help the students in imparts the knowledge they learned in class to consolidate, and can improve the efficiency of face-to-face classroom teaching. Give student's communication and practical applications of activity the opportunity; tutoring class can improve the students' interest in learning English and the ability of autonomous learning.

3 CONCLUSIONS

Application in English teaching mode of distance open education theory research, summarizes the teaching experience, this paper holds that the English of distance open education is to learners as the main body of learning, teachers to guide the way of teaching, and through the let students autonomous learning and face-to-face teaching and tutoring class teaching way, USES the effective teaching resources, to improve the students' English level. But now, the students' autonomous learning way of evaluation and examination is not comprehensive enough, and the application of network information in teaching is not enough skilled, students in English listening and speaking learning also faces some difficulties, therefore, the current implementation of remote opening education English teaching model also cannot achieve the best effect, but through constant research and exploration in the future, believe in open and distance education English teaching pattern is bound to achieve the best teaching effect.

REFERENCES

Vivian lulu red, XingFa countries. Remote opening education English teaching pattern reform and try [J]. Journal of Shijiazhuang school of economics, 2005 (2).

Dan-dan liu, xiaojian. Two kinds of English teaching in modern distance education mode comparative study [J]. Foreign language world, 2003 (6).

Investigation of [3]. Build socialism learning theory and college English teaching mode selection [J]. Journal of BBS of trade unions, 2007 (1).

Liu Xiping. Modern open and distance education teaching and learning mode study [J]. China distance education, 2002 (6).

Dai Jinrong. Theory of modern open distance education in teacher's leading role [J]. Journal of Zhuzhou institute of technology, 2006 (1).

Jian-xin Zhao. Try to talk about the function of the distance education teaching materials and writing characteristics [J]. China distance education, 2001 (1).

Li Baochuan. The functional change of open and distance education grassroots teachers think [J]. Open education research, 2002 (6).

XingFa. Open education for English majors of formative assessment practices and thinking [J]. Journal of hebei radio and television university, 2002 (4).

Liu Yanxiu. Computer-assisted autonomous learning + class mode exploration and learners study [J]. Foreign language world, 2008 (1).

Information, Computer and Application Engineering – Liu, Sung & Yao (Eds)
© 2015 Taylor & Francis Group, London, ISBN 978-1-138-02717-6

The exploration and practice of continuing education for professional and technical personnel for mechanical manufacturing industry

Tie Jun Zhang
Research Institute of New Materials Technology, Chongqing University of Arts and Sciences, Chongqing, China

Ying Guan
Library, Chongqing University of Arts and Sciences, Chongqing, China

Yan Zhao
School of Foreign Languages, Chongqing University of Arts and Sciences, Chongqing, China

ABSTRACT: Mechanical manufacturing industry is one of the main developments of Chinese economic growth. Scientific and technological innovation driven manufacturing upgrading is the key for Chinese manufacturing industry from large to strong, and more importantly, it is to train the professional and the technical personnels in this industry. Through the teaching practice of undertaking municipal and national level continuing education for professional and technical personnels, the working experience of continuing education can be summarized at Chongqing University of Arts and Sciences as follows: first, strenuously serve for professional and technical personnel's knowledge update project, departments at all levels catch condominium together and make unity and cooperation. Second, elaborately design the content on the basis of transformation of enterprises and application of new materials and technologies. Third, catch organization, management and service work. Last but not the least, timely collection of trainees' feedback and evaluation information. These provide some new thoughts to improve the quality of training and teaching professionals and technical personnels in mechanical manufacturing industry.

KEYWORDS: Mechanical manufacturing industry; Knowledge Upgrading project; Professional and technical personnel; Continuing education

1 INTRODUCTION

In addition to the acceleration of the globalization process, the relationship between economy and education becomes closer day by day. China is now at a crucial point of building a well-off society, quickly transforming the pattern of economic development and comprehensively deepen its reform. As a result, we need higher education to supply the frontline service of modern production with applied, inter-disciplinary and innovative talents, who not only master modern science technology, but also has received a system of skill training, especially the professional and technical personnel on the top of the industry chain. The decision at the 3rd plenary session of the 18th central committee of CP is to comprehensively deepen its reform. So during the process of innovation driving and transformation development in China, the Research Institute of New Materials Technology, Chongqing University of Arts and Sciences is not only a research center, but also an education center. The key thing which we have explored for a long time is how to play a further role in servicing the local

economic society. In recent years, the awareness of the new universities to run schools locally is weak. Influenced by the experience and thought of the traditional universities, new universities pursue high level, large scale and overall field schools. Generally, they get far away from the thinking of the local. Therefore, the time for discussiing and re-understanding of the local is not long.

First, it is the objective demand of the rapid development of knowledge economy for the local universities to train professional and technical personnel. With the rapid development of the knowledge economy, the economic development of our country has turned to depend on the intensive technology and intensive knowledge instead of the natural resources. Also, colleges are now cut themselves from the economic development, but face the economic development of the dominant economy. This is a new era, where the development of the higher education and the economic society are combined together firmly. The development of economic society does not just rely on researchers, but more on those people, who apply, implement and realize the modern technique. The needs of the society for talents are always changing, so the quality of the professional and technical personnels now has already slow down the speed of the economic development. According to their self-position, local universities must adjust themselves to adapt to the changes of the economic development and the industrial structure, and explore the new ways of training application-oriented personnels. From then on, we conducted the exploration and practice of updating the professional and technical personnel' knowledge.

Second, training high quality technical personnels for regional economic development is the mission of application-oriented university. As our higher education has changed from elite education to mass education, application-oriented university conform to the trend of higher education and create a huge room for the growth of school transformation. Therefore, to study the concept of scientific development to meet the needs of the economy and society development, and to seize the opportunity to realize the characteristic development of the school, it is an accurate positioning and a smart move to conduct the engineering of updating the professional and technical personnel' knowledge.

2 THE OBJECTIVE DEMAND OF TRANSFORMATION FOR LOCAL APPLICATION-ORIENTED UNIVERSITY

2.1 Background

China has stepped into a period of mass education. In 2020, the gross matriculation rate of higher education will be 40%. The book, Outline of the National Medium-and Long-term Program for Education Reform and Development, says that we should establish a system of classified management for higher education, speed up to establish a system of modern vocational education, and focus on expanding the scale of the training for application-oriented professional and technical personnels. So, for China's modernization drive, we need a large number of innovative personnels and also millions of skilled personnels. Those personnels can apply important science achievements to the fields of production and daily life to promote industrial transformation upgrading and the development of economic society. Therefore, mass higher education needs to speed up the transfer, application and accumulation of advanced technology. One of the major tasks for mass higher education is to train high quality technical talent in front line of production. The construction of university has been on the agenda of China. To conduct Chinese characteristic application-oriented university, we should learn from foreign universities, accumulate technological innovation and integrate them into the development of regional industry, based on the needs of real economy development.

That is to conduct the system of training personnel, an overpass to conduct the growth of personnel, and an essential part to create a room for the growth of front line labors. The higher education system and structural reform must focus on the close relationship between the drive of training talent in colleges and universities and the needs of the economic society.

2.2 Industry-academy-research cooperation

It is bound to pass a long and hard time to conduct local application-oriented university to promote the development of transformation in local higher education institutes. Certainly, it will also meet many problems, difficulties and challenges. So our government, industry and community need to reach a consensus and gather together, with more courage, confidence and determination, to reform local higher education institution, to integrate education with production and to conduct transformation development. We are having some effect at present. For example, we have provided CSIC Red Oil Machinery Co., Ltd., with PVD commodity supply coating, which prolongs the piston pump life. We also have done the work for the failure analysis as well as the diagnosis and countermeasures on that the key part roller fractured during production, such as providing China Merchants Aluminum (Chongqing) Co., Ltd., with the direct evidence of large roller's breaking, which helps the company claim successfully and avoid a great economic loss, and so on.

2.3 The demand of regional industry development

The automobile and motorcycle industry is the main industry in Chongqing. As a district far from main city, Yongchuan is the place where plants in main city have been relocated. According to the survey in Bishan, Qinggang, Dazhu, Da'an and Shanjiao, which are towns in Yongchuan, by the machinery industry and dozens of companies, we know that from 2010, most of the new graduates from universities are unwilling to work at these local companies. Medium and small private companies or joint-stock materials processing companies badly lack relevant technical staff. In these companies, the detection technology is poor, and the current technicians cannot update their knowledge as soon as possible. As the only local university in Yongchuan, our school has made a target for establishing a regional, application-oriented and multi-subject university, and taken a developmental strategy for "holding up the heavens and supporting the earth" development. "Holding up the heavens" is to set up excellent characteristic subjects, based on the needs of the local economic and cultural development. "Supporting the earth" is to root in the cultural development to train application-oriented personnel for local economic and cultural development. It is our goal to root in Yongchuan, serve Chongqing, and face with the whole China. Since 2010, Academician Tu has been giving lectures on new technologies in materials in Yongchuan and Dazhu. Under the leadership of the academician, Tu Mingjing, the Research Institute of New Materials Technology has done a lot of preliminary survey and preparation work to implement the Mid-to-Long Term Development Plan for Training Professional and Technical Personnel in Chongqing (2010–2020), to train high level personnels for machinery manufacturing, to strengthen the ability of independent innovation and the ability of comprehensive competition in manufacturing organizations in Chongqing, to promote the improvement of industrial technology, and to speed up the transformation and upgrading of companies.

3 LAYOUT OF THE EXPLORATION AND PRACTICE IN UPDATING THE PROFESSIONAL AND PERSONNEL'S KNOWLEDGE

Under the rapid development of knowledge economy, we have explored the training mode of the engineering for updating the professional and technical personnel' knowledge, deepen the security of training system and the reform of teaching content, strengthen the training of practical ability and improving the occupation quality of the talent. We have held the senior seminars for updating the professional and technical personnel' knowledge successfully, which are shown in Table 1.

Among them the paper of technology study was collected into the Collection of outstanding papers written by four students, Huang Li from CSIC Chongqing Hydraulic Machinery Co., Ltd., Zhao Yong from Huayi Machinery Casting Co., Ltd., and Liu Liang from Xintai Machinery Co., Ltd. This seminar plays an active role in training high level talent for machinery manufacturing, strengthening the ability of independent innovation in manufacturing organizations in Chongqing, promoting the improvement of industrial technology, speeding up the transformation and upgrading of companies, and promoting the industry group evelopment. At the same time, it broadens the companies' awareness of our newly founded local university, Chongqing University of Arts and Sciences. According to the plan for the senior project of the engineering for updating the professional and technical personnel' knowledge of Chongqing ongqing Human Resources and ocial Security Bureau and Chongqing Economic and Information Commission has been authorized Chongqing University of Arts and Sciences is to host the 2014 advanced training class for the popularization and application of the whole city advanced Engineering materials and Manufacturing technology in mechanical manufacturing industry. This training class has been approved to become the base of the engineering for updating the professional and technical personnel' knowledge of Chongqing, and has won the approval of Human Resources and Social Security to undertake the project of 2014 professional and technical personnel of national advanced training. Next year, we go on to hold the senior seminar.

Table 1. Continuing education for professional and technical personnel for mechanical manufacturing industry.

No. Time	Participants	District distribution	Subjects
1 2013Dec	58	Bishan, Dadukou, Jiangjin, Rongchang, Dazhu, ect*	Advanced materials engineering, Enterprise management
2 2014Jun	60	Yubei, Dianjiang, Qijiang, Wanzhou, Yuzhong, etc**	Advanced engineering materials, Manufacturing technology
3 2014July	85	Beijing, Tianjing, Heilongjiang, Guangdong Guizhou, etc***	Advanced materials and new technologies

*8 districts altogether.
**18 districts altogether.
***13 province and municipality cities.

4 CONCLUSIONS

The conclusion section should state concisely the most important propositions of the paper as well as the author's views of the practical implications of the results. It is an important trend toward the development which is environmentally friendly to the industry, seeks characteristics, and strengthens the application, for both the domestic and the foreign, to establish the senior project of the engineering for updating the professional and technical personnel' knowledge and construct the training system for high level application-oriented talents. Our university has conducted a theoretical study and a practical exploration of the engineering for updating the professional and technical personnel' knowledge, preliminarily and effectively, which laid a foundation for the research.

The transformation has just begun. We have to face the history and reality, breaking throughthe trouble, developing in reform and innovation. As positioning and directioning are clear, we should expand the senior project of the engineering for updating the materials engineering technicians' knowledge, following our own path. We will finally realize the vision that we can try our best to do something for the progress and prosperity of the mechanical manufacturing industry.

ACKNOWLEDGMENTS

This work was supported by Science Research Foundation Program (Grant No. R2012CJ19) and Teaching Reform Project (Grant Nos. 130334 and 140348) of Chongqing University of Arts and Sciences. The authors also thank the Chongqing Human Resources and Social Security Bureau and Ministry of Resources and Social Security of the Peoples Republic of China for the financial support in the year 2013, 2014 and 2015. Corresponding author: Zhao Yan.

REFERENCES

Zeping Sun and Guixin Qi. 2013. Practicing and exploring of applied talent cultivation system building, taking Chongqing University of Arts and Sciences as an example. *Chongqing Higher Education Research* 1(1):54–58.(in Chinese).

Mingjing Tu and Di Xu. 2013. The strategy of low-competition and non-competition areas in science and technology. *Chongqing Higher Education Research* 1(1):32–35. (in Chinese).

Maoyuan Pan. 2010. What is application-orented higher education. *Higher Education Exploration* 19(1):10–11. (in Chinese).

Yongyao Su and Dongping Shi. 2013. Boost practice to enhance the quality of the postgraduate students of material engineering. *Chongqing Higher Education Research* 1(5):68–71 (in Chinese).

Liang Liu and Tiejun Zhang. 2014. Failure analysis of fracture roller in 700 mm for a rolling mill. *Journal of Chongqing University of Arts and Sciences* 38(4): 105–108. (in Chinese).

Information, Computer and Application Engineering – Liu, Sung & Yao (Eds)
© *2015 Taylor & Francis Group, London, ISBN 978-1-138-02717-6*

MALL via mobile applications: Evaluation of an electronic English dictionary on smartphones

Y.L. Liao & L.J. Gao
College of Information, Yunnan Normal University, Kunming, Yunnan, China

ABSTRACT: There is no doubt that a dictionary is the most important and necessary reference for language learning and an appropriate dictionary would enhance language learning outcome. Compared with paper dictionary, E-dictionary with the advantages of light, portability, costless, rich content has become an upcoming trend. This paper is to make an effort by picking out essential factors via two typical electronic English dictionaries, analyze these features according to the applications, and finally offer an available evaluation template for language learners. This study is also supposed to be helpful for language instructors, dictionary editors, publishers, and syllabus designers.

KEYWORDS: Mobile applications; E-dictionary; Mobile phone; MALL (Mobile-assisted Language Learning); Merriam-Webster; Youdao

1 INTRODUCTION

To date, MALL (Mobile-assisted Language Learning) has extended the path traced by CALL (Computer-assisted Language Learning). Meanwhile, incredible progress has been made in mobile phones when MALL began. With prices falling and functions increasing, the actual number of mobile phone users is up to 4.3 billion, over 60% of the entire world population. Meanwhile, in developed countries mobile phone subscriptions now represent over 128% of the population (Bursto, 2014). Even in developing countries, mobile phones are virtually common, especially among adolescents and young adults. According to the China Internet Network Information Center, Internet users have increased to 632 million in China by the end of June, 2014, and Internet users by mobile phone account for 83.4% of the total Internet users.

The ubiquity of mobile phone ownership and the available access to Internet have motivated the mobile applications developers. Bestuniversities.com listed 100 best iPhone applications for college students by emphasizing mainly on study help with foreign languages and math, newspaper sites, shopping deals, local guides, and references (Truong, 2014). On the basis of Li Lan's (2006) investigations in Hong Kong Polytechnic University, above 70% of students interviewed claimed that they used electronic dictionary more often than the traditional bulky paper products (Zarei & Gujjar,2012). What's more, the mobile phone has become a complicated device with the incorporation of multiple functions, which was earlier found only on MP3, PDAs, and laptop, such as high quality audio recorder and playback, graphics and video playback, digital still and video camera, accompanied by large data storage, programmability, and access to the Internet (Bursto, 2014).

While mobile applications are increasing in diverse area and the hardware is upgraded on the mobile devices with more advanced technologies every year, there is still a lack of analysis and evaluation to help instructors and students in choosing suitable applications for their needs, such as in language learning. Mass production of electronic dictionaries and human inclination towards convenience have urged students to use electronic dictionaries more than ever before owing to their various advantages, such as portability as they are set-up in a smart phone, low cost (some of them even cost free), compact and faster than any paper dictionary, easily updated by using Internet and software, many volumes compressed into one electronic dictionary, large memory that contain large volumes of vocabulary, authentic recorded voice facilities for better pronunciation, cross-referencing by linking to the Internet, etc.

Most of the researches are concerned with survey, development, and investigation, for instance, contribution of paper and electronic dictionaries to EFL learners' vocabulary learning (Zarei & Gujjar, 2012); development of E-dictionary (Omar &Dahan, 2011); empirical evidence regarding the effects of using different forms of dictionary, including printed dictionaries, pocket electronic dictionaries, and online type-in dictionaries, on English vocabulary retention (Chiu & Liu, 2013).

However, few studies focus on the evaluation of electronic dictionary. Diversity in characteristics of mobile applications, such as cost, content, functions, and access to the Internet, will affect learners' decision when they need to choose appropriate electronic dictionaries fit for their mobile phone and individual language level. The purpose of this research is to present an evaluation template of electronic dictionaries in MALL (Mobile-assistive Language Learning) and develop criteria for the decision of language reference applications based on CALL. Moreover, this research will help to to identify the usability issues and potential problems of the use of various dictionary types with students at different educational levels.

2 EVALUATION OF ELECTRONIC DICTIONARIES: MERRIAM-WEBSTER VS. YOUDAO

Generally speaking, evaluating software and websites for reading and writing in CALL consists of the following factors: cost, structure, technical feature, goals, presentation, appropriateness, outcomes, evaluations, and notes (Egbert, 2005). The evaluation is usually made by teachers and help teachers to meet their classroom instruction. Nevertheless, it seems not sufficient for language learners due to individual demand and personal mobile device once we are engaged in a mobile learning environment. Hence, more factors should be considered, such as presentation (interface, content design, media format), appropriateness, outcomes, feedback, and Internet connection. Moreover, it is worth to notice that electronic dictionary, to some extent, is more likely sort of e-book instead of a common software.

Two electronic English dictionaries are selected for the analysis based on their characteristics, such as content, interface, media format, update frequency, access to Internet, and improvement of learning skill, etc.

The first choice is Merriam-Webster electronic English dictionary. As a well-known publisher for more than 150 years, Merriam-Webster offers all kinds of language dictionaries whatever in print, online and in mobile products. Merriam-Websters' electronic English dictionary on smart phone with Android operating system is one of open English learning resources. The second is Youdao English-Chinese electronic dictionary, published by the website of cidian.youdao.com and developed by NetEase corp., it is free to download and use on smart phone with Android operating system. They are both running on smart phones with Android operating system, and are able to download and use for free. However, there are a number of varieties worth

deeper comparison and analysis. For convenience of comparison, Merriam-Webster electronic English dictionary will be replaced by Merriam-Webster e-dictionary, and Youdao English-Chinese electronic dictionary will be replaced by Youdao e-dictionary in the following of this paper.

2.1 Content

Merriam-Webster e-dictionary is a conventional English reference in content. It supplies common items for vocabulary learning like attributes, spelling, pronunciation by authentic recorded voice, English explanation, example sentence, origin, first use, synonyms, and antonyms.

Youdao e-dictionary might represent a new trend in the electronic publishing. Its content is a rich and colorful combination that embed electronic dictionary and online-journal, including not only universal items similar with Merriam-Webster, but also more bilingual columns, such as daily proverbs, sayings, essays, jokes, scientific news, informations of study abroad, and English slangs.

2.2 Presentation format

As for the presentation format, text is mainly presented in Merriam-Webster e-dictionary while multiple media formats, such as text, image, online photo, online audio, are introduced in Youdao e-dictionary.

One-fold media and multiple media have their own advantage. English-to-English Text explanation in Merriam-Webster e-dictionary is helpful for the learners' concentration on English vocabulary retention when the learner has hold enough storage of vocabulary; rich media formats, like colorful image, picture, recorded essay online-audio with bilingual text presented in Youdao e-dictionary, would improve learner's comprehension of target language and motivate learner' interest.

2.3 Accessibility

Most functions of Merriam-Webster e-dictionary can be used straight after acquisition in offline access, while Youdaoe-dictionary, as a typical online e-dictionary, is supposed to get access to the Internet once learners long for more details than the elemental meaning.

2.4 Frequency of update

Merriam-Webster e-dictionary can be updated by users in an uncertain frequency, depending on the download version; Youdaoe-dictionary is daily updated automatically by the online system.

2.5 *Personalization settings*

There are distinct settings for individual learners in pop-up menu of Merriam-Webster electronic dictionary, for instance, *Recent* can mark out recent searching track of the vocabulary; *Favorites* is a record of learner's favorite words.

Similarly, such functions are set-up in Youdaoe-dictionary, like *My Youdao Notebook*, which mark each word you feel difficult to remember, and count the words; besides, you can catalog the words by their order of character, marked date, and review a certain number of words according to your learning pace.

These personalization settings would assist learners to make up their learning plan and check their own progress day-by-day, step-by-step.

2.6 *Interactivity*

Interactivity might be the most important difference between paper dictionary and electronic dictionary. Both e-dictionaries, Merriam-Webster e-dictionary and Youdaoe-dictionary provide users and publishers with multiple communication channels, for example, *Feedback*, which promote users to mail the publisher through the Internet; *Rate This App*, which allow users to grade the e-dictionary according to their own experience and submit the online survey to the publisher; *Share This App,* which encourage learners to recommend this app to other learners in order to enhance their common learning experience.

2.7 *Advertisements*

It might be a problem for a devoted learner when a number of advertisements are planted into the free e-dictionary. Both e-dictionaries, Merriam-Webster e-dictionary and Youdaoe-dictionary, also face this problem, especially in Youdaoe-dictionary, there are more advertisements than Merriam-Webster e-dictionary because of its integration of e-book and online e-journal, which is updated daily. Obviously, advertisements, the most popular commercial element, would be attached for free download. We take it for granted that there is no way to have a free lunch in the world.

However, everything always has both sides. On the one side, these advertisements would distract the learner's concentration on language learning; on the other side, learners might get a chance to acquire more related applications and information of language learning from these advertisements planted in offline e-dictionary or poppedup in online e-dictionary. This depends on learners' reasonable choice.

2.8 *Distinct features*

Word of the Day, as a distinct feature in Merriam-Webster e-dictionary, provides daily word where it derive from and its history automatically via online system.

Online Instant Translation, which offer a timely platform once learner need to translate target language into their native language or vise versa, is a distinct feature in Youdaoe-dictionary. Moreover, Error-tolerant input (Zarei & Gujjar, 2012) helps users to look up words with wrong spelling; cross-referencing (Zarei & Gujjar, 2012) with relevant Website link offers the users an opportunity to get rapid access to other references.

3 CONCLUSIONS

Electronic dictionary, as a kind of reference electronic resource, or an integration of e-book and online e-journal, should have its certain factors to consider. That characteristic makes it slightly different from the evaluation of language learning software. In this paper, an evaluation template is provided for language learners by utlizing two types of e-dictionaries, Merriam-Webster e-dictionary and Youdaoe-dictionary. As a result according to the evaluation, Merriam-Webster e-dictionary fits for those who desire to improve and polish English in fluency and accuracy; Youdaoe-dictionary is more effective for those who are in between elementary stage and medium level of language learning, need to review new words every day, and enhance their motivation in English.

ACKNOWLEDGEMENTS

This research was financially supported by China National Science Foundation Project (No. 61262071); Key Laboratory Program of China Ministry of Education (EIN2011C001); China Scholarship Council Award.

REFERENCES

Bursto, J. 2014. MALL: The Pedagogical Challenge. Retrieve from: http://dx.doi.org/10.1080/09588221.2014.914539.

Chiu, L.L. and Liu, G.Z. 2013. Effects of Printed, Pocket Electronic and Online Dictionaries on High School Students' English Vocabulary Retention. *Asia-Pacific Education Resources* 22(4):619–634.

Egbert, J.2005. Developing and Practicing Reading and Writing Skills. In P. Gibbs (eds), *CALL Essential: Principles and Practice in CALL Classroom*: 25–26.

Alexandria, Virginia, USA: TESOL (Teachers of English to Speakers of Other Language).

Omar, C.A.M... & Dahan H.B.A.M. 2011. The Development of E-dictionary for the Use with Maharah Al-Qiraah Textbook at a matriculation Centre in a University in Malaysia. *TOJET: The Turkish Online Journal of Education Technology* 10(3): 255–264.

Truong, D. 2014. How to Design a Mobile Application to Enhance Teaching and Learning? Retrieve from: http://dx.doi.org/10.3991/ijet.v9i3.3507.

Zarei, A.A. & Gujjar, A.A. 2012. The Contribution of Electronic and Paper Dictionaries to Iranian EFL Learner's Vocabulary Learning. *International Journal Social Science& Education* 2(4): 628–632.

Information, Computer and Application Engineering – Liu, Sung & Yao (Eds)
© 2015 Taylor & Francis Group, London, ISBN 978-1-138-02717-6

The analysis of enterprise value drivers based on the EVA discounted taking the State Grid provincial company as example

Hai Peng Sun, Li Ping Lei & Ju Zhang
State Grid Shanxi Electric Power Company, Shanxi, China

Yu Luan Huang
North China Electric Power University, School of Economics and Management, Beijing, China

ABSTRACT: Because the perspectives of understanding the value of the enterprise are different, the value drivers have different classifications. This passage proposed classification of the form of the enterprise value based on the EVA discounted, thinking that the enterprise value can be divided to the current trading value and future growth value, analyzed the drivers of two values, respectively and took the provincial company for example to analyze the drivers in its operating activities combining the power industry characteristics.

KEYWORDS: EVA discounted; Corporation value; Value drivers; State Grid provincial companies.

1 THE ENTERPRISE VALUE AND THE VALUE CONSTITUTION

Enterprise value is the meaning of the existence of the corporation. The corporation owners invest the capital for the purpose of obtaining higher return than the cost of capital.[1] Thus the value of the business lies in creating a steady flow of investment that is higher than the opportunity cost of capital gains (that is EVA). On the basis of this concept, we use EVA discounted method to represent the corporate value. In this way, the enterprise value is equal to the total amount of capital investment plus the discounted present value of the future phases of the EVA, the discount rate is a weighted average cost of capital. The value expression based on the assumption of going concern enterprise is:

$$EV = TC_0 + \sum_{t=1}^{\infty} \frac{EVA_t}{(1+WACC)^t} \qquad (1)$$

In the equation, EV represents the enterprise value; TC_0 represents the beginning of business investment capital; EVA_t represents economic added value of the first period t; WACC represents the weighted average cost of capital. Invested capital represents the core business activities of enterprises (fixed assets, intangible assets and operating liquidity capital) on the cumulative amount that has invested.[2]

When assuming that the business annual EVA fully coming from the beginning of the invested capital, then the above formula is:

$$EVA_t = NOPAT_t - TC_{t-1} \times WACC$$

Capital investment is the base of increasing profits of an enterprise, and is the key of enterprise value creation. The assets of the enterprise are the capital invested in the past. To improve the income of enterprises, on the one hand, improving the operational efficiency of existing assets, fully utilizinating idle assets or inefficient assets; on the other hand is to increase capital investment and new assets.[3]

Capital investment will increase the EVA of enterprises during the future. [4] On the one hand, the current EVA is the capital invested by the past (shown as existing assets) in the current operation, and the capital will constantly produce EVA during the period of the future through the continued business activities of enterprises; on the other hand, the new capital such as the construction of fixed assets will be converted to the assets of the enterprise after a certain process, breaking through the assets operation efficiency bottleneck or opening up new markets to gain more profits by optimizing the existing portfolio, brings the enterprise with the addition value of EVA. [5]

According to the enterprise value expression before, the enterprise value of zero growth model is as follows:

$$EV = TC_0 + \sum_{t=1}^{\infty} \frac{EVA_1}{(1+WACC)^t} = TC_0 + \frac{EVA_1}{WACC}$$

That is the enterprise of the current operating value (COV) [6], representing the value created by the existing capital scale of continuing operations.

Tthat $COV = TC_0 + \dfrac{EVA_1}{WACC}$ is,

Using the constant growth model it assumes that the enterprise of EVA is at constant growth rate g, then the enterprise value is:

$$EV = TC_0 + \sum_{t=1}^{\infty} \frac{EVA_1(1+g)^{t-1}}{(1+WACC)^t} = TC_0 + \frac{EVA_1}{WACC - g} \quad (2)$$

The difference between the enterprise values according to the constant growth model minus the calculation at zero growth model of enterprise is the enterprise's future growth value (FGV) calculated on the same growth model, representing future value growth ability of the enterprise, namely

$$FGV = TC_0 + \frac{EVA_1}{WACC - g} - (TC_0 + \frac{EVA_1}{WACC})$$

$$= EVA_1 \times \frac{g}{WACC(WACC - g)}$$

The derivation of a more general expression of enterprise value:

$$EV = TC_0 + \sum_{t=1}^{\infty} \frac{EVA_t}{(1+WACC)^t} = TC_0 + \frac{EVA_1}{(1+WACC)} + \frac{EVA_2}{(1+WACC)^2} + \dots$$

$$= TC_0 + \frac{EVA_1}{(1+WACC)} + \frac{EVA_1 + \Delta EVA_2}{(1+WACC)^2} + \frac{EVA_1 + \Delta EVA_2 + \Delta EVA_3}{(1+WACC)^3} + \dots$$

$$= (TC_0 + \frac{EVA_1}{WACC}) + \sum_{t=1}^{\infty} \frac{\Delta EVA_{t+1}}{(1+WACC)^t}$$

$$= COV + FGV$$

$\Delta EVA_{t+1} = EVA_{t+1} - EVA_t$ representing the first $t+1$ period of EVA value.

So we get the expression of enterprise's overall value:

$$EV = (TC_0 + \frac{EVA_1}{WACC}) + \sum_{t=1}^{\infty} \frac{\Delta EVA_{t+1}}{(1+WACC)^t}$$

$$= COV + FGV$$

2 THE DECOMPOSITION OF VALUE DRIVERS

The above distinguished the enterprise value to the current operation of the value and future growth value, so the overall enterprise value is the sum of the two parts. [7] The factors influencing on the value of the current operating mainly is the invested capital, capital cost and the current EVA. When other conditions remain unchanged, the more the capital,

the greater is the current value of the operations of the enterprise and that the cost of capital is smaller. The factors that influence the value of the enterprise future growth is the EVA increment and the cost of capital, when other conditions remain unchanged, the smaller cost of capital of the enterprise value, the greater the future growth. Enterprise managers should be prepared for long term value, reducing the cost of capital and making the balanced choice between the current value of the operations and future growth, making the enterprise current EVA as well as the future periods EVA is the largest, when the two cannot be taken into account, choosing the overall value of the enterprise maximization as the standard. [8]

For a certain period of the enterprise, the initial capital investment and the cost of capital are certain, the enterprise current operating value drivers is the enterprise current EVA, drivers on the value of enterprise future growth is ΔEVA, namely each period increment of EVA. The economic behavior that can bring the current and future ΔEVA is valuable.

Further derivation of the current operating value expression is:

$$COV = TC_0 + \frac{EVA_1}{WACC} = TC_0 + \frac{NOPAT_1 - TC_0 \times WACC}{WACC}$$

$$= \frac{NOPAT_1}{WACC} = \frac{TC_0}{WACC} \times ROIC$$

When TC_0 and WACC remains unchanged, the ROIC is the core drivers of enterprise current operating value.

Future growth of the value of the drivers ΔEVA can be divided into two parts: one part is generated by the increase of existing operating capital levels, another part is produced by the new capital. Then the future growth value drivers can be divided into raising the level of the existing assets operating factors and the new asset factor.

$$\Delta EVA_{i+1} = EVA_{i+1} - EVA_i$$

$$= TC_i \times (ROIC_{i+1} - WACC) - TC_{i-1} \times (ROIC_i - WACC)$$

$$= (TC_i - TC_{i-1}) \times ROIC_{i+1} + TC_{i-1} \times (ROIC_{i+1} - ROIC_i)$$

$$-(TC_i - TC_{i-1}) \times WACC$$

$$= \Delta TC_i \times (ROIC_{i+1} - WACC) + TC_{i-1} \times \Delta ROIC_{i+1}$$

ΔTC_i represents the stage i of newly increasing capital, that is the increase of the investment of stage I. $\Delta ROIC_{i+1}$ is the increase of the return on investment of phase $i+1$. $TC_{i-1} \times \Delta ROIC_{i+1}$ is the increase of returns of the stage I beginning total capital on the increasing of the operating level of the period of $i+1$.

3 ANALYSIS OF THE VALUE DRIVERS OF STATE GRID PROVINCIAL COMPANY

The enterprise value based on EVA discounted can be decomposed into the current trading value and future growth value. [9] Take the provincial company in power industry for example, segmenting the current trading value and future growth drivers, combining with power management features of enterprise value, discussing the control of the power grid enterprise value drivers in daily management specifically.

3.1 Long-term value analysis

The so-called long term value is that enterprise managers reasonably plan for the future periods of EVA on the basis of the enterprise strategy inorder to make whole enterprise value to the maximum.

Long term value is the basis of current analysis of enterprise operating value drivers and future growth value driving factors, it mainly has two purposes, one is to influence the current value of the operations and the value of future growth factors of management as a whole, the second is the plan of the future periods of EVA, making the enterprise operating current value and future growth value maximum to realize the goal of whole enterprise value maximization. [10]

The factors in enterprise economic activity can be divided into the following three categories, according to the relationship with current EVA and future ΔEVA: only influencing current factors; only influencing the future factors; influencing current and future factors. Reclassifying the activities and elements in the operating of the provincial company according to this standard:

Table 1. The reclassification of the elements in the provincial company operating activities.

Impact term	Business	Activities	Factors
Only influence the current period	Buy and sell electricity	Sell electricity	Revenue from electricity sales
		Buy electricity	Power purchase costs
		Management activities	Employee benefits
			Other operating expenses
Influence the current and the future	Electricity transmission and distribution	Fixed asset management and maintenance	Depreciation fee
			Material fee
			Maintenance fee
		Others	R & D expenses
			Training expenses

Combining the classification of the enterprise value and the reclassification of State Grid provincial company operation elements can analyze specific impact of provincial company operating current value and future growth value driving factors, targeting for the current operating value and future growth value for the concrete control to realize the value management goals of the whole enterprise value maximization.

3.2 Analysis of the drivers of the current operating value

From the above analysis, the current operating value of a company is equal to the EVA of current period divided by the cost of capital, coupled with the current total capital invested. [11] The core driving factor of the current operating value is return on invested capital ROIC, And the return on invested capital can be decomposed into the rate of return on sale and the asset turnover ratio, which each measure the performance of the enterprise in sales and assets. For a provincial Power Grid Corporation, when we subdivide the factors affecting the rate of return on sale and the asset turnover ratio in daily management activities, we can get further indicators of the current operating value driving factors.

3.3 The rate of return on sale

In the guidance of EVA, the rate of return on sale is one of the main drivers of improving and enhancing the current operating value. The rate of return on sale measures the corporate performance in sales, and it is made up of operating income and operating costs. These two aspects mainly reflect the operating activities of the enterprise, and is the main activities to maintain and improve the current operating value.

1 Analysis of operating income. Most of income of a provincial Power Grid Corporation is from electricity selling income, which depends on two aspects of sales price and electricity sales. Under the existing system of electricity market, the price of electricity is not a controllable factor for electric power enterprises. In the case of the electricity price policy unchanged, the influence of the electricity price factor is constant. Therefore, operating income mainly depends on the electricity sales.

2 Analysis of operating costs. The costs of a provincial Power Grid Corporation are mainly composed of electricity purchasing costs, employee compensation, materials expenses, depreciation costs, repair costs and other operating expenses. Among these expenses, electricity purchasing cost, labor costs and other operating costs only affect

the current EVA. However, materials expenses, depreciation costs and repair costs are the comprehensive factors which affect both the current EVA and the future EVA. In the process of execution, the enterprise controls the current amount of sales expenses and management expenses according to the sales expenses and management expenses budget, and evaluates the cost center's performance according to the cost budget, thus to reduce the unreasonable expenses.

For the costs which only affect the current, they should be strictly controlled. Given the income isn't changed, cutting the cost can increase the current EVA. For the costs which affect both the current and the future, they cannot just be controlled. we should consider the impact on the future EVA and make the necessary inputs. If the occurrence of some costs in the enterprise cannot produce EVA neither in the current period nor in the future period; the costs are unnecessary and should be avoided.

3.4 The asset turnover ratio

The asset turnover ratio measures the efficiency of assets in the enterprise. Enterprise assets are divided into current assets and fixed assets.

1 Analysis of fixed assets. The provincial Power Grid Corporation is a typical asset-intensive enterprise. The value of its fixed assets is huge. The driving of fixed assets to value creation is mainly manifested in two aspects: one is to create business value through the use of fixed assets; the other is to improve the using efficiency of the assets, reduce capital occupation of idle assets and reduce the value of damage, thus increasing the value.
2 Analysis of current assets. Current assets mainly include the monetary funds, accounts receivable, inventory, etc.

Strengthening the management of funds and improving the turnover speed of inventory and accounts receivable and other current assets can reduce the occupancy of the capital, thus reducing the cost of capital and increasing the enterprise value.

Generally speaking, for a provincial Power Grid Corporation, the driving factor of operating income is less controllable. The main emphasis of control during the period should be focused on reducing costs and increasing the service efficiency of assets. In order to reduce the cost effectively, we can reduce the unnecessary spending and put money into the projects which can generate large profits without affecting the income. Reasonable allocation of resources, reasonable arrangement of the structure of current assets and fixed assets can improve the total asset turnover ratio.

4 ANALYSIS OF THE DRIVERS OF THE FUTURE GROWTH VALUE

The drivers of the enterprise future growth value are mainly the factors which can increase the rate of return on capital and future assets. Combined with the specific situation of a power grid enterprise, the drivers of the future growth value in a provincial Power Grid Corporation are the investment on fixed assets and the costs which influence both the current costs and the future costs.

1 Fixed assets investment. For a provincial Power Grid Corporation, Long-term investment is mainly the productive long-term investment, the performance of fixed assets investment. Fixed assets investment, especially productive fixed assets investment, are essential for the long-term value creation capacity of a provincial Power Grid Corporation, and are the core driving factor of the future growth value, even of the overall value of the enterprise.
2 Costs which influence both the current and the future. As a deduction item of the current EVA, these are included in the current cost. These costs do not work in the current period and do not create any value for the enterprise. However, they work in the future and are long-term and they contribute to increase the future and long-term value of the enterprise.

5 SUMMARY

Overall, this article, based on the enterprise value model of EVA discount, deduces the expression of the current operating value and the future growth value. In addition, it conducts further description by combining the production and operation feature of a provincial Power Grid Corporation. We hope that this article will be useful for the future research on the value drivers of the future power grid enterprises.

REFERENCES

[1] Chang Cai. 2003. Value Based Management: Improve Cash Flow and Enhance Enterprise Value. Haitian Press.
[2] Applied Investment & Finance Analyst Institute. 2011. Valuation Modeling. China Finance Publishing House: 39–40.
[3] Martin J.D. and Petty J.W. 2005. Value Based Management: The Company's Response to the Transform of Shareholders. Shanghai University of Finance and Economics.

[4] James I. Grant. 2005. The Foundation of Economic Value Added. Dalian: Dongbei University of Finance and Economics Press.

[5] Tom Copeland, Tim Koller Companies, Jack Murrin. 2002. Valuation: Measuring and Managing the Value of Companies. Beijing: Electronic Industry Press.

[6] Shanghai National Accounting Institute. 2011. Value management. Economic Science Press, P. 267.

[7] Ying Zhang. 2013. Research on the Value Drivers of Thermal Power Enterprise. North China Electric Power University.

[8] James A. Knight. 2002. Beijing Tianze Economic Research Institute Translate. Business Based on the Value. Yunnan Press.

[9] A. Eibar. 2001. Economic Value Added – How to Create Wealth for Shareholders. Beijing. China CITIC Press.

[10] Alfred Rapport. 1986. Creating Shareholder Value. Free Press:102–110.

[11] Ran He and Yunsi Xie. 2010. Market Value Management: The unity from Value Creation to the Value Realization. *Financial Supervision* (24):47–48.

Information, Computer and Application Engineering – Liu, Sung & Yao (Eds)
© 2015 Taylor & Francis Group, London, ISBN 978-1-138-02717-6

Preliminary analysis on the application of two-dimensional Gaussian spline function in geomagnetic navigation

Zhi Gang Wang & Chun Sheng Lin
Department of Weapon Engineering, Naval University of Engineering, Wuhan, China

ABSTRACT: A main aspect of geomagnetic navigation is to identify the carrier's location on an existing geomagnetic map, and several navigation algorithms as relative matching and Kalman algorithms are the most prevalent methods that many scholars are using. However, conventional relative matching algorithm, with discrete geomagnetic data used as database, is usually derived from terrain navigation; as a result, its navigation performance depends heavily on the accuracy and resolution of geomagnetic map. Due to that background, a novel algorithm that is different from relative matching algorithm for geomagnetic navigation is developed in this paper. The algorithm implements geomagnetic navigation by directly estimating the carrier's location through Kalman filter algorithm and the key part of this implementation is a 2-D Gauss spline function. First, the principle of local geomagnetic model based on 2-D Gauss spline function is given in this paper; whereafter, theoretical derivation and calculation are done, and from which we can see the given local geomagnetic model, with its good performance in accuracy and analyticity, can effectively establish the observation equation required in the implementation of geomagnetic navigation by Kalman filter algorithm. Finally, simulation test is done on actual grid geomagnetic data with encouraging results. From the results we can see that the 2-D Gauss spline function can be used in geomagnetic navigation, and with which the navigator error can be reduced markedly. Although this algorithm is developed for aided navigation using geomagnetic data, it is equally applicable to other domains, for example, aided navigation on gravity or terrain data.

KEYWORDS: Geomagnetic navigation; Gauss spline function; Local geomagnetic field; Extended Kalman filter; Inertial navigation system.

1 INTRODUCTION

Carrier's aided navigation is an area of research with broad commercial and military application. Recently, there has been greater interest in using geophysical maps (for example gravity and terrain) to amend errors inherent in inertial navigation system. These navigation methods, as gravity and terrain navigation[1–4], can bound errors inherent in traditional navigation systems based on dead-reckoning or inertial navigation to a certain extent. However, a gravimeter with high precision is very expensive, and just as terrain navigation, gravity navigation also has limited applicable field, if data variation is not obvious, navigation accuracy will drop greatly. In this case, we introduce geomagnetic navigation[5–7], which is an autonomous navigation system, and has many excellent features such as cost-effective, undetectable, high-accuracy, continuous navigation. With these excellent features, geomagnetic navigation can cover the shortage of gravity and terrain navigation, and is very suitable for long distance and timely use.

Geomagnetic navigation, for its merits of best economy and reliability, in the west, especially the USA, has been developed for a long time, and many research findings have been applied inengineering[8,9]. However, what the authors feel pity is that the civil research findings, which only date from the turn of this century, just stay in the stage of simulation. In order to practically apply this navigatuin technology in engineering; three key problems should be solved first, that is the research of real-time navigation algorithm, compensation algorithm of magnetic measurement and the algorithm of background mapping[10,11]. In these three problems, many significant research findings have achieved in navigation and compensation algorithm, only left is the algorithm of background mapping a field of less concern.

A magnetic background map with high accuracy is the data base of geomagnetic navigation, and decides whether this technology can be applied in endineering or not. However, from the available documentations we can see that civil researches on background mapping method are insufficient and most of those researches are elated to various world geomagnetic models as IGRF or WMM[12], which are unsuitable for geomagnetic navigation because of their low precision and resolution. Moreover, as to navigation algorithm, which is another key problem of geomagnetic navigation, borrow ideas mainly from terrain navigation, while the fact is that, because of the difference in data form and measuring method, geomagnetic navigation should develop its own navigation algorithm. Due to that background, in the following parts of this paper, we will illustrate in full a new algorithm of background mapping, and based on that a navigation algorithm is developed; those works are significant to the engineering application of geomagnetic navigation.

2 PRINCIPLE OF GEOMAGNETIC NAVIGATION

The configuration of geomagnetic navigation is shown in Fig.1. As indicated, a magnetometer provides the measured geomagnetism of the sample at the carrier's actual position. Then we find those we name indicated geomagnetism on background map according to INS indicated positions. Sending this indicated geomagnetism and measured geomagnetism, together with the INS navigation parameter to a central solver. At the existence of navigation algorithm then the corrected positions can be computed eventually.

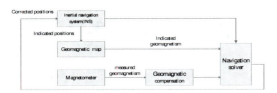

Figure 1.   Sketch of geomagnetic navigation system.

There is an issue which we must remind the reader, as the measured geomagnetism are measured with several disturbances which are different from the background map data, in this case a compensation operation should be applied to these measured geomagnetism to put them to the same level with background map data.

According to the configuration mentioned above, it is easy to see that geomagnetic background map is the base of geomagnetic navigation, and the map accuracy and resolution have significant influence

on the performance of geomagnetic navigation. Nowadays, conventional background maps include those established by various world geomagnetic models and those established by local geomagnetic models that are build on interpolation formulas, and the navigation algorithms include relative matching algorithm and Kalman filtering algorithm. However, relative matching algorithm, in essence, is one kind of graph matching algorithm, and the local background map it uses is discrete instead of analytic. In this case, this relative matching algorithm can only deal with measuring data in batch processing instead of real-time.

Different from relative matching algorithm, Kalman filter is a real-time algorithm, which has higher precision and has been applied in engineering in abroad. A main aspect of implementing geomagnetic navigation by Kalman filter is how to establish the required Kalman observation equation. To achieve these ends will depend heavily on the analytical relationship between the geomagnetic data and their geographic coordinates, namely the confirmation of geomagnetic model. Various world geomagnetic models are fit of this requirement; however, as indicated in part.1, those world geomagnetic models, in essence, are one kind of harmonic analysis under a spherical coordinates system. Due to those complicated Legndre functions with double indexes, when geomagnetic models with a higher resolution are expected, one faces two-sided difficulties that is the complexity of calculating large quantity Legndre functions and the complexity of linearization required in filtering management. In this case, considering the characteristic of local application, a geomagnetic model with simple and analytic form is expected. To implement this idea, we introduced a 2-D Gauss spline function based on local geomagnetic model, and the use of this model in geomagnetic navigation will illustrate in full in the following parts of this paper.

3 2-D GAUSS SPLINE FUNCTION BASED LOCAL GEOMAGNETIC MODEL

As indicated above, background map is playing a dominate role in geomagnetic navigation, to estimate carrier's position directly by geomagnetic data, the geomagnetic field model we use should be an analytical expression like global geomagnetic model, and at the same time need not be necessary to calculate a large number of complex Legndre functions, to fill those two qualifications we consider 2-D Gauss spline function to be a good choice.

Recently, Gauss spline function has been extensively applied in fields like medical-image, digital-terrain and geodesy. In document [13], how to establish covariance function by Gauss spline function is discussed. And document [14] discusses how to reconstruct local gravity data by Gauss spline function. In this paper, 2-D Gauss spline function is introduced to geomagnetic navigation for the first time, to establish a local geomagnetic model required in Kalman filter. On this model, the geomagnetic data can be expressed by a continuous Gauss spline function, and any point of the earth's geomagnetic data can be computed once the local geomagnetic model is determined.

3.1 The principle of approximated 2-D Gauss spline function

The Gauss spline function can be expressed as $G(x) = \exp(-\dfrac{x^2}{a_x^2})$, assume that $\{(x_i\ y_i)|i = 1,...,m;$ $j = 1,...,n\}$ is a grid point set distributed on the frame of XY, and $z_{i,j}(i = 1,...,m; j = 1,...,n)$ the corresponding value set. Starting from the Gauss spline function, the single-dimensional Gauss spline function in X direction can be denoted as $L_x = span\{G_x((x - x_i)/\Delta x)\}$; in Y direction, we have $L_y = span\{G_y((y - y_i)/\Delta y)\}$; considering L_x and L_y, then the following 2-D Gauss spline equation can be deduced subsequently

$$L(x,y) = L_x \otimes L_y = span\left\{\begin{matrix} G_x((x - x_i)/\Delta x) \times ... \\ G_y((y - y_i)/\Delta y) \end{matrix}\right\} \quad (1)$$

Eq. (1) also holds

$$L(x,y) = \sum_{i=1}^{m}\sum_{j=1}^{n} c_{i,j} G_x((x - x_i)/\Delta x)G_y((y - y_i)/\Delta y) \quad (2)$$

assume that

$$\left\{\begin{matrix} \vec{L}_x = \begin{bmatrix} G_x((x - x_1)/\Delta x), G_x((x - x_2)/\Delta x), \\ ..., G_x((x - x_m)/\Delta x) \end{bmatrix} \\ \\ \vec{L}_y = \begin{bmatrix} G_y((y - y_1)/\Delta y), G_y((y - y_2)/\Delta y), \\ ..., G_y((y - y_m)/\Delta y) \end{bmatrix} \end{matrix}\right. \quad (3)$$

then Eq. (3) can be simplified as

$$\left\{\begin{matrix} L(x,y) = \vec{L}_x * C * \vec{L}_y^T \\ C = (c_{i,j})_{m \times n} \end{matrix}\right. \quad (4)$$

Considering, $\{L(x_i, y_j) = z_{i,j}|i = 1,...,m; j = 1,...,n\}$, following linear equation can be obtained

$$XCY^T = Z$$

$$X = \begin{bmatrix} G_x(0) & G_x(1) & ... & G_x(m-1) \\ G_x(-1) & G_x(0) & ... & G_x(m-2) \\ . & . & & . \\ . & . & & . \\ . & . & & . \\ G_x(1-m) & G_x(2-m) & ... & G_x(0) \end{bmatrix}$$

$$Y = \begin{bmatrix} G_y(0) & G_y(1) & ... & G_y(m-1) \\ G_y(-1) & G_y(0) & ... & G_y(m-2) \\ . & . & & . \\ . & . & & . \\ . & . & & . \\ G_y(1-m) & G_y(2-m) & ... & G_y(0) \end{bmatrix} \quad (5)$$

$$Z = (z_{i,j})_{m \times n}$$

Note that X, Y are non-singular matrixes, and Eq. (5) has unique solution; solving Eq. (5), then coefficient matrix c can be obtained; in this case, after substituting in c Eq. (2), approximated analytic expression of local geomagnetic data on 2-D Gauss spline function can be finally deduced.

According to Eq. (5), matrix c only depend on X, Y, Z; and X, Y, depend on a_x, a_y. In document [14], the influence factors on inverse operation errors are illustrated in detail, through which, we can see that the higher the order of matrix X, Y and the smaller the number of a_x, a_y, the larger the value of inverse operation errors will be. Specifically, when the orders of matrix X, Y or the value of a_x, a_y are less than 5, the value of inverse operation errors will be quite small. As to the optimum value of a_x, a_y, we invite the reader to consult document [15] for Fibonacci sequence to optimize $a_x \in [1,5], a_y \in [1,5]$.

3.2 The test on approximation accuracy of 2-D Gauss spline function

According to above-mentioned theoretical analysis, approximated local geomagnetic model on 2-D Gauss spline function is one Gauss fitting method in essence. And the fitting accuracy directly determines whether local geomagnetic model can be applied in geomagnetic navigation or not. Once the model accuracy meets our expectation, and considering its analytical expression, then it can be applied in carrier geomagnetic navigation.

To verify the accuracy of this local geomagnetic model, a numerical computation is done on one test

897

area $i:117^{\circ}E-121^{\circ}E,j:21^{\circ}N-25^{\circ}N$. Test data consist of geomagnetic anomalies with a resolution of $2'\times2'$. First, the base data with a resolution of $4'\times4'$ are extracted from actual test data; starting from base data, then the approximated analytic expression of local geomagnetic data on 2-D Gauss spline function is deduced. Finally, on the base of approximated expression, comparison is done between $2'\times2'$ approximated data and actual test data to verify the accuracy of this local geomagnetic model.

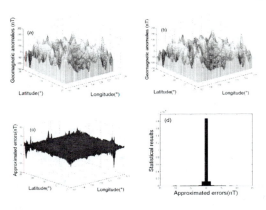

Figure 2. The simulation results of approximated local Geomagnetic anomaly map with 2-D Gauss spline. (a) The actual $2'\times2'$ geomagnetic anomalies;(b) The approximated $2'\times2'$ geomagnetic anomalies;(c) The approximation errors;(d)The statistical histogram of approximation errors.

Computation results are shown in Fig. 2(a–d), Fig. 2(a) shows the actual $2'\times2'$ test geomagnetic anomalies data, Fig. 2(b) shows the $2'\times2'$ approximated data computed on $4'\times4'$ extracted base data. Comparing Fig. 2(a) and 2(b), then we can see that the approximated and actual geomagnetic anomalies data perform quite similarly in appearance (i.e. 2-D Gauss spline function based local geomagnetic model is practical).

Fig. 2(c) shows the difference between approximated and actual geomagnetic anomalies data, from which we see the high accuracy of 2-D Gauss spline function based local geomagnetic model. Fig. 2(d) and Table 1 denote error statistics of the approximated geomagnetic anomalies.

Table 1. Precision comparison of computation and approximation (Unit: NT).

Items	Maxmum	Minmum	Mean	MSE
$\left|\Delta g_t-\Delta g_a\right|$	10.81	$1.10\cdot10^{-6}$	0.26	0.41

Δg_t stands for the actual test geomagnetic anomalies;

Δg_z stands for the approximated geomagnetic anomalies.

4 THE 2-D GAUSS SPLINE FUNCTION BASED OBSERVATION EQUATION

In this paper, Kalman filter algorithm is used to implement geomagnetic navigation. And the navigation equation that includes carrier's uniform speed and acceleration mode of motion is as follows:

$$x(k+1)=F(k)x(k)+\Gamma(k)v(k) \qquad (6)$$

in which

$$x(k)=\begin{bmatrix}\varphi(k)\\V_{\varphi}\\\lambda(k)\\V_{\lambda}\end{bmatrix},\; F(k)=\begin{bmatrix}1&T&0&0\\0&1&0&0\\0&0&1&T\\0&0&0&1\end{bmatrix},$$

$$\Gamma(k)=\begin{bmatrix}T^{2}/2&0\\T&0\\0&T^{2}/2\\0&T\end{bmatrix}$$

where stands for the system matrix at time; and the system control matrix; and denote the carrier position in latitude and longitude; V_{φ} and V_{λ} denote the carrier velocity in latitude and longitude; T is sample interval; $v(k)$ is zero-mean, white system noise vector.

The observable equation is expressed as

$$z(\varphi,\lambda)=\Delta g(\varphi,\lambda) \qquad (7)$$

where

$$\begin{cases}\Delta g(\varphi,\lambda)=\vec{L_{\varphi}}*C*\vec{L_{\lambda}}^{T}\\\vec{L_{\varphi}}=\begin{bmatrix}[G_{x}((\varphi-\varphi_{1})/\Delta\varphi),G_{\varphi}((\varphi-\varphi_{2})/\Delta\varphi),\\...,G_{x}((\varphi-\varphi_{m})/\Delta\varphi)\end{bmatrix}\\\vec{L_{\lambda}}=\begin{bmatrix}[G_{\lambda}((\lambda-\lambda_{1})/\Delta\lambda),G_{\lambda}((\lambda-\lambda_{2})/\Delta\lambda),\\...,G_{\lambda}((\lambda-\lambda_{m})/\Delta\lambda)\end{bmatrix}\\C=\left(c_{i,j}\right)_{m\times n}\end{cases}$$

$z(\phi, \lambda)$ denotes the measurement form geomagnetometer; other variables mentioned above are detailed in Part. 3. Then we have the discrete-time observation equation.

According to Eq. (7), these observation are nonlinear in their analytical expression, since the extended Kalman filter[18,19] is the most popular approach to nonlinear estimation, a linear approximation is used in modeling this observable as follows:

$$\begin{cases} z_\varphi(k) = \Delta g_\varphi(\varphi, \lambda) \\ z_\lambda(k) = \Delta g_\lambda(\varphi, \lambda) \end{cases} \quad (8)$$

where

$$\begin{cases} \Delta g_{L_\varphi}(\varphi, \lambda) = \overrightarrow{} * C_\varphi * C * \overrightarrow{L_\lambda}^T \\ \Delta g_{L_\lambda}(\varphi, \lambda) = \overrightarrow{} * C * C_\lambda * \overrightarrow{L_\lambda}^T \\ C_\varphi = diag\left(\left[\dfrac{-2(\varphi - \varphi_1)}{\Delta\varphi^2 * a_\varphi^2}, \dfrac{-2(\varphi - \varphi_2)}{\Delta\varphi^2 * a_\varphi^2}, \cdots, \dfrac{-2(\varphi - \varphi_m)}{\Delta\varphi^2 * a_\varphi^2}\right]\right) \\ C_\lambda = diag\left(\left[\dfrac{-2(\lambda - \lambda_1)}{\Delta\lambda^2 * a_1^2}, \dfrac{-2(\lambda - \lambda_2)}{\Delta\lambda^2 * a_1^2}, \cdots, \dfrac{-2(\lambda - \lambda_m)}{\Delta\lambda^2 * a_1^2}\right]\right) \end{cases}$$

Applying Taylor expansion to Eq. (7), then the linearized observation equation is as follows:

$$z(\varphi(k), \lambda(k)) \approx H(k)x(k) + W(k)$$

$$= \begin{bmatrix} z_\varphi(k) \\ 0 \\ z_\lambda(k) \\ 0 \end{bmatrix}^T \cdot \begin{bmatrix} \varphi(k) \\ V_\varphi \\ \lambda(k) \\ V_\lambda \end{bmatrix} + W(k) \quad (9)$$

where $H(k)$ is the linearized observation matrix; $W(k)$ is the measurement noise.

By now, the carrier position can be estimated on navigation equation Eq.(6) and observation equation Eq.(9) through the extended Kalman filter.

5 SIMULATION

In this section we will apply the 2-D Gauss spline function to geomagnetic navigation to estimate the carrier's position. Simulation is done on one test ground with a dimension of 100 m × 70 m, a caesium optical pump, that is, G858 is used as magnetometer to measure the total magnetic intensity of test ground. Simulation is done according to the following two steps:

1 Measuring the geomagnetic background data

In order to decrease the measuring interference in full, some arrangements are made in advance; first, the test ground we choose is open enough to assure the uniform distribution of magnetic intensity; second, ground calibration with resolution 5 m × 5 m is done to fix measurement point; final, measuring procedure is implement statically in morning when the sum is inactive. After that, final measurements are got; these conclude 14 measuring lines with 20 data each; and other three testing lines with 60 data each. The sketch of measuring and testing lines is illustrated in Fig. 3. At the existence of measurements, then starting from the local geomagnetic model given in this paper, a geomagnetic background map is given in Fig. 4.

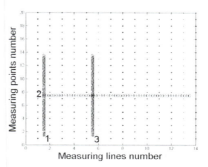

Figure 3. Sketch of measuring and testing lines.

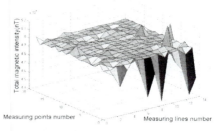

Figure 4. Map of geomagnetism background.

2 Geomagnetic navigation simulation

To verify the feasibility of our navigation method, simulation is carried out on three testing paths with typical motion quality. As shown in Fig. 3, three testing paths whose initial positions according to map grid are (1.5, 1.5), (1.5,2.5), (5.5,1.5) respectively, consists of 60 testing data sampled in measuring point or line direction with a sample interval 1 meter (0.2 map grid). Compare to 1 meter sample interval, we assume each paths with 50% position error (05 m) in measuring point and line direction. Note that the measurements and geomagnetic map we use in simulation are derived from actual data; the results will be more credible than pure simulation method.

The results of geomagnetic navigation simulation are shown in Figs. 5–7 and Tables 2–3.

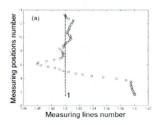 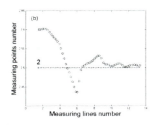 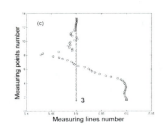

Figure 5. Simulation results of geomagnetic navigation. (a) The result of testing path; b) the result of testing path 2; (c) the result of testing path 3.

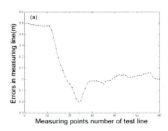 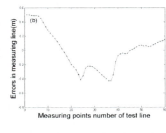 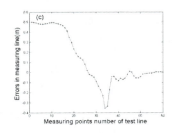

Figure 6. The position errors in measuring line. (a) the errors of testing path 1 (b) the errors of testing path 2; (c) the errors of testing path 3.

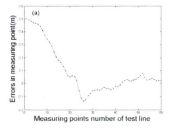 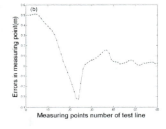 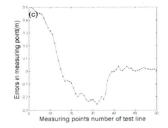

Figure 7. The position error in measuring point. (a) The errors of testing path 1; (b) the errors of testing path 2; (c) the errors of testing path 3.

Table 2. Error statistics of geomagnetic navigation in measuring line (Unit:mile).

Items	Maximum	Minimum	Mean	MSE	Corrected rate
Testing path1	0.502	0.002	0.145	0.181	71%
Testing path2	0.501	0.007	0.216	0.156	57%
Testing path3	0.501	0.002	0.201	0.195	60%

Table 3. Error statistics of geomagnetic navigation in measuring point (Unit:mile).

Items	Maximum	Minimum	Mean	MSE	Corrected rate
Testing path1	0.501	0.010	0.133	0.154	73%
Testing path2	0.511	0.008	0.160	0.167	68%
Testing path3	0.500	0.000	0.151	0.150	70%

Fig. 5 shows the geomagnetic navigation results of three testing paths. Fig. 6 shows the estimation of carrier position errors in measuring line; and Fig. 7 shows the estimation of carrier position errors in measuring point; as shown in Figs. 6 and 7, after few steps of operation, the errors are held to low values, and in Fig. 5 the estimated path matches well with the actual testing paths, which demonstrates the effectiveness of geomagnetic navigation algorithm developed in our paper, it is feasible to extract the carrier position contained in geomagnetic measurement.

Tables 2 and 3 show error statistics of geomagnetic navigation in measuring line and point, from which we can see the accuracy of our geomagnetic navigation algorithm. As indicated, with geomagnetic navigation, after 60 steps of operation, the average navigator errors in measuring line have been reduced more than 71%, 57%, 60% of the initial navigator error respectively, meanwhile, errors in measuring point have been reduced more than 73%, 68%, 70% of the initial navigator error, respectively.

For the Kalman filter tuning used, and we have reasons to believe that with the increase of samples (i.e. more than 60samples) the corrected rate will also increase.

6 CONCLUSIONS

Since geomagnetic map is a key element in geomagnetic navigation, the quality of the maps, available in an operating area, affects system performance. Conventional geomagnetic navigation algorithm, with discrete geomagnetic data used as database, is usually derived from terrain navigation; as a result it cannot provide real-time navigator results. To meet the challenge to provide accurate and real-time navigation and at the same time to enhance the quality of the maps, a 2-D Gauss spline function is introduced in this paper. With this function, we develop a continuous local geomagnetic model (geomagnetic map) and further develop a novel navigation algorithm different from conventional relative matching algorithm, for it is continuous (real-time) instead of being discrete. Theoretical derivation and simulation are done, which show the effect of 2-D Gauss spline function based local geomagnetic model, and demonstrate that our navigation algorithm provides high-accuracy navigation. Although this algorithm is developed for aided navigation using geomagnetic data, it is equally applicable to other domains, for example, aided navigation on gravity or terrain data.

REFERENCES

[1] Zhigang W, Shaofeng B. 2008. A local geopotential model for implementation of underwater passive navigation. *Progress in Nature Science*, 18 (9):1139–145.
[2] Tong Yu-de, Bian Shaofeng, Jiang Dongfang, et al. 2011. Gravity matching aided navigation based on local continuous field. *Journal of Chinese Inertial Technology*, 19(6) 611–615.
[3] Yang Yong,Wang Kedong, Wu Zhen, et.al. Evaluation of performance of ICCP algorithm with different parameters [J]. ACTA AEROM AUTICA ET ASTRONAUTICA SINICA, 2010, 31(5):996–1003.
[4] Yan Li, Cui Chenfeng, Wu Hualing. The gravity matched navigation on TERCOM algorithm[J]. Geomatics and Information Science of Wuhan university, 2009, 34(3): 261–264.
[5] Xiao Sheng-hong, Bian Shao-feng. Research on Regional Model of Continuous Fourier Series of Marine Magnetic Anomaly Field Using for the Geomagnetic Navigation[C]. International Conference on Industrial and Information System, 2010: 437–440.
[6] ZHAO Jianhu, ZHANG Hongmei, WANG Aixue, et al. Underwater Geomagnetic Navigation Based on ICCP[J]. Geomatics and Information Science of Wuhan University, 2010, 35(3):261–264.
[7] HAO Yanling, ZHAO Yafeng, HU Junfeng. Preliminary analysis on the application of geomagnetic field matching in underwater vehicle navigation.[J] PROGRESS IN GEOPHYSICS, 2008, 23(2):594–599.
[8] Goldenberg F. Geomagnetic navigation beyond the magnetic compass [J]. Position Location and Navigation Symposium, 2006:684–694.
[9] GUO Caifa, HU Zhengdong, ZHANG Shifeng, et al. A Survey of Geomagnetic Navigation [J]. Journal of Astronautics, 2009, 30(4):1314–1319.
[10] Huang Xuegong, Fang Jiancheng, Liu Gang, et al. Geomagnetic mapping and validity estimation[J]. Jounal of Beijing University of Aeronautics and astronautics, 2009, 35(7):891–894.
[11] Xu Daxin, Wang Yong, Wang Hubiao, et al. Statistical analysis of regional gravity anomaly aided navigation and positioning[J].Journal of Chinese Inertial Technology, 2009, 17(4):444–448.
[12] Huan Xiaoyin, Bian Shaofeng. The research evolution of international high-resolution geomagnetic models[J]. HYDROGRAPHIC SURVEYING AND CHARTING, 2010, 30 (30):79–82.
[13] BIAN Shaofeng, JOACHIM M. determining the parameter of a covariance function by analytical rules[J]. ZfV, 1999(7): 212–216.
[14] TONG Yude, BIAN Shaofeng, JIANG Dongfang, et al. The Reconstruction of Local Gravity Anomaly Field Based on Gauss Spline Function[J]. ACTA GEODAETICA et CARTOGRAPHICA SINICA, 2012, 41(5): 754–760.
[15] Overholt K J. Efficiency of the Fibonacci search method[J]. BIT, 1973, 13(1): 92–96.

Information, Computer and Application Engineering – Liu, Sung & Yao (Eds)
© 2015 Taylor & Francis Group, London, ISBN 978-1-138-02717-6

Ceramic art design, design America

Xiu Mei Wu

Jingdezhen Ceramic Institute, College of Art and Design, China

ABSTRACT: One of the design of aesthetic content created by the United States as human beings, especially the increasing development of modern society, so that the results of scientific research constantly produces new technologies, new materials, and thus the production of handicrafts are gradually being replaced by machines. Ceramic products as people indispensable part of daily life, but also continue to reflect the importance of design beauty. Ceramic products design reflects the beauty of practical value and aesthetic value of the depth of mutual unity, indicating that the wisdom and strength and emotional attitudes of designer material and spiritual factors that combine to create a new humanity, reflects the human need to be more healthy, perfect the new ceramic design products. Ceramic design beauty is natural and science combined, is people-oriented, and to humanity as a preconditional wisdom.

KEYWORDS: Ceramic, Design, Design aesthetics

Design in people's daily life everywhere, ceramic design is part of the design, but also the oldest human design. With the advent of large-scale machine Industrial Revolution, new technologies, machine production to replace manual labor produced products formed a new design beauty. A variety of new design also appears appropriate, product utility function and mental function, practical value and aesthetic value to get a more unified, reaching US combination of art and technology. Design can be said to involve the United States in all aspects of design. Of course, the field of ceramics is no exception, the emergence of new technologies and new materials takes ceramic design to a new level of development. Ceramic design because of the particularity of its materials, processes, so the aesthetic ceramic design has the same design aesthetics and other features of the design has its own unique beauty. Below ceramic design will be to analyze the design of America.

1 Development of ceramic product made from hand to machine design beauty

Pottery is the earliest human, the most practical creation. Painted pottery first appeared can be said is derived from practical and eventually beyond practical sense of beauty. It fully reflects the design of human consciousness and the establishment of the creation and pursuit of beauty, its shape and decoration has reached a perfect degree, this pursuit of practical and aesthetic unity, laid the foundation for the development of ceramic design beautiful. After entering class society, the meaning of the United States and ceramic design will not only reflect the practical and aesthetic unity, as the ruler of a small number of the aristocracy

aesthetic ideas and ideology also directly affects the design of the formation of the United States, and its integration into a richer spiritual content. According to the original styling and decorative porcelain, are carefully designed, with high aesthetic value and practical value, while the value of the purpose of creation is a symbol of sacrifice in order to meet the needs or power, has some religious significance and symbolism.

Emergence and development of handicraft era ceramic design beautiful, laid the factors and elements of its composition: the purpose of the function embodied in Art Deco style to convey the beauty and formal beauty, and the impact of the development of ceramic technology and a variety of social factors, economic factors that the United States has become different from other ceramic design art design design sisters beauty and the creative process in a highly uniform pursuit of ceramic products functional and aesthetic function of the state of mind exists in the design object.

The emergence and development of modern industrial revolution, breaking the era of low labor efficiency handicraft defects. And the new ceramic design must be completed by the production of new machines. Practical function and aesthetic functionality has been further improved and unified. Kind of handicraft era "Hand and art" produced by combining a unified state, can not meet the needs of fast-paced people. So the machine instead of manual production of ceramics is the trend of social development. Ceramic design must also be designed to meet the machinery products. At this time reflects the beauty of it is different from the US ceramic design hand-crafted ceramics.

Ceramic art and design innovative beauty techniques is dependent on the development of a ceramic material. Yangshao, Majiayao cultural painted pottery. Replace the engraved lines early Neolithic pottery. Shuai was made due to the master potter pottery color material and firing techniques. Longshan Culture period pottery art was very popular, was structural improvements and pottery kiln firing technology to improve device quality inevitable result: Song before. To rely on Chinese porcelain glaze win, its peak is the Chinese Song Dynasty, "brother", "official", "Ru", "scheduled", "Jun" five famous kilns. They use glaze charm that caused all sorts of wonderful artistic image of colorful ceramic, porcelain same time, aggravation and glaze firing skills to adapt. Ming and Qing porcelain colorful, partly because the fine white porcelain firing success, potters provide useless; on the other hand is highly developed technology glaze firing crystallization. Therefore, ceramic design beauty, is the study of ceramics production process, through scientific and technological means to achieve the desired design and natural beauty of the formation.

2 The nature and characteristics of the design of ceramic art and design in the United States is reflected in design beauty, which is the mutual penetration of art and technology, combined with another product; is to study how to use the technology in product design aesthetics of science. With the continuous development of human society, the results of scientific research into technology increases constantly, handicraft production is gradually being replaced by machines produced. Performance design aesthetic is gradually becoming evident. Design aesthetic reflects the fundamental characteristics of human-made tools of production activities. Design America is the "beauty" of the scale, to measure the effectiveness of product function and mental function, practical value and aesthetic value of mutual unity; then it can be said that the United States is actually designed for the handicraft era to exist, just design and handicraft era is the era of industrialization and it differs because of different production depends on both carriers. The handicraft era is designed mainly for the craftsmens US labor skills and experience, they are materialized as a sophisticated form of beauty in the processing of products; and in the era of mechanized production, technical beauty is manifested in specific manufacturing processes and technology the formation of a unified, high-volume production of exquisite artistry process.

The nature of the design can be seen from the United States, on ceramic design is concerned, whether it is made of hand-painted pottery Neolithic wheel or modern machine production of household ceramic products, they have designed the United States. And the United States is not only reflected in the designin of ceramic gc products is also reflected in the external technical ceramics. The new ceramic technology, the use of new materials is directly related to the new forms of ceramic design.

Design of the United States contains the material and spiritual factors. It is a combination of technology and art, it has both the characteristics common aesthetic beauty and artistic beauty of technology.

Technology refers to the United States and the United States mainly focuses on mechanical industrial technology. Fron the "technical beauty" point of view full meaning of human technology forms are varied, we shall include Mimi 's technique of manual techniques and mechanization. US manual techniques beauty and mechanical technology compared to manual techniques in the United States, often with personal taste, permeated with the spirit of the individual, to maintain the experience and emotional characteristics. In the field of art, such as technical ceramic handmade, often direct nature of the arts, such as ceramics in the mud plate construction is both a technology, but also a style of art. Thus in the era of large-scale machine production, the technology can put into the field of arts and crafts and handmade plastic arts, and mechanical technology, mechanical constraints, making this technology products show little taste of the human spirit, it is more of a mechanical process the technical performance of its technical beauty is very clear. In ceramic design technology, the ceramic materials are mainly from the United States and they create ceramic products functional beauty.

Ceramic material texture, is the basis of Tao Pirates artistic image. Texture of ceramic materials, including porcelain and enamel gloss, transparency and moisturizing sense they are ceramic skin, flesh color. A wide variety of ceramic materials are as follows: celadon porcelain class, black ceramic, porcelain, ivory porcelain, bone china, steatite, magnolia porcelain. Lu Yu porcelain; pottery was divided into fine pottery, stoneware, pottery purple; there ranged between porcelain and earthenware stoneware and so on. Porcelain aggravation of acid, alkali resistance, sudden heat, quench, firm texture consistent performance, namely chemical and thermal stability, and the materials used are inseparable. Manufacture of ceramic glaze porcelain variety is dazzling, there are nearly a hundred. Different texture characteristics of the material, only with the shape and decoration for their identity before they can give full play to the advantages of its intrinsic beauty. Jingdezhen porcelain is hard to nourish and it is smooth, with this material design style, type body structured, turning accountable clear, clear detail treatment, with a bright, beautiful, robust, compact, light very artistic style. Longquan celadon glaze thickness, use it to produce containers should highlight its solemn, deep, subtle, moist, fresh and artistic features. Potters used if not familiar with their performance and specialty ceramic materials that are not performing at artistic creation. Bound thankless, and even diametrically opposed to failure, accomplished potters not only are able to correctly understand and adapt to the properties of the material but also using artistic ideas, good use and give full play to the unique beauty of ceramic material texture. Like Jingdezhen who produced the kind of high eggshell white glaze bottle, which is perfectly clean, shiny and transparent shadow. If the decor is overfilled, ornamentation becomes too heavy, it will superfluous, conceal their beauty device quality while reducing its artistic appeal; if then massage pastel, only to control the quality and ornamentation off each other and greatly enhance its aesthetic value.

Ceramic products functional beauty is reflected in the practicality of its functions. Although modern glass, plastic, metal and other new materials, products continue to emerge, ceramic products are still the main items of daily life. So the usefulness of ceramic products is crucial, especially for the design of ceramics for daily use. Its usefulness is more obvious, such as ceramic tableware design, not only it is simple and neat design, it is decorated according to the latest fashion inorder to keep up with with the fast pace of modern life. Ceramic functional beauty is mainly reflected in the ceramic shapes, different shapes have different practical applications, such as bowls, plates, dishes reflect the different functions in the ceramic tableware; bottles, cans, pots also have different purposes. Therefore, the designing of ceramic product is done to meet different practicality. Different shapes give a different aesthetic value. It can be said that the ceramic design is the primary function of the United States, to meet the practical premise to reconsider artistic decoration.

Beautiful ceramic products have not only practical function, but also has aesthetic beauty of art. It is a manifestation of spiritual products, ceramic products, artistic beauty subtle cultivate people's sentiments and thoughts and feelings, ceramic design of artistic beauty is manifested in its decorative beauty and style beauty. Ceramic decoration reflects the different characteristics of each era style and also contributes to the development of other designs. Ceramic art and design is part of the design. Therefore, the United States and other styling ceramic art artistic beauty, as derived from the natural beauty and the beauty of life, from the objective material world, with a particular social and historical basis, but it is not the original beauty of natural forms, but after the artist's subjective process, which permeates the artist's aesthetic point of view and aesthetic ideals and artistic originality (creative personality, artistic style, artistic skills), so it is more than natural beauty, the beauty of life is more concentrated, and more typical. Similarly ceramic art to showcase their beauty is grace charm through specific image, and the image of this art has a certain aesthetic value, it can cause people's beauty, people enjoy the beauty, embodies a certain aesthetic ideals and aesthetic emotion, it is a concentrated expression of human social life of aesthetic judgment, is the artist's aesthetic sense materialized. Ceramic art is the creation of beauty, but also like other plastic arts pay attention to subtle social function, and follow the vivid aesthetic principles.

Ceramics not only meet the daily necessities, but also make the people to enjoy the arts and crafts. It has not only apractical function, but also has aesthetic value, so it cannot be like the other which are purely used for plastic arts appreciation, as the blind pursuit of a spectacular and pleasant cardiopulmonary specific artistic image, but the premise of upholding the practical by harmonious proportions, smooth lines, undulating contours of such a sleek, decorative patterns and reattachment in body and color contrast of the organic unity, To convey a general thought, emotion, mood, and show some style, fashion, and even makes the atmosphere idle. This aesthetic characteristics of ceramic artis to create a unique form of expression it was designed beauty – through its own texture, shape and decoration of organic integration, to show their graceful look.

Technical ceramic products beauty (ceramic texture) and artistic beauty (shape and decoration), although each has its own independent aesthetic connotations, but both of them are interdependent, mutual restraint, melt is body. Texture of ceramic materials, ceramic art images constitute fleshy skin; ceramic art pottery owned styling impart the image of the soul; it is creating ceramic art decoration a heroic image of beauty. Therefore, in the design of ceramic artit has owned the pottery, ceramics and ceramic decoration in the shape of a unified image of the background art and the artistic treatment takes into consideration, in order to be expected out of people heartstrings brewing art to share.

3 Ceramic art and design factors in the design of beauty

Ceramic products design quality and design of the house is not only relevant but also dependent on the progress of science and conservation technologies, such as furnace structure and kiln technology, ceramic decoration technology, ceramic decorative pigments invention and so on. Forming a porcelain glaze beauty and decorative patterns and whether they can achieve the desired effect of the design, which depends mainly on the quality of the kiln technology. Because of the firing temperature, the cooling method kiln gas composition changes, and their concentrations are relying on kiln technology to master. Porcelain kiln placed in position, firing atmosphere is oxidation or reduction, sintering temperature, vitrification range and kiln technology are inseparable. Furnace structure is also very important and it has gone through a variety of transformations, which species are numerous, there are bread kiln, horseshoe kiln, the development of tunnel kiln and so on. Puzzle kiln refractory bricks and refractory kiln constructed in cotton is the most important. Kiln firing is directly related to the quality of porcelain. Furnace produced its good balance kiln firing temperature at various locations, porcelain glaze hair color from different angles are almost the same. So as to achieve a uniform translucent glaze, there is no good light, good kiln, and kiln technology is not enough, the master control of the kiln, kiln temperature, when heated, are there times when cooling control. Good kilns and good technical ceramic kiln firing both musts.

Development of ceramic decoration technology for ceramic design is also essential. Such as the use of ceramic decals ceramic technology has dramatically changed the decor, especially the ceramic decals used in the design of household ceramic products, which is both simple and elegant and neat unity. Ceramic decal it relates to the use of decal production, decal printing process by use of the design patterns printed on the ceramic pigment particular paper, depending on the rapid development of today's printing industry

Ceramic decorative ceramic pigment of the present invention is also a great influence on the design. Decorative ceramic pigment chemical formula is developed from the scientific method. Each of the emergence of new decorative paint will appear a new ceramic design style, such as the new color pigments can be repeated several times due to having fired feature allows creators to be free of paint until it reaches satisfied; ceramic decals can be color pigments printed on paper and then pasted onto the porcelain like structured to meet the unified design of consumer groups.

4 Ceramic art and design in the United States and the role of ceramic design throughout the United States, it is in common with other designs, it is inseparable from the development of science, but it has its own uniqueness. Today's ceramic products

still occupy the major part of people's daily lives, it can be said that the United States would have to play an important role in ceramic design.

4.1 To improve the quality and market competitiveness of ceramic products, advances in modern technology, the design of ceramic products requires more emphasis on functionality and aesthetic combination. NA, not beautiful ceramic technology products will be eliminated in the market competition. New materials, new styling ceramic products continue to market. The emergence of new technologies on the one hand to improve the quality of ceramic products designed to promote the favorable development of the market economy, on the other hand it also brings convenience to people's lives, for example, bone china can be put in the microwave, which greatly respect the people's lives.

4.2 Promote the development progress and modern design era. Advances in technology itself is social progress, technology products improve the lives of people with unprecedented ease and convenience, especially the development of the use of modern high-tech computers, mobile phones and other IT can be said to promote the era progress, and to narrow spatial distance. Improving the design and technology development is a process of interaction techniques for designing and providing advanced materials. On the contrary, the design also provides advanced technical tools. Both combined with each other and promote each other, and jointly promote the development of the society forward. Ceramic production areas are also increasingly embarked on a high-tech production. This leads to the emergence of a new ceramic product.

REFERENCES

[1] Kai Zhuge forward. 2006. Ten Lectures on Art of Design, Shandong Pictorial Publishing House, Shandong.
[2] Chen Wang Heng. 1993. Technological Aesthetic Principles, Shanghai Science and Technology Press, Shanghai.
[3] Xu Heng alcohol editor. 1995. Practical Technical Aesthetics, Space Law Science and Technology Press, Tianjin.
[4] Liao Xiong, 1989. With Ethnic Characteristics of Ceramic Aesthetics and Aesthetic of Chinese Ceramics, Zhejiang Academy of Fine Arts Publishing House, Zhejiang.
[5] Li Chaode. 2004. With, Design Aesthetics, Anhui Fine Arts Publishing House, Anhui.

Information, Computer and Application Engineering – Liu, Sung & Yao (Eds)
© 2015 Taylor & Francis Group, London, ISBN 978-1-138-02717-6

Research on simulation for auto of border fairing technology stamping

Rui Jun Liu & Xiang Wen Dang

College of Automobile & Civil Engineering, Beihua University, Jilin, China

ABSTRACT: Liquidity and ability to plastic deformation of the material is an important factor stamping effects, studies show stamping formability with the complexity of the boundary shape stampings magnitude decrease exponentially. To make the design of stamping die with good formability, die face design phase requires border stamping CAD model appropriate fairing, which aims to make the design of stamping die forming the boundary substantially beneficial "convex" polygons. In addition, after the fairing parts overlapping boundaries and circumstances discount does not appear when generating addendum section line.

KEYWORDS: Automobile; Border fairing; Simulation

1 INTRODUCTION

At present, domestic and foreign aid roller basic algorithm to determine the area to be fairing stampings border, using methods of fairing cover boundary surfaces, the algorithm can accurately determine the fairing area and generate consistency with the requirements of the smooth surface simulation. However, the above boundary smoothing algorithm stampings still possess the following disadvantages: 1) determine the criteria to judge the roller with a series of discrete parts of the border to determine the intersection point is the low efficiency of the method fairing area. 2) Covering method to generate smooth surfaces and stamping of the original boundary is difficult to achieve a continuous C1. In order to achieve efficient automatic stamping boundary for smooth operation, chapter proposed boundary line fairing part feature-based ring intersection judgment algorithm, and the development of automatic stamping boundary smoothing module is based on ACIS modeling system. Algorithm to calculate the size of the bounding box stamping to obtain judgment segment intersection of optimum length to the line intersects the projection plane parts group judged the situation to be fairing area fairing area boundary line to meet the requirements of the smooth region and generate comply with C1 continuous smooth surface.

2 STAMPING AUTOMATIC IDENTIFICATION AND DISCRETE BOUNDARY

ACIS modeling uses a boundary representation (B-Rep) to achieve a solid model description, boundary representation in order to distinguish between the two phenomena in the above share one side there is

the concept put forward edge (COEDGE), there is the ability to record to the side edge (EDGE) occurs at the boundary surface of one. Side relationship and directed edges are shown in Fig.1.

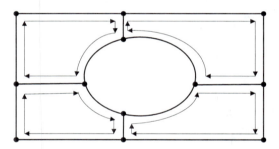

Figure 1. Stamping surface containing bore, side, directed graph edges.

According to the definition of edges, any one of the top surface has a directed edge connecting ring that is made between the surface and the surface have a common edge that belongs to the edge is not unique. Stampings for complex are spatial three-dimensional surface, not only for the process and the subsequent installation requirements in the design of the surface but also the corresponding holes. Stamping surface containing bore, side, there's the relationship side, belongs to the interior side of stamping directed edge is 2, and the boundary edge (boundary edge and the outer boundary edge of the hole stamping) have to belong side is 1, so you can determine the edge belongs to the edge if there is a way to find out all the boundary edge stampings model.

CAD model after stamping model transformations have the complete topology information, according to the topology information to complete the part boundary of automatic recognition, automatic recognition process is described as follows:

Step1: According topology traversal parts stamping models constitute all surfaces, and then traverse the edge of each side of the composition, while the edges have to belong to the number of edges to determine if there are sides thatbelong to the number of edges is greater than a description of this edge for the internal parts of the edge, the edge belongs if there is an equal number of edges to illustrate this edge is part of the boundary edge, the boundary edge storage container and related information to develop (vector1).

Step2: optional vector1 in from one side edge as the initial seed, according to the principles of the endpoint coordinates attached on the surface side coincides with the counter-clockwise direction to find the edge of the terminal point of the initial seed edges overlap, you will find that the side connected with the initial seed counterclockwise edge termination point as seeds continue to look for the next level of connected edges, and so on until you find the initial seed side counterclockwise beginning until end. These are referred to herein constitute the boundary edge end to end ring is characterized by a ring, with endpoints coincide way to find all the features of the boundary edge vector1 ring composed of these features constitute the outer ring and the inner boundary of the hole parts.

Step3: All the features of the ring plane are perpendicular to the pressing direction projected to generate a feature ring projected minimum binding rectangle (Minimum Bounding Rectangle, called MBR), the smallest rectangular area is relatively binding minimum binding characteristics of ring projection area of a rectangle, the maximum projection characterized in that the ring corresponding to the outer boundary of a boundary part, the rest of the inner ring are both characterized hole part.

To determine the fairing area judging line in stamping model boundary reasonable step to move the line, you need to determine the boundaries of discrete parts. Typically, such as a discrete way the curve is divided into two kinds of chord length and arc length, etc., taking into account the possible existence of complex stampings boundary changes, such as the arc length is determined using the discrete parts of the outer boundary of the way. Results chord is like dispersion curve of the arc length, etc., shown in Fig. 2. Looking at Fig. 2 shows that when large parts of the border region of the curvature transformation exists, such as chord length of discrete parts boundary arc quite different, the second arc discrete components to achieve boundary are associated with discrete purposes.

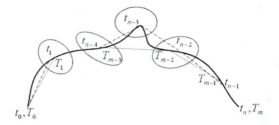

Figure 2. Results of the chord and arc length of dispersion curves, etc.

And other discrete arc will first need to boundary curve parameterization, and then determine the equation of the curve based on parameters such as coordinate information of discrete point's arc, arc length, etc. While discrete judgment to ensure the accuracy of fairing area, but there are shortcomings arc tedious calculation one of which is the focus of future research.

Boundary curve discrete point arc curve parametric equations need to get back calculation, derivation Inverse Discrete point formula is as follows:

Set parameters curve equation is:

$$p(u) = [x(u) \quad y(u) \quad z(u)] \tag{1}$$

$$\dot{p}(u) = \frac{dp}{du} = [\frac{dx(u)}{du} \quad \frac{dy(u)}{du} \quad \frac{dz(u)}{du}] \tag{2}$$

$$\dot{p}(u_0) = \lim_{\Delta u \to 0} \frac{p(u_0 + \Delta u) - p(u_0)}{\Delta u} \tag{3}$$

The adoption curve parameter indicates that there is a direction. Direction of the curve is corresponding to the curve parameter's increasing direction, the arc length of the curve and so that the direction of the discrete curve (positive or negative) is carried. Parts located within the boundaries of the curve in the first derivative of the parameter vector everywhere nonzero vector, then the derivative of formula 1 was:

$$dp = [dx \quad dy \quad dz] \tag{4}$$

$$ds^2 = dp^2 = dx^2 + dy^2 + dz^2 \tag{5}$$

$$ds = \sqrt{\left(\frac{dp}{du}\right)^2} du = \left| \dot{p}(u) \right| du \tag{6}$$

$$\dot{s} = \frac{ds}{du} = \left| \dot{p}(u) \right| > 0$$

The remaining arc segments do not need any treatment; storage can be used directly as a discrete arc. Stamping model discrete boundary flow chart shown in Figure 3.

Figure 3.   Stamping a discrete model boundary effect.

3   STAMPING MODEL PROJECTION MATRIX

Segment intersection algorithm is the premise that stamping surface models and boundary point to the pressing direction of discrete vertical projection, so to get the part surfaces projected polygon and the corresponding projection of discrete points. Stamping surface model to the plane perpendicular to the direction of the projection transformation matrix is derived as follows:

$$\begin{bmatrix} x' & y' & z' & 1 \end{bmatrix} = \begin{bmatrix} x & y & z & 1 \end{bmatrix} \begin{bmatrix} 1 & 0 & 0 & 0 \\ 0 & \cos\theta & \sin\theta & 0 \\ 0 & -\sin\theta & \cos\theta & 0 \\ 0 & 0 & 0 & 1 \end{bmatrix} \quad (7)$$

After mapping the boundary point "convex" after adjustment package has a good "convex" package, and therefore only need to use the control points on the curve approaching the impact of the larger spline map these points are connected with a smooth continuous C1 curves. Based on the above considerations, choose quadratic B-spline curve as the boundary curve fairing area. In order to make the second B-spline curve through the control polygon endpoint, you need to repeat the endpoint node vector equal to 2.

CAD model after stamping model transformations have the complete topology information, according to the topology information to complete the part boundary of automatic recognition, automatic recognition process is described as follows:

Step1: Topology traversal parts stamping models constitute all surfaces, and then traverse the edge of each side of the composition, while the edges have to belong to the number of edges, determine if there are sides that belong to the number of edges are greater than a description of this edge for the internal parts of the edge, the edge belongs if there is an equal number of edges to illustrate this edge is part of the boundary edge, the boundary edge storage container and related information to develop (vector 1).

Figure 4.   Parts boundary smoothing curve.

Auto cover internal model contains a large number of holes, which is basically a function of the hole, mounting holes and structural lightweight holes. In front panel forming simulation requires internal holes to fill, cover model purpose is mainly to fill the whole punch to achieve generation and die by means of bias. Currently, the fill hole operation process is divided into following steps: 1) determine the type of the hole to be filled, the so-called type is the hole, the hole or holes border, which is mainly disadvantage to different types of algorithms fill holes caused by different holes. 2) by way of a mouse click to select the boundaries that need to be filled in the hole, requiring the operation of the selected border is closed "ring."

3) Click the OK button to complete the filling of the selected bore. The traditional method has many drawbacks to fill holes, 1) encounter irregular bore, the operator is difficult to determine the type of hole, only through the fill hole after observing the effect to determine the type of selection is correct, which increases the operator's work intensity. 2) Each of the hole to be filled is required for type selection and border options, for there are many different types of inner hole cover models, fill holes operation workload will be very large. 3) Sometimes overlook the smaller bore diameter, resulting in the forming simulation process to fail. In order to solve the drawbacks of the conventional algorithm fill holes, this chapter presents a targeted car models that cover the hole filled automatic algorithm, the algorithm can automatically identify all of the hole, the hole does not need artificial selection border, once completed all bores are filled. By C ++ programming language features fill holes in the three-dimensional modeling ACIS platform, tests show that the algorithm is accurate, and efficient.

4   EXAMPLES OF VERIFICATION

In order to validate the model boundary stampings fairing area, correctness of the algorithm using eta/DYNAFORM boundary smoothing algorithm, AUTOFORM boundary smoothing algorithm and boundary smoothing algorithms presented in this chapter for a judge stampings border region fairing and fairing surface testing. The test results, shown in Fig. 4, eta/DYNAFORM software and AUTOFORM software cannot accurately determine the need fairing area, let alone meet the requirements to generate smooth surfaces. Boundary smoothing algorithms presented in this chapter (KMAS / Die_Face) can accurately determine the needs of the region and the fairing can be generated to meet the requirements of the smooth surface, Figs. 5 and 6 show the KMAS/Die_Face of two models fender border fairing and fairing surface generated results.

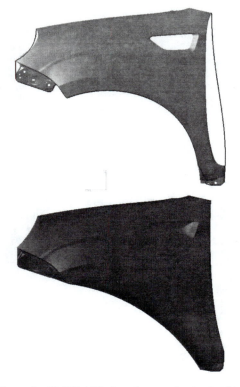

Figure 5.   KMAS / Die face for stamping border fairing results.

5   CONCLUSION

Boundary smoothing algorithm stamping is one of the important functions of stamping die face design parametric design, this chapter describes the line of intersection of the fairing method for determining the area and meet the requirements generated by B-spline surfaces smooth, this chapter contains the following:

1   The boundary maximum principle to determine the bounding box part boundary.

2   The derivation of the discrete arc length, etc., on the part of the arc discrete boundaries.
3   To achieve stampings model derivation based on the projection of the direction of the vertical projection stamping.
4   The proposed line segment intersection algorithm to determine the functional parts fairing area.
5   The proposed boundaries of discrete points mapped parts fairing plane algorithm to achieve a smooth plane boundary mapping parts of discrete points.
6   With such a smooth adjustment approximation of boundary points "convex" package.
7   Generate a good continuity of B-spline smoothing the boundary line and the B-spline smoothing surfaces.

Corresponding Author: Dang Xiangwen

REFERENCES

Kim H S, Hong S K, Hong S G, Huh H. 2003. The evalua2tion of Crashworthiness of vehicles with forming effect, 4th European LS2DYNA Users Conference, 25–33.
Lee S H, Han C S, Oh S L, Wriggers P. 2001. Comparative crash simulations incorporating the result s of sheet metal analyses. Engineering Computations 20(526):744–758.
Lee S H, Han C S, Oh S L, Wriggers P. 2001. Comparative crash simulations incorporating the result s of sheet metal analyses. *Engineering Computations* 20(526):744–758.
Hu Ping, Bao Yidong, Robert Huth, Bo, et al. 2006. Car body crash simulation into the process factors analysis. Journal of Solid Mechanics 27 (2):148–158.
Bao Yidong. 2004. Automobile body parts of one step inverse forming FEM and lean collision simulation of, PhD thesis, Jilin University.
Huh H., Kim K.P., Kim, S.H., Song J.H., Kim H.S., Hong S.K. 2003. Crashworthiness assessment of front side members in an auto2body considering the fabrication histories. *International Journal of Mechanical Sciences*, 2003 , 45:1645–1660.
uh H., Kim K.P., Kim, S.H., Song J.H., Kim H.S., Hong S.K. 2003. Crashworthiness assessment of front side members in an auto2body considering the fabrication histories. *International Journal of Mechanical Sciences* 45:1645–1660.
For many years, Liu Li, Li Hongjian, et al. 2008. Study on automobile technology application of FFS method to the analysis of vehicle bumper crash simulation (3):27–29.
He Wen, Zhang Weigang, Zhong Zhihua. 2005. Finite element simulation. *Chinese Journal of Mechanical Engineering, Welding Connection Simulation of Vehicle* 41(9):73–77.
Shi Yuliang, Zhu Ping, Shen Libing, et al. 2007. The relationship between the finite element simulation method for spot welding of car crash simulation based on. *Journal of Mechanical Engineering* 43(7):227–230.

Information, Computer and Application Engineering – Liu, Sung & Yao (Eds)
© 2015 Taylor & Francis Group, London, ISBN 978-1-138-02717-6

Research on ideological and political education of college students based on new media-based network platform

Da Ming Liu

No. 1 Clinical College, Beihua University, Jilin, China

ABSTRACT: With the rapid development of computer network technology and the popularity of the Internet and the daily life of contemporary college students learn to produce a profound impact, which greatly changed the college students' ideology, morality and values and so on. Analyze and develop a network of new media technology background coping strategies for ideological and political education of college students has important significance. Under the new media for ideological and political education in the problems raised reflections on the ideological and political education.

KEYWORDS: New media and networks; Ideological and political education; Students

1 INTRODUCTION

With the rapid development of computer network technology and the popularity of the Internet and the daily life of contemporary college students learn to produce a profound impact, which greatly changed the college students' ideology, morality and values and so on. College students are the future hope of the motherland, how to make full use of network dissemination of socialist core values, promote socialist theme, to overcome the negative impact of new media that brought the community to occupy the ideological positions of mainstream thinking in front of college students to become political workers for important issues. Therefore, strengthening the ideological and political education under the new media, for us to carry out ideological and political education has a strong guiding significance.

2 IDEOLOGICAL AND POLITICAL EDUCATION OF COLLEGE STUDENTS' PROBLEMS THAT EXIST UNDER THE NEW MEDIA

Since the reform and opening up, some Western countries tend to use their influence and ability to control the dissemination of information through the Internet, the new media means to transport their decadent ideology and culture of the socialist countries. This new channel of communication networks for some ulterior motives of Western countries ideological and cultural infiltration facilitated to our country, they use network technology advantage in foreign scientific and technical output, while output and decadent ideological and cultural values. Backward countries are relatively slow in the introduction of advanced technology; it is often affected by decadent ideology and culture. Because young students at a crucial stage of life, values, world view, which is often prone to be hidden behind the technology, ideological and cultural identity and values, leading to the blind worship of Western capitalist culture.

With the rapid development of computer popularity of the network, ideological and political education website becomes an important educational resource university. Many universities have established online school which has its own characteristics, network Party School, a large auditorium and other characteristics of the network of ideological and political theme of the site, but in the process of building the application, there is still a lack of attractiveness, click on the site utilization rate is not high, lack of education and teaching resources, interactive network platform construction and poor management and other issues. Because of these problems makes it difficult to attract students to browse the ideological and political sites, thus losing the use of advanced network means the value of ideological and political education.

Ideological and political education of employees is to optimize the ideological education of college students in the new media organizations reliable protection. They themselves often have a direct impact on the quality of the effect of ideological and political work. This new media on the Internet ideological and political education practitioners of traditional ideas, tools and other proposed new challenge, especially when they are faced with the network information analysis and processing, not only do ideological and political work, more important is the need to master certain network technology, only a good combination of these two points helps to handle the new issues facing the new media age.

Thus, the network era of ideological and political education of employees proposed new requirements, they need to be with the times, constantly update their knowledge structure, learning new knowledge. But in reality the majority of employees are not ideological and political ideological and political education graduates, and the actual work needs continued exploration of ideological and political work law, but also a lack of expertise in the work system of learning opportunities for further study, and the other due to some schools for students ideological and political work is not enough investment, often resulting in the treatment of the ideological and political work of employees is relatively low, which leads to the ideological and political staff mobility is relatively large, the objective existence of these factors to carry out ideological and political work of the network played a certain impediment. Ideological and political education for employees' enthusiasm and initiative had not affected, some of these problems exist shows the insufficiency of current ideological and political ranks Universities. Network World has a massive information, which will inevitably there are some harmful information, in addition to network control laws and regulations is not perfect, who need access the self-awareness of their online behavior constraints, which often lead to college students in the face there will be more and more psychological conflicts network temptation.

Free and open network features, on the one hand to achieve a high degree of freedom of speech, it also makes pornographic, reactionary, and so some of the harmful impact on the dissemination of information and college students on the network in the next package of modern technology. This harmful information on the student's original ethics forms the impact, resulting in a dilution of some students' morality. In addition, college students can easily make laws contrary to moral behavior, such as the manufacture and dissemination of false information, such as the spread of the virus.

3 THOUGHTS ON NEW MEDIA NETWORK IDEOLOGICAL AND POLITICAL EDUCATION OF COLLEGE STUDENTS

Network provides a variety of information resources to the people, but these information resources in different ways, and some are even harmful to us instill negative information and ideas through a variety of forms, so during the ideological and political education work, we must adhere to the correct theoretical guidance, arm themselves with scientific theory. We should adhere to under the guidance of the important thought of Marxism-Leninism, Mao Zedong Thought, Deng Xiaoping Theory and the "Three Represents" and the scientific concept of development, from the reality of the students' actively socialist core values education. Through these education to help college students to establish a correct outlook on life and values, thus improving their ability to resist corruption.

First, the ideological and political education "Red Website" message content must be in accordance with the needs of living close to the students, close to the students thought the pursuit of growth closer to the principle of student needs to build, in order to attract students of resonance, thus creating favorable conditions for ideological and political education. Second, an interactive network platform between student teachers to create a good atmosphere for the network of ideological and political education work. We want to take full advantage of the interactive network to serve the ideological and political work of our colleges and universities. College To take full advantage of the interactive features of the network, to build an interactive platform for exchange between students and teachers, the use of micro-Bo, QQ and other carriers to facilitate communication and exchange between teachers and students.

Ideological and political workers should also take the initiative to participate in the discussion forum on campus, conscious appropriate guidance for students, for students to use the Internet to solve problems encountered in learning and life. Only fully use the network to provide services to students, in order to attract the majority of students come to the ideological and political education website. Third, we must do the work of public opinion analysis and opinion leaders. At present, many colleges and universities have attached great importance to the Internet public opinion analysis work, to collect and analyze information on student interest in various forms, for college students to carry out the work provided the basis for, and achieved good results.

Ideological and political education practitioners should strengthen political theory of Marxism-Leninism in order to continuously improve their theoretical knowledge. Second, the ideological and political education workers should strive to broaden their knowledge in order to adapt to online media under the ideological and political work. Third, the ideological and political education of employees must master certain network communication technology and network knowledge. Good skills and knowledge as a means of ideological and political education of employees work, is an important way to interact with college students. Ideological and political education of employees through Fetion, QQ chat, microblogging and other exchanges with students on an equal footing, to fully understand the students' interests, targeted work, so that the network becomes the assistant to carry out ideological education.

Colleges and universities to strengthen publicity and education on the one hand the network of laws and regulations, norms network behavior of college students, advocating civilized it have a good awareness of the law, efforts to improve college students the ability to identify harmful information network. On the other hand you want to enhance students 'self-restraint to help them establish the correct network values and ethics, improve college students' self-awareness and self-control, to enable students to be responsible for their own network behavior, consciously resist the intrusion of bad information, do propriety civilized and law-abiding Internet users, and consciously safeguard the network order.

4 UNDER THE NEW MEDIA ENVIRONMENT, THE IDEOLOGICAL AND POLITICAL EDUCATION OF COLLEGE STUDENTS AND INNOVATIVE STRATEGIES

Use new media tools for the ideological and political education of college students to develop a new space, so the ideological and political education workers must grasp this opportunity to improve the way of thinking and ways of working to achieve in the new media environment for ideological and political education innovative work.

4.1 Emphasis on the use of new media, old media, give full play to their respective roles

In the ideological and political education of college students, we want to use on the basis of the traditional media, the importance of new media and new technologies used, that old and new media can complement each other, in order to give full play to their role in old and new media, all to strengthen and improving the ideological and political education. For example, mobile media, online media in the dissemination of time-sensitive information, the diverse interaction, we can open up a "Mobile Forum" and "Network interviews" and other columns, on a hot and difficult issues of concern to students in such forums or interviews the discussion; newspapers, magazines and other sharp-oriented, credibility and strong, we can use it to report figures advanced deeds, using it to guide the correct guidance of public opinion on campus.

4.2 Strengthen the campus network culture; grasp the ideological and political education initiative

In the new media age, online forums, blog, SMS, flying letters, chat and other new media tools in new ways to spread the message, but also gradually formed naturally, but also a new form of Campus Network Culture . Therefore, in the new media environment, to grasp the initiative in the ideological and political education, we must pay attention to these changes in the new media; new technologies in the conditions of Campus Culture must further strengthen the cultural construction of campus network. Strengthen campus network culture, we must pay attention to its advanced nature, must be advanced ideological theories to lead the campus network culture. To promote socialist ideology, use scientific, public, national and socialist advanced culture to capture the Campus new media positions. Therefore, we should make a lot of vivid and intuitive new media clips, select the ideological and political education should be the theme of the website theme set ideology, knowledge and humanity in one, to try to make the operation of the site can be done fast, massive and interactive as far as possible so that it can close the campus, close to the college students' learning, real life and thought.

Ideological and political education workers to take full advantage of new media fast, superb illustrations and audio-visual features, trying to improve the campus network positions attractive and appealing in order to attract more students into the construction of Campus Network Culture come, so seize the initiative in ideological and political education work.

4.3 Innovative tools and support, strengthen ideological and political education of penetration

In the new media age, SMS, blog and QQ, have been infiltrated into all aspects of college students' learning and life. Ideological and political education workers must face up to this situation, and try to play a new carrier network media, mobile media and other functions, to interactive, experiential, guided and infiltrating other ways to carry out ideological and political education, virtual space and real space of harmony and unity. Ideological and political education workers must comply with the requirements of the times and make full use of the means of communication and exchange of college students are widely used to promote the theme of education into college students to study and daily life, thus enhancing the ideological and political education understanding force.

Ideological and political education workers are the backbone of the effective implementation of the ideological and political education work. In the new media environment, the ideological and political education workers universities must continue to improve their understanding and use of new media level, must try the network technology and mobile communications technology to be applied to the daily work of ideological and political education among the students. Should pay attention to building up their network platform on ideological and political education, guiding students to use the Internet to learn; to be concerned about on-line information and statements, timely comments on it as a guide for students; should pay attention to understand the idea of dynamic on the network, timely implementation of the relevant public opinion research; should use new media to effectively carry out the theme campaign. In addition, the ideological and political education workers universities should closely follow the trend, pay attention to understand the new media in particular, on the network student favorite language and communication, in a timely manner have content targeted filled into the ideological and political education, and improve the way of education to make their own ideological and political education of college students can be closer to life, make it easier for college students to understand and accept.

5 OPPORTUNITIES IN THE NEW MEDIA ENVIRONMENT FACED BY IDEOLOGICAL AND POLITICAL EDUCATION

Seen from the preceding analysis, new media technology is a double-edged sword, it is to the students learning and life that brings convenience and positive impact, it would also give them a negative impact. This feature of new media technologies, to make the ideological and political education work to strengthen and improve both opportunities, and challenges, but overall, the opportunities outweigh the challenges. The new media has carrying capacity, speed, multimedia, three-dimensional, covering a wide range of interactive and other advantages. Thus, we have ideological and political education of college students which not only can make use of the new media on both the rich educational resources, but also greater scope to take the initiative to quickly correct the student promotion and dissemination of ideas, theories and policies that can overcome the traditional limitations in education, providing a new, unprecedented broad platform for ideological and political education of college students.

In the new media age, SMS, blog, online forums, are flexible and fast and so is increasingly becoming a new ideological and political education of new carriers and new tools, and shows its unique advantages. Through new media, and in some cases, college students do not have the traditional way at a specified time to a specified place to receive education, access to knowledge and information they may need, such as mobile phones via SMS and network channels, which greatly facilitates the students, but also greatly enriched the means of ideological and political education work.

In the new media environment, college students may change from passive learning to active learning and change is "indoctrination" for independent reading. Through the Internet and other new media, college students are free to choose what content they want to learn or access the information they want, and you can take a different approach to reply to the information source, can easily participate in timely feedback and re-creation. Thus, the two-way interactive communication and exchange of information on ways to make educated by passively accept change in order to actively participate, you can make the effect of ideological and political education that has been greatly improved.

6 CONCLUSION

New media technology is to break the boundaries between the real world and the virtual world, and changing the way the people exchanges fundamentally. In the new media, interpersonal communication between the conditions, everyone can hide their true situation; you can speak freely with impunity, to express their views. Thus, it is conducive to the education of college students who learn about the true thoughts, so that their ideological and political education are targeted; also conducive to a more in-depth discussion of the relevant issues, ideological and political education should produce some of its practical effect.

In the ideological and political education, the degree of trust between teachers and students is an important factor in the impact and effectiveness of education and quality of education. In traditional teacher–student relationship, both teachers and students are always in a state of inequality, which makes the student and the teacher reluctant to tell the truth, it prevents to enhance the effect of ideological and political education. And with the help of SMS, blog, forums and other new media, the exchange between education and the educated have some hidden, and thus bring a sense of equality of the two sides in the personality, rights and status, that is conducive to the formation of a relaxed and harmonious atmosphere, thereby eliminating the gap between teachers and students, thereby enhancing the degree of trust between teachers and students, ideological and political education to have good teaching.

REFERENCES

Xiao Xuebin, Julie. 2009. The impact of new media on the ideological and political education and its response. *Ideological Education Research* (7).

Wang Huancheng. 2010. New trends in the new media environment of ideological and political education of college students. *Contemporary Education Forum (Management)*, (8).

Lai Yong. 2010. Ideological and political education in the new media environment exploration. *Net Wealth* (7).

Li Yan, who Velen, Haitao. 2010. Ideological and political education of college students new carrier of the new media environment. *Chongqing University of Posts and Telecommunications: Social Sciences* (5).

Information, Computer and Application Engineering – Liu, Sung & Yao (Eds)
© *2015 Taylor & Francis Group, London, ISBN 978-1-138-02717-6*

Thinking of ideological and political education of university students in the internet era

Bao Sheng Guo
College of History and Culture, Beihua University, Jilin, China

ABSTRACT: Modern universities are faced with the dual pressures of university reform and international competition, how to overcome difficulties to play the role of modern ideological and political education in colleges and universities to enhance the core competitiveness of universities, colleges and universities modern ideological and political education problems to be solved. Based on the current situation of modern universities in the premise of the Internet platform fully demonstrated the characteristics and viability of the Internet ideological and political education, the necessity of modern ideological and political education of college Internet. Discusses the Internet targeting ideological and political education and important role is in the ideological and political education to guide the development of modern universities Internet.

KEYWORDS: Internet era; Ideological and Political Education; University Students

1 INTRODUCTION

Rise and fall of colleges and universities, a direct impact on livelihood issues of national energy security, economic development and the people, plays an important role in the process of building socialism with Chinese characteristics. Therefore, how to deposit is the university sustained, rapid and efficient development of great economic significance and far-reaching strategic implications. Ideological and political education in colleges and universities in the development process has been to protect the university sustained, healthy and rapid development has made a significant contribution. As the market economy continues to develop, in-depth reform of state-owned university, college party mobility, independence, increasing dispersion, coupled with the Internet age young party members active thinking, resulting in ideological and political education faced unprecedented difficulties and challenges. Therefore, to explore the use of the Internet platform to carry out effective ideological and political education, and promote the healthy and orderly development of universities, colleges and universities to strengthen cultural cohesion, enhance the international competitiveness of modern universities, colleges and universities has become a new topic of modern ideological and political education research needs.

2 THE IMPORTANCE OF MODERN IDEOLOGICAL AND POLITICAL EDUCATION

As a large state university college of energy is the basis for the development of the national economy,

it is to enhance China's comprehensive national strength, reflecting the superiority of the socialist market economy, the window, I was an important foundation for a stable ruling party, but also to realize the important position of our Party's mass line. Modern ideological and political education of college will be directly reflected in the central role of political party organization in colleges and universities, and is an important aspect of management and operation mechanism. With the oil industry's market economic internationalization, universities have from the previous administration into a national monopoly to participate in international competition in the market economy subject. Therefore, improving the modern ideological and political education of college based work, for speeding up the economic development of the country's construction and reform of decisive significance, is a modern development of university's important organizational guarantee. Ideological and political education of the guiding ideology of "economic catch around ideological and political education, a good job of ideological and political education to promote economy," which indicates that the promotion of economic development is the central task and the goal of ideological and political education work, and ideological and political education but also for the economy spiritual foundation for development, promote market is economy healthy, orderly and rapid development.

Dialectical relationship between ideological and political education and the economy shows the ideological and political education as an indispensable position and role, only in this dialectical relationship guidance to determine the exact ideas and methods of modern ideological and political education in order from a strategic height strengthen the party's base, and promote sustained universities, stability and development. In the full international competition in modern society, we should take the new situation and problems, "eighteen" spirit study ideological and political education is facing, so the modern college party organizations to maintain the advanced nature forever, purity, and full combat effectiveness and creative and cohesion, to enhance the core competitiveness of universities to contribute to a new era.

3 PRACTICAL NECESSITY OF THE IDEOLOGICAL AND POLITICAL EDUCATION

Ideological and political education reform after the existence of "serious economic, ideological and political education of light", staff mobility and strong ideological and political education less attractive forms and other issues, needs to carry out new forms, content diversity and effective ideological and political education form, solving complex problems of modern ideological and political education colleges face. The rise of the Internet as a twenty-first century ideological and political education of modern college gives new vitality, by the group continues to explore the network platform for network platform party activities to promote party exchange in the form of new programs to enhance the Party's theoretical study, exercise of party members party, provide for the development of the core competitiveness of universities. The necessity of the ideological and political education is in the following areas:

3.1 Emerging "network politics" is the leading ideological and political education of new ideas

Development and use of Internet technology is changing the way the work of ideological and political education. January 5, 2010, Xi Jinping, ideological and political education through national grassroots work phone information system, to the national grass-roots party secretary, students "Village" issued a greeting message. This regards access to a vast grassroots ideological and political educator's enthusiastic response, become my party grassroots advantage of emerging Internet media an important manifestation of the ideological and political education. February 22, 2010, People named "Hu" microblogging Home Information bar shows: the CPC Central Committee General Secretary and State President and CMC Chairman. Users quickly add attention, fans in one day, "Hu microblogging" on the super-people, and the number of fans still rising. Party and state leaders to convey decree through SMS, microblogging

and other network form, send messages, listen to suggestions, as a new form, try new ideas and new Chinese Communist Party's ideological and political education.

3.2 A large number of "network citizens" is the Internet ideological and political education of the new object

Since 2009, the cumulative incidence of "Network mass incidents" show, "the network platform in the field" has been formed, "network citizen" has grown, matured. "Statistical Report of the 32nd China Internet Development" China Internet Network Information Center (CNNIC) 2013 released in July showed that as of the third quarter of 2013, Chinese Internet users reached 608 million, the size of mobile phone users reached 464 million people. Above tall, the ideological and political education can be the party's theory with practice combined, through a modern Internet platform, party and government activities continue to explore new forms of ideological education, improve grassroots party organizations. Promote the adoption of the ideological and political education platform party activities to strengthen party building theory, improving their party level.

4 FEATURES INTERNET PLATFORM

Development of Internet technology and the user, the network communication into public acceptance of information exchange. Internet is increasingly becoming an important means of business, education, cultural exchanges, news media, and other areas of ideological propaganda. Internet efficiency is to achieve real-time characteristics of the information capacity, transmission speed, time-sensitive functional requirements, and direct or indirect impact on people's thinking through different forms, values and spiritual pursuits. Internet platform has the following features:

1 Openness and sharing of information in the Internet platform for knowledge acquisition and exchange of information to facilitate the condition, a new way to acquire new knowledge and communicate with each other.
2 Internet platform to ensure the transmission of information regardless of time, geographical restrictions, information convenient, fast and widely disseminated.
3 Internet platform with strong real-time interaction, the user can place unlimited, you can anytime, anywhere to communicate, interact, express ideas and communicate accordingly.

5 THE MAIN PURPOSE OF LOCATING IDEOLOGICAL AND POLITICAL EDUCATION

The main purpose of ideological and political education is the use of the Internet platform, according to the status of ideological and political education to upgrade to the traditional forms of ideological and political education, so as to realize the party organization at the university to join the international competitive environment to play their own public authority in charge of the construction and management of social functioning the role and effectiveness of the positive energy generated by other aspects. Ideological and political education is the new era of the Chinese communists to actively adapt to changes in the social environment and working methods based on the current situation of innovation and network environment, and the basic objectives include.

5.1 Close liaison party unity

Internet ideological and political education can be achieved "were scattered parties together," the purpose of using the Internet to break the geographical limits of the original platform for ideological and political education, and extends the work of the geographical space between regions, between departments, the lateral interaction between party members, the advantages of the network open to get the effective development, college party anywhere in the world, you can get relevant information and assistance through the network. Communication and interaction platform are Internet ideological and political education, ideological and political education to achieve unity of party members and liaison purposes.

5.2 Strengthening democratic life

Internet ideological and political education platform is an easy, fast, real-time, efficient and understanding, and supervise the work of the party platform. Internet platform for ideological and political education should focus on the protection and implementation of the party and the people's right, to know the majority of party members and cadres to strengthen the implementation of the right of expression, the right to effectively implement and safeguard the democratic participation of party members and cadres to strengthen the implementation and protection of the majority of party members and cadres of supervision. Through the platform to enhance people's awareness of the work of the party, strengthen the supervision of the work of the party sensing capabilities, the real purpose of the network guarantee their democratic political power.

5.3 Strengthen the cohesion of the party organization, serving the people

New forms of organization are covering the full features of the Internet and ultra-strong space, developing a network of grassroots party organizations, online organizational life and online services platform. Entirely new ideological and political education of life, create new operational and organizational models and innovative party service system, develop and maintain the Party's advanced nature, thus ensuring the vitality and energy of party organizations, strengthen party cohesion. By adapting to the new situation and changes in the modern college party members, were the party organization and network service system, the establishment of a full range of ideological and political education service platform, forming a real network platform to serve the masses.

6 ROLE OF THE INTERNET IN MODERN IDEOLOGICAL AND POLITICAL EDUCATION

Dissemination of information on the Internet as a way of being permeated to all areas of social life, quickly changing the way people work, lives and ways of thinking. Advantages of using the Internet more and more to the ideological and political education of college administration get the practice and promotion in the ideological and political education in different forms, to play its role to protect the core competitiveness of universities.

The Internet has become the new carrier to spread advanced ideas. With the rapid development of the Internet, the university can build the unit's website. By ideological and political education of the comprehensive section of the site, such as "ideological and political education Expo", "Friends of party members" and other party knowledge platform, the party's political theory classics, the Party's basic line, principles and policies, the latest of the party and state leaders speech and the ideological and political education and other dynamic data sharing. Provide all party workers and members abreast of the latest information on the ideological and political education, the exchange of ideas and political education work experience, explore the carrier of ideological and political education theory, easy to learn about the latest current political party, for the majority of party members to provide a new theoretical ideas of the carrier.

The Internet has become a new model of Party work. College party management through the establishment of an integrated electronic system for party work can improve efficiency and reduce the cost of party work. By using ideological and political education of electronic accounting system can achieve a paperless office, party workers can be recorded via the network party work, access to documents on previous work subordinates affairs and transport, which saves the cost of office, but also improves the work efficiency. Internet-based digital party management and daily work, improve the efficiency of party work, to achieve a comprehensive new model of party building and run party affairs management services.

The Internet has become the new position of party members in management education. You can use the Internet features across time and space, construction of ideological and political education information dissemination, open Internet Party, online ideological and political work, honest government, mass organizations and other online ideological and political education positions. You can also take advantage of the openness of the characteristics of the Internet, the ideological and political education platform links to other units, and learn from the best practices and advanced experiences of other units ideological and political education, expanding the learning dimension of party members, improve the demonstration effect of advanced ideological and political education. By constantly enriching the content of the ideological and political education network, enhancing the depth and breadth of the party of education, the ideological and political education created a good atmosphere of culture, and strengthened the network of ideological and political education. Make the Internet platform to become the party's knowledge mastered college party, the party's theory, become party to enhance the sense of responsibility, improving their quality of new political positions.

The Internet has become a new platform for interaction between the parties. Open interactive use of the Internet, the majority of party members through the "ideological and political education exchange," "Hot Topics" and other Internet platforms, ideological report, comments, ideas and other work of the party. University Party platform through the Internet can grasp the ideological trend of the masses, for the masses of party members or share those problem-solving. In order to establish a university with a typical propaganda, you can put a candidate deeds, photo sharing to the Internet platform, organize the masses to participate in discussions and voted, this work is in the form of not only to improve the participation and awareness of the public, but also for the excellent typical extensively propaganda in order to create a new platform for the exchange of university work of the party.

7 CONCLUSION

New media technology to break the boundaries between the real world and the virtual world, changing the way people exchanges fundamentally. In the new media, interpersonal communication between the conditions, everyone can hide their true situation; you can speak freely with impunity, to express their views. Thus, it is conducive to the education of college students who learn about the true thoughts, so that their ideological and political education targeted; also conducive to a more in-depth discussion of the relevant issues, ideological and political education should produce some of its practical effect.

Ideological and political education is the degree of trust between teachers and students and it is an important factor in the impact and effectiveness of education and quality of education. In traditional teacher–student relationship, both teachers and students are always in a state of inequality, which makes students and teachers reluctant to tell the truth, it prevents to enhance the effect of ideological and political education. And with the help of SMS, blog, forums and other new media, the exchange between education and the educated have some hidden, and thus bring a sense of equality of the two sides in the personality, rights and status, there is conducive to the formation of a relaxed and harmonious atmosphere, thereby eliminating the gap between teachers and students, thereby enhancing the degree of trust between teachers and students, ideological and political education to have good teaching.

REFERENCES

Liu Hongjun. 2009. Several colleges and universities scientific development thinking. Learning and exploration, (5).

Cold Dives. 2006. Adapt to the new situation and to explore new ways – on the petroleum and petrochemical *Ideological and Political Education. Theoretical Front*.

Chen Li. 2006. Ways to enhance the effectiveness of grassroots ideological and political education in colleges and universities. Party School.

Mr. Zhao. Universities in New ideological and political education of ideological and political work study. Sichuan People's Publishing House, 2009.

Jiang Tiezhu. 2008. Famous Ideological and Political Education [M]. Wenhui Press.

Sun Chuanming. Ideological and Political Education Problems and Solutions. Coastal Universities and Science and tTechnology, 2005.

Xiao Xuebin, Julie. The impact of new media on the ideological and political education and its response. *Ideological Education Research*, 2009 (7).

Wang Huancheng. New trends in the new media environment of ideological and political education of college students. *Contemporary Education Forum (Management)*, 2010 (8).

Lai Yong. Ideological and political education in the new media environment exploration. *Net Wealth*, 2010 (7).

Information, Computer and Application Engineering – Liu, Sung & Yao (Eds)
© 2015 Taylor & Francis Group, London, ISBN 978-1-138-02717-6

Study on data mining for customer relationship management

Li Ping Zhang
School of Business Administration, Chongqing University of Science & Technology, China

ABSTRACT: Customer relationship management is an important aspect of the business. Using ID3 decision tree to analyze the losing customers, we explain how to extract the basic characteristics of the losing customers, and provide some help to the telecom operators in the customer relationship management.

KEYWORDS: Data Mining; Customer Relationship Management; Decision Tree

1 INTRODUCTION

With the entering into the WTO, and China's telecommunications market is gradually opening up to the outside world, international telecom operators gradually penetrate into China's telecommunications market through various means. Domestic telecom operators are facing an intense market competition environment. In the high-growth of increasingly competitive and complex environment, the mature telecom operators are concerned on how to improve the mobile user' satisfaction, how to reduce the loss of customers, and how to mine potential value of the user and profit growth, as well as the stable operation of its own network, improvement of customer perception. These have become the key to maintain a competitive advantage and to compete for market leadership in the future.

The telecommunications industry is a typical data-intensive industry. Compared with other industries, the telecommunications industry grasps more users in information. The ones who can properly analyze these data to obtain useful knowledge can provide better services for the user and find more business opportunities, so that they can win in the competition. Telecommunication companies must save user's call data for accounting and monitor status of network running and network planning, and the operators need to analyze these data in order to find useful patterns to optimize the network.

Data mining is a process to auto-obtain useful information or knowledge from a large amount of data, which belongs to the BI (Business Intelligence). BI is a very important branch of Artificial Intelligence and Computational Intelligence Science. Inorder to have an access to new knowledge, data mining is to put various data generated in the process of business operations and unify them together, and data warehouse provides an effective method of data centralization and unified mechanism. Therefore, data mining and data warehousing have important application value in the telecommunications industry.

2 PROBLEMS AND RESEARCH OBJECTS

Currently, operators' basic information system is fairly complete, but the data of each system is relatively independent. With the change of business development innovation and user' behavior, data mining faces the following status quo.

Supporting system is relatively isolated and different data in different formats are stored in different systems, all these can easily lead to an isolated island of information, and business and system construction plan lacks reliable decision to support data analysis.

Operator's mobile data center construction is gradually mature, unified storage, cloud-oriented platform service has become increasingly evident, and vendors' follow-up cooperation model is to provide development services and solutions based on the vast amounts of data.

Network user's behavior and network business have become more diversified, end-user-oriented, understanding of customer perception, positioning user consumption habits, and mining user potential.

This thesis focuses on the use of decision tree analysis, which gets a tendency of customer losss by establishing a customer value model through the analysis of customers' monthly ARPU (Average Revenue Per User), online time, payment of premiums. Customer is the profit of the operators, and through the analysis of customer loyalty, providing service of high quality to those high valued customers, retaining the customers have losing tendency ,operators can increase the market share and provide better service to the customers.

3 KEY STEPS OF DATA MINING

3.1 Determine the boundary of the system

For data mining and establishment of a data warehouse, the most important thing is to identify the target.For example, the analysis of results will lead to which decisions and which types that decision-maker is interested in, and what info needs to be provided for decision.In some sense, defining the boundary of the system can also be demand analysis of data warehouse system design, for it reflects data analysis requirement of decision-maker inform of the definition of system boundary. To analysis the tendency of customer value and losing, we need to get the distribution of various packages in the market, customer satisfaction and behavioral characteristics of losing customer, and then we could provide support for those decision-makers.

3.2 Determine analysis subject field

The second step is to determine the theme of the analysis and build analysis model based on these themes, master the association between the subject fields, and determine the subject domain attribute group. Customers and domain-experts combine with industry analysis itself andmining knowledge information to provide data for decision-making. The subject of this study is customer value and losing tendency analysis. For this theme, we need to invite the customers, customer manager from the operator, marketing manager, the experts from various fields, such as accounting and so on, to discuss and determine the original basis data from various production systems for analysis. Table 1 is the call and payment information of losing customer.

Table 1. Customer's call information and payment information.

ARPU	Lease Months	Call Len(Sec)	Over Due	Call Number	Cancel Lease
150	36	300	No	100	No
120	30	280	No	80	No
100	20	270	No	80	No
90	18	250	No	100	No
80	25	280	No	85	No
78	22	165	Yes	10	Yes
70	20	150	Yes	10	Yes
60	30	200	No	20	No
50	25	150	No	50	No

3.3 Data warehouse logical model design

Logic model is an important part in the implementation of data warehouse, because it can directly reflect the requirement of business and it plays an important role in guiding physical implementation of the system.The main tasks of logic model:

Analyze subject field and determine the the meneeds be loaded into the data warehouse;

Determine hierarchical division of granularity, hierarchical division of data granularity needed to complete this theme;

Determine data partitioning strategy;

Definition of relation mode and recording system,-determine data extraction model and data sources, etc.

After completing logical design of the data warehouse, it needs physical design of data warehouse and ETL work, and this article will not describe them indetail.

3.4 Data mining prediction and regression

Data mining technique develops rapidly in recent years, and algorithms for data mining research is in continuous innovation. IEEE International Conference on Data Mining (ICDM) selects the 10 classical algorithms in the field of data mining in December 2006: C4.5, k-Means, SVM, Apriori, EM, PageRank, AdaBoost, kNN, Naive of Bayes and CART. C4.5 is one of decision tree algorithms, which has been improved from ID3 algorithm. This article focuses on the ID3 algorithm constructing decision trees.

Based on ID3 algorithm of information theory, use Information Gain as indicator of measuring the quality of node splitting to select the splitting attribute. Highest information gain split will be a division program. In 1948, Shannon (CE Shannon) brought forward information theory. Below are the definition of Information and Entropy:

$$\text{Information} = -\log_2 P_i$$
$$\text{Entropy} = -P_i \log_2(P_i)$$

Actually, entropy is the weighted average of system information; also it is the average information of the system. Information gain is the information obtained by the system due to the classification. Entropy is a statistics which measures the degree of system confusion. The greater the entropy, greater the systems' confusion. The purpose of the classification is to extract system information, and allow systems to a more orderly, regular, organized direction. So the best split program is to reduce entropy to the maximum amount (i.e. information gain).

After pre-processing of data, we begin to summarize the decision tree. Because ID3 algorithm is a recursive process, we only discuss a particular node split.

Information entropy calculation.

First, Information entropy calculation

Information entropy:

$$I(U) = -\sum_i P(u_i) \log_2 P(u_i)$$

Probability of type u_i

$$P(u_i) = \frac{|u_i|}{|S|}$$

$|S|$ is the sum of all examples, $|u_i|$ is the count of type u_i.

Second, Conditional entropy calculation
Conditional entropy

$$I(U/V) = -\sum_j P(v_j) \sum_i P(u_i/v_j) \log_2 P(u_i/v_j)$$

Probability of v_j when property X gets value v_j:

$$P(v_j) = \frac{|v_j|}{|S|}$$

Conditional probability of type u_i

$$P(u_i/v_j) = \frac{|u_i|}{|v_j|}$$

Third, Information gain calculation
Information gain Value:

Gain(X, S) = I(U)–I(U|V)

The definition of information gain is the difference between the two information amounts: one is required to determine the amount of information S and another is to be determined after the attribute value X, that is, the information gain is related to the attribute X.

Using the definition of function Gain to arrange the attribute, the property with highest information gain is selected as a test property to the given set of S. First, we create a root node, mark with the property, create branches for each value of the property, and then create the tree recursively, finally it constructs a decision tree. Each node is the property with maximum gain in properties. Below is the generated decision tree:

Fig. 1 is a decision tree for judgment on losing of mobile customer possibility. It points out that whether a mobile customer will be lost. Each internal node (square gbox) is a detection of a property. While

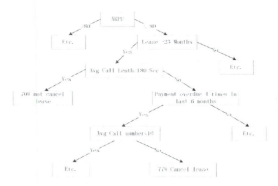

Figure 1. Loss of Mobile Customer Probability Analysis Decision Tree.

enterin the new decision record, we can predict its class that the record belongs to.

We analyze losing customer from different operators and different regions to get their common characteristics. The example applies losing customer characteristic analysis, as shown in Table 2.

Table 2. The characteristic of Loss Customer.

No	Name	Value	Memo
1	ARPU	Less than 80 Yuan	
2	Month of Account Active	Less than 23 months	
3	Avg seconds per call	Less than 180 sec	
4	Payment overdue	4 times in last 6 months	
5	Number of call per month	Less than 10 times	

4 CONCLUSION AND FUTURE RESEARCH

Through data mining analysis, we could master the new customer behavioral characteristics of different periods and understand the losing customer cyclical characteristics, and provide an objective scientific basis for development of methods and strategies of customer retention and caring. With the multi-dimensional characteristics the analysis of new customer and losing customer, we could master new customer retention and cyclical behavior characteristics of losing customer. The purpose is to reduce the loss rate and improve the retention rate of new customer.

All in all, the data mining technology has broad application prospects in the field of operator decision management. The decision tree analysis method is fast and comprehensible, so it plays a special role and great potential in the telecom data mining application. It should be pointed out, in addition to the data mining based on decision tree analysis method, the implementation of data mining of telecom also needs the integration of a variety of data mining methods, such as association rules, neural networks, Bayesian belief networks and so on. If someone can be in the lead in the field of data mining application, he would win the market opportunities and effectively enhance the core competitiveness of its enterprises. Because of the diversity of data mining tools, all the operators should choose the type of model and process carefully. These tools do not produce the data flow or system flow and the role of data mining tools have to rely on commercial data collection and accuracy, as well as the complete establishment of production and operation system, especially the establishment of data warehouse systems and processes. Conclusion for the application of a variety of tools also needs the confirmation and evaluation from the business experts. We believe that the correct

use of data mining technology will play a bigger role, and it will become a powerful weapon to improve the core competitiveness of the telecom operators.

REFERENCES

[1] Chen Wen Wei, Wang Jin Cai. 2004. Data Mining and WareHouse, The People's Posts and Telecommunications Press.
[2] Shao Fengjing, Yu Zhongqing. 2003. Data Mining-Principle and Algorithm, China Water Power Press.
[3] Liao Li, Yu Yingze. 2001. Data Mining and Data Warehouse with Their Application in Telecommunications, Institute of Computer Science and Technology, Chongqing.
[4] Zhao Hongbo, Meng Yaling, Data mining for customer relationship management in telecom, *Telecommunications Technology*, 2012.

Information, Computer and Application Engineering – Liu, Sung & Yao (Eds)
© 2015 Taylor & Francis Group, London, ISBN 978-1-138-02717-6

Thinking of the art design teaching in higher vocational colleges

Xin Lian Zhu

Jiangxi Tourism and Commerce Vocational College, Jiangxi, Nanchang, China

ABSTRACT: Art design is, a comprehensive subject, a set of art, culture, science and technology in a body. Many of the art design teaching in Higher Vocational Education in our country has the following drawbacks: the existence of the teaching goal setting is unreasonable, monotonous teaching content, teaching methods and teaching evaluation are not comprehensive, and other mechanical defects, because of these reasons it could not adapt to the needs of occupation education in the new period. The teaching objectives, good teaching content, the teaching process of autonomy, diversity of the teaching evaluation are the four aspects that are proposed to improve the higher vocational art design teaching related countermeasures.

KEYWORDS: Higher vocational education; Art design teaching; Disadvantage; Countermeasure

Art design is a set of art, culture, science and technology in an integrated disciplines, at present, many higher vocational colleges opened artistic design major, with its relatively low entry threshold, better employment prospects, attracting a large number of students. But throughout many art design in Higher Vocational Teaching in our country, there are still many shortcomings, it cannot adapt to the needs of occupation education in the new period. With the deepening of higher education reform, today there is an urgent social demand for compound talents, how to improve the teaching of art design in higher vocational education quality has become an important subject in front of us.

1 TEACHING OF ART DESIGN VOCATIONAL DISADVANTAGE

One is the teaching goal setting is not reasonable, on the other hand, it does pay attention to the individual differences of students. Some student's painting foundation is better, and some students have no basis, but in actual teaching, the teacher cannot take into account the comprehensive ability, cognitive, psychological emotional factors, the teachers cannot teach students in accordance of their aptitude. According to different levels of students, set the same goal, no individual difference methods reflect the students' capability. Lead to some basic painting better students "food", the basis of poor students "do not eat" situation. On the other hand, because of the transformation of Higher Vocational Art Design Specialty in our country which is mostly derived from traditional art education, discipline orientation is not clear, the

teaching system is not mature, hence the deviation of the teaching goal. Many of the simple superposition of Higher Vocational Art Design Speciality for traditional art teaching and information technology can achieve the perfect combination of technology and arts and culture, and thus become a microcosm of the ordinary undergraduate education, uniqueness cannot reflect the teaching goal and forward-looking, lack of professional characteristics, cannot adapt to the modern society in urgent need of compound talents training requirements.

Second is the monotonous teaching content. On the one hand, the contents of the current teaching of art design specialty of some higher vocational college is still focused on technology, which belongs to the "elite" education, not to stimulate the enthusiasm of the art design of all the students, to stimulate students interest in learning. Although the reform of higher education has been put forward for many years, but most of the artistic design specialized reform just in order to adapt to the trend of Higher Vocational Teaching Reform of "image" change, content than the design course of the form. China's Ministry of Education put forward clearly "orientation of higher vocational education is the employment education", but a lot of art design in higher vocational teaching content with teaching materials, the disconnection between theory and practice, practice teaching project less, cannot adapt to the employment market. On the other hand, art design teaching must take the student'strong humanistic quality as the basis, only in a heavy cultural guide, in order to design a successful works. At present, higher vocational art design speciality is generally related to lack of humanistic quality training. Some vocational colleges uses the ordinary teaching theory with more utilitarian trends, ignore the cultivation of comprehensive quality of art design talents, pay no heed to the comprehensive development of students, violated the original intention and essence of art design education.

The three is the inflexible teaching methods. Although most of the colleges are in constant exploration of new art design teaching means, but in practical applications are often unable to highlight these art design course in Teachers' personality, oral teaching and demonstration of guiding, "stressing skills", "light participation", "heavy teaching", "lack of practice", "valuing commonness and personality characteristics of light" still exists. To realize the organic combination of "teaching, learning and doing", They are unable to reflect the "double main body" teaching, ignoring students' main body status in teaching, how to learn how the teacher taught, teaching, unable to mobilize to explore the knowledge and skills of students' subjective initiative and the independence of the project teaching method, task driven method and research-based learning new teaching method is not widely used in art design teaching.

The teaching evaluation system is not comprehensive. The art design of the learning process is a long-term accumulation, repeated training process, and in many colleges and universities teaching, teachers still importance to outstanding students based on make "qualitative evaluation", but ignore the daily learning" quantitative assessment", not to promote the formation of students' comprehensive art design in the improvement of quality and the concept of lifelong learning. As the evaluation targets, excessive focus on results and ignoring the process. At the same time, different design level students use the same evaluation methods; the art basis of poor students cannot see their own progress, to the professional to lose interest in.

2 COUNTERMEASURES TO IMPROVE THE TEACHING OF ART DESIGN IN HIGHER VOCATIONAL COLLEGES

2.1 The teaching goal hierarchy

First of all, in the teaching of art design in higher vocational education, formed to positioning in the professional ability, the promotion, the student painting technical level and quality of art students interested in art and design, to develop the habit of lifelong learning as the final goal, setting the goal of teaching more scientific, reasonable, goal setting is no longer simple art class in ordinary colleges and universities continue, but to promote students in art design on the road to go further, and promote comprehensive development of students. To the students'

artistic perception, technical level, behavior and other aspects of the target to quantify, refine, for students to better target evaluation. Second, the difference should pay attention to the teaching goal, according to the different levels students set up their own accord goal, let each of the students to meet their learning needs, "the successful design" and "design", "active design" concept to guide teaching, stimulate the student to the art design of confidence.

2.2 Teaching content rich

The art of teaching design in higher vocational college should change the single teaching mode in teaching materials, making the art more close to the lives of students, make students willing to accept, to explore a more suitable teaching mode for students. On the one hand, development should pay attention to the conventional classroom teaching. Fully consider students' personal feeling, stimulate students to participate in artistic design motives, be based on conventional design courses on a variety of elective courses to vigorously expand the new mode, so that students can learn according to their own preferences, characteristics, leading to the development of art teaching platform. On the other hand, attention should be paid to the classroom teaching and student's extracurricular life by combining the students' extracurricular activities for teaching the students of art design of the second class, the implementation of the "integration" inside and outside the course of art design teaching mode, take the employment as the guidance, extensive contact with the design studio, outside enterprises, provide more training opportunities for students, training students' practical ability. Can also guide the students to become design studio, to undertake the design task outside the school, advance business simulation. Make good use of the Internet as a platform and encourage students to actively participate in the network academic forum, established QQ group, the establishment of BLOG, is also available for students to apply for "pig", "Witkey China" power guest account, let the students finish or bidding in power guest website design works through group cooperation, realize their own value, advanced technology and successful learning the concept of excellent works, make the student extracurricular activities more personalized.

2.3 The process of teaching autonomy

USA educator Polya once said, "any knowledge of the best way to learn is by himself found. Because of this discovery, the most profound understanding, but also the most easy to grasp the inherent law, nature and contact." The teacher wants to use "meta cognitive learning theory as the support, to fully appreciate the significance of" two subjects teaching, give full play to students' subject status, guide students through the "task drive" active thinking, active exploration. On the one hand, to the creation of self-exploration learning environment for students, arouse students' interest. The creation of a strong learning atmosphere for students, stimulate students interest in learning is the first step. If the students for learning objects lack the interest, they will not take the initiative to explore knowledge. In the process of teaching, teachers should carry out the creation of situations before class, by guiding the interest, questioning, and mobilize all positive factors of students. For example, learning advertising design, assuming you have graduated for students to simulate the situation – to do self-promotion, for personal "VI design" a vivid image of the, or let the students for their own good friend to design a meaningful birthday gift, etc. Let the students with questions, into the design practice. Questioning stimulates students to explore the motivation of knowledge of art and design. On the other hand, guide the students to explore independently. Sue Home Linsky also said that "in the depths of the human mind, have a deep-rooted need, I hope he is a discovery and a researcher, explorer." Once the students' learning motivation is stimulated it results in effective inquiry and cooperative learning.

2.4 *Diversified teaching evaluation*

On the one hand, the content of evaluation should be diversified. By the quantitative evaluation of simple to quantitative and qualitative evaluation of the combination way. To quantify the evaluation of students' learning attitude, psychological, emotional, cognitive and so on, the evaluation should not only focus on the design of technical control, but also should pay more attention on the overall development of body and mind. By combining the objective evaluation and the formative evaluation, the evaluation of the teaching of art design should not only be limited to the end of the year, and during the course of learning of students, real-time diagnostic, formative evaluation of students' problems, timely guidance, promote the forming of student's basic quality. Proper attention should be given to individual differences.

On the other hand, the diversification of the valuators, the traditional teacher evaluation of single, interactive evaluation into the teacher–students, students, through student self-assessment, peer assessment, peer evaluation are the ways to make students to encourage each other, and common progress. Self-assessment evaluation combined with him, teaching evaluation is the essential stage of design course. Through the formation of teaching evaluation it can promote students' self-recognition.

Information, Computer and Application Engineering – Liu, Sung & Yao (Eds)
© 2015 Taylor & Francis Group, London, ISBN 978-1-138-02717-6

The quality structure of English teachers in higher vocational colleges from the perspective of "planning of excellent engineers"

Xiang Ying Cao & Xiao Li Leng
Nanchang Institute of Technology, Nanchang, China
Jiangxi Vocational College of Economic Management, Nanchang, China

ABSTRACT: Implementing "Planning of Excellent Engineers" is one of the important approaches of embodying the characteristics of education in higher vocational colleges. The teacher's quality is the key to the success of the education for all-around development. The researcher investigated fifty English teachers (including the deans of the Foreign Department) and 200 students from Nanchang Institute of Technology from October to December in 2013. The investigation aims in knowing about the quality structure of English teachers in higher vocational colleges from the Perspective of "Planning of Excellent Engineers". On the basis of the results of the investigation, the research puts forward some strategies to improve their qualities.

KEYWORDS: Planning of Excellent Engineers; English teachers in higher vocational colleges; Quality structure; Strategies

Higher vocational education shifted its focus from scale expansion to improving teaching quality and promoting connotative construction. One of the important approaches to enhance the student's all-around quality and its competitiveness in future employment is to implement "Planning of Excellent Engineers". The teacher's quality is the key to improve the students' quality and the rate of the future employment; and therefore, it's quite necessary and essential to investigate the quality structure of English teachers in higher vocational colleges from the Perspective of "Planning of Excellent Engineers".

1 THE MEANING OF TEACHERS' QUALITY

The Teacher's quality is a decisive factor to the success of educational effect. (Shao Weiping& Li Fangfeng,2012:31-32). Teachers should possess the following qualities: self-identity of faith in professional value, gracious moral cultivation, perception of modern education, optimized knowledge structure, the development and improvement of teaching ability, the establishment and consolidation of the image of researcher, favorable somatopsychic qualities, team orientation, co-operative, specifically with other divisions (Wang Bo,2001:37). Teachers' qualities include teachers' belief, knowledge structure and ability structure (Du Hebing, Qin Xuhong, 2002:23).

2 EMPIRICAL RESEARCH ON TEACHERS' QUALITY STRUCTURE

The subjects of the research are 200 three-year-term students from Nanchang Institute of Technology who are randomly drawn and 50 English teachers (including the deans of the faculty) from five different higher vocational colleges. The questionnaire to the students is divided into two stages. The first one consists of an open-ended question: "What qualities should your ideal college English teachers possess". The purpose of the study is to investigate the key qualities of college English teachers expected by the students. After the students finished their questionnaire, the researcher sorted out, classified and analyzed their answers and then presented the results of the ideal higher vocational college English teachers' qualities as follows:

1 Appearance and image: good image, fashion, presentable appearance, good temper, elegant, dressed properly;
2 Character: compatible, optimistic, patient, amiable, responsible, having a sense of humor, strict, vigorous;
3 Professional intension: a lover of teaching, devoted to teaching career, enthusiastic;
4 Communication skills: concerning for every student, being friends with students, keeping eye contact;

5 Proficiency in English: standard pronunciation, fluent oral English, good expressive ability in English;
6 Professional knowledge: wealthy in knowledge, profound subject-matter knowledge;
7 Teaching skills: creating interesting and happy class, stimulating initiative, denying copying teaching materials and program, providing students with knowledge out of class, discouraging learning by rote, motivate students' interest;

As for the second-stage questionnaire to the students, the researcher listed 7 categories by summarizing the results of the first-stage questionnaire. The open-ended question in this questionnaire is "please choose the first three important qualities that higher vocational college English teachers should possess from the above 7 categories". The results show that the order of the most important qualities is as the following: proficiency in English (89%), communication skills (72%), the characters (63%), teaching skills (50%), professional knowledge (37%), professional intension (25%), appearance and image (16%). Thus it can be seen that the patient teachers, who are good at interpersonal communication, and have clear and fluent oral English are most welcomed by the students.

The question in the interviews to the 50 higher vocation college English teachers (including the deans of the faculty) is "what are the most important three qualities that the ideal higher vocational college English teachers should possess". After sorting the results of interviews, the results follow: 86% teachers hold that the most important quality is standard pronunciation and oral expression; 78% teachers take profound professional knowledge (consisting of linguistics, second language acquisition and business English, etc.) and various teaching methods as the second important quality; 65% teachers regard research ability as the quality that English teachers should possess; 53% teachers consider advancing with times and being bold in innovation are important qualities, with which teachers are capable of attracting students' attention and motivating students; and 46% teachers hold the view that loving educational career and students, and the willingness to communicate with them are the essential qualities. Thereby it can be concluded that proficient English, profound professional knowledge and research ability are the three most essential qualities that the excellent English teachers should have in the eyes of higher vocational college English teachers as well as the deans of the faculty.

3 THE CONCLUSION OF THE EMPIRICAL INVESTIGATION

After the investigation, the researcher has a comprehensive understanding of the quality structure that higher vocational college English teachers should possess. The following are the quality structure that 50 higher vocational college English teachers (including the deans of the faculty) and 200 students expect: (1) proficiency in English, such as, having standard pronunciation as well as clear and fluent oral expression; (2) loving students, and willing to communicate with them; (3) profound professional knowledge and interesting ways of teaching; (4) research ability; (5) patient and optimistic; (6) advancing with time and devoting themselves to development and innovation. But it also shows that some problems do exist, for example, the poor ability of oral expression as well as research ability, and lack of creative teaching methods and professional knowledge bother college English teachers a lot.

4 THE STRATEGIES TO PROMOTE THE QUALITIES

4.1 Establishing a new teaching notion

The traditional teaching lays stress on explanation of knowledge and imparting of technical skills. The classroom is the only teaching environment, in which students passively and negatively accept the infusion of the knowledge. However, the teaching environment under the Mode of Combining Learning with Working is an open classroom, where enterprises and companies can also be regarded as teaching classroom, as a result, the studying enthusiasm of students is aroused and favorable teaching effect is obtained. College English teachers should make the transition from the old conception to the new teaching methods, in which they attach importance both to the instruction of knowledge and to the improvement of the students' practical use of the knowledge.

4.2 Developing the double-professional teachers

Double-professional teachers require English teachers own favorable professional ethic and faith, reflect the characteristics of "skills" and "profession", both thoroughly understand the relevant theory concerning the subject and master professional skills. The strategies to develop the double-professional teachers are as the following: first, the leaders in higher vocational colleges organize series of lectures, inviting translators and interpreters from different enterprises and institutions, and salesmen from foreign trade companies to the lectures, in this way, English teachers can have face-to-face communication with the experts and obtain the opportunities to improve their technical skills; second, the colleges can co-operate with foreign trade companies and foreign affair organizations, providing English teachers with the chance to put their knowledge into practice. Double-professional teachers usually try more interesting ways of teaching, for example, they turn the traditional classroom into a more authentic teaching environment, such as, imitating activities that the translators or foreign trade salesmen conduct everyday, in such way, students can improve their abilities of practical skills.

4.3 Combining teaching with research

One of the efficient approaches to promote teachers' qualities is by combining teaching with research. The research should be based on college English teaching practices, from which the researcher discover, study and then solve problems, as a result, the teaching quality improves. Higher vocational colleges should increase money for research and conduct academic exchanges, for instance, inviting famous teachers and experts to introduce the advancing front. At the same time, other activities, such as teaching observance, exchange among teachers can create various opportunities for them to involve in researching. Teachers question their own teaching objective and activities, reflect their teaching, and write down teaching diaries, all of which can help the teachers pursue their professional development.

5 CONCLUSION

In short, through the research on the higher vocational college English teachers' qualities, a conclusion can be drawn as the following: the qualities, such as, high level of proficiency in English, clear and fluent oral English, good at communicating with students,

patience and interpersonal ability that excellent higher vocational college English teachers own, are most favored by the students, while profound professional knowledge, various teaching skills and research ability are the three qualities that English teachers including the deans of the faculty count much. English Teachers' optimizing their own quality structure is of great value to develop more high-quality and competent students.

REFERENCES

[1] Du Hebing, Qin Xuhong. The Requirement of Teachers' Qualities in the New Century. The Journal of Neimenggu Oil and Chemical Engineering, 2002(4).23.
[2] Qiao Xiaoliu. 2012. The Conception of College English Teaching Reform from the Perspecitve of "Planning Excellent Engineers. *Shangdong Foreign Language Teaching Journal*,(3):69–70.
[3] Wang Bo. 2001. A research on main factors of teachers' quality structure. *The Theory and Practice of Education* (4):37.

ABOUT THE AUTHORS

Cao Xiangying(1980-), female, Han nationality, from Jiujiang of Jiangxi province, Lecturer in Nanchang Institute of Technology, Research Interests: Professional development and second language teaching.

Leng Xiaoli(1965-), Male, Han nationality, from Jiujiang of Jiangxi province, professor in Jiangxi Vocational College of Economic Management, Research Interests: English Language and Literature.

Information, Computer and Application Engineering – Liu, Sung & Yao (Eds)
© *2015 Taylor & Francis Group, London, ISBN 978-1-138-02717-6*

Research and countermeasures of English vocabulary teaching in higher vocational colleges

Mu Yu Lin

Shaoyang Medical College, Hunan, Shaoyang, China

ABSTRACT: The vocabulary teaching is an important part of business in English teaching. In this paper, taking the English Teaching of Higher Vocational Education "practical, good enough for the degree of" principle as the guidance, analyzes the characteristics of the current situation of business English Majors in Vocational Colleges and business English vocabulary and discusses how to take corresponding strategies of effective vocabulary teaching methods to improve vocabulary learning efficiency, fully mobilize the enthusiasm of the students study business English, good students of business language and communicative competence, and then improve the quality of teaching.

KEYWORDS: higher vocational education; Business English; lexical features; teaching strategies

1 HIGHER VOCATIONAL BUSINESS ENGLISH PROFESSIONAL STATUS

1.1 *The present situation of the students*

Because of China's current education system, overall, the quality of students' language and culture in higher vocational colleges are slightly worser than undergraduates, especially the English foundation is weak. But in recent years with the expansion of college enrollment and students shrinking, this issue is more prominent. Differences in Business English of vocational college students are larger, learning objective and learning atmosphere are not the same. There are different degrees of psychological problems in learning English, mainly in the following aspects: first, the lack of learning motivation. English learning purposes is not clear, do not know what is the use and how to learn the English. Some students learn business English majors just to cope up with the job or temporary major adjustment and the choice of professional training, so do not understand, with blindness. Second, learn the negative emotions. Some students with poor English foundation, especially the part adjustment, heavy burden, in anxiety and learning achievement is not good cycle and unable to extricate themselves. Also some students' independent consciousness is strong, not willing to accept other people's arrangement and guidance, so as to form a psychological confrontation emotionally negative, action, produce "dislike" for learning English, aversion to learn Eenglish. Third, lack confidence in learning. Some of the students to understand the importance of business English, have to seriously study with will and determination, and once when they face a specific problem like failing in an English exam, they start doubting their ability to learn which automatically leads to loss of confidence and development of inferiority complex.

1.2 *The professional characteristics*

First of all, the business English major in Higher Vocational College is the first English majors, we should adhere to the English for this principle, training of personnel must have wide English foundation and cultural knowledge, with good basic skills of English listening, speaking, reading, writing and translating, based on language knowledge and skills training, Teachers should focus on the cultivation of students' practical ability and the ability of applying English. Second, higher vocational college English teaching and English for specific purposes ESP certain (English for SpecialPurpose) uses English occupation scope of EOP (English for Occupational Purpose) teaching features, teaching is more inclined to linguistic functions and activities, its purpose is to make the learner acquire its social purpose terminal behavior related "". Again, higher vocational business English majors is a composite type cross disciplinary, the professional training of personnel should not only have a solid foundation in English, and must master a wide range of business knowledge. The requirements of "practical give priority to, good enough for the degree of", make the business knowledge and business English language application ability and communication ability of combining.

Vocabulary is an important part of business English teaching, students are also one of the most headache. The students' vocabulary learning mainly include the following: first is to belittle or barely learning vocabulary, do not pay attention to memorizing words, resulting in less vocabulary; two is the learning method is improper or heavy meaninglight pronunciation, although efforts but the effect is not obvious; the three is the large vocabulary, but the actualapplication ability is poor. All these will cause bad influence to further enhance the students' comprehensive language ability.

In addition, the hardware and software of Higher Vocational Business English teaching and practice in general is relatively backward, the overall quality of teachers, teaching mode, teaching method and scientific research abilityand other aspects of undergraduate college teachers still have many gaps.

2 CHARACTERISTICS OF BUSINESS ENGLISH VOCABULARY

Business English (Business English) from the ordinary English (English for General Purposes, namely EGP), it is the same with the ordinary English, plays an important role in language acquisition and language communication. But business English belongs to English for specific purposes (ESP), compared with the common English has a strong professional in vocabulary, is a comprehensive application of professional knowledge and english. Therefore, business English in addition to have a common linguistic characteristics of English has its unique characteristics.

2.1 Rich professional terminology

Professional terminology is applied to different fields of professional word, is used for the right words to express scientific concepts, has a rich connotation and extension, professional term requires the monosemy, exclusion polysemy and ambiguity, and vocabulary are fixed, not free to change. The term of business English major with international universal, its precise sense, a single, unambiguous, and not emotional, vocabulary is rich, professional vocabulary of which possess strong interdisciplinary characteristics, the professional range is very wide, involving the subject knowledge to marketing, economics, finance, accounting and management. Business activities involving foreign trade, technology import, investment, foreign labor contract, business negotiation, contract, bank collection, international payment and settlement, foreign insurance, overseas investment, international transport etc.. Every fieldhas its own specialized vocabulary, terminology so business

English exist lots of different areas. For example: marketing mix (Marketing Mix), the forward contract (futures contract law), documentary credit (documentary letter of credit), financial instruments (financial instruments), bill of exchange (Draft), currency futures (currency futures), option (option), claim settlement (Li Pei) etc.. Do not understand the terminology, is unable to carry out business activities.

2.2 The common lexical specialization

Common words are commonly used in general English vocabulary. Business English is English based, in addition to its terminology, but also from the common words derived from the basic meanings, have special meaning in Business English vocabulary, namely common lexical specialization, which accounted for the vast majority of business English vocabulary, the use of frequent. For example: Appreciation (appreciation), collection (Collection), balance (balance), delivery (delivery), margin (profit, margin), maturity (maturity), settlement (settlement), returns (profit), Imports exports (import and export of goods or the import and export volume), liabilities (liabilities), customs (customs), bonds (bearerbonds) etc.

2.3 Application of words

The ancient words (archaism) are generally not common, British linguist Leech (1998) pointed out that in the classification theory of word meaning in English, professional words, archaic words and foreign words have to belong to a formal style vocabulary (words with formal stylistic meaning), seriousness and formality of the use of ancient words can reflect the business letters, business contract, legal documents. The ancient words are often used in Business English for some compound adverbs, often by here, there, where as the root, respectively, plus after, at, as, from, to, in, by, of, with, under, etc. constitute the synthesis adverbs prepositions. For example: whereby (according to, with), herein (=in which here, in...... In hereby (=by), which thereof (thus), thus), whereas (Jian Yu), etc. These words with legal means, rarely used in everyday English, but in the text of the contract but the meaning expressed more clearly, but also make the whole discourse formal.

2.4 The use of abbreviations

Abbreviation is characterized by concise language, convenient use, large amount of information. There are a lot of abbreviated terms in Business English, and is an important part of business English vocabulary. This reflects the "economy principle of language use, time-saving and labor-saving needs" – the French linguist Maltin Hei (A. Martinet). Forms of abbreviations used in Business English mainly includes the following several points.

2.4.1 The acronym (initialisms), such as B/L (bill of lading bill of lading), L/C (letter of credit of credit), D/A (documents against acceptance D/a), D/P (documents against payment D/P), FOB (free on board FOB shipment), etc.

2.4.2 Semi acronyms, such as e-business (Electronic Business Electronic Commerce (Treasury), the T-bond bond treasury bond), etc.

2.4.3 Truncated word (clipped word), such as biz (business business), Corp (Corporation), PLS (please), RPY (reply), Exp. (exit export), T.S. (transshipment transshipment), etc..

2.4.4 Blends (blending word), such as FOREX (foreign exchange exchange), hi-tech (high technology advertics (advertising statistics Tech), advertising Statistics), workfare (work welfare workfare) etc..

2.4.5 The first letter of Pinyin word (acronyms), such as ISO (International Standardization Organization International Organization for Standardization (Organization), OPEC of Petroleum Exporting Countries, the organization of Petroleum Exporting Countries), etc.

2.4.6 Hybrid acronyms, such as B-to-B or B2B (business-to-business business to business), etc.

3 BUSINESS ENGLISH VOCABULARY TEACHING STRATEGY

The above analysis and business English in Higher Vocational Business English major students and the present situation of the lexical features of Colleges Based on depicting, in the teaching process in addition to the basic skills to use such as association, word formation, comparison method to master common words, but also practical to adopt more flexible methods. We can try to improve the traditional vocabulary teaching method, applied to business English vocabulary teaching. Of course, vocabulary learning is a long-term accumulation process; only through classroom learning is not enough. Master degree students vocabulary and inevitably there are individual differences, classroom vocabulary teaching unified is difficult to meet the needs of students themselves.

3.1 *Perfect English learning psychology of students*

According to the psychological characteristics of the students in higher vocational colleges, teachers should develop a good attitude of students learning English through teach by precept and example, strengthen students' learning motivation, interest in learning, try various devices to help students overcome negative emotions, and enhance the confidence of the students. Let the freshmen to adapt as soon as possible after learning English, do the positive psychological preparation for vocabulary learning.

3.2 *Combination of vocabulary teaching and business knowledge*

Business English is a wide-ranging expertise that covers trade, financial and legal fields, and is very comprehensive. Therefore, students will continuously engage in professional knowledge of these disciplines, pay attention to the latest business information, the accumulation of solid business background knowledge and professional knowledge, cultivating business thinking, so as to enrich vocabulary, and broaden the knowledge.

REFERENCES

[1] Chen Ling J. 2008. The lexical features of business English. *China Electric Power Education*, (2).
[2] Zai Shu Mo J. 2003. Lexical features of business contract English. *Shandong Foreign Language Teaching*, (4).

Information, Computer and Application Engineering – Liu, Sung & Yao (Eds)
© 2015 Taylor & Francis Group, London, ISBN 978-1-138-02717-6

The current situation and strategy of China's new energy vehicle development

De Jun Xue & Peng Zhou

Nanchang Institute of Technology, Jiangxi, Nanchang, China

ABSTRACT: This paper discusses the new energy vehicles. Pure electric vehicles, fuel cell vehicles, hybrid electric vehicles, advanced diesel powered vehicles and the use of bio liquid fuel cars belong to the category of energy saving automobiles. So called pure electric vehicle refers to the motor vehicle which is power driven by battery or fuel cell; there are different kinds of fuel cell cars, at present the optimal way is recognized in proton exchange membrane fuel cell. The characteristics of this technology are to use hydrogen as fuel, generate electricity through the electronic movement, storage and use; the hybrid electric vehicle is defined by a conventional gasoline engine or diesel engine and electric motor combination powered automobile.

KEYWORDS: New energy automobile; development;

1 THE GREAT SIGNIFICANCE OF THE DEVELOPMENT OF NEW ENERGY VEHICLES

The automobile industry is the pillar industry of the national economy, and it is closely related to people's lives, and it has become an indispensable part of modern society. However, the traditional oil fuel automobile industry provides people with fast, comfortable means of transport and at the same time increases the national economy's dependence on fossil energy; deepen the contradictions between energy production and consumption. Along with the continuous increase of the double pressure of resources and environment, the development of new energy vehicles has become the future direction of the development of the automotive industry.

1.1 The development of new energy automobiles are important measures to reduce oil shortage

The development of new energy vehicles is to reduce the dependence on foreign oil, the route one must take to solve the contradiction between energy demand and the rapid growth of oil resources will be exhausted. In recent years, China's auto market is developing rapidly and has become the world's second largest car market. On 2009, Chinese car retains the quantity breaks through 70,000,000, production and sales both exceeded 13,000,000, annual oil consumption will reach 1.2 tons. At present, China's per capita GDP has exceeded 3000 US dollars, the upgrading of consumption structure is the inevitable trend, China is at an important stage of industrialization, urbanization and motorization, the rapid growth of demand for cars is difficult to avoid. Every 1000 people car owners in China only 38 cars, 139 cars and there is a large gap between the average level of the world, the automobile consumption market has large development space. It is expected that by 2020, car ownership in China will reach 150,000,000 units, and the use of fossil fuels and oil consumption will reach 2.5 tons, accounting for about China's total oil consumption by 55%. Therefore, the development of new energy vehicles is to ease China's oil shortage and reduce the dependence on foreign.

1.2 The development of new energy vehicles is an effective way to reduce environmental pollution

New energy vehicles do not burn gasoline and diesel, the lithium battery is used by the internationally recognized environmental protection battery. And compared to the traditional vehicle, electric vehicle had no pollution at startup, environmental performance is also excellent. For efficiency, the energy conversion efficiency of conventional cars is only 17%, the electric vehicle efficiency is 90%, even taking into account the loss of efficiency of coal-fired power generation, the total efficiency of electric cars is greater than 30%, about two times as much as that of conventional cars, the energy saving effect is very obvious. Especially in recent years, countries around the world attach attention to greenhouse gas emissions and climate change issues, although China is a developing country, the level of per capita greenhouse gas emissions is low, but because of China's large population, years of sustained and rapid economic development, energy consumption has been ranked second in the world, the future face international pressure will gradually increase. A survey shows that about 25% of the carbon dioxide is globally exhausted from cars. Our country can be the first to achieve a breakthrough in the field of new energy vehicles, will change the passive position of China on climate change, and to make contribution to the global energy to solve the increasingly serious environmental problems.

1.3 The development of new energy vehicles is the route one must take the development of the automotive industry

A revolution of new energy vehicles will be the birth of vehicle power technology, and will drive the automobile industry upgrading, the establishment of industrial model of national economic strategy, is the route one must take the development of the automotive industry. New energy vehicle use cost is very low. The 100 km with electricity costs of conversion, the cost of electricity is only 20% of the cost of oil, that is to say, the use of new energy vehicles only spend 1/5 of the money and it can be run for quite a few kilometres.

Ordinary cars, whether manual or automatic, are used, gearbox, transmission is electric motor drive, no gear box, and also very strong. In addition, the electric vehicle four wheels drive, simple principle, easy realization, and convenient operation and maintenance, no need to change the oil. The characteristics of new energy vehicles, determines that it has strong vitality and broad market prospect. Assume that in 2020, China's new energy vehicles accounted for 20%, about 30000000 vehicles a year can save 50,000,000 tons of oil, equivalent to a year for Daqing oil field oil production.

1.4 The development of new energy vehicles is an important part of smart grid construction

In traditional power system, there is an inherent conflict between the actual volatility of electricity load and generator rated conditions required for electric load for stability, how to deal with the power system peak valley difference has been a headache problem for power grid enterprises. China's power installed has exceeded 800,000,000 kilowatts, and will continue to grow rapidly, but the current power station's annual utilization hours is only 5000 hours, that is to say, many units to the peak load short time to deal with the power system and the construction measures, if successfully implemented, installed capacity of 600,000,000 kilowatts enough construction.

Just think, if the government strongly advocated the development of new energy vehicles, all city residents to buy electric cars, the evening with low-cost electricity trough for electric charge, the daytime peak can also sell electricity to the grid by higher prices on peak effect, so the peak valley difference problem can be readily solved. Established in accordance with this assumption of power grid, will have a certain self adjustment ability, generating electricity, transmission, distribution and sale, will each link and scheduling to form an effective interaction, become an organic whole intelligent, thereby greatly improving the security and reliability of power system. It can be expected that, as an important part of smart grid construction, a revolution of new energy vehicles will bring power system.

2 INTERNATIONAL NEW ENERGY VEHICLE STATUS AND TRENDS

In developed countries, Japan, Germany and America attache great importance to the new energy automotive technology, starting from the automotive technology transformation and industrial upgrading strategy, promulgated and formulated preferential policies and measures, and actively promote the national new energy automobile industry development, in order to enhance the international competitiveness of domestic automobile industry in the global auto industry, a new round of competition to occupy a favorable position. According to incomplete statistics, before the outbreak of the international financial crisis, developed countries each year for research and development of new energy vehicles and industrial development fund of not less than US $1,000,000,000, has invested a total of $about 10,000,000,000.

In the policy incentives and guidance, the world's major automobile manufacturers are speeding up the industrialization of new energy vehicles. At present, "hybrid vehicles with low emissions" has entered the large-scaleindustrialization phase, has more than 1,000,000 vehicles worldwide cumulative sales, to occupy the dominant position in the current new energy automobile market. "Mass production time zero emission electric vehicles have ahead to 2015, 10 years to 15 years earlier than the expected time. According to the forecast, before and after 2012will usher in the new energy automotive industry of the climax, the next 10 to 20 years will be a crucial period for new energy automobile industry pattern formation, new energy vehicles will be pulling a new growth point of economicdevelopment.

At present, the country's largest global hybrid car sales to the United States. In 2008 the United States hybrid car sales of 3,14,000 vehicles, 10.3% year-on-year decline. In 2007 3,50,000. Although the United States is the global sales of hybrid cars the biggest countries, but the hybrid cars in the United States auto sales proportion is very low, in 2008 accounted for about 2.4% of the overall car sales. After President Obama took office, the plug-in hybrid electric vehicle as a trump card to stimulate the economy and save the auto industry. At his initiative, the federal government for plug-in hybrid electric vehicle plans to promote, within months of being issued a series of strong measures to spend $14000000000 to support the key parts, power battery R & D and production, support the charging infrastructure construction, consumer subsidies and government procurement. The United States also has established a total of $25,000,000,000 fund, with cheap credit way to support the manufacturers of energy-saving and new energy automotive R & D and production, the target is an annual automobile fuel economy doubled. It is expected that by 2012, half of the United States federal government car is a plug-in hybrid vehicles and pure electric vehicles, by 2015, the United States will have 1,000,000 hybrid cars put into use.

3 THE CURRENT SITUATION OF THE DEVELOPMENT OF NEW ENERGY VEHICLES IN CHINA

In recent years, China has accelerated the pace of the new energy vehicle development, at present has the basic condition to realize industrialization development, initially formed a relatively complete system of key parts and components, the development of a number of new energy vehicles, realize small batch production capability of the whole vehicle. To carry out small-scale demonstration application in Beijing, Tianjin, Shanghai, Wuhan, Zhuzhou and other cities, as the new energy vehicles to further promote provides the technology, product and operation experience of the security.

4 THE EXISTENCE OF CHINA'S NEW ENERGY VEHICLE DEVELOPMENT ISSUES

New energy automobile industry in China is still at the starting stage, there are still many problems to be resolved, mainly include the following aspects:

（一）The level of technology needs to be further improved.

（二）Business investment is obviously insufficient. New energy vehicles in China is still in the making "prototype" level, to make it become the mainstream commodity users satisfied, regardless of whether the vehicle or parts are still a large number of subsequent development work.

（三）The construction of industrial system is not perfect.

（四）All kinds of factors and resources need to be further integrated.

（五）The lack of a national strategy clear and forceful policy measures. New energy vehicle is a new thing, in the current technology level condition, the cost of production is still not completely have the ability to compete with conventional cars, need the government to formulate preferential policies to support to the healthy development of. At present, China's new energy vehicle strategy is not clear, the state has not yet developed the new energy automobile strategic development planning, encourage new energy automobile production and consumption policy is not the system, lack of coordination between them, although the country has formulated the subsidy policy of new energy vehicles, but the relevant implementation details have not been announced, many subsidies actually difficult to obtain in addition to the local government, prison, for foreign enterprises to enter the local market to take the rejection policy, lead to local protectionism, so that our development of new energy automobile enterprise market environment deteriorated further narrow the space for survival.

Information, Computer and Application Engineering – Liu, Sung & Yao (Eds)
© *2015 Taylor & Francis Group, London, ISBN 978-1-138-02717-6*

Application of circular vortex pattern of the painted pottery

Xiu Mei Wu
Jingdezhen Ceramic Institute, College of Art and Design, China

ABSTRACT: This paper leads round WHORL start from Painted Pottery and its origin from three original worship WHORL initially formed a circle that nature worship, relationship reproductive worship, totem worship and original aesthetic vortex pattern in a circle to discuss the reason why earthenware is widely used.

KEYWORDS: Circular vortex pattern; Painted Pottery

As human beings prehistoric earthenware is a most systematic and complete material and a cultural creation, in addition to functions other than the fact it can be said is a most aesthetic sense of creativity. This aesthetic sense, on the one hand, reflected in their particular mix of shapes, the other, in their colorful colors and a great variety of ornamentation. These work together to create a rich connotation "significant form" where ornamentation is in the form of painted pottery of variety. They all have varied characteristics, its meaning also includes an extremely wide range of content. Different types of painted pottery culture pattern-based broadband lines. Miaodigou type multi-purpose combination of straight lines and curves, Majiayao type circular vortex pattern is popular on the outside in a spiral pattern around the point, there is the feeling of movement, mid-type circular vortex and sawtooth pattern is combined with decorative patterns Machang type polyline based. Featuring the use of a straight line. These ornamentation, the application of the impact circle WHORL curve is maximum. The main reason is that the original worship and original aesthetic are inseparable. Here's to circle the origin and formation of vortex lines and primitive worship, and to elaborate aesthetic relationship.

1   ORIGIN OF THE CIRCLE FORMED VORTEX PATTERN

Circular vortex pattern is circular and its variants are geometric patterns. Their dot pattern, concentric pattern, semicircular pattern, circle pattern, multiple concentric circle pattern, big circle pattern, dot hook leaf pattern, spiral pattern, four circle pattern, WHORL and other variations are widely used in Majiayao culture. This initial formation of ornamentation is that people in long-term productive labor formed, in many Neolithic sites and found snail shells and clam shells.

With the apparent snail shell spiral pattern; the rules on solitary line clamshell arrangement, others such as the sun's circle, fire light, undulating water fluctuations and mountains, primitive life is in regular contact In nature, in addition to the sun and the moon, people see things that were sharp circular rotation of activities around the axis sputtered objects, due to centrifugal force, will form a vortex-like trajectory of some such as: stone, jade perforation, log fires, spinning wheel fuel lines, and wheel manufacturing and running and so on. These are round source WHORL pattern. Another reason constitute a pattern of circular vortex pattern, it may be due to the direct production of pottery tools and methods or caused by legacy. Making pottery was hand-made and wheel system is divided into two into two different kinds: hand made earlier, wheel system in the post. Clay plate construction method which is a hand-made law. The strip is growing mud system to spiral approach plate built into the shapes up. Another is circled clay, which is made up of layers of shapes being piled up. Spiral plate construction method is made up of shapes, there is a clear spiral pattern, made up of pile construction method shapes, face lined for the same interval. If it is a large or small mouth bottom floor of the mouth of a small abdominal shapes, looking to become concentric circle pattern. In addition, hand nie tao is leaving fingerprints when the finger and be inspired to form a circle pattern, and so on. These are the body's awareness of long-term production practice, through a combination of reasoning and art evolved. Is continuing to create a sense of rhythm practice. Rhythm and regularity of reproduction, 1956 Hubei Jingshan, Qujialing Xichuan unearthed cultural relics show that the spinning wheel color pattern, and many more arcs, mostly spinning wheel to rotate a mimic (by the test : no color pattern on a spinning wheel with color grain or around on-line, in the rotation, and more will produce concentric circles, and other spin -shaped or curved linear), is a symbolic pattern was spinning, but also the spiritual life of the people was a direct reflection. When the spinning wheel rotate the color pattern, not only can produce a variety of beautiful melodies, but also for the observed rotation rate. Conducive to the production and easier to use. It can be said, is a circular vortex pattern varied, decorative patterns can get a better artistic effect. By coincidence line, repeated and density arranged in a row, etc., made with a rhythm and rhyme of decorative patterns, because of its rich beauty of the arc, it is rounded and smooth, soft and uniform and is widely decorated in painted pottery.

2 THE RELATIONSHIP BETWEEN THE ORIGINAL CIRCULAR VORTEX PATTERN OF WORSHIP

Generally round WHORL widely bear the following reasons: the primitive worship and aesthetic are closely linked. The original is a certain phenomenon that primitive worship of nature cannot be explained, cannot overcome a certain type of difficulty or inability to conquer a certain class of things, etc., to find a settled under or in spirit. And all about the idea that everything in the world of images, you have to show through certain material carrier out of these substances carrier, in a sense, when people constitute the object of worship, and these things and their lives are closely linked. The evolution of primitive worship is roughly to nature worship, reproductive worship, totem worship and other concrete manifestations. WHORL formed a circle with these worship has a direct link.

Nature worship is the worship of man's most primitive, primitive thinking as the initial thinking human beings, it is extremely simple, and they worship, as well as a natural phenomenon caused by the psychological fear of the outside world and their own lives most closely related. They generally worshiped by natural objects have the earth mountains, the sun and the moon Morning Star, thunder and lightning storm, rocks, water and fire, etc. Circular vortex pattern is the nature of many natural phenomena variant abstraction, such as day, month, water and so on. As Machang type of pottery in a spiral pattern as the sun maneuver between sports, continuous cycle of meaning; Qujialing culture earthenware pottery spinning wheel has many vortex pattern, helical form, the sun and the moon and downs roundabout, cold and heat alternating imagery inherent in reincarnation among them, thus linked to the need to divide the season, so the motion imagery in a circle WHORL also reflects the rhythm and continuity of time. Polka dot image with its sun sign, continuous arrangement of the sun's movement pattern is the point, the sun

maneuver and plot points into a line, the line will contain the movement of the sun rise and downs imagery. Such as: There are many forms of painted pottery culture dawenkou vortex pattern, consisting also richly colorful pattern with circumferential ring having a dynamic rhythm. Pattern circle dot pattern combinations for sun image reproduction, it forms an arc -shaped ornamentation, showing the sun rises off the roundabout symbolic meaning. Mid-type of pottery prevalent form of a continuous pattern of four circles. Corrugated curve 1973 Lanzhou almond ears station unearthed painted pottery bottle rotation pattern, graceful, tactful, with a strong dynamic, a few dots on the midpoint of the line parallel to the wave, like water on a vibrant wave beads, smooth rich artistic beauty, show when people rely on the water and nostalgic feelings. Pottery excavated from the ruins can be seen essentially distributed in the Yellow River, Yangtze River and its tributaries, as people familiar with the maintenance of the water of life and worship, so in painted pottery to water as a prototype as ornamentation. Round WHORL use the most, the most widely distributed, the largest number of artistic implication most abundant. Obviously circular vortex pattern is when people of their living environment and objective description of life worship.

Human worship is mainly reflected in the reproductive life of worship. The original human reproductive worship is a historical phenomenon throughout the world. Hegel have pointed out that "In the discussions we have already mentioned the symbolic Oriental art and reverence is often emphasized universal life force of nature, not ideology but the spirit and the power of creativity especially reproductive. In India, this religious worship is universal, it also affects the Folijiya and Syria, the reproductive performance of huge image of the goddess and later even the Greeks have accepted this concept. More specifically, the natural fertility of the general view is that the shape of male and female genitalia to use performance and worship." (Hegel "aesthetics" Vol. 3, Commercial Press, 1979 edition) This worship of reproductive performance in the original painted pottery art is particularly prominent. Painted with naked women on such Qinghai Ledu Willow Bay excavations portrait painted pottery pot like and exaggerated dough the vulva. Description of the primitive in class on reproductive worship, exaggeration and distortion used by painted pottery decoration. Circular vortex pattern is female genital abstraction. As 1958 Lanzhou color pottery unearthed Majiayao type that is a symbol of female genital shape round shape WHORL. On the other hand, primitive that day, month, water, fire constitute life on Earth, without which there will be no life. Circular vortex pattern is the day, month, water, fire symbol shape. And also the offsprings of female genital mutilation. By varying the shape circular vortex pattern has become a symbol of female genital decorative ornamentation on painted pottery as a symbol of our ancestors desire for more children, so the circular vortex pattern that is the symbol of life, is a reproduction of worship. In addition, the proliferation of life is also reflected in the totem worship.

The original totem worship is the most important hominid reproductive ceremony. There will be any impact on the totem' ancestor species reproduction. Totem worship is primitive for some natural phenomena, and material life of the people is related to the survival of certain animals and plants, which are regarded as tribal ancestors or gods, becoming the tribal fetish worship together, as well as becoming the eyes of most tribal peoples sacred, the best symbol of the tribal society is to maintain collective spiritual pillar. So, people put it on the pottery decorated with tables worship. Start with realism, to the later development of abstraction. Circular vortex pattern is a kind of snake-like variants. Snake coiled posture of a circle, forming a circular vortex pattern. Wave pattern that is serpentine line should be wriggling snake. It is said that the snake is a tribal totem. They snake as a totem worship, think that the snake is a symbol of male roots. The formation of a snake wrapped around a circle symbolizes the vulva, so the snake is male and female combination of performance, are symbols of human offspring. Snake is considered tribal totem object of worship. By depicting abstract deformation on painted pottery as a decorative form an aesthetic sense of the totem.

3 ROUND WHORL RELATIONSHIP WITH THE ORIGINAL AESTHETIC

The original inhabitants of the aesthetic sense WHORL round except from totem worship, but also because of its repetitive sense of movement and continuity, smooth lines, curves mildly sexual. Circular vortex pattern of circular shape and balance between having a distinctive artistic effect. Curve has a beautiful aesthetic quality that is lively and varied and so subtle dynamic performance and strong. As the four circle pattern depicts Majiayao culture period, full of get rid of "framework" after the sense of freedom and spiritual movement, arbitrary and there are laws, flying smooth and harmonious nature, just right. Full round schemata, flying fluency displayed a certain degree of mental health with people being optimistic, enthusiastic, reflects a person's cycle, the cycle is a repeated nature of mental schema changes. Round WHORL application reflects a human emotion ripples, and is the soul of flying is flying lines beating heart strings. Should be more moderate but continuous curve, which both stimulate people's attention and not dispersed to their attention, so people feel warm and pleasant, with a gentle nature, sensuality and bite.

4 CONCLUSION

Because of the characteristics of the above circular vortex pattern, only to large number of applications, until modern still used in all types of decorative, the large number of applications in round bronze WHORL is following the pattern of circles painted pottery WHORL variants, as well as Vatan, bronze mirrors and others have decorated with circular vortex pattern pattern pattern, modern types of packaging also has applications in the form of circular vortex pattern, dynamic, gives a youthful, exuberant sense of the rhythm of life. Visible, circular vortex pattern is decorative patterns in enduring, widely used as a decoration.

REFERENCES

[1] Guohua Achillea. 1990. "Reproductive Worship Culture", China Social Sciences Press, 1990 edition.
[2] Zheng was with: "Chinese Painted Pottery Art", Shanghai People's Publishing House, 1985 edition.
[3] A Holy with: "Religious Art" theory, Jinan University Press, 1988 edition.
[4] Niu Jun with: "Minority Religious Culture and Aesthetic Yunnan", China Social Sciences Press, 2001 edition.
[5] Xiaohui Hu with: "Mother's Song – Chinese Painted Pottery and Rock Art Motif of Life and Death," Shanghai Culture Publishing House, 2001 edition.
[6] Wushan ed: "Chinese Neolithic Pottery Decorative Arts", Heritage Press.

Information, Computer and Application Engineering – Liu, Sung & Yao (Eds)
© *2015 Taylor & Francis Group, London, ISBN 978-1-138-02717-6*

Neuroesthetics—the neural basis of art

Jian Feng Hu

Institute of Information Technology, Jiangxi University of Technology, Nanchang, China

ABSTRACT: What is art? Why can it not become a common culture that is acceptable to all the society? With the rapid development of neuroscience, a new field – neurasthenics is used to explain the questions. The research goal of neuroesthetics is to explore the neural basis of creating, understanding and aesthetic appreciation of the arts. The study of neuroesthetics unite art, music and other art forms and means of scientific research to explore the mysteries and will certainly make revolutionary contributions to the development of aesthetics.

KEYWORDS: Neuroesthetics; Neuoscience; Art

1 INTRODUCTION

Why great art works are eternal? What prompted the prehistoric humans to create art? If someone says, "maybe I don't understand art, but I know what I love". In this case, he knows much more than what he realizes. More and more people hold such a view: all human activities – from the art and music to the language, literature and architecture are the products of human brain tissue, and follow the neural rules. It is precisely this point that makes great art works unfading. Mozart can accurately characterize human psychology in his opera performance of their own; Da Vinci reveals the essence of human with a scalpel subtly just like he used the same brush.

Similarly, prehistoric art is not "original", but "represents significant characteristics in the mind of human in the exploration process". This means that, art and science link together at the beginning of human form, at that time, the most effective way to obtain knowledge is vision – which is why about 1/3 part of the brain is related to vision. Research on the visual system of Semir Zeki shows Mondrian, Picasso and other artists also should be viewed as neuroscientists, because they have to study the brain imperceptibly when they create art works. Zeki describes two brain rules for art: one is "eternal law", the roles of the brain and the great art works, in the contact information changing, are to understand the nature of the object of constant exploration. Second is "abstraction law". Zeki thinks, "Art is abstract, and thus make the brain internal working specific". Zeki proposed a new research field – neuroesthetics, is used to study the neural basis of artistic creation and aesthetic. Also, as a common human language – neural mechanisms are also included in the study of music. That is why Beethoven created the Ninth Symphony, causes the people all over the world for having heard it many times. Discordant concert activated brain region for specific unpleasant emotions, only certain areas make a pleasant reaction on harmonious and pleasant music.

2 BACKGROUND

What is art? Why can't it become a common culture that all society can accept? Over the years, the society discussed this question but had not obtained a satisfactory answer. In fact, this is not surprising, because this discussion is usually carried out beyond the brain research domain, however, only in the brain, art can be conceived, created and appreciated. The art activity is an advanced form of conscious activity, involving our perception, emotion, imagination, memory, value judgments, and other advanced cognitive processes. Art is a human activity and includes moral, legal and religious policies, abided by the laws of the brain just like other human activities. Aesthetics research mainly involves metaphysical discussion until twenty-first century, then the aesthetic research began to focus on the research of aesthetic cognitive mechanism. With the vigorous development of neuroscience, especially the great progress of the brain's visual system; we began to study neural aesthetics.

Neuroscientists explore the neural mechanism, and the artist also explore its own visual brain potential and the ability, because all the art works will follow the visual brain rules, so they can tell us the results of their research through the art works. At this point, Paul Klee believes that art is not just copy of the visual world; it also makes some invisible nature being revealed. Only by understanding neural rules of all areas of human activity (aesthetic, moral, legal, religious and political and economic, art), we will hopefully get a right understanding of human nature.

Of course, the study of neuroesthetics has just started, the current study focused on the following aspects: (1) research of aesthetic activity in the visual art works, and the purpose is to perceive and integrate the brain for all factors of visual stimuli in the line, shape, color, position, and movement. (2) Research of aesthetic perception in the auditory art works. The aim is to reveal brain activity mechanism related to music aesthetic perception. (3) Research of the different artistic styles, such as Zeki's study to portrait painting, abstract expressionism, realism painting, impressionism, and fauvism works. (4) Research for other aesthetic activities (such as movies, novels, and so on) to explore the neural mechanisms. (5) Research of the history of art, thus revealing the aesthetic changes in the brain in the long-term evolution. (6) Research of aesthetics.

3 NEUROSCIENCE METHODS FOR NEUROESTHETICS RESEARCH

At present, the research method of neuroscience is becoming an important method to explore the basic theory of aesthetics. Three kinds of brain imaging techniques, including EEG record, PET scanning technique and fMRI, are very popular tools of neuroscience research at present. Each technique has its advantages and disadvantages.

1 EEG (electroencephalography) recording method directly reflects the electrical activity of the nerve, have a high time resolution (in milliseconds), and low cost, no trauma, operation and maintenance are more convenient. However, the spatial resolution is low; the reliability of location algorithms has not yet been confirmed.
2 PET (positron emission tomography technology) of the recorded neurons was sufficient to identify the cortex region; the spatial resolution is higher than EEG. However, it needs longer imaging time, and the scope of application is limited by the radioactive substances; limits of metrology, the same subjects should not be frequently participated in the PET experiment, and the cost is also very high.
3 fMRI (functional magnetic resonance imaging) has very high spatial resolution, convenient for brain localization; but the time resolution (second level) is lower than EEG, the cost of the system is also very high.

4 THE FUSION OF NEUROSCIENCE AND ART

Most of the current vision researchers have given up on Helmholtz's theory, they no longer talk about sense, but believe that the different parts of the brain perform a variety of highly specialized tasks at the same time, respectively deal with the shape, or sports, or color, and, these processes can be combined with each other, to form a unified visual perception. However, to give up the Helmholtz vision theory does not mean to deny this fact by artists to explore on the visual system of creation. The different art styles are corresponding to different parts of visual neural system. For example, the dynamic art is corresponding to visual cortex V5, because the area of the visual cortex has the response to exercise stimulation; Fauvism art corresponds to the V4 zone, because here is sensitive to colors; the creation of Mondrian painting will significantly stimulate the V1 zone, because it is sensitive to the line, and so on. That is to say, different types of art will stimulate different groups of neurons in the brain.

Mild Brain vascular damage (such as stroke) may cause a profound influence in the painter's painting style imperceptibly. After a stroke, the painter can be more bold in the usage of color and composition, and his artistic style of impressionist is suddenly changed into a style of abstract. Researchers at the University of Lausanne in Switzerland think that this may be due to the inhibition of many regions in the human brain, which are related to certain emotions, when these regions were injured, these emotions can be liberated. By scanning the brains of two painters with minor stroke, the injured area of one painter is related to the process of visual imaging, stroke caused mild visual defects to him, and his mood also becomes more difficult to control, but his work with the color became brighter and the theme becomes more abstract. The occurrence of stroke of another painter is in the back of the brain in the left thalamus is related to attention and emotion. Compared to his previous level, the impressionism, smooth style becomes bright, bold, lines and contrasts, composition become simple and abstract, more bold style, in contrast to the more intense, the symbolic weakened, realistic means thick.

5 COMBINATION OF NEUROSCIENCE AND MUSIC

Listen and match performances involve thinking. Areas of the brain for controlling the language are used for understanding rhythm, and the areas controlling the visual part are used for imagining tone. The right side of the brain is a regional, whose task is to help us identify the melody or tone. The brain has an unpleasant emotional area, which is not affected by the harmonious music; on the contrary, there are specific regions for the pleasure of the musical response. According to this universal way of working, we naturally explain the reaction with emotion and down when listening to music. As Wagner said – you don't need to understand an opera libretto, because music has played most incisive. The music has really touched our string of emotion: fear, joy and sadness caused the corresponding psychological change of pulse increase, respiratory increase and heart rate variation.

6 NEUROESTHETICS FOR HISTORY OF ART

What is the origin of art? Neuroesthetics for history of art will make the origin of art as the research content. The works of art appeared in Cave of Chauvet in France before 32,000 shocked us more. In addition to neuroesthetics, no one can explain why the first piece of art naturally recorded physiological and psychological situation of bears and lions.

Neuroesthetics for history of art also explained to us why Florence painters will use more of the lines and the Venice painter who uses more color. The reason for that is the plasticity of neural center , which makes the difference effected by natural or human environment, leading to the differences of the visual preference. Similarly, it also shows that European artists like standing work in vertical front of the canvas, and China artists prefer sitting in the square in front of the paper writing, the reason why there is this difference, is due to the presence of mirror neurons.

7 FUTURE PROSPECTS

Research and achievements of neuroesthetics, for the development of today's aesthetic theory and the practice, have extremely vital significance. Although it not enough to build the aesthetic theory based on it, but it provides a new possibility for the development of aesthetics; mining the neural mechanism of aesthetic theory will make a revolutionary contribution to the development of aesthetics.

ACKNOWLEDGEMENTS

This work was supported by Natural Science Foundation of Jiangxi Province [No. 20142bab207008], project of Science and Technology Department of Jiangxi Province [No. 2013BBE50051] and project of Jiangxi [No. XL1406].

REFERENCES

[1] D Freedberg, V Gallese. 2007. Motion, emotion and empathy in esthetic experience. *Trends in Cognitive Sciences* 11(5):197–203.
[2] M Iwase, Y Ouchi, H Okada, C Yokoyama, S Nobezawa et al. 2002. Neural substrates of human facial expression of pleasant emotion induced by comic films: a PET study. *NeuroImage* 17, 758–768.

Information, Computer and Application Engineering – Liu, Sung & Yao (Eds)
© *2015 Taylor & Francis Group, London, ISBN 978-1-138-02717-6*

Research on control method of retail chain enterprises' logistics cost

Xiao Yong Li

Weifang University of Science and Technology, Shou Guang, China

ABSTRACT: In this paper, the concept, attributes and characteristics of logistics cost are analyzed. We obtain the logistics cost composition and influence factors of logistics cost (including competitive factors and product factors). On the basis of this, we developed a basic method of logistics cost control; the flexible budget method and the zero base budget method are mainly introduced. It is implemented into the logistics cost control of retail chain enterprises, analyze and optimize the control of logistics cost.

KEYWORDS: Retail enterprises; Cost control; Algorithm; Influencing factors

1 INTRODUCTION

With the growth of awareness for logistics management and development of modern logistics industry, there is a concern for the growth of enterprise for logistics costs. Enterprises that are in the practice of logistics management, reducing logistics cost is their top priority. Modern logistics is called after the labor, natural resources, "the third profit source", and realizing the profit source is the key to reduce the logistics cost. How to reduce the enterprise huge logistics costs, eliminate the logistics iceberg, excavate the third profit source, but also to our country enterprise has the vital practical significance, which is peculiar to the Chinese retail market.

2 LOGISTICS COST OVERVIEW

2.1 *The concept of logistics cost and properties*

2.1.1 *The concept of logistics cost*
Logistics cost shows the organizations implementation and management of various costs that are incurred in the logistics activity and material consumption of currency performance, it also include the items in the packaging, loading and unloading, transportation, storage, distribution processing, other entities in the process of flow spending by the summation of manpower, financial and material resources. Logistics costs account for the proportion of the retail price which is 20% to 30%. As a share of GDP, 10% in developed countries; 18% to 20% in China.

2.1.2 *The characteristics of logistics cost*

1 Accounting difficult: "the tip of the iceberg"
   "Iceberg theory" logistics is a theory of Professor at Washed University West Ze Xiu, he found that when the logistics cost advances the financial accounting system and the accounting method cannot grasp the actual situation of the logistics cost, and thus the people understanding of the logistics cost is blank, and mostly false. He compared the situation to "Logistics iceberg", when the most heavy iceberg under the water of the black area can be seen, but what people see is a small part of the logistics cost.

2 Between the costs of project "benefit against" phenomenon
   The theory of benefit refers to the logistics of several functional elements that exist between profit and loss of contradictions, in the optimization of functional factors and interests at the same time, on the other hand, there are one or several different functional elements of the profit loss. The most typical is the logistics of each link "benefit against". It refers to the enterprises that take measures to decrease the cost of a logistics link, because the influence of benefit against another link cost will increase, so that the logistics generally tend to rise.

2.2 *The composition of logistics cost*

Composition of logistics cost includes transportation, warehousing, packing, handling charges, handling fees, circulation processing, Transportation cost and storage cost are in a dominant position in the logistics cost, transportation cost is the goods in the logistics transportation, warehousing costs is to store the goods that need space, manpower, equipment cost, packing, loading and unloading, handling, distribution processing are arising in the course of logistics operation link cost, reducing the cost of those links can greatly improve the logistics costs, thereby creating greater profits for the company.

2.3 The main factors influencing the logistics cost

2.3.1 Competitive factors

1 The order cycle. Enterprise logistics system can efficiently and inevitably shorten the order cycle, reduce the inventory of the customer to reduce the inventory cost of the consumer, improve the level of enterprise customers, and enhance the enterprise' competitive ability.
2 Inventory level. Production enterprises and circulation enterprises, the inventory of strict control, strict grasp of replenishing one's stock, the number of times and varieties, can reduce capital take up, spending of the interest payments, reduce inventory, storage, and maintenance costs.
3 bTransportation. Different means of transport and the cost of high and low, are also sizes.

2.3.2 Product factors

1 Product value. Product value height will directly influence the size of the logistics cost, with the increase of product value, the cost of each logistics activity is increased.
2 Product density. The product density, the greater the transportation units on the same goods more, the lower the shipping costs.
3 Vulnerability. Item's vulnerability to the influence of the logistics cost is the most direct, the vulnerability of the products at each link of logistics is put forward higher request, the cost will be high.

2.4 The basic method of the logistics cost control

Flexible budget law and the zero-base budget law

1 Flexible budget law
   Flexible budget, also called a variable budget or sliding budget. It is part of the relatively fixed budget for budget method. Basic principle is: the costs according to cost behavior can be subdivided into variable cost and fixed cost of two parts. Due to fixed costs in its range, it generally does not change along with the business increase or decrease in total, so to adjust the budget is in accordance with the actual business, only need to adjust the change fee.
2 The zero-base budget law
   Zero-base budget, otherwise known as the "zero based planning and budgeting," in the preparation of indirect costs or a fixed expenses budget, the traditional method is: in the past various fees project based on the number of actual spending, given the budget period business change, past spending for the appropriate adjusted to determine the increase or decrease. Shortcoming of the traditional method in: historical spending will

have unreasonable charges, if only on this basis to increase or decrease in general, is likely to make these unreasonable expenses continue to exist, can't make budget play its proper role.

3 RETAIL CHAIN LOGISTICS COST ANALYSIS

3.1 The logistics cost of retail chain enterprises in China, the status quo

1 Low efficiency of logistics operations, high rate of out of stock
   The outstanding performance of the low efficiency of logistics in the retail enterprise is out of stock rate is extreme. According to Roland burger and China association of chain operation in September 2003 for the domestic retail chain three cities five a dozen hypermarket chain retail enterprises in the investigation, the result shows: China's terminal stockout rate at about 10%, which is far higher than the international level of about 7%. For retail enterprises, goods shortage situation will cause consumer reactions and eventually lead to losing sales of the enterprise.
2 Low degree of information sharing, low level
   Retail enterprises and suppliers and different departments within the retail enterprise information sharing are very important. Many retail businesses as like private property, the messages have not been shared with the suppliers, the suppliers do not have access to retail enterprises inventory and the sales information, the results to the rise in the cost of goods, goods of retail enterprises competitiveness is also decreased. Within the enterprise, single operations and procurement, the necessary information is not entirely shared, communication also led to a serious shortage. In commodity management, purchasing department did not well attend to the needs of different stores, make some poor sales of goods which are stranded on the shelf for a longer time, but also hindered the more popular products for sale. In the data management, because of the lack of communication between the headquarters and stores, the database got affected seriously due to inaccurate date. The store managers for headquarters and distribution system of replenishment have no confidence.

3.2 The necessity of our retail chain enterprises logistics cost control

1 Pull the important channel that domestic consumption
2 The main channels for necessities of emergency supplies

3 Farmers and ", "" of thousands of villages and townships in the main contract
4 Make outstanding contributions to environmental protection and energy saving

4 RETAIL CHAIN LOGISTICS COST CONTROL STRATEGY

4.1 *Optimize the organization*

Establish a clear division of responsibilities for organizations, setting the specialist for supplier quality management. Clear division of responsibility, supplier development engineer is responsible for product conformity and the technical process of the ability to conduct a preliminary assessment of the quality engineer in the process of production using the system tools to the supplier's quality performance for effective management and assessment on a regular basis. Such controls prevent the happening of the procurement quality risk management structure.

4.2 *Establish perfect logistics information system, unified logistics platform*

Informatization is one of the principal reasons for success. In China 80% of large- and medium-sized chains adopted computer management in different degrees, the international retail giant is above 2%. From the level of technology application, the chains of the lower level of informationization in our country such as sales management, finance, customer relationship management and applications of data mining system lags behind that of the foreign enterprises, and enterprises lack the overall market, cash flow, and logistics control ability. This suggests that chain informatization strategy in our country is not perfect; it has yet to be an increasing investment.

4.3 *Joint distribution*

Centralized distribution helps to reduce logistics to strengthen the construction of distribution center. Some large chain enterprises in China have their own logistics distribution center, although in recent years our country's chain enterprise expansion space is very big, but there is a certain gap compared with foreign chains.

4.4 *Perfect supplier*

1 Perform supplier certification program clearly
2 Evaluate supplier performance on a regular basis
3 Set clear quality standards in a timely manner to the supplier
4 Help suppliers improve the quality of the product

5 CONCLUSION

We want to realize the strong competition of retail enterprises, which will be a very good control of the logistics cost. Enterprise can combine their own authentic environment by appropriate and reasonable cost control, so as to realize the maximization of enterprise profit. With the advent of the era of the new economy and information, the competition among chains has become a fast response and low price competition strategy, reducing the retail enterprise logistics cost control the profits of retail enterprises can be greatly improved.

REFERENCES

[1] Maolei Hu. 2008. How to dance with Wolf. Journal of Chinese retail price monthly, (7):38–39.
[2] Zhengfei Lu. 2009. Financial management (3rd edition) [M]. Nanjing University Press, 2.
[3] Yanli Li. 2006. Our country enterprise logistics cost control problems research review. Journal of Financial and Economic Politics and Law, (4).
[4] Daliang Zhang, Xiaoping Fan. 2009. Marketing Management: Theory, Application and Case. Science Press, 8.
[5] Fucheng Liu, Hui Zhao. 2009. Just try to talk about logistics enterprise marketing operation. *Journal of Changchun University*, 2009 (9).

Information, Computer and Application Engineering – Liu, Sung & Yao (Eds)
© 2015 Taylor & Francis Group, London, ISBN 978-1-138-02717-6

Study on measures of risk prevention in engineering projects

Shou Sheng Zhang
Weifang University of Science and Technology, Shou Guang, China

ABSTRACT: The concept of project risk is pointing to: internal and external interference factors in the project that cannot be determined in advance. We mainly study the preventive measures of engineering project risk. Process of risk analysis system involves four steps: engineering project risk analysis is based on the following points: technical and environmental risks. The risk of economic aspects involve the risks of contract signing and performance. Ways of risk prevention includes monitoring the risk and transfer risk. Risk prevention measures include a claim for compensation in accordance with the law; activist of rights based on the contract law; the risk of guard against illegal construction.

KEYWORDS: engineering projects; bidding risk; prevention awareness

1 INTRODUCTION

Project bidding for getting contract is a risky business. The contractor's low price will win at all costs, the situation of the high price and settlement in the bidding work is increasingly impossible. The competition is high because of the increased build value in engineering., the bidding is more and more standard; the winning is no longer the ultimate goal. Risk is objective existence, uncertainty factors, how to effectively prevent the bid of the project risks, guarantee the bid and gain reasonable profits, bidding and construction enterprise need to face the problems in the current project.

2 ENGINEERING PROJECT RISK PREVENTION

2.1 The concept of project risk

There is a danger in any project. Project is a set of economic, technology, management and all aspects of the comprehensive social activities; it exists in every aspect of uncertainty. These cannot be determined in advance of the internal and external interference factors. People call it a risk. Risk is an unreliable factor in project system, it will cause the engineering project implementation out of control, such as time limit for a project delay, cost increase, modify, eventually led to the economic benefit is reduced, clung even project failure. The characteristics of modern engineering project are a large scale, new technology, long duration, to participate in the unit, the interface with the environment become more complex, in the

process of project implementation in crisis, so to speak. Many fields, because of its high project risk, the harm is large. Many of the projects in China, the losses caused by risk are shocking. So, in the modern project management, risk control has become one of the research hot spot.

2.2 Engineering project risk analysis theory

What is the risk analysis

In a broad sense, the risk analysis is a way of qualitative and quantitative estimation of risk for the impact of decisions. Use of technology is a mixture of two kinds of qualitative and quantitative technology. Given the likely outcome is a better understanding, is the purpose of these methods in any kind to help decision makers to choose an action. Risk analysis system of risk analysis is a method of enumerative, this kind of method to explore the decision-making results identified as a probability distribution. In general, the process of risk analysis system includes the following four steps:

Development model, defined in the format of the project encountered problems and situations.

Identify uncertainty: number and probability distribution in the project to specify their possible values, to analyze the uncertainty of project results.

Using the Monte Cargo simulation model: the scope of your project all the possible results and probability.

Decision making: based on the results of the offer.

Risk analysis system has first three steps, which are provided with the project work as a powerful and flexible tool. They make it easy to model the risk analysis. Decision makers can then use the result of a risk analysis system to help choose an action plan. Fortunately, adopted technology of the risk analysis system is very intuitive. Therefore, need not on faith accept the role of risk analysis.

3 ENGINEERING PROJECT RISK ANALYSIS

Construction project management factors to the risks of construction enterprises

3.1 Technical risks to the environment

1 Geological foundation conditions. The developer shall generally be able to provide the corresponding engineering geological data and foundation technique requirements, but these data and the actual discrepancy are large, sometimes we need to handle exceptions like geology or encounter other obstacles that will work more and extend the time limit for a project.
2 Hydrology meteorological conditions. Mainly manifested in the emergence of abnormal weather, such as typhoon, rainstorm, snow, flood, debris flow, soil slip of force major such as the construction of natural phenomena and other natural conditions, can cause the construction period delay and the loss of the property.
3 Construction preparation. Because the landlord will provide the construction site surrounding environment and other aspects of natural and man-made obstacles or "tee – equal" preparation is insufficient, lead to the early stage of the construction of construction enterprise can't be ready for work, brings to the normal operation of engineering construction difficulties.
4 Design changes or drawing supply not in time. Design changes will affect the construction schedule, resulting in a series of problems; Design drawing supplied not in time, lead to the construction schedule delay, due to the contractor construction period delay and economic losses.
5 Technical specification. Especially the technical specification outside of the special process, because the developer is not well adopted the standards, specifications, in the process of working procedure and not well coordinated and unified, project acceptance and settlement after impact.
6 Construction technology of coordination. Engineering construction process does not match with its technical expertise of the engineering technical problems, among professional and not timely coordination difficulties, etc; Because of the developer's management engineering technology level being poor, for the contractor to the employer to solve technical problems, and not made a reply in a timely manner.

3.2 The risk of economic aspects

1 The bidding documents. This is the main basis of the tender, tenderer guidelines, in particular, design drawings, engineering quality requirements, the terms of the contract and the bill of quantities, etc. There is potential economic risk, and must be careful in analysis research.
2 The market price. Factor market such as the Labour market, materials, equipment market, the market price changes, especially the prices, directly affects the project contract price.
3 Money, materials, equipment supply. Main show is the money supply by contractor, materials or equipment give rise to poor quality or untimely supply.
4 National policy adjustment. State of macroeconomic regulation and control, such as salary, tax and tax rates will bring to the construction enterprises must be risked.

3.3 Risk and performance of the contract signing

1 Defective and unfair contract. The terms of the contract are not comprehensive, not perfect, the text is not careful, not tight, loopholes in the contract. As on the terms of the contract, it is not perfect or not transfer risk guarantee, claims, insurance and other relevant provisions, the lack of economic losses caused by third party influence the construction period delay or the terms of the existence of unilateral constraints, such as too harsh right imbalance terms.
2 The developer credit factors. The developer's economic deterioration, lead to poor performance ability, inability to pay the payment; client credit is poor, not good faith, not by the contract for project settlement, and is related to all overdue payments.
3 The subcontract. Selection of subcontractor is undeserved, meet the subcontractor default, not quantity and timely completion of the subcontract works, which affect the entire project schedule or economic losses.
4 The performance aspect. Process of the contract, due to the developer in site representative or supervisory engineer's work efficiency is low, can't timely solve the problems, and even send the wrong instruction, etc.

The above according to the main aspects of risk factors, the risk can be divided into technical, environmental risks and economic risks as well as the contract signing and performance of three kinds of risks, etc. All kinds of risks are the threat of building enterprise survival and development. Therefore, must carry out effective risk prevention.

4 ENGINEERING PROJECT RISK PREVENTION MEASURES

4.1 Construction enterprise risk prevention way

1 Control risk
Familiar with and master the relevant laws and regulations of engineering construction stage. Involvement in the construction phase of the laws and regulations is legal according to protecting the interests of the project by contracting parties, construction enterprises are only familiar with and master the laws and regulations, according to the laws and regulations, to enhance the consciousness of using laws to protect their interests, effectively and to control the project risk in accordance with the law.

2 Transfer of risk
By adopting the system of claim for compensation. Because it is unpredictable, there is always some risk, the occurrence of risk events is the root cause of economic loss or time loss, contract between both sides is expected to transfer risk. So in the contract, claim for compensation system is tantamount to transfer risk. Nonetheless, engineering claim system in our country has not yet become universal, construction companies are not very understanding of the claim, the claim of the specific practice is still very unfamiliar, therefore, we must understand the significance of the claim system shifting risk, learn to claim for compensation method, make the transfer project risk legal, reasonable claim system like health, gradually to the international engineering practices.

4.2 Construction enterprise risk prevention measures

1 Claim for compensation in accordance with the law
Claim is in the implementation of the contract, the parties pursuant to the provisions of the law, and management is not due to its fault, but should be caused by the contract of each other's responsibility, and the actual loss, to the other party to ask for compensation, it is the main way to transfer risks. Construction claims are the loss of rights and law, is a kind of protection for construction enterprises,

protect their legitimate rights and interests, avoid a loss, and increase profit. Although in cultivating and developing the construction market in China still exists in the relevant laws and regulations is not perfect, not enough strict problem, but to gradually implement construction claim work, objectively has certain conditions. The general principles of the civil law", "contract law" and "construction and installation enterprises owned by the whole people convert the measures for the implementation of operational mechanism has the engineering claim terms, such as can be used as the legal basis of implementation of engineering claim.

2 To the "contract law" article 286 and its judicial interpretation based on human rights
"Contract law" the 286th regulation: "The developer failed to pay the price in accordance with the contract, the contractor may demand that the payment from the developer within a reasonable period. The employer fails to pay, except in accordance with the nature of the construction project should not be auctioned, outside, the contractor can deal with the developer of the project at a discount, can also apply to the people's court will auction the project in accordance with the law. The construction project price shall be paid in priority out of proceeds from the liquidation or auction of the project." The supreme people's court and in June 2002 to the problem of "the priority of compensation with respect to price," explains. Rules "construction engineering contractor of the priority of compensation is better than that of mortgages and other debts". "Contract law" Article 286 and its judicial interpretation for construction enterprises to maintain their legitimate rights and interests, is "a shot."

3 Guard against the risk of illegal construction
China's "building law" regulation, allows the construction of the project should meet the following conditions: building land approval formalities; within the urban planning of construction projects, has made planning permit; Need for demolition, the demolition schedule meet the construction requirements, etc, otherwise, the law to the project.

5 CONCLUSION

As long as we study the risk of timely countermeasures, reference and summarizing various experiences, and put forward measures and closely linked with the national policy, at the same time be lawful, reasonable, reasonable, no matter what kind of risks are encountered in the construction corporation, can dissolve or reduce the risk to a lower level.

REFERENCES

[1] Wei Liu, Jianbo Yuan. Construction project risk coping strategies apply. *The Chinese and Foreign Road*, 01, 2006.

[2] Hao Wu. Engineering project risk factors and counter-measures. *Shanxi Building*, 26, 2007.

[3] Yuanfang Xie. Construction project cost risk analysis and prevention. *Shanty Institute of Economic and Trade Journal*, 2000, (6).

[4] Dianfen Han g. International engineering contracting project risk analysis. *Western China Science and Technology*, 2009, (13).

[5] Qingjun Meng. Construction projects risk analysis and prevention [J] installation, 2005, (4).

Information, Computer and Application Engineering – Liu, Sung & Yao (Eds)
© 2015 Taylor & Francis Group, London, ISBN 978-1-138-02717-6

A study on tourism motivation of college students—the case of college students in the city of Wuhan, China

Shu Zhang
Wuhan Buiness University, Wuhan, China

ABSTRACT: Why the college students go for traveling has become a problem that needs to be solved urgently. This paper focuses on a research of college students' tourist motivation in the city of Wuhan, and then got the following results: the traveling motivation of college students consists of impulsive travel motivation, knowledge and experience travel motivation, dissimilation and self-improvement travel motivation, pleasure-seeking travel motivation, and other travel motivations. The strategies of stimulating college students' motivation to travel are based on these factors.

KEYWORDS: College students; Tourism motivation

1 THE PROPOSED PROBLEM

In ancient China, the emperor made tour activities, Confucius traveled far and wide, and Li Bai, the great poet of China, would like to put their creative inspirations on the basis of mountains and rivers. As an old saying, "learn knowledge from thousands of books and accumulate experience by traveling thousands of miles," which has encouraged the young people to explore and research the great rivers and the mountains of their motherland. When the wheel of history goes on, and the people gradually improved to become a leisure class, travel changes from a minority activity into a mass-based one. At present, why do so many young people and college students like to travel? This is an urgent issue which the author aims to solve.

Despite its decisive role in decision-making process about the travel behavior, studies and researches about travel motivation remain a relatively new field.

Motivation can have inner and direct impact on the individual behavior (Iso-Ahola, 1982), and throughout the whole decision-making process. Scholars also tried to give travel motivation a scientific definition, John L. Crompton & Stacey "cKay (1997) said that"travel motivation can be defined as a dynamic process that makes the individual create internal psychological factors like tension and imbalance". Juergen Gnoth (1997) proposed that travel motivation encourages tourists to seek symbols existing in things, scenes and events, and these symbols imply commitment to reduce the current tension. In order to better illustrate travel motivations, scholars have constructed different theoretical models such as the theoretical model of Maslow's hierarchy of needs,

Iso-Ahola Model, and the travel motivation construction proposed by Cromptom that broke the routine (Hyounggon Kim, 2006), which has boosted the research of this field markedly. The follow-up that the scholars have adopted in various empirical study methods and they attempted to examine and correct the theoretical models. The early scholar Nesbit (1973) divided the tourism market into four categories: personal business travel, government or corporate business travel, visiting friends and relatives, and pleasure vacation travel. In 2012 Tze-Jen Pan explored the motives of the volunteered student tourists from Taiwan who participated in overseas travels, whereas Songee Kim and other researchers (2013) selected 161 South Korean families for investigation and analysis. After carding the research achievements of foreign scholars, they recognized the existence of multidimensional travel motivation and used the empirical research methods to study the different regions and different groups of tourists. But what is interesting is that the number of researches on college students is relatively less.

2 RESEARCH METHODS AND OBJECTS

2.1 *Research methods*

This research collected data by means of questionnaire survey; it dealt with and analyzed data mainly through statistical software of SPSS18.0 by adopting the study method of factor analysis.

The basic idea of factor analysis is to find out several random variables which can control all variables to describe the co-relativity among variables by studying the interior structure of variable correlation coefficient matrix and to classify them according to how they are correlated with each other so that variables with high co-relativity can fall into the same group and variables of different groups have low co-relativity. Each group represents a basic structure which can be called public factors. Factor analysis can compress information, lower dimensional index and simplify the structure of index to make it easier, more straight and effective for us to analyze problems. Undergraduates have diverse motivations for traveling around, which can be clearly reflected through factor analysis.

2.2 Questionnaire design

This research is done mainly to study the traveling motivation of college students, based on reference of researching processes conducted by both domestic and foreign experts and scholars. The first section is the main body of the sheet which is designed by adopting Likert-type scale, involves 27 subjects on traveling motivation of college students; the second section is more specifically involved with traveling time, traveling approaches, preferred traveling types and many other items.

2.3 The basic situation of the sample

The Questionnaire adopts to take sample by random sampling method to ensure that each sample unit is pumped to the equal opportunity; at the same time, on research methods, street intercept interview-based survey is chosen in order to obtain higher questionnaire recovery. The author carried out the questionnaire survey in Wuhan University, Huazhong Normal University, Hubei University, Jianghan University, Wuhan Business University from March 1 to 15 in 2014. Of 550 questionnaires, 493 are valid questionnaires, and 57 are invalid questionnaires. The recovery rate of the questionnaire was 89.64%.

There are 231 males and 262 females which respectively accounted for 46.86% and 89.64%; Looking from the perspective of grade distribution, freshmen were 126, sophomores were 131, junior students were 123, senior students were 98, other people were 15, accounting for 25.56%, 26.57%, 24.95%, 19.88%, 3.04%. From the perspective of monthly living expenses, students who have less than $500 are 18 people, the number of students in $500–$700 range is 140, 700$–1000$ is 233, more than 1000 yuan is 1000, this shows that the current college students have certain consumption ability and travel possibility. It can be seen from the statistic–results that the questionnaire has the content validity and reliability from the surface, the selection of sample also has certain rationality.

3 ANALYSIS OF COLLEGE STUDENTS' TRAVEL MOTIVATION

3.1 Adaptability test of factor analysis

Adaptability test of factor analysis is the premise condition of factor analysis; KMO (Kaiser – Meyer – Olkin) test and Bartlett's spherical test are appropriate test tools for adaptability test. In Kaiser's view, it is very suitable to do factor analysis when the KMO is > 0.9; it is quite suitable to do factor analysis under the condition of 0.8 < KMO < 0.8, it is very suitable for factor analysis under the condition of 0.7 < KMO < 0.7, it is less suitable for factor analysis under the condition of 0.6 < KMO < 0.6, it is not suitable for factor analysis when under the condition of KMO < 0.6. By KMO and Bartlett's test we found that when the KMO value is 0.913 > 0.9, it is very suitable for factor analysis; at the same time, significant probability of the Bartlett's sphere test is 0.000, it shows that the correlation coefficient is not unit matrix which has good construct validity for factor analysis.

3.2 Factor analysis

After the variable through appropriate test, we use principal component analysis and Varimax orthogonal process to extract the factor. In extracting factors, determining the number of factors on the principle of eigenvalues greater than 1, and therefore, should extract six corresponding common factors. As you can see in the cumulated variance contribution ratio, the first six factors have explained variance of 63.491% of the variation, which contains most of the information. Therefore, 27 variables can be divided into six groups by conducting further analysis of sample variables. In order to determine which index all the six common factors have, we need to rotate maximum variance of factor loading and name the factors by the meaning of high load that public factor include in order to make clear information was contained by the public factors.

Factor 1 incorporated six items of motivation, identified aesthetic pleasure travel motivation, this factor accounted for approximately 15.522% of the variance in the data. Factor 2 incorporated five items of motivation, identified impulsive travel motivation, this factor accounted for approximately 12.956% of the variance in the data. Factor 3 incorporated six items of motivation, identified knowledge and experience travel motivation travel motivation, this factor accounted for approximately 12.105% of the variance in the data. Factor 4 incorporated four items of motivation, identified dissimilation and self-improvement travel motivation, this factor accounted for approximately 9.559% of the variance in the data.

Table 1. Factor analysis of motivation of college students.

Variables Item	Factor					
	Factor 1	Factor 2	Factor 3	Factor 4	Factor 5	Factor 6
1. Opportunities for rest and relaxation	0.839					
2. Beautiful natural scenery	0.864					
3. Relax away from the ordinary	0.780					
4. See nature all around the country and human geography	0.694					
5. Together with friends and family members, fell happy	0.621					
6. Cultivate one's taste	0.555					
7. During the holidays, friends are going to travel, so I will go		0.706				
8. To pass the time		0.743				
9. Special ticket		0.859				
10. The worship of idols		0.781				
11. Religious		0.736				
12. Escape from learning stress			0.607			
13. New experience			0.559			
14. Travel to historical heritage sites			0.613			
15. Rafting, hiking and so on, accept the challenge,			0.689			
16. Scientific tourism			0.577			
17. Make friends			0.542			
18. To understand the customs of minority areas				0.565		
19. A force or person that inspires to write, paint, etc.				0.814		
20. Practicing sports or exhibitions				0.714		
21. Simple folkway				0.661		
22. Attracted by the movie, TV, and magazines describes					0.651	
23. Enjoy life					0.671	
24. Get fun form adventure sports, such as bungee jumping					0.578	
25. Theme park					0.538	
26. In the case of depressed, can adjust the mood						0.770
27. Visiting friends						0.754

Table 2. The results of exploratory factor analysis.

Factor	Factor 1	Factor 2	Factor 3	Factor 4	Factor 5	Factor 6
Factor name	Aesthetic pleasure travel motivation	Impulsive travel motivation	Knowledge and experience travel motivation	Dissimilation and self-improvement travel motivation	Pleasure-seeking travel motivation	Other travel motivations
Eigenvalues variance	4.346	3.628	3.389	2.676	2.026	1.712
contribution rate	15.522	12.956	12.105	9.559	7.234	6.116

Factor 5 incorporated four items of motivation, identified pleasure-seeking travel motivation, this factor accounted for approximately 7.234% of the variance in the data. Factor 6 incorporated two items of motivation, identified other travel motivation, this factor accounted for approximately 6.116% of the variance in the data.

From the perspective of variance contribution rate, the factor 1 has the maximum influence. So to say

the college students' aesthetic pleasure travel motivation has occupied a major portion in the traveling process; followings are impulsive travel motivation, knowledge and experience travel motivation, dissimilation and self-improvement travel motivation, pleasure-seeking travel motivation, other travel motivations. It also provides the development of tourism products with a certain solution as there are differences among these factors.

4 THE DEVELOPMENT SUGGESTIONS FOR THE UNIVERSITY STUDENTS' TOURISM PRODUCTS

First, tourism products with much aesthetic pleasure are useful for increasing tourism development. From the results of exploratory factor analysis (EFA), aesthetic pleasure factor accounts for a large proportion of university students' motivation for traveling. The experience of esthetic pleasure factor refers to sight-seeing and leisure activities, making tourists relaxed and delighted, so the development of tourism products should focus on the ones involving aesthetic pleasure. Therefore, more efforts should be put into the tourism products that are dominated by scenery with mountains and rivers with much aesthetic pleasure in the development. Another issue of concern is that it does not equal the development of other tourism types, since the scenery type of tourism is mainly a low-level tourism product, whose tourist route is simple and the agency quoted price is low, which is suitable only for university students' consumption level but it is not good for travel agencies profit. To benefit the agency more and trigger the agency to concentrate on the university students' tourism market, experiencing factors should be permeated into aesthetic pleasure during the development. For example, developing a tourist route, which integrates sight-seeing and experience, such as drift and mountain-climbing, helps digging into the connotation of tourism resources and thereby advancing the economic value of tourist routes.

Second, expanding tourist routes' promotional channels is a good choice. From the investigation, there are several ways for the respondents to collect the tourism. Nevertheless, a majority of respondents stem from friends' recommendation, Internet and TV programs, and only 145 respondents get information directly from the travel agency. In this situation, the travel agency should do the propagandas according to the special audience group of university students: first, conducting agent services on campus helps to introduce products face to face, so the direct propagandas through campus agency can be more broad and effective; second, supporting the campus activities properly gains reputation to the travel agency and influences the customers' choices; third, it is a good idea to advertise online through campus forum, Baidu Post Bar, BBS and so on, using promotion signs, pictures and travel guides to catch students' eyes and to

stimulate their motivation for traveling, since students are easily attracted by the amazing scenery on magazines, TV programs and movies according to the survey; henceforth, the travel agency can set up specialized tourism-consulting websites for university students, providing any information demands and various services.

The third step is to strengthen the promotion of travel products. As shown in the investigation, university students sometimes travel based on impulse, because of the discount airfare, idolatry, religious cult, etc. These are irrational travel behaviors, so the travel agency can make promotion accordingly by adopting a price strategy, in other words, providing a reasonable low price, will attract the attention of the students and stimulate them to travel. The summer vacation is a worth effort, because the respondents' spare time is centralized during this period and longer trips can be planned. Third, through digging and utilization of idolatry and religious cult, students' motivation could be stimulated while their concerns are understood completely.

5 CONCLUSION

To understand university students' motivation for travelling, I designed the research method and the research process, including collecting data from questionnaire survey and obtaining the results from descriptive statistics method and exploratory factor analysis. However, some deficiencies may occur in the research process, such as the limited quantity of questionnaires and deficient places to hand out the questionnaires. Increasing the quantity of questionnaires may improve the accuracy of the results. Unfortunately, these deficiencies may leave some regret. But surely, with the help of this research, further discovery of university students' travel behavior will becomes my follow-up research.

REFERENCES

[1] John L. Crompton, Stacey L. McKay. 1997. Motives of visitors attending festival events. *Annals of Tourism Research*, 24(2):425–439.
[2] Tze-Jen Pan. 2012. Motivations of volunteer overseas and what have we learned – The experience of Taiwanese students. *Toursim Management*, 33:1493–1501.

Information, Computer and Application Engineering – Liu, Sung & Yao (Eds)
© 2015 Taylor & Francis Group, London, ISBN 978-1-138-02717-6

Construction and research on application of electronic technology in higher vocational linking "3 + 2" staging training courses system

Min Li
Huanggang Polytechnic College, China

ABSTRACT: In our educational system, the vocational convergence "3 + 2" segment on vocational and vocational education health, coordinating role in promoting the development of a comprehensive training and economic and social development of skilled personnel. In high school, secondary vocational education is the key component for skilled personnel to focus on training, and give full play to the basic role. Electronic technology professionals in vocational convergence "3 + 2" segmented training, focus on training highly skilled personnels to create and build a modern vocational education in the curriculum, according to "unified school system, school management unified, unified personnel training programs and training objectives unity "principle analysis and Discussion on Curriculum for construction staging.

KEYWORDS: "3 + 2" segment, in vocational convergence, electronic technical expertise, personnel training

For the development of education for the National Party to be the full implementation of the spirit of eighteen, systemic reform of vocational examination and enrollment system, focusing on training highly skilled personnel to create and build a modern vocational education in the curriculum, comprehensive training highly skilled personnel, to the past, "3 + 2" Careers mode comprehensive reform, according to the "training objectives unified unified school system, school management unified, unified personnel training programs and personnel," principles of analysis and Discussion Curriculum for construction staging. This paper focuses on the application of electronic technology professionals in vocational convergence "3 + 2" system construction staging training courses for analysis and research.

1 THE PROBLEMS OF ELECTRONIC TECHNOLOGY PROFESSIONALS IN VOCATIONAL CONVERGENCE "3 + 2" STAGING TRAINING

1.1 Target cultivation problems

"Education Dictionary" vocational training objectives of education is to make the following explanations: the culture at all levels of management, technical staff, and rural laborers and skilled workers. Vocational education has a consistent educational purpose, but there are some differences between levels. Vocational education for the third-level technical education, and its main goal is to cultivate applied talents for SMEs;

and vocational education training for the community mainly intermediate or junior technical workers. At this stage, many vocational colleges specializing in electronic technology has no clear training objectives positioning. Tandaqiuquans' vocational training model to imitate undergraduate goal, the goal of vocational training model to imitate vocational, vocational classroom curriculum to follow vocational continuous compression, inaccurate positioning of vocational training objectives, resulting in personnel training and scientific convergence is difficult, the quality of teaching has a direct impact.

1.2 Teaching problems plan

Fundamentally speaking, the vocational convergence "3 + 2" is more than vocational colleges docking corresponding professional vocational colleges. Vocational colleges differ due to differences in talent needs and educational resources, making their teaching plans different. The difference for the vocational convergence "3 + 2" segment of assessment and training has caused great difficulties, there are many in vocational colleges have duplicate bridging courses, for example, electronic technical expertise in basic courses in high selected class vocational schools, the use of teaching materials and teaching plans are the same, the students have learned in these two stages of the course, too, a serious waste of teaching resources for students interested in a serious impact. Their cohesion curriculum is extremely lacking coherence, with great differences in course content, not effective convergence of vocational education programs have an impact on the quality of teaching.

c.Science marital problems

2 MEASURES AND RECOMMENDATIONS ABOUT STRENGTHENING TECHNICAL EXPERTISE IN ELECTRONICS VOCATIONAL CONVERGENCE "3 + 2" SYSTEM CONSTRUCTION STAGING CLASSROOM TRAINING

2.1 Draw up integrated "3 + 2" segmented training objectives

In high school, secondary vocational education is the key component for skilled personnels to focus on training, and give full play to the basic role. Higher education, vocational education is the key component of the training to be focused primarily on high-skilled personnels, their education will give full play to the leading role. In vocational convergence "3 + 2" system of staging classroom training, vocational colleges should lead created by vocational colleges teaching with electronic technology management team. Fundamentally speaking, vocational education should serve as a regional industry, which requires teaching management team of local talent demand and economic development of in-depth investigations, and electronic technology industry talent analysis and discussion of electronic technology professional positions together, and in accordance with the "Electronic Technology professional positions capacity analysis "to develop vocational school for two years with the segment having accurate positioning and learn three years of vocational integration of modern electronic training plan.

Professional integration of electronic technology training programs, vocational school students in the main section of the basics of electronic technology, electronic applications and other skills, skills-based and can be engaged in moderate activities related to the application and the electronic payment and settlement, online marketing and advertising design publicity talents. On the basis of the vocational school segment, and learn not only with the electronic technology vocational skills, but also have the electronic equipment maintenance, electronic building, site system management skills, after graduation generally engaged in the process technician, product maintenance staff, hardware development staff and service engineers and other skilled qualified professionals.

2.2 Development of "dual-card fusion" of comprehensive teaching plan

The so-called "double certificate fusion" refers to the process of vocational education in academic education, effective docking courses or types of professional and vocational qualifications, the training content, the realization of vocational qualification standards and vocational curriculum standards effective fusion fundamentally speaking, the vocational qualification certificate is not only a direct proof of professional knowledge workers, but also the level of professional skills of workers facade, is the employment of workers "stepping stone."

Professional training programs is in accordance with the integration of electronic technology, the convergence of "3 + 2" staging training curriculum system based on the "double certificates fusion" requirement of unity, curriculum standards converge with each other to be developed to ensure that vocational and vocational school section and learn teaching plans to cover different levels of vocational qualification certificate to the specific requirements. In addition, the vocational convergence "3 + 2" system should be in accordance with subparagraph training courses and learn the difference, you need to learn section vocational curriculum content of electronic technology to be deepened and broadened, and vocational school opened this section between teaching level, so that the course content convergence and continuity can be truly integrated. Also avoid duplication fees bridging courses that appear to effectively link different levels of vocational education teaching programs. For example, vocational school integration is the main section of skills and knowledge required by electronic and technical workers, grassroots electronic technology personnel for training to learn how to do electronic work and obtain a vocational diploma and e-commerce operator's certificate, and to the vocational stage, we need to integrate the skills and knowledge required electricity consultant, for middle-level electronic personnel for training to understand what electronic talent can do, and to qualify the electronics industry qualifications, national vocational qualifications and professional skills certificate. In vocational convergence "3 + 2" system implemented staging training courses and diploma qualifications blending pick, is to make the students' employment competitiveness and professional quality which can be improved to achieve employment and vocational education effective docking of key measures.

3 CONCLUSION

All in all, a professional electronic technology in vocational convergence "3 + 2" system of staging training courses and just a simple docking time, you also need effective convergence of educational content. In the "3 + 2" convergence in vocational training, and the key is in accordance with the level of talent demand and economic development of local enterprises, the

Chinese and vocational colleges and technical personnel to the electronic specific training objectives were clear positioning, jointly develop a comprehensive, integrated electronic training objectives in order to achieve effective docking of Higher Education and two school vocational education segment.

REFERENCES

[1] Xia Tang, Xinqiao Wang. 2012. Logistics Management in Higher Vocational interface issues discussed three two segments. *Communication of Vocational Education*, (23): 120–121.

[2] Huilan He,Haofeng Wu. 2012. Reflections on Vocational convergence model in Guangdong Province three two segments. *Modern Enterprise Education*, 2012 (14): 147–148.

[3] Zihua Liu, Guanghua Tang. 2010. Discussion on the application of electronic expertise innovative talents. *China Electric Power Education*, (01): 122–123.

Information, Computer and Application Engineering – Liu, Sung & Yao (Eds)
© 2015 Taylor & Francis Group, London, ISBN 978-1-138-02717-6

Elementary analysis of management innovation of higher education teaching

Ming Hui Zhang
Beihua University, China

ABSTRACT: With the continuous development of China's economy and society, education reform to the deepening of the rapid development, China's higher education environment changes, and the teaching management becomes innovative. As a key content in colleges and universities, teaching management level directly affects the quality of university teaching and quality of personnel training. However, the current university teaching management is still having some problems and it needs to actively innovate, to break the inherent ideas, and to improve the management level. This paper focuses on the current status of the teaching management and to analyze and proposeinnovative measures in teaching management.

KEYWORDS: Colleges and universities; Teaching management; Issues; Innovation

The main task in college is teaching. The main content management system of colleges and universities are teaching management. In recent years, the reform of higher education in China has increased. Under the background of deepening reform, teaching management is faced with new content and tasks, the traditional teaching management system has been unable to fully meet the teaching needs, and the management level needs to be improved. Therefore, universities should use modern management tools, change the traditional teaching of management thinking, improve teaching management mechanism, innovation management, and constantly improve the quality of university teaching and training of qualified personnels.

1 AT PRESENT, CHINA'S HIGHER EDUCATION TEACHING MANAGEMENT PROBLEMS

1.1 *Teaching management concept lag behind, the way of management is single*

Teaching management philosophy is behind restricting the level of teaching management and hinders management level. At present, in the management of higher education, the managers lag behind in the following areas: understanding of management concepts, not comprehensive, rational understanding of the importance of teaching management, meaning, and failed to establish a sense for teachers, students and services [1]. In particular in teaching management, the managers failed to understand the students, teaching, teacher's problems, lack of reality based, multi issued a notice based on subjective judgment,

experience, conveying the document, the lack of "people-oriented" management philosophy, thereby reducing student teachers' enthusiasm for teaching activities, which restricts the development of universities. In terms of teaching management, teaching management focus on teaching activity itself, limit the scope of teaching affairs management, ignoring the student management, personnel training and management. At the same time, management is simple, single means, the lack of specific management, ignoring the students own concerns. Under this management, teaching management is unfocused, the quality of teaching is not improved, and teaching management has not played its role.

1.2 *Teaching scientific management system is unscientific, weak enforcement of the management system*

Object of teaching is people; teaching should be people-centered, attention, understanding. Then, in the current teaching management in universities, the presence of a strong management system "official standard" colors, heavy "system" light "body" does not listen to goodwill students, the teacher's voice [2]. In the specific management, first, training objectives and not out of ideological elite education, teaching content, curriculum design does not need a good combination of students and society. Second, the management system should be rigid, lack of humanity, for example, too much emphasis on the management of teachers normative, syllabus, planning, assessment standards too much emphasis on unity, lack of flexibility. Finally, the relationship between managers and teachers, students discord, management existence of errors, managers in an absolutely dominant position, autocratic right against the student, the teacher's participation. In addition, the implementation of the management system is not optimistic, some universities did not establish a sound system of accountability, and did not go to perform in accordance with the relevant system; management system and the others cannot play an ideal role in managing the effect.

1.3 Lower overall quality of management staff

Colleges teaching managers' quality level directly affects the level of teaching management. Currently, many managers do not realize the quality of colleges and universities and the importance of the ability to work, the lower the overall quality of the management team, management needs to be improved. In the management team, some managers have not received systematic training in management, teaching management of unskilled work requirements, methods, management methods are simple, low quality of management. Meanwhile, in recent years, with the enrollment of major colleges and universities, more and more college students, university professionals increases, teaching management faced with more content, task management becomes more complex. In this context, work stress management personnel increases, job slack, lower management levels.

2 COLLEGE TEACHING MANAGEMENT INNOVATION STRATEGY

2.1 Improve teaching management concepts, implementation of modern management tools

Here the managed objects are human, and therefore, in teaching management, we should establish a people-oriented management philosophy, focussing heavily on students, teachers, who take reasonable management measures according to their specific circumstances. First, attention, respect for the students to understand and respect the students' ideas, lifestyle, personality traits, etc., to carry out targeted teaching, creating a vibrant, positive campus environment. And people-oriented determine the syllabus, teaching content, flexible use of teaching methods, curriculum. Second, strengthen ideological connotations schools will be teaching management to academic heights, constantly sum up innovation management, constantly updated management concepts, so that the management moves toward a healthy direction [3]. Meanwhile, colleges and universities teaching management innovation is based on the implementation of modern management tools, the use of computer information technology to build information management platform for teaching, teaching management

of information technology. For example, in favor of computer and networking technology, the establishment of teaching information management system to collect, record teaching data, information, comprehensive, unified processing, management information data to improve management efficiency, and for storing information data, query, to facilitate the sharing of resources.

2.2 Improve teaching management system, increase enforcement

To improve the teaching management system for personnel training provides an effective incentive and restraint mechanisms, management, is a standard and effective educational resources [4]. Therefore, in teaching management in universities, managers should establish a flexible, strict management system, fully mobilizing the students, teachers involved in teaching are enthusiastic, take initiative and give students ample space for development, teachers play a leading role. At the same time, managers should clean up; improve the existing teaching management rules and regulations; highlighting the diversity of management; and the implementation of flexible management. At the same time, correct positioning of outstanding teaching characteristics. In each professional colleges are different and have different strengths, so in teaching management, managers should reflect the school's training model, combined with social needs, from the school's actual teaching situation, their own advantages and formulate reasonable management system. In addition, to strengthen the supervision system, reward and punishment, the establishment of an accountability system, a clear division of labor jobs, increase efforts to develop the system.

2.3 Improving quality of management

3 CONCLUSION

As the core of the work of university management, teaching management has a positive role in improving the quality of teaching and training of qualified personnels and other aspects. Therefore, in the current teaching administration in universities, schools should actively innovative teaching management, depending on the existence of problem management, change management philosophy of teaching, improve teaching management system to improve the quality of the management team to continuously improve the management level, to carry out teaching provide protection and improve the quality of university teaching, the output of high-quality personnel for the society.

REFERENCES

[1] Yimin Hong. 2010. US undergraduate teaching and management experience and enlightenment. *Chinese University Teaching* 4:93–96.

[2] Liping Shi. 2010. Teaching management system innovation particularity, the body and the path. *Modern Education Management* 4:71–73.

[3] Shimeng Guo. 2013. Elementary analysis of college teaching management innovation. *Modern Communication* 9:190–191.

[4] Shize Li, Huanmin Yang, Chunren Wang. 2012. Management innovation and performance practice teaching colleges – analysis of veterinary medicine practice teaching management construction Heilongjiang Bayi Agricultural University as an example. *Era of Education (Education Edition)* 1 10–11.

[5] Qiuying Liu. 2009. Innovative teaching modern management model. *Changchun University of Traditional Chinese Medicine* 25(6):981.

Information, Computer and Application Engineering – Liu, Sung & Yao (Eds)
© 2015 Taylor & Francis Group, London, ISBN 978-1-138-02717-6

Kinematic analysis on Tkatchev stretched action of elite Chinese male gymnast Chenglong Zhang

Hui Liu
Institute of Physical Education, Guiyang University, Guizhou, China

ABSTRACT: Aim: the horizontal bar is one of the China men's gymnastics team advantage project, Tkatchev stretched belongs to "flight action" of the horizontal bar. We have tested on the Tkatchev stretched action of world elite gymnast Chenglong Zhang at the game site, and have kinematics analysis on the complete movement, establish kinematic model of the technology, thus provide theoretical basis and technical reference to the development of the action. Method: By using the 3D video analysis method. Result: Zhang has successfully completed the Tkatchev stretched action.

KEYWORDS: Elite athlete, horizontal bar, Tkatchev stretched.

1 STUDY OBJECTIVES

Tkatchev stretched action belongs to D group, its score is 0.4; the action was innovated by Tkachev. Earlier our athletes have successfully completed this action in the World Gymnastics Championships and the Olympic Games, at present it is still one of the difficult tidal flow of horizontal bar. The whole process of the competition was recorded at the game site of 2013 Dalian National Games; we took Chenglong Zhang (hereinafter referred to as "Zhang") action. By the vertical and horizontal three-dimensional kinematics comparative analysis, find the key elements of his action, so as to provide reference for the technical training of coaches and athletes.

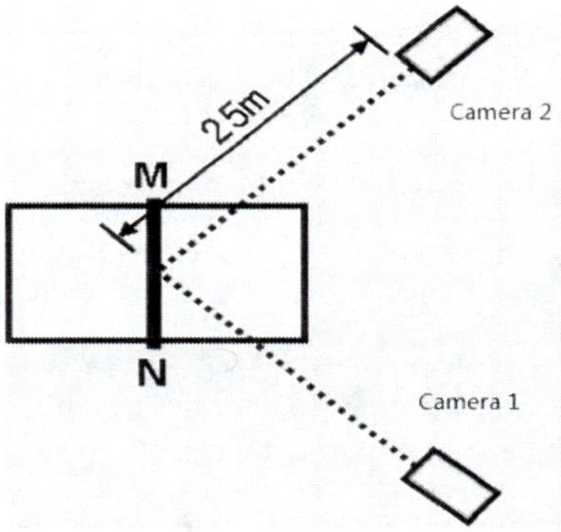

Figure 1. Schematic diagram of cameras.

2 STUDY METHODS

The main method is a three-dimensional video analysis. The whole process of the competition was recorded by two JVC cameras at 50 Hz from different angles (the angle of the two cameras was about 90°). (Figure1). The shooting frequency is 50 frames per second. The record analysis used 3-D SignalTec system and series analysis. The anthropometric dummy is Japanese Songjingxiuzhi phantom (22 articulation points, 16 segments). It passed the original data filter and the cut off frequency is 8Hz.vStudy objects, Zhang, The champion of the 42nd world gymnastics championships of the horizontal bar in 2010, The 2012 London Olympic Games men's team title. (Figure2)

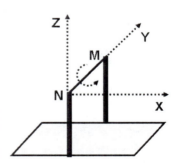

Figure 2. Schematic diagram of coordinate system.

3 STUDY RESULTS

In order to facilitate the analysis of Tkatchev stretched action (Figure3), according to its operating characteristics, we divided it into three phases:

A: Giant hem stage: refers to the inverted moment movement to the body center of gravity moves to the lowest center of gravity moment. B: Giant swing phase: refers to the human body from under lowest center of gravity to the time of his hands off the movement process. C: Somersault stage: two hands pushing off the bar moment after flying to re grasp the bar end time.

Figure 3. Diagrammatic sketch of Tkatchev stretched action.

3.1 Giant hem stage

The stage is inverted moment movement to the body center of gravity moves to the lowest center of gravity moment.

Along with the body of the hem, the whole body should be straight stem head out top, shoulder angle fully open top, shoulder and hip angles increase gradually, in order to increase the gravity torque, gravity short more big momentum is greater, round bar loop ability is strong, which is one of the main power to complete the action [1].

According to the experimental data, the shoulder angle of Zhang in center of gravity to lever for the front horizontal moment is 167.0°, the hip angle is 178.0°, the knee angle is 176.0°, the velocity of center of gravity in the Z direction is 3.90 m/s, gravity center distance average value is 1.11 m. All these data show that Zhang completed the action successfully. Knee angle of Zhang is close to 180°, but the hip angle is obviously too large, a stretch of booth, the body does not reach the ideal attitude. The lower limb of Zhang also fully extended, but the shoulder angle is relatively small, the body basically reaches the ideal position.

3.2 Giant swing phase

Giant swing phase refers to the body from under lowest center of gravity moment after horizontal bar surface before to hands pushed away from the moving process.

When the body is put to the bar sagging face, legs should be rapid upward around the legs, the shoulder sink, which can make the bar maximum bending deformation, bar elastic potential energy for the body to provide more energy movement. In the process, arms should press bar, shoulder open the top as far as possible, most likely to prolong the action time push bar, the purpose of this is to make the supporting force to do more work, thus to gain bigger flip angle velocity and flight height. [2]

According to the experimental data, lowest center of gravity moment: the distance of Zhang of the center of gravity to bar is 1.10m, the velocity of center of the horizontal direction is 5.23 m/s, the value of shoulder, hip angle and angle respectively: 157.0°, 148.0° and 177°. When the center of gravity to the lowest point, the value of shoulder and hip angles were reduced by 10° and 30.5°, the hip angle decreases more, the body gradually fold to increase the moment of inertia to prepare, the hip angle of Zhang is a little smaller, and the center of gravity in the horizontal velocity of center of gravity speed is 0.18 m/s faster than Zhang, which shows Zhang completed the action successfully.

3.3 Somersault stage

This stage is from hands pushing off the bar moment after flight to the holding time at the end of the motion bar again.

Push hands off the bar into the somersault stage, this time the body into the non support state, by the influence of gravity, the height of flight and flight time is limited. According to the law of conservation of angular momentum theorem, to better complete the action, it is necessary to increase the turning angular velocity [3]. In this stage, the release timing is very important. In theory, the focus placed on the phase trajectory can be approximated as composed of a plurality of different curvature of the arc tangent direction, speed up time center of gravity and the corresponding end point of arc consistency. So, early release may result in smaller center of gravity flight angle, thus it is impossible that the body to re grasp the bar in flight phase from the horizontal bar; once release late, the jumping angle is too large, the flight phase body off the bar too close and make bar. Therefore, the appropriate release angle is the key factor in this stage.

According to the experimental data, the release angle of Zhang is 54.8°, the value of shoulder, hip angle and knee angle is respectively: 101.3°, 247.3° and 175.5°, velocity in the vertical direction is 4.21 m/s, we find the hip angle of Zhang is bigger, from the image can also be seen in a two hands pushing off the bar body already formed the bow that moment, hip and shoulder over expansion. Zhang in the release instantly, hip, and shoulder were launched reasonably. The release angle of Zhang is bigger, according to the mechanics analysis, bigger release angle and higher release point can obtain higher flight parabola and body motion.

After releasing his legs to brake, and upper body upwards, the whole body to do the reverse movement, the transition from level one when the vertical state, arms by the lift to front movement, when the center of gravity moving to the highest point, the upper part of the body close to the vertical position, the hip angle of Zhang decreases from 247.3° to 143.2°, the body center of gravity away from the bar height is 1.50 m.

The body's center of gravity over the highest point, his legs continue to back and crossed the bar, arms reach to grasp the bar, when the upper part of the body the reverse movement to the horizontal position with both hands to grasp the bar down. Grab bar when the arms should be straight, the body center of gravity in a higher position.

According to the experimental data, flight time of Zhang is 0.72 s, the value of shoulder, hip angle and knee angle are respectively: 158.9°, 167.9° and 179.0°. The velocity in the vertical direction is 3.86 m/s. We can see Zhang had a longer flight time, this may be due to the bigger release angle, increase the residence time in the air, notably in there grasp the bar phase, the hip angle of Zhang is 167.9°, the shoulder angle is 158.9°, which explained Zhang is conducive to complete the action, because a grab bar when the position of the body is far higher shoulder angle is larger, can obtain the gravitational potential energy of larger and moment of momentum.

4 STUDY CONCLUSIONS

Through the comparison analysis of kinematics parameters of Tkatchev stretched action of elite Chinese gymnast Zhang, we can draw conclusion as follows:

Giant hem stage: In the center of the bar before the plane moment, the knee angle of Zhang is close to 180°, but the hip angle is obviously too large, a stretch of booth, the body does not reach the ideal attitude. A center distance of 1.11m, relative bigger, and this is because of the higher height.

Giant swing phase: Center of gravity to the lowest point, the shoulder and hip angle reduced by 10° and 30°, the center of gravity in the bar before level time, the hip angle decreases more, at the same time, the hip angle of Zhang is 148°.

Somersault stage: Zhang's flight time is 0.72s, a prolonged suspension piece. Two hands pushing off the bar time Zhang's hip angle is 247.3°, the shoulder angle is 201.3°, release angle of Zhang is 51.7°, which shows he can obtain higher flight parabola. The highest point of the body center of gravity, the hip angle of Zhang is 143.2°, off the bar height 1.50m. In this stage, the athlete completes the technical movement. Regrasp the bar moment, the hip angle of Zhang is 167.9°, the shoulder angle is 158.9°, explained Zhang is conducive to complete the action.

REFERENCES

[1] Jun Tsuchiya. Koichiro Murata (2004). Tetsuo Fukunaga. Kinetic Analysis of Backward Giant Swing on Parallel Bars, *International Journal of Sport and Health Science*, 211–221.
[2] L Chen, X W Yao, Y H Wen. LiYa uneven bars hang kinematics forward somersault leg to 51-re grasp the bar. *Sports Science and Technology*, 2007, 5:56.
[3] J H Zhou, X Li, the parallel world elite circle piked two weeks into the hanging arm kinematics analysis. *Journal of Chengdu Sports University*, 2012, 04:63–64.

Information, Computer and Application Engineering – Liu, Sung & Yao (Eds)
© *2015 Taylor & Francis Group, London, ISBN 978-1-138-02717-6*

Research on network model of fresh agricultural products cold chain logistics in Yunnan province

Xiu Ying Tang, Lin Lin Yang, Li Chang Chen, Jie Shi & Yu Li
Faculty of Mechanical and Electrical Engineering, Yunnan Agricultural University, Kunming, Yunnan, China

Bin Cheng & Jing Wang
Kunchuan Design Institute, Kunming, Yunnan, China

ABSTRACT: On the basis of thethe survey of fresh agriculture products, cold chain logistics and collection of documentation, we sum up the existing model of cold chain logistics in Yunnan province, which is domestic cold chain logistics system and abroad cold chain logistics system. After that, we proposed the integrated model which includes three types which is the regional cold logistics system, the inter-provincial cold logistics system and abroad cold logistics system.

KEYWORDS: Fresh agriculture produces; Cold chain logistics; Network model

1 INTRODUCTION

Yunnan is in the southwest of China, and is the important production base of fresh agricultural products which has wonderful weather and abundant rainfall. On one hand, the output of agricultural products increases rapidly in recent years, the agricultural total output value reached 268 billion in 2012 [1]. On the other hand, the demand for chilled and frozen agricultural products is also growing sharply with the improvement of people living level and consumption concept. The continued growth of output and consumption need the support of cold chain logistics system. But the cold chain logistics is in initial stage form that is the current development status of Yunnan province. According to statistics, more than 90% of vegetables are still in normal temperature transportation, storage, processing, and more than 30% loss rate in those stages, while only 5% loss rate in developed countries [2]. So it is necessary to research the cold logistics system to decrease the loss rate and the cost of the whole system.

2 EXISTING PATTERNS OF COLD CHAIN LOGISTICS SYSTEM

We learn the existing patterns of cold chain logistics system through investigation for three months, the places include Kunming, Anning, Songming, Dali, etc. It is divided into two patterns, one is domestic cold chain logistics system and other abroad cold chain logistics system.

2.1 *Domestic cold chain logistics system*

It includes two sub-systems, which are sub-systems inside and outside of province, its specific logistics process is shown in Fig. 1:

1 Sub-system in the province : the agricultural products, which are fror the production base, plantation, farmers and etc., are sold to chain supermarket, agriculture products sales points, the local and other wholesaler and, etc., the local wholesaler sell the agriculture products to consumers through the local agriculture products market, while other wholesalers sell the agriculture to all parts of the province through the normal temperature transportation. Take Jinning vegetable production base as an example, some vegetables are directly supplied to supermarkets such as Wall-Mart, Carrefour, another part of vegetables are supplied to vegetable direct-sale stores, the rest of the vegetables are sold to wholesales and then are distributed to each city in the province.

2 Sub-system out of province: the agricultural products, which are form the production base, plantation, farmers and etc., are sold to local or out-of-town cold storages. Another situation is that the owners of cold storages purchase the agriculture products from surrounding area. Next, the agriculture products are put in cold storage for pre-cooling, processing and packing. Then, the agriculture products are transported out of province by long-distance truck under normal temperature. After that, the products are delivered to hotel, supermarkets, dining room and, etc., by

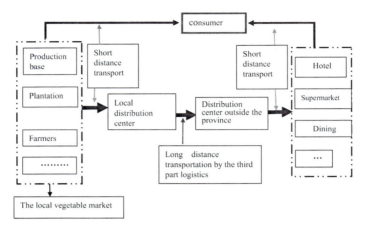

Figure 1. Flow chart of domestic cold chain logsitics system of fresh agricultural products.

short distance transportation. Finally, the products are reached to customers. Take Yanglin vegetable as an example, the cold storages purchase the vegetables from surrounding town. Next, the products are put in cold storage for pre-cooling, which maybe half an hour or four days. Then, workers pick the leaves of vegetables which are affected in quality, such as the leaves with the soil or the old leaves (Seeing Fig. 2). After that, the vegetables are put in bubble chamber with simple insulation measure (Seeing Fig. 3), two liquid bottles are put in the diagonal of bubble chamber. Later, the chambers are transported to different distribution centers of China. Finally, the customers get the vegetables from supermarkets, agriculture products sales points and, etc.

Figure 3. Insulation measure.

Figure 2. Processing in cold storage.

Figure 4. Foreign transport vehicle.

2.2 Abroad cold chain logistics system

Its specific logistics process is shown in Fig. 5. It is mainly divided into two types:

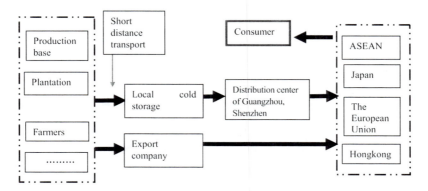

Figure 5. Flow chart of abroad cold chain logsitics system of fresh agricultural products.

1 Export by Export Company in Yunnan Province. These export companies purchase high-quality fresh agricultural products. After pre-cooling, processing and packaging, the part of agricultural products export to ASEAN by freight car under normal temperature through international road, while, the others export to Japan, the European Union and other countries by air under normal temperature.

2 Export by distribution center of Guangzhou and Shenzhen. The front part of process is the same; the difference is that agricultural products export to Hong Kong, Japan, the European Union and other countries by truck or air.

From the current cold chain logistics mode, there are mainly three types, which is the regional cold logistics model, the inter-provincial cold logistics model and abroad cold logistics model. There are many forms of three types, so the cold chain logistics system is complex. At the same time, agricultural logistics has many links; the logistics channel is not smooth, which cause the high cost for system. We also learn that only in the stage of cold storage the fresh agricultural products are in low temperature environment of the whole cold chain logistics system, the refrigerated trucks are scarcer and almost all the transportation is under the normal temperature, which cause that the decay rate of fresh agricultural product is high and the loss is very serious. So it is necessary to research on the cold chain logistics model of fresh agricultural products which could decrease the links, decrease the decay rate and the total cost of logistics system.

3 PROPOSED COLD CHAIN LOGISTICS SYSTEM

We propose an integrated model in view of the existing problems of cold chain logistics and combined with the actual situation in Yunnan province. The cold chain logistics system almost is in the suitable low temperature from pre-cooling after field picking to customers, which could make the products in preservation station, low loss and decrease the cost of the whole system. The cold chain logistics model is shown in Fig. 6.

The model is also divided to three types which is the regional cold logistics system, the inter-provincial cold logistics system and abroad cold logistics system.

1 The regional cold logistics system. There are two situations; one is that the whole process can make sure the agricultural products keep quality in certain time which is not necessary to put the products in cold chain stage, it could be in normal temperature as much as possible. The another is the time is more than the normal time for preservation of agricultural products, the products from the production base and plantation should precool after picking in the field. Then, they are transported to local cold storages under the normal temperature. Next products are in processing, storage, and packing in the cold storage. Finally, they are transported to different places by refrigerated truck under suitable low temperature in the province and reached the consumers.

2 The inter-provincial cold logistics system. The products from the production base and plantation should precool after picking in the field. Then, they are transported to local distribution center under the normal temperature. After that products are done for processing, packaging and storage in distribution center in suitable low temperature. Next, the refrigerated trucks are used to transport the products to different distribution center of demand location. Finally, the products are distributed to customers.

3 Abroad cold logistics system. The export companies purchase the qualified fresh agricultural

975

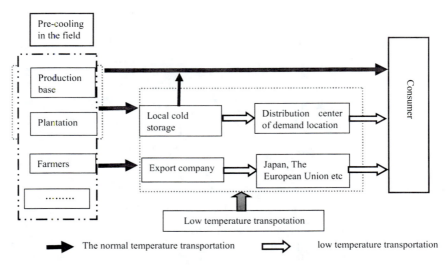

Figure 6. Flow chart of an integrated cold chain logistics model.

products from supply center, which have been precooled in the field. Then, processing, storage and packaging are done in the cold storage. According to the different countries, we could choose different mode of transportation. If they export to ASEAN, the refrigerated truck could be choose, and if the export to Japan, the airline is used for keep fresh.

4 CONCLUSION

Through the literature review and spot investigation, we sum up the existing model of cold chain logistics of fresh agricultural products in Yunnan province, which is domestic cold chain logistics system and abroad cold chain logistics system. After that,

we proposed the integrated model which includes three types which is the regional cold logistics system, the inter-provincial cold logistics system and abroad cold logistics system. The new model makes the agricultural products almost in suitable low temperature, which could keep the products in low loss and decrease the cost of the whole system.

REFERENCES

[1] Yunnan provincial bureau of statistics. 2013. Yunnan statistical yearbook, China Statistics, Inc., Beijing, 156.
[2] Hu Min. 2013. Facing the opportunities and challenges of Yunnan_ the development of vegetable cold chain logistics. *Logistics Technology (Equipment)*(7):22–25.

Information, Computer and Application Engineering – Liu, Sung & Yao (Eds)
© 2015 Taylor & Francis Group, London, ISBN 978-1-138-02717-6

Research and design on universities electronic information management system based on ASP.NET

Xin Zhang

Jilin Business and Technology College, Changchun, China

ABSTRACT: Information management system was designed based on the actual situation of the management of university in order to improve the university management. In the development process of the system, Microsoft's .NET architecture was used as a system development framework, and Visual Studio.NET was adopted as environment of the system. In the design process of the database, SQL Server 2005 was adopted as the database management platform, database interaction and access to technology in ASP.NET was applied to achieve pages and databases interactive each other. This system was designed in this paper has strong stability, reliability and strong practical value, which can basically meet the needs of information management.

KEYWORDS: In order to; MS Word

1 INTRODUCTION

With increasing depth of higher education development and the gradual increase in the level of information, how to implement information management level institutions of higher education, is the great challenge facing managers. Through the application of network communication technology, all of the teachers and students of the transaction can be done through the electronic information system network, which greatly improves the efficiency of management work [1]. In the development process of this paper, the combination of theory and computer IT college information management, design of an electronic information management system based on ASP .NET, thus improving the efficiency of student management personnel management for institutions of higher education students to science to lay the foundation of normalization [2].

2 SYSTEM REQUIREMENTS ANALYSIS

After research on the specific college, college student information management involves the following aspects: management of teachers, student management, curriculum management, and performance management. The specific content of these management work are summarized below. Teacher management is the teacher's personal archives information management, information includes the teacher's name, sex, age, where the college ...... and so on. When personal information was changed, it should be allowed to carry out their request to modify and expand the like. Student management is to manage the student's personal information which includes student information, gender, where classes have been revised credits and so on. When personal information was changed, it should be allowed to carry out their request to modify and expand the like too. Course management is provided to the college curriculum management, such information includes course number, course name, course category and course credits and the like. Teaching administrator has modified permissions. For course category, provide management functions, curriculum categories of information, including public courses, specialized courses, professional electives and other information. Performance management is to manage student achievement, such information including course number, course name, student number, student name and so on. Teachers and teaching administrator has modified permissions. As a student permission to provide the results of the query function [4].

3 INFORMATION SYSTEM DESIGN

3.1 *Overall flow of information systems*

The overall flow of information system design is shown in Fig. 1:

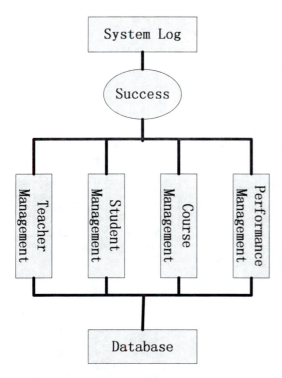

Figure 1. Overall system flow chart.

3.2 Overall information system architecture

According to the needs of the above analysis, the electronic information system is divided into five sub-modules in this paper which is shown in Fig. 2:

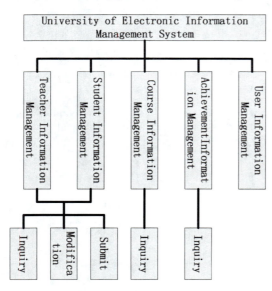

Figure 2. System function block diagram.

Profile information for teachers to manage information, including the teacher's name, sex, age, where the academy and other information. On the student's personal information management, such information includes student's information, gender, where classes have been revised credits and so on. The module is divided into two parts: the first part of the student information management, mainly in the form of a list showing all the student information, both modify and delete functions; The second part is the new section, click on the module, the new student information to fill in the pop-up window [5]. College courses on management, information including course number, course name, course category and course credits and the like. Student performance management module to complete the function is on student achievement management, information includes the student selected course number, course name students selected, and students learn numbers, the student's name and so on. The module is divided into two parts, the first part of the query results, according to the input query conditions, the query results are displayed in a list being; the second part is the score entry, click on the module, the new pop-up window filled course grade information [6].

3.3 Overall information system deployment

The system for the convenience of anytime, anywhere access operations can be performed, preclude the use of B/ structure, the specific deployment structure is shown in Fig. 3:

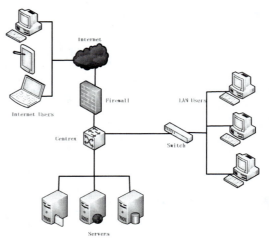

Figure 3. Actual system deployment diagram.

3.4 Database design

The Universities of electronic information management system used by the database management platform for SQL Server 2008, the database name is "db-GXM". Teachers table includes four fields, namely number of teachers, teacher's name and department number and gender of teachers, their physical structure as shown in Table 1.

Table 1. Teacher data sheet (part).

Item-name	Data-Type	Data-Length	Premary-key	Memo
id	Varchar	20	Yes	Teacher Number
Name	Varchar	100	No	Teacher Name
Department	Varchar	100	No	Department Number
Sex	Char	2	No	Teacher Sex

Student table includes six fields, namely student number, the name of the student, the student gender, physical structures such as shown in Table 2.

Table 2. Student data sheet (part).

Item-name	Data-Type	Data-Length	Premary key	Memo
id	Varchar	20	Yes	Student Number
Name	Varchar	100	No	Student Name
Class_id	Varchar	100	No	Class Number
Sex	Char	2	No	Student Sex
Birth_data	Varchar	100	No	Student birthday
Address	Varchar	100	No	Student Address

Curriculum includes six fields, respectively, course number, course name and semester and other information, its physical structure as shown in Table 3.

Table 3. Course data sheet (part).

Item-name	Data-Type	Data-Length	Premary key	Memo
id	Varchar	20	Yes	Course Number
name	Varchar	100	No	Course Name
Term	Char	1	No	School Term
Examtype	Char	10	No	Exam Type
Total	Number	100	No	Phase
Score	Number	100	No	Credit

Exam table includes seven fields, respectively Course ID, student ID number and other information and teachers, as shown in Table 4.

Table 4. Examination data sheet (part).

Item-name	Data-Type	Data-Length	Premary-key	Memo
Course_id	Varchar	20	Yes	Course Number
Student_id	Varchar	20	No	Student Number
Teacher_id	Varchar	20	No	Teacher Number
Term	Char	1	No	School Term
Examtype	Char	100	No	Exam Type
Examtime	Varchar	20	No	Exam Time
Address	Varchar	100	No	Address

4 ACHIEVEMENT OF ELECTRONIC INFORMATION SYSTEM

4.1 System log interface

Login window of universities electronic information management system is shown in Fig. 4:

Username: hello

Password: •••

Submit   Cancel

Figure 4. Login interface.

4.2 Teacher management information interface

Teacher Management Information page of college's electronic information management system is shown in Fig. 5.

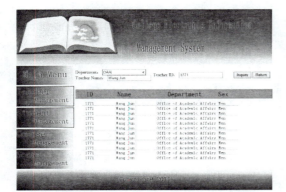

Figure 5.   Teacher management interface.

4.3 *Student management information interface*

Student management information page of college's electronic information management system is shown in Fig. 6:

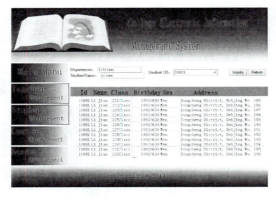

Figure 6. Student management interface.

5   CONCLUSION

There is a great practical significance about information management system that was designed in this paper, but some are not technically perfect and mature. It also needs to be refined and improved, and then it will gradually become powerful and easy tools of electronic information management of universities.

REFERENCES, SYMBOLS AND UNITS

[1] W.l.iang.Human Resources Management of University Research, Anhui University of Technology, 2005.
[2] S.T.Zhao, X.L. Chen. Visual Studio 2005 + SQL Server 2005 Database Application Development, Electronic Industry Press, 2007.
[3] J.Chen, W.B.Wang. Achievement of Enrollment Management System based on B/S and C/S Structure, Computing Technology and Automation, 2005.
[4] F.Guo. Enterprise Personnel Management Information System Based on B/S, Xiamen University, 2014.
[5] Y.F. Lou. Design and Implementation of Educational Administration System Based on B/ mode,Jilin University, 2014.
[6] M.Wang. Analysis and Design of Student Information Management System, Xiamen University, 2014.

Information, Computer and Application Engineering – Liu, Sung & Yao (Eds)
© 2015 Taylor & Francis Group, London, ISBN 978-1-138-02717-6

Decision and analysis of liquid cargo replenishment

Wei Deng, Duan Feng Han & Wei Wang
Harbin Engineering University, Harbin, Helongjiang, China

ABSTRACT: Replenishment ship is a unique type of ship, which keeps other ships functioning well by providing fuel, food, fresh water and other supplements. This kind of ship plays a crucial role in increasing the operation time and radius of other ship. The supplement determinable system is established based on the multi-objective decision theory. The TOPSIS algorithm modified by permutation and combination is used to solve supply plan. In addition, artificial selection and modification module are combined with the existing example to analyze the results, and thus indicating the practicability of this system.

KEYWORDS: Replenishment ship; Supplement; Multi-Objective Decision; TOPSIS;

1 INTRODUCTION

Along with the enhancement of China's military power, more multi-function ships began to execute different tasks means replenishment for a large number of reserve consumption should be more accurately and quickly[1,3]. In this paper, liquid cargo replenishment schemes are analyzed; the simulation modeling about the discrete event is discussed and determines the original replenishment schemes, and then uses TOPSIS approach ideal solution sorting algorithm to find the optimal solutions.

2 ANALYSIS OF LIQUID CARGO REPLENISHMENT SCHEMES

2.1 Search the replenishment schemes

Assuming that there are 4 kinds of liquid cargo tanks, which include 8 fuel oil tanks, 6 diesel oil tanks, 2 fresh water tanks and 1 jet fuel tank (for air craft) which are shown in Fig. 1.

Two symmetrical tanks are considered as a whole to be a rule for searching replenishment schemes, so following the symmetry principle, make a serial numbers to each kind of liquid cargo tank and make a same series number to each kind of symmetrical tanks which are shown in Fig. 2.

The residual amount of each symmetrical tanks are known, in this addition the objective of course is to find a best replenishment scheme from all schemes which are calculated by algorithm. In order to achieve this goal all possible permutation and combination schemes which are suitable for liquid cargo demand should be found.

According to permutation and combination theory[4], the method to determine replenishment schemes of fuel oil and diesel oil is same. In this additions which residual amount of each symmetrical tanks is known all possible schemes can be got. And there is only one fresh water tank, according to symmetric supply rule only one scheme is found without using permutation and combination theory. Jet fuel tank is as same as fresh water tank.

Based on the above mentioned method, if different kinds of oil are needed to replenish at the same time, we also should use permutation and combination to find the replenishment schemes. Assuming a ship which needs to be supplied with both fuel oil and diesel oil together, and there are M schemes for supplying fuel oil and N schemes for diesel oil. So M×N schemes are found.

2.2 Search reasonable solution with adaptive step traversal algorithm

For increasing the precision of calculation and speed, adaptive step traversal algorithm is used. In the actual situation we set traversal step to be transformable. In the process of the searching schemes, first big step length and then small step length for searching till the best solutions are found.

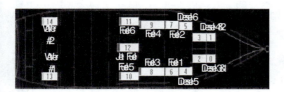

Figure 1. Oil tanks arrangement plan

Figure 2.   Oil grouping schematic
Green: fuel oil tanks; Yellow: diesel oil tanks
Purple-blue: fresh water tanks; Pink: jet fuel tank

First, to process every scheme which is found with permutation and combination, the amount of each oil tank is divided into 10 parts and every part is set to be a step length for replenishment. There are 11 kinds of oil tank residual quantity state: 0; 1/10;2/10...;1, 0 represents oil completely empty and 1 represents oil is not need to supply. At the same time another kind of symmetrical tank state can be determined.

Assuming the amount of port and starboard oil tank are 1000t and 600t. Each tank is divided into 10 parts, then all 11 kinds of state are calculated. The best solution of heel angel is smallest, and the solution is used as first state for small step length binary search.

In binary search, based on the best solution which is found in first search the step length is changed to 1t, then continue searching with the replenishment quantity is already determined in the first search. Searching will not stop until better solution was found, so the solution is one of the best solutions that the step length is 1t.

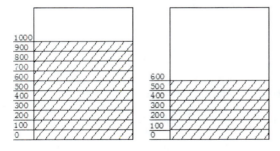

Figure 3.   Separation of oil tanks

3   OPTIMIZATION CHOICES OF LIQUID CARGO REPLENISHMENT SCHEMES

3.1   *Mathematic model of multi-objective decision*

Decision and analysis about Liquid Cargo Supply's Project is a multiple attribute decision making (MADM) problem. It is one of optimization controlling projects.

Assuming x_j represents the jth attribute a_j represents the i_{th} scheme $u_{ij} = f(a_i, x_j)$ represents preference value of the i_{th} scheme which is under the j_{th} objective attribute, so the mathematic model of MADM is shown as follows:

$$\begin{cases} DR[a_1(u), a_2(u), ... a_m(u)], u \in U \\ U = [u | u_{ij} = f(a_i, x_j)], i = 1,2, ... m; j = 1,2, ... n \end{cases} \quad (1)$$

Before calculating the MADM's mathematic model, we should know the constrains in the process of supplying float state and stability of ship. The constrains are shown as follows:

1   The absolute value of heeling angel is less than 0.5^0;
2   $\overline{GM} > 0.75$m;
3   Trim angel is less than <0.4%L (ship length), and as big as better. In this paper, the value ranges from −0.9 to −0.

3.2   *Seeking answer for multi-objective optimization problem.*

3.2.1   *Determine feasible scheme with TOPSIS method*

In TOPSIS theory, first step is to find out the best and worst schemes which is called ideal solution and negative ideal solution[4].

Second make all various measures of each scheme compare with measures of ideal and negative ideal solution. If one scheme's measures which is the closest to ideal solution and the farthest to negative ideal solution at the same time is called optimal decision for supplying. If not it is called the worst decision.

According to the constrains of supplying, data structure of decision and analysis about Liquid Cargo replenishment should be built[5]. Then use the TOPSIS theory to determine all possible schemes which is called evaluation unit (EU), and determine the evaluation attribute (EA) together. The two elements which make up the data structure is shown in Table 1.

Table 1. Data structure of decision analysis

EU \ EA	Heeling angel	$\overline{GM}$	Trim angel
Scheme 1	y_{11}	y_{12}	y_{14}
Scheme 2	y_{21}	y_{22}	y_{24}
...	...	...	...
Scheme N	y_{m1}	y_{m2}	y_{m4}

3.2.2 The improvement of TOPSIS decision algorithm

TOPSIS algorithm is improved in this part based on case of final value is close to target value. We changed MOP (multi-objective problem) model to SOP (single-objective problem) model[6], the mathematic model is as follows:

$$\min_i F(f_i) = w_1 f_i(\alpha) + w_2 f_i(t) + w_3 f_i(\overline{GM})$$

$$\text{subject to} \begin{cases} |\alpha| < 0.5^0 \\ -0.9m \leq t \leq 0m \\ \overline{GM} > 0.75m \end{cases} \quad (2)$$

$$f_i(\overline{GM}) = \frac{\overline{GM}_{max} - \overline{GM}_{min}}{\overline{GM}_i}$$

$$f_i(t) = \frac{|t_i|}{\max\{|t_i|\} - \min\{|t_i|\}}$$

$$f_i(\alpha) = \frac{|\alpha_i|}{\max\{|\alpha_i|\} - \min\{|\alpha_i|\}}$$

$F(f)$ ——objective function
f_i——ith possible scheme in data structure
$\overline{GM}$——initial stability height of replenishment ship
α ——heeling angle of replenishment ship
t ——trim angle of replenishment ship
w_1——weight value of objective function's heeling angle
w_2——Weight value of objective function's trim angle
w_3——Weight value of objective function's initial stability height
Judgment matrix of 3 subgoals is:

$$A = \begin{cases} 1 & 2 & 2 \\ 1/2 & 1 & 1 \\ 1/2 & 1 & 1 \end{cases} \quad (3)$$

Normalize characteristic vector of matrix above is $w = (w_1, w_2, w_3) = (0.5, 0.25, 0.5)$ So (2) can be shown as:

$$\min_i F(f_i) = 0.5 f_i(\alpha) + 0.25 f_i(t) + 0.25 f_i(\overline{GM})$$

$$\text{subject to} \begin{cases} |\alpha| < 0.5^0 \\ -0.9m \leq t \leq 0m \\ \overline{GM} > 0.75m \end{cases} \quad (4)$$

Formula (4) is the improved mathematic model based on objective theoretical value, it can solve the evaluation of all indexes[7]. The improved multi-objective decision formula is:

$$Object[i] = \frac{1}{Ya[i]+1}$$
$$+ \frac{Max_{GM} - Min_{GM}}{Max_{GM} - Min_{GM} + |GM[i] - GM_{goal}|}$$
$$+ \frac{Max_t - Min_t}{Max_t - Min_t + |t[i] - t_{goal}|} \quad (5)$$

$Ya[i]$ represents the ith possible plan;
Max_{GM} represents max $\overline{GM}$ value of all schemes;
Min_{GM} represents min $\overline{GM}$ value of all schemes;
$GM[i]$ represent the ith plan of all schemes;
GM_{goal} represents target value of all schemes;
Max_t——the max value of trim angle;
Min_t——the min value of trim angle;
$t[i]$ ——value of the ith plan.
t_{goal} ——target value of trim angle

4 LIQUID CARGO REPLENISHMENT DECISION SCHEMES CALCULATION EXAMPLE ANALYSIS

4.1 Caculation example

In a case of full load departure, the needed amount of replenishment is shown in Table 2.

Table 2. The replenishment quantity of ship

Kinds	Amount
Fuel oil	1000t
Diesel oil	500t
Jet fuel	800t
Fresh water	200t

To calculate the stability of replenishment ship, the solution is: the ship's displacement is 48,000 t, initial heel angle is -0.180, trim angle is -0.507 m, GM is 1.553 m, forward draft is 10.516 m, aft draft is 11.023 m and the barycentric coordinate of the ship is (−2.381, −0.005, 12.092).

The barycentric coordinate of the ship is (−2.381, −0.005, 12.092).

4.2 Analysis of calculation result

Calculating with multi-objective decision based on ship's work condition to find the possible schemes which satisfy constraints. Finally, 3 possible schemes are found as follows:

Take Plan 1 as real replenishment scheme and the state curve of float and stability are shown in Fig. 4.

Table 3. Analysis of picked schemes

	Plan 1	Plan 2	Plan 3
Whether meet the replenishment demand	YES	YES	YES
Number of fuel oil tank for replenishment	5, 6	5, 6	1, 2
Number of diesel oil tank for replenishment	5, 6	3, 4	5, 6
$\overline{GM}$	1.575	1.616	1.602
Heel angle	−0.083	−0.095	−0.087
Trim angle	−0.754	−0.853	−0.622

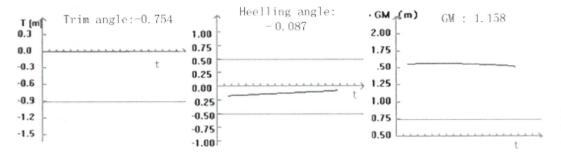

Figure 4. State curve of float and stability

From figure, in the process of supplying the range of trim angle value is(−0.9, 0), and heel angle value change in range from −0.5 to 0.5 and $\overline{GM}$ is more than 0.75 m. The status in whole process is suitable.

After replenishment, the state of ship is changed and the updated state is shown in Table 4

Table 4. Updated state of ship's state

Displacement(t)	Fore draft(m)	Aft draft(m)
45500	9.944	10.699
Heel angle (°)	Trim angle (m)	GM (m)
−0.083	−0.853	1.575

4.3 Analysis of calculated solution

In order to verify the 3 optimal schemes' superiority, we pick other 3 schemes (nos. 5, 62, 243) from alternative schemes to compare with. The result of comparing is shown in Table 5.

The values of θ and $\overline{GM}$ in Plan 1, 2 and 3 are better than nos. 5, 62, 243 schemes. So Plans 1, 2 and 3 are optimal under constrains.

5 SUMMARY

In this paper, the stress characteristics are analyzed when load are changed at any position of replenishment ship, and the calculation method of floating state, stability and strength are summarized. One basis, which constraints of supply ship during operation, should be definite. The feasible strategy is regarded as the evaluation unit and constraints of supply ships are considered as evaluation standards, which contribute to establish multi-objective determinable matrix. Based on the application of TOPSIS, the algorithm alternatives are solved in the same trend and the dimensionless standard matrix is established. Then the relatively optimal replenishment solution can be obtained by checking the Euclidean distance

Table 5. Comparison result

		Plan 1	Plan 2	Plan 3		No.5 scheme	No.62 scheme	No.243 scheme
θ	Picked schemes	−0.083	−0.095	−0.087	Phased out schemes	−0.056	−0.074	−0.082
t		−0.754	−0.853	−0.622		−0.221	−0.241	−0.876
$\overline{GM}$		1.575	1.616	1.602		1.676	1.627	1.684

between each scheme and ideal solution. Examples of special replenishment quantity calculation are analyzed, and the result verifies the algorithm in paper is good and suitable.

REFERENCES

[1] Chapter XII Additional safety measures for bullk carriers amendments to the international convention for the safety of life at sea. *IMO*, 1997, MSC68.
[2] Shields J J. Containership stowage: a computer-aided preplanning system. *Marine Technology*, 1984, 21(4): 370–383P.
[3] John J.D.Expert System Applications to Ocean Shipping-A Status Report. Marine Technology, 1990, 27(5):265–284P.
[4] Botter R C,Brinati M.A.Stowage Container Planning: A Model for Getting an Optimal Solution. Computer Applications in the Automation of Shipyard Operation and Ship Design, IFIP Transactions B(Application in Tech.) 1992:217–229P.
[5] Dubrovsky O,Levitin G,Penn M.A Genetic algorithm with a compact solution encoding for the container ship stowage problem. *Journal of Heuristics*, 2002, 8: 585–599P.
[6] Xiong Sheng-wu, Li Feng. Parallel strength pareto multi-objective evolutionary algorithm, The 2003 Congress on Evolutionary Computation, CEC' 2003, IEEE 2003:223–232.
[7] Lu Hai-ming,Yen G G. Rank-density-based multi-objective genetic algorithm and benchmark test function study. IEEE Trans on *Evolutionary Computation*, 2003, 7(4):56–71.

Information, Computer and Application Engineering – Liu, Sung & Yao (Eds)
© 2015 Taylor & Francis Group, London, ISBN 978-1-138-02717-6

Kinematics analysis of Tkatchev stretched action of gymnastics horizontal bar champion Kai Zou in the 12th national games

Xiu Ling Bian, Ji He Zhou & Zhan Le Gao
Chengdu Sport University, Chengdu, China

ABSTRACT: By using the 3D video analysis method, we have tested on the Tkatchev stretched action of gymnastics horizontal bar champion Kai Zou at the game site; obtain the relevant kinematics' parameters through analysis on the complete movement. The results show that Zou Kai had a good finish of the action, but the hip was not straight when the hem to the lowest point and the center of gravity away from the bar. In addition, the legs were not fully close together throughout the course of action.

KEYWORDS: Horizontal bar; Kinematics analysis; Tkatchev stretched

1 STUDY OBJECTIVES

Kai Zou is an elite athlete of China, and his height, 1.58 m; weight, 47 kg. He is the The horizontal bar champion of the 2008 Beijing Olympic Games; The 43rd World Gymnastics Championships horizontal bar title in Tokyo, 2011; The 12th national games men's horizontal bar title in 2013.

The horizontal bar is one of the China men's gymnastics team advantage project, Tkatchev stretched belongs to "flight action" of the horizontal bar, at present it is still one of the difficult tidal flow of horizontal bar. This paper aims to find the key elements of their action, so as to provide reference for the technical training of coaches and athletes.

2 STUDY METHODS

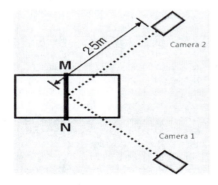

Figure 1. Schematic diagram of cameras.

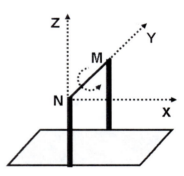

Figure 2. Schematic diagram of coordinate system.
Note: N is the origin of coordinates, line MN expressed high bar.

The main method is a three-dimensional video analysis. The whole process of the competition was recorded by two JVC cameras at 50 Hz from different angles (the angle of the two cameras was about 90°) (Figure 1). And shooting frequency is 50 frames per second. The record analysis used was 3-DSignalTec system and series analysis. We used Japanese Songjingxiuzhi phantom as the anthropometric dummy (22 articulation points, 16 segments). It passed the original data filter and the cut off frequency is 8 Hz

3 STUDY RESULTS

In order to facilitate the analysis of Tkatchev stretched action, according to its operating characteristics, we divided it into two phases: Giant swing phase (refers to the highest point of the inverted moment to the moment that his hands off the bar.); somersault stage (two hands pushing off the bar moment after flying to re grasp the bar end time).

4 ANALYSIS OF THE GIANT SWING PHASE

There is a very important technical aspect called "vibration wave", and it consists of two parts, hem and pocket the legs and the body can get the maximum kinetic energy of the stage [1].

Table 1. Kai Zou Giant swing phase parameters.

Average parameters of A phase	Shoulder joint angle (°)	147.45
	Hip joint angle (°)	152.08
	Knee joint angle (°)	167.10
	Ankle joint angle (°)	149.42
Distance between gravity center and the horizontal bar when swing lowest		1.64
Hip angle when swing lowest (°)		162.85
Pocket leg time hip angle range (°)		63.80
Pocket leg end time center of gravity flight speed (m/s)		3.11
Swing stage time (s)		1.46

Note: A phase refers to the bar on the highest point to the lowest point of the hem.

Table 1 shows Zou Kai Giant swing stage time is 1.46 s, the average value of shoulder joint angle is 147.45°, the average value of hip joint angle is 152.08°, the average value of knee joint angle is 167.10°, the average value of ankle joint angle is 149.42°. The hem of the lowest point, the body's center of gravity from the bar distance is 1.64m. Data reflect Zou Kai body basically straight into a state in the process of hem, conforms to the maximum torque principle. From pocket leg angle range 64.87°, it can be seen that Kai Zou had a large margin in vibration wave technology. Hip angle in the time hem lowest is 162.85°, failed to

fully straighten. Pocket leg end time center of gravity flight speed is 3.11 m/s, This ensures the flight time and without the influence of grasping the horizontal bar technology below

5 SOMERSAULT STAGE

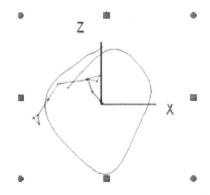

Figure 3.   Schematic diagram of the trajectory of gravity center.

Somersault stage is from hands pushing off the bar moment after flight to the holding time at the end of the motion bar again. Push hands off the bar into the somersault stage, this time the body into the non support state, by the influence of gravity, the height of flight and flight time is limited. According to the law of conservation of angular momentum theorem, to better complete the action, it is necessary to increase the turning angular velocity [2]. In this stage, the release timing is very important. In theory, the focus placed on the phase trajectory can be approximated as it comprises the plurality of different curvature of the arc tangent direction, speed up time center of gravity and the corresponding end point of arc consistency [3]. So, early release may result in smaller center of gravity flight angle, thus it is impossible for the body to re grasp the bar in flight phase from the horizontal bar; once release late, the jumping angle is too large, the flight phase body off the bar too close and make bar. Therefore, the appropriate release angle is the key factor in this stage.

According to the mechanics analysis, bigger release angle and higher release point can obtain higher flight parabola and body motion [4]. In terms of speed, the loss of vertical velocity of Zou is large, the reason may be Zou in braking legs, chest out, pretty hip movements, because hip flexion angle, pretty hip need longer, resulting in brake leg late timing. After releasing his legs to brake, and upper body upwards, the whole body to do the reverse movement, the transition from level one when the vertical state, arms by the lift to front movement, when the center of gravity moving to the highest point, the upper part of the body close

Table 2. Kinematics parameters in Somersault stage.

phase	Gravity center speed		Shoulder angle (°)	Hip angle (°)	Knee angle (°)	Release angle (°)	Flight height (m)	Flight time (s)
	Vx	Vz						
B	0.21	3.52	182.5	210.5	147.6	51.5	----	----
C	0.18	1.47	42.8	119.6	179	----	0.88	----
D	1.62	3.89	130.2	132.5	179	----	----	0.60

Note. B: The kinematics parameters in the hands off the bar phase, C: The kinematics parameters in the highest center of gravity phase, D: The kinematics parameters in the re-grasp bar phase. The release angle is defined as release instant the body center of gravity to shake hands with the horizontal plane angle. Flight height refers to the distance between the highest point of the center of gravity and the horizontal bar. Flight time refers to time from the release moment to re grasp the bar moment.

to the vertical position, the hip angle of Zou decreased from 210.5° to 129.6°, the body center of gravity away from the bar height is 0.98 m. In this stage, Zou completed the technical movement, and had a beautiful air gesture. From Table 2, we can see Zou's flight time is 0.60 s, the shoulder angle is 158.9°, hip angle of Zou is 132.3°, and the shoulder angle is 131.2°.

6 STUDY CONCLUSIONS

Through the analysis of kinematics parameters of Tkatchev stretched action of elite Chinese gymnasts Kai Zou, we can draw conclusion as follows:

1 Giant swing phase: Zou Kai Giant swing stage time is 1.46 s, the average value of shoulder, hip, knee joint angle is 147.45°, 152.08° and 167.10°. The hem of the lowest point, the body center of gravity from the bar distance is 1.64 m. Data reflect Zou Kai body basically straight into a state in the process of hem, conforms to the maximum torque principle.
2 Somersault stage: In this stage, the body center of gravity away from the bar height is 0.98 m. Zou's flight time is 0.60 s, the shoulder angle is 158.9°, hip angle of Zou is 132.3°, and the shoulder angle is 131.2°. Zou completed the technical movement, and had a beautiful air gesture.

Kai Zou had a good finish of the action, but the hip was not straight when the hem to the lowest point and the center of gravity away from the bar. In addition, the legs were not fully close together throughout the course of action.

REFERENCES

[1] Jingguang Qian. Study on the straight Tkatchev vault innovation action. Master's graduate thesis.
[2] Zhou Jihe, Li Xi, Cheng Qingqing, Cheng Liang, Lai Qicai. 2012. Parallel bars world excellent athletes circle two weeks into the kinematics hanging arm movement analysis. *Journal of Chengdu Sport University*, 04:63–66.
[3] Li Xi. 2012. The kinematics analysis of excellent gymnastic athletes bar straight Eger and swivel 360 technical movement. Chengdu Sport University.
[4] Yao Xiawen, Wang Zheng, Wang Rong. 2010. Analysis of the Xiao Qin, Teng haibin horizontal bar and the leg Tkatchev vault kinematics. *Journal of Beijing Sport University*, 02:99–101.
[5] Dallas G et al. 2011. *Influence of angle of view on judges' evaluations of inverted cross in men's rings. Perceptual and Motor Skills*, 112(1):109–121
[6] Maurice R.Yeadon. 1994. Twisting techniques used in dismounts from the rings. *Journal of Applied Biomechanics*, 10(2):178–189.

Information, Computer and Application Engineering – Liu, Sung & Yao (Eds)
© 2015 Taylor & Francis Group, London, ISBN 978-1-138-02717-6

Analysis of landscape elements' application in landscape architecture

Xiao Hua Wu, Jin Ying Liu & Li Xia Song
Langfang Polytechnic Institute, China

ABSTRACT: Landscape design is an integral part of urban greening, and also the embodiment of a low-carbon city, eco-city ideas. This paper describes the importance and application principles of the application landscape elements in landscape design, and also analyzes the problems in the current urban landscape design. At last some corresponding countermeasures are put forward to optimize the city landscape.

KEYWORDS: Scenic Landscaping; Important significance; Principle; Countermeasures

With the continuous improvement of China's urbanization, people require high living environment requirements increasingly, so does scenic elements used in landscaping design more widely. The core of the landscaping projects is the garden design and the landscape construction, so enough attention must be given in actual construction. In order to enhance the building greening effect, improve the urban ecological environment and to provide a better living environment, landscape elements need to be used fully to optimize the landscaping configuration.

1 THE IMPORTANCE OF THE LANDSCAPE ELEMENTS' APPLICATION IN SCENIC LANDSCAPING

1.1 Scenic landscaping can maintain the ecological balance of the city

Landscaping is an artificial ecosystem reconstruction, which simulates the natural way to preserve the natural environment, suburban scenic spots, nature reserves of the city. It is also to maintain the ecological diversity of the city, to create a good living environment for a variety of animals and plants and even maintain the ecological balance of the city through the artificial facility. Designers can take advantage of landscape elements, such as garden vegetation, to promote the natural environment for better development.

1.2 Scenic landscape can effectively alleviate the problem of urban pollution and be eco-efficient

The ratio of greening space is increased by some greening ways, such as planting trees, grass, flowers and so on and the greening blind spots are lay out as reasonable as possible. The integrated functions of the stereoscopic landscape elements can be fully realized urban landscape elements in the ecosystem of the nutrient cycle, resulting in considerable ecological benefits but can also serve as noise reduction dust, wind and sand, micro-climate, adjust the temperature and humidity. Besides that urban pollution problems will be alleviated, the urban air quality will be improved urban, and it will bring considerable ecological benefits.

1.3 Scenic landscape can improve the living environment and be social-efficient

Reasonable application of landscape elements in the landscape is conducive to implementation of landscape greening works, to help realize the sustainable development of eco-environment between human and the city, is conducive to boosting quality of urban residents and the construction of the urban civilization, as well as optimization of the quality of the urban environment to provide a healthy living environment for urban residents, can to some extent improve the residents' quality of life. Urban landscape construction is not only able to provide a suitable place for the residents to do all kinds of outdoor activities, to enrich their daily lives, but also can attract tourists to promote tourism. People will feel the charm of nature in the landscape, and then leading them to the revered nature for nature conservation.

1.4 The application of landscape elements can meet the needs of residents for green plants and be landscape-efficient

The plants in scenic landscape are structured well and extremely beautiful, which can not only increase green areas, but also enhance the viewing pleasure.

Different colored and shaped vegetation in different places is more artistic and ornamental than littering individuals. And then can be a great extent to meet the viewing needs of the residents.

2 THE PRINCIPLES OF THE LANDSCAPE ELEMENTS' APPLICATION IN SCENIC LANDSCAPING

2.1 *Application of landscape elements should adjust to local conditions*

From the characteristics of climate and the city's architectural style, the landscape designers need to configure the landscape plants scientifically in light of the local conditions, select the plants whose growth habit can meet the demand of the landscaping needs, while also ensuring selection of plants are adapted to the local climate and environment. From the perspective of the overall landscape design, the application of landscape elements should pursuit not only higher ornamental but also ecological value, with the consideration of rational combinations of color and variety of plants, and other buildings, rockery fountains.

2.2 *Application of landscape elements should follow the natural law and overall priority*

Landscape configuration must be consistent with the nature of landscape development, besides that overall priorities and the appropriate configuration are also important principles of landscape elements. The designer should comprehensively take the surrounding natural landscape, other buildings, the construction environment into account for the rational allocation of landscape greening. As the seasons change, growth of the plants exhibits different characters, and with the consideration of the color characteristics, designer may mix the landscape elements together, such as, flowers, grass, tress, which present differently in different seasons, so reasonable mix can extend the viewing period of landscape vegetation.

2.3 *Focus on landscape artistically and functionally*

In order to make people feel good, visual landscape and beauty, the designers combine the natural form, artificial form and their inspiration, and make full use of landscape elements to design the landscape greening. They should focus on landscape artistically and functionally, but without flashy design. For ex-ample, design the garden rode, besides the plane ge-ometry, the strict control of its safety must be taken into account. Also it should be in strict accordance with criterion for building materials, attaching im-portance to the road's structural strength standards, safety regulations, use of convenience, and both use-ful and beautiful landscaping is the only successful landscaping design.

2.4 *Stick to the sustainable introduction of multi-party system and clear green targets*

Besides the sustainable ways at technical level are proposed, the design of the landscape greening need to be perfected at system level, which have to introduce the multi-parties and systems, follow the fair principles and ensure the reasonable allocation, so as to let all kinds of factors in the design appears to complement each other. The use of specific digital indicators needs measuring the ecological benefits in the landscape design. For example, green design of lawn can be selected, through the analysis of greening of specific indicators, such as per capita green rate, green coverage rate, to choose a simple lawn mode or a variety of plant combination model.

2.5 *Respect the local context to choose the landscape elements*

Not only the ecological benefit should be taken into account, but also the inheritance cultures should be taken into during the city landscape design. The modern landscape design can mix the reinforced concert glass and other new building materials with some traditional landscaping elements, such as , bridges, water, lotus and bamboos. In the ancient landscape garden it also can show a perfect interpretation of the new era of vitality and the realize the inheritance of the traditional culture.

3 THE PROBLEMS IN THE APPLICATION OF LANDSCAPE ELEMENTS IN GREENING

3.1 *The greening structure is single, and the space utilization is not high*

At present, the landscape greening is mainly grassland, and the kinds of vegetation is relatively less, so the structure of greening is too simple. Landscape elements are not coordinated to guarantee diversity of landscape plants. Widely cultivated grassland expanded the green area and enhanced the greening effect, but did not achieve the desired goal of ecological benefit. There are more grasses than trees, so the single space does not highlight the landscape greening of space utilization level and does not fully reflect the artistic landscape.

3.2 The application of landscape elements is lack of flexibility and the landscape is lack of vitality and is in single color

There are some errors in the selections of landscape vegetation diversity, because the designers do not fully understand the diversity of vegetation. Though there are a wide variety of plants, but it lacks the vitality. Designers do not use the landscape elements in the landscape design, but simply list them together, without optimal landscape configurations. Nowadays many designers often ignore the layers and colors in landscape greening. When there is no novelty with the seasonal color transformation and landscape is in single color, only the green plants are selected. As it may make viewers aesthetically fatigue.

3.3 The application of landscape elements is lack of local characteristics and is lack of the local heritage context

The greening design of many cities in China is too dependent on foreign case design, and the designs of landscaping do not adjust to local conditions. It also lacks the awareness of innovation. The traditional culture is not continued and inherited by the application of landscape elements. Only with the lawn combined with the fountain, the pattern of landscape greening is the same without new ideas. The landscape elements with traditional characteristics are not applied in the landscape greening, so as it is lack of oriental characteristics.

3.4 The design of landscape is not environmental and is high consumption

Advancing urbanization in China seek economic benefits excessively and ignore the environmental protection, which is not only at expense of environment and ecological benefits to some extend, but also lack of long term vision for overall planning. Landscape design is aesthetic than being ecological, ignoring the ecological benefit. The application of landscape elements, which is low benefit and high consumption, into the landscape design so it will not fully take the ecological balance.

3.5 The management of landscape construction is inadequate

When constructing the landscape greening, the construction units lack enough management and the refined attitude of excellence. Data management and project quality cannot be ensured, the management of maintenance management of construction and acceptance is often overlooked by the construction units. There is no timely and careful maintenance, which increases the difficulty of landscape plant survival.

4 THE CORRESPONDING COUNTERMEASURES TO OPTIMIZE THE LANDSCAPE DESIGN

4.1 Designers should start from the whole, and use the landscape elements scientifically and reasonably

Designers need to inherit the regional context, based on the local social and cultural background, and mix the design of landscape greening, regional characteristic with the building environment around the garden skillfully, and then to design the landscape elements which cannot only meet the functional requirements functionally and aesthetically, but also fully show the local culture and unique artistry. In the landscape configuration, the designers should take the plant growth and local climate into account, and instead of the random plant configuration and blindness of design, they should realize the practical design scientifically and to enhance the ecological benefits.

4.2 Landscape greening should take advantage of the local tree species

Local plants are not only the landscape elements with excellent visual and ecological effect, but also can enhance the visual sensory effects and increase the ecological effects of the landscape. In the local environment, compared to other alien species, the local trees species are with better economy, adaptability, stress-resistance and ornamental. Besides that it can adapt to the local climate and soil environment, with higher survival, with which more comprehensive functions of vegetation are ensured to environment protection. Then not only the ecological benefits of landscape is increased, but also the art of garden ornamental and entertainment is fully embodied.

4.3 The green space should be ensured as the foundation for the landscape greening

The designers should mix the plants reasonably according to its growth habit and seasons change, and also should consider the effective protection of the landscape plants. After taking into account of the people's full range of requirements to the green space and other landscape greening, the designers can apply the open, ornamental and tread lawn and greening facilities into the landscape construction. The administrators of the garden should improve their management consciousness, strengthen the management of plans, protect the landscape soil and plants health as environmentally as possible, cure plant diseases with less chemical agent, achieve the sustainable development of ecological environment.

4.4 The project quality, data management and maintenance management after the construction check should be guaranteed during the garden construction process

In the process of landscape construction, the responsibilities should be defined clearly, and the importance of qualitative control also should be defined, and the quality, schedule and the cost of construction should be directed carefully. At the same time, the safety and the civilization of construction should be guaranteed. In the construction stage, the construction units should supervise the engineering conscientiously, keep the construction records strictly and objectively, manage the each project approval procedures timely, manage the engineering material carefully, make sure the use of engineering materials to implement and the data management without error. The maintenance management of construction after acceptance is a key and a long-term task, and the most critical maintenance period is several months after the completion of the construction. During this time, enough water should be provided timely in order to make sure the relative balance of soil moisture above ground and underground, and to improve the plants survival rate and accelerate plants growth, to avoid water logging and drainage timely as to protect the plants. Besides the consideration of the age of the planted trees; it should timely cultivate and loosen the soil, weed, increase soil permeability and thermal insulation effect, enhance the soil temperature, encourage fertilizer dissolved, and promote plant root development.

5 CONCLUSION

Designers should give full play to the landscape elements, landscape ecological benefit, and social benefit. Although in the current landscape design, the structure of the greening is single and the space utilization is not high, the landscape is lack of vitality and local features, which is not environmental protective but high energy consumption, with inadequate management of garden construction and so on. The designers, however, can take the green area as the basic configuration of the green vegetation instead of random design, and take the advantage of the local trees, strengthen the construction management, optimize the landscape design, improve the city environment, and the living standards of the residents.

REFERENCES

[1] Hu Zexing. *Problems of the Garden Design Planning in the Small and Medium Cities* [J] *Modern Gardens*, 2014, (2):103–103.

[2] Gu Wenwei. *The Design Principles and Construction Schedules of the Landscape Greening* [J]. *Engineering and Architectural Design*, 2013, 980:31–32.

[3] Chen Hanfang, Lou Wei. Construction management problems and solving measures of the current landscape greening project. *Investment and Cooperation*, 2012, (6):274–275.

Information, Computer and Application Engineering – Liu, Sung & Yao (Eds)
© 2015 Taylor & Francis Group, London, ISBN 978-1-138-02717-6

Research on the relationship between technical & knowledge and cluster innovation performance

Peng Zhao & Fu Zhou Luo
School of Management, Xi'an University of Architecture & Technology, Shaanxi, Xi'an, China

ABSTRACT: With the rapid development of science and technology, the breadth, depth, and the efficiency of technological innovation and knowledge innovation have become two driving forces of cluster innovation performance in knowledge economics era. The article sets up the path model taking technology and knowledge as the explanatory variables and cluster innovation performance as explained variable on the basis of diffusion mechanism study of technology and knowledge. Then the theoretical model is fitted with a confirmatory analysis based on Baoji equipment manufacturing industrial cluster. The article concluded that the speed of technological innovation diffusion, knowledge spillover scale and knowledge integration efficiency have significant positive influence on cluster innovation performance. The ultimate goal of this research is to provide theoretical foundation for cluster innovation factor identification and innovation ability promotion.

KEYWORDS: industrial cluster; innovation performance; technology innovation and diffusion; knowledge spillover and integration; path fitting.

1 INTRODUCTION

Innovation is not only the decisive force in support of cluster sustainable development but also the important factor and inexhaustible power for cluster to gain competitive advantage. Under the background of knowledge economy, the open innovation environment brought by world's industrial division of labor and the global manufacturing network put forward higher requirements on the cluster innovation. People pay more attention to the accumulation of the technology, resource acquisition and smooth information channel. Then cluster innovation network model emerged. Cluster nodes realize resource sharing, complementary advantages and risk-sharing. Innovative resources, information, technology and knowledge flowing are flowing in cluster network organization, which are blood flowing in the skeleton of cluster network organization.

2 VARIABLE DESIGN AND MODEL BUILDING

Cluster innovation is essentially innovative behavior of network nodes using cluster network relationships. Network nodes achieve innovation value through knowledge integration, thereby improving innovation performance. The most profound impact factors of cluster innovation performance are technology innovation diffusion and knowledge spillover integration. In fact, technology innovation is also the innovation process of knowledge.

2.1 Technology innovation diffusion mechanism

Technical potential difference caused by innovation enterprises using new technology and new knowledge and the same social and cultural background caused by close geographical position[1].

All of these inspire the catch-up effect among cluster enterprises. High liquidity of technical personnel increases network density and leads to technology diffusion. Universities and research institutes in the cluster provide targeted research to effective dissemination of new knowledge. In turn, new knowledge will be converted into final product and benefits by companies and companies achieve the transformation and application of knowledge. The value of knowledge can be achieved. Technology innovation and diffusion is generally issued by the innovation source enterprise, and then innovations are used in the production of products by technical invention, and finally spread to other companies by means of certain policies or intermediary organizations[2].

2.2 Knowledge spillover integration mechanism

Under the background of knowledge economy, ability to obtain, use and create new knowledge has become decisive factors in the survival and development of enterprise. Theory and practice studies have shown that: network mechanism of cluster innovation is closely connected with knowledge spillover integration. Knowledge spillover and integration is the process of knowledge spillover, share, access, absorption, integration, transformation and reconstruction. In the process of knowledge overflow and transfer, companies will capture knowledge with their own knowledge after recombination, then output new knowledge. Network not only improve the efficiency of the use of knowledge, but also promote the restructuring and optimizing the allocation of knowledge. So we can think that the scale and efficiency of knowledge spillover and integration directly affect the speed of the industrial cluster innovation and performance[3]. Innovation is competed by knowledge transmission, sharing, integration and turning into new knowledge. Asymmetry of knowledge is fundamental driving force for knowledge spillover and integration.

2.3 Model construction

According to the above analysis, we assume that the model is as follows:

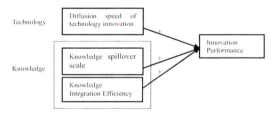

Figure 1.   Basic theoretical assumption and path diagram.

3   RESEARCH METHODS AND EMPIRICAL TEST

3.1 Questionnaire design and data collection

Questionnaire is distributed mainly through paper and electronic version using Likert Scale 5 segments except of firm size, ownership category and business life. Every latent variable are measured by 3–5 observed variables. Baoji equipment manufacturing industrial cluster is chosen research object for two reasons. First, as the western industrial city, Baoji is a typical industrial city with relatively strong industrial

foundation. Second, most enterprises of Baoji equipment manufacturing industrial cluster focused on the Economic and technological development zone and industrial park, which is convenient for sample data collection.

3.2 Data analysis

3.2.1 Reliability analysis
Whether a survey is carried out more than a single person or a single person repeated, the result should be roughly the same. We think that it has credibility[4]. In this paper, Cronbach's Alpha coefficient is used to measure the internal consistency of measuring items. Cronbach's Alpha coefficients generally can be expressed as:

$$\alpha = \left(\frac{n}{n-1}\right)\left(1 - \sum \frac{S_i^2}{S^2}\right) \qquad (1)$$

WU Minglong thinks it is acceptable for reliability at least 0.7 in the exploratory research generally[5].

Table 1.   Total reliability table.

Reliability Statistics	
Cronbach's Alpha	N of Items
.952	18

Reliability of total table is 0.952 (>0.8) and the minimum reliability of subscales is 0.959 (>0.8). It is believed that the accuracy of the scale is quite high.

3.2.2 Validity analysis
Validity Analysis refers to the degree of authenticity and accuracy of a study. This paper adopted confirmatory factor analysis to test validity.

Table 2.   KMO and Bartlett sphere validity.

KMO and Bartlett's Test		
Kaiser-Meyer-Olkin Measure of Sampling Adequacy.		.899
Bartlett's Test of Sphericity	Approx. Chi-Square	5940.275
	df	153
	Sig.	.000

It is generally believed that factor analysis will get better effect when KMO value above 0.7. KMO value of this case is 0.899(>0.7), so we think it's very suitable for factor analysis.

From table load factor, we see that each facet is formed different corresponding factors and factor loadings are greater than 0.6. Cumulative variance explained rate reached 90.452% and sufficient amount of information has been extracted, so structure validity passed by the test.

3.3 Path simulation and evaluation

According to 3 latent variables and 13 Observable Variables, we use Structural Equation Model (SEM) for data detection and path simulation With SPSS22.0 and AMOS 21.0. Path coefficients calculated as Fig.2.

If significance probability P value is less than 0.05, we think the path is significant. Its P value is greater than 0.05, so the effect could be insignificant as supposed.

Scoring and index standard comparison results show that model fit meets the requirements and model fit well.

4 CONCLUSIONS

Model fitting results show that every assumption on the path gets the support. Three paths have been empirically supported. Technology and knowledge innovation are significant positive correlated with cluster innovation performance. It will have a significant effect if we accelerate technology innovation diffusion and increase the size of the knowledge spillover and improve the efficiency of knowledge integration. Cluster innovation performance is a multidimensional concept based on network organization and affected by network, technology and knowledge.

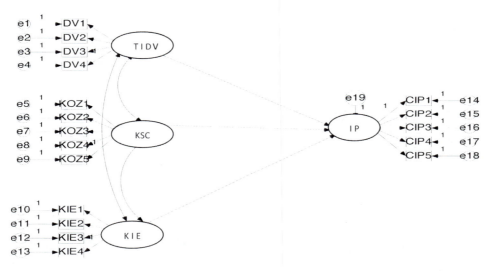

Figure 2.   Initial path.

Table 3.   Model fitting path coefficient.

			Estimate	P
Innovation Performance	<---	Technology innovation diffusion velocity	.213	.017
Innovation Performance	<---	Knowledge spillover scale	.421	***
Innovation Performance	<---	Knowledge Integration Efficiency	.187	.013

Table 4.   Model fitting standard and index scores.

Index		GFI	RMSEA	IFI	TLI	CFI	NFI
Standard	Fair	>0.7	<0.1	>0.7	>0.7	>0.7	>0.7
	Better	>0.9		>0.9	>0.9	>0.9	>0.9
Score		6.103	0.736	0.154	0.888	0.869	0.887

REFERENCES

[1] WANG Bo. Study on Efficiency of high-tech industry innovation activities based on new network model DEA [D]. University of Science and Technology of China, 2014: 41–42.

[2] LIU Yajun. Research on influence among Enterprise intellectual capital, absorptive capacity and innovation culture and technology innovation performance based on the research of manufacturing[D]. Tianjin University, 2010:77–79.

[3] YUAN Zhigang. Research on network, social capital and cluster life cycle—A new economic sociology perspective[M]. Shanghai: Shanghai people's publishing house, 2005.

[4] LING Wenxuan, Fang Liluo. Psychological and Behavioral Measures [M]. China Machine PRESS, 2003: 111–113.

[5] WU Minglong. SPSS Statistical Applications of Practice [M]. Beijing: science press, 2003: 237–239.

Information, Computer and Application Engineering – Liu, Sung & Yao (Eds)
© 2015 Taylor & Francis Group, London, ISBN 978-1-138-02717-6

Risk analysis of Guizhou wind power project based on low-carbon economy

Li Min Jia & Xun Cao

College of Economics and Management, China Three Gorges University, Yichang China

ABSTRACT: With the development of social economy, energy shortage and environmental pollution problems have become increasingly serious, so the development mode of low-carbon economy emerges, as the times require. As a kind of renewable clean energy, the wind power has become an important means of developing low-carbon economy. From the perspective of low-carbon economy, this paper expounds the connotation and signification of low-carbon economy, then analyzing the present situation of wind energy resources and wind power development in Guizhou province, and evaluates the Risk of Guizhou wind power development project. The conclusion of this paper can provide a basis for the feasibility study on wind power.

KEYWORDS: low-carbon economy; wind power development; risk analysis

1 THE CONNOTATION AND SIGNIFICATION OF LOW-CARBON ECONOMY

Low-carbon Economy is a kind of economic developing mode which takes low energy consumption, low environmental pollution and low greenhouse gas emissions as its foundation. It appears in the context that the problems facing mankind such as energy shortage and environmental damage are becoming increasingly serious.

At the very critical juncture of building a moderately prosperous society in all respects, China has been facing serious energy demands and greenhouse gas emissions problems. In 2007, Hu Jintao clearly put forward several suggestions to promote the development of low-carbon economy at the meet of the Asia Pacific Economic Cooperation (APEC). Then in the year 2009, Chinese State Council proposed Action Plan of controlling CO2 emissions by 2020. A year later, in 2010, the pilot work of low-carbon economy development was officially started by the National Development and Reform Commission, which aimed to lay a foundation for reducing greenhouse gas emissions in China.

The development of low carbon economy has a very important significance on transforming the economic development mode, promoting the ecological civilization construction, and realizing the sustainable development of economy and society. First of all, China's resources and energy utilization efficiency is low, and the pattern of economic growth is inefficient, developing low-carbon economy helps to promote the optimization and upgrading of industrial structure, to quicken the development of emerging industries. The second, the development goal of

modern civilization is to realize a win-win ecological modernization with the coordination between economic development and environment protection, the development of low-carbon economy is the basic and powerful way to resolve the contradiction between the socio-economic development, the energy shortage crisis and the eco-environmental protection. Finally, clean energy such as wind and solar power is almost inexhaustible, developing low-carbon economy and invigorating new energy industries has been the foundation for the survival and development of human beings.

2 THE CURRENT SITUATION OF WIND ENERGY RESOURCES IN GUIZHOU

China is rich in wind energy resources. It has a total installed capacity of 1 billion kW, according to conservative estimates. The wind energy resources of China are widely distributed, but mainly concentrated in the southeast coast and the other areas, which are the northeastern, northern, and northwestern China. Located in the Yunnan-Guizhou Plateau, Guizhou has relatively poor wind energy resources, of which the total installed capacity is estimated to be 0.123 billion kW, but still has great value of exploitation and utilization.

In general, wind energy resources are relatively abundant in plateau area, as the wind speed, wind power density and effective utilization hours increase along with the altitude increases correspondingly. Guizhou is located in low latitude of the Yunnan Guizhou Plateau, mountainous areas accounts for 94% of the whole province area, with an average elevation of more than 2,000 meters. In addition, Guizhou has obvious monsoon climate characteristics, the temporal and spatial distribution of wind energy is rather complicated, leading wind directions are southwest wind in winter and marine monsoon in summer. But in general the planning wind farm of Guizhou belongs in low-speed wind farm. The distribution of wind energy resources in Guizhou is imbalanced that there are better resources in the west than in the East, while it's better in the midland than in the north and south. What's more, the distribution of High value area is relative scattered and complicated. To be detailed, the areas that rich in wind energy resources are the west, south and north-central of Bijie District; the middle and south of Liupanshui City; the north-central of Zunyi City; the middle of Guiyang City; partial of the central-eastern Qiandongnan Prefecture; the special borderland between Rongjiang County and Libo County; the northern Qiannan Prefecture; partial of the central Qianxinan Prefecture; and partial of Tongren City.

Because of the scarcity of resources, the severe weather conditions and the natural rugged terrain, Guizhou was once thought to be a province without wind energy resources. It was not until the year 2011 that wind power emerged in this area. In just over two years, the wind power industry of Guizhou province has developed sharply. By the end of 2013, the total on-grid wind power installed capacity of Guizhou reached 1,023,600 kW, not including the capacity under construction of more than 1,864,500 kW.

Table 1. On-grid wind power installed capacity of Guizhou in 2013.

Number	Name of the area	installed capacity (10 thousand kW)
1	Bijie District	54.3
2	Liupanshui City	9.5
3	Qiandongnan Prefecture	4.96
4	Qiannan Prefecture	25.65
5	Guiyang City	7.95
	Summation	102.3

3 THE RISK ANALYSIS OF GUIZHOU WIND POWER PROJECT BASED ON LOW-CARBON ECONOMY

3.1 *The risk identification for Guizhou wind power development project*

Due to the complex physical geographic environment, the Guizhou wind power development project is faced with many risks. Here we mainly analyze it from four aspects: first ones, policy risks, including carbon sequestration trade policy risk, power interconnection policy risk, electricity price policy risk and so on; the second, technical risks, including construction risk, operation risk, personnel risk and target risk; the third, economic risks, such as cost budget risk and interest rate risk; the last ones are the unpredictable risks, for example, force majeure risk, default risk.

1 Policy Risks

At the present stage, the government has enacted a great deal of policy to support the wind power project. Such as, Fifty percent VAT rebate policy, take the year the project starts making profit as the first taxation year, its income tax shall be exempted in the first 3 years, and be levied by 50% from the 4th year to the 6th year. In addition, there are preferential policies in many areas, including wind subsidies, power interconnection, electricity price and credit. However, when enjoy the benefits of the policies, we should take the potential risks into consideration, once the policies changed, the impact on the wind power development project would be severe, and even fatal.

2 Technical Risks

As the development of wind power energy started late in china, it is still at an early stage compared with developed western countries, and relies heavily on foreign technology. The increase of "equipment localization rate" does not solve the problem essentially. The Only way to get rid of the passive situation is to master the core technology.

In the design stage of wind power project, the main technical risk is the target risk. The aspects of technical design, the Micro site of wind power farm and equipment selection are critical to successful project, which determine whether the wind power project can achieve the stated objectives. In the construction period of wind power project, the main technical risks it faced are equipment installation and personnel safety. While in the operational phase of wind power project, the main risks are the safe operation of equipment, personnel safety and technological upgrading.

3 Economic Risks

The main risks faced by the wind power farm are the cost budget and the risk and the interest rate risk.

The decentralized development model of Guizhou wind power project determines that we cannot copy the cost budget of the centralized development model, as the increased cost due to the dispersion of wind farms should be taken into consideration.

The major source of wind power project financing is mainly from the loan. The volatility of interest rates determines the uncertain amount of project investment. If we didn't take preventive measures for potential changes in interest rates, the normally construction and operation of the project would be directly affected, thus the cost of construction, operation and maintenance would increase.

4 Unpredictable Risks

Besides, there are many unpredictable risks faced by Guizhou wind power development project, such as, force majeure risk, default risk and so on. Being placed in the plateau mountainous area, the fan equipment would be frozen or even collapse due to force majeure such as the snowstorm disaster.

3.2 *The risk analysis for Guizhou wind power development project based on low-carbon economy*

In this paper, we firstly construct the evaluation index system of wind power development project risks by using the fault tree analysis (FTA), then with the method of fuzzy comprehensive evaluation, we analyze the risks of the project. According to the maximum membership degree principle, we determine the risk rank of the project, and identify the serious risks that should be focused on.

3.2.1 *The construction of wind power development project risks evaluation index system*

The core of wind power project is the economy. All of the safety problems can be solved through technical means, while the balance of the interests of various parties requires higher-level strategic considerations. We determine the intermediate event (A1~A4) and the bottom event (B1~B16) of wind power development project risks (W) through the method of FTA, as shown in Table 2.

3.2.2 *The fuzzy comprehensive evaluation method based on fault tree*

First, we construct the pairwise comparison judgment matrix of the intermediate event according to the expert evaluation method. Then, using MATLAB software we calculate the index weight vector of the intermediate event with normalization processing. $\omega=(\omega1,\omega2,\omega3,\omega4)$ Table 3 shows the result of consistency check.

In the same way, we can check the consistency of the bottom event's judgment matrixes. The result of its combined weight coefficients is shown in Table 4.

Explanation of the scale: For W =1,3,5,7,9 ,the "1" represents "factor Ai is as important as factor Aj", "3" represents "factor Ai is slightly more important than factor Aj", "5" represents "factor Ai is obviously

Table 2. The evaluation index system of wind power development project risks.

index 1	index 2	index 3
project risks W	Policy risks A1	Industrial policy B1
		Trading mechanism B2
		Power interconnection B3
		Electricity price B4
		Tax policy B5
		Loan policy B6
	Economic risks A2	Interest rate risk B7
		Electricity price risk B8
		Cost budget risk B9
	Technical risks A3	Technical design B10
		Equipment installation B11
		Personnel safety B12
		Technological upgrading B13
	Unpredictable risks A4	Target risk B14
		Default risk B15
		Force majeure risk B16

more important than factor Aj", "7" represents "factor Ai is far more important than factor Aj", and"9" represents "factor Ai is extremely more important than factor Aj". RI=0.9

Under the consideration for the occurrence and consequences of wind power development project, we divide the risks into 5 levels: very low risk, low risk, moderate risk, high risk and very high risk. According to the risk level, we evaluate the probability of the bottom event, and draw the probability evaluation matrix, as shown in table 4.

Then by synthesis of normalization Matrix W and Matrix Q, we get Matrix A, the comprehensive evaluation matrix of the intermediate event (See table 5.).

Above all, we composite the Matrix A with the index weight vector of the intermediate event and finally we get the comprehensive evaluation matrix of the project, Matrix O.

$$O=\omega \cdot A== (0.2825\ 0.2368\ 0.2381\ 0.1837\ 0.0589)$$

Table 3. The judgment matrixes of the intermediate event.

W	A1	A2	A3	A4	ω	consistency check
A1	1	4	3	4	0.5304	λmax=4.2381 CI=0.0794 CR=0.0882<0.1
A2	1/4	1	1/3	1/4	0.0755	
A3	1/3	3	1	1/2	0.1606	It meets the requirement of consistency
A4	1/4	4	2	1	0.2334	

Table 4. The combined weights and probability evaluation matrixes of the bottom event.

Q		combined weight coefficients	the value of risk evaluation				
			very low	low	moderate	high	very high
A1	B1	0.04609176	0.4	0.3	0.2	0.1	0
	B2	0.05951088	0.2	0.3	0.2	0.2	0.1
	B3	0.08231808	0.3	0.2	0.2	0.2	0.1
	B4	0.15233088	0.3	0.3	0.3	0.1	0
	B5	0.05738928	0.5	0.2	0.2	0.1	0
	B6	0.13275912	0.4	0.3	0.2	0.1	0
	B7	0.0471875	0.2	0.2	0.3	0.3	0
A2	B8	0.01030575	0.2	0.3	0.2	0.2	0.1
	B9	0.01800675	0.2	0.2	0.3	0.2	0.1
	B10	0.0166221	0.2	0.1	0.3	0.3	0.1
	B11	0.01071202	0.3	0.2	0.3	0.1	0.1
A3	B12	0.09173472	0.3	0.2	0.2	0.2	0.1
	B13	0.02490906	0.1	0.2	0.3	0.2	0.2
	B14	0.0166221	0.1	0.1	0.4	0.3	0.1
A4	B15	0.0777222	0.2	0.2	0.3	0.2	0.1
	B16	0.1556778	0.2	0.2	0.2	0.3	0.1

Table 5. The comprehensive evaluation matrix of the intermediate event.

A	A1	(0.34414,0.27366,0.22872,0.12674,0.02674)
	A2	(0.20000,0.21365,0.28635,0.26250,0.03750)
	A3	(0.23793,0.179300,0.25323,0.21403,0.11551)
	A4	(0.20000,0.20000,0.23330,0.26670,0.10000)

We can see that for the Guizhou wind power development project, the probabilities of a very low, low, moderate, high risk and very high risk are 0.2825, 0.2368, 0.2381, 0.1837 and 0.0589, respectively. We can determine that the project is a less risky one, according to the principle of maximum membership degree. In addition, we can see the values of B2, B3, B4, B5, B6, B12, B15 and B16, the combined weight coefficients of the bottom event, are all bigger than 0.05 from table 4. That's to say, they are the serious risks that should be focused on.

3.3 The risk management for Guizhou wind power development project based on low-carbon economy

The risk analysis above shows that the serious risks are mainly concentrated in the aspects of the policy and unpredictable factors. For the unpredictable risks, not only should we strengthen preventive measures, but also avoid risks by transferring them to insurance companies.

In policy terms, first of all, the trading system and compensation mechanism of carbon sequestration should be established and refined as soon as possible, to develop the higher ability of Guizhou's carbon sequestration adequately. Carbon trading has two forms: One is that the government imposes caps on the emissions and charging CO_2 emission tax from carbon emissions units, then the revenue will be given to the carbon sink units as compensation. The other one is that the total amount of emissions is determined by the government, while the market determines carbon price, and through the establishment of climate exchange, the carbon sink units can trade directly with the carbon source unit. Secondly, it is necessary to establish an eco-compensation mechanism for wind power connection, as it can inspire the enthusiasm of them. Last but not the least, the government should continue to implement other preferential policies of wind power, to ensure the long-term development of wind power enterprises.

Moreover, in core technology terms, we should enlarge the research and development efforts and speed up the industrialization progress, to reduce technology dependence on developed countries. And in economic terms, we should strengthen market research, cost control and financial management, so as to improve our competitiveness.

4 SUMMARY

In this paper, based on the low-carbon economy, we firstly introduce the situation of wind energy resources and wind power development in Guizhou province, then discuss the LEBs of Guizhou wind power development project, lastly, we analyze and evaluate the project's potential risks in detail. The study can provide decision-making references for Guizhou wind power project.

Information, Computer and Application Engineering – Liu, Sung & Yao (Eds)
© 2015 Taylor & Francis Group, London, ISBN 978-1-138-02717-6

The research on competitive and complementary of equipment and rubber plastics manufacturing industry

Qiang Feng
School of Management, Donghua University, Shanghai, China
School of Management, Ningbo Institute of Technology, Zhejiang University, Ningbo, China

ABSTRACT: Competitive cooperation theory as a concept has become a model and dynamic mechanism, and to combine competition with cooperation in order to enhance the enterprise the competitive advantage. In this paper, the formation of industrial clusters and 20 manufacturing industries as the analysis object, the data are collected from every region of the calendar year statistical yearbook. Selection of industrial enterprises above designated size output between 2000 and 2012, considering the price factor, calculated using the 2000 constant prices. Use of D–S model, using Eviews6.0 software processing. Choose the Shanghai area as the denominator. Study on the competitive and complementary relationship of the Yangtze river delta and the Zhujiang river delta region of communication equipment, computer and other manufacturing and electrical machinery and equipment manufacturing industry.

KEYWORDS: Competitive and complementary, General Equipment, Special Equipment, Rubber Plastics, Manufacturing Industry.

1 INTRODUCTION

Dendrinos–Sonis model (hereinafter referred to as D–S model) application is more mature in abroad. By the earliest, Dendrions and Sonis (1988) used to study the population of the competitive and complementary problem [1]. Subsequently, Hewings et al. (1996) also used to study the population problem in the United States, and use the model to predict the regional differences. Thus the model gradually came to be known as D–S model [2].

Due to the regional competition process that can be seen as a market, in the process, economic activity or factors of production are distributed in each area. Therefore, some scholars introduced the model in the analysis of competition and complementarity of economic activity, and is used to measure the competitiveness of a regional system. Magalhaes [3] (1999 used the model study the North-East Brazil and the United States the great lakes region of the economy. Jackson and Sonis [4] (2001) on the analysis of the share in the process of social economy, introduce the random item forecasting model, the further development of D D–S model. Jaime Bonet [5] (2003), Suahasil Nazara et al. (2000) [6] and Richard Healy and Randall w. Jackson [7] (2001), Yiannis Kamarianakis, agelis Kaslis [8] (2005), Miguel a. Marquez Geoffery J.D.H ewings [9] (2003) and then also USES the model to study economic problems at home and abroad.

The application of D–S model in domestic also belongs to the early stage. Zhu-lie (2008) based on D–S model to guangxi as an example to study the regional competition and complementary relationship in the process of regional economic growth [10]. Zhao-yu (2009) based on D–S model with five major economic zones in guangxi as an example, the performance of regional innovation competition and complementary relationship between the study of [11]. Su-fanglin [12] (2010), such as D–S model is used to analyze the competition in the complementary relationship between industry of zhongyuan urban agglomeration. Subsequently, and the model was applied to China's 30 provinces domain competitive and complementary of low carbon consumer behavior research.

2 THE THEORY METHOD D–S MODEL

D–S model for the system as the research object, the basic idea is to assume that a regional system consists of several mutually dependent child areas, a sub area of share change under the influence of other areas, all child area share changes to zero. Areas within the system, one is the regional share growth may lead to another is the region's share of the growth or decline. If in two children these variables are negative correlation between areas is considered to be competitive relationship between two sub areas. If these variables are related, there is a complementary relationship.

Assume that a large area economy is made up of n independent sub area. Definition $y_i(t)$ is moment area in the economy as a whole share of the industry, so the whole regional economy system the industry share of the distribution function is:

$$Y(t) = [y_1(t), y_2(t) \cdots, y_n(t)], \; i=1,2 \ldots, n, \; t=1,2 \ldots, T \quad (1)$$

Formula (1) can be sees as a distribution dynamics of discrete systems, the available:

$$y_i(t+1) = \frac{F_i[y(t)]}{\sum_{j=1}^{n} F_j[y(t)]}, i, j = 1, 2, \cdots n; t = 1, 2 \cdots T \quad (2)$$

Among them, $0 < y_i(0) < 1$, $F_i[y(t)] > 0$, $\sum_i y_i(0) = 1$, $F_i[y(t)]$ is a positive definite function. It means t time regional comparative advantage in Time and space.

Choose a reference area as the denominator, remember the first area to get:

$$G_j[y(0)] = \frac{F_j[y(0)]}{F_1[y(0)]}, j = 1, 2, \cdots n$$

Formula (2) can be rewritten as

$$\begin{cases} y_1(t+1) = \dfrac{1}{1 + \sum_{j=2}^{n} G_j[y(t)]} & j = 1, 2 \cdots, n \quad (3) \\ y_j(t+1) = y_1(t+1) G_j[y(t)] \end{cases}$$

The definition $G_j[y(0)]$ for the following Cobb–Douglas functional form:

$$G_j[y(0)] = A_j \prod_k y_{kt}^{jk}, j = 1, 2 \cdots, n; k = 1, 2 \cdots n$$

And then make it for the log-linear form, get the following models:

$$\ln y_j(t+1) - \ln y_1(t+1) = \ln(A_j) + \sum_{k=1}^{n} a_{jk} \ln y_k(t), j = 1, 2 \cdots n \quad (4)$$

Among them: $y_j(t+1)$ is a period j t moments after regional industry index of relative share of the gross

weight in all area index; $y_1(t+1)$ refers to the area as the denominator in t time lag of industry index; A_j means J region in t time location advantage; A_{jk} is elastic coefficient, namely for correlation coefficient.

Formula (4) can use the maximum likelihood method to estimate the parameter values. When the elastic coefficient of symbol timing, it means regional I and k is a complementary relationship. K industry share increased by 1% of the region also will lead to increased regional j industry share in percentage points. When the elastic coefficient of symbol is negative, the said regional I and k is a competition. Share increase k industry of the region will lead to a drop in share area j industry. Mathematically, the greater the relationship between two regions the more stronger it is.

3 CALCULATION AND ANALYSIS

3.1 Data selection and processing

In this paper, the formation of industrial clusters and 20 manufacturing industries as the analysis object. All the data are from every region of the calendar year statistical yearbook. Selection of industrial enterprises above designated size output between 2000 and 2012, considering the price factor, calculated using the 2000 constant prices. Specific calculation was carried out using Eviews6.0 software processing.

3.2 Calculation and analysis

By theoretical analysis that, in the use of D–S model, needs to select a reference region as the denominator, for the first area. The choice of the denominator is very important, different ways of choosing can directly affect the whole calculation results. In theory and practice both at home and abroad at present has not unified standard, the selection of common reference objects have to share is the largest, smallest or share in the middle. In this paper, considering the Shanghai area being in an important position in manufacturing, and its development direction is in the process of manufacturing industry development in the country has an indicator function, so choose the Shanghai area as the denominator. Horizontal column in each table for the independent variable, transverse bar as the dependent variable, and lists only by 10% significance level value of the test. In the concrete analysis of each industry, the first is to analyze the reliability of the calculation results, and then a concrete analysis of the industry competition and the complementary relationship between the regions. Then the formation of industrial cluster region, according to the geographical area of Yangtze river delta and pearl river delta were analyzed.

4 MAIN IMPLICATIONS

Table 1 shows the General Equipment Manufacturing industry's most competitive is between Shanghai and jiangsu, the elastic coefficient reached 1.5133. Complementary to the strongest is between Shanghai and guangdong, the elasticity coefficient is 1.3152. Yangtze river delta region, is a kind of competition. Strong competition between Shanghai and jiangsu, between Shanghai and zhejiang, jiangsu and zhejiang is a kind of weak competition. The formation of industrial cluster region is Shanghai, jiangsu and zhejiang, which suggests that the industry competition is mainly embodied in the Yangtze river delta region.

Table 2 shows the competitive and complementary to each other of the Special Equipment Manufacturing industry's are not strong, and of the elastic coefficient of mutual between the two regions are between 1 and 1. In the Yangtze river delta, competitive relationship exists between Shanghai and zhejiang, Shanghai and jiangsu, jiangsu and, but

Table 1. The analysis results on d-s model of general equipment manufacturing.

Region	Shanghai	Jiangsu	Zhejiang	Guangdong
Shanghai		−0.1839	−0.254	0.3339
Jiangsu	−1.5133		−0.8097	−0.7318
Zhejiang	−0.8019	−0.6093		−0.2682
Guangdong	1.3152	−0.3907	−0.1903	

Table 2. The analysis results on d-s model of special equipment manufacturing industry.

Region	Shanghai	Jiangsu	Zhejiang	Guangdong
Shanghai		−0.3349	−0.6759	0.0603
Jiangsu	−0.6909		−0.2866	−0.8543
Zhejiang	−0.4363	−0.0641		−0.1457
Guangdong	0.1272	−0.9359	−0.7135	

Table 3. The analysis results on d-s model of rubber and plastic products industry.

Region	Shanghai	Jiangsu	Zhejiang	Guangdong
Shanghai		0.7400	−0.2041	0.3679
Jiangsu	1.2081		0.0326	−0.1163
Zhejiang	−4.5628	0.3149		−0.8837
Guangdong	2.3547	−1.3149	−1.0326	

the intensity of the competition is not large. Form of industrial clusters in the industry in Shanghai and jiangsu, which means that the industrial cluster is a kind of weak competition, mainly reflected in the Yangtze river delta region.

Table 3 shows the Rubber and Plastic Products industry's competition between Shanghai and zhejiang, and the elastic coefficient reached 4.5628. The second is between jiangsu and guangdong, and the elasticity coefficient reached 1.3149. Then is between zhejiang and guangdong, its elastic coefficient reaches 1.0326. Complementary to the strongest is between Shanghai and guangdong, elasticity coefficient is 2.3547. In Yangtze river delta region, strong competition is between Shanghai and zhejiang, is a strong complementary relationship between Shanghai and jiangsu, jiangsu and zhejiang is weaker, the complementary relationship between the elastic coefficient is 0.3149. Formation of industrial clusters is the industry of Shanghai, zhejiang and guangdong, Shanghai and guangdong is strong complementary relationship, zhejiang and guangdong is strong competitive relationship. In other words, the internal competition in the industry in Yangtze river delta is mainly between Shanghai and zhejiang, between the Yangtze and pearl river deltas, chiefly competition between guangdong and zhejiang.

REFERENCES

[1] Dendrinos D, Sonis M. 1988. Nonlinear Discrete Relative Population Dynamics of the US Regions, *Journal of Applied Mathematical Computations* 25:265–285.

[2] Hewings G.J.D., Sonis M., Cuello F.A, and Mansouri F. 1996, The role of regional interaction in regional growth:competition and complementarity in the US regional system, *Australasian Journal of Regional Studies* 2:133–149.

[3] Magalhaes, A., M. Sonis, and G.J.D. Hewings. Regional Competitionand Complementarity Reflected in Relative Dynamics and Growth of GSP: A Comparative Analysis of the Northeast of Brazil and the Midwest States of the U.S. REAL Discussion Paper 99-T-8. Urbana, Illinois, University of Illinois.

[4] Jackson,R. and M. Sonis. 2001. On the spatial decomposition of forecasts, *Journal of Geographical Analysis*, 33:58–75.

[5] Jaime Bonet. Colombian regions: competitives or complementaries? J. REAL Discussion Paper 03-T-25. Urbana,Illinois,University of linois.

[6] Suahasil Nazara, Michael Sonis and Geoffrey J.D. 2000. Hewings. Interregional competition and complementarity in Indonesia. REAL Discussion Paper.

[7] Richard Healy and Randall W. Jackson. 2001. Competition and Complementarity in Local Economic

Development: A Nonlinear Dynamic Approach.Studies in Regional & Urban Planing.

[8] Yiannis Kamarianakis,Vagelis Kaslis. 2005. Competition-complementarity relationships between Greek Regional Economies.ERSA Conference Papers.

[9] Miguel A. Marquez, Geoffrey J.D. 2003. Hewings. Geographical Competition Between Regional Economies: The Case of Spain. The Annals of regional Science, Springer-Verlag.

[10] Zhu-lie. 2008. An empirical study on the regional competition and complementary relationship of Regional GDP share in the process of evolution, *Journal of The Development of Society and Economy.* 8:77–80.

[11] Zhao-yu. 2009. Complementary competition analysis of regional innovation performance, *Journal of The Northern Economy.* 10:19–20.

[12] Su-fanglin,Hou-xiaobo, 2010.Based on D - S model of the central plains urban agglomeration industry development, *Journal of Hunan College of Finance and Economics* 10:89–92.

Information, Computer and Application Engineering – Liu, Sung & Yao (Eds)
© *2015 Taylor & Francis Group, London, ISBN 978-1-138-02717-6*

The expression and analysis of C- fos protein, Sp1 protein in liver cirrhosis, liver cancer tissue

Jian Guang An

Baicheng Medical College, Baicheng, China

ABSTRACT: C-fos protein, Sp1 protein may play a key role in the occurrence and development of liver cirrhosis and liver cancer, they may also play a vital role in the regulation of gene transcription level, C-fos protein may also closely related to the pathogenesis of liver cirrhosis.

KEYWORDS: Liver cirrhosis, Liver cancer, C-fos protein, Sp1 protein

1 INTRODUCTION

The occurrence and developmental mechanism of liver cirrhosis and liver cancer are complex. It involves multiple gene regulation and multiple signal pathway, gene transcription regulation which may play a key role. Over the years, we observed the expression of transcription factor of C-fos protein, Sp1 protein in liver cirrhosis and hepatocellular carcinoma tissue. Now let us report the results and analyze the significance.

2 MATERIALS AND METHODS

Homochronous puncture or surgical obtained 141 pathologic specimens, among which, 53 specimens of hepatocellular carcinoma tissue, 53 specimens of tissue adjacent to carcinoma, 25 specimens of cirrhotic tissues, 10 cases of normal liver tissue. Candidate standard: clinical and pathological data integrity; the preoperative treatment of tumor such as radiotherapy and chemotherapy was not given. Sp1 protein and C-fos protein expression of various kinds of tissue samples were evaluated by immunohistochemical SP method, specific steps were taken strictly following the kit instructions. Determination standard of protein expression results: if light yellow particles, brown yellow particles, brown particles are observed in the cytoplasm, the result is positive, according to the proportion of positive cells, it is divided into four levels:< 10% is 0 point, 10%–30% is 1 point, 30%–60% is 2 points, > 60% for is 3 points, ; Based on the staining degree, it is divided into three levels: from light yellow to tan, becoming brown, it is, in turn, rated as 0–2 points. Total score = positive cells percentage grade ×staining degree scores, ≤3 points is determined negative (−), 3–6 points is determined positive(+), >6 points is determined strong positive(++). Using SPSS13.0 statistical software for statistical processing, adopting two or more nonparametric independent samples and t test between groups, it has significant difference when $p \leq 0.05$.

3 RESULTS

The expression of C-fos protein, Sp1 protein in normal liver tissue, cirrhosis of the hepatic tissue, the tissue adjacent to carcinoma, tissue of liver cancer are shown in Table 1.

4 DISCUSSION

Liver fibrosis and liver cirrhosis, liver cancer are continuous progress of developing illness .Ap-1 is a kind of immediate early genes encoding's transcription factor, its constituent members include C-fos family's C-fos, FosB, Fra-1, Fra-2 and C-Jun family's C-Jun, junB, JunD. Proto-oncogene c-fos positioning in human long arm of chromosome 14, the oncoprotein which share a common encoding forms nucleoprotein complex. It has specific binding for gene regulation sequence. Yingchao Zhang and other researchers' study showed that the C-fos in cirrhosis hyperplastic nodule and in liver cancer tissue are excessive expression, and its level is associated with liver cancer cell differentiation degree, it can be used as an evaluation index of malignant degree of liver cancer. C-fos expression positive rates in liver cancer and tissue adjacent to carcinoma in this data are significantly higher than that of normal liver tissue, prompt: C-fos may participate in the disease process of liver cancer; C-fos protein expression positive rates in the tissue adjacent to carcinoma are slightly higher than that of liver cancer tissue, the reason may be that the C-fos expression is an early event in liver cell cancer; Liver cells canceration process activate other cancer genes at the same time, it restrains C-fos transcriptional expression. From this data, C-fos protein's positive rate in cirrhosis of the

Table 1.   The expression of C-fos protein, Sp1 protein in hepatic tissue.

Tissue types	n	Negative		Positive		Strongly positive	
		E.g.	%	E.g.	%	E.g.	%
Normal liver tissues	10						
C-fos protein		8	80.0	1	10.0	1	10.0
Sp1 protein		10	100.0	0	0.0	0	0.0
Cirrhotic tissues	25						
C-fos protein		8	32.0	9	36.0 *	8	32.0*
Sp1 protein		22	88.0	3	12.0	0	0.0
Pericarcinous tissues	53						
C-fos protein		6	11.3	25	47.2 *	22	41.5*
Sp1 protein		32	60.4	13	24.5*	8	15.1*
HCC tissues	53						
C-fos protein		9	17.0	27	50.9*	17	32.1*
Sp1 protein		14	26.4	8	15.1*	31	58.5*

liver tissue is also significantly higher than that of normal liver tissue, prompt: C-fos may be involved in the pathogenesis of liver cirrhosis process. Jinjun Zhao and other researchers thinks that C-fos gene was undetectable during cell stationary growth period, but in this data C-fos gene also has a certain expression in normal liver tissue, which may be associated with liver tissue hyperplasia. The study of Jie Liu and other researchers also shows that c-fos gene has involved in the whole process of hepatocirrhosis turn to cancer.

Sp1 is the first cloned factor in the family of Sp/Smy transcription factors, which is a sequence specific DNA binding protein and can regulate GC sequence gene transcription of some promoters. It's important and essential transcription factors that could regulate cell survival, growth and angiogenesis of functional genes. So Sp1 is a kind of nuclear transcription factor which is closely related to the development of tumor, the excessive expression has the functions of positively regulating tumor proliferation and metastasis potential role. The data shows that Sp1 protein expresses slightly in liver cirrhosis tissues, there is no statistically significant difference compared with the normal tissue, prompt: Sp1 has no obvious correlation

with the cirrhosis of the liver. Tissue adjacent to carcinoma cancer has the risk of cancerization.

5 CONCLUSIONS

To sum up, C-fos protein, SP1 protein may have a certain role in occurrence and development of liver cancer, C-fos protein may be associated with the onset of cirrhosis of the liver.

REFERENCES

Iredale JP, Benyon RC, Rickering J, et al. 1998. Mechanisms of spontaneous resolution of rat liber fibrosis, hepatic stellate cell apoptosis and reduced hepatic expression of metalloproteinase inhibitors [J]. J Clin Invest,120(3):538. Wang X. Kan, B.Q. Wu, et al. 1995. Molecular Pathology Progress of Hepatitis Virus and Cancerization of Liver. Chinese Journal of Pathology, 24 (6)/;393–395 (in Chinese).
J. Feng, Z.H. Tang, R.G. Zhang, G.Q. Zang and Y.S. Yu, the study of expression of C-fos.EGFR in liver cancer and liver cirrhosis, Journal of Clinical Hepatology, 24(3):201–212 (in Chinese).

Information, Computer and Application Engineering – Liu, Sung & Yao (Eds)
© *2015 Taylor & Francis Group, London, ISBN 978-1-138-02717-6*

Evaluation of technological innovation capability of equipment manufacturing industry in Liaoning province

Yuan Yuan Guo & Xian Chen Jiang
School of Management, Shenyang Normal University, Shenyang, China

Jie Bai
School of Business English, Dalian University of Finance and Economics, Dalian, China

ABSTRACT: Through empirical analysis of technological innovation capability of equipment manufacturing industry in Liaoning Province, the paper reveals the basic situation of technological innovation capability of major equipment manufacturing industries in Liaoning Province, which can provide the basis and foundation for policies and recommendations that are used to develop and upgrade technological innovation capability of equipment manufacturing industry in Liaoning Province.

KEYWORDS: Liaoning Province, Equipment manufacturing industry, Technological innovation capability

The equipment manufacturing industry provides guarantee for national economic development and national security by manufacturing technology and equipment. The development of equipment manufacturing industry is the basis for industrialization and the key to improve the country's comprehensive competitiveness. Liaoning province has always been an important research and production base of national equipment manufacturing technology industry, which has a number of industries with huge scale of production and high-tech. Share of industrial output accounted for a larger proportion of the province's total economy. The equipment manufacturing industry is an important pillar industry of the province's economy, which plays an important role in the development of Liaoning province's industrial economy. However, there are many problems in the equipment manufacturing industry of Liaoning province, such as reliance on imports of core technology and key equipment, poor high-end products, industrial innovation backward, and weak market competitiveness, etc. The root is that the technological innovation capability of equipment manufacturing industry in Liaoning is weak and new products and new technology development cycle is too long. Simultaneously, the industrial development is still mainly relies on traditional equipment manufacturing industry with low technology and low end and the profit margins gradually shrank, which has seriously hampered the healthy and sustainable development of equipment manufacturing industry of Liaoning. For Liaoning, a province who treats industry as a leader, the weak technological

innovation capability is stagnating the province's economic development, which has a negative impact on the revitalization of old industrial bases in Northeast China

1 THE CONSTITUENTS OF TECHNOLOGY INNOVATION CAPABILITY OF EQUIPMENT MANUFACTURING INDUSTRY

Technological innovation capability is an important part of the enterprise's core competitiveness. Technological innovation capability is the more specific, more object-oriented theoretical extension of technology innovation. In 1980s, the concept of technological innovation capability was first explicitly proposed. Hereafter, the scholars from various countries have never stopped the research and study for its connotation and extension.

In the field of technical innovation capability, the two most representative foreign scholars are Larry and Barton. Both of them, from a different perspective, made specific and elementary study on the technological innovation capability. Larry (1981), from the perspective of organizational behavior, believes that the technological innovation of enterprises is constituted by organizational capacity, adaptability, innovation and the ability to obtain information technology. Barton (1992) by analyzing the subject of technological innovation activities considers a core element of enterprise technology innovation, which includes skilled personnel, technological innovation system, enterprise management capabilities and corporate culture. The research production of domestic scholars on technological innovation capability is also very rich. Zhou Yuping (2000) analyzes the implementation process of technological innovation so that she divides technological innovation capability into investment capacity in R & D, the ability to allocate researchers, the ability to achieve innovation and innovation output capacity. Liu Haitao (2007) argues that enterprise technology innovation capability is the comprehension of R & D, management, production, marketing and many other capabilities. Gao Cuijuan and Yin Zhihong (2012) divides the independent innovation capacity of SMEs into R & D capability of independent innovation, transforming ability of independent innovation and management capabilities of independent innovation.

Throughout the related researches of domestic and foreign scholars on technology innovation capability, despite different emphasis, most of the researches focus on technology innovation process functionality and try to make concrete and elements of technical innovation capability from the perspective of process, system, etc. The author believes that technological innovation is a dynamic evolution process in which its constituent elements are under the control of innovation dynamic mechanism. Technological innovation capability can be broken down into three major elements, namely, investment ability in technological innovation, the ability to achieve technological innovation and transforming ability of technological innovation. These three factors are respectively covering financial management capacity of corporate R&D funds, the ability to implement R&D activities including R&D personnel and the overall R&D capabilities and the ability to fund operations including market research for new product and new technology, production, marketing strategies and the flow of capital.

Table 1. Innovation evaluation index system of Liaoning's equipment manufacturing industry.

Technology Innovation Process	Internal Factors	External Factors
Technological innovation investment	Investment capacity of R&D funding	Government's scientific investment in the proportion of research funds
	The proportion of R&D personnel	
		Financial institutions' scientific investment in the proportion of research funds
	The proportion of R&D equipment	
Technological innovation achievement	Patent number	The participation of university research institutes and intermediary service organizations
	Market development capability of new products	
	The proportion of new products	Conversion rate of scientific and technological achievements of university research institutes
	The overall level of R&D personnel	
Technological innovation transformation	New product margins	
	New product sales rate	The support of government fiscal policy
	New product export rate	
	New product innovation rate	

2 INNOVATION EVALUATION INDEX SYSTEM DESIGN OF LIAONING'S EQUIPMENT MANUFACTURING INDUSTRY

According to the selection principles of technical innovation evaluation index, combining evaluation index system design of relevant research literature and the current development status of Liaoning's equipment manufacturing industry, in this paper the architecture of evaluation index system is designed for three one-level indicators, eleven two-level internal evaluation indicators and five two-level external evaluation indicators. As shown in Table 1:

3 EMPIRICAL ANALYSIS OF TECHNOLOGICAL INNOVATION CAPABILITY IN LIAONING'S EQUIPMENT MANUFACTURING INDUSTRY

According to variable values from factor score coefficient matrix and standardization, we can draw the factor scores of $6 \times 7 = 42$ sample data that come from the seven sub-sectors of Liaoning's equipment manufacturing industry in six years. Its mathematical expression and the coefficient matrix are shown in Table 2:

According to statistical measurement, to make weighted summary in accordance with corresponding weight designed by variance contribution rate of

Table 2. The average score and ranking of seven industry factors.

	F1		F2		F3		F4		F5		F6	
	SCore	N	SCore	N	SCore	N	SCore	N	SCore	N	SCore	N
H1	-9.75002	7	-8.31644	7	-1.64233	4	-2.48226	6	4.09598	1	1.13327	4
H2	-3.64801	5	-0.77924	4	-3.47556	6	0.59037	3	-0.69472	5	0.90282	5
H3	1.78159	2	8.01947	1	-1.1702	3	-0.95072	5	2.99834	2	3.68091	1
H4	0.47595	3	-3.75696	5	6.22153	2	1.51546	2	-2.95041	6	1.9831	2
H5	-4.16089	6	5.12384	3	-2.24702	5	9.35149	1	-4.79192	7	-1.35147	6
H6	15.00914	1	-6.61966	6	-7.44364	7	-0.31101	4	0.49902	4	1.8457	3
H7	0.29225	4	6.329	2	9.75722	1	-7.71335	7	0.84372	3	-8.19433	7

factors, the composite score and the final ranking of the unit can be calculated according to the following formula. As shown in Table 3:

$$S=0.1608F1+0.1472F2+0.1269F3+0.1089F4+0.080F5+0.062F6 \qquad (1)$$

Table 3. The composite score and the final ranking of seven industries.

Industry	Composite score	Ranking
Special equipment manufacturing	2.10	1
Instruments and cultural and office machinery manufacturing	1.74	2
Communication equipment, computers and other electronic equipment manufacturing	1.06	3
Electrical machinery and equipment manufacturing	0.85	4
Transportation equipment manufacturing	0.29	5
General equipment manufacturing	-0.11	6
Fabricated metal products manufacturing	-1.07	7

4 CONCLUSION

First, the input factor "F_2" of technological innovation resources from the three manufacturing – electrical machinery and equipment manufacturing, instruments and cultural and office machinery manufacturing and special equipment manufacturing – has a higher score with score>5, which illustrates that these three sectors have attached great importance to the input of technological innovation resource and they are stronger than the other four sectors in terms of the input of technological innovation resource.

Second, the conversion factor of technological innovation research belonging to instruments and cultural and office machinery manufacturing and transportation equipment manufacturing is positive and >5, which shows higher efficiency in research transformation of technological innovation process. The conversion factor "F_3" of technological innovation research scores<0, which belongs to communication equipment, computers and other electronic equipment manufacturing, general equipment manufacturing, special equipment manufacturing, electrical machinery and equipment manufacturing and fabricated metal products manufacturing. Among the several industries with positive score, the instruments and cultural and office machinery manufacturing has the highest score, which indicates that this sector has a higher technological R&D efficiency and is able to achieve more scientific research achievements and put them into practice. The last ranking is communication equipment, computers, and other electronic equipment manufacturing. Its factor score is far lower than other sectors, indicating that this sector has great difficulty in scientific research transformation and the less efficiency.

Third, the external support factor "F_4" of technological innovation attaching to general equipment manufacturing, electrical machinery and equipment manufacturing and transportation equipment manufacturing scores>0. Among them, electrical machinery and equipment manufacturing has the highest score. Compared to the other two sectors with positive scores, it is far ahead, which reflects the external power of dynamic mechanism of technological innovation. For example, the government agencies and financial service institutions provide strong support for the sector. The factor score values – communication equipment, computers and other electronic equipment manufacturing, instruments and cultural and office machinery manufacturing, fabricated metal products manufacturing and special equipment manufacturing – are<0, which illustrates the external dynamic of technological innovation is weak in these four sectors. Among them the lowest score is instruments and cultural and office machinery manufacturing, reflecting efforts to support this sector from the relevant government agencies and technical innovation service industry are very weak. However, the research conversion factor of instruments and cultural and office machinery manufacturing scores are the highest, indicating the strong support of technological innovation within the industry and the higher eventual conversion efficiency.

Fourth, the Non-R&D factor "F_5" of technological innovation – communication equipment, computers and other electronic equipment manufacturing, special equipment manufacturing, fabricated metal products manufacturing, instruments and cultural and office machinery manufacturing – scores>0. The factors of special equipment manufacturing and fabricated metal products manufacturing have higher scores, indicating that in the process of technological innovation these two sectors are not very high reliance on research capacity and rely more on the two indicators of product volume and market share. However, the Non-R&D factor of technological innovation scores<0, belonging to transportation equipment manufacturing, electrical machinery and equipment manufacturing and general equipment manufacturing. Combining with other factors' scores, it tells that they lay less emphasis on end links of technological innovation activities and put too much emphasis on R&D links, which is also unreasonable.

Fifth, the R&D factor "F_6" of technological innovation – communication equipment, computers and other electronic equipment manufacturing, fabricated metal products manufacturing, transportation equipment manufacturing, general equipment manufacturing and special equipment manufacturing –scores>0. And this factor of these sectors score or less. It shows that they all focus on R&D factor of technological innovation to the same extent. The special equipment manufacturing focuses most on investment of R&D resources and gets the highest score. The R&D factor of innovation – electrical machinery and equipment manufacturing and instruments and cultural and office machinery manufacturing – scores<0. The lowest score is instruments and cultural and office machinery manufacturing. Its effort to invest in the resources of technological innovation R&D activities are very low, indicating that its emphasis on technological innovation is not enough and should be corrected.

REFERENCES

[1] Wang Kun, Yuan Jing, Hotspot analysis and related revelation of innovation evaluation. Technical Economics and Management Research, 2012 (02):35–38.
[2] Li Qinqin, Liu Zhiying, Technological innovation index research on chinese provinces, *Science and Technology Progress and Decision-making*, 2012 (19):47–50.
[3] Liu Chunzhi, Nie Ying, Statistics and analysis of the technological innovation situation of liaoning's equipment manufacturing, *Journal of Shenyang Normal University (Social Sciences)*, 2006 (04):6–9.

Information, Computer and Application Engineering – Liu, Sung & Yao (Eds)
© *2015 Taylor & Francis Group, London, ISBN 978-1-138-02717-6*

The synergy mechanism research on diversified coal enterprises based on grounded theory

Xi Min Sun & Xi Luo
Xi'an University of Architecture and Technology, School of Management, Xi'an, China

ABSTRACT: In the past ten years, along with the rapid growth of diversification in China's large coal enterprise groups, problems like unclear industry synergy motivation, path and target also appear. To solve these problems, a case study of a Coal & Chemistry Industry Group is used to analyze the synergy mechanism of its diversified industries based on grounded theory. The study shows the association between the main and expanded coal enterprises business, containing three forms of association, which are business association, function association and political association, respectively. Different methods like risk diversification, economy of scope, economy of scale and market force are also used to promote the enterprise image, achieve excess profit and lower the cost of opportunity, production and operation.

KEYWORDS: Coal Enterprises; Diversification; Industry Synergy; Mechanism model; Grounded Theory

1 INTRODUCTION

Since 2003, along with rapid economic development of the huge demand for energy, China's coal enterprises through the "golden decade" and coal production from 1.667 billion tons in 2003 up to 3.66 billion tons in 2012. Not only the enterprise scale increases exponentially, business model from the past to the "coal mining, selling coal" specialized operation fission as "coal and non-coal" simultaneously diversified operation. According to 2013 statistics, the number of coal enterprises in China's top 100 enterprises is five, there are 13 coal enterprises in China's top 500 enterprises and they all implement diversified operation. The Coal Industry Association statistics show that the proportion of non-coal value up to 60% and the business scope covers coal, chemical industry, electric power, construction, machinery, building materials, metallurgy, transportation, infrastructure and many other downstream industries. The vertical integration development model taking coal as main has become the strategy choice of many coal enterprises.

2 RESEARCH METHODS AND DATA

The grounded theory was proposed by Glaser and Strauss in the 1960s, a kind of more effective research methods of quantitative theoretical analysis based on qualitative data, aimed at based on detailed information, from down to up, spiral circulation gradually raise the level of abstraction of concepts and its relationships through scientific logic, induction, comparison and analysis, eventually establish a substantive theory model [1]. Its outstanding advantage is not to make logical deduction to the assumption that researchers predetermined, but to summarize and analysis the data and naturally present to construct theory. So it is widely used in the field of qualitative modeling.[2] The research process is shown in Figure 1:

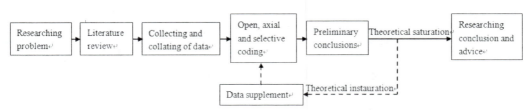

Figure 1. Research process of grounded theory.

Table 1. Categories in Open Coding.

No.	Category	Concept
1	Internal Market	mainly including: sales industry (1-6-4-5), sales way (1-7-4-6), sold to electric power plant (1-10-6-2) internal market ratio (3-10-3-2), etc.
2	Circular Utilization	specifically including: coal tar hydrogenation (2-6-3-1), coal tar processing (2-14-8-3), circulating economy chain (2-18-9-1), utilization of waste gas (2-5-2-3), etc.
3	Transaction Cost	specifically including: settlement business (4-5-4-3), cost calculation (6-5-9-1), etc.
4	Merge and Reorganization	specifically including: holding open securities (8-1-2-1), setting up Chang'an bank (8-3-3-1), sharing Happy Insurance (8-6-5-1), setting fund company (8-7-6-1), building Hanzhong Iron &Steel Company (9-3-2-1), etc.
5	Scarce Resources	specifically including: resources reserve (1-5-3-1), iron ores (9-9-7-1), mines resources (9-12-7-4), etc.
6	Risk Diversification	mainly including: anti-risk capacity (1-11-4-4), taking into account the multi-aspect (3-9-2-5), financing capacity (10-4-2-1), etc.
7	Economy of Scope	mainly including: coal chemical industry strategy (2-24-14-1), professional support (3-9-2-5), integration of coal and electricity (6-1-1-1), cement for building (9-7-5-2), etc.
8	Economy of Scale	specifically including: the increasing of economic indicator (1-12-4-7), key technology (1-13-7-1), scale merit (5-6-4-5), external market development (5-10-6-1), etc.
9	Market Force	mainly including: market concentration ratio (1-1-1-1), principal market (1-8-4-7), bargaining capacity (5-7-5-1), etc.
10	Resources Sharing	specifically including: natural resources (2-3-2-1), local resources (2-4-2-2), coal resources (2-9-4-3), informatization platform and logistics park (5-11-7-1), etc.
11	Differential Land Rent	specifically including: the coordination of motor transport (9-16-8-4), railway construction (10-7-8-1), etc.
12	Management and Control Mode	specifically including: unified management (1-3-2-4) and No.3-4-6-1, the coexistence of multi-equity (3-3-1-3) and (6-2-2-1), basic management (5-2-8-1), progressively responsibility and poles focus (5-4-4-2), etc.
13	Social Responsibility	specifically including: safe clean (1-13-7-1), clean utilization (1-13-7-2), etc.
14	Production Cost	specifically including: cost control (2-17-8-6), cost of purchasing (5-5-4-3), etc.
15	Stable Market	specifically including: advantage of region (5-9-5-3), Northern Shaanxi market in Xi'an (7-7-4-1), the base of building materials (9-6-5-1), etc.
16	Service Transformation	specifically including: new type of coal chemical (2-10-5-4), maintenance, transformation and manufacturing (3-6-2-2), etc.
17	Profitability	specifically including: loss status (2-16-8-5), (2-20-10-2) and(8-13-10-2), break-even (3-11-5-1), profitability (11-6-3-3), etc.
18	Business Association	specifically including: the incorporation of chemical companies (2-1-1-1), the development of coal and chemical (2-2-1-2), etc.
19	Function Association	specifically including: production and operation (1-4-2-2), internal maintenance (3-14-8-3), logistics channel (9-11-7-3), research platform (11-7-5-1), etc.
20	Political Association	specifically including: government requirement (3-1-1-1), the credit of state-owned enterprises (5-8-5-2), the cooperation between coal and electricity (6-6-8-2), etc.

3 RESEARCH PROCEDURE

3.1 *Open coding*

Data selection and analysis technology in grounded theory is critical to satisfy the promotion, reproducibility, accuracy, rigor and verifiability of research

Table 2. Four Relationships Based on Axial Coding.

Serial-number	The kind of relationship	The category of affecting the relationship
1	Diversified motivations	Risk Diversification, Economy of Scope, Economy of Scale, Market Force
2	The pathways of industry synergy	Internal Market, Resources Sharing, Circular Utilization, Transaction Cost, Merge and Reorganization, Scarce Resources,
3	The ways of industry association	Business Association, Function Association, Political Association,
4	The efficiency of industry synergy	Stable Market, Production Cost ,Profitability, Management and Control Mode, Service Transformation, Differential Land Rent, Social Responsibility,

result [3]. Therefore, at work complying with Strauss' [4] and others coding technology procedures rigorously is to ensure the research reliability and model validity.

3.2 Axial coding

Axial coding is to establish relationships among different categories which are divided in the open coding, through the type analysis. When associating, we need to analyze each category on the conceptual level whether there is a potential connection relations or not, so as to search for clues [5].

3.3 Selective coding

Selective coding is a process of comparing the core of selective category with the other categories to check their relationships and complete the category which conceptual development has not yet completed. [6]

3.4 The test of the theory saturation

The test of the theory saturation is a touchstone used to decide the time to stop sampling. Through the coding analyses for the reserved 26 key events, they find that the consequences are still according to the relationship of "Synergy Mechanism of Diversified Coal Enterprises Model" and new categories and relation have not been discovered, so we can think that the above-mentioned model is saturate.

4 CONCLUSION AND REVELATION

Synergy Mechanism of Diversified Coal Enterprises law the rules of "way-pathway-benefit". The business of coal have three ways of association that are business, function and political with others. In addition to this, with the help of the internal market, circular utilization, reducing transaction cost, resources sharing, merge and reorganization, acquiring scarce resources and other measures, coal business, to achieve synergistic effects, such as risk diversification, economy of scope, economy of scale and market force, etc, which will give some concrete benefits to the enterprise in the aspects of reduce the cost, obtain the profit, improve their image, etc.

REFERENCES

[1] Glaser B G, Strauss A. L. 1967. The discovery of grounded theory: Strategies for qualitative research. Chicago: Aldine.
[2] Glaser B G, Strauss A. L. 1967. The discovery of grounded theory: Strategies for qualitative research. Chicago: Aldine.
[3] Hesth H, Cowley S. 2004. Developing a grounded theory approach: a comparison of Glaser and Strauss. International Journal of Nursing Studies, 41(2):141–150.
[4] Strauss A L, Corbin J M. 1990. Basics of Qualitative Research: Techniques and Procedures for Developing Grounded Theory. Los Angeles: Sage Publications, Inc., 1–45.
[5] TAO Hou-yong, LI Yan-ping, 2010. LUO Zhen-xin. Formation mechanism of shanzhai model and its implications to organizational innovation. China Soft Science 11:123–135, 143.
[6] LI Zhi-gang, LI Xing-wang. 2006. Applying grounded theory in the models and determinants study of Mengniu's fast growth. Management Sciences in China 19(3):2–7.

Information, Computer and Application Engineering – Liu, Sung & Yao (Eds)
© 2015 Taylor & Francis Group, London, ISBN 978-1-138-02717-6

Research on students' innovative entrepreneurial training practice under the tutorial system

Zhe Xin Liu
Zhuhai College, Jilin University, China

Bing Han
Zhuhai Nanping Middle School, China

Zhe Li Liu
Nankai University, China

ABSTRACT: To carry out the tutorial system under the college students' innovative entrepreneurial training project is important for cultivating innovative talents in colleges and universities. In view of the present college education, creative ability training problems under the tutor system is discussed and several problems of cultivating the ability of college students' innovative undertaking. We learn from the problems and solutions of training mode and the practical forms, results and so on. Four aspects are analyzed and summarized respectively.

KEYWORDS: College students; Tutorial system; Innovative undertaking training.

1 INTRODUCTION

College students' innovative entrepreneurial training program includes three categories: training project innovation, entrepreneurial training projects and entrepreneurship practice. undergraduates in individual or team, independently complete the innovative research project design, the research condition of preparation and project implementation (academic) exchange, research report, achievements, etc., under the guidance of a tutor. Entrepreneurship training program in the undergraduate group, each student in the team is the process of project implementation the role of one or more specific, through the preparation of a business plan, feasibility study, simulation enterprise operation, participate in practice, writing a business report, etc. Entrepreneurship practice project is that the student team, supervisor in school under the guidance and mentor utilize the former innovation training program (or innovative experiment), which has a market prospect of innovative product or service to promote entrepreneurship practice activities on this basis.

2 PROBLEM STATEMENT

2.1 *The lack of perfect supervision and appraisal system and inadequate incentives*

Tutorial system under the current status of college students' innovative entrepreneurial training ability training mode is: students who can find a good topic selection determine the guiding teacher or teacher according to their own research projects or life after good research topic selection, to find a student organization team. Mentor work arrangement, students start training project of innovation and entrepreneurship, tutor job is in a state of free from the management, institute for mentor work no specific assessment and evaluation system, to student's training schedule is only a mid-term examination, lack of good supervision and evaluation mechanism. Most teachers do not like working as a tutor, and a portion of the teacher to the student guide is a state of casual, it seriously restricts the tutorial system implementation and development of colleges and universities.

2.2 *Propaganda work is not enough*

It is mainly reflected in the universities with not enough attention for training innovative undertaking projects, publicity is not in place. Tend to be college students' innovative entrepreneurial training work of project application, symbolic of the relevant documents sent to the department, teachers' box, interested in or responsible teachers will take the initiative to contact and communicate the relevant document to the students, but most of the teachers wouldn't ignore these things. Most of the students for the project of training is not very understanding and even do not know at all. No special project presentation and guidance, student participation and guidance teacher's enthusiasm is not strong, students do not know much about the related data of each teacher, do not even know each mentor research direction or expertise, thus affecting the enthusiasm of the students.

2.3 Lack of funding, limited experimental conditions

Although the documents stated in each college must have a special project grants, only three thousand Yuan of funds is used to support every project. It is not enough to complete a training of college student' innovation and entrepreneurship. Project, directed by himself, for example, the project of 5 students, need for market investigation, the completion of the project from Zhuhai to Guangzhou trip back and forth once you need eight hundred Yuan, other experiment material, such as textiles face, art supplies such as test materials, shortage of funds, and hard to due to a serious shortage of funds. And specialized laboratory construction is also relatively lags behind, which seriously limits the college students' innovative entrepreneurial training activities.

3 SOLUTIONS

3.1 Emphasis on the training program to promote the significance of talent training mode reform

Participate in the tutorial system under the various colleges and universities should fully attaches great importance to the college students' innovative entrepreneurial training program of talent training mode reform of the important function and meaning. Universities should be formulated feasible measures for the administration and supporting policies, will the daily management of college students' innovative entrepreneurial training program in undergraduate teaching management system.

3.2 College students' innovative entrepreneurial training program towards the talent training scheme and teaching plan

Participating college teaching management from the curriculum construction, student course selection, test, the results identifying, and credits should give policy support, flexible management, etc. To the training of innovative entrepreneurial project, elective courses at the same time to the organization construction and innovation training courses on innovative thinking and innovative methods, and project

management related to entrepreneurship training courses, business management, and risk investment.

3.3 Emphasis on the college students' innovative entrepreneurial training plan, the construction of the contingent of tutors

Participating colleges and universities should formulate the relevant incentives, encourage the school teachers in college students' innovative entrepreneurial training plan mentor, teacher guide students actively recruit enterprise entrepreneurship training and practice.

3.4 Emphasis on the construction conditions on the implementation of college students' innovative entrepreneurial training program

Attaches great importance to participating universities demonstration experiment teaching center, all kinds of open LABS and key laboratories at all levels should be provided free of charge to participate in the project students experimental field and experimental instruments and equipment. Participating college of university science park to actively take college students' innovative entrepreneurial training mission, for students to provide technical, sites involved in the planning, policy, management and other support and business incubation services.

3.5 Participating colleges and universities to build innovative entrepreneurial culture atmosphere

Students build project communication platform, to carry out communication activities on a regular basis. Encourage performance outstanding students and support project students to attend academic meetings, face-to-face communication for students' innovative entrepreneurial experience, show achievements, share resources and the opportunity. The school will organize regular project of communication between teachers.

4 PRACTICE FORM

4.1 Build new multidimensional teaching training course system

To college students' innovative entrepreneurial training and teaching and talent training scheme, combine with the industry interaction to increase practice teaching content and class, and a separate class, practice the modular teaching process. Formed by the direction of basic skills, professional basis, and comprehensive training module design new training curriculum system composed of four modules. In order to guarantee the students' training effect, the laboratory shall practice a system of all-weather open; meet the demand of the implementation of the new training curriculum system, teacher's leading role in the process of teaching, and the students as the main body of teaching activity. Training system is still at the actual demand in industry as the guidance, the discipline competition at home and abroad, the introduction of university–industry cooperation teaching, improve the openness of specialized courses and practical, formed by competition, virtual topic to drive innovation, the virtuous cycle of innovation into practice, and promote the patent of incubation.

4.2 Implementation of "studio learning style" to drive students towards scientific research activities

With the "professional design studio" based on laboratory in various professional direction, students go into the teacher's studio, combined with the studio learning experience research content, in scientific research through the two-way choice between teachers and student, promote the teachers' leading role in education, cultivate and improve students' scientific research consciousness and research ability, so as to improve the students' ability of innovation education. Combined with college students' innovative entrepreneurial training projects, each project item must carry on the detailed division of labor, and according to the students' scientific research team, each team 3-5 people, actively guide the student to carry on the innovation practice activity. Practice has proved that this form is very effective.

4.3 Build a multi-level production practice base and business incubators, guides the student to self-employment

Through setting up practice base of production and business, production, collaborative innovation platform, implementation technology, knowledge, talent, information and other resources of each flow, it provides platform for student internships, in the real environment for the cultivation of professional quality and training, enables students to course learn professional knowledge, skills, combined with the actual production, enhances students' practical ability. Through the open teaching in the teaching, visits the internship, short-term internships and graduation practice by guiding teachers and the professional on-site guidance and guidance of the enterprise, lets the student in learning and market, and jobs during the period of zero distance contact, make them to quickly set up their career ideal, and form a good habit and professional ethics, practice solid high vocational skills, implement the work and learning.

5 ACHIEVEMENT

For example, in the department of art, including the author, the current college students' innovative entrepreneurial training activities under the department tutorial system has been in the third stage. Training effect is becoming more and more obvious. In the class, they are ordinary college students but when in their spare time, they become "inventor" and "small boss" under the guidance of teachers in innovation and entrepreneurship. So far, the student to obtain all kinds of awards at or above the provincial level of more than 200 items, all kinds of certificates from more than 100; entrepreneurship student success more than 50 cases, such as personal guidance student Wu Shenglan classmates founded "longitudinal era culture communication co., LTD.", since it was established in August 2012, the cumulative turnover reached 600000 Yuan, has been the opening of a branch office in shenzhen, in guangdong college students business competition "second prize, was convicted by the silk business incubator, named" new star ", and was used as a typical college students undertaking in the zhuhai special economic zone and each big media propaganda.

6 CONCLUSION

The development of the college students' innovative entrepreneurial training activities based on tutorial system, aims to "cultivate creative thinking, spurt to create passion and provide innovative platform" has obtained certain achievements. Though there are many shortcomings, this is a continuos process to explore and improve. Through college students' innovative entrepreneurial training activities, it will arouse the enthusiasm of the students' learning and initiative, cultivate the students' innovation ability and practice ability, promote our country's higher education personnel training mode reform is of great significance.

ACKNOWLEDGMENTS

This work is supported by The National Natural Science Foundation of China (No. 61300241) and the project consists of Zhuhai College of Jilin University teaching quality project provincial project "Baoli Garment Co. Ltd. of art department internship teaching base".

REFERENCES

[1] Regarding the ministry of education, The ministry of education "teaching of undergraduate course project" national college students' innovative entrepreneurial training plan implementation notice [Z]. Beijing, 2012:1–1.

[2] zhang Jinqiu. 2010. And the study of the theory of the tutorial system [J]. Journal of teaching: higher education BBS, 6:30–32.

[3] Song Yang, Zhao jing, Liu Chenguang, 2010. Honor student tutor system effective realization forms of the tutorial system. *Journal of North China Coal Medical College* 12(1):122–124.

[4] Ding Lin. 2009. Tutorial system: meaning, and mess. *Journal of Heilongjiang Province Higher Education Research*, (5):74–77.

[5] Sun Haozhe. 2011. Based on the perspective of human resource development of college students' innovative quality training mode study. PhD thesis: The Capital University of Economics and Trade, 10.

[6] Ren Yang. 2013. Undergraduate tutor support behavior: an empirical study of influence on college students' creativity tendency. Ph.D. dissertation, University of Electronic Science and Ttechnology 5.

Information, Computer and Application Engineering – Liu, Sung & Yao (Eds)
© 2015 Taylor & Francis Group, London, ISBN 978-1-138-02717-6

The international competition of innovative propaganda poster design for agricultural environmental protection

Rui Lin Lin

*Department of Commercial Design, Chienkuo Technology University,
Changhua City, Taiwan*

ABSTRACT: This study adopted the team teaching method with two teachers guiding two students who were willing to participat in the innovative propaganda poster design competition of the 2011 International Charity Advertisement Festival for University Students held by the Beijing Union University with the topic of air pollution in the agricultural environment. The satirical method was applied to present how the nature has been damaged by humans, resulting in global warming, seriously influencing the production in the agricultural environment and the ecological balance of the nature, brining great worries to humans. The students won one silver medal and one bronze medal in the competition. It was hoped that, through this study, the concepts of maintaining agricultural environment and energy saving & carbon reduction can be promoted.

KEYWORDS: Agricultural environment, Energy saving and carbon reduction, Propaganda poster

1 INTRODUCTION

In the recent years, due to the influence of global environmental changes, humans have been facing a lot of difficult environmental issues, such as global warming, disappearance of biological diversity, and insufficiency of water resources. More attention needs to be paid to these issues. The generation and solution of environmental issues are related to knowledge from various fields. It is necessary to promote concepts of environmental protection through education, in order to improve people's environmental literacy and knowledge regarding ecological conservation. There are many kinds of air pollution.

The main pollutants in the air include solid particles, VOCs, oxides of sulfur, nitrogen oxides, and hydrocarbons. Solid particles may cause serious damage to humans' respiratory systems. Oxides of sulfur may influence health and cause acid rain, acidifying water and soil. The influences on the agricultural environment, ecological environment, existences and balance of animals and plants may last very long [4][1]. Furthermore, agriculture includes businesses in the agricultural environment related land, water, sunlight, temperature, and humidity, and various production businesses, depending on the nature. Agriculture may provide foods, raw materials for the industry, energy, ecological environment, tourism, and job opportunities. People should place more importance on it [8].

2 LITERATURE REVIEW

2.1 *Agricultural environment*

Environmental pollution has caused serious heavy metal pollution in soil. In the recent years, there have been a lot of literatures related to environment risk assessment of heavy metals and applications of follow-up coping methods. Researchers have had not only find out the total concentrations of these heavy metals, but also explore their distributions, especially, in cases where farms have been polluted by heavy metals from, as inferred, irrigation ditches. Heavy metals are released to ditches from sources and then go in to farms through irrigation. According to the research results, the ditch sample and the farm soil sample were highly related in the concentration of aqua regia at the polluted site and heavy metal pollutants, such as Cr, Cu, Zn, Ni, Pb, and Cd. Longer irrigation distance and deeper cross-section depth would lead to lower concentration of heavy metal. Although the concentration reduced as distance increased, the distribution of the total concentration of heavy metal was not a normal distribution. It was concluded that the pollution of farm soil might have been caused by solid particles carried to farms through irrigation by bottom mud or suspended particles. This was why the concentration of pollution differed with the irrigation distance [7][5].

Some scholars applied a new technology based on the coordinate algorithm, combining differential GPS (DGPS) and DEC to develop a stable and reliable guiding control system. When the system was installed in a car which was driven outdoors in agriculture, the location of the car could be determined correctly. And through the cubic spline function, the manual guiding route could be built. With this new control algorithm, the system could automatically guide the car moving toward the target route. The system was satisfyingly applied in the automatic sprinkling irrigation in the agricultural environment [9]. A study found that farms may have damaged the environmental resources due to limited current conditions or lack of professional guidance and suggestions. Therefore, the study suggested making an overall plan, with consideration of ecological design, with respect to ecological environment, and laid emphasis on sustainable development of local resources, using the methods and the principles of leisure agricultural environmental design, in hope of achieving better results [2].

3 RESEARCH METHODS

3.1 *Designing propaganda posters*

This study applied the team teaching methods, with the teachers guiding two freshmen with outstanding design talents who entered the university in 2010 to participate in the international competition of innovative propaganda poster design for agricultural environmental protection during their after-school leisure time. The teachers applied innovative thinking and teaching methods to guide the students individually to discuss and modify their innovative ideas, encourage them to participate, give them inspiration, and develop their unique creativity.

3.2 *Creative thinking and teaching*

Creativity is the result of thinking. Creativity is about innovative, novel, and unique thinking and new inventions. Creativity is to come up with new ideas, solutions, or plans in the face of difficulties. It can be seen as ability. The ability of creativity includes: (1) fluency: having a quick mind and being able to come up with various ideas in a short time; (2) flexibility: having unpredictable thoughts and being able to act according to circumstances; (3) originality: think differently and being able to come up with unique solutions when encountering problems; and (4) elaboration: being able to provide insights and conduct in-depth analyses when encountering problems. Creativity is a thinking process. The stages of the process include: (1) preparative stage: exploring the problem found; (2) incubational stage: a difficult period when no innovative ideas can be generated immediately during the process of

creation; (3) illumination stage: coming to an understanding all of a sudden; and (4) verification stage: verifying obtained new ideas or inspiration. The factors which may influence creativity are intelligence, way of thinking, personality, social culture, education method, and etc [3].

To cultivate students' creativity through education, in the aspect of teaching strategies, suggested methods to be adopted include: (1) brain storming: through group thinking, members of the group can storm each other to inspire creative thinking; (2) encouraging association: teachers encouraging students to make associations to related things or concepts; (3) partial changing: making partial changes to the original; (4) checklist method: making a checklist of many related items for students to consider, producing new creative concepts; (5) combination method: combining some concepts or things to further produce creativity; (6) property listing: listing features, advantages, disadvantages, etc. of each item and coming up with possible improvements; (7) encouraging students' divergent thinking: teachers asking students questions without standard answers so that they can consider various ways to solve those questions; and (8) cultivating students' personality with creativity [6].

4 DESIGN RESULTS

The first work was designed based on the idea of global warming, with other design elements including polar bears, penguins, and a refrigerator, creating the visual feeling of living creatures disappearing, delivering the message of gradual extinction. In the aspect of color arrangement, the refrigerator was dark brown, making it look hard and solid. The bright pink was applied to present the theme of the design, with a warning flavor. The dark gray was used to represent environmental pollution. The different shades of the skin color showed the warmth of life and added more variation to the entire poster (figure 1). The theme of this work was to "Protect the Forest, Maintain Ecological Balance", and the slogan was "Would an icehouse be the last home of polar bears?" This poster won the silver medal in the competition.

The second work was designed based on the idea of air pollution. The pure images of students were added to deliver the message of cherishing our living environment and working on the plans of green energy to leave a clean space to the next generation. The gas masks were used to mock the fact that if we continuously pollute the earth, the result of the greenhouse effect would be humans living their daily lives with gas masks on their faces all the time. In the aspect of color arrangement, the cream color was a hint of the polluted environment. And the color red was a warning, a metaphor of danger and the harm the pollution has done to the environment. The dark green was applied to deliver the message of plants being polluted and losing their gloss of life (figure 2). The theme of this work was "Save the Petroleum to Reduce Air Pollution", and the slogan was "I do not want to go to school wearing a gas mask." This poster won the bronze medal in the competition.

| Figure 1. | Figure 2. |

5   CONCLUSION

The topics chosen by this study included cultural environment, cherishing forest, and energy saving and carbon reduction. Through the satire design method, the posters deliver the messages regarding the greenhouse effect caused by air pollution, which has seriously influenced ecological balance. The slogans on the posters were designed with warning, with a hope of achieving the goal of promoting environmental protection.

6   DISCUSSIONS AND SUGGESTIONS

The teachers guided the students during their leisure time to participate in the international competition and won some prizes. The influence on the students' learning motivation is very positive. And this result may increase other students' willingness to participate in competitions to gain more field experiences, which are very useful for their further education and future career, especially, the two teachers with different design specialties worked together to guide the students in the competition, providing the students a more diversified space to improve their professional knowledge in different aspects in design. This is a goal most activities in curriculums can hardly achieve. It is suggested to work toward this direction in the future by choosing an international competition which is held annually or periodically and forming a design team to participate in the competition

to stimulate students' creative thinking and improve their design abilities.

Second, in the teachers' teaching process and the students' learning process, the teachers learned from each other through the team teaching and expanded their professional knowledge, while the students improved their professional skills through discussions and modifications of their works. The school should cite the students for their achievements or offer a prize to encourage more teachers and students to participate in competitions. In addition, the teachers spent their leisure time guiding the students. The students should cherish this rare opportunity, especially, the two teachers have different specialties in design, making this the best chance for the students to improve their professional abilities. Of course, the teachers had to not only make time to guide the students to participate in the international competition, but also be passionate enough about teaching to be willing to do this. And their willingness came from the students' intention to participate in the competition. Therefore, the teachers and the students were closely related.

In the process of teachers guiding students to participate in competitions, students' professional skills can be improved, more experiences can be gained, and there are chances to win awards. Also teachers and students can develop a good relationship to increase students' interest in learning and sense of achievements. Teachers' retire rate may also be reduced. Therefore, schools should provide software and hardware equipment for teachers and students to use, as well as administrative supports, such as planning for spaces for discussions, opening discussion rooms, managing spaces, and making opening hours more flexible, in order to encourage more teachers and students with willingness to participate in competitions to devote themselves in accumulating personal experiences and building good reputations for schools.

REFERENCES

[1] B. F. J. Jojo, T. Hodgkin, M. Jones, Introduction to special issue on agricultural biodiversity, ecosystems and environment linkages in Africa, *Agriculture, Ecosystems & Environment*, 157, August (2012), pp. 1–4.

[2] C. L. Liu, The Research of Leisure Agriculture Environmental Design from the Viewpoint of Ecological Design-Two Cases Study in Yi-Lan, Master dissertation, Graduate Institute of Environmental Policy, National Dong Hwa University, 2005.

[3] C. X. Ye, Educational Psychology, Psychological Publishing Co., Ltd., Taipei, 2011.

[4] H. C. Zhang, Introduction to Damaging Effects of Environmental Pollution, New Wun Ching Developmental Publishing Co., Ltd., Taipei, 2012.

[5] J. Wang, J. Liu, The Change of Agricultural Policy's Effect and the Importance of Developing Environment-friendly Agriculture-A case in republic of Korea, Energy Procedia, 5 (2011), pp. 2246-2251.

[6] L. A. Chen, Theories and Practices of Creative Thinking Teaching, Psychological Publishing Co., Ltd., Taipei, 2008.

[7] W. C. Fu, A Study of the Distribution and the Transformation of Metal Speciation in Contaminated Agro-environment Exemplified by Cr, Cu, Zn, Ni, Pb, and Cd, Master dissertation, Department of Soil and Environmental Sciences, National Chung Hsing University, 2007.

[8] W. F. Fu, Theories and Practices of Agricultural Cooperation, Harvest Farm Magazine, Taipei, 2003.

[9] Y. T. Qiu, Development of a Guiding Control System for Autonomous Vehicle Working in Agricultural Environment, doctoral dissertation, Graduate Institute of Bio-Industrial Mechatronics Engineering, National Chung Hsing University, 2011.

Information, Computer and Application Engineering – Liu, Sung & Yao (Eds)
© *2015 Taylor & Francis Group, London, ISBN 978-1-138-02717-6*

A study of the innovative design of knowledge management implementation case

Jing Chen Xie
Department of Digital Media Design and Management, Far East University, Taiwan
Department of Industrial Education and Technology, National Changhua University of Education, Tainan City, Taiwan

ABSTRACT: This study applied the innovative design of knowledge management implementation case, to perform team creative thinking and product innovative R&D for a chessboard gaming method to play chess. Through the establishment, use, and sharing of an online knowledge base, following the knowledge management life cycle, this study performed knowledge retrieval, knowledge creation, knowledge circulation, knowledge storage, knowledge learning, and knowledge value chain, to help the case to come up with creative ideas and implement its design plans.

KEYWORDS: Knowledge management, Innovative design, Gaming method.

1 INTRODUCTION

Knowledge is the experiences, values, and abilities under operation of information. It must be operated through a process to create innovative and networking effects. Knowledge is intangible, thus it can achieve great creative abilities. Humans are the most fundamental and primary vehicles of knowledge. Due to the property of accumulativeness, creation can be performed based on predecessors' and ones' own achievements. Knowledge can be generated at once and applied in an organization for unlimited number of times. Especially, in the new economic age when the Internet is very popular, with the innovation of technology, apparently economic activities with intangible values have caused unprecedented impacts on all industries. And thus the business model of knowledge management has been implemented by enterprises as the optimized tool for resource integration [3].

Knowledge management is a continuous process to offer correct knowledge promptly to members who need it, to help them take correct actions to improve performances of an organization. The influential constructs of knowledge management include: (1) knowledge workers, (2) data, information, knowledge, wisdom, and sharing, and (3) information technology which helps to establish knowledge management. To survive in an environment with fierce competition, an enterprise must continuous learn and create new knowledge, sharing knowledge throughout the entire enterprise or even the entire industry and rapidly integrating knowledge into its technologies and products

to improve its competitiveness and advantages in future survival. In the knowledge management lifecycle of generation, propagation, application, storage, value, creation, and re-production, an enterprise can begin its integration and transformation. In other words, the e-operation of the knowledge industry is to establish and produce a structure of strategy innovation, organization innovation, and process innovation [1] [2].

The effects of knowledge management are based on the performances of an enterprise's business management growth, including (1) introducing a new enterprise model, (2) increasing advantages in competitions and values of enterprise, (3) improving ability to process information technology, (4) forming an organizational culture, (5) transforming workers into knowledge workers, (6) increasing enterprise's intangible assets, and (7) improving learning ability of organization. Through the implementation of knowledge management information platform, other information systems can be efficiently integrated, to maximize the effectiveness of technology innovation and technique innovation. In the aspect of applying knowledge management, implementation must be conducted from the viewpoints of core elements. And knowledge management must be promoted to case members, through trainings and communication to reduce differences in communication among members. In addition, when an enterprise develops strategies for knowledge management, it is important to pay attention to the breadth and features of product lines, segmentations and choices of target markets, degrees of vertical integration, limited scales and scale economics, geographic coverage, and weapons for competitions [12].

2 LITERATURE REVIEW

If an organization wants to establish a new innovative platform, it shall implement its main innovative acuities every day. Breakthrough points include solving problems together, using and integrating new technologies and tools, doing experiments and tests, absorbing external knowledge, and learning from markets. Knowledge transformation include: (1) communization: which is the process to transform an individual's tacit knowledge into tacit knowledge of another person (group of persons); (2) externalization: which is the process to transform tacit knowledge into explicit knowledge through metaphors or analogies; (3) combination: which is the process to summarize explicit knowledge into a set of systemized knowledge; and (4) internalization: which is the process to internalize systemized knowledge an individual obtained through contact and combination into his tacit knowledge [7].

In the aspect of knowledge base, it can be established using the methodologies such as search engine, data mining, agent, and knowledge portal. The key factors of successful implementation of knowledge management include: (1) culture: case company must be able to propose a correct method to stimulate member interactions and knowledge sharing; (2) evaluating knowledge: in the beginning of implementation of knowledge management, costs and benefits from investments shall be calculated; (3) processing knowledge: in the case company, there must be a leader finding out how knowledge can be gathered, stored, processed, and distributed through technologies; and (4) action knowledge: there must be a driving force from a clear concept. And only through identification with the organization, knowledge management can lead the learning innovation of the organization and improve its competitiveness. Furthermore, factors which may cause failures shall be avoided, such as issues related to humans and institutions, humans and systems, and system technologies [10].

3 RESEARCH METHODS

Experiences are knowledge one has obtained before. Knowledge is the product of a series of catalyses of data and information. This study implemented the idea of knowledge management according to the demand of the case company for innovative design, to help the case company perform team brain storming to gather important data to build a knowledge base of creative thinking. Before the case company implemented knowledge management, first it had to find knowledge composition, understand

keywords, management meanings, and methods to look for knowledge. Secondly, according to the properties of knowledge, which documents belong to knowledge composition could be determined. Then, knowledge objects were categorized to speed up knowledge access, which was good for creation, usage, storage, propagation, and formation of re-use life cycle of knowledge objects [11].

The new product development procedure is to first analyze customers' demands, propose a conceptual design, then start product prototype design and submit the design to the engineering department for evaluation and technical modification. If a problem is found during testing, the design department must modify the design and then the testing will be conducted by the production department. In the process of knowledge management implementation for a case company, all the members of the case company must consider themselves as knowledge workers, so that each individual can show his self potentials, control his work directions, and complete his tasks. And their abilities to achieve work effectiveness and enterprise goals must be consistent to create benefits for the company [4].

According to Arthur Andersen's viewpoint on knowledge management model, it shall include management process and knowledge management enablers. Through scientific paradigm and social paradigm knowledge construction can be provided, so that knowledge embodiment can perform knowledge dissemination for users to use. The innovative design of the case company of this study adopted knowledge innovation as the key to its core competitiveness and sustainable development. Innovative design includes, from small to large scale, product improvement, new product development, and breakthrough or discontinuous innovation. Thinking is required for product innovation. Products must be customized for customers to meet their demands according to their lifestyles [5] [8].

In order to systematically summarize obtained data, the case study method was adopted. The unit can be a person, a family, a group, or an institution. The purpose of the case study method is to understand the unit being studied or repeated life cycles, or to further explore and analyze important parts of an event, in order to explain current status and assess performance, or to describe interactive situations which may lead to changes or influences in the long run. It is a vertical research method used to explore developmental phenomenon in a certain period of time [9].

4 EXPERIMENTAL RESULT

Knowledge management is, like knowledge, difficult to be defined specifically. With the advancement of information technology and the internet, the concept of knowledge management has been practically applied in many fields. This study introduced the concept of knowledge management to help the case company to perform innovative design for the dual chessboard game rules and chessmen. This study first collected currently game rules of related products, and discussed and analyzed their advantages and disadvantages. Then, through teamwork and division of works, new game rules were designed, discussed, and modified. Finally, a new set of game rules for players of all ages was proposed. In a game, both players can choose roles and a scenario they prefer and may take a card according to the settings. There are different ways to play chess according to ranks of chessmen.

The chess game was designed for two players. One of them uses red chessmen while the other uses black ones. They take turns to place one chessman at a time. Chessmen include humans and zombies. Ranks include king, general, warrior, raid, dog, and soldier. Human roles: a survivor represents king, a lady with double blades represents general, a bulky fellow represents warrior, a sniper represents raid, a patrol dog represents dog, and a private represents soldier. Zombie roles: a carrier represents king, a butcher represents general, a tank represents warrior, a crow represents raid, a zombie dog represents dog, and a zombie represents soldier. As for obstruct settings, there are two-grid building, one-grid building, phone booth, and underground passage. Roles of the warrior rank can move to two-grid buildings and one-grid buildings, kings and soldiers can use phone booths or exchange places. All the roles can use underground passages. Chessmen are placed in a wood box. The box can be turned upside down and placed on a table as the chessboard for the game. There are 4 colors on the chessboard to mark the grids based on the rules. The gray grids are used for underground passages only. The light gray ones are the starting points for zombies and privates. The black ones are general grids, where chessmen and buildings can be placed. And when a chessman stops on a grid with the special mark, the player can take a chance card.

Chance card settings: when a chessman stops on a grid with the card mark, the player can take a chance card. There are 25 cards in total, with functions such as moving any one of opponent's chessmen, reviving the dead, moving a chessman back to the starting point or an underground passage, moving at will for 3 times, pausing for 1 time, and exchanging positions with one of opponent's chessmen with the same rank. Moving settings: a king can move 1 grid at a time within a 6-grid gray square. To move to other grids, the king must use a phone booth. However, if there is no obstruct on the straight line between two kings, a king can directly take out the other. A general can move 1 grid at a time. It cannot take out the king. It can move straightly or diagonally. A warrior can

move 1 grid at a time. It cannot take out the king. However, it can move on 1-grid and two-grid buildings. A raid can move 1 grid at a time. However, when there is no obstruct, it can attack any one of opponent's chessman within the 5-grid range. A dog can move any number of grids, but straightly only. A solder can move 1 grid at a time.

5 CONCLUSIONS

A case study was performed for the theories related to knowledge management. Innovative rules of a two-player chess game and designs for chessmen were proposed. This study first collected data regarding types of chess games, then summarized data, combined them with the creative thinking from the team, and categorized them, as the foundational data to establish a creative design knowledge base. Then, in the aspect of the main body of the design, although there are too many chess games available, they are largely identical but with minor differences. Innovations of game rules and chess designs are very rare. Thus, the achievements of this study are very novel. An application for a patent has already been submitted.

This study established a database for innovative design and applied the optimized innovative decision strategies to simulate game rules, select roles, design looks, and arrange colors. Finally a new set of game rules was developed which can test two players' wisdom. The chessmen were made like action figures, increasing fun and true feelings during the game competition. However, the updates of innovative design knowledge base and system maintenance also need to be considered. The only way to make the best out of knowledge management in the present economic age is for team members to share their professional abilities and pass on their experiences.

REFERENCES

[1] C. W. Holsappleed, Handbook on Knowledge Management, New York : Springer-Verlag, 2003.
[2] E. Pasher, T. Ronen, The Complete Guide to Knowledge Management: A Strategic Plan to Leverage Your Company's Intellectual Capital, Hoboken, N.J.: John Wiley, 2011.
[3] Findbook, Knowledge Management: Implementation and Case Analyses, Dr Master, Taipei, Taiwan, 2008.
[4] K. T. Ulrich, S. D. Eppinger, Product Design and Development, Boston: McGraw-Hill, Irwin, 2004.
[5] M. Kao, Research in Education, Ding Mao, Taipei, 2007.
[6] T. H. Huang, S. D. Wu, Knowledge Management Theories and Practices, Ch Wa, Taipei, Taiwan, 2003.
[7] T. Knight, T. Howes, Knowledge Management: A Blueprint for Delivery, Oxford: Butterworth-Heinemann, 2003.

[8] T. Schipper, M. Swets, Innovative Lean Development: How to Create, Implement and maintain a learning culture using fast learning cycles, Boca Raton: CRC Press, 2010.

[9] W. K. Wang, C. H. Wang, Research in Education, Wunan, Taipei, 2008.

[10] Z. X. Wu, J. B. WANG, Knowledge Management: Strategy and Practice, Lin King, Taipei, Taiwan, 2001.

[11] Z. X. Yang, Knowledge Management: Theories and Practice, Yang Chih, Taipei, Taiwan, 2005.

[12] Z. Y. Hu, Technology Innovation Management, New Wun Ching, Taipei, 2005.

Information, Computer and Application Engineering – Liu, Sung & Yao (Eds)
© 2015 Taylor & Francis Group, London, ISBN 978-1-138-02717-6

Current status and future strategy for the application of new technologies in agricultural mechanization

Yun Yu Li
Department of Physics and Electronic Engineering, Guangxi Normal University for Nationalities, Chongzuo, Guangxi, China

Ming Li
College of Mechanical Engineering and Transportation, Southwest Forestry University, Kunming, Yunan, China

ABSTRACT: This paper summarizes current status and issues on the application of new technologies in agricultural mechanization in China. The authors propose strategic advice in five aspects for the development of future application of new technologies in agricultural mechanization. The authors also summarize factors that impact the technological innovation in this field. These insights may be used for supporting decision-making in agricultural mechanization and associated technological innovation and promotion.

KEYWORDS: New technology, agricultural mechanization, application, problems, strategy

1 INTRODUCTION

Since the begging of the 21st century, the amount of farm land in China has declined continuously. Urbanization has accelerated and rural agricultural population has declined. To ensure food security in China in the new century, agricultural technology, agricultural high-tech popularization and application are particularly important. This paper presents a comprehensive analysis on the present application of new technologies in the domestic agricultural mechanization and existing problems. This paper proposes the corresponding strategies that are in emergent need. These strategies provide support for decision-making in China's agricultural mechanization and associated new technology innovation and promotion.

2 PRESENT STATUS OF THE APPLICATION OF NEW TECHNOLOGY IN INTERNATIONAL AND DOMESTIC AGRICULTURAL MACHINERY

2.1 *The international aspect [1]*

The United States of America, Germany, Israel and other countries are leading in the development of agricultural mechanization and technology: ① In USA, the technology is trending towards high speed automation stage, large power, high speed technology, satellite global positioning system for monitoring operation. ② In European countries,

precise machineries equipped with GPS systems are becoming more and more popular. ③ The Japanese agricultural technology is highly advanced. It has realized indoor and outdoor mechanization, light-weighted, comfortable, and helicopter operations. ④ Other countries such as South Korea have made great progress in the mechanization level after importing new technologies or with technical assistance from other countries. The directions of new technology and new process in South Korea are in the design, manufacture, and use. It realizes and understood the innovation after importing new agricultural technology.

2.2 *Domestic technology has the following characteristics[1]*

2.2.1 *Computer-aided technology in agricultural machinery design, manufacturing, application got fast popularization*

In China in 1990s, computer started to be really used in mechanical design and manufacturing industry and developed rapidly in the past thirty years. Computer-aided technology was first applied in the manufacturing of aerospace, electronic technology, and automobile design industry in our country, but also promoted the innovation and application of new technology in agricultural machinery. A development climax will also appear in the future, the computer-aided technology is widely used in CAD, CAM, CAPP, and CAE technology.

2.2.2 Application of hydraulic technology with good sensor in agricultural machinery

At present in our country, large cultivator, combine harvester, rapeseed production and other machineries use this technology. The development of standardization of hydraulic components, integrated mechanical, electronic and hydraulic systems are playing a powerful performance advantage.

2.2.3 The use of automation, network technology

In our country, intelligence and automation started to be used in variable sowing machine, spraying machine, fertilizing machine, and combines. For example, Wu Weixiong [2] studied automatic control of the depth of tillage using mounted plough set, which can be adjusted by pressing a button control.

Self-innovation and importing of technology are complementary. Network technology exists in agricultural machinery in three aspects: the mechanical local network control, network control between machinery, and remote control network technology between agricultural machinery and control center. Our country is concentrated on the study and application of these aspects, such as the Shanghai Jiao Tong University Electrical Control Research Institute Zhou Guoxiang developed a digital remote communication controller GSM [3]. At the same time China has accelerated the construction and promotion of supply chain.

2.2.4 Unmanned machine has become a reality

Agricultural unmanned technology has been the research focus at present. The emergence of robot are: self walking tillage, fertilization, weeding, picking robot, grafting, seedling, although the performance is not stable enough, but it represents the direction of theoretical research and product application. Not only unmanned machinery technology has made rapid progress, the UAV technology has made a breakthrough. As of 2011 June, China Agricultural University hosted the "low altitude low amount unmanned aerial spray vehicle" entered the demonstration stage [4]. At the same time unmanned technology obtained the global positioning system (GPS), geographic information system (GIS), remote sensing (RS), data acquisition sensor (CDS), and decision support system (DSS) these five breakthroughs in technology and application. No fuel power equipment such as solar energy and wind energy powered agricultural machinery increased a sense of mystery. Developed countries such as USA, etc. have started to conduct research in this area.

Since 21st- century, our country has made rapid progress in agricultural industry in the application of new technologies. The total output value of the agricultural machinery industry exceeded 300 billion Yuan and China is now a major producer of agricultural machinery in the world [5]. However, the research and development to promote the use of new technology of mechanization of farming still has many problems and difficulties that need to be solved.

3 ANALYSIS OF DIFFICULTIES IN AGRICULTURAL MECHANIZATION TECHNOLOGY DEVELOPMENT AND ITS COUNTERMEASURES

3.1 The development, application and extension of new technology suffers from existing difficulties

Although the application and popularization of agricultural mechanization and its related technology in China has made considerable achievements in the past, but compared with developed countries there is still a big gap. Technology is not advanced and not stable, which restricts the innovation and application of new agricultural technology. These restrictive effects are mainly reflected in six aspects:

3.1.1 Strong policy-oriented and low profits causes slow enterprise development and weak technical innovation

At the present stage, technical innovation is weak; the present situation of enterprise is big but not strong: 2008 agricultural machinery manufacturing enterprises in China has more than 9000 companies. Among them 2349 are large scale enterprises (sales of more than 5 million Yuan), accounted for 26.1%, the average sales revenue of 79.36 million Yuan [6].

3.1.2 Investment in agricultural mechanization is still at a relatively low level. Therefore, the agricultural mechanization in China is lagging behind developed countries, so the popularization and application of new agricultural technology still faces difficulty

From 2001–2010 years, our country agricultural mechanization investment scale increased, but the input intensity did not change much, around 0.2% of GDP. It is well below the agricultural mechanization industry contribution of gross national product (GNP), which is 1.736% [7]. This suggests that financial investment in agricultural mechanization is still in a relatively low level, the investment in research into new technology is even lower. Therefore, our country agricultural machinery technology promotion and innovation still face difficulties and challenges.

3.1.3 The national standardization, serialization technical guidance document is still incomplete, which lacks leading power

At present, China is formulating quality standards, but for parts design, specifications and standards are not mature. So agricultural enterprises in our country design their products independently, parts are not compatible. This incompatibility increases the cost of the design and manufacturing at individual enterprise, and increases the cost of replacement and maintenance. Some standardization has been realized within an enterprise, or within a brand, but there is no standardization between enterprises. This is also one of the reasons why Chinese agricultural machinery enterprises cannot grow bigger and stronger. For technical requirements for serialization, our country still has no uniform standard. In order to keep competitiveness, each enterprise has its own respective series product, so the standardization and serialization degree is not high. The leading strength of standardization and serialization is limited.

3.1.4 Number of students in colleges and universities who major in new technology in agricultural machine is limited and enthusiasm is not high. Agricultural and mechanical colleges are not strong enough; R & D innovation still has much room for improvement

In the recent years, university students majoring in Agricultural Mechanization in China did not increase, but the number of students was reduced. Overall increase of enrollment of college students did not increase the enrollment of students majoring in agricultural mechanization. This trend has tremendous adverse impact on China's Agricultural Mechanization in teaching, scientific research and personnel training. These engineering majors obtained huge development, but agricultural undergraduate majors had no rapid development. The number of students decreased, agricultural engineering and agricultural mechanization major shrink [8]. Among the universities that have agricultural mechanization major, only China Agricultural University and Zhejiang University ranked in top tier among universities in China, the quality of other schools have a lot of room for improvement. The innovation efforts still have much room for improvement.

3.1.5 Labor force migration increased. The desire and demand by farmers for purchasing high power, and advanced automatic machineries are not high

The gap between urban and rural areas is large and labor transfer willingness is increased. By early 2013,

China had approximately 260 million migrant workers. Because the scale of cities is bigger and bigger, city education facilities are better than those in rural areas. There are more employment opportunities in cities. Labor force relocation desire was increasing, especially among the generations born in 1980s and 1990s. Labor force transfer increase leads to the fact that only elderly, youngsters, women and the weak are staying in the countryside. The average education level among those staying is low; this weakens the ability to use machinery. Agricultural cost will increase with the decrease of labor force staying in the rural area. Labor force reduction leads to low degree of intensive, large scale, industrialization. So farmers have low incentives to farm, cost for land transfer is high and the transfer speed is not fast enough. The purchasing desire for large power and advanced automatic machinery is not high and the demand is reduced.

3.1.6 Social service system for agricultural machinery operation, maintenance, rental and leasing is not well established. Service quality is low. It restricts the development of new technologies and applications

The overall level of Agricultural Mechanization in China is not high. The repair and maintenance service system in agricultural machinery in China was established during the initial stage of Chinese reform and opening up. Each village and town set up an agricultural station, which is responsible for the management, promotion, and repair work. But now these service stations are not fully functional. Their service is limited in the management aspect because of the existence of the gap between technician and advanced technologies. In some townships Agricultural Stations have no technicians, and therefore, repair and maintenance skills have lagged behind. The licensing and qualification process for service technicians are difficult. The investment on maintenance equipment is limited. Therefore, the maintenance and service model for agricultural machinery in China cannot fulfill the needs in the new period for the promotion and application of new agricultural technology. This resulted in the situation that companies only sell agricultural machinery and no or few companies specialize in agricultural machinery repair and service business. It is possible that a specialized agricultural machinery repair shop cannot be found in many villages and towns. The rental market development lags behind. Agricultural machinery leasing business is developing in some local areas this year. Because this business needs large amount of investment, the promotion is difficult. towing to the above-mentioned reasons, the maintenance and service system of agricultural machinery promotion, management, maintenance are not sound enough. This poor maintenance and service system restrict the development and application of new technology.

3.2 Countermeasures to promote rapid development of new technology in agricultural mechanization

3.2.1 Policy support and legal protection to promote the sustainable development of agricultural mechanization technology

Since 2004, The Chinese Communist Party Central Committee and the State Council issued a series of executive orders regarding agriculture, which covers the construction of infrastructure, financial investment, insurance policies, tax breaks, financial credit, government responsibility, agricultural machinery safety supervision and management, agricultural cooperatives, fuel supply and other aspects of support. These executive orders provide policy support for agricultural mechanization development in our country [6].

Increase financial support, support and encourage the development of new technology, improve benefits and social status of technical professionals in agricultural machinery, promote the agricultural industrialization and large scale operation, improve the support for key agricultural machinery types and new technology and new projects. Clearly define targeted timeline milestone. Realize the goal that a full range of technology in agricultural machinery can be developed and produced in China. Agricultural machinery technology can be applied in all aspects of agriculture including the whole production, custody, harvesting, and storage.

Subsidies for agricultural machinery upgrade, subsidies on fuel should cover all agricultural production, custody, harvest, and storage. Support priority should be given to independent research and development of agricultural machinery products. Breakthrough in weak technical areas in a short time should focus on research and development in rice planting, sugar cane harvesting, cotton harvesting, energy conservation and environmental protection of agricultural mechanization technology. Key technology should also be introduced to promote independent research and development. The development speed needs to be improved to catch up with other existing technologies.

All functional departments have to do a good job of planning, guidance, service, good publicity and implement the "Agricultural Mechanization Promotion Law", agricultural machinery management regulations in provincial levels, and 2014 No. 1 document of central government. Make sure financial support policy is established for short-term

interests. Legal documents protection is not a mere formality, indeed, create a good environment for new agricultural technology innovation and application. Accelerate agricultural mechanization development of new technology and improve the level of agricultural mechanization, to promote the long-term sustainable development of new technology in agricultural mechanization.

3.2.2 Increase training intensity to improve agricultural machineries talents. Increase returns for technology so the talents love agricultural machinery industry

First, optimize the configuration of rural school resources in villages and towns, focus on improving the teaching quality, cultivate new farmers and retain new knowledge-based farmers. This will enable new farmers to accept new technology application and popularization of such knowledge, skilled farmers to accept new technology and the corresponding technical training. This will in turn promote farmers and agricultural enterprises to adopt new technology and equipment.

Second, the government, colleges and universities should collaborate, support, and expand agricultural machinery related technical level and quality of professional education, cultivate innovative talents, high quality farm machinery technology professional talents, high employment rate, employment quality is high. Increase the support from the society for agricultural technology extension related professional and increase their enthusiasm and desire. This will also attract more excellent talents to enrich the industry team, to reverse the trend of de-agriculturalization in technical school, vocational schools, vocational and technical college, and agriculture and forestry colleges and universities. Agricultural colleges and universities are not only for training skilled farmers, but also the promoters and participants of technology innovation and technology teaching.

Third, if few employment opportunities exist for agricultural machinery technical personnel after graduation and the work condition is hard, the wages and benefits are low, there is little room for career development, less entrepreneurial opportunities and profit is low, then these conditions lead to a low social status and low revenue. Incomplete survey in other industries such as automotive, machinery manufacturing, electronics manufacturing industry showed that salary of technical personnel is twice as much as farm machinery industry. Automobile drivers and agricultural machinery driver salary differs even more. If income does not increase after obtaining the professional qualification certificate of senior agricultural machinery operators, this greatly dampened the enthusiasm of personnel engaged in agricultural machinery industry and this will lead to talent career change or loss. Countermeasure is to consider the whole value chain of new technology in farm machinery, planning and implementation from design, production, management, promotion, and maintenance. Each dot in the value chain should be supported and improve benefits, increase technical innovation encouragement. This can actually remove the inferiority mentality of the employees.

Finally, speed up the education in the construction of new countryside, let the people in the agriculture industry has a perfect living environment, children can receive good education. The technical personnel will settle down easily and love the agriculture industry.

3.2.3 *Emphasize agricultural technology promoting, let people who make use of new agricultural technology taste the sweetness of technology investment*

Promotion of agricultural machinery is one of the ways to enhance people understanding and usage motivation to improve, also one of the measures to improve agricultural machinery sales. Agricultural extension workers should not only say how good a machine is when they are doing new technology promotion. They should present corresponding actual data or live demonstration, and compare with human labor to calculate the cost savings. Also product upgrade and maintenance costs, how long to recover costs, labor cost savings need to be presented so that the cost-saving is more persuasive.

Let buyers or owners be willing to accept new things. When the mechanical failure needs repair, maintenance should be trouble-free. We should draw lessons from the car industry experience and practice; make the farm machinery supermarket and 4S stores and various repair shops available across all medium cities and in each township and village farm. This will make new agricultural technology users to taste the sweetness of new technology, make people in the industry to obtain higher profits. This will make wider application of new technology.

Attract social capital investment and learn from automobile, real estate industry experience. Reflections on the automobile industries rapid development, why real estate industry is developing rapidly, a main reason is that sales are big, output is high, and profit is high, therefore the investment in the industries are high. This attracts more people to join the industry and promotes the technology development.

3.2.4 *Pay attention to scientific and technological innovation, strengthen the technical support system, and enhance national competitiveness of agricultural machinery*

The Ministry of Science and Technology, the Ministry of Agriculture should work together to formulate a national standardization and serialization technology guidance documents, which must be forward-looking, comprehensive, objective, guiding documents. For instance, to make the development of the total agricultural production and new technology catalogue, the application type products statistics, recommend public promotion, and the product list also should learn from the example of the car industry practices, to publish directory and its technical indicators in a timely manner, so as to guide and accelerate the development of innovation and promotion of speed.

We should promote the collaboration between agricultural universities, agricultural enterprises, agricultural scientific research institutions, and Agricultural Mechanization Promotion and Service Organizations. A new science and technology innovation system should be established which is based on the production, study, research and promotion of the agricultural machinery. Enterprises are the development main body, colleges and scientific research institutions are for technical support, promotion agencies are the backbone, agricultural machinery service organization are the foundation [9].

In order to promote enterprise technological innovation, to complete the large group and large enterprises competition pattern and the elimination of small workshop type backward mode of production, we should accelerate to support the expansion and merger of agricultural enterprise. This is also good for us to invest much more capital in new technology, to promote the level of science and technology, and to invent new technologies. Enterprise will have more money and human resources for the development of new technologies and products after its growth.

Encouraging enterprises to expand and grow, which is conducive to improve the benefits of employees and improve staff's working environment, to invest large capital in high-tech manufacturing technology, to make the production processes more automated and intelligent. It is also good for controlling the quality of products, establishing professional laboratory, attracting global technical and management elite, and benefiting the global design and collaborative manufacturing.

We make the tactic to boom agricultural and national subsides in agricultural machineries as a driving force, to make the expansion of the popularization and application in new technology and machinery as an opportunity, and to put the education of scientific research institution and enterprise R&D as the main body, to accelerate the production of products with new technology, expanding the enterprise market and increasing its benefit. Accordingly, we can achieve the goal of improving the technology's international image and the international competitiveness of export products, and this will in turn feed the innovation of new technologies.

Straighten out the relationship between the government departments, scientific research institution and enterprise, increase support for inputs and services, reduce the corresponding enterprise tax, enlarge the propaganda promotion, and expand financing channels and income, and effectively improve the development of talent, capital, environment and subsidies support of agricultural new technology and utilization, Accordingly, enhance the level of international agricultural extension, new technology and equipment, and the production and sales of new products.

3.2.5 *Adjust measures to local conditions, combine the reality in our country, and develop appropriate new product and technology*

China has large mountainous and hilly areas. The economic development lags behind in these areas due to poor natural conditions and high labor costs, The agricultural products are hard to sell for a good price, the transportation cost in the remote area is costly, it is common that farmers purchasing power is low, the farm field is small, contiguous operation efficiency is not high, and the number of types of crops are high.

At present, in the western area of China, especially in the mountains area agricultural mechanization develops slowly and lags those in the rest of other areas. The human resources, especially skilled technical personnel, are in short of supply. The maintenance and service market have not established, and the maintenance costs is high, the difficulty in the implementation of mechanized transitions, the lag in contiguous land transfer, and the poor quality of the land, all of these factors limit the application and popularization of agricultural mechanization and new technology and technology upgrade. How to adjust measures to local conditions and develop the agricultural mechanization and new technology, innovative research, design and develop new technology products, and how can we make the rapid popularization and application, all these questions need systematic and strategic thinking.

To speed up the development of mechanization, consider elevation, topography and soil structure and other factors, a reasonable allocation of resources, optimizing the mechanical structure; actively guide the agricultural households to face the market, participate in the purchase and lease, both market benefit and win–win principle, develop the market and

increase sales; construct beneficial model of service system for agricultural machinery operation, management, technical repair, promotion. It is the key to expand enterprise and stimulate circulation. It is the primary contributor to promote new technology and innovation and application. The use of administrative, economic, policy and other means, to mobilize all levels of government and households, agricultural cooperatives, and agricultural enterprises, implement layer by layer, and expand the application type and market. This can promote the rapid development and prosperity in agriculture mechanization, which can promote the birth of new technologies and applications.

4 CONCLUSION

In short, to encourage scientific research innovation, improve benefits of workers, enhance the attractiveness, incentives for the application of new technology or new machinery, are also an important aspect of economic development in the present stage in our country. Chinese modern science and technology development momentum enabled the innovation and application of new agricultural technology. This also makes enterprises more dynamic, expanding, paying attention to scientific research, cultivation of talent, and enhancing the speed of technological innovation. Government, propaganda, and urbanization provided guarantee and guiding force for a good incubation environment for new agricultural technology. The status for realizing agricultural mechanization in the areas of automation, comfort, and large scale is promising and has a long way to go. Only when government, enterprises, scientific research units, promotion department and users form a unified force, the modernization of our country's agriculture machinery and new technology application and innovation will accelerate rapidly, leading the world trend of agricultural technology.

ACKNOWLEDGEMENT

This research project has been 2013 year Guangxi University of science and technology of funded research projects, project number (2013LX164)

REFERENCES

[1] Li yunyu, Li ming. Application of novel technologies in agricultural mechanization and its new directions [J]. Journal of Chinese Agricultural Mechanization, 2014(3).
[2] Wu Wei-xiong, Ma Rong-chao. Design of the automatic control of the plowing depth of the integrated plowing set [J]. Journal of Agricultural Mechanization Research, 2007(9).

[3] Zhou Guoxiang, Zhou Jun, Miao Yubin, Liu Chengliang. Development and application on GSM-based monitoring system for digital agriculture[J]. Transactions of The Chinese Society of Agrcultural Engineering, 2005, 21(6).

[4] Wang Minghui. Low altitude unmanned aerial spray amount of technology exchange will be held in Heilongjiang[J]. Xinjiang Agricultural Mechanization, 2011(5).

[5] The 2012 farm machinery industry top 10 news[J]. Agricultural Science & Technology and Equipment, 2012(12).

[6] Chen JingHu. Analysis and Countermeasures on the problems existing in the development of agricultural mechanization Chinese[D]. Master's thesis of Yangtze University, 2012 May.

[7] Li Zehua, Ma Xu. An empirical study on structure of financial resource and spatial distribution characteristics of agricultural mechanization investment[J]. Journal of Chinese Agricultural Mechanization, 2013(4).

[8] Zhang Jianjun, Ma Yongchang. Agricultural mechanization training mode and practice of undergraduate talents[C]. Proceedings of the 2008 academic annual meeting of Chinese society of agricultural machinery, in 2008 September 19 ~ 22 days, Ji'nan Shandong china.

[9] "Twelfth Five Year" development planning of Agricultural Mechanization of Shanxi province [EB\OL].

Information, Computer and Application Engineering – Liu, Sung & Yao (Eds)
© 2015 Taylor & Francis Group, London, ISBN 978-1-138-02717-6

Discussion on management innovation and patterns of the real estate's construction costs

Li Xia Song, Jin Ying Liu, Gui Zhen Ji & Qin Qin Qiao

Langfang Polytechnic Institute, China

ABSTRACT: With an increasing number of populations in the society, it urges the development of the real estate. And construction costs run through the entire real estate business. How to work out the cost of development and how to make the real estate keep pace with the times have become problems worth exploring. This paper analyzes existing problems in the management of the real estate's construction costs and studies the measures to improve cost management in order to promote the development of the real estate.

KEYWORDS: Construction costs; Real estate; Cost management

At present, the real estate's cost management has three problems: cost consciousness weakens, cost management weakens, and cost behavior softens. So high costs and high house prices have restricted and affected the development of the industry. With the development of the society, the previous old cost management methods can no longer deal with the fierce competition in the market today, and there are a series of problems. Therefore, we must adopt a series of measures to reduce costs, save money, and win the maximum profit with minimum input, which are the purposes of the real estate business.

1 PROBLEMS EXISTING IN COST MANAGEMENT

1.1 *The less strict job of cost management at early stage*

The positioning at the early stage is very important for real estate enterprises. If the initial stage is not ready, then the project may fail. If the attention is only focused on the job of the construction costs management, the project contract price, the project price change, and the project settlement at the construction phase, and some owners even at the decision-making phase try to get approval who will deliberately estimate the investment less, it will emerge "angling engineering projects". At the designing stage, some of the designers' informal cost indicators, conservative designing ideas, and the above mentioned problems make many new expanded projects extend the time limit for the projects, gain a poor investment benefit, have unqualified estimate results, and default on project funds. So by making the positioning on the cost proper and paying attention to the various stages,

a well graded and well arranged project can be constructed which can take a very good control of the cost and meet the requirements of the maximum cost target.

1.2 *Cost management is of low professional level*

Firstly, due to the imperfect construction market, cases of administrative intervention occur frequently. Cost consulting service is poor. In most cases, project settlement is only on the side of the quota party. Secondly, whether the project can be continued smoothly is directly determined by the strictness of the contract terms. If the terms of the contract are too general, it may cause the two parties to have different opinions which will lead to the project can not be completed on schedule. Besides that, another reason for the unsatisfactory cost management in the real estate is that construction costs personnel have narrow and a lower level of knowledge; they can only understand the drawings or edit construction budget or either of them. Those construction personnel with a low professional level can not meet the requirements of the modern construction costs management "to calculate construction costs and to effectively control the cost".

1.3 *To reduce costs and organize management improperly*

The construction costs personnel in the real estate enterprises always reduce costs blindly and ignore the project's positioning and grade. It causes many companies to establish an office building that is originally positioned as class A, but when it is completed, it can't even up to class B which will cause a lot of trouble for sale. When selecting the projects, the project grades and cost objectives need to be decided first. To improve the level of the project will increase sales, satisfy consumers' demand, and promote the development of the real estate industry. So the blind reduction of the cost can't bring profits to the real estate developers, but play an opposite role. As construction costs personnel, they should be aware of that cost management is not only about blindly lowering costs, but also about improving the level of the project. The cost needs to be within a suitable scope, getting maximum profits with minimum costs. In order to achieve the purpose of decentralizing the authority to restrict each other, real estate enterprises divide cost control departments into purchasing, contract, and pre settlement. The departments restrict each other so that none of them take the responsibility for the cost and even shirk their responsibilities which results in the cost control in a state of chaos. The construction belongs to one project, so various departments need to be connected despite of the division of labor. If various departments can't be harmoniously connected which will disrupt the circulation of information and result in information isolated island phenomenon, construction costs management is not truly implemented and this is not successful. Cost control is a systematic project and runs through every stage in the construction process. Various departments are in charge of individual responsibilities and result in the failure of completion. So we can learn from the cost management in other countries.

2 MEASURES OF CONSTRUCTION COSTS MANAGEMENT IN THE REAL ESTATE

2.1 Construction costs management must realize informatization and form a network

Everyone must have heard the term neural network. Construction costs management can also use the theory of neural network. This method by simulating human brains to search for information about the project and relying on experience to determine the construction cost is similar to the pattern of Hong Kong. If there is a very rich project database, some advanced technologies can be applied to the construction costs management. We lack them because of lower social average level and enterprise individual level, but at the same time, it also avoids a complicated problem, that is, during the evaluation of bid, the enterprise's quote is lower than the cost quote. In order to make the personnel actively make decisions for the project instead of passively reporting cost results, the problem which should be solved first is to train a batch of project cost management personnel who can satisfy the needs of socialist modernization to solve the contradiction between the

quality of talents and the enterprise's needs. Carrying out certified qualification system of registered cost engineer can help solve this problem. Chief economist and senior management can be trained to meet the need of socialist modernization. By strengthening the construction of College engineering cost management discipline, cost management talents who have knowledge on skills, economy, laws, management, computer, and foreign languages can be trained. Today is the information age, computer network technology is very common; software and network have become the best. So cost management personnel must learn the advanced technology to accept the challenge of the modern world. Nowadays, network has a very strong potential. The introduction of advanced technologies reflects that the project in the real estate industry has made some innovations, which can break down the traditional management and operation patterns and methods bringing incalculable benefits for the enterprise. The application of advanced technologies reflects that we can keep pace with the world and is also the outcome of keeping pace with the times.

2.2 Construction costs control and management at the decision-making stage

Any investment decision made at the decision-making stage is very important for the whole project because it is the begging of the cost control. The procedure is to submit project proposals, carry out studies on the feasibility of the project, draw up the investment budget, and make a detailed assignment book. The last one is a crucial stage, so at this stage construction costs control (capital operation, debt repayment, etc.) should assign responsibilities to each individual and make a strict system. The government or the supervisor can strictly supervise and manage the project by establishing supporting measures. When carrying out supervision and management or making market analysis, we should follow the principle of seeking truth from facts. Carping will make decision-making on the project investment with blindness and risk. We should try to consider the project which can increase the market competitiveness and implement the project legal person system.

2.3 Strengthening costs control at the designing stage

Firstly, nowadays, technical design has a lot of design quality problems. For example, the design quality is poor and can't live up to standards. We should promote quota design and can't allow designers to change randomly, which will result in the great loss of construction costs control. The establishment of check-up and reward systems can inspire designers to rise with spirit. If we analyze the problem when designing a construction only from the perspective of drawing but not considering its true value, the construction design has no meaning at all. We should make a comprehensive analysis of the design, optimize the design scheme, and reduce the cost of the project in order to make the real estate industry achieve greater benefits. Secondly, some designers randomly change sizes and standards of the design without developers' permission which may bring cost losses to them. Generally speaking, it's not allowed to change the design. If the design must be modified, it must get the developers' permission. Important control measures of cost management include calculating the cost brought by design changes and redesigning when problems arise because of the changes. Problems like arbitrarily changing the design and poor economic consciousness should be handled by adopting systems to prevent any arbitrary change and signing contracts and terms. For example, design directors need to deduct certain design fees if the change causes economic costs. Therefore, we should enact binding terms to regulate design standards, project quantities, budgetary estimates and budget indicators.

2.4 Dynamic tracking at construction stage and settlement auditing at completion stage

In the process of construction, the construction cycle, the prices of materials and equipment, the market supply and demand fluctuations, policy adjustment and other factors need to be considered. We should make monthly construction budget analysis to explain the reasons for the project changes if the amount increases or decreases, so that the deviation is corrected and the project dynamic change can be controlled.

At settlement stage, we should seek truth from facts, make strict examinations, and check every system in the contract to check whether the project is qualified. Auditors should not only master the rules to calculate the project quantity, but also know construction design procedures and skills. The calculation is based on as-built drawings, design change orders and on-site visas. Project quantities can be calculated in accordance with national rules. The project must conform to the local government's relevant regulations and then the project can be accepted as qualified.

3 CONCLUSION

With constant developments of society, the real estate industry develops rapidly; people's living standards have made some improvements and there is a soaring demand for houses. The price in the future real estate industry is not clear. The enterprise which can better control the cost will possibly gain a foothold in the competitive market and get the right to use and develop the land. As a person in charge of cost, cost management is very important. We should strictly control the designing process to control and save costs, and improve the quality of the project. Cost management control is the real estate's indispensable stage.

REFERENCES

[1] Zhao Lingzhi. A brief discussion on innovation patterns of project management in the real estate [J]. Value Engineering, 2013, (25):76–76, 77.
[2] Mo Ning. Construction costs management in the real estate under macroeconomic control [J]. Value Engineering, 2014, (4):78–79.

Information, Computer and Application Engineering – Liu, Sung & Yao (Eds)
© *2015 Taylor & Francis Group, London, ISBN 978-1-138-02717-6*

A game theory based analysis on the supervision over a public company's accounting fraud

Xiao Zhou Xu & Zhi Wei Zhang
Sichuan TOP Vocational Institute of Information Technology, Chengdu, Sichuan, China

ABSTRACT: With the development of capital market, listed companies in our country have made great progresses and developments, no matter in quantity, size, strength, or the operation and management. But the public company accounting fraud has brought serious damages to social and economic development. This paper uses the game theory analysis method analyses the problem of public company accounting oversight, according to complete information static supervision game point of view. The author analyzes the game equilibrium results and put forward some countermeasures of our country public company accounting oversight.

KEYWORDS: Game theory; the listed company; Accounting fraud; Accounting oversight

1 INTRODUCTION

Accounting fraud is a worldwide issue, recently, we witnesses a lot of case involving public company accounting fraud. These frauds have strong negative impact on social and economic credit and stability. The consequence is severe, it will weaken government's efficiency in macro management over economy, and it also will mislead investors and creditors' choice making, finally, it will increase the exchange rate and stop the transactions to be done. The impact of accounting fraud has aroused a lot of attention and supervision not only from abroad, but also inside the country. Considering this background, this thesis uses relevant theories of game policy to draw a static game policy model with incomplete information in exploring of how to enhance the supervision over accounting fraud.

2 BACKGROUND AND LITERATURE

Accounting fraud is not a new issue; it has its foot stage not only in overseas countries but also in Chinese companies. Many scholars have a large amount of literature on this issue.

The literatures are always focusing on accounting information; there are little literature on exact item such as accounting fraud. They tend to use the items of financial statement fraud or earnings management. Meanwhile, the literatures have great amount of portion focusing on fraud behaviors and motivations, they consider the fraud as a dynamic and organic system that cannot be fully eliminated. Their solutions to fraud are strengthening inner and outer supervision.

Bayou considers fraud system as an open system, it will occur since the build-up of this system. Even though the cost issue is not the restrain, fraud will not disappear. Zabihoolah Rezaee analyszes systematically the motives and reasons why fraud occurs. He thinks that the motives are tax evasion, catering to analyst's expectation of company profit's increase, abusing of company's assets and hiding management's improper behaviors. Furthermore, he points out that the inefficiency of corporate governance is the key reason of accounting fraud.

According to the existed information and literature mentioned above, this thesis will use the static game policy model with incomplete information to draw the analysis of public company's accounting fraud. Through the perspective of economics, the thesis analyzes both parties' profit model, and brings out some solutions to solve the problems of accounting fraud.

3 GAME POLICY AND ITS BASIC HYPOTHESIS

3.1 Game policy and the research chart

Game policy is an economic theory to research decision maker's behavior and its equilibrium when behaviors have relationship and cross effects.

Static game policy with incomplete information means that all players join into the game simultaneously without observing other players' choices. Supposing other players' decision, every player's decision is completely subjected to his own type.

Because every player only knows the probabilities and types of other players, and he has no information about the real type of other players, so he has no possibility of knowing other players' decisions. The only information he can get is to predict other players' decision according to their types.

According to static game policy with incomplete information, the thesis builds up a relevant model

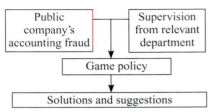

3.2 Basic hypothesis of game policy

In the real world, it is not deterministic whether companies do accounting fraud or not, relevant department take supervision or not, some people report or not. Public companies' accounting fraud is one of the various marketing behaviors. All players joining into the game include public companies and supervision departments, and all players are not capable of knowing others' decision. The game policy model is here to analyze their behaviors and equilibrium.

Hypothesis 1: There are only two players in the game: public company and supervision department. Both parties know each other's game policy structure and his own payoff.

Hypothesis 2: The supervision department has the liability to audit public company's financial statement. There are two choices of the supervision department: conspiracy and no conspiracy.

Hypothesis 3: Public companies have the motive and chance to make accounting fraud. There are two decision portfolio: fraud and no fraud.

According to the real environment, further hypothesis are:

First, supposing company's fraud profit is a (a>0), where tax evasion, share allocation and other risk avoidance could make positive profit to a company. If not, the profit is -a

Second, supposing company's penalty of being discovered of fraud is f (f>0)

Third, supposing the cost of supervision department is c(c>0) under the condition of no conspiracy. The cost of c is the total cost of the supervision department in planning and implementation of the supervision. If not, the supervision department will have a cost reduction of c.

Suppose the company makes accounting fraud, and the supervision department has no supervision over the company. In this circumstances, there are two occasions: some people report to make the company

discovered of being fraud (not being penalized yet), then the supervision department will be penalized as d (d>0); the second situation supposes that no one reporting, and the company's fraud is not being discovered, so the supervision department also has no profit loss.

4 GAME POLICY MODEL AND ANALYSIS

4.1 The game policy model

According to the aabove-mentioned hypothesis, the game policy model of accounting fraud is displayed below

Table1. Game policy model for public company's accounting fraud.

		Supervision dept.			
		conspiracy		No conspiracy	
		Being discovered	No being discovered	Being reported or unveiled	Not being reported or unveiled
Public company	fraud	(-f, f-c)	(a, -c)	(a, -d)	(a, 0)
	No fraud	(-a, -c)	(-a, -c)	(-a, c)	(-a, c)

Note: payoff is in parenthesis: the first digit is the payoff of the public company; the second digit is the payoff of the supervision department. From the chart above, apparently, there is no advantageous decision or pure Nash equilibrium.

This thesis supposes that the game policy is incomplete information game policy, so mixed strategy Nash equilibrium can be solved. Furthermore, a stepped hypothesis is drawn: whether public make fraud or not, whether supervision being taken or not, whether supervision works or not, whether people report or not, all possibilities mentioned above are unsure.

To make these more feasible, this thesis supposes: company making fraud's probability is p, not making fraud's probability is 1-p; supervision department's supervision probability is q, 1-q to the counter case; supervision department's finding fraud's probability is r, 1-r to the counter case; and supervision department's being reported of breach of duty's probability is w, 1-w to the counter case.

Supposing company's expected payoff is π_1, so the company department' payoff is

$$\pi_1 = p \cdot \{q \cdot [r \cdot (-f) + (1-r) \cdot a] + (1-q) \cdot [w \cdot a + (1-w) \cdot a]\}$$
$$+ (1-p) \cdot \{q \cdot [r \cdot (-a) + (1-r) \cdot (-a)] + (1-q) \cdot [w \cdot (-a) + (1-w) \cdot (-a)]\}$$

Supposing supervision department's payoff is π_2, so the supervision department's expected payoff is

$$\pi_2 = q \cdot \{r[p \cdot (f-c) + (1-p) \cdot (-c)] + (1-r) \cdot [p \cdot (-c) + (1-p) \cdot (-c)]\}$$
$$+ (1-p) \cdot \{w \cdot [p \cdot (-d) + (1-p) \cdot c] + (1-w) \cdot [p \cdot 0 + (1-p) \cdot c]\}$$

Whereas the first order condition of payoff maximization of company and supervision department is

$$\frac{\partial \pi_1}{\partial p} = 0 \quad \text{and} \quad \frac{\partial \pi_2}{\partial q} = 0$$

Therefore, this game policy's mixed strategy Nash equilibrium solution is

$$p^* = 2c/(rf + wd + c) \quad \text{And} \quad q^* = 2a/[r(f+a)]$$

According to the above equation, it can be concluded that company will take P^* as their hypothesis to make fraud or not and supervision department will take q^* as their hypothesis to take supervision or not.

4.2 Model analysi

Some conclusion can be drawn from the above analysis.

4.2.1 Game policy analysis from the angle of public company

When a company makes fraud, the probability of discovering company's fraud increases (r) will decrease the probability of company's making fraud (p) for low level of supervision, bad operation environment or low corporate ethics. On the contrary, the decreasing probability of discovering company's fraud (r) will increase the probability of company's making fraud (p).

If supervision department will not take any action to companies, if the probability of company's being reported (w) increases, the probability of company's making fraud (p) will decrease. That means that the inner-company supervision and public surveillance has efficient function to supervise. If the penalty to supervision (d) is increased, it will burden up their liability to take efficient actions to supervise the public company's fraud. Finally, it will decrease the probability of company's making accounting fraud.

The increase of supervision department's cost (c) will increases the probability of company's making fraud (p* is the increasing function of c). That will weaken the motives of the supervision department.

4.2.2 Game policy analysis from the angle of supervision department

Increasing the penalty to company's making fraud (f) will decrease the probability of supervision department's supervision (q). That will increase 1-q because it will have the motive to abuse their rights and lose their power. The payoff of company's making fraud (a) will increase the probability of supervision (q). That is to say, if the company has great willingness and motive to make fraud will increase the burden and liability of the supervision department.

The increase of probability of company's fraud being discovered by the supervision department (r), the probability of supervision department's being not supervise (1-q) will increase for the level up of surveillance and risk of penalty coming from the public company. The probability of company's making fraud (p) will decrease, so the supervision is not always needed in this circumstances.

5 CONCLUSION AND SUGGESTION

Through the game theory analysis of public company's accounting fraud, it can be concluded that there is apparently a relationship between public company's accounting fraud and the check and penalty from the supervision department. Considering the circumstances, this thesis will provide some solutions and suggestions to decrease public company's accounting fraud from three perspectives: inside company, supervision department and economic environment.

5.1 Enhances the intra-company control

China has taken a new intra-company control system, but the stable and efficient intra-company control system can supervise all the circulation of accounting information's generation and discovery to eliminate the accounting fraud from its origins. Stable public company intra-company control not only requires company to discover information about intra-company control but also requires company's board of supervisors and committee on public accounts to give measurement and suggestion. In doing so, it will let the board members to shoulder more burden of surveillance and restrain to the general manager and find out quickly the improper behaviors of the general manager. Meanwhile, independent director plays an important role in the generation and discovery of accounting information, they have the liability to check company's accounting rules and most importantly the connected transactions. Intra-company auditors should take responsibility of measure and sign signature on intra-company report on audition, which will drive them to be more conscientious and finally improve company's competitive power.

5.2 Urges the supervision department taking efforts necessary to supervise and penalize the improper deeds by public companies

Surveillance should be powered up to penalize public companies which have improper behaviors and keep balance between rewarding and penalizing. Meanwhile, it is very important to realize that the penalty is not unlimited. Government should cut the cost of surveillance, innovate the surveillance system and strategy and optimize personnel and process to cut the cost of supervision departments. If the cost has been lowered in some degree that public company's accounting fraud is easily been discovered, rationally speaking, most public companies will not make accounting fraud for cost reasons.

5.3 Construct sound social system and environment

The management and accountants of a public company are the direct involvers and generators of accounting information. To take thorough solutions to avoid accounting fraud and enhance the personnel's business ethics, it is essential to let these professionals feel that the real terror comes from fraud. The construction of accounting confidence and credibility system is not only the requirement of market economy but the requirement of basic social ethics. Everyone in a society must take the rule of game play, social channels are needed to provoke and publicize this idea and make the whole society realizing the importance of credibility and the hazard of not doing so.

Publicity is another important channel; mass media plays an important role in surveillance and supervision. Mass media can play the role of "rangers" using its keen insight, unique professional advantage and the empowerment of law to supervise and report any information about public company's improper behaviors. The prompt report from the mass media can offer the information to every corner of the society to function the system by the effect of restrain and control. From a great deal of events involving accounting fraud, mass media's surveillance is of great importance. Finally, a sound environment will be founded.

REFERENCES

Safeguard and protection of migrant workers' interests from the perspective of game theory. *Asian Agricultural Research*, 2010, 11:20–25.

Bracken T. Wimmer,Ivan G. Krapac,Randy Locke,Abbas Iranmanesh. Applying monitoring, verification, and accounting techniques to a real-world, enhanced oil recovery operational CO_2 leak. *Energy Procedia*, 2011, 4:3330–3337.

Ursula F. Ott. International business research and game theory: looking beyond the prisoner's dilemma. *International Business Review*, 2012:480–491.

Anna Jadlovská,Kamil Hrubina. Algorithms of optimal control methods for solving game theory problems. *Kybernetes*, 2011, 401:290–299.

Chun-Keung Hoi,Ashok Robin. Labor market consequences of accounting fraud. *Corporate Governance*, 2010, 103:113–143.

Ratul Lahkar. Evolutionary game theory: an exposition. *Indian Growth and Development Review*, 2012, 52:202–213.

Svitlana Kuznetsova. The transformation of accounting systems in the chaotic economy structuring: The synergetic approach. *Risk and Decision Analysis*, 2011, 23:509–525.

Information, Computer and Application Engineering – Liu, Sung & Yao (Eds)
© 2015 Taylor & Francis Group, London, ISBN 978-1-138-02717-6

Developing mode and strategic research for education in universities with industry characteristics—exemplified by Jingdezhen Ceramic Institute

Xin Feng & Jing Yi Sun
Jingdezhen Ceramic Institute, Jingdezhen, Jiangxi, China

ABSTRACT: Innovation has become a main driving force for economic and social development in today's world. Under such background, it is of great significance for universities with industry characteristics and significant education characteristics to elevate their original innovation, integrated innovation and the capability of re-innovation, as well as actively promote collaborative innovation.

KEYWORDS: University with Industry Characteristics; Collaborative Innovation

1 INTRODUCTION

Universities with industry characteristics are the main force of Chinese higher education, but presently those universities are generally faced with new challenge and developing bottleneck of technological innovation. Although with such difficulties, universities having industry characteristics still bear obvious advantages over others, because they possess deep industry background, unique subject priority, profound industry influence, and same cultural feelings.

2 THE INHERENT INDUSTRY BACKGROUND

With a long period of practicing education, universities with industry characteristics have come into being to promote the development of their industries. So they have formed their educational advantages and subject features. These universities are important forces to push forward the industry development. Together with the industry development, they have emerged, survived, and developed. The rise and fall of the universities are closely connected with the vicissitude of industries. The developing level of universities with industry characteristics, to some degree, represents the industry level and industry competitive power in China.

Jingdezhen Ceramic Institute has always been regarded as the birth place of ceramic talents. The origin of the institute was China Pottery Industry School founded in 1910. In 1927, it began to be managed by Jiangxi Province, and in 1947 it was named Ceramic Industry College. In 1958, it got its present name and began to recruit undergraduates. Formerly, it was under the jurisdiction of National Ministry of Light Industry; but in 1998, it started to be constructed by both central and local governments. Now, Jingdezhen Ceramic Institute is the only college with ceramic characteristics and multi disciplines in China. The ceramic characteristics especially distinguish it from other colleges; therefore, it takes the lead in ceramic industry and field.

3 THE DISTINCTIVE SUBJECT CHARACTERISTICS

Universities having industry characteristics take advantages in exploring the leading edge of the discipline, grasping discipline direction, crossing and fusing related disciplines to promote co-development. Such universities will not set the aim to make them huge and comprehensive; instead, they focus on the growth of their advantageous features. Surrounding the specific industry and service object, they develop forward with clear and definite development orientation.

For years, Jingdezhen Ceramic Institute shoulders the responsibility of advancing and spreading Chinese ceramic culture, and reviving Chinese ceramic industry. As a result, to walk the way of characteristic development is the consistent guideline of its educational theory and practice. The institute sticks to the idea of rooting in ceramic industry, serving local economy, catering to the nation, and walking towards the world. As the base of ceramic education, it has accumulated amounts of experience for talent cultivation, characteristic subjects, scientific research, artistic creation, and community service. With these orientations of running a college, Jingdezhen Ceramic Institute has shaped its unique educational characteristics, and enjoyed preferable competitive advantage and power. On July 11th, 2013, The Academic Degrees Committee of the State Council formally approved that Jingdezhen Ceramic Institute is authorized to grant doctorate. Material science and engineering, together with design science, is the level discipline to have the authorization for doctor's degree, which shows that the institute has obtained another break-through for its level of higher education. To establish a characteristic college with both domestic and for-eign influence, it has made a key step forward.

4 THE CULTIVATION OF CHARACTERISTIC INDUSTRY TALENTS

For the subjects and specialties, universities with industry characteristics closely centre on their related industries and industry chains. Therefore, these uni-versities have both their characteristic core disciplines and other related disciplines to meet the demands of their industry development. The arrangement of sub-jects makes their particularity and applicability prom-inent, and emphasizes on forming a talent-educating idea of combining learning with research and produc-tion, so as to establish an integrative teaching system with a good combination of theory and practice.

Take Jingdezhen Ceramic Institute as a typical example: with years of development, the institute has become a characteristic college for its principal subjects of ceramic engineering; besides, it has established other related subjects such as arts, economy, and management. Thus, it has developed into a well-known college for its integrative education system and professional ceramic subjects. For decades, the college has cultivated thousands of ceramic talents to promote the development of Chinese ceramic industry. Guided by the idea to take lead in ceramic industry, serve local economy, cater to the nation, and walk towards the world, the institute has trained numerous graduates and undergraduates to serve the country nationwide, many of whom have become the navigator of ceramic factories and institutes. At the same time, a large number of famous artists, scholars, executives who graduated from Jingdezhen Ceramic Institute are contributing to advancing Chinese ceramic culture and reviving Chinese ceramic industry.

5 THE NEW STRATEGIES FOR COLLABORATIVE INNOVATION

5.1 *To set discipline construction as the core*

In these years, universities with industry characteristics have quickened their pace toward multi disciplines. Although the disciplines have been

greatly enriched, the characteristic disciplines should never be undervalued or replaced, because they are the root and base of these universities. Presently, there are some universities that have shifted their attention from characteristic disciplines to others (mainly because of financial temptations); as a result, these universities are in lack of energy and money to be invested into their characteristic disci-plines. Therefore, their characteristic disciplines will become bankrupt for their features, insufficient in innovation, and short of forward planning. Gradually, their characteristics will disappear, and the devel-opment of the universities will be greatly hindered. Hence these universities must have an overall outlook for their development: to scientifically rearrange the distribution of disciplines, and to continuously con-centrate on their characteristics. They should resist the temptations of perfection and finance, and cultivate the distinct research direction of their characteristics based on the present foundation and condition. They should also elevate their competitive power to better serve the local economy and social development.

This perspective can also be exemplified by Jingdezhen Ceramic Institute. After having got the authority to grant doctorate, the institute pours most of its energy and finance to construct the level disciplines (material science and engineering, design science), in order to unswervingly stick to and emphasize its characteristic subjects. On the basis of this, it aims at producing a series of landmark fruits and cultivating a group of influential leading fig-ures and outstanding talents in ceramic industry. The institute tries to perfect the present characteristic dis-cipline system and, at the same time, boost the fuse of different disciplines. As for its characteristic subjects, the institute is now engaging in forming a discipline group made by subjects like art design, ceramic cul-ture, ceramic material engineering and machinery, ceramic economy and management. The purpose is to promote the combination of different subjects, to optimize the mode of discipline construction, and to establish sound teams of characteristic disciplines. Through the formerly mentioned means, Jingdezhen Ceramic Institute wishes to set up a scientific con-structing and managing mode which is made up of perfect disciplines, various degrees, academic eche-lons, and developing platforms.

5.2 *To regard technological innovation as the symbol*

The level of scientific research symbolizes the comprehensive power of a university, and it is also related to the enhancement of teaching and talent-cultivating qualities. Technological inno-vation demands universities to bring about more original, creative, high-level research findings to meet the requirements of economic construction and social development. It is high time that universities changed the singular research unit of project group or laboratory, which is like the primitive manual workshop; on the contrary, they should spare no effort to promote collaborative innovation and to enhance the capability of technological innovation. Meanwhile, universities ought to integrate their resources of scientific research, form and take advantage of all kinds of advanced scientific research bases and platforms, select and cultivate potential technological talents, and elevate their scientific research level and academic status with no stop.

In 2003, Jingdezhen Ceramic Institute, Ceramic Research Institute of National Ministry of Light Industry, Jiangxi Ceramic Research Institute, Jingdezhen Special Ceramics Research Institute began to integrate for better resources of ceramic technology. After having been approved by Ministry of Science and Technology, "The National Daily and Architectural Ceramics Engineering Technology Research Center" came into being in Jingdezhen Ceramic Institute. It is the first engineering technology research center authorized by Ministry of Science and Technology in Jiangxi Province, and Jingdezhen Ceramic Institute becomes one of the few universities that are qualified for such a center.

Under the background of new times, Jingdezhen Ceramic Institute should especially depend on its characteristic disciplines; along with the perfect platform of The National Daily and Architectural Ceramics Engineering Technology Research Center, it should also actively cooperate with domestic and foreign high-level universities and institutes, and famous large enterprises. In addition, the platform for technological innovation ought to be perfected; the pace of systematic reform and innovation for scientific research ought to be quickened; the collaboration of production, education, research ought to be improved; the commercialization of research findings, especially industrialization development, ought to be vigorously promoted. These steps are to enable the institute to become the promoter of new industries and the booster of local development.

5.3 To count serving the society as the aim

At present, universities with industry characteristics have the inclination to break away from regional economy and social development to some extent: the collaboration of production, education, research is in a low level; the conversion of technological findings is at a low rate. Therefore, quantities of technological

fruits cannot be converted into practical productivity. These universities are supposed to vigorously strengthen its function of serving the society, and actively borrow from both domestic and foreign universities the successful experience for collaboration of production, education, research. Moreover, they must promote the cooperation with industries, and establish a strategic alliance which regards industries as main body, the market as orientation, projects as the carrier, and the capital as the link. This alliance is characterized by its long term, tightness and marketization.

In these years, based on serving the local economy and industry development, Jingdezhen Ceramic Institute has put more energy in collaboration of production, education, research. Altogether, it has supplied almost 200 scientific fruits to the related ceramic factories, and set up long-term cooperation with over 30 factories from Foshan (Guangdong Province) and Gao'an (Jiangxi Province). Through the transfer of scientific fruits, technological development and service, the institute has achieved economic benefits of nearly 10 billion; for this reason, the First Chinese Collaboration of Production, Education, Research awarded the institute a prize for innovation. Under the circumstances of national innovative strategy, Jingdezhen Ceramic Institute should make full use of its advantages to face the challenge, utilize its mature research and development to unite ceramic factories, institutes and platforms. With the mode of collaborative innovation, the institute is surely to develop vigorously and successfully.

6 CONCLUSION

To construct a famous, characteristic, influential university nationwide and worldwide means to continuously enhance its core competitive power, academic influence, and contribution to the society. Universities with industry characteristics, serving as the driving force to carry out national innovative strategy, are supposed to grasp the opportunity for collaborative innovation, seek and highlight their advantages, and achieve prosperity based on their characteristics.

ACKNOWLEDGMENTS

This article is the stage achievements of Jiangxi Social Science Planning Project, NO. 13JY53. It is also the stage achievements of Jiangxi Art Science Planning Project, NO. 2014–169.

REFERENCES

[1] Chen Lin, Wu Guilong. Advantages and means of collaborative innovation for universities with industry characteristics. *College Education Management*, 2013, (03).
[2] Guo Guangsheng. the difficulties and measures for constructing high-level universities. *Chinese University Science and Technology*, 2011, (12).
[3] Yin Xiangwen. The fixed position and value pursuit of collaborative innovation in universites. *Chinese University Science and Technology*, 2012, (07).
[4] Zhan Mingrong. The origin of Jingdezhen ceramic history and culture. *Ceramic Science and Art*, 2012, (03).

BRIEF INTRODUCTIONS TO THE AUTHORS:

Xin Feng (1983–), Han nationality, female lecturer, Master's degree; her main research field: theory of literature and art, ceramic art criticism.

Sun Jingyi, (1977–), Han nationality, female lecturer, Master's degree; her main research field: ceramic culture, ceramic intercultural communication.

Information, Computer and Application Engineering – Liu, Sung & Yao (Eds)
© 2015 Taylor & Francis Group, London, ISBN 978-1-138-02717-6

On the everlasting charm of *A Rose for Emily*

Hong Chun Deng

Jingdezhen Ceramic Institute, Jingdezhen, China

ABSTRACT: *A Rose for Emily* is the most famous short story by William Faulkner, the representative of American southern writers and the Nobel Prize laureate in literature. Since its publication, this story has been popular among readers and critics as well as the topic of great interest for American literature research. Based on the reading of the story and related research articles, the everlasting charm of the story is discussed from three aspects: its unique narrative skills, wonderful gothic elements and profound themes.

KEYWORDS: *A Rose for Emily*; charm; narrative skills; gothic elements; themes

1 INTRODUCTION

William Faulkner, the famous American writer in the 20th century and the Nobel Prize laureate, was born in a declining southern plantation family in Mississipi in 1897. In total, he composed 19 novels and 125 short stories, most of which are set in the fictional Yoknapatawpha County, the prototype being his hometown, and depict the history of the south of America. Thus he is considered as the representative of American southern writers. *A Rose for Emily* is his most famous short story. In 1930, it was published and attracted considerable publicity. Eighty-four years passsed, yet this story still boasts great charm for readers. The author of this article intends to make an analysis of the everlasting charm from three aspects: its unique narrative skills, wonderful gothic elements and profound themes.

2 EVERLASTING CHARM OF A ROSE FOR EMILY

2.1 *Unique narrative skills*

Faulkner is honored as "a great experimentalist on novel skills among novelists in the 20th century." The unique narrative skills in *A Rose for Emily* perfectly reflect his talents as a great writer.

This story gives up the traditional narrative mode of natural time sequence, and adopts the mode of time disorder which means the disagreement between the natural time of the story and that in narration. The natural time of the story cannot be altered, while the time in narration can be rearranged by the narrator. This story falls into five sections. The first section tells about Emily's funeral, then recalls her refusal to pay the tax when the officials come to her home ten years before her death. The second section tells us the efforts made by officials to deal with the decayed smell in Emily's house 30 years ago, and her father's death 32 years ago. The third section tells us Emily falls in love with Homer after her father died, and a year later, she buys arsenic and her two sisters-in-law come to visist her. The fourth section explains the reason why her sisters-in-law come and tells us Emily orders a man's toilet set and clothes, Homer disappears, and Emily teaches children how to panit on the ceramics when she is 40. The fifth section tells us Emily dies at 74 and people in the town find Homer's corpse. It can be seen that the story begins and ends with Emily's funeral. With her father's death as the turning point, the plots are arranged with two clues. The first clue begins with Emily's death with flashbacks of the tax paying issue 10 years ago, then the smell affair 30 years ago and her father's death 32 years ago. The second clue adopts a natural time order: her father's death, buying arsenic, the visit of sisters-in-law, the order of toilet set and clothes, Homer's disappearance, Emily's death and the finding of Homer's corpse. Faulkner employs the narrative mode of time disorder as if he is building blocks with time. At first glance, readers will feel that they are lost in the maze of time and become confused. However, after careful reading, they will suddenly see the light and admire Faulkner's elaborate arrangement. Just because of this special narrative mode, the story avoids the boredom which the traditional narrative mode will cause, while it increases the reading difficulty and suspense and involves readers to guess what will happen next with great interests.

2.2 Wonderful gothic elements

Gothicfiction is a new type of fictions which developed in the late 19th century. It usually chooses old castles, wasteland or wilderness as the background. The plots are horrible and thrilling with violence, murder, death and revenge, even accompanied by ghosts and transnatural phenomena, thus creating a misterious and ghastly atmosphere. *A Rose for Emily* is universally recognized as a classical gothicfiction, because it embodies all the wonderful gothic elements: a gloomy high mansion, an eccentric spinster, a dumb balck servant, a decayed male corpse and horrible murder.

Most of the plots are set in Emily's torn-out mansion, which is the last symbo of her noble identity. After her father's death and her lover Homer's disappearance, Emily lives a solitude life, and few people can enter her prison-like big house. In the tax paying issue, Faulkner describes the gruesome scene of the house: the aged black servant ushers the government officials into the dim hall and through a dark stairway. Because the room has been vacant for a long time, it smells of dust and disuse. The furniture is covered by leather which is cracked. When the black servant opens the blinds of one window and the officials sit down, a faint dust rises sluggishly about their thighs, spinning with slow motes in the single sun-ray......
At the end of the story, the town residents enter the room upstairs which no one has seen in forty years. The room is furnished as for a bridal, yet a thin and acrid pall as of the tomb seems to lie everywhere: curtains of faded rose color, the rose-shaded lights, the dressing table, the man's toilet things backed with tarnished silver, the man's collar and tie, the carefully foded suit, the two mute shoes and the discarded socks, the man's skeleton in the bed...... All the things make us recall ghosts and feel absolutely terrified.

The characters in the story give us a feeling of eccentricity and terror. As a member of the declining noble family, Emily still thinks highly of herself. She vanquishes the goverment officials, horse and foot. She was arrogant and impolite and refuses to tell what it is for when she buys arsenic. She repels the progressive social change, refuses to let people fasten the metal numbers above her door and attach a mail box to it, and lives all alone. In the tax paying issue, "she looked bloated, like a body long submered in motionless water, and of that pallid hue. Her eyes, lost in the fatty ridges of her face, looked like two small pieces of coal pressed into a lump of dough as they moved from one face to another while the visitors stated their errand.", impressing us as a bloodcurdling living dead human being. Her black servant, Tobe, is also like a walking corpse, silent all day long, takling to no one, walking in and out with a basket in his hand. The eccentric, unsociable and escaping characters make us feel extremely terrified as if falling into a hell.

Faulkner is also good at setting up foreshadowing and suspense to create a horrible atmosphere. When her father dies, Emily insists that he is not dead and refuses to bury him for three days. She does not compromise until she is to face the penalty by law. In fact, her abnormal obsession with her dead father establishes an idication to her sleep with Homer's corpse for as long as forty years. What is more, Homer once says that he likes men and he is not a marrying man, which establishes a foreshadowing for Emily's murder of Homer. Besides, the strange smell from Emily's house, which vanishes after officials sprinkle lime, serves as a suspense for readers to guess the real source of the unpleasant smell. Another suspense is Emily's refusal to tell her purpose for buying arsenic. Does she buy it to kill rats, kill herself or kill other people? Till the end of the story, when we see the male skeleton in the bed, the whole truth of murder comes out, leaving readers a great shock.

2.3 Profound themes

On the surface, *A Rose for Emily* tells a gothic love tragedy, in which a declining noble, Miss Emily, murders her lover and sleep with his corpse for forty years. However, after careful reading, we will find that the story contains more profoud themes and expresses Faulkner's deep concern with human beings.

After the American civil war, the plantation-based southern economic system and slavery collapsed, and the southern noble class suffered an unprecedented fatal failure. However, the noble tradition and philosophy are still deep-rooted in the declining southern noble class and general public, incapable of being removed in a short time. Emily was born and brought up in a noble family, thus forming a personality of arrogance and pride. Even if her family has lost power and her father dies, she stilll thinks highly of herself and tries every means to maintain the noble tradition so as to reveal her noble identity. She refuses to pay the tax to the government and tell the real usage of the arsenic to the druggist. She rejects the metal numbers above her door and mail box and lets the unpleasant decay spread to the neighborhood. All these evolve from her noble birth, while few people in the town can deal with her.

On the other hand, as a member of the noble family, she is the victim of the old southern patriarchy society and puritan doctrine of women. In the Bible, the god creates a man, Adam, with the earth according to his own look, and then makes a woman, Eve, with Adam's ribs. This order and way of making of a man and a woman have a great influence on the general American public who are immersed with Christianity, making females subordinate to males insistently, which is perfectly reflected in the patriarchy in the family relationship, with the father being absolutely dominant. With the plantation as its economic base, the south of America has a very strict hierarchy. As the owner of a plantation, Emily's father drives away all the young people who propose to her with a whip in order to maintain the dignity of his noble family, because he thinks they can not match Emily's noble origin. As a result, Emily is still unmarried in her thirties and she is robbed of the right to pursue her own happiness. After her father's death, Emily at long last can have the freedom to choose her own lover, but she is interfered with by all the residents in the town. According to the puritanism, a woman's virginity must not be contaminated and she should keep pure without any desire. In the eyes of the town dwellers, Emily is the monument of the southern tradition, the symbol of the noble identity and the representative of American southern virtuous ladies. Therefore, she is supposed to stay pure and noble, to supress her sexual desire. She can only marry a noble man but not an ordinary citizen. As a consequence, it is a disgrace to the town and a bad example to the young people for Emily to fall in love with Homer, a northern day laborer. What is more, it is her own degeneration and a disgrace to the noble family if she is to marry a civilian. Owing to this deep-rooted philosophy, pople in the town force the Baptist minister to dissuade her, and wirte to her two cousins to interefere with her marriage with Homer. The poor Emily has no right to pursue her own happiness. After her father's death, she plucks up her courage to fight for her happiness regardless of her own noble identity, but she still can not break away from the imprisonment of traditional women doctrines and the worldly oppression. In the end, she becomes mentally distorted and kills her lover to live her rest of life solitarily with Homer's corpse. On one hand, Faulkner mercilessly discloses

the stubbonness and arrogance of Emily, the member of the old declining nolbes. On the other hand, he sings highly of Emily's great courage to fight for her own happiness, strongly citicizes the destruction of the southern patriarchy and puritan women doctrines to the human nature and shows deep sympathy to Emily's misfortune with the presentation of a rose.

What is more, Faulkner discloses the evil of the southern slavery and racism mercilessly in the story. Emily's black servant, Tobe, serves her faithfully from a young guy to an aged man as a slave. Even after the slavery has been aboished because of the civil war, he is unwilling to leave. Tobe disappears silently only after his master, Emily, dies. Subconsciously, he has acceptped the unequality between the white and the black and become contented to be a slave for the southern noble family, which perfectly embodies the deep oppression and bondage which the slavery and racism impose on the black, and the disappearance of Tobe also indicates the complete collapse of the slavery and racial discrimination.

REFERENCES

[1] William Faulkner. *Collected Short Stories of William Faulkner*[M]. Trans. by Tao Jie. Beijing: Foreign Literature Press, 2001.

[2] Xiao Minghan. *Why Does Faulkner Present Emily with a Rose?*[J]. Masterpieces Review, 1996(6).

[3] Xiao Minghan. *The Gothic Tradition in American and British Literature*[J]. Foreign Literature Review, 2001(2).

[4] Wang Minqin. *On the Narrative Features of A Rose for Emily*[J]. Journal of Foreign Languages, 2002(2).

[5] Introduction to the author: Deng Hongchun, male, 35, teaches at Jingdezhen Ceramic Institute. Address: Jingdezhen Ceramic Institute, Jingdezhen City, China.

Information, Computer and Application Engineering – Liu, Sung & Yao (Eds)
© 2015 Taylor & Francis Group, London, ISBN 978-1-138-02717-6

The educational management that mainly centers on teachers

Jing Yi Sun, Lun Wang & Xiao Ping Yu
Jingdezhen Ceramic Institute, Jingdezhen City, P. R. China

ABSTRACT: As long as the university wants to achieve successful education, it must center on the teachers, because teachers are the main force for the development of the university; so the administrators for education are supposed to discard the old-fashioned management notion and practices, and try to concentrate on the all-side development and service for all the teachers. The university leaders and managing staff should try their best to elevate the status of the teacher, create an excellent academic environment for the teachers. In addition, as the departments serving the teacher, the educational administrations should try their best to solve the difficulties that hang around the teacher, including life problems. If these are successfully solved, teachers are sure to work with more energy and enthusiasm, so as to enhance the level of education for the university.

KEYWORDS: Educational Management; Center; Teachers

1 INTRODUCTION

The teacher, the student, and the school are generally regarded as the three indispensable elements for education. As for the three elements, the teacher is most significant. Teachers are the main body of teaching, research, and even the whole education; they should be the most essential and necessary treasure for universities and the whole educational cause. Teachers are the least easily got and least replaceable generative factor, so all members of society ought to respect the hard work of teachers. Although China's educational departments are now emphasizing human-oriented management in universities, actually they exclude all the teachers in the classroom; it is actually student-oriented. As all know, the main goal for education is to raise the quality of teaching, to improve the abilities and grades of students. To reach these goals, the all-round development of the teacher is the major premise. Without this premise, the main educational goals would not be realized. The educational management that centers on teachers asks the university administrators to regard the teacher as the root of education. The administrators ought not to impersonally use their power in hand to mange or supervise the teacher, but to place the teacher at the top of education. Anyway, it is teachers who are always studying and staying together with students, and it is they who know their students well, not the university administrators. Therefore the administrators should convert their bureaucratic practices, truly respect the teacher's value and dignity, and fully exert the teacher's potentiality. In a democratic, free, equal, just, and harmonious working environment, the teacher's

comprehensive qualities will be elevated, and they will be and should be the main body for educational development. The teachers' dedication to educational career will surely be promoted; the teachers' position in the university will also be greatly enhanced; and they will make more contribution to the development and innovation of the university.

2 DEFECTIONS OF THE CURRENT EDUCATIONAL MANAGEMENT

2.1 Low status of the teacher

Universities are part of governmental institutions, so the leaders and managing staff in universities are endowed with certain administrative ranks, which directly results in the fact that the notion of bureaucracy is extended till every corner of management. Thus, the administrative power is superior to teaching work. As a result, the educational administrators are the absolute controller of education. The consequences of this management system are that the administrators who do not know how to teach are managing teaching and controlling the teachers who know how to teach. The teachers' contribution and hardworking to teach are not highly thought of. As a result, most teachers are not active in teaching, and the teacher's zeal for education will decrease gradually, even may fade away.

Therefore, ordinary teachers in most Chinese universities can only do the jobs like teaching, correcting mistakes in students' homework, and scoring the papers. Except all these, the teacher is sheer "being controlled". In fact, it is the teacher who knows how to improve the management of the student and the university, because they are much closer to the students than the administrators. Their suggestion, however, are always neglected by the administrators, so the teacher's enthusiasm for enhancing educational qualities will gradually diminish and vanish. Many Chinese universities have the yearly convention to listen to the teacher's suggestions, but the role of the convention is minor. Most of the participants are not teachers, but administrators or leaders. Convention is only a form, not of practical value. Ordinary teachers' desires and appeals cannot be put forth in the meeting. By and by, the authority of the university, of the whole education, and even of the government will be greatly affected.

2.2 *Insufficient respect to the teacher*

From the year 1985, September 10th has become Chinese Teachers' Day; however, with the elapse of almost thirty years, the teacher's status has not been improved too much since. Many schools only regard the very day as the day for teachers' rights. As to other days left in the whole year, teachers are still being supervised or managed by the administrators. The university administrators are not willing to let the teacher become the manger of teaching or let the teacher control the education totally, because they want to grasp the right in their hands. This idea is so deeply rooted in their mind that they do not understand what teacher-centered management really means. In their mind, the teacher should only stand on their platform to teach the student; as for the matters beyond the classroom, the administrators are the king. The teacher, however, is the most principle part for teaching and research, the most essential element for the development of the university, and the most valuable treasure for education. The university administrators ought to recognize the importance of the teacher and mobilize the teacher into an active and creative working state, and they ought to establish a teacher-centered notion indeed, so as to show enough respect to the teachers, because it is teachers who are the main force of the university.

2.3 *Heavy teaching and research requirements to the teacher*

In the recent years, almost all universities have set up a series of criteria to access the teacher, but the process of carrying it out needs improving. Some criteria are so high that they have deviated from the reality and become impossible to reach. For instance, one college requires its assistant teachers to accomplish 290 hours of teaching each year, lecturers 380,

and professors 400; and requires assistant teachers to publish one paper in ordinary magazines every year, lecturers one paper in core magazines, professors two in core magazines. We all know that professors are the main force for research, but they are asked to complete so many hours of teaching at the same time. How can they give consideration to research as a result of so many burdens in teaching? This kind of criteria for assessment will easily lead to an impetuous style in academy, which is very disadvantageous to the teacher's wholehearted devotion to teaching and research, and is thus disadvantageous to the development of education. So the universities should lessen the workload of the teachers, making them teach happily and positively.

3 STRATEGIES TO ACHIEVE TEACHER-CENTERED EDUCATIONAL MANAGEMENT

3.1 *Discarding the current managing ways*

3.1.1 *Establishing a teacher-centered managing notion*

With the development of society and education, it is high time that the administrative departments established a really teacher-centered management notion at present. Only by doing this will the teacher be more enthusiastic in educational management, will they exert their utmost in the teaching career. As long as the environment for respecting the teacher and their work is formed in the university or in the society, the teacher's full potentials will be made use of to the highest degree. So the prosperity of the university can be anticipated.

3.1.2 *Making clear the roles played by administrative departments*

In fact, members of administrative departments are only the helper and servant of university education; but because they are grasping the right to manage the university, they traditionally regard themselves as the supervisor of teachers and facilities. They will supervise the teacher not to do this or not to do that. Actually, in their mind, they are superior to teachers, as they have certain rights in their hand. Nevertheless, the teacher-centered management should endeavor to change this concept. The administrators, in fact, are the service suppliers in the university. They should be supervised by the service receivers (teachers). What they have done for the teacher, the quality of their service, whether teachers are satisfied or not with their service are the proofs for their performance. With this means, members of administrative departments will surely behave themselves.

3.1.3 Innovating the methods of administration

In order to achieve the effect above, the university should train the administrative staff to acquire the relevant knowledge, to enhance their ability of service, and therefore, force them to discard the old notion and practices. The modern management of human resources is to convert the experiential and administrative management modes into human-oriented, scientific, and standard modes. Meanwhile, some management experts can be invited to evaluate the management and service of the administrators, so as to guarantee the smooth progress of administration. This is exactly the right way to carry out administration for universities.

3.2 Elevating the teachers' status

To elevate the teachers' status, the teacher-centered educational management is to start from the need of the teacher. The teacher's teaching and research fruits should be fully respected. First of all, the university ought to make clear that the teacher is the principle part of education, and then the main-body status of the teacher will be ensured. In the system of teacher-centered educational management, teachers and administrators are cooperative mates, not the controlees and controllers. Teachers, not the administrators, are the core for education; teachers' devotion to teaching and the university is of essential significance to education. It is their dedication to education that makes the universities run on.

3.3 Awakening the master consciousness of the teacher

The master consciousness asks the teachers to actively participate in the management of the universities. The teacher is the source of educational development, and the progress of education needs the co-work of all teachers'. The administrators should spare no effort to encourage teachers to participate in the university management. Through this, the teachers' master consciousness is greatly raised, so they will trust the university more, they will regard the university as their home, and therefore they will devote all their efforts to the prosperity of education and make more contributions to the country.

3.4 Setting up perfect academic environment

The university is the place for learning and researching, so an excellent academic atmosphere is the key to cultivating excellent faculty and students and the key to improving the quality of education. Besides teaching, the teacher has to research in various professional fields. The administrators are to form a good academic environment for the teacher; in addition, they should provide all kinds of opportunities for the teacher's development. Therefore, teachers can do their research conveniently, efficiently, and happily. With such efficiency for work, more research finding made by the teachers will come into being.

3.5 Being considerate for the life of the teacher

As the staff to serve the teachers, the administrators should do everything at the teacher's convenience, including caring for their life. The university should make all efforts to raise the teachers' living standards, to improve their housing conditions, and to increase their pay according to the economic development. Thus, the teachers can comfortably commit themselves to teaching and research. In addition, the university should aid the teacher for publishing essays or books, and for further study. All human beings have feelings. As long as the teachers have felt the deep love and care from the university, they will surely make full efforts to bring about more teaching and researching fruits.

4 CONCLUSION

The teachers are the center and main force of university education, so the educational administrators should stick to the guideline to serve teaching and teachers. All the administrators ought to strengthen the notion of service and discard the old practices. Everything must be done to meet the needs of teachers, to help them accomplish teaching and research. Anyway, it is teachers who are the main body of education. Without teachers' hard work, education will go no further. Therefore, it is time to enhance the status, awaken the master consciousness, and create an excellent career environment for the teacher. Through these methods, the level of university management will be profoundly enhanced, and the teachers' commitment and enthusiasm to education will be greatly increased; thus the teaching effects will be substantially improved.

REFERENCES

[1] Chen Changhua. On Establishing a stimulating mechanism by teacher orientation. *Journal of Southwest Agricultural University*, 2007, (3):182–184.
[2] Huang Zhicheng. Theories on Modern Educational Management[M]. Shanghai Education Press, 1999.
[3] Liu Xuequn. The thought on human-oriented management of college teachers. *Education and Teaching Forum*, 2011, (14):142–143.

[4] Yu Wensen, Lian Rong, Hong Ming. Teachers' Professional Development. Fujian Education Press, 2007.

[5] Zhang Yujiao. On the approaches of conducting human management in colleges. *Reform and Opening*, 2009, (12):238–240.

AUTHORS IN BRIEF

Sun Jingyi (1977-), female, Han nationality, master of literature, lecturer, mainly engaged in college education, teaching methods, and teaching management.

Wang Lun (1977-), male, Han nationality, master of literature, vice-professor, mainly engaged in English teaching theories and practices.

Yu Xiaoping (1965-), male, Han nationality, master of literature, professor, mainly engaged in higher educational management and development.

Information, Computer and Application Engineering – Liu, Sung & Yao (Eds)
© 2015 Taylor & Francis Group, London, ISBN 978-1-138-02717-6

The application of MOOCs to teacher training in China: Potential, possible problems and countermeasures

Lan Wei & Deng Long Fan
Chongqing University of Science and Technology, Chongqing, China

ABSTRACT: This paper focuses on the application of MOOCs to teacher training in China. The potential, possible problems and countermeasures are especially explored. MOOCs in teacher training have many advantages: they can meet the need for large-scale teacher professional development; they are convenient and ensure self-directed learning with high-quality resources, much adaptability and rewarding peer learning. However, their application to teacher training in China may encounter difficulties: lack of technological infrastructure, lack of good command of computer skills and online learning skills, lack of learning autonomy and inadequacy of MOOCs in Chinese language. Countermeasures for applying MOOCs to teacher training in China are proposed: a blended training model, administrative intervention and leadership are suggested in promoting teacher training; more high-quality MOOCs for teacher training in Chinese language should be designed; teachers should try to improve language ability, computer skills and online learning ability; the governments and schools should improve technological infrastructure.

KEYWORDS: MOOCs; Online learning; Teacher training; Teacher professional development

1 INTRODUCTION

MOOCs (massive open online courses), representing a new stage in distance learning, are growing in popularity in the recent years. They are actually hot issues in the field of education. And it is even a tendency to learn certain MOOCs. Meanwhile, the rapid development of information technology offers many new opportunities and approaches for teachers' professional development. Many teachers choose distance learning which provides much convenience. Attending MOOCs is becoming a new approach of teacher professional development. In November, 2014, U.S. President Obama announces free verified certificates for MOOCs from edX and Coursera for teacher professional development, which marks a new step toward offering more opportunities for teacher education. As said by President Obama: "EdX is offering more than a dozen training courses to teachers nationwide for free." EdX and its partner institutions will offer free verified certificates to U.S. teachers for one year. Coursera's university and nonprofit partners will provide 50 high-quality teacher professional development (TPD) courses to U.S. teachers for free throughout a two-year partnership. Courses for teacher professional development provided by these top two MOOC providers can be available on their websites. U.S. teachers will be able to earn free verified certificates upon successful

course completion. MOOCs are leading a new revolution in education for both teaching and learning. The application of MOOCs in China should also be explored. This paper mainly focuses on the application of MOOCs to teacher training in China. The potential, possible problems and countermeasures are especially explored.

In MOOC College by Guokr, a leading company offering MOOCs in China, some courses designed for teacher education or professional development are provided. The data from that website show that some teachers in China have already participated in learning certain courses, and they have successfully completed course learning and got a certificate. A conclusion can be drawn that it is possible for teachers in China to learn MOOCs for professional development. From the comments on the courses, it can be seen that most courses are popular among teachers and considered much helpful and valuable. Courses like Surviving Your Rookie Year of Teaching: 3 Key Ideas & High Leverage Techniques, History and Future of (Mostly) Higher Education, Task-based Language Teaching with Digital Tools and so on are highly evaluated and recommended by teachers who have learned the courses. But the number of teachers who learn MOOCs is still very small, though comparatively much more courses are offered. There is still a long way to go for the application of MOOCs to teacher training in China. Therefore, it is necessary to explore the possibilities of applying MOOCs to teacher training in China and possible problems should also be taken into account. Corresponding countermeasures also need to be proposed to promote the application of MOOCs to teacher training in China.

2 ADVANTAGES OF MOOCS IN TEACHER TRAINING

Continuing education for teachers is necessary but often time, energy and resource consuming. If it is conducted well, it can promote both teaching and learning, and be beneficial for teachers' professional development. But if it is poorly done, it can be ineffective and worthless. MOOCs are a good option for teacher's professional development.

There are several factors which make MOOCs a good option for teacher training in China.

First, they can meet the need for large-scale teacher professional development, as they are available to anyone who wants to learn online. As You (2014) mentions in his paper, using MOOCs for teacher training in China has a unique advantage of ensuring large-scale learning. Compared to the ratio of teachers to students in distance learning in foreign countries (1:6 or 1:12), the ratio of teachers to students in distance learning in China is 1:100. And the course completion percentage is up to 80%–90%.

Second, they are convenient, as teachers can learn without space and time limitations. Most courses are offered for free or at a nominal fee. It can obviously cut down the cost of teacher training if teachers can learn in their districts and even without leaving home. It can also save much time. Much flexibility fits in with teacher's busy lives.

Third, they ensure self-directed learning. When learning MOOCs, teachers can definitely have more freedom of choice, by choosing what they like and what they need. They can personalize their own goals, select among a large number of resources, and decide whether, when, and how to engage in projects and discussions to further their own professional learning. (Glenn et al., 2013).

Fourth, they are high-quality resources which are provided by the world's best universities, colleges and organizations. Andrew Ng, the co-founder of Coursera, believes that there is a huge need for high quality teacher professional development. High quality is one of the advantages MOOCs have. Teachers can not only learn the updated knowledge or content based on the courses, but also be inspired by the teaching methods and strategies of the excellent teachers in the world. Materials, resources and strategies in some MOOCs can be employed by teachers to develop effective pedagogy and facilitate their own teaching.

Fifth, teachers can benefit a lot from peer interaction. Peer learning can make teachers feel more connected to others outside their institutions through participation in MOOCs. As Manning, C., Morrison, B. R., & McIlroy, T. (2014) mention in their paper, interactivity of the peer-reviewed assignment feedback was encouraging and motivating. Teachers can learn original ideas from peer interaction, which also suggests the potential for collaboration and cooperation with colleagues overseas. Justin (2013) holds that MOOCs also provide teachers with a medium for discussing and sharing their PD challenges and triumphs with others in the field.

Last, not only do MOOCs provide flexibility, in many cases, the content that they provide is also adaptable to different circumstances. Justin (2013) mentions edX and Coursera are two prime examples of free content that is available to be modified and remixed in different ways to meet the specific needs of a particular audience. This feature allows those responsible for managing professional development for teachers to customize the experience for their audience.

3 POSSIBLE PROBLEMS OF APPLYING MOOCS TO TEACHER TRAINING IN CHINA

Though there are excellent reasons to believe MOOCs are a potentially valid avenue to help teachers with their professional development needs, but problems still remain. The application of MOOCs to teacher training in China may encounter some barriers. Here are some possible problems which may exist.

First, lack of technological infrastructure like necessary computers and access to the Internet is a big problem for high school and elementary school teachers in small cities and especially in rural areas. A survey (Fan et al., 2013) conducted among preschool teachers in Tianjin Municipality shows that 30% preschool teachers lack necessary online learning facilities. Another survey (June, 2012) among high school and elementary school teachers in a small county in Jiangxi Province shows that 35.3% teachers surveyed do not have computers, so they cannot conduct online learning. These statistics show that MOOCs are still unavailable to a large number of teachers in China, especially to those preschool and elementary school teachers in rural areas.

Second, lack of good command of computer skills and online learning skills is another obstacle, which may cause ineffective and inadequate communication and interaction online. A study (Fan et al., 2013) shows that only 15.5% preschool teachers in Tianjin Municipality have mastered online learning skills which just include manipulating common office soft wares and downloading resources. Only 14.2%

teachers know how to make course wares. 70.8% teachers surveyed have never received distance education before. Jun (2012) finds 49% teachers do not master network skills like searching for materials, downloading resources, participating in forum discussion, etc.

Third, lack of learning autonomy can be a problem which cannot be ignored. From previous surveys (Jun, 2012; Fan et al., 2013), some teachers do not have an active attitude toward online education or a strong autonomous online learning awareness. They just try to finish the learning tasks assigned by schools. Some even ask colleagues to help complete the learning assignments. Therefore, when learning MOOCs without administrative intervention, teachers' learning motivation may be a challenge.

Fourth, most courses designed for teacher developments are offered by teachers of English speaking countries. English is the main language used in class. Few MOOCs targeting teacher training in Chinese language are offered at present. It is difficult for teachers in China to understand the contents if they do not have a good mastery of English. Some teacher learners have already reported their trouble in understanding some courses in English. And they hope the courses could at least offer Chinese subtitles in order to help their understanding.

4 COUNTERMEASURES FOR THE APPLICATION OF MOOCS TO TEACHER TRAINING IN CHINA

In light of the above problems, some countermeasures are put forward as follows.

First, a blended training model should be employed to optimize teacher training in China. The traditional face-to-face training approach and MOOCs should be combined in teacher training to have complementary advantages. Feedbacks from learners show that they need face-to-face communication in addition to online learning. A learner at Guokr who learns Teaching Character and Creating Positive Classrooms reports: "Although many kindhearted staff and peer learners answer all questions and I also can get help from the forum discussions, I still feel the inconvenience when I want to ask for help. I hope there is another platform to provide more "face-to-face" communications with peers and teachers." So, after watching videos and finishing online assignments on their own, teacher learners need to come together to share and discuss in their district with teachers and peer learners, which can also compensate the incapability of teachers who have limited exposure and experience of online learning.

Second, necessary administrative intervention for MOOCs in teacher training is needed at present.

Measures can be taken to enhance teachers' learning autonomy. For example, teacher development centers in charge of teacher development can be established to support, monitor and evaluate teachers' continuing education including online learning. MOOC completion can be incorporated into teacher evaluation structures with scientific evaluation systems. Anissa (2014) holds learning credits for MOOC completion may provide the necessary incentives. Yanching Institute of Technology, an undergraduate institute in China, in 2014, has already required every teacher learn at least one MOOC, which stands for a good try of MOOCs in teacher development in China.

Third, leadership should play a crucial part in promoting teacher training from MOOCs. Leaders should help to establish professional learning networks, and teams of teachers or even entire faculties can take a MOOC together. School leaders or experts should also help to seek out useful and appropriate MOOCs for teachers, as not every MOOC is worth teachers' time. The encouragement and support from leaders can contribute to the successful implementation of learning MOOCs.

Fourth, more high-quality MOOCs for teacher continuing education should be specifically designed by Chinese top universities and institutions. In the field of teacher training, currently there are only a limited number of excellent courses in China. The overall quality of courses is not satisfactory with various problems, and most of them cannot meet the requirement of being open to large audiences (You, 2014). Clifford (2013) points out that, when designing MOOCs for teacher education care must be given to selectivity regarding what is most effective to be taught and its fit with existing teacher education programs. Consideration would also need to be given to the appropriateness of MOOCs for different kinds of teacher education.

Fifth, a suitable operating system for teacher training MOOCs should be established in China. You (2014), an expert in education in China, points out a suitable operating system for teacher training MOOCs is needed in China to ensure sustainable development. Coursera and Udacity are run by companies; although Edx is free, it can get financial support from universities. You (2014) expresses his worry: how to guarantee the funds for designing excellent courses and the fees supporting learning, and how to support continuing hard and soft ware maintenance should be carefully considered. The motivations to develop MOOCs and expand their scale need a scientific operating system.

Sixth, teachers in China should try to improve language ability, especially English proficiency if they want to learn MOOCs, as English is the main medium of language in MOOCs at present. Also, they need to improve their computer skills and enhance online learning ability to ensure active and successful participation in MOOCs. Schools should help to enhance teachers' skills and abilities concerned.

Last, the governments at all levels and schools should try to provide and improve technological infrastructure for teachers' learning MOOCs, as for active participation in MOOCS, teachers must have a fairly up-to-date computer device and access to some form of broadband Internet.

To summarize, MOOCs in teacher training in China have great potential because of their unique advantages. However, the application of MOOCs in teacher training in China has difficulties as mentioned above. To apply MOOCs to teacher training in China, we have to make them adapted to local circumstances, innovate the system design and maintain existing advantages to ensure and keep sustainable growth. Some corresponding countermeasures are proposed in the hope of helping to promote the application of MOOCs to teacher training in China.

ACKNOWLEDGMENT

This paper is supported by the project "Research on quality development in foreign language education in multi-learning environments" sponsored by Chongqing Municipal Education Commission, China.

REFERENCES

Anissa, L. 2014. Grab a MOOC by the horns. *Educational Leadership* 71(8):61–65.

Clifford, O. F. 2013. Teacher education MOOCs for developing world contexts: issues and design considerations. Sixth International Conference of MIT's Learning International Networks Consortium.

Glenn, M. K, Mary, A. W, & David, F. 2013. The digital learning transition MOOC for educators: exploring a scalable approach to professional development. http://www.mooc-ed.org/wp-content/uploads/2013/09/MOOC-Ed-1.pdf.

Jun, Y. S. 2012. Problems and countermeasures on teachers' distance training in the rural primary and middle schools. *Education Research Monthly* (7):58–60.

Justin, M. 2013. Why MOOCs are good for teacher professional development! http://www.onlineuniversities.com/blog/2013/05/why-moocs-are-good-for-teacher-professional-development/

Manning, C., Morrison, B. R., & McIlroy, T. 2014. MOOCs in language education and professional teacher development: possibilities and potential. *Studies in Self-Access Learning Journal* 5(3):294–308.

Ping, F., Li, N. L. & Zhen, Z. L. 2013. Rural preschool teachers' training needs and solutions in Tianjin Municipality. *Distance Education in China* (9):52–59.

You, Q. R, 2013. How does teacher training join hands with MOOCs? http://www.ict.edu.cn/news/n2/n20131108_5482.shtml

Information, Computer and Application Engineering – Liu, Sung & Yao (Eds)
© 2015 Taylor & Francis Group, London, ISBN 978-1-138-02717-6

Factor analysis of the development of low-carbon economy in China

Xiao Ling Wang & Xue Chi Zhao

Harbin University of Science and Technology, China

ABSTRACT: With the increasing emphasis on climate changes, Low-Carbon Economy (LCE) is becoming the strategic focus of future development globally. As a nation undergoing rapid urbanization and accelerating industrialization with ever increasing demand for energy, China faces severe challenges in developing LCE. This paper selects 17 provinces as the sample, quantitatively analyses the variables using factor analysis method in SPSS 17, and computes the comprehensive score of LCE for each province respectively. The results show that social support, local development, ecological index and environmental protection are the determining factors in influencing LCE development. Lastly, relevant countermeasures are proposed based on the results to provide valuable references for future studies in the subject area.

KEYWORDS: Factor analysis; Low-carbon economy; Energy

1 INTRODUCTION

Low-carbon economy (LCE) is a new economic development pattern aiming to actualize sustainable economic development, while the survival and progress of human beings are seriously threatened by the emerging problems of climate change, environmental pollution and energy deficiency. Rapid developments in urbanization and industrialization lead to ever increasing demand for energy. It makes reducing carbon emissions while ensuring the steady development of Chinese economy one of the most important tasks that we are facing at present.

Pertaining to LCE related research and studies, Soytas et al. (2007) implement VAR model into the study of causal relationship among energy consumption, GDP and carbon emission in the US, and find energy consumption to be the primary cause of carbon emission[1]; Puliafito and Grand (2008) utilize the Lotka Volterra model and prove that population, GDP and energy consumption are all determining factors of carbon emissions[2]; in search of means of reducing carbon emissions, Dagoumas and Barker (2010) apply the Energy-Economy-Environment Model in analyzing the situation of carbon emission in the UK[3].

Studies on LCE theories and appraisal system are also conducted in China in recent years. Tao (2010) uses the grey correlation theory in investigating the determinants of LCE development, and proposes quantitative and qualitative indices for evaluating LCE levels [4]. Zheng, Fu and Li (2011) construct the index system of LCE development levels, comprehensively assessing the Chinese LCE development level

with analytical hierarchy process[5]. Liang and Han (2012) quotes carbon productivity as the core indicator of LCE development, and conduct empirically analyses with economic development level, industrial structure, energy structure, carbon financing and carbon sink construction as influencing factors[6].

This paper analyses and appraises the development level of LCE in China using factor analysis method, and subsequently proposes the relevant countermeasures to the further development of LCE.

2 CONSTRUCTION OF LCE DEVELOPMENT APPRAISAL MODEL

2.1 *Variables and data selection*

We choose the following 10 index variables for the purpose of our study: carbon emission per capita (X1), carbon emission intensity (X2), energy consumption per unit of GDP (X3), key pollutants in exhaust emission (X4), local environmental protection expenses (X5), urban per capita disposable income (X6), forest coverage rate (X7), investment in scientific research (X8), high school or higher education attainment rate (X9), household waste decontamination rate (X10).

The raw data used in this paper is adopted mainly from the China Statistical Yearbook 2012, with the remaining quoted from the China Energy Statistical Yearbook 2012, Communiqué on the 2nd Local R&D Census, and Communiqué on the 6th National Population Census etc. Factor analysis method is then applied to the 10 variables to yield 4 factors. The comprehensive score of LCE development in each province is then calculated and finally an overall assessment on LCE development level is provided accordingly.

2.2 Factor analysis on LCE development levels

2.2.1 KMO (Kaiser-Meyer-Olkim) measure of sampling adequacy and Bartlett's test of sphericity

Before conducting the factor analysis, we first test the variables for correlations using the KMO measure and the Bartlett's Test of Sphericity. The KMO statistic is used in the test of correlations. A KMO value close to 1 indicates that patterns of correlations are relatively compact so factor analysis should yield reliable factors. KMO values below 0.5 suggest that either the data is insufficient or the variables are inappropriate to run factor analysis. Bartlett's measure tests if there are relationships between variables. A significant test with a significance value less than 0.05 would suggest factor analysis to be appropriate. We run the raw data through KMO and Bartlett's tests in SPSS17.0, and yield the following results as shown in Table 1.

Table 1. KMO and Bartlett's tests.

Kaiser-Meyer-Olkin Measure of Sampling Adequacy		.674
Bartlett's Test of Sphericity	Approx. Chi-Square	168.067
	df	45
	Sig.	.000

According to the test results, the KMO value is 0.674, proving that the data collected are acceptable for factor analysis. Meanwhile the Bartlett's test gives an approximate chi-square value of 168.067 with the degrees of freedom being 45 and significance value being 0.000, therefore factor analysis is appropriate.

2.2.2 Factor extraction

Factors are extracted from the original observed variables to reflect the joint variations in the variables. Generally speaking, a resulted cumulative percentage extraction sums of squared loadings greater than 85% is considered significant and can explain most of the variability in the original variables. Extract the factors through principal component analysis in SPSS17.0, and the results are shown in Table 2.

As seen in Table 2, the extraction of components with eigenvalues greater than 1 yields 4 factors, with a cumulative sums of squares loadings

of 88.318%. It shows that these 4 factors accounts for 88.318% of the variability in the original variables in depicting the development of LCE of the 17 chosen provinces.

Table 2. Total variance explained.

	Initial Eigenvalues			Extraction Sums of Squared Loadings		
Factor	Total	% of Variance	Cumulative %	Total	% of Variance	Cumulative %
1	4.838	48.378	48.378	3.895	38.945	38.945
2	1.829	18.290	66.668	2.287	22.867	61.812
3	1.136	11.364	78.032	1.339	13.387	75.199
4	1.029	10.285	88.318	1.312	13.119	88.318

2.2.3 Component rotation

Factors in component matrix are usually not sufficiently differentiative. Hence we rotate the component matrix through varimax rotation to get the rotated component matrix as shown in Table 3.

Table 3. Rotated component matrix.

	Factor			
	1	2	3	4
X_1	.948	-.144	.008	-.077
X_2	.943	.264	.070	.138
X_3	.767	.433	.267	.285
X_4	.608	.539	.282	-.278
X_5	-.947	-.239	-.064	-.155
X_6	.265	.790	.165	.390
X_7	.399	-.256	.675	-.272
X_8	.054	-.217	-.829	-.257
X_9	.084	.958	-.070	-.015
X_{10}	.117	.078	.065	.901

It can be seen from the loadings for each variable onto each factor that Factor 1 is strongly loaded by variables including per capita carbon emission, emission intensity, energy consumption per unit GDP, key pollutants in exhaust emission, and local environmental protection expenses, at 0.948, 0.943, 0.767, 0.608 and 0.947 respectively. Factor 1 reflects information such as carbon dioxide emission and local financial support etc., thus can be called the LCE social support factor.

Factor 2 has relatively high loadings of 0.790 and 0.958 for urban per capita disposable income and high school or higher education attainment rate respectively. This factor reflects regional economic development and population quality, we therefore address it as the LCE local development factor.

With loading values of 0.675 in forest coverage and 0.829 in scientific research funding respectively, Factor 3 mainly reflects the situation of regional ecological construction and development so is also known as the LCE ecological index factor.

Factor 4 has a high loading of 0.901 in household waste decontamination rate, hence is called the LCE environmental protection factor.

2.2.4 Factor score

We first compute the scores of each factor based on its scoring coefficient and the standardized values of the original variables, then perform weighted summation based on factor weights (calculated by dividing each eigenvalue by the sum of all 4 eigenvalues) to get the comprehensive scores of the 17 sample objects, as detailed in Table 4.

2.2.5 Conclusion

According to the results above, the LCE development levels differ greatly among provinces, gradually decreasing from coastal regions to inland regions. Out of the 17 provinces, Inner Mongolia, Hebei, Liaoning and Heilongjiang are at the bottom level of LCE development. This may be primarily attributed to their energy distribution, consumption pattern and excessively coal-reliant energy structure. The distribution characteristics exhibited by LCE development are the synthesized outcome of various factors such as energy production, energy consumption structure, technology development, population quality, carbon sink construction etc, among which energy distribution and consumption structure, as well as technology development are the most important elements.

3 COUNTERMEASURE SUGGESTIONS TO ACCELERATE LCE DEVELOPMENT

Based on the empirical analysis above, the following measures need to be implemented for the acceleration of LCE development:

The economic development modes need to be transformed, and the industrial structures upgraded and optimized. Firstly, high carbon industries are to be low-carbonized. Industries with high pollution, high energy consumption and high emission are to be closely monitored and the economic emphasis on them gradually shifted. The government needs to increase its inputs in research for low-carbon technologies to increase the efficiency of energy sources. Meanwhile, advanced pollutant treatment facilities are to be adopted to enhance the efficiency of recycling industrial waste water and comprehensive utilizing sewage, and the development path of low-carbonize high-carbon industries also need to be gradually explored. Secondly, new energy industries

Table 4. Comprehensive score on provincial (municipal or district) LCE development.

Region	Factor1	Factor 2	Factor 3	Factor 4	Score	Rank
Beijing	0.79824	2.38242	-0.24239	0.19154	0.96055	1
Zhejiang	0.67529	-0.06561	1.36669	0.79103	0.60545	2
Shanghai	0.04307	2.48124	-0.49329	-0.07097	0.57611	3
Fujian	0.78692	-0.37415	1.72626	0.21246	0.54335	4
Guangdong	0.80002	-0.04844	0.87609	-0.1829	0.44586	5
Jiangsu	0.23457	0.31503	-0.61739	1.38803	0.2976	6
Hunan	0.64456	-0.65113	0.78403	0.18137	0.26142	7
Anhui	0.32607	-0.71147	0.32651	0.41569	0.07081	8
Sichuan	1.05255	-1.00703	-1.26534	-0.29797	-0.03266	9
Hubei	0.61196	-0.06813	-1.26909	-1.46357	-0.15755	10
Shandong	-0.59989	-0.44929	-0.10744	1.60465	-0.15879	11
Shanxi	-0.54293	-0.14455	0.84576	-0.16123	-0.17259	12
Henan	0.34509	-0.93571	-1.43874	0.31362	-0.26159	13
Heilongjiang	-0.48874	-0.08537	0.97772	-2.81672	-0.50782	14
Liaoning	-0.63078	-0.09609	-1.06375	-0.39279	-0.52262	15
Hebei	-1.04047	-0.86719	-0.83599	-0.03213	-0.81483	16
Neimenggu	-3.01553	0.32546	0.43036	0.31988	-1.13272	17

are to be actively developed. China is in the phase of rapid industrialization and urbanization, and there exists a rigid demand for energy. In order to achieve a low-carbon economy, the only way is to reduce our reliance on carbon-based energy, actively develop and utilize solar, wind, biomass, ocean energy, nuclear energy and other clean energy sources, promote the development of new energy industries, increase its share in the economy, and achieve continuous upgrading of industrial structure[7].

Population quality is to be improved, and attention to be paid to nurture talents to provide intellectual support for low-carbon economy. Talents are an essential stimulus to the development of low-carbon economy. However due to the large population base and the relatively low population quality, the development of low-carbon economy is restricted to some extent. Therefore we must continue to increase inputs in education and to place emphasis on both general and vocational education, so as to improve the level of education of the general population. Meanwhile, we also need to pay great attention to the training of low-carbon talents. Companies can establish training facilities to provide their employees with on job trainings. Universities can collaborate with research institutes in R&D to nurture specialists with low-carbon ideology and relevant skills to boost the development of low-carbon economy.

Expand afforestation to realize forest carbon sink potential. Carbon sink refers to the process of which plants absorb carbon dioxide from the open atmosphere through photosynthesis and store it in vegetation and soil, thus reducing the atmospheric carbon dioxide concentration. Forests are the largest carbon stocks in the terrestrial ecosystem, and play vital roles in reducing greenhouse gas concentrations in the atmosphere to mitigate global warming. Scientific research shows that the growth of an additional 1m3 of forest trees can help absorb 1.83 tons of carbon dioxide. So the forest resources have significant impacts in addressing climate change and promoting low-carbon economy. Therefore, the development of low-carbon economy requires us to expand the area of forests and increase forest coverage. At the same time, we also need to strengthen management to the existing forests, artificially promote forest growth, improve forest reserves, and realize forest carbon sequestration potentia.

Set up a special purpose fund for energy conservation, and increase financial support. First, allocate funds in the government's budget for energy saving and low-carbon economy development, and ensure a certain annual growth rate. Budget expenditure is mainly used for R&D in energy saving technology and application promotion. Second, increase the level of financial support. Low-carbon economy depends on scientific and technological progress. R&D in energy saving technology, solar energy, wind energy, tidal energy and any other new energy technologies all require substantial investment that enterprises cannot afford on their own. This is where both central and local governments need to increase research funding and environmental protection investment. Third, provide additional government subsidies. To be more specific, one is to subsidize the energy-saving infrastructure alterations; the other is to provide interest subsidies price subsidies, investment subsidies and others to energy saving enterprise and project loans.

REFERENCES

[1] SOYTAS U, SARI R, EWING T B. Energy consumption,income,and carbon emissions in the united states[J]. Ecological Economics, 2007(62):482–489.
[2] PULIAFITO S, PULIAFITO J, GRAND C M. Modeling population dynamics and economic growth as competing species: An application to CO_2 global emissions[J]. Ecological Economics, 2008(65):602–615.
[3] DAGOUMAS A S, BARKER T S. Pathways to a low-carbon economy for the UK with the macro-econometric E3MG model[J]. Energy Policy, 2010(38):25–41.
[4] TAO Ai xiang.AnAnalysisonChina's Low—carbonEconomicDeveloPment BasedonGrayRelationTheory[J] Journal of Linyi Normal University, 2010(9):24–29
[5] ZHENG Lin-chang,FU Jia-feng,LI Jiang-su.Evaluation on the Development Level and Spatial Progress of Low-carbon Economy atProvincial Scale in China[J]. China Population, 2011(7):70–85.
[6] LIANG Shu—yi, HAN Shui—Hua. Factors Analysis of Low—carbon Economic Development Level in China[J]ChinaSoftSciencesupplement, 2012(9): 29–37.
[7] Cong Mingyao, Yun Miaoying, XiaJianxin. Index system of low carbon city based on factor analysis in northeast China[J].Environmental Science & Technology, 2014, 37, 2, :195–198.

Information, Computer and Application Engineering – Liu, Sung & Yao (Eds)
© *2015 Taylor & Francis Group, London, ISBN 978-1-138-02717-6*

The study of rural primary and middle school students' ideological and moral construction

Jian Ping Gong
Weifang University of Science and Technology, Shou Guang, China

ABSTRACT: First this article analysis the current situation of rural primary and middle school students' ideological and moral construction. Second analyzed the rural primary and middle school students in school, such as social environment, family factors, the influence of ideological and moral level changes, and a declining trend. Last this paper expounds the countermeasures to strengthen the construction of rural primary and middle school students' ideological and moral.

KEYWORDS: The countryside; Primary and middle school students; Ideological and moral; Genesis; countermeasures.

1 THE CURRENT STATUS OF PRIMARY AND MIDDLE SCHOOL STUDENTS' IDEOLOGICAL AND MORAL CONSTRUCTION

1 Subjective identity and maintain social morality, but the initiative is not strong, and on the basic morality accomplishment there is a decreasing tendency with the increase of age. The middle school students' daily behavior standards as the current society advocated the most basic standards of public life, get the majority of the students. Many parents are greatly influenced by the feudal hierarchical, use authoritarian way of schooling, they can't accept their children to be "equal", especially the grandparents every generation of education, the feudal thought is more serious. Or for the child's care and love, think the child's first task is to learn. Healthy body is the most important thing, and not willing to let the child to participate in family related affairs. Thus affect the behaviour of middle school students' family virtues practice, in the long run, will weaken the nurturance of middle school students' family virtues.

2 Ideal diversification, "selfish" and "utility" stand out. One is an ideal, life values to the contemporary middle school students has shown a trend of diversification, pragmatic and utilitarian tendency to start on them. They are undeveloped, but in his heart with "self-interest" has to "the public" is just as important as the understanding of the position. When talking to students, there are students who talked about "money makes the mare go," money can wear name brand clothes, etc. The second is the contemporary rural middle school students in the choice of values, based on the self, both to others and society, behavior more driven by the interests of the reality, the city they work during the holiday season to parents, realized the reality of the disparity is too big gap between the urban and rural and cruel, one-sided that money and wealth is the greatest pursuit and goal in life, so put personal interests before the main location, the purpose is to want to like the city of the same age to live a rich life, can we truly achieve economically equal political in the sense of equality.

3 Accept the fair competition, pay attention to the unity and cooperation. In the face of increasingly fierce competition, rural middle school students to more positive team spirit to career success. Many rural middle school students, from a young age because of family economic conditions, such as a variety of reasons, cannot be like city peers during the holiday season when they attend all kinds of training, the city children go for vacation, parents then sent them to all kinds of training classes, while the rural children cannot have such an opportunity and conditions of the rural middle school students is autistic, art special features for rural students is a luxury, not confident on group activities, they fear make a fool of yourself, be laughed at by the people, and for a variety of artistic activities not very actively attend the performance.

4 Collective evaluation and exchanges show that the rural middle school students plain view of interpersonal communication himself was born in the countryside, the life in the countryside, the rural middle school students' attitude is calm. Reflect the rural middle school students in real life are pure, enthusiastic and sincere.

5 Also exists some problems that cannot be ignored: the majority of the rural students' parents are working outside, the influence of the society, the status of migrant workers in cities is relatively low, cannot get the respect they deserve, they see is money determine social status and human dignity of the individual, some parents also produce eyes only for money, ignore the good morality education of their children, some parents to give their children money directly on the values of life. With the development of economy and the progress of the society, the change of the rural migrant farmers thought, the divorce rate has obvious rising trend of rural households, rural middle school students in single-parent families are more and more, and rural middle school students for parents remarried children born again after the divorce, child neglect education and management of the original, the students became grandfathers discipline for like, specifically, the student's family education is almost a blank, family education is far behind; Rural school students, although the city is better than the students' information sources, but under the major premise of balanced development of education, the use of distance education in rural schools, and the construction of digital campus, many rural students have access to the Internet, in the school, teacher management and constraints can be civilized or healthy on the Internet, the quickening pace of urbanization, the popularity of Internet, Internet cafes, little house, and so on the rapid development of all kinds of cultural entertainment, some contain pornography, violence, gambling, ignorance, superstition backward decadent culture and take advantage of a loophole to spread harmful information, to a certain extent, corrupt the minds of some minors; rural students a home over the weekend, don't know where his forefathers to network, in the rural middle school students to the Internet cafes, secretly or use his cell phone to the Internet, some decadent culture and the decadent lifestyle has different degrees of damage to part of the rural middle school students; full, comprehensive care and support rural ideological and moral construction of minors atmosphere has not yet formed. Under the influence of these negative factors, a few minor's world outlook, the outlook on life and values distorted, and dislocation, become the hidden trouble of the minor crime and the main reason.

2 ANALYSIS OF THE CAUSES OF THE CURRENT SITUATION

Reasons for these problems are various, in particular has the following several aspects:

2.1 The school education

School education in formalism. School is for students to carry on the educational institutions, and the teacher is playing a leader role in student education. The ancients cloud; "Teacher, so preaching, putting to reassure also." That is to say, teachers by using their own teaching practice teach students knowledge, cultivate the students' practical ability, causes the student to understand all kinds of reason, which is the main mission. There is only one standard score society sees in a school. Therefore, in a unity of parking, in order to get a good ranking, a school's collective cheating. Ask such education to children is what? As a teacher, has put all the time in teaching, in the naval battle of the students' questions, actively responds to the call, symbolic undertake to the student ideological education, education of a float to the surface. Thus naturally produced a kind of "teaching is tough, moral education is soft task" thoughts, disdainful of subject knowledge education, ignores the form of students' ethical personality education.

2.2 Family influence

Soon after birth the Children first saw the faces of their parents, the initial environment the children face is the "family", while at the same time of his birth parents become educators and children's example. Therefore parents became the first teachers of children, family birth place is the child's former school.

Due to the influence of the market economy, the material and spiritual life has become increasingly improves and changed, combined with various forms of cultural life, leaving some families dissociation and recombination. This kind of special children being raised in the family, and school education more attention to the object. According to the city for minors illegal crime situation, according to the results of a survey of exceptional family minor illegal crime rate is higher. Parents divorced or separated structure such as incomplete children crime rate is much higher than that of ordinary families with children of the family. And broken family, these minors may not be due a father or a mother's love, enjoys the warmth of family, become warped personalities, extremely easy on the road of crime.

2.3 The social environment

The story of "mentions's mother three times," tell us a good environment for the child the importance of the education effect. In the information age today, the Chinese on the material, improve the economy, but can't keep up with the improvement on the heavenly civilization. Internet, TV, movies, magazines and other convey information in their respective economic purpose. Often combined with adult entertainment and learning to show the spread of the liberal way in front of the students. Make students blindly chasing singers, movie stars, and the indifference of great men and heroes, scientists, writers. Network of ineffective supervision, which many commercial Internet cafes, specifically with violence, pornography and other harmful information to attract students, become a terrible minors' growth path "network trap".

2.4 Their own reasons

The ideological and moral development of primary and middle school students receives the restriction of its own development. Elementary student moral consciousness is in the phase of mere cage, physiology and psychology is not yet mature, intellectual curiosity, strong imitativeness, curiosity and vanity, hostile, rebellious attitude, easy to tell right from wrong and self-discipline, control ability is weak, easily affected by the social bad style or instigated astray.

3 COUNTERMEASURES OF PRIMARY AND MIDDLE SCHOOL STUDENTS' IDEOLOGICAL AND MORAL CONSTRUCTION

3.1 Perfect school assessment mechanism, the class management, work to promote the foundation of moral education innovation in place.

The class is the essential foundation of the school andis the main carrier of moral education. School moral education work in the every grade and class to refine the objectives, clear requirements, strict examination, adhere to the moral education as a primary content of class management, and as the main basis evaluation to judge the teacher in charge work quality, and strengthen supervision to guide and promote their moral education research and innovation, improve and optimize the class management. Emphasis on strengthening process supervision and examination: one is beginning to urge and guide all class study close to the actual and effective moral education work plan, and supervise the implementation, tracking, unplanned or a plan is not practical to overcome the moral education work in the form of socialist practice. Second it is usual to supervise and guide the class good theme party or in the morning and evening meeting, with specific and series education, do a around a theme, select a point of view, inspire and guide students to conduct a self education, ZhaJiu a bad performance. At the end of three is at the end of the semester and learn, but also urge and guide the moral education work project summary for your class. School organization special evaluation at

the same time, the class of civilization and the moral education work advanced individual award, and help them summarize experience; not value, work does not reach the designated position of moral education, is seen as incompetent, implementation of admonishing the conversation, and promote its efforts to improve.

3.2 Perfect joint collaboration mechanism, school, family and community education works in place.

Suggest grass-roots government will this work into three civilization construction, does planning, deployment, inspection, examination, together. We need to improve the community education institutions, enrich the community education, and to give the necessary funds to safeguard. School with the parents of the rural women's federation, closes working the committee, Committee jointly does a good job in school, parents and the parent, community education institutions and relevant departments to strengthen the contact and communication. Can be led by the township women's federation and closes working committee, organize relevant personnel research teenagers education work regularly, so as to give full play to their respective functions, be truly, should form a trinity education network, multi-party cooperation, mutual cooperation, let students on campus have been concerned about outside, someone care, someone is discipline. Such, teenagers can be comprehensive progress, healthy growth.

3.3 Will care for left-behind children in school education

Left-behind children are a vulnerable group, age is generally small, curiosity is strong, adapt to the environment, poor ability of self-control is weak, lack of foresight to danger, susceptible to radio and television, and the influence of gangs and produce some bad behavior. For a long time, school, family security cohesion between vacuum, school is not comprehensive, guardian and general lack of safety consciousness, hurt or being hurt accidents occurred frequently in children. Schools have a certain level of legal system education, can make the children know what their lawful rights and interests, on the behavior of which is prohibited by law, thereby significantly reducing all kinds of illegal crime the happening of the accident.

4 CONCLUSION

Primary and middle school students are the future of the motherland, the nation's hope, their healthy growth, is a direct result of the future of the country and the future of the nation. To strengthen and improve the rural middle school students in the construction of the ideological and moral education is a long-term project, but also reasonably event. Society, school and family should unite as one, manages concertedly, constantly sum up experience, efforts to explore, bold innovation, for the country, cultivating qualified talents for the society.

REFERENCES

[1] Chao Zhang. Primary school moral education series. Gumshoe Education, 2008 (9).

[2] Jiefang Huang. "Change the two key points of the school moral education", "*The ninety Education*" 2011, (4).

[3] Shouyou Huang. Ideological education strategy for teenagers. *Education and Occupation*, 2006 (21): 162.

[4] Chaorong Li. Discuss the influence factors of youth ideological and moral education and countermeasures. *Secondary Vocational Education*, 2004 (14): 38.

[5] Degang Wu. Thinking of strengthening teenagers' ideological and moral education. *Education Research*, 2008 (7).

Information, Computer and Application Engineering – Liu, Sung & Yao (Eds)
© *2015 Taylor & Francis Group, London, ISBN 978-1-138-02717-6*

Rehearsal features of movement structure in difficulty-beauty projects

Guang Hua Ren & Bei Wang
Harbin University of Science and Technology, Weihai, Shandong Province, China

ABSTRACT: This thesis discusses and studies movement structure in difficulty-beauty projects from perspective of biomechanics and rehearsal features and focuses on investigating rehearsal features of movement structure to conclude similarities. It is a kind of innovative content that enriches rehearsal theory theoretically and provides theoretical reference for exercise training and competition activities.

KEYWORDS: Difficulty-beauty; Movement structure; Rotation axis

1 INTRODUCTION

Improving quality and value of rehearsal about movement structure in difficulty-beauty projects of China has become one of the critical problems that need be solved urgently. This thesis carries out exploration from perspectives of sports biomechanics, athleticism and aesthetics, hoping to disclose basic features and internal laws of rehearsal about movement structure in difficulty-beauty projects and provide theoretical and application reference for improvement in quality of rehearsal about movement structure in difficulty-beauty projects.

2 RESEARCH OBJECT AND METHODS

Research object: movement structure in difficulty-beauty projects

Research methods: document literature method, video analysis and logical analysis

3 RESEARCH RESULTS AND ANALYSES

A difficulty-beauty project shows beauty of human body, where technical training is a core. It is mainly reflected by accurate judgment on space and time, ability to control body postures, skillful mastery about all professional instruments and coordination with partners. In tactics application, disadvantages of movement rehearsal are avoided, advantages are adopted and layout of movements is reasonable. In the process of rehearsal, individual movements, composite ones, linking ones and difficulty ones are rehearsed according to features and nature of specific sport events and competition environment.

3.1 Individual movement structure

Arms and legs, trunk, head and expressive force constitute elements of an individual movement. Such elements changes in time and space, which makes movement structure vary. Individual movements with similar nature and gestures will not be placed together. Indeed, different types of movements with various natures are connected together to show rehearsal level, such as a whole set of movements, rhythm and style, and reflect difficulty in routine movements and individual technique style. An individual movement structure is not single but various and changes with movement forms. Movement technique of all difficulty-beauty projects is divided in to 'plain displacement' and 'three-dimensional displacement'. Statistic movements are routine modeling, while dynamic ones are methods and approaches of movement process.

3.2 Composite movement structure

Composite movements contain movement pattern combination, difficulty movement combination as well as transition and linking combination, etc. Technical combination is a set that is formed by linking several independent technical movements. Selection about content of composite movements is much flexible, including all kinds of basic movements, elements and movement patterns. A complete action contains combination of several individual movements. A whole element is composed of a series of individual movements, such as preparatory phase, theme movements and ending movements. Thus, when it ensures movements are finished, skills are highlighted to a lager extent. As shown in Fig. 1, considering composite structure of movements, basic movements are changed into the ones with single movement structure and then a complete movement combination is formed further. Therefore, structure of each movement cannot be neglected. In the process of movement rehearsal, it is essential to consider whether the rehearsed movement combination is single movement structure or multiple movement structure as well as nature, whether forms of movements are statistic or advancing composite changes and whether there are changes in formation or not as well as think about attribute of their sport events.

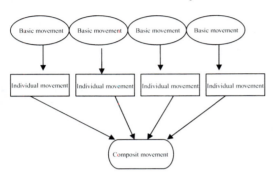

Figure 1. Structural representation about composite movements.

3.3 Difficulty movement structure

Difficulty movements are an important component of a whole set of movements and significant content of judges' grading. Rules vary largely in the aspect of total quantity of difficulty, difficulty type and value of difficulty movements, so we should pay attention to combining with competition rules and rehearsing movements scientifically and reasonably in the process of rehearsal.

3.3.1 Difficulty rehearsal accords with basic axis of human body

Axes of movement in human body contain coronal axis, sagittal axis and vertical axis. Movement planes involve frontal plane, sagittal plane and horizontal plane. Movements of human body should be developed in basic axes of movement and movement planes even if they are quite splendid. Speed of reversal movements, such as pushing hands and jumping, mainly derives from horizontal velocity of barycenter. Besides, the horizontal velocity plays an important role in improving reversal speed of the horizontal axis. Turn may be generated in the air by two mechanisms. First, hip joints bend and stretch and circle rings. Second, overall angular momentum generates component in the vertical axis of human body by asymmetric arm swing Complicated difficult movements are realized by surrounding several

axes of human body rather than on rotation axis only. Value of movements grows as difficulty increases.

3.3.2 Difficulty movements and sports biomechanics analysis

Difficulty movements are a bright spot for a whole set of movements. When novelty of movement is kept, rehearsal should accord with features of sports biomechanics. Human body is affected by the outside during movements, which drives our body to suffer displacement. The force human body suffers is divided into two kinds, i.e., internal and external. Only external can make movement status of human body change.

Take diving for example. It contains run-up, take-off, flying and entry. A key deciding whether diving can be successful or not lies in the force applied to a take-off surface and direction during take-off. With respect to flip turn, a diving platform has counter-force which tends to one or several directions to generate angular momentum of flip, depending on the direction of acting force which is relative to an active line of barycenter. Here, we introduce the concept of angular momentum.

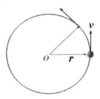

Under the effect of centripetal force, a mass point has uniform circular motion. In this case, the momentum may be expressed as $P = mv$. Since direction of speed has been changing rather than constant but $r \times mv$ radius of the mass point multiplied by the momentum is a constant vector which equals mvr, the direction is perpendicular to and outward all the time. $r \times mv$ is just angular momentum of the mass point. Thus, the angular momentum remains constant.

If we want to generate rotation, there must be a distance between the active line of the diving platform's counterforce and the barycenter. Arm of force is the vertical distance from the active line of force to a rotation axis or a fixed point when an article rotates around a fixed axis or point. Here, the arm of force refers to the distance from the barycenter to the active line of force. The larger the force or the arm of force is, the larger the angular momentum of an athlete is and the more easily the rotation will be generated. The angular momentum depends on rotational inertia and angular velocity.

We introduced the concept of the rotational inertia in this thesis. It only relies on the shape of rigid bodies, mass distribution and position of rotation axis but is not related to rotational state when rigid bodies rotate around an axis, for instance, the angular velocity. As attributes m and r of an objective are constant in movements, we replace variables m and r with K (K = mr2). K is the rotational inertia.

According to mechanics principle, stability of movements may be improved by reducing the barycenter. With reduction in the barycenter, difficulty in movements will also be decreased accordingly. In movements, for instance, flip and turn, angular momentum of both coronal axis and vertical axis of our body are needed.

3.4 Transition and linking movements

Transition refers to flexible and smooth conversion among movements in the air and on the ground and linking movements mean basic pace, movement combination, difficulty and lift are linked smoothly. Transition and linking movements show movements among key movements or key sections. Link-up forms of them contain difficulty movement + linking + difficulty movement + movement pattern + difficulty movement + movement pattern. Difficulty movement combination is divided into dynamic–dynamic linking combination and dynamic–static linking combination. The dynamic–dynamic linking combination reflects an athlete's ability to finish different difficulty movements continuously and connect movements, while the dynamic–dynamic linking combination shows the athlete's ability to control his or her body after finishing difficulty movements. Ways to connect movements directly show rationality of rehearsal and technical level of athletes.

3.5 Movement patterns

A movement pattern refers to technical movements except difficultly movements, transition and linking movements. It is an important component of a whole set of movements, acting as a bridge and link connecting difficulty movements and movements on the ground. Movement patterns tend to take less time and spend more time on difficulty movements, transition and linking movement. Dissymmetry of movements finished by upper and lower limbs reflects complexity and diversity of movement patterns and changes in body direction is also an important reflection of movement diversity. Even, smooth and complicated movement paths as well as changes in formation and rhythm are bright spots that enhance a whole set of actions score. As time passes by, complete movements will be updated regularly and movements will be upgraded based on original ones.

REFERENCES

[1] Theories of Sport Training. General Teaching Materials for Physical Cultural Institutes in China, People's Sports Publishing House.
[2] Chen Zhiqiang. Research on athletes' scientific material selection and methods in freestyle skiing aerials of China. A master's thesis at Northeast Normal University, 2006.
[3] Sports Biomechanics. General Teaching Materials for Physical Cultural Institutes in China, People's Sports Publishing House.

Information, Computer and Application Engineering – Liu, Sung & Yao (Eds)
© 2015 Taylor & Francis Group, London, ISBN 978-1-138-02717-6

Location choice of China's banking sector in the Middle East

Yu Hua Zhao & Tian Jiao Feng

School of Economics, Trade and Event Management, Center for Financial Management of High-end Industry, Beijing International Studies University, Beijing, China

ABSTRACT: The Middle East has become a new investment highlight for China's banking sector in recent years. The key issues here are how to identify the preferred countries and choose the suitable business forms in the above region. This paper tries to analyze these issues by using the clustering method. Four variables are employed including the export, outward FDI, GDP per capita and the position in Global Financial Center Ranking. The results show that 16 sample countries can be classified into five categories: The United Arab Emirates and Qatar are the most preferred regions, branches can be established; Bahrain is also a good choice, but the form of representative office is better; Saudi Arabia is the best candidate for further investment; the next suitable regions are Kuwait and Oman, either branches, subsidiaries or representative office can be established; Egypt, Turkey and Algeria come last for representative offices; Syria and other 6 countries are not suitable for outward investment until now.

KEYWORDS: China's banking sector; The Middle East; Location choice of investment

1 INTRODUCTION

The internationalization of China's banking sector has started 90 years ago. However, until now it is still in the initial stage. One striking improvement is the investment locations are much wider than before since 2008. The business operations are not heavily focused on Asia-Pacific areas or global financial centers in Europe and America. They have gradually extended to the emerging markets such as Africa, South America, Eastern Europe and the Middle East. For example, during the period 2008–2012, China's banks have established 49 overseas agencies, 7 of which are located in the Middle East, indicating that the Middle East is becoming the new investment highlight for China's banking sector.

The Middle East is a region that roughly encompasses a majority of Western Asia and northern Africa. According to the common thoughts in China's academic field, it includes 14 countries in Western Asia, such as Saudi Arabia, The United Arab Emirates, Kuwait, Bahrain, Qatar, Oman, Jordan, Syria, Lebanon, Iraq, Iran, Palestine, Israel, Yemen, and 6 countries in Northern Africa like Egypt, Sudan, Morocco, Algeria, Tunisia and Libya. These countries have significant differences in politics, economy, law and the foreign policy with China. How to identify the preferred location and choose the suitable operational form are the issues for research.

2 LITERATURE REVIEW

The basic framework for analyzing outward FDI in banking sector still follows Dunning's eclectic paradigm. It is also called OLI-Model, showing ownership advantages, location advantages and internalization advantages as 3 important factors for deciding the corporation's international activities. As far as the banking sector is concerned, this model can be explained as two motives: following the customer and seeking market (Erranmilli, 1990).

The hypothesis of following the customers are proposed by Brimmer & Dahl (1975), Grubel (1977), Gray (1981) and Kindleberger (1983). They argue that the banks can serve the banking needs of their home country clients more competitively than other banks (both local and foreign) due to their pre-established relationship with and better access to information on their clients at home. In addition, these banks may need to accompany their clients abroad in order to ensure a continuing business relationship with the foreign subsidiaries. Failure to do so would force foreign subsidiaries to turn to foreign banks or domestic rivals with offices there, thus the motivation of following the customers can be seen as a defensive strategy. Cho (1985,1986) proposed banks' experience in internationalization and capacity for product differentiation are the other sources for ownership advantages. Many empirical studies found evidence for these arguments. The research made by Hultman & McGee (1989),

Grosse & Goldberg (1991), Heinke & Levi (1992), and Brealey et al. (1996) showed that the business connection between the host country and the home country and the host country's open-up policy for foreigners are the key factors for a country's banking sector internationalization. Some researchers like Ball & Tschoegl (1982) and Ursacki & Vertinsky (1992) use the data concerning individual bank's information found the bank size and internationalization experience also affects banks' internationalization activity.

Another motivation of the banking internationalization is market seeking. Jones (1993) noticed that the establishment of some British bank in 19th century is for serving the foreign markets, rather than domestic customers. Tripe (2000) pointed out that the main motivation of the banking internationalization is for profit. Williams (1997) further proposed what the banks care is the global profit maximization, not the individual foreign market. Lensink and Hermes (2004) suggested that the host country's profitability potential, imitating their competitors, decreasing the cost of capital and risk diversification also lead to banks' internationalization. Yamori (1998) found seeking markets is also one of the motivation of Japanese banks' internationalization by using the per capita GDP as the indicator of host countries' financial market size.

Many researches find out that the customers following is also the motivation of Chinese banking internationalization. Aili Liu (2007), Yixing Deng (2008), and Qirui Huang (2010) pointed out that both outwards FDI of manufacturing sector and bilateral trade volume have a positive correlation with Chinese banking internationalization, however, the motivation of seeking market still lacks empirical support.

So far, most of the research about China's banking internationalization is to study its motivation by using the sample countries already invested. A more urgent issue in practice is to tell which country banks should enter or call for more attention in the near future. The answer of these questions undoubtedly needs all the samples both countries invested and countries without investment. Unfortunately, to our knowledge, the similar studies are not found. As far as the area of the Middle East is concerned, the studies are even more limited. Most of studies concerning this area are just description of the present condition, lacking for quantitative study and deep research. This paper will try to answer these questions by using clustering method and the research framework for choosing the variable will still follow Dunning's OLI-Model.

3 RESEARCH METHOD AND DATA

3.1 Data description

On the basis of the previous literature review, two kinds of indicators are chosen. One kind of business

connection indicators is to represent the motivation of "following the customer". The variables of average FDI and export of China with the countries in the Middle East during the period of 2005–2012 were adopted. All the data come from WIND. The other kind of economic development and financial market development indicators is for representing the motivation of seeking market. The variables such as the average GDP per capita from 2005 to 2012 and the ranking of global financial markets are chosen. The data of GDP per capita comes from World Bank. The rankings are from GFCI (global finance city index). In the recent 5 years, only Bahrain, Qatar and Dubai rank top 100 in the GFCI, thus these 3 countries are set 1, the other countries are set 0 for this dummy variable.

There are 20 countries in the Middle East. However, Turkey, Palestine, Israel and Sudan are deleted from the samples due to lack of data of outward FDI and export. Iran is also dropped since it is under international sanctions. Eventually 16 countries are kept, including Saudi Arabia, United Arab Emirates, Kuwait, Bahrain, Qatar, Oman, Jordan, Syria, Lebanon, Iraq, Yemen, Egypt, Morocco, Algeria, Tunisia and Libya.

3.2 Research method

Cluster analysis or clustering is the task of grouping a set of objects in such a way that objects in the same group (called a cluster) are more similar (in some sense or another) to each other than to those in other groups (clusters). The basic cluster analysis includes: Hierarchical cluster analysis and K-means clustering methods. K-means clustering methods choose K points randomly, and then calculate the distance to central points separately. And the samples are assigned to the nearest central points. Repeat the process again, if there is no different central points, we can adjust for the next iteration. This paper uses K-means clustering methods and classifies the 16 samples into five types. The tree of Hierarchical cluster analysis is also supplemented for the more clear presentation.

As four parameters differ in the units and scales, standardization is preprocessed before clustering. The results after standardization is showed in Table 1.

Table 1. Standardization of variables.

	n	min	max	average	s.d
Z(FDI)	16	-.68110	1.3594	-.17995	.6924
Z(GDPPER)	16	-.6996	2.75362	.03367	.0228
Z(EXPORT)	16	-.62840	3.21178	-.093202	.9535
Z(CENTER)	16	-.44909	2.09575	.028068	.0259

4 RESULTS

4.1 K-means clustering

Using K-means clustering, the 16 countries in the Middle East can be divided into 5 categories based on outwards FDI, export, GDP per capita and rankings of global financial centers. Variance analysis shows that the differences among these groups' mean are significant. The results are shown in Tables 2 and 3.

Table 2. Results of K-means clustering.

n	FDI	Per capita GDP	Export	Ranks	Country
1(mean)	6393	5.8779	519599.33	1	Qatar, UAE
2(mean)	68	1.9657	23921.03	1	Bahrain
3(mean)	7501	1.6165	3.75E+06	0	Saudi Arab
4(mean)	544	3.062	951130.53	0	Omen Kuwait
5(mean)	2285.11	0.3636	167302.9	0	Syria. etc*
total	2768.475	1.5686	524514.8	0.1875	

* Syria, Lebanon, Tunisia, Jordan, Libya, Morocco, Yemen, Algeria, Iraq, Egypt.

Table 3. ANOVA.

Setting	clustering Mean square	df	errors Mean square	df	F	Sig
Zscore(FDI)	3.157	4	0.281	12	11.2	.001
Zscore(GDPPER)	2.865	4	0.378	12	7.6	.003
Zscore(EXPORT)	3.071	4	0.31	12	9.9	.001
Zscore(CENTER)	4	4	0	12		

The first category includes Qatar and the United Arab Emirates. They are the best choices for banking internationalization, and branches or subsidiaries are suitable. First, both of them have global financial center, developed financial markets and highest GDP per capita in the Middle East, so the banks can seek local profitability here. Second, China has the closest business connection with these countries, which were reflected as the highest export and FDI. China's banks can follow their customers to set up branches or subsidiaries here. At first, they can provide wholesale banking or international settlements. Later they can start to carry out retail banking.

The second category only includes Bahrain. It is also the preferred choice, since it is the traditional global financial center in the Middle East. But the form of representative office is better considering its comparatively low GDP per capita and not very close

business connection with China's economy. The form of representative office can satisfy the needs for keeping the latest developments in the Middle East in a more economical way.

Saudi Arab is the only country in the third category. Though it is not a financial center, its government is actively promoting developments of financial markets and attracted foreign Banks to set up agency there. The Saudi Arab is also the most developed economy in the Middle East. Its business relation with China is also close, which are shown as the highest export and FDI. In brief, Saudi Arab can be taken as one of the preferred location just after the above 3 countries, and branches or subsidiaries are more suitable forms.

4.2 Hierarchical clustering

Based on the four indicators above, the Hierarchical cluster analysis is also adopted. The result is as follows.

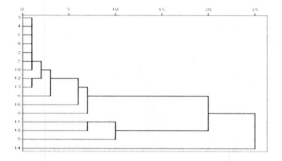

Figure 1. Hierarchical clustering.
* 14:Saudi Arab; 5,15,11:Bahrain, United Arab Emirates, Qatar; 9:Kuwait; 16:Algeria; 6:Omen; 13:Iraq; 12:Egypt; 3, 4, 1, 8, 2, 7, 10:Syria, Lebanon, Tunisia, Jordan, Libya, Morocco, Yemen.

According to the hierarchical clustering, the samples can be divided into 8 categories. The result is similar as the K-means clustering except Omen and Algeria. It also verifies that the fifth category mentioned in the K-means clustering can be further divided into two types. One is Syria, Lebanon, Tunisia, Jordan, Libya, Morocco, Yemen, and the other is Algeria, Iraq, and Egypt.

5 RESULTS VERIFICATION AND FORECAST

5.1 Profile of China's banking in the Middle East

Table 4. Chinese banks' branches in the Middle East.

	2007	2008	2009	2010	2011	2012	2013
Number	1	3	7	7	7	9	10

Table 5. Forms of business organization in 2013.

Setting	ICBC			BC			ABC			CCB			CDB		
	B	S	A	B	S	A	B	S	A	B	S	A	B	S	A
UAE(2008)	1			1					1			1			
Qatar(2008)	1														
Bahrain(2008)						1									
Egypt(2009)	1														
Turkey(2012)						1									
Ttoal	2	1	0	0	1	3	1	0	0	0	1	0	0	0	1

* ICBC: Industrial & Commercial Bank of China; BC: Bank of China; ABC: Agricultural Bank of China; CCB: China Construction Bank; CDB: China Development Bank;B represents branches, S represents subsidiary, R represents representative office.

Bank of China sets up a representative office in Bahrain in 2007, which was the first bank in China to enter the market in the Middle East. Then in 2008, Industrial & Commercial Bank of China Ltd., opened the branches in Dubai and Doha, representing the first Chinese bank to set up branches or subsidiaries in the Middle East. In 12th 2009, ICBC was approved to operate wholesale service in the form of branches in Abu Dhabi, United Arab Emirates. It is also the first Chinese bank to get license from the central bank in the Middle East. By 2013, both Agricultural Bank of China and China Construction Bank established branches or sub-branches in the Middle East. Different from the above big-four banks, China Development Bank chooses the Cairo in Egypt as the target country, setting up the representatively office in 2009, because its overseas investment and corporation focuses on Africa. Until now, China Development Bank has supported 27 projects in 18 African countries.

Besides the form of representative office, branch, sub-branch, Bank of China also strengthens its agency form by opening Chinese service counter. Bank of China sends salesman to the correspondent bank in the Middle East for providing financial service related with China's business. Until now, then Bank of China has set up counters in the country like Omen, Turkey and Egypt.

5.2 Conclusion and suggestion

The current development of China's banking in the Middle East basically corresponds with the results of clustering analysis. China has already established business operation in Dubai of United Arab Emirates, Qatar, and Bahrain, and Dubai is the most preferred area for China's banks. Based on their different level with China's economy, branches, subsidiaries and representative offices were set up separately.

The result of clustering analysis also predicts that Saudi Arab, Kuwait and Omen can be the best candidates for China' s further investment. The forms of business operation can choose among the representative office, branch, subsidiary and agent bank depending on local country's economy and business relation with China.

ACKNOWLEDGMENTS

Beijing Municipal Commission of Education (Grant No. 356005), Beijing Municipal Party Committee (Grant No. 2011D005008000003), China Scholarship Committee (No. 201307127016), Beijing International Studies University (BISU)

REFERENCES

Jones. G.1993. British multinational banking strategies over time. In The growth of global business. H. Cox, J. Clegg & G. Ietto-Gilles(eds). London and New York:Routledge.

Erranmilli, M.K. & Rao. C.P. 1990. Choice of foreign market entry modes by service firms: role of market knowledge. Management International Review 30(2):135–150.

Ursacki, T. & Vertinsky I. 1992. Choice of entry timing and scale by foreign banks in Japan and Korea. Journal of Banking and Finance16:405–421.

Ball,C.A.&Tschoegl,a.T. 1982.The decision to establish a foreign bank branch or subsidiary: An application of binary classification procedures. Journal of Financial and Quantitative Analysis 17:411–424.

Cho,K. 1986. Determinants of international banks. Management International Review 26:10–23.

Williams, B. 2002. The defensive expansion approach to multinational banking: evidence to date. Financial Markets, Institutions and instruments 11(2):127–203.

Williams, B. 1998. Factors affecting the performance of foreign-owned banks in Australia: a cross-sectional study. Journal of Banking and Finance 22(2):197–219.

Kindleberger, C.P. 1983. International banks as leaders or followers of international business. Journal of Banking and Finance 7(4):583–595.

Tripe, D. 2002. The international expansion of Australian banks. 13th Australian Finance and Banking Conference, Sydney:Dec.2002.

Gray, J.M.& Gray, H. P. 1981. The multinational bank: A financial MNC. Journal of Banking and Finance 5:33–63.

Cho,K. 1985. Multinational banks: their identities and determinants. Michigan: UMI Research Press.

Williams, B. 1997. Positive theories of multinational banking: eclectic theory versus internalization theory. Journal of Economic Surveys 11(1):71–100.

Grubel, H.A. 1977. Theory of multinational banking. Banca Nazionale del Lavor Quarterly Review 30:349–364.

Lensink, R & Hermes, N. 2004. The short-term effects of foreign bank entry on domestic bank behavior: Does economic development matter. Journal of Banking and Finance 28:553–568.

Information, Computer and Application Engineering – Liu, Sung & Yao (Eds)
© *2015 Taylor & Francis Group, London, ISBN 978-1-138-02717-6*

Internet financing: An interpretation from the perspective of management theories and managerial innovation

Zhi Wei Zhang & Xiao Zhou Xu
Sichuan TOP Vocational Institute of Information Technology, Chengdu, Sichuan, China

ABSTRACT: Internet financing is enjoying its prosperity in china; a lot of companies dealing with mixed businesses competes with each other in the new market. to understand this new phenomenon better, this thesis draws a review and interpretation of internet financing in depth. the thesis concludes that internet financing can find its supports in classic management theory, customer acceptance theory and the theory of business process reengineering. finally, this thesis takes an outlook of internet financing.

KEYWORDS: Internet financing, classic management theory, customer acceptance theory, BPR

The year 2013 witnessed the infancy of China's Internet Financing. Stepped with the reformation of interest rate liberalization, the internet companies are aggressive in penetrating into financing industry, Yu Ebao, P2P credit and crowdfunding companies are enjoying their first coin of money in China. In the Internet financing environment, how to respond to new challenges and opportunities to realize their own management innovation is becoming a hot topic in today's society.

This thesis is based on the characteristics of Chinese Internet financing, by interpreting the classic management theories, it provides a strong support to Internet financing from three perspectives: the classic management theories, customer perception and business process. Meanwhile, in-depth, systematic management of business enterprises to promote innovation in the financial environment of Internet strategies and measures.

1 THE CLASSIC MANAGEMENT THEORIES

1.1 *Schumpeter's creative destruction*

Creative destruction is among the most famous theories by the great economist Schumpeter. It constructed his early theoretical foundation. From Schumpeter's view, creative destruction is the key and critical matter of capitalism. It is very important to know how capitalism create and destroy economic structure. The way to create and destroy economic structure is innovation rather than price or competition. Every innovation will eliminate the old technology and production system to erect the new ones.

The theory shocked the theoretical circle, now the global economic recession and its value perfectly proved his profound thought. It can be said that the power of creative destruction is still enhancing, and it has become the critical idea of mainstream economic thesis.

Schumpeter has said that in stochastic unbalanced economy, the way to make profit and create new economic point of growth is to destroy the original structure and build up the new Eco-System. Silicon Valley embedded his great idea, in shortly 2 to 3 decades, Silicon Valley became the benchmark of IT industry all over the world, and their magazine *business 2.0* also advocates new economy.

Schumpeter's creative destruction precisely offered traditional financing industry a new way to develop, namely, it is to creatively destroy banks' traditional business with low efficiency and build up a new "Bank 2.0" taking customers as their business center.

Researchers have a large range of literature to support this idea, creativity has strong impact on every new industry, especially the IT industry; destruction has all the way to build up a new one, the suitable one and the profitable one. We can see some of these rich literatures below.

1.2 *Dynamic capabilities theory by Teece et al.*

There were lot of theories and models since 80s last century such as Micheal Porter's industry organization theory; Penrose's resource basics theory; Prahalad's core competitiveness theory and Nelson's evolution theory. All the above theories have finally contributed to the dynamic capabilities theory by Teece et al.

Teece (1997) defined dynamic capabilities as companies integrate, build, reconfiguration and competence to cultivate their capabilities to adapt to the new changing and unstable environment.

Table 1. Literatures on creative destruction.

Researchers	Year	Research and Standpoint
Dosi	1990	Financial innovation in industry transformation
Brown&Perterson	2009	Research on cash flow, and outside equity and financial innovation's impact on profit growth
Cosh&Hughes	2009	Research on post-crisis multinational corporations; theory on how to eliminate crisis and recover from losses
Mazzucate	2013	Research on stochastic relationship; analysis on how financing industry serves innovation
Schubert	2013	Bringing out how to measure Schumpeter's creative destruction

After Teece, a lot of scholars led researches on dynamic capabilities from different angles. Among the most famous are: Teece (1994, 1997), Pisano (1994) and Seung Ho Park's (2001) angle of strategic management, they defined dynamic capabilities as a special ability, namely, the ability of a company to build, reconstruct and combine inside and outside resources to adapt to the changing environment. Nonaka (1994), Ken Kusunoki (1998), Eisenhardt & Martin (2000), Wang & Ahmed (2007) and Schreyogg & Kliesch-Eberi's (2007) angle of organization capabilities, they defined dynamic capabilities as the ability of organization study and knowledge creation. They also classified organizational adaptation capability as low-level capability, and absorb capability as high-level capability. Zolo & Winter (2002) and Zahra's (2006) angle of group study. They classified dynamic capabilities as study mechanism, dynamic capabilities and operation routine.

The key to acquire dynamic capabilities is to cultivate the power and create new core competitive. This strategy is dynamic and agile, and all organization structure must adapt itself to this principle.

The traditional banks are facing complex environment, under the impact of Internet financing and interest rate liberation, the critical point is how to create their own dynamic capabilities.

From the bringing-out of this theory, scholars have researched a lot of achievements and large amount of literature has strong relevance with combination between dynamic capabilities and Internet financing, the main literature picked below are among the typical ones.

Table 2. Literatures on dynamic capabilities.

Researchers	Year	Research and Standpoint
Calantone&Cooper	1979	The reason why new products failed in new industry: using the bias model
Cepeda&Vera	2007	Analysis on dynamic and operating capabilities from the perspective of knowledge management
Ambrosini&Bowman	2009	Research on dynamic capabilities' impact on corporate's strategic management
Hao., et al.	2010	Research on the route of achieving dynamic capabilities from the perspective of entrepreneurship and organization study

2 THE RESEARCH ON CUSTOMER'S ACCEPTANCE OF INTERNET FINANCING

Scholars have many theories on the research of the acceptance of new technology, especially the information technology. There are mainly five models: the theory of reasoned action, diffusion of innovation, technology acceptance model, social cognitive theory and technology rejection model. Internet financing is strongly powered by information technology, all five models mentioned above have concrete proofs, and we just list them below

Table 3. Literatures on customer's acceptance of technology.

Researchers	Year	Research and Standpoint
Fishbein&Ajzen	1975	Theory of reasoned action: analysis on how attitude influences people's behaviors deliberately
Puschel&Afonso	2010	Analysis the mobile bank's feasibility and characteristics using general acceptance inclination framework
Rogers	1962	The five key factors whether new technology is accepted: relative advantage, compatibility, complexity, trialability and observability
Polatoglu&Ekin	2001	Empirical study on Turkey commercial bank's internet financing
Davis et al	1989	The two critical factor determining customer's acceptance of new technology: the perception of the availability and the ease of use
Tan & Teo	2000	Analysis on factors of the acceptance of internet financing
Durkin&Jennings et al	2004	Analysis the main barriers to use internet financing

3 BUSINESS PROCESS REEINGEERING BY HAMMER AND CHAMPY

Business process reeingeering (BPR) was brought out by MIT's professor Micheal Hammer and CSC Consulting's James Champy in 1990s. They pointed out in their book that people always build up and manage their companies based on the idea of specialization in Adam Smith's book over 200 years. That idea focus on dividing the job into simplest and most basic steps, but now people should regroup these specialization into unified process.

Hammer and Champy defined BPR as to leapingly enhance the main basics of operation such as cost, quality, service and speed, the work process should be fundamentally rethink and totally revolved. BPR broke the traditional way to set the work process according to specialization since the industrial revolution. BPR is totally customer-oriented.

For small commercial bank, BPR has its own advantage. Small banks take relatively lower cost, and they have strong desire to take BPR for their service diversification.

4 ENTERPRISE MANAGEMENT INNOVATION AND INTERNET FINANCING ENVIRONMENT

4.1 Innovation and enterprise management concept: grasping the corporate strategic management direction

The current financial and economic environment has provided the Internet a new ear to invade the traditional industry of corporate management. The traditional management over hierarchical class in company has aroused a sense of crisis, so the process of innovation in the management of enterprises should first understand the concept of innovation, and ad-vance with the times.

In the current economic environment, companies should also promptly develop forward-looking and strategic goals, and grasp opportunities for develop-ment under the new trends in Internet banking, to avoid the external environment, potential risks and threats.

4.2 Actively promote innovative organizational structure

The organizational structure: Companies should try to establish a flat organizational structure, through the traditional multi-layered structure, companies can break, achieve reduction and compress levels of management functions of the body, so that they can get feedback and deliver information more quickly; At the same time, they can create a virtual organizational structure, combined with network technology for organizational restructuring and innovation to have a comprehensive knowledge of the dynamic unfolded human resources, information and data in order to establish a shrewd and efficient management organization.

4.3 Actively promote innovative enterprise system

Enterprise system is the main constraint of economic management, but most of the current economics of the business management system is not able to develop with the times, and therefore not well adapted to the need for new financial and economic development.

Therefore, in order to promote enterprise management innovation, companies must perform management system innovation and adjustment to build a reasonable and effective management system.

5 CONCLUSION

Overall, the new trend in Internet financing, actively promotes enterprise management innovation of great value and significance. However, the management innovation of enterprises needs to go through a long stage, not overnight. In addition, it is also important to increase investment and attention on the management philosophy, management direction, human resource management, organizational restructuring, management system of innovation. At the same time, more importantly, in the promotion of enterprise management innovation process, we still need to fully and timely understand the Internet, based on the new trends in finance; we have to make new strategies, new initiatives, new enterprise management innovations in order to achieve a new Leap.

REFERENCES

Allen et al, E-Finance: An Introduction, *Journal of Financial Services Research*, 2002.

Andersen, E. S., Schumpeter's Evolutionary Economics, *Anthem Press, London*, 2009.

A Nation of Shoppers, *Economist*, 21 May, P.40, 2011.

Barreto, Ilídio. Dynamic Capabilities: A Review of Past Research and an Agenda for the Future. *Journal of Management*, Vol. 36 No. 1, January 2010.

Brown, J., S. Fazzari and B. Peterson, Financing innovation and growth: cash flow, external equity, and the 1990s R&D boom, *Journal of Finance*, 2009.

CK Prahalad and Gary Hamel, The Core Competences of the Corporation. *Harvard Business Review*. May 1990.

Christensen C, S Anthony, A Roth, SeeingWhat's Next: Using the Theories of Innovation to Predict Industry Change[M], *Harvard Business School Press*, 2004.

Christian Schubert, How to evaluate creative destruction: reconstructing Schumpeter's approach, *Cambridge Journal of Economics*, 2013.

Christine D. Reid, *Internet Finance Review* 2000–2001, 2000.

David J. Teece, Gary Pisano, and Amy Shuen, Dynamic Capabilities and Strategic Management, *Strategic Management Journal*, 1997.

Davis F. Perceived usefulness, perceived ease of use, and user acceptance of information technology[J]. *MIS Quarterly*, 1989.

Davenport T H, Grover V.General perspectives on knowledge management:Fostering a research agenda[J]. *Journal of Management Information Systems*, 2001.

Mark Durkin et al, Key influencers and inhibitors on adoption of the Internet for banking, *Journal of Retailing and Consumer Services*, 2008.

Information, Computer and Application Engineering – Liu, Sung & Yao (Eds)
© 2015 Taylor & Francis Group, London, ISBN 978-1-138-02717-6

Cultural similarities and differences between Chinese and Korean pottery

Zhao Hui Zhang

Jingdezhen Ceramic Institute, College of Art and Design, China

ABSTRACT: In the 21st century, significantly accelerate the process of economic globalization, showed more obvious economic integration in East Asia. Development and cultural exchanges between China and ROK are closely linked to each other. China and South Korea are geographically adjacent, similar culture, aesthetic taste is similar for both countries that belong to the Confucian orthodoxy Confucian culture. It is more similar to the promoters of the further development of bilateral cultural exchanges, especially the exchange of ceramic culture has its similarities, but there are differences. Ceramic culture stems from the similarity of common traditional culture, and reflects the impact of the difference in the alienation of modern Western culture. It is the existence of commonality and difference, pottery culture, only China and South Korea were a constant exchange in order to further innovation and development.

KEYWORDS: South Korea; Pottery culture; Exchange

In ancient China and South Korea, exchange of ceramics has been a lot of records, and Chinese celadon Korea that is the exchange of crystallization, we know the history of Korean celadon source of China, after decades of accumulation, after the formation of roots and shine in their own style world. In South Korea, today pottery has not only developed into a separate category of art, but also plays a critical role in the world. Especially after entering the 21st century with the international standards, significantly accelerated the process of economic globalization, economic integration in East Asia also showed more and more obvious development, and cultural exchange between China and Korea ceramic closely linked to each other, China and South Korea because of its geographical proximity they share a common culture and asthetics and have more exchanges of contemporary ceramics culture which has its similarities, but there are differences, common history and culture from the commonality, the difference is the subject of modern Western culture the impact produced alienation. We can clearly see its development through exchanges; learn skills in communication with each other constantly innovative development of ceramic art is undoubtedly of great significance.

1   SOUTH KOREA CULTURAL SIMILARITIES POTTERY

Turning to the similarity of pottery in South Korea, we have to talk about the similarities and South Korean culture. The history of Korean culture by the Chinese influence is obvious, as early as the Tang Dynasty, Silla Korean Peninsula on specially sent to China to learn Chinese culture and rule strategy. In South Korea, Confucianism, and Buddhism have its orthodoxy, and culture of the Korean Confucianism is as follows: first, the pursuit of the content and form of harmonious beauty; second, focus on simple beauty; third, that everything begin with human-centered culture. It is this pursuit of simple, frugal Confucianism that promoted the rapid development of Korean celadon, especially under the influence of China Koryo celadon and celadon trend developed, forming a golden age in the history of Korean ceramics. This is the oldest large-aperture Yun unearthed in the kiln bottom wall carved jade porcelain parrot patterns relics that can get a glimpse of its great similarity with the Chinese celadon from the 8th century to the Five Dynasties period.

It is the similarity of the ancient aesthetic ideology and culture, in the traditional ceramic decoration reached its commonality, Chinese and Korean celadon celadon still favored until today, and developed an enduring more innovative products.

2   SOUTH KOREA CULTURAL DIFFERENCES POTTERY

South Korea pottery difference comes from the impact of Western modern ceramics, pottery face of Western modern, exhibit different attitudes between the two countries.

South Korea is an advantage of easily absorbed by foreign culture for its own purposes in the nation, it is this spirit of innovation that made the Korean culture vigorously developed. Since the 17th and 18th century, South Korea began to Seoul and Kaesong as the center of the development of national commercial networks, independent artisans have been actively developed, they strive to explore the nation's history and tradition, and open up a lot of very unique and innovative products. But the history of this nation's tradition of their products is relatively small, the historical development of shorter, therefore, must be absorbed by the innovative use of foreign cultures. Performance ceramics since the mid-12th century AD to 14th century AD is reflected in this stage of the Korean celadon mosaic that is the foundation for learning to develop innovative products on the Chinese celadon. Especially today with the highly developed science and technology, in the shape of porcelain patterns and have gained a lot of development, there have been a number of highly dynamic new design that reflects the new era of artistic charm, they draw on the western ceramic art the use of language to be innovative, which had significant differences with the Chinese ceramic art. To imitate and absorb Western ceramic art and pottery, Korea can be said to some extent at the forefront of China.

On the one hand, China is affected by the deep-rooted Confucian culture, has always been a relatively conservative country. On the other hand, the history of Chinese ceramics in the world ever created a brilliant achievement, therefore, in the ceramic production has unique skills, especially with ceramic decoration in high technology, we can only inherit the tradition of innovation on the basis of this nation, so as to easily form their own specialty products.

3 CHINA AND SOUTH KOREA EXCHANGE SIGNIFICANCE POTTERY

South Korea is the presence of pottery more commonality and differences, pottery culture between the two countries were only a constant exchange in order to further innovation and development, visibility, communication is a necessary prerequisite for the development of ceramic culture, which obviously have the following meanings:

First, the exchange of pottery culture can provide a clear direction for the future development of their pottery. Pottery is the quintessence of Chinese culture position in the world cannot be questioned, however, since entering the modern Chinese ceramics gradually lagged in the world, has continued the tradition combined with modern Western techniques no innovation and development in the traditional basis, especially contemporary pottery and ceramics industry, the lack of high-quality work. On the contrary,

due to the highly developed modern Western technology, the quality of life has been significantly improved, the requirement of ceramic utensils also have higher requirements, especially in Europe and America, many families are handmade ceramic ware made up of high-quality products. The ceramic utensils our restaurant or household use or slip casting most popular goods, not the market is not in demand, but we do not offer, hand casting of ceramic tableware in the market almost did not, some seemingly painted very fine tea, Grouting of its mold and also book tire, although good painter, because it is not a return to nature and the lack of manual tire artistic taste, which is obviously not high. In the ceramic industry, our volume is increasing, but the design quality of Italy and other countries there are still gaps, often become domestic ceramic processing, the lack of their own design brand products, and the lack of design innovation. Exchanges between the two countries will help to see their own shortcomings and deficiencies, in order to learn the advantages of others to provide innovative use of clear development goals for their own development.

Second, pottery and cultural exchanges between the two can be seen in the characteristics of their pottery making specialty products, development of the national culture. Although the two ceramics are similar, but the influence of Western culture since modern times by a careful look is very different. China is more interested in decorative ceramics and crafts, exquisite style and techniques, giving a sense of dignified, pottery works either shape, glaze, and so there is still a shadow form of traditional Chinese culture.

And see from the visual point of view, the Korean ceramics is relatively simple, more casual, at the same time focus on practical performance is very imagery and personalization, which is affected by the western countries and out of the national character. The inheritance of traditional Chinese ceramics and development better, we know we must do Featured Product innovation on the basis of ethnic characteristics, we have to do ceramics on the basis of traditional Chinese culture, if simply to imitate his country's unconventional away work is meaningless, there is no cultural heritage. We should extend our Chinese culture, while breaking the shackles of creative thinking, with Chinese characteristics of ceramic art road.

Third, cultural exchanges between the two pottery ceramic art education also has a great role in promoting. For example, in setting employment professionals solve ceramics, ceramic courses, and industrial ceramics education on how to communicate more effectively, AC apply their knowledge and other specific issues. In South Korea, the university engaged in ceramic education teacher taught most modern styling is based ceramics education, therefore, the lack of systematic majors pottery heritage and teach traditional ceramic styling education awareness. Education focused on learning and technology-specific wheel molding, ceramic culture of the nation's lack of understanding of the theory, blindly follow the West, so the lack of innovation and diversity of pottery theory and humanities education, which is not a good thing pottery heritage. On the contrary, China 's ceramics education, pottery professional students receive hands-on education is a model that is imbued with the apprenticeship, therefore, accepted in education is more traditional ceramic art.

Korean celadon source for China, after the formation of their own roots style and shine in the world, after the international ceramics community standards, the creation of ceramic raw materials, firing artwork, ahead of South Korea. In the 1950s, South Korea's major universities have created pottery department. After decades of accumulation, and now 'pottery', not only in South Korea has developed into a separate category of art, but also plays a critical role in the world.

Jingdezhen porcelain in Korea Koryo era that incoming native South Korea and North Korea accelerated era ceramics exchange between the two countries through bilateral trade and personal selling and other means. Therefore, the exchange between the two countries is a source of promoting the development of ceramics, and in the development process, but also according to their own distinctive culture.

4 CONCLUSION

Overall, the development of bilateral pottery is different, whether it is to learn from the West or the absorption of the national traditional pottery ceramic art, as long as they can push forward the development of ceramic art is beneficial exchanges and in the ceramic art between China and South Korea promote the cooperation of the national culture on the one hand; on the other hand you can see the advantages of self-development and inadequate, so as to enhance the level of international potters to make their own ceramic art development towards international arena.

REFERENCES

[1] A "relatively modern pottery Education – China, Japan, the United States and ceramic art education more" Shanghai Academia Press, March 2008, 150 000 words.
[2] Chen before rain, the "Eleventh Five-Year" national key audio and video electronic publications, "China Jingdezhen Ceramic Culture Series feature films – China Jingdezhen porcelain", Huazhong University of Science and Technology Electronics Audiovisual Press, 2008.06.
[3] Chen before rain, on the "Jingdezhen Tao Song" literary beauty, Chinese ceramics, 2009.12, core.
[4] Zhou Si, the study of modern folk pottery "living map", art, 2006.09.

Information, Computer and Application Engineering – Liu, Sung & Yao (Eds)
© 2015 Taylor & Francis Group, London, ISBN 978-1-138-02717-6

Early 20th century American modern advertising research (1902–1929)—a case study of a century of advertising company Mccann

Zhi Yang
Zhengzhou College of Fine Arts, China

ABSTRACT: In this paper, commercial and cultural context of the background, analyzes the centuries McCann in the early 20th century, a number of advertising works, to extract the typical case, discusses the featured ad in the social and cultural, economic and political background rendering, and then analyze some typical features of advertising design language development. On this basis, refining, summed up the causes of advertising to generate visual images as well as the characteristics of the strategy was to sell advertising, and ultimately to the development trend of the early advertising of the 20th century, the West has made a broad summary.

KEYWORDS: Business culture; Advertising visual image; McCann Company

1 INTRODUCTION

In the early 20th century, numerous medium-sized advertising professional company appeared in the United States, which then produced a large number of well-known development impact advertising companies so far. Erickson (Alfred W. Erickson) advertising companies and advertising agencies McCann (The HK McCann Company), although only one of the representatives of it but they should not be overlooked in the contribution of the advertising industry, we can pass these two companies (not pre-merger) advertising works glimpse of the early 20th century the entire US advertising design development of the situation.

2 THE CHANGES OF THE EARLY MODERN ADVERTISING DESIGN LANGUAGE

(A) The design gradually shift the center of gravity. In the late 19th century and 20th century, the time before advertising cannot be called a real sense of modern advertising, because both the theoretical level or skill level, mostly selling, notices inform, advertising and other properties. From the existing data collected early advertising point of view, the basic text-based advertising content advertising space filled with dense text of course, there are many combinations of text plus pattern appears in the display form of advertising, but still a large proportion of the text.

The rise of the advertising industry was the direct cause of the commodity market saturation and economic maturity. In the late 19th and early 20th century's time, there has been a lot for people to choose the merchandise, and many household items are produced in this period of time, such as ovens, safety razors, cereal, milk bottles, Coke, canned fruit and other food and beverage on the market, telephone, telegraph, typewriters and other office supplies, as well as movies, phonograph emergence and establishment, such as department store chains. People around the emergence of new facilities to make life chores, office process becomes easy for them; increased number of live entertainment, and enrich people's spiritual life. People enjoy the benefits of all this at the same time, advertising is also endlessly to demonstrate the merits of these public goods. Then sell rational theory Albert Lasker • John • Kennedy • Claude Hopkins and a number of advertising representative for the formation theorist, identified as modern advertising "paper salesmanship" status. Advertising in this period for the transition from simple to inform manufacturers and commodity-centric salesmanship. Due to the theoretical guidance of advertising, this time advertising has had the prototype of modern advertising, merchandising combine theory, detailed insight into the market began, with the planning, the initial creation of the concept of marketing and advertising activities have become increasingly specialized. The overall state of the US's major advertising agency is: just completed sales from a simple magazine advertising space for agents to their own product design and creative advertising copywriting service providers transition customers

Early promotional advertising design, focus mainly on the manufacturers, on the advantages to explore product features, with plenty of copy elaborate product features, copywriting rather long explanation advertising theme meant emphasis on graphics performance is less likely exaggerated patterns, the majority of them was based on the product itself and truthfully depict the appearance of products and details, overall advertising based on objective facts selling products to sell to consumers . So, the basic design of the early advertising paradigm case of text with illustrations and written language of high status in the ad. This phenomenon has been recognized by many well-known advertising and advertising industry theorist George • Lewis said that advertising is a text of the original media, the industry pioneer in the creative copy writer. Lier Si also believe that the early stages of the entire 20th century, the main center of advertising but on the text, rather than the visual arts. The advertising industry has been hailed as the creation of modern advertising, one of the six giants • Albert Lasker (Albert Lasker) saying the advertising industry wisdom: Pictures should be applied only in the case of telling a story better than words . From these arguments, we see that the text was dominant in advertising design.

Rise (b) visual images

Text language holds a strong position in advertising, and in the 1920s was completely broken, the proportion of newspaper advertising text gradually becomes narrow, four-color magazine ads exquisite graphics are also in large numbers, and then well-known illustrator to help the advertising industry, to create a a number of well-known illustrator advertising, as well as humorous comics art form to join, the United States has witnessed a boom in advertising.

Why, then, has been the dominance of text-based advertising industry will be gradually replaced by the image? Observe its long evolution, the following are the main reasons:

First, the technical level of influence. First, print advertising is directly related to printing technology upgrades. An early form of propaganda advertising is the main content of the text, there is an image data shortage, the problem was associated with the level of productivity and printing technology is limited. Lots of color image data due to the development of printing technology is imperfect, forming a direct impact on the color image, not to mention beautifully printed images, thus greatly limiting the use of images in advertising. At the same time, one of the essential elements of the text as an advertising component, it is advantage in printing technology than the image clear, with a significant improvement in the technical aspects of typesetting. Hot metal typesetting machine for mass printing production basis, this comparison, greatly enhance the productivity of the printed text, fonts, layout quickly and conveniently. So naturally the advertising agency focusing on advertising copywriter elaborated by attractive, novel, suspense exaggerated headline to attract consumers, the hard sell-style language,emphasis on product and price advantage as the best aspirations points. However, this situation lasted until the 20th century, 10 years, then it began to change . This is

something we can see from the 1902 to 1910 years ago, the company and the McCann Erickson advertising company confirmed cases, especially color image has been used freely in ad creation, combine to create an atmosphere of mass entertainment depicting like color lithographs of the 19th century in general, colorful, well-made, popular consumer popularity. Second, the invention of photography for the rise of visual images full of plans. From 1839, the invention of the camera, the film appears to live volume, and then in 1888, a lot of the market Kodak camera, photography has become the most realistic image acquisition, the most convenient means of hand-drawn illustrations can save time and labor costs, Therefore, its popularity for advertising creative material to provide a good image began to increase the proportion of advertising.

Second, a large number of issuing newspapers and magazines for the development of the visual image advertising has provided a broad space. These large-scale printing carrier issues led to a sharp rise in the number of print advertising, commercial advertising for survival-based newspapers and magazines, in order to draw a number of businesses in the creative advertising can be described as painstaking. Due to the large number of newspapers and magazines writing itself, if the ad text-based screen then appears in front of the reader, not bound to get their attention, so many newspapers, magazines and advertisements are interspersed with illustrations, a larger share of the proportion of the layout, the magazine also appeared in the form of a full-page ad, plus on the use of color printing technology matures, when the advertising works, although copywriting words really reduced, but compared to the previous advertising designs, began showing increasing trend in the proportion of the image, breaking the text with illustrations of fixed paradigms.

3 ADVERTISING EMPHASIS ON GRAPHIC DESIGN

In 1914, The American Institute of Graphic Arts founded, the largest number of ads in the form of still print ads, people began to learn the way to the creation of graphic design advertising works, creative forms inspired graphic design elements constitute innovative advertising works of art, advertising appears in the novel font design and layout, picture composition no longer monotonous, Varied forms, so treat advertising as graphic arts, proposed this concept can be said was a big step forward advertising designer, has laid a good foundation for the future development of advertising design . Subsequently, the dominant image advertising works have become popular.

Third, the image type of early advertising summary

Early 20th century advertising works, with the gradual increase in the proportion of visual image component, advertising images appear in the form of expression of the following types:

(A) "Illustration plus testimony " type of advertising design language

Due to the amount of creative text advertising Chinese case has not plummeted, still writing than the major cases, so the text to explain the role of advertising is still in the advertising value, but it is a prominent feature, copy the text from the literary sensational speech gradually, encouraging atmosphere into a plain text description of credible testimony style. This trend has been applied to many ad creations, so that consumers feel the earth feeling from hype clouds, so that they generate a sense of trust for advertising.

Illustration of a large number of applications and image scale is consistently in a increasing trend, more and more advertising companies to increase image picture in the ad creation, intuitive feel to the consumer, such as McCann was also hired well-known American painter for them to advertise illustration, commercial propaganda . Rennes as Dirk (FC Leyendecker), Norman • Rockville (Norman Rockwell), have their hand surgeon commercials, style and delicate, real scene, coupled with artistic text, if that is a painting.

(B) images show life situations story

By 1902 to 1929 statistics between the company and the McCann Erickson advertising company advertising case, it can be seen, when the ad content more publicity proportion living everyday products: gasoline fuel, construction materials, cosmetics and medicine . These ads with the daily lives of the American people solidarity, so advertising design process, designers focused on the advertising scene is set in the environment, consumers are familiar with family life, and put any merchandising are seen as constituting the consumer part of life in the scene . Advertiser in the early 20th century saw that the creativity should be based on the consumer's life scenes. Because advertising is the consumer' public face of the group, in the ad creation process, only considering the consumer's life scene as a theme, it can make them resonate. Designed to give consumers the content as the protagonist, and then create an event around, such as time, location, and other factors of things together constitute a combination of advertising image narrative presented to them . When deliver advertising messages to consumers, these visual scenes so that they can be combined to form their own experience preset scenes influence consumers' purchasing decisions, and guide consumers. So the story to life situations for the content of advertising images abound, and this trend has been in use since retained many of today's creative content or material excavated from the consumer's life scene.

Transition (three) hand-painted advertising to advertising photography

As mentioned above it has affected the rise of visual images of the technical level, including the birth of photography and invention. McLuhan said: people from printing (Typographic Man) This step toward the era of images of people (Graphic Man) times is due to the invention of photography. Towards late 19th century, it was the emergence of photography for the advertising industry has injected vigor, so that gradually get rid of the traditional text-based advertising creative mode, updated in real clear picture of the direction of the leading advertising design, advertising, visual ingredients gradually increase understanding weakened textual representation of the error, intuitive visual communication is more people receive information, it also shows that advertising language and culture to the visual culture of the early stages of the formation of the conversion.

Widely used photo-shooting technology, design images for advertising material provides another way. The use of photographs in advertisements for visual images brought to life flourished, many advertising works appear in the pictures are similar to the style of news photos, the photos feature is objectively true, the viewer can feel accurate and timely information, there is a certain the scene documentary sense, so that consumer confidence increased advertising . Extensive use of images for advertising artistry indeed increased considerably, In addition to art, but its advantage is gradually revealed, not only by advertising people pay attention it also has been recognized by the manufacturer. It is because of the picture in the ad during the transmission of information gradually showing much greater advantage than the text, photography was able to actively develop advertising. But this time, did not completely abandon advertising works hand-painted creative way, but in the conversion process, hand-painted and photographic images in the form of numerous creative advertising is still a situation of alternately.

4   CONCLUSION

From the above discussion it can be seen as advertising images played an important role in the later stages of evolution, especially opened up an important channel in terms of creative expression. This phase of advertising design techniques increasingly focus on the effect of rendering the visual arts, is no longer confined to a single model with a copy of the illustration, began to pursue new forms of expression, was also hired well-known contemporary artists for advertising illustrations. Advertising creative intent expressed by the image began to focus on individual feelings passed on to consumers, advertising as a product of modern art, industrial technology and popular culture and shine, art and business of beginning an intimate combination of different ways of artistic expression by the public authorized to accept.

REFERENCES

[1] Stewart Alter. Truth Well Told, US: Mccann-Erickson Worldwide, 1995.

[2] G.Lois P & B.Pitts The Art of Advertising.US:. Harry N. Abrams, Inc., 1996.

[3] (US)Er SI Li with Hai Long Ren translation abundant fable : Cultural History of American Advertising Shanghai People's Publishing House, 2005.

[4] (plus) • Marshall McLuhan with, what channel width translated Understanding Media On the extension of the human [M] . Nanjing: Yilin Press, 2011.

Information, Computer and Application Engineering – Liu, Sung & Yao (Eds)
© 2015 Taylor & Francis Group, London, ISBN 978-1-138-02717-6

Challenges faced by teachers in web-based English teaching context

Y. Zhou

English Teaching Section, Guangzhou Civil Aviation College, China

ABSTRACT: The web-based English teaching brings conveniences as well as challenges for English teachers in China. In this paper, the author lists the advantages web-based English teaching has and explores the challenges faced by English teachers and reflects on the technical challenge and methodological challenge brought by web-based English teaching.

KEYWORDS: technical; teachers' role; motivation; strategy

1 INTRODUCTION

1990s witnessed the rapid development of internet. In August 1991, Tim Berners_Lee publicized his new World Wide Web project and laid a foundation for today's internet. In the year of 1993, HyperText Markup Language (HTML) and HyperText Transfer Protocol (HTTP) were created by Tim Berners_Lee, which made the creation of web pages and transmission of information on the WWW possible. Since then, the modern Internet began to have a impact on every aspects of life including EFL learning and teaching.

CALL, an acronym for computer-assisted language learning, has begun to be used by teachers as an important addition to the classroom teaching in 1980s. It is used in order to increase the learner's ability to interactive situation or as a replacement for teachers. With the booming of internet in 1990s, CALL has rapidly shifted to web-based teaching and learning.

At present, one of the important tasks of the ongoing college English teaching reform in China is to reform the traditional teaching model and establish the web-based multimedia English teaching model. Lots of colleges in China have been doing this. Take Guangzhou Civil Aviation College as an example, the college set up its own self-access learning center in 2009 to facilitate its English language teaching. Both students and teachers are actively involved in a web-based English teaching and learning context. Besides the traditional classroom teaching, teachers assign homework online and give feedbacks to the students online. Each term, the students are supposed to finish certain amount of reading exercise online and their online learning is an import part of final score. The outcome of doing this is encouraging and inspiriting.

2 THE ADVANTAGES OF WEB-BASED ENGLISH TEACHING AND LEARNING

The advantages of web-based English teaching and learning are obvious.

First, it provides a multimedia presentation. In a web-based context, students have easy access to audios, animations and songs. They can view the latest news, get information from the advertisements or listen to an English song. Authentic videos bring students all kinds of voices in all kinds of situations, with full contextual back-up. Authenticity itself is attractive. There is a special thrill when the students are able to understand and enjoy the real things. Authentic video is a window on English-language culture. It shows how people work, think and behave in different situations. After viewing the videos, students will have a deeper understanding of western culture. A minute of vivid showing on a video is worth hours of telling from a teacher in a classroom. That is the profound impact that internet has.

These materials online provide a vast linguistic resource of accents, words, new expressions, grammar and syntax. These are something that neither textbook nor classroom or teacher can do.

Besides, Web-based English teaching provides the most desirable context for students to study a foreign language.

In the constructivism learning theory, people produce knowledge and form meaning based upon their experiences. Two of the key concepts which create the construction of an individual's new knowledge are accommodation and assimilation. Assimilating means that an individual incorporates new experiences into the old experiences, develop new outlooks, rethink what were once misunderstandings, and finally change their perceptions. Accommodation is reframing the world and new experiences into the mental capacity already present. The student learns and comes to the conclusions on their own instead of being told.

According to constructivism learning, students should be exposed to data, primary sources, and have the opportunities to interact with other students and teacher. From this learning and pedagogical perspective, we can say that web-based English learning does provide a platform for the students to be exposed to real authentic language environment and for the students to interact with classmates and teachers dynamically.

3 CHALLENGES FACED BY TEACHERS

3.1 Technical challenge

To begin with, Teachers are technically challenged in a web-based English teaching. In China, most of the English teachers are English majors or liberal arts graduates. Their computer literacy can't be as good as the science graduates. Although they can turn to the assistants experienced with computers in schools' self-access learning center for help, they still need the basic ability to integrate web-based stuff into their language teaching. And sometimes students may depend on their English teachers to provide technical support when they have trouble solving computer problems.

In-service technical training is quite necessary for English teachers. After learning the basic programming language, teachers can add discussion forums or chat sessions to their online courses and receive feedback from their students through CGI, Javascript, PHP, ASP and some other computer programs.

3.2 Renewing conception about teachers' role

Teachers are supposed to renew his or her conception about teachers' role. Holec (1997) listed the three necessary conditions for autonomy. One important condition is to train teachers in a new role (the other two are to ensure that learners have appropriate resources at their disposal and to ensure that learner acquire the ability to take on responsibility.)

In the traditional classroom teaching, teachers are usually privileged as leaders or lecturers whose authorities could not be offended. They argue that language is a grammatical system, if students grasp the grammar rules through teachers' explanation, they will surely acquire the language competence.

However, role of teachers is quite different within the constructivism learning theory. Instead of giving a lecture, the teachers in the web-based environment function as facilitators whose role is to aid the student

when it comes to their own understanding, or function as collaborators during performance. Teachers are supposed to treat their students as friends with equal social status and work as their helpers whenever needed. Teachers are also supposed to encourage students to learn as hard as possible. Just as H.D. Brown in *Teaching by Principles* (1994) suggested, "Think of yourself not so much as a teacher who must constantly 'deliver' information to your students, but more as a facilitator of learning whose job is to set the stage for learning, to start the wheels turning inside the heads of your students, to turn them to their own abilities, and to help channel those abilities in fruitful direction."

In addition to being a facilitator and a collaborator in virtual world, teachers should be continually in conversation with the students in real life, finding the needs of the student as the learning progresses. Teachers following Piaget's theory of constructivism must challenge themselves by making students effective critical thinkers instead of a passive reproducer of knowledge. Teachers' role is more challenging and demanding. They are not merely a "teacher" but also a mentor, a consultant, and a coach.

3.3 Stimulating the students' motivation

The next big challenge is how to motivate the students to learn in a web-based context. In a traditional classroom, teachers sometimes tell them "be attentive" or "listen carefully" when the students' minds are absent. In a web-based environment, how can the teachers really motivate the students to learn?

According to the cognitive theory, the learners will learn when they actively think about what they are learning. But the willingness or the enthusiasm counts a lot. Before learners can actively think about something, they must have a strong desire to think about it.

This brings us to a matter which has been one of the most important factor in foreign language teaching—the learner's motivation. Gardner and Lambert (1972) identified two forms of motivation: instrumental and integrative. Instrumental motivation is the reflection of an external need. The need may vary. One may have the need to speak with a foreign customer; one may need to read some English materials for a test. Their needs are different while their motivation is the same external one. Integrative motivation derives from a desire on the part of the learners. Integrative motivation is based on interest in learning a foreign language because of a desire to know about or communicate with the people who use it. It is internally generated instead of being externally imposed.

Large quantities of evidence have shown that most Chinese students' English learning is impelled by the College Entrance Examination and job application, instead of merging into the target culture. It is a typical instrumental one. But motivation sometimes is a complex and highly individual thing. The author has made an investigation among the 160 students majoring in aircraft maintenance engineering in Guangzhou Civil Aviation College about their purpose of learning English. More than 60% of the students tick the item of making friends with English —speaking people and merging into the target culture. The investigation indicates the changing trend that the number of learners with integrative motivation is increasing. That is a good phenomenon.

Motivation stimulation is a detailed, time-consuming and challenging work. In order to stimulate favorable motivations for their learning, the learners' needs should be analyzed. If the teachers know the learners' learning purpose, they can decide what will be acceptable as reasonable content on the web-based English course. The physical setting where the language will be used; how the language will be used; with whom the learner will use the language; when the language will be used. These are all the elements that teachers should consider carefully. Then he or she knows the content to which he or she should pay special attention. The content may be a text, dialogue, video-recording, diagram or any piece of communication data, depending on the needs' analysis that the teacher have done.

3.4 *Fostering students' learning strategy*

Fostering students' learning strategy in web-based teaching and learning context is another big challenge for English teachers. In order to motivate the students to learn, teachers must know students' learning strategies and help them to foster appropriate learning strategies. The fostering of learning strategy takes away focus from the teacher and lecture and puts it upon the student and their learning.

Learning strategy is closely related to one's learning style. A learning style refers to one's natural, habitual, and preferred ways of absorbing, processing, and retaining new information and skills which persist regardless of teaching methods or content area. Everyone has his or her unique learning style.

A learning style involves perception, cognition, conceptualization, and behavior and many other elements. Teachers usually refer to the sensory channels through which perception occurs as modalities: auditory (hearing), visual (seeing), tactile (hands-on), and kinesthetic (whole-body movement). Students vary considerably with respect to their modality strengths. We may notice that some students absorb information most effectively by silent reading and some prefer to absorb information by listening.

Students with different learning styles usually choose different learning strategies. Learning strategies are the particular approaches or techniques that learners use in learning a foreign language. Since the 1970's, studies on learning strategies used by L2 learners have been growing. Their general assumption the studies tried to test is that poor learners use fewer varieties of effective strategies and use them less frequently while effective learners use more and frequently.

O.Mally and Chamot (1990) identified 3 kinds of learning strategies: cognitive strategies, metacognitive strategies and social/affective strategies.

Cognitive strategies are those involved in the analysis, synthesis, or transformation of learning materials. Cognitive strategies are limited to the specific learning. It has something with the learning material directly.

Metacognitive strategies are those involved planning, monitoring, and evaluating. Among three types of strategies in the classification of O' Malley and Chamot, metacognitive strategies may be regarded as core learning strategies because they are applicable to a variety of learning tasks. The metacognitive strategies allow learners to control their own study process by planning what they will do, checking how it is going, and then evaluating what they have done.

Social/affective strategies concern the ways in which learners choose to interact with other speakers. The strategies are for regulating emotions, motivation, and attitudes and for self-encouragement. They include studying, clarifying questions and asking for help.

English teachers should allocate some time to develop students' learning strategies to promote their self-autonomy and produce autonomous lifelong learners. Strategy training, especially the training in metacognitive strategies, is a must in today's English teaching. How do teachers incorporate the strategy component into a daily lesson? How do teachers incorporate strategies training into a web-based course? This is another big challenge.

Some strategies are double edged and some strategies do not function well individually. In ancient China, great educator Confucius strongly advocates that teachers should teach students in accordance with their aptitude. Teachers are supposed to spend lots of time on knowing the students. Knowing their learning style, preference in learning, learning purpose and even their personality will help teachers a lot. Based on the investigation, teachers may find it much easier when training students' learning strategy.

The demanding and challenging web-based teaching context together with the lack of teaching methodology contribute to the current English language teaching situation in China. Promoting autonomous foreign language learning in China is our long term pedagogic goal and is of great importance. Teachers' facing challenges in web-based English teaching context is an indispensible part of the teaching reform. In-service training including some professional development courses and proper guidance is the top priority for teachers. Teachers should be confident that they will play an even more important role in the web-based English teaching while the technology demonstrates its significant function in English teaching.

REFERENCES

Brown H. Douglas. 1994. *Teaching by Principles: An Interactive Approach to Language Pedagogy*. Beijing: Foreign Language Teaching and Research Press.

Gardner, R.C. and Lambert, W.E. 1972. *Attitude and Motivation in Second Language Learning*, Newbury House.

Holec, H. 1997. *Learner Autonomy* in Holec, H & Huttunen, I. (Eds.).

O.Malley. J M & Chamot. A U. 1990. *Learning Strategies in Second Language Acquisition*. London: Cambridge University Press.

Warschauer, M., & Whittaker, F. 1997 *The internet for English teaching: Guidelines for teachers*. TESL Report, 30(1), 27–33.

Information, Computer and Application Engineering – Liu, Sung & Yao (Eds)
© *2015 Taylor & Francis Group, London, ISBN 978-1-138-02717-6*

Behaviors of controlling shareholders and corporate governance

Lin Pan

China West Normal University, Nanchong, China

ABSTRACT: Concentrated ownership structure view has been widely recognized by the academia; the behaviors of controlling shareholders in the concentrated ownership structure are getting attention, this paper reviews related researches. Controlling shareholders mainly through the pyramid structure to realize separation of the control rights and cash flow rights, resulting in negative and positive effects on enterprises, namely tunneling and propping.

KEYWORDS: Controlling shareholders; Corporate governance; Tunneling; Propping

1 INTRODUCTION

Since Berle and Means (1932), there have been considerable researches on modern corporations, and most of them are based on a common assumption, that is, shares are dispersed. The dispersed ownership structure view has received severe challenges in recent years, a representative research is from La Porta et al. (1999). Through the study of 27 rich countries' listed companies, they found that except for a few countries such as USA and England, concentrated ownership structure is very common in the world. In addition, even in the United States, there are evidences that some large domestic enterprises have concentrated ownership structure (Demsets, 1983). Significant changes in view of ownership structure make research focus transfer, the traditional theory holds that the core issue of corporate governance is the agency costs generated by the conflict of interests of owners and operators under dispersed ownership structure (Jensen and Meckling, 1976), which was replaced by the another agency costs, that is conflicts of interest between shareholders and small shareholders under concentrated ownership structure (Shleifer and Vishny, 1997).

2 CONTROLLING SHAREHOLDER'S CONTROL RIGHTS AND CASH FLOW RIGHTS

La Porta et al. (1999) found that the separation of control rights and cash flow rights of controlling shareholders is very common. Previous studies show that the control rights and cash flow rights are usually separated by three kinds of methods: pyramidal structure, cross-shareholdings and dual-class shares.

In these three ways, pyramid structure is the most common one. Why pyramid structure prevail? La Porta et al. (1999) pointed out that the existence of concentrated ownership structure revealed lack of effective legal protection of investors, they also used this view to explain the existence of pyramid structure. However, they failed to explain why in perfect legal system countries pyramid structure exist. Following traditional view, pyramid structure enable the controlling shareholders to obtain greater control power with less cash flow rights, and under the separation of cash flow rights and control rights, controlling shareholders have the motivation to tunnel. The view that the construction of pyramid structure is to realize the separation between cash flow right and control right is challenged by some studies. Almeida and Wolfenzon (2006) argued that the traditional view could not explain at least two realistic phenomena. First, controlling shareholders under the pyramid structure may have higher cash flow rights, which leads to a smaller deviation between cash flow rights and control rights. Second, since the issue of dual-class shares can also cause separation, why pyramid is more popular? One possible explanation is that the issuance of dual-class shares is banned in some countries, but in some Western Europe countries, offering dual-class shares is legitimate, why pyramid in these countries has not been completely replaced by dual-class shares?

3 TUNNELING

Generally, the separation of control rights and cash flow rights will lead to tunneling of controlling shareholders. "Tunneling" is put forward by Johnson et al. (2000), which means transferring the assets and profits by various legal or illegal behavior. Johnson et al. put the tunneling behavior into two categories: one includes fraud, theft and sale of company assets at low price, setting company assets as collateral loan guarantees, plunder of lower companies' investment opportunities. The other is through the dilution to expand their holdings of equity share, so as to achieve the expulsion of medium and small shareholders for the purpose of related party transactions. Lots of researches on the controlling shareholders' tunneling give corresponding evidences. Claessens et al. (2002) found that the tunneling is common in East Asian countries. Faccio et al. (2001) found controlling shareholders use dividend policy to tunnel the companies on the bottom of pyramid. Bertrand et al. (2002) pointed out that the India family group tunneling is serious, and gave evidence company resources move from lower to upper in pyramid structure, tunneling non operating profit is the main mean of India family group's controlling shareholders. Many researches found the controlling shareholders' tunneling from related party transactions and the M & A events in evidence. Cheung et al. (2006) studied 375 related party transactions of Hong Kong listed companies in 1998–2000, found outside shareholders suffer the loss in related party transactions, confirming the existence of tunneling. They also found companies that the ultimate controllers traced back to mainland China have more related party transactions, because the Chinese mainland and Hongkong have a huge difference in the legal system, this finding confirms the legal opinions and tunneling (La Porta et al., 1998; Johnson et al., 2000). What is worth mentioning is that, due to the supervision and controlling shareholders tend to take covert means to escape the public and regulators, to directly measure the degree of controlling shareholders' tunneling is very difficult, so that most researches adopt indirect ways to measure tunneling behavior in the empirical studies.

4 PROPPING

The traditional view is challenged by another question: if controlling shareholders always tunnel lower companies in pyramid structure, why outside shareholders still invest in these companies? Almeida and Wolfenzon (2006) suppose upper in pyramid structure, outside shareholders have only bounded rationality, their interests are tunneled by controlling shareholders. However, this hypothesis is not convincing from a long-term perspective, standing in the long-term perspective, outside shareholders of the upper companies could also stop investment because of long-term tunneling from controlling shareholders. Obviously, only using the traditional tunneling theory to explain the existence of pyramid structure is not enough.

Friedman et al. (2003) pointed out that controlling shareholders sometimes inject personal private property or capital of other companies in control to save companies in dilemma, for the first time these behaviors was called propping. Friedman et al. (2003) believed that controlling shareholders have tunneling motivation, but when the lower companies are in financial crisis, they prop up. Tunneling and propping are symmetrical; propping may be in order to obtain a higher return on investment, or likely to further tunneling. In emerging markets with non-effective law system, tunneling and propping are easier. They believe that in the emerging markets, because of legal and institutional environment for the protection of investors is relatively weak, corporate lendings are usually supported by banks or the government, so the company's liability is under the soft budget constraint. Although the legal protection is poor, and lenders cannot pay the debt while the debit is often not effective access to collateral, thus unable to recover the debt, but the soft budget constraint of debt can reduce the controlling shareholders' debt evasion, when companies with poor performance controlling shareholders will prop up to help company get out of the financial crisis, so as to maintain the continued occupation of the interests of small shareholders or option to obtain legal returns from listing corporations. Therefore, the existence of debt for the controlling shareholder has propping motivation, this effect also causes the liabilities as a signal of propping to outside investors, in order to attract outside investors, which partly explains why emerging markets have high debt rate. Through empirical research, Friedman et al. (2003) found during 1997–1998 Asian financial crisis, the high debt ratio companies' stock price is relatively low, but the pyramid companies with high debt rate have higher stock price compared with non-pyramid companies, which shows propping of controlling shareholders, there is positive relationship between the debt ratio and propping, while negative correlation with tunneling. In these countries, the high debt ratio is an alternative to poor corporate governance.

Because of strong confidentiality, propping is difficult to measure, for the studies in propping, especially empirical studies are rare. Baek et al. (2004) found that Korean family shareholders prop up companies in financial crisis. Cheung et al. (2006) also found that the Hong Kong listed companies have propping related party transactions. To avoid creditors occupation, Indian business groups often prop up lower companies (Gopalan et al., 2004).

5 DOMESTIC RESEARCHES

China as an emerging market country, the degree of legal protection of the interests of small investors is weak, the tunneling and propping of controlling shareholders in recent years have aroused wide interest of scholars. Tang Zongming and Jiang Wei (2002) using block of shares transfer pricing, studied 88 listing companies from 1999 to 2001 in Shanghai and Shenzhen with 90 large state-owned shares and legal person shares as the sample of intrusion events, to study the private benefit of control. In their sample, listing corporation block transfers average price higher than its net assets per share of 27.9%, the premium level reflect degree of tunneling in China listing companies. To study the influence factors of large shareholders expropriation degree, they found that the proportion of the shares to be transferred with the premium level is related to the scale and level of premium is negatively related to listing corporation.

As the use of the research method is the same, Tang Zongming and Jiang Wei (2002) compared Chinese expropriation degree with Dyck and Zingales (2004), results show that protection of rights and interests of investors reduce tunneling. Compared with other countries, China listing corporations' expropriation degree is not high, but a few large shareholders' expropriation is serious. Su Qilin and Zhu (2003) studied 128 family listing corporations in 2002 as samples, found that China's family controlling shareholders and other countries, through the separation of cash flow right and control right to expropriate the interests of minority shareholders.

In terms of research relates to propping, Jiang Wei (2005) used cash flow sensitivity to analyse, the results support the holding that tunneling and propping are symmetric. Li Zengquan (2005) studied 1998–2001 China's capital market 416 acquisition of listing Corporation, it is found that when the company has the rights or avoiding loss motivation controlling shareholders or local government will support the merger and acquisition, and without "protect the allotment of shares" and "security shell", the motive of M & A often leads to tunneling. Their empirical results also show that propping in the short term can significantly enhance the company's accounting performance, tunneling will damage the companies' value.

6 A BRIEF REVIEW ON THE RESEARCH STATUS AND PROSPECTS FOR FUTURE RESEARCH

Concentrated ownership structure has been widely recognized in the academic circles, the concentrated ownership structure hold the problem of corporate governance of shareholders control rights and cash flow rights caused by the separation is getting attention. The existing researches focused on the implementation of controlling shareholders' control rights and cash flow rights separation

mode and degree, which generally believed that the separation of controlling shareholders is mainly through the pyramid structure, and analysis of problems in the cause of widespread existing in pyramid structure, the existing literature mostly from the point of view of private benefits of control degree, investor protection and internal market and gives theoretical explanation; on the other hand, behavior and the related companies governance issues under shareholder control right and cash flow separation are receiving more and more attention. The existing literature focuses more on the negative effect of the controlling shareholders, namely tunneling. In tunneling, incentive mechanism analysis of theoretical studies focus on the pyramid structure encroachment, empirical researches pay more attention to the possibility of tunneling, and influence degree, etc. In contrast, researches on the controlling shareholders' propping are rare, but the tunneling and propping view is gradually widely accepted in academic circles. The theoretical studies on the support behavior is mainly used to explain the existence of the pyramid structure, empirical researches are mostly from two aspects, distinguishing the different nature of the related party transactions or merge motivations, and to explore its influence on the enterprise, related to a relatively limited research, an important reason is that propping of controlling shareholders is difficult to measure directly, makes the empirical studies uneasy to promote.

In Chinese enterprises, concentrated ownership structure is a very common phenomenon. In this context, governance problems caused by controlling shareholders behavior have more practical significance. The author believes that, in the study of behaviors of controlling shareholders and corporate governance, the existing literature for further study left a lot of space: first, analysis of influence factors of systematic controlling shareholder's tunneling and propping is uncomplete. Internal factors analysis of existing researches mostly focus on traditional (such as the company's characteristics and the agency cost), involves no more on external factors. In the aspect of external factors, law and finance is a very active research branch of the corporate governance in the current literature, such a common conclusion is, the national legal system to protect the rights of investors in the extent exist differences, causes the financial development level of all countries and the corporate governance model is not the same. In addition, system of foreign law plays in corporate governance role in recent years has been more attention, such as ethics, culture, media and the internal constraints of taxation implementation aspects (Zheng Zhigang, 2007). Then, when the analysis of factors influencing the behavior of controlling shareholders, mechanism of reasonable introduction of law and the legal system outside the is a direction; secondly, as the controlling shareholder act of concealment, for its direct measurement is often more difficult, many studies have used indirect methods to measure. For example, in the measure of the tunneling behavior, some studies use external benefits of different company impact (performance shock) reaction degree to describe the marginal value of outflow and calculation of tunnel mining group company internal resources (Bertrand et al., 2002), otherwise a lot of research on proxy for cash flow rights and control rights in proportion to the degree of deviation as holding shareholder tunneling behavior (Claessens et al., 2002; Faccio and Lang, 2002), in addition to the use of bulk stock transfer pricing (Barclay and Holderness, 1989), dividend payout (Faccio et al., 2001) and the fluctuation of stock price (Cheung et al., 2006) and other methods to measure the tunneling behavior. In propping, the existence of many studies use the stock market reaction to test support behavior, there is no consistent measure. Due to the selection of indicators are not the same, between different research results comparable and therefore reduce, how to innovate in the measure of controlling shareholders behavior indicators, so as to improve the empirical study on the comparability and persuasion, is a problem to be solved; Finally, can transfer part of attention from the controlling shareholders behavioral consequences and implications to explore the different mechanism of corporate governance impact on the behavior of controlling shareholders, etc.. These have the potential to become the research direction in the future.

ACKNOWLEDGEMENTS

I would like to express my gratitude to all those who have helped me during the writing of this paper. I gratefully acknowledge the help of Professor Cao. I do appreciate his encouragement and professional instructions. This paper is supported by MOE (Ministry of Education in China) Project of Humanities and Social Sciences (No. 14XJC630004), Scientific Research Fund of Sichuan Province Education Department (No. 14SB0094), and Fundamental Research Funds of China West Normal University (No. 13D037).

REFERENCES

[1] Berle, A. A., and G. C. Means. The Modern Corporation and Private Property[M]. New York: MacMillan, 1932.
[2] La Porta, Rafael, Florencio Lopez-de-Silanes, and Andrei Shleifer. Corporate ownership around the World[J]. Journal of Finance, 1999, 54(2):471–518.
[3] Demsetz, H. The structure of ownership and the theory of the firm[J]. Journal of Law and Economics, 1983, 26(2):375–390.
[4] Jensen, M. and Meckling, W. Theory of the Firm: Managerial Behavior, Agency Costs and Ownership Structure[J]. Journal of Financial Economics, 3(4):30–360.
[5] Shleifer, Andrei, and Vishny, Robert. A Survey of Corporate Governance[J]. Journal of Finance, 1997, 52:737–83.
[6] Almeida, H. and Wolfenzon, D. A theory of pyramidal ownership and family business groups[J]. Journal of Finance, 2006, 61(6):2637–2680.
[7] Claessens, Stijn, Djankov, Simeon, and Lang. The separation of ownership and control in East Asian Corporations[J]. Journal of Financial Economics, 2000, 58:81–112.
[8] Franks, Julian, and Mayer, Colin. Ownership and Control of German Corporations[J]. Review of Financial Studies, 2001, 14(4):943–978.
[9] Lefort, Fernando, and Walker, Eduardo. Ownership and Capital Structure of Chilean Conglomerates: Facts and Hypotheses for Governance[J]. Revista Abante, 1999, 3(1):3–27.
[10] Faccio, Mara, and Lang, Larry H. P. The ultimate ownership of Western European corporations[J]. Journal of Financial Economics, 2002, 65:365–395.
[11] Bianchi, Marcello, Bianco, Magda, and Enriques, Luca. Pyramidal groups and the separation between ownership and control in Italy[Z]. Bank of Italy: Working paper, 2001.
[12] Valadares, S. M., and Leal, R. P. C. Ownership and control structure of Brazilian companies[J]. Abante, 2000, 3(1):29–56.
[13] Volpin. Governance with poor investor protection: evidence from top executive turnover in Italy[J]. Journal of Financial Economics, 2002, 64(1):61–90.
[14] Claessens, Stijn, and Fan, Joseph P. H. Corporate governance in Asia: A survey[J]. International Review of Finance, 2002, 3:71–103.
[15] Johnson, Simon, Rafael La Porta, Florencio Lopez-de-Silanes, and Andrei Shleifer. Tunneling[J]. American Economic Review, 2000, 90(2):22–27.
[16] Faccio, Mara, Lang, Larry H. P., and Young, Leslie. Dividends and expropriation[J]. American Economic Review, 2001, 91:54–78.

Information, Computer and Application Engineering – Liu, Sung & Yao (Eds)
© 2015 Taylor & Francis Group, London, ISBN 978-1-138-02717-6

The design and application of the basketball teaching CAI course ware

Han Jiang

College of Physical Education, Wuhan Sports University, Hubei, China

ABSTRACT: Combined with the actual case of physical education of normal university and in the guidance of physical education and education and technology theory and methods, for the research of the design and application of the basketball teaching CAI course ware. For this, this paper put forward a design principle of student-centered, then analyze the structure design of the basketball CAI course ware and teaching content and function design and so on, and discussion for the applications of the multimedia technology and the inadequate of the multimedia teaching.

KEYWORDS: the Basketball Teaching, CAI Course ware, Design and Application

1 INTRODUCTION

Combined with the true situation in physical education at college, this paper makes research on some design and application of CAI lecture software under the physical educational teaching and theory. It studies student's center teaching, analyzes the frame, contents and function of CAI lecture software for basket-ball. It also points out the shortcomings of the media technology and teaching.

CAI is short of Computer-Aided Instruction. For now, many experts use Author ware, Director, Founder Oswald, Course ware Masters with Photoshop, word. The production of CAI course should be consistent with the teaching goal and be centered of students [1].

2 THE PRODUCTION OF CAI COURSE WARE SHOULD BE CONSISTENT WITH THE TEACHING GOAL

The production of basketball teaching course must be based on course outline of basketball teaching course, and supplemented something related to get the goal of the expectancy of results.

3 THE PRODUCTION OF CAI COURSE WARE SHOULD BE CENTERED OF STUDENTS

According to the different ages, sex and psychological characteristics, and in accordance with the cognitive of students and personality traits. Conceiving and designing CAI course ware form the angle of education and psychology, from this vision, we can see the features and the developing trends of basketball. First, situations take an important role of acquiring the knowledge. The level to knowledge for students is just to accept, but CAI course ware teaching stressed the construction of knowledge. Second, learning circumstances also can promote the initiative of study of students. Learning circumstance is a place of freedom to explore and study for students, CAI is important for the understanding of learning content. Last but not the least, multimedia course ware contribute to transform old knowledge to a new one.

4 THE SELECT OF LEARNING CONTENT OF CAI COURSE WARE

Making sure to complete the content of basketball teaching outline, we can introduce a large amount of related knowledge and information [2]. Fig. 1 is a structure of basketball general teaching content. This part focus on basic technology, basketball teaching and training, and the same time, basketball appreciation also can improve the interesting in the study in basketball.

Basketball CAI course ware can supplement the theory of basketball which the lack of textbook, then also can help students to master basketball technology and theoretical knowledge.

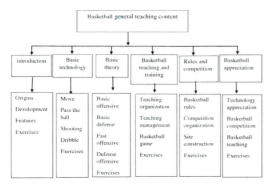

Figure 1. A structure of basketball general teaching content.

5 THE MULTIMEDIA OF CAI COURSE WARE

We can use voice, animation and video expect to words and picture to get the level of teaching form diversification.

According to the requirement of the total structure of CAI course ware, we can design the interface which is divided into three parts. That is the main interface, the interface of content select and the interface of content commentate. In the light of the content of basketball teaching, we can set a wide variety of teaching means. For example, the introduction of words, picture, fragment of video and the commentate of flash.

We must pay attention to the preparation of a course. Not every part has its style, and we need to set buttons of selection and switching for reaching a goal of reducing the number of times.

6 THE APPLICATIONS OF BASKETBALL CAI COURSE WARE

Basketball CAI course ware is good to improve teaching efficiency. Using basketball CAI course ware teaching and a large of information, students can improve their skills themselves and get a good teaching result.

Basketball CAI course ware is good to master technology and theory. Basketball CAI course ware imparts same teaching content in using different forms. In this way, technology would express better and understand easier [3]. We can improve the quality of teaching in using basketball CAI course ware teaching.

Basketball CAI course ware is good to improve the initiative of study. The advantage of CAI course ware is expressive abounding, interactive power and sharing. The useful of CAI course ware plays an important role in initiative of study.

7 THE PROBLEMS OF THE BASKETBALL TEACHING

The effect of teachers in basketball teaching. Though CAI course ware contains a lot of information and data, the information can only read. Animation and image are two-dimensional, and if the animation and image are three-dimensional, it also cannot be an alternative to the teachers. The basketball teaching course ware is not having key action highlights, but the convergence of different action and the tactical coordinate opportunity in the basketball teaching. So the basketball teachers must explain these contents [4]. And only in these ways, can the students learn the basketball action.

Consider the limitation of the basketball teaching CAI course ware. The development of every sports is very rapid. And the CAI course ware is not including basketball technology and theory knowledge [5]. So we must correct and adjust the basketball teaching CAI course ware to meet the need of the students. But the basketball teaching CAI course ware has many advantage, and it can help students to learn basketball action quickly, for example, use the CAI course ware in center of basketball teaching. And you can find more information in the Fig. 2.

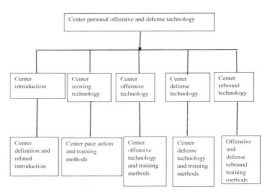

Figure 2. In center of basketball teaching use the CAI course ware.

Perfect the teaching facilities. As a result the restriction of the basketball space and equipment, the basketball teaching is insufficient, and we cannot launch local area network teaching widely. Multimedia teaching includes a little of theory knowledge. And the college should put in lot of funds to build multimedia classroom [6] Nowadays, there are many universities which had build mobile multimedia classroom that have a laptop, a digital camera and a portable digital projectors to fulfill the need of the practical teaching.

8 CONCLUSIONS

Using the basketball CAI can solve the problem that the teachers have little time to explain the basketball theory knowledge in the class better, and it is a perfect teaching process of basketball action. From the above advantage, you can find that the basketball teaching CAI course ware make the basketball teaching more specific. The using of multimedia teaching promotes the transformation of teaching method and teaching artifice and the development of the teaching theory and teaching pattern. Universities should increase the study of multimedia of teaching theory and practice to adjust to the need of modern teaching.

REFERENCES

[1] Sanding Luo, Li Sha sanding, Development of hypermedia authoring tool for CAI course ware, 10(2007)993–999.
[2] Hu wanliang,Liu Dan,Experimental Research on the Multimedia CAI Course ware in the University Tennis Teaching,Procedia Engineering, 2011, 23:339–344.
[3] Yuan, Guicai, Construction of fuzzy-mathematics-oriented comprehensive evaluation in aerobics CAI Course ware, 2012, 7(8):185–192.
[4] Billings D M,An instructional design approach to developing CAI courseware. Computers in Nursing, 1995, 3(5):217–23.
[5] Yonde HE,Approaches for environmental education as reflected in CAI, 2006, 7(1).
[6] Derong Fu, Wu Liu, Guangming Huang, Study on Multimedia CAI Authoring Systems, Educational Media International, 1998, 35(4).

Information, Computer and Application Engineering – Liu, Sung & Yao (Eds)
© 2015 Taylor & Francis Group, London, ISBN 978-1-138-02717-6

The research of building RBAE adult education based on the role model

Mei Bai

Tianjin Open University, Tianjin, China

ABSTRACT: Adult education is an important part of lifelong education system and learning society into an important way. The traditional model of adult education regards the education entity as the center, which neglects the disparities in education of the individual. It cannot meet the diverse needs of educating a body, not reasonable, efficient use of educational resources. Therefore, the construction of an adult education RB AE model based on role is very necessary. It explores the feasibility of the model, and has important significance for the prospect of its application.

KEYWORDS: The role of vision; Adult education; RBAE model

1 INTRODUCTION

In the twenty-first century, it is a production, distribution, which uses knowledge and information is the base of the era of knowledge economy. As an important part of lifelong education system, the adult education will be an important way of learning type. It will be its development pattern, which is facing severe challenges. As the Poland famous educational philosopher Succo Dos Ki said, "we should focus on the future of education, which expressed such a doctrine: the current reality is not the only reality. Therefore, we cannot constitute the only requirement of education. Focusing on the future education spirit, we should be beyond the scope of the present, to create tomorrow's reality as the goal. This thesis starts with the analysis of the current model of adult education in our country, which makes some preliminary exploration on the learning type society China adult education development model.

2 ANALYSIS OF THE CURRENT MODEL[1–2]

Adult education in China mainly include: job training, adult basic education, adult higher education, in the social and cultural life of teaching

They are the main mode of conduct: school based model, enterprise based mode; the occupation education based model. These patterns are the essence of adult education model of the entity as the center, just as shown in Fig. 1:

Adult education model reflects the entity as the center, which is a one of many relationships. It is namely education entities and educated individuals directly linked model. That is to say of the adult education enrollment, professional setting, curriculum,

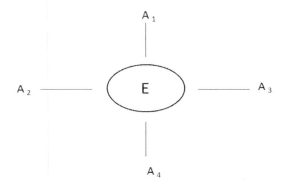

Figure 1. The education entity as the center of the adult education model.

teaching organization form, the student studies appraisal. The degree of granting and a series of work are determined by the education entities E and are responsible for implementation. I can only passively accept the specific services provided by the educational entity E.

3 THE REQUIREMENT OF THE CONSTRUCTION OF NETWORK COURSE OF ADULT EDUCATION

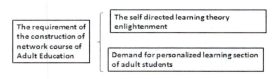

Figure 2. The requirements of the construction.

4 THE SELF-DIRECTED LEARNING THEORY ENLIGHTENMENT

Adult learners have a clear purpose, most of which are brought to solve the occupation's actual needs and problems in work into learning. Therefore, their learning is very specific, hopes that the teaching content and their actual work are related. In the network learning process, they usually choose learning content, which is closely related with their work, life, and other information filtering with its little relationship. That is to say, adult students will learn through the network in the aspect of learning objective, learning content and learning methods. They have shown a very strong self-orientation, openness, sharing.

The interactive network provided for adult students is to fully demonstrate. The ability of self-oriented provides technical support. Thus, self-directed learning ability is the adult learning in network essential ability. Self-directed learning theory defines the learning characteristics of adult students, which plays a fundamental role in the theoretical research on adult education.

5 DEMAND FOR PERSONALIZED LEARNING SECTION OF ADULT STUDENTS

Self-directed learning theory clearly points out that adult students have been mature in psychology, physiology and so on. They have strong consciousness of independence, self-control, self-management, to make their own decisions. They will take responsibility for your own life. So the adult in the learning process are eager to play the initiative and independence. The self-diagnosis learning needs to establish learning objectives, independently for learning resources and learning strategies.

6 THE BASIC REQUIREMENTS OF THE CONSTRUCTION OF NETWORK COURSE OF ADULT EDUCATION[2]

For the construction of network course, the adult education has its particularity, which must follow the law of adult learning. They embody the characteristics of adult learning, and adult students meet the personalized learning needs. At present, adult education network education courses still has general school" characteristics, the content of many, subject centered, emphasizing academic, lack of practicability, Therefore, the construction of network course

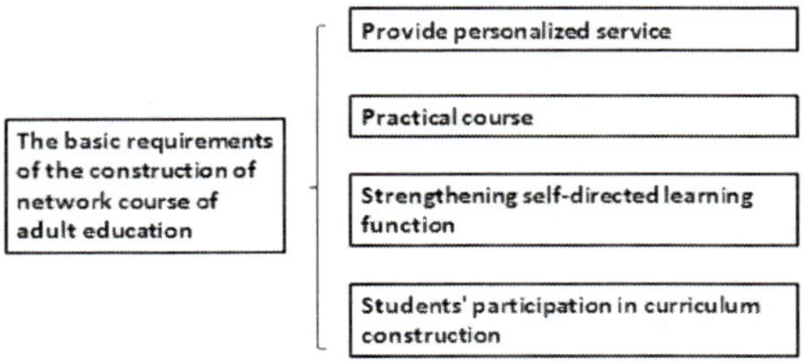

Figure 3. The basic requirements of the construction of network course of adult education.

of adult education should follow the following few requirements:

7 PROVIDE PERSONALIZED SERVICE

The scope of network course of adult education service object is very extensive, including all social identity for adult learners, regardless of age, gender, education level, occupation and experience what's the difference. Even with a course of study, the differences are between them in age, life experience, social stratum, social role, occupation, education background. The background knowledge structure of solid is unthinkable. There are complex differences between adult students, requirements of the construction of network course. The adult oriented is not only to meet the general needs of students, but also to meet the needs of students' personality.

8 PRACTICAL COURSE

Adult students are charged with the work, life, learning and other aspects of the pressure. They can be used to study time and energy, which is limited. In addition, adult students' learning purpose is very clear, the lack of what school does. In order to better meet the needs of adult students, it will update their knowledge structure in the short term to improve the practical ability of the learning requirements. In the course of the selection, design, layout of the network course of adult are oriented in follow the principle of fewer but better, highlighting the practical. "Little" that requires a lot of cut excess in curriculum, unnecessary knowledge; "fine" refers to the basic structure of knowledge and the basic principle to structure. At the same time, the network curriculum design should be based on the occupation, and the practice is regarded as the direction, showing its practicability.[3–4]

It constitutes a connection already, special module and relatively independent. The construction of realistic situation will make use of virtual technology, which enables the student to be personally on the scene to learn. For example, learners can be according to the lectures or presentations steps, simulation provided in accordance with the interactive program. It guides the simulation program, which can complete the operation process simulation.

9 STRENGTHENING SELF-DIRECTED LEARNING FUNCTION

Network course is on computer technology, network technology support, which can break through the traditional boundaries of time and space. So that the separation of teachers and students, the relative separation of teachers' teaching and students' learning process, which requires the participation of network course learning of students with self-directed learning ability. According to the analysis of the self-directed learning theory, adult is with self-directed learning. The so-called adult self-directed learning is the "individual learners in others or without the help of others self-diagnosis, clear learning objectives. The learning needs identified for study of human resources and materials, process selection and implementation of appropriate learning strategies and to assess learning results.

10 STUDENTS' PARTICIPATION IN CURRICULUM CONSTRUCTION

Adult education experts Knowles think that adult self-concept is from childhood "dependent individuals" to "self can guide individuals". This means that the study "manipulation" is the learners and adult can be independent, self-discipline, and control the learning process itself.

11 THE FEASIBILITY OF RBA MODEL AND PRACTICAL SIGNIFICANCE

The most essential feature of knowledge economy era is knowledge and information changes rapidly. The educated adult education and general education are different in that they face the dual pressures of work and study. They should pay the time and energy of each, which is not identical. While the difference exists in their knowledge structure, the ability structure, employment background, which determines the diversity, complexity of the demand by the education and individual education. Adult education must meet the personalized needs of education model by the individual.

12 CONCLUSION

Educational entity has different courses of each professional attribute analysis. The element service is secondly according to these courses which provide specific educational activities. The assessment of learning outcomes is through the completion of the base element. RBA E model will make education physical work fully simplified education individual learning full freedom. Their maneuverability is strong. The learning type society have application prospect.

REFERENCES

[1] Liu Xuanwen. Humanism learning theory review. *Journal of Zhejiang Normal University*, 2002 (1).
[2] Tang Aimin. Twenty-first Century adult education development model on the. *Beijing Adult Education*, 2000 (9).
[3] Tang Xue, From now on to 2000, the content of education development in global perspective [M]. Beijing: Educational Science Publishing House, 1997.
[4] Tian Lin, Floor of a peak at China transcentury thinking the development mode of human education. *Adult Higher Education Journal*, 2000 (5).

Information, Computer and Application Engineering – Liu, Sung & Yao (Eds)
© 2015 Taylor & Francis Group, London, ISBN 978-1-138-02717-6

The reasonable control analysis of architectural engineering cost

Li Xia Song, Xiao Hua Wu, Li Jun Song & Jun Jie Wang
Langfang Polytechnic Institute, Song Lijun, Hebei University of Science and Technology, China

ABSTRACT: With the rapid development of the economy, the speed of the urban construction becomes faster and faster, which requires a reasonable cost control during the process of construction, to control it within the cost range, which can be approved, guaranteed for the smooth progress of the project. To control the cost of all the aspects of construction work, thus reducing the construction cost effectively, to maximize the resource utilization. This thesis is based on the present situation of the construction project cost control, to analyze the problems of the construction cost control and propose appropriate measurements.

KEYWORDS: Construction cost; Control; Status; Problems; Measurements

Construction cost mainly refers to the cost which spends in the process of the beginning to the completion of the construction work, which includes the cost of materials of the construction, installation engineering fees, rental fees, equipment and apparatus acquisition costs, taxes and other construction costs. Reasonable control of the construction cost can greatly improve the utilization of manpower, material and financial and other resources of the architectural engineering. This paper analyzes the cost control problems and proposes measurements for improvement, thus ensuring the efficiency of cost control.

1 THE STATUS OF CHINA'S CONSTRUCTION COST CONTROL

Construction Cost Management System in China is formatted in the 1950s of the last century, and has been improved in the eighties of the last century. China's construction cost management is participated directly by the state. China's relevant departments of construction cost management make a more complete overview of the budget quota management system, make strict provisions of the budget, the principles of establishment, fixed, fixed costs, methods, approaches and budget approval and other equipment and materials. In the recent years, along with the deepening of China's reform and opening up, the past construction cost management system which is directly involved by the country cannot meet the current needs of the actual situation of the construction cost, construction cost management has more new requirements.

In the modern society, the nature of the project has undergone great changes, thus the demand of improving project management is increasing, and the importance of construction project cost control has become increasingly prominent. The effective control of the construction cost is to control the overall construction costs, thereby making it effective measures to stay within the budget range. Traditional method of construction cost control in our country cannot meet the current development, the relevant departments also improve the concepts and methods of construction project cost control according to the modern situation of the construction project cost, to make it better adapt to the development of China's construction cost at the present stage.

At this stage, the engineers of construction cost is still in a relatively short state, the number of registered cost engineers in the country-wide cannot meet the relevant needs of the industry, in addition, to improve this phenomenon still needs more time, making the development of China's construction cost related industries severely constrained. This requires all colleges and universities to strengthen the construction cost of talent cultivation, thus contributing to the healthy development of the entire construction cost of related industries.

2 CONSTRUCTION COST CONTROL'S PROBLEMS

2.1 *Construction project cost control' control concepts backward*

As the economy continues to develop, building construction is also undergoing tremendous changes. But most of nation's construction project's cost control concept not change with these developments, traditional construction cost's control concept still has a significant impact on present construction cost, management and decision-making government departments still have a serious impact on the construction cost. Leading in the serious shortage of market of many construction cost's sensitivity and awareness, do not pay attention to the great role of the market, resulting in serious mistakes in estimating, budgeting and settlement process of construction cost.

For many years, our nation controls a serious shortage in pre-construction cost, while focusing on the late stage of the project or the account stage, resulting in project expenditures exceed the planned investment seriously. The backward construction cost's control concepts lead to construction cost's supervision and regulation out of place, so that construction cost increased, money and building resources wasted, resulting in reducing the overall cost of construction project.

2.2 Construction cost's practitioners lack of professionalism

Construction cost industry's comprehensive is very strong, which involves the country's policies, economic and financial management expertise, technology and knowledge of construction cost, and so on and has a strong practicality, policy, economy and technicality. At this stage, the employees' professionalism of cost industry of construction project is still not enough, can not meet the social needs for a large number of relevant professionals, it's a severe shortage. Although many professional talents related to project cost from various universities go into the industry in recent years, to a certain extent, improved the construction cost industry practitioners' professionalism, but compared to the entire industry needs of construction cost, these improvements also seem so insignificant, and the role is rare.

Because of of the construction of cost industry has always been practiced with estimate budget management system, resulting in a lot of practitioners' knowledge of construction cost industry is too single, they can only use the national's fixed amount to review and budget, and lack knowledge for building other relevant industries, it is difficult to adapt to the needs of modern construction cost of the industry. In addition, due to the practitioners' professional quality of construction cost industry is not enough, leading to a serious impact on the economic benefit of the entire industry, and even some practitioners lack necessary work ethic, make use of improper means to operate, causing a bad society impact on the entire industry.

2.3 The advice and practice system of construction cost is not perfect

Our construction cost's consultancy industry is still in the primary stage of development, its market hosting is short. The purpose of construction cost consultancy is to give contractors, owners, and other related aspects' management and control services of construction project cost. The relevant intermediary services are required to have strict independence and impartiality, but many organizations have better sources of business are affiliated institutions relevant to government departments, their dependence is very strong. While many construction cost of enterprises can solve their sources problems of business, but its corporate impartiality can not get good protection.

3 REASONABLE CONTROL MEASURES OF CONSTRUCTION COST

3.1 Complete construction project's cost control system

In the process of building construction projects, for the construction and engineering design and the process of construction, implementing strict supervision in the period of pre-project, and engineering design optimization, and then control the construction cost within the relevant financial limits, so that the investment profits reached the maximization, and economic benefits of companies optimal. Also this limited design method can be applied in the whole construction projects, and so as to control the entire building project cost more efficiently. The relevant departments of construction cost may also design a variety of programs to cost control, and do a detailed comparison for the various programs, make the entire cost of the design process more perfect, so that there are more economy and practicality in constructing cost's design.

3.2 To change the concept of construction cost control

The concept of controlling the construction cost should be changing and making progress with the establishment and reform of the domestic market economic system by newly making the orientation for construction cost, accepting new concepts of cost control and forming new methods of cost control. The control of construction cost includes maintaining the cost within the investment during the stages of strategic discussion, decision-making, design and implementation of the construction project, and making timely correction for the deviations existed in the process of implementation. In this way, the construction project can be guaranteed to develop toward the objective designed. It can greatly promote the efficiency of human, material and money to update the concept of cost control so that the loss in project can be minimized and better economic efficiency of the enterprise could be ensured.

3.3 *To improve the professional qualification of construction cost staff*

At present, staff in the construction cost related industry are usually with lower qualities, which demand that colleges and universities, combining the social reality with industry needs, increase the training of professionals related with construction cost into those who are qualified with professional knowledge, like financial management, economic management and construction cost, etc. and with the knowledge of machinery and construction skills and who are able to better adapt to the need of comprehensive abilities required by the construction cost industry. In addition, working staff should receive regular training of relevant professional knowledge and skills in order to improve their professional qualification and better adapt to the need in construction cost industry. Organization working in construction cost should set relevant courses in accordance with the professional requirements in this industry so as to provide the society with more professionals.

3.4 *To strengthen the contract management in construction cost control*

At the stage of project implementation, contract is an important basis for controlling the construction cost, which requires the enterprise to thoroughly consider any possible disagreement and claim of expenses in implementation, avoiding some disputes due to undefined contract items in financial accounting when it signs the contract. Besides, the contract should definitely show the standard of charging and the like to prevent unnecessary disputes.

During site-management process, workers of budget and project manager should be familiar with the contract, but also the financial and technicians understand it well. In this way, the contract can be ensured to honor thoroughly, some unnecessary phenomenon can be reduced, like changing the design or approval on site, and the high efficiency of construction cost control can be guaranteed.

4 CONCLUSION

As a comprehensive project managerial work, construction cost control involves social, managerial and economic knowledge, etc, and reaches out every stage of the project. Highness or lowness of construction cost has direct influence on the return of investment in a project. So enough importance must be attached to the control of construction cost. Taking into consideration of the problems that existed in the construction cost, conduct overall control and management in the whole process and try to create the best project by using the minimum investment so as to maximize the returns of the construction enterprise.

REFERENCES

[1] Yan Rong. The analysis of reasonable control of construction cost. *Science and Wealth*, 2012, (4): 496–496.
[2] Shen Qiaohong. The analysis of reasonable and effective control of construction cost *Science and Wealth*, 2011, (6): 151.
[3] Yang Hua. The analysis of difficulties in construction cost budget and how to control *Low-carbon World*, 2014, (10): 157–158.

Information, Computer and Application Engineering – Liu, Sung & Yao (Eds)
© *2015 Taylor & Francis Group, London, ISBN 978-1-138-02717-6*

Some thoughts about information commons—taking Jilin Agricultural University as an example

Qi Wang & Wen Yong Chen
Library of Jilin Agricultural University, Changchun, China

ABSTRACT: Information Commons is a new library information service model in a new technology environment, and also the future development trend of university library. Taking Jilin Agricultural University library as an example, lists the basic elements of building information commons, hoping to provide a template for future library information commons construction.

KEYWORDS: Information commons; IC; Information service; JLAU

1 INTRODUCTION

During 20 years' development, "Information Commons" (IC) has become increasingly mature and widely concerned in foreign countries. In China, IC gradually becomes one of the conduction of university library commons design. IC redefines the connotation of library as service commons and learning places, emphasizing the organic integration of information technology and library service in digital environment, with central task is to provide integrated space, resources and services which has positive meaning to support researching universities' goal.

2 DEFINITION OF INFORMATION COMMONS

The basic theory of information commons formed in the 90's of last century. Now have shaped two distinctions of information commons concept:

First, a unique online environment. Under this environment, users through a graphical user interface achieve various digital service, and search engines which installed on net workstations search the collections and other digital resources.

Second, a new type of physical facility or commons, which can manage commons and provide service in the integration digital environment. This commons can be a department, a floor or an independent physical facility of library, it constitutes a new information environment, and increases librarians and new services based on the first mode.

3 THE COMPOSITION OF INFORMATION COMMONS

There are three level modes of the most typical information commons studies: physical layer, virtual layer and support layer.

3.1 Physical layer

Physical layer comprises physical commons, hardware instruments and service facilities. It is the main part of IC building space, IC core area (reference table, multi-function computer area, and open resource access area), electronic classroom, collaborated discussion area, independent study room, leisure area, etc.

3.2 Virtual layer

It contains information resources, software instruments and virtual environment. Information resources include all printing and digital library resources; software instruments include various software of terminal; virtual environment includes web-based cooperative learning community, online courseware and various training courses. This layer mainly provides digital resources environment, which user through a friendly graphical user interface (GUI), use search engines to obtain digital information service from each workstation. The service's content includes library bibliographic information and more digital information resources.

3.3 Support layer

Support layer consists of service organization, service specification and service evaluation system. Service organization is composed by some department staff; service specification and service evaluation system are the operation guarantee of information commons.

4 IC CONSTRUCTION OF JLAU LIBRARY

4.1 Physical layer construction

Currently, it plans to build seven regionals of JLAU library information commons:

1 Individual learning areas: library provides a number of separate small reading rooms for readers, which have network interface and comfortable personal learning space.
2 Multi-media reading area: providing a plurality of notebook computer interfaces to facilitate autonomous learning and high-grade integrated machine for readers.
3 Collaborative research area: meeting the needs between different professional teachers and students' communication in the opening teaching mode, providing a number of independent study rooms, and equipping with a large LCD screen display, projector and network interfaces.
4 Academic report hall: it can hold all kinds of meetings and lectures and play films for college and student societies.
5 Multi-functional training rooms: equipping with a projector and computers for readers' training, database lectures, humanistic information literacy training and so on.
6 Self-service printed text: readers log into the homepage of library self-service page to upload the required documents and print output, it is suitable for nontradable literature.
7 Other regions: the library will improve the hardware instruments to satisfy readers, such as relaxation space and communication space.

4.2 Virtual layer construction

IC construction should not only have physical commons support but also the construction of virtual layer. University library information commons charges with the responsibility of school teaching and scientific research to help users' access of information resources to complete knowledge transformation, which requires virtual layer to provide support.

4.2.1 Virtual layer framework
Information commons virtual layer is composed by information resources, software instruments and virtual environment, the actual framework.

4.2.2 Information resource database construction
Information commons' service concept and mode determine its particularity of information resources construction (i.e. virtual). The information resource database construction of information commons cannot break the construction process of library digital resources, but integrate the mass digital information to support users' research and study. In this regard, we can integrate multi-level digital resources; build index database and place in a conspicuous position of IC web. Such as bibliography database (OPAC resources), network database, full text electronic literature database (including electronic books, electronic journals), database navigation system, CDs database, characteristic collection database, and network free resources. And the most important work is CDs database building, whose operation is difficult.

4.2.3 IC (virtual environment) site construction
In the IC virtual layer construction, there must be a network platform to carry various kinds of information resources, through which we can provide service and propaganda, so IC site construction is particularly important. IC web construction includes homepage design, function design, and program design. From the research of foreign and domestic college IC page, the design must be concise, with clear level and a large amount of information, well reflecting the IC's resources, services and management, knowing the IC's layout, computers number and software and hardware configuration, setting individual study area, online courses and teaching materials, teaching reference books, and also service contents, contacts introduction and facilities guide.

Based on the principles and combining with the web of Jilin Agricultural University library, we design the information commons homepage divided into two modules: the service module and virtual learning module. Service module contains reservation service, subject service, technology services and reader training. Virtual learning module includes course management system, science and technology exchange platform, online virtual cooperative learning area, information syndication push and instant communication. They enable the reader to realize convenient scientific learning and make the traditional classroom teaching into online learning resources.

4.3 Support layer construction

Combined with our library practice, support layer is composed of senior reference librarian, senior technical personnel and managing personnel, in addition, we also invite professional teachers and high grade students with certain professional basis as subject guiders.

Senior reference librarians can be the subject librarians or information literacy training staff in the library information service center, who is responsible for the higher level information retrieval, reference, subject navigation, information literacy training and information push service. Senior technical staff held by library technical department personnel, who is responsible for guiding users to use the hardware and software facilities, all machinery and equipment management and maintenance, integration of library digital resources and operation automation system management. The managers should have a keen information consciousness, independent information processing capability, perfect business skills, strong communication and promotion ability. The management personnel can be from the elite group of reader service sectors.

In order to make the information commons integrates into the teaching and researching process, we use collaborative learning room as the teaching classroom, which teachers guide students to carry on inquiry teaching by internet, computer software and hardware facilities and rich library resources. It aims to make up the professional shortage of library staff and make better frontier discipline guidance to improve students' study ability.

5 CONCLUSIONS

Information Commons is a new library service mode, a kind of "active service", and also a developing concept. With the development of higher education, readers' information resources requirements are much higher, which requires us "as the users for center" to explore new service concepts and methods to improve the service of our library information commons.

ACKNOWLEDGMENTS

The corresponding author of this paper is Wenyong Chen. This paper is supported by the Youth Foundation of Jilin Agricultural University (Grant no. 201337), CALIS Programs of the National Agronomy Literature Information Center(Grant no. 2014026).

REFERENCES

[1] Huiru Wu. The practice situation of university library information commons. *Information and Documentation Services*, 2009 (03):68–71.
[2] Yao Xie. Construction status and characteristics analysis of 211 university library iinformation commons. *Research on Library Science*, 2013 (08):54–57.
[3] Jintao Liu. From supporting with information resources to involving in learning process—construction of information commons in Hong Kong university libraries. *Library Tribune*, 2014 (04):135–140.
[4] Hui Li,Cheng Yi. Research on the construction of information sharing space at college library—taking the information sharing space of the library of haikou college of economics as an example. *Journal of Hainan Radio & TV University*, 2013 (02):155–158.
[5] Baojin Fnag. Comparative Spatial Study on Information Commons of China and Abroad. *Research on Library & Information Work of Shanghai Colleges & Universities*, 2008 (1):4–9.

Information, Computer and Application Engineering – Liu, Sung & Yao (Eds)

The quality management of landscaping projects

Xiao Hua Wu, Li Xia Song & Jin Ying Liu
Langfang Polytechnic Institute, China

ABSTRACT: With the acceleration of China's modernization process, people's living and working environment is undergoing tremendous changes, they require the quality of the environment that around them have become increasingly demanding, therefore, the amount of landscaping projects becoming more and more, how to ensure the quality of landscaping projects become the focus project. The article discussed the significance of landscaping project management, the quality issues and status of landscaping project, and proposed the measures of conducting the landscaping project quality management.

KEYWORDS: Landscaping project; Quality management; Countermeasure

1   INTRODUCTION

With the rapid development of the cause of landscaping projects, our nation's landscaping projects quality has also been greatly improved with people's increasing demands for the urban environment. Now, the quality management of landscaping project becomes one of the project' core work, good quality management landscaping project is not only to meet the people's needs, but also play a crucial role in urban construction, so the article will focuses on the quality management of landscaping project's strategies.

2   THE CONCEPT AND CHARACTERISTICS OF LANDSCAPING PROJECT

Landscaping project is the work carried out by landscape construction item, through application engineering technology and other advanced technology, make people's living or working's building or structure be harmony with the landscape, so give people a comfortable living or working environment, make the city's landscaping environment improved, and then improve the urban landscape effectively. Landscaping projects included many works, mainly landscaping, landscape architecture, landscape water management projects, garden paving projects and other projects. Because the object of landscaping projects is a living body, which different from some other project, so landscaping project has some its own engineering features:

First, landscaping construction and management's objects are living. Green garden project's construction is a project that relies on green plants landscape design, in the construction and management's process, through reasonably arranged some artificial cultivation of plants; it presents a certain aesthetic effect. On the other hand, in the process of landscaping construction, some green plants can be planted, which remove the dust and purify the air, so as to reduce the pollution of the city effectively.

Second, there are many projects in landscaping projects. As mentioned, there are many landscape construction projects, and the quality management is difficult, so the expertise and the technical requirements of the construction for workers who are working on the landscaping is relatively high, so the staff in landscaping project's construction must have good skills and knowledge to ensure the project quality effectively.

Third, landscaping projects require high artistry. The purpose of the landscaping projects construction is in order to better beautify the urban environment, and therefore, the landscaping project requires high demand for the plant flowers, from plants' colors to morphology, it focused on the art form of expression more. On the other hand, in the process of building landscaping project's construction, better to prune plants regularly to enhance their artistic.

Fourth, the price of materials used in landscaping projects in the market changes very frequently. The plant used in landscaping works strongly influenced by geographical requirements, the types used in the construction are more than the others, in addition, the period of time of landscaping project's construction is longer, and therefore it's difficult to control the cost used in the landscaping project investment, and finally bring a lot of inconvenience to project cost.

Fifth, landscaping construction has its own characteristics, such as seasonality. As mentioned above, the period of landscaping project's construction is longer; mainly due to the growth of plants have their own different growth cycles, of cause, the landscaping construction time will be affected, leading to the landscaping project's construction more difficult. On the other hand, in order to ensure the final design requirements for landscaping projects, need to maintain different plants reasonably; which will directly lead to the cycle of the building landscaping project's construction becomes longer.

3 THE SIGNIFICANCE OF LANDSCAPING PROJECT'S QUALITY MANAGEMENT

As people's requirements for landscaping project's quality continue to increase, our nation's landscaping project's quality management level has been greatly improved, but in the actual process, there are still a lot of quality issues in landscaping projects, the consciousness of some management or construction staff in the landscaping project for quality management of the whole process is not strong enough, eventually leading to the quality problems of landscaping project or the unsatisfied results from the project quality management.

The purpose of the project quality management is to ensure the construction quality of the project, for the actual problems of construction, it can response and rectify timely. Landscaping object is plant with the characteristic of long-term and cyclical, and the work of the whole project contains much content, and complex, which is a highly comprehensive project. Through effective quality control to landscaping works, it plays a significant role in improving the social and economic benefits.

Landscaping not only beautify the urban environment, also played an effective role in purification. If the urban landscaping projects can not be managed with quality management effectively, it will waste resources; therefore, the meaning of landscaping project's quality management is significant.

4 THE STATUS AND QUALITY ISSUES OF LANDSCAPING PROJECT

The ecological environment is not only closely related to social development, also still affect the survival and development of the city to a large extent. But in recent years, the city's population and industry is too concentrated, and they destroyed the urban ecological environment. Therefore, in order to achieve sustainable development of the city, we must strengthen the construction of the urban landscape, improve the ecological environment of the city, create a vibrant eco-city. Now, the city's landscape construction projects including many aspects, for example, green squares,

eco-parks, parks, urban landscape avenues, and so on. While on the surface, the status of urban landscape's construction is extremely prosperous, but in fact there are still some problems. For example, some green area just to cope with the government's request to simply plant a few trees, without any design, lead to poor visual effects; Again, the quality of some green construction team is not high, resulting in the construction of landscaping design effects can not reach the request; also, because lack of oversight mechanisms, some construction units are shoddy, resulting in phenomenon of "green in the first year, yellow in the second year".

5 THE SPECIFIC MEASURES OF LANDSCAPING PROJECT'S QUALITY MANAGEMENT

5.1 The principles of landscaping project's quality management

According to the characteristics of landscaping projects, landscaping projects have to adhere to the quality management in the whole process, we must develop the project management and control the current landscaping projects from project construction in preparatory work to the project during construction and to the end stage. The whole process of quality management can monitor and control the entire project quality effectively, and ensure the implementation of the project smoothly. On the other hand, the landscaping project's construction must adhere to the principle of objective management. Principles of objective management require managers to adhere to the goal-oriented, by refining the project, work out the overall quality management goals and periodical quality management, and make the periodical work to staff with effective implementation, and clear the responsibility. Because any project has the presence of uncertainties, and therefore the manage workers should clear the risk to deal with milestones while developing the stage goals. Except for the above two principles about quality management, there are many other principles that the landscaping projects need to adhere to, clearing the principle is good for working, and protecting the quality of the project.

5.2 Preparatory phase for quality management

In the landscaping project's preparation phase, the selection requirements for Project Management Department staff is must strict, choosing experienced quality control engineers is also very important for the project's quality management. After the formation of a good management team and construction workers for landscaping project, need to plan landscaping construction project's management objectives. After constructing the drawings and surveying the construction site by actual, make specific planning for overall goals and periodical goals of the quality management's project, so proceeds the management of the next quality objectives smoothly. After conducting the goal of landscaping project's management, in order to guarantee the work's proceeding, it is need to establish detailed work system and criterion for the management team and the construction team, so that can clear responsibilities.

5.3 Quality management in the construction phase

Landscaping projects construction has long cycle, and much project work, to do the quality management of the construction well mainly from the following aspects:

First, enhance the staff's awareness of quality management

Landscaping construction workers play an important role in the quality of the project, every project work proceeds in the hands of constructors, therefore, the construction workers' responsibility of landscaping project and the knowledge of project are directly related to the ultimate goal of the project results. In the process of landscaping project's construction, the managers can use rewards and penalties to fully mobilize the enthusiasm of the project's constructors, and enhance staff's awareness of quality management. Adhere to the ideology of "safety first, quality first", to ensure the smooth realization of the landscaping project's quality management objective.

Second, the quality of construction materials proceeds engineering quality management and control work. The quality of the landscaping work's material impact the quality of the project, so in the actual process, must ensure the quality of construction materials to meet the requirements, for example, whether the soil quality standards, whether the plant conform to local ecological environment, and whether the pipeline quality is qualified. Through strictly controlling the quality of landscaping project's construction materials, can ensure the overall quality of the project. On the other hand, landscaping works are also affected by the climate, in the actual project, must ensure the quality of construction materials during storage, master the growth pattern and the growth cycle of all types of plants, in order to find the most suitable time of all kinds of plant types.

Third, the quality management of construction process focuses on processes quality management.

Landscaping project is a highly integrated project, in the process of constructing projects, it requires itemized the entire project, detailed specific tasks and goals, ensuring the quality of every process favors for controlling the overall quality of the project. In the specific implementation of the project, the manager required to inspect each procedure quality strictly, when quality problem appear, should take effective measures to correct timely, if not qualified, the next process can be carried out. When inspecting the quality during sub-project, control the quality of assessment work strictly, when the quality of each of the sub-items or the quality of every step work past, the quality of the entire project can achieve the desired goal. On the other hand, landscaping projects must adhere to the principles of prevention, that may arise for effective prevention well in advance of the risk factors, prepare preventive work well, and this is one of the measures of landscaping project's quality management.

Fourth, do landscaping project's supervision well. One of the effective implementation of any of the essential work of the project is to carry out project supervision. Carrying out landscaping project supervision work effectively can ensure the overall quality of the project. As a third-party engineering, the supervision unit's task is to protect the engineering's quality, the interests of owners and the interests of the construction unit, to be honest, fair and legal, so as to guarantee the quality of landscaping projects effectively.

5.4 Construction end phase and the following-up quality management

One of the important work of the construction end phase is to proceed plants' maintenance and management in landscaping projects and others. Conservation of plants is an ongoing process, at the end of the landscaping project's construction, in order to ensure the survival of plants, improve the visual effects of landscaping projects, need to do some regular post-maintenance landscaping project's work. On the other hand, must constantly improve accept information of project's completion, the information of project's completion has an important role in the detection of landscaping project, the managers should pay attention to the importance of the accept information; ensure the quality of the project qualified.

6 CONCLUSION

Landscaping construction is an important work in the process of building the cit. Only it follows the entire process of quality management principles, can the final quality management's objectives of landscaping projects be ensured in long-term project. Landscaping project is a complex and comprehensive project, each project team should based on local conditions, and plan the architecture, landscape and lighting in the landscaping project's construction rationally, at the same tome, increase instructors' professional tectonics, strengthen the control for quality management, so as to achieve local social benefits and project's economic benefits effectively.

REFERENCES

[1] Zhu Dejun. A brief talking of quality management of landscaping projects[J]. The Guide of Building Materials Development (below), 2012,10 (1):157–158.

[2] Wang Aijun. Quality management of landscaping project [J]. The Quality and the Goods· construction and development, 2014, (6):366–366.

[3] Yang Daosong, Cao Yongjun. Landscaping project's quality management [J] Chinese and Foreign Entrepreneurs, 2013, (20):82–83.

[4] Wang Yijie. A brief talking of quality management of landscaping projects [J] Technology Rich Wizard, 2014, (18):242–242.

Information, Computer and Application Engineering – Liu, Sung & Yao (Eds)
© *2015 Taylor & Francis Group, London, ISBN 978-1-138-02717-6*

To explore the use of financial analysis in enterprise reorganization

Xiao Liang Xu

Weifang University of Science and Technology, Shou Guang, China

ABSTRACT: To explore the application of financial analysis in enterprise restructuring. The concept of corporate restructuring is analyzed from two aspects, content and method. The study of financial analysis includes: cash flow analysis, profitability, debt paying ability analysis, analysis of financial situation and its importance in the corporate restructuring. The specific uses of financial analysis in corporate restructuring take CAAC as an example, to explore the cause of corporate restructuring, financial analysis method and the existing problems.

KEYWORDS: Financial analysis; Corporate restructuring; Importance; Strengthen measures

1 INTRODUCTION

In the recent years, due to the continuous development of a market economy in our country, the financial analysis of the importance of corporate restructuring is becoming more and more prominent. Effective corporate restructuring to enhance enterprises to adjust to market economy, the survival and development ability has a very important role. The key to financial analysis of corporate restructuring is the judicious choice of the mode of corporate restructuring and reorganization. Financial analysis of corporate restructuring mode and the choice of the ways of the standard in corporate value creation. Corporate restructuring value source channels, to determine the level of corporate restructuring value creation and clarify the beneficiaries of corporate restructuring, and so on for the financial analysis of corporate restructuring have a very important role.

2 AN OVERVIEW OF CORPORATE RESTRUCTURING

2.1 *The content of the corporate restructuring*

Corporate restructuring is to increase the capital value as the goal, using the assets restructuring, debt restructuring and the property right restructuring mode, optimize the structure of enterprise assets and liabilities structure and the property right structure, in order to make full use of existing resources, and achieve optimal allocation of resources. Corporate restructuring, it is according to the characteristics of the general strategy of the enterprise restructuring and capital operation, can continue to take the original mode of enterprise restructuring, enterprise merger

and reorganization model and discrete enterprise restructuring, etc. The key to enterprise restructuring is the reasonable choice of the mode of corporate restructuring and reorganization. Restructuring and structural adjustment of model selection criteria is to create enterprise value, to achieve capital appreciation.

2.2 *Corporate restructuring*

Adjustment of our country's current enterprise restructuring practice there are usually two questions: one is the partial enterprises reorganization merger or expand, and ignore its sale, stripping and other enterprises to the capital operation of contraction mode; the second is its merger and acquisition, and division. Source of value analysis the connotation of corporate restructuring, corporate restructuring, defines corporate restructuring way is very necessary. The major methods are: Merger; Merger and Acquisition (m&a); Acquisition; take over or receive; tender offer; stripping; selling; Division; bankruptcy.

3 AN OVERVIEW OF FINANCIAL ANALYSIS

3.1 *The content of the financial analysis*

Financial analysis is based on the content of the information users are divided into external analysis and the analysis of internal content, can be set to a thematic content analysis. Enterprise financial analysis is based on the financial analysis enterprise's financial statements and other accounting documents, use special accounting techniques and methods of financial activity, risk and performance analysis. It is the enterprise production, management, an important part of management activities, mainly including the following:

1 Financial situation analysis

With the company's assets, liabilities and owners' equity, to reflect the different production scale of enterprises, liquidity conditions and stability of the business. This paper analyzes the financial condition of enterprises which includes: analysis of the capital structure, capital efficiency and asset use efficiency. Efficiency analysis using the usual efficiency of funds and assets of the operation ability of the composition of the enterprise, which is the focus of the financial situation analysis.

2 Profitability analysis

Profitability analysis is a kind of professional tools and methods, to determine whether there is a certain part of the operating profit or loss of management tools.

3 Solvency analysis

Enterprise's solvency is using its assets to repay its long-term debt and short-term debt ability. Enterprises have the ability to pay in cash and debt the ability is the key to enterprise's survival and healthy development. Enterprise debt paying ability is the reaction ability of enterprise financial status and an important symbol of vitality.

4 Cash flow analysis

Cash flow analysis refers to the difference between the net cash flow of cash inflows and cash outflows. The expected net cash flow may be positive, sometimes may be negative. Is a positive number, for a net; if it is negative for the net. Net cash flow is to reflect the cash flow of the final result of the formation of the various activities.

3.2 The importance of financial analysis in enterprise restructuring

In the increasingly competitive market economy, the enterprise and competition between enterprises are increasingly fierce, maintain long-lasting competitive advantage, in the competition through the integration of existing resources, optimize the allocation of resources, improve the overall strength is a good method.

Financial analysis is very important, because the company's main financial indicators such as assets, liabilities, equity, cash flow from operating activities such as financial indicators, adopt scientific evaluation standard and applicable analysis method to analyze and evaluate, and compared with the overall level of enterprise industry enterprise's financial indicators, analysis their advantage, find out deficiency, learning and absorbing others' experience and teach, to prevent the risk of the enterprise, to develop suitable for enterprise development decision.

4 FINANCIAL ANALYSIS OF EXAMPLES USED IN THE CORPORATE RESTRUCTURING

4.1 The basic situation of China civil aviation enterprise restructuring is introduced

After China's entry into the world trade organization, China's civil aviation as a major industry of the national economy, the opportunity at the same time, also faces many challenges of the international aviation industry. The so-called restructuring, is six group of companies in China, directly under the general administration of civil aviation of nine airlines and four established on the basis of service enterprises. They are respectively: Air China Group co., LTD., China Eastern Airlines corporation, China southern airlines group company of three air transport group and Chinese civil aviation information group co., LTD., China Aviation oil corporation, China aviation supplies import and export group co., LTD. Three major aviation service groups. After restructuring, three major aviation transportation group total assets accounted for 80% of the total assets of the civil aviation, a total of nearly 500 have aircraft.

4.2 China civil aviation enterprise financial analyses

4.2.1 China's civil aviation enterprise restructuring of financial analysis

1 Access to financial analysis of strategic opportunities. China's civil aviation enterprise restructuring is one of the motive is to buy the future development opportunities. When a company decided to expand its business in a particular industry, is an important strategic merger in the existing enterprises in the industry, rather than relying on their own internal development.

2 The synergistic effect plays a financial analysis. The corporate restructuring of synergistic effect refers to the restructuring can produce $1 + 1 > 2$ effect.

3 Improve the management efficiency of financial analysis. The value of the financial analysis of the CAAC another source is to improve the management efficiency. Through the case analysis, now operating in a nonstandard way, therefore, more efficient business acquisition, managers will be replaced, so as to improve the efficiency of management.

4 The capital market mispricing found the financial analysis, more efficient business acquisition, managers will be replaced, so as to improve the efficiency of management.

4.2.2 The method of Chinese civil aviation enterprise restructuring of financial analysis

1 Classification analysis

Is actually a method of empirical analysis, classification analysis reveals the characteristics of the enterprise restructuring, rather than a merger.

2 Forecast analysis

China civil aviation enterprise reorganization of the candidate's motivation has always been making investment strategy, can predict the candidate or the possibility of merger.

4.3 The problems existing in the Chinese civil aviation enterprise in financial restructuring

1 Insufficient development of the civil aviation market, mergers and acquisitions of financial intermediaries is not developed. China's civil aviation market mechanism is still not perfect, especially in the development of capital market; it is not mature, lack of financial instruments and the shackles imposed by the financing bottleneck. Property rights trading market is not healthy, on the other hand, would have resulted in many mergers and acquisitions take the non-market way, make the property right transaction unfair trade program is incomplete.

2 Civil aviation assets evaluation, and unclear property rights definition.

In mergers and acquisitions in the operation of the civilian aviation transport enterprise, how to determine the value of enterprise property rights is an important problem. Of assets evaluation link at present stage, however, many problems have not been resolved effectively.

3 Government intervention in financial analysis of a civil aviation enterprise m&A activity. China's civil aviation industry is in a particular historical period, a unique background of system transformation, presents the complexity of the airlines merger and acquisition is very big. Obviously, in this phase the airline is not the only mergers and acquisitions. The government's preference to a large extent affected the motivation of enterprise merger and acquisition. This undoubtedly greatly weakened the rules of the market.

4.4 Strengthen the CAAC financial analysis of thinking used in corporate restructuring

As a result of the three airlines group restructuring process has ended, airlines in our country are faced with the foreign to cope up with the great challenge of competition and the internal resources integration. Group, in fact, after the completion of the reorganization, is only part of the relative legal system idea, from the point of view of management, the whole process of restructuring is far from complete. Especially China's civil aviation restructuring is not normative and particularity, the transformation of modern enterprise system in China's airlines has not yet been established, there is a lot of can't ignore the management complexity of problem, such as the relationship between the management is fuzzy, not balance the interests of the enterprise culture conflict, conflict and cooperation partners, as well as some companies did not officially implemented restructuring, the mechanism and the organization structure is no big change, some only simple addition of the number or size of the company, but there is no true advantages of group.

5 CONCLUSION

Enterprise in the process of restructuring if only limited to the physical resources reorganization, and ignore the cultural fusion, cannot reflect the assets reorganization, resource relocation biggest advantage. Cultural integration and integration of the project such as financial, organization, personnel, are to ensure that mergers and acquisitions one of the means to success in the end. Only by adopting leading cultural fusion, attaches great importance to in the merged enterprise into the enterprise culture and advanced management mode, mergers and acquisitions to make be enterprise mergers and acquisitions to grow vitality.

REFERENCES

[1] Xianzhi Zhang.Modern financial analysis procedure and method of system reconstruction. *Journal of Qiushi Journal*, 2008, (4).
[2] Jing Zhu.Try to talk about mergers and acquisitions in the financial analysis and function. *Journal of Heilongjiang Science and Technology Information*, 2007, (13).
[3] Xiali Ni.What should be paid attention to financial analysis. *China's Statistics*, 2005, (4).
[4] Yu Dong.Enterprise financial analysis problems and their improving. *Journal of Business Accounting*, 2005, (24).
[5] Yanfen Wang.On the discussion of strengthening our country enterprise restructuring. *Journal of Business Review*, 2010, (4).

Information, Computer and Application Engineering – Liu, Sung & Yao (Eds)
© 2015 Taylor & Francis Group, London, ISBN 978-1-138-02717-6

The analysis of landscape design in housing districts

Jin Ying Liu, Xiao Hua Wu, Yuan Ya Guo &Yuan Yuan
Langfang Polytechnic Institute, China

ABSTRACT: With the constant development of science and technology and the improvement of people's living standards, people have higher requirements for their living environment. Hence, when the real estate enterprises develop housing property, they should pay attention to the landscape design in the housing district in order to create a more beautiful living environment for the residents. The landscape design in the housing district must take many factors into consideration. This paper covers many aspects, including the significance of landscape design, its principles, existing problems, measures of optimized design, etc.

KEYWORDS: Housing Districts; Landscape; Design; Principle; Significance; Optimized measures

Sustainable development is the trend of today's society. The pursuit of a harmonious relationship between man and nature has become the goal pursued by all walks of life. Because people's environmental awareness is improving, real estate enterprises should pay attention to landscape design in the housing district during the process of the residential development in order to better improve the quality of the housing district.

1 THE SIGNIFICANCE OF LANDSCAPE DESIGN IN HOUSING DISTRICTS

Landscape design in the housing district is of great significance to improve the residents' feelings and artistic effect in the housing district.

1.1 To improve the residents' feelings

Landscape design can improve the ecological environment of the housing district during the process of its construction. Diversified green plants in the residential area can make the residents feel closer to nature and let them live in a natural and relaxed environment. Nowadays, the pace of life quickens, so a good housing district with good landscape designs can make the residents feel the joy of life after stressful work and let them improve their living-in feelings which can meet the requirements of the residents' living conditions.

1.2 To make the housing district have more artistic features

Landscape design not only means planting flowers and trees, but also means that through the designers'

ingenious vegetation arrangement, their scientific and rational design of the landscape structure and other embellishment, they can make the whole residential area have living-in and more artistic characteristics, making it like a picture scroll, which will fulfill the residents' aesthetic needs.

1.3 To improve the quality of living conditions

Landscape design in the housing district mainly focuses on vegetation. Well-designed vegetation in the district can not only reduce noise pollution and air pollution in the community, but also can realize the internal adjustment of the temperature and humidity which will make the residents' living conditions more comfortable.

1.4 To create a harmonious living environment for the residents

Landscape design can not only provide residents with a good living atmosphere, but also a leisure place with beautiful environment. Designers put greening the land and leisure together by designing, which not only can make the leisure place bring people a more comfortable environment, but also can better promote the communication between residents.

2 PRINCIPLES OF LANDSCAPE DESIGN IN THE HOUSING DISTRICT

Landscape design in the housing district should follow the following principles: firstly, adjust measures to local conditions; secondly, the designing process should emphasize humanization; thirdly, the designing process should fully respect the growth rules of the plant.

2.1 Adjust measures to local conditions

Adjust measures to local conditions is the most important principles in landscape design, and at the same time it is also a basic principle. This principle requires designers when designing the landscape in the housing district to make full use of its terrain features to reduce earthwork, thereby reducing the capital and personnel investment in constructing the garden. On the other hand, making the best use of terrain features can make the garden more natural, more poetic. The principle also requires designers to give priority to select indigenous plants and then choose a certain number of exotic plants with strong adaptability and of ornamental values. Through scientific arrangement of vegetation, designers make ecological environment of the garden and the nature bear more similarities and create a distinctive natural environment for residents at all seasons.

2.2 The designing process should emphasize humanization

Designers should also follow the principle of humanization when designing the landscape in the housing district. The green land, bonsai and flowers can provide the residents with a better environment for leisure, sports and talk. So the scenic design must have human touch. Only putting scenery and people's lives together can the garden create a maximum value. Designers must take the needs of playing for children, walking, playing chess and chatting for old people, alleviating the pressure of work for adults into account. The designing process should emphasize the principle of humanization which also requires the designer to green the residence as well as every household as much as they can so that people can best enjoy green and the landscape.

2.3 Follow the growth rules of the plant

Designers should also follow the growth rules of the plant when designing the landscape. Once the residents live in the housing district, it's rather difficult to make some changes for the landscape. So in the selection of plants, the growth habit and rules must be fully respected. There's no need to make frequent replacement of the plant in the garden. At the same time, fully taking the living habits of plants into consideration can also make the ecological value of district to the full.

3 CURRENT PROBLEMS EXISTING IN THE LANDSCAPE DESIGN IN THE HOUSING DISTRICT

Although at present are paying more and more attention to their living environment, there are still many problems existing in the landscape design in the housing district. First of all, the landscape design in the housing district is lack of humanistic concern. The designers should follow the principle of humanization, but actually the designing process is influenced by various factors. The landscape design is still a lack of humanistic concern. Many housing districts have taken the needs of leisure and entertainment into account in the process of environmental construction, but the research on the residents' behavioral rules and psychological demand doesn't carry out further. Designers in the designing process put more emphasis on greening, on the form and scale rather than how to make people feel comfortable. The landscape design in some housing districts even appeared the phenomenon of formalism which emphasizes the changeable presentation and the mix of colors but lacks emotional investment; Secondly, some designers blindly pursue hot spots in the landscape design process. In order to facilitate the sale of houses and have more unique features, many designers emphasize more about visual effect and the modern trend of landscape in their design plans. The excessive pursuits of ornament and the composition of a picture in the landscape design usually increase hard landscape and ignore natural plants landscaping functions. In addition, the blind pursuit of the hot spot is also reflected in being blindly keen on foreign styles. As people's living standards improve, people yearn for the foreign way of life and they crave for houses like garden villas. So when the real estate enterprises develop residential areas, they will usually introduce western styles to the residential design. More western designing principles are introduced by designers in the landscape design. Western elements like pillar fountains emerge from residential gardens which make the residential district become neither a Chinese one nor a western one; Finally, the relationship between different species is overlooked in the landscape design process which will play a negative role in biodiversity. Designers need to take biodiversity between species into consideration. However, designers are lack of the knowledge about the relationship between species. The mix of different plants will form different effects. Symbiotic plants or plants having no competitive relationship can grow together for a long time. Due to the designer's lack of consideration on the interaction between species, the plants in many housing districts even have a competitive relationship which is not good to the growth of the plants. Some residential areas have even emerged plant waste problems.

4 MEASURES OF STRENGTHENING THE LANDSCAPE DESIGN IN THE HOUSING DISTRICT

Designers should do the following work in the process of landscape design in order to make it more effective: first, establishing modern concepts of landscape design;

second, designing based on the positioning of the housing district and the residents' demand; third, coordinating the relationship between ecological benefit and economic effect; fourth, deepening the understanding of plants.

4.1 Establishing modern concepts of landscape design

As people have higher requirements for their living environment, designers should establish modern concepts of landscape design which doesn't mean that designers need to blindly cater to the needs of modern people and blindly worship the so-called European style and tropical landscape, but means that they should give full consideration to the diversity of urban culture and ecological environment. The modern landscape design in the housing district should pursue its ultimate goal, that is, the harmonious relationship between man and nature. Designers are required to give serious considerations to the key element, that is, the landscape design will luster people's lives.

4.2 Designing based on the positioning of the housing district and the residents' demand

Designing based on the positioning of the housing district and the residents' demand can make the greening achieve the best effect. Different residential areas with various positioning target people at different levels who will have different requirements for their living environment. Hence, the landscape design in the housing district should be based on its positioning. Generally speaking, crucial elements of the landscape design include green lands, flowers and trees. Green lands close to residents' lives. The design of green lands should not only conform to the balanced composition of buildings but also consider the arrangement of the plant in the residential area; the arrangement of flowers should be lively, flowers need to bloom in three seasons and remain green throughout the year; the selection of trees should also be rich, so that the housing district can have distinctive features all year round. In addition to some ordinary plant species, designers should add distinctive landscape according to the positioning of the area. Take high-grade residential areas as an example, the positioning targets senior white-collar workers, gold collar workers and professional managers. So when the designers design the landscape in such kind of residential areas, they should pay more attention to the comfortableness and the aesthetic effect. By doing so, they can have a more suitable positioning on high-grade residential areas.

4.3 Coordinating the relationship between ecological benefit and economic effect

Landscaping in residential areas is also one of the factors determining house prices. So designers need to coordinate the relationship between ecological benefit and economic effect. Ecological benefit can not be ignored in the pursuit of economic benefit. If the landscape in four seasons shows different characteristics and there are a variety of fine plants for greening and garden ornaments which can promote the quality of the whole housing district, the housing district will have more economic benefit. In this way, designers take both ecological value and economic value into consideration which can make the greening plan more effective.

4.4 Deepening the understanding of plants

Because different plants have different living habits, different species will survive in different geographic environment. So designers for landscape design must constantly improve their understanding of the plants. First of all, designers should have a clear understanding of the common plants in the housing district and get familiar with a variety of common plants; Secondly, designers need to know the living habit of the plant, understand the relationships among plants and grasp the factors that affect the growth of plants, which can help them better arrange the plant in the designing process and make the arrangement of the plants more scientific and more reasonable.

5 CONCLUSION

Landscape design in the housing district has an important impact on residents' daily lives. So designers should fully realize the great significance of landscape design in the area, figure out the problems existing in the landscape design, focus their attention on the use of urban and vegetation resources, and follow the designing principles in the landscape design process. And then they can create a livable living environment for residents where they can close to nature and develop a harmonious relationship between man and nature.

REFERENCES

[1] Wu min. A brief analysis of landscape design in housing districts [J]. Shanxi Science and Technology, 2013, 3(3) 11–12.
[2] Qiu Rongsheng. A brief analysis of landscape design in housing districts [J]. Technology and Design of Constructional Engineering, 2014, 24(22) 30–31.
[3] Huang Yongwen. Study on the landscape design strategy in communities [J]. City Construction, 2012, 34(34) 19–20.
[4] Wang Yanqiu. The growing trend of the landscape design in housing districts [J]. Heilongjiang Science and Technology Information, 2012, 35(32) 11–12.

Information, Computer and Application Engineering – Liu, Sung & Yao (Eds)
© 2015 Taylor & Francis Group, London, ISBN 978-1-138-02717-6

Research of driving force model of city tourism development

Yi Lin Mao

Zhengzhou Tourism College, Zhengzhou Henan, China

ABSTRACT: Tourism development is driven by a driving force constituted by the push from tourist consumption and the pull from the attraction of tourist products and they are related with tourism medium and conditions for development. Tourism development not only refers to the specific issues such as the development of tourism re-sources, construction of tourism projects and other service facilities, but also is an important part in the development of economy and society as a whole. The article tries to break away from the traditional mode of thought that tourism relies on developing resources and building projects, put forward the driving force system and its structure model of tourism development in accordance with the relation between supply and demand. Try to find out the main factors and auxiliary factors that drive tourism development from the combination of driving force of tourism development and set up driving force model for different types of regions to facilitate the working out of the strategy for the driving force mechanism of concerted development of regional tourism and economy and society.

KEYWORDS: City; Tourism; Driving force model

1 INTRODUCTION

As the city to become a strong point of modern tourism, urban tourism research, much attention has been paid to have become one of the major hot spots of tourism research in recent years. Many scholars carry on the research in this field, but about the Driving Mechanism of urban tourism (Driving Mechanism) research is still relatively rare, in particular, there is lack of Driving Mechanism of urban tourism, special discourse system. However, the theory and practice are calling the further deepening of city tourism development motivation research. Carrying on the research of urban tourism development momentum is not only beneficial to correct the city managers eager for tourism and the blind action, guide urban tourism shape characteristic, in order to achieve the ultimate goal of sustainable development, and to improve the international competitiveness of China's urban tourism, strive to gain a foothold in the international tourism market. Therefore, deepening the research on city tourism motivation has important theoretical and practical significance.

Development of tourism resources and tourism project construction is an important way of promoting the development of tourism, and in the "resource driven" destination is particularly important. Because of varied modern tourism attraction elements, the traditional sense of "tourism resources" is difficult to cover travel appeal. Lack of high strength of tourist attractions, such as Guangdong, but to become

a big tourist province, its reality well above target market tourist spots. It proves that tourist market of Guangdong tourism to attract much more than in the traditional sense of the following places. As a result, tourism development is not only a tourism resource development, tourism project construction and service facilities such as tourism department usually focus on specific issues, but is an important part of regional economic and social development. Tourism market is not pure, it is the market of tourist spots, but the area and the common market of the city. Therefore, the author thinks that the tourism development, starting from the comprehensive needs of regional and urban development, is committed to creating sustainable development of tourism and the dynamic mechanism of sustainable economic and social development as a whole.

2 CONSTRUCTION OF THE INDEX SYSTEM DYNAMIC MODEL OF URBAN TOURISM DEVELOPMENT

To establish index system is the precondition of judging dynamic model of urban tourism development, and the requirements of the selected indicators must be able to reflect the change rule of characteristics and urban tourism development. According to certain principles of constructing the index system, this paper uses the AHP method and Delphi method to determine the index and its weights.

The principle of index selection:

- Systemic: city tourism by the urban system and tourism system and so on, should with the method of system analysis to determine the power of the development of urban tourism.
- Objective: index system should objectively reflect the scientific connotation of urban tourism development.
- Hierarchy: Index system level must be clear, clear logical relations, making the discriminant index organically linked, forming a distinct as a whole, should not only fully embodies the core index, and auxiliary evaluation indexes, give attention to two or more things to ensure the rationality of the index and representation.
- Operability: indicators should be easy to get, meaning scientific and accurate, and data collection and convenient measurement, standards and specifications, can not only reflect the index system of quantitative results, and can explain the qualitative state of the index system, so as to ensure the simplicity of index system and implementation of sex.
- Comparison: the selected indicators and their weights can not only reflect the common features of tourist inner cities, and the horizontal comparison between the cities.

Index system is used for the construction of urban tourism development dynamic model of judging whether science is accurate, index selection is critical. In accordance with the above index selection principle, through the expert decision-making consultation, for all involved in the city tourism drive index on the basis of comprehensive consideration, using AHP method and Delphi method to dynamic model of urban tourism development of index analysis, building dynamic model of the urban tourism development of index system and its hierarchy model (Figure 1).

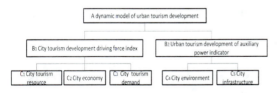

Figure 1. The index system of city tourism development model.

3 THE FORMATION MECHANISM OF THE DYNAMIC DEVELOPMENT MODEL OF URBAN TOURISM

In the process of urban tourism development practice, mutual influence, interaction between driving force

can form a dynamic portfolio model, dynamic model namely. Whereas dynamic model refers to the city tourism subject under certain conditions, by using various driving force for organic combination, to promote the development of city tourism specific mode of operation to promote. Relying on the dominant power of urban tourism development can be summarized as resources, economic power and social forces, with resources, economic power and social forces of the interaction between the dominant powers became means, dynamic model. Leading power of urban tourism development in the process – "resources, economic power and social forces" interaction, mutual influence, and deduce the dynamic pattern of diversification. According to the stage of development roughly, there are four kinds of models, such as traditional system dynamics model, dynamic model and image of modern market power and even social dynamics model from the side of the force (Fig. 2).

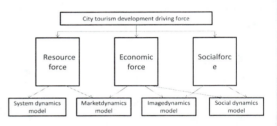

Figure 2. The dynamic model of urban tourism development mechanism.

Because of the difference of different urban development condition, in the different developing stages of the city, dynamic state is different, and different dynamic model is usually administered "catalytic media" to speed up the power output or improve power utility. What power a city tourism development mode, has his own development regularity and mechanism, and the existence of self-reinforcing mechanism, this is the path dependence. Whatever the motivation mode should be suitable for the local environment, mode of variation for the normal reflection of regional differences. So-called "power mode" refers to "in a certain city, a certain historical condition, the characteristic of tourism development path", which is the generalization of the characteristics of the particular urban tourism development. Environment and conditions variability determines the dynamic mode of variability, it cannot be fixed. This is the pattern of innovation. When we say that a dynamic model is innovative, actually contains two meanings, one is the pattern has innovation in new stage of development, the reasonable connotation of original pattern in the new pattern of inheritance. Second, many cities beyond the conventional power, with a unique power mode, drive the rapid development of urban tourism. Stage is the time property of urban tourism development dynamics model. Mode is the formation of various natural factors, economic factors and human factors, the result of the interaction with the evolution of the natural condition and social economy development, the three factors will be constantly changing, the result will inevitably lead to the power mode in the goal and the content, structure and function, the diversity and complexity, and soon have the corresponding change, thus present a certain stage. Different cities or urban tourism development in different stages of economic development level can be found that the dominant power, USES the main power to the rational allocation of power resources, promoting the healthy and rapid development of urban tourism. Different phase power model, dynamic model is suitable for different periods (Fig. 3).

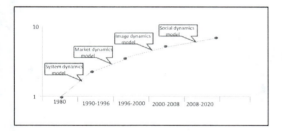

Figure 3. City tourism motivation mode to adapt to the development stage.

4 CONCLUSION

Driving mechanism of tourism development is the coordination interaction of each element of tourism development program. Tourism to sustainable development, first, it has continuous demand and continuing to attract, and has a good environment to support and active agents. Tourism development decision must be from the big background of regional development, tourism industry distribution and regional division of labor can be considered, with the concept of big tourism, based on the build tourism environment, committed to creating tourism sustainable development and benign mechanism for the sustainable development of regional economy. Urban tourism development dynamic model for the development of urban tourism has important guiding significance.

REFERENCES

[1] Peng Hua. Create excellent tourist city of thinking: introduction to tourism development and urban construction [J]. *Travel Journal*. 1999, 14(2).
[2] Gu Shi-yun. Urban tourism research progress. *Travel Journal*. 2001, 11(3).
[3] Myriam J. Inner-citytourism: resources. Tourists and promoters. *Annals of Tourism Research*, 1986, 13(1):79–100.
[4] kevin M. Consuming (in) the civilized city. *Annals ofTourism Research*, 1996, 23(2):322–340.

Information, Computer and Application Engineering – Liu, Sung & Yao (Eds)
© 2015 Taylor & Francis Group, London, ISBN 978-1-138-02717-6

Study on ceramic industry cluster for Guangdong province

Geng Zhong Zheng
Department of Computer Science and Engineering, Hanshan Normal University, Chaozhou, Guangdong, China

Qiu Mei Liu
Library, Hanshan Normal University, Chaozhou, Guangdong, China

ABSTRACT: Industrial cluster is a powerful model of economic development in today's world, which has great market competitiveness. In this paper, through study of the industrial cluster and the effect of industrial clusters competitiveness, we summarize the factors affecting the competitiveness of industrial clusters. Based on these, according to the characteristics of the ceramic industry cluster, we establish the evaluation system of the competitiveness for industrial cluster, discusses the status quo of Guangdong province of ceramic industry cluster, analyzes the existing problems in its development process, and gives the corresponding countermeasures, put forward the sustainable development strategy for Guangdong Province.

KEYWORDS: Industrial cluster; Clusters competitiveness; Ceramic industry; Strategy

1 INTRODUCTION

Industry cluster is a group of geographically adjacent, interconnected enterprises and relevant institutions, and connected by common and complementary for each other. Industry cluster is an outcome of market economy under the conditions of industrialization, and is an important phenomenon in economic development. The competitiveness of industrial cluster is the embodiment of comprehensive ability of industrial clusters, which reflect on the development of industrial clusters.

Guangdong provincial ceramic industry has a long history and has two major ceramic producing areas Foshan and Chaozhou. During 20 years development, Foshan has a new production line of ceramic production of nearly 1000, relates to the building ceramics, sanitary ceramics, art ceramics, ceramics for daily use and many other varieties. Chaozhou ceramic industry scale is the first during the Song Dynasty, after the reform and opening up, Chaozhou ceramic industry has become China's largest ceramics production and export base. In the paper, we study and analyse the current situation of ceramic industry cluster in Guangdong, discusses how to optimize Guangdong ceramic industry cluster, provide some suggestions for the development of Guangdong ceramic industry.

2 OVERVIEW OF DEVELOPMENT OF GUANGDONG CERAMICS INDUSTRY

2.1 Characteristics and distribution of Guangdong province ceramics

Foshan and Chaozhou is the representative of Guangdong ceramic industry cluster. Foshan is known as 'Chinese ceramic capital', the famous ceramic producing areas, mainly in building ceramics, sanitary ceramics, art ceramics production, ceramics etc. Chaozhou has grown rapidly in recent years, great strength, known as the "Chinese porcelain". In 2004, output value of Chaozhou ceramic industry is $20 billion, with exports amounting to $55.9billion, the annual export volume is increasing by 20% per year. Chaozhou is the largest daily ceramic and sanitary ware export base and daily-use ceramics in China, sanitary ware and ceramics production base, sanitary ware output accounts for about 57% of total output; electronic ceramic matrix, the annual output of 1200 million, accounting for 70% of national output, accounting for about 50% of global output. In the national ceramic producing a most complete, and has formed a complete industrial chain, involving product and craft porcelain, antique porcelain, Christmas environmental bone porcelain and many other categories of the series, has become the regions' export volume of China's largest ceramic, ranking first in the ceramic producing areas of the first. The strong radiation by Foshan and Chaozhou ceramic producing areas, ceramic industry in other areas such as bamboo shoots after a spring rain like Guangdong are booming, such as Qingyuan, Meizhou etc.

2.2 Advantages of Guangdong province ceramics industry

2.2.1 Resources and varieties

Guangdong ceramic not only have production and exports, but also have complete varieties. Only the building ceramics have glazed tile, stone imitation ceramic tile, mosaic brick, split brick (including patent technology "McDonald's" brick), architectural glass products, glass ceramics composite ceramic tiles and other products. According to reports, the current building ceramics in China has more than 2000 kinds of varieties, almost can produce all kinds of international. As an important ceramic producing area, ceramic raw materials demand is huge. Production of raw materials, ceramics are generally required to use high-quality non renewable production building, ceramics for daily use, this will inevitably lead to high quality non renewable ceramic raw materials are reduced. Guangdong ceramic raw material consumption mainly distributed in Foshan and Chaozhou. Qingyuan, Meizhou, Heyuan will become the ceramic raw material supply source for Foshan. Chaozhou china clay is rich in resources. According to incomplete statistics, Chaozhou's clay resources is more than million tons, and it can be used as the raw material of high quality. The "Flying Swallow" porcelain clay mine reserves most, porcelain clay optimal, storage capacity of 32,190,000 tons, ranking the exploration has been the porcelain clay mine of China clay resources reserves second.

2.2.2 Overseas market

Guangdong province building ceramics production accounted for more than 60% of the country, sanitary ceramics production accounted for about 25% of the country, domestic ceramics production accounted for about 30% of the country. Its' domestic market share are respectively 60%, 65%, 25%. Guangdong ceramic not only occupy most of the domestic market, but also have influence on foreign markets. Guangdong is adjacent to Hong Kong, Macao and Southeast Asia, the geographical position is superior, the foreign exchange is carried out frequently. China's ceramics city, Huaxia ceramic exhibition city has become a foreign exchange Foshan ceramic products window, at the same time

as the Foshan annual spring, autumn two season ceramic products fair well in the "exhibition platform". Information exchange and trade talks create lot of achievement by Foshan and the Guangdong ceramic brilliant. Ceramic products export market from Hong Kong, Macao and Southeast Asia, Middle East, extended to West Asia and Europe and America market. According to reports, Guangdong building wall tile exports mainly is the EU, Hongkong, South Korea, Singapore, Saudi Arabia and American; sanitary ceramics export is the main target of EU, Hongkong, Spain, South Korea and Saudi Arabia, USA; ceramics export is the main target of Southeast asia. According to the geographical division of Guangdong, the ceramic tile exports object 70%–80% in Asia; sanitary ceramics export region dominated by European and American countries and rich area, these areas of export exports account for about 70%–80% of all sanitary ceramics sanitary ware accessories; the main export destination is the United States and Europe and the Middle East markets in Latin America and Africa.

2.3 Disadvantage of Guangdong province ceramics industry

2.3.1 Production technology

In Guangdong ceramic industry, small and medium-sized ceramic enterprises accounted for the majority, to their own development, a lot of workshop production use traditional way to product, they have not enough money to introduce advanced technology and equipment and to carry out technology research and development; production technology is backward, such as ceramic raw materials production. Because of technology, equipment behind, the mining of raw materials, it is difficult to determine the level by dressing, and different region, its chemical composition fluctuation, which directly affect the quality of the products. In addition, the raw material is not standardized, and there are lot of raw materials which cannot meet the requirements for the advanced equipment, resulting in tremendous waste. In contrast, in foreign countries, advanced porcelain country such as Germany almost all the ceramics raw material standardization and specialization of production, production line by computer operation, of roughing by separation technology, ultrafine grinding, high gradient magnetic separation, centrifugal separation, the microcomputer burden, eddy mixing and modern analysis and testing technology and equipment realize the standardization, series production, product quality and stability.

2.3.2 Innovation

Imitation has been accompanied by Guangdong ceramic industry, which leads to decades of devel-opent. From the first automatic grouting line, mechanical line, the first roller kiln, the first print-ing machines, there is no existing traces of imitation. The imitation for the new enterprises is essential, but the enterprise development to a certain extent, inno-vation has become the enterprise winning the market power and booster. Internationally famous ceramic enterprises have attached great importance to inno-vation. At present, the majority of ceramic enter-prises concepts have not changed, are often used to imitate and follow the trend, always follows behind others run, and even some bad plagiarism. Many so-called innovative products and energy-saving technology are introduced from outside. Therefore, Guangdong ceramic industry but also in innova-tion, especially independent innovation more efforts, transfer of technology and the two innovation organ-ically, form their own energy-saving technology sys-tem has characteristics of the enterprise, increase the level of comprehensive utilization of the enterprise "three wastes".

3 EVALUATION AND EMPIRICAL ANALYSIS OF GUANGDONG CERAMICS INDUSTRY CLUSTER

3.1 Factors affecting the competitiveness of industrial clusters

Hardware factor refers to the industrial cluster com-petitiveness external, intuitive, directly measurable factors, industry clusters are reflected in the market competition to meet the needs of the market and the competition for market share, hardware factor is the direct result of the soft power of industrial cluster. Soft ware refers to the competition ability factors in industrial cluster internal, it cannot be directly meas-ured and represents the ability of sustainable develop-ment of industrial clusters, a potential development, on behalf of the industry clusters' competitive in the future.

3.2 Competitiveness evaluation system of ceramic industry cluster

Based on the analysis of industrial cluster competi-tiveness evaluation for current stage, in combination of factors affecting the competitiveness of industrial clusters, a number of indicators are selected to con-struct the evaluation index of competitiveness for ceramic industry cluster.

Table 1. Evaluation index.

One level index	Two level index	Three level index
Hardware	Scale	number of enterprises in cluster
		enterprises within cluster density
		value of industrial output
		total sales
		market share of products
		fixed assets
	Economics Benefits	sales profit
		net profit
		industrial added value
		means of production
	Resource	labor resources
		land resources
		technical resources
Software	Transaction Network	enterprise subcontracting outsourcing
		clusters of industry chain length
	Innovation Network	technical cooperation
		cooperative conditions
		innovation environment
	Brand	brand reputation
		brand awareness
	Administration	government put into force
		intellectual property protection
Cluster environment		attitude of the government information update
	Association	Association
		guide association
	Finance	invested capital
		policy

4 STRATEGY TO IMPROVE CLUSTER COMPETITIVENESS FOR GUANGDONG PROVINCE CERAMICS INDUSTRY

4.1 Production process

Through reorganize production system and intro-duce more advanced technology to improve the input–output ratio, improve the efficiency of process in the value chain to reduce production costs, and improve the economic efficiency of enterprises. Extend the industrial china chain, the links will be more detailed, improve the product-on process each link. Division of labor and cooperation to improve the efficiency.

4.2 Cooperation among enterprises

Industrial cluster is in a certain geographical area, specialized division of some network support ing the production of a product of the industrial chain and related support services system gather together and form, is the organic connection and cooperation of enterprises in the region. Division of labor and specialization determines the formation and production of industry cluster, and then promote the rapid development of industry cluster of regional division of labor and specialization of the formation of industrial clusters. Specialization is conducive not only to improve the technological level and innovation capacity of the enterprises within the cluster, enhance the competitive advantage but also conducive to extensive cooperation, complementary advantages between enterprises in the region, to create regional brand, to prevent excessive competition. By strengthening the enterprises and other institutions of cooperation network and the network of interpersonal relationships, the cluster network, and promote the upgrading of clusters.

4.3 Capability of independent innovation

Innovation is the key to the development of cluster and cluster competitiveness. Actively promote the independent innovation of enterprises in the cluster, encourage enterprises to increase R & D investment, research and development and transformation of achievements and encourage the advanced textile technology, adopt various forms to establish R & D center, design center and engineering center. Small and medium-sized science and technology enterprises and encourage the development of specialized, attract Multi-National Corporation set up R & D center in the cluster, to encourage local enterprises and foreign joint venture established R & D center. Through acquisitions or build overseas R & D center to strengthen the capability of independent innovation of enterprises.

4.4 Service function

The key industry associations work is the development of industry to provide information resources for the enterprise, the establishment of industry self regulatory mechanism, coordinate on relations of enterprises, promote fair competition, participation in the formulation of industry standards, to promote the healthy development of the industry. Through investigation and research, the industry of new industrialization overall conception, policy advice and work plan, carry out enterprise oriented information, consulting, training, as well as international platform and market development wor k, the implementation of a comprehensive tracking, guidance and services to the industry of new industrialization process.

5 CONCLUSION

In Guangdong Province of ceramic industry, ceramic industry cluster from the time the successsfu l experience should not apply mechanically, to in-depth analysis of specialized industrial clusters need conditions, is not only the establishment of Industrial Park, the company cited in, but a careful analysis of its internal relations in the ceramic industry, enterprises have synergistic effect in accordance with the on a certain scale in the industrial park, in order to obtain economies of scale in industrial cluster brings, promote the development of regional economy. The kind of "bags of potato" of the enterprise cluster will not have cluster effects, it will bring serious consequences to area economy. Guangdong Provincial ceramic industry cluster according to their actual situation, to play their own advantages. Such as resource advantage, talent advantage, traditional industry, industrial base and cernaic cultural differences, to create its own characteristics and advantages of professional ceramic industrial district. We believe ceramics in Guangdong Province will be better tomorrow tomorrow.

ACKNOWLEDGMENT

This research was financially supported by the Breeding project of Education Department of Guangdong Provincial (No. 205020513)

REFERENCES

[1] Weilin Zhaoa, Chihiro Watanabea, Charla Griffy Brownb. Competitive advantage in an industry cluster: The case of Dalian Software Park in China. *Technology in Society*. 2009, 31(2): 139–149
[2] Jane Zheng, Roger Chan. The impact of 'creative industry clusters' on cultural and creative industry development in Shanghai. *Culture and Society*.2014,5(1):9–22
[3] Yung-Lung Lai, Maw-Shin Hsu, Feng-Jyh Lin, Yi-Min Chen, Yi-Hsin Lin .The effects of industry cluster knowledge management on innovation performance. *Journal of Business Research* 2014,67(5): 734–739
[4] H. Celik. Technological characterization and industrial application of two Turkish clays for the ceramic industry. Applied Clay Science.2010,50(2):245–254
[5] Zsuzsanna Turóczy, Liviu Marian. Multiple Regression Analysis of Performance Indicators in the Ceramic Industry. Procedia Economics and Finance. 2012(3): 509–514

Information, Computer and Application Engineering – Liu, Sung & Yao (Eds)
© 2015 Taylor & Francis Group, London, ISBN 978-1-138-02717-6

A modeling for reconfigurable control II: Procedures of aggregation

H.X. Hu
Agricultural and Animal Husbandry College, Tibet University, Tibet, P. R. China
Energy and Electrical Engineering College, Hohai University, Nanjing, P. R. China

Y.F. Sheng & Y.Q. Zheng
Energy and Electrical Engineering College, Hohai University, Nanjing, P. R. China

Y. Zhang
Computer and Information Engineering College, Hohai University, Nanjing, P. R. China

C. Shan
Business English Department, Shen-Yang Institute, Engineering, Shen-Yang, P. R. China

ABSTRACT: In this paper, we propose an aggregation modeling approach for reconfigurable control and illustrate the main result by two tank system. In this aggregation modeling, the reconfigurability aspect is analyzed and the aggregation modelling procedure is revealed how to construct a hierarchical system and to assure the coherences between specification and realization at each level and between levels.

KEYWORDS: Model aggregation; Service; Operating mode; Reconfigurable control

1 INTRODUCTION

In this paper, we continue to present the part of the model aggregation modeling for reconfigurable control. As in its sister paper "A Modeling for Reconfigurable Control I: Generic Component", reconfiguration control considers the problem of automatically changing the control structure and the control law after a fault has occurred in the plant (Blanke, M. et al. 2003). The Generic Component Model (GCM) describes components from a point view of the users, who receive services and can use them differently in different operating modes (Staroswiecki, M. & Bayart, M. 1996). A formal analysis of the reconfigurability has been described by using GCM (Staroswiecki, M. & Gehin, A.L. 1998). Component interconnections are taken into account by considering higher level components, which result from the aggregation of lower level ones. Choukair and Bayart present an approach of modeling of a distributed architecture based on the Cartesian products of components (Choukair, C. & Bayart, M. 1999). The model aggregation is a useful method by which a large-scale system is built as a hierarchical system composed of interacting subsystems (Aström, K. et al. 2001 & Hu, H.X. et al. 2006).

In this paper, our goal is to realize model aggregation for reconfigurable control and to present a systematic procedure for the model aggregation of a hierarchical system. This procedure finds its justification from the coherence it achieves in a level and between levels, unifies the two elementary notions service and the operating mode during the model aggregation, and produces the important information about reconfigurable control.

The paper is organized as follows. The example used to illustrate an aggregation procedure for a hierarchical system in section 2. In section 3, we illustrate the main result by two tank system. In section 4, we provide a brief summary of the main results of this paper.

2 BUILDING SYSTEMS FROM COMPONENTS

System architectures can be described at different hierarchical levels. The decomposition of a system to several subsystems and so on until the elementary components can be represented by a pyramidal architecture. Since it may be advisable to allow some components to belong to several subsystems, the physical structure is not purely hierarchical but is a pyramidal one (figure 1).

In a pyramidal architecture, each component of level *l-1* belongs to at least one component of level *l* and any component of level *l* includes at least one component of level *l-1*.

To establish the pyramidal structure of a system, the functional viewpoint may be adopted. For example, the TTS will be decomposed into two subsystems, each subsystem grouping the elementary components allowing the level regulation in one of the tank (figure 2).

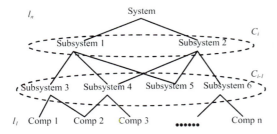

Figure 1.   Physical pyramidal structure of a system.

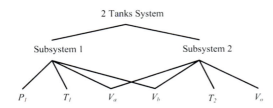

Figure 2.   TTS pyramidal structure.

2.1   Procedure of aggregation

Let $S_\alpha(a)$, $O_\alpha(a)$ (resp. $S_\beta(b)$, $O_\beta(b)$) be the services and the objectives associated to the component a being in the mode α (resp. the component b being in the mode β). $S_\alpha(a)$ (resp. $S_\beta(b)$) allows the realization of $O_\alpha(a)$ (resp. $O_\beta(b)$).

Let c be a component of level l which aggregate the components a and b of level l-1. The combination of operating modes α and β, if it is consistent from a view point of user, provides a service of the component c. More generally, we will say that the set φ of the potential services of the component c is the Cartesian product of the operating modes offered by a and b.

$$\varphi = \{\Delta = \alpha \times \beta, \text{ such as } \alpha \in M(a), \beta \in M(b)\} \quad (1)$$

Of course, not every combination of lower level OMs is significant or allowed in real practice, for example associate a *Test_OM* with a *Regulation_OM* may be not relevant for the application realization, and such a combination should be removed from φ. Moreover there may exist combinations which have the same functional interpretation, they constitute versions of the same service and provide fault tolerant control perspectives.

Removing irrelevant services from φ allows specifying the set of relevant services φ_r which has to be organized into consistent OMs. Note that it is designers who indeed determine φ_r from φ and structure φ_r into OMs.

Defining φ and φ_r and structuring φ_r into OMs constitute the three steps of the aggregation procedure.

This procedure finds its justification from the coherence it achieves in a level and between levels. Removing irrelevant services from the set of possible combination of lower level components' OMs and organizing the set of relevant services into OMs allows guaranteeing the coherence between the available services and the objectives to achieve. In other words, it is a mean to express the relation of "what the aggregated component could do" and "what it should do".

Defining the services of a higher level component from the OMs of the components it aggregated and then structuring them into OM through the objective notion allows taking into account a specification given in a hierarchical way, assuring a coherence between levels and decreasing the number of combinations at each level of aggregation. Relations between OMs, objectives and services through the different levels are expressed by the figure 3.

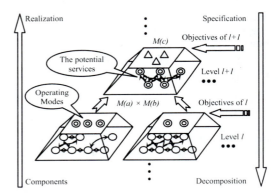

Figure 3.   The aggregation of a hierarchical system.

2.2   Aggregation of services

On figure 3, we can see that an OM is a set of services allowing the realization of the objectives associated to this OM and a service can be provided by a combination of the OMs associated to the components of lower level. Each of them can transform his role into another one in different levels.

How to succeed this transformation? Firstly, We note the tth service of the component k at level l as $S_t(k)\_l$, which is a combination of the operating modes associated with n components at level l-1. It is represented as an element of $\varphi(k)\_l$, where $\varphi(k)\_l$ is the production, such as $M(1)\_l$-$1 \times M(2)\_l$-$1 \times \ldots \times M(n)\_l$-$1$ and $M(k)\_l$-1 is a set of $S_t(k)\_l$-1.

As the definition in the sister paper "A Modeling for Reconfigurable Control I: Generic Component" section 3.1, $S_t(k)\_l = <cons_t(k)\_l,\ prod_t(k)\_l,\ proc_t(k)\_l,\ rqst_t(k)\_l,\ res_t(k)\_l>$, if let $M_i(k)\_l$-1 is the ith operating mode of $M(k)\_l$-1, $|M(k)|$ is the number of OM in this set and $/M_i(k)/$ is the number of services of ith mode in $M(k)$, then we can obtain a service $S_t(k)\_l$ from the following procedure:

For $i\_1 = 1$ to $/M(1)\_l\text{-}1/$
For $i\_2 = 1$ to $/M(2)\_l\text{-}1/$
......
For $i\_n = 1$ to $/M(n)\_l\text{-}1/$

(There are n components at level $l\text{-}1$ forming one component at level l and several modes in each component.)

$\{Cons_{i_1 i_2 \ldots i_n}(k)\_l = 0$
For $j=1$ to n
$\{m = i\_j$
For $s=1$ to $Mm(j)\_l\text{-}1/$
$Cons_{i_1 i_2 \ldots i_n}(k)\_l = cons_{i_1 i_2 \ldots i_n}(k)\_l \cup cons^s_m(j)\_l\text{-}1\}\}$

(Here, $cons^s_m(j)\_l\text{-}1$ is the variables consumed by the sth service in mth OM of jth relevant component at level $l\text{-}1$.)
End

As the same procedure, we obtain $prod_t(k)\_l$, $proc_t(k)\_l$, $rqst_t(k)\_l$ and $res_t(k)\_l$. In fact, expect for level 0, the services of level l result from the Cartesian production of the operating modes at level $l\text{-}1$ as mentioned above. So we can infer the signification of this service from variables it consumes and produces.

3 APPLICATION TO THE TTS

3.1 Model aggregation

The specification of TTS is firstly given in a hierarchical way by figure 4.

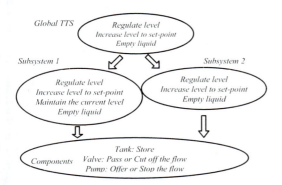

Figure 4. The hierarchical specification for TTS.

The operating modes of elementary components are given by table 1.

The elementary components are those that can not be decomposed into other components. Each operating mode in these components has only one service

Table 1. The operating modes of elementary components.

Component	OM	Services	Objectives
Tank	1	T_i_store	Store the flow
Valve	1	V_i_open	Pass the flow
	2	V_i_close	Cut off the flow
Pump	1	$Deliver\_Q_i$	Offer the flow
	2	$Stop\_Q_i$	Stop the flow

Table 2. Services combinations of subsystem 1.

T_l	P_l	V_a	V_b	Service
T_l_store	$Stop\_Q_l$	V_a_close	V_b_close	$H_l \rightarrow$
		V_a_close	V_b_open	$H_l \downarrow$
		V_a_open	V_b_close	$H_l \downarrow$
		V_a_open	V_b_open	$H_l \downarrow$
	$Deliver\_Q_l$	V_a_close	V_b_close	$H_l \uparrow$
		V_a_close	V_b_open	$H_l \downarrow$
		V_a_open	V_b_close	$H_l \downarrow$
		V_a_open	V_b_open	$H_l \downarrow$

Table 3. The versions of services for subsystem 1.

Versions		Lower OMs	Resources
$H_l \rightarrow$	V_0	$T_l\_store, Stop\_Q_l$ V_a_close, V_b_close	T_l, P_l V_a, V_b
$H_l \uparrow$	V_0	$T_l\_store, Deliver\_Q_l$ V_a_close, V_b_close	T_l, P_l V_a, V_b
$H_l \downarrow$	V_0	T_l_store V_a_open, V_b_close	T_l V_a, V_b
	V_1	T_l_store V_a_close, V_b_open	T_l V_a, V_b
	V_2	T_l_store V_a_open, V_b_open	T_l V_a, V_b

and is associated with an objective. For simplicity reasons, the same names for these services and OMs are chosen.

Following the pyramidal decomposition of the TTS given by figure 2, the Cartesian product of the OMs of the elementary components P1, T1, Va and Vb is calculated to define the potential service set of the subsystem 1 (see Table 2). The three relevant services H1→ (maintain), H1↑ (increase) and H1↓ (decrease) can then be defined for the subsystem1 and H1↓ is defined under three versions (see Table 3). The relevant service set is then organized into OMs which are consistent with the objectives to achieve in this subsystem 1 level. These operating modes are shown in table 4.

As the same procedure, we can also obtain the operating modes of subsystem 2 (see tables 5, 6 and 7).

Table 4. The operating modes of subsystem 1.

OM	Services	Objectives
Empty	$H_l\downarrow$, $H_l\rightarrow$	Empty liquid
Preparation	$H_l\uparrow$, $H_l\rightarrow$	Increase level to Set-point
Regulation	$H_l\uparrow$, $H_l\rightarrow$, $H_l\downarrow$	Regulate level
End	$H_l\rightarrow$	Maintain the level current

Table 5. Services combinations of subsystem 2.

T_2	V_o	V_a	V_b	Service
T_2_store	V_o_open	V_a_close	V_b_close	H_2_below
		V_a_close	V_b_open	H_2_above
		V_a_open	V_b_close	H_2_above
		V_a_open	V_b_open	H_2_above

Table 6. The versions of services for subsystem 2.

Versions		Lower OMs	Resources
H_2_below	V_o	T_2_store, V_b_open V_a_close, V_b_close	T_2, V_o V_a, V_b
	V_o	T_2_store, V_b_open V_a_close, V_b_open	T_2, V_o V_a, V_b
H_2_above	V_1	T_2_store, V_b_open V_a_open, V_b_close	T_2, V_o V_a, V_b
	V_2	T_2_store, V_o_open V_a_open, V_b_open	T_2, Vo V_a, V_b

The aggregation of the subsystems 1 and 2 allows defining the TTS OMs (see Tables 8, 9 and 10). Note that combinations which are not significant in real practice are rejected.

Table 7. The operating modes of subsystem 2.

OM	Services	Objectives
Empty	H_2_below	Empty liquid
Preparation	H_2_above	Increase level to Set-point
Regulation	H_2_above, Hv_below	Regulate level

Table 8. The table of services combinations of TTS.

Subsystem 1	Subsystem 2	Service
Regulation	Regulation	Regulation
	Empty	Rejected
	Preparation	Rejected
Preparation	Regulation	Rejected
	Empty	Rejected
	Preparation	Preparation
Empty	Regulation	Rejected
	Empty	Empty
	Preparation	Rejected
End	Regulation	Rejected
	Empty	Empty
	Preparation	Rejected

Table 9. The versions of services for TTS.

Versions		Lower OMs	Resources
Regulation	V_0	Regulation_1 Regulation_2	Subsystem 1 Subsystem 2
Preparation	V_0	Preparation_1 Preparation_2	Subsystem 1 Subsystem 2
Empty	V_0	Empty_1 Empty_2	Subsystem 1 Subsystem 2
	V^1	End_1 Empty_2	Subsystem 1 Subsystem 2

Table 10. The operating modes of TTS.

OM	Services	Objectives
Empty	Empty	Empty liquid
Preparation	Preparation	Increase level to Set-point
Regulation	Regulation	Regulate level

3.2 Fault scenario

To illustrate the reconfiguration on the TTS, we consider a fault scenario. Let suppose, the current OM be the Regulation one and the necessary services and OMs of nominal version given in a hierarchical way as figure 6. Scenario: V_a blocked in the closed position. Service V_a_close gets permanent in time and service V_a_open becomes unavailable. Therefore, the nominal version of $H_l\downarrow$ and $H_{2-above}$ becomes unavailable, but the degraded version, $\{H_l\downarrow$: T_l_store, V_a_close, $V_b\_open\}$ and $\{H_{2-above}$: T_2_store, V_a_close,

$V_h\_open\}$, remains available (ref. table 4 and 7). The OMs, Regulation_1 and 2, are not affected and the mission of regulation TTS can still be achieved using these degraded versions of services.

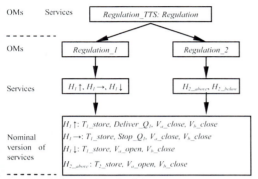

Note: Level 0 is omitted for simplicity reason.

Figure 6. The hierarchical description for regulation mission.

4 CONCLUSION

In this paper, a model aggregation procedure based on generic component model to reconfigurable control has been proposed. Two elementary notions, the service and the operating mode have been introduced to construct a hierarchical system and to assure the coherences between realization possibilities and given specifications not only at each level of the system decomposition but also between the levels of the decomposition. It is a mean to expose the relation of "what the aggregated system could do" and "what it should do". The important benefit of this procedure is that we can describe a system at any hierarchical level in a systematic way. Moreover it provides a unified framework for facilitating the knowledge acquisition of reconfiguration.

ACKNOWLEDGMENT

This work is supported by "The Natural Science Funds of Tibet", "the Fundamental Research Funds for the Central Universities".

REFERENCES

Aström, K., Albertos, P., Blanke, M., Isodori, A., Schaufelberger, W. & Sandz, R. 2001. *Control of Complex Systems.* Springer.
Blanke M., Kinnaert M., Lunze J. & Staroswiecki M. 2003. *Diagnosis and Fault-Tolerant Control.* Springer.
Choukair C. & Bayart M. 1999. Application of External Model of intelligent Equipment to Distributed Architectures. *In ISAS'99, Orlando (U.S.A.), pp. 329–335, Juillet 1999.*
Hu, H. X., Gehin, A.L. & Bayart, M. 2006. Model Aggregation for Reconfigurable Control Based on Generic Component Model. *In ICSSSM'06, Troyes, France, 2006.*
Staroswiecki, M. & Bayart M. 1996. Models and Languages for the Interoperability of Smart Instruments. *Automatica* 32(6): 859–873.
Staroswiecki, M. & Gehin, A. L. 1998. Analysis of System Reconfigurability using Generic Component Models. *In Control'98, Swansea (UK), 1998.*

Information, Computer and Application Engineering – Liu, Sung & Yao (Eds)
© 2015 Taylor & Francis Group, London, ISBN 978-1-138-02717-6

The study on brand development of sports events of artistic gymnastics in China

Yong Sheng Sun, Jia Hui Zou & Wen Sheng Wang
Capital University of Physical Education and Sports, Beijing, China

ABSTRACT: The results of this article show that as an event of winning gold medal, artistic gymnastics is on the road of marketization and socialization, and only first-class brand competition can promote the development of gymnastics in China. And this is an effective way to realize the value of the artistic gymnastics and promote gymnastics public participation.

KEYWORDS: Brand; Artistic gymnastics; Sports events

1 SUBJECTS AND METHODS

1.1 Subjects

The brand development of sports events of artistic gymnastics in China.

1.2 Methods of study

The diagnostics was got with documents, expert interview, and logical analysis methods.

2 RESULTS AND ANALYSIS

2.1 Strategic significance of the brand development of sports events of artistic gymnastics in China

2.1.1 Comply with the market rule and cater to the market demand

As for the movement of artistic gymnastics is beautiful and full of thrilling, the sports event has the very high appreciation value for the audience. Under market economy, the question of sports consumption put to sports business units by external environment is mainly showed through the performance of sports market, including material object and labor sports consumptions. Therefore, the events of artistic gymnastics own big market which is helpful to enter into the industrialization market. The successful events could not only meet the demands of the public, but also provide generous interest. It also complies with the development of market, and promotes the sports industrialization effectively.

2.1.2 To maintain the sustainable development of artistic gymnastics needs

As an event of gold medal capable of winning, although artistic gymnastics will get much more support from the government, it will be difficult to maintain the development of artistic gymnastics in future. Therefore, from the aspect of survival and development, we should seek a path of marketization and the brand development instead of relying on government. With the development of economy, the proportion of the consumption of cultural, entertainment and fitness products to the total consumption expenditure continues to grow, which provides a wide space for the development of marketization of the athletic gymnastics. The marketization of artistic gymnastics started late, which is somewhat weak compared with other sports events. In the process of marketization, to maintain the survival of gymnastics and the development of events, and participate in the competition of other sports events, it needs to improve the quality of the gymnastics competition and the level of marketing operation. Through the brand operation, we can expand the market share; get more profits, so as to achieve the sustainable development.

2.1.3 Boost the sports industry development

The successful hosting of the Olympic Games in Beijing has the milestone significance, which marks the new position of Chinese sports in the contemporary international sports. From then on, the sports industry entered into our vision. As an advantage event of China, artistic gymnastics events play an important role in numerous sports in the transformation from the athletic powerhouse to sporting powers. The competitive sports are at the early stage, and the branding of artistic gymnastics sports events have not developed. But its development space is huge. Successful brand operations will boost the development of sports industry, which is of great strategic significance.

2.2 The currency of artistic sports events in China

2.2.1 The unitary competition system

There is not so many artistic gymnastics competition in China, and these events are in strict accordance with the international competition in the form of game operation, so the form is single. The species is less and the content is simple from the national to regional competitions. And there is seldom some exhibition, match play, friendly match and success brand events. The basketball events learned from the NBA game mechanism, and introduced a new basketball super league, which works well now. The simple event leads to less attention and severely restricted the benign development of the artistic gymnastics; meanwhile, it is not conducive to the realization of its economic value.

2.2.2 Lack of publicity, insufficient competition influence

There are seldom reports and TV shows before the artistic gymnastics competition in China. With the rapid development of modern science and technology, media has penetrated into all walks of life, and the effect of media propaganda is much more important for sports promotion and operation. Media is the most direct measure to promote globalization of an event. The event could eliminate the number limitations of audiences, and further expand the number of viewers. As for some managers and coaches lack of the realization of marketization of gymnastics, the event propaganda is poor. Therefore, the operation of the event is poorer, which leads that the audience is hard to learn more about gymnastics. They always think that gymnastics is a high risk project that only few people can participate in, thus the number of people involved in gymnastics is few.

2.2.3 The pressure from other brand competition

As the market development of artistic gymnastics in China is late, the current results are preliminary. Compared with basketball, the participants of artistic gymnastics are less. With the development of economy and the opening up of people's ideas, many foreign sports already have a great influence on Chinese sports consumers. In terms of sporting events, such as European Super League, attracted most of TV viewers in China. The good mass base and mature market development worldwide are the main reasons. So there is a great impact and influence on the promotion of artistic gymnastics competitions.

2.2.4 The imperfections of laws and regulations of marketization and industrialization

The market playing a role in the allocation of resources and the protection of every subject's interests all depend on a sound system of laws and regulations. At present, as for the development of marketization and industrialization is still at the exploratory stage, a lot of laws and regulations are not perfect. So we need to improve relevant laws and regulations, standardize the management and operation behavior of social groups and organizations, clear each participator's responsibilities and rights which is particularly important to speed up the brand operation of artistic gymnastics.

3 CONCLUSIONS AND SUGGESTIONS

3.1 Brand culture building

From their own brand in many parallel events competition, brand culture must be set up. The development of potential culture of Chinese artistic gymnastics competition needs to make clear of the essence of the brand culture of Chinese artistic gymnastics. The artistic gymnastics culture includes event product conception, design, brand, advertising and so on, which stains with the cultural quality, culture personality, and aesthetic consciousness of athletics gymnastics.

3.2 Promote the reform of mechanism of artistic gymnastics competition

The reform of competition system and transformation of operation mechanism could promote the socialization and the industrialization of the competition by the combination strength of country and society instead of a single pattern of country-oriented. The market system of sports competition is full of vigor and in line with the law of sports competition, which makes the industry of competition as the pillar industry. The artistic gymnastics athletes started the hard training when they were young, but the time of achieving result is very short, in which time, competition able to participate in is very limited, and this resulted in the waste of human resources. Therefore, the current competition system has limited the athletes' chances to compete. The system must be reformed and the athletes could also get the equal rewards.

3.3 Reinforce the medium publicity

Mass media is a brand window of expanding event influence, shaping a powerful game brand. According to World Snooker China Open Report, there are more than 20 Chinese and foreign media that make the reports of events, and the number of reports by media publicity is rising year by year. The central and local major sports TV media for events were widely reported. Meanwhile, innovation can also be conducted to highlight events from the content. Artistic gymnastics in essence is the combination of difficulty, innovation and beauty. So in artistic gymnastics programs, the most entertaining elements of gymnastics events can be mixed into within the limits of entertainment elements.

3.4 *Take advantage of star athletes*

The core factor of brand sports events is the quality of competition, which is the foundation of brand competition. High quality of competition appeals to the audience and the media most. The audience is eager to enjoy the high quality and competitive game, not only by watching the game, but also in the media. The most important element of the high quality of the competition is the participation of star athletes. Without them, an event is difficult to become brand sports event. There are rich resource of star and excellent athletes in China, which leads to the intense competition and high competitive view. The participation of star athletes will greatly improve the visibility of the event, namely the deep influence of brand shaping and expansion. From the perspective of main supply body of competition market, they are the premise of improving high quality of products and service.

ACKNOWLEDGMENT

The Young Talent Program of BeiJing Municipal Education Commission (YETP1712).
Corresponding author: WenSheng Wang.

REFERENCES

Zhang Shiwei. Study of China's sports personnel training in period after olympics. *Journal of Xi'an Institute of Physical Education*, 2009 (5).
Tang Hao, Wei Nongjian. The Study on the Market of Artistic Gymnastics Industry in China.Shanghai: Xue Lin Publishing, 2005.
Li Fang.Study on Artistic Gymnastics Reform and Development in China. The 7th National Sports Science Conference Assembly 1st, 2004.
Jiang Jiazhen, Zhong Bingshu.The Study on Evaluation System of Principle and Method of Brand Sports communication value. *Journal of Beijing Sport University*, 2008, (2).

Author index